ENCYCLOPEDIA OF
BIOLOGICAL INVASIONS

ENCYCLOPEDIA OF BIOLOGICAL INVASIONS

EDITED BY

DANIEL SIMBERLOFF

University of Tennessee, Knoxville

MARCEL REJMÁNEK

University of California, Davis

UNIVERSITY OF CALIFORNIA PRESS

Berkeley   Los Angeles   London

University of California Press, one of the most distinguished university presses in the United States, enriches lives around the world by advancing scholarship in the humanities, social sciences, and natural sciences. Its activities are supported by the UC Press Foundation and by philanthropic contributions from individuals and institutions. For more information, visit www.ucpress.edu.

Encyclopedias of the Natural World, No. 3

University of California Press
Berkeley and Los Angeles, California

University of California Press, Ltd.
London, England

Library of Congress Cataloging-in-Publication Data

Encyclopedia of biological invasions / edited by Daniel Simberloff and Marcel Rejmánek.
    p. cm. — (Encyclopedias of the natural world ; 3)
    Includes bibliographical references and index.
    ISBN 978-0-520-26421-2 (cloth)
    1. Introduced organisms—Encyclopedias. I. Simberloff, Daniel. II. Rejmánek, Marcel.

QH353.E53 2011
578.6'2--dc22—dc21   2010010391

Manufactured in China.
19 18 17 16 15 14 13 12 11
10 9 8 7 6 5 4 3 2 1

The paper used in this publication meets the minimum requirements of ANSI/NISO Z39.48-1992 (R 1997) (*Permanence of Paper*). ∞

Cover photograph: Large, pink masses of eggs laid by golden apple snails, courtesy N.O.L. Carlsson. Insets, from left: *Pheidole megacephala* tending aphids, courtesy Alex Wild; European starling, courtesy Michelle St. Sauveur, Coventry, Rhode Island; oral disc of a parasitic-phase sea lamprey, courtesy Ted Lawrence, Great Lakes Fishery Commission; flowering tropical weed *Turnera subulata*, courtesy Spencer C.H. Barrett.

Title page photo: An observer dwarfed by an overgrowth of giant hogweed (*Heracleum mantegazzianum*) invading the Slavkovský les Protected Landscape Area, Czech Republic, Central Europe. Introduced to Europe from the Caucasus, invasive giant hogweed can grow to over four meters in height. Its phototoxic sap blisters human skin and can cause serious scarring. Photograph by Petr Pyšek. From *Ecology and Management of Giant Hogweed* (Heracleum mantegazzianum), edited by P. Pyšek, M. J. W. Cock, H. P. Ravn, and W. Nentwig (2007, CAB International, Wallingford, UK).

CONTENTS

CONTENTS BY SUBJECT AREA

CONTRIBUTORS

CHRISTOPHER B. ANDERSON
University of Magallanes
Punta Arenas, Chile
Mammals, Aquatic

LARS W. J. ANDERSON
USDA–Agricultural Research Service
Davis, California
Freshwater Plants and Seaweeds

GREGORY P. ASNER
Carnegie Institution for Science
Stanford, California
Remote Sensing

PRZEMYSLAW BAJER
University of Minnesota, St. Paul
Carp, Common

SPENCER C. H. BARRETT
University of Toronto, Ontario, Canada
Reproductive Systems, Plant

DEVIN M. BARTLEY
California Department of Fish and Game, Sacramento
Aquaculture

ARIJANA BARUN
University of Tennessee, Knoxville
Carnivores

JACQUELINE BEGGS
University of Auckland, New Zealand
Wasps

ROGER A. BERGSTEDT
U.S. Geological Survey
Millersburg, Michigan
Sea Lamprey

ROSS A. BICKNELL
Plant and Food Research
Christchurch, New Zealand
Apomixis

BERND BLOSSEY
Cornell University
Ithaca, New York
Enemy Release Hypothesis

PIETER BOL
Delft University of Technology
Delft, The Netherlands
Pathogens, Human

LEONARD A. BRENNAN
Texas A&M University, Kingsville
Game Animals

FRED C. BRYANT
Texas A&M University, Kingsville
Game Animals

JENNIFER L. BUFFORD
University of Hawaii, Honolulu
Life History Strategies

STANLEY W. BURGIEL
Global Invasive Species Programme
Alexandria, Virginia
Black, White, and Gray Lists

MARC W. CADOTTE
University of Toronto, Ontario, Canada
Darwin, Charles

RAGAN M. CALLAWAY
University of Montana, Missoula
Novel Weapons Hypothesis

JAMES T. CARLTON
Williams College
Mystic, Connecticut
Ballast
Invertebrates, Marine

TED CENTER
U.S. Department of Agriculture
Ft. Lauderdale, Florida
Water Hyacinth

MATTHEW K. CHEW
Arizona State University, Tempe
Invasion Biology: Historical Precedents

MICK N. CLOUT
University of Auckland, New Zealand
New Zealand: Invasions
Predators

PETER COATES
University of Bristol, United Kingdom
Xenophobia

JULIE COETZEE
Rhodes University
Grahamstown, South Africa
Water Hyacinth

ANNA COHUET
Institut de Recherche pour le Développement
Montpellier, France
Malaria Vectors

ROBERT H. COWIE
University of Hawaii, Honolulu
Snails and Slugs

JEFFREY A. CROOKS
Tijuana River National Estuarine Research Reserve
Imperial Beach, California
Lag Times

CURTIS C. DAEHLER
University of Hawaii, Honolulu
Life History Strategies

CARLA D'ANTONIO
University of California, Santa Barbara
Grasses and Forbs

ADAM S. DAVIS
USDA-ARS Invasive Weed Management Unit
Urbana, Illinois
Agriculture

MARK A. DAVIS
Macalester College
St. Paul, Minnesota
Invasion Biology

PETER DE LANGE
New Zealand Department of Conservation, Auckland
New Zealand: Invasions

MARIE-LAURE DESPREZ-LOUSTAU
Institut National de la Recherche Agronomique (INRA)
Villenave d'Ornon, France
Fungi

CHRISTOPHER R. DICKMAN
University of Sydney, New South Wales, Australia
Competition, Animal

JOSEPH M. DITOMASO
University of California, Davis
Herbicides

ANDY DOBSON
Princeton University
Princeton, New Jersey
Rinderpest

IAN C. DUGGAN
University of Waikato
Hamilton, New Zealand
Aquaria

JEFFREY S. DUKES
Purdue University
West Lafayette, Indiana
Climate Change

RICHARD P. DUNCAN
Lincoln University, New Zealand
Propagule Pressure

JOAN G. EHRENFELD
Rutgers University
New Brunswick, New Jersey
Transformers

EMIEL ELFERINK
Centre for Agriculture and Environment
Culemborg, The Netherlands
Ecoterrorism and Biosecurity

KATHARINA A. M. ENGELHARDT
University of Maryland Center for Environmental
Science, Frostburg
Eutrophication, Aquatic

ROBERT E. EPLEE
USDA APHIS PPQ
Whiteville, North Carolina
Early Detection and Rapid Response

FRANZ ESSL
Environment Agency Austria
Vienna, Austria
Bryophytes and Lichens

EDWARD W. EVANS
Utah State University, Logan
Ladybugs

MARA A. EVANS
University of California, Davis
Herbivory

DAVID M. FORSYTH
Department of Sustainability and Environment
Heidelberg, Victoria, Australia
Grazers

J. HOWARD FRANK
University of Florida, Gainesville
Biological Control, of Animals

JASON D. FRIDLEY
Syracuse University, New York
Invasibility, of Communities and Ecosystems

BELLA S. GALIL
National Institute of Oceanography
Haifa, Israel
Mediterranean Sea: Invasions

RACHAEL GALLAGHER
Macquarie University
Sydney, New South Wales, Australia
Vines and Lianas

EMILI GARCÍA-BERTHOU
University of Girona, Spain
Fishes
Rivers

PIERO GENOVESI
Institute for Environmental Protection and Research
(ISPRA)
Ozzano Emilia, Italy
Eradication

FRANCESCA GHERARDI
University of Florence,
Florence, Italy
Crayfish

STEPHAN GOLLASCH
GoConsult
Hamburg, Germany
Canals

DAVID GOULSON
University of Stirling, Scotland, United Kingdom
Bees

CHARLES L. GRIFFITHS
University of Cape Town, South Africa
South Africa: Invasions

EDWIN GROSHOLZ
University of California, Davis
Crabs
Seas and Oceans

JESSICA GUREVITCH
Stony Brook University, New York
Competition, Plant

CARLA J. HARRIS
Macquarie University
Sydney, New South Wales, Australia
Seed Ecology
Vines and Lianas

ALAN HASTINGS
University of California, Davis
Epidemiology and Dispersal

WARREN HAYS
Hawaii Pacific University, Honolulu
Small Indian Mongoose

HENNING S. HEIDE-JØRGENSEN
University of Copenhagen, Denmark
Parasitic Plants

DANIEL A. HERMS
Ohio State University, Columbus
Pesticides for Insect Eradication

GRAHAM J. HICKLING
University of Tennessee, Knoxville
Pathogens, Animal

MARTIN HILL
Rhodes University
Grahamstown, South Africa
Water Hyacinth

RICHARD J. HOBBS
University of Western Australia, Crawley
Disturbance
Land Use

MARK S. HODDLE
University of California, Riverside
Biological Control, of Animals

RICARDO M. HOLDO
University of Florida, Gainesville
Rinderpest

KRISTEN T. HOLECK
Cornell University, Ithaca, New York
Great Lakes: Invasions

JODIE S. HOLT
University of California, Riverside
Weeds

ROBERT D. HOLT
University of Florida, Gainesville
Rinderpest

CAROL C. HORVITZ
University of Miami
Coral Gables, Florida
Demography

CHO-YING HUANG
National Taiwan University
Taipei, Taiwan
Remote Sensing

PHILIP E. HULME
Lincoln University, New Zealand
DAISIE Project

INDERJIT
University of Delhi, India
Allelopathy

SAMUEL W. JAMES
University of Kansas, Lawrence
Earthworms

BENJAMIN R. JANES
University of California, Davis
Tolerance Limits, Plant

VOJTĚCH JAROŠÍK
Charles University
Prague, Czech Republic
CART and Related Methods

PETER T. JENKINS
Defenders of Wildlife
Washington, DC
Pet Trade

JONATHAN M. JESCHKE
Ludwig-Maximilians-University Munich
Planegg-Martinsreid, Germany
Range Modeling

ALBERTO JIMÉNEZ-VALVERDE
Universidad de Málaga, Spain
Tolerance Limits, Animal

LADD ERIK JOHNSON
Université Laval
Québec, Canada
Zebra Mussel

STEPHEN JOHNSON
Industry and Investment New South Wales
Orange, Australia
Lantana camara

MIC JULIEN
Commonwealth Scientific and Industrial Research
Organisation (CSIRO)
Indooroopilly, Queensland, Australia
Water Hyacinth

JITKA KLIMEŠOVÁ
Czech Academy of Sciences, Třeboň
Vegetative Propagation

ROBERT C. KLINGER
U.S. Geological Survey–BRD
Bishop, California
Fire Regimes

ANNA M. KOLTUNOW
CSIRO Plant Industry
Adelaide, South Australia, Australia
Apomixis

FRED KRAUS
Bishop Museum
Honolulu, Hawaii
Reptiles and Amphibians

PHIL LAMBDON
Royal Botanic Gardens
Kew, United Kingdom
Bryophytes and Lichens

LOUIS LAMBRECHTS
Institut Pasteur
Paris, France
Malaria Vectors

DOUGLAS A. LANDIS
Michigan State University, East Lansing
Agriculture

CAROL EUNMI LEE
University of Wisconsin, Madison
Evolution of Invasive Populations

WILLIAM G. LEE
Landcare Research
Dunedin, New Zealand
Islands

MICHELLE R. LEISHMAN
Macquarie University
Sydney, New South Wales, Australia
Seed Ecology

DAVID LE MAITRE
Council for Scientific and Industrial Research (CSIR)
Stellenbosch, South Africa
Hydrology

CHRISTOPHER LEVER
University of Cambridge, United Kingdom
Acclimatization Societies

ANDREW M. LIEBHOLD
USDA Forest Service
Morgantown, West Virginia
Forest Insects
Gypsy Moth

JORGE M. LOBO
National Museum of Natural Sciences
Madrid, Spain
Tolerance Limits, Animal

JULIE L. LOCKWOOD
Rutgers University
New Brunswick, New Jersey
Taxonomic Patterns

W. M. LONSDALE
CSIRO Entomology
Canberra, Australian Capital Territory, Australia
Risk Assessment and Prioritization

LLOYD LOOPE
U.S. Geological Survey
Makawao, Hawaii
Hawaiian Islands: Invasions

L. PHILIP LOUNIBOS
University of Florida, Vero Beach
Disease Vectors, Human
Mosquitoes

TIM LOW
Invasive Species Council
Chapel Hill, Queensland, Australia
Australia: Invasions

HUGH J. MACISAAC
University of Windsor, Ontario, Canada
Lakes

RICHARD N. MACK
Washington State University, Pullman
Cheatgrass

JOHN S. MACKENZIE
Curtin University of Technology
Perth, Western Australia, Australia
Flaviviruses

EARL D. MCCOY
University of South Florida
Biological Control, of Animals

DEBORAH G. MCCULLOUGH
Michigan State University, East Lansing
Forest Insects
Pesticides for Insect Eradication

R. M. MCDOWALL
National Institute of Water and Atmospheric
Research
Christchurch, New Zealand
Dispersal Ability, Animal

ARTHUR C. MEDEIROS
U.S. Geological Survey–BRD
Maui, Hawaii
Melastomes

SCOTT J. MEINERS
Eastern Illinois University, Charleston
Succession

JEAN-YVES MEYER
Government of French Polynesia, Papeete
Melastomes

LAURA A. MEYERSON
University of Rhode Island, Kingston
Databases

MARC L. MILLER
University of Arizona, Tucson
Laws, Federal and State

EDWARD L. MILLS
Cornell University
Bridgeport, New York
Great Lakes: Invasions

CHARLES E. MITCHELL
University of North Carolina, Chapel Hill
Pathogens, Plant

NICOLE MOLINARI
University of California, Santa Barbara
Grasses and Forbs

CAROLINA L. MORALES
Instituto de Investigaciones en Biodiversidad y
Medioambiente, CONICET
Comahue, Argentina
Pollination

PETER B. MOYLE
University of California, Davis
Fishes
Rivers

WOLFGANG NENTWIG
University of Bern, Switzerland
DAISIE Project

ROBERT F. NORRIS
University of California, Davis
Integrated Pest Management

STEPHEN J. NOVAK
Boise State University, Idaho
Geographic Origins and Introduction Dynamics

TAKASHI OKADA
CSIRO Plant Industry
Adelaide, South Australia, Australia
Apomixis

JENNIFER O'LEARY
University of California, Santa Cruz
Nitrogen Enrichment

DAVID A. ORWIG
Harvard Forest, Harvard University
Petersham, Massachusetts
Hemlock Woolly Adelgid

JAE R. PASARI
University of California, Santa Cruz
Nitrogen Enrichment

MICHEL PASCAL
Institut National de la Recherche Agronomique (INRA)
Rennes, France
Rats

ANNE M. PERRAULT
Center for International Environmental Law
Washington, DC
Black, White, and Gray Lists

CHARLES PERRINGS
Arizona State University, Tempe
Invasion Economics

GAD PERRY
Texas Tech University, Lubbock
Brown Treesnake

ANDREA J. PICKART
U.S. Fish and Wildlife Service
Arcata, California
Mechanical Control

STEWARD T. A. PICKETT
Cary Institute of Ecosystem Studies
Millbrook, New York
Succession

MICHAEL J. PITCAIRN
California Department of Food and Agriculture, Sacramento
Biological Control, of Plants

C. W. POTTER
University of Sheffield, United Kingdom
Influenza

RICHARD B. PRIMACK
Boston University, Massachusetts
Endangered and Threatened Species

ANNE PRINGLE
Harvard University
Cambridge, Massachusetts
Mycorrhizae

ROBERT M. PRINGLE
Harvard University
Cambridge, Massachusetts
Nile Perch

PETR PYŠEK
Academy of Sciences
Pruhonice, Czech Republic
DAISIE Project

WOLFGANG RABITSCH
Environment Agency Austria
Vienna, Austria
Bryophytes and Lichens

JOHN RANDALL
The Nature Conservancy
Davis, California
Protected Areas

JAMIE K. REASER
Ecos Systems Institute
Stanardsville, Virginia
Agreements, International

ROBERT N. REED
USGS Fort Collins Science Center
Fort Collins, Colorado
Burmese Python and Other Giant Constrictors

SARAH REICHARD
University of Washington, Seattle
Horticulture

MARCEL REJMÁNEK
University of California, Davis
Eucalypts
Invasiveness

ANTHONY RICCIARDI
McGill University
Montreal, Quebec, Canada
Crustaceans (Other)

JAMES H. RICHARDS
University of California, Davis
Tolerance Limits, Plant

DAVID M. RICHARDSON
Stellenbosch University
Matieland, South Africa
Eucalypts
Forestry and Agroforestry
South Africa: Invasions
Trees and Shrubs

DAVID M. RIZZO
University of California, Davis
Fungi
Phytophthora

VINCENT ROBERT
Institute de Recherche pour le Développement
Montpellier, France
Malaria Vectors

GORDON H. RODDA
USGS Fort Collins Science Center
Fort Collins, Colorado
Brown Treesnake
Burmese Python and Other Giant Constrictors

MEGAN A. RÚA
University of North Carolina, Chapel Hill
Pathogens, Plant

JENNIFER RUESINK
University of Washington, Seattle
Ostriculture

JAMES C. RUSSELL
University of California, Berkeley
Predators

KRISTIN SALTONSTALL
Smithsonian Tropical Research Institute
Balboa, Panama
Genotypes, Invasive

NATHAN J. SANDERS
University of Tennessee, Knoxville
Ants

KRISTINA A. SCHIERENBECK
California State University, Chico
Hybridization and Introgression

EUGENE W. SCHUPP
Utah State University, Logan
Dispersal Ability, Plant

PAUL C. SELMANTS
University of California, Santa Cruz
Nitrogen Enrichment

TAMARA SHIGANOVA
Russian Academy of Sciences, Moscow
Ponto-Caspian: Invasions

DANIEL SIMBERLOFF
University of Tennessee, Knoxville
Carnivores
Elton, Charles S.
Kudzu
"Native Invaders"
Rodents (Other)
SCOPE Project

AJAY SINGH
Gorakhpur University, India
Pesticides (Fish and Mollusc)

JENNIFER E. SMITH
University of California, San Diego
Algae

WILLIAM E. SNYDER
Washington State University, Pullman
Ladybugs

NAVJOT S. SODHI
National University of Singapore
Birds

PETER W. SORENSEN
University of Minnesota, St. Paul
Carp, Common
Sea Lamprey

JOHN J. STACHOWICZ
University of California, Davis
Mutualism

KAREN STAHLHEBER
University of California, Santa Barbara
Grasses and Forbs

SCOTT STEINMAUS
California Polytechnic State University
San Luis Obispo, California
Habitat Compatibility

THOMAS J. STOHLGREN
U.S. Geological Survey
Fort Collins, Colorado
Landscape Patterns of Plant Invasions

SHARON Y. STRAUSS
University of California, Davis
Evolutionary Response, of Natives to Invaders

DONALD R. STRONG, JR.
University of California, Davis
Herbivory

BERND SURES
University of Duisburg-Essen, Germany
Parasites, of Animals

PATRICK C. TOBIN
U.S. Department of Agriculture
Morgantown, West Virginia
Gypsy Moth

DAVID R. TOWNS
New Zealand Department of Conservation, Auckland
New Zealand: Invasions

ALEJANDRO E. J. VALENZUELA
Centro Austral de Investigaciones Científicas,
(CONICET)
Ushuaia, Tierra del Fuego, Argentina
Mammals, Aquatic

WOUTER VAN DER WEIJDEN
Centre for Agriculture and Environment
Culemborg, The Netherlands
Ecoterrorism and Biosecurity

ROY G. VAN DRIESCHE
University of Massachusetts, Amherst
Biological Control, of Animals

KURT J. VAUGHN
University of California, Davis
Restoration

DIEGO P. VÁZQUEZ
Instituto Argentino de Investigaciones del las Zonas Áridas,
CONICET
Mendoza, Argentina
Pollination

ELSE VELLINGA
University of California, Berkeley
Mycorrhizae

MONTSERRAT VILÀ
Estación Biologica de Doñana, Spain
DAISIE Project

THOMAS VIRZI
Rutgers University
New Brunswick, New Jersey
Taxonomic Patterns

BETSY VON HOLLE
University of Central Florida, Orlando
Invasional Meltdown

DAVID A. WARDLE
Swedish University of Agricultural Sciences, Umeå
Belowground Phenomena

RANDY G. WESTBROOKS
U.S. Geological Survey
Whiteville, North Carolina
Early Detection and Rapid Response

OLAF WEYL
South African Institute for Aquatic Biodiversity
Grahamstown
South Africa: Invasions

DESLEY A. WHISSON
Deakin University
Melbourne, Victoria, Australia
Pesticides (Mammal)

DAVID T. WILLIAMS
Curtin University of Technology
Perth, Western Australia, Australia
Flaviviruses

JOHN R. WILSON
South African National Biodiversity Institute
Matieland
South Africa: Invasions

PHYLLIS N. WINDLE
Union of Concerned Scientists
Washington, DC
Regulation (U.S.)

BENJAMIN WOLFE
Harvard University
Cambridge, Massachusetts
Mycorrhizae

RAM PRATAP YADAV
Gorakhpur University, India
Pesticides (Fish and Mollusc)

HILLARY YOUNG
University of California, Santa Cruz
Nitrogen Enrichment

TRUMAN P. YOUNG
University of California, Davis
Restoration

ERIKA S. ZAVALETA
University of California, Santa Cruz
Nitrogen Enrichment

JOY B. ZEDLER
University of Wisconsin, Madison
Wetlands

The *Encyclopedia of Biological Invasions* is a comprehensive and authoritative reference dealing with all the physical and biological aspects of invasive species and invasion biology and theory. The articles are written by researchers and scientific experts and provide a broad overview of the current state of knowledge with respect to the patterns and processes of invasion, the theories associated with invasion, and particular accounts of organisms that have become invasive. Biologists, ecologists, environmental scientists, geographers, botanists, and zoologists have contributed reviews intended for students as well as for the interested general public.

To aid the reader in using this reference, the following summary describes this *Encyclopedia*'s features, reviews its organization and the format of the articles, and is a guide to the many ways to maximize the utility of this *Encyclopedia*.

SUBJECT AREAS

The *Encyclopedia of Biological Invasions* includes 153 topics that review the various ways scholars have studied invasive species. The *Encyclopedia* comprises the following subject areas:

· Invader Attributes
· Ecosystem Features
· Processes
· Impacts
· Notable Taxa
· Pathways to Invasion
· Management and Regulation
· History
· Notable Invasions

ORGANIZATION

Articles are arranged alphabetically by title. An alphabetical table of contents begins on page v, and another table of contents with articles arranged by subject area begins on page ix.

Article titles have been selected to make it easy to locate information about a particular topic. Each title begins with a key word or phrase, sometimes followed by a descriptive term. For example, "Genotypes, Invasive" is the title assigned rather than "Invasive Genotypes," because *genotypes* is the key term and is thus more likely to be sought by readers. Articles that might reasonably appear in different places in the *Encyclopedia* are listed under alternative titles—one title appears as the full entry; the alternative title directs the reader to the full entry. For example, the alternative title "Coccinellidae" refers readers to the entry entitled "Ladybugs."

ARTICLE FORMAT

Because the articles in the *Encyclopedia* are intended for the interested general public, each article begins with an introduction that gives the reader a short definition of the topic and its significance. Here is an example of an introduction from the article "*Phytophthora*":

> *Phytophthora* is a genus of approximately 100 species of fungal-like, plant pathogenic organisms classified in the kingdom Stramenopila. Often known as "water molds," *Phytophthora* species have a swimming spore stage (when they are known as zoospores). *Phytophthora* species are well known as pathogens of agricultural, ornamental, and forest plants. Across the genus, individual *Phytophthora* species may infect roots, stems, leaves, flowers, or fruits of susceptible plants and cause dieback, decline, or death. Some *Phytophthora* species are specialists (infecting only one or a few plant species), and others are generalists (infecting many plant species across several or many plant families). The diversity of *Phytophthora* is quite astounding, and many new species have been described in the past ten years. In addition, *Phytophthora* species have become one of most important causes of emerging diseases in native plant communities.

Within most articles, and especially the longer articles, major headings help the reader identify important subtopics within each article. The article "Melastomes" includes the following headings: "Naturalized and Invasive Melastomes," "Diversity of Life Forms and Habitats," "Reasons for Success," "Impacts and Control Methods," and "Conclusions."

CROSS-REFERENCES

Many of the articles in this *Encyclopedia* concern topics for which articles on related topics are also included. In order to alert readers to these articles of potential interest, cross-references are provided at the conclusion of each article. At the end of "Disturbance," the following text directs readers to other articles that may be of special interest:

> SEE ALSO THE FOLLOWING ARTICLES
>
> Fire Regimes / Invasibility, of Communities and Ecosystems / Land Use / Restoration / Succession / Transformers

Readers will find additional information relating to Disturbance in the articles listed.

BIBLIOGRAPHY

Every article ends with a short list of suggestions for "Further Reading." The sources offer reliable in-depth information and are recommended by the article's author or authors as the best available publications for more lengthy, detailed, or comprehensive coverage of a topic than can be feasibly presented within this *Encyclopedia*. The citations do not represent all of the sources employed by the contributor in preparing the article. Most of the listed citations are to review articles, recent books, or specialized textbooks, except in rare cases of classic, ground-breaking scientific articles or articles dealing with subject matter that is especially new and newsworthy. Thus, the reader interested in delving more deeply into any particular topic may elect to consult these secondary sources. This *Encyclopedia* functions as ingress into a body of research only summarized herein.

GLOSSARY

Almost every topic in the *Encyclopedia* deals with a subject that has specialized scientific vocabulary. An effort was made to avoid the use of scientific jargon, but introducing a topic can be very difficult without using some unfamiliar terminology. Therefore, each contributor was asked to define a selection of terms used commonly in discussion of their topic. All these terms have been collated into a glossary at the back of the volume after the last article. The glossary in this work includes over 600 terms.

APPENDICES

Owing to the great number of currently invasive organisms, it is not possible to include an account of every invasive species in this *Encyclopedia*. Therefore, the editors have included, as an appendix, the IUCN list of the world's worst 100 invasive species. (Note that many other species have also been documented as invasive.) A second appendix lists key references—important book/volume references on biological invasions.

INDEX

The last section of the *Encyclopedia of Biological Invasions* is a subject index consisting of more than 3,800 entries. This index includes subjects dealt with in each article, scientific names, topics mentioned within individual articles, and subjects that might not have warranted a separate, stand-alone article.

ENCYCLOPEDIA WEBSITE

To access the *Encyclopedia of Biological Invasions* website, please visit

> http://www.ucpress.edu/books.php?isbn=9780520264212

This site provides a list of the articles, the contributors, several sample articles, published reviews, and links to a secure website for ordering copies of the *Encyclopedia*. The content of this site will evolve with the addition of new information.

As you read this, thousands of species of plants, animals, fungi, and microbes have been or are being transported by humans to new locations, whether deliberately or inadvertently. This geographic rearrangement of the earth's biota is one of the great global changes now underway. Of course, species have always managed to spread, even without human assistance, but much less often, much more slowly, and not nearly so far. Over the last 150 years, beginning with the advent of steamships and accelerating with air travel, the rate of movement of some organisms has increased many-fold. Any distance can be quickly spanned by a plane; a hitchhiking seed, spore, or insect can be transported from Asia to South America, or from Africa to Australia, in a day. Although many introduced species fail to establish populations or remain restricted to the immediate vicinity of the new sites they land in, others establish populations and invade new habitats, spreading widely and sometimes well beyond the initial point of introduction. Their interactions, both within native communities and with other introduced species, have been noticed by biologists and, increasingly, by governments and the lay public.

Ecologists, evolutionists, economists, geneticists, agronomists, fisheries and forestry scientists, and many others study biological invasions for a variety of reasons. Some invaders are enormously costly, damaging agriculture, forestry, fisheries, and other human endeavors as well as natural areas. Other invaders are pathogens that cause human, animal, or plant disease; yet others are vectors that carry these pathogens. Many invaders depress populations of native species, and even threaten some with extinction, by preying on them, competing with them, hybridizing with them, infecting them with disease, or changing their habitat. The cost of biological invasions to the economy of the United States alone is estimated at over $100 billion annually.

However, trying to understand, estimate, and mitigate the damage caused by some invaders is only one reason for the rapid growth of research on biological invasions. Invasions are, in a sense, unplanned experiments in population, community, and ecosystem ecology; sometimes there are even forms of experimental control and replication—as, for example, when a species is introduced to certain islands in an archipelago but not to others. For legal, ethical, and logistic reasons, most if not all such introductions would be impossible to perform as planned scientific experiments, so it is not surprising that scientists rush to investigate them when such opportunities arise. Initially, most such "purely" scientific research was done by ecologists interested in such questions as what impact a wholly new species has on a native community and ecosystem, and to what extent a similar native species can impede invasion by a newly introduced species. More recently, recognition that an invasion constitutes an experiment in both evolution and ecology has led to an explosion of research on the evolutionary consequences of invasions for both invaders and residents. How, and how quickly, do invaders evolve adaptations, how do native communities adapt to an invader, and how does the small number of invading individuals affect the genetics of the expanding population of invading organisms?

Biological invasions have attracted attention for other reasons. The variety of impacts of introduced species is astounding. Many invasions have such idiosyncratic and bizarre effects that they cannot fail to arouse our curiosity simply as fascinating tales of natural history. For example, who would have predicted that introducing kokanee salmon to Flathead Lake, Montana, and, many years later, opossum shrimp to three nearby lakes would ultimately have led to population crashes of grizzly bears and bald eagles through a complicated chain reaction? Or that introducing myxoma virus to Great Britain to control introduced rabbit populations would have led to the extirpation of the large blue butterfly there? Who would have suggested that introducing a particular grass species would lead to hybridization with a native congener, subsequent polyploidization, and origin of a new vigorous invasive species that would change entire intertidal ecosystems? Teasing apart such intriguing causal chains is a scientific accomplishment of the first order. But the variety and idiosyncrasy of invasion effects also challenges attempts of invasion biologists to produce general laws or rules and to be able to explain why some introductions have no major impacts, yet others lead to huge invasions. Being able to predict which species will fall in the latter category if introduced, and which in the former, is the elusive holy grail of invasion biology.

Choosing topics for this work was difficult; even a large encyclopedia cannot contain every possible topic of interest relating to biological invasions. Had we restricted ourselves simply to species, in addition to the famed "100 of the World's Worst Invasive Alien Species" (see Appendix), we would still have had to consider several hundred other noteworthy invasive species. Many kinds of habitats, and several regions, have been particularly afflicted by invasions. Certain groups of organisms are especially well represented among invasive species. As the science of invasion biology and technologies of managing invasions have matured, myriad subjects can now be treated in some depth. We could not cover all relevant topics, but we attempted to choose those that, in total, would treat the broadest possible range of subjects related to biological invasions. The additional readings suggested with the articles and at the end of this volume will help the reader further pursue individual topics and may provide an entree into topics that are not covered individually.

We are indebted to the 197 authors who contributed articles to this encyclopedia. Experts are always busy people, yet these made time to produce novel syntheses for a publication quite different from the scientific journals for which they usually write. Gail Rice of the University of California Press assisted us enormously in all facets of producing this encyclopedia—helping us locate experts and photographs, editing articles, keeping track of revisions, reminding tardy authors (and editors!), and even providing aesthetic advice on layout and formats. Her organizational and substantive skills were crucial to the project. Chuck Crumly of the University of California Press suggested the need for this encyclopedia, convinced us that we were the right people to edit it, and encouraged us as impediments arose. Finally, our families (Mary, Tander, and Ruth and Eliška, Honza, and Daniel) were enthusiastic, patient, and encouraging even as the encyclopedia, at times, dominated our lives.

Daniel Simberloff
University of Tennessee, Knoxville

Marcel Rejmánek
University of California, Davis

ACCLIMATIZATION SOCIETIES

CHRISTOPHER LEVER

University of Cambridge, United Kingdom

Acclimatization societies were organizations formed by groups of like-minded but otherwise diverse individuals (aristocrats, landowners, biologists, agriculturalists, sportsmen, and others), whose mutual interest was the introduction of exotic animals and plants. These societies were formed to improve domestic stock, to supply additional food, to provide new game animals, to satisfy nostalgic yearnings, to control pests, and (in Russia) to substantiate the claims of evolutionists. They died out due to declining and unscientific membership, apathy by the public and scientific bodies, inadequate funding and dwindling revenues, increasingly strict legislation on the introduction of exotic species, and the growing realization that such introductions were ecologically unsound.

ACCLIMATIZATION SOCIETIES IN FRANCE AND ITS COLONIES

From the 1780s onward, Louis Jean Marie Daubenton (1716–1799) was responsible for the increasing involvement of the *Jardin des Plantes du Roi* in Paris in the study of economic zoology, including the acclimatization of exotic species for commercial and agricultural purposes. In 1793 the Jardin was converted into the *Muséum National d'Histoire Naturelle*, which in turn was succeeded by the *Jardin Zoologique d'Acclimatation*.

In February 1854 a group of savants, under the chairmanship of then director Isidore Geoffroy Sainte-Hilaire (1772–1844), founded La Société Zoologique d'Acclimatation for the introduction, acclimatization, and domestication of animals and for the cultivation of plants. Subsequently, satellite societies were formed in Grenoble, Nancy, Algeria, French Guyana, Guadeloupe, Martinique, and Réunion.

The most important plant introduced by the Société to France was a new variety of potato from Australia, to combat the impact of the same blight, *Phytophthora infestans*, that had caused the potato famine in Britain and Ireland in the 1840s. The most potentially valuable animal importations were Chinese silkworms and various species of fish.

The principal achievements of the Société were the formation of a menagerie in Paris and the development of agricultural crops in metropolitan France and Algeria.

By the late 1860s, membership of the Société and visitors to the menagerie were in decline, and government subsidies and other revenue were dwindling. In 1901 the Société was declared insolvent.

ACCLIMATIZATION SOCIETIES IN BRITAIN

The prime influence behind the acclimatization movement in Britain was Francis Trevelyan Buckland (1826–1880). At the time of his birth, Britain was still suffering from the economic consequences of the Napoleonic Wars of 1792–1815 and the Industrial Revolution. During this period, the corn harvests were exceptionally poor, and the wars hindered the importation of corn from abroad. The population, and the price of food, increased dramatically, and the rising labor pool helped to lower wages.

It was against this background that Buckland (Fig. 1) began to develop his interest in acclimatization. This

FIGURE 1 Frank Buckland physicking a porpoise. "There was only one way; so I braved the cold water and jumped into the tank with the porpoise. I then held him up in my arms (he was very heavy), and, when I had got him in a favourable position, I poured a good dose of sal-volatile and water down his throat with a bottle." From *The Curious World of Frank Buckland* by G.H.O. Burgess, quoting from F.T. Buckland's *Curiosities of Natural History* (4 vols.) 1857–1872. Richard Bentley, London.

concept was not new; a similar policy had been declared by the Zoological Society of London (ZSL), founded in 1826.

The trigger for the founding of the Society for the Acclimatization of Animals, Birds, Fishes, Insects, and Vegetables within the United Kingdom was a dinner held in London in 1860, attended by Buckland and presided over by the distinguished zoologist Sir Richard Owen (1804–1892), at which eland *Taurotragus* (*Tragelaphus*) *oryx* from London Zoo was the principal dish. So impressed by the eland were Owen, Buckland, and the other guests that later that same year, the Society was founded, with Buckland as honorary secretary. Also in the same year (1860), a branch was formed in Scotland (Glasgow), and in 1861 in the Channel Islands (Guernsey).

In 1865, the Society, clearly in financial straits, merged with the Ornithological Society of London. A decline in membership and an apparent lack of interest by the council led in 1868 to the Society's demise.

The principal reasons for the ephemeral life of the Society were its failure to attract enough scientific members, most of whom were drawn from the aristocracy and gentry; its inability to gain adequate government funding; and a lack of facilities for keeping exotic species, most of which were entrusted to the care of individual members. These factors collectively jeopardized the Society's ability to differentiate itself from such competitors as the ZSL. Nor was the Society's progress furthered by the apathy on acclimatization shown by the public and the prestigious British Association for the Advancement of Science. In contrast to the French Société, which examined the commercial and economic benefits of acclimatization to all classes, the Society inclined to the introduction of species to benefit only the upper class. Furthermore, most of the species considered for acclimatization by the Society were wholly unsuitable for that purpose.

ACCLIMATIZATION SOCIETIES IN AUSTRALIA

Acclimatization societies in Australia were formed in 1879 in New South Wales (in Sydney, having evolved from a society founded in 1852); in 1861 in Victoria (Melbourne); in 1862 in South Australia (Adelaide) and Queensland (Brisbane); in 1895 and 1899 in Tasmania (Hobart and Launceston, respectively); in 1896 in Western Australia (Perth); and at various provincial centers.

As in Algeria, the activities of acclimatization societies in Australia reached their zenith during the final days of protectionism, especially in such colonies as Victoria, which possessed the most important such society and which, even in the 1860s, depended on protective tariffs. A close parallel can be detected between the position and status of Victoria in the British Empire and Algeria in its French counterpart. The economy of both colonies was remarkably similar, and favorable tariffs on imports and government grants resulted in increasing interest in acclimatization. Algiers and Melbourne were both centers of rapid demographic growth.

In Australia, the acclimatization movement met with the same apathy as in Britain, based on the belief that the societies were acting in the interest of the privileged minority (Fig. 2).

The societies claimed that their introduction of insectivorous birds increased crop production, while pastoralists claimed that they consumed crops and displaced native birds. Deer provided sport and venison but damaged crops and trees. Eventually, many societies degenerated into importing species solely as curiosities or for ornamental purposes, and some metamorphosed into menageries. Few, if any, attempted to "improve" domestic stock or cultivars on which the prosperity of Australia depended. Those that survive are involved mainly with the introduction of fish.

FIGURE 2 Cartoon from the *Melbourne Punch*, May 26, 1884.

ACCLIMATIZATION SOCIETIES IN NEW ZEALAND

The thirty or so acclimatization societies that were formed in New Zealand between the 1860s (the first in Nelson in 1861) and the early 1900s had, as in the case of many of those elsewhere, two principal objectives: the introduction of game animals for sport and insect-eating birds to control pests.

As in other countries, some of the New Zealand societies eventually failed due to lack of support and falling revenues, coupled with increasing public criticism. Founded and run, as in Britain, by enthusiastic amateurs, they were managed unprofessionally, and they failed to keep adequate records that would have shown their critics the value of revenue derived from visiting sportsmen and the benefit to crops.

After the Second World War, control of the societies by the government increased, and their operations became mainly confined to conservation; the promotion of sport; and, in a complete role reversal, the prevention of further introductions of nonnative species.

The main income of acclimatization societies in New Zealand today comes from the sale of sporting licenses; part of this income is used to acquire wetland habitats, fund research, and educate the public on conservation issues: the balance funds the societies' own conservation programs.

ACCLIMATIZATION SOCIETIES IN RUSSIA

An interest in the acclimatization and domestication of nonnative animals and plants existed in Russia since at least the early 1840s and was led by the distinguished biologist Karl Frantsevich Rul'e (1814–1858).

The primary topic among scientists at the time was the immutability or mutability of species. Rul'e used the transformation of species through acclimatization, domestication, and cultivation to support the theory of mutability (evolution).

Under the leadership of Rul'e, the Imperial Russian Society for the Acclimatization of Animals and Plants was formed in Moscow in January 1864; branches were later established in St Petersburg, Khar'kov, and Orel.

After Rul'e's death, his successors, led by his protégé Anatoli Petrovich Bogdanov (1834–1896), continued his work. As early as 1856, Bogdanov and his colleagues had formulated the idea of establishing a scientifically based zoo in Moscow. Almost from the start, however, dissent broke out between those who gave preference to pure research and those who favored applied research in acclimatization, domestication, and hybridization. This controversy was soon overshadowed by the zoo's financial failure, though it rumbled on well into the twentieth century and had a profound effect on the development of Russian science. Thereafter, the Society began to stagnate, although outside its ranks, the interest in acclimatization actually increased, in particular in the translocation of native fur bearers and the widespread formation of many research *sad* (gardens).

By the early 1900s, it had become accepted within the Society that acclimatization must give way to conservation, and that the introduction of exotics could be actually harmful. By 1930 the Society had ceased to exist.

Although during its 65 years the Society failed to acclimatize (naturalize) any alien species in Russia, it did encourage local attempts in acclimatization of a wide range of, albeit as in Britain, wholly unsuitable, species.

ACCLIMATIZATION SOCIETIES IN THE UNITED STATES

Although since as early as 1846, songbirds, including the house sparrow *Passer domesticus*, were successfully released in the United States, the founding father of the acclimatization movement was a New York pharmacist, Eugene Schiefflin, who, with John Avery, in 1871 founded the American Acclimatization Society, which in 1877 successfully released the first European starlings *Sturnus vulgaris* in Central Park.

In 1873 Andrew Erkenbrecher formed the Cincinnati Society of Acclimatization, which in 1873–1874 unsuccessfully (except in the case of house sparrows) released in the city 21 alien bird species.

At about the same time, the Society for the Acclimatization of Foreign Birds was founded in Cambridge,

Massachusetts. In 1872–1874, it freed large numbers of goldfinches *Carduelis carduelis*, some of which survived until at least the turn of the century.

In Portland, Oregon, in 1880, C. F. Pfluger founded the Society for the Introduction of Useful Songbirds into Oregon (the Portland Songbird Club), which in 1889 and 1892 unsuccessfully (except in the case of the European starling) released 15 species.

These societies spawned several others throughout the United States, of which the Country Club of San Francisco was formed mainly to introduce brown trout *Salmo trutta* to California. It also dispatched chinook salmon *Oncorhynchus tshawytscha* ova to New Zealand, where, under the name quinnat salmon, it became a popular game fish.

In 1884 the Cincinnati Society of Natural History rightly stated that the introduction of alien species was ecologically unsound, a pronouncement that seems to have sounded the death knell for acclimatization in the continental United States.

ACCLIMATIZATION SOCIETIES IN THE HAWAIIAN ISLANDS

Although since 1865 private individuals had released in the Hawaiian Islands (then a territory of the United States) a number of bird species with varying degrees of success, it was not until 1930, under the presidency of Mrs. Frederick J. Lowery, that the Hui Manu (Hawaiian for "bird society") was formed, for the introduction of songbirds to the islands. In the same year, immigrants from Japan founded the Honolulu Mejiro (the national name of the Japanese white-eye *Zosterops japonica*) Society, specifically for the introduction of Japanese songbirds. Among the species successfully freed by these two organizations were northern mockingbirds *Mimus polyglottos*, white-rumped shamas *Copsychus malabaricus*, Japanese bush-warblers *Cettia diphone*, varied tits *Parus varius*, Japanese white-eyes, red-crested cardinals *Paroaria coronata*, northern cardinals *Cardinalis cardinalis*, red avadavats *Amandava amandava*, and black-headed mannikins or munias *Lonchura malacca*.

In 1968, diminishing funds and increasingly strict regulations about importing and releasing alien birds in the islands caused the Hui Manu to disband.

ACCLIMATIZATION SOCIETIES IN GERMANY AND ITALY

In 1858 the *Akklimatisations-verein* was formed in Berlin, and in 1861 the *Società di Acclimazione* in Palermo, Sicily.

SEE ALSO THE FOLLOWING ARTICLES

Australia: Invasions / Birds / Game Animals / Hawaiian Islands: Invasions / New Zealand: Invasions / Xenophobia

FURTHER READING

Jenkins, C. F. H. 1977. *The Noah's Ark Syndrome: One Hundred Years of Acclimatization and Zoo Development in Australia.* Perth: Zoological Gardens Board of Western Australia.

Lever, C. 1992. *They Dined on Eland: The Story of the Acclimatization Societies.* London: Quiller Press.

McDowall, R. M. 1994. *Gamekeepers for the Nation: The Story of New Zealand's Acclimatisation Societies.* Christchurch: Canterbury University Press.

Osborne, M. A. 1993. *The Société Zoologique d'Acclimatation and the New French Empire: Science and Political Economy during the Second Empire and Third Republic.* Bloomington: Indiana University Press.

ADELGID

SEE HEMLOCK WOOLLY ADELGID

AGREEMENTS, INTERNATIONAL

JAMIE K. REASER

Ecos Systems Institute, Stanardsville, Virginia

International agreements are used between or among national governments in order to establish mutual understanding, shared objectives, and, if legally binding, common law. Nearly 50 international agreements address some aspect of invasive species management, although the explicit prevention and control of invasive species is a relatively recent objective. International agreements focused on such issues as trade, agriculture, transportation, and energy have, however, inadvertently forged pathways for the spread of invasive species—likely for thousands of years. There is considerable need to strengthen the capacity of governments to implement international agreements on invasive species, as well as to raise awareness of the invasive species issue within the context of those international agreements that have substantial influence on the pathways of biological invasion.

CHARACTERISTICS OF INTERNATIONAL AGREEMENTS

International agreements take many forms:

- *Bilateral agreements* exist between two governments, while *multilateral agreements* are made by three or more governments.

- *Legally binding agreements* (generally referred to as treaties or conventions) must be observed and met in good faith; in contrast, *nonbinding agreements* (generally called "soft law"—e.g., codes of conduct) provide guidance but are not enforceable. *Protocols* are supplementary, often more specific, guidance within the context of legally binding agreements.
- *Regional agreements* are made among neighboring countries and may include the distant protectorates of those neighboring countries.
- *Nongovernmental organizations* (e.g., the International Union for Conservation of Nature [IUCN]) may also develop guidelines or policy positions to inform negotiating parties.
- *The focus* of international agreements may be relevant to a specific driver of biological invasion (e.g., climate change, trade, agriculture), a region (e.g., country or set of countries), an ecosystem (e.g., wetlands), or a species (e.g., migratory wild animals), or it may broadly encompass multiple dimensions of the issue.

UTILITY OF INTERNATIONAL AGREEMENTS

International trade, travel, and transport greatly facilitate the movement of species around the world. Organisms are also inadvertently relocated as "hitchhikers" through international military activities, famine and disaster relief, development assistance, and financing programs. Once an invasive species becomes established in one country, it can threaten the entire region, as well as every country along the network of intersecting pathways. Thus, no country can effectively prevent biological invasion without engaging in international dialogue and cooperation, as well as helping to raise the capacity of other countries to effectively manage invasion pathways and invasive species within their own borders.

Although aspects of the invasive species issue have been a topic of international agreements since the 1950s, national and international responses to the invasive species problem as a whole are very recent. The Global Invasive Species Programme (GISP) has played a significant role in advancing international agreements on invasive species by producing numerous documents (e.g., *The Global Strategy on Invasive Species*), as well as by hosting national and regional workshops to help governments understand and constructively engage in this technically complex issue.

The effectiveness of international agreements, however, largely depends upon the will and capacity of member governments to enforce their provisions. Despite the recent promulgation of mandates and "soft law" tools

aimed at invasive species through various international agreements, relatively few governments are investing in the development of well-coordinated policies and programs across relevant sectors (e.g., agriculture, environment, trade, transport, defense). Many countries are constrained by lack of implementation capacity (especially financial, technical, and informational). Furthermore, competing political priorities (e.g., trade expansion versus invasion prevention) exist within and among governments and often hamper international negotiations. Governments routinely find themselves challenged by the perceived need to support economic growth while simultaneously protecting their natural environments and domestic industries from potentially harmful imports. For this reason, there is increasing interaction between the World Trade Organization (WTO) and international organizations with an invasive species mandate.

LEGALLY BINDING INTERNATIONAL AGREEMENTS

A multiyear negotiation process is standard for binding treaties, conventions, and associated protocols. The process of agreement is commonly reached through consensus and results in general, broadly interpretable guidance. Separately negotiated, detailed rules can be developed in associated annexes, but this is rare given the length of overall negotiating time required. In most cases, the agreements must be signed and ratified by the cooperating governments in order to bind them to the provisions.

The following are examples of legally binding agreements that have an explicit focus on invasive species:

- *International Plant Protection Convention* (IPPC, 1951 with revisions entering into force in 2005): Applies primarily to pests of plants that occur in international trade ("quarantine pests"). Member countries must implement a series of "phytosanitary measures" to prevent the spread of organisms potentially harmful to plants and plant products. Regional plant protection organizations (e.g., the North American Plant Protection Organization [NAPPO]) exist to facilitate implementation of the IPPC.
- *Convention on Biological Diversity* (CBD, 1993): Article 8(h) calls on member governments to prevent the introduction of, or to control or eradicate, those alien species that threaten ecosystems, habitats, or species. The CBD has negotiated guiding principles and programs of work focused on invasive species, and the invasive species issue is also addressed as a topic under other thematic areas.

- *International Maritime Organization* (IMO): In 2005 the Marine Environmental Protection Committee (MEPC) adopted formal guidelines for the implementation of the 2004 International Convention for the Control and Management of Ship's Ballast Water and Sediments.
- *Convention on the Conservation of European Wildlife and Natural Resources* (Bern Convention, 1979): Requires member governments to strictly control the introduction of nonnative species (Article 11.2.b). This single legal provision has been used to develop a pancontinental strategy (*European Strategy on Invasive Alien Species*), as well as many species-specific recommendations.
- *Convention for the Protection of the Natural Resources and Environment of the South Pacific Region* (SPREP, 1986): Among other things, Article 14 calls for member governments to take all actions necessary to protect rare and endangered species in the convention area, including the regulation of activities (e.g., trade) that could negatively impact them. The *Invasive Species in the Pacific: A Regional Strategy* has been adopted. The *Protocol to the Antarctic Treaty on Environmental Protection* (1991) prohibits the introduction of nonnative species to the Antarctic Treaty Area without a permit.

"SOFT LAW" INSTRUMENTS

Nonbinding agreements can often be reached in a much shorter timeframe than treaties or conventions and do not require a ratification process. "Soft law" resolutions are generally adopted within the context of intergovernmental organizations and may be produced as forward-looking guidelines, codes of conduct, recommendations, programs of work, or declarations of principles. These nonbinding instruments often serve as precedents or complements to binding agreements.

The following are examples of "soft law" approaches that focus on the invasive species issue:

- *Food and Agriculture Organization of the United Nations* (FAO): Addresses invasive species through a variety of economic sectors and cooperates with other international instruments. For example, codes of conduct relevant to fisheries (1995) and biocontrol agents (1995) recommend actions that member governments can take to limit the introduction of harmful nonnative species.
- *International Union for the Conservation of Nature* (IUCN): Adopted *IUCN Guidelines for the*

Prevention of Biodiversity Loss caused by Alien Invasive Species as drafted by its Invasive Species Specialist Group (ISSG). These guidelines had substantial influence on the development of GISP's *Global Strategy* and the CBD's *Guiding Principles* on invasive alien species.
- *International Civil Aviation Organization* (ICAO): Three resolutions (A32-9, A33-18, and A35-19) request the ICAO Council to work with other United Nations organizations to identify and report on approaches it might take to reduce the risk of introducing potential invasive species through civil air transportation and urge member governments to support each other's efforts to reduce the risk of invasive species transport.

LOOKING AHEAD

International bodies are making substantial strides in increasing understanding and synergies among the various international agreements, even across sectors (e.g., cooperative work between the CBD and IPPC). This will help reduce the likelihood of gaps, inconsistencies, and duplication in the future and provide the clarity and consistency needed for effective implementation.

Globalization and large-scale environmental changes require that even more attention be given to the application of international agreements as a fundamental tool in the prevention and control of invasive species. The alternative energy sources being fostered in the context of climate change (e.g., biofuels) are already creating political conflicts under agreements such as the CBD, while the invasive species issue is not yet being adequately acknowledged under international agreements focused on climate or energy.

The decline in the global economy will hinder the ability of many governments to participate effectively in international negotiations and make their implementation of existing agreements even more challenging. International funding agencies (e.g., Global Biodiversity Facility [GBF]) need to support these governments by making invasive species a higher funding priority.

International governing bodies (e.g., the United Nations) are increasingly engaging nongovernmental organizations and the private sector in supportive roles. For example, the CBD recently called on the Pet Industry Joint Advisory Council (PIJAC) to work with GISP to create a toolkit of best management practices for reducing risks associated with the pet trade as a release pathway. If this trend continues, the private sector and nongovernmental organizations will be better poised to help raise the capacity of governments to participate in and enact both legally binding and "soft law" tools aimed at minimizing the impact of invasive species.

SEE ALSO THE FOLLOWING ARTICLES

Black, White, and Gray Lists / Laws, Federal and State / Regulation (U.S.)

FURTHER READING

McNeely, J. 2001. *Global Strategy on Invasive Species*. Gland: Global Invasive Species Programme/IUCN – World Conservation Union.

Reaser, J. K., E. E. Clark, and N. M. Meyers. 2008. All creatures great and minute: A public policy primer for companion animal zoonoses. *Journal of Zoonoses and Human Health* 55: 385–401.

Reaser, J. K., B. B. Yeager, P. R. Phifer, A. K. Hancock, and A. T. Gutierrez. 2004. Environmental diplomacy and the global movement of invasive alien species: A U.S. perspective (362–381). In G. Ruiz and J. Carlton, eds. *Invasive Species: Vectors and Management Strategies*. Washington, DC: Island Press.

Shine, C. 2008. *A Toolkit for Developing Legal and Institutional Frameworks for Invasive Alien Species*. Nairobi: The Global Invasive Species Programme.

Shine, C. 2007. Invasive species in an international context: IPPC, CBD, European Strategy on Invasive Alien Species and other legal instruments. *OEPP/EPPO Bulletin* 37: 103–113.

Shine, C., N. Williams, and L. Gündling. 2000. *A Guide to Designing Legal and Institutional Frameworks on Alien Invasive Species*. Gland: IUCN – The World Conservation Union/The Global Invasive Species Programme.

AGRICULTURE

ADAM S. DAVIS

USDA-ARS Invasive Weed Management Unit, Urbana, Illinois

DOUGLAS A. LANDIS

Michigan State University, East Lansing

Agricultural production of food, feed, fiber, or fuel is a local human activity with global ecological impacts, including the potential to foster invasions. Agriculture plays an unusual role in biological invasions, both because it is a source of nonindigenous invasive species (NIS) and because it is especially susceptible to invasions. A formerly innocuous species may become invasive when its environment no longer constrains its population expansion, either due to geographic dislocation or because of local environmental change. Agriculture is associated with both types of environmental alterations, triggering invasions by NIS and causing other species to become "native invaders."

AGRICULTURE AS A SOURCE OF INVADERS

Intentional cultivation and dispersal of plant and animal species in an agricultural context is an ancient human activity with a modern twist. Historically, the growing of food, fiber, and fuel was fundamentally a site-specific practice; over millennia, crop cultivars and animal breeds were selected for their ability to thrive in certain environments. Long-term cultivation of specific crops in an area results in stable management practices and agroecosystem community composition. In contrast, modern agriculture frequently includes regional and international trade, with traded species changing rapidly in response to varying market demands.

Agriculture and agricultural trade may promote biological invasions in several different ways (Fig. 1). First, intentionally transporting species, allowing them to "sample" different environments, results in high propagule pressure. Trade can introduce commercially valuable species to areas that may benefit economically from their production, yet in a new environment, these species may become invasive. Second, the transfer of a crop species may result in the unintentional transfer of close associates, such as pathogens, weeds, or insect pests of the crop from its source range. Third, the cultivation of species in different environments promotes invasions by reducing stochastic effects on survival normally associated with founder events. Rather than one low-probability chance

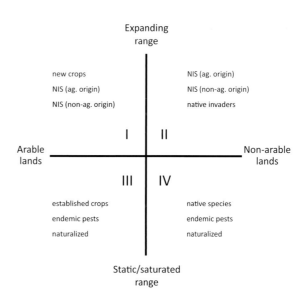

FIGURE 1 The relationship of agricultural systems to the spread of nonindigenous invasive species (NIS) can be defined by (1) the type of land management in the introduced range (arable or nonarable lands), and (2) whether the population is expanding or not at a given place and time, either due to saturation of the new range or due to characteristics of the species or environment that hinder invasion.

of establishing in a new environment, as might happen with unintentional species transfer (often resulting in very high mortality), cultivated populations are maintained under favorable conditions, with multiple chances to disperse beyond cultivated boundaries.

AGRICULTURE AS A SINK FOR INVADERS

Several features of agricultural management make agroecosystems particularly susceptible to biological invasions. First, all agroecosystems are intentionally simplified in time and space: one or two (at most, several) species are reared in a designated area during a defined time period. At its most extreme, this results in continuous monoculture, in which a single crop is grown repeatedly. Historically, cropping systems were more diversified, in the sense of including both plants and animals grown in complementary temporal sequences and featuring spatially heterogeneous arrangements. However, at present, cropping system diversity is low throughout many of the world's productive agricultural regions. Just as increasing ecosystem diversity can provide biotic resistance to invasions, decreasing diversity may increase vulnerability to invasion by pest organisms.

Second, all agroecosystems are subjected to repeated, predictable disturbances in the form of agricultural management, including tillage, planting, soil fertility management, and harvest. These disturbances collectively exclude many organisms, opening up biological space for colonization by those species able to exploit these conditions.

Third, resource availability, especially that of essential nutrients, in agricultural ecosystems is often much greater than in nonarable systems. Many invasive species thrive in nutrient-rich environments, outcompeting natives that are adapted to lower nutrient levels.

Finally, the high levels of connectivity in the agricultural landscape facilitate dispersal of invaders. In the northern corn belt of the United States, is it not uncommon for states to have more than two-thirds of their land area devoted to crop production. In more heterogeneous agricultural landscapes, transmission of invasive organisms from one field to another may be reduced, but the potential for ongoing host–pest cycles between agricultural hosts and overwintering hosts in nonarable lands, as with soybean aphid and buckthorn (see below), may increase.

A BESTIARY OF AGRICULTURAL INVADERS

The difference between an organism termed an "agricultural pest" and one labeled as an "invasive species" is primarily one of geographic origin. Native organisms that reduce agricultural productivity, such as the redwing blackbird (*Agelaius phoeniceus*) in North American maize production, are considered pests. Weeds are also considered pests, although many weeds are not native to the crop production areas they infest and therefore could be considered "invasive." Here, we present examples of a variety of invasive and exotic pest species associated with agricultural systems (Table 1).

Plants

WEEDS Plants that compete with crops or other desired species for light, nutrients, and water are termed weeds. They may be of native origin or may have been brought to the production field as contaminants of soil or seed from an exotic source. Most long-term intensive agricultural systems are characterized by a relatively stable weed flora with several dominant species. Over time, these species disperse throughout the crop production range

TABLE 1

Examples of Invasive Species Associated with Agriculture

Organism	Invader Type[a]	Native Range[b]	Invaded Range
Aphis glycines (soybean aphid)	I	A	soybean production worldwide
Harmonia axyridis (multicolored Asian lady beetle)	I	A	United States, Europe
Phakopsora pachyrhizi (soybean rust)	P	A	soybean production worldwide
Arundo donax (giant reed)	C	A	riparian ecosystems, southwestern United States
Abutilon theophrasti (velvetleaf)	W	A	agronomic crops worldwide
Kochia scoparia (kochia)	C	EA	Australian rangeland (eradicated)
Amaranthus rudis (common waterhemp)	HR	NA	soybean production, north central United States
Sorghum halapense (Johnsongrass)	H, C	A	agronomic crops, United States
Sus scrofa (feral swine)	L	EA	worldwide, many ecosystem types

[a] Invader type abbreviations: C = crop introduction, H = alternate host for pest organism, HR = herbicide resistant biotype, I = insect pest, L = livestock, P = plant pathogen, W = weed

[b] Native range abbreviations: A = Asia, EA = Eurasia, NA = North America

and become so highly represented in the soil seedbank that even exotic weeds can saturate the invaded range. Velvetleaf (*Abutilon theophrasti*) and common lambsquarters (*Chenopodium album*) were both brought to North America in early stages of European colonization. Their range expansion followed the plow, and both species can be found in all U.S. states and Canadian provinces.

CROP ESCAPEES The great majority of exotic invasive plant species were transported to their invaded range intentionally, often as ornamentals or food crops. Johnsongrass (*Sorghum halapense*) was imported to the United States in the early 1800s as a forage crop. Less than a century later, the United States Department of Agriculture organized a control program for this species, now found throughout North America. Ironically, although *S. halapense* produces large amounts of forage, it can be toxic to ruminants in early growth stages, or when trampled or stressed, severely limiting its use as a forage. Giant reed (*Arundo donax*) is another species disseminated for commercial purposes that later escaped cultivation. At the turn of the twentieth century, *A. donax* was brought to the southwestern United States for a variety of purposes, including woodwind instrument reed production and streambank stabilization. The plant proliferated rapidly in riparian areas (Fig. 2), choking out native vegetation and reducing streamflow. More recently, *A. donax* has been proposed for use as a biofuel feedstock, a move likely to result in further expansion of its invaded range.

FIGURE 2 *Arundo donax* (giant reed) is a warm-season perennial grass that spreads primarily via clonal expansion and fragmentation. Native to the Indian subcontinent, it has invaded riparian areas with Mediterranean climates worldwide. It is currently being considered as a biofuel feedstock crop in the southeastern United States. (Photograph courtesy of Dr. John Goolsby, USDA-ARS Beneficial Insects Research Unit, Weslaco, Texas.)

HITCHHIKERS OF GLOBAL TRADE Not all invasive plant species associated with agriculture are disseminated intentionally. Some are dispersed as byproducts of global agricultural trade. Seed sold for use in planting crops is generally certified as being nearly weed-free, but feed grains are not subject to the same level of cleaning or scrutiny. The seeds that move with grains tend to be of a similar size and density, thus avoiding separation through sieving or forced-air cleaning. One plant invader that has gained notoriety recently via this pathway is giant ragweed (*Ambrosia trifida*), a native of North America and a common weed of corn and soybeans. It has recently invaded agricultural lands in both Europe and Asia, bringing with it the allergenic pollen that causes hay fever.

EVOLUTION OF INVASIVE GENOTYPES With the development of herbicide-tolerant crop cultivars at the close of the twentieth century, agricultural impacts on weed communities changed profoundly. Previously, weed management had varied considerably over time, with one technique having been the use of herbicides with different modes of physiological action; now, stresses have become highly predictable and uniform. Unsurprisingly, weeds have adapted to these stresses, with over 180 weed species having become herbicide resistant. Glyphosate-tolerant soybean and corn cultivars dominate crop sequences throughout much of the field crop production area of the Americas. An increasing number of weed species, including marestail (*Conyza canadensis*), common waterhemp (*Amaranthus rudis*), and Palmer amaranth (*Amaranthus palmeri*), all of which were once thought to be immune to the development of herbicide resistance, have now become resistant to glyphosate. These invasive biotypes often start out being highly localized, but seed and pollen dispersal via wind and human activities help them rapidly to invade new territory, leaving behind obsolete herbicide modes of action.

Animals

INSECTS Similar to weeds, insects can be native pests, escapees from cultivation, or exotic invaders. In North America, the Colorado potato beetle, or CPB (*Leptinotarsa decimlineata*), is an example of an endemic insect pest. The CPB is believed native to central Mexico but is now the principal pest of cultivated potatoes throughout the continent. In the past, CPB fed on native Solanaceae, but it switched to cultivated potato when this crop was introduced into the beetles' native range. Following this host switch, CPB expanded its range, eventually cooccurring with potatoes throughout much of the northern hemisphere.

While insects are seldom managed as strictly agricultural species, the gypsy moth (*Lymantrea dispar*) was originally brought to the United States from Europe for experiments in silk production. In its native range, it naturally feeds on a variety of deciduous trees. It escaped and is now a major pest of forests throughout the northeastern United States west to the Great Lakes region.

New exotic insects also continue to colonize agricultural systems. Soybean aphid (*Aphis glycines*) and associated species provide excellent examples of how agricultural- and natural-area invaders can interact. Soybean aphids were first discovered in the United States in the summer of 2000, when they were found feeding on cultivated soybeans in southern Wisconsin. Currently, the aphids occur from the northeastern United States west to the Dakotas and south to Missouri, and they can reach populations of up to 60,000 aphids or more per plant. The rapid expansion of the soybean aphid was fostered by the prior establishment of its primary overwintering host in North America, common buckthorn (*Rhamnus cathartica*). Common buckthorn was originally imported into the United States for use as a hedging plant. It escaped cultivation, colonizing fencerows, woodlots, and natural forests throughout much of the eastern and north central United States. Thus, the widespread cooccurrence of the soybean aphid's overwintering host (common buckthorn) and summer host (cultivated soybean) provided an ideal situation for rapid landscape spread of this exotic insect. Moreover, outbreak populations of soybean aphid favor another exotic, the multicolored Asian lady beetle (*Harmonia axyridis*). The *Harmonia* beetle is a natural predator of the soybean aphid (Fig. 3) and can build to large numbers in aphid-infested soybeans. It is also an intraguild predator, known to eat the eggs and larvae of other lady beetles, and it has been associated with declines in native lady beetle species. Finally, *Harmonia* has the habit of overwintering in large numbers in homes and other buildings, where it can cause human allergies.

VERTEBRATES Rodents, including rats, mice, and rabbits, are serious invasive pests in cropping systems throughout the globe. Rats and mice have been widely transported via human activities, including agricultural trade. These species may cause losses in the growing crop, or more frequently, in stored agricultural products. It is estimated that rats cause losses of $25 billion per year in India alone. Endemic vertebrates are a common cause of crop losses as well. In the United States, whitetail deer (*Odocoileus virginianus*) and Eastern wild turkey (*Meleagris gallopavo*) and geese are common direct pests in agricultural crops. Feeding in agricultural fields subsidizes their populations

FIGURE 3 *Aphis glycines* (soybean aphid) is a phytophagous insect pest of soybean fields that originated in East Asia. In both its native and exotic ranges, it is frequently attacked by *Harmonia axyridis*. (Photograph courtesy of Kurt Stepnitz, Michigan State University.)

and can heighten their impacts on natural ecosystems. For example, snow geese (*Chen caerulescens*) in North America spend the winter feeding in agricultural fields in their overwintering range. Given this highly nutritious food source, they return to their breeding grounds in better condition, where they then produce more offspring. In some areas, they have increased to the point that they are causing ecological damage to the tundra.

Agricultural livestock, including feral swine, goats, and rabbits, cause damage in natural ecosystems. Feral swine (*Sus scrofa*) are generalist feeders and have been reported to cause harm to native plant, insect, amphibian, reptile, and vertebrate communities. The European rabbit (*Oryctolagus cuniculus*) was imported into Australia by European settlers. In the absence of natural predators, populations exploded, causing widespread damage in crop and natural habitats. Extensive feeding on native plants left soil unprotected and exposed to erosion.

Pathogens

Plant pathogens may invade a new geographic region through several pathways, including transportation of infected plant tissue, wind and rain dispersal, and movement of animal vectors. Late blight of potato, caused by the pathogenic fungus *Phytophthora infestans*, requires overwintering potato tubers to survive. When infected tubers are brought to a new location, this devastating

FIGURE 4 *Phakopsora pachyrhizi* (soybean rust) is a soybean pathogen that originated in East Asia and is increasingly common in soybean production worldwide. (Photograph courtesy of Dr. Glen Hartman, USDA-ARS National Soybean Research Laboratory, Urbana, Illinois.)

disease follows, wiping out entire crops. The primary defense against this, and many other plant pathogens, is to breed resistant crop varieties.

Prevailing winds can carry fungal spores hundreds of miles. Soybean rust (*Phakopsora pachyrhizi*; Fig. 4), native to East Asia, was first identified in the U.S. soybean production region in 2004. Soybeans are the main economic species affected by this fungus, but there are many alternate hosts, including another invader, kudzu (*Pueraria montana*), and so soybean rust can overwinter easily and infect new plants. Wind then disperses new spores to uninfected regions. Fortunately, fungicides have proven very effective against this pathogen.

Sometimes, one pest may facilitate another. In the case of maize dwarf mosaic virus (MDMV), three pests interact to complete the disease cycle. Aphids that feed on infected maize can be infected within minutes; they then carry it to other maize plants or *S. halepense*, an alternate host. Because *S. halepense* is a perennial, it provides an excellent overwintering location for MDMV to begin the infection cycle anew the following growing season.

PREVENTING FUTURE AGRICULTURAL INVASIONS

Preventing agricultural invaders is considered the best strategy for reducing their impacts on both agriculture and natural ecosystems. This should involve careful screening of new crop species and varieties for potential invasiveness and to ensure that insects and diseases are not introduced along with crop material. Increased interest in biofuel production means active interest in new crop introductions in many parts of the world. Biofuel crops share many characteristics with invasive plants and should be subjected to particularly stringent review.

There is much scope for managing agricultural systems to reduce the opportunity for invasion, the impacts of invaders, and the invaders that agricultural systems supply to natural areas. These include improved sanitation to prevent transport of propagules from one field to another, improved monitoring to facilitate early detection and eradication or containment of invaders, and increased crop and landscape diversity, which can improve biotic resistance to invaders.

SEE ALSO THE FOLLOWING ARTICLES

Genotypes, Invasive / Herbicides / Integrated Pest Management / "Native Invaders" / Parasitic Plants / Pathogens, Plant / Seed Ecology / Weeds

REFERENCES

Buddenhagen, C. E., C. Chimera, and P. Clifford. 2009. Assessing biofuel crop invasiveness: A case study. *PLoS ONE* 4(4), e5261: 1–6.

Clements, D. R., A. DiTommaso, N. Jordan, B. D. Booth, J. Cardina, D. Doohan, C. L. Mohler, S. D. Murphy, and C. J. Swanton. 2004. Adaptability of plants invading North American cropland. *Agriculture, Ecosystems and Environment* 104: 379–398.

Mack, R. N., and M. Erneberg. 2002. The United States naturalized flora: Largely the product of deliberate introductions. *Annals of the Missouri Botanical Garden* 89: 176–189.

Margosian, M. L., K. A. Garrett, J. M. S. Hutchinson, and K. A. With. 2009. Connectivity of the American agricultural landscape: Assessing the national risk of crop pest and disease spread. *BioScience* 59: 141–151.

Pimentel, D., ed. 2002. *Biological Invasions: Economic and Environmental Costs of Alien Plant, Animal, and Microbe Species.* Boca Raton, FL: CRC Press.

Raghu, S., R. C. Anderson, C. C. Daehler, A. S. Davis, R. N. Wiedenmann, D. Simberloff, and R. N. Mack. 2006. Adding biofuels to the invasive species fire? *Science* 313: 1742.

Rossman, A. Y. 2009. The impact of invasive fungi on agricultural ecosystems in the United States. *Biological Invasions* 11: 97–107.

Seward, N. W., K. C. Ver Cauteren, G. W. Witmer, and R. M. Engeman. 2004. Feral swine impacts on agriculture and the environment. *Sheep and Goat Research Journal* 19: 34–40.

Smith, R. G., B. D. Maxwell, F. D. Menalled, and L. J. Rew. 2006. Lessons from agriculture may improve the management of invasive plants in wildland systems. *Frontiers in Ecology and the Environment* 4: 428–434.

AGROFORESTRY

SEE FORESTRY AND AGROFORESTRY

ALGAE

JENNIFER E. SMITH

University of California, San Diego

Algae are an exceptionally diverse group of generally autotrophic marine, freshwater, and terrestrial organisms that lack true tissues and organs and are thus not typically considered true plants. They comprise a group of eukaryotic uni- or multicellular organisms that possess nuclei and chloroplasts bound with one or more membranes. While cyanobacteria (blue-green algae) were once lumped into the group "algae," they are not eukaryotes and so are not treated as algae here. Small single-celled algae are often described as phytoplankton, while larger multicellular species are described as macrophytes, if growing in freshwater, or seaweeds, if growing in the ocean. In this article "invasive introduced algae" are treated as any species of marine, freshwater, or terrestrial algae that has been introduced into a region where it does not naturally occur, has become highly successful, and is causing ecological or economic harm.

ALGAL INVASIONS

Algae can take on numerous growth forms and can exist as single cells, colonies, or more complex multicellular forms (e.g., filamentous forms, cylindrical forms, sheets, crusts, etc.). These primary producers are common components of most ecosystems around the world and are responsible for producing more than 70 percent of the world's oxygen. Algae can form blooms in their native environments, where they may grow excessively in response to some biotic or abiotic trigger (e.g., nutrient enrichment, pollution), and they may form blooms associated with being introduced into new environments.

While many species of algae have been introduced around the world to new environments, marine algae appear to be the most invasive. Marine algal invasions have occurred and continue to occur around the world as a result of many vectors, including intentional aquaculture introductions as well as accidental introductions associated with shellfish aquaculture and ship traffic. While few introduced algal species become invasive, those that do have been shown to have large negative impacts on associated species and communities. Control or eradication of algal invaders is extremely costly and has been successful in very few cases.

GLOBAL PATTERNS

The majority of documented cases of invasive introduced algae involve marine seaweeds. Out of some 277 species of marine algae that have been introduced around the world, only 13 can be considered invasive. However, many of these species are invasive in more than one locale, region, or ocean basin. Invasive introduced marine algae include members of the Chlorophyta (3), Phaeophyta (3), and Rhodophyta (7), with the number of invaders being proportional to the size of the taxonomic group. While many phytoplankton species can form massive blooms in their native ranges and cause adverse ecological impacts, very few species have been definitively identified as being invasive, probably because of difficulties in taxonomy and incomplete distributional data. Among freshwater taxa, many flowering plants are known to be invasive, but only a few species of algae, specifically some diatoms, have become successful invaders.

Despite the large number of algal species that are considered to be nonnative introductions, very little is known about most species. In fact, the majority of studies that have examined invasive algae have focused on a few high-profile species. Furthermore, many algal invasions are likely to have occurred prior to scientific study or the emergence of the field of "invasion biology." Thus, the results and summaries presented below should be treated with caution, as it is difficult to make generalizations about the effects of invasive introduced algae on invaded communities.

SPECIES OF CONCERN

Among the marine algae or seaweeds, only two species are currently listed on the IUCN Invasive Species Specialist Group's list of the 100 worst invaders in the world (see Appendix); these are the green alga *Caulerpa taxifolia* and the brown alga *Undaria pinnatifida* (Figs. 1A and B, respectively). *Caulerpa taxifolia*, otherwise known as the "killer alga" for its reputation in the Mediterranean Sea, has been introduced to three oceanic basins: the Mediterranean, Australia, and the west coast of North America. In all cases, it has become highly invasive. *Caulerpa* is native to tropical waters around the world, but a cold water–tolerant strain was first introduced to the Mediterranean via accidental release from the Monaco Aquarium in the 1980s. This aquarium strain grows nearly ten times larger than *Caulerpa* in its native habitats and is known for monopolizing soft-bottom and seagrass habitats, where it outcompetes native species and impacts the livelihoods of fishermen. The kelp *Undaria pinnatifida* is native to Asia and has been introduced to Europe, the Mediterranean, Australia, New Zealand, and the west coast of North America, where it has become highly successful in most cases. This species is cultivated in Japan (where it is known as *wakame*) for human consumption, and it was intentionally introduced to parts of Europe. Causes of *Undaria* introductions in other parts of the world remain unknown but are believed to be associated with ship traffic (hull fouling and ballast water). Impacts of *Undaria* on native communities remain undocumented in many cases, but it has been shown to alter community structure and reduce native species diversity.

FIGURE 1 Photographs of some of the most invasive introduced algae from around the world. (A) *Caulerpa taxifolia* (the killer alga) dominating habitat in the Mediterranean; (B) the invasive kelp *Undaria pinnatifida* from Monterey Bay, California; (C) *Kappaphycus* spp. in Kane'ohe Bay, Hawaii; and (D) the invasive freshwater diatom *Didymosphenia geminata* in lakes of the northeastern United States. (Photographs courtesy of the author.)

Numerous other species of marine algae have been considered to be invasive based on studies that have explicitly examined the ecological impacts of the invaders on communities in which they have been introduced. These include (1) the red algae *Gracilaria salicornia*, *Dasya sessilis*, *Acrothamnion preisii*, *Womersleyella setacea*, *Kappaphycus alvarezii* (Fig. 1C), and *Eucheuma denticulatum*; (2) the brown algae *Fucus evanescens* and *Sargassum muticum*; and (3) the green algae *Codium fragile* and *Caulerpa racemosa*.

The freshwater diatom *Didymosphenia geminata*, otherwise known as Didymo or "rock snot," is native to the cool temperate regions of the northern hemisphere (Fig. 1D). However, it was recently found in New Zealand and appears to be spreading out within its native range in North America and Europe; it is now considered to be one of the worst freshwater invasive introduced algal species. Although it is a single-celled alga, it can form large colonies that attach to the bottom of both lakes and streams

where it smothers native biota including fish, plants, and invertebrates. It is described as having an unpleasant appearance and may cause adverse effects to the fishing and tourism industry. Didymo is believed to spread to new locations though human activity, primarily as a result of algal cells hitchhiking on footwear or fishing gear.

VECTORS OF INTRODUCTION

Based on a recent review of seaweed invasions around the world, a number of vectors were identified and ranked according to prevalence. Interestingly, the mode of introduction for the majority of introduced seaweeds has been undocumented. Of those introductions where the vectors have been identified, ship traffic (including both ballast water and hull fouling) was the most significant source of algal introductions. Because algae are photosynthetic organisms and require sunlight to grow, ballast introductions are not as likely as hull fouling to transport algal propagules. A number of algal introductions have also occurred as a result of aquaculture. Several seaweed species are cultivated around the world in open water cultures both for human consumption and for production of colloids (agar and carrageenan), which are used as thickening agents in a number of human products such as toothpaste, shaving cream, hair products, low-fat foods, ice cream, and even beer. In most cases, these seaweed introductions have not resulted in severe invasions. However, in the Hawaiian Islands, a number of seaweeds introduced for experimental aquaculture in the 1970s have become invasive on the coral reefs, where they cause large ecological and economic impacts (Fig. 1C; see section below on impacts of seaweed invasions). A number of other invasive introduced algae arrive in new locations indirectly through shellfish aquaculture. Algae can "hitchhike" on the shells of oysters and other shellfish, and as these species are transported around the world for human consumption, the algae go with them. Algae have also been commonly used as packing material to pad shellfish when they are transported from one location to the next. Other common vectors for the introduction of algae to new locations include fishing gear, SCUBA gear, and clothing, as small algal fragments can become entangled in these items when they move from location to location. Aquarium introductions have also occurred accidentally though discharge pipes or intentionally when people want to free or discard the organisms living in their home aquaria. Lastly, in the past researchers introduced algae to new environments for the purpose of research, but today, intentional introductions are no longer common outside of the aquaculture industry.

IMPACTS OF INVASIVE INTRODUCED ALGAE

Despite the large number of algae (277 species of marine algae and an unknown number of freshwater algae) that are known to have been introduced to new locations around the world either intentionally or accidentally, very little is known about the impacts these species are having since introduction. A recent review of the scientific literature revealed a total of 68 published studies that have explicitly examined the ecological impacts of seaweed invaders on the invaded communities. A variety of response variables have been used to assess impacts, including changes in the abundance and diversity of the native biota and changes in productivity, community structure, community function, and feeding and performance of native species. The majority of these studies reported that the seaweed invaders caused negative impacts or changed community structure in the invaded communities. However, in some

cases, seaweed invaders have been shown to increase the abundance, diversity, and feeding rates of native species. It is important to note here that although the majority of studies have documented negative impacts of invaders on the native biota, only a very small number of seaweed invaders have been studied in any detail. Of the 277 species of introduced marine algae, only a small number of these species are truly considered invasive, but those that are invasive have had large negative impacts. *Caulerpa taxifolia* is the best studied of the invasive seaweeds, and it has been shown to negatively affect performance, feeding, diversity, community function, and abundance of native species. Other species that have been shown to have negative effects on invaded communities in at least ten different cases include *Codium fragile* ssp. *tomentosoides, Sargassum muticum, Caulerpa racemosa,* and *Euchema denticulatum.* In the case of freshwater algal invasions, few studies have quantified impacts, but anecdotal evidence suggests that when the diatom Didymo blooms, it may reduce oxygen levels in water bodies at night while the algae are respiring and cause hypoxia and subsequent harm to native species. Didymo blooms also reduce light levels and may negatively affect other plant or algae species growing nearby. Didymo blooms, along with many other algal blooms, appear unpleasant to tourists and may negatively impact recreational activities.

At least one study has documented the economic costs associated with an invasive introduced alga. The red seaweed *Hypnea musciformis* was first introduced to the island of Oahu in the Hawaiian Islands in the 1970s for experimental aquaculture. Over the next several decades, this species spread to a number of other islands in the main Hawaiian Islands, and it began forming extensive, large blooms on the island of Maui. The blooms, which are believed to be associated with nitrogen and phosphorus pollution occurring in the nearshore environment, result in large amounts of rotting algal biomass that accumulates on the beaches and creates an unpleasant environment and a very foul odor. In a formal economic analysis, Van Beukering and Caesar (2004) determined that the city and county of Maui were losing an estimated $20 million per year in costs associated with beach cleanups, reduced property values, and reduced occupancy rates in hotels and condominiums in bloom-affected areas. The authors concluded that the costs associated with preventing the introduction of this invasive alga would likely have been far less than the costs associated with managing the current problem. Several other studies have anecdotally noted economic impacts of seaweed invaders as a result of colonization and fouling of aquaculture facilities, but no quantitative data are currently available.

MANAGEMENT AND ERADICATION

Invasive introduced algae can be managed in a number of ways and during a number of different stages (i.e., transport, establishment, and spread) of the invasion process. By far, the best form of managing invasive introduced algae is to prevent introductions from occurring altogether. This is especially true for algal species that have proven to be invasive in other locations around the world or for species that are closely related to known invasive species. Taxonomy has in some cases been used to predict species that are more likely to become invasive than others, and taxonomic risk assessments are often used for other taxa (e.g., flowering plants) as a screening mechanism to prevent introductions. Species traits have also been used to predict invasive algal species for 113 species of macroalgae in Europe. Trait categories were grouped according to dispersal ability, establishment, and ecological impact, and introduced species were compared with native species to determine whether some traits were good predictors of invasiveness. Certain characteristics, such as an alga's ability to disperse through vegetative propagation, were common among many invasive introduced algal species. Other mechanisms that can help to prevent the introduction of algae into new locations include strict guidelines for the shipping industry to manage both ballast water and hull fouling introductions; these regulations will of course need to operate at an international level. Aquaculture activities should be strictly regulated to prevent algae from (1) "hitchhiking" on the shells of invertebrates, and (2) escaping from farm plots in open water systems. Finally, educational programs can help to inform people of the dangers of introducing nonnative species and can help to prevent intentional dumping of aquarium organisms into the natural environment.

Once an alga has been transported to a new location, it may either persist and establish or die. The "tens rule" has been used to describe the invasion process, whereby only one in ten species introduced to an area typically successfully grows, and of these succeeding species, usually only 10 percent become prolific and spread. If an introduction is detected early enough, then a rapid response activity can result in complete eradication. After the killer alga *Caulerpa taxifolia* was found in waters near San Diego, California, a rapid response was initiated, and using a combination of reduced light and chlorine, the Southern California *Caulerpa* Action Team was the first to eradicate a marine alga. The eradication was declared in 2006, some five years after the infestations were initially detected, and the estimated cost upon completion was approximately $8.3 million. The success of the *Caulerpa* story in San Diego was likely due to the quick response and the dedicated effort by a number of individuals representing state and federal agencies as well as consulting firms and nongovernmental organizations. A number of eradication attempts have been made with other species of algae, but none has been successful.

In most cases, by the time an introduced alga has been detected and declared invasive, it is too late for successful eradication. In these cases, control is really the only management option. A variety of control strategies have been used and include simply removing the plants manually, by hand or underwater suction device (see information on the "super sucker": www.nature.org/wherewework/northamerica/states/hawaii/press/press2376.html); controlling them with chemicals, such as algicides or bleach; changing water temperatures; reducing the light; and even employing biological control methods by using native or nonnative herbivores. For many of the most invasive algal species, control strategies are used either to reduce the impacts of the invaders on the native communities or to prevent further spread of algal invaders into new areas.

SEE ALSO THE FOLLOWING ARTICLES

Aquaculture / Aquaria / Ballast / Eutrophication, Aquatic / Freshwater Plants and Seaweeds / Invasion Economics / Seas and Oceans / Taxonomic Patterns

FURTHER READING

Beukering, P. J. H. Van, and H. S. J. Cesar. 2004. Ecological economic modeling of coral reefs: Evaluating tourist overuse at Hanauma Bay and algae blooms at the Kihei coast, Hawai'i. *Pacific Science* 58: 243–251.
Hewitt, C. L., M. L. Campbell, and B. Schaffelke. 2007. Introductions of seaweeds: Accidental transfer pathways and mechanisms. *Botanica Marina* 50: 326–337.
Inderjit, D. Chapman, M. Ranelletti, and S. Kaushik. 2006. Invasive marine algae: An ecological perspective. *Botanical Review* 72: 153–178.
Johnson, C. R., and A. R. O. Chapman. 2007. Seaweed invasions: Introduction and scope. *Botanica Marina* 50: 321–325.
Nyberg, C. D., and I. Wallentinus. 2005. Can species traits be used to predict marine macroalgal introductions? *Biological Invasions* 7: 265–279.
Ribera Siguan, M. A. 2003. Pathways of biological invasions of marine plants (183–226). In G. M. Ruiz and J. T. Carlton, eds. *Invasive Species: Vectors and Management Strategies*. Washington, DC: Island Press.
Schaffelke, B., and C. L. Hewitt. 2007. Impacts of introduced seaweeds. *Botanica Marina* 50: 397–417.
Schaffelke, B., J. E. Smith, and C. L. Hewitt. 2006. Introduced macroalgae—a growing concern. *Journal of Applied Phycology* 18: 529–541.
Williams, S. L., and J. E. Smith. 2007. A global review of the distribution, taxonomy, and impacts of introduced seaweeds. *Annual Review of Ecology, Evolution and Systematics* 38: 327–359.

ALLELOPATHY

INDERJIT

University of Delhi, India

In the year 1937, Hans Molísch coined the term "allelopathy" to describe the effect of ethylene on fruit ripening. Elroy L. Rice (1974) defined allelopathy as the effect of one plant (including microorganisms) on another plant through the release of chemical compounds into the environment. Because Rice's definition includes both positive and negative effects and also includes the influences of soil microorganisms, it is considered too broad in its scope by many researchers. Limiting its meaning to the suppression of seed germination and growth of one plant species by another through the release of chemical compounds is more acceptable to most ecologists.

MAJOR SHORTCOMINGS OF ALLELOPATHY RESEARCH

During the 1960s through the 1980s, Elroy Rice, Cornelius Muller, and A. M. Grodzinsky investigated allelopathy in many natural and agricultural communities, proposing that allelopathy is a major factor in community organization. Their work, however, was severely criticized by John Harper on methodological grounds, and much work from this era (1960–1980) fell into disrepute. From 1980 onward, numerous studies have shed light on the complexities of identifying allelopathy as a mechanism for plant growth inhibition. These include investigations on bracken fern, crowberry, *Sorghum bicolor, Pluchea lanceolata,* the Florida scrub, and more recently invasive species such as *Centaurea maculosa* and *Ageratina adenophora* (Fig. 1). Does any of this work provide conclusive evidence that release of chemicals influences the growth and establishment of neighbors? Probably not. Are there still questions to be answered? Yes. But allelopathy as a science has certainly progressed far beyond the days of Harper's criticisms.

Many allelopathy studies are limited by a focus on in vitro circumstances, which do not explain field patterns. Inadequate methodology remains a major concern in allelopathy research, including (1) use of artificially sensitive species and unnatural growth medium in laboratory bioassays, (2) neglect of ecological factors, and (3) lack of appreciation of field conditions in the laboratory or greenhouse bioassays. Allelopathic effects are often modified by additional biotic and abiotic stress factors, regional

FIGURE 1 *Ageratina adenophora,* a native of central Mexico, is expanding its range aggressively in the foothills of the northwestern Himalayas, and forms thick monocultures. *Ageratina* produces millions of viable seeds and has tremendous potential for vegetative reproduction. (Photograph courtesy of the author.)

agroecological practices, uncertain climatic events, and physical, chemical, and biological edaphic factors, all of which can influence the residence time and persistence, concentration, and fate of allelopathic compounds in the environment. There is a need to generate ecologically relevant data on the residence time and fate of allelochemicals in soil. However, this has proven difficult in a system as dynamic as soil, in which several chemicals may be present in minute and constantly fluctuating quantities. Recently, there has been progress in quantifying chemicals from the intact rhizosphere. Jeff Weidenhamer and his colleagues at Ashland University have developed soil probes to quantify the dynamics of nonpolar chemicals in soil. Such efforts would certainly further our understanding on the ecological roles of chemicals in the soil environment.

MAJOR CHALLENGE

The major challenge in allelopathy research is to establish that the release of chemicals does influence the growth and establishment of neighboring plants. Ian Baldwin at Max Planck Institute for Chemical Ecology, Jena, Germany, suggested that the use of silenced plants that have been altered for their ability to synthesize and release or perceive specific compounds could generate conclusive evidence for allelopathy. Using *ir-aco* mutants of *Nicotiana attenuatta* silenced for its ability to produce ethylene, Inderjit and coworkers showed that the release of ethylene from *Nicotiana attenuata* seedlings does influence neighboring siblings. The use of silenced plants to establish the allelopathic potential of a chemical would be an indispensable tool for putative allelopathic crops such as *Sorghum bicolor*.

ALLELOPATHY AND PLANT INVASIONS

Allelopathy has been investigated as a probable cause of invasion success of nonnatives. However, the traditional approach that examines nonnatives in the same way as other native plants also suspected of allelopathic activities was taken. Ragan Callaway and colleagues at the University of Montana compared allelopathic effects of exotic invasives on species from their native and invaded communities, which provides stronger evidence than the traditional approach for whether or not allelopathy actually contributes to invasive success. The term "novel weapons hypothesis" was proposed to describe the general effects of *Centaurea* species, and chemicals contained in their root exudates, which were more effective against species in invaded regions than against related species in native regions. *Centaurea diffusa* and *C. maculosa* roots exude 8-hydroxyquinoline and (±)-catechin, respectively, in both their native lands in Eurasia and in their naturalized ranges, but in some experiments, these chemicals inhibited North American species more than Eurasian species. Some of this work, however, was criticized due to inadequate methodology for soil extraction and conflicting data on recovery of catechin from *C. maculosa* soils. This may be the result of climatic and edaphic variation that creates a great deal of conditionality, which influences the exudation and availability of chemicals in soil.

CONCLUSION

A better knowledge of the biochemical pathways (enzymes and genes) involved in the production of putative allelochemicals, their accumulation and transport to the soil, and the potential in vivo interactions of these compounds will provide the physiological basis for improved understanding of the role of allelopathy in both agricultural and natural ecosystems. Allelopathy is a conditional and species-specific effect, and the term "allelopathy" may be misleading unless this conditionality is addressed. In addition to investigating the roles of allelochemicals at population level, there is a need to explore temporary versus long-term ecological changes caused by allelochemicals, and to define changes at population and community levels. Allelopathy needs to be conceptualized and investigated in terms of soil chemical ecology, which can further allelopathy research and reduce some of the less fruitful controversy surrounding this science.

SEE ALSO THE FOLLOWING ARTICLES

Competition, Plant / Novel Weapons Hypothesis / Seed Ecology / Vegetative Propagation

FURTHER READING

Callaway, R. M., and W. M. Ridenour. 2004. Novel weapons: Invasive success and the evolution of increased competitive ability. *Frontiers in Ecology and the Environment* 2: 436–443.
Inderjit. 1998. Influence of *Pluchea lanceolata* (Asteraceae) on selected soil properties. *American Journal of Botany* 85: 64–69.
Inderjit, and R. M. Callaway. 2003. Experimental designs for the study of allelopathy. *Plant and Soil* 256: 1–11.
Inderjit, R. M. Callaway, and J. M. Vivanco. 2006. Plant biochemistry helps to understand invasion ecology. *Trends in Plant Science* 11: 574–580.
Inderjit, C. C. von Dahl, and I. T. Baldwin. 2009. Use of silenced plants in allelopathy bioassays: A novel approach. *Planta* 229: 569–575.
Inderjit, and J. Weiner. 2001. Plant allelochemical interference or soil chemical ecology? *Perspectives in Plant Ecology, Evolution and Systematics* 4: 3–12.
Inderjit, K. M. M. Dakshini, and C. L. Foy, eds. 1999. *Principles and Practices in Plant Ecology.* Boca Raton, FL: CRC Press.
Rice, E. L. 1974. *Allelopathy.* Orlando, FL: Academic Press.
Weidenhamer, J. D., P. D. Boes, and D. S. Wilcox. 2009. Solid-phase root zone extraction (SPRE): A new methodology for measurement of allelochemical dynamics in soil. *Plant and Soil* 322: 177–186.

AMPHIBIANS

SEE REPTILES AND AMPHIBIANS

ANTS

NATHAN J. SANDERS

University of Tennessee, Knoxville

Five of the world's 100 worst invasive species (see Appendix) are ants, although ants make up only a minuscule fraction of the Earth's biodiversity. There are over 12,000 described ant species, yet in some places, it is possible to spend entire days (if not entire field seasons) studying the ant fauna without seeing any native ants. Almost without fail, when populations of invasive ant species become established, they dramatically affect populations of native ants, other arthropods, and in some cases populations of vertebrates and plant communities.

INVASIVE ANT SPECIES

Approximately 150 ant species have become established outside of their native ranges, but not all of them have become invasive—that is, have established in and disrupted native ecosystems outside of their native range.

The five most widespread and problematic invasive ant species are *Anoplolepis gracilipes* (yellow crazy ant), *Linepithema humile* (Argentine ant), *Pheidole megacephala* (big-headed ant), *Solenopsis invicta* (red imported fire ant), and *Wasmannia auropunctata* (little fire ant).

Though these species are globally widespread and recognized as pests, tremendous variation exists in how much is known about each. There are over 1,300 publications about *S. invicta*, 467 about *L. humile*, 116 about *P. megacephala*, and 85 about *W. auropunctata*, but only 33 about *A. gracilipes*. This article will, by necessity, focus on species for which adequate information is available. But that is not to deny that many other potentially problematic species might threaten native communities and ecosystems, or at least become pests, in their introduced ranges.

NATIVE AND INTRODUCED RANGES

Introduced invasive ants originate in both the Old and New Worlds. With few exceptions (e.g., *Myrmica rubra*), most introduced invasive species have originated in the southern hemisphere. Of the five most widespread species, three are South American: *Linepithema humile*, *Solenopsis invicta*, and *Wasmannia auropunctata*. While the native ranges of *L. humile* and *S. invicta* include Argentina, Uruguay, and Paraguay, *W. auropunctata* occurs from Argentina north to Central America, and the origins of invasive populations are still unclear. *Pheidole megacephala* and *Anoplolepis gracilipes* originated in Africa (or Asia for *A. gracilipes*).

These species now occur on most continents and many isolated oceanic islands. The introduced range of *A. gracilipes* includes southern Africa, southern Asia, the Caribbean, and several islands in the Indian and Pacific oceans. *Linepithema humile* has become established in mild Mediterranean climates globally (e.g., California, the Mediterranean coast of Europe, coastal Australia, parts of New Zealand). *Solenopsis invicta* became established in the southern United States in the 1930s and has since spread as far west as Texas and California and as far north as Maryland and Delaware in the United States. It has also been detected in China, New Zealand, Hawaii, Australia, Malaysia, and Taiwan. *Pheidole megacephala* is from southern Africa, and it now occurs in many of the world's temperate and tropical regions. *Wasmannia auropunctata* has been introduced to parts of Africa, North and South America, and Israel, as well as to islands in the Caribbean and the Pacific Ocean (New Caeldonia, Vanuatu, Tahiti, Galapagos, Hawaii, and the Solomon Islands).

Most research on invasive ant species has occurred in the introduced ranges of the species (with one exception; more is known about *Myrmica rubra* in its native range than in its introduced range). This paucity of research is unfortunate because studying species in their native ranges might reveal important differences in the behavior, life histories, population biology, and ecology of invasive species that could help explain the success of invasive ant species in their introduced ranges. Of course, the native ranges of some invasive species remain unknown, and it is often difficult to do fieldwork in remote locations. However, several recent studies, especially with *L. humile*, have shown that the population genetic structure, competitive interactions with the rest of the ant community, and trophic ecology can differ dramatically between native and introduced ranges. These differences have illuminated, at least for this species, some of the mechanisms that might account for its success as an invasive species.

Modeling the Potential Introduced Range of Invasive Species from the Native Range

Environmental niche models aim to predict the potential introduced ranges of invasive ant species based on information about environmental conditions in their native ranges. The general approach of environmental niche modeling is to obtain information on climatic conditions in the native range of the species and then to find locations with the same, or at least similar, conditions in other parts of the globe. Places outside the native ranges with similar environmental conditions to those in the native range are considered susceptible to invasion. However, this approach generally ignores three important facts. First, both biotic and abiotic factors limit the distributions of ant species, and most environmental niche models are not equipped to include the influence of factors such as competition or disturbance. Second, in order to model the potential range of an introduced species, the species must be at equilibrium with climate in its native range. That is, it has to occur everywhere it possibly can, at least as determined by climatic conditions, and it has to have been sampled well enough in its native range to know where it occurs and where it is absent. And third, species evolve and undergo niche shifts. If either evolution or niche shifts occur, perhaps because of genetic bottlenecks common to many invasions, then the environmental niche models will fail to predict accurately the potential ranges of introduced ant species.

Several recent studies have modeled the potential invasive range of *Solenopsis invicta* in North America based on its apparent distribution in its native range. However, the models frequently fail. That is, in some cases, the predicted invasive range is substantially smaller than the actual range of *S. invicta*. But in other cases, the models undoubtedly overestimate the potential range of *S. invicta* in North America. For example, it is highly unlikely that *S. invicta* will become established in Canada or points northward, though some models have predicted such a scenario.

Species Traits That Might Correlate with Success

One early goal of invasion biology was to identify traits of particular species that might promote their success as invaders. Most investigators would argue that searching for traits alone as a key to invasion success oversimplifies the issue—that the interplay between the traits of the invader and the characteristics of the native community and environment, along with the opportunity to invade, are the key determinants of invasion success. The same is likely true for invasive ants, but are any life history traits shared among the most successful invasive ant species?

TROPHIC ECOLOGY The five worst invasive ant species are all omnivores. They prey on small arthropods, scavenge dead animals, and tend hemipterans for their carbohydrate-rich secretions. But many, if not most, ant species are omnivorous, so omnivory alone cannot account for the success of invasive species. There is some evidence that the ability of some invasive species to exploit hemipterans more efficiently than do native ant species might fuel their success.

Stable isotope analyses have proven useful in characterizing diets of many native and nonnative ant species. Although more studies are needed, recent evidence suggests that the diet of invasive ant species, at least for *Linepithema humile*, may be more flexible than that of native ant species. Such dietary flexibility could promote invasion success by allowing invasive ants to exploit a wide variety of food resources in order to use the most abundant or nutrient-rich resources.

BODY SIZE Body size is an important life history trait for most species. In ants, body size spans several orders of magnitude, both among species and even within the same colony of a single species. Edward O. Wilson was among the first to argue that within-colony variation in body size, or polymorphism, might allow workers to specialize and thereby increase colony efficiency. Is there evidence that the most successful invasive species are polymorphic? Some are (*Solenopsis invicta*, *S. geminata*, and *Pheidole megacephala*), but others (*Linepithema humile*, *Anoplolepis gracilipes*, *Wasmannia auropunctata*) are not. Similarly, many noninvasive species exhibit extreme polymorphism, so polymorphism alone cannot account for invasion success.

It has been argued that successful invasive species are smaller than native species in the communities in which they invade, or are smaller than congeners that are not invasive. Indeed, some invasive ant species are small relative to noninvasive ants and to native species in the communities they invade. But many are not. *Anoplolepis gracillipes*, for example, can be over 5 mm in length. Additionally, previous global analyses of body size of invasive ants did not consider polymorphic species (e.g., *Pheidole megacephala*, *Solenopsis invicta*, *Solenopsis geminata*). Whether body size is an important predictor of invasion success remains an open question.

COMPETITIVE ABILITY Some ant species are notoriously good at quickly discovering food resources, but those species are often quickly displaced by more aggressive species. Other species are good at displacing competitors from resources but slow to discover resources. Thus, a tradeoff exists: some species are good at discovering resources, but others are good at displacing their competitors from resources, and few species excel at both. These dominance–discovery tradeoffs might promote coexistence in native ant communities. But the ability to break the tradeoff—to be able both to discover resources faster than your competitors and to displace competitors once they do discover the resources—could be an important aspect of invasion success for some invasive species. In fact, many invasive ant species do seem to break this tradeoff. When an ephemeral resource becomes available, invasive ant species are likely to discover it before any native ant species does, to recruit nestmates to the resource in higher numbers than native species, and to initiate aggressive interactions before native ant species do. The ability to excel at both exploitative and interference competition undoubtedly confers a competitive advantage on invasive ant species.

It might be that invasive ants not only engage in competition with native species but also prey on them. Distinguishing between competition and predation among ants is challenging. For example, when a group of *L. humile* workers is seen dismembering a worker of a native ant species, is that competition or predation? However, there are some clear-cut instances of predation. For instance, both *Linepithema humile* and *Solenopsis invicta* are known to raid nests of native ant species, where they aggressively attack and kill workers and remove brood (Fig. 1). In actuality, some combination of predation and superior competitive ability probably accounts for the success of invasive ant species. It would be interesting to know more about the relative effects of exploitative competition,

FIGURE 1 *Linepithema humile* workers attacking a *Pogonomyrmex subdentatus* worker near Davis, California. (Photograph courtesy of Alex Wild.)

interference competition, and predation by invasive ants on native ant biodiversity.

COLONY SIZE One reason some invasive ants may be able to break the dominance–discovery tradeoff and successfully engage in colony raids of other species is that they tend to have large colonies. Because colonies are large, invasive ant species can invest in scouts (workers that search for food resources), recruit in high numbers, and overwhelm native species in interference interactions for food resources.

But this begs a question: how do colonies of invasive species become large when they presumably begin with, at most, a few queens and workers? One intriguing potential mechanism is that colonies of invasive species may become large because they are unicolonial. Unicoloniality is the condition in which colonies consist of connected networks of nests and exchange workers freely among nests. Because nests exchange workers and presumably do not compete with one another, colonies are free to invest all of their energy in interspecific competition rather than intraspecific competition. Unicoloniality can confer ecological dominance by enhancing colonization ability or resource exploitation. It can also shift the outcome of interference interactions. Indeed, unicolonial colony structures have been documented or inferred for *Anoplolepis gracilipes*, *Linepithema humile*, *Pheidole megacephala*, *Solenopsis invicta*, and *Wasmannia auropunctata*.

Most research on unicoloniality in invasive species has been on the Argentine ant, *Linepithema humile*. Early work suggested that the Argentine ant perhaps passed through a genetic bottleneck upon introduction in California, and so as the species expanded its range, nests over large areas remained highly related, and intraspecific aggression was absent over large areas, resulting in the formation of expansive "supercolonies." As a result of

a lack of intraspecific aggression at the population level, Argentine ants are free to focus entirely on interspecific competition. In contrast, the biology of the Argentine ant appears different in the native range; colonies are often aggressive toward one another within populations (e.g., scales of tens to hundreds of meters), which could explain why the Argentine ant is not as ecologically dominant in its native range. More recent work, however, has documented variation in the scale at which intraspecific aggression occurs in Argentine ants in both their introduced and native ranges, so while unicoloniality is the norm in introduced populations, expansive supercolonies extending hundreds of meters still occur in native populations. Research on other unicolonial invasive ant species has shown that most exhibit variation in the degree of intraspecific aggression, particularly between native and introduced ranges, and that introduced populations may often result from multiple introductions from the native range. Together, detailed studies of Argentine ants in their native and introduced ranges and a handful of studies on other invasive species suggest that unicoloniality alone is an insufficient explanation for ecological dominance by invasive species.

It is important to note that unicoloniality is also common for many noninvasive species across the ant phylogeny. A key distinction is that many unicolonial noninvasive species form large, long-lasting nests in mature habitats, whereas unicolonial invasive species form small, transient nests in disturbed habitats. Perhaps it is this distinction between the nature of nests that allows invasive species to be transported, to become established, and to increase when rare.

What else might account for the success of invasive ant species? As with other invasive taxa, studies have indicated that escape from natural enemies (e.g., interspecific competitors, disease, and parasitoids) may account for some of the success of invasive species outside of their native ranges. It is most likely that a combination of rapid colony development, coupled with the ability to maintain small, ephemeral networks of nests and with the benefits that large colony size confers (e.g., efficient location and retrieval of food resources), has led to the broad distributions of several introduced invasive species, as well as their impacts on biodiversity and native ecosystems.

IMPACTS

The effect of invasive exotic ant species on populations and communities of native ants and arthropods is not a new problem. In fact, effects of invasive ants were documented as early as the early sixteenth century in the West Indies.

E. O. Wilson surmised that *Solenopsis geminata* destroyed crops and invaded homes on the island of Hispaniola in the early 1500s and that *Pheidole megacephala* ravaged sugarcane fields throughout the Lesser Antilles the late 1700s. These two species are still pests throughout many parts of the world.

Impacts on Ants

A large and growing body of literature indicates the dramatic effects that invasive ant species can have on native ant species. In invaded areas, both richness and abundance of native ants is substantially lower than in the absence of invasive ants. Examples abound on nearly every continent of the diversity and abundance of native ants and invertebrates being lower in the presence of *Linepithema humile* than in its absence. Similarly, the effects of *Solenopsis invicta* on native invertebrates, especially in the southern United States, are well documented. *Anoplolepis gracilipes* greatly reduces the abundance of ants and other invertebrate species on the islands in the South Pacific and in Australia. *Pheidole megacephala* has displaced native ants and other invertebrates in northern Australia and threatens several taxa in Florida and Mexico. *Wasmannia auropunctata* has reduced diversity in the Galapagos, Central Africa, and New Caledonia.

Although most studies have documented how invasive ants reduce richness and abundance, a handful of studies have illustrated more subtle effects. For example, *Solenopsis invicta* may shape biogeographic patterns like the latitudinal gradient in diversity of native ants in the eastern United States. And both *Solenopsis invicta* and *Linepithema humile* disrupt co-occurrence patterns of native species (the tendency of species to occur together more or less than expected by chance).

LONG-TERM IMPACTS Most ecological studies are short—at most, covering a few field seasons. Similarly, most studies of the impacts of invasive ants on native ant communities last only a few field seasons. A couple of notable exceptions exist. A team of researchers at Jasper Ridge Biological Preserve in northern California has tracked the spread and impact of the Argentine ant since 1993. Early work at the site showed that native ant species richness was much lower in the presence of Argentine ants than in its absence. But a recent study has shown that the impact of *L. humile* tends to dissipate with time. The number of native ant species in plots that had *L. humile* in the early stages of the invasion has increased. At Brackenridge Field Station in Texas, one of the first thorough studies of the impact of an invasive ant species showed that native ant

diversity and abundance were dramatically lower in the presence of *S. invicta* than in its absence. However, when the invaded sites were resampled at Brackenridge 12 years later, the ant community had apparently rebounded, and many species had returned. Two counterexamples are worth noting. In southern California, the richness of native ant species has yet to rebound at many sites invaded by *L. humile*, almost ten years after it first invaded intact native ant communities. Similarly, native ant diversity has yet to rebound in areas infested with *Pheidole megacephala* in northern Australia. In fact, native ant diversity was reduced effectively to zero and remains near zero in what were once species-rich sites.

EXPERIMENTS (OR LACK THEREOF) Experiments are the lifeblood of ecology. Without them, separating causation from correlation is challenging. However, most studies of the impact of invasive ants on native ant communities have been based on correlations and have compared native arthropod communities in the presence and absence of the invasive ant species of interest. Of course, there are potential problems with such correlational studies. Namely, the invaded and intact sites may differ in subtle environmental characteristics (e.g., soil moisture, temperature, habitat type, disturbance history) that could actually be driving native ant community structure. Or perhaps invasive ants become established only after native ant diversity has been reduced by some environmental disturbance. But in many cases, evidence that invasive ants affect native ant communities is clear. For example, detailed pre-invasion and post-invasion comparisons, especially with *Linepithema humile*, show that communities can change within a year of *L. humile* becoming established in a community. Well-replicated long-term experimental removals of entire colonies could be an ideal way to assess impacts of invasive ant species. Addition experiments, in which colonies of an invasive ant species are deliberately introduced to a previously uninvaded community, are ethically questionable and probably should not be undertaken.

Impacts on Non-ants

The most obvious impacts of invasive ants are on other ants. But other taxa, ranging from other ground-foraging arthropods to charismatic vertebrates and plants, can also be affected by the spread of invasive ants. These effects result from competition and predation by invasive ants and from indirect factors, such as the potential displacement of other ant species that play key roles in ecosystem function.

IMPACTS ON NON-ANT INVERTEBRATES The effects of *Anoplolepis gracilipes*, *Linepithema humile*, *Pheidole megacephala*, *Solenopsis invicta*, and *Wasmannia auropunctata* on non-ant invertebrates are all well documented in a variety of ecosystem types. The list of affected taxa includes, but is not limited to, spiders, springtails, flies, beetles, yellowjackets, honeybees, ticks, mosquitoes, tree snails, apple snails, butterflies, mites, scorpions, pseudoscorpians, and hemipterans. The impacts of predation by invasive ants on other invertebrates are especially striking in agroecosystems and on oceanic islands. For example, *Pheidole megacephala* apparently led to the extinction of a ground-foraging fly on Oahu. On Maui, the presence of *L. humile* has led to reductions in total arthropod richness of more than 32 percent and to reductions in endemic arthropod richness of over 50 percent.

One particularly striking example comes from several islands in the South Pacific, where *Anoplepis gracilipes* has displaced crabs by both preying on them and competing with them. On Christmas Island, the presence of *A. gracilipes* has led to a severe decline in the population of the endemic red land crab, *Gecarcoidea natalis*. The loss of crabs has led to dramatic changes in the plant community and has increased the abundance of at least one nonnative species on the island.

Invasive ant species can also favor other invertebrates. This usually occurs when ants engage in a protective mutualism with hemiptera, whereby the ants protect the aphids from predators and parasitoids, and in return, the aphids provide honeydew, a sugary excretion rich in carbohydrates and amino acids. In the presence of the ants, aphid abundance is usually higher than in their absence (Fig. 2). Few studies have compared the relative impact of invasive ants and native ants on the hemiptera they tend, but the few that have have generally found that the invasive ants have much stronger positive effects than the native ants. One consequence of the mutualism between ants and the hemipterans they tend is that the ants remove or prey on the other herbivores on the host plant of the hemiptera. Thus, by engaging in a mutualism with hemiptera, invasive ants can affect many other herbivorous invertebrate taxa. And by reducing the abundance of herbivores and presumably the damage those herbivores cause to the plants, introduced ants can have positive effects on some plant species. This research area deserves more attention.

IMPACTS ON VERTEBRATES Not only do invasive ants affect invertebrates, but they can also have surprisingly big effects on growth or survival of vertebrates. The list of affected species includes about 20 bird species, more than 10 mammal species, and more than 15 reptile and amphibian species. The impacts of *Solenopsis invicta* on vertebrate species are the best studied, but there are documented effects of each of the five worst invasive ant species on the survival, growth, behavior, and in one case morphology, of some vertebrate species.

Predation by invasive ants can affect survival and success of bird nestlings. When exposed to high densities of *Solenopsis invicta*, chicks of bobwhite quail gained weight more slowly than did chicks exposed to lower densities of *S. invicta*. Additionally, long-term bird surveys have linked declines in bobwhite quail populations to infestations of *S. invicta* at several locations in the southeastern United States. *Pheidole megacephala* and *Solenopsis geminata* attack nestlings of seabirds and may have short-term effects on fledging success. Generally speaking, the effects are usually greatest on species that cannot escape attack or predation by the ants (e.g., chicks unable to leave the nest).

There are other means by which invasive ants might affect vertebrate species. For example, invasive ants may reduce the availability of nesting sites or alter the behavior of individuals. Not surprisingly, these effects might act as strong agents of selection on vertebrate populations. For example, the presence of *Solenopsis invicta* has led to adaptive responses in the eastern fence lizard in the southern United States. Fence lizards that have co-occurred with *S. invicta* for a longer time are more likely to exhibit defensive behaviors (body twitching and fleeing) when confronted with *S. invicta*, and they are more likely to have relatively longer hind limbs that are important for removing attacking *S. invicta* workers.

FIGURE 2 *Pheidole megacephala* tending aphids. (Photograph courtesy of Alex Wild.)

Ant invasions can also have indirect effects on vertebrate populations. In Florida, *Pheidole megacephala* appears to disrupt sea turtle and seabird nestlings. *Wasmannia auropunctata* has reduced the diversity and abundance of birds and reptiles on the Galapagos and has affected lizard populations in New Caledonia. In southern California, *Linepithema humile* has displaced many native ant species, and this might have indirectly led to population declines of coastal horned lizards. One candidate explanation for the decline in horned lizard populations is that *L. humile* displaces the native ants, which make up 90 percent of the diet of coastal horned lizards. Indeed, when individual horned lizards are offered *L. humile* as food, growth rates decline. In addition, continued habitat alteration in southern California has negative impacts on coastal horned lizards but positive effects on *L. humile* by making some sites more susceptible to invasion. Thus, it appears that habitat alteration combined with invasion by *L. humile* could be responsible for declines in coastal horned lizard populations.

IMPACTS ON PLANTS Ant–plant mutualisms are common in nature and range from tightly co-evolved interactions to more diffuse interactions such as seed-dispersal mutualisms. When native ants are displaced by invasive ant species, more often than not, the invasive ants do not adequately fill the roles played by the displaced native species.

Many plant species rely on ants to disperse their seeds, and some of these plants have a clever adaptation to encourage seed dispersal by ants: their seeds have a lipid-rich appendage called an elaisome. Ant species disperse the seed away from the maternal plant, remove the elaisome, and then may bury the seed or deposit it in the refuse pile near the nest. Studies on the effects of ant invasions on individual plants or plant communities have yielded mixed results, with some showing that invasive ants effectively replace native ant species without major impact on the plants and others documenting negative effects of ants on plant populations and communities. For example, in many instances, invasive ant species collect and disperse fewer seeds or smaller seed than do native species, disperse the seeds only a short distance, or fail to bury the seeds effectively. Invasive ants may not only be poor seed dispersers: they may in fact be seed predators. Ultimately, these differences in seed dispersal may affect plant community structure, if seeds preferred by invasive ants successfully germinate while seeds not preferred by invasive ants do not. It is important to note that not all studies have found negative effects of ant invasions on seed dispersal mutualisms. However, a recent quantitative review explicitly compared the effects of invasive ants and native ants and found that the effects of ant invasions on seed dispersal mutualisms are, on average, negative.

In other instances, invasive ants can have net positive effects on plants. In the case of plants that provide extrafloral nectar (EFN) to ants and in return receive protection from herbivores or pathogens, there are several documented cases of invasive ants conferring more positive benefits than do native ants. However, a handful of studies have shown either negative or no effects of invasive ants on the EFN-providing plants.

Generally speaking, the overall effects of invasive ants on seed-dispersal mutualisms is likely to be negative, while the effects of invasive ants on EFN-producing plants is likely to be positive. Too few studies have been conducted over sufficiently long periods to generalize about the magnitudes of the effects of invasive ants on plant communities.

IMPACTS ON HUMANS Invasive ant species also seriously threaten human well-being, and the economic impacts of invasive ant species are likely to be substantial. For instance, *Solenopsis invicta* can inflict painful stings (Fig. 3), which in some cases have led to anaphylaxis and death. One recent study estimated that more than 10 million people are stung each year in the United States by *S. invicta*. In California, for example, it is estimated that the economic impact of *S. invicta* could range from $3 to $9 billion over the next ten years if *S. invicta* is left unchecked. The costs in Texas are estimated to be $300 million per year. Most of the costs come from applying pesticide treatments, restoring and replacing property and equipment, and providing medical and veterinary treatment for stings.

Other invasive ant species also have important impacts on humans. The best-documented impacts

FIGURE 3 *Solenopsis invicta* (red imported fire ant) stinging. (Photograph courtesy of Alex Wild.)

on human well-being are from *L. humile*. Although *L. humile* does not pose serious threats to public health, it has substantial economic impacts, mostly by disrupting biocontrol efforts and facilitating hemipteran pests on crops. The agricultural impacts of *W. auropunctata* are also well documented in many parts of its introduced range. The total costs of control and eradication efforts for invasive ant species likely total tens of millions of dollars, but the outcomes of several successful eradication and control programs indicate that the costs are worth the investment.

There have been failed attempts at controlling invasive ant species. *The Fire Ant Wars* (Blu Buhs, 2004) provides a thorough and compelling review of the U.S. government's failure to control the spread and impact of *Solenopsis invicta*. Although there have been failures, it is worth noting that there have been some successes as well. For example, New Zealand has successfully prevented establishment by several invasive ant species, and the Australian government has had some success controlling *A. gracilipes* on Christmas Island and potentially several other species on mainland Australia. Together, these successes, however moderate, illustrate that control and even eradication is possible. However, it does seem highly unlikely that some firmly established species with broad introduced ranges (e.g., *L. humile* and *S. invicta* in the United States) will ever be totally eradicated.

CONCLUSION AND FUTURE DIRECTIONS

This review has focused especially on a limited number of invasive ant species: *Anoplolepis gracilipes*, *Linepithema humile*, *Pheidole megacephala*, *Solenopsis invicta*, and *Wasmannia auropunctata*. Even within this already limited list, *S. invicta* and *L. humile* are the most thoroughly studied species. The paucity of information on many problematic and potentially problematic species (e.g., *Pachycondyla chinensis*, *Paratrechina longicornis*, *Paratrechina fulva*, *Pheidole obscurithorax*, *Lasius neglectus*, *Monomorium pharoensis*, *Monomorium sydneyense*, *Myrmica rubra*, *Technomyrmex albipes*, *Tetramorium tsushimae*, *Doleromyrma darwiniana*, *Ochetellus glaber*, and several species in the genus *Cardiocondyla*) provides a clear impetus for more research on the basic biology (in both the native and introduced ranges) and potential impact of these species. Additionally, native ant species can have far-reaching effects on ecosystem processes. To date, too few studies have addressed whether and how invasive species might disrupt ecosystem processes provided by ants. More experiments, especially long-term removal experiments, aimed at understanding the effects of invasive species are

an important research aim. Finally, habitat alteration and climate change continue to alter terrestrial ecosystems and affect biodiversity. Understanding how these global change factors interact to mediate the impacts of invasive ants on biodiversity will be a challenging but important area for future research.

SEE ALSO THE FOLLOWING ARTICLES

Competition, Animal / Habitat Compatibility / Invasion Economics / Life History Strategies / Mutualism / Pesticides for Insect Eradication / Predators / Range Modeling

FURTHER READING

Blu Buhs, J. 2004. *The Fire Ant Wars*. Chicago: The University of Chicago Press.

Bond, W., and P. Slingsby. 1984. Collapse of an ant-plant mutualism: The Argentine ant (*Iridomyrmex humilis*) and myrmecochorous Proteaceae. *Ecology* 65: 1031–1037.

Helanterä, H., J. E. Strassmann, J. Carrillo, and D. C. Queller. In press. Unicolonial ants: Where do they come from, what are they, and where are they going? *Trends in Ecology and Evolution* 24: 341–349.

Hölldobler, B., and E. O. Wilson. 1990. *The Ants*. Cambridge, MA: The Belknap Press of Harvard University Press.

Holway, D. A., L. Lach, A. V. Suarez, N. D. Tsutsui, and T. J. Case. 2002. The causes and consequences of ant invasions. *Annual Review of Ecology and Systematics* 33: 181–233.

Human, K. G., and D. M. Gordon. 1996. Exploitation and interference competition between the invasive Argentine ant, *Linepithema humile*, and native ant species. *Oecologia* 105: 405–412.

Ness, J. H., and J. L. Bronstein. 2004. The effects of invasive ants on prospective ant mutualists. *Biological Invasions* 6: 445–461.

Rodriguez-Cabal, M., K. L. Stuble, M. A. Nuñez, and N. J. Sanders. 2009. Quantitative analysis of the effects of the exotic Argentine ant on seed dispersal mutualisms. *Biology Letters* 5: 499–502.

Tschinkel, W. R. 2006. *The Fire Ants*. Cambridge, MA: Harvard University Press.

APOMIXIS

ANNA M. KOLTUNOW AND TAKASHI OKADA

CSIRO Plant Industry, Adelaide, South Australia, Australia

ROSS A. BICKNELL

Plant and Food Research, Christchurch, New Zealand

Apomixis is the asexual propagation of a plant through seed without meiotic reduction and fertilization. Seedlings that develop via apomixis are genetically identical to the maternal plant and to each other, forming uniform populations that can persist for many years and over large land areas. Most apomicts are also capable of a limited amount of sexual reproduction (facultative apomixis), ensuring genetic changes over time, and therefore displaying adaptive radiation. Occasional hybridization and/or mutation events can lead to the formation of genetic variants, many of which can be perpetuated by apomixis as a "true breeding" line. As a result, apomictic genera frequently occur as a patchwork of distinct "apo-species," displaying considerable taxonomic complexity. Some, but not all, apomicts are invasive weeds. Here, we consider characteristics of apomictic plants that may have facilitated their spread and assisted some in becoming invasive.

BIOLOGICAL AND GEOGRAPHICAL DISTRIBUTION OF APOMIXIS

Apomixis is known in approximately 40 flowering plant families of monocotyledonous and eudicotyledonous taxa. Intriguingly, about 75 percent of known apomicts belong to only three families—the *Asteraceae*, *Rosaceae*, and *Poaceae*—which collectively comprise 10 percent of flowering plant species. The reason for the overrepresentation of apomixis in these families is unknown. It is speculated that it reflects predispositions for the trait due to unique developmental or genetic factors that characterize these families.

Apomixis is represented in plants with a broad range of physical forms, life histories, and adaptive strategies. One conspicuous exception to this is its almost complete absence among annual herbaceous species. Some apomicts are adapted to the colonization of disturbed habitats, and, as seen among sexual plants, these are the species most likely to be implicated in the aggressive invasion of nonnative habitats and cultivated lands. Notable weedy apomictic species include members of the genera *Taraxacum* (the common dandelion), *Hieracium* (hawkweed), *Rubus* (bramble), *Cortaderia* (pampas grass), *Chondrilla* (skeleton weed), *Opuntia* (prickly pear), and *Erigeron* (fleabane).

Apomicts are often reported to predominate at higher latitudes or at higher altitudes than related sexual species. Many also exist over a much greater range than their sexual counterparts, such as is the case with *Rubrus* in Europe, *Crepis* in North America, and *Dicanthium* in India and Africa. Often, the sexual form occupies a central region of the geographic range of the apomict.

APOMICTS AS INVASIVE WEEDS

One way that apomixis may favor an alien species is through the facilitation of successful founder events. For example, *Hieracium* (hawkweed) species, which originated in Europe, are regarded as invasive and noxious weeds in North America, Australia, and New Zealand. *Hieracium*

FIGURE 1 (A) *Hieracium caespitosum* plant and inflorescence (inset). Bar = 10 cm. *Hieracium aurantiacum* (B) flower head and (C) single floret. (D) Seed head and single seed (inset). Bars in (B), (C), (D), and inset in (D) = 5, 2, 5, and 2 mm, respectively. Abbreviations in (C): a = anther, o = ovary, p = petal, pa = pappus, s = stigma. (Photographs courtesy of T. Okada.)

invades highlands, roadsides, forest openings, fields, and pastures, and reduces the forage value of these lands for grazing. It also displaces native plants due to its aggressive growth and capability to spread asexually via runners. Each flowering shoot consists of 5 to 30 flower heads (Fig. 1A), and each flower head contains 30 to 50 individual florets (Figs. 1B and 1C). Seeds have a tuft of bristles (Fig. 1D) that enable them to stick to hair, feathers, clothing, and vehicles, and thus to be carried long distances. Seeds can also be dispersed by wind and water. A high fecundity and an efficient dispersal mechanism are typical characteristics of any plant adapted to the colonization of dispersed or disturbed habitats. In addition, however, many *Hieracium* species are fully autonomous apomicts, producing seeds without pollination. This assists reproduction in conditions where pollination is restricted or potential pollination partners are remote. Apomicts may therefore act as better colonizers than their sexual counterparts, as they are more able to successfully colonize remote sites with single individuals. In addition, apomicts are less likely to suffer from genetic bottlenecks following a founder event. Conversely, however, following establishment, the clonal nature of an apomict may restrict its ability to adapt to a new environment. The apomict hawkweed *H. pilosella* was recorded as a minor species in the pastures of New Zealand for more than 120 years before it underwent a rapid expansion of its range in the 1970s. In this case, rare hybridization events are implicated in the eventual adaptation of this facultative species to a new environment.

MECHANISM OF APOMIXIS FOR CLONAL SEED FORMATION AND ADVANTAGES FOR AGRICULTURE

Apomixis appears to have evolved polyphyletically. The developmental mechanisms involved have been grouped into two types, termed gametophytic and sporophytic. Sporophytic apomixis, common in citrus and mango, defines the formation of embryos from somatic cells adjacent to a sexually formed embryo sac. These embryos take advantage of the nutrient endosperm formed during fertilization of the adjacent sexual embryo sac.

In gametophytic apomixis (in the *Asteraceae*), unreduced embryo sacs form, where, unlike in sexual reproduction, meiotic reduction is bypassed. Gametophytic apomixis subdivides into diplospory and apospory. In diplospory, the cell starting meiosis fails to complete the process and undergoes mitosis. In apospory (Figs. 2A–C), one or more embryo sacs can develop mitotically from neighboring somatic cells. Depending on the apomictic species, the sexual embryo sac may or may not persist. The egg cell formed in unreduced embryo sacs develops into an embryo without fertilization. In *Asteraceae*, endosperm development is fertilization independent (Fig. 3), but in many apomict species, fertilization is required for endosperm formation. Apomixis

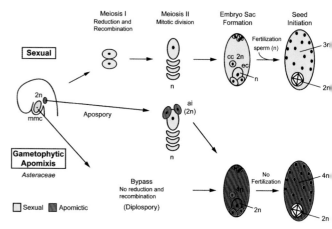

FIGURE 3 A schematic representation of sexual and apomicitic seed development. In sexual plants, the diploid (2n) mmc undergoes meiotic reduction and only one of four haploid (n) megaspores survives and forms an embryo sac after three rounds of mitotic division. The embryo sac contains a haploid (n) egg cell (ec) and a diploid (2n) central cell (cc). The egg cell fuses with a haploid sperm cell giving rise to an embryo (2n) and fusion of the central cell and a sperm cell develops endosperm (3n) after double fertilization. In plants with gametophytic apomixis, meiosis is avoided by diplospory or apospory (see text), and the diploid cell develops the unreduced embryo sac containing a diploid egg cell and a 4n central cell. In *Asteraceae*, seed initiation occurs without fertilization and an embryo (2n) and endosperm (4n) are developed autonomously. Cells in sexual and apomictic reproductive pathways are indicated in yellow and red, respectively. Abbreviations: ai = aposporous initial, cc = central cell, ec = egg cell, mmc = megaspore mother cell. (Schematic prepared by T. Okada.)

appears to be controlled by a few (one to five) dominant genetic loci in currently examined species. Because apomixis avoids meiosis and fertilization to form clonal seedling progeny, it may have the potential to fix hybrid vigor. If apomixis could be harnessed as a controllable breeding tool, then it could assist the accumulation of desirable traits in plants, increase food yields, and enable the improvement and development of more locally adapted plant varieties that cope with changing climate pressures.

SEE ALSO THE FOLLOWING ARTICLES

Genotypes, Invasive / Habitat Compatibility / Hybridization and Introgression / Pollination / Vegetative Propagation

FURTHER READING

Bicknell, R. A., and A. M. Koltunow. 2004. Understanding apomixis: Recent advances and remaining conundrums. *Plant Cell* 16(Suppl): S228–S245.

Hoerandl, E., U. Grossniklaus, P. J. van Dijk, and T. S. Sharbel, eds. 2007. *Apomixis: Evolution, Mechanisms and Perspectives.* Koenigstein: Koeltz Scientific Books.

Koltunow, A. M., and U. Grossniklaus. 2003. Apomixis: A developmental perspective. *Annual Review of Plant Biology* 54: 547–574.

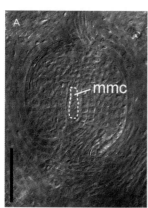
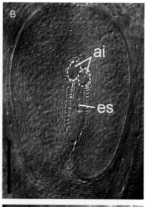
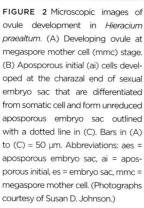
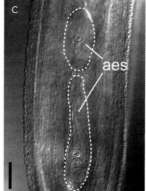

FIGURE 2 Microscopic images of ovule development in *Hieracium praealtum*. (A) Developing ovule at megaspore mother cell (mmc) stage. (B) Aposporous initial (ai) cells developed at the charazal end of sexual embryo sac that are differentiated from somatic cell and form unreduced aposporous embryo sac outlined with a dotted line in (C). Bars in (A) to (C) = 50 μm. Abbreviations: aes = aposporous embryo sac, ai = aposporous initial, es = embryo sac, mmc = megaspore mother cell. (Photographs courtesy of Susan D. Johnson.)

Ozias-Akins, P., and P. J. van Dijk. 2007. Mendelian genetics of apomixis in plants. *Annual Review of Genetics* 41: 509–537.

Ravi, M., M. P. Marimuthu, and I. Siddiqi. 2008. Gamete formation without meiosis in Arabidopsis. *Nature* 451: 1121–1124.

AQUACULTURE

DEVIN M. BARTLEY

California Department of Fish and Game, Sacramento

Aquaculture, the farming of aquatic organisms in inland and coastal areas, has become the fastest growing food production sector and now accounts for nearly half of all aquatic species consumed by humans. The global value of aquaculture products in 2007 was approximately $85.9 billion. Aquaculture has also been identified as the main reason for the deliberate introduction of aquatic species to areas outside of their native range. While the use of nonnative species is a vital component of international trade and aquaculture development, it also poses significant risks to native biodiversity, as farmed aquatic species may become invasive in certain environments. Although it is in the aquaculturists' best interest to keep organisms from escaping, experience has shown that species in fact do escape. Additionally, "unwanted travelers" and disease organisms may be inadvertently introduced along with the farmed species.

GLOBAL PATTERNS

The use of introduced species in aquaculture is not new, although modern technologies have greatly facilitated the process. European oysters (*Ostrea edulis*) were transplanted around the Greek Isles in the Golden Age of Greece. Ancient Romans introduced common carp (*Cyprinus carpio*) and perch (*Perca fluviatilis*) in rudimentary aquaculture systems to many parts of Europe; the practice was continued by monks, clerics, and lay people in the Middle Ages. In the thirteenth and fourteenth centuries, Vikings transported the soft-shell clam (*Mya arenaria*) to Europe, and in the fifteenth century, Portuguese explorers returned to Europe with a Japanese oyster (*Crassostrea angulata*). Nineteenth-century Europeans migrating to Africa and Latin America brought with them familiar species of trout and salmon in hopes of establishing fisheries in the new lands. Some of these introduced species (e.g., the common carp) are so well established that humans forget they are not native. These early introductions were to primitive aquaculture facilities, usually a small bay, pond, or reservoir, where animals were simply held, and no controlled breeding or feeding was practiced. As aquaculture technology and husbandry improved in the mid- to late twentieth century, it became easier to move species around the world and to farm them in new environments.

Aquaculture is the primary reason for the deliberate movement of aquatic species outside of their range, and much of modern aquaculture is based on introduced species. Species have been moved to and from all continents except Antarctica, with Asia and Europe having received most of the introduced species. Tilapia (*Oreochromis* spp., *Tilapia* spp.), a group of cichlid fishes native to Africa, are now farmed in greater quantities in Asia and the Americas than on their native continent; Chile has become the world's second leading producer of salmon, based on the farming of salmon introduced from Europe and North America; the Pacific oyster (*Crassostrea gigas)* is the mainstay of oyster production in Europe and on the Pacific coast of North America, where native oysters have been reduced by habitat loss and disease.

The most commonly introduced aquatic species for aquaculture include herbivores, omnivores, and carnivores from tropical and temperate environments (Table 1). They are easy to grow and breed, they grow fast, or they have wide environmental tolerances. These qualities also contribute to their potential for becoming invasive.

The species in Table 1 are primarily freshwater vertebrate species with one marine bivalve and a brackish-water prawn. With increased consumption and farming of marine shrimp, introductions of several

TABLE 1

The Ten Most Commonly Introduced Aquatic Species for Aquaculture Purposes

Species	Trophic Status[1]
Rainbow trout (*Oncorhynchus mykiss*)	4.5
Common carp (*Cyprinus carpio*)	3.1
Nile tilapia (*Oreochromis niloticus*)	2.0
Grass carp (*Ctenopharyngodon idella*)	2.0
Mozambique tilapia (*Oreochromis mossambicus*)	2.0
Silver carp (*Hypophthalmichthys molitrix*)	2.0
Big head carp (*Aristichthys nobilis*)	2.3
Giant river prawn (*Macrobrachium rosenbergii*)	NA
Blue tilapia (*Oreochromis aureus*)	2.0
Pacific oyster (*Crassosrtea gigas*)	<2.0

NOTE: From Food and Agriculture Organization of the United Nations Database on Introductions of Aquatic Species.

[1]From www.FishBase.org: 2.00–2.19 = herbivores; 2.20–2.79 = omnivores; >2.80 = carnivores.

shrimp species are increasing. The white-leg shrimp (*Penaeus vannamei*), native to the Americas, is now being widely farmed in China and Southeast Asia because it is easier to breed than native Asian species and resists several viruses that impact native species. It may be expected that, with changes in climate, domestication, improved breeding technologies, and consumer preferences, other nonnative species may become important in aquaculture production.

Farmed aquatic species enter aquatic ecosystems either through deliberate release by stocking (i.e., the release of hatchery-produced organisms in order to create added value to a fishery), or from accidental escapes from aquaculture facilities. Millions of introduced Pacific salmon escaped in the mid 1990s in Chile as a result of severe storms. In the North Atlantic, Atlantic salmon of farmed origin compose significant components of the salmon fishery and the spawning population; approximately 90 percent of Atlantic salmon in the Baltic Sea are believed to have originated on farms.

TYPES OF IMPACTS

Introduced species in aquaculture can impact native biodiversity through interactions between species at different trophic levels, competition, genetic interactions, disease transmission, habitat alteration, and the inadvertent introduction of associated species.

Interactions between Trophic Levels

The introduction of top-level predators has often been cited as the most dangerous or invasive type of introduction, as in the infamous example of the Nile perch introduction into Lake Victoria. Although such introductions may occasionally present a spectacular example of impact, the more common and pervasive threat appears to be from species that have a range of feeding habits and that feed lower on the food chain. Examination of a species' trophic status (i.e., where it feeds on the food chain) and the number of reported adverse impacts from an introduction revealed that omnivores and herbivores were more likely to have adverse environmental impacts than were predatory species.

PREDATION Significant reduction in native fishes in Chile, South Africa, Australia, and New Zealand has been attributed to the introduction of several species of salmonids from the northern hemisphere. In Chile, trout and salmon were introduced to establish fisheries at the end of the nineteenth century, and stocking of Pacific salmon was undertaken in the late twentieth century. However, it

was not until the expansive growth in farming of Atlantic (*Salmo salar*), Chinook (*Oncorhynchus tshawytscha*), and coho (*O. kisutch*) salmon in the 1980s that large quantities of salmonids were produced. Through escapes and deliberate releases, introduced salmonids have reduced the occurrence of native fishes, primarily galaxiids (*Aplochiton spp. and Galaxias spp.*), in coastal and inland waters in the southern hemisphere. In South Africa, predation by brown trout (*Salmo trutta*) has eliminated local fish species. In Patagonian lakes and streams, there appears to be a partitioning of the habitat between native fishes and introduced salmonids. Whether this segregation is a mechanism to reduce species interactions or is a direct result of the interaction is unclear. Predation by North American crayfish (*Pacifastacus leniusculus*) on young of the native European crayfish (*Astacus astacus*) is one of several mechanisms responsible for reduction of the native crayfish in most areas where the nonnative species is farmed.

HERBIVORY The golden apple snail (*Pomacea canaliculata*), a native of South American floodplains, was introduced to many Asian countries in the 1980s to provide a new source of food protein for humans and a potential export for Asian aquaculturists. Markets for the snails never developed, and they were discarded into local waterways. The snails eventually became established in rice fields, where they became a voracious pest, feeding on young rice plants. In the Philippines, economic losses to rice production range from 20 to 40 percent of the value of rice. The snails reduced food security by consuming large quantities of rice and were implicated in reducing native biodiversity such as the species of snail Filipinos preferred for food, *Pila luzonica*.

Grass carp have been introduced in many parts of the world for both food and aquatic vegetation control. Grass carp eliminated submerged aquatic plants in Donghu Lake, China. This produced other ecological changes, leading to the disappearance of most of the lake's fish fauna (as described in Habitat Alteration, below). Nile tilapia that escaped from cages in Lake Apoyo, Nicaragua, preferentially feed on the macroalgae *Chara*, which native cichlids used for nesting and refugia.

Competition

Competition between farmed and native species can occur over food, habitat, spawning sites, mates, and other essential resources. Pacific oysters introduced to Australia in the mid-1900s have established reproducing populations that are competing for growing space with

FIGURE 1 Pacific oyster, *Crassostrea gigas*, (A) bag-culture in northern California (B). Good growing conditions, but unsuitable conditions for reproduction and settlement, prevent establishment of the nonnative oyster in most growing areas of California. (Photographs courtesy of the author.)

the native oysters, *C. commercialis*, and the Sydney rock oyster (*Saccostrea commercialis*). The Pacific oyster (Fig. 1) has higher fecundity and a higher growth rate than the natives and is crowding them out of their native habitat. The Pacific oyster has been widely introduced around the world, and Australia is one of the few areas where competitive interactions have caused problems. This could be due to the fact that the Pacific oyster was often introduced into areas where native oysters had declined because of disease or habitat degradation and thus were not capable of producing viable populations.

Introduced tilapia compete successfully with many native species because of their short generation time, fast growth rate, wide environmental tolerances, aggressive behavior, and omnivorous feeding ability. In the Philippines and Pacific islands, brackish-water populations of Mozambique tilapia (*O. mossambicus*) have displaced milk fish (*Chanos chanos*), mullet (*Mugil cephalus*), and shrimp (*Fenneropenaeus merguiensis*) species through competition for food resources, primarily algae. Escaped Nile tilapia's aggressive behavior during breeding, which allowed them to occupy prime nesting sites, reduced abundance and diversity of native fishes in the southern

United States. In Nicaragua, introduced Nile tilapia have reduced the abundance of local cichlids through competition for nesting sites.

Genetic Impacts

Genetic impacts occur when farmed fish interbreed with local species and when interactions with farmed species reduce a local population to the point that inbreeding and loss of genetic diversity become problematic. Genetic impacts from introduced species may include loss of species integrity from mixing with introduced genotypes, reduced reproductive efficiency from hybridization that results in unviable offspring or lack of gonadal development, decreased fitness from incorporation of introduced genes, or the loss of coadapted gene complexes.

Hybridization between escaped farmed Atlantic salmon (*Salmo salar*) and native brown trout (*S. trutta*) in Scotland has been shown to reduce the reproductive efficiency of both species. Rainbow trout (*Oncorhychus mykiss*) have been widely introduced for aquaculture, but they also are regularly stocked for recreational fisheries. Stocked nonnative rainbow trout have hybridized with threatened Apache trout (*O. apache*) and endangered Gila trout (*O. gilae*) in the United States causing a loss of pure populations of native trout.

Although tilapia have been widely introduced abroad, they have also been moved extensively in Africa to areas where other native tilapia exist. Most species of tilapia are interfertile, and many escapes of farmed tilapia throughout Africa have resulted in the loss of pure strains in many parts of their native ranges. Escape of nonnative Nile tilapia into the Zambezi and Limpopo river systems of southern Africa and subsequent hybridization have caused the loss of pure strains of local Mozambique and three-spot tilapia (*O. andersoni*) and have reduced catch of the local species.

In Thailand the North African catfish (*Clarius gariepinus*) is routinely hybridized with the native Thai bighead catfish (*C. macrocephalus*). This hybrid comprised approximately 80 percent of Thai catfish production. However, the widespread use of the introduced African catfish has resulted in the loss of many pure populations of wild Thai catfish to the point that native clariiad catfish are becoming species of special concern or are threatened. Hybrid catfish (*Clarias batrachus x C. gariepinus*) are also impacting local catfish in Bangladesh.

ALIEN GENOTYPES Aquaculture may involve the controlled breeding of aquatic species, and this process can lead to genetic changes in the farmed species relative to its wild relative. This is the domestication process, and it is accomplished through both directed genetic improvement programs (e.g., selective breeding) and inadvertent selection brought about by cultural practices. Thus, a species farmed within its native range may be genetically different from native conspecific populations. Furthermore, farmed species are routinely moved between countries in standard aquaculture business practices (e.g., farmed Atlantic salmon in Scotland probably originated in Norway).

These genetically differentiated groups are often called "alien" genotypes, and they can adversely impact native populations in a similar manner to nonnative species. Numerous genetic, biological, and behavioral differences between farmed and wild populations have been documented, mainly in salmonids, and it is presumed that many of these differences are related to fitness. Survival to smoltification is less in farmed Atlantic salmon than in wild salmon, but farmed fish grow faster; survival of wild brown trout in Norway is higher than that of escaped farmed fish; farmed Atlantic salmon spend more time at sea before returning to spawn. Farmed Atlantic salmon are usually larger and more aggressive than their wild relatives, which gives the farmed salmon a temporary competitive advantage early in life. Competition between escaped Atlantic salmon and local stocks has led to a reduced production of wild smolts in parts of Scotland and Ireland.

When alien genotypes breed with native genotypes, it is usually to the detriment of the natives. Besides the loss of the wild genotype, interbreeding has led to reductions in wild populations. Farmed and hybrid rainbow and steelhead trout have lower survival, are more risk prone, and are more likely to be taken by predators than are wild trout. Farmed and hybrid Atlantic salmon outcompete wild salmon for territory and food, leading to displacement of the wild population.

Baltic strains of Atlantic salmon resist the parasite *Gyrodactylus salaris,* but Norwegian strains are highly susceptible. Introductions of the alien genotype from Sweden on the Baltic to Norway in the early 1970s resulted in the reduction of many wild salmon populations (as described in Disease).

Disease

The spread of pathogens is not unique to nonnative species; pathogens can spread from the farming of virtually any species through improper farming techniques. However, nonnative species present special challenges.

A major problem in preventing the spread of disease is that often the pathogen does not cause problems in

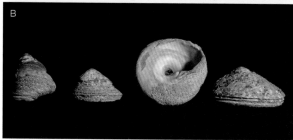

FIGURE 2 (A) Left, Adult red abalone, *Haliotis rufescens*, infected with a sabellid worm parasite accidentally introduced into California with abalone from South Africa. Right, Normal red abalone. (B) Native California gastropods that became infested with the parasite in an aquaculture facility (left to right, *Chlorostoma brunnea, Lottia digitalis, Chlorostoma monterey,* and *Lottia pelta*). (Photographs by M. Newnham (A) and C. Culver (B), courtesy of California SeaGrant.)

its native environment or under nonfarming conditions. California abalone farms and some wild gastropods were infected with a sabellid worm (*Terebrasabella heterouncinata*) inadvertently introduced with abalone (*Haliotis* spp.) from South Africa. This parasite was not pathogenic under South African conditions but caused widespread devastation in California farms and spread to native gastropods (Fig. 2). European oysters were introduced to North America where they became infected with a blood parasite, *Bonamia* spp. When these oysters were reintroduced to Europe for aquaculture, the parasite was also introduced and led to the collapse of European flat oyster populations in the Mediterranean.

Aeromonas salmonicida, which causes furunculosis disease in salmonids, was introduced into Norwegian salmonid farms through infected stocks of rainbow trout that had been imported from Denmark in 1966. The pathogen spread to over 500 fish farms and 66 salmon streams by 1991. The spread of both *Gyrodactylus* (as described in Alien Genotypes, above) and *A. salmonicida* was probably facilitated by stocking programs that inadvertently used infected fish.

The culture of marine shrimp has been fraught with viral disease problems brought about by the introduction of farmed shrimp and by poor farming practices. Introduction of diseased shrimp containing *Penaeus monodon*–type baculovirus, yellow-head virus, and other newly discovered viruses caused financial losses of over a billion dollars in Asia in the early 1990s and led to the collapse of the Taiwanese shrimp aquaculture industry. Infectious hypodermal and haematopoietic necrosis virus in white-leg shrimp can devastate *P. stylirostris*, which is cultured in several tropical countries. White-spot and yellow-head viruses, which have caused multimillion dollar losses in Asian shrimp farms, have been found in the Americas. These viruses may have been introduced through dumping of untreated waste from processing plants for imported giant tiger prawn (*Penaeus monodon*) and from transport of infected white-leg shrimp.

The introduction of crayfish (*Pacifastacus leniusculus*) from North America to Europe also introduced the crayfish fungus (*Aphanomyces astaci*) to European and Scandinavian crayfish species (*Astacus astacus*). The North American crayfish species is a resistant carrier that also outcompetes native European crayfish, owing to their higher reproductive and growth rates; the plague gives the invaders an additional competitive advantage by weakening European crayfish.

Habitat Alteration

Some common aquaculture species can significantly alter their new habitats if they escape. Grass carp are effective at removing aquatic vegetation that may provide food, shelter, and reproductive habitat for local species. Such removal can lead to plankton blooms and decreases in invertebrates and many species of fish (e.g., pike [*Esox* spp.], blue gill [*Lepomis macrochirus*], crappie [*Pomoxis* spp.], and largemouth bass [*Micropterus salmoides*]) that rely on vegetated habitats as forage areas or nursery and spawning habitats. However, for other species such as channel catfish (*Ictalurus punctatus*) and tench (*Tinca tinca*), stocks have remained unaffected or have improved following grass carp introduction because of the removal of nuisance aquatic vegetation and the reduction of refugia for prey species. Common carp feeding remobilizes benthic sediments, thereby reducing water clarity, and has led to loss of aquatic macrophytes and phytoplankton that form the basis of the aquatic food chain.

The invasive smooth cordgrass (*Spartina alterniflora*) was inadvertently introduced (as described in Fellow Travelers, below) in packing material with shipments of oysters (*C. virginica*) from the eastern United States to coastal areas of Washington State in the late 1800s. The cordgrass became established and transformed thousands of square kilometers of mudflats, which now contain a completely different aquatic community.

Fellow Travelers

Fellow travelers are species inadvertently introduced along with desired aquaculture species. Fellow travelers have been found in packing material, as epibionts on shells and surfaces of fish, and in transport water.

One of the major risks of bivalve mollusc introduction is fellow travelers. Twenty-nine species, including algae, diatoms, protozoans, and invertebrates, were found in the transport water of oyster shipments. The Manila clam (*Ruditapes philippinarum*) was accidentally introduced to the west coast of the United States with shipments of Pacific oysters from Japan. A slipper shell (*Crepidula fornicata*) that competes with local oysters and a predatory oyster drill (*Urosalpina cinerea*) were fellow travelers in shipments of *C. virginica* from the east coast of the United States and have now become established on the west coast of North America.

The invasive zebra (*Dreissena polymorpha*) and quagga (*D. bugensis*) mussels, primarily spread through water transport systems, ballast water, and recreational boats, have infested water sources used by aquaculturists in North America. In some cases, farms have been infested and have been closed down. Although aquaculture has not been responsible for the introduction of these mussels to new areas, and many types of aquaculture-growing facilities and farmed species appear to preclude the establishment of the mussels, the potential does exists for dispersal through aquaculture. Farm and water management protocols and monitoring are used by aquaculturists in high-risk areas to prevent further spread of the invasive mussels.

CONCLUSION

Although it is difficult to predict precisely when a species will become invasive, guidelines, risk assessment methods, and comparative information are available that can help farmers and policymakers to make informed decisions on using nonnative species. Evaluation of the impacts of farming nonnative species is also difficult, owing to a lack of good baseline information and to the confounding influence of other factors such as overfishing, water diversion, loss of habitat, or pollution. A species may become invasive in one area but not in another. For example, Pacific oysters grown in northern California bays have not become invasive after several decades of cultivation. However, in Australia, Pacific oysters have displaced native oysters. The culture of introduced tilapia by urban and peri-urban

FIGURE 3 Nile tilapia, *Oreochromis niloticus*, taken from Chinese rice fields and sold in local markets. (Photograph courtesy of FAO/D. Bartley.)

small-scale farmers in Asia (Fig. 3) has had minimal adverse impacts while providing a viable fishery product in highly modified aquatic habitats where few other native species would survive. In Nicaragua, however, tilapia have been shown to reduce both native plants and fishes.

It is acknowledged that aquaculture uses many species that can adversely impact certain environments. However, proper siting of aquaculture facilities, wise choice of species to farm, and good farming practices can greatly reduce the risk of farmed species becoming invasive, while at the same time allowing aquaculture to produce needed food resources and to encourage economic development.

SEE ALSO THE FOLLOWING ARTICLES

Carp, Common / Crayfish / Genotypes, Invasive / Hybridization and Introgression / Ostriculture / Parasites, of Animals / Pathogens, Animal

FURTHER READING

Bartley, D. M., ed. 2006. *Introduced Species in Fisheries and Aquaculture: Information for Responsible Use* [CD-ROM]. Rome: Food and Agriculture Organization of the United Nations.

Bert, T. M., ed. 2007. *Ecological and Genetic Implications of Aquaculture Activities.* Dordrecht: Springer.

Database on Introductions of Aquatic Species. Food and Agriculture Organization of the United Nations. www.fao.org/fishery/dias

FAO. 2009. *State of World Fisheries and Aquaculture.* Rome: Food and Agriculture Organization of the United Nations. ftp://ftp.fao.org/docrep/fao/011/i0250e/i0250e.pdf

FishBase. www.fishbase.org

Gozlan, R. E. 2008. Introduction of non-native freshwater fish: Is it all bad? *Fish and Fisheries* 9: 106–115.

Holmer, M., K. Black, C. M. Duarte, N. Marbà, and I. Karakassis, eds. 2008. *Aquaculture in the Ecosystem.* Dordrecht: Springer.

National Research Council. 2004. *Nonnative Oysters in Chesapeake Bay.* Washington, DC: National Academic Press.

Naylor, R. L., R. J. Goldberg, J. H. Primavera, N. Kautsky, M. C. M. Beveridge, J. Clay, C. Folke, J. Lubchenco, H. Mooney, and M. Troell. 2000. Effects of aquaculture on world fish supplies. *Nature* 405: 1017–1024.

Sunderland, O. T., P. J. Schei, and Å-e. Viken. 1999. *Invasive Species and Biodiversity Management.* Dordrecht: Kluwer.

AQUARIA

IAN C. DUGGAN

University of Waikato, Hamilton, New Zealand

Keeping home aquaria is a popular hobby worldwide. Millions of specimens from thousands of species are transported annually via the aquarium trade to areas where they are not native. While fish are generally the central focus, a variety of aquatic plants and snails are commonly kept in freshwater aquaria, while macroalgae and invertebrates, including corals, are also used in marine aquaria. Invasions from home aquaria are a multistep process, requiring collection of species in a native range, transportation to stores in other regions (sometimes via aquaculture/breeding facilities), transportation to homes, and finally release into natural waters (Fig. 1).

HISTORY OF THE AQUARIUM TRADE

The potential risk of invasions from home aquaria has increased through time. The first evidence of fish being kept indoors is from Asia in the mid-1500s. At this time, fish were kept in earthen or jade vessels, and comprised locally collected native fish. Home aquaria, commonly defined as vessels for housing fish with at least one glass side, are more recent inventions. Aquarium-keeping first became popular in Britain and Germany in the 1850s, and soon after in North America. Initially these amateur aquarists also generally kept native fish, although goldfish (*Carassius auratus*) were also used. Greater availability and ease of keeping of species from abroad ultimately increased the popularity of home aquaria through the twentieth century. The first tropical freshwater fish became available in the United States, for example, in the early 1900s; these now dominate the trade. Fish were initially carried by cargo ships, but the development of airline cargo

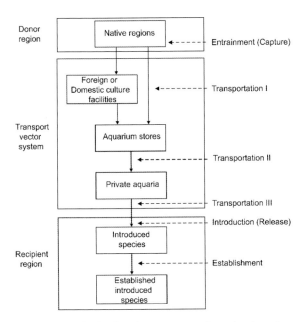

FIGURE 1 Steps in the invasion sequence of species from home aquaria. The transportation phase can be divided into multiple substeps, although care is generally taken to ensure survival in the stages leading up to taxa entering home aquaria. Only a small proportion of those kept in home aquaria will be introduced to natural waters, and only a small proportion of these will establish populations. Although this figure illustrates introductions from home aquaria, escapes from aquaculture facilities are also recognized through this trade.

allowed more fish to be successfully imported from distant regions. Houses became supplied with electricity throughout the early 1900s and were commonly so by the mid-1900s, permitting aquaria to have heating, artificial light, aeration, and filtration. Improved packaging techniques, including plastic bags and Styrofoam containers, have also allowed easier transportation. Amateur saltwater fish keeping gained popularity in the 1950s and has increased since the 1970s with improvements in artificial seawater and the development of protein skimmers. All of these factors have led to the hobby increasing in popularity, raising both the numbers and variety of species available to hobbyists but also the number and variety of nonnative fish potentially available as invaders.

SPECIES IN THE AQUARIUM TRADE: ENTRAINMENT AND TRANSPORTATION

Invasions begin with the capture of species and their movement to a new geographical area (Fig. 1). Species in the aquarium trade are not a random subset of global taxa. For example, fish that dominate both marine and freshwater aquarium trades typically originate from tropical regions, are colorful, and are small. Freshwater fish supplied to the global trade are mostly farm-raised

in southeast Asia (e.g., Hong Kong, Singapore), but this is commonly not the origin of the species themselves. However some imports, including freshwater fish from South America, are wild-caught native species. Fish and invertebrates carried for the marine aquarium trade globally are also almost exclusively wild-caught from their regions of origin. Domestic production also supplies species in some regions. In North America, for example, the cultivation of goldfish began in the 1880s, and facilities in Florida and Hawaii now supply some tropical species.

Species available for sale vary globally. North America and Europe are relatively unregulated with respect to what can be imported, as tropical aquarium species have been considered to pose a negligible risk there. Canada and various U.S. states use "blacklists," restricting importation of species deemed to have a high probability of invasion success, or potential effects. Snakeheads, walking catfish, and piranha, for example, are common listings. However, this leaves numerous species that may be imported, with some estimates putting the figure available at over 5,000. Fish imported into the United States, for example, can number over 1,000 species in any given month. Some countries have more restricted access, using "whitelists" to reduce the probability of importing species that can survive or have significant impacts. New Zealand, for example, currently lists as permissible around 900 freshwater aquarium fish, all named to species level; this has recently been altered from a list of about 280 names, 178 of which were genera that comprised multitudes of species (e.g., *Barbus*). Australia lists 225 taxa, with many identified to genus only. Around 1,500 fish species and over 600 invertebrate species are estimated as being used in the marine aquarium trade globally. At least 150 freshwater aquatic plant species are available in most regions.

Despite many species potentially being imported, only a small number dominate the trade. For example, North American store surveys and importation records show fish species including goldfish, guppy (*Poecilia reticulata*), neon tetra (*Paracheirodon innesi*), swordtails (*Xiphophorus helleri*), and platy (*X. maculatus*) to be very common. Only several hundred other fish species are likely in common use; far fewer than that are potentially available. For example, only about 300 species were observed from a survey of 20 stores in southern Ontario, Canada, while over 60 species of aquatic plant were found in the same survey.

The volume of fish moved by the marine aquarium trade is small relative to that in the freshwater trade.

Damselfish (family Pomacentridae) make up almost half of the fish trade, with species of angelfish (Pomacanthidae), surgeonfish (Acanthuridae), wrasses (Labridae), gobies (Gobiidae), and butterflyfish (Chaetodontidae) accounting for approximately another 25–30 percent. The most traded species include the blue-green damselfish (*Chromis viridis*) and clown anemone fish (*Amphiprion ocellaris*).

From stores, species are returned to homes. Survival of species in these early phases of the invasion sequence (i.e., entrainment, transport I and II; Fig. 1) is likely high relative to other aquatic vectors (e.g., ballast water).

INTRODUCTION TO NATURAL WATERS AND ESTABLISHMENT

Despite the movement of numerous aquarium fish (and other aquatic life) being allowed globally, laws exist in many countries to discourage their release into natural waterways (e.g., Invasive Species Acts). However, it is clear that fishes, and other taxa, are released through this vector. Fish are commonly cited as being released because owners tire of them, or they become too aggressive, large, or prolific for their tanks. Release into natural waters is seemingly perceived as more humane than killing. As these introductions are deliberate, release sites hinge on social factors (e.g., ease of getting to a location, perceived probability of a former pet surviving). In a study of restored ponds in England, for example, goldfish were recorded as being preferentially released into ponds close to roads, footpaths, and houses.

Some fish introductions are suggested to result from breeders wanting to return later and harvest the progeny, while in Florida in 1992 Hurricane Andrew reportedly led to an aquarium containing six lionfish (*Pterois volitans*) to break, flushing the fish into the ocean (these were seen alive nearby several days later). Escape from breeding facilities is often responsible for the introduction of species considered aquarium releases.

Introduction effort through high release sizes or frequencies, known as propagule pressure, is increasingly recognized as one of the most important determinants of invasion success. Most fundamentally, fish need to find mates and reproduce, and greater (or more frequent) releases increase the chances of this occurring. In comparisons of data on freshwater aquarium fishes in stores and records of the same species in Canadian and U.S. waters, clear relationships exist between the "popularity" of fish and those recorded as introduced or established (i.e., the more readily available the fish, the more likely it is to be released and subsequently establish). Goldfish, the most commonly used species in the aquarium trade, has invaded most widely, as have guppies and swordtails. Only a subset of taxa recorded as introduced establish populations. For example, of the 94 freshwater fish species recorded as potentially introduced into Canada and the United States via the aquarium trade, only 34 have established populations. Additionally, a number of those not established were recorded as single individuals only. Many more species have been introduced than have been recorded.

Records of introductions and successful establishment of marine fishes are relatively rare. However, a link has also been made between their popularity (based on numbers imported) and their occurrence off the coast of subtropical Florida. Sixteen nonnative aquarium fishes, all among those imported in greatest numbers into the region, have been observed. Ten of these were in more than one location, although only two were seen in schools of two or more individuals: the orbicular batfish (*Platax orbiculus*) and the Indo-Pacific lionfish (*Pterois volitans*). Of these the lionfish, at least, is considered established. Emperor angelfish (*Pomacanthus imperator*), sailfin tang (*Zebrasoma veliferum*), and other tang species (*Zebrasoma* spp.) were also commonly observed.

It is unlikely that plants or macroalgae are released as a humane means of disposal. These are more likely released incidentally with fish (or invertebrates). Taxa with larger propagule supplies likely also have the greatest chance of success.

Besides propagule supply, environmental conditions in the recipient regions are critical. As most aquarium fish are tropical, few are able to survive or breed in temperate regions. Latitude therefore plays an important role. For instance, in contrast to the northern United States, where shipping is seemingly the dominant vector for aquatic invaders, aquarium species are among the dominant invaders in subtropical regions (e.g., Florida and Hawaii). Pacu (*Colossoma* sp.) and piranha (*Serrasalmus natterreri*), for example, have been recorded from the North American Great Lakes, but even if propagule supplies were great enough, these could not survive the cooler temperatures. Goldfish is the most notable exception among aquarium invaders, having established widely in temperate areas. This trend applies to marine systems also, with a surprising number of nonnative aquarium fish recorded from the southern Florida reefs.

Another hotspot for freshwater aquarium fish is hot springs. For example, in Cave and Basin Hot Springs,

Alberta, Canada, sailfin molly (*Poecilia latipinna*) and African jewelfish (*Hemichromis bimaculatis*) have established, while others have been recorded as introduced. Similarly, in New Zealand geothermal springs sailfin molly, guppy (*P. reticulata*) and swordtail (*Xiphophorus helleri*), three of the most common species in the aquarium trade, have established. Spread of these species from these springs is usually impossible.

Diminishing the probabilities of establishment, bird predation (e.g., by herons) is more likely on conspicuous or colored fish, which will be taken preferentially over more cryptic prey. Factors important in the entrainment and transport phases, such as being colorful and warm-water species, may thus be detrimental to establishment and spread (Fig. 1).

Besides fish, a large number of plants and snails have established populations in freshwaters through the aquarium trade. Plant invasions are numerous, including *Egeria densa* (Brazilian waterweed), native to South America, which has invaded widely (e.g., Canada, United States, France, Germany, Australia, New Zealand, and at least 14 other countries). Once extremely common in the aquarium trade, its use has been banned in many areas (e.g., New Zealand, some U.S. and Australian states). Fanwort (*Cabomba caroliniana*), native to the subtropical–temperate regions of America, has invaded Japan, China, Peru, Malaysia, Australia, and Canada. Snails also establish—for example, *Marisa cornuarietis* (Giant Rams Horn) sold in North America has invaded Florida. Many snail invaders are commonly incidental components of aquaria (i.e., not kept intentionally). One such species, *Physella acuta* (European Physa), has invaded Australia, New Zealand, Hawaii, and several continental U.S. states.

IMPACTS AND ADDITIONAL CONCERNS

Most invasions from aquaria are considered fairly benign. However, predatory or competitive interactions of fish are inevitable, while plants may clog waterways and hinder recreation (e.g., *Egeria densa*). Some invasions though are noteworthy. Declines in fish with restricted distributions due to aquarium releases have been documented. For example, the Central American convict cichlid is considered partially responsible for declines in native speckled dace, and shortfin molly for decline of the endangered Moapa dace, at some localities in Nevada. Lionfish, established in Florida, is also a voracious predator on smaller fish. This marine species can also be a direct danger to people, as its spines deliver venom that can cause extreme pain, sweating, and respiratory distress.

Although marine macroalgal invasions are apparently uncommon, that of *Caulerpa taxifolia,* native to the Indian Ocean, is notable. This species was first recognized in the Mediterranean Sea from an aquarium release in 1984. It rapidly spread along the Mediterranean coast and now covers thousands of hectares of near-shore waters, displacing native species (e.g., other algae, sessile invertebrates).

Surveys of parasites and pathogens of aquarium fish entering various countries have recorded bacteria, protozoans, monogeneans, cestodes, and nematodes that could pose a risk to indigenous fish populations. For example, the Asian nematode *Camallanus cotti* was seemingly introduced into New Caledonia with nonindigenous poeciliid fishes and has since been recorded from two native species. A number of aquarium snails may also act as intermediate hosts for parasites.

Live food sold for aquarium fish can also produce a risk. In New Zealand, the North American calanoid copepod *Skistodiaptomus pallidus* has been found among live *Daphnia* sold as food for aquarium fish, and has established two nonindigenous populations. Live freshwater oligochaetes for aquarium fishes in Germany have also been found to contain parasitic infections of actinosporean myxozoans; this has the potential to introduce both parasites and suitable hosts to new areas.

SEE ALSO THE FOLLOWING ARTICLES

Algae / Aquaculture / Fishes / Parasites, of Animals / Propagule Pressure / Snails and Slugs

FURTHER READING

Courtenay, W. R., Jr. 1990. The introduced fish problem and the aquarium fish industry. *Journal of the World Aquaculture Society* 21: 145–159.

Crossman, E. J., and B. C. Cudmore. 1999. Summary of North America fish introductions through the aquarium/horticulture trade (129–133). In R. Claudi and J. H. Leach, eds. *Nonindigenous Freshwater Organisms: Vectors, Biology and Impacts.* Boca Raton, FL: Lewis Publishers.

Duggan, I. C., C. A. M. Rixon, and H. J. MacIsaac. 2006, Popularity and propagule pressure: Determinants of invasion success in aquarium fish. *Biological Invasions* 8: 377–382.

Mackie, G. L. 1999. Mollusc invasions through the aquarium trade (135–147). In R. Claudi and J. H. Leach, eds. *Nonindigenous Freshwater Organisms: Vectors, Biology and Impacts.* Boca Raton, FL: Lewis Publishers.

Meinesz, A. 1999. *Killer Algae: The True Tale of a Biological Invasion.* Chicago: University of Chicago Press.

Semmens, B. X., E. R. Buhle, A. K. Salomon, and C. V. Pattengill-Semmens. 2004. A hotspot of non-native marine fishes: Evidence for the aquarium trade as an invasion pathway. *Marine Ecology Progress Series* 266: 239–244.

Wabnitz, C., M. Taylor, E. Green, and T. Razak. 2003. *From Ocean to Aquarium.* Cambridge, UK: UNEP-WCMC.

AUSTRALIA: INVASIONS

TIM LOW

Invasive Species Council, Chapel Hill, Queensland, Australia

Islands are often more susceptible to invasive species than continents, and in this respect, Australia—known as the "island continent"—has proved more like an island than a continent. Its diverse biota, which evolved in isolation, has proved very vulnerable to introduced predators and pathogens. Many Australian species are now endangered or extinct as a consequence.

MAMMALS

Fox and Cat

Across much of Australia, all native mammals weighing between 35 g and 5.5 kg have disappeared. Complete mammal faunas can be found only in some of the wetter coastal and near-coastal forests in the east, and in parts of the far north. Of the 22 completely extinct marsupials and rodents, the red fox (*Vulpes vulpes*) and cat (*Felis catus*) have possibly contributed to the disappearance of all but two, although their culpability in many extinctions remains speculative. Foxes and cats are also blamed for the loss from the mainland of another eight species that survive on islands, sometimes precariously. Foxes are also blamed for the massive range contractions of several marsupials that survive on the mainland only where foxes are baited.

Most of the rodents that disappeared did so earlier than the larger marsupials, consistent with feral cats spreading across Australia from European settlements during the nineteenth century. The later advance of foxes, after their release in the 1860s and 1870s by sport hunters, was followed by the disappearances of several bandicoots and small wallabies weighing 200 g or more. Foxes are more strongly implicated in extinctions than cats because their arrival in different regions can be matched with species disappearances. Also, fox poisoning in nature reserves has led to substantial increases in 12 marsupial species. Tasmania, which until recently was fox free, is the only state not to have lost small mammals. Threatened mammals have been reintroduced to several sites on mainland Australia inside vast predator-proof enclosures.

Overgrazing by livestock and feral herbivores, changed fire regimes, and nutrient depletion resulting from landscape degradation have been proposed as alternative explanations or contributing factors for some or most of the extinctions. In some arid areas, mammals became extinct soon after the Aboriginal inhabitants left, suggesting a role for changed fire regimes or for predation by cats, whose numbers rose when they were no longer hunted for food.

Dingo

The dingo (*Canis lupus*), a breed of dog from Southeast Asia, is Australia's first known introduction. It arrived 3,500 to 4,000 years ago, long after the Aboriginal colonization of Australia more than 40,000 years ago. Sometimes called Australia's "native dog," the dingo has unique status today as the only introduced species featured in wildlife books and widely protected inside national parks. Conservationists campaign for protection of the last genetically pure populations, as hybridization with domestic dogs has occurred over most of Australia. Dingoes contribute to ecosystem stability by preying on kangaroos and wallabies, which otherwise overgraze vegetation, and by suppressing introduced foxes, cats, pigs, goats, and rabbits. A 5,614 km–long dingo fence divides the sheep lands of southeastern Australia, from which dingoes have largely been eliminated, from northern and western Australia where dingo–dog hybrids are common and possibly increasing in numbers.

Rabbit and Hare

European rabbits (*Oryctolagus cuniculus*) are ranked with foxes as perhaps Australia's worst pest. After their release by sport hunters in 1858, they spread across southern Australia as a "grey blanket," consuming native vegetation, pastures, and crops, and even killing small trees by chewing their bark. They now occupy 4 million km². By consuming seedlings of dominant plants such as mulga (*Acacia aneura*), they are preventing woodland regeneration across vast areas. Rabbits are considered a threat to many rare plants, and to several threatened animals whose habitats they degrade. Two biological control agents, myxomatosis and rabbit hemorrhagic disease, proved very effective at first, but rabbits have developed some resistance. Rabbits are now a major food for birds of prey, in place of extinct marsupials.

European brown hares (*Lepus europaeus*) are widespread, but their impacts are relatively minor.

Ungulates

In northern Australia, the combined biomass of introduced ungulates sometimes matches that found in African national parks.

Pigs (*Sus scrofa*) escaping from farms have become major agricultural and environmental pests, found across 38 percent of the mainland. They grub up damp soil for roots and invertebrates and raid crops at night. They are considered a threat to many native animals and plants. Fences are maintained to keep them out of important wetlands and away from stands of grass used by endangered Gouldian finches. A recent problem is pigs on remote northern beaches digging up and consuming eggs of threatened marine turtles.

Goats (*Capra hircus*) denude and erode stony slopes and inhibit woodland regeneration when they reach high numbers, as they often do on degraded sheep lands. High densities are reached only where dingoes are scarce.

Water buffalo (*Bubalus bubalis*) in the Northern Territory have severe impacts on floodplains and adjoining forests. In 1985, before a major culling program, their population reached 340,000, with densities of up to 34 per square kilometer. The swim channels created when they force their way through muddy plains have allowed seawater ingress into freshwater wetlands up to 35 km away, resulting in the death of vegetation.

Cattle (*Bos taurus*) have formed wild herds in many national parks in the north and east. In the Kimberley region, they damage scarce rainforest tracts and facilitate invasion by weeds. Banteng (*Bos javanicus*), which are Asian rainforest cattle, are degrading Garig Gunak Barlu National Park in the Northern Territory, to which they are confined by a fence.

The one-humped camel (*Camelus dromedarius*) became extinct in the wild in prehistoric times, and the Australian Outback now has the only wild populations, which date back to pioneering days when they were valued pack animals. Their impacts were thought to be minimal until the 1990s, when numbers were found to be increasing substantially, and they are now known to foul and deplete desert oases and reduce densities of favored food plants.

Donkeys (*Equus asinus*) are widespread in inland and northern regions, sometimes reaching densities as high as ten per square kilometer. In the Northern Territory, they have damaged hundreds of Aboriginal cave paintings by sheltering from rain under sandstone overhangs.

Horses (*E. caballus*) are considered an environmental and pastoral pest when they foul watering points, consume vegetation, and damage fences.

Deer, escaping from poorly managed or failed deer farms or from inadequately fenced paddocks, or having been freed by hunters to create future sporting opportunities, have rapidly increased in numbers and range since deer farming was popularized in the 1970s by books such as *Gold on Four Legs*. Six species occur in the wild: red deer (*Cervus elaphus*), sambar (*C. unicolor*), rusa deer (*C. timoriensis*), fallow deer (*Dama dama*), chital (*Axis axis*), and hog deer (*A. porcinus*). Sambar and rusa severely damage small rainforest tracts, and the other species damage other habitats and compete for food. Deer are expected to increase greatly in range and environmental and economic impacts.

Rodents

House mice (*Mus musculus*) are not excluded from natural habitats by a diverse native rodent fauna, often reaching high densities in arid shrublands and grasslands after rain. They are important prey for native owls, kites, snakes, and introduced cats. Mouse plagues are a problem for crop farmers.

Black rats (*Rattus rattus*), like house mice, now occupy natural woodlands as well as urban and rural areas, but they are not considered a conservation problem, except on islands.

BIRDS

Australia has 27 species of introduced birds, although seven of these are confined to islands. Introduced birds cannot convincingly be blamed for severe ecological impacts except on a local scale, a situation that partly reflects their failure to colonize intact habitats. Indian mynas (*Acridotheres tristis*) aggressively displace parrots from nest sites in tree hollows, but only where forest abuts farms and towns. Common blackbirds (*Turdus merula*) and common starlings (*Sturnus vulgaris*) spread the seeds of fruit-bearing weeds.

Superb lyrebirds (*Menura novaehollandiae*), taken from mainland Australia to Tasmania in a misguided conservation measure in the 1930s, may be causing greater ecological impacts than any bird introduced from overseas. By raking up large volumes of soil and litter while searching for prey, they erode slopes, alter soil dynamics, and destroy understory plants.

REPTILES

Introduced reptiles have as yet had little impact on mainland Australia. Quarantine regulations forbidding their importation as pets have restricted opportunities for introductions. The Asian house gecko (*Hemidactylus*

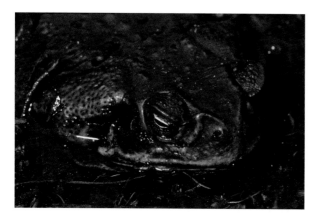

FIGURE 1 The introduced cane toad (*Rhinella marina* [*Bufo marinus*]) is now the biggest amphibian found in Australia, with large individuals reaching lengths of 15–20 cm. (Photograph courtesy of the author.)

frenatus) has become abundant in urban areas and some nearby forest edges in the tropics and subtropics, having rapidly expanded its range since the early 1980s, long after it first appeared in the Northern Territory in the mid-nineteenth century. Red-eared slider turtles (*Chrysemys scripta*), imported illegally, have established wild populations around Sydney and very recently around Brisbane, although the latter are targets for eradication using a sniffer dog. Flowerpot snakes (*Ramphotyphlops braminus*) are an accidental introduction to northern Australia.

AMPHIBIANS

The only introduced amphibian, the cane toad (*Rhinella marina* [*Bufo marinus*]) (Fig. 1), has become abundant in the northern half of Australia since its introduction to control a sugar cane pest in 1935. Toads and their tadpoles and eggs are toxic to most animals that try to eat them. Cane toads continue to spread west across northern Australia, causing mass deaths of sensitive vertebrates when they reach new areas. In 2005 the northern quoll (*Dasyurus hallucatus*), a marsupial carnivore, was declared endangered after a sharp decline attributed mainly to toad poisoning. Other animals killed include freshwater crocodiles, monitor lizards, and snakes. Many birds have learned to exploit cane toads by consuming their nontoxic parts. Cane toads have been introduced to many countries, but the most dramatic impacts have been recorded from Australia, where they often reach very high densities.

FISH

Introduced fish have major impacts on biodiversity, greatly outnumbering native fish along large sections of some catchments. Of at least 42 introduced species, seven were introduced for fishing, 30 were introduced for the aquarium trade, and four are marine species that arrived with ballast water. New species continue to become established.

Brown trout (*Salmo trutta*) and rainbow trout (*Oncorhynchus mykiss*), stocked for fishing, have caused catastrophic declines of some native fish, including four galaxias (*Galaxius* species) that are now endangered. Trout also consume tadpoles of endangered frogs. Redfin perch (*Perca fluviatilis*) are also implicated in various fish declines. Mosquitofish (*Gambusia holbrooki*) are locally abundant in many catchments following their release for mosquito control, and they are thought to threaten many small fish and some frogs, although some of the evidence is ambiguous. Several small fish, including the endangered red-finned blue-eye (*Scaturiginichthys vermeilipinnis*), seldom cooccur with mosquitofish. Common carp (*Cyprinus carpio*) now dominate large stretches of Australia's largest river, the Murray, making up 80 percent or more of biomass along some stretches; they increase turbidity and damage aquatic plants. The rapid spread of tilapia (*Oreochromus mossambicus*) in northern Australia is raising concerns. Other widespread fish include goldfish (*Carassius auratus*) and guppies (*Poecilia reticulata*).

More than 50 Australian fish have been translocated, mainly to increase fishing opportunities, with potentially serious but poorly documented impacts. Lake Eacham rainbowfish (*Melanotaenia eachamensis*) disappeared from Lake Eacham after the introduction of mouth almighties (*Glossamia aprion*), but they survive sparingly elsewhere.

INVERTEBRATES

A wide range of introduced invertebrates occurs. Two groups of particular significance are mentioned here. Other noteworthy invaders include Portuguese millipedes (*Ommatoiulus moreletii*), earthworms such as *Pontoscolex corethrurus*, European wasps (*Vespula germanica*), and Collembola that displace native decomposers. The quarantine service has successfully kept out giant African snails (*Lissachatina fulica* [formerly *Achatina fulica*]), Asian gypsy moths (*Lymantria dispar*), and various timber borers, but new crop pests periodically become established.

Ants and Bees

Tramp ants were recognized as a problem only in the twenty-first century. Red imported fire ants (*Solenopsis invicta*) raised alarms when large populations were detected in Brisbane in 2001, and a major eradication campaign, one of the world's largest, was mounted, and appears to be succeeding. Little fire ants (*Wasmannia auropunctata*), detected in Cairns in 2006, are also subject to eradication.

Yellow crazy ants (*Anoplolepis gracilipes*) have been detected at many sites in the tropics and subtropics, and although they are eradication targets in Queensland, a large population in the Northern Territory, and one on Christmas Island, are beyond eradication. Australia also has big-headed ants (*Pheidole megacephale*) invading forests, Argentine ants (*Linepitheme humile*) in urban areas, and small populations of tropical fire ants (*S. geminata*) in the Northern Territory.

Honeybees (*Apis mellifera*) are abundant almost throughout Australia, and competition with birds and native insects for floral resources, and with hole-nesting birds for tree hollows, are sometimes considered significant issues. Honeybee impacts vary locally from minor to serious, depending on the circumstances, but they are not convincingly blamed for any species declines. In fragmented landscapes, where native pollinating insects may be scarce, they are important pollinators of rare plants.

Asian honeybees (*Apis cerana*) are periodically detected in the tropics after entering on ships, and they may become established.

The bumblebee (*Bombus terrestris*) appeared in Tasmania in 1992 and has since become common and widespread in that state. It now competes with birds and native insects for nectar, and it pollinates weeds, increasing their seed output.

Marine Invertebrates

More than a hundred introduced marine invertebrates have established in Australian waters, and possibly more than twice this number, but the correct status (native or introduced) of many species is difficult to determine. The cooler waters of Victoria and Tasmania harbor the most introductions, which are attributed mainly to transport via ships' hulls or ballast water, although the Pacific oyster (*Crassostrea gigas*) was deliberately introduced.

The Northern Pacific seastar (*Asterias amurensis*) (Fig. 2), a large (0.5 m) starfish, is a voracious predator of molluscs and other invertebrates that reached very high numbers in Tasmanian waters soon after it was detected in 1986. New Zealand screwshells (*Maoricolpus roseus*) reach densities of up to 1,500 per square meter on sea floors in southeastern waters, and the large accumulations of shells, both occupied and abandoned, have altered ecosystem dynamics. The European shore crab (*Carcinus maenas*) is a major predator of native crabs and molluscs. The Asian bag mussel (*Musculista senhousia*) can dominate benthic communities in bays by forming dense aggregations that exclude other species, and a polychaete worm (*Euchone limnicola*) can reach densities of more than 2,000 per

FIGURE 2 The Northern Pacific Seastar (*Asterias amurensis*) is one of several marine pests indicative of maritime trade between Australia and Japan and Korea. (Photograph courtesy of the author.)

square meter; the European fan worm (*Sabella spallanzanii*) also achieves very high densities. An introduced chiton (*Chiton glaucus*) and barnacle (*Notomegabalanus algicola*) have become common in Tasmania.

WEEDS

Weeds have been a national concern since prickly pear (*Opuntia stricta*) engulfed large tracts of land in the nineteenth century, forcing farmers off the land. Australia now has more than 2,700 introduced weed species, with another ten species establishing each year. In addition, Australian plants grown as ornamentals outside their original ranges readily invade forest remnants, and they now represent a significant category of environmental weeds. Garden escapees make up 65 percent of new weeds. In response to its weed problems, Australia has become a world leader in biological control and weed risk assessment.

Grasses

Livestock grazing is the main land use in Australia, with stock often grazed on introduced grasses, many of which have become very invasive. Favored because they produce more biomass than native pastures, these grasses fuel intense fires if they are not grazed, invading woodlands via a grass-fire cycle. African gamba grass (*Andropogon gayanus*) fires are eight times as hot as native savanna grass fires, and several such fires can kill eucalypt trees (Fig. 3). Australian vegetation is very fireprone, and African grasses are increasing fire intensity, frequency, and reach, with profound consequences for ecosystem functioning and composition. Buffel grass (*Pennisetum ciliare* [*Cenchrus ciliaris*]) covers vast tracts of inland Australia, and the fires

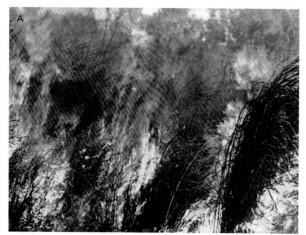

FIGURE 3 (A) Gamba grass (*Andropogon gayanus*) fires burn much hotter than fires fueled by native Australian grasses. (B) Fires fueled by gamba grass (*Andropogon gayanus*) are often so intense that eucalypt trees are killed. (Photographs courtesy of Michael Douglas.)

it fuels are reducing dry rainforest remnants. Other flammable grasses of concern include mission grass (*Pennisetum polystachion*), molasses grass (*Melinus minutiflora*), green panic or guinea grass (*Panicum maximum*), and perennial veldtgrass (*Ehrharta calycina*).

The aquatic pasture grasses hymenachne (*Hymenachne amplexicaulis*), para grass (*Brachiaria mutica*), and aleman grass (*Echinochloa polystachia*), which are grown in large seasonal ponds, have proved invasive in waterways, as has ricegrass (*Spartina anglica*) in southern estuaries. Hymenachne was declared one of Australia's 20 Weeds of National Significance only 11 years after it was released for sale to farmers in 1988.

Invasive grasses introduced to southern beaches to control erosion have altered dune structure and sediment movement, with profound consequences for native back dune vegetation. Marram grass (*Ammophila arenaria*) produces steep, high dunes, and sea wheatgrass

(*Thinopyrum junceiforme*) grows closer to the sea than any native plant.

Aquatic Weeds

On what is the driest habitable continent, aquatic weeds prove a blight when they choke dams and waterways. Water hyacinth (*Eichhornia crassipes*) and salvinia (*Salvinia molesta*) were introduced as ornamental plants, cabomba (*Cabomba caroliniana*) was introduced for fish aquaria, and alligator weed (*Alternanthera philoxeroides*) entered accidentally with dry ship ballast.

Japanese kelp (*Undaria pinnatifida*) is highly invasive along the coast of Tasmania, overgrowing and excluding native algae. *Codium fragile tomentosoides* is another invasive marine alga. The toxic dinoflagellates *Gymnodium catenatum* and *Alexandrium minutum* cause toxic blooms in Tasmanian waters, leading to closures of oyster farms.

Vines

Vines often invade habitats of high conservation value, notably rainforest remnants and riparian forests within savanna. Rubber vine (*Cryptostegia grandiflora*), bridal creeper (*Asparagus asparagoides*), and cats-claw creeper (*Macfadyena unguis-cati*) (Fig. 4) are especially serious, but there are many others. Most of Australia's 179 introduced vines are escaped garden plants.

Shrubs

Introduced shrubs dominate large areas of farmland and disturbed forests. Prickly pear occupied 24 million ha of land by 1924, almost half of it so densely that farming became impossible. Biological control employing the cactus moth (*Cactoblastis cactorum*) proved highly successful. More serious today, as reflected by their government status as

FIGURE 4 Cat's claw creeper (*Macfadyena unguis-cati*) is one of several invasive vines that can dominate riverine vegetation in Australia. (Photograph courtesy of the author.)

Weeds of National Significance, are mimosa (*Mimosa pigra*), mesquite (*Prosopis* species), lantana (*Lantana camara*), bitou bush/boneseed (*Chrysanthemoides monilifera*), gorse (*Ulex europaeus*), and blackberry (*Rubus fruticosus*). Blackberry and gorse provide cover for rabbits, foxes, and other pests, but they also protect threatened mammals from fox predation. Most of the weedy shrubs invading eucalypt forests have succulent fruits attractive to birds.

Trees

After floods in 1974 and 1988, Athel pine (*Tamarix aphylla*) colonized 600 km of the Finke River in central Australia, forming monocultures along the banks up to 30 km long. The tree was not previously considered a problem in Australia, but it is now a target of major control operations. Other invasive trees that form monocultures include prickly acacia (*Acacia nilotica*), parkinsonia (*Parkinsonia aculeata*), willow (*Salix* species), pond apple (*Annona glabra*), camphor laurel (*Cinnamomum camphora*), olive (*Olea europaea*), Chinese elm (*Celtis sinensis*), pine (*Pinus* species), and date palm (*Phoenix dactylifera*). In the northern rainforests, weed trees such as miconia (*Miconia calvescens*), harungana (*Harungana madagascariensis*), cecropia (*Cecropia peltata*), and pond apple colonize when cyclones create openings in intact forest.

PATHOGENS AND PARASITES

The disappearance of six frog species from eastern Australia, mainly from upland rainforest, and declines of other species, are attributed to chytrid fungus (*Batrachochytrium dendrobatidis*), a pathogen blamed for many frog extinctions in Latin America.

The plant pathogen *Phytophthora cinnamomi* is found worldwide, but its impacts are nowhere greater than in Australia (Fig. 5). In southwestern Australia, it has caused widespread deaths of jarrah (*Eucalyptus marginata*), a dominant forest tree, and banksias (*Banksia*) have been lost from some woodlands they once dominated. In Western Australia, *Phytophthora* poses a threat to 800 endemic plant species, including several endangered species that it threatens with extinction. Animals are affected by plant deaths, as nectar-rich shrublands give way to understories dominated by sedges. Plants are also killed in eastern Australia, within heathlands, woodlands, and rainforests, but not on the same scale.

Pathogens and parasites such as goldfish ulcer disease (*Aeromonas salmonicida*), the parasitic copepod *Lernaea cyprinacea*, and the tapeworms *Ligula intestinalis* and *Bothriocephalus acheilognathi* have entered on aquarium fish and now infect declining native fish.

FIGURE 5 Signs warning about *Phytophthora cinnamomi* dieback can be found in many Western Australian national parks. (Photograph courtesy of the author.)

An introduced disease may have contributed to the extinction of the Tasmanian thylacine and caused other marsupial declines, but proof is lacking.

ISLANDS

Australia has jurisdiction over several remote islands, where the impacts of invasive species have been catastrophic.

Two endemic rodents on Christmas Island, an Indian Ocean territory, became extinct by 1901, after contracting a trypanosome blood parasite from introduced black rats. Tens of millions of endemic red land crabs (*Gecarcoidea natalis*), which dominate the island, were killed by yellow crazy ants, whose populations expanded dramatically in the 1970s. Ongoing ant control is necessary to prevent ecosystem "meltdown." The island shrew and endemic bat are endangered, and most of the native reptiles have become very rare, with invasive species, which include the wolf snake (*Lycodon acaulis*), a centipede (*Scolopendra morsitans*), the feral cat, the black rat, and the Asian house gecko, attracting the blame.

On Lord Howe Island, 600 km east of New South Wales, black rats became established in 1918, and five birds, including two endemic species, subsequently disappeared. Rat impacts on the island's endemic invertebrates are very serious today. Pigs and cats posed a dire threat to the endangered Lord Howe woodhen but were removed. Weeds such as strawberry guava (*Psidium cattleianum*) are major concerns.

FIGURE 6 On subantarctic Macquarie Island, European rabbits (*Oryctolagus cuniculus*) have caused severe erosion of steep slopes. (Photograph courtesy of the author.)

On Norfolk Island, 1,600 km northeast of Sydney, black rats, probably in concert with cats, caused the extinction of four birds and today threaten the endangered Norfolk Island Green parrot. Philip Island nearby was almost entirely denuded by pigs, goats, and rabbits, but these have been removed.

On subantarctic Macquarie Island, slopes have been destabilized and vegetation denuded (Fig. 6) by European rabbits, which have increased in numbers since cats were exterminated to stop their preying on seabirds. Eradication of the rabbits, and also black rats, which prey on seabirds, is planned.

CONCLUSIONS

The impacts of invasive species have been greater in Australia than on any other continent. Australia has lost far more mammals to extinction in modern times than any other landmass, with introduced predators implicated in many, if not most of the extinctions. The introduced rabbit is now considered Australia's most abundant mammal, and introduced foxes, cats, and dingoes are the dominant predators. Feral ungulates sometimes achieve densities rivaling those found in Africa. Introduced mammals are a significant presence in most habitats, very few of which retain a complete native mammal fauna.

Many introduced species besides mammals have become abundant and ecologically damaging. Chytrid fungus is blamed for the extinction of six frog species since the 1970s. The plant pathogen *Phytophthora cinnamomi* has transformed habitats by killing susceptible shrubs and trees, and it threatens the survival of many plants. Introduced fish now dominate major sections of inland catchments, and cane toads have become one of the most abundant vertebrates in northern Australia. Introduced starfish, molluscs, and algae are now dominant in some temperate marine habitats.

More than 2,700 introduced plants have naturalized, and many are highly invasive. African pasture grasses are a special concern because they fuel intense fires that can kill trees and alter ecosystem functioning and composition.

This article and its author sponsored by the Wettenhall Foundation.

SEE ALSO THE FOLLOWING ARTICLES

Acclimatization Societies / Algae / Ants / Carp, Common / Fire Regimes / Grazers / Invertebrates, Marine / Islands / *Phytophthora* / Rodents

FURTHER READING

Groves, R. H. 1998. *Recent Incursions of Weeds to Australia 1971–1995*. Adelaide: CRC for Weed Management Systems.

Johnson, C. 2006. *Australia's Mammal Extinctions: A 50,000 Year History*. Melbourne: Cambridge.

Lach, L., and M. L. Thomas. 2008. Invasive ants in Australia: Documented and potential ecological consequences. *Australian Journal of Entomology* 47: 275–288.

Lintermans, M. 2004. Human-assisted dispersal of alien freshwater fish in Australia. *New Zealand Journal of Marine and Freshwater Research*. 38: 481–501.

Listed Key Threatening Processes. www.environment.gov.au/cgi-bin/sprat/public/publicgetkeythreats.pl

Low, T. 1999. *Feral Future: The Untold Story of Australia's Exotic Invaders* Melbourne: Penguin.

Rolls, E. C. 1969. *They All Ran Wild: The Story of Pests on the Land in Australia*. Sydney: Angus and Robertson.

AVIAN FLU

SEE INFLUENZA

B

BALLAST

JAMES T. CARLTON

Williams College, Mystic, Connecticut

Ballast is any material that is used to help stabilize a vessel at sea. Adding weight into a ship increases its draft, meaning that it lowers the ship into the water. When ballast is properly loaded, the vessel's trim—its fore and aft leveling—can also be changed, propeller immersion increased, and stress loads adjusted. Ballasting a ship is thus a fundamental part and principle of sailing, and, therefore, vast amounts of ballast have been moved around the world for centuries. Solid ballast, which typified vessels until the middle of the twentieth century, included soil and rocks, often rich with associated animal and plant life. Water ballast, which began to be used in the nineteenth century, similarly engages all the life in it at the time the water is boarded into the ship. Both ancient and modern forms of ballast have thus been some of the most profound vectors for moving species across and between oceans. It is probable that careful analysis of historical shipping lanes, ballast records, and the disjunct distributions that characterize so many animals and plants would reveal that the movement of ballast has played a far greater role in mediating invasions than has been previously suspected.

SOLID BALLAST

That ballast played a fundamental physical role in coastal seaport communities is reflected in names that remain today along coastlines worldwide: Ballast Point, Ballast Cove, Ballast Island, Ballast Reef, Ballast Lane, and Ballast Hill. A full-time career in the 1700s and 1800s was that of "ballast master," a port officer, one of whose duties was to oversee where ballast was jettisoned so as not to fill up harbor channels. Ballast that inadvertently engaged living organisms typically consisted of soil, rocks, gravel, and sand; other materials used as ballast that may have also borne animals and plants included coral, oyster shells, and cowry shells. These materials were transported between all continents of the world and then dumped both near and on the shore. Ballast was placed in the bottom of ships' holds; the environmental conditions in these holds in the days of solid ballast are poorly known.

Plant Introductions

Ballast loading and unloading were sufficiently conspicuous and well-known enterprises that they did not escape the attention of early naturalists, who began to ascribe to ballast traffic an increasing number of exotic plants, insects, and seashells discovered far from their known or presumed homes. Plants dispersed by solid ballast became known as "ballast waifs," and by the mid-1800s, botanists were actively seeking out alien plants in seaports. In the summer of 1877, the naturalist Isaac Martindale wrote that "A few days ago I paid my first visit of the season to the ballast ground near Philadelphia," in a search for European plants. His note in the *Botanical Gazette* concludes with an early statement about the fascination of biological invasions: "This transportation of seeds and consequent introduction of new plants . . . is an interesting matter for consideration, and as I review these ballast deposits, and detect so many strangers, I feel a re-awakening of that interest which a ramble about our fields and woodlands fails to create."

So thoroughly did these Victorian era naturalists and botanists work the Philadelphia ballast grounds that a rare data set was yielded documenting the number of species that were inoculated by seed, and then found growing, but which failed to colonize the continent. No fewer than 81 species of plants known only from ballast heaps are recorded for the flora of Pennsylvania; none has been seen since the 1800s. These plants represented Mediterranean, African, and South American species, testimony to the scale of traffic arriving in Philadelphia harbor in the late nineteenth century. Ballast plants can reveal the exact nature of the ballast used: one British species found growing on ballast in the harbor was the wall-inhabiting sticky-weed *Parietaria judaica*, indicating the use of wall rubble as ballast.

Today, on all continents but Antarctica, a wide variety of species of plants, worms, insects, and isopods, many of which have now moved far inland, are thought to have been introduced by ships dumping foreign soil, rocks, and gravel. In North America, one of the most famous of these is purple loosestrife (*Lythrum salicaria*), although this plant and many other species are demonstrably polyvectic, having also been potentially introduced with livestock feed, in forage crops, in grain shipments, in mattress stuffing, through the horticulture trade, and so on. The type locality of the native North American saltwater cordgrass, *Spartina alterniflora*, is France, where this plant was introduced with ballast probably in the 1700s. *Spartina alterniflora* then crossbred with the native *Spartina maritima* in England, giving rise to a new species, *Spartina anglica*.

The concern for the introduction of noxious and harmful weeds by solid ballast inspired discussion of control regulations in the 1890s, which included quarantine and chemical treatment possibilities and the observation that ballast released at sea (so as to not deposit plant seeds on land) would only result in the widespread distribution of exotic species all along a coast. These concerns foreshadowed nearly identical discussions about ballast water that commenced 100 years later.

Ballast Beetles

A great many coastal arthropods were moved by solid ballast over the centuries, a record that has not been worked out with nearly the historical and biogeographical sophistication of the plant dispersal record. One of the most well-investigated histories of ballast-mediated insect dispersal is the classic work of Carl Lindroth (1911–1975), Professor of Entomology at Lund University, whose 1957 *The Faunal Connections Between Europe and North America* is one of the foundation works of invasion biology. Lindroth sought to understand the biogeographic history and origins of carabid beetles common to Europe and Newfoundland and in so doing established the long history of ballast traffic across the North Atlantic. Lindroth defined criteria for recognizing introduced species, speculated about the minimum size of a viable population as applied to invasion ecology, and analyzed a suite of factors (including soil type, moisture requirements, level of food specialization, and reproductive biology, among others) that influenced which beetles could have been transported (174 species) compared to those that were in fact introduced (68). Lindroth noted that "it can be concluded that the selection of animal species which managed to cross the Atlantic by ships, first and foremost in [shore] ballast, has not been a random one."

Sand, Beach, and Rocky Shore Ballast Introductions

Many strictly shore-dwelling species now owe their distribution to the movement of sand and rocks as ballast. Sand and associated beach debris have been used for thousands of years as ballast: the *corpus saburrariorum* was the guild of sand ballast heavers in ancient Rome. Along with sand, beach plants, insects, isopods, and amphipods—the last commonly known as beach-hoppers or shore-hoppers—were widely dispersed, with European plants appearing on Australian beaches, and European beach-hoppers colonizing South African shores. Such far-flung populations were often redescribed as native species, and not a few may still hide under such names around the world as pseudoindigenous species. Thus, the North Atlantic shore snail *Truncatella pulchella* was cast ashore on the Pacific coast of Panama and described as a native species, *T. bairdiana*, in 1852; the European shore snail *Myosotella myosotis* was renamed no fewer than eight times around the world, including from Peru as early as 1837 and from Australia in the 1870s. Numerous coastal isopods were renamed again and again as they were spread around the world.

Many meiofaunal species, including protozoans, ostracods, copepods, worms, and other taxa, are said to be naturally cosmopolitan, but the use of sand as ballast calls into question how natural many of these distributions are.

A famous introduction, associated in part with the movement of sand ballast, was the invasion of Africa by the South American sand flea (jigger or chigoe flea) *Tunga penetrans* (Siphonaptera, Hectopsyllidae), which burrows into the feet of humans. Infested sand ballast (and infested crew) from Rio de Janeiro aboard the ship *Thomas Mitchell* came ashore in Angola in September 1872. The sand fleas soon spread in the coastal area. *Tunga* causes tungiasis, a serious human skin disease, which now plagues people throughout sub-Saharan Africa and India.

In the same vein, a blood-sucking sand fly, the white no-no, *Styloconops albiventris* (Diptera, Ceratopogonidae) was introduced in sand ballast from Papua New Guinea to the Marquesas by German ships at the start of World War I. Marquesans named it *no-no purutia*, or Prussian no-no.

That sand could be infested with living organisms did not pass without notice by ships' crews: in the 1860s, a ship that had taken aboard sand in New Jersey was found within a few days to be contaminated with such huge numbers of newly hatched horseshoe crabs (*Limulus polyphemus*) that the entire load had to be thrown overboard.

The role of ballast moved between rocky shores in influencing the distribution of intertidal species is not well known and remains a rich area for study. The European periwinkle *Littorina littorea* and the European rocky shore seaweed *Fucus serratus* are both thought to be ballast rock introductions to North America, although *L. littorea* may have arrived as a popular food item as well. The presence of the African air-breathing limpet *Siphonaria pectinata* in the Gulf of Mexico and Caribbean may be a remnant of ballast traffic.

WATER BALLAST

In contrast to the multimillennial history of solid ballast, the transoceanic movement of water as ballast did not come into common use until the 1870s and 1880s, with the widespread conversion from wooden to steel ships, with the transition from sail to steam, and with the design of watertight bulkheads. Ballast water may be freshwater, brackish (estuarine) water, or fully marine water. The water is held in ballastable cargo holds, in port and starboard ballast tanks, in fore and aft tanks, in double-bottom tanks, and in other spaces. The water can be from various geographic sources within one ship, and the environmental conditions may vary widely between ballast tanks and ballasted holds (Fig 1.).

FIGURE 1 (A) Sampling a topside ballast tank aboard a bulk cargo vessel with a zooplankton net. (B) Sampling ballast water in a flooded cargo hold of a bulk carrier. (Photographs courtesy of the author.)

The organisms in the ballast are whatever species are in the water column at the time the water is gravitated or pumped into a vessel. These include nekton (including fish up to 30 cm in length), holoplankton, and vertically migrating plankton. Also transported are viruses and bacteria, including cholera bacteria. Benthic species are boarded as eggs or larvae (meroplankton) (Fig. 2) and as juveniles and adults attached to small pieces of floating wood, seaweed, seagrass, plastic, or other material. Thus, benthic organisms with nonplanktonic larvae may also be transported. Infaunal benthic species with nonplanktonic larvae may also be caught up in ballast water as tychoplankton. Larvae may also settle on tank bottoms (forming benthic communities) or on tank or hold walls (forming fouling communities); the populations so created may then reproduce, producing more larvae. Scores of species may occur in one ship. In the ballast water of just nine ships (from Japan, Europe, the Caribbean, and elsewhere in North America) entering the Port of Morehead City, North Carolina, 342 different phytoplankton species were found. Ballast water thus successfully transports around the world a staggering variety of marine life representing a wide variety of habitats and trophic guilds.

Marine, estuarine, and freshwater organisms may have been distributed in water inside ships long before ballast water came into use, but little is known about the potential of nonballast water in vessels to transport living organisms. Bilge water, in particular, is a combination of ambient water, rain water, and general seepages that accumulate in the bilges, or the lowest internal parts of a ship, above the keelson. Ambient water with living organisms, including larvae, may have entered the bilges of old sailing ships through holes created by shipworms and gribbles.

FIGURE 2 Polychaete worm larva intercepted alive in ballast water arriving from Japan and released in Oregon. (Photograph courtesy of J. Goddard.)

Crews drew this water out by hand-operated pumps; the water would then be brought up to deck level and would flow overboard through scupper holes or freeing ports.

Ballast Water Introductions

Hundreds of invasions of nonindigenous species around the world have been attributed to ballast water transport; examples are in Table 1. The biogeographic and historic resolution of this vector is coarse grained: thousands of species may have been introduced by ballast water, especially many smaller taxa that are often mistakenly presumed to be naturally widespread (a bias in invasion diversity assessment known as the "smalls rule"). Detecting the full extent of ballast water invasions is further challenging because many species are polyvectic: barnacles, for example, may be transported by ships both externally (in hull fouling) or internally (as planktonic larvae), and the same is true for a great many other fouling taxa that could, as noted above, alternatively be transported either in the planktonic stage or as juveniles or adults on small pieces of floating material.

With the appearance of the Asian diatom *Odontella sinensis* in the North Sea in the early 1900s, interoceanic transport of seawater came to the attention of a few marine biologists. Less than a decade later, the Asian mitten crab *Eriocheir sinensis* appeared in Germany, an event also linked to ballast water transport. Ballast water as a possible, probable, or certain vector was then steadily recognized with increasing frequency for the rest of the century. The relatively few invasions shown in Table 1 prior to the 1960s, compared to the number of invasions shown from the 1970s and after, is approximately representative of the relative numbers of invasions accredited to ballast water over time.

The fewer ballast-mediated invasions reported in the first half of the twentieth century may be due to the general failure to recognize the widespread existence of this vector; as with solid ballast, careful study will reveal more examples of earlier introductions mediated by ballast water. Nevertheless, ballast water appears to have played an increasingly important role in the second half of the century. The number, size, routes, and speed of oceangoing vessels have increased steadily, concomitant with the economic forces driving increasing global trade. In turn, the stage was set for both an increased scale of inoculation (due to more water being moved) and an increased survival (because of faster ship speeds) of propagules of nonnative species. An increase in water quality both outside and inside the ship may have further increased invasion potential. Coastal water quality began to improve in the

TABLE 1
Examples of Ballast Water–Mediated Invasions around the World

Species	Introduced			Era
	From	⇒	To	
Odontella sinensis diatom	Asia		Europe	1900s
Eriocheir sinensis mitten crab	China		Europe	1900s
Acartia tonsa copepod	North America		Europe	1910s
Eurytemora americana copepod	North America		Europe	1930s
Omobranchus punctatus fish (blenny)	Indo-West Pacific		Caribbean	1930s
Attheya armatus diatom	Australasia		North America (Pacific)	1940s
Oithona oculata copepod	Asia		Caribbean	1950s
Praunus flexuosus mysid (opossum shrimp)	Europe		North America (Atlantic)	1950s
Palaemon macrodactylus shrimp	Asia		North America (Pacific)	1950s
Alexandrium catenella dinoflagellate	Asia		Australia	1950s
Pleurosigma simonsenii diatom	Indian Ocean		Europe	1960s
Acanthogobius flavimanus	Asia		North America (Pacific)	1960s
Tridentiger trigonocephalus fish (goby)	Asia		North America (Pacific)	1960s
Theora fragilis clam	Asia		North America (Pacific)	1960s
Odontella sinensis diatom	Europe		North America (Atlantic)	1960s
Coscinodiscus wailesii diatom	North Pacific		North America (Atlantic)	1960s
Coscinodiscus wailesii diatom	North Pacific?/North America (Atlantic)?		Europe	1970s
Ensis directus razor clam	North America		Europe	1970s
Ilyocryptus agilis cladoceran (water flea)	Europe		North America (Atlantic)	1970s
Neomysis americana mysid (opossum shrimp)	North America (Atlantic)		South America (Atlantic)	1970s
Limnoithona sinensis	Asia		North America (Pacific)	1970s
Oithona davisae	Asia		North America (Pacific)	1970s
Sinocalanus doerii copepod	Asia		North America (Pacific)	1970s
Nippoleucon hinumensis cumacean	Asia		North America (Pacific)	1970s
Deltamysis holmquistae mysid (opossum shrimp)	Asia		North America (Pacific)	1970s
Tridentiger trigonocephalus	Asia		Australia	1970s
Acanthogobius flavimanus fish (goby)	Asia		Australia	1970s
Gymnodinium catenatum dinoflagellate	Asia		Australia	1970s
Centropages abdominalis copepod	Asia		Chile	1970s
Hemigrapsus takanoi crab	Asia		Europe	1980s
Marenzelleria spp. polychaete worm	North America		Europe	1980s

(Continued)

TABLE 1
(Continued)

Species	Introduced		Era
	From ⇒	To	
Mnemiopsis leidyi comb jellyfish	North America	Black Sea	1980s
Coscinodiscus wailesii diatom	Europe/North Pacific	Brazil	1980s
Hemigrapsus sanguineus crab	Asia	North America (Atlantic)	1980s
Membranipora membranacea bryozoan	Europe	North America (Atlantic)	1980s
Dreissena polymorpha	Eurasia	North America (Great Lakes)	1980s
Dreissena bugensis zebra mussel	Eurasia	North America (Great Lakes)	1980s
Apollonia melanostomus	Eurasia	North America (Great Lakes)	1980s
Proterorhinus semilunaris	Eurasia	North America (Great Lakes)	1980s
Gymnocephalus cernuus fish (goby; ruffe)	Eurasia	North America (Great Lakes)	1980s
Bythotrephes cederstroemi cladoceran (water flea)	Europe	North America (Great Lakes)	1980s
Thalassiosira baltica diatom	Europe	North America (Great Lakes)	1980s
Rangia cuneata clam	southern United States	northern United States (Hudson River)	1980s
Acartia omorii copepod	Asia	Chile	1980s
Pseudodiaptomus forbesi	Asia	North America (Pacific)	1980s
Pseudodiaptomus inopinus	Asia	North America (Pacific)	1980s
Pseudodiaptomus marinus copepod	Asia	North America (Pacific)	1980s
Corbula amurensis clam	Asia	North America (Pacific)	1980s
Palaemon macrodactylus shrimp	Asia	Europe	1990s
Rapana venosa Asian whelk	Mediterranean	North America (Atlantic)	1990s
Convoluta convoluta flatworm	Europe	North America (Atlantic)	1990s
Potamopyrgus antipodarum snail	Europe	North America (Great Lakes)	1990s
Echinogammarus ischnus amphipod	Europe	North America (Great Lakes)	1990s
Cercopagis pengoi cladoceran (water flea)	Europe	North America (Great Lakes)	1990s
Eurytemora americana copepod	northern hemisphere	Argentina	1990s
Hyperacanthomysis longirostris	Asia	North America (Pacific)	1990s
Acanthomysis hwanhaiensis	Asia	North America (Pacific)	1990s
Acanthomysis aspera mysid (opossum shrimp)	Asia	North America (Pacific)	1990s
Acartiella sinensis	Asia	North America (Pacific)	1990s
Limnoithona tetraspina	Asia	North America (Pacific)	1990s
Tortanus dextrilobatus copepod	Asia	North America (Pacific)	1990s
Nuttallia obscurata clam	Asia	North America (Pacific)	1990s
Philine auriformis sea slug	New Zealand	North America (Pacific)	1990s
Neomysis japonica mysid (opossum shrimp)	Asia	North America (Pacific)	2000s
Assiminea parasitologica snail	Asia	North America (Pacific)	2000s

FIGURE 3 Oceangoing vessels carry abundant marine life in ballast water, on the hull, in sea chests, and in other compartments and spaces. (Photograph courtesy of D. Reid.)

last few decades of the twentieth century, due to increasingly aggressive water pollution legislation in the United States and elsewhere. Concomitantly, tanker vessels were required to switch from transporting ballast water in oily cargo holds to separate dedicated (or so-called segregated) ballast tanks. The increased number of invasions over time is not due to increased search activity for exotic species: rather, both coastal exploration by zoologists, botanists, and naturalists in general and the number of expert taxonomists able to identify organisms have declined steadily and dramatically over these same decades.

Today, billions of gallons of ballast water are moved around the world annually, transporting on any one day thousands of species in tens of thousands of vessels (Fig. 3).

The Management of Ballast Water as a Vector of Alien Species Invasions

The invasion of Japanese dinoflagellates in Australia in the late 1970s and of Eurasian zebra mussels (*Dreissena polymorpha* and *D. bugensis*, the latter also known as the quagga mussel) in the Laurentian Great Lakes in North America in the 1980s planted the seeds for renewed attention in the early 1990s for the need to address global ballast water management at the United Nation's International Maritime Organization (IMO) in London. Precedent had been set with the IMO's earlier formulation in the 1970s of "Marpol," the International Convention for the Prevention of Pollution from Ships.

After many years of discussion, the IMO issued the International Convention for the Control and Management of Ships' Ballast Water and Sediments in February 2004. This convention calls for ballast regulation with the goals of preventing or minimizing the introduction of nonnative species. The convention enters into force 12 months after ratification by 30 nations, representing 35 percent of global shipping tonnage. Around the world, individual nations, and in some cases individual jurisdictions (such as states within the United States), have promulgated separate federal or lower-level ballast management regulations.

Short-term ballast water management strategies globally focus on "ballast water exchange": fresh, estuarine, or coastal ballast water is deballasted in the open ocean, followed by reballasting. The original neritic ballast life is thus exchanged for open ocean marine life, which, when later released in ports and harbors, cannot survive. However, complete deballasting is often not mechanically possible and may further be unsafe due to sea conditions. Long-term ballast management solutions thus focus on shoreside or shipboard treatments to remove most living organisms from the water prior to its release elsewhere in the world. Potential mechanisms to achieve this goal include mechanical treatment (such as filtration, centrifugation, or cavitation), chemical treatment, physical treatment (heat, ultrasound, ultraviolet irradiation, or deoxygenation), or combinations thereof.

SEE ALSO THE FOLLOWING ARTICLES

Algae / Canals / Crabs / Crustaceans / Fishes / Invertebrates, Marine / Seas and Oceans

FURTHER READING

Carlton, J.T. 1985. Transoceanic and interoceanic dispersal of coastal marine organisms: The biology of ballast water. *Oceanography and Marine Biology, An Annual Review* 23: 313–371.

Carlton, J.T., and J.B. Geller. 1993. Ecological roulette: The global transport of nonindigenous marine organisms. *Science* 261: 78–82.

Drake, L.A., M.A. Doblin, and F.C. Dobbs. 2007. Potential microbial bioinvasions via ships' ballast water, sediment, and biofilm. *Marine Pollution Bulletin* 55: 333–341.

Lindroth, C.H. 1957. *The Faunal Connections between Europe and North America.* New York: Wiley.

Mack, R.N. 2003. Global plant dispersal, naturalizations, and invasion: Pathways, modes, and circumstances (3–30). In G.M. Ruiz and J.T. Carlton, eds. *Invasive Species: Vectors and Management Strategies.* Washington, DC: Island Press.

McCarthy, H.P., and L.B. Crowder. 2000. An overlooked scale of global transport: Phytoplankton species richness in ships' ballast water. *Biological Invasions* 2: 321–322.

National Research Council. 1996. *Stemming the Tide: Introductions of Nonindigenous Species by Ships' Ballast Water.* Washington, DC: National Academy of Sciences.

Wonham, M.J., J.T. Carlton, G.M. Ruiz, and L.D. Smith. 2000. Fish and ships: Relating dispersal frequency to success in biological invasions. *Marine Biology* 136: 1111–1121.

BEES

DAVID GOULSON

University of Stirling, Scotland, United Kingdom

Bees (superfamily Apoidea) belong to the large and exceedingly successful insect order Hymenoptera, which also includes wasps, sawflies, and ants. There are currently approximately 25,000 known species of bee, and undoubtedly many more remain to be discovered. All bees feed primarily on nectar and pollen throughout their lives, with the adults gathering food for their sedentary larvae. The life history of bees varies across a spectrum, from solitary species (which make up the vast majority of species) to those that are highly social, living in large colonies with sometimes hundreds of thousands of individuals. The most familiar bee is of course the domesticated honeybee (*Apis mellifera*; Fig. 1), highly valued for its role as a crop pollinator and source of honey.

BEES AS INVASIVE SPECIES

Because of the obvious benefits they provide to humankind, various bee species have been deliberately introduced in parts of the world to which they are not native.

FIGURE 1 The honeybee, *A. mellifera*, now perhaps the most widespread insect on Earth. (Photograph courtesy of the author.)

The honeybee is thought to be native to Africa, western Asia, and southeastern Europe, although its association with humans is so ancient that it is hard to be certain of its origins. It has been domesticated for at least 4,000 years. Because of its economic value, the honeybee has been introduced to every country in the world (being absent only from the Antarctic). It is now among the most widespread and abundant of insects on Earth.

Various bumblebee (*Bombus*) species have also been deliberately introduced to new countries. The earliest successful bumblebee introduction was to New Zealand in 1885 (no bumblebees naturally occur in Australia or New Zealand). Four species became established and survive to this day. During the late 1980s, the commercial rearing of bumblebees was developed, primarily for pollination of glasshouse tomatoes, and this has since developed into a worldwide trade with in excess of 1 million nests per year being exported from Europe to at least 19 countries around the globe. The main trade is in the European species *Bombus terrestris*. As a result of this trade, *B. terrestris* became established in the wild in Japan in the 1990s. In 1992 *B. terrestris* arrived in Tasmania from New Zealand, perhaps having been accidentally transported in cargo. Recently, *B. terrestris* was deliberately introduced to Chile. This is the second U.K. species to arrive in Chile, for *B. ruderatus* was introduced in 1982; it had spread across the Andes to Argentina by 1993, and *B. terrestris* arrived there in 2006.

Other bee species were deliberately introduced far from their native range in the twentieth century. The alfalfa leafcutter bee (*Megachile rotundata*), a native of Eurasia, has been introduced to North America, Australia, and New Zealand for alfalfa pollination. At least six other leafcutter bee species (Megachilidae) have been introduced to the United States for pollination of various crops, mainly from Europe and Japan. The alkali bee (*Nomia melanderi*), a native of North America, was introduced to New Zealand for pollination of alfalfa and has become established.

Because bees are generally regarded as beneficial organisms, should these introductions be a cause for concern? There are a number of possible undesirable effects of exotic bees, including (1) competition with native species, (2) introgression with native species, (3) transmission of parasites or pathogens to native species, (4) pollination of exotic weeds.

Competition with Native Species

The two bee species that are most widespread outside their native range, the honeybee and the bumblebee *B. terrestris*, are both generalists. Honeybees usually visit a hundred or more different species of plant within any one region and in total have been recorded visiting nearly 40,000 different plant species—and *B. terrestris* is similarly polylectic, having been recorded visiting 419 plant species in New Zealand alone. Honeybees and bumblebees differ from many other flower visitors in having a prolonged flight season; honeybees remain active for all of the year in warmer climates, while bumblebees commonly forage throughout the spring and summer in the temperate climates where they naturally occur. Thus, the wide distribution, broad diet, and long field season of these two bee species mean that their niches overlap with many thousands of different native species.

Of course, demonstration of niche overlap is not proof of competition. If flowers are abundant, there may be plenty of nectar and pollen to go around. In fact it is notoriously difficult to provide unambiguous evidence of competition, particularly in mobile organisms such as bees. Both bumblebees and honeybees begin foraging earlier in the morning than many native insects; they are able to do so because their nests are kept warm through the night. Studies in Tasmania suggest that the combined action of nonnative bumblebees and honeybees removes 90 percent of the available nectar before native bees have begun to forage. This could give these nonnative organisms a competitive advantage over most native insects. This and other studies demonstrate that the presence of high densities of either honeybees or bumblebees can depress availability of floral resources, and there is good evidence that this can displace native organisms from the most profitable flowers. For example, the presence of honeybees on particular nectar sources has been found to deter foraging by hummingbirds.

Asymmetries in competition may also occur because of the ability of honeybees and bumblebees to communicate the availability and/or location of valuable food sources with nestmates, thus improving foraging efficiency. In contrast, the majority of other flower visitors are solitary, and each individual must discover the best places to forage by trial and error. Thus, social species are collectively able to locate new resources more quickly, which again may enable them to gather the bulk of the resources before solitary species arrive.

Asymmetries in competition may not be stable, because the relative competitive abilities of bee species are likely to vary during the day, according to temperature and resource availability, and are likely to vary spatially according to the types of flowers available. Bumblebees and honeybees are large compared to most of the native species with which they might compete. Large bees are at a competitive advantage in cool conditions because of their ability to maintain a body temperature considerably higher than the ambient air temperature. However, large bees are not always at an advantage. The energetic cost of foraging is approximately proportional to weight; thus, large bees burn energy faster. As nectar resources decline, the marginal rate of return will be reached more quickly by large bees. Thus, large bees are likely to be at a competitive advantage early in the day and during cool weather, but small bees can forage profitably even when rewards per flower are below the minimum threshold for large bees; at these times, honeybees and bumblebees may survive by using honey stores. Small insects are also able to maintain activity in high ambient temperatures, when bumblebees would swiftly overheat. Thus, the relative competitive abilities of different bee species are not consistent, and the strength of competition is likely to vary with time of day, season, and according to what types of flower are available.

Is there evidence that competitive effects of nonnative bees reduce the populations of indigenous species? Studies in Argentina, Israel, and on islands near Japan have all found that native flower-visitor abundance was lower in places where either nonnative bumblebees or honeybees were more abundant. In Japan, the arrival of *B. terrestris* has led to declines in the native bumblebee *B. hypocrita*, and this is thought to be at least in part due to competition for nest sites. However, such studies can be criticized on the grounds that the relationship between invasive bee abundance and declining native bee populations (if found) need not be causative. Increasing invasive bee populations are often associated with increased environmental disturbance by humans, which may itself explain declines in native bees.

The only way to test unequivocally whether floral resources are limiting and competition is in operation is to conduct experiments in which the abundance of the introduced bee species is artificially manipulated, and the population size of native species is then monitored. Removal of feral honeybee nests and domesticated hives from part of Santa Cruz Island in California resulted in marked increases in numbers of native bees and other flower-visiting insects. Similarly, a decrease in abundance of native insects was found when hives of Africanized honeybees were placed in forests in French Guiana. Native bumblebee (*B. occidentalis*) nests placed near honeybee hives in California brought back less food to the nest and produced fewer offspring than those that were not near honeybee hives.

To summarize, it seems almost certain that abundant and widespread exotic organisms that singlehandedly use a large proportion of the available floral resources do impact the local flower-visiting fauna. Consider, for example, the Australian native bee community. Australia has over 1,500 known bee species, and many more probably exist. Nowadays, by far the most abundant flower-visiting insects throughout most of Australia are honeybees, often outnumbering all other flower-visiting insects by a factor of ten or more. In Tasmania, the second most abundant flower visitor is usually the bumblebee *B. terrestris*. The majority of floral resources are gathered by these bees, often during the morning before native bees have become active. It is hard to imagine how the introduction of these exotic species could not have substantially altered the diversity and abundance of native bees. Unfortunately, we will never know what the abundance and diversity of the Australian bee fauna was like before the introduction of the honeybee. The same applies to most other regions such as North America, where the honeybee has now been established for nearly 400 years. It is quite possible that some, perhaps many, native bee species were driven to extinction by the introduction of this numerically dominant species or by exotic pathogens that arrived with it. Even if it were practical or considered desirable to eradicate honeybees from certain areas, it would be too late for such species.

Introgression with Native Species

The global trade in bumblebees poses a threat to genetic diversity that has received very little attention. The trade is largely in *B. terrestris dalmatinus* from southeastern Europe, which are shipped throughout the range of *B. terrestris*, which consists of a number of distinct subspecies: *B. terrestris terrestris* in much of Western Europe, *B. t. audax* in Great Britain and Ireland, *B. lusitanicus* in Iberia, and various named subspecies on different islands in the Mediterranean and Canary Islands. In a laboratory setting, the subspecies readily interbreed, but this does not necessarily mean that they will interbreed in a natural setting. The transport of *B. t. dalmatinus* throughout Europe poses the threat that the distinct local races will be lost through introgression, resulting in an overall loss of genetic diversity within the species. However, there has been no attempt to ascertain whether this is happening.

Nonnative bees also pose a different threat through interspecific matings. In 2007, 30 percent of queens of the native Japanese bumblebee *B. hypocrita* were found to have mated with *B. terrestris* males, matings that result in no viable offspring and so effectively sterilize the queens. Such interspecific mating is to be expected among closely related species and is probably contributing to the decline of *B. hypocrita*.

Transmission of Parasites or Pathogens to Native Organisms

Bees and their nests support a diverse array of predatory, parasitic, and commensal organisms, including viruses, bacteria, protozoans, mites, nematodes, fungi, and parasitoid wasps and flies. There is no doubt that many bee parasites have been transported to new regions with their hosts, particularly where introductions were made many years ago when awareness of bee natural enemies was low. Thus, for example, the honeybee fungal disease chalkbrood, foulbrood (a bacteria disease), the microsporidian *Nosema apis*, and the mite *Varroa destructor* now occur throughout much of the world. Similarly, bumblebees in New Zealand are host to a parasitic nematode and three mite species, all of which are thought to have come from the United Kingdom with the original introduction of bees. Studies in Japan have demonstrated that *B. terrestris* imported from Europe are frequently infested with tracheal mites. Exposure of hosts to novel strains of mite can have dramatic consequences, as demonstrated by the recent spread of *V. destructor* in honeybees. There is strong circumstantial evidence that the most dramatic declines that have been observed in any bumblebee species are the result of exposure to a nonnative pathogen. In the 1990s, queens of various North American species were taken from North America to Europe and reared in factories alongside the European *B. terrestris*. The established nests were then returned to North America. Shortly afterward, *B. occidentalis*, *B. terricola*, and *B. affinis*, all widespread and abundant species, disappeared from much of their range. These three species, which are all closely related, belong to the subgenus *Bombus*. The only other North American member of this subgenus, *B. franklini*, was always very rare but has recently disappeared from former localities and may be extinct. Thus, an entire subgenus has been devastated across a continent in the space of a few short years. It is hard to conceive of an explanation for this decline that does not invoke a disease outbreak. Anecdotal evidence suggests that a nonnative strain of the disease organism *Nosema bombi* was transported to North America with the commercial colonies, but in truth we shall probably never know.

It is hard to exaggerate our ignorance of the natural enemies of most bee species, particularly their pathogens. We do not know what species infect them or what the host ranges of these pathogens are. The natural geographic range of bee pathogens is almost wholly unknown. Given the current collapse of honeybee populations in North America and perhaps also in Europe, thought to be driven by one or more viral diseases perhaps interacting with parasitic mites, there is an urgent need to improve our understanding of the biology of bee diseases. In the meantime, legislation to enforce strict quarantine of all bees prior to transportation would seem to be sensible.

Pollination of Exotic Weeds

As we have seen, both honeybees and invasive bumblebees visit a broad range of flowers. They appear to prefer to visit nonnative flowers; for example, in New Zealand, *B. terrestris* has been recorded visiting only 19 native species but 400 exotic plants, almost all from the natural range of *B. terrestris* in Europe. These preferences presumably occur because the bees tend to gain more rewards by visiting flowers with which they are coadapted.

Do visits by exotic bees improve seed set of weeds? By virtue of their abundance and foraging preferences, they often make up a very large proportion of insect visits to weeds. In a site dominated by European weeds in Tasmania, honeybees and bumblebees were found to comprise 98 percent of all insect visits to the problematic weed creeping thistle (*Cirsium arvense*). In North America, honeybees increase seed set of the yellow star thistle (*Centaurea solstitialis*) and are the main pollinators of the invasive weed purple loosestrife (*Lythrum salicaria*). Of the 33 worst environmental weeds in New Zealand, 16 require pollination and are visited by honeybees, and

one is pollinated more or less exclusively by them (the barberry shrub, *Berberis darwinii*). In addition, the tree lupin (*Lupinus arboreus*), broom (*Cytisus scoparius*), and gorse (*Ulex europeaus*) are self-incompatible and rely on pollination by bumblebees (Fig. 2).

The tree lupin is currently a minor weed in Tasmania. However, seed set in areas recently colonized by *B. terrestris* has increased, and it is likely that *L. arboreus* may become as problematic in Tasmania as it is in New Zealand, now that it has an effective pollinator. And *L. arboreus* is only one of many weeds in Tasmania, New Zealand, and southern Australia that originated in the temperate northern hemisphere and that are coadapted for pollination by bumblebees.

At present Australia alone has 2,700 exotic weed species, and the costs of control and loss of yields due to these weeds is an estimated 3 billion Australian dollars per year. The environmental costs are harder to quantify but are certainly large. The majority of these 2,700 exotic weeds are, at present, scarce and are of trivial ecological and economic importance. The recent arrival of bumblebees in Tasmania may awaken some of these "sleeper" weeds, particularly if they are adapted for bumblebee pollination. Positive feedback between abundance of weeds and abundance of bumblebees is probable, because an increase in weed populations will encourage more bumblebees, and visa versa. If even one new major weed occurs in Australia owing to the presence of bumblebees, the economic and environmental costs could be substantial.

CONCLUSION

It must be remembered that introduced bees provide substantial benefits to humans in terms of pollination of crops, and in providing honey. These quantifiable benefits need to be weighed against the likely costs. In areas where weeds pollinated by exotic bees are a serious threat or where native communities of flora and fauna are particularly valued, it may be that the benefits provided by introduced bees are outweighed by the costs. Further investigation of the potential of native bees to provide adequate crop pollination is needed. A ban on the import of *B. terrestris* to North America led to the swift development of the native *B. impatiens* as an alternative pollinator for tomatoes. Most parts of the world probably have native bee species that could be exploited. For example, there are native Australian bee species that are able to pollinate tomatoes, but adequate means of rearing these bees for glasshouse use have not yet been developed.

The precautionary principle argues that, in the meantime, we should prevent further deliberate release of exotic

FIGURE 2 A nonnative bumblebee *B. terrestris* worker pollinating nonnative lupins in New Zealand. (Photograph courtesy of the author.)

bee species. Unlike many of the other human activities that have an impact on the environment, introduction of exotic species is usually irreversible, and this is almost certainly true of bees. Similarly, if an exotic pathogen escapes into wild bee populations, there is no way it can be eradicated. If bees are to be moved between countries (regardless of whether or not they are native to the country of importation), rigorous screening should be used to ensure that they are not carrying parasites or pathogens.

Given the many potential interactions between alien bees and their pathogens, on one hand, and native flower visitors, native plants, and nonnative weeds, on the other, it seems almost certain that introducing new bee species has had serious impacts on natural ecosystems that we have not yet begun to appreciate.

SEE ALSO THE FOLLOWING ARTICLES

Competition, Animal / Hybridization and Introgression / Pollination / Wasps / Weeds

FURTHER READING

Goulson, D. 2003. Effects of introduced bees on native ecosystems. *Annual Review of Ecology, Evolution and Systematics* 34: 1–26.

Goulson, D. 2009. *Bumblebees: Their behaviour, ecology and conservation.* Oxford: Oxford University Press.

Ings, T. C., N. L. Ward, and L. Chittka. 2006. Can commercially imported bumble bees out-compete their native conspecifics? *Journal of Applied Ecology* 43: 940–948.

Inoue, M. N., J. Yokoyama, and I. Washitani. 2008. Displacement of Japanese native bumblebees by the recently introduced *Bombus terrestris* (L.) (Hymenoptera: Apidae). *Journal of Insect Conservation* 12: 135–146.

Kato, M., A. Shibata, T. Yasui, and H. Nagamasu. 1999. Impact of introduced honeybees, *Apis mellifera*, upon native bee communities in the Bonin (Ogasawara) Islands. *Researches on Population Ecology* 2: 217–228.

Roubik, D. W. 1996. Measuring the meaning of honeybees (163–172). In A. Matheson, S. L. Buchmann, C. O'Toole, P. Westrich, and I. H. Williams, eds. *The Conservation of Bees.* London: Academic Press.

Thomson, D. M. 2006. Competitive interactions between the invasive European honey bee and native bumble bees. *Ecology* 85: 458–470.

BELOWGROUND PHENOMENA

DAVID A. WARDLE

Swedish University of Agricultural Sciences, Umeå

All terrestrial ecosystems consist of primary producers (plants), aboveground consumers (herbivores and their predators), and belowground consumers (bacteria, fungi, and soil animals). These organisms are all directly or indirectly interlinked: plants determine the amount and quality of food resources for both the aboveground and belowground consumers, aboveground consumers affect the availability of resources to belowground consumers, and belowground consumers regulate decomposition and nutrient cycling processes that affect plants and aboveground consumers. Because of the degree of linkage between the aboveground and belowground ecosystem components, invasive plant and aboveground consumer species have the potential to greatly alter the belowground subsystem, while invasive belowground organisms can greatly influence what we see aboveground. The many spectacular examples of invasive organisms influencing biota on the other side of the aboveground–belowground interface include plants such as legumes, C4 grasses, and coniferous trees; aboveground consumers such as deer, rats, and ants; and belowground consumers such as earthworms, flatworms, and fungal pathogens. The effects of these invaders in turn alter ecosystem properties such as ecosystem production, nutrient cycling, and soil fertility, and can thus greatly transform the functioning of the ecosystem.

INVASIVE PLANTS

Invasive plant species exert their most dramatic effects on the belowground subsystem when they differ in some fundamental way from the native species present. Classic studies on the island of Hawaii have shown that the invasive shrub faya (*Morella [Myrica] faya*), native to the Azores and Canary Islands, can greatly transform native montane forest ecosystems. This is because faya differs from all the native species present in being able to form root nodules that convert atmospheric nitrogen to potentially biologically available forms. As a consequence of invasion by this shrub, ecosystem nitrogen input is increased by over fourfold, greatly transforming the fertility of the soil. Another example involves the invasion of northern hemisphere pine (*Pinus*) tree species in many southern hemisphere ecosystems (Fig. 1). In New Zealand grasslands and forests, invasion by North American lodgepole pine (*Pinus contorta*) results in large reductions of soil animals such as nematodes that are responsible for regulating nutrient cycling and decomposition processes. This situation likely arises because of the low quality, highly acidic litter produced by the invader. Invasive plants also have important belowground effects when they alter the ecosystem's disturbance regime. For example, invasive C4 grasses in forests in Australia and Hawaii are highly flammable and greatly increase the fuel load for fires at the ground level. This switches the ecosystem to a new stable state, which

FIGURE 1 Invasive coniferous tree species in New Zealand (foreground), such as lodgepole pine (*Pinus contorta*) and Douglas fir (*Pseudotsuga menziesii*) from North America, can greatly reduce densities of belowground fauna such as nematodes that cycle nutrients, relative to native tree species such as mountain beech *Nothofagus solandri* var. *cliffortioides* (background). This probably arises through the poorer quality acidic litter produced by the conifers. (Photograph courtesy of the author.)

is dominated by C4 grasses and which burns frequently, causing losses of soil organic matter and impairment of belowground organisms and ecosystem processes.

While the above examples represent especially dramatic belowground effects of invasive plant species, most invasive plant species have lesser effects, either because they become only a minor component of the flora, or because they do not differ greatly in terms of key attributes or traits from the native plant species present. However, there is some evidence that native and exotic components of floras do show some consistent differences in key traits; separate studies performed on the Hawai'ian Archipelago, in eastern Australia, and in New Zealand have shown that, on average, invasive plant species have higher specific leaf areas (i.e., leaf area divided by leaf mass) and concentrations of leaf nutrients than do coexisting native species. These attributes are, in turn, well known to lead to greater quality and decomposability of leaf litter, suggesting that, in many cases, invasive plant species should have positive effects on soil biota and the belowground ecosystem processes that they regulate. Consistent with this prediction, the majority of studies have found invasive plant species to produce more rapidly decomposing litter than that of coexisting native species with similar life forms, and to enhance soil nutrient flux rates and availability of nutrients. However, there are many exceptions to these generalizations, which are most pronounced when the invasive species

have fundamentally different life forms than those of the native species (e.g., when pine trees invade grasslands).

Soil organisms are important drivers of plant growth, and invasive plant species' effects on the soil biota can, in turn, feed back to influence the success of both the invasive and associated native plant species in the community. For example, a landmark study performed in Canadian grassland and meadow ecosystems showed that invasive plant species consistently entered positive feedbacks with their associated soil biota, while rare native plant species entered negative feedbacks. Another study showed that black cherry (*Prunus serotina*) was negatively influenced by the soil community that formed under it in its native range in North America, but was positively affected by its soil community when it invaded northwestern Europe. Such studies provide evidence that plant species may escape soil antagonists (e.g., pathogens) that regulate their densities in their native ranges when they invade new ecosystems. Conversely, other studies point to invasive plant species being adversely affected by the absence of necessary belowground mutualists in their new environment. For example, invasion of pine trees in New Zealand and South Africa can be impaired by the lack of compatible mycorrhizal fungal species (i.e., fungi that form mutualisms with plant roots and assist plant nutrient uptake) and will spread only once associated fungi also arrive. Similarly, invasive nitrogen-fixing plant species may spread only when compatible strains of nitrogen-fixing bacteria are also present.

INVASIVE ABOVEGROUND CONSUMERS

Aboveground primary consumer organisms (pathogens and herbivores) can potentially exert major belowground effects through promoting the replacement of host plant species with other plant species, thus altering the species composition of the vegetation. Compositional shifts of this type are well known to influence belowground organisms and processes, especially when the host plants and species that replace them differ in the quality and quantity of organic materials (as dead leaves, wood, and root materials) that they input to the soil. As such, invasive aboveground consumers can have important indirect effects on soil organisms and processes, especially when they undergo outbreaks in the absence of regulation by their natural enemies and cause widespread losses of particular plant species. For example, the pathogenic chestnut blight (*Cryphonectria parasitica*) from Asia has resulted in widespread loss of chestnut (*Castanea dentata*) in eastern U.S. forests and its replacement by other hardwood tree species. Because chestnut produces poorer quality wood that is less favorable for decomposers than does the species that it replaces, it has been proposed that invasion by this blight has greatly altered belowground organisms and processes, promoting changed nutrient cycling rates. Similarly, the widespread loss of hemlock (*Tsuga* spp.) in the northeastern United States, following invasion by the hemlock woolly adelgid (*Adelges tsugae*, an aphid-like insect), has led to replacement by other tree species that produce far superior quality litter. This shift in litter inputs is likely to have important positive consequences for decomposer organisms and nutrient cycling processes.

Mammalian herbivores, for example deer, goats, rabbits, and pigs, have been introduced to many parts of the world where they sometimes greatly transform natural ecosystems. For example, several species of deer, along with feral goats, were introduced to New Zealand between the 1770s and 1920s; prior to that time, New Zealand had no native browsing mammals. Long-term deer exclusion plots throughout New Zealand forests have shown that these invaders have caused widespread replacement of plant species that produce high-quality litter by species that produce low-quality, poorly decomposing litter (Fig. 2). They have also revealed that deer and goats can either promote or reduce soil microorganisms and certain soil processes depending on environmental factors, while they usually adversely affect larger-bodied soil organisms as a result of physical disturbances including trampling. Disturbance resulting from the physical effects of mammalian herbivores may be particularly acute when

the type of disturbance is novel to the ecosystem. For example, the feral pig (*Sus scrofa*), which is invasive in many ecosystems worldwide, often causes substantial soil disturbance through cultivating the soil in search of roots and invertebrates. This activity has wide-ranging and usually adverse effects on forest vegetation and on soil organisms. Another example is widespread tree felling and death caused by invasion by introduced beavers (*Castor canadensis*) in southern beech (*Nothofagus* spp.) forests in southern South America. Here the forests are replaced by meadows, and the belowground consequences of this transformation, while unexplored, are likely to be substantial.

Some of the strongest belowground effects of aboveground invasive consumers involve invasive predators, and these effects are the most acute when predators remove other animals that themselves perform important roles in the ecosystem. For example, on Christmas Island in the Indian Ocean, the red land crab (*Gecarcoidea natalis*) plays an important role in consuming and breaking down leaf litter. The island has recently been invaded by the yellow crazy ant (*Anoplolepis gracilipes*), which serves as an aggressive predator of the crabs and leads to their elimination. The net effect of this loss of crabs is reduced litter decomposition and accumulation of litter on the soil surface. Other examples involve nesting seabirds, which perform important ecological roles in many coastal regions worldwide, both through transferring nutrients from the ocean (as fish) to land (as guano) and through increasing soil disturbance. Recent studies have shown that invasion by predators of nesting seabirds (and their chicks and eggs), such as foxes on the Aleutian Islands and rats on offshore islands in New Zealand, has severely reduced seabird densities, and therefore their effects on the ecosystem (Fig. 3). The net effect is that transport of nutrients from the ocean to the land by the birds is thwarted, leading to greatly reduced soil fertility and availability of nutrients for plants and invertebrates. In the New Zealand example, seabird elimination by rats also adversely affects most components of the soil invertebrate community, soil biological activity, and the decomposability of litter produced by the dominant vegetation.

INVASIVE BELOWGROUND ORGANISMS

In general, little is known about the ecological significance of invasions by soil microbes (i.e., bacteria and fungi). This is especially true for microbes that decompose dead plant material—most decomposer microbial species have not been taxonomically described, so if any species were to become invasive it would almost certainly

FIGURE 2 Deer exclusion plot in northern New Zealand. On the right-hand side of the fence, vegetation is free of deer, and there is a rich forest understory. On the left-hand side, where the deer can access, the understory is removed and replaced by unpalatable ground-level plant species. These effects of deer cascade belowground, influencing soil decomposer microbes and invertebrates, and belowground ecosystem functioning (Photograph courtesy of the author.)

FIGURE 3 Soil burrowing activity caused by nesting seabirds on an offshore island in northern New Zealand (A). These seabirds have major effects in promoting soil fertility, enhancing densities of soil invertebrates, and promoting breakdown and burial of leaf litter. (Photograph courtesy of the author.) (B) A nearby offshore island has been invaded by rats, in which the rats have largely eliminated seabirds through predation of their eggs and chicks. This invasion has resulted in the reversal of seabird effects, leading to substantially reduced soil fertility and soil organisms and to accumulation of leaf litter on the soil surface. (Photograph courtesy of T. Fukami. From Fukami et al. 2006 *Ecology Letters* 9: 1299–1307.)

remain undetected. Furthermore, it is likely that any effects of invasive decomposer microbes on other components of the ecosystem would be insignificant. However, there are some well-documented examples of invasive pathogenic soil microbes adversely affecting natural vegetation. Prominent examples include dieback of natural vegetation in Australia due to invasion by root pathogenic oomycetes and fungi, such as *Phytophthora cinnamonii* and *Armillaria luteobubalina*, and sudden oak death syndrome caused by invasion of the pathogenic oomycete *Phytophthora ramorum* in California. In addition, there are several examples of invasive mycorrhizal fungi that

form mutualistic associations with native tree species, a conspicuous one being the fly agaric (*Amanita muscaria*, characterized by its red coloring with white spots), which occurs commonly in some natural forests in New Zealand and Australia. However, the effect of invasive mycorrhizal fungi on the native trees that host them, or on the native mycorrhizal fungal community, remains largely unexplored.

Invasive belowground animal species also have the potential to influence the invaded ecosystem, especially when they perform a novel function. The best-studied example involves invasive earthworms, which can have wide-ranging effects in ecosystems that lack a functionally comparable native earthworm fauna. In particular, invasion of earthworms in North American forests that do not otherwise contain earthworms (because of elimination of their previous earthworm faunas through Pleistocene glaciations) has been shown to cause large losses in soil organic matter and forest floor material, alterations of densities of soil microbes and other soil invertebrates, increased availability of soil nutrients for plant growth, and large shifts in the species composition of the understory vegetation. Furthermore, invasion of the earthworm *Pontoscolex corethurus* in many tropical regions worldwide can either enhance or reduce soil compaction and thus greatly affect soil porosity; this may in turn impact on both other soil organisms and plant growth. Other belowground decomposer animals are also known to be invasive in many ecosystems worldwide, for example, springtails, isopods, millipedes, and larval stages of beetles and true flies. While invasions by these organisms certainly have the potential to substantially alter decomposer processes, soil nutrient availability, and plant growth, these effects have, to date, attracted relatively little attention, and they remain little understood.

Finally, belowground predators of decomposer animals can themselves exert major effects both above and below ground. One convincing example is the predatory New Zealand flatworm (*Arthurdendyus triangulata*), which has invaded grasslands in the British Isles and the Faroe Islands. This flatworm serves as an aggressive predator of earthworms, thus reducing the effects of the earthworms on soil properties. In the Scottish highlands, this invasion has been shown to lead to reduced soil porosity, greater waterlogging, domination of the vegetation by rushes (*Juncus* spp.), and reduced densities of burrowing moles. Another example involves invasion of subantarctic Marion Island by the house mouse (*Mus musculus*). Here, predation by mice severely reduces densities of the endemic flightless moth *Pringleophagus marioni*, which otherwise serves as the main faunal agent of litter breakdown; this in turn impairs litter breakdown by the moth.

SEE ALSO THE FOLLOWING ARTICLES

Earthworms / Fungi / Grazers / Mycorrhizae / New Zealand: Invasions / Nitrogen Enrichment / Transformers

FURTHER READING

Allen, R. B., and W. G. Lee, eds. 2006. *Biological Invasions in New Zealand.* Berlin: Springer.

Bardgett, R. D., and D. A. Wardle. 2010. *Aboveground-Belowground Linkages: Biotic Interactions, Ecosystem Processes and Global Change.* Oxford: Oxford University Press.

Ehrenfeld, J. G. 2003. Effects of exotic plant invasions on soil nutrient cycling processes. *Ecosystems* 6: 503–523.

Lovett, G. M., C. D. Canham, M. A. Arthur, K. C. Weathers, and R. D. Fitzhurgh. 2006. Forest ecosystem responses to exotic pests and pathogens in eastern North America. *BioScience* 56: 395–405.

Van der Putten, W. H., J. N. Klironomos, and D. A. Wardle. 2007. Microbial ecology of biological invasions. *The ISME Journal* 1: 28–37.

Wardle, D. A., and R. D. Bardgett. 2004. Human-induced changes in densities of large herbivorous mammals: Consequences for the decomposer subsystem. *Frontiers in Ecology and the Environment* 2: 145–153.

Wolfe, B. E., and J. N. Kliromonos. 2005. Breaking new ground: Soil communities and exotic plant invasion. *BioScience* 55: 477–487.

BIOGEOGRAPHY

SEE GEOGRAPHIC ORIGINS AND INTRODUCTION DYNAMICS

BIOLOGICAL CONTROL, OF ANIMALS

J. HOWARD FRANK

University of Florida, Gainesville

ROY G. VAN DRIESCHE

University of Massachusetts, Amherst

MARK S. HODDLE

University of California, Riverside

EARL D. MCCOY

University of South Florida, Tampa

Biological control, often called biocontrol, is the deliberate use of a living species (biocontrol agent) against a target (the pest). Its origin is in the observation that populations of many animal species may be lowered by adversaries (predators, parasites, parasitoids, and pathogens), collectively called natural enemies and deemed to be beneficial when they control a pest. The four major ways in which biological control can be used are (1) augmentation, (2) introduction, (3) conservation, and (4) biopesticide-based control. Biological control is a powerful weapon in the battle against biological invasions by animals. As with all things that have the ability to transform ecosystems, it must be used carefully and is not appropriate in many instances, and not feasible in others, for the management of invasive animals.

FORMS OF BIOLOGICAL CONTROL

Augmentation (augmentative biocontrol) is the recognition that a natural enemy already in place, either introduced or native, is having some effect on a pest population, but not an adequate effect to reduce the population of that pest to the desired level. More individuals of that natural enemy are therefore added to enhance that effect. Those extra individuals are typically brought in from elsewhere, such as from a commercial vendor that sells mass-reared natural enemies. Only a few dozen species are produced in the United States, including some ladybird beetles and lacewings. The expectation following release is a short-term reduction of the pest population, usually to protect a crop for a single season. The population of the biocontrol agent subsequently falls to near its previous level. Fig. 1 shows an example of an augmentative biocontrol agent.

Introduction (classical biocontrol) is the introduction of a coevolved and preferably highly specific natural enemy from the native range of an invasive species. It is the form of biocontrol that has produced the most spectacular successes. The objective is to import a natural

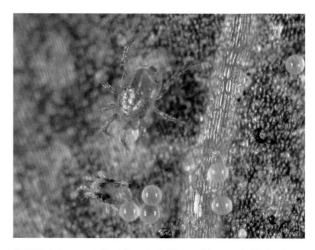

FIGURE 1 Augmentative biocontrol: *Phytoseiulus persimilis* (a phytoseiid mite) is effective against spider mites. (Photograph by Jack Kelly Clark, courtesy of the University of California Statewide IPM Program. Copyright: Regents of the University of California.)

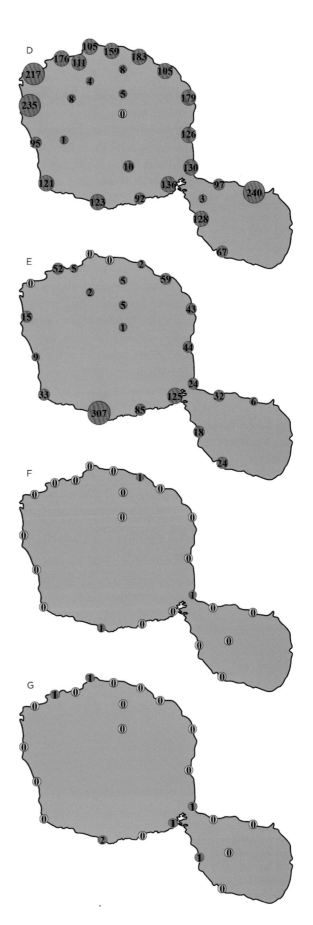

FIGURE 2 Classical biocontrol: (A) Adult glassy-winged sharp-shooter (GWSS), *Homalodisca vitripennis* (a cicadellid bug). This invasive pest was not only a menace to agriculture, but also threatened natural areas because of its ability to acquire and transmit a lethal plant pathogen. It posed a continuous invasion threat to other South Pacific nations. (B) Egg parasitoid, *Gonatocerus ashmeadi* (a myrmarid wasp), of GWSS. This was the key natural enemy that dramatically reduced GWSS populations in Tahiti over a seven-month period. (C) GWSS egg mass showing *G. ashmeadi* emergence holes. (D–F) Reduction of GWSS populations on Tahiti following the release of *G. ashmeadi*. (D) GWSS densities on Tahiti from one-minute sweep net sampling of hibiscus bushes before the release of *G. ashmeadi* (April 14, 2005). (E) GWSS densities five months after parasitoid release (October 1, 2005). (F) GWSS densities seven months after parasitoid release (December 12, 2005). (G) GWSS densities two years after parasitoid release (April 4, 2007). (Photographs by Jack Kelly Clark, courtesy of the University of California Statewide IPM Program. Copyright: Regents of the University of California. Maps courtesy of Mark S. Hoddle.)

enemy that seems important in regulating the pest in its native range and help it to establish a population in the new range of the invasive species. Ideally, the established natural enemy will increase and spread in time throughout the invasive range of the pest and permanently reduce its population density. Thereafter, the biocontrol agent and its target will continue to exist but at lower population levels. This method is the most likely of the four to produce permanent control of an invasive species without need for further human action. Classical biocontrol has sometimes been called inoculative biocontrol because it usually inoculates (introduces) rather small numbers of biocontrol agents into a new habitat to establish a permanent breeding population that is self-dispersing. Fig. 2 shows an example of a classical biocontrol project.

Conservation biocontrol is the organized attempt to make crop fields or other sites better habitats for natural enemies. This is done by enhancing resources important to key natural enemies, either introduced or native. The expectation is of a greater reduction of the pest population so long as the habitat remains altered; if the habitat reverts to its previous condition, the benefits attained are lost. Examples are the provision of nectar sources for some beneficial insects, or nesting boxes for insectivorous birds, or refuges for key predacious insects (e.g., "beetle banks" in grain fields). Fig. 3 shows a nectar-seeking wasp.

Biopesticides are living viruses, bacteria, entomopathogenic nematodes, or fungal pathogens produced typically in industrial quantities for broadcast application into

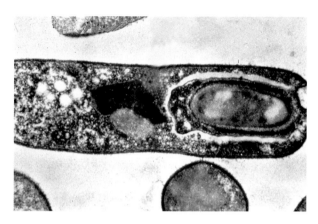

FIGURE 4 Biopesticides: *Bacillus thuringiensis*—a transmission EM picture of *Bt* var. *kurstaki*, a bacterial variety specialized to lepidopterous larvae and used as a biopesticide. Another variety, *Bt* var. *israelensis*, is effective against mosquito larvae. (Photograph courtesy of Drion G. Boucias, University of Florida.)

the environment. Such biopesticides are used much like chemical pesticides in sprays or baits but are selected for use because they are much more target specific, and thus safer for nontarget organisms. The expectation is that the populations of these pathogens decline to zero or almost so shortly after use and that reapplication will be needed if the pest population should later resurge from a low level. Fig. 4 shows a bacterium used as a biopesticide.

None of these four biocontrol methods is designed to eradicate an invasive species. Instead, they control it by reducing its population density.

BIOLOGICAL CONTROL AS APPLIED POPULATION ECOLOGY

Biocontrol began as an attempt to mimic ecological phenomena such as predation. Augmentative biocontrol was practiced in Chinese citrus groves at least 1700 years ago and involved relocating native ant colonies to control pests. Classical biocontrol, involving moving insect predators and parasitoids from one country to another, began in the late nineteenth century in response to invasions of pests into crops in New Zealand and North America—California's citrus industry benefited spectacularly in the 1880s. Biopesticidal control had its beginnings with the study of honey bee and silkworm diseases in the nineteenth century, while conservation biocontrol emerged as a subdiscipline in the 1960s. Soon after ecology became a recognized science in the 1920s, biocontrol was developed as an applied branch of population ecology. Biocontrol practitioners discovered that successful biocontrol agents should be capable of responding at the population level to

FIGURE 3 Conservation biocontrol: *Larra bicolor* (a crabronid wasp) taking nectar from a flower. Populations of this introduced wasp can now be assembled in Florida at patches of some nectar source plants such as *Spermacoce verticillata*, shown here. This functions like butterfly gardening but with the objective of suppressing populations of invasive *Scapteriscus* mole crickets, which are hosts for larvae of the wasp. (Photograph courtesy of Lyle Buss, University of Florida.)

changed pest densities (density dependence). Biocontrol studies were in the forefront of the development of ecological life table studies, beginning in the 1950s, because the quantified causes of population regulation are the essence of interpreting biocontrol outcomes.

Biocontrol practitioners recognized early that specialist biocontrol agents, as contrasted with generalists, were more likely to be successful when introduced to new areas. Specialists were selected because they would be better at seeking out their target host or prey in a complex environment, and thus would likely be more able to reduce the population density of that target. The practice of using specialists for introduction likely reduced potential harm to nontarget species by natural enemies.

TARGETS AND SUCCESSES OF BIOLOGICAL CONTROL

Potentially, a wide range of insects might be managed with biocontrol, but groups for which this approach has been especially successful include scale insects, whiteflies, mealybugs, psyllids, leaf-feeding beetles, spider mites, sawflies, and leafminers. Invasive species that are pests in crops may be controlled permanently over wide areas with classical biocontrol or may be managed temporarily in crops or greenhouses by integrating pesticides with natural-enemy conservation and augmentation. For example, the invasive soybean aphid (*Aphis glycines*) in the United States is partially suppressed by native generalist predators, especially ladybird beetles, whose presence can be enhanced by reducing the use of pesticides. Similarly, many pests of greenhouse vegetable crops, including invasive whiteflies and thrips, can be managed with releases of parasitoids (e.g., aphelinid wasps [*Encarsia* and *Eretmocerus*] for whiteflies) or predators (e.g., species of phytoseiid mites for thrips).

The interface between ecology of invasive species and biocontrol, however, is greatest when classical biocontrol is applied against invasive species, especially in wild lands. Such projects may be directed against either agricultural or environmental pests. The successful suppression of damage from the invasive cassava mealybug (*Phenacoccus manihoti*) in Africa through introductions of South American parasitoids from the pest's native range, for example, prevented hunger and restored food security to several hundred million farmers and their families in tropical Africa. Similar protection of stored maize in West Africa from the ravages of the invasive larger grain borer (*Prostephanus truncatus*) was achieved with the introduction of a predacious beetle, also from the pest's native range. More than 200 invasive agricultural pests have been suppressed in this manner with classical biocontrol, making it perhaps the most effective and ecologically sensitive form of insect pest management.

In addition to controlling agricultural pests, classical biocontrol can be employed to protect native plants and their associated ecosystems. Restoration of hemlock (*Tsuga canadensis*) forest communities in affected parts of the eastern United States, for example, will depend on introduction of predators capable of suppressing the invasive Japanese hemlock woolly adelgid (*Adelges tsugae*). Similarly, on St. Helena in the Atlantic Ocean, the native gum tree *Commidendrum robustum* was rescued from extinction (threatened by the invasive scale *Orthezia insignis*) by the introduced lady beetle *Hyperaspis pantherina*. In Florida, ten species of threatened or endangered bromeliads plus the aquatic invertebrates that depend on some of these bromeliads as their only habitat are being destroyed by the invasive weevil *Metamasius callizona*, against which a Central American tachinid (*Lixadmontia franki*) parasitoid has recently been released (in 2007). These and other examples indicate the importance of classical biocontrol as a primary response against damaging invasive species of insects. While the method has often worked well against insects, classical biocontrol against invasive animals other than insects and mites has not typically been attempted, and such "nontraditional" programs are discussed below. In an extensive analysis of 441 biocontrol projects conducted worldwide and published in 1988, 13 percent provided complete control, 25 percent substantial control, 6 percent partial control, and 56 percent no control.

BIOLOGICAL CONTROL OF NONTRADITIONAL TARGETS, INCLUDING VERTEBRATES

Most animal biocontrol uses have been against insect or mite pests of agriculture, pastures, horticultural crops, or forests. Biocontrol has been applied much less often against other kinds of pests. However, the increasing awareness of the environmental effects that invasive animals cause in natural areas has increased interest in the potential for biological control of such groups. Among these emerging projects is one assessing the feasibility of controlling a marine crustacean, the European green crab (*Carcinus maenas*), with a parasitic barnacle. Other novel targets (realized and potential) for biocontrol include terrestrial snails and slugs, invasive frogs, and snakes. For example, the cane toad (*Rhinella marina [Bufo marinus]*) is a serious invasive pest in Australia and has been evaluated as a possible target for control with helminth lungworms. This parasitic nematode reduces toad longevity, vigor, and fecundity of adults but has a sufficiently high level of host specificity to present a negligible threat to native amphibians. Highly host-specific parasitic flies have been established in parts of Australia for the classical biocontrol of invasive land snails that originated from the Mediterranean region. *Pasteuria* spp. (bacteria) are under investigation for their potential in controlling plant-parasitic nematodes.

Efforts to achieve biocontrol of vertebrates have focused on rabbits, cats, mice, and to a lesser extent the European red fox (*Vulpes vulpes*) and brush tail possum (*Trichosurus vulpecula*). Feral cat populations have been controlled successfully on some remote oceanic islands through the introduction of feline-specific viruses. Early attempts at biocontrol of rabbits in the 1880s in New Zealand and Australia were poorly thought out and simply employed the importation and release of generalist European predators. The catastrophic effects of the establishment of stoats, weasels, and ferrets were greatest in New Zealand, and populations of many native species of birds were dramatically reduced as a result. Greater sophistication in concept and program development resulted in the use of host-specific rabbit pathogens in Australia. The myxoma and rabbit hemorrhagic disease viruses provided substantial suppression (over 50 percent reduction in pest populations) in many areas. European rabbits in Australia and elsewhere developed a genetic resistance to the myxoma virus, reducing its effectiveness, but resistance to the rabbit hemorrhagic virus is not yet evident.

Immunocontraception, the use of self-disseminating genetically engineered microorganisms that cause sterility in infected hosts, has been assessed for the control of rabbits, mice, and fox populations in Australia but has failed to lead to the development of a successful means of control.

THE ETHICS OF BIOLOGICAL CONTROL

The various forms of biocontrol raise several ethical issues. Conservation biocontrol raises few issues, being minor in scale compared to other human-induced changes to wild habitats. Augmentative releases of insects and mites pose a potential allergy risk to applicators and some risk of permanent establishment of the released species in a new area without consideration of that outcome. Biopesticides pose minor allergy risks but produce few lasting effects and so are rather like chemical pesticides in the social concerns they raise—unless the pathogen is genetically

modified. If so, the concerns raised depend on the kinds of genetic modification to the organism. The use of any of these three forms of biocontrol obliges practitioners to minimize harm to the environment.

For some, introduction (classical) biocontrol raises an ethical concern. A strict conservationist viewpoint mandates no intentional introductions of alien species. The principal fear, apart from potential harm to the environment, is the deliberate homogenization of the world's biota. This is an important concern that should factor into decisions to release biocontrol agents in the same way that it should factor into decisions to introduce any species. Other persons, who might not worry about introductions in principle, often are concerned with the possibility of ensuing environmental damage. "Nontarget effects" of biocontrol agents have been the main ethical issue surrounding biocontrol in recent years. Egregious errors in judgment in the past, which led to introductions of environmentally damaging biocontrol agents such as the small Indian mongoose (*Herpestes auropunctatus*) and the rosy wolf snail (*Euglandina rosea*), have prompted much of the justifiable concern about nontarget effects. Many other ill-advised introductions fortuitously have had little apparent environmental effect, and the prior bad practice of introducing generalist predators is now encouragingly rare.

Practitioners who introduce biocontrol agents are obliged to minimize harm to the environment. In particular, they are obliged first to ascertain whether the harm caused by a target species warrants biocontrol. They are then obliged to determine whether introduction of a biocontrol agent is the best means of control. For example, introduction of a biocontrol agent may be appropriate for widely established pests or for those beyond the point of eradication that have the potential to become widespread pests. Finally, practitioners are obliged to make every effort to anticipate potential problems engendered by a biocontrol agent and to determine whether the problems are likely to materialize. They must avoid the temptation to dismiss the importance of potential nontarget effects occurring just because the evidence for particular effects is scant. Currently, the overall importance of nontarget effects is difficult to evaluate, and caution is warranted. At another level, there are ethical issues surrounding the decision not to engage in biocontrol of important invasive environmental pests. For example, if an invasive species has caused great harm to natural systems and we choose not to attempt biocontrol, we accept by default some responsibility for the damage the pest does to the affected natural system.

SEE ALSO THE FOLLOWING ARTICLES

Agriculture / Biological Control, of Plants / Integrated Pest
Management / Ladybugs / Wasps

FURTHER READING

Bigler, F., D. Babendreier, and U. Kuhlmann, eds. 2006. *Environmental
   Impact of Invertebrates for Biological Control of Arthropods*. Wallingford,
   UK: CABI Publishing.
Clausen, C. P. 1978. *Introduced Parasitoids and Predators of Arthropod
   Pests and Weeds: A World Review*. Agricultural Handbook No. 480.
   Washington, DC: USDA.
Greathead, D. J., and A. H. Greathead. 1992. Biological control of insect
   pests by parasitoids and predators: The BIOCAT database. *Biocontrol
   News and Information* 13(4): 61N–68N.
Gurr, G., and S. Wratten, eds. 2000. *Biological Control: Measures of Suc-
   cess*. Dordrecht: Kluwer.
Gurr, G. M., S. D. Wratten, and M. A. Altieri. 2004. *Ecological Engineer-
   ing for Pest Management: Advances in Habitat Manipulation for Arthro-
   pods*. Ithaca, NY: Cornell University Press.
Heinz, K. M., R. G. Van Driesche, and M. P. Parrella, eds. 2004. *Biocontrol
   in Protected Culture*. Batavia, IL: Ball Publishing.
Neuenschwander, P., C. Borgemeister, and J. Langewald, eds. 2003.
   Biological Control in IPM Systems in Africa. Wallingford, UK: CABI
   Publishing.
Van Driesche, R., M. Hoddle, and T. Center, eds. 2008. *Control of Pests
   and Weeds by Natural Enemies: An Introduction to Biological Control*.
   Oxford: Blackwell.
Waage, J. K., and D. J. Greathead. 1988. Biological control: Challenges
   and opportunities. *Philosophical Transactions of the Royal Society of
   London B* 318: 111–128.
Wajnberg, E., J. K. Scott, and P. C. Quimby, eds. 2001. *Evaluating Indi-
   rect Ecological Effects of Biological Control*. Wallingford, UK: CABI
   Publishing.

BIOLOGICAL CONTROL, OF PLANTS

MICHAEL J. PITCAIRN

California Department of Food and Agriculture, Sacramento

Plant populations are limited by many factors, including abiotic conditions, resource limitations, germination safe sites, plant-to-plant competition, predation (herbivory), pollination, and plant disease. The large and various groups of herbivores and diseases that consume or infect a particular plant are called its natural enemies, and the damage they impart due to their feeding or infection works together with the other limiting forces to maintain a plant's population density around some reduced level. Biological control of invasive plants is a pest control method where the natural enemies of an organism are intentionally manipulated to further reduce its abundance.

Usually, this is achieved by the introduction of additional natural enemies (either new species or more of the same species) or by encouraging an increase in the abundance of local natural enemies by habitat modification.

BIOLOGICAL CONTROL STRATEGIES

Weed biological control efforts can be grouped into three different strategies: introduction, augmentation, and conservation. Introduction, or classical biological control, is the movement of selected natural enemies of a targeted plant species from its native range into the new area invaded by the weed. It is common for exotic weeds to lack natural enemies in their new area of invasion. When a plant is imported as an ornamental or crop plant, effort is taken to ensure it is free of insects, mites, and disease. Similarly, plants accidentally introduced are usually transported as seeds or pieces of stem or rhizome, plant parts too small to include or support natural enemies that require leaves or larger pieces of the plant to survive. If the plant escapes cultivation or is accidentally introduced and begins to spread, the reason for its success as an invader may be due to the difference in the level of herbivory it receives compared to plants native to the area, which are damaged and infected by their own group of natural enemies. This difference in the level of herbivory or disease has been proposed as one of the reasons that exotic plants become invasive and is called the "enemy release hypothesis." Classical biological control involves the discovery of specific natural enemies in a plant's native range, an evaluation of their safety (through host-specificity testing) and efficacy, and the study of their release and establishment in the invaded range. The objective is for the exotic natural enemies to permanently reduce the weed population. It is generally accepted that the weed will not to be eradicated and that both the weed and biological control agents will permanently persist, but at densities below economic or ecological threshold levels where the weed is no longer problematic. Classical biological control is the most common biological control method used against plants and should generally be part of an integrated pest management program.

Augmentative biological control is the addition of natural enemies, either native or exotic, to provide a temporary boost to the background level of herbivory. Natural enemies released in an augmentative program are usually not expected to survive past their life spans or the growing season and often do not become permanently established. The released organisms are mass reared in laboratory cultures so that thousands are released at a time.

Augmentative biological control was originally developed against insect pests in greenhouses and field crops, where it is economically feasible to produce thousands of parasitic or predatory insects on high value crops, especially if the pests are resistant to insecticides. There are very few examples of augmentative biological control being used on plants, probably because it is not cost-effective compared to herbicides, which are able to control broad classes of weeds. For invasive plants that infest large areas, augmentative strategies are not likely to be cost-effective. Some have called the use of sheep or goats an augmentative control activity, but the use of grazing is traditionally considered a cultural control method because the animals must be herded. Some plant diseases have been developed for use as bioherbicides and can be classified as augmentative control. Unlike insect agents, impacts from a bioherbicide may occur for several generations of the pathogen, but they usually do not extend beyond a single field season. Eight pathogens worldwide have been registered for use as bioherbicides against weeds.

An example is Collego, a commercial product consisting of spores of the fungus *Colletoctricum gloeosporioides* f. spp. *aeschynomene*, for control of northern jointvetch, a native leguminous weed in rice and soybean crops in the southeastern United States. Most of these products have not been economically viable because they are effective against a single weed species and must compete in a marketplace with broad-spectrum herbicides effective against many weed species. The use of plant pathogens as an augmentative biological control tool has great potential and needs to be explored further.

Recently, a stem-boring wasp and a scale insect have been proposed for use in augmentative control releases against *Arundo donax*, a giant reed that has invaded the riparian community along the Rio Grande, the river that serves as a border between Mexico and the United States. Both insects are exotic and were obtained from Spain, where *A. donax* is native. The proposed objective is to rear hundreds of thousands of these species in a mass-rearing facility and then release them early in the growing season

STEPS IN A CLASSICAL WEED BIOLOGICAL CONTROL PROGRAM

1. Target selection. Identify weed species using morphological and molecular techniques and identify area of origin. Resolve conflicts regarding the commercial or environmental value of the target weed. Perform cost–benefit analysis.

2. Foreign exploration in weed's area of origin. Examine literature and explore target weed's native range to discover and collect potential biological control agents. When extensive, native areas with the most similar climate to the invaded range should be identified as priority. Correct identification of all collected material is critical for purposes of safety and project success. Plants closely related to the target weed should be examined in the native range to see if they are damaged by candidate control agents.

3. Host specificity studies. All potential biological control agents should be subjected to a series of choice and no-choice tests. Results will be used to predict field host range and potential risks to nontarget species after release of a control agent in the invaded range.

4. Approval of agents by government regulatory agencies. Most major countries have enacted laws to regulate the introduction of exotic plants and animals. These laws also regulate the introduction of biological control organisms. Results of host specificity tests must be summarized and submitted for consideration to the regulatory authority before a permit will be issued. The review process can take months to years.

5. Implementation. Upon approval for introduction, initial release and establishment of the biological control agent will occur in field nursery sites, areas with high densities of the target weed located in climatic areas deemed optimal for the control agent. Usually, only a few organisms (usually fewer than 1,000) are available for initial releases. Once they are established and their numbers increase, collections of surplus agents will be used to redistribute them throughout the invaded range. Regional redistribution can be facilitated through outreach events, such as "field days" where local land managers and property owners are invited to visit the nursery site, learn the biology of the target weed and control agent, and receive a small quantity of the agent for release on their property.

6. Post-release monitoring. Following their initial release, nursery sites should be monitored to determine whether the agent establishes, populations begin to increase, spread occurs into nearby plant infestations, and the new populations support collections for further redistribution. Generally, an agent is considered established once it has survived at least two consecutive years after release. Monitoring should be performed to examine impact of the agent on the target plant population and the occurrence of any nontarget effects. This kind of monitoring is more detailed and should occur for several years following release.

along the critical areas of the Rio Grande, in an attempt to suppress growth of the reed.

Conservation biological control is the manipulation of the habitat to encourage more activity by the local community of the pest's natural enemies. This could include providing resources needed by the natural enemies (e.g., overwintering sites) and reducing interference by other control strategies, such as mowing or applying herbicides when they are likely to harm the weed's natural enemies. This method is better developed for use against insect pests, such as the attraction of syrphid flies (hover flies) using certain flowering plants. Syrphid larvae are predators on aphids, and some control of aphids is achieved by the interplanting of *Alyssum* as a nectar subsidy among row crops such as lettuce. The use of conservation biological control needs to be more fully explored for control of weeds.

CLASSICAL BIOLOGICAL CONTROL OF PLANTS

The objective in a classical biological control program is to locate host-specific natural enemies from a plant's area of origin and introduce them into the plant's invaded range. Those natural enemies identified for use in a plant's invaded range are called biological control agents. All potential control agents are tested for safety prior to introduction, and a permit for their introduction is required from the appropriate regulating agency of the recipient country (see box). The goal is to establish permanent, self-sustaining populations of the biological control agent that increase to critical population levels and reduce the abundance or impact of the target plant. Control of a target weed can result from death of the attacked plant, but this is unusual. More commonly, control is achieved by the cumulative stress from nonlethal impacts that reduce

a plant's reproduction, competitive ability, and growth rate (Fig. 1).

It is common for a complex of natural enemies to be introduced against a target weed. Approximately 40 percent of past weed control projects introduced at least three exotic natural enemy species. A complex of agents is especially needed for weeds that are widely distributed throughout different climatic and geographic regions in the invaded range, are genetically variable, and have several modes of reproduction.

HISTORY

The first recorded use of a natural enemy to control a plant population was the introduction of a cochineal insect to control the prickly pear cactus, *Opuntia vulgaris*, in Sri Lanka in 1863. Carmine red dye was produced using the cochineal insect, *Dactylopius coccus*, which can be grown on several species of *Opuntia* cactus. Both the cactus and insect are native to the tropical regions of North and South America and were introduced in the early 1800s to India and Australia for commercial production of carmine dye. Some cactus species became troublesome weeds, and it was discovered that another cochineal insect, *Dactylopius ceylonicus*, which had been mistakenly introduced as *D. coccus* into southern India, appeared to suppress populations of *Opuntia vulgaris*, one of the cactus species that had become a weed. In 1863, prickly pear pads infested with *D. ceylonicus* were transported from India to Sri Lanka where it established and substantially reduced the weedy cactus populations.

The first classical biological control program in which a weed's area of origin was explored for natural enemies was the program against *Lantana camara* in Hawaii in 1902. This woody shrub is a native of Mexico and was introduced into Hawaii as an ornamental. It built up

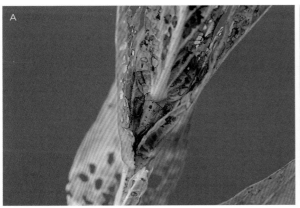

FIGURE 1 (A) Adult *Galerucella calmariensis*, a leaf beetle released as a biological control agent on purple loosestrife in northern California. (B) Purple loosestrife plants damaged from feeding by the leaf beetle. (Photographs courtesy of Baldo Villegas, California Department of Food and Agriculture.)

dense populations in a wide variety of habitats, displacing native vegetation. In 1902, the entomologist Albert Koebele, who was previously involved in the introduction of natural enemies for the control of the cottony cushion scale in California, searched the native range of *Lantana* in Mexico for natural enemies. His exploration efforts resulted in shipping 23 insect species to Hawaii, of which 14 were released and eight became established. These introductions resulted in successful control of *Lantana* in the drier lowland habitats of Hawaii; however, *Lantana* continues to be a problem in the wetter, upland habitats.

The next major program was against a complex of *Opuntia* cacti that were rapidly invading rangelands in Australia. In 1912, two Australian scientists explored some of the native range of prickly pear cacti in South America and introduced five insects into Australia between 1913 and 1914. The cochineal insect, *D. ceylonicus*, successfully controlled *O. vulgaris*, but other species of *Opuntia* continued to spread, especially *O. stricta* and *O. inermis*. In 1920, a renewed effort was initiated with scientists searching Mexico, the southwestern United States, and Argentina for new natural enemies. The efforts resulted in the discovery of 48 different species of insects, 12 of which were released and established in Australia. The most important of these was the moth, *Cactoblastis cactorum*, which was introduced from Argentina in 1925. At the time, it was estimated that over 60 million acres were infested with prickly pear cactus. By 1933, less than ten years following its introduction, all of the large stands of cactus had been destroyed (Fig. 2). Control of prickly pear cactus by *C. cactorum* in Australia continues today.

In the United States, the first major weed biological control program was against Klamath weed (*Hypericum perforatum*, St. John's wort) in California. Klamath weed is a European herbaceous perennial that became an invasive weed in Australia, South Africa, Chile, New Zealand, Hawaii, and western North America. It is toxic to livestock, and invaded rangelands and pastures could no longer be used for cattle and dairy production. Previously, scientists had introduced six insects into Australia, some of which proved successful in controlling Klamath weed. Bolstered by the success in Australia, scientists in California introduced four of the six insects between 1940 and 1950. The two most important insects were the chrysomelid beetles *Chrysolina quadrigemina* and *C. hyperici*. Larvae of these beetles burrow through the stems, and the adults feed on the foliage. Following their introduction, Klamath weed populations were reduced to less than 5 percent of their former abundance. Most local landowners at the time were involved in dairy production and credit the *Chrysolina* beetles with having saved their livelihood. In gratitude, they erected a monument to the beetles that sits in front of the Agricultural Commissioners Office in Humboldt County in northern California.

HOST SPECIFICITY TESTING

The risks associated with attack on nontarget plant species are reduced by selecting organisms with high host specificity—that is, those that damage only the target plant. Many insects and pathogens have long, coevolved

FIGURE 2 Before (A) and after (B) photographs of *Opuntia inermis* following release of *Cactoblastis* moths at a location in Queensland, Australia. The "before" photograph was taken in October 1926; the "after" photograph was taken in October 1929. (From Dodd, A.P. 1940. *The Biological Campaign against Prickly-Pear*. Brisbane, Australia: Commonwealth Prickly Pear Board.)

associations with their hosts and have developed high levels of host specificity. They search out and reproduce on only a few plant species and have proven to be safe when introduced into new areas.

Prior to introduction, potential biological control agents are subjected to a series of tests to examine their host specificity. Upon establishment in a new habitat, a biological control agent will be exposed to hundreds of new plant species. It is not possible to test all plant species, and so methods have been developed to select a sample of plant species to best predict the host range of a potential biological control agent once it is released in the new region. A plant list is created, consisting of related genera, crop plants, and native species. Observations of control agents released in past programs have shown that those plants most closely related to the target weed are the most vulnerable to damage, and emphasis is placed on testing representatives of these species. This method of constructing a test list is called the phylogenetic method.

Several types of feeding and oviposition tests are performed using all of the plant species listed for testing. If possible, it is important to reject a potential agent as early in the process as possible, to reduce costs and lost time. As a result, the most conservative tests (no-choice tests) are usually performed first. No-choice tests consist of enclosing an organism with a test plant, where it must either use the plant or die. Those plant species that are not fed on or in other ways damaged are considered to be unusable by the organism and are removed from further testing. The remaining plants are exposed to the potential control agent in a series of choice tests. In a choice test, the control agent can choose on which plants to feed, deposit eggs, or infect. Choice tests are performed in cages in a quarantine laboratory or in outdoor gardens in the native range of the target weed. Upon completion, the results of all tests are summarized and submitted to the appropriate governmental regulatory agency as a petition requesting permission to field release the organism.

The costs of developing a classical biological control program are very high and can exceed $1 million and take five or more years per agent. The highest costs occur during the exploration and safety testing of potential biological control agents. Because of the high development costs, classical biological control is usually directed at weeds that infest large regions and produce significant negative impacts such as reducing forage plants in rangelands, poisoning livestock, displacing native plant and animal species by dominating

TABLE 1

Weed Species under Complete Biological Control in All or Part of Invaded Range

Acacia saligna
Alternanthera philoxeroides
Carduus nutans
Centaurea diffusa
Chondrilla juncea
Chromoleana odorata
Cordia curassavica
Eichhornia crassipes
Euphorbia esula
Hypericum perforatum
Lythrum salicaria
Mimosa invisa
Opuntia spp.
Pistia stratiotes
Salvinia molesta
Senecio jacobaea
Sesbania punicea
Sida acuta
Tribulus terrestris
Xanthium occidentale

habitats, clogging water flow in irrigation canals, and modifying ecosystem services (Table 1).

TYPES OF ORGANISMS USED FOR CLASSICAL BIOLOGICAL CONTROL OF PLANTS

Worldwide, there have been over 1,100 releases of biological control agents against 365 weed species in 75 countries. Most releases have occurred in five countries: the United States, Australia, South Africa, Canada, and New Zealand. Insects are the most common type of organism used, accounting for 98 percent of the species used. The other organisms consist of mites, nematodes, and plant pathogens. As a group, beetles (Coleoptera), moths (Lepidoptera), true bugs (Hemiptera), and flies (Diptera) are the most commonly used organisms, making up 65 percent of all species released.

BENEFITS OF CLASSICAL BIOLOGICAL CONTROL OF WEEDS

Classical biological control is an attractive control method because, when successful, control is permanent and requires little human input thereafter. Ongoing expenditures for pesticides, labor, and specialized equipment are significantly reduced or removed altogether, saving enormous amounts of time and money. Over time, these cost savings accrue and can become substantial. For example, Klamath weed in the western United States has been successfully controlled since the 1940s following introduction of four exotic insects. The savings from not using herbicides or other control efforts was estimated at $5 million annually. Economic analysis of the biocontrol of leafy spurge, a noxious weed of rangeland in the north central United States, by introduction of several species of exotic *Aphthona* flea beetles, estimated the cost–benefit ratio to exceed 150:1, despite the high prerelease costs for the program's development.

Environmental benefits of classical biological control can include the reduction of pesticide use, an increase of biodiversity, a reduction of the occurrence of fire, and an increase in ecosystem services, such as increased water flow in irrigation canals and access to fishing in lakes and rivers. For some weeds, the immense size and spread of weed infestations preclude the use of chemical or other traditional control methods as both logistically and economically impossible. In such cases, classical biological control is especially valuable.

Evaluation of success rates in classical biological control programs worldwide has been largely subjective. For those programs where the target weed is reduced to a fraction of its former abundance, such as occurred with Klamath weed in the western United States and *Opuntia* cactus in Australia, the level of success is obvious. However, for many programs, reduction of the target weed is not dramatic, and other control methods are still needed. Still, a reduction in control effort, such as using less herbicide or less frequent applications, may be considered a partial success. When programs identified as having had complete or partial success are combined, the success rate can be very high (Table 2), exceeding 80 percent in Australia, South Africa, and New Zealand.

DISADVANTAGES AND RISKS OF CLASSICAL BIOLOGICAL CONTROL

There are several disadvantages in a classical biological control project: high development costs, risks of damage to nontarget plant species, the lack of a guarantee of ultimate success, uncertain food web effects, and the inability to remove a control agent once it has been released. Classical biological control targets only one species and is not efficacious in habitats (e.g., cultivated crops) where many weed species need to be controlled.

The introduction of a new exotic organism has the potential to cause both direct and indirect nontarget impacts. Direct impacts consist of feeding by a control agent on nontarget native, economic, or other desirable plant species. The risk of direct nontarget damage is reduced through the host-specificity tests that are performed prior to their introduction. Predictions of the realized host range from these tests have been mostly accurate. Reviews of past weed biological control programs report that, of nearly 400 species of control agents released worldwide, 12 (3%) have been recorded attacking nontarget plants. Of these, most are transitory, short-term episodes of exploratory feeding on nearby plants that sometimes occurs when an outbreak population of a biocontrol agent has destroyed its local host population.

However, there are two examples of potentially significant effects on nontarget plant species from biological control agents: the thistle seed head weevil, *Rhinocyllus conicus*, on native *Cirsium* spp. in the United States, and the cactus moth, *Cactoblastis cactorum*, on native *Opuntia* cacti in Florida and Mexico. The seed head weevil, *R. conicus*, was introduced from Europe to the United States during 1968 and 1969 for control of musk thistle (*Carduus nutans, C. thoermeri*). While there are no native North American *Carduus* species, there are two native genera (*Cirsium* and *Sassaurea*) in the same tribe Cardueae. Field host records in Europe and results from the prerelease host-specificity testing suggested that four genera (*Carduus, Cirsium, Sylibum,* and *Onopordum*), all within the tribe Cardueae, can be used as hosts by *R. conicus*. Because none of the ornamental or agricultural crop species was damaged in the host-specificity studies,

TABLE 2
Successful Weed Biological Control Programs

Location	Total Weed Targets	Complete Success	Partial Success	Proportion of Total (%)
South Africa	23	6	13	83
Hawaii	21	7	3	48
Australia				
completed	15	12	0	80
ongoing	21	4	3	33
New Zealand	6	1	4	83

NOTE: Reported number of weed biological control programs that have achieved complete (no other control options needed) or partial (reduced weed densities, some control options needed) success.

and because many thistles (both native and exotic) were considered at the time to be weeds, release of this insect was approved. The target weed was successfully controlled at many locations; however, attack by *R. conicus* has been reported on over 20 native *Cirsium* species. The impact of this weevil on one native thistle, *Cirsium canescens*, has been studied in detail, and results showed that populations of this native thistle are seed-limited and that population densities are reduced following attack by this weevil.

What has been learned from this program is that the error in the decision to release *R. conicus* was not due to a misjudgment of its predicted host range but to the societal view of the time that native species need not receive the same level of protection as agricultural crops and ornamental species. Social views on biodiversity have changed since the 1960s, and native plants are now seen as a valued resource worth protecting. Consistent with this change, both Canada and the United States have revoked all permits to collect and distribute *R. conicus* within their country's borders.

The second example of direct attack of a weed biological control agent on a new host is the cactus moth (*C. cactorum*) on native *Opuntia* cacti in southern Florida. In 1957 through 1960, the cactus moth was introduced into several islands in the Caribbean to control native *Opuntia* cacti that had infested pastures due to overgrazing. Several decades later, the moth accidentally spread into southern Florida, where it has been observed attacking native *Opuntia* cacti, including the endangered cactus *Consolea* (formerly *Opuntia*) *corallicola*. An even greater threat is the potential for attack on the much larger *Opuntia* flora of Mexico.

The decision to release the cactus moth in the Caribbean was primarily economical, not ecological. Its release in Australia was ecologically sound, as there were no native cacti that were vulnerable to attack. In the Caribbean, the cactus moth was released into an ecological region that has a high diversity of *Opuntia* cacti, and, once it was established, was able to spread. The lesson learned from this project is that the degree of host specificity required of a biological control agent is dependent on the ecology of the region of introduction.

Indirect nontarget impacts can occur through changes in the food web. If a biocontrol agent builds up high populations on a target weed but fails to control its host, then the biocontrol agent becomes an abundant new resource that can be exploited by generalist predators in the community. As a result, higher abundance of generalist predators can lead to increased predation on desirable native species unrelated to the target weed system. For example, the fruit fly, *Mesoclanis polana* was introduced in 1996 as a seed predator of *Chrysanthemoides monilifera* (bitou bush), a perennial shrub that invaded the bush country of Australia. After introduction, *M. polana* developed moderate population levels, but seed destruction was too low to cause a decline in the host plant's abundance. A few years after initial release, field observations found high rates of parasitism of *M. polana* by native parasitoid wasps, which normally attack several species of native insect seed predators. Recently, field observations have shown that species richness and abundance of the local native seed predators declined where the introduced biological control agent increased. Ineffective biological control agents that remain abundant in the community are most likely to have persistent, indirect negative effects. However, eventual achievement of successful control of the target weed will reduce populations of ineffective biological control agents, thereby reducing indirect nontarget impacts to acceptable levels.

Biological control practitioners have developed an International Code of Best Practices, which provides a set of guidelines for individuals engaged in the biological control of invasive weeds. The code consists of 12 guidelines (Table 3) and covers all aspects of classical biological control, including prerelease and postrelease activities. The code attempts to incorporate the lessons learned from past projects and to help identify actions that reduce risk and enhance effectiveness. By following this code, it is hoped that the practice of biological control will continue to improve and remain a viable control option for invasive weeds in the future.

TABLE 3

The International Code of Best Practices for Classical Biological Control of Invasive Weeds

1. Ensure that the target weed's potential impact justifies the risk of releasing non-endemic agents
2. Obtain multi-agency approval for the target weed
3. Select agents with potential to control the target weed
4. Release safe and approved agents
5. Ensure that only the intended agent is released
6. Use appropriate protocols for release and documentation
7. Monitor impact on the target weed
8. Stop releases of ineffective agents, or when control is achieved
9. Monitor impacts on potential nontarget species
10. Encourage assessment of changes in plant and animal communities
11. Monitor interactions among biological control agents
12. Communicate results to the public

SEE ALSO THE FOLLOWING ARTICLES

Biological Control, of Animals / Enemy Release Hypothesis / Freshwater Plants and Seaweeds / Herbicides / *Lantana camara* / Pathogens, Plant / Weeds

FURTHER READING

Carvalheiro, L. G., Y. M. Buckley, R. Ventim, S. V. Fowler, and J. Memmott. 2008. Apparent competition can compromise the safety of highly specific biocontrol agents. *Ecology Letters* 11: 690–700.

Coombs, E. M., J. K. Clark, G. L. Piper, and A. F. Cofrancesco Jr. 2002. *Biological Control of Invasive Plants in the United States*. Corvallis: Oregon State University Press.

Denoth, M., L. Frid, and J. H. Meyers. 2002. Multiple agents in biological control: Improving the odds? *Biological Control* 24: 20–30.

Hoffman, J. H. 1996. Biological control of weeds: The way forward, a South African perspective (77–89). In C. H. Stirton, ed. Weeds in a Changing World International Symposium, November 20, 1995, Brighton, England. British Crop Protection Council Monograph no. 64. Farnham: British Crop Protection Council.

Julien, M. H. 1997. *Biological Control of Weeds: Theory and Practical Application*. Canberra: Australian Centre for International Agricultural Research.

Julien, M. H., and M. W. Griffiths. 1998. *Biological Control of Weeds: A World Catalogue of Agents and their Target Weeds*, 4th edition. Wallingford: CABI Publishing.

McFadyen, R. E. C. 1998. Biological control of weeds. *Annual Review of Entomology* 43: 369–393.

Rose, K. E., S. M. Louda, and M. Rees. 2005. Demographic and evolutionary impacts of native and invasive herbivores on *Cirsium canescens*. *Ecology* 86: 453–465.

Van Driesche, R., M. Hoddle, and T. Center. 2008. *Control of Pests and Weeds by Natural Enemies: An Introduction to Biological Control*. Oxford: Blackwell Publishing.

BIOSECURITY

SEE ECOTERRORISM AND BIOSECURITY

BIOTIC RESISTANCE HYPOTHESIS

SEE INVASIBILITY

BIRDS

NAVJOT S. SODHI

National University of Singapore

Humans are responsible not only for creating a conducive environment for invasive species but also for introducing many of these species themselves. Worldwide, 1,400 attempts to introduce 400 bird species have been recorded. Approximately 70 percent of bird introductions have been to islands, although islands make up only about 3 percent of Earth's ice-free land area. Just over half of global bird introductions have been to Pacific islands (notably to Hawaii) and to Australasia. Bird introductions do not always result in an established population. A case in point is the introduction of more than half a million common quail (*Coturnis coturnix*) to more than 30 states in North America between 1875 and 1958. Incredibly, this experiment failed. In the case of Australasia, of the 242 bird species introduced in the past two centuries, 32 percent established viable populations.

BIRD INVADERS

Most bird introductions took place in the eighteenth and nineteenth centuries, during a period of major European colonization. Because of this, a large proportion of invasive (alien, exotic, or introduced) bird species originated from temperate regions. For instance, 60 percent of introduced birds in New Zealand originated from the Palaearctic and Australasian regions. Birds were introduced by settlers to various countries, predominantly for aesthetics, hunting, and biocontrol. Two-thirds of the species chosen for introduction have been from 6 (of 145) bird families: Anatidae (ducks), Columbidae (pigeons and doves), Fringillidae (finches), Passeridae (sparrows), Phasianidae (pheasants), and Psittacidae (parrots). This overrepresentation of certain families points to the reasons for these introductions. For example, ducks and pheasants were introduced for hunting, while finches, sparrows, and parrots were introduced as pets. Three bird species included among the 100 worst invasive species (see Appendix) are the European starling (*Sturnus vulgaris*), red-vented bulbul (*Pycnonotus cafer*), and common myna (*Acridotheres tristis*) (Fig. 1).

The negative effects of invasive bird species on native biodiversity, ecosystems, and humans have been widely recognized and are briefly discussed below. Most long-distance bird introductions to new areas are the direct or indirect result of human activities, and social and economic factors are often as critical as biological factors in the introduction of exotic species. Activities such as logging and grazing further enhance establishment of exotics by creating optimal habitat for colonization (e.g., through range expansion). Agriculture can facilitate bird species invasions when pests in agroecosystems are exposed to agricultural practices for many generations, resulting in selection for characteristics that make them persist. It is possible that bird invasions will increase in the future owing to increase in human trade and traffic, and "global warming" may further facilitate species movements to new locations.

CHARACTERISTICS OF SUCCESSFUL BIRD INVADERS

Life history and ecological attributes influence introduction success, and an understanding of them can assist in identifying "potential invaders." Successful invading bird species should be able to maintain high survival and reproductive success in new locations. Breeding early and throughout the year may result in high productivity in some invasive birds. For example, the feral pigeon (*Columba livia*) can breed at the age of 1, and throughout the year. The latter may be because they feed on seeds, which are generally available year-round.

Many invasive bird species exhibit social behavior, such as facultative colonial nesting, communal roosting, and flocking at feeding sites. The likely benefits of this social behavior include efficient predator avoidance and possible increase in foraging and reproductive success. Among colonists in anthropogenic habitats, granivores seem most successful, suggesting abundance of seeds in such areas. All successful invasive bird species show high behavioral flexibility. For example, invasive species can use a variety of nest sites. The case in point is the common myna, which can nest in holes in the trunk of trees, drainage holes in retaining walls, holes in buildings, bridges, and crevices in

FIGURE 1 The three worst invasive bird species, according to the list of 100 worst invasive species (see Appendix): (A) European starling, (B) red-vented bulbul, (C) common myna. (Photographs courtesy of Michelle St. Sauveur, Coventry, Rhode Island [A] and J.M. Garg [B, C].)

other structures such as lamp posts, air conditioners, ventilators, and disused vehicles. Helping them is the fact that hole nests generally suffer lesser nesting mortality than do open-cup nests. Hole nests are also protected from severe climatic changes such as storms. Therefore, this habit may have assisted in the successful reproduction of some hole-nesting species in newly colonized areas.

Success of an invasive bird also depends upon the size of the founding population. Larger populations can have higher success than smaller ones. Smaller populations have a greater risk of extirpation because of factors such as loss of genetic diversity. Two other hypotheses have been proposed regarding the success of invasive species. According to the "enemy release hypothesis," the lack of native predators and parasites assists invasive species to flourish in newly colonized areas. The "climate-matching hypothesis" suggests that invasive populations have a higher chance of becoming established if the climate between native and exotic locations closely matches. The results from the rose-ringed parakeet (*Psittacula krameri*) seem to support both these hypotheses. These parakeets suffer heavier nest predation in their native India than in the invaded UK. Similarly, their reproductive success is higher in warmer Mediterranean areas than in the colder UK; both are invaded locations. It has also been hypothesized that in some areas, invasive bird species can exploit vacant niches left by extinct or declining native species. If this is true, we should be expecting more invasive bird species in deforesting tropics.

INVASIVE BIRDS AND NATIVE BIODIVERSITY

Scientists have observed that invasive species can wreak particular havoc on oceanic islands. This is because life on such islands is generally less diverse and likely to have evolved without any threat from nonnative predators. Several ecological and life history attributes of island species, such as their relatively small ranges and population sizes, make island biotas vulnerable to displacement by invading species. Indeed, Birdlife International's statistics in 2004 reveal that alien invasive species threaten 67 percent of endangered bird species on oceanic islands.

Invasive bird species can depress the reproductive success of native birds through predation and competition. In Kenya, house crows (*Corvus splendens*) have been observed raiding the nests of weaver and other small bird species. European starlings were introduced in New York in 1890 and have spread throughout much of North America since then. Many scientists thought that starlings

harmed native birds, especially cavity-nesters. However, inspection of long-term data (since 1900) on population trends of cavity-nesting birds showed that starlings may have had little impact on them.

Invasive bird species can also hybridize with native species. Studies show that where the feral pigeon is abundant, the feral form and the wild form (i.e., the rock pigeon) interbreed, and the wild pigeon genotype eventually is genetically replaced by that of the feral pigeon.

In the United States, 75 exotic bird species have established themselves. In addition, many native species have expanded their ranges thanks to humans altering the habitat to their advantage. Some exotic birds produce mixed emotions in people. For example, some people love to watch exotic mute swans (*Cygnus olor*), while others fear that these aggressive new arrivals may displace native waterfowl. The Hawaiian Islands have suffered the highest rates of bird introductions in the world, with 145 introduced bird species brought in from a wide variety of countries since the 1800s. Of these, 54 have now established breeding populations. Introduced species there include the barn owl (*Tyto alba*), Japanese white-eye (*Zosterops japonicus*), Kalij pheasant (*Lophura leucomelanos*), and house finch (*Carpodacus mexicanus*). These exotic birds compete for resources with the endangered native Hawaiian avifauna, disperse nonnative plants, and act as reservoirs for avian parasites and diseases (e.g., avian pox).

Overall, more research is needed to understand the effects of invasive birds on native biodiversity.

INVASIVE BIRDS AND HUMANS

Invasive birds may impact humans, since many of these are found in human-altered habitats, but the large majority are not inherently harmful to humans. For example, only 10 percent of all invasive species in Britain have become "pests." Worldwide, the bill for bird damage to agriculture amounts to billions of dollars. Some species, such as munias and sparrows, parrots, crows, geese, and cranes can damage crops, vegetables, and fruits. Invasive pest birds have been reported as reducing the yield of rice (*Oryzis* sp.) by 50 percent in Thailand. Across Europe, sparrows and doves are pests of the sunflower fields. The greater flamingo (*Phoenicopterus ruber*) is reported from France as a pest threatening rice farming. In the United Kingdom, the wood pigeon (*Columba palumbus*) damages rapeseed (*Brassica* sp.), also known as canola, and the bullfinch (*Pyrrhula pyrrhula*) raids orchards. During the British winter,

FIGURE 2 Canada goose in the United Kingdom. (Photograph courtesy of Darwin Sodhi.)

brant geese (*Branta bernicla*) graze in agricultural fields, as do invasive Canada geese (*B. canadensis*; Fig. 2). Across Europe, starlings, pigeons, and gulls are a nuisance in urban areas (see below).

Communally roosting invasive bird species such as crows and mynas foul gardens, pedestrian footpaths, buildings, and vehicles beneath or near their roosts. Excessive noise from roosts, especially in early morning, can cause annoyance. Pigeons and starlings alone are estimated to cost the United States almost $2 billion a year in damage and control costs. Feral pigeons have become a nuisance in most of the world's great metropolises because they foul buildings, statues, cars, paths, and street furniture with their droppings. Pigeon droppings not only deface but also accelerate the deterioration of buildings, statues (Fig. 3), and cars,

FIGURE 3 Feral pigeon droppings on a statue in Italy. (Photograph courtesy of Darwin Sodhi.)

render fire escapes hazardous, are sometimes deposited on unwary pedestrians, and produce objectionable odors, especially when deposited in ceilings and on window sills. Sometimes invasive birds can cause human fatalities; for example, in 1960 an aircraft taking off from Logan Airport in Boston crashed when it hit a flock of European starlings. All 62 people onboard perished.

Some invasive species may spread disease to humans. For example, West Nile virus was first discovered in Uganda in 1937. West Nile virus mainly infects birds but can be transferred to humans through infected mosquitoes. West Nile virus can cause mortality in humans and was first discovered in the United States in 1999 in the New York area. It is speculated that this virus was introduced to New York by a migratory or invasive bird species. Invasive species have also sometimes been found to carry the bird flu virus (H5N1)—for example, the house crow (*Corvus splendens*) in Hong Kong.

BENEFITS OF INVASIVE BIRDS

Rarely, invasive bird species can carry on the ecosystem processes disrupted by extinction of native species. For example, in Hawaii, a vine (*Freycinetia arborea*) that was originally pollinated by an extinct bird now survives owing to pollination services provided by the introduced Japanese white-eye (*Zosterops japonicus*). Similarly, in Hong Kong, the greater necklaced laughingthrush (*Garrulax pectoralis*) is now the largest-gaped fruit-eating bird capable of swallowing and dispersing certain large fruits in shrublands and secondary forests. Until this bird arrived, there was no other animal around to disperse these native fruits. Likewise, the red-whiskered bulbul is a pest in many other places, but because it eats fruit, it is now possibly one of the most important seed dispersal agents in Hong Kong. In addition to helping forest regeneration through seed dispersal, bulbuls can remove insect pests in agricultural areas. Other insectivorous invasive bird species may also be providing the latter benefit. Some invasive bird species such as crows can be useful scavengers in urban areas and may keep rats (*Rattus rattus*) in check.

EXTINCTIONS CAUSED BY INVASIVE SPECIES

Avian malaria is another example of invasions causing biodiversity catastrophe on islands. The mosquito (*Culex quinquefasciatus*) that carries the avian malaria parasite (*Plasmodium relictum*) was inadvertently introduced to Hawaii in 1826, with the parasite arriving thereafter in introduced birds. Since then, avian malaria has been responsible for the decline and extinction of some 60 endemic forest birds on the Hawaiian Islands. Without having evolved in the presence of avian malaria, Hawaiian bird populations were unable to cope with the novel disease. However, since the establishment of the disease 100 years ago, some native thrushes (*Myadestes* spp.) are now showing resistance, although many other species such as endangered honeycreepers (e.g., Hawaii creeper, *Oreomystis mana*) remain vulnerable. The most feasible method of reducing transmission of malaria seems be to eliminate vector mosquito populations through chemical control and elimination of larval habitats.

MANAGING INVASIVE BIRDS

Attempts have been made around the world to disperse or eradicate invasive bird species by shooting, poisoning, or scaring them (e.g., with balloons). The use of short-term control measures (e.g., bioacoustics, chemical repellents, shooting) may well be successful in certain sites or over the short-term. Problems such as bird habituation to these measures make them less useful for long-term bird control. In Singapore, shooting has not been successful until recently because the biology of house crows was poorly understood. Once the biology of the species became better known, it was easier to decide on the numbers needed to be culled in order to reduce the population. Currently, many wildlife management programs have shifted their emphasis toward ecological approaches to bird control, based on understanding the ecology of invasive bird species, especially their habitat and food requirements. Once ecological knowledge is

FIGURE 5 A poster in Hong Kong informing the public of fines that may be levied for feeding feral pigeons. (Photograph courtesy of the author.)

sound, the hope is that long-term control will be possible. Integrating sound science and habitat management into the long-term control of invasive bird species is a good wildlife management strategy and should be practiced. In every invasive bird species control program, public education is necessary, because people can release birds or feed them (Figs. 4 and 5), thus facilitating the establishment of these species.

SEE ALSO THE FOLLOWING ARTICLES

Agriculture / Competition, Animal / Disease Vectors, Human / Malaria Vectors / Predators

FURTHER READING

Blackburn, T. M., J. L. Lockwood, and P. Cassey. 2009. *Avian Invaders: The Ecology and Evolution of Exotic Birds*. Oxford: Oxford University Press.

Brook, B. W., N. S. Sodhi, M. C. K. Soh, and H. C. Lim. 2003. Abundance and projected control of invasive house crows in Singapore. *Journal of Wildlife Management* 67: 808–817.

Duncan, R. P., T. M. Blackburn, and D. Sol. 2003. The ecology of bird introductions. *Annual Review of Ecology and Systematics* 34: 71–98.

Lever, C. 1987. *Introduced Birds of the World*. Harlow, UK: Longman Scientific and Technical.

Long, J. L. 1981. *Introduced Birds of the World: The Worldwide History, Distribution and Influence of Birds Introduced to New Environments*. New York: Universe Books.

Lowe, S., M. Browne, S. Boudjelas, and M. De Poorter. 2000, updated 2004. *100 of the World's Worst Invasive Alien Species: A Selection from the Global Invasive Species Database*. Auckland, New Zealand: Invasive Species Special Group of the World Conservation Union.

Schwartz, A., D. Strubbe, C. J. Butler, E. Matthysen, and S. Kark. 2008. The effect of enemy release and climate condition on invasive birds: A regional test using the rose-ringed parakeet (*Psittacula krameri*) as a case study. *Diversity and Distributions* 15: 310–318.

Sodhi, N. S., and I. Sharp. 2006. *Winged Invaders: "Pest" Birds of Asia-Pacific*. Singapore: SNP.

FIGURE 4 People feeding feral pigeons in Venice, Italy. (Photograph courtesy of the author.)

BLACK, WHITE, AND GRAY LISTS

STANLEY W. BURGIEL

Global Invasive Species Programme, Alexandria, Virginia

ANNE M. PERRAULT

Center for International Environmental Law, Washington, DC

Black, white, and gray lists are used to simplify the process of evaluating and addressing the risks of invasiveness of species that are to be intentionally introduced into a new environment. When an organism is proposed for introduction, it is placed on one of these lists based on its risk of invasiveness as well as the acceptability of this risk. A white list indicates that risk of invasiveness is low and the species is acceptable for introduction, a black list indicates that risk of invasiveness is high and introduction of the species is unacceptable, and a gray list indicates that risk of invasiveness is indeterminate and more research is required prior to a decision on introduction of the species. These lists can then be used as a reference for exporters considering shipments of new species into a country, as well as for customs officials reviewing shipments at a country's border control facilities. A clear and better understanding of risks allows decision-makers to develop appropriate means to manage those risks and to decide whether to ban imports or to require particular conditions for the treatment of imports.

LISTING SPECIES

The risk of invasiveness of a species includes consideration of that species' ability to survive, reproduce, and adversely impact native species or preexisting biota, as indicated by available scientific data. A species' potential risk is generally established through a process of risk analysis, which includes risk assessment (an evaluation of the potential threat that a species may pose within a particular set of environmental conditions or habitats), risk management (development of sanitary and phytosanitary measures that can be taken to reduce potential risks), and risk communication (conveyance of information from the risk assessment and management measures to policymakers for a decision and to other stakeholders for their use).

The threshold of acceptable risk is generally based on the perceived needs and views of society and key stakeholders. Governments vary widely on the level of risk they are willing to accept in exchange for some additional benefit. A low level of acceptable risk might be established to prevent the introduction of pests and diseases that might have severe impacts on the environment, agricultural production, or human health. Establishing a low level of acceptable risk is likely to involve more costly and time-consuming inspection and management procedures. A higher level of acceptable risk could be adopted to help expedite the movement and volume of goods and passengers into and out of a country, while recognizing that there may be potential introductions of unwanted species from this freer exchange. Conventional wisdom and practice holds that it is impossible to establish zero risk (moreover, this would require a halt to all movement of goods, vehicles, and individuals into a country).

Frequently, listing systems are established as one component of a broader biosecurity system that may also use other quarantine, border control, and screening processes.

LISTS AND TRADE RULES

Decisions to ban or place conditions on imports can restrict international trade. These decisions can, therefore, run afoul of international trade agreements (such as those under the World Trade Organization), unless the regulations have been transparently developed and embody a scientific rationale and procedure in their formulation and application. The WTO's Agreement on the Application of Sanitary and Phytosanitary Measures (SPS Agreement) sets out the general framework for such regulations, particularly with regard to the application of measures that would restrict trade for the benefit of animal, plant, or human health; appropriate risk analysis processes; transparency and consistency in the application of restrictive measures; the use of existing international standards; and the ability to take provisional measures in cases of uncertainty.

LISTS IN DETAIL

In some cases, black or white lists are employed individually and not in tandem. A more comprehensive system would use all three to manage imported species. The section below provides a more detailed description of each list and closes with some thoughts about how they relate in a single risk analysis system.

Black Lists

Black lists (also referred to as "dirty lists") identify those species whose introduction is prohibited due to their potential adverse effects on the environment or on human, animal, or plant health. Such lists can be a significant component of an invasive species prevention regime because they clearly state which species are banned from import. Black lists are the most common type of listing mechanism and are found in a range of countries, including the United States, Australia, New Zealand, and Poland. To add a species to a black list, a risk analysis needs to gauge that species' potential invasiveness and impacts. Black lists are often reactive tools; a species is most often listed only after it has become invasive within the country. For this reason, black lists may not be sufficient to manage the full set of risks presented by international trade and imports; they should be considered in conjunction with other species lists. In the United States, the injurious wildlife provision of the Lacey Act serves as a mechanism to restrict imports of listed species. It should be noted that this mechanism has come under criticism for its reputed inability to review and list new injurious species in a timely fashion (i.e., before they are established).

The efficacy of black lists depends largely on their scope and on a country's ability to add and amend them in a timely fashion. Governments are often notably reluctant to adopt stringent measures that would severely constrain or ban trade in a particular good or species given impacts on trade and scrutiny by other governments and the private sector alike. For example, the U.S. government maintains a list of noxious weeds that are banned from import, but few species have been added since the list's inception. In 1993, there were approximately 93 taxa listed, and a decade later there were only about 96 (two removed and five added), despite a number of new introductions of invasive species and a backlog of data on other potentially harmful noxious weeds.

Black lists and their bans on species can be fully consistent with international trade rules and are extremely effective when countries have the flexibility to add and remove species as available information and necessary precaution dictate. Finally, attention needs to be given to the process of how species are removed from a black list, particularly in the case of temporary bans. Experience with national measures to prevent entry of mad cow disease or avian flu in livestock and poultry has highlighted the need for countries to have transparent processes for reassessing risk and reopening trade.

White Lists

While black lists identify species that are denied access or import into a country, white lists (also referred to as "clean lists") identify species that are low risk and approved for introduction. The key idea behind a white list approach is that intentional introduction of an organism should be authorized only if the species is on this list of safe organisms. The presence of a given species on a white list reflects that a risk assessment has generally determined the species to be safe. White lists can be used on their own or in tandem with black lists as a means to increase transparency for potential importers and, ultimately, to facilitate trade in those species. White lists are commonly associated with agricultural and animal crops and are found in New Zealand, Argentina, and the Australian states of the Northern Territory and Western Australia.

Two other subcategories of species can also be included in white lists: species that can be imported under certain conditions and species that may already be widespread within a country. During the risk analysis of a species, regulators may identify particular risk management practices that can reduce the risk of invasiveness posed by a species to a manageable level. Such measures could range from sterilization of seeds and organisms to particular sanitary practices to reduce the potential transmission of diseases and pests associated with a species. The conditions or particular sanitary and phytosanitary requirements for import would then be clearly identified within the listing process for use by exporters and border officials.

There may also be cases where a particular species is already widespread within a country and beyond the reasonable hope of control or eradication. In such cases, those species can be included on a white list despite their inherent risk, as they may not pose any additional risks to those already extant within the country. For example, regulations targeted at preventing the import of purple loosestrife (*Lythrum salicaria*) into the United States would likely do little to limit the existing impacts of the species. However, listing of such species should be taken with care, as further introductions will continue to increase propagule pressure and could thereby lead to further adverse impacts or spread. For white-listed species, it is also important to consider inclusion of new genotypes that might have a higher tendency toward invasiveness.

Gray Lists

Finally, there may be instances in which the risk associated with an alien species cannot be determined conclusively, and hence the species cannot yet be listed on either a white or a black list. These species can be put on an intermediate "gray" or "pied" list, subject to further review. Sophisticated versions of gray lists place the species within a spectrum or matrix according to the likelihood of invasiveness based on presently available information. Gray lists are frequently developed as potential importers submit requests to a government to see whether their product or organism is eligible for import. The list serves as a virtual holding area as the appropriate regulatory agency makes the relevant determination. Gray lists thus serve a precautionary function by temporarily preventing the import of a species until further assessment can be conducted to determine whether a species is safe or potentially invasive. In some senses, gray lists function as provisional black lists. For example, the U.S. Department of Agriculture is considering guidelines on a category of imported plants intended for planting that is termed "not authorized for import pending risk analysis" (NAPRA). By its nature, such a list would bar the import of plants specified on the list until further scientific review can establish their safety or potential risk of invasiveness.

In the context of the international and regional trade agreements (particularly the WTO's SPS Agreement), gray lists can be viewed as provisional measures, which require a process seeking to obtain further information for making a risk assessment. Within its risk analysis system, the regulatory authority of the importing country can request that the exporter facilitate this process by providing scientific evidence and analysis for the assessment, thereby decreasing the amount of work by the importing country (it is also possible to require administrative fees to recoup costs incurred by the evaluation process). It generally is in the exporter's interest to assist, because doing so may facilitate a timely response to its import request. If there is still insufficient evidence to make a definitive assessment of risk, then the species can be left on the gray list pending the collection of additional information and a subsequent reassessment.

INCREASING UTILITY OF LISTS

The success of a system using these three types of lists is inherently related to its adaptability and flexibility, particularly with regard to processing new submissions and proposals for movement from one list to another. The development of such lists also requires consideration of other issues such as the scope of listings (e.g., designation of an entire genus or taxon versus identification of particular genotypes or species within a

genus) as well as species identification (e.g., if species look alike but have different potentials for invasiveness). While each of these lists can stand alone, they are most effective when all three are combined into a unified multiple-listing approach. The use of such a system does not have to catalog all known species, which would be impossible in practical terms. Instead, it codifies allowed and banned species that are already known, and then places requests by other governments or importers of unlisted species onto the gray list for case-by-case evaluation.

Current application of this multiple-listing approach is limited to a handful of countries, most specifically New Zealand and Australia. However, its use can be one of the most effective mechanisms to assess intentional introductions and to define clearly where a particular species falls in the regulatory process. More exchange of experiences with the application of the multiple-list system, as well as sharing of data from assessments of particular species, can certainly help to promote and widen its application. As a tool, it does entail a commitment of resources to provide the institutional, regulatory, and scientific processes necessary to assess and list species. Such listing systems are also limited to intentional introductions and thereby require associated instruments to address the threat of unintentional introductions. Despite these limitations, multiple-list systems are an increasingly important tool for facilitating scientific evaluations and decision making in order to manage the threat of invasive species.

SEE ALSO THE FOLLOWING ARTICLES

Agreements, International / Invasiveness / Risk Assessment and Prioritization

FURTHER READING

Burgiel, S., G. Foote, M. Orellana, and A. Perrault. 2006. *Invasive Alien Species and Trade: Integrating Prevention Measures and International Trade Rules.* CIEL, Defenders of Wildlife, TNC.

Christensen, M. 2004. Invasive species legislation and administration in New Zealand (23–50). In M. L. Miller and R. N. Fabian, eds. *Harmful Invasive Species: Legal Responses.* Washington, DC: Environmental Law Institute.

Di Paola, M. E., and D. G. Kravetz. 2004. Invasive alien species: Legal and institutional framework in Argentina (71–88). In M. L. Miller and R. N. Fabian, eds. *Harmful Invasive Species: Legal Responses.* Washington, DC: Environmental Law Institute.

Fowler, A., D. M. Lodge, and J. H. Hsia. 2007. Failure of the Lacey Act to protect US ecosystems against animal invasions. *Frontiers in Ecology and the Environment* 5(7): 353–359.

Krzywkowska, G. 2004. The Polish invasive species experience, legislation and policy (109–124). In M. L. Miller and R. N. Fabian, eds. *Harmful Invasive Species: Legal Responses.* Washington, DC: Environmental Law Institute.

BROWN TREESNAKE

GAD PERRY

Texas Tech University, Lubbock

GORDON H. RODDA

USGS Fort Collins Science Center, Colorado

The brown treesnake, *Boiga irregularis,* was transported to Guam following World War II. A nocturnal, arboreal, and cryptic species, it initially escaped detection. Within a few decades, however, it reproduced, spread, and devastated the terrestrial vertebrate fauna of the island, causing economic damage and cultural disruption. The species is an excellent disperser; brown treesnakes originating from Guam have since been found as near as the island of Rota and as far away as Spain and Diego Garcia atoll. Research on control and interdiction methods has been extensive and productive, but eradication remains improbable, and the risk of further dispersal continues.

GUAM: GEOGRAPHIC AND HISTORICAL SETTING

Guam is a long way from everywhere. All islands within 1,500 km in every direction are even smaller than Guam's 550 km^2, and even these are few and far between. Despite this, humans first reached the island some 4,000 years ago. Although some nonnative species possibly arrived in pre-European times, many of those that have exacerbated the brown treesnake problem arrived with the Spanish colonialists, starting in the mid-1600s. Even more arrived with the Americans, who took possession of Guam after the Spanish-American War of the late 1800s. Invasive rodents, shrews, deer, feral hogs, Eurasian sparrows, and skinks either provide food for the snake (Fig. 1) or compound its negative impacts on the ecosystem, causing invasional meltdown.

THE BROWN TREESNAKE ARRIVES: GUAM, 1950–1980

In the wake of WWII, Guam served as a regional military base for the U.S. military. Movement of salvaged equipment resulted in the arrival of the snake around 1950. Details of this period are sketchy, and most of what little we know about spread of the snake on Guam emerged from the work of Julie Savidge, who reconstructed the process from interviews held in the early 1980s. As is the case with many invaders, the period between arrival

FIGURE 1 A brown treesnake containing three introduced Eurasian sparrows (*Passer montanus*). Common introduced species thought to be benign, such as the sparrow and the curious skink (*Carlia ailanpalai,* formerly referred to as *C. fusca*), can subsidize snake populations and enhance their impact on native species. (Photograph courtesy of G. Perry.)

and irruption ("lag period") was characterized by a slow buildup in brown treesnake numbers and effects. With abundant food, few predators, and no known diseases or parasites on Guam, snakes grew up to 3 m long. Early reports attached little importance to the snake's arrival, predicting that it would be beneficial by reducing rat populations.

Lack of species on Guam that feed on or parasitize the snake, as well as abundance of naive prey, helped brown treesnake populations to explode. By the 1970s, brown treesnake numbers were high, their distribution included most of the island, and native birds were in clear decline. Initial thoughts on the cause of the bird decline, based on avian diseases in Hawai'i, turned out to be wrong; no explanation except that of the brown treesnake was supported. Nonetheless, Savidge faced considerable skepticism when she identified the brown treesnake as the culprit, since there was no previous example of a snake causing such ecosystem-wide impact.

BROWN TREESNAKE IMPACTS: 1980 ONWARD

Considerable work has focused on documenting brown treesnake impacts on Guam. Human impacts have taken three forms. Venomous snake bites to humans, and especially to infants, have not resulted in fatalities but have produced some cases of respiratory arrest. Economically, power outage caused by the brown treesnake is at the top of the list. Snakes climb into the transmission system, seeking food or simply moving along. Whenever they short the system, damage that ranges from purely local to islandwide can ensue, causing damaging power outages and requiring costly repairs. Lost tourist revenues resulting from bad publicity are also a concern. Culturally, the impact has been loss or massive decline of native species that were part of folktales and traditional lifestyles, such as the Mariana fruit bat (*Pteropus mariannus*, locally known as fanihi, and an important food source) and the Mariana fruit dove (*Ptilinopus roseicapilla*, tottot). Ecologically, the impacts have been some of the most extreme seen in any invasion, primarily as a result of direct predation. Native species had not evolved with a snake predator, and they had few defenses. Snake populations at the height of the irruption were higher than those for comparable snakes measured elsewhere, compounding the problem. Of the three native bat species, two are extinct and the third is barely holding on, despite considerable conservation effort. Practically all native forest birds—nine out of eleven, some of them species or subspecies unique to Guam—have become locally or globally extinct. Native reptiles have fared little better, with most species either gone or in decline. With most bird and mammal prey gone, large snakes are no longer common on Guam, and most adults are about 1.5 m in length.

Some of the extirpated species, such as the fantail (*Rhipidura rufifrons*, chichirika) were insectivorous, and their loss has resulted in changes in invertebrate populations. Other, perhaps more extensive if still unfolding, cascading effects resulted from the snake-caused extinction of important pollinators and seed dispersers such as the Mariana fruit dove and the Micronesian honeyeater (*Myzomela rubrata*, egigi). In an example of how invasive species can have synergistic effects, reduced pollination and seed dispersal are exacerbated by the invasive feral pig (*Sus scrofa*) and Philippine deer (*Cervus mariannus*) grazing on young plants. As a result, old-growth forest is not regenerating after natural or anthropogenic loss.

BROWN TREESNAKE DISPERSAL FROM GUAM

The fate of Guam is an alarming demonstration of the extensive damage that an invasive species can cause when conditions are right. Unfortunately, the same basic conditions exist on many Pacific islands, making them highly susceptible to invasion from a brown treesnake–like species. Even more unfortunately, high snake numbers, combined with the position of Guam as a civilian and military transportation hub, have allowed repeated human-aided dispersal of snakes to a remarkable diversity of locations (Fig. 2). Although some are relatively close (Fig. 2A), perhaps within the capacity of eventual natural dispersal for the brown treesnake, many are considerably further away (Fig. 2B), and a large number (Fig. 2C) would be considered long-distance dispersal by any standard.

BROWN TREESNAKE CONTROL EFFORTS ON AND OFF GUAM

More than anything else, it is the risk of further invasion that has prompted policymakers to fund brown treesnake interdiction efforts on Guam. These have focused on two primary goals. The first is to eliminate snakes from the transportation network. The second, discussed below, centers on understanding the biology of the snake on Guam, and on devising methods to control populations there. Guam's geographical isolation is an advantage in that snakes can leave the island only on aircraft or sea vessels. Indeed, brown treesnakes originating from

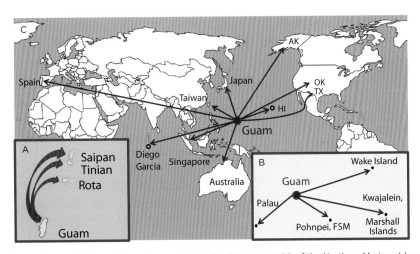

FIGURE 2 Documented brown treesnake dispersal from Guam. (A) Into the Commonwealth of the Northern Mariana Islands (CNMI; scale: tens of kilometers). (B) Within the region (hundreds of kilometers). (C) Globally (thousands of kilometers). Island locations and sizes are approximate. Some sites, such as the CNMI and Hawaii, received multiple snakes over the years, but most reported only one documented arrival.

Guam have been found on, or associated with, both. Although one might think that interdiction at two airports and two sea ports (one military and one civilian of each) would be easy, such has not been the case. Operational procedures, some local and others determined by agencies far away, limit operational access to sites and what may be done while there. Short-term and narrow economic interests also limit what can be done. Finally, much of the cargo shipped from Guam is prepared off-site, in a shifting number of privately owned facilities. Control staff have spent considerable effort identifying these facilities and gaining access to their operational areas so that snake education and inspections can be provided, with variable success.

Three primary operational tools are used on Guam. Snake traps are installed around the perimeters of ports and airports and trap hundreds of snakes annually as they approach the facility, but their success turns out to be surprisingly sensitive to details such as trap placement and the weight and material of the flap used to allow snakes in but prevent their exit. Both small and large snakes are relatively unlikely to be caught by such traps and require alternative methods to interdict. Barriers, either permanent or temporary, block snakes from entering specific areas. Although expensive in the short-term, they offer a savings over the long-run because they require relatively little maintenance. Detector dogs provide a last line of defense, inspecting both cargo that is ready to load and vessels. Research has focused on fine-tuning the efficacy of each of these methods to determine when they are most helpful and under what conditions they are ineffective.

Although brown treesnake interdiction operations on Guam have become increasingly more efficient as a result of lessons learned and research conducted, no system is perfect. Snakes are still occasionally sighted at other locations, especially those that have regular transportation links with Guam. Several locations, most notably Hawai'i and the Commonwealth of the Northern Mariana Islands, have established their own standing interdiction efforts, relying on one or more versions of the three tools described above. Because snake damage has not yet occurred at these locations, policy impediments tend to be greater than on Guam; budget levels fluctuate, and short-term economics are more likely to interfere with snake interdiction. In addition, a rapid-response team has been assembled, with trained members and at least some equipment available on multiple islands, which responds to new sightings and attempts to quickly capture and remove any snakes seen off of Guam.

ERADICATING THE BROWN TREESNAKE FROM GUAM

The argument has been made that brown treesnake damage on Guam is as bad as it is likely to get, and therefore interdiction should be the only concern. This view is short-sighted for two reasons. First, so long as the snake remains on Guam, expensive interdiction operations will be required and occasional escapes will occur. Since establishment of invasives is often tied with propagule pressure, the risk of eventual brown treesnake establishment elsewhere is unacceptably high. Second, with increasing success of island eradications and restoration efforts and the availability of some extirpated species in captive colonies, much can be done to improve things on Guam itself. Although Guam is larger than sites of most successful eradication efforts, the Oriental fruit fly (*Dacus dorsalis*) has been eradicated on Guam, showing that the process may be possible.

One of the most commonly asked questions about the brown treesnake is why the small Indian mongoose has not been released on Guam to control it. Unfortunately, this mongoose has caused more harm than good when introduced elsewhere, is not adept at climbing trees, and seems unlikely to be effective against an arboreal snake. Other biological control agents, such as diseases, currently also seem unlikely to be effective. However, research has identified a number of possible toxicants that are effective against the brown treesnake and suggests that aggressive application can drastically reduce, and with repeated coverage perhaps even eradicate, the snake from modest areas. Applying existing tools would be very difficult on Guam, most of which is privately owned and much of which is topographically rugged—but perhaps not impossible. However, the likely cost—perhaps several hundred million dollars—is likely to remain prohibitive for the foreseeable future.

SEE ALSO THE FOLLOWING ARTICLES

Biological Control, of Animals / Eradication / Invasion Biology / Islands / Lag Times / Reptiles and Amphibians / Restoration / Small Indian Mongoose

FURTHER READING

Brown Tree Snake Control Committee. 1996. *Brown Tree Snake Control Plan: Report of the Aquatic Nuisance Species Task Force, Brown Tree Snake Control Committee.* Honolulu: U.S. Fish and Wildlife Service.

Nathan, R. 2001. Dispersal biogeography (127–152). In S.A. Levin, ed. *Encyclopedia of Biodiversity*, Vol. 2. San Diego: Academic Press.

Perry, G., and J.M. Morton. 1999. Regeneration rates of the woody vegetation of Guam's northwest field following major disturbance:

Land use patterns, feral ungulates, and cascading effects of the brown treesnake. *Micronesica* 32: 125–142.

Perry, G., and D. Vice. 2009. Forecasting the risk of brown treesnake dispersal from Guam: A mixed transport-establishment model. *Conservation Biology* 23: 992–1000.

Perry, G., E. W. Campbell III, G. H. Rodda, and T. H. Fritts. 1998. Managing island biotas: Brown treesnake control using barrier technology. *Vertebrate Pest Conference* 18: 138–143.

Rodda, G. H., T. H. Fritts, and D. Chiszar. 1997. The disappearance of Guam's wildlife: New insights for herpetology, evolutionary ecology, and conservation. *BioScience* 47: 565–574.

Rodda, G. H., T. H. Fritts, and P. J. Conry. 1992. Origin and population growth of the brown tree snake, *Boiga irregularis*, on Guam. *Pacific Science* 46: 46–57.

Rodda, G. H., Y. Sawai, D. Chiszar, and H. Tanaka, eds. 1999. *Problem Snake Management: The Habu and the Brown Treesnake.* Ithaca, NY: Cornell University Press.

Savarie, P. J., J. A. Shivik, G. C. White, J. C. Hurley, and L. Clark. 2001. Use of acetaminophen for large scale control of brown treesnakes. *Journal of Wildlife Management* 65: 356–365.

Savidge, J. A. 1987. Extinction of an island forest avifauna by an introduced snake. *Ecology* 68: 660–668.

BRYOPHYTES AND LICHENS

FRANZ ESSL AND WOLFGANG RABITSCH

Environment Agency, Vienna, Austria

PHIL LAMBDON

Royal Botanic Gardens, Kew, United Kingdom

Globally, invasions of bryophytes and lichens are strongly underrecorded; the best data exist for temperate regions with a strong tradition of floristic and taxonomic research. Compared to other taxonomic groups, numbers of alien bryophytes are rather low. In Europe, there are 45 bryophyte species that are considered to be alien in at least some parts of Europe. On this basis, only 1.8 percent of all European species are certainly alien; if cryptogenic species (i.e., species that are assumed, but not known with certainty, to be alien) are included, then the estimate rises to 2.5 percent. The cumulative number of alien bryophytes in Europe, and probably worldwide, has increased exponentially in recent decades. Countries and regions with humid climates are most heavily invaded. In comparison with other taxonomic groups, the contribution of distant regions (especially from the opposite hemisphere) to alien bryophyte floras is remarkable. The dominant pathway is unintentional introduction with ornamental plants. Alien bryophyte species display a strong affinity for human-made habitats. Within lichens, only a very few alien species have been recorded, and these are mainly restricted to human-made habitats in urban areas in the northern hemisphere.

GLOBAL PATTERNS

Invasions of bryophytes are strongly underrecorded, and the spatial distribution of data is very skewed toward temperate regions with a strong tradition of floristic and taxonomic research. Hence, for most (sub)tropical regions, even approximate numbers of alien bryopyhtes are currently impossible to estimate. However, one globally valid pattern is their low number of alien species. One explanation for the paucity of alien bryophytes is the lack of distribution data and historical knowledge, so some alien bryophytes (especially inconspicuous species) might well have been overlooked and therefore be wrongly considered to be indigenous. Spores of bryophytes are very efficient at long-distance dispersal, which means that human activities play a much less prominent role in overcoming geographic barriers than with vascular plants. In fact, many bryophytes appear to have colonized both hemispheres by natural means. Of those species considered to be native to the United Kingdom, 75 percent are also known from North and Central America, and 14 percent from Australia; 3 percent are even known from Antarctica. Although their biogeographic history remains largely unknown, many appear to be widespread and ecologically well integrated across their range, with little evidence to suggest recent arrival. Furthermore, bryophytes are only rarely transported for economic purposes; hence, intentional introduction—the prevailing pathway for vascular plants, for example—is of little importance.

The pattern of bryophyte invasions in the temperate regions of the northern hemisphere is best known for Europe due to the DAISIE project. Patterns emerging from this data set are presented below and supplemented by case studies from other continents. For alien lichens, the data situation is woefully incomplete, which limits analyses of invasion patterns. Checklists are available for only a few countries (e.g., Austria, Czech Republic, United Kingdom). However, this appears to genuinely reflect the rarity of alien lichens.

SPECIES NUMBERS AND INVASION HOTSPOTS

Globally, numbers of alien bryophytes are rather low. In Europe, there are 45 bryophyte species (excluding greenhouse species) that are considered to be alien at least in

TABLE 1
Alien Bryophytes in Europe

Status	Sub-phylum	Family	Species	Number of Countries or Regions
Alien	Bryophyta	Dicraniaceae	*Campylopus introflexus*	21
	Bryophyta	Orthodontiaceae	*Orthodontium lineare*	15
	Bryophyta	Pottiaceae	*Didymodon australasiae*	11
	Hepaticopsida	Ricciaceae	*Ricciocarpos natans**	8
	Bryophyta	Pottiaceae	*Leptophascum leptophyllum*	6
	Bryophyta	Pottiaceae	*Hennediella stanfordensis*	4
	Bryophyta	Pottiaceae	*Tortula bolanderi*	4
Cryptogenic	Hepaticopsida	Lunulariaceae	*Lunularia cruciata**	12
	Hepaticopsida	Ricciaceae	*Riccia rhenana**	12
	Bryophyta	Pottiaceae	*Scopelophila cataractae*	7
	Bryophyta	Rhabdoweisiaceae	*Dicranoweisia cirrata**	4

NOTE: Ordered by decreasing number of invaded countries. Only species invading more than three countries or regions (e.g., large islands) are shown.
* = native in some parts of Europe (from Essl and Lambdon, 2009).

some parts of the continent. These comprise 21 alien mosses and 11 liverworts, but no hornworts. This figure includes 13 cryptogenic species (11 mosses and 2 liverworts), which are strong candidates to be alien, although there is insufficient evidence to be certain. Eight of the aliens are native to Europe as a whole but alien to some countries, while the remainder are alien to the entire continent. Most species are rare; only 11 species have been recorded as alien or possibly alien from more than three countries (Table 1). Very few of these (especially *Campylopus introflexus* and *Orthodontium lineare*) are widespread. The ruderal thalloid liverwort *Lunularia cruciata*, which is native to the Mediterranean region, has also greatly expanded its range northward in recent decades.

In North America, the few rather widespread bryophytes are mostly of European (e.g., *Brachythecium* spp., *Lunularia cruciata*, *Pseudoscleropodium purum*, *Thuidium tamariscinum*) and southern hemisphere

(e.g., *Campylopus pyriformis*, *C. introflexus*) origin. In temperate regions of the southern hemisphere, naturalized introduced bryophytes include the pleurocarp European mosses *Pseudoscleropodium purum* (Australia, New Zealand, South America, St. Helena), *Rhytidiadelphus squarrosus* (Tasmania, New Zealand), *R. triquetrus* (New Zealand), and *Sphagnum subnitens* (New Zealand). Significant examples of bryophytes assumed to be alien in (sub)tropical regions include *Jungermannia sphaerocarpa* (tropical Africa, South America, Borneo, New Guinea), which is widespread in human-made high mountain habitats. For Australia, 12 mosses are recorded as alien. Concordantly to alien vascular plants, some well-studied tropical islands boast a rather high number of alien bryophytes (e.g., 14 alien mosses in Hawaii and 25 possible introductions on St. Helena, equivalent to 23% of the known bryoflora). In the whole of Europe, there are 1,292 species of mosses and 474 species of liverworts (excluding subspecies and varieties). On this basis, only 1.8 percent of all European species are certainly alien; if cryptogenic species are included, the estimate rises to 2.5 percent. This relationship strongly contrasts with the much higher proportion of aliens among vascular plants and seems to be valid globally.

TEMPORAL AND SPATIAL TRENDS

The first records of alien bryophytes in Europe were made in the nineteenth century (e.g., 1828: *Lunularia cruciata* in Karlsruhe, Germany). The cumulative number of alien bryophytes in Europe has increased exponentially (Fig. 1). This temporal picture seems to be valid for other

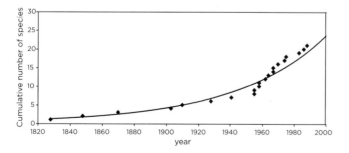

FIGURE 1 Temporal trends in invasion of 25 alien and cryptogenic bryophytes in Europe, for which data on introduction dates are available (from Essl and Lambdon, 2009). Shown is the cumulative number of recorded species ($R^2 = 0.98$; $y = 2E\text{-}14e0.0174x$).

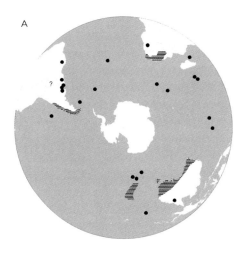

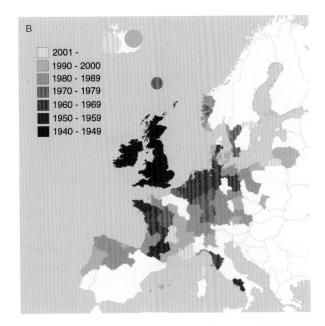

FIGURE 2 Native range of *Campylopus introflexus* and the spatiotemporal pattern of its invasion in Europe (Hassel and Söderström, 2005).

continents. Due to its rapidity, the colonization of Europe by the southern hemisphere moss *Campylopus introflexus* represents one of the rare well-documented bryophyte invasion events (Fig. 2). This species has also successfully invaded the western United States and Canada since the first record in 1975.

The countries with the most alien bryophytes in Europe are the United Kingdom, followed by France and Ireland, while in North America, states in the Pacific West and in the Northeast are most invaded. In the southern hemisphere, New Zealand (14 alien mosses) and Tasmania show the highest numbers of alien bryophytes. Regions with prevailing humid climates are invasion hotspots, while countries with drier and warmer climate are poor in alien bryophytes.

No lichens are unequivocally known to be alien in Europe, although a few may be considered cryptogenic. The best known of these is *Lecanora conizaeoides*, which has expanded to become one of the commonest species encrusting trees with acid bark throughout Europe in recent decades. It is strongly pollution tolerant and may have spread naturally from small relict populations following increasing acidification of rainwater by industrial emissions. *Lecanora conizaeoides* has also spread considerably in the western United States. A few other pollution-tolerant lichens are currently spreading within Europe (e.g., *Anisomeridium polypori*, *Buellia punctata*, *Hypocenomyce scalaris*, *Physcia biziana* var. *aipolioides*, *Scoliciosporum chlorococcum*).

REGIONS OF ORIGIN

Compared to other taxonomic groups, the important contribution of distant regions (especially from the southern hemisphere) to the alien bryophyte floras of North America and Europe is remarkable. This result is linked to the excellent dispersal capacities of bryophytes, which makes it easier for species of adjacent regions to overcome geographical barriers without the assistance of humans. However, for southern hemisphere species, crossing the equator by natural means presents a barrier because of the prevailing wind directions in the intertropical convergence zone.

In Europe, the most important donor regions in decreasing order of importance are South America, Australasia, North America, and Africa. Another 7 percent of the species are native to Oceanic islands. Only 7 and 8 percent of the species are native to parts of Europe or temperate Asia, respectively.

PATHWAYS

The most important introduction pathway for bryophytes is with ornamental plants. In Europe, 15 species have been introduced this way (e.g., in association with Australasian tree ferns). The only other significant introduction pathway is unspecified accidental import as a hitchhiker on ships and planes, perhaps on clothing or goods.

The negligible contribution of deliberate introduction strongly contrasts with vascular plants. In Europe, only one species (*Ricciocarpos natans*), which is floating in water and used for garden ponds and aquaria, is known to be introduced deliberately. In North America, *Pseudoscleropodium purum* has been introduced intentionally in packing material for nursery stock and is likely to be occasionally transported as a contaminant of peat-based compost.

Secondary spread often mainly depends on the dispersal capacity of the species. However, human activity may enhance it. Spread may be favored by anthropogenic changes to existing habitats (e.g., by airborne pollutants), by the creation of new habitats, or by unintentional transport. In Europe, some alien bryophytes are known to have expanded due to acid rain (e.g., *Campylopus introflexus, Orthodontium lineare*). The spread of *Campylopus introflexus* in Europe is fostered by mechanical disturbance (e.g., caused by rabbits or trampling by people), which breaks up the moss mat and subsequently disperses propagules to initiate new populations. The formation of specialized structures for vegetative reproduction is a potential mechanism to aid invasion, via vegetative fragments, gemmae (simple asexual buds shed from the leaf tips or other organs), bulbils, or tubers (larger gemmae-like structures borne respectively in the leaf axils or on rhizoids), although most species are probably only moved short distances by such means. Fifty-four percent of U.K. alien bryophytes possess these structures, compared with only 28 percent of natives. However, many bryophytes are dioecious, and because there is often only one sex in the founder population (e.g., *Leptophascum leptophyllum, Lunularia cruciata* in its alien European range), they can only survive via vegetative propagation. This feature is therefore likely to be partly a consequence of their colonizing ability rather than a cause of invasion.

For lichens, introduction pathways are poorly understood. However, it is known that *Lecanora conizaeoides* has increased in Europe as a result of acidification of precipitation, and *Parmelia submontana* seems to be associated with alien trees and may thus be an accidental garden import.

INVADED ECOSYSTEMS

Alien bryophytes display a strong affinity for human-made habitats, especially gardens, roadsides, and walls. In temperate regions, a considerable number of (sub)tropical species are restricted to glasshouses (in Europe, e.g., *Marchantia planiloba, Vesicularia reticulata, Zoopsis liukiuensis*). In North America, several bryophytes have successfully colonized frequently cut lawns (e.g., *Brachythecium* spp., *Pseudoscleropodium purum*). This is in strong contrast to the patterns displayed among native bryophytes. Few bryophytes have

so far widely established themselves in near natural vegetation. In Europe and western North America, *Campylopus introflexus* colonizes mainly coastal dunes, heathlands, and bogs. In Australia and New Zealand, some northern hemispheric species (e.g., *Pseudoscleropodium purum, Rhytidiadelphus squarrosus, R. triquetrus*) have locally invaded native bushland and forest vegetation, while *Sphagnum subnitens* is invading wetlands on the west coast of New Zealand. The few cryptogenic lichens in Europe mostly colonize acidic bark of trees.

ECOLOGICAL AND ECONOMIC IMPACTS

The impacts of alien bryophytes and lichens are less obvious than in vascular plants. Their small size means that they are rarely strong competitors for light, nutrients, or space. However, they may compete with native bryophytes and lichens, or with germinating seedlings. In Europe, *Campylopus introflexus* (Fig. 3) is the only species known to cause strong impacts of this type. It is capable of forming dense mats, significantly reducing the species diversity of other lower plants and lichens. A few threats posed by other alien bryophytes have been documented, but so far the implications have not been well studied. In the United Kingdom, *Orthodontium lineare* competes with a rare relative, *O. gracile*, on sandstone rocks, and has caused the loss of this species from many localities. In northwestern Europe, *Lophocolea semiteres* is displacing the native congener *L. heterophylla* in acid heaths and pine forests. In the southern hemisphere, few European pleurocarp mosses (*Pseudoscleropodium purum* in Australia and St. Helena, *Rhytidiadelphus triquetrus* in New Zealand) are considered to outcompete native bryophytes.

FIGURE 3 Dense mat of *Campylopus introflexus* in a coastal dune near Bremen, Germany. (Photograph courtesy of Maike Isermann.)

Aside from competition, it is also likely that abundant alien bryophytes can alter ecosystem functioning occasionally by stabilizing soils, binding leaf litter, altering decay rates, and creating humid microhabitats which affect the composition of microfaunal communities, but the potential consequences of invasions at this microenvironmental level remain almost unexplored.

MANAGEMENT OPTIONS

Deriving feasible management strategies for bryophytes and lichens is very difficult for several reasons. Introduction is difficult to control, identification of species needs expert knowledge, long-distance dispersal and thus reimmigration is likely to be frequent, and most management measures are difficult and costly to apply because of the small size of the invaders. In Europe, the only bryophyte perceived as a threat to biodiversity is *Campylopus introflexus*. Tested management methods include liming dunes, burning, treating with herbicides, and introducing grazing animals to trample the mats. These have generally met with only limited success. Efforts are also under way to eradicate *Rhytidiadelphus triquetrus* in New Zealand, and the invasion of *Pseudoscleropodium purum* in urban lawns in the western United States has led to a thriving "moss killer" industry based on herbicide application.

EXPECTED TRENDS

Human-made modifications of existing habitats and creation of new habitats, along with increasing intercontinental trade, are the dominant drivers for the spread of alien bryophytes. Airborne acidification has been strongly reduced in most regions in the last few decades, so it is likely that in the future this factor will lose relevance. However, nitrogen deposition in large parts of Europe exceeds critical loads and is still increasing. As several alien bryophytes can take advantage of airborne nitrogen, it is expected that some species will make gains from this trend.

Climate change and increasing temperatures may foster range expansions of alien bryophytes in the future. In Europe, some (sub)tropical bryophyte species formerly restricted to glasshouses have started to establish outdoors in recent years (e.g., *Didymodon australasiae*), which might reflect the recent warming trend.

As intercontinental trade increases further, and especially with the increasing popularity of exotic ornamental plants, there is great potential for increased movement of alien bryophyte and lichen species across the world. Thus, we can probably expect that numbers, abundance, and associated impacts of alien bryophytes and lichens will increase in the future considerably.

SEE ALSO THE FOLLOWING ARTICLES

DAISIE Project / Dispersal Ability, Plant / Disturbance / Geographic Origins and Introduction Dynamics / Invasibility, of Communities and Ecosystems / Nitrogen Enrichment / Weeds

FURTHER READING

Australian National Botanic Garden. 2009. *Bryogeography: Imports and Exports.* www.anbg.gov.au/bryophyte/bryogeography-imports-exports .html

Essl, F., and P. Lambdon. 2009. The alien bryophytes and lichens of Europe (29–42). In *The Handbook of Alien Species in Europe.* Berlin: Springer.

Gradstein, S. R., and J. Váňa. 1987. On the occurrence of Laurasian liverworts in the tropics. *Memoirs of New York Botanical Garden* 45: 388–425.

Hassel, K., and L. Söderström. 2005. The expansion of the alien mosses *Orthodontium lineare* and *Campylopus introflexus* in Britain and continental Europe. *Journal of the Hattori Botanical Laboratory* 97: 183–193.

Hedenäs, L., T. Herben, H. Rydin, and L. Söderström. 1989. Ecology of the invading moss *Orthodontium lineare* in Sweden: Spatial distribution and population structure. *Holarctic Ecology* 12: 163–172.

Holyoak, D., and N. Lockhart. 2009. Australasian bryophytes introduced to South Kerry with tree ferns. *Field Bryology* 98: 3–7.

Miller, N. G., and N. Trigoboff. 2001. A European feather moss, *Pseudoscleropodium purum,* naturalized widely in New York State in cemeteries. *The Bryologist* 10: 98–103.

Pfeiffer, T., H. J. D. Kruijer, W. Frey, and M. Stech. 2000. Systematics of the *Hypopterygium tamarisci* complex (Hypopterygiaceae, Bryopsida): Implications of molecular and morphological data. *Journal of the Hattori Botanical Laboratory* 89: 55–70.

Schofield, W. B. 1997. Bryophytes unintentionally introduced to British Columbia. *Botanical Electronic News* 162. www.ou.edu/cas/botany -micro/ben/ben162.html/

Söderström, L. 1992. Invasions and range expansions and contractions of bryophytes (131–158). In J. W. Bates and A. M. Farmer, eds. *Bryophytes and Lichens in a Changing Environment.* Oxford: Clarendon Press.

BURMESE PYTHON AND OTHER GIANT CONSTRICTORS

ROBERT N. REED AND GORDON H. RODDA

USGS Fort Collins Science Center, Colorado

The ecology of invasive reptiles has historically received little attention compared to that of birds and mammals. A turning point was the recognition that the disappearance of birds from Guam resulted from introduction of the brown treesnake (*Boiga irregularis*). Having learned what damage can be caused by invasive snakes, wildlife authorities have become very concerned about a recent irruption of Burmese pythons (*Python molurus bivittatus*) in southern Florida. Florida is also home to at least two other introductions of exotic constrictors, one of which has also invaded islands in the Caribbean. This article focuses primarily on the Burmese python, because more data are available on the biology and impacts of this species.

SPECIES CHARACTERISTICS AND BASIC BIOLOGY

Identification

Several species of giant constrictors are of concern as established or potential invasive species, but only three are known to have established invasive populations. Of these, the best known by far is the Burmese python (*Python molurus bivittatus*), which is established over thousands of square kilometers of southern Florida (Fig. 1A). The species *P. molurus*, native to Asia, has traditionally included the Burmese python in Southeast Asia and the Indian python

FIGURE 1 (A) An adult female Burmese python brooding her eggs in Everglades National Park, Florida. Up to 85 eggs have been recovered from a single female python in the Everglades. (B) A boa constrictor from a long-established population in southern Miami, Florida. (Photographs courtesy of M. Rochford.)

(*P. m. molurus*) in south-central Asia. More recently, some authorities have considered Burmese and Indian pythons to be separate species (i.e., *P. molurus* and *P. bivittatus*). The Burmese python exhibits a darker color pattern and an arrowhead-shaped mark on the top of the head that does not fade toward the snout.

The northern African python (*Python sebae*), also referred to as the African rock python, also appears to be established in southern Florida, but in a smaller area of central Miami–Dade County. This species is superficially similar to Burmese pythons, and distinguishing between the species usually hinges on a combination of color pattern and scale counts. Burmese and northern African pythons hybridize readily in captivity, and introduced populations are found in the same area of southern Florida; the fertility and fitness of hybrids are largely unknown, as are the potential effects of hybridization on invasiveness.

The boa constrictor (*Boa constrictor*) has established several invasive populations in Florida and the Caribbean (Fig. 1B). This is a wide-ranging and variable species, with at least seven commonly recognized subspecies. Some authorities consider this taxon to be composed of several species, but there is as yet little agreement on the distinctiveness of these subgroups. Boas have no enlarged scales on the top of the head and no obvious heat-sensing pits on the labial (lip) scales, allowing them to be easily distinguished from pythons.

Hundreds of extralimital individuals of these and several additional species of giant constrictor snakes have been found in the United States and other countries, typically due to escapes or releases from captivity, but thus far, none of the other species are confirmed to have established reproductive populations. Aside from the three species mentioned above, large snakes including reticulated pythons, green anacondas, and yellow anacondas are regularly reported as having been introduced.

Body Size and Reproduction

The Burmese python is one of the four "true giants" among snakes, a group in which individuals can reach or exceed 6 m in total body length; the other members of this group are the green anaconda (*Eunectes murinus*), northern African python, and reticulated python (*Python* [or *Broghammerus*] *reticulatus*). In all of these species, females attain larger body sizes than do males. Fantastically exaggerated tales of huge body sizes are the rule, rather than the exception, for these giant snakes, but most such tales do not survive careful scrutiny; while there are fairly reliable records of free-ranging reticulated pythons over 8 m in length, individuals over 6 m long appear to be rare in any population of giant snakes. Burmese pythons over 5 m in length are reported to be scarce in the species' native range, but well-fed captives have reached 8 m and 180 kg. Females collected from the introduced population in Florida have occasionally exceeded 5.2 m and 75 kg, and median body size of Florida specimens hovers around 3 m and 10–15 kg. Female Burmese pythons can produce clutches of up to 107 eggs, although 25–50 is more typical. Females from the Everglades have contained as many as 85 shelled eggs in their oviducts. Hatchlings emerge at lengths of 50–75 cm, and growth appears to be rapid.

The northern African python is closely related to the Burmese python and exhibits a similar range of body and clutch sizes in its African distribution. Of about a dozen northern African pythons found in Florida to date (January 2010), adults have typically exhibited superlative body condition, sometimes with visible rolls of fat along their bodies. One male captured in early 2010 may be larger both in length (4.4 m) and mass (63 kg) than any male ever recorded in the native range (Fig. 2). Such unusually large body sizes may be common among some populations of invasive snakes during early stages of invasion, especially when such snakes represent ecologically novel predators on naive prey.

Although they are very large and heavy-bodied, boa constrictors do not attain truly giant body sizes. Their maximum size is probably a bit over 4 m, but individuals larger than 3 m are uncommon in most areas of the native distribution. A very large female boa (about 3.5 m and 43 kg, close to the maximum known body mass for free-ranging boas) was captured in 2009 on a small island in the Florida Keys, and close examination of its color

FIGURE 2 Employees of the U.S. National Park Service and South Florida Water Management District with an extremely large (440 cm, 63 kg) male northern African python captured in January 2010 during a multiagency search effort. (Photograph courtesy of N. Yglesias. Reprinted by permission from the South Florida Water Management District, a public corporation of the State of Florida.)

pattern and scalation allowed it to be conclusively identified as the same individual that had been photographed on the island almost three years prior. The ability of this snake to attain such a large body size on a small human-occupied island while evading capture attests to the highly cryptic nature of large snakes. As with anacondas, and unlike pythons, boas are live bearing, with maximum litter sizes of about 60.

Diet

Almost without exception, the majority of the diet of giant constrictors consists of warm-blooded prey, although reptiles including iguanas and crocodilians are also taken. To this smorgasbord, anacondas add fish, turtles, and snakes. A wide range of prey including porcupines, primates, leopards, and flying foxes has been recovered from Burmese and Indian pythons in the native range, and other giant constrictors appear to consume a similarly broad range of prey.

The large number of Burmese pythons collected from southern Florida has allowed a more thorough dietary analysis than is available from the entire vast native range. Nearly three dozen species of native vertebrates are known to have been consumed by Burmese pythons in Florida, including deer, bobcats, wrens, herons and egrets, alligators, and raccoons. These have included federally endangered species such as the wood stork and Key Largo woodrat, as well as species of conservation concern such as white ibis, round-tailed muskrats, and limpkins. Burmese pythons may already have reduced populations of mid-size mammals such as raccoons, round-tailed muskrats, and marsh rabbits in Everglades National Park. Invasive Burmese pythons appear to readily eat any bird or mammal of suitable size, along with the occasional reptile. Some media sources have sensationalized the risk of attacks on humans by giant constrictors—although attacks on humans are well documented for some species in their native ranges (especially reticulated pythons) and snake keepers are occasionally killed by their pets, there have been no fatalities from any free-ranging introduced giant constrictor, and unbiased assessments typically conclude that the risk to humans is minuscule.

GEOGRAPHIC DISTRIBUTIONS

Native Distributions

The native distribution of the Burmese python includes much of Southeast Asia including most or all of Vietnam, Cambodia, and Thailand, portions of Myanmar,

Bangladesh, and Bhutan, and some parts of Indonesia. Burmese pythons also occur to the west along the border of India and Nepal. Indian pythons occupy Sri Lanka and the remainder of the subcontinent of India. The range of the Burmese python in China is surprisingly poorly understood, perhaps owing to thousands of years of persecution by humans; the status of current or historical populations between Sichuan and more southerly Chinese locales is particularly poorly documented.

The native distribution of the northern African python includes most of central Africa south of the Sahara, from southern Mauritania across to Somalia, with a northerly extension of the range through central Ethiopia into Eritrea and northeastern Sudan. In arid regions, pythons appear to be limited to the vicinity of watercourses. In southern areas of the range, the species may be found in close proximity to the closely related southern African python (*Python natalensis*).

The native distribution of the boa constrictor spans a vast range in the New World, from Argentina to northern Mexico. The species is distributed more or less continuously at elevations below about 1,500 m through Central America and south through the Amazon basin. The Argentine boa, a subspecies, is found to at least 30 °S latitude, inhabiting areas that experience colder winter conditions than more northerly forms of the species.

Introduced Distributions and Establishment History

Burmese pythons are now distributed across thousands of square kilometers of southern Florida. The origin of their invasive distribution is Everglades National Park, and pythons have been collected from every major habitat type in the park, including sawgrass prairie, cypress domes, hardwood hammocks, mangroves, and human-made canals. In the last few years, increasing numbers of pythons have been collected north of the park; numerous individuals have been collected, and there is evidence of reproduction (i.e., recent hatchlings). Multiple Burmese pythons have been observed or collected outside of their known range in Florida, and it is difficult to delineate clearly the extent of the current population. As an example, several observations of Burmese pythons east of the city of Sarasota could represent independent releases or escapes from captivity, a second incipient population, or an extension of the known population. Rumors about a second introduced population of Burmese pythons in the United States territory of Puerto Rico have circulated for several years but have proven difficult to verify.

The origin of the established population of Burmese pythons has been hotly debated. Overall, the spatiotemporal distribution of the earliest pythons collected from Everglades National Park appears to support the hypothesis that the population originated in the extreme southwestern corner of the park near the end of the main park road. This pattern of observations in a remote area at the end of a road suggests that pythons may have been intentionally released, perhaps by pet owners wishing to find a "natural" home for an unwanted pet or by python aficionados attempting to establish a population. Conversely, many pythons were known to have escaped from poorly secured commercial animal-holding facilities along the eastern margins of the park during Hurricane Andrew in 1992. Although it is unlikely that the origin of the population will ever be known with certainty, all parties appear to agree that the ultimate source of the population is the pet trade, and in terms of ecological impacts, it is immaterial whether pythons were released intentionally or unintentionally.

Northern African pythons have only recently been confirmed to be established in Florida. The population almost certainly has its ultimate origins in the pet trade, but it is unknown whether it resulted from escapes or intentional releases. Thus far, the population is known from a small (perhaps 20–40 km²) marshland area west of Miami. A hatchling was collected in summer 2009, prompting a shrewd National Park Service biologist to solicit additional records from other agencies that occasionally respond to snake sightings. The results of these inquiries yielded five more records of the species from the same small area, suggesting the existence of a population. A concerted multiagency and volunteer effort to confirm the presence of a population and delineate its geographic extent occurred in early 2010. The search yielded five more adult northern African pythons, a Burmese python, and a boa constrictor (the last was probably a former captive). Events associated with the discovery of this population point to the importance of centralized record-keeping for early identification of "hotspots" of observations of invasive species, as well as to the potential efficacy of high-intensity searches to identify the boundaries of incipient populations. Two additional northern African pythons have been collected in the last few years from a relatively small area east of Sarasota, Florida.

The boa constrictor has established more invasive populations than any other large constrictor snake and is confirmed to be established in southern Florida and on the Caribbean islands of Aruba (Netherland Antilles) and Cozumel (Mexico). Populations on additional islands off the coasts of Mexico and Colombia may also be the result of human introductions, and recent evidence strongly suggests an established population in western Puerto Rico. The population in southern Florida has been present for several decades and may be confined to the vicinity of a county park in the city of Miami. Boas are commonly found in outlying areas of Miami, however, and few attempts have been made to survey outside the known area of occupancy or to determine whether outliers represent independent introductions from captivity, dispersing waifs from the known population, or members of a geographically expanding population. The Miami population appears to consist of boas of southern Colombian stock, which were popular in the pet trade several decades ago. The origin of the high-density boa population on Aruba is not known with certainty; most authorities consider the pet trade to be the most likely source, but unintentional introduction from the mainland via horticulture imports or other means is also a possibility. The Cozumel population stands apart as having become established under faintly bizarre circumstances. A small number of boas (reported as two to six individuals) were imported to Cozumel and used during the filming of a movie in 1971, after which they were released and became established on the island.

PATHWAYS AND TRADE

Extralimital individuals of various giant constrictor species have been reported from all over the world for over a century; for example, a northern African python was collected in the state of Pennsylvania in 1901. The pathways by which these snakes arrive and become introduced are numerous. Some individuals have been unintentionally transported with cargo or agricultural products (in previous decades, boa constrictors regularly entered the United States in banana shipments from Central America), but advances in shipping and inspection methods have decreased the frequency of unintentional transport. Unintentional transport relies on stowaways escaping detection, an improbable prospect for very large snakes.

In the last few decades, most introductions of giant snakes have been associated with the international trade in reptiles as pets. This trade includes wild-caught snakes, captive-bred or captive-hatched juveniles from areas within native ranges, and domestically captive-bred animals. It is no surprise that most of the invasive populations are in the United States, as by some estimates, over 80 percent of the global trade in live reptiles passes through the United States. In an average year over the last few decades, tens of thousands of large constrictors

(primarily boa constrictors, with smaller numbers of Burmese and reticulated pythons and even fewer northern African pythons and anacondas) entered the United States. The keeping of hundreds or thousands of imported individuals of the same species provides the opportunity for multiple escapes after natural disasters or via insecure housing. Captive propagation, chiefly in the United States and northern Europe, adds many thousands of snakes to the overall trade (especially Burmese pythons, reticulated pythons, and boas), although the size and species composition of the domestic trade in any given country is unknown.

RISKS AND MANAGEMENT

Risk Factors

The giant constrictors share a number of ecological traits that either increase their probability of establishment or magnify the potential ecological consequences of their establishment. Such traits include large body size, high reproductive output, rapid growth, a broad diet (Fig. 3), broad habitat tolerance, and the capability to attain high population densities. The primary impact of established populations is likely to be predation on native species, with the most severe impacts being likely to fall on prey species that are already threatened by habitat alteration or other environmental problems. Secondary impacts are also possible, as these snakes are capable of hosting parasites (especially ticks) that can vector pathogens of economic, human health, and agricultural importance.

Non-ecological factors also have a good deal to do with the probability of establishment of giant constrictors. These snakes are widely available, and prices are often

FIGURE 3 American alligators and Burmese pythons now represent the largest predators in the Everglades (both attain larger sizes than the Florida panther). Both species regularly consume the other; in this case, a large alligator has captured a midsized python. (Photograph courtesy of L. Oberhofer.)

low (hatchling Burmese pythons sell for as little as $25 each). While buyers may be attracted by the thought of an impressively large pet, the costs of feeding and housing a very large snake are considerable. Pythons are incredibly strong, and they commonly escape from poorly constructed cages. Moreover, imported wild-caught snakes can be riddled with parasites and are prone to biting, rendering them undesirable as pets. Zoos and animal shelters typically refuse offers of large snakes owing to the physical space and expense associated with their care, so snake owners often find themselves without an obvious outlet for disposing of an unwanted python. All of these factors converge to increase the odds of intentional or unintentional release, as evidenced by hundreds of extralimital observations of giant constrictors around the globe.

Climate Matching

Invasion biologists often attempt to use the climate experienced by a species in its native range to predict where it might invade. "Climate" is a complicated mix of many factors, however, and uncertainties in matching climates include a number of methodological, philosophical, and statistical concerns, including such basic decisions as which areas of the native range should be treated as likely source areas (Fig. 4). Several climate-matching scenarios for Burmese pythons in the United States have been proposed; most of these predict suitable climates in Hawaii,

Florida, the island territories, the southeastern Coastal Plain, and in some cases even larger areas.

A rare cold snap in southern Florida in early 2010 resulted in the deaths of large numbers of Burmese pythons and at least some northern African pythons, along with other introduced species such as green iguanas and armored catfish and native species such as manatees and American crocodiles. It is as yet difficult to assess the demographic impact of this event or interpret its ecological or evolutionary implications for the northerly spread of the Florida population. Mortality of pythons may have been due to cold intolerance, inappropriate behavioral responses to sudden and prolonged decreases in temperature, or low availability of thermally suitable belowground refugia in Everglades wetlands.

Management Opportunities

Giant constrictors share a suite of traits that render management difficult. Their camouflaged patterns, immobility for long periods of time, and regular use of vegetated or aquatic habitats render even the largest snakes extremely cryptic, thus reducing the efficacy of visual searches. As ambush predators, these snakes may move infrequently between ambush sites; this habit reduces the probability that snakes will regularly encounter control tools such as traps. Most pythons in Florida are found by driving roads at night during the warm summer months or by walking along levees searching for basking sites in the winter.

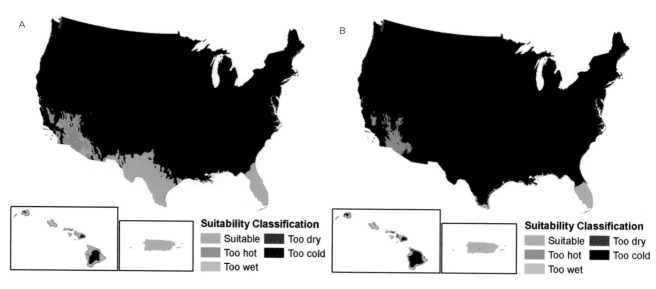

FIGURE 4 Climate-matching extrapolations to the United States for *Boa constrictor*, based on mean monthly rainfalls and temperatures from the native range. (A) is based on localities for the entire species including the cool-climate Argentine subspecies; (B) excludes Argentine localities. (Reproduced from Reed and Rodda, 2009.)

While these methods have yielded over 1,000 individuals, roads and levees comprise a minuscule proportion of the total occupied area. Traps have proven capable of catching pythons, but capture rates are low. Chemical or pheromonal control, biocontrol, and other "high-tech" tools have been proposed for investigation, but all either have a very poor track record for control of introduced vertebrates or constitute a major risk to nontarget species. Bounty systems and citizen involvement have been suggested as possible solutions, but these tend to be subject to the limitations of the existing methods (limited access to occupied habitat and low detection probabilities of the target species). Overall, there appear to be no control tools currently available that would suffice to eradicate a widespread population of giant constrictors. However, available tools may allow protection of small areas with high ecological value (e.g., bird rookeries) and prevention of spread of existing populations to islands (e.g., prevention of establishment in the Florida Keys). Developing an effective network of rapid responders might also increase the odds of eradicating incipient populations of giant constrictors.

SEE ALSO THE FOLLOWING ARTICLES

Brown Treesnake / Climate Change / Early Detection and Rapid Response / Pet Trade / Predators / Reptiles and Amphibians

FURTHER READING

Everglades National Park. Burmese python: Species profile. www.nps.gov/ever/naturescience/burmesepython.htm

Harvey, R. G., M. L. Brien, M. S. Cherkiss, M. Dorcas, M. Rochford, R. W. Snow, and F. J. Mazzotti. 2008. Burmese pythons in south Florida: Scientific support for invasive species management. University of Florida IFAS publication WEC242. http://edis.ifas.ufl.edu/uw286.

Kraus, F. 2009. *Alien reptiles and amphibians: A scientific compendium and analysis.* New York: Springer.

Murphy, J. C., and R. W. Henderson. *Tales of Giant Snakes: A Historical Natural History of Anacondas and Pythons.* Malabar, FL: Krieger.

O'Shea, M. 2007. *Boas and Pythons of the World.* Princeton: Princeton University Press.

Reed, R. N., and G. H. Rodda. 2009. Giant constrictors: Biological and management profiles and an establishment risk assessment for nine large species of python, anacondas, and the boa constrictor. United States Geological Survey Open-File Report 2009-1202. Reston, VA: U.S. Geological Survey. Available at www.fort.usgs.gov/products/publications/pub_abstract.asp?PubID=22691

Snow, R. W., K. L. Krysko, K. M. Enge, L. Oberhofer, A. Warren-Bradley, and L. Wilkins. 2007. Introduced populations of *Boa constrictor* (Boidae) and *Python molurus bivittatus* (Pythonidae) in southern Florida (418–438). In R. W. Henderson and R. Powell, eds. *Biology of the Boas and Pythons.* Eagle Mountain, UT: Eagle Mountain Publishing.

CANALS

STEPHAN GOLLASCH

GoConsult, Hamburg, Germany

Similarly to other species introduction vectors (e.g., ballast water, hull fouling, and species introductions for aquaculture purposes), canals enable organism transfers between the waters they connect. Canals overcome biogeographic barriers, thereby providing opportunities for species to migrate to waters from which they were formerly completely isolated. This is not only the case for canals that enable the passage of oceangoing vessels, but also for inland waterways, where freshwater and brackish biota migrate through canals connecting inland waters and seas.

CANALS FOR INLAND SHIPPING

Canal construction has resulted in extensive linkages of rivers and lakes in the Americas, Asia, and Europe. In order to facilitate cargo shipments, navigable canals have been constructed since antiquity, and some of them are engineering masterpieces. In the sixth century BCE, a canal was built to connect the Nile with the northern Red Sea. Another example is the Chinese Grand Canal, which was built in the fourth century BCE. This canal of almost 1,000 km in length stretched from Peking to Hangzhou. In Europe, a canal network connects the North, Baltic, White, Black, Azov, and Caspian Seas. These connections have been in place for more than 100 years, linking previously isolated, predominantly brackish water bodies. Commercially, the most important European inland waterway is the Ponto-Caspian–Volga–Baltic canal network; annually, more than 300 million tons of cargo are shipped through it, as of the turn of the twenty-first century.

Species sometimes spread through canals by natural range expansion. In other cases, sudden and isolated appearances of alien biota have been observed. This latter phenomenon has been termed "jump dispersal." Vessels are an important vector of alien species spread in inland waterways.

CANALS FOR OCEANGOING VESSELS

As a result of technological innovations and of increasing demand for intercontinental shipping, beginning in the eighteenth century, two major interoceanic canals were completed, the Suez and the Panama. Such canals for oceangoing vessels serve as the world's greatest short-cuts, and so they are heavily used shipping routes. The Suez Canal, opened in 1869, connects the Mediterranean Sea with the Red Sea and saves ships up to 8,500 nautical miles, or 25 days at sea. Since 1914 the Panama Canal has enabled passage between the Atlantic and the Pacific Oceans, cutting the route between the two oceans by up to 8,100 nautical miles and saving ships approximately three weeks of sailing time (Table 1). In addition to shortening sailing times between the oceans, the Suez and Panama Canals improved safety in shipping; the passages around South America and South Africa are dangerous shipping routes that have wrecked many vessels, especially prior to the twentieth century.

The elimination of natural land barriers between the oceans as a result of the construction of these canals has led to the dispersal of marine organisms that otherwise would not have been able to colonize the other side of the canal. In addition to the natural migration of organisms

Canal	Opening	Length (km)	Features	Alien Species Movements/ Migrations	Number of Ships in Transit	Cargo in Transit (mt)
Panama	1914	81.5	locks, marine–freshwater–marine	moderate	13,154	191,301,069
Suez	1869	161.0	no locks, marine–saline–marine	extensive	15,667	457,965,000

SOURCES: Panama Canal www.pancanal.com/eng/maritime/reports/table01.pdf, and Suez Canal Institute of Shipping Economics and Logistics (ISL) Bremen, ISL Shipping Statistics and Market Review (SSMR), Volume 48 (2004). (Modified after Minchin, et al., 2006.)

through the canal, ship traffic also aids this biological dispersal, as organisms may cling to the hull of vessels or move in the ballast water of ships. These two invasion vectors are always relevant in shipping, but when ships move in canals, the organisms spend remarkably little time in transit, thus increasing their survival probability. This is especially true for ballast water transports. The longer organisms are kept inside ballast tanks, the fewer the individuals that will survive. Because a canal passage usually takes no longer than a day, organism survival through canals between the two oceans is much higher than when ships travel around the southern tip of the two continents.

INCREASE OF GLOBAL SEA TRADE

In addition to their role as corridors for ship-mediated alien species invasions, canals, in combination with globalization, facilitate aquatic invasions globally by increasing the overall volume of ship-borne trade and also by changing the routing patterns of vessels. World seaborne trade increased in 2004 to 6.76 billion

FIGURE 1 Construction works in the Suez Canal. (Photograph courtesy of the author.)

metric tons, mostly driven by the Asian, American, and European economies, and this resulted in a substantial traffic increase through the Suez and Panama Canals. As shipping demand reaches capacity, the Suez Canal Authority has been expanding the channel to accommodate larger vessels, and the Panama Canal Authority has started to construct another channel with time-efficient water-recycling locks to enable canal passage by larger ships. The canal enlargement projects will likely increase their role as invasion corridors. The construction of new docks and berthing facilities and the modernization of ports may further enhance the spread of biota (Fig. 1).

CANALS AS INVASION CORRIDORS

The impact of the two principal maritime canals as invasion corridors for aquatic species differs. More than 50 possible species introductions through the Panama Canal have occurred from the Atlantic to the Pacific Ocean, while only about 25 have occurred in the reverse direction. The Suez Canal has facilitated a massive spread of species. More than 300 species occurring in the Red Sea have now been found in the Mediterranean Sea. This species movement is termed "Lessepsian migration," after Ferdinand de Lesseps, who built the canal. The differing numbers of alien biota believed to have migrated through the canals may be related to the absence of locks in the Suez Canal, which results in an open gateway for organisms to migrate. However, the Suez Canal had, for the first half of its existence, a hypersaline species migration barrier in the form of the Great Bitter Lakes. The number of Erythrean invaders migrating through the canal is directly related to the decline in salinity in the Bitter Lakes and to the rise of salinity in the northern half of the canal since the Nile was dammed.

FIGURE 2 The tight Miraflores Locks in the Panama Canal. (Photograph courtesy of the author.)

By contrast, the Panama Canal includes a locked freshwater corridor (Fig. 2) between the two oceans it connects. Thus, much of the length of the Panama Canal consists of Gatun Lake, which provides a freshwater barrier between the Caribbean Sea and the Pacific Ocean (Fig. 3). This freshwater barrier accounts for the relatively small numbers of alien species that have migrated through the canal and established populations at either end.

SPECIES MIGRATION BARRIERS

Migration barriers, including multiple air-bubble curtains or electric barriers in freshwater, may reduce the movement of biota through canals. Air-bubble curtains are in use, with or without sound emissions, to reduce the unwanted entrainment of fish at urban or industrial water intake

FIGURE 3 Vessels anchoring in Gatun Lake. (Photograph courtesy of Mariusz Slotwinski.)

points. Larger-scale air-bubble systems may prove suitable for canals. One example of an electric barrier was established in the Chicago Sanitary and Ship Canal to prevent the Asian silver carp and bighead carp from reaching the Chicago River and thus the Great Lakes. This barrier has been in operation since 2002. These migration barriers, however, also affect the movement of native species, and this may result in impacts on fisheries and other recreation areas. Furthermore, these barriers are not effective against species in ballast water tanks or fouling ship hulls.

The most efficient method, and at the same time the most technologically challenging option, would be the inclusion of an environmental barrier in the canal. As studies of alien biota have shown, species migration through the freshwater Gatun Lake in the Panama Canal is relatively rare. This may indicate that a salinity barrier would reduce species movements between oceans.

SEE ALSO THE FOLLOWING ARTICLES

Ballast / Invertebrates, Marine / Mediterranean Sea: Invasions / Ponto-Caspian: Invasions / Rivers / Seas and Oceans

FURTHER READING

Bij de Vaate, A., K. Jazdzewski, H. A. M. Ketelaars, S. Gollasch, and G. Van der Velde. 2002. Geographical patterns in range extension of Ponto-Caspian macroinvertebrate species in Europe. *Canadian Journal of Fisheries and Aquatic Sciences* 59: 1159–1174.

Cohen, N. C. 2006. Species introductions and the Panama Canal (127–206). In S. Gollasch, B. S. Galil, and A. N. Cohen, eds. *Bridging Divides—Maritime Canals as Invasion Corridors.* Dordrecht, The Netherlands: Springer.

David, M., S. Gollasch, M. Cabrini, M. Perković, D. Bošnjak, and D. Virgilio. 2007. Results from the first ballast water sampling study in the Mediterranean Sea: The Port of Koper study. *Marine Pollution Bulletin* 54(1): 53–65.

Galil, B. S. 2006. The marine caravan: The Suez Canal and the Erythrean invasion (207–301). In S. Gollasch, B. S. Galil, and A. N. Cohen, eds. *Bridging Divides: Maritime Canals as Invasion Corridors.* Dordrecht, The Netherlands: Springer.

Gollasch, S., J. Lenz, M. Dammer, and H. G. Andres. 2000. Survival of tropical ballast water organisms during a cruise from the Indian Ocean to the North Sea. *Journal of Plankton Research* 22(5): 923–937.

Hewitt, C., D. Minchin, S. Olenin, and S. Gollascg. 2006. Epilogue: Canals, invasion corridors and introductions (302–306). In S. Gollasch, B. S. Galil, and A. N. Cohen, eds. *Bridging Divides: Maritime Canals as Invasion Corridors.* Dordrecht, The Netherlands: Springer.

Minchin, D., B. S. Galil, M. David, S. Gollasch, and S. Olenin. 2006. Overall introduction (1–4). In S. Gollasch, B. S. Galil, and A. N. Cohen, eds. *Bridging Divides: Maritime Canals as Invasion Corridors.* Dordrecht, The Netherlands: Springer.

Panov, V. E., B. Alexandrov, K. Arbaciauskas, R. Binimelis, G. H. Copp, M. Grabowski, F. Lucy, R. Leuven, S. Nehring, M. Paunovic, V. Semenchenko, and M. O. Son. 2009. Assessing the risks of aquatic species invasions via European inland waterways: From concepts to environmental indicators. *Integrated Environmental Assessment and Management* 5(1): 110–126.

Ruiz, G. M., J. Lorda, A. Arnwine, and K. Lion. 2006. Shipping patterns associated with the Panama Canal: Effects on biotic exchange?

(113–126). In S. Gollasch, B. S. Galil, and A. N. Cohen, eds. *Bridging Divides: Maritime Canals as Invasion Corridors.* Dordrecht, The Netherlands: Springer.

Slynko, Y., L. Korneva, I. Rivier, V. Papchenkov, G. Scherbina, M. Orlova, and T. Therriault. 2002. The Caspian–Volga–Baltic invasion corridor (399–411). In E. Leppäkoski, S. Gollasch, and S. Olenin, eds. *Invasive Aquatic Species of Europe: Distribution, Impact and Management.* Dordrecht, The Netherlands: Kluwer Academic Publishers.

CARNIVORES

ARIJANA BARUN AND DANIEL SIMBERLOFF

University of Tennessee, Knoxville

The Carnivora are a diverse order of placental mammals: almost all Carnivora are primarily meat eaters, though some species (such as the small Indian mongoose and the brown-nosed coati) are omnivorous. Many predatory species other than mammals are colloquially termed "carnivores," but in this entry the word refers to a member of the Carnivora. Carnivores range in size from the least weasel through the southern elephant seal and include dogs, bears, raccoons, weasels, mongooses, hyenas, and cats, among many other creatures. Many global declines and extinctions can be wholly or partially attributed to introduced carnivores. Carnivores have been most often deliberately introduced to prey on pest animals, but many have also been either escapees or intentional releases from fur farms. Predation by introduced carnivores is a major current threat to several species, but carnivores have other impacts as well: they affect human health and economies, and they hybridize with native species. Long-term carnivore control is required to prevent the declines and possible extinctions of some endemic species. Successful eradication campaigns are increasingly being undertaken, although these have largely been restricted to islands to date.

GLOBAL PATTERNS

The earliest introduced carnivore was probably the dog, brought to the Americas by Paleoindians and to Australia by Aboriginal explorers as early as 3000–5000 BC. Most carnivore introductions were for the fur industry and occurred between 1850 and the early twentieth century, while accidental introductions of cats and dogs peaked during World War II as a result of military activities.

Carnivores were introduced for many different reasons. Some escaped from captivity, such as from fur farms, but many were deliberately released for economic gain, recreational hunting, or biological control of introduced pests such as rats and rabbits. During the early stages of colonization of many parts of the world, many domestic animals, including cats and dogs among the carnivores, turned feral after arriving with humans. Cats were often brought on ships as companions or for rodent control, and many were introduced unintentionally during stopovers. Arctic and red foxes, sable, and American mink were introduced to Europe, Asia, and many islands in the Pacific by the fur industry. They were kept either in enclosures or cages, but free-living populations soon arose. Some introductions in mainland Europe and Great Britain resulted from "animal liberation" activities. Hunters and trappers introduced large numbers of carnivores, such as the 19,000 mink, 10,000 raccoon-dogs, and 1,200 raccoons that were released on hunting grounds throughout the former USSR. Most species of *Mustela* and the Viverridae (mongoose and civet) family were introduced as biological control agents in attempts to reduce rabbit or rat populations, but in many cases the introduced carnivore became a more consequential pest itself.

Details on the majority of individual introductions are lacking, but because almost all introduced carnivore species are conspicuous, we have relatively good accounts of their presence. Globally, a minimum of 29 carnivore species have been introduced. Some populations have dwindled and disappeared without apparent reason, but many species have become serious threats. Four carnivores are listed among the IUCN's list of 100 of the world's worst invasive alien species (see Appendix): the feral cat (*Felis catus*), the small Indian mongoose (*Herpestes auropunctatus*) (q.v.), the stoat (*Mustela erminea*), and the red fox (*Vulpes vulpes*). Many others vie for positions on this list, including the raccoon (*Procyon lotor*), the raccoon-dog (*Nyctereutes procyonoides*), the feral dog (*Canis familiaris*), and the brown-nosed coati (*Nasua nasua*). Several mustelids introduced as a result of fur farms are also notorious: these include the weasel (*Mustela nivalis*), the ferret or polecat (*Mustela putorius*), and the American mink (*Neovison vison*).

The rapid expansion of some native species beyond their usual ranges is sometimes viewed as an invasion. A good example is that of the coyote (*Canis latrans*), which until 1900 was present only west of the Mississippi River in the United States and west of Ontario's Lake Nipigon in Canada. Coyote populations have expanded eastward, helped by the disappearance of wolves and by habitat modification. In Europe, a similar expansion of the golden jackal (*Canis aureus*) is occurring into the Balkans.

NOTABLE EXAMPLES

Cats

Cats (*Felis catus*) were domesticated from the Eurasian wildcat (*Felis silvestris*) in the eastern Mediterranean around 3,000 years ago. Because cats were good at controlling rats, they traveled around the world on ships. During stopovers, some escaped, but many were also intentionally introduced to control rodents near newly established colonies. Domestic cats are very adaptable and have survived in inhospitable conditions on many remote oceanic islands. Wherever cats are present, they have immense impacts on wildlife, preying on small mammals, birds, reptiles, and amphibians. Cats are touted as being by far the most dangerous introduced carnivore for native prey, because they have been introduced to many islands worldwide. They are responsible for 26 percent of all predator-related island bird extinctions; possibly the most famous example was that of the Stephen Island wren (*Xenicus lyalli*), the only flightless songbird in the world, whose extinction was caused by one lighthouse keeper's cat in 1894. In subsequent years, cats caused 12 more extirpations of native birds from this island. Stomach contents of a single feral cat caught in New Zealand included at least 34 native skinks (*Leiolopisma* spp.). Unlike some predators, a cat's desire to hunt is not suppressed by adequate supplemental food. Even when fed regularly by people, a cat's motivation to hunt remains strong, and so it continues hunting. In addition, hybridization and disease transmission between domestic cats and wildcats is by far the greatest threat to the existence of wildcat subspecies all over their range of distribution. Feral cats act as reservoirs in the transmission of many diseases, creating a health hazard affecting both wildlife and human populations. In the United States in 2000, 249 of the 509 cases of rabies detected in domestic animals were found in cats.

Stoats

The stoat, or ermine or short-tailed weasel (*Mustela erminea*), is native almost everywhere throughout the northern temperate, subarctic, and Arctic regions of Europe, Asia, and North America. The stoat is an intelligent, versatile predator specializing in small mammals and birds. It is fearless in attacking animals larger than itself and is adapted to surviving periodic shortages by storage of surplus kills. Stoats have been introduced for small mammal control to several Scandinavian islands, mainland Shetland Island, and the north of Scotland. In an unsuccessful attempt to control introduced rabbit populations, hundreds of stoats were introduced to New Zealand in the 1880s despite objections by ornithologists (Fig. 1A). The success of stoats in New Zealand is likely at least partly related to their capacity to survive in any habitat, from sea level to elevations well above the tree line. In New Zealand they are responsible for significant damage to populations of native species, including two threatened endemic birds, the yellowhead (*Mohoua ochrocephala*) and takahe (*Porphyrio hochstetteri*), which still exist on the New Zealand main islands, but only in protected areas where stoats are controlled or from which stoats have been eradicated (Fig. 1B). Two other native bird species, the kakapo (*Strigops habroptila*) and saddleback (*Philesturnus carunculatus*), are found only on offshore islands as a consequence of predation by the stoat and several other introduced predators. The stoat and the ship rat (*Rattus rattus*) contributed to the extinction of at least five endemic bird subspecies. Although stoat populations in New Zealand have declined from a peak in the 1940s, stoats are still abundant on the two main islands

FIGURE 1 (A) Stoat (*Mustela erminea*, also known as short-tailed weasel), Kanuti National Wildlife Refuge, Alaska. (Photograph by Steve Hillebrand, courtesy of U.S. Fish and Wildlife Service.) (B) Takahe (*Porphyrio hochstetteri*) killed by a stoat while on the nest. (Photograph courtesy of Department of Conservation, New Zealand.)

and several of the nearer small fringing islands, which they reached by swimming.

Red Foxes

The red fox (*Vulpes vulpes*; Fig. 2) is native to Europe, Asia, North Africa, and boreal regions of North America. It has been introduced to Australia and many regions of North America (multiple times to many eastern U.S. states, the lowlands of California, and the Aleutian Islands of Alaska). It is now the most widely distributed carnivore in the world, mostly because it can colonize very rapidly when prey are abundant. The rate of spread in Australia was 160 km/year and can be closely linked with the spread of the introduced rabbit. Foxes have often been imported by hunt clubs (Alaska) and, even more frequently, have escaped from fur farms (California, Canada). From 1650 to 1750 European foxes were introduced many times to eastern states and have possibly hybridized with local populations. Red foxes negatively affect many native species. The spread of the fox in Western Australia appears to coincide with the disappearance or population decline of several small and medium-sized rodent and marsupial species, but their true impact is masked by agricultural development and other introduced species (cats, dogs, sheep, and cattle). The Aleutian Canada goose (*Branta canadensis leucopareia*) and other ground-nesting birds have been severely reduced in numbers as a result of red fox translocations. For example, on Shaiak Island, two red foxes devastated a colony of 156,000 nesting seabirds when all eggs and nestlings were killed and cached all over the island. The red fox is an important wildlife vector of rabies in Europe, the United States, and Canada. Millions of dollars are spent each year on bounties to reduce numbers and to vaccinate foxes. On the other hand, introduced sterilized red foxes were used successfully as biological control agents to eliminate introduced Arctic foxes (*Alopex lagopus*) from two Arctic islands.

Raccoons

The raccoon (*Procyon lotor*; Fig. 3) is native to North America, but as a result of fur farm industries it was introduced to islands off Alaska, Canada, and the continental United States. In the mid-twentieth century raccoons were deliberately introduced by hunters or fur industries to France, Germany, the Netherlands, and Russia. In Japan, up to 1,500 raccoons were imported as pets each year after the success of the 1960s anime series *Rascal the Raccoon*. They are now widely distributed across the European mainland, the Caucasus region, and Japan. For many years it was believed that an indigenous species of raccoon inhabited the Bahamas, but recent morphological and genetic analyses show that Bahamas raccoons are recent descendants of raccoons from North America. Owing to their adaptability and increased habitat availability, raccoons have extended their native range from deciduous and mixed forests to mountainous areas, coastal marshes, and even urban areas, where some homeowners consider them pests. They are one of the major wildlife vectors of rabies in the United States, and restocking of raccoon populations by hunting clubs in the 1970s led to the spread of rabies from the southeastern states to the mid-Atlantic states. Raccoons plague game management by preying on waterfowl, quail, and many other ground-nesting birds. On the Queen Charlotte Islands and other islands off the coast of British

FIGURE 2 Red fox (*Vulpes vulpes*), Cape Newenham State Game Refuge, Alaska. (Photograph by Lisa Haggblom, courtesy of U.S. Fish and Wildlife Service.)

FIGURE 3 Raccoon (*Procyon lotor*), Lower Klamath National Wildlife Refuge, California. (Photograph by Dave Menke, courtesy of U.S. Fish and Wildlife Service.)

Columbia, introduced raccoons are responsible for the destruction of 95 percent of seabird colonies. The raccoon is the most economically important furbearer in the United States. Over 5 million raccoons were harvested per year in the early 1980s in the United States alone.

NEGATIVE IMPACTS

Impacts of introduced carnivores have ranged from almost none to major economic, health, ecological, and cultural losses.

Human and veterinary health problems caused by wild carnivore populations have been a major concern for public health departments and international organizations. Several introduced carnivores are important reservoirs of rabies, such as the small Indian mongoose in the West Indies and feral dogs and cats in many parts of the world. Salmonella may be transmitted from dogs to humans via flies feeding on feces. Dogs, through their urine, have been implicated in spreading leptospirosis to people. Feral cats serve as a reservoir for many wildlife and human diseases, including toxoplasmosis, mumps, cat scratch fever, leptospirosis, distemper, histoplasmosis, plague, rabies, ringworm, salmonellosis, tularemia, and many endo- and ectoparasites.

Many economic costs are generated by introduced carnivores, particularly feral dogs and cats. The direct costs of managing populations of introduced carnivores to acceptable levels can be huge. Millions of dollars in the United States were paid out in bounties in the last 30 years to reduce red fox populations, but with little success. Many other indirect costs accrue over time. For example, the small Indian mongoose will kill every chicken in a coop in broad daylight, so small-scale chicken farming is completely absent in areas where the mongoose is present, unless the chickens stay constantly in well-built enclosures.

The ecological impacts of introduced carnivores are varied, including those derived from their roles as predator, as competitor of biologically similar species, and as potential hybridizer with native congeners. The best-known impact of introduced carnivores is predation of native animals. The population-level impact of this predation can be none, coexistence in an equilibrium, or extinction of the prey species or population. Empirical evidence of the first two impacts is scant—not only because a stable relationship between an introduced predator and native prey is probably uncommon, but also because of the difficulty in demonstrating prey regulation. One notable example might be the reported inability

of the small Indian mongoose to reduce populations of introduced rat species on some islands where it was introduced. There are many examples of major declines, local extirpations, and island extinctions of native prey owing to the introduction of carnivores. Many introduced carnivores have become notorious solely because of this impact. For instance, the small Indian mongoose has been responsible for many extinctions, extirpations, population reductions, and range restrictions of birds, amphibians, and reptiles on islands. It is not uncommon for many species to exist on mongoose-free islands but to be absent or to exist only in low numbers on nearby islands where the mongoose is present. The extinction of the Stephen Island wren by a housecat, mentioned previously, is another example. Introduced American mink are implicated in the decline of many seabirds and inland waterfowl in Great Britain, as well as of the water vole (*Arvicola terrestris*).

Competition with native species occurs when individuals of native species suffer reduced abundance, fecundity, survivorship, or growth as a result of resource exploitation or interference with introduced species. On the Kerguelen Islands, where cats are present, there are not enough petrels for the native skuas to eat to reproduce, and the skua population has plummeted. The presence of an introduced congener might prevent the establishment of a subsequently introduced species. For example, introductions of Arctic foxes (*Alopex lagopus*) to islands where red foxes were absent were successful, but where red foxes occurred, the Arctic species disappeared.

Hybridization involving introgression of introduced species with natives is an even subtler impact, because it leads gradually to the loss of genetic integrity of native species and to their extinction as separate species. If interbreeding has occurred for a long time, there may be no reliable methods for phenotypic or genetic comparison, and so the precise history and impact of this process will not be able to be described. This is the case with dingoes and wild domestic dogs in Australia, and feral cats and wildcats in Scotland. Hybridization with dogs has also led to the introduction of dog genes into gray wolves (*Canis lupus*) and the endangered Ethiopian wolf (*Canis simensis*). There may even be impacts when interspecific matings do not lead to genetic introgression. For example, the larger American mink males mate with European mink (*Mustela lutreola*) females, which then do not permit other males to approach them. The embryos resorb, and the female leaves no offspring for that year, while the American mink females reproduce. This removal of females from the breeding population must exacerbate the imperilment of the European mink.

In addition to direct effects on prey populations, introduced carnivores can generate a trophic cascade strong enough to alter the abundance and composition of entire plant communities. The introduction of arctic foxes to the Aleutian archipelago induced strong shifts in plant productivity and community structure. Foxes reduced nutrient transport from ocean to land by preying on seabirds, affecting soil fertility, and transforming grasslands to dwarf shrub– or forb-dominated ecosystems.

In some locations, many different species of carnivores have been introduced, and they may interact with one another and with other species (e.g., rats) to modify food web structure, making it difficult to characterize the impact of a single introduced predator on native species. For example, the Hawaiian Islands have no native mammals, but several introduced carnivores (cats, dogs, mongooses) have devastated populations of native birds. Rats can also prey on some of the same species that introduced carnivores consume. In addition, introduced prey species (rats, mice, and rabbits) are probably supplementing the diet so that predators can increase their numbers and maintain pressure on even low numbers of native prey, eventually leading to extirpations of native fauna. The Macquarie Island parakeet (*Cyanoramphus novaezelandiae erythrotis*) was unaffected by cat predation until rabbits were introduced. Rabbits provided the cats with a food supply year-round, allowing cat numbers to multiply, which drove the parakeet to extinction. Often interactions between introduced species have a synergistic effect on local species. For example, the construction of a tourist hotel on Caicos Island led within three years to the near extirpation of the 5,500 endemic West Indian rock iguanas (*Cyclura carinata*) that were hunted by introduced cats and dogs. Most likely, the cats preyed on the young and the dogs on larger adults of iguanas, but it is difficult to disentangle the different effects of many carnivore species combinations.

Many introduced carnivores have more severe impacts on prey than native predators do, because in communities where predators and prey have coexisted for long periods, prey species have evolved behaviors and morphologies that reduce the chance of encounters with predators or that increase the likelihood of escape once predators are detected. In contrast, naive prey in communities with novel introduced carnivores lack those avoidance behaviors. For example, Australia never had placental carnivores until they were introduced by humans, and these new predators (cats and red foxes) have different hunting and tracking tactics than do native predators. The best-known impact of alien carnivores, elimination of native birds and other vertebrates on oceanic islands, occurs mostly because of native avifaunal and herpetofaunal naïveté.

MANAGEMENT AND ERADICATION

To alleviate problems caused by established introduced carnivores, we can exclude, control, or eradicate them. Exclusion is done in a localized area from which the target species is being removed—but outside the exclusion area, the invader probably thrives. In New Zealand several predator-proof fences have successfully excluded many introduced carnivores (cats, stoats, ferrets) and other introduced species. Once introduced predators have been removed, it is possible to restore areas to nearly its condition prior to human habitation of New Zealand. In Australia, fewer kangaroos and emus are found on the northwestern side of the dingo-proof fence, where dingoes are present, suggesting that the dingoes' presence depresses their populations. However, fencing has had a limited effect, so other forms of control (trapping, poisoning) are necessary.

Control usually means reducing the size of the pest population to acceptable levels. Because control is not complete removal of the invasive species, a constant or repeated effort is needed to keep the population at the desired level. The ultimate goal of many efforts to control introduced carnivores is eradication, but this is in many cases an impossible task, so the control must be done constantly, or only during periods when the native species are at most risk. The small Indian mongoose is trapped on beaches on several islands in the West Indies during the peak of sea turtle reproduction. Such control temporarily reduces predation pressure on young turtles until they move to the sea. The drawback is that this procedure must be repeated every year. Many such control efforts are undertaken for other species of introduced carnivores in Hawaii, New Zealand, Australia, and many other islands.

Unlike control, eradication should have to be performed only once. Eradication is the complete removal of all individuals of the target. This is difficult to achieve because it is usually very challenging to remove the last individual of a population, and eradication, even where technically feasible, is often limited by prohibitively high costs. Nevertheless, introduced carnivores have been eradicated from many islands, some of which are quite large. For example, the Arctic fox was eradicated from Attu Island, Alaska (905.8 km²), cats from Marion Island, South Africa (190 km²), and the red fox from Dolphin Island, Australia (32.8 km²). Overall, at least 75 feral cat, 42 fox, 5 feral dog, 35 mustelid, and 4 raccoon populations have been eradicated from islands worldwide. Three main eradication techniques are chemical (poisoning), physical (fencing, shooting, and trapping), and biological (introduction of a competitor or pathogen, or immuno-contraception). The most difficult part of any method is removing individuals when low densities are reached, because even a single pregnant female can initiate a population resurgence. For example, the attempt to eradicate the small Indian mongoose from Amami-Oshima, Japan, has been unsuccessful particularly because of the difficulties of removing the mongoose at low densities.

The upshot of carnivore eradications has often been an improvement in the status of the species under threat. But it is not always enough simply to eradicate the top predator. Eradication of cats from Little Barrier Island, off the coast of New Zealand's North Island, led to a decrease in breeding success of a resident seabird, Cook's petrel. The reason for this decline was an explosion in numbers of rats, which prey on the seabirds. Rat eradication was followed by a rise in petrel productivity. In addition, recolonization by local native species is not always possible following removal of an introduced carnivore, both because some extirpated species were endemic to islands and lack neighboring populations that can act as recolonization sources and because introduced species may have irreversibly damaged the environment.

Long-term carnivore control will be required to stop the declines and possible extinctions of some endemic species. Widespread control of carnivores (such as immuno-contraception) is needed to aid eradication over large areas. There are also ethical considerations; biological control (particularly of cats) may prove unacceptable to the general public, so extensive public outreach campaigns must be conducted prior to control efforts.

SEE ALSO THE FOLLOWING ARTICLES

Eradication / Hybridization and Introgression / New Zealand: Invasions / Predators / Rats / Small Indian Mongoose

FURTHER READING

Courchamp, F., J.-L. Chapuis, and M. Pascal. 2003. Mammal invaders on islands: Impacts, control and control impact. *Biological Review* 78: 347–383.

Gittleman, J.L., S.M. Funk, D.W. MacDonald, and R.K. Wayne. 2001. *Carnivore Conservation*. Cambridge: Cambridge University Press.

Johnson, C.N., J.L. Isaac, and D.O. Fisher. 2007. Rarity of a top predator triggers continent-wide collapse of mammal prey: Dingoes and marsupials in Australia. *Proceedings of the Royal Society B* 274: 341–346.

King, C. 1984. *Immigrant Killers*. Auckland: Oxford University Press.

Kinnear, J.E., N.R. Summer, and M.L. Onus. 2002. The red fox in Australia: An exotic predator turned biocontrol agent. *Biological Conservation* 108: 335–359.

Letnic, M., F. Koch, C. Gordon, M.S. Crowther, and C.R. Dickman. 2009. Keystone effects of an alien top-predator stems extinctions of native mammals. *Proceedings of the Royal Society B* 276: 3249–3256.

Lever, C. 1985. *Naturalized Mammals of the World*. London: Longman.

Long, J.L. 2003. *Introduced Mammals of the World: Their History, Distribution and Influence*. Collingwood: CSIRO.

CARP, COMMON

PETER W. SORENSEN AND PRZEMYSLAW BAJER

University of Minnesota, St. Paul

The common carp is one of the world's most widely introduced and invasive species of fish. This large minnow presently dominates the fish biomass of many shallow lakes, rivers, and wetlands in North America, Australia, and New Zealand, where it is largely responsible for poor water quality and a massive loss of waterfowl habitat. Ironically, it is also highly valued for sport and food in Europe and China, where it is not invasive. Invasiveness appears to be caused by location-specific interactions of abiotic (environmental instability) and biotic factors (life history). Efforts to control the common carp have been ongoing for a century, focus on poisons and barriers, and have had only moderate success. New research and technologies offer promise.

TAXONOMY AND BIOLOGY OF THE COMMON CARP

The common carp (*Cyprinus carpio*) is one of the largest of the minnows (family Cyprinidae). It can live to be over 50 years old and can reach a length of over 70 cm and a weight in excess of 10 kg. Most believe that the common carp (hereafter "carp") originated in large rivers and estuaries of the Ponto-Caspian region of Europe (the Danube and Volga Rivers and the Caspian, Aral, and Black Seas). Indeed, a gold–bronze-colored, fully

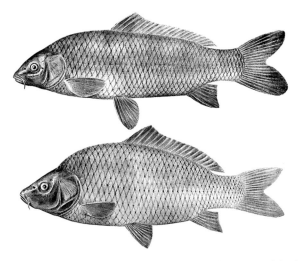

FIGURE 1 A wild common carp (top) and its domesticated feral form (bottom) from the Danube delta. (From Antipa, G. 1909. *Fauna Ichtiologică Ramâniei.* Bucharest: Carol Gobl.)

scaled, torpedo-shaped "wild" fish is still found in this region, albeit in low numbers. Many other forms of the carp are found in other locations, including a "German" or "European" type that appears to reflect the effects of domestication and that has a deeper body and a distinctive hump behind its head (Fig. 1). Varieties of this form are invasive across North America and Australia. Other domesticated forms include the scaleless "leather" carp and the "mirror" carp, which has only a few large scales. An ornamental form, the koi carp, originated in Japan and is characterized by brilliant red, white, black, and orange pigmentation; it is invasive in New Zealand. The carp is closely related to both the goldfish (*Carassius auratus*) and Crucian carp (*C. carassius*) but not to the "Asian carps" (*Hypophthalmichthys* sp.), which also can be invasive.

The common carp is renowned for its omnivorous benthic diet, fast growth, large size, high fecundity, and physiological resiliency, all of which appear to contribute to (but not fully explain) its invasiveness. This species has a well-developed sense of taste mediated by two sets of facial barbels and an internal palatine food-sorting mechanism that allows it to locate and extract food from deep in the sediments in a way that few fish can, but that also embroils the water with expelled mud. Its dietary habits are remarkably omnivorous, and it will consume all nature of plant and animal material, which it digests without the aid of a true stomach. So great is its initial growth rate (~14 cm/yr) that within months it faces little predation pressure and can live long. While the carp grows fastest in warm temperatures (25–30 °C),

it can tolerate severe winters and can survive under the ice, where it tends to aggregate for unknown reasons and consequently can be targeted for removal. The carp is remarkably tolerant of low oxygen, high turbidity, and modest salinities—such as are found in the Ponto-Caspian. It also has an excellent sense of hearing (mediated by a large swim bladder) and sense of smell that serve it well in turbid waters. Finally, various experiments and the experiences of fishermen suggest that the carp is quite intelligent. In sum, the common carp is a long-lived generalist with remarkable physiological abilities.

LIFE HISTORY

The carp has a poorly understood but fascinating life history that appears to contribute to its invasiveness. Carp mature at a few years of age, and females can carry up to 3 million eggs. Once mature, adults of both sexes migrate far into shallow wetlands to spawn each spring. Migrations can range from 100 m to 300 km, depending on local conditions. Spawning is mediated by an elaborate and well-understood multicomponent sex pheromone system that triggers final maturation and then permits males to locate females. Males follow and nudge and push females until they release their eggs, which are immediately fertilized and stick to submerged vegetation—a spectacular and well-described event (Fig. 2). Carp eggs are tiny (~1 mm) but hatch and develop within a week. It appears that this strategy evolved to swamp predators and, in North America, regions with shallow wetlands that suffer seasonal hypoxia and lack predators are particularly prone to carp invasions, especially if also connected to deeper waters that can function as refuges for adults (Fig. 3). Carp populations can reach extreme densities in just a few years, although this phenomenon has not been well studied.

FIGURE 2 Common carp spawning in a shallow lake in Minnesota. (Photograph courtesy of David Florenzano.)

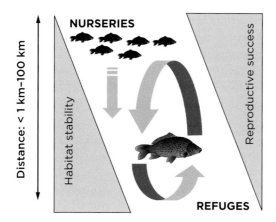

FIGURE 3 Schematic figure of the life history of common carp, showing how carp alternatively exploit shallow, unstable basins for spawning and nurseries and then use deeper, more stable connected waters as adults.

INVASIVE PATHWAYS AND MECHANISMS

The carp has been repeatedly introduced by humans to all continents except for Antarctica. The Romans were responsible for the initial spread of the carp approximately 2,000 years ago, when they developed simple carp culture in the Danube River and then moved the species across Europe to feed legionnaires and their families. Later, the Catholic Church assumed this role and domesticated it further for use in its monastery ponds. By the late 1800s, the common carp was found across all of Europe but rarely (if ever) described as undesirable or invasive; indeed, carp culture had become a fine art, and the British stocked it as a prized game fish. Naturally, European emigrants at the turn of the last century wanted this species in their new homes. There are many accounts of small-scale private introductions across the globe at this time, but few seemed to succeed, so settlers turned to the government for help.

The presence of the carp in North America is attributed to the United States Fish Commission, which by the 1880s was receiving thousands of letters every year demanding its introduction. Carp were subsequently imported from Germany in 1887 to the Washington, D.C., area, bred, and distributed widely across the country and Canada (at Canada's request) for nearly two decades, after which they were found in most inland waters. Undoubtedly, much of this spread was supplemented by the carp's natural tendency to move through interconnected rivers. In any case, the introduction proved disastrous: feral carp proved to be less flavorful than cultured individuals, North American marine fish were found to be surprisingly good-tasting, and the carp was linked with severe declines in water quality. Carp stocking ended by the turn of the century,

and control efforts commenced. By the 1920s, significant government-led carp removal programs dominated most inland fisheries programs (see below). However, success proved elusive, and by the 1950s most systematic efforts had stopped, although substantial local efforts still continue today. Currently, carp densities across mid-North America appear to reach 400 kg/ha commonly, and carp comprise the majority of fish biomass in many locations.

Carp in Australasia followed a different course than in North America, and this fact brings with it a lesson on the need to regulate industry and pay attention to genetic strains. Although initial efforts to introduce carp across Australia in the 1800s failed, in the 1960s a fish farm in Boolarra (southern Australia) imported a new strain of European carp that thrived. A wary government agency noticed and banned their culture, but not before carp escaped into the Murray-Darling River. Within only three decades, carp had spread throughout this highly interconnected and hypoxia-prone system, reaching a biomass of up to 1,000 kg/ha. Today carp are associated with serious water quality issues and a loss of native fishes. A somewhat different scenario played out in neighboring New Zealand, where initial introductions also failed. In the 1960s large areas of the Waikato River on the North Island were found to be infested with koi carp, apparently reflecting accidental releases of contaminated goldfish shipments. The spread continues, and fines for transporting live koi range up to NZ$100,000.

DAMAGE CAUSED BY CARP

Although the presence of high densities of carp consistently correlates with poor water quality and a loss of native fish and waterfowl, the precise role played by carp is not well understood. Likely carp are both the cause and effect of degraded waters, and their effects are complex. Nevertheless, it is well established that carp uproot massive amounts of rooted vegetation in shallow waters and release vast amounts of sediments, which destroys habitat, suppresses plant recovery, and permits wind mixing (causing further deterioration). In the past century, countless reports describe clear lakes and wetlands in North America with abundant vegetation turning into devegetated and permanently turbid systems after the arrival of carp (Fig. 4). Likely tens to hundreds of thousands of hectares of waterfowl habitat in North America, and probably Australia, have been devastated in this manner. In addition, it is now becoming increasingly apparent that by feeding in the bottom and growing rapidly, carp often serve as significant nutrient "pumps," leading to algal blooms and other problems that are exacerbated

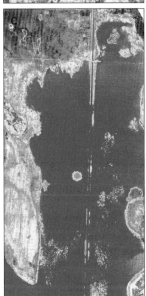

FIGURE 4 Infrared photographs showing the disappearance of aquatic vegetation (shades of red and yellow) versus open water (blue) in Hennepin and Hopper Lakes, Illinois. This set of shallow lakes was invaded by the common carp in 2001 after restoration. Photographs were collected in (A) 2004, (B) 2006, and (C) 2008, when the biomass of carp in the lake was 10, 110, and 250 kg/ha, respectively. (Satellite images courtesy of the Wetlands Initiative, Digital Globe, Inc., and Sidwell Co.)

by the loss of plant cover. Notably, many of these ecological changes likely cause reductions in native fishes and further deterioration in water quality, promoting a self-reinforcing cycle.

CARP CULTURE AND FISHING

The factors that make carp invasive continue to make it a good fish for culture. Currently, the carp is widely cultured in Asia, Eastern Europe and Africa, where it is an important source of animal protein, and it is one of the top cultured fish in the world. Ornamental forms of carp (koi) are also revered in Japan, where these fish can fetch thousands of dollars each. Carp fishing is still valued in the United Kingdom and other areas of Europe. Thoughts of controlling carp in Australia using pathogens and genetic technologies are complicated by concerns for these industries.

CARP CONTROL

Carp control has been ongoing across North America for a century. Initial efforts focused on constructing fish barriers and large-scale removal using seining and, while likely moderately successful, were not documented and were soon abandoned because of lack of obvious success. Efforts by local government agencies, waterfowl protection groups, and watershed districts continue at great expense. Many of these claim modest success, especially with severely impacted waterfowl lakes and wetlands that are often drained in winter or poisoned using rotenone, procedures that kill all fish. Native fish and plants are then often restored at great expense, and electrical barriers typically built to suppress reinfestation. Large-scale winter seining under the ice is also employed by lake groups concerned with water quality and can be effective, as carp form dense winter aggregations. However, positive effects are usually short-lived, as ecosystems are often reinvaded. A success story has recently been reported from Lake Sorrel in Tasmania, where carp have been extirpated using systematic removal by exploiting their tendencies to aggregate using radio-tagged "Judas" fish to find and remove groups. This process took many years and cost millions of dollars.

There has been a recent surge in interest in using integrated pest management strategies to suppress carp. These schemes exploit recruitment (survival of young to adulthood), reproduction, and removal of adults. The recent discovery that carp recruitment is often restricted to shallow nursery habitats in North America offers hope. Pheromones are also being developed as attractants while "carp separation cages" are being deployed in the Murray-Darling Basin to capture carp and leave native fish. "Daughterless" carp, the result of genetic technology that seeks to introduce gene silencers into the population to cause a bias toward the production of males, is being explored in Australia. Statistical management models for carp also now exist. Given our understanding of the common carp and its worldwide importance as a damaging invasive, there is good reason to believe that with the right support, these efforts can succeed.

FURTHER READING

Balon, E. K. 1995. Origin and domestication of the wild carp, *Cyprinus carpio*: From Roman gourmets to the swimming flowers. *Aquaculture* 129: 3–48.

Bajer, P. B., and P. W. Sorensen. 2009. The superabundance of common carp in interconnected lakes in Midwestern North America can be attributed to the propensity of adults to reproduce in outlying habitats that experience winter hypoxia. *Biological Invasions* 12: 1101–1112.

Billard, R., ed. 1995. *Carp Biology and Culture.* Paris: Springer-Verlag.

Buffler, J. R., and T. J. Dickson. 1990. *Fishing for Buffalo.* Minneapolis: Culpepper Press.

Cooper, E. L. 1987. *Carp in North America.* Bethesda, MD: American Fisheries Society.

Haas, K., U. Köhler, S. Diehl, P. Köhler, S. Dietrich, S. Holler, A. Jaensch, M. Niedermaier, and J. Vilsmeier. 2007. Influence of fish on habitat choice of water birds: A whole system experiment. *Ecology* 88: 2915–2925.

Koehn, J. D. 2004. Carp (*Cyprinus carpio*) as a powerful invader in Australian waters. *Freshwater Biology* 49: 882–894.

Koehn, J., A. Brumley, and P. Gehrke. 2000. *Managing the Impacts of Carp.* Canberra, Australia: Bureau of Rural Sciences.

McCrimmon, J. R. 1968. *Carp in Canada.* Bulletin 165. Ottawa: Fisheries Research Board of Canada.

Sorensen, P. W., and N. E. Stacey. 2004. Brief review of fish pheromones and discussion of their possible uses in the control of non-indigenous teleost fishes. *New Zealand Journal of Marine and Freshwater Research* 38: 399–417.

CART AND RELATED METHODS

VOJTĚCH JAROŠÍK
Charles University, Prague, Czech Republic

Classification and regression trees (CART) are a computer-intensive data-mining tool originally designed for analyzing vast databases of often incomplete data, with an aim to find financial frauds, suitable candidates for loans, potential customers, and other uncertain outputs. Searching for potential invasive species and their traits responsible for invasiveness, predicting their potential distributions in regions where they are not native, or identifying factors that distinguish invasible communities from those that resist invasion are similar risk assessments. This is perhaps one reason why CART and related methods are becoming increasingly popular in the field of invasion biology. Identifying homogeneous groups with high or low risk and constructing rules for making predictions about individual cases is, in essence, the same for financial credit scoring as for pest risk assessment. In both cases, one searches for rules that can be used to predict uncertain future events.

PRINCIPLES OF CART

CART models use decision trees to display how data may be classified. Their method is technically known as *binary recursive partitioning*: the data are successively split along coordinate axes of the explanatory variables so that, at any node, the split is selected that maximally distinguishes the response variable in the left and right branches. If the response variable is categorical, the tree is called a *classification tree*; if continuous, it is called a *regression tree*. Explanatory variables can be both categorical and continuous. The process is binary because parent nodes are always split into exactly two child nodes by asking questions that have a "yes" or "no" answer, and it is recursive because the process can be repeated by treating each child node as a parent.

Making a decision when a tree is complete is best achieved by growing the tree until it is impossible to grow it further and then examining smaller trees obtained by gradually decreasing the size of the maximal tree in a process called pruning. A single *optimal tree* is then determined by testing for misclassification error rates of candidate trees. When the data are sufficiently numerous (i.e., greater than 3,000 records), they are divided into a learning (also called training) sample and a test sample, created by a completely independent set of data or a random subset (e.g., 20%) of the input data. To calculate the misclassification error rate, the model is fitted to the learning sample to predict values in the test sample. The error rate is calculated for the largest tree as well as for every smaller tree. When the data are not sufficiently numerous to allow for separate test samples, cross-validation is employed. Cross-validation involves splitting the data into a number (e.g., 10) of smaller samples with similar distributions of the response variable. Trees are then generated, excluding the data from each subsample in turn. For each tree, the error rate is estimated from the subsample excluded in generating it, and the cross-validated error for the overall tree is then calculated. The cross-validated or test sample errors are then plotted against tree sizes (insets in Fig. 1). The optimal tree is the one with the lowest error rate (SE rule 0 in the inset) or that is within one standard error of the minimum (SE rule 1 in the inset). A series of 50 or 100 validations of error rates for both rules is recommended, of which the modal (most likely) optimal tree is chosen for description. This tree is then represented graphically (Fig. 1), with the root

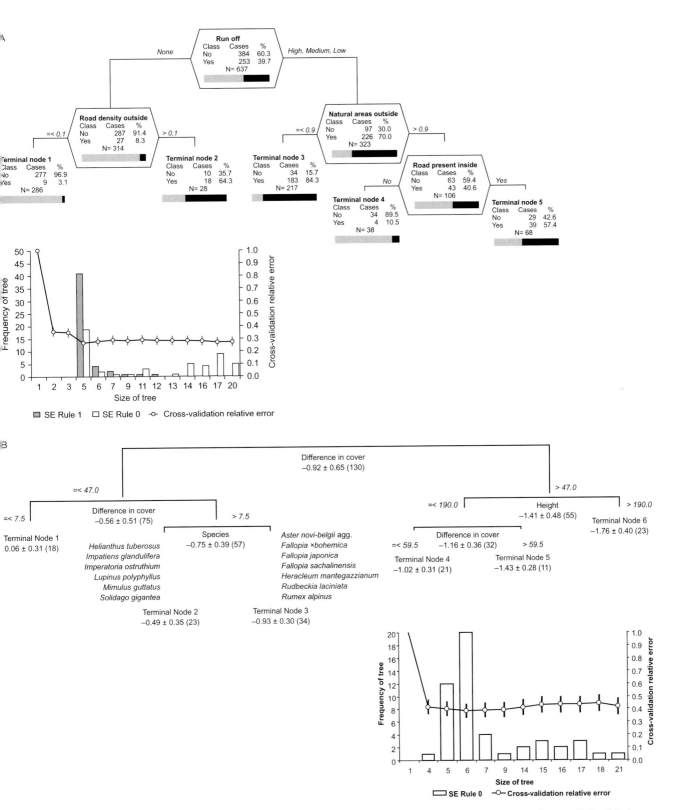

FIGURE 1 (A) Classification tree describing the probability of an alien plant presence (No/Yes) in boundary segments of Kruger National Park (KNP) based on explanatory variables from within the KNP and areas adjacent to the park (after Foxcroft et al., Protected Area Boundaries as a Natural Filter of Plant Invasions from Surrounding Landscapes, under review). (B) Regression tree describing the impact of individual invading plant species on diversity of native plant communities (Shannon's index of diversity *H*') based on absolute and relative population performances of the invaders. (After Hejda et al. 2009. *Journal of Ecology* 97: 393–403.) For both trees, the vertical depth of each node is proportional to its improvement value that corresponds to explained variance at the node. *(Continued on next page)*

standing for undivided data at the top, and the terminal nodes, describing the most homogeneous groups of data, at the bottom of the hierarchy.

The quality of each split can be expressed based on deviance explained by the split and visualized as a vertical depth of the split. The overall quality of the best classification tree (Fig. 1A) can be expressed as its misclassification rate by comparing the misclassification rate of the optimal tree with the misclassification rate of the null model (e.g., 50% null misclassification rate for a presence–absence response variable) and with the misclassification rate for each category of response variable. The overall quality can be also described as specificity and sensitivity of cross-validated or test samples. Specificity is defined as the true positive rate, or the proportion of observations correctly identified (for instance, the ability of the model to predict that a species is not invasive when it is not). Sensitivity is defined as the true negative rate (e.g., the ability to

predict that a species is invasive when it is). The overall quality and appropriateness of the optimal regression tree (Fig. 1B) can be expressed similarly to analogous features of a linear model. We can express explained variance r^2 of the tree, and because we know observed and predicted values for each terminal node, we can also calculate residuals as a difference between observed and expected values and use them as a diagnostic check of the model.

RELATED METHODS AND THEIR COMBINATIONS WITH CART

When explanatory variables have no missing values and either have or can be transformed to an approximately normal distribution, CART models can be replaced by linear or multivariate statistics. Linear models can then substitute for regression trees. If we have a categorical response variable with two classes and at least

FIGURE 1 *(Continued)* Insets: Cross-validation processes for the selection of the optimal trees. The lines show a single representative tenfold cross-validation of the most frequent (modal) optimal tree with standard error (SE) estimate for each tree size. Bar charts are the numbers of optimal trees of each size (frequency of tree) selected from a series of 50 cross-validations based on the minimum-cost tree, which minimizes the cross-validated error (white, SE rule 0), and 50 cross-validations based on SE rule 1 (gray, SE rule 1), which minimizes the cross-validated error within one standard error of the minimum. The most frequent (modal) classification tree, established based on both SE rules, has five terminal nodes (A), and the modal regression tree, established based on SE rule 0, has six terminal nodes (B).

(A) Each node (polygonal table with splitting variable name) and terminal node with node number shows table with columns for class (No/Yes), number of cases and percent of cases for each class, total number of cases (*N*), and graphical representation of the percentage of No (gray) and Yes (black) cases in each class (horizontal bar). Except for the root node standing for undivided data at the top, there are the splitting variable name and split criterion above each node. "Run-off" is the categorical measure of mean annual runoff from the surrounding watershed (million m³), "natural areas outside" refers to the percentage of natural areas within a 5 km radius outside the KNP boundary, "road density outside" refers to the density of major roads within a 10 km radius outside the KNP boundary, and "road present inside" refers to the presence or absence of all roads in the segment inside the KNP. In segments with no river, the probability of being invaded depended on the density of major roads within a 10 km radius outside the KNP boundary. If there was a river, invasion was unlikely only in segments with more than 90 percent of protected natural areas with natural vegetation in the adjacent 5 km radius outside the KNP, and with roads absent within the park. The overall misclassification rate of the model is 13.5 percent, compared to 50 percent for the null model; the sensitivity (true positive rate, defined as proportion of observations correctly identified as suitable) is 0.90; the specificity (true negative rate) is 0.80; the misclassification rate for the presence of an alien species is 0.10; the misclassification for the absence is 0.20. This model is an alternative tree after dropping a primary splitter, the continuous explanatory variable "mean annual runoff," from the optimal tree. The categorical surrogate "run-off" appeared at Node 1 of the optimal tree with an association value of 0.86, and it explained 86.8 percent of the variability of the primary split. The optimal tree had a bit higher misclassification rate (14%), but higher sensitivity (0.92) and specificity (0.81), and a lower misclassification rate for the presence (0.075) and absence (0.19) of alien species, with only three compared to five terminal nodes for the alternative tree. The chosen continuous explanatory variables of the optimal tree, "mean annual runoff" and "road density outside," lacked collinearity and could be used as explanatory variables in a logistic regression.

(B) Each node shows the splitting variable, splitting criteria, mean ± standard deviation of the difference in species diversity between invaded and uninvaded pairs of plots (negative value indicates a decrease due to invasion), and number of plots in brackets. "Difference in cover" is a cover difference between an invading species and the dominant native species in uninvaded plots (in percent), "height" is height of the invading species (in centimeters), and "species" are the scientific names of the invading plants. To reduce the splitting power of the high categorical variable "species" (13 factor levels), the species were adjusted to have no inherent advantage over continuous explanatory variables. The impact was first divided based on the cover difference of approximately 50 percent (≤47%). The group with the small differences in cover exhibited no impact on diversity if the cover of the invading and native dominant species differed by less than or equal to 7.5 percent; for cover differences between 8 and 47 percent, the impacts were species-specific. The group with differences in cover above 47 percent indicated the absolutely highest impact on diversity if the invading species was taller than 190 cm. If the invading species was shorter than 190 cm, then the impact on diversity was further divided based on differences in cover. The tree explains 74 percent of the variance. The alternative linear model explained 76 percent of the variance, but all the variance was included in interactions in a way that rendered the model noninterpretable.

one explanatory variable is continuous, then a suitable method to replace a classification tree is a binary logistic regression; if all explanatory variables are continuous, then a suitable alternative is also multivariate discriminant function analysis. For missing values, CART and a related technique called random forests (RF) can be applied. RF gives, at least for small samples, more robust results than CART and allows for a ranking of explanatory variables, but it does not have the CART virtue of easily followed graphic presentation. The ability to treat missing values, however, makes both RF and CART invaluable tools when one tries to identify and rank traits associated with invasiveness, impact of invasive species, and similar tasks in which we usually deal with very incomplete data.

When classification trees are compared with logistic regressions or discriminant analyses, the optimal tree often performs better on the learning sample, and because of the CART model verifications on validated samples, the tree is usually more accurate on new data. The same is usually true for RF, owing to its self-testing procedure based on an extension of cross-validation. When we compare linear models and regression trees on data determining susceptibility to invasions of different habitats by plant invaders, we also find slightly higher explanatory power for regression trees, and the trees are much easier to interpret than linear models. The reason is that, unlike linear models, which uncover a single dominant structure in the data, CART models are designed to work with data that might have multiple structures. The models can use the same explanatory variable in different parts of the tree, dealing effectively with nonlinear relationships and higher-order interactions. In fact, provided there are enough observations, the more complex the data and the more variables that are available, the better tree models will do compared to alternative methods. With a complex data set, understandable and generally interpretable results often can be found only by constructing trees (Fig. 1B).

However, trees are also excellent for initial data inspection. CART models are often used to select a manageable number of core measures from databases with hundreds of variables. A useful subset of predictors from a large set of variables can then be used in building a formal linear model (Fig. 1A). CART can also suggest further model simplification by converting a continuous variable to a categorical one at cut-points used to generate splits and by lumping together subcategories of categorical variables that are never distinguished during splitting.

IMPORTANT PROPERTIES OF CART

CART models are nonparametric. Consequently, unlike with parametric linear models, nonnormal distribution and collinearity do not prevent reliable parameter estimates. Because the trees are invariant to monotonic transformations of continuous explanatory variables, no transformation is needed prior to analyses. Outliers among the response variables generally do not affect the models because splits usually occur at non-outlier values. However, in some circumstances, transformation of the response variable may be important to alleviate variance heterogeneity.

Surrogates of each split, describing splitting rules that closely mimic the action of the primary split, can be assessed and ranked according to their association values, with the highest possible value 1.0 corresponding to the surrogate producing exactly the same split as the primary split. Surrogates can then be used to replace an expensive primary explanatory variable by a less expensive, although probably less accurate, one, and then to build alternative trees on surrogates (Fig. 1A). Unlike in a linear model, a variable in CART thus can be considered highly important even if it never appears as a primary splitter. Surrogates also serve to treat missing values, because the alternative splits are used to classify a case when its primary splitting variable is missing.

However, as it is easier to be a good splitter on a small number of records (e.g., splitting a node with just two records), to prevent missing explanatory variables from having an advantage as splitters, the power of explanatory variables can be penalized in proportion to the degree to which they are missing. High-level categorical explanatory variables have inherently higher splitting power than continuous explanatory variables and therefore can also be penalized to level the playing field (Fig. 1B). Finally, proportions calculated from larger samples give more precise estimates, and therefore, proportional response variables can be weighted by their sample sizes; a similar approach can be applied to stratified sampling on strata having different sampling intensities. CART models can also accommodate situations in which some misclassifications are more serious than others. For instance, if invasion risks are classified as low, moderate, and high, it would be more costly to classify a high-risk species as low-risk than as moderate-risk. This can be achieved by specifying a differential penalty for misclassifying high, moderate, and low risk.

LIMITATIONS

CART models are good, but not as good to solve completely all problems with data that violate a basic assumption of the independence of errors of observations due to temporal or spatial autocorrelation, or due to a related problem of phylogenetic relatedness. Fortunately, as we do not need formal parametric tests of statistical significance, spatial autocorrelations cannot prevent correct use of CART and related nonparametric methods (e.g., for prediction of species distributions). However, CART cannot distinguish fixed and random effects and thus cannot be used with mixed effect and nested statistical designs. Trees thus do not allow for phylogenetic corrections using mixed-effect general linear models in which taxonomic hierarchy is included as nested random effects. They are also inefficient when used on principal coordinate axes derived from phylogenetic trees to account for relatedness. The reason is that the principal coordinate axes are orthogonal, and trees exhibit their greatest strengths with a highly nonlinear structure and complex interactions. Their usefulness decreases with increasing linearity of the relationships, and consequently, on mutually independent principle coordinates, no trees are built. Because, in invasion biology, we usually need values for each individual species, phylogenetic contrasts derived from related species are also usually not helpful. The only way to include phylogeny in CART models seems to be to use the hierarchical taxonomic affiliations of the individual species. Considering phylogenetic effects is important not only to treat a lack of statistical independence, but also to solve practical implications (e.g., when one predicts whether a species belonging to a particular family, order, or class would be more predisposed to invasion than other species belonging to other taxa at the same hierarchical level).

The tree-growing method is data intensive, requiring many more cases than classical regression. While for multiple regression it is usually recommended to keep the number of explanatory variables six to ten times smaller than the number of observations, for classification trees, at least 200 cases are recommended for binary response variables and about 100 more cases for each additional level of a categorical variable. Efficiency of trees decreases rapidly with decreasing sample size, and for small data sets, no test samples may be available.

SEE ALSO THE FOLLOWING ARTICLES

Invasiveness / Life History Strategies / Remote Sensing / Risk Assessment and Prioritization

FURTHER READING

Bourg, N. A., W. J. McShea, and D. E. Gill. 2005. Putting a CART before the search: Successful habitat prediction for a rare forest herb. *Ecology* 86: 2793–2804.

Breiman, L., J. H. Friedman, R. A. Olshen, and C. G. Stone. 1984. *Classification and Regression Trees*. Pacific Grove: Wadsworth.

Cutler, D. R., T. C. Edwards Jr., K. H. Beard, A. Cutler, K. T. Hess, J. Gibson, and J. J. Lawler. 2007. Random forests for classification in ecology. *Ecology* 88: 2783–2792.

De'ath, G., and E. Fabricius. 2000. Classification and regression trees: A powerful yet simple technique for ecological data analysis. *Ecology* 81: 3178–3192.

Křivánek M., P. Pyšek, and V. Jarošík. 2006. Planting history and propagule pressure as predictors of invasions by woody species in a temperature region. *Conservation Biology* 20: 1487–1498.

Kumar, S., S. A. Spaulding, T. J. Stohlgren, K. A. Hermann, T. S. Schmidt, and L. L. Bahls. 2009. Potential habitat distribution for the freshwater diatom *Didymosphenia geminata* in the continental US. *Frontiers in Ecology* 7: 415–420.

Reichard, S. H., and C. W. Hamilton. 1997. Predicting invasions of woody plants introduced into North America. *Conservation Biology* 11: 193–203.

Rejmánek, M., and D. M. Richardson. 1996. What attributes make some plant species more invasive? *Ecology* 77: 1655–1661.

Steinberg, G., and P. Colla. 1995. CART: *Tree-Structured Non-Parametric Data Analysis*. San Diego, CA: Salford Systems.

Vall-llosera, M., and D. Sol. 2009. A global risk assessment for the success of bird introductions. *Journal of Applied Ecology* 46: 787–795.

Venables, W. N., and B. D. Ripley. 2002. *Modern Applied Statistics with S*, 4th ed. New York: Springer.

CHEATGRASS

RICHARD N. MACK

Washington State University, Pullman

Cheatgrass (*Bromus tectorum*), or downy brome, is a cleistogamous (i.e., almost totally self-pollinating) annual grass that occupies an enormous native range in Eurasia and the northern rim of Africa. In the past 200 years it has been transported worldwide, almost always as an accidental introduction, and it now has a naturalized range that includes North and South America, Australia, and temperate environments in Oceania. It has become a widespread invader in arid North America, especially in the largely treeless region between the Rocky Mountains and the Cascade and

Sierra Nevada ranges (the Intermountain West). In this region, it causes enormous damage as a persistent weed in cereal agriculture and a strong competitor of native steppe species in many other habitats. Its most costly economic and environmental damage is inflicted by providing highly combustible fuel for vast fires that destroy vegetation and thereby trigger massive erosion, soil runoff, and increased sedimentation in the regional watersheds.

NATIVE AND INTRODUCED RANGES

In its native range across Western Europe, especially in countries surrounding the Mediterranean Sea, *Bromus tectorum* (Fig. 1) is at least common, although rarely abundant. The grass's range extends into Eastern Europe, including the Ukraine. In Asia it occurs in a wide swath, from Turkey to Rajasthan in northern India. The grass is a community dominant in the steppe in Turkmenistan and Uzbekistan. Tibet and Xinjiang apparently form its eastern Asiatic range boundary.

Cheatgrass has been accidentally dispersed worldwide and has become naturalized in temperate regions in Australia, New Zealand, North America, and South America. Although it now occurs throughout the United States, including Alaska and Hawaii, it reaches its greatest prominence as a destructive plant invader in the steppe in the Intermountain West of the United States and adjacent western Canada. The grass appears to have increased in prominence in the Great Plains (the midcontinental United States) during the last 50 years, suggesting that its range expansion in North America is

FIGURE 1 *Bromus tectorum* (cheatgrass, downy brome), a native annual grass of Eurasia that has become highly invasive in much of arid North America. (Photograph courtesy of the author.)

not complete. Nevertheless, it is its role as an invader in the Intermountain West that has attracted so much deserved attention.

INTRODUCTION AND SPREAD OF *B. TECTORUM* ALONG THE COASTS OF EASTERN AND WESTERN NORTH AMERICA AND INLAND

Given this grass's common occurrence in its native range in association with agriculture, the modes by which it could have arrived in the United States are readily envisioned: it could have come as a seed contaminant in crop seed (especially the seeds of wheat, oats, and barley) or as a contaminant in any form of dry cargo or ballast. Its long awn (>1 cm) adheres readily to fur or clothing, so it could have arrived even as a hitchhiker on livestock or human immigrants, although the number of individuals arriving in a single event via this mode would seem unlikely to have produced a founder population. At any rate, the first record of *B. tectorum* in the United States is around 1790, in Lancaster, Pennsylvania. Whether any descendants from this early population persisted is uncertain as there are no reliable records of the grass for the next 70 years.

The oldest surviving herbarium specimen was collected in 1859 in West Chester, Pennsylvania, about 80 km from the 1790 collection site in Lancaster. By 1870, *B. tectorum* had been collected repeatedly along the eastern seaboard (e.g., Providence, Boston, Philadelphia, and Long Island). If *B. tectorum* did arrive or persist between about 1790 and 1859 on the U.S. east coast, it apparently remained in small, isolated populations. Local botanists and naturalists in that era collected assiduously, and a new occurrence of a nonnative species in the mid-nineteenth century would have had a much higher chance of early detection than a similar event would have today!

Curiously, *B. tectorum* was not reported in the Midwest until 1886. This occurrence was soon joined, however, by populations that stretched across the United States in the last decade of the nineteenth century. By 1900 cheatgrass had been found in Iowa, Colorado, Nebraska, and Missouri. Early collection records in Colorado and Iowa are coupled with the invaluable observations that the grass was associated with packing straw—the likely mode of transportation.

Cheatgrass was first detected in the Intermountain West in 1889 in the Okanogan Valley, British Columbia. It was found soon thereafter (1893) in Ritzville,

Washington. Clusters of other collection dates also occur in eastern Washington and northeastern Oregon in the years bracketing 1900. These early records are significant for at least two reasons. Although they are contemporaneous with the earliest collection dates in the Great Plains, the single pre-1900 collection record for the grass in the 1500 km between eastern Colorado and eastern Washington is from northern Utah. The geographic isolation of these early collection records, coupled with the prominence today of a unique genotype (Got-4c) among populations in the Intermountain West, strongly suggests that cheatgrass entered the Pacific Northwest from a route other than across the continent (i.e., entry likely also occurred on the west coast, and the grass was transported inland). Cheatgrass also appears to have first entered Canada almost simultaneously in the 1880s on the west coast and in Ontario.

Spread of *B. tectorum* in the West, including the Great Basin, moved swiftly after 1900; pre-adaptation of the grass to the region's arid environments (see below) would have greatly facilitated this invasion. By 1915 cheatgrass was being reported in numerous locales as a serious pest in wheat fields. Extensive plowing and livestock grazing aided the invasion; both activities create soil disturbance that fosters cheatgrass establishment. In their zeal to claim free farmland from the government, many new farmers attempted to plow soils far too saline or alkaline to support crops. But by destroying the thin cryptogamic crust in these sites, experiencing crop failure, and then abandoning the land, these would-be farmers inadvertently set the stage for the rapid range expansion of *B. tectorum*.

The invasion of *B. tectorum* in the Intermountain West followed a common trajectory of a lag phase (about 20–25 years), in which range expansion stemmed from a few isolated foci and the area occupied was restricted. This period was followed by a log phase (15–20 years) in which range expansion accelerated, such that most habitats in which *B. tectorum* can persist were occupied by about 1930. Given the grass's attributes (successively germinating cohorts, enormous phenotypic plasticity, ability to usurp soil resources, tolerance of grazing and trampling, and tolerance of frequent fire, all of which are discussed below), its spread is not surprising. Collectively, these traits allowed *B. tectorum* to transform the steppe in the Intermountain West in about 45 years (roughly from 1890 to 1935), and probably in less than 25 years (from around 1910 to 1935), into a virtual monoculture of one invasive plant species.

PRE-ADAPTATION TO ENVIRONMENTS IN THE U.S. INTERMOUNTAIN WEST

In retrospect, the massive invasion of cheatgrass into the Intermountain West is predictable, given the grass's ecology. *Bromus tectorum* is well suited to the region's arid environments, with their prolonged dry periods and soils with pH that may exceed 8.4. The timing of precipitation on these sites is often erratic, beyond a general seasonal pattern (i.e., periodic rains in the autumn and winter, with little rain in the summer). In these environments, *B. tectorum* displays enormous plasticity for a wide range of morphological and ecological traits, such as germination time, rate of growth, number of tillers, and fecundity. Late spring flowering is one of the few phenological events for which the grass displays little site-to-site variation.

Broad phenotypic plasticity for many traits has a direct bearing on the extent and persistence of this plant invasion in the Intermountain West. With the resumption of rain in late August to early September of each year, the first cohort of *B. tectorum* seedlings can emerge. Other bouts of germination can join this first cohort, with each being triggered in as little as 72 hours by a precipitation event. Thus, a series of cohorts can emerge from late summer into winter and onward, as long as the soil surface is not snow covered. This pattern continues until mid-March. As a result, by May, there may be plants side by side that vary in lifespan by as much as seven months. This difference in age translates directly into plant size, into the number of tillers, and most importantly into the number of seeds produced. Old (germinating in late August) plants may produce more than 500 seeds; small plants, such as those emerging in spring, may produce only a single seed. But practically every plant alive on the first of May will produce at least one seed. Alternatively, germination may be seasonally restricted (e.g., where autumn precipitation is nil, only spring germination occurs).

The remarkably long period of potential germination means that autumn-germinating plants can die before spring but that later emerging individuals will take their place—the basis for considerable population resiliency to environmental stochasticity, given the gauntlet of environmental hazards plants may experience between September and June. Seed survival in or on the soil also accounts for the grass's persistence: even if a site burns during summer, some seeds survive in the shallow seed bank.

Other aspects of cheatgrass biology, in the context of the Intermountain West, explain its proliferation.

In autumn, before seedlings of most of the native species have germinated, cheatgrass seedlings rapidly extend their root systems deep into the soil. Establishment of a root system at depth sets the stage for events during warmer conditions in late winter and early spring, when these roots undergo further rapid growth and usurp soil water and nutrients, especially nitrogen, from competitors. Although *B. tectorum* may offer little competition to adults of the large plant species in these communities (e.g., *Artemisia tridentata, Festuca idahoensis, Poa secunda*), it eventually dominates the site by usurping soil resources from successive generations of these species' seedlings. Over time, as recruitment among these species is effectively truncated, cheatgrass becomes dominant. Furthermore, cheatgrass seedlings are much more tolerant of repeated grazing than are the seedlings of the native perennial grasses. This invasive grass can tolerate not only periodic (but not prolonged) drought, but also snow cover and a wide range of soil textures, salinity levels, and nutrient levels.

Despite tolerance of an impressive array of environmental forces, cheatgrass mortality can be extensive. Plants that germinate in late summer or autumn run the risk of desiccation in September or October, if precipitation is infrequent. Chief mortality agents in winter are grazing by voles and frost heaving, the latter of which effectively isolates the desiccating plant on a small soil ped. Although cheatgrass can tolerate prolonged snow cover, it is vulnerable on thoroughly frozen, bare ground. Competition for water and nutrients can greatly affect plant vigor but does not usually cause plant death. However, *B. tectorum* is highly vulnerable to competition for light. This vulnerability is evident in its native and introduced ranges: the grass dominates arid sites, which usually have less than 100 percent plant cover. As a plant community's canopy coverage exceeds 100 percent, there is increasing likelihood that insufficient light will reach *B. tectorum* for its survival. For example, within the Intermountain West, cheatgrass probably reaches its highest abundance in the arid steppe, once dominated by *A. tridentata* (big sagebrush), or on other habitat types that once supported communities dominated by the shrubs *Purshia tridentata* or *Sarcobatus vermiculatus*, as well as somewhat less arid sites co-dominated by the perennial grasses *Agropyron spicatum* (*Pseudoroegneria spicata*) and *Festuca idahoensis*. Competition for light is much greater within forests, where *B. tectorum* can occur in low numbers under the open canopy within *Pinus ponderosa* forests, but the grass is rare or absent in forests upslope (e.g., dominated by *Pseudotsuga menziesii, Abies grandis*).

Shade intolerance emerges as a major environmental limitation to its invasion.

Any ruptures or cracks in the soil (e.g., frost-heaved soil, hoof prints, tire tracks, animal burrows) provide an appropriate seed bed for the grass; these sites are often unsuitable for many other species. In contrast, many of the native angiosperms in the Intermountain West flower infrequently, are poor competitors, do not produce viable seeds every year, and do not tolerate livestock grazing and trampling. Not only does cheatgrass appear pre-adapted to the Intermountain West habitats circa 1850, but its attributes are also supremely well adapted to the environments created with European settlement in the last 150 years.

THE ROLE OF CHEATGRASS IN STEPPE COMMUNITIES

The steppe communities of the state of Washington illustrate the environmental spectrum of sites in the Intermountain West across which *B. tectorum* is widespread. In general, *B. tectorum* is most abundant on those sites that are most arid (<250 mm annual precipitation). As a result, the grass is the dominant in the arid steppe (e.g., *A. tridentata/A. spicatum* zone) and is common but usually not abundant within the meadow steppe (*Symphoricarpos albus/Festuca idahoensis* zone), the steppe zone that receives the most precipitation (>350 mm) and that forms the lower timberline ecotone with *Pinus ponderosa* communities. Further west and especially south into the Great Basin (*A. tridentata/A. spicatum* and *A. spicatum/F. idahoensis* zones), farming is less prevalent, and much more of the landscape is devoted to livestock. The immense areas in the Intermountain West with the arid steppe environment, once dominated by big sagebrush, present the most difficult areas in which to control cheatgrass and are the sites on which the grass inflicts the greatest ongoing damage.

CONTROLLING CHEATGRASS: A CHALLENGE WORTHY OF NATIONAL ACTION

Transformation of much of the Intermountain West by cheatgrass in the last century is readily apparent in comparing even landscape views of big sagebrush communities with and without invasion by this grass (Fig. 2). Loss of the multitiered structure of the sagebrush community has obviously affected the environment not only for native plants but also for many native vertebrates, including the potentially threatened sage grouse. Conservation of the native biota is threatened from the ratchet-like ability of cheatgrass to usurp water and nutrients from the

A

B

FIGURE 2 Comparison of the striking effects of the invasion by *Bromus tectorum* in western grasslands. (A) A mature stand supporting the native *Artemisia tridentata/Agropyron spicatum* community. (B) A site for which the native vegetation seen in (A) has been completely replaced by cheatgrass after its invasion and accompanying fires. (Photograph courtesy of the author.)

seedlings of native species. Sharp decline in the prominence of several native annual grasses that once played the role of colonizers on sites of disturbance (e.g., *Festuca octoflora*), coincident with the arrival of cheatgrass, is but a small part of the alteration of the native flora.

The role cheatgrass has had in altering the frequency and character of fire in this region is clearly of paramount economic and environmental importance. As an annual, the vegetative plants die in summer and become a ready source of fuel. Although fire occurred in this steppe before the entry of cheatgrass, fires were apparently less frequent, and the fuel was not evenly distributed across sites. With a blanket of dead cheatgrass straw across a site, a fire advances quickly, igniting any remaining shrubs and totally consuming the aboveground biomass. These fires can destroy even adults of the perennial caespitose grasses, such as *A. spicatum*. The resulting environment is nonetheless suitable for cheatgrass recruitment, beginning as soon as the following autumn.

Damage to property (fences, buildings, livestock) in the path of a cheatgrass-fueled fire can be extensive, especially because some fires have consumed more than 500,000 ha of grassland. Much more serious, however, are the consequences with resumption of precipitation: with little live or dead plant material to check erosion, sheet-wash becomes extensive. This soil-laden water eventually flows through a watershed until it reaches a river, where rates of sedimentation can be much altered. Perhaps most serious in economic terms are consequences of these sediments in river systems now dotted with hydroelectric dams. Along the Snake and Columbia rivers in the Pacific Northwest, there are 18 public and privately operated dams erected to provide a reliable supply of water for irrigation, channels for barge navigation, energy for power generation, and flood protection. The environmental chain reaction caused by the widespread dominance of cheatgrass upstream greatly complicates maintenance of these river courses, quite aside from the negative consequences the sediment addition has for the freshwater biota, including migrating salmon.

Controlling an invader that has now occupied such a large, sparsely populated region gives new meaning to the term "daunting," especially because the grass appears to be still increasing its range in the Great Plains and Rocky Mountains. Nevertheless, even the crudest of estimates of its cost to the national economy, in terms of the aforementioned damage (in addition to the costs of annual fire fighting and frequent loss of human life), illustrate that it merits a much more science-based and concerted effort at control than has so far been mounted. Policymakers and the public need to be alerted to the immense shared cost of this pest. One useful step would be a comprehensive economic tally of damage, quite aside from an assessment of the loss of the biological heritage of the Intermountain West's grasslands.

SEE ALSO THE FOLLOWING ARTICLES

Fire Regimes / Grasses and Forbs / Lag Times / Landscape Patterns of Plant Invasions / Land Use / Seed Ecology / Transformers / Weeds

FURTHER READING

Haubensak, K., C. D'Antonio, and D. Wixon. 2009. Effects of fire and environmental variables on plant structure and composition in grazed salt desert shrublands of the Great Basin (USA). *Journal of Arid Environments* 73: 643–650.

Mack, R. N. 1989. Temperate grasslands vulnerable to plant invasion: Characteristics and consequences (155–179). In J. Drake, ed. *Biological Invasions: A Global Perspective*. New York: Wiley and Sons.

Mack, R. N., and D. A. Pyke. 1984. The demography of *Bromus tectorum*: The role of microclimate, predation and disease. *Journal of Ecology* 72: 731–748.

Novak, S. J., and R. N. Mack. 2001. Tracing plant introduction and spread into naturalized ranges: Genetic evidence from *Bromus tectorum* (cheatgrass). *BioScience* 51: 114–122.

Schachner, L., R. N. Mack, and S. J. Novak. 2008. *Bromus tectorum* (Poaceae) in mid-continental United States: Population genetic analysis of an ongoing invasion. *American Journal of Botany* 95: 1584–1595.

Smith, S. D., R. K. Monson, and J. E. Anderson. 1997. *Physiological Ecology of North American Desert Plants.* New York: Springer.

CHEMICAL CONTROL
SEE HERBICIDES; PESTICIDES

CLASSIFICATION AND REGRESSION TREES
SEE CART AND RELATED METHODS

CLIMATE CHANGE

JEFFREY S. DUKES
Purdue University, West Lafayette, Indiana

As human activities release increasing amounts of carbon dioxide (CO2) and other greenhouse gases into the atmosphere, more infrared radiation that would otherwise escape from Earth's atmosphere gets trapped. As a consequence, climate models (and basic physics) suggest surface temperatures will rise, and precipitation patterns will change. These changes will have consequences for living organisms around the globe, including invasive species. Many plants and animals will respond to climate change by migrating, evolving, or adapting. Species that are slow to shift ranges, adapt, or evolve will be disadvantaged. The abundance and distribution of many invasive species depends largely on human actions, whether intentional or otherwise. As humans react to the changing climate, societal responses will disproportionately affect the species that are most influenced by people. Several lines of reasoning suggest that many invasive species will benefit from climate change, but many of these ideas are difficult to test experimentally. This article details the environmental changes that are expected during this century and the likely responses of invasive species to these changes.

EXPECTED CHANGES IN THE ATMOSPHERE AND CLIMATE

Before the onset of the Industrial Revolution, the amount of CO2 in the atmosphere had remained quite stable for about 10,000 years. Since then, as society has burned fossil fuels and converted forests to agricultural landscapes, the CO2 concentration has risen to a level not seen in at least 650,000 years (and probably much longer than that). By 2008, concentrations had increased by 38 percent over their preindustrial levels, reaching 385 parts per million (ppm). This change, in concert with increases in concentrations of methane, nitrous oxide, and other greenhouse gases, is causing the oceans to become more acidic, and causing heat to be trapped in the atmosphere that would otherwise have escaped to space.

As the atmosphere more effectively insulates our planet from space, the climate on Earth's surface is warming. The rate of future warming will depend largely on societal decisions about how fast to use fossil fuels and how best to use land, but a likely range of average warming lies between 1.1 and 6.4 °C for Earth's surface by the year 2100. This average can be misleading, though—large spatial variation is expected in the warming, with less warming over the oceans. Greater-than-average warming is likely on land and in the northern boreal and arctic regions. In some northern locations, climate could warm by more than 7.5 °C.

Changes in precipitation patterns will accompany warming. At high latitudes, winters are very likely to become wetter, while many parts of the subtropics, which are already relatively dry, are likely to receive less rain. Warming will change the nature of water inputs in colder areas; a smaller fraction of their precipitation will arrive as snow, and snow cover will disappear more quickly. In addition, storms are expected to intensify, with more precipitation coming in large storms and less in small events. Accordingly, the length of time between rain events is likely to increase in many systems.

Warming will lead to many changes in the oceans. Acidification, driven by carbon dioxide in the atmosphere, will threaten marine organisms such as corals and shellfish that build structures out of calcium carbonate. Warming waters expand, and this, in combination with melting ice, will lead to rising sea levels. Some computer models also predict the complete melting of summertime arctic sea ice before century's end.

Other more localized changes in the atmosphere could also affect interactions among species. For instance, industrial activity and fertilizer use are increasing the amount of nitrogen pollution that falls out of the atmosphere and onto ecosystems. This unintentional fertilization varies regionally, and is increasing in many locations. Levels of ground-level ozone also vary by region, with consequences for human health as well as plant growth and competition.

Rapid changes in the atmosphere, climate, and ocean will directly affect the survival and competition of invasive species in a variety of ways, also altering the transport and management of these species.

EFFECTS OF CLIMATE CHANGE ON PATHWAYS TO INVASION

Invasive species typically arrive in new locations with the help of humans. A variety of the pathways for invasive species may change in the future as a result of climate change. For instance, patterns of commerce or transport could change. This could change the number of arriving propagules at a given site. To reduce the rate of climate change, among other reasons, many governments have mandated or incentivized biofuel production. Nonnative, weedy biofuel species such as giant reed (*Arundo donax*) and jatropha (*Jatropha curcas*) have been planted in or proposed for use in many regions, from which they may escape into natural systems. In areas where warming reduces cold stress, new suites of ornamental species might be sold, with the potential for some of them to become invasive. Similarly, suitable zones for aquaculture will change, with larger areas becoming susceptible to escapes of warmwater fish.

Warming could also change transport routes: If a reduction in sea ice allows ships to save time by navigating the Arctic Ocean, organisms transported unintentionally in ballast water or by other mechanisms will arrive at new ports more quickly than they previously had. Since many organisms perish during transport, faster voyages would likely increase survival and lead to the introduction of more individuals to the new site.

EFFECTS OF CLIMATE CHANGE ON THE ESTABLISHMENT OF INVASIVE INTRODUCED SPECIES

The change in atmospheric CO_2 concentrations will directly affect the growth and tissue chemistry of plants, and will affect terrestrial animals primarily through changes in plant chemistry and plant community composition. In general, plants grow faster when CO_2 is more abundant, although some plants are more responsive than others. Some basic traits such as the photosynthetic pathway of plants can be used to generate rough predictions of "winners" and "losers"; fast-growing C_3 plants are typically responsive to CO_2 increases, for instance, while C_4 plants and slow-growing C_3 plants respond less vigorously. However, these predictions have not always held up in field experiments, where plants are grown in competition. Some invasive species, such as red brome (*Bromus madritensis* ssp. *rubens*) and Japanese honeysuckle (*Lonicera japonica*), have been demonstrated to respond strongly to CO_2 in the field, at least in some years. It is not yet clear, however, how frequently invasive species will benefit from the changing atmosphere at the expense of native species.

Climate change will shift the potentially suitable ranges for all species, typically moving them poleward and up in elevation. Many studies have already observed shifts in species' ranges that are consistent with recent shifts in climate. The mechanisms behind such range shifts have also been observed: On the shores of New England, recruitment patterns of native and invasive sea squirt species fluctuate by year, with the invasive species being favored by years with warmer waters.

Several researchers have used projections of future climate in conjunction with climatic tolerances of particular invasive species (often inferred from current distributions of the species) to predict changes in areas susceptible to invasion in the future. These analyses sometimes predict increases in the range of the invaders, and sometimes predict range decreases. The inexact nature of climate projections, particularly with respect to precipitation, loads any particular prediction of a species' future range with uncertainty. However, these predictions remain illustrative of potential changes in the coming century.

Climate envelope studies of invasive plant species suggest some range expansions and some contractions. Several of these studies have been carried out, looking at invasive plants on many continents. An analysis of South African plant invaders suggests that an increase of 2 °C would reduce suitable range for five species, to 63–92 percent of their currently suitable habitat area. For all of these species, soil types restricted potential ranges, and were jointly responsible for the predicted range reductions. A similar analysis that examined five invasive plant species in North America (but that did not incorporate soil type) suggested future climates would lead to range expansions of two species, shifts of two others, and range contraction for one species. Where ranges of invasive plant species contract, restoration opportunities may await.

Climate envelope studies have also been carried out for animals, suggesting, for instance, that the distribution of cane toads (*Rhinella marina* [*Bufo marinus*]) in Australia will expand southward under climate change. A biophysically based model has also been used to infer possible range shifts in this species, predicting a southward range expansion by about 100 km.

While a new, stable climate may favor some invasives and limit the ranges of others, a *changing* climate could favor species having certain traits—what's more, traits that are commonly found among invasive species. In particular, mobility, short generation times, and broad environmental tolerances—all traits of many invasive species—may prove advantageous. Species such as slow-growing trees may be disadvantaged in a changing climate if they are slower than competing species to extend their ranges into newly suitable habitat. Slow-spreading species may gradually become less competitive with their neighbors as conditions change away from those to which they are adapted. Plant species with short generation times and effective long-distance seed dispersal mechanisms are likely to quickly exploit newly suitable areas. Comparisons within some genera show that invasive plant species are more likely to have traits that allow them to disperse over long distances (e.g., *Pinus*), and to have broader climatic tolerances than noninvasive species, which should allow them to better tolerate changes in climate.

Because species respond to different environmental cues, shifting climates may disrupt some mutualistic relationships between plants and pollinators or seed dispersers. While some native species rely on specialized mutualistic relationships that could be subject to disruption, invasive plant species typically do not depend on specialists, and so will avoid these potential problems.

Invasive plant species tend to have traits associated with fast growth, and many thrive in areas with high resource availability. These include areas that are inadvertently fertilized, such as from nitrogen pollution from the atmosphere in industrial regions (a large fraction of which comes as a byproduct of fossil fuel combustion) or in heavily agricultural landscapes, and areas that have received additional water. In addition, areas in which plant communities have been disturbed have high resource availability, plant demand having been artificially suppressed. One hypothesis for the success of invasive plant species in areas of high resource availability holds that natural enemies have the most suppressive effects on native species growing in high resource areas. Thus, plant species that are adapted to conditions of abundant resources benefit more from enemy release than other species. There is good circumstantial support for this hypothesis; native species in resource-rich areas host more pathogen species than those from resource-poor areas. Evidence from observed patterns of invader abundance, experimental nutrient manipulations, and this hypothesis all suggest that increases in resource availability, whether via nitrogen deposition, increases in growing-season water availability due to climate change, or both, may lead to increased abundance of invasive plant species.

EFFECTS OF CLIMATE CHANGE ON THE IMPACT OF INVASIVE INTRODUCED SPECIES

The impact of an invasive species can be conceptualized as a product of the species' range and abundance within that range, and the effect that each individual has on the ecosystem process or property of interest. Climate change can affect an invader's impact by altering any of the components of this equation (Fig. 1).

New climates will alter constraints on invasive species' distributions, sometimes dramatically altering their impacts. In eastern North America, for instance, the impact of the hemlock woolly adelgid (*Adelges tsugae*; an insect pest of hemlock trees) is likely to increase as the climate warms, because the adelgid will expand its range into areas where its populations were previously limited by cold winters. In the United States, this species has already largely eliminated eastern hemlock (*Tsuga canadensis*) from mid-Atlantic states and is now moving into the Northeast. Similarly, on Hawaii, the introduced avian malaria parasite *Plasmodium relictum* has led to the elimination of Hawaiian honeycreepers from forests at lower elevations. Remaining bird populations survive primarily in the colder upper elevations, where the parasite's vector, the introduced mosquito *Culex quinquefasciatus,* is rare and where *Plasmodium* development is slow. In a warmer climate, the impact of malaria will increase as its range extends upwards into some of the few remaining areas of safe habitat for honeycreepers.

Other species will become more (or less) competitive within their ranges, altering their impact. For instance, competition studies and natural distributions indicate that, while brown trout and brook trout compete equally for food in colder temperatures, brown trout dominate in warmer water. Climate change may thus increase their impact on other species. Experimental manipulations similarly suggest that climate and atmospheric change will cause some invasive plant species to become more competitive.

In other cases, climate change may alter the per capita impact of an invader. Ectothermic species such as fish require more food as their body temperatures increase. In the Bonneville Dam area of the Columbia River, in

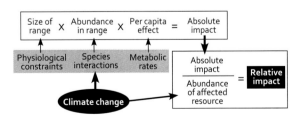

FIGURE 1 Conceptual diagram of factors affecting the impact of an invasive species, and some of the mechanisms through which climate change can affect those factors. (Figure adapted from Hellmann et al. 2008.)

the western United States, an increase in water temperature of 1 °C could increase the rate at which introduced smallmouth bass and walleye eat native salmonids by 4–6 percent.

Finally, in some cases the relative impact of an invasive species may change simply because the affected species or resource has responded to climate change. For instance, the impact of the invasive salt cedar tree (*Tamarix* spp.) could change in the future; these trees are thought to use more water than native tree communities, reducing streamflows in much of southwestern North America. Some climate models project decreases in precipitation in that region by the end of this century. If water became scarcer in this already dry region, each liter of water would become more valuable, increasing the relative impact of salt cedar on water availability.

EFFECTS OF CLIMATE CHANGE ON THE MANAGEMENT AND CONTROL OF INVASIVE INTRODUCED SPECIES

A variety of the techniques used to combat invasive species may be affected by climate and atmospheric change. The effectiveness of biocontrol species is likely to change for a variety of reasons: For instance, ranges of climatic suitability of the target and biocontrol species may shift, leading to different amounts of overlap between the two. Also, the effectiveness of a given biocontrol species varies with climate, and so there may be shifts in the success of this technique within a given target species' range. Finally, the palatability of plant species can be affected by elevated CO_2, which could affect some biocontrol interactions.

The increase in atmospheric CO_2 is likely to increase tolerance of some invasive plant species to herbicides. Results of a field trial with Canada thistle (*Cirsium arvense*) suggest that CO_2 enrichment will allow this agricultural weed to recover more quickly after spraying, presumably because an observed faster accumulation of belowground resources leads to greater vigor in resprouting.

Warmer temperatures will allow many species to grow faster, including many aquatic and terrestrial invasive species. This is likely to lead to increased control costs in settings in which a maximum amount of the invasive species may not be exceeded, such as where zebra mussels (*Dreissena polymorpha*) have colonized water intake pipes that must remain open.

For species that currently cannot overwinter but that will be able to in the future, management strategies may change; such species may include ornamental plant species or tropical aquarium species that are currently unable to survive outside their artificial environments. Climate change may lead to new recommendations for acceptable species in gardens and aquaria.

For land managers, the most useful tools for predicting future effects of invasive species will be large databases that accurately capture ranges of native and nonnative species, as well as the environmental conditions in which they exist, over large areas and at high resolution. These databases could be used with climate projections to identify areas that will become vulnerable to new invasions. For most areas of the world, such databases do not currently exist, although relevant data are being collected in many regions.

OUTLOOK

For any particular invasive species, changes in climate and atmosphere could be, on balance, positive or negative. The effects are typically not easy to predict. In general, invasive species are likely to benefit from a combination of factors because of the suites of traits that they are likely to possess. For instance, invasive plant species tend to have broad environmental tolerances (and accompanying large ranges), fast growth, and little dependence on other species for survival or reproduction. In addition, invasive species often thrive in disturbed areas and areas with high resource availability, which are likely to be more common in the future. In areas where climate change results in more frequent fires, for instance, some fire-adapted invasive plant species could become more widespread, potentially accelerating the fire cycle even further, to the detriment of native species.

How should land managers address potential new threats under climate change? Essentially, land managers will need to do more of what they are doing already: keeping vigilant for the arrival of new invasive species, identifying and removing new or outlying populations of nonnative species that are or could become invasive, and controlling the species already present, particularly those that have the greatest potential impacts.

Policymakers can address this issue by crafting policies that address climate change with an awareness of potential consequences for invasive species, for instance by mandating use of only noninvasive plant species as bioenergy crops. Policymakers also could direct some climate-related revenue, such as that derived from sales of carbon allotments, toward invasive species control, with the recognition that some invasive species problems will become worse because of climate change.

SEE ALSO THE FOLLOWING ARTICLES

Enemy Release Hypothesis / Evolution of Invasive Populations /
Hydrology / Invasiveness / Land Use / Life History Strategies /
Propagule Pressure / Range Modeling / Vines and Lianas

FURTHER READING

Bradley, A.B., M. Oppenheimer, and D.S. Wilcove. 2009. Climate
change and plant invasions: Restoration opportunities ahead? *Global
Change Biology* 15: 1511–1521.

Dukes, J.S., and H.A. Mooney. 1999. Does global change increase the
success of biological invaders? *Trends in Ecology and Evolution* 14:
135–139.

Hellmann, J.J., J.E. Byers, B.G. Bierwagen, and J.S. Dukes. 2008. Five
potential consequences of climate change for invasive species. *Conservation Biology* 22: 534–543.

Invasive Species Council. Double trouble. http://doubletroublebulletin
.wordpress.com

Low, T. 2008. Climate change and invasive species: A review of interactions. Report of the Biological Diversity Advisory Committee for the
Department of the Environment, Water, Heritage and the Arts. Commonwealth of Australia, Canberra.

Mooney, H.A., and R.J. Hobbs, eds. 2000. *Invasive Species in a Changing
World.* Washington, DC: Island Press.

Pyke, C.R., R. Thomas, R.D. Porter, J.J. Hellmann, J.S. Dukes, D.M.
Lodge, and G. Chavarria. 2008. Current practices and future opportu-
nities for policy on climate change and invasive species. *Conservation
Biology* 22: 585–592.

Rahel, F.J., and J.D. Olden. 2008. Assessing the effects of climate change
on aquatic invasive species. *Conservation Biology* 22: 521–533.

COCCINELLIDAE

SEE LADYBUGS

COMMON CARP

SEE CARP, COMMON

COMPETITION, ANIMAL

CHRISTOPHER R. DICKMAN

University of Sydney, New South Wales, Australia

Competition occurs when two or more organisms require
similar resources to grow and reproduce, and then harm
each other while seeking or using those resources. The
effects of competition are often detectable on individual
organisms but more commonly are measured at the level
of species populations. Thus, species can be said to be in
competition if they have mutually depressive effects on
each other's population size, performance, reproduction,
growth rate, survival, and ultimately, fitness. Strong competitive ability is an attribute of many invasive animals

and is probably a key reason for their success in new
environments.

IDENTIFYING COMPETITION

Competition takes a very wide variety of forms and can be
difficult to identify and distinguish from other processes.
Consider the situation where two species have similar
population densities in places where they occur on their
own, but densities that are much reduced where they
occur together. Competition may be suspected, especially
if the two species use similar resources. However, the situation could arise if the two species are suppressed by a
common predator where they cooccur (apparent competition), if each species acts as vector for the transmission of
disease or parasitic organisms that reduce populations of
the other, if the habitat where the species cooccur is suboptimal for both, or even simply by chance. The situation
could arise also if the two species hybridize and produce
fewer young from hybrid matings than from within-species matings. These possibilities can be difficult to disentangle, especially in reconstructions of past events. For
example, the extinction of two species of native rodents
on Christmas Island, in the Indian Ocean, has long been
attributed to competition from the black rat (*Rattus rattus*) after the latter species was introduced in 1899. However, recent evidence suggests that both the bulldog rat
(*R. nativitatis*) and Maclear's rat (*R. macleari*) succumbed
to trypanosome parasites carried by the black rat, with
R. macleari also disappearing due to hybridization with
the invader. Competition can be confirmed more reliably
in contemporary invasions, especially if investigators are
able to identify the type of competition that is occurring
and determine which species and resources are involved.

TYPES OF COMPETITION

Two broad types of competition are usually recognized:
interference competition and exploitation competition.

Interference Competition

In interference (also encounter or contest) competition,
animals harm one another while seeking shared resources,
even if those resources are not in short supply. Harm
can be inflicted directly by fighting, chasing, or other
forms of aggression, or indirectly by the production of
poisons and other toxic compounds that constitute a
form of chemical warfare. This type of competition is
seen in many of the most successful invasive animals. In
Australia, for example, the European red fox (*Vulpes vulpes*) is an aggressive interference competitor with native

FIGURE 1 Spotted-tailed quoll, *Dasyurus maculatus*. (Photograph courtesy of Dr. Alistair Glen.)

species of quolls (*Dasyurus* spp.; Fig. 1) and limits quolls to small areas where they can find exclusive refuge. The invasive American mink (*Neovison vison*) likewise aggressively outcompetes its European relatives such as the polecat (*M. putorius*) and European mink (*M. lutreola*), with the latter species having disappeared from more than 80 percent of its former range in central and northern Europe. In both cases, intense aggression by the dominant competitor can result in the death of subordinate animals; this interaction becomes "intra-guild predation" if the subordinates are then eaten. Many social insects also are effective interference competitors. The Argentine ant (*Linepithema humile*) has been introduced to many parts of the world and is outcompeting native species of ants over large geographical areas. Another South American native, the red fire ant (*Solenopsis invicta*), has a similarly large invasive range and, like *L. humile*, is indiscriminately aggressive toward other species of ants. It is spreading in the southern United States, parts of China, and Southeast Asia, but it has been contained following control actions in eastern Australia.

Although overt aggression is often characteristic of interference, species engaging in this type of competition can harm each other in much more subtle ways. Sponges, colonial ascidians, and soft corals, for example, produce many biologically active compounds that deter predators and reduce fouling but that also may negatively affect competitors. The Indo-Pacific soft coral *Stereonephthya* aff. *curvata* has been expanding its range off the southeastern coast of Brazil since its introduction there in the mid-1990s; its production of secondary

metabolites causes necrosis in the tissues of the local gorgonian coral *Phyllogorgia dilatata* and appears to play an important role in the invasion process. Chemical warfare may be assisting another Indo-Pacific species, the sponge *Chalinula nematifera*, in its invasion of coral communities on the shores of the eastern Pacific Ocean. The production of chemical defenses against competitors, sometimes termed allelopathy, is particularly widespread in marine organisms and also in green plants, but it has yet to be confirmed as important in vertebrate animals.

Another subtle but effective form of interference occurs in sedentary animals such as mussels, barnacles, encrusting bryozoans, and ascidians that can reach such high local densities that other species are crowded out or overgrown. Introduced to the Laurentian Great Lakes in the 1980s, the Eurasian zebra mussel (*Dreissena polymorpha*) is expanding rapidly into drainage systems across much of North America. Local densities of this species are commonly 2,000–4,000 animals per square meter but can reach 100,000 animals per square meter, allowing the invasive mussel to overgrow and outcompete its native counterparts, as well as large snails, caddisfly larvae, and other fauna that need smooth surfaces for attachment. Similar patterns of overgrowth have been recorded in the nearshore environment, with North Atlantic marine mussels such as *Mytilus edulis* being overwhelmed by the European sea squirts *Ascidiella aspersa* and *Ciona intestinalis*. These invaders grow quickly and can outcompete resident species by growing over them if they arrive and settle later.

Exploitation Competition

In exploitation (also resource or scramble) competition, animals harm one another by using common resources that are in short supply. Immediate physical harm does not occur in this type of competition; indeed, exploitation competitors may not even meet or be active at the same time, but they will still fare poorly because each species depletes resources that could have been used by the other.

Humans provide some excellent examples of exploitation competition, using a wide range of land and freshwater resources so efficiently that other organisms can gain little access to them; reducing fish stocks so far that populations of seals, dolphins, and other fish-eaters are in decline; and outcompeting many species for space. In Australia, the invasive European rabbit (*Oryctolagus cuniculus*) has been an effective competitor since its establishment in 1858, reducing the food available to wallabies, rat-kangaroos, and probably many species of native rodents, and helping to drive declines in their populations as a result. In New Zealand, introduced wasps (*Vespula* spp.) can attain densities of 10,000 workers per hectare in native beech forests and reduce the standing crop of honeydew by over 90 percent for several months of the year. Competition from these voracious wasps reduces local populations of native insects and birds that use the honeydew resource.

In the marine environment, a well-studied exploitation competitor is the comb jellyfish *Mnemiopsis leidyi* (Fig. 2). Introduced from the western Atlantic to the Black Sea in 1982, this species had reached peak densities of 400 animals per cubic meter of water before the end of the decade. Its varied diet of zooplankton, fish eggs, and larvae reduced the food base for many species of fish and caused reductions in fishery catches of 75 to 90 percent. The species has spread subsequently to the Caspian, North, and Baltic seas, reducing fisheries in the Caspian Sea at least. Although *Mnemiopsis leidyi* is a strong competitor, its consumption of fish eggs and larvae likely contributes also to fishery depletion.

WHICH SPECIES COMPETE?

A common observation is that competition occurs more frequently and intensely between members of the same species than between those of different species, even if the type of competition involved is the same. This is not surprising; individuals of the same species can be

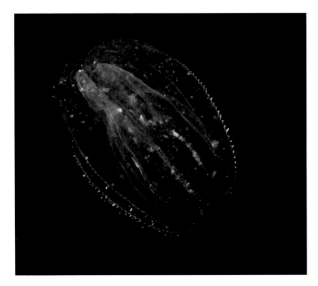

FIGURE 2 Warty comb jellyfish, *Mnemiopsis leidyi*. (Photograph courtesy of Dr. Marco Faasse, Acteon Marine Biology Consultancy.)

expected to use very similar resources, and hence to compete if these resources are scarce. If the intensity of competition increases with resource overlap, then closely related species should be more likely to compete strongly among themselves than with distant relatives. This is often the case, and most studies documenting competition have focused on the interaction between sibling species or members of the same genus. However, competition need not be confined to related species and potentially may occur among any that use similar resources in similar ways. Species that make similar use of resources are sometimes said to belong to the same guild and may be termed collectively as fruit-eaters, wood-borers, cave-dwellers, and so on, depending on the type of resource that they exploit. In Australia, for example, the introduced honeybee (*Apis mellifera*) can be placed into two guilds, the first grouping it with species that use tree hollows for nests, and the second with species that use nectar as their primary food source. Mounting evidence suggests that honeybees use interference behavior to evict bats, parrots, and marsupials such as brush-tailed possums (*Trichosurus vulpecula*) from nest hollows in trees, and exploitation to usurp floral resources from native insects, honeyeaters, and small nectar-feeding marsupials such as pygmy-possums (*Cercartetus* spp.).

Competition is often thought to involve pairs of species, but the guild concept makes it clear that many species may be involved simultaneously. The situation where a species is competitively affected by suites of other species has been termed "diffuse competition." Diffuse interactions are more difficult to disentangle and quantify than those between pairs of species but provide explicit recognition that natural communities usually contain many species.

WHICH RESOURCES ARE AT STAKE?

Among mobile animals, competition often occurs for food or shelter, or for the habitats where these and other scarce resources can be found. If the resources are ephemeral or distributed patchily, such as in the case of fallen fruit, carcasses, or temporary ponds, competition among the resource consumers can be brief but intense. Thus, invasive cane toads (*Rhinella marina* [*Bufo marinus*]) quickly deplete invertebrate food resources that could be used otherwise by native frogs on the floodplains of northern Australia; African clawed toads (*Xenopus laevis*) appear to reduce food resources similarly around water bodies in their new range in parts of Europe and

North America. In sedentary or sessile animals, by contrast, space itself may be the scarce resource for which competition occurs.

Food and habitat, or space, are usually the most important resources for which competition occurs, but time is sometimes also considered to be important. This is especially so if competitors mutually restrict the time that each has available to find or harvest other essential resources. Competing desert rodents provide good examples of temporal competition, with an extreme example recorded in spiny mice (*Acomys* spp.) of the Negev Desert in Israel. Here, the dominant species *A. cahirinus* is nocturnal and forces the subordinate *A. russatus* to be active during the risky hours of daylight; its removal allows the subordinate species to revert to nighttime activity. There is little evidence that invasive species cause deleterious shifts in the activity of resident species, but this topic has so far attracted little research attention.

THE EFFECTS OF COMPETITION

By definition, competing organisms reduce each other's fitness. However, the extent of fitness reduction will often differ between species, times, and places, and also be evident in different responses that are shown by the competitors themselves. Species' responses to competition can be divided into two groups.

Ecological Responses to Competition

One of the most common findings of competition studies is that species are seldom equally matched; dominant ("winner") and subordinate ("loser") species usually can be identified. Such competitive asymmetry can arise in many ways. In interference competition, dominant species often are larger or more aggressive than subordinates and may exert their effects even when they are present at low density. In exploitation competition, by contrast, dominant species simply gain access to common resources more quickly or in greater numbers than subordinates and use them more efficiently. Exploitation competitors often cooccur at the same times and places, with dominance sometimes shifting between species depending on their relative numbers. For example, in eastern Australia, the native New Holland mouse (*Pseudomys novaehollandiae*) is competitively superior to the invasive house mouse (*Mus musculus*) when the latter is at low density but not when house mice reach high numbers. Dominance switches from one species to the other depending on which has the numerical ascendancy. Similar dominance switching occurs between

the house mouse and gray-bellied dunnart (*Sminthopsis griseoventer*) on Australia's west coast and may be a common phenomenon between house mice and other exploitation competitors.

The effects of competition are manifold. At a broad scale, competitors may restrict each other to particular types of habitat or even exclude each other from regional areas. If such areas are small or restricted to begin with, such as for species confined to islands, the arrival of competitively superior invaders may cause extinctions. The invasion by black rats of María Madre Island, Mexico, and the Mediterranean islands of Corsica and Sardinia appear to provide good examples of competitive exclusion, with the resident Nelson's oryzomys (*Oryzomys nelsoni*) and the Tyrrhenian field rat (*Rhagamys orthodon*), respectively, disappearing in the wake of the black rat's introduction. At finer scales, competitors may confine each other to particular components of the habitat while still cooccurring more broadly.

If competitors persist in each other's presence, they may exert a range of effects that are manifest at the both the individual and population levels. Following the removal of invasive rodents from islands, for example, resident small mammals in several studies have displayed increases in survival, growth rate, body mass, or reproduction, and population increases ranging from 2- to 32-fold have been documented. Removal of black rats from one of the Galápagos Islands further allowed resident Santiago nesoryzomys rats (*Nesoryzomys swarthi*) to increase immigration and recruitment, reducing seasonal declines in their population size. These effects can be interpreted as being largely due to competitive release, although release from predation is also possible in some cases. Similar effects have been documented in a wide range of other taxa following the removal of invasive competitor species, suggesting that competition can have pervasive effects on the individual fitness and population performance of resident animals.

Evolutionary Responses to Competition

As competition is an energetically expensive process that negatively affects the species involved, we might expect selective benefits to accrue to animals that use different resources from their competitors. Furthermore, as many competitive interactions are asymmetric, with larger or more aggressive species being dominant, selection could be expected to be especially strong on competitive "losers." In consequence, differences in resource use, size, shape, or other attributes between cooccurring species have commonly been interpreted as evolutionary responses to competition, especially if such differences are accentuated in places where species cooccur compared to places where the species occur on their own. If competition is sufficiently intense, it is even possible that gene flow could be disrupted between those parts of a species' range where it occurs with, versus without, its competitor, giving rise to a sympatric speciation event. Structured patterns in species' distributions and ecology often have been attributed to the "ghost of competition past." However, as differences in resource use or morphology between species may arise for many reasons, the work of the ghost should be accepted only when alternatives have been scrutinized.

Competition-mediated evolutionary shifts have been described in a small number of case studies of invasive species and have not always conformed to expectation. In Canadian lakes invaded by benthic-feeding white suckers (*Catostomus commersoni*) and creek chub (*Semotilus atromaculatus*), native brook char (*Salvelinus fontinalis*; Fig. 3) show predictable shifts from benthic to pelagic feeding and also exhibit shifts in color, fin shape, and diet that are consistent with use of the pelagic niche. Benthic-feeding char remain prevalent only in lakes that have not been invaded by white suckers or chubs. By contrast, American mink decreased in size after their invasion into Eastern Europe, while competing polecats and European mink grew larger; as a result of these shifts, all three species unexpectedly converged in body size.

Because so many invasions have taken place relatively recently, it may be too early to expect consistent patterns or large competitively induced evolutionary shifts. The fossil record suggests that, in general, the numbers of species driven to extinction by invaders have been less than the numbers generated by subsequent evolutionary

FIGURE 3 Brook char, *Salvelinus fontinalis*. (Photograph courtesy of Biopix.)

diversification and adaptive radiations of the invaders themselves. Of course, this is not always obvious during contemporary timeframes, when losses of native species to invaders can be catastrophic. The unprecedented range of species being introduced and the rapidity with which they are establishing make the current situation difficult to compare with invasion scenarios in the past. Nonetheless, prediction of the diverse effects of invasive species may require both short- and long-term perspectives to be considered.

ARE INVASIVE SPECIES GOOD COMPETITORS?

The foregoing discussion provides many examples of introduced species that have succeeded in competition to become invasive in their new homes, but this does not mean that introduced species are always good invaders or that they necessarily use competition to achieve success. In Australia, for example, 23 of 40 introduced species of mammals have become established, and only 11 of these have expanded to occupy large areas (≥8%) of the continent's land surface; six of these species likely achieved invasion success via competition. Similarly, in the United States, some 1,500 species of insects had become established by the mid-1980s out of a much larger but unknown number that had been introduced. Just 235 of these established insects are considered to be invasive pests, with a fraction of these achieving success via competition. Despite the apparently low numbers of species that achieve invasion success by competition, those that do can have dramatic and negative effects on populations and communities of resident species. Witness the plight of flightless rails on islands of the Pacific region: some estimates suggest that many hundreds of species were driven to extinction by the shocks of human colonization, with competition and predation from introduced rats being key factors in their demise.

Successful invaders are often species that are successful in their "home" environments, in the sense that they occupy extensive geographical ranges or achieve large population sizes. Their competitive abilities may be enhanced further if they leave behind their natural competitors, predators, parasites, or other enemies when they move to new environments, or if their invasions are assisted by human agency. In Australia, for example, rabbits were introduced persistently until their numbers became self-sustaining around 1858 and were aided by intensive campaigns mounted by settlers to poison both native predators and potential competitors. Rabbits were assisted also by the creation of large areas of pasture for foraging, the felling of trees for cover, and the introduction of edible plant species. These changes to the environment, as well as the presence of the rabbit, contributed to successful efforts to introduce the European red fox in the early 1870s and precipitated an "invasional meltdown" that is still continuing. These examples suggest that invasive species have varied competitive abilities and that further experimental quantification is needed to identify the traits that best predict competitive success.

FURTHER READING

Duncan, R. P. 1997. The role of competition and introduction effort in the success of passeriform birds introduced to New Zealand. *American Naturalist* 149: 903–915.

Harris, D. B. 2009. Review of negative effects of introduced rodents on small mammals on islands. *Biological Invasions* 11: 1611–1630.

Mooney, H. A., R. N. Mack, J. A. McNeely, L. E. Neville, P. J. Schei, and J. K. Waage, eds. 2005. *Invasive Alien Species: A New Synthesis.* Washington, DC: Island Press.

Pimm, S. L. 1991. *The Balance of Nature?* Chicago: University of Chicago Press.

Sax, D. F., J. J. Stachowicz, J. H. Brown, J. F. Bruno, M. N. Dawson, S. D. Gaines, R. K. Grosberg, A. Hastings, R. D. Holt, M. M. Mayfield, M. I. O'Connor, and W. R. Rice. 2007. Ecological and evolutionary insights from species invasions. *Trends in Ecology and Evolution* 22: 465–471.

Strauss, S. Y., J. A. Lau, and S. P. Carroll. 2006. Evolutionary responses of natives to introduced species: What do introductions tell us about natural communities? *Ecology Letters* 9: 357–374.

Vellend, M., L. J. Harmon, J. L. Lockwood, M. M. Mayfield, A. R. Hughes, J. P. Wares, and D. F. Sax. 2007. Effects of exotic species on evolutionary diversification. *Trends in Ecology and Evolution* 22: 481–488.

Vermeij, G. 2005. Invasion as expectation: A historical fact of life (315–339). In D. F. Sax, J. J. Stachowicz, and S. D. Gaines, eds. *Species Invasions: Insights into Ecology, Evolution, and Biogeography.* Sunderland, MA: Sinauer Associates.

COMPETITION, PLANT

JESSICA GUREVITCH

Stony Brook University, New York

Competition is a negative interaction between plants using the same resources, in which one or more of the competing plants suffers a reduction in growth, chance of survival, or reproductive output as a result. Competition among plants in nature is very common, and its effects are ubiquitous. It may be important at any stage of an individual plant's life history—from its start as an embryo inside of a seed, to the seedling stage, as a juvenile or mature reproductive individual—and may hasten senescence and death. Competition can also affect plant populations and alter their abundances in communities, can affect species' distributions at both small and large spatial scales, and can change the direction and course of evolution. Not surprisingly, competition can affect plant invasions in a number of important ways. How important is competition from native plants in limiting the success of invasive species? What different types of evidence can be brought to bear on the role of competition from natives in affecting the invasion process? Do invasive species have characteristics that make them competitively superior, or do they evolve such characteristics in their novel environment? What are the impacts of competition from invasive plants on native plants, and what are the consequences for biodiversity? Many of these questions, and interpretation of the evidence to date, remain controversial.

BIOTIC RESISTANCE AND PLANT COMPETITION

Charles Elton, in his landmark book on invasions, *The Ecology of Invasions by Animals and Plants* (1958), hypothesized that one of the major factors limiting invasions is *biotic resistance*: the ability of native species to prevent an invader from becoming established by their interactions with propagules of the invader. These negative population interactions that might reduce the likelihood of successful invasion include competition, predation and herbivory, and disease. Competition from native plants is believed to have a major effect on limiting invasion of exotic species. Competition may serve as an initial barrier to invasion, or it may reduce the growth and reproduction of invasive individuals after they become established. Competition from native plants may also have effects on invaders at other scales, limiting their rate of population growth and range increase, or constraining their effects on native communities at later invasion stages.

Elton predicted that communities that had greater numbers of species—that is, that were higher in native species diversity—would have higher biotic resistance than those that had fewer species, and would therefore be better able to resist invasion by exotic species. However, even if a greater number of species resulted in greater biotic resistance (not necessarily the case), other factors also vary as the number of species varies. Recent work has shown that especially at larger spatial scales, the greater the number of native species, the larger the number of exotic species that invade and become established. It is likely that the same factors that promote native species diversity are also favorable for invasive species diversity. At smaller spatial scales, Elton's prediction is more likely to hold true, particularly in experimental tests, where plots with greater numbers of species tend to resist invasion better than those that are poorer in native species.

Many more species are introduced to novel environments than become established, and many more species establish populations in nonnative environments than become invasive. To what extent is competition from native species responsible for preventing the initial establishment of aliens, or for limiting noninvasive aliens and preventing them from spreading or becoming dominant? How important is competition relative to abiotic factors, enemies (herbivores, predators, or pathogens), or lack of mutualists, in these "biological filters"? Because there are limited data addressing these questions, we can only speculate that competition is likely to play a role. The available evidence suggests that competition from native species on invasive species is common and important but may not prevent them from becoming established or successful as invaders. However, we do not know about the effects of competition on immigrants that arrived but failed to become established, and we have very limited information on those that are present in low numbers in the novel environment but are not invasive.

A related issue is whether invasion is facilitated by the existence of empty ecological niches in a community. For example, if a native plant community does not include a species of a particular functional type, it might be particularly vulnerable to invasion by exotic species with the characteristics of the functional type that is "missing" in that community. That is because species with similar ecological roles might be expected to exert greater competitive effects on one another. An exotic species functioning in a role that is very different than that of other species in a community would be likely to experience less competition from native species. Some examples of this are the invasion of understory shrubs, such as *Lonicera maackii* and *Berberis japonica* in forest communities that previously did not have understory shrubs in the eastern and southern United States. Another example is the invasion of *Morella (Myrica) faya,* a nonnative nitrogen-fixing small tree or shrub, into plant communities in Hawaii where no native nitrogen-fixing tree species exist. Such invasions can cause dramatic changes in community structure and function.

Another factor that is generally associated with "opening the door" to invasive plants is disturbance of the native community. The problem with this explanation is that the term "disturbance" encompasses a wide range of different phenomena, with different effects on both native and invasive species. However, if the nature of the disturbance is that it reduces the competitive effects exerted by the native species, thereby releasing resources that are made available to the invader, then it is reasonable to expect that this would afford an opportunity for exotic species that were previously kept out by competition from the natives to gain a foothold in that community. As an example, in forests in the northeastern United States (and undoubtedly elsewhere) in which gaps were created by either natural (e.g., hurricanes) or human disturbance (e.g., logging) in the recent past after invasive species were present in the area, exotic species have become established and may be important components of the regenerating forest community. In similar local forests without such disturbances, most invasive species remain largely absent. In forests that experienced such disturbances long ago, before invasives were common in the region, forest regeneration occurred without invasives becoming established, and so invasives are likewise absent or present in only small numbers.

INVASIVE PLANTS AND COMPETITIVE SUPERIORITY

Various hypotheses have been proposed for the success of invasive species that invoke competition. It has been suggested that plants that succeed as invasives are inherently superior competitors (that is, inferior competitors do not become invasive), that they are larger in size, that invasive plants grow more rapidly or respond to resource fluctuations more vigorously, or that they evolve greater competitive ability in the novel environment due to reallocation of resources available after release from predation and herbivory by the enemies they have left behind. We do not have definitive general tests of the relative importance of these hypotheses, and it is not always clear whether the relevant comparisons are to native plants in the invaded environment (congeners or ecologically functional equivalents), to noninvasive alien species in the novel environment, or to conspecifics in the home environment. Quantitative syntheses of the limited available information suggest that some invasive plants are larger or grow more rapidly than congeneric natives, or than the same plants in their home environment, and

FIGURE 1 (A) *Celastrus orbiculatus*, an invasive vine, overgrowing and covering a small native tree, *Acer rubrum*, in New York State. (B) The small *Acer rubrum* tree (approximately 7 m tall), with the vine completely removed. (Photographs courtesy of the author.)

appear to allocate less to reproduction than do congeneric natives. There is some evidence that invasive plants are more responsive (phenotypically plastic) to enhanced or fluctuating resources, particularly light, than are noninvasive species. Invasive species appear to have somewhat higher competitive effects on natives than vice versa, but results are highly heterogeneous among sites and species.

Not surprisingly, invasive species also appear to have higher rates of population increase than natives, with plant growth and fecundity being particularly important for invasive population growth rates. Population growth rates of invasive plants vary depending on whether the population is at the invasion front (likely to be higher) or in a well-established part of the new distribution (likely to be lower), and such rates may or may not be higher in the invaded than in the native range. Considerable effort has been devoted to attempting to identify traits associated with invasion success. Some of the suggested traits are associated, of course, with rapid growth and superior competitive ability. The results of these efforts are complex and controversial. It has often been pointed out that invasion is not merely a result of the properties and characteristics of invaders but is the product of the interaction between invading species and their novel environments. This is certainly the case for rapid growth rates, phenotypic plasticity, and competitive ability.

Invasive plants clearly compete with other invasive plants in all of the same ways that they compete with native species, but curiously, competition between invasive alien plants has rarely been studied. Consequently, we have little idea of how this might influence the success of invasions.

Do invasive plants ever have beneficial, facilitative effects on native plant species? There are many different ways that plants can have beneficial effects on one another—to some extent the inverse of a competitive effect—although these tend to be far less well studied than competitive effects. There is some evidence that invasive species in some circumstances do benefit native plants, although the generality and importance of facilitative interactions between native and invasive plants are not known.

IMPACTS OF INVADERS AS COMPETITORS WITH NATIVE PLANTS

Once established, invasive species have competitive effects on the plants that are already present in the environments that they invade. They may compete with natives for various resources, with sometimes devastating negative impacts on the native species. Species of various growth forms may compete for light by overtopping natives, including, for example, invasive vines such as *Pueraria lobata* (kudzu) in the southern United States and *Celastrus orbiculatus* (Asian or oriental bittersweet; Fig. 1) in the northeastern United States, trees such as *Acer platanoides* (Norway maple) in the midwestern and eastern United States, and grasses such as *Imperata cylindrica* (cogon grass, which also dominates space and has other effects) in Florida. Other species compete with native species for water, such as *Tamarisk ramosissima* (tamarisk or saltcedar trees) in the southwestern United States. *Acacia mearnsii* and *Pinus radiata*, among other invasive plants, similarly reduce available water in South Africa. Invasive plants may compete with native plants for nutrients, pollinators, and dispersers, or space for seeds to germinate and seedlings to become established. Some invasive species are hypothesized to exert negative impacts through allelopathic interactions (Fig. 2), although unambiguous

FIGURE 2 *Centaurea stoebe* (violet-pink flowers), a plant that has been hypothesized to have allelopathic effects on its neighbors. Interpretation of evidence for allelopathy in this species has been highly controversial. (Photograph courtesy of the author.)

demonstration of these effects in natural populations in the field has thus far been limited.

The consequences of competitive effects of invasive plants on native plant species at local scales may be a decline in the abundance of natives and in reduced diversity of natives, although it may be difficult to determine whether the decline in the natives is being caused by the invasive plants, or whether other factors (e.g., habitat degradation, nitrogen deposition, grazing) are simultaneously leading to the decline of the natives and offering opportunities for the proliferation of invasive plants. At somewhat larger spatial scales, it is unclear whether invasive plants cause, or are even associated with, biodiversity loss for native plants, and in some cases, it has been shown that invasions on islands do not lead to loss of native species but rather to an increase in total species numbers. This would suggest that those communities are not saturated with species, but are able to absorb additional plants without resulting in local extinction of natives. At the largest spatial scales, it seems highly likely that the proliferation of invasive plants, particularly if they competitively exclude natives in some habitats, may result in the decline of global biodiversity. However, we do not yet have a clear picture of the extent to which this is true, nor do we know under what conditions it is true. At all spatial scales, better evidence is needed, but the forensic picture may be difficult to reconstruct once invaders have become well established and natives have declined.

SEE ALSO THE FOLLOWING ARTICLES

Allelopathy / Competition, Animal / Disturbance / Invasibility, of Communities and Ecosystems / Invasiveness / Life History Strategies / Succession / Transformers

FURTHER READING

Bruno, J.F., J.D. Fridley, K.D. Bromberg, and M.D. Bertness. 2005. Insights into interactions from studies of species invasions (13–40). In D.F. Sax, J.J. Stachowicz, and S.D. Gaines, eds. *Species Invasions: Insights into Ecology, Evolution and Biogeography.* Sunderland, MA: Sinauer Associates.

Freckleton, R.P., A.R. Watkinson, and M. Rees. 2009. Measuring the importance of competition in plant communities. *Journal of Ecology* 97: 379–384.

Hawkes, C.V. 2007. Are invaders moving targets? The generality and persistence of advantages in size, reproduction and enemy release in invasive plant species with time since introduction. *The American Naturalist* 170: 832–843.

Levine, J.M., P.B. Adler, and S.G. Yelenik. 2004. A meta-analysis of biotic resistance to exotic plant invasions. *Ecology Letters* 7: 975–989.

Levine, J.M., M. Vila, C.M. D'Antonio, J.S. Dukes, K. Garigulis, and S. Lavorel. 2003. Mechanisms underlying the impacts of exotic plant invasions. *Proceedings of the Royal Society of London B* 270: 775–781.

Ramula, S., T.M. Knight, J.H. Burns, and Y.M. Buckley. 2008. General guidelines for invasive plant management based on compara-

tive demography of invasive and native plant populations. *Journal of Applied Ecology* 45: 1124–1133.

Rejmánek, M. 2002. Intraspecific aggregation and species coexistence. *Trends in Ecology and Evolution* 17: 209–210.

Vila, M., and J. Weiner. 2004. Are invasive plant species better competitors than native plant species? Evidence from pair-wise experiments. *Oikos* 105: 229–238.

CRABS

EDWIN GROSHOLZ

University of California, Davis

Crabs are among the most familiar members of the crustacean order Decapoda. Many crabs of economic and ecological importance have been moved around the planet by humans and have quickly established new populations. Because of their comparatively large size and their often great impact on their surroundings, crabs are among the best known and most thoroughly described group of introduced marine invertebrates. Unsurprisingly, we know quite a lot about a handful of well-studied introductions and much less about the majority of introductions. Thus, our understanding about the means of dispersal, ecological and economic consequences, and other features of crab introductions is strongly influenced by a small number of well-known introductions. Nonetheless, our knowledge of introduced crabs does permit some analysis of the causes and consequences of introductions of this important group.

PHYLOGENETIC DISTRIBUTION OF INVASIVE CRABS

Two infraorders of Decapoda, Anomura and Brachyura, comprise all species of crabs. Both groups contain species that have contributed to the growing list of introduced species; however, introduced crabs are found in a comparatively small number of families. Among the 14 families of anomuran crabs, only two contain established introductions: Lithodidae and Porcellanidae. Of the 74 families of brachyuran crabs, only 18 contain species that have established populations after introduction. Several families have multiple introduced species, including Leucosiidae, Majidae, Pilumnidae, Panopeidae, Porcellanidae, Portunidae, Varunidae, and Xanthidae. The swimming crab family Portunidae contains the most introduced species, with 13 of the 47 introduced crabs in Table 1, although these are represented by just six genera.

TABLE 1
Crab Invasions of the United States and the World

Genus species (Family) (Native Range)	Introduced Region	Genus species (Family) (Native Range)	Introduced Region
Acantholobus pacificus (Panopeidae) (Indo-Pacific)	Hawaii	*Ixa monodi* (Leucosiidae) (Indo-Pacific)	Mediterranean
Atergatis roseus (Xanthidae) (Indo-Pacific)	Mediterranean	*Leucosia signata* (Leucosiidae) (Indo-Pacific)	Mediterranean
Calappa pelii (Calappidae) (Tropical Atlantic)	Mediterranean	*Libnia dubia* (Majidae) (eastern United States)	Mediterranean
Callinectes bocourti (Portunidae) (Caribbean)	Southeastern United States	*Metopograpsus oceanicus* (Grapsidae) (Indo-Pacific)	Hawaii
Callinectes exasperatus (Portunidae) (Caribbean)	Southeastern United States	*Micippa thalia* (Majidae) (Indo-Pacific)	Mediterranean
Callinectes sapidus (Portunidae) (eastern United States)	Western Europe, Mediterranean	*Myra subgranulata* (Leucosiidae) (Indo-Pacific)	Mediterranean
Cancer novaezelandiae (Cancridae) (New Zealand)	Australia	*Nanosesarma minutum* (Sesarmidae) (Indo-Pacific)	Hawaii
Carcinus aestuarii (Portunidae) (Mediterranean)	Japan, South Africa	*Pachygrapsus fakaravensis* (Grapsidae) (Indo-Pacific)	Hawaii
Carcinus maenas (Portunidae) (Atlantic Europe)	Eastern North America, Japan, western North America, , Australia, South Africa, South America	*Panopeus lacustris* (Panopeidae) (southeastern United States, Caribbean)	Hawaii
Cardisoma guanhumi (Gecarcinidae) (South and Central America)	Southeastern United States	*Paralithodes camtschaticus* (Lithodidae) (northwestern North America)	Barents Sea, Scandinavia
Charybdis hellerii (Portunidae) (Indo-Pacific)	Florida, South America, Mediterranean	*Percnon gibbesi* (Plagusiidae) (southeastern United States, Caribbean)	Mediterranean
Charybdis japonica (Portunidae) (Indo-Pacific)	New Zealand	*Petrolisthes armatus* (Porcellanidae) (South America)	Southeastern United States
Charybdis longicollis (Portunidae) (Indo-Pacific)	Mediterranean	*Petrolisthes elongatus* (Porcellanidae) (New Zealand)	Australia
Chionoecetes opilio (Oregoniidae) (western North America, Greenland)	Scandinavia	*Pilumnus oahuensis* (Pilumnidae) (Indo-Pacific?)	Hawaii
Dorippe quadridens (Dorippidae) (Indo-Pacific)	Mediterranean	*Pilumnopeus vauquelini* (Pilumnidae) (Indo-Pacific)	Mediterranean
Dyspanopeus sayi (Panopeidae) (eastern United States)	Mediterranean	*Platychirograpsus spectabilis* (Glyptograpsidae) (Mexico, Caribbean)	Florida
Eriocheir sinensis (Varunidae) (China, Yellow Sea)	Western United States, western Europe	*Portunus pelagicus* (Portunidae) (Indo-Pacific)	Mediterranean
Eucrate crenata (Euryplacidae) (Indo-Pacific)	Mediterranean	*Pyromaia tuberculata* (Inachoididae) (western North America)	Japan, Korea, New Zealand, Australia, Brazil, Argentina
Glabropilumnus seminudus (Pilumnidae) (Austalia)	Hawaii	*Rhithropanopeus harrisii* (Panopeidae) (eastern United States)	Western United States, Central America, Mediterranean, Black Sea, Europe, Japan
Hemigrapsus takanoi (Varunidae) (Asia)	Europe		
Hemigrapsus sanguineus (Varunidae) (Asia)	Northeastern United States, Europe	*Schizophrys aspera* (Majidae) (Indo-Pacific)	Hawaii
Herbstia nitida (Majidae) (tropical Atlantic)	Mediterranean	*Scylla serrata* (Portunidae) (Indo-Pacific)	Hawaii
Heteropanope laevis (Panopeidae) (Indo-Pacific)	Mediterranean	*Thalamita gloriensis* (Portunidae) (Indo-Pacific)	Mediterranean
Hyas araneus (Majidae) (North Atlantic)	Antarctica	*Thalamita indistincta* (Portunidae) (Indo-Pacific)	Mediterranean
Hyastenus hilgendorfi (Majidae) (Indo-Pacific)	Mediterranean	*Thalamita poissonii* (Portunidae) (Indo-Pacific)	Mediterranean
Hyastenus spinosus (Majidae) (Indo-Pacific)	Hawaii		

NOTE: The list is incomplete for areas outside the United States.

DISPERSAL VECTORS

Ship traffic and trade is only one of several vectors that are likely responsible for the introduction of crabs to new regions. Larval crab stages (zooea, megalopae) have been found in ship's ballast but are by no means common, and crabs are rarely on hulls. Adult crabs have been found in bottom sediments in ballast tanks and in sea chests and other areas not routinely affected by ballast water management. *Charybdis hellerii* is a likely example of an introduction due to ship traffic. Other vectors such as the live bait and live seafood trades have likely contributed to many significant crab invasions, such as those of the European green crab (*Carcinus maenas*) and the Chinese mitten crab (*Eriocheir sinensis*). A few invasions, such as those of the red king crab (*Paralithodes camtschaticus*) and snow crab (*Chionoecetes opilio*) in the Barents Sea, were probably intentional introductions to create new fisheries. Once introduced to a new continental margin, crabs may frequently expand their range through dispersal of planktonic larval stages by advection of ocean currents. Several introduced crab species have been rapidly dispersed by ocean currents along a coastline following an initial human-mediated introduction to a new continental region (see below).

IMPACTS ON NATIVE SPECIES

Introduced crabs constitute some of the best examples of introduced marine and estuarine species that have had significant impacts on coastal habitats and economies. Among the best studied is the European green crab *Carcinus maenas* (Fig. 1), for which ecological and economic impacts have been demonstrated on several coasts. Green crabs contributed substantially to the demise of the commercial soft-shell clam fishery in the northeastern United

FIGURE 2 An adult Chinese mitten crab (*Eriocheir sinensis*) from the San Francisco Bay-Delta region of central California. (Photograph courtesy of Lee Mecum.)

States during the 1940s and 1950s. During the early stages of their introduction in the western United States during the 1990s, green crabs caused significant but smaller-scale losses in clam fisheries. There have also been long-term changes in the benthic communities in bays and estuaries in central California because of green crab predation on small native crabs and clams. Green crabs have also had long-term impacts on shell shape and morphology of herbivorous snails in New England.

The Chinese mitten crab *Eriocheir sinensis* (Fig. 2) is another species for which impacts have been quantified fairly extensively. In Europe during the early twentieth century, population explosions of the catadromous mitten crab resulted in extensive efforts to mitigate their impacts in rivers, canals, and associated municipal facilities. In the San Francisco Bay following population outbreaks in the late 1990s, substantial investments by state and federal agencies were required to prevent clogging of water pumps and associated fish salvage facilities. Major losses were experienced by commercial and sport fishing interests as well in central California.

The grapsid crab *Hemigrapsus sanguineus* has been shown to significantly affect both native and other introduced species in its invaded range in the northeastern United States. It has restricted the distribution and abundance of the European green crabs in areas where they overlap. This species also significantly affects populations of native blue mussels.

The red king crab *Paralithodes camtschaticus* has also been the focus of recent study regarding its impacts on benthic communities in its introduced range in the Barents Sea off the Norwegian coast. These crabs can be seasonally abundant in shallow water during spring breeding periods, and they are very opportunistic consumers,

FIGURE 1 An adult European green crab (*Carcinus maenas*) from Bodega Harbor in central California. (Photograph courtesy of the author.)

so there is the potential for significant impacts on benthic communities. Feeding rate studies suggest that introduced king crabs could significantly affect native scallop populations in this region, although impacts in the field remain to be assessed.

RANGE EXPANSION AND INTRODUCTION

Among factors that color our interpretation of what is and what is not native is the tendency of some species to expand their native range significantly into areas that are considerably outside of their typical range. This is probably especially true for crabs, which in many cases have long planktonic development periods, during which developing larvae may be carried many hundreds of miles by ocean currents. Among the species that best exemplify this pattern are the swimming crabs in the genus *Callinectes*, which include the commercial blue crab *Callinectes sapidus*. On the eastern coast of the United States, in addition to the native *C. sapidus*, there are other native species including *C. similus* and *C. ornatus*. However, there are several species that are typically listed as subtropical with distributions in the Caribbean or further south that are routinely captured in bays and estuaries of the southeastern United States, where they establish temporary populations. These include *C. bocourti*, *C. larvatus*, and *C. exasperatus*, which are listed in records from South Carolina, Georgia, Florida, and the Gulf.

In bays and estuaries of the western United States, another portunid crab, *Callinectes arcuatus*, shows a similar pattern. It is native to Mexico and regions further south of the U.S. border, but in some years it can be found abundantly in southern California estuaries and then be entirely absent in other years. The pattern seems to be consistent with a larval dispersal event that results in a single cohort of *C. arcuatus* becoming temporarily established in a southern California estuary for several years and then rapidly disappearing and remaining absent for several years afterward.

Once introduced to a new region, many crabs can rapidly expand their range along the coastline. These dispersal events include some of the fastest range expansions recorded for any introduced species. Among the most rapid range expansions is that of the European green crab, which spread along the western coast of the United States at rates of 200 km per year. Also, the crab *Hemigrapsus sanguineus* experienced rapid range expansion on the eastern coast of the United States. In Europe, *Eriocheir sinensis* also spread rapidly along the coast and throughout many river systems during the early twentieth century.

Rapid spread is by no means the rule or even the norm for the same species. For instance, *Carcinus maenas* has spread very little in southern Australia and Tasmania, and the rates of spread in eastern North America were very minimal for long periods. Also, several crabs introduced into the San Francisco Bay, including *Rhithropanopeus harrisii* and *Eriocheir sinensis*, have not spread at all along the California coast since their initial introduction.

ABUNDANCES IN DIFFERENT REGIONS

Of course the list of introductions is strongly influenced by the level of taxonomic effort in addition to characteristics of propagule pressure, the recipient community, and so forth. Given the strong reporting biases in Table 1, it is problematic to attach much biological significance to the patterns that emerge. Nonetheless, particular recipient regions in warm temperate and subtropical areas such as the Mediterranean region and Hawai'i have the greatest number of established introduced crab species. Among the regions that could potentially provide the source for introduced crabs, the most common source region is overwhelmingly the Indo-Pacific region, which in particular is the source of a few well-described subtropical faunas. The vast majority of crabs are intertidal or shallow-water (bay, estuary) species. By comparison, there have been far fewer documented introductions of coldwater or deepwater crabs. Factors such as propagule pressure, habitat matching between source and recipient communities, and other general processes known to influence species introductions have clearly influenced patterns of crab introductions as well.

SEE ALSO THE FOLLOWING ARTICLES

Ballast / Crustaceans / Dispersal Ability, Animal / Invertebrates, Marine / Mediterranean Sea: Invasions

FURTHER READING

CIESM. 2002. *Atlas of Exotic Species in the Mediterranean.* Monaco: CIESM Publishers.
Gollasch, S., B. Galil, and A. N. Cohen. 2006. *Bridging Divides: Maritime Canals as Invasion Corridors.* Berlin: Springer.
Grosholz, E. D. 2002. Ecological and evolutionary consequences of coastal invasions. *Trends in Ecology and Evolution* 17: 22–27.
Leppäkoski, E., S. Gollasch, and S. Olenin, eds. 2002. *Invasive Aquatic Species of Europe: Distribution, Impacts, and Management.* Dordrecht: Kluwer.
National Management Plan for the Genus *Eriocheir* (Mitten Crabs). 2003. Federal Aquatic Nuisance Species Task Force.
Rilov, G., and J. A. Crooks, eds. 2009. *Biological Invasions in Marine Ecosystems: Ecological, Management, and Geographic Perspectives.* Berlin: Springer.
Yamada, S. B. 2001. *Global Invader: The European Green Crab.* Corvallis: Oregon Sea Grant.

CRAYFISH

FRANCESCA GHERARDI
University of Florence, Italy

There are over 600 species of crayfish in the world, taxonomically organized into two superfamilies, Astacoidea in the northern hemisphere with the two families of Cambaridae and Astacidae, and Parastacoidea in the southern hemisphere with the single family of Parastacidae. Although quite large, crayfish native diversity is unequally distributed across the world. Over 380 species occur in the nearctics, and 150 are found in the Australasian ecoregion, but only about 60, 30, and 10 occur in the neotropical, palaearctic, and Afrotropical ecoregions, respectively. Crayfish are naturally absent from the Antarctic continent, continental Africa, and much of Asia and South America, while Europe has only five native species. Superimposed on this natural distribution, however, is one orchestrated by humans, the product of a massive translocation of species between and within continents.

A HISTORY OF CRAYFISH INTRODUCTION

Crayfish introduction has a long history: even the distribution of species of current conservation concern seems to owe much to human intervention. For instance, as suggested by historical records and genetic studies, the European *Austropotamobius pallipes*, now protected under the EU Habitats Directive, was introduced into Ireland from France by monastic orders in the twelfth century. During the last few decades, however, the human-aided movement of nonindigenous crayfish species (hereafter referred to as NICS) has increased as a result of the exponential growth in the volume and complexity of international trade. Ten and 21 crayfish species have been introduced outside their native ranges between and within continents, respectively (Table 1), but most introductions are of six species now widely diffused in the world (*Astacus leptodactylus*, *Cherax destructor*, *Orconectes limosus*, *Orconectes rusticus*, *Pacifastacus leniusculus*, and *Procambarus clarkii*); one of them, *P. clarkii*, accounts for over 40 percent of all introduction events recorded (56).

In some cases, introductions have been accidental (e.g., through ballast, canals, escapes from holding facilities), but the majority of them has been deliberate (for aquaculture and both legal and illegal stocking, for live food trade, as aquarium pets, as live bait, for snail and weed control, and as supplies for science classes). Some were driven by narrow objectives: *P. clarkii*, for instance, was imported to Japan in 1927 as a food for the bullfrog, *Rana catesbeiana*. Sometimes, on the contrary, introductions were intended to ameliorate human conditions: in Africa many releases into the wild of *P. clarkii* from the 1960s onward were aimed at controlling freshwater snails that carry human schistosomiasis. Most often, however, crayfish introductions were motivated by our desire to eat them, which in turn generates economic interests and stimulates further human-assisted dispersal; *C. destructor*, for instance, after its first introduction to Western Australian farm dams for aquaculture in 1932, rapidly spread, now threatening the over ten endemic crayfish species of this state. A handful of other species are highly valued as gourmet food, and, in locations such as Scandinavia and Louisiana, feasting on them has become a cultural icon. As a consequence, some NICS are commercially harvested from wild stocks or have become aquaculture commodities in countries such as Australia, China, and the United States. The aquarium and pond trade has been another powerful vector of NICS; it represents a great threat to regions, such as southern Europe, where the climate is today amenable to species from tropical regions—and it will be more so in the near future with the projected global warming. Finally, the use of live bait has also aided the spread of NICS, particularly in North America in the case of *O. rusticus*.

Indeed, some introductions may have provided socioeconomic benefits. However, once introduced for stocking and aquaculture and kept in outdoor ponds, crayfish almost inevitably escape and easily establish self-sustaining populations in the colonized habitats. Because of their ability to integrate into the food web at many levels and to travel long distances even overland, a large majority of the naturalized populations have the potential to become noxious and to spread widely from their points of introduction. Additions of crayfish species may significantly alter the structure of freshwater food webs, which ultimately affect ecosystem services and, as a consequence, human well-being.

INVASIVE CRAYFISH OF THE WORLD

Crayfish species that today cause major concern in the recipient areas include the Ponto-Caspian *A. leptodactylus* in some European countries, the Australian *C. destructor*

TABLE 1

Crayfish Species Moved between or within Continents outside Their Native Ranges

Family	Species	Common Name	From	To
Astacidae	*Astacus astacus* (Linnaeus)	Noble crayfish	Europe	Africa, Europe
	Astacus leptodactylus Eschscholtz	Narrow-clawed crayfish	Asia, Europe	Asia, Europe
	Austropotamobius pallipes (Lereboullet)	White-clawed crayfish	Europe	Europe
	Pacifastacus gambelii (Girard)	Pilose crayfish	N. America	N. America
	Pacifastacus leniusculus leniusculus (Dana)	Signal crayfish	N. America	Asia, Europe, N America
	Pacifastacus leniusculus trowbridgii (Stimpson)	Columbia River signal crayfish	N. America	N. America
Cambaridae	*Cambarus longirostris* Faxon	Atlantic slope crayfish	N. America	N. America
	Cambarus rusticiformis Rhoades	Depression crayfish	N. America	N. America
	Orconectes causeyi Jester	Western plains crayfish	N. America	N. America
	Orconectes immunis (Hagen)	Calico crayfish	N. America	Europe, N. America
	Orconectes juvenilis (Hagen)	Kentucky river crayfish	N. America	Europe
	Orconectes limosus (Rafinesque)	Spinycheek crayfish	N. America	Africa, Europe, N. America
	Orconectes neglectus neglectus Faxon	Ringed crayfish	N. America	N. America
	Orconectes palmeri creolanus (Creaser)	Creole painted crayfish	N. America	N. America
	Orconectes rusticus (Girard)	Rusty crayfish	N. America	N. America
	Orconectes sanbornii Faxon	Sanborn's crayfish	N. America	N. America
	Orconectes virilis Hagen	Virile crayfish	N. America	Europe, N. America
	Procambarus sp.	Marbled crayfish	N. America	Africa, Europe
	Procambarus acutus (Girard)	White river crawfish	N. America	N. America
	Procambarus clarkii (Girard)	Red swamp crawfish	N. America	Africa, Asia, Europe, C., N. and S. America
	Procambarus zonangulus Hobbs and Hobbs	Southern white river crawfish	N. America	N. America
Parastacidae	*Cherax destructor* Clark	Yabby	Oceania	Europe, Oceania
	Cherax quadricarinatus (von Martens)	Redclaw	Oceania	Africa, Asia, C., N. and S. America
	Cherax tenuimanus (Smith)	Marron	Oceania	Oceania

NOTE: After Hobbs, et al., 1998; Souty-Grosset, et al., 2006; Taylor, et al., 2007; and Gherardi, 2010.

in Africa and Western Australia, and four North American species: *O. limosus* in Europe, *O. rusticus* in the midwestern United States and Canada, *P. leniusculus* in California, Europe, and Japan, and *P. clarkii* in Africa, California, Europe, and Japan (Figs. 1 and 2). Other species, such as the Australian *Cherax quadricarinatus* in South America and the parthenogenetic North American "marbled crayfish" (*Procambarus* sp.) in Europe and Madagascar, are expected to generate problems in the near future.

Illustrative in this respect is the story of the human-assisted movement of the red swamp crayfish, *P. clarkii*, one of the 100 worst invasive species in Europe. This species occurs naturally in northeastern Mexico and in the south-central United States, extending westward to Texas, eastward to Alabama, and northward

to Tennessee and Illinois. It has been extensively cultivated since the 1950s in the southern United States, reaching a maximum production of 3,000 kg/ha. Because of its commercial value, it has been introduced into several U.S. states; its current range now includes the east and west coasts and extends northward into the states of Idaho and Ohio. Outside the continental United States, *P. clarkii* has been successfully introduced to Hawaii, western Mexico, Costa Rica, the Dominican Republic, Belize, Brazil, Ecuador, Venezuela, Japan, mainland China, Taiwan, the Philippines, Egypt, Uganda, Kenya, Zambia, South Africa, Israel, and several European countries. Strong economic and social reasons apparently led to its first introduction to a European country, Spain, in 1973. This introduction

FIGURE 1 Invasive crayfish species of the world: (A) *Astacus leptodactylus*, (B) *Cherax destructor*, (C) *Cherax quadricarinatus*, and (D) *Orconectes limosus*. (Photographs courtesy of Chris Lukhaup.)

was even solicited by local institutions striving to satisfy, with a "plague-resistant" species, the large demand for crayfish by the European market. The native crayfish production, in fact, had been drastically reduced since the mid-1800s by the spread of the oomycete *Aphanomyces astaci*, the causative agent of a lethal disease for indigenous species, the "crayfish plague." All the legal procedures were followed and respected; there was even the consensus of American experts who had previously visited Spain to identify zones appropriate for crayfish introduction. Because the native *A. pallipes* had never been present in nor was suited to the areas of introduction, and because its potential to transfer *A. astaci* was unknown, there was confidence that *P. clarkii* would be innocuous to the native stocks and would provide great economic benefits to the local population. The habit of selling it alive as a food item or as an aquarium pet accelerated the successful invasion of this species into natural waters. However, the exact trajectory of its rapid spread across Europe is still unknown; recent genetic studies even suggest that multiple introductions have occurred since the 1970s from source areas other

than Spain or Louisiana, including the Far East and Kenya.

THE SUCCESS OF CRAYFISH INVADERS

Once added to a system, some NICS have the potential to pose considerable environmental stress, and in most instances, they may induce irreparable shifts in species diversity. In areas without any native ecological equivalent, the changes caused by NICS introductions usually affect all levels of ecological organization. The modes of resource acquisition by crayfish and their capacity to develop new trophic relationships, coupled with their action as bioturbators, may lead to dramatic direct and indirect ecosystem effects.

When NICS replace native crayfish, their ecological effects should not be novel to the colonized community, and so the resulting impact is expected to be weak. But their overall effect can be strong if, once introduced, they are capable of building high densities or achieving large size. Several NICS often reach much higher densities than native crayfish: more than 70 m^{-2} for *O. limosus* in Poland, over 20 m^{-2} for *O. rusticus* in North America,

FIGURE 2 Invasive crayfish species of the world: (A) *Orconectes rusticus*, (B) *Pacifastacus leniusculus*, (C) *Procambarus clarkii*, and (D) the "marbled crayfish" (*Procambarus* sp.). (Photographs courtesy of Chris Lukhaup.)

and 30 m^{-2} for *P. leniusculus* in the United Kingdom. On the contrary, reported densities of the native species range from 1 m^{-2} for *Pacifastacus fortis* in California to 3 m^{-2} for *Paranephrops planifrons* in New Zealand, 4 m^{-2} for *Cambaroides japonicus* in Japan, and 14 m^{-2} for *Astacus astacus* in Sweden.

Several biological traits contribute to the achievement by crayfish of high densities and large size. Compared to native crayfish, some NICS are characterized by higher fecundity (more than 500 pleopodal eggs in *P. clarkii*), protracted spawning periods, faster growth rates (50 g in three to five months in *P. clarkii*), and maturity reached at relatively small size (10 g in *P. clarkii*). They are also extremely plastic in their life cycle and are better at coping with changes induced by human activities that cause pollution and habitat destruction. For instance, *P. clarkii* is a good colonizer of disturbed aquatic habitats and can survive in anoxic and dry conditions in burrows; it tolerates elevated turbidity and a wide range of water temperatures and salinities. A higher survival rate is also expected when a species is introduced without a full complement of specific parasites, pathogens, and enemies. And large sizes, in turn, make crayfish both resistant to gape-size limited

predators (such as many fishes) and agonistically superior in resource fights. As a consequence, NICS exert a greater direct (through consumption) or indirect (through competition) effect on the other biota, particularly on other crayfish species, benthic fishes, mollusks, and macrophytes. Invasive crayfish seem also to be affected by the so-called aggression syndrome that makes them highly abundant and active at the same time, despite the elevated intraspecific aggression exhibited. Obviously, large size usually translates into a higher energy and nutrient demand, but NICS may also be more efficient energy converters and may display higher metabolic rates when compared with similarly sized native species.

ECOLOGICAL IMPACTS

The negative impact that invasive crayfish inflict on the environment occurs at multiple levels of ecological organization. NICS may outcompete or prey upon native species, eventually leading to their extirpation and, in at least one case, extinction. For instance, the ability of *P. leniusculus* to outcompete fishes by expelling individuals from their shelters, making them more vulnerable to predators, has contributed to the drastic reduction in abundance of *Cottus gobio* and *Noemacheilus barbatulus* in some rivers of England. Similarly, its heavy predatory pressure upon the eggs of the newt *Taricha torosa*, despite their antipredator chemical defense (tetrodotoxin), allowed *P. clarkii* to extirpate this species from some streams in southern California. Competition and predation, coupled with reproductive interference, enhance the effects of habitat loss, overexploitation, and pollution in inducing a dramatic decline of crayfish diversity. Of the 67 threatened crayfish species in North America, for instance, over 5 percent are subject to interference competition by NICS. Along with urbanization and overexploitation, *P. leniusculus* has contributed to the global extinction of the crayfish *Pacifastacus nigrescens*, once common in the creeks of the San Francisco Bay area; in northeastern California, it is now displacing the Shasta crayfish, *P. fortis*. Similarly, the European native species *A. astacus*, *A. pallipes*, and *A. torrentium* are threatened by *A. astaci*, introduced to Europe via the North American NICS. To make things worse, the parasite does not require its host in order to spread; the spores can be transported on damp surfaces (e.g., on fishing equipment), as is thought to have triggered the crayfish plague outbreak in central Ireland in 1986. Hybridization with invaders is an additional threat for native crayfish species. In Wisconsin, hybrids between the invader *O. rusticus* and the native *O. propinquus* were found to mate with pure *O. rusticus*, which leads to a massive genetic introgression of nuclear DNA from the native to the invasive species and thus to the gradual elimination of *O. propinquus* genes from the population.

At the community level, NICS exert direct and indirect effects on invaded ecosystems. Their intense grazing on aquatic macrophytes, coupled with nonconsumptive plant clipping and uprooting, induces a significant decline in plant abundance. Macrophyte destruction is generally followed by a switch from a clear to a turbid state, dominated by surface microalgae growth with consequent reduction in light penetration and decrease in primary production of benthonic plants.

The biomass and species richness of macroinvertebrates are altered by NICS as the result of consumption, increased drift through prey escape, incidental dislodging by their foraging, and possible inhibition of invertebrate colonization. Molluscs are the taxon most affected: some gastropod species, particularly thin-shelled snails, have sometimes been extirpated. Crayfish predation is weak only on species that move quickly enough to escape tactile-feeding crayfish (e.g., amphipods) and that live in cases (e.g., Trichoptera) or in the sediment (e.g., some Diptera). Through consumption of macrophytes and detritus, crayfish may also lead, indirectly, to the decline of macrophyte-associated taxa, particularly collector-gatherers, while their predation upon other zoobenthic predators such as Odonata larvae causes an increased abundance of their prey. Finally, NICS can be prey items for a large number of fish, bird, and mammal species, such as eels, storks, herons, egrets, and otters, thus representing a new resource for higher trophic levels in the areas of introduction.

At the ecosystem level, NICS may alter pathways of energy flow by augmenting connectance by feeding at several trophic levels and increasing the availability of autochthonous carbon as a food source for higher trophic levels. The intense NICS burrowing activity and locomotion often result in bioturbation: water quality is impoverished, light penetration and plant productivity are reduced, and the benthic community is affected by changes in the riverbed substrate.

EFFECTS ON HUMAN WELL-BEING

Introduction of NICS has sometimes been assumed to have benefited human societies by, for example, restoring cultural traditions such as crayfishing (in Sweden), producing economic benefits in poorly developed areas (in southern Spain), inducing the development of extensive or semi-intensive cultivation systems (in People's Republic of China), and increasing the volume of international trade (in Spain).

Several examples, however, show that often the introduction of commercially valuable crayfish has also led to negative results in the marketplace. In Scandinavia, Germany, Spain, and Turkey, the plague led to a loss of over 90 percent in the production of *A. astacus* and *A. leptodactylus*, with considerable economic damage. For instance, when the plague spread to Turkey in the 1980s, the annual catch of *A. leptodactylus* plunged from 7,000 to 2,000 tons, nearly eliminating exports from Turkey to Western Europe. In Africa, very few of the several projects that led to crayfish importations have been successful: in Lake Naivasha, Kenya, less than 40 metric tons of *P. clarkii* are now fished annually for exclusive local consumption (mainly tourism), while crayfish spoil valuable fish species and damage fish nets.

NICS may affect other human activities: *P. clarkii* is a recognized pest in rice cultures in various parts of the world, causing a decrease in profits that amounts to over 6 percent in Portugal, for example. Burrowing by several species can be a problem in agricultural and recreational areas, such as lawns, golf courses, levees, dams, dykes, and canal irrigation systems, and in rivers and lakes, where NICS may destabilize the banks. The nonmarket economic damage of NICS, owing to their impact on biodiversity, seems to be enormous. For instance, the reintroduction of *P. fortis* in California cost $4.5 million, and the (unsuccessful) eradication of *P. leniusculus* in Scotland cost about £100,000.

Little attention has been paid until now to the harm that NICS pose to human health. Invasive crayfish often live in areas contaminated by sewage and toxic industrial residues and have high heavy metal concentrations in their tissues. Their potential to transfer contaminants to their consumers, humans included, is obviously high. The finding that *P. leniusculus* and *P. clarkii* may also accumulate toxins produced by cyanobacteria is of increasing concern for human health; *P. clarkii* is also suspected to be an intermediate host for many helminth parasites of vertebrates and a vector of transmission of the bacterium *Francisella tularensis*, the causative agent of human tularemia.

On the other hand, *P. clarkii* may control snails known to host *Schistosoma* spp., the agents of human schistosomiasis. Owing to the quick spread of this crayfish in African water bodies, the epidemiology of schistosomiasis is expected to be significantly altered with time, although the possibility remains that African snails will soon evolve measures to avoid crayfish predation or that the parasite will change its host.

PREVENTION, MANAGEMENT, AND EDUCATION

The "three-stage hierarchical approach" (prevention, early detection/rapid response, and containment/control) recommended by the Convention on Biological Diversity for the management of invasive species applies well to NICS. The first stage, prevention, is particularly critical in this case. NICS, in fact, can be hard to detect and can disperse rapidly, making eradication or control extremely difficult and expensive. Much effort should therefore be directed at minimizing the risks of intentional and unintentional introductions, as is attempted by the current legislation of some countries. For instance, in the United Kingdom, *A. astacus*, *A. leptodactylus*, and *P. leniusculus* have been designated as pests under the Wildlife and Countryside Act; much of Britain has been declared a no-go area for keeping *P. leniusculus*, and the whole of Britain for keeping all other NICS (except the tropical *C. quadricarinatus*). Similarly, in Japan, *Astacus* spp., *Cherax* spp., *O. rusticus*, and *P. leniusculus* have been classified as invasive alien species under the Invasive Alien Species Act; their import and live keeping are banned except for scientific purposes.

In the EU, Council Regulation No 708/07 "concerning use of alien and locally absent species in aquaculture" has been in force since 2007; its novelty is to take a "white list" approach: only the importation of species that have been appropriately screened after a thorough risk assessment analysis can be allowed. This contrasts with the analogous regulation in the United States, which permits the importation of species unless they are on a "black list" (classified as "injurious wildlife species" by the U.S. Fish and Wildlife Service). The above restrictions on import of NICS for aquaculture use, however, seem not to be well harmonized with the legislation concerning the aquarium trade. NICS, such as the parthenogenetic marbled crayfish, are easy to buy for ornamental use, particularly via e-commerce. Finally, illegal importations of NICS are very difficult to police, and their accidental introductions as, for instance, contaminants of fish stockings are common.

Several attempts have been made in different countries to eradicate or to control invasive populations of NICS, and experimentation on different methods is under way. Independently of the method adopted, however, a high rate of failures has generally been lamented. Mechanical removal using baited traps of various design or electrofishing has had some effect only when conducted for an extended period of time, which means considerable cost and human-power; besides, the prevalent removal of large and dominant individuals from the population might reduce their pressure on juveniles that are usually trap-shy, thus allowing them to grow and to give rise to even larger populations. Drainage of ponds, diversion of rivers, and construction of barriers (either physical or electrical) may also be used in the case of confined crayfish populations, but very little is known about their efficacy. Biocides, including organophosphate, organochlorine, pyrethroid insecticides, rotenone, and surfactants, lack specificity: other invertebrates may be eliminated along with crayfish, and toxin bioaccumulation and biomagnification in the food chain are likely. Other solutions lie in recourse to two autocidal methods already used with success against insect pests and now under investigation for the control of the *P. clarkii* populations in Italy—the use of sex pheromones and the sterile male release technique (SMRT); both, although expensive, cause no environmental contamination or nontarget impacts. Traditional biological control methods include the use of fish predators, disease-causing organisms (e.g., engineered strains of *A. astaci*), and microbes that produce toxins, such as strains of the bacterium *Bacillus thuringiensis*. Only the introduction of predaceous fish species, however, has provided some positive results, for example in Switzerland; eels, burbot, perch, and pike are well known predators of crayfish, but they are usually gape-size limited, preying only on small crayfish; in some instances, the presence of fish predators induces a change in the behavior of crayfish by reducing their trophic activity and increasing the time spent in shelter. But it is the combination of different methods, such as intensive trapping and an induced increase in predation pressure, that can be followed by some success against invasive NICS. In Sparkling Lake (Wisconsin), *O. rusticus* was mechanically removed from 2001 to 2005, and the harvest of fish species known to consume crayfish was restricted. As a result, crayfish catch rates declined by 95 percent, from 11 crayfish per trap per day in 2002 to 0.5 crayfish in 2005, and the native community showed a slow but steady recovery.

Whatever method is used, however, any intervention against NICS should be based on a thorough understanding of their threats by the general public, decision-makers, and other stakeholders. On the contrary, as recently shown in southern Spain, most stakeholders seem to have a limited knowledge of the nature of invasive species, and have varied perceptions of their impacts and different attitudes toward their introduction or eradication. Education and public awareness campaigns seem thus to be indispensable prerequisites for developing shared responsibility in solving the problems generated by these invaders.

SEE ALSO THE FOLLOWING ARTICLES

Aquaculture / Black, White, and Gray Lists / Crustaceans / Eradication / Lakes / Pathogens, Human / Rivers

FURTHER READING

Ackefors, H. 1999. The positive effects of established crayfish introductions (49–61). In F. Gherardi and D. M. Holdich, eds. *Crayfish in Europe as Alien Species*. Rotterdam, Netherlands: A. A. Balkema.

Aquiloni, L., A. Becciolini, R. Berti, S. Porciani, C. Trunfio, and F. Gherardi. 2009. Managing invasive crayfish: Using X-ray sterilization of males. *Freshwater Biology* 54: 1510–1519.

Gherardi, F. 2007. Crayfish invading Europe: The case study of *Procambarus clarkii*. *Marine and Freshwater Behaviour and Physiology* 39: 175–191.

Gherardi, F. 2007. Understanding the impact of invasive crayfish (507–542). In F. Gherardi, ed. *Biological Invaders in Inland Waters: Profiles, Distribution, and Threats*. Dordrecht, The Netherlands: Springer.

Gherardi, F., C. Souty-Grosset, G. Vogt, J. Diéguez-Uribeondo, and K. A. Crandal. 2010. Infraorder Astacidea, chapter 67. In F. Schram, ed. *Treatise on Zoology: Decapoda*, Vol. 9A. Leiden, The Netherlands: Brill.

Hein, C. L., M. J. Vander Zanden, and J. J. Magnuson. 2007. Intensive trapping and increased fish predation cause massive population decline of an invasive crayfish. *Freshwater Biology* 52: 1134–1146.

Hobbs, H. H. III, J. P. Jass, and J. V. Huner. 1989. A review of global crayfish introductions with particular emphasis on two North American species (Decapoda, Cambaridae). *Crustaceana* 56: 299–316.

Holdich, D. M., R. Gydemo, and W. D. Rogers. 1999. A review of possible methods for controlling nuisance populations of alien crayfish (245–270). In F. Gherardi and D. M. Holdich, eds. *Crayfish in Europe as Alien Species*. Rotterdam, The Netherlands: A. A. Balkema.

Lodge, D., C. A. Taylor, D. M. Holdich, and J. Skurdal. 2000. Nonindigenous crayfishes threaten North American freshwater biodiversity: Lessons from Europe. *Fisheries* 25: 7–20.

Nyström, P., C. Brönmark, and W. Granéli. 1996. Patterns in benthic food webs: A role for omnivorous crayfish? *Freshwater Biology* 36: 631–646.

CRUSTACEANS (OTHER)

ANTHONY RICCIARDI

McGill University, Montreal, Quebec, Canada

Crustaceans (phylum Arthropoda: subphylum Crustacea) are among the aquatic animals most widely introduced by humans. Their geographic spread is driven by activities such as local and transoceanic shipping, the construction of canals, aquaculture, and the aquarium trade. Although the impacts of most crustacean invasions are unknown, dozens of nonindigenous species have been observed to be ecologically or economically disruptive—particularly in freshwater and estuarine systems, where they can reduce native biodiversity, alter food webs, and modify the physical structure of habitats. Economic impacts have also been incurred due to biofouling (e.g., by barnacles), damage to river banks and pier pilings by burrowing species (e.g., crayfish, the Chinese mitten crab *Eriocheir sinensis*, the Australasian isopod *Sphaeroma quoyanum*), declines of commercially important fisheries resulting from food web disruptions (e.g., by mysids), and direct predation on cultured bivalves (e.g., by green crabs).

GLOBAL PREVALENCE OF CRUSTACEAN INVADERS

Most large freshwater and coastal marine systems worldwide have been invaded by crustaceans such as amphipods (Amphipoda), mysid shrimp (Mysida), waterfleas (Cladocera), and copepods (Copepoda), among other groups. At least 40 nonindigenous crustaceans have invaded European inland and coastal waters, 70 have invaded North American coastal areas, and 30 have invaded New Zealand coastal areas. Intercontinental transfers of species are common; for example, an African freshwater cladoceran (*Daphnia lumholtzi*) has invaded North American lakes, North American crayfishes have invaded European and African rivers, and Eurasian amphipods, cladocerans, and copepods have invaded the North American Great Lakes and connecting waterways. A few species, such as the barnacle *Balanus glandula* and the marine isopod *Synidotea laevidorsalis*, have achieved global distributions due to human activities. Moreover, some crustaceans (e.g., the copepod *Eurytemora affinis*) that were previously confined to marine and estuarine environments are being discovered more frequently in freshwater lakes and rivers across the northern hemisphere, apparently indicating a recent adaptation to low-salinity waters.

FACTORS CONTRIBUTING TO INVASION SUCCESS

Nonindigenous crustaceans often share a suite of life history traits that predispose them to being successful invaders: short generation times (rapid growth, early maturity, short life span), high rates of reproduction, broad environmental tolerances, and broad diets. Their large-scale dispersal is often aided by transoceanic shipping. They are among the most abundant macroscopic invertebrates in the ballast tanks of ships because (1) many species are euryhaline and thus are abundant around ports and can survive fluctuating salinities within ballast tanks during long voyages; (2) they possess planktonic life stages that are easily taken up and released during ballasting/deballasting procedures; and (3) planktonic crustaceans such as cladocerans and copepods possess resting eggs that allow them to survive periods of unfavorable conditions—these eggs occur commonly in the bottom sediments of the ballast tanks of ships from freshwater and estuarine ports, in numbers that reach tens of millions of eggs per ship. Desiccation tolerance of the resting eggs creates opportunities for short-distance overland dispersal between water bodies. Furthermore, the ability to alternate between phases of sexual and asexual reproduction allows rapid population growth from a very small number of individuals.

ECOLOGICAL IMPACTS

In marine systems, introduced crustaceans other than crabs are rarely associated with strong ecological impacts. By contrast, freshwater systems are more vulnerable; for example, 20 percent (4 of 20) and 25 percent (5 of 20) of the nonindigenous crustaceans in the Great Lakes and the Rhine, respectively, have been implicated in native species declines.

The greatest impacts are caused by species (particularly predators) that are ecologically unique in their new environments. However, impacts can vary markedly along environmental gradients (e.g., temperature, oxygen, conductivity), such that within a heterogeneous aquatic system, a nonindigenous crustacean may extirpate a native or another nonindigenous species at one site, but may coexist with or be dominated by the same species at another site.

Mysids

Predatory mysids (*Mysis*, *Hemimysis*, *Limnomysis*, *Neomysis*, and others) have been stocked widely as a supplementary food source for fishes in North American and European lakes. A common consequence of these intentional introductions is the rapid reduction of large zooplankton, which in turn often causes declines in the growth and abundance of pelagic fishes, as well as other food web repercussions. In a classic case, the opossum shrimp *Mysis diluviana* (formerly *M. relicta*) was introduced as a prey item for kokanee salmon (*Oncorhynchus nerka*) in Flathead Lake in Montana. A nocturnal species, it avoided predation by the kokanee, which is most active during daylight hours. Moreover, *M. diluviana* is an efficient planktivore, and it outcompeted kokanee for dwindling food resources, which resulted in a precipitous crash in the kokanee population. This event was followed by declines in local populations of grizzly bears and bald eagles, which feed on kokanee during salmon spawning runs. Mysid introductions in other lakes have caused (1) severe reductions in zooplankton; (2) enhanced bioaccumulation of contaminants (mercury and polychlorinated biphenyls); (3) increased parasitism of fishes by nematodes, cestodes, and acanthocephalans; and (4) declines in the growth, abundance, and productivity of pelagic fishes.

Cladocerans

The Eurasian spiny waterflea *Bythotrephes longimanus* can cause rapid reductions in the abundance and species richness of zooplankton in invaded lakes in Europe and North America. It suppresses native cladocerans through competition and predation. Similarly, the fishhook waterflea *Cercopagis pengoi*, a Ponto-Caspian species, is suspected to have caused the decline of some dominant species of zooplankton in Lake Ontario. Both invasive cladocerans may themselves escape substantial predation because they possess a long spine that renders them unpalatable to small planktivorous fish. Furthermore, both species are generalist predators, which are able to feed on prey of a broad range of sizes and swimming abilities.

Amphipods

Amphipod invasions have contributed to declines in native and exotic invertebrates in European inland waters and in the North American Great Lakes. In North

FIGURE 1 *Dikerogammarus villosus*, a predatory freshwater amphipod known as "killer shrimp." (Photograph courtesy of D. Platvoet.)

America, the Ponto-Caspian amphipod *Echinogammarus ischnus* has replaced the native *Gammarus fasciatus* as the dominant amphipod on rocky substrates in Lake Erie and Lake Ontario. Most European invasions involve Ponto-Caspian species that were deliberately introduced as prey resources to enhance fisheries, or that were dispersed along canals. Interactions between native and exotic amphipods commonly involve intraguild predation, such that a superior ability to prey upon recently molted individuals while resisting predatory attacks often governs which species dominates and which is excluded. Species that are dominant intraguild predators and have high reproductive output compared with native species are more likely to replace the natives. A noteworthy example is *Dikerogammarus villosus* (Fig. 1), a voracious predator known as "killer shrimp," which is larger than many other amphipods. In laboratory experiments, it has been shown to attack a wide variety of invertebrates and, in particular, to rapidly eliminate large numbers of other

crustaceans. It has replaced native and exotic amphipods in various rivers as it has expanded its range into northwestern Europe.

Another Ponto-Caspian species, the filter-feeding amphipod *Chelicorophium curvispinum*, constructs muddy galleries of tubes on stones, thereby altering bottom habitat and reducing the diversity of epibenthic communities. In the 1980s, *C. curvispinum* invaded the Rhine River, colonizing hundreds of kilometers and establishing enormous densities (greater than 100,000 m^{-2}) within only a few years. By altering rocky substrate and competing for suspended food particles, *C. curvispinum* nearly extirpated populations of filter-feeding caddisflies (*Hydropsyche* sp.) and the zebra mussel from the lower Rhine.

SEE ALSO THE FOLLOWING ARTICLES

Ballast / Crabs / Crayfish / Invertebrates, Marine / Ponto-Caspian: Invasions / Rivers

FURTHER READING

Dick, J. T. A., I. Montgomery, and R. W. Elwood. 1993. Replacement of the indigenous amphipod *Gammarus duebeni celticus* by the introduced *Gammarus pulex*: Differential cannibalism and mutual predation. *Journal of Animal Ecology* 62: 79–88.

Dick, J. T. A., and D. Platvoet. 2000. Invading predatory crustacean *Dikerogammarus villosus* eliminates both native and exotic species. *Proceedings of the Royal Society of London B* 267: 977–983.

Kestrup, A., and A. Ricciardi. 2009. Environmental heterogeneity limits the local dominance of an invasive freshwater crustacean. *Biological Invasions* 11: 2095–2105.

Lee, C. E. 1999. Rapid and repeated invasions of fresh water by the copepod *Eurytemora affinis*. *Evolution* 53: 1423–1434.

Spencer, C. N., B. R. McClelland, and J. A. Stanford. 1991. Shrimp stocking, salmon collapse, and eagle displacement. *BioScience* 41: 14–21.

Van den Brink, F. W. B., G. Van der Velde, and A. Bij de Vaate. 1991. Amphipod invasion on the Rhine. *Nature* 352: 576.

Yan, N. D., R. Girard, and S. Boudreau. 2002. An introduced invertebrate predator (*Bythotrephes*) reduces zooplankton species richness. *Ecology Letters* 5: 481–485.

DAISIE PROJECT

PETR PYŠEK
Academy of Sciences, Pruhonice, Czech Republic

PHILIP E. HULME
Lincoln University, New Zealand

WOLFGANG NENTWIG
University of Bern, Switzerland

MONTSERRAT VILÀ
Estación Biologica de Doñana, Spain

A milestone in European knowledge of alien species has been achieved with the Delivering Alien Invasive Species Inventories for Europe (DAISIE) project, a major portal for information on biological invasions that was established through the Sixth Framework Programme of the European Union. The rationale was to develop a pan-European inventory of invasive alien species by integrating existing databases to describe patterns and evaluate trends in biological invasions in Europe, identify priority species, and assess their ecological and economic impacts. Although still an ongoing process, the foundation, scope, and technological architecture of DAISIE was established through a consortium of leading researchers of biological invasions in Europe from 18 institutions across 15 countries. Results were delivered through the cooperation of experts in ecology and taxonomy from throughout Europe, which in total amounted to 182 contributors. The inventory, accounts, and distribution maps, today provide the first qualified reference system on invasive alien species for the European region.

The information presents an outstanding resource to synthesize current knowledge and trends in biological invasions in Europe. The data have helped to identify the scale and spatial patterns of invasive alien species in Europe and led to greater understanding of the environmental, social, economic, and other factors involved in invasions. DAISIE has also been used as a framework for considering indicators for early warnings.

OUTPUTS OF DAISIE

The main project outputs available to the public on the DAISIE portal (www.europe-aliens.org) are (1) the European Alien Species Database, (2) the European Invasive Alien Species Information System, and (3) the European Expertise Registry.

The European Alien Species Database is an inventory of all alien species known to occur in Europe and resulted from compiling and peer-reviewing national and regional lists of alien fungi, bryophytes, vascular plants, invertebrates, fish, amphibians, reptiles, birds, and mammals. Data were collated for all 27 European Union member states (and separately for their significant island regions), for other European states (Andorra, Iceland, Liechtenstein, Moldova, Monaco, Norway, the European part of Russia, Switzerland, Ukraine, the states of the former Yugoslavia), and for Israel. Marine lists are referenced to the appropriate political region with administrative responsibility. To have full coverage of the Mediterranean Sea, data for the marine regions of North African and Near East countries are included. The coverage of the database currently comprises 71 terrestrial and nine marine regions. Although for some regions and taxa the database includes information on some casual (not established) alien species, the focus is on naturalized or established taxa. For the sake of comparability across regions, established species were defined as aliens that form free-living populations persisting in the wild, unsupported by and independent of humans, that were introduced to Europe after 1500. For each species, an attempt was made to gather information on the date of introduction, habitats invaded, pathways of introduction, known impacts in a region, and native geographical distribution.

The European Invasive Alien Species Information System contains detailed accounts of 100 of the most invasive alien species in Europe, selected so as to cover a range of life forms, functional types, invaded ecosystems, and impacts on biodiversity, environment, economy, and human health. For each species, information on biology, ecology, impact, and management is included, with key references, images, and distribution maps at approximately 50 × 50 km grid cells, using the Common European Chorological Grid Reference System. The current distribution is plotted, but where a species previously occurred but was eradicated is also indicated. Mapping of aquatic invaders is based on linear distributions along the coast or in maritime areas.

The European Expertise Registry represents the first effort to identify a critical mass of expertise in alien species research. The information includes the field of expertise (distribution, conservation, ecology, economy, genetics, legislation, management, pathways, physiology, risk assessment, taxonomy), systematic focus, and geographic area studied. Although primarily focused on Europe, the registry currently contains information on 1,700 experts from more than 90 countries across the world, and for over 3,400 higher taxa of plants and animals.

In addition, the information accumulated by DAISIE has been summarized in the *Handbook of Alien Species in Europe* (DAISIE 2009), which contains an analytical chapter on each taxonomic group as well as fact sheets of the 100 most invasive alien species in Europe, with distribution maps and images. The book also lists all alien species recorded, ranked taxonomically; this list can be used as a reference for future assessment of trends in biological invasions in Europe.

TABLE 1

Alien Species in the European Union

Region	Fungi	Bryophytes	Plants	Terrestrial Invertebrates	Fishes	Amphibians and Reptiles	Birds	Mammals	Total
United Kingdom	42	22	1,779	924	34	16	62	31	2,910
Belgium	22	10	1,969	318		10	19	11	2,359
France	48	15	1,258	709	1	9	52	17	2,109
Czech Republic	26	8	1,378	528		6	14	16	1,976
Germany	45	11	851	686	22	27	58	31	1,731
Austria	30	5	1,086	459	28	12	25	13	1,658
Sweden	16	8	1,201	308	15	5	15	13	1,581
Denmark	19	8	978	446	16	6	11	18	1,502
Spain	25	13	933	417		28	51	6	1,473
Italy	41	10	557	714	38	11	47	15	1,433
Finland	12	4	918	201	17	1	9	8	1,170
Latvia	8	4	886	136	23		9	6	1,072
Bulgaria	17	4	708	310		1	1	8	1,049
Lithuania	19	4	827	152	20	1	4	8	1,035
Ireland	20	13	734	219	17	4	9	13	1,029
Hungary	30	6	711	265		1	6	8	1,027
Slovakia	23	6	741	194		3	3	8	978
Portugal	19	9	547	272		3	24	2	876
Romania	24	5	435	236	31	1	2	7	741
Poland	25	6	300	284	23	2	7	12	659
Netherlands	21	9	232	323	1	8	19	10	623
Greece	17	5	315	226	21	6	6	5	601
Estonia	7		416	134	9	1	2	5	574
Malta	2		195	179		8	2	1	387
Cyprus	3	2	223	90			9	2	329
Slovenia	24		87	161	16	3	5	9	305
Luxembourg		1	118	67			3	6	195

NOTE: Numbers of alien species recorded in 27 member states of the European Union, according to taxonomic groups. Note that the figures reflect not only the actual levels of invasion load but also research intensity. For plants, the focus among regions varied, and some contain data on casual (not established) species. Empty cells indicate that data were not available.

SCIENTIFIC RESULTS OF DAISIE: THE STATE OF THE ART OF BIOLOGICAL INVASION IN EUROPE

Species Numbers

By November 2008, the database included records of 10,771 alien species belonging to 4,492 genera and 1,267 families. Plants are most represented, accounting for 55 percent of all taxa (5,789 species), with terrestrial invertebrates at 23 percent (2,477 species), followed by vertebrates (6%), fungi (5%), molluscs (4%), annelids (1%), and rhodophyts (1%). In total, the database includes records of 45,211 introduction events to particular regions (plants: 28,093; terrestrial invertebrates: 11,776; aquatic marine species: 2,777; terrestrial vertebrates: 1,478; aquatic inland species: 1,087). The regions considered differ in the number of alien species recorded (Table 1). Due to unprecedentedly thorough assessment, DAISIE substantially improved the accuracy of estimates of alien species numbers derived from previous datasets. For example, from the dynamics of introduction of alien plants it can be inferred that about 38 percent of the species present in Europe in the 1980s were not captured by the authoritative Flora Europaea completed then. The increase in the numbers of species is thus partly due to the quality of data collated by DAISIE, and it partly also reflects new introductions.

Dynamics of Introductions

The detailed information about invasive species from a wide range of groups in Europe makes it possible to evaluate the dynamics of introductions of alien species to Europe over the last century. The numbers of newly recorded naturalized taxa on the continent are generally increasing, in both terrestrial and aquatic environments, and this increase exhibits accelerating trends (Fig. 1).

Origin

Although there are differences in the frequency of species coming from other continents, Europe receives most of its alien species from the Americas and Asia. A large proportion of aliens recorded in particular European regions are also species native to other parts of Europe: for example, among plants, these species account for 44.7 percent of the total number, with the remaining 55.3 percent being alien to Europe as a whole (i.e., coming from other continents) (Fig. 2).

Geographic Distribution

The most invasive species are unevenly distributed among European regions. The western part of Europe, including the Mediterranean region, harbors higher numbers of the most serious terrestrial and aquatic inland invaders than parts of the continent located further to the east (Fig. 3). Of the three marine regions, the Mediterranean basin harbors a higher number of alien species (501) than the Atlantic coast of Europe (359) and the Baltic Sea (112).

Habitats

The highest numbers of alien plants and insects occur in human-made habitats (ruderal areas, cultivated land, parks, and gardens); riparian habitats also support high numbers of alien plant species. Unlike the case of plants and insects, where invasions are concentrated in these few highly invaded habitats, invasions by vertebrates are more evenly distributed among habitats, with aquatic and riparian areas,

A

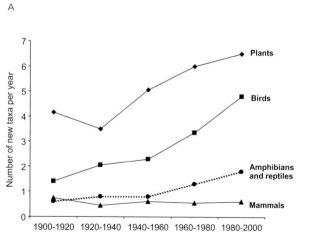

B

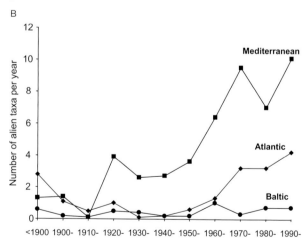

FIGURE 1 Dynamics of introduction of alien plants and vertebrates in terrestrial environments (A) and that of multicellular aquatic species in three marine areas (B) in Europe. Alien taxa newly recorded as naturalized (A) and the total number of aliens recorded (B) are shown per annum for the time periods indicated. (Based on data from Hulme et al. 2009. *Science* 324: 40–41 and DAISIE 2009.)

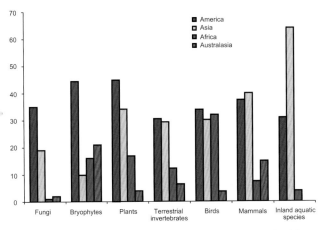

FIGURE 2 Regions of origin of species alien to Europe, arrived from other continents.

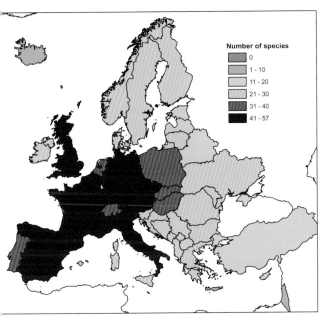

FIGURE 3 Geographical distribution of the 66 invasive terrestrial and aquatic inland species in European regions (based on 3 fungi, 18 plants, 16 invertebrates, 15 vertebrates, and 14 aquatic inland taxa among the 100 worst species addressed by DAISIE, 2009).

as contaminants of commodities or as stowaways on transport vectors [i.e., independently of a commodity]). This contrasts with the case of terrestrial insects, of which 90% of aliens arrived unintentionally (75% associated with a commodity, 15% as stowaways). Only 0.5 percent of alien plants can be attributed to deliberate releases (planted or sown in the wild, not escaped), a pathway that is responsible for the many vertebrate introductions (e.g., 38% of 193 bird species and 35% of all known 411 cases of introduction events of mammals). The majority of aquatic inland species introductions were escapees from aquaculture, deliberate releases for stock enhancement, or stowaways on shipping vessels. In terms of marine invasions, most organisms spread along canal corridors as stowaways on the hulls and in the ballast water of vessels or as escapes from aquaculture.

Impact

DAISIE has allowed the first assessment of the extent of the impact of alien species in Europe. There are 1,094 species with documented ecological impacts and 1,347 with economic impacts. The two taxonomic groups with the most species causing impacts are terrestrial invertebrates and terrestrial plants. The North Sea is the maritime region that suffers from the most impacts. Across taxa and regions, ecological and economic impacts are highly correlated. Terrestrial invertebrates create greater economic than ecological impacts, while the reverse is true for terrestrial plants. Many of these species impact on a variety of ecosystem services in several European countries. Vertebrates (*Branta canadensis, Cervus nippon, Myocastor coypus, Salvelinus fontinalis*) are most represented among the top ten alien species exerting the most diverse impact on ecosystem services in Europe; other species represent plants (*Oxalis pes-caprae*), freshwater invertebrates (*Dreissena polymorpha, Procambarus clarkii*), marine algae (*Codium fragile, Undaria pinnatifida*), and marine invertebrates (*Balanus improvisus*). The monetary costs of biological invasions in Europe are likely underestimated because, despite the efforts of DAISIE, the impacts of the majority of invasive species remain unknown or have not been translated into economic terms.

THE ROLE OF DAISIE

As a result of DAISIE, Europe is today the continent with the most complete information on its alien biota. The continent has been working toward implementing an effective strategy on invasive alien species, and DAISIE is considered to be one of the major instruments for achieving this goal. An Internet-accessible knowledge base such as DAISIE can provide crucial information for the

wooded and cultivated lands, and parks and gardens being the most invaded. Coastal habitats, mires, bogs and fens, grasslands, heathlands, and scrub are generally less invaded.

Pathways

Alien organisms are being introduced to Europe by a variety of pathways. About two-thirds of plants (63%) were introduced intentionally (with ornamentals most represented, at 39%), 37 percent were introduced unintentionally (mostly early detection, eradication, and containment of invasive aliens—outcomes that are most possible for species that have just arrived. Through DAISIE, managers and policymakers addressing the invasive alien species challenge can easily obtain data on which species are invasive or potentially invasive in particular habitats, and then use this information in their planning efforts. Agencies responsible for pest control can quickly determine if a species of interest has been invasive elsewhere in Europe. Importers of new alien species can access data to make responsible business choices. Land managers can learn about control methods that have been useful in other areas, reducing the need to commit resources for experimentation and increasing the speed at which control efforts can begin. DAISIE is potentially a model for other continents with currently less detailed information on their alien biota available.

SEE ALSO THE FOLLOWING ARTICLES

Databases / Early Detection and Rapid Response / Geographic Origins and Introduction Dynamics / Mediterranean Sea: Invasions / Risk Assessment and Prioritization

FURTHER READING

Chiron, F., S. Shirley, and S. Kark. 2009. Human-related processes drive the richness of exotic birds in Europe. *Proceedings of the Royal Society B London* 276: 47–53.

DAISIE. 2009. *Handbook of Alien Species in Europe*. Berlin: Springer.

Desprez-Loustau, M. L., C. Robin, M. Buee, R. Courtecuisse, J. Garbaye, F. Suffert, I. Sache, and D. M. Rizzo. 2007. The fungal dimension of biological invasions. *Trends in Ecology and Evolution* 22: 472–480.

Lambdon, P. W., P. Pyšek, C. Basnou, et al. 2008. Alien flora of Europe: Species diversity, temporal trends, geographical patterns and research needs. *Preslia* 80: 101–149.

Hulme, P. E., W. Nentwig, P. Pyšek, and M. Vilà. 2009. Common market, shared problems: Time for a coordinated response to biological invasions in Europe? *Neobiota* 8: 3–19.

Hulme, P. E., P. Pyšek, W. Nentwig, and M. Vilà. 2009. Will threat of biological invasions unite the European Union? *Science* 324: 40–41.

Hulme, P. E., D. B. Roy, T. Cunha, and T.-B. Larsson. 2009. A pan-European inventory of alien species: rationale, implementation and implications for managing biological invasions (1–14). In *Handbook of Alien Species in Europe*. Dordrecht: Springer.

Nentwig, W., E. Kühnel, and S. Bacher. 2009. A generic impact-scoring system applied to alien mammals in Europe. *Conservation Biology* 24: 302–311.

Pyšek, P., S. Bacher, M. Chytrý, V. Jarošík, J. Wild, L. Celesti-Grapow, N. Gassó, M. Kenis, P. W. Lambdon, W. Nentwig, J. Pergl, A. Roques, J. Sádlo, W. Solarz, M. Vilà, and P. E. Hulme. 2010. Contrasting patterns in the invasions of European terrestrial and freshwater habitats by alien plants, insects and vertebrates. *Global Ecology and Biogeography* 19: 317–331.

Vilà, M., C. Basnou, P. Pyšek, M. Josefsson, P. Genovesi, S. Gollasch, W. Nentwig, S. Olenin, A. Roques, D. B. Roy, P. E. Hulme, and DAISIE partners. 2010. How well do we understand the impacts of alien species on ecosystem services? A pan-European cross-taxa assessment. *Frontiers in Ecology and the Environment* 8: 135–144.

DARWIN, CHARLES

MARC W. CADOTTE

University of Toronto, Ontario, Canada

Charles Darwin (1809–1882), best known for articulating the process of natural selection, was an astute student of life whose numerous and multifaceted observations still inform our understanding of the operations of nature. Most if not all of the life sciences subdisciplines mark their intellectual beginning with Darwin. Collectively, his works span all major taxonomic groups (plants, mammals, invertebrates, birds, etc.), as well as many major biological processes (evolution, parasitism, plant reproduction, competition, predation, biogeography, species invasions, etc.). By far, his most important and influential work was *The Origin of Species*, published in 1859. In *The Origin*, Darwin includes a plethora of natural world observations for the purpose of either supporting his theory of species descent by modification or of documenting patterns that require an explanation, which his theory supplies. Throughout *The Origin*, there are a number of empirical threads Darwin repeatedly uses (e.g., domesticated species), but the most pervasive is his use of species that have been introduced to new regions.

FIGURE 1 Charles Darwin (1809–1882) in his later years. Photograph by J. Cameron, 1869.

INVASIONS IN *THE ORIGIN*

Although observers (e.g., Alphonse de Candolle) had commented on biological invasions before the publication of *The Origin*, Darwin used invasions as a sort of natural experiment, a test of his ideas about the origin of species and the outcomes of ecological interactions. Thus, *The Origin* employed a two-sided use of species invasions, as support for Darwin's theory and as a process explained by his theory. With the latter, Darwin actually created an informative set of explanations for understanding invader success.

The first, and most obvious, observation of biological invasions is that they involve individuals of a species that have been introduced by humans into a new place, where the species did not occur historically. That is, even though species are often adapted to widely available habitats, they are not found in all places, even those places to which a species may appear perfectly adapted. Why is this an important observation? The pre-Darwinian explanation that seemed to resonate with most thinkers was that an omnipotent creator independently and perfectly created species for their particular habitats. However, if species were not found in all those places where they were best suited, or if they readily gave way to better-adapted nonnative species, then how could this explanation make sense? Darwin commented on the evidence against special creation several times. In one example, he wrote:

> (The) general absence of frogs, toads and newts on so many oceanic islands cannot be accounted for by their physical conditions; indeed it seems that islands are peculiarly well fitted for these animals, for frogs have been introduced into Madiera, the Azores and Mauritius, and have multiplied so as to become a nuisance . . . . But why, on the theory of creation, they should not have been created there, it would be very difficult to explain (p. 382). (Quotations throughout correspond to the 1979 Gramercy Books reprint of the first edition.)

In fact, for Darwin, this was such a crucial point that he reiterated the evidence as the focus of an entire chapter (Chapter 11: Geographical Distribution). Furthermore, in a passage later on the same page as the above quote, he stated that, in his research, he had not found a case where terrestrial mammals occupied islands greater than 300 miles from the nearest continental landmass. Again, however, he noted that these islands readily receive and support terrestrial vertebrates introduced by humans.

WHY CERTAIN SPECIES ARE SUCCESSFUL

The fact that so many species readily invade new regions when dispersal barriers are overcome begs the question of why and how success is attained. According to Darwin, invader success or failure comes down to two things: the place and the species (which are still major demarcations in studies of invasions today). According to Darwin, those places that seem to readily give way to introduced species tend to be rather species poor. He noted that isolated islands generally have few species relative to comparable mainland areas, and he thus observed:

> We have evidence that the barren island of Ascension aboriginally possessed under half-a-dozen flowering plants; yet many have become naturalised on it as they have on New Zealand and on every other oceanic island which can be named (p. 379).

Again, Darwin saw this as evidence in favor of his theory over that of species creation, because on islands "man has unintentionally stocked them from various sources far more fully and perfectly than has nature" (p. 379). While the place is important because of native species richness and the forms that have evolved there, what is it about particular species that allow them to be successful invaders, while others fail? Darwin's general response is that, because species evolve to local conditions, they may have unique attributes allowing success, when species in other regions may not have attained those attributes.

The ability for introduced species to rapidly increase in range and numbers begs explanation. Darwin gives several examples, from horses in South America to plants in India, and remarks:

> In such cases, and endless instances could be given, no one supposes that the fertility of these animals or plants has been suddenly and temporarily increased in any sensible degree. The obvious explanation is that the conditions of life have been very favourable, and that there has consequently been less destruction of the old and young, and that nearly all the young have been enabled to breed. In such cases the geometrical ratio of increase . . . simply explains the extraordinary rapid increase and wide diffusion of naturalized productions in their new homes (p. 118).

For Darwin, the reason for the success of these invaders is uniqueness. While it may be intuitive to think, "the plants which have succeeded in becoming naturalised in any land would generally have been closely allied to the indigenes" (p. 158), the reality is that "floras gain by naturalization, proportionally . . . far more in new genera than in new species" (this is often referred to as Darwin's naturalization hypothesis). Because Darwin believed that two species from unique genera differ more than two species from the same genus, he argued that successful invaders are generally the more different species. What this observation meant to Darwin is that, from those successful species, we can gain some crude idea in what manner some of the natives would have to be modified, in order to have gained an advantage over other natives . . . [and] that diversification of structure, amounting to new generic differences, would have been profitable to them (p. 158).

THE LIMITS OF UNIQUENESS: NAIVETÉ

Conversely, according to Darwin, failure is likely because the introduced species are less perfectly adapted or are naive to local conditions and interactions. Populations face numerous checks to the growth of their numbers, but why do these checks differentiate some introduced species from others? Darwin explicitly considered three checks that are likely important for limiting success. First, even though many introduced species can survive climates in the new range successfully, as evidenced in gardens, they "cannot compete with our native plants, nor resist destruction by our native animals" (p. 122). Thus, in many cases, there is some benefit to adapting to local communities and to the new suite of interactions. The second interaction he considered important is naiveté to parasites. He offers as evidence the lack of successful mammal invasions into Paraguay because "of a certain fly, which lays its eggs in the navels of these animals when first born." If fly numbers were reduced "then cattle and horses would become feral" (p. 124). Finally, the third check is a lack of coevolved positive interactions, especially pollinators for plants and that, for example, "the exotic *Lobelia fulgens*, in this part of England, is never visited by insects, and consequently, from its peculiar structure, never can set seed" (p. 125). Thus, it appears as though there may be a cost to uniqueness: namely, not being the best suited competitor, being susceptible to new pathogens and parasites, and lacking required coevolutionary partners. Regardless, Darwin viewed success of introduced species as being related to having unique traits or occupying unique niches.

SUCCESS AND IMPACT

A potential problem resulting from Darwin's hypothesis that species uniqueness influences success is that, because success is determined by niche uniqueness, successful invaders should have minimal impact on the native community because they are not competing for the same resources. Darwin observed that there are likely always open niche opportunities and that in "the Cape of Good Hope, where more species of plants are crowded together than in any other quarter of the world, some foreign plants have become naturalised, without causing, as far as we know, the extinction of any natives" (p. 154). Thus,

invasions do not necessarily entail negative impacts, especially if successful invaders are occupying unique niches.

Logically then, it should be the invaders most similar to natives that have the greatest impact. Darwin concluded this when concluding his thesis on relatedness, uniqueness, and competition:

> As species of the same genus have usually, though by no means invariably, some similarity in habits and constitution, and always in structure, the struggle will generally be more severe between species of the same genus, when they come into competition with each other, than between species of distinct genera (p. 127).

He cited several examples, including the spread of swallows, thrushes, cockroaches, and charlock and their impacts on close relatives, as evidence of the greatest impact actually coming from invaders with closely related natives. While most successful invasions are the result of having unique requirements relative to the natives, those that are successful because of competitive superiority have strong impacts on closely related species. Thus, to Darwin, success and impact were not synonymous, but rather were the outcomes of two potentially different processes.

This dichotomy between the causes of success and impact has important repercussions for management decisions about which nonnative species may become problematic. While Darwin's naturalization hypothesis has long been tested and debated in the scientific literature, this success–impact dichotomy is only now being developed and tested. Given the number and scope of Darwin's insights about evolutionary and ecological forces influencing the success of introduced species, *The Origin* has been, and will likely continue to be, an important source of ideas and hypotheses about introduced invaders.

SEE ALSO THE FOLLOWING ARTICLES

Competition, Animal / Competition, Plant / Elton, Charles S. / Endangered and Threatened Species / Invasion Biology / Invasiveness / Taxonomic Patterns

FURTHER READING

Cadotte, M. W. 2006. From Darwin to Elton: Early ecology and the problem of invasive species (15–33). In M. W. Cadotte, S. M. McMahon, and T. Fukami, eds. *Conceptual Ecology and Invasion Biology: Reciprocal Approaches to Nature.* Dordrecht, The Netherlands: Springer.
Cadotte, M. W., M. A. Hamilton, and B. R. Murray. 2009. Phylogenetic relatedness and plant invader success across two spatial scales. *Diversity and Distributions* 15: 481–488.
Daehler, C. C. 2001. Darwin's naturalization hypothesis revisited. *American Naturalist* 158: 324–330.
Darwin, C. 1859. *The Origin of the Species by Means of Natural Selection.* London: Murray.

Diez, J. M., J. J. Sullivan, P. E. Hulme, G. Edwards, and R. P. Duncan. 2008. Darwin's naturalization conundrum: Dissecting taxonomic patterns of species invasions. *Ecology Letters* 11: 674–681.

Proches, S., J. R. U. Wilson, D. M. Richardson, and M. Rejmánek. 2008. Searching for phylogenetic pattern in biological invasions. *Global Ecology and Biogeography* 17: 5–10.

Richardson, D. M., and P. Pyšek. 2006. Plant invasions: Merging the concepts of species invasiveness and community invasibility. *Progress in Physical Geography* 30: 409–431.

DATABASES

LAURA A. MEYERSON
University of Rhode Island, Kingston

Databases are repositories of information collected and organized to serve a specific purpose. Because of the global nature of invasive species, and because invasions occur over both space and time, many types of quantitative and qualitative databases have been developed to help scientists, managers, and policymakers better understand and manage invasions and prevent future introductions. For example, some invasive species databases focus on specific species, locations, or impacts of invaders, while others record invasions of multiple species at regional or global scales (Fig. 1). Ideally, the information held in all of these databases would be integrated to create a single large database with global coverage that included invasions by all taxa at all spatial and temporal scales. In reality,

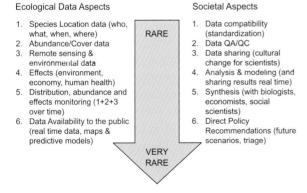

FIGURE 1 While great progress has been made, invasive species databases that include all of the information that scientists and policymakers need to be able to address invasive species issues have yet to be fully developed. This figure depicts the kinds of data that are needed to better understand biological invasions, and the extent to which they are currently available. (Figure courtesy of Tom Stohlgren, USGS.)

however, a significant portion of existing data cannot be easily harmonized in this way, because the data were collected using incompatible methods and approaches. Furthermore, many databases on invasive species are not readily available to the public, or even to scientists studying biological invasions, but are instead privately owned and controlled by individuals or organizations.

INVASIVE SPECIES DATABASES, DATA PORTALS, AND CLEARINGHOUSES

In general, invasive species database Web sites fall along a continuum in terms of the kinds of information that they include. Some are highly focused on particular taxa or geographic regions, such as NIMPIS (National Introduced Marine Pest Information System), which provides data and information on marine invasive species in Australia, and DAISIE (Delivering Alien Invasive Species Inventories for Europe), which provides an online inventory of invasive species in Europe. Some sites serve as "distributed databases" that facilitate the simultaneous search of multiple independent invasive species databases. For example, NISbase (Non-Indigenous Species Network) allows the user to search the records of up to ten different databases by group (such as molluscs or algae), genus, species, or geographic area. Other Web sites serve as "clearinghouses"—as does, for example, GISIN (Global Invasive Species Information Network). These kinds of databases provide links to online invasive species databases that cover particular regions or taxa. They encourage database owners to submit their data for public use and can also facilitate searches of submitted databases.

Biological invasions have ecological, evolutionary, economic, and social aspects, and because of this, gathering information to prevent new invasions and manage existing ones increasingly requires reaching out beyond ecological data. Databases that focus on genetics, trade, and transport pathways and the costs of invasions are increasingly being mined for information related to the introduction and success of invasive species—a powerful approach that is yielding new insights. For example, NCBI (the National Center for Biotechnology Information) is a public repository for molecular biology information ranging from DNA and protein sequences to reference literature. This database can be mined for basic genomic information on an invasive species or used to enhance or expand another data set. Rapid and frequent transport and international trade have facilitated the spread and survival of invasive species by substantially shortening the time an organism spends in transport (thus increasing the chances of its survival) and in many cases increasing both the frequency and number of locations to which they will be introduced. The global airline transport network records all scheduled commercial and charter flights for large and small aircraft. This is a significant and growing network, and one that is vitally important, given that travel by air can move organisms around the world in less than 24 hours. Although not as rapid as air travel, container ships can inadvertently carry much larger numbers of organisms (e.g., fish, larvae, bacteria, and viruses) to new locations across the globe in ballast water, attached to the ship's hull, and in the containers and merchandise that they deliver. Many maritime databases track ships by route, port of origin, destination, type of ship, and cargo, among other attributes. These kinds of transport databases related to the invasion process can be used to track invasion pathways, to predict likely future invasions, and to strengthen biosecurity efforts, when coupled with other ecological and economic data.

STRENGTHENING INVASIVE SPECIES DATABASES

While much information is currently available on the Internet, there are significant gaps in the focus and scope of data that are collected and published. For example, many of the publicly available regional or global scale ecological invasive species databases include information on species identity and location, but somewhat less frequently include information on species abundance. Knowing how many individuals are present (e.g., density or percent cover) can provide an indication of factors such as the degree of impact or the invasion stage (e.g., early or late). Some groups of species are well represented by invasive species databases, while others are not. For example, many databases cover invasive plant, bird, fish, mammal, insect, and mollusc taxa. Many fewer databases exist for invasive parasites, viruses, and fungi, among other groups. These gaps are due in part to the ease of studying and identifying particular organisms and the availability of scientists with the appropriate taxonomic expertise. For example, plants are not generally mobile and usually have readily visible characteristics, while identification and tracking of microbial invaders may require sophisticated and relatively costly genetic techniques. The databases that do exist for these groups generally include information on the locations where the invasive organisms have been reported to occur but often do not include information on the geographic origin of the pathogens, because these are often unknown. One critical piece of information for pathogen databases is to report the range of the host (e.g., plant or animal host) for the invasive

pathogen. This provides data on the potential spread of the invasive organism in its introduced range.

Invasive species databases may also be biased geographically, with some parts of the world having significantly greater representation than others. For example, wealthier or more developed nations may be able to devote more resources to invasive species data collection than less developed regions, including in terms of the frequency of data updates over time. The disparity in data collection and availability can make comparisons of invasions across regions difficult or impossible. A further challenge to widespread use occurs when databases are published on the web in the native language of the country of origin without translation into the standard scientific language. Ideally, databases could be published in both the native and standard language.

Information on the effects of invasive species on natural and human ecosystems and economies, and how to mitigate those effects, is often of particular interest to policymakers. To that end, gross estimates of economic impacts due to the costs of managing or eradicating invasive species have been prepared for many regions, and in some cases for individual species. Data collection of local impacts has become more common and in a few instances is occurring at the regional level. However, quantifying the effects of invasive species over large areas can pose a formidable challenge. For example, declines in plant diversity or other ecosystem functions due to the indirect effects of an invader may not be readily apparent, or may vary over space and time. Quantifying such relationships is critical, but devising approaches to practically measure and monitor them can be difficult and expensive. Although the challenges are significant, comprehensive databases on invasive species and their impacts are essential to help inform policymakers and natural resource managers to make the best decisions. This was recognized by the U.S. government when it established the U.S. National Invasive Species Council (NISC) in 1999. NISC was charged with creating an up-to-date and comprehensive invasive species database for scientists, policymakers, managers, and the public. However, this goal has not yet been accomplished and taxon- or habitat-specific databases are still managed by individual federal and state agencies. Nonetheless, invasive species databases continue to evolve and improve at all spatial scales. Efforts to integrate disparate data sets and to make invasive species data more widely accessible are ongoing, and as a result, a better understanding of the environmental and social effects of invasions is beginning to emerge, and predictive models and risk assessments are becoming more robust.

SEE ALSO THE FOLLOWING ARTICLES

DAISIE Project / Range Modeling / Risk Assessment and Prioritization

FURTHER READING

DAISIE European Invasive Alien Species (Delivering Alien Invasive Species Inventories for Europe) Gateway. www.europe-aliens.org.

GISIN (Global Invasive Species Information Network). www.gisinetwork.org.

NCBI (the National Center for Biotechnology Information). www.ncbi.nlm.nih.gov.

NIMPIS (National Introduced Marine Pest Information System). www.marine.csrio.au/crimp/nimpis.

NISbase (Non-Indigenous Species Network). www.nisbase.org.

Tatem, A. J. 2009. The worldwide airline network and the dispersal of exotic species: 2007–2010. *Ecography* doi: 10.1111/j.1600-0587.2008.05588.x.

USDA Fungus-Host Database. http://nt.ars-grin.gov/fungaldatabases/fungushost/fungushost.cfm.

VIDE (Virus Identification Data Exchange) plant virus database. http://md.brim.ac.cn/vide/refs.htm, http://micronet.im.ac.cn/vide/index.html.

DEMOGRAPHY

CAROL C. HORVITZ

University of Miami, Coral Gables, Florida

Demography is the population-level study of vital rates—birth, death, growth, and migration—and of how these rates vary across the life cycle by genotype, phenotype, age, size, and stage. Understanding how vital rates determine emergent properties of populations, including changes in population size, density, structure, and spatial extent, comprises the fundamental issue of demography. This is also the fundamental issue of invasion biology: understanding why some species increase in number and spread more rapidly in populations that are introduced into a new region of the world than they do in populations located in their native range. Management of invasive species also depends upon understanding how different components of the life cycle impact the dynamics of population growth and spread. Models of population dynamics (Fig. 1) and their sensitivity analyses are key for addressing this issue.

FUNDAMENTAL ISSUE

Invasive species are found across diverse taxa, including ferns, forbs, grasses, shrubs, trees, birds, small mammals, amphibians, reptiles, fishes, snails, shellfish, algae, kelp, insects, marine invertebrates, and microorganisms.

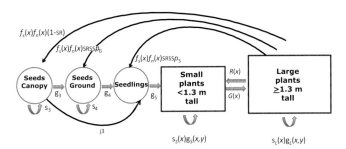

FIGURE 1 Life cycle diagram for the population projection model of paperbark *Melaleuca quinquenervia* (Myrtaceae) a tree that is native to Australia and invasive in southern Florida, Puerto Rico, and the Bahamas. The round nodes represent different discrete stages, and the rectangles represent two blocks of continuous stages. Arrows represent transitions among stages or blocks of stages over one time step (Table 1), but within each rectangle, transitions occur among continuous sizes over one time step. This model is an integral projection model; it also includes three discrete stages: seeds held in the capsules in a canopy seed bank, seeds in a soil seed bank, and tiny seedlings. Continuous size is used to model fates in two distinct phases of the life cycle: plants under 1.3 m in height, in which diameter is measured at the base of the stem, and plants at least 1.3 m in height, for which diameter is measured at breast height. The regression relationships of the size-dependency of growth and survival for these two groups of plants are estimated separately. Only the large plants reproduce; seed production depends upon size, which influences both the probability that a plant will reproduce and the number of seeds that will be made by a plant that does reproduce. (Diagram courtesy of Lucero Sevillano and Carol Horvitz.)

A species is considered as invasive when its populations increase more rapidly in numbers and spread more quickly across the landscape in the introduced range than in the native range, particularly when it reaches such high density that it has negative effects in the introduced range. The issue is this: what are the demographic features that are most critical to the success of populations in their native versus novel ranges, and can these features be used to characterize invasiveness?

Species that are well adapted to disturbance and have good colonizing ability (those that have many, highly dispersed offspring with rapid growth), such as the tree of heaven (*Ailanthus altissima*), are often noted as being preadapted to becoming invasive. This general paradigm predicts adaptation to disturbance across all ranges, however, and does not predict a difference in demographic behavior in native versus novel ranges. Also, this paradigm fails to account for the fact that there is another suite of invasive species that can successfully recruit in relatively closed habitats; examples include the shade-tolerant seedlings of Norway maple (*Acer platanoides*) in temperate forests and the fleshy fruited shoe-button Ardisia (*Ardisia elliptica*) in subtropical forests.

TABLE 1
Integral Projection Model of Paperbark

Parameter	Meaning
s^3	probability that seeds in the canopy remain in the canopy
g^3	probability that seeds in the canopy fall to the ground and remain as seeds
j^3	probability that seeds in the canopy fall to the ground and become seedlings
s^4	probability that seeds on the ground remain as seeds on the ground
g^4	probability that seeds on the ground become seedlings
g^5	probability that seedlings become small plants
$s_S(x)$	size-dependent probability of survival for small plants
$g_S(x,y)$	conditional probability that small plants of size x become plants of size y
$G(x)$	size-dependent probability that small plants become large plants
$s_L(x)$	size-dependent probability of survival for large plants
$g_L(x,y)$	conditional probability that large plants of size x become plants of size y
$R(x)$	size-dependent probability that large plants become small plants
$fs(x)$	size-dependent probability of being reproductive
$fn(x)$	size-dependent conditional seed production
SR	proportion of seeds that rain out of the canopy on a regular basis
SS	proportion of seeds that survive on the ground
p_G	conditional proportion of seeds that remain as seeds
p_S	conditional proportion of seeds that become seedlings

NOTE: Transitions among stages or blocks of stages over one time step (and other vital rates) for the integral projection model of paperbark, *Melaleuca quinquenervia* (Myrtaceae).

What is needed are comparative data on what drives population dynamics and determines population densities in the native versus introduced ranges. There are few such data. Existing studies on this issue differ in their findings. For example, in mile-a-minute weed *Polygonum perfoliatum* (native to Japan and an invasive weed in the northeastern United States), projection matrix models showed that the population growth rate in the introduced range was lower and much less sensitive to adult survival than in the native range. In some invasive species, density is higher even though the population growth rate is lower in the introduced range than in the native range, as in biennial houndstongue *Cynoglossum officinale* (native to Europe and invasive in western North America), which was studied with integral projection models. On the other hand, in the thistle *Carduus nutans* (native to Eurasia and invasive in the Americas, Australia, and New Zealand), projection matrix models showed that the population growth rate in the native range is sensitive to seed production and is kept low (populations are declining) especially by insect herbivores, which feed on seed heads, and because of the additional effects of sheep grazing and drought. In the introduced ranges, where populations are free from these pressures, the population growth rate is high (populations are increasing and spreading); but the thistle's sensitivity structure varies across its introduced range.

ENEMY RELEASE AND BIOLOGICAL CONTROL

A commonly proposed hypothesis is that populations in the introduced range do better because there they are free from (coevolved) enemies that are present in the native range. The idea is that not only are they free from the effects of direct damage (tissue removal, disease transmission), but also that they are free from having to invest resources in expensive defensive measures. Part of the enemy release hypothesis is that saved resources are then reallocated to increase reproduction and growth. Thus, this hypothesis essentially predicts measurable differences in demographic rates induced by the presence or absence of enemies. Such differences should be detectable by enemy-exclusion types of experimental manipulations in the native range, in which the entire life cycle is the response variable. A predicted outcome for a life table response experiment (LTRE) is that population growth and spread rates should be altered by the presence or absence of enemies and that those specific parts of the life cycle impacted by the enemy should make large contributions to any observed variation in the population growth rate between treatments. Such data from the native ranges of invasive species are very scarce; there are relatively more data on the effects of simulated herbivory on individual plants in the introduced range. One example is *Lespedeza cuneata*, for which it was found that population growth was relatively insensitive to herbivory.

An important corollary of the enemy release hypothesis is the adoption of biological control as a method to manage invasions. The idea is centered on bringing some of the natural enemies from the native range over to the introduced range. Here again, the appropriate metric for the success of the endeavor is demographic: will the introduced putative enemy actually slow the rate of growth or rate of spread of an invasive population? Most biocontrol programs to date have not included the necessary demographic data on the invasive species before and after the introduction of the enemy to be able to answer this question. Instead, the focus and funding have been targeted on ensuring that introduced agents will not harm native or crop species and on monitoring whether introduced agents have naturalized.

A few studies have found that biocontrol agents have had effects at the population level, including the examples of the successful control of *Senecio jacobaea* and *Cirsium nutans* in their introduced ranges by insects introduced from their native ranges. A few other studies have focused on evaluating the demographic sensitivity structure of an invasive species in its introduced range(s) and have made specific recommendations about which biocontrol agents should be used based on that analysis, most notably for *Carduus nutans* (for which different insects are recommended for New Zealand, Australia, and the Americas to correspond to distinct sensitivities of population dynamics), and more recently for selecting the appropriate weevils as biocontrol for garlic mustard *Alliaria petiolata* (native to Eurasia, invasive in temperate forests of North America).

DISPERSAL AND SPREAD

Changes in population number and local density are only part of the dynamic picture of invasive species. The spatial population dynamics are also of interest. For structured populations, every life cycle transition is potentially associated with individuals moving out across the landscape. Generally, however, not all stages move. For example, plants move at the seed stage, while birds may move at the fledgling or adult stages. Recently, models have been developed that can track both demography and dispersal at different life history transitions of structured populations. Dispersal is measured by the probability that individuals of a particular stage move a particular distance each time step. When dispersal data are combined with demographic data, new parameters emerge including the structure of the population at the dispersal "wavefront" and the speed at which the population is moving across the landscape, the rate of spread.

Although rapid spread across the introduced range is one of the defining features of an invasive species, no studies have yet compared the rates of spread of populations in native and introduced ranges at the population level. Some dispersal may continue to be mediated by humans even after initial introduction to a new region. For example, it is known that one way Caribbean anoline lizards move northward in Florida is by "hitchhiking" in cars. Another example is that ornamental plants continue to be intentionally moved and planted. Nevertheless, in many cases, following the initial human-mediated translocation of a species to a new part of the globe, dispersal proceeds by natural (i.e., nonhuman) means in the new range. At the landscape level, rates of spread have been estimated by tracking when species first appear in places they did not previously occupy, as in the classical studies of the American muskrat invading Europe and the African/Spanish cattle egret invading America. However, it remains largely unknown how dispersal dynamics at the population level compare between introduced and native ranges. For biotically dispersed plants, for example, little is known about the acquisition of dispersal agents in the introduced range.

REGENERATION NICHE DYNAMICS

Another potentially important difference between the native and the introduced range is the temporal patterning of variability in the environment. There may, by chance, be more abundant opportunities (i.e., favorable variants of the environment) in the introduced range. In their native ranges, species do not generally recruit at random, and often, for plants at least, the recruitment niche for juveniles is distinct from the adult niche. Populations may be recruitment-niche limited in the native range. As an example, imagine an animal that needs a certain type of nest cavity, or a plant that needs a certain microhabitat. It is possible that such a patch of resource is limiting in the native range but turns out to be abundant in the introduced range in either space or time. In a way, the enemy release hypothesis is a special case, in which the limiting factor in the native range is the enemy-free space. Other cases could be adaptation to a particular frequency of fire or other type of natural disturbance. For example, time since fire in English temperate heathlands determines whether *Pinus sylvestris* can successfully invade. The spatiotemporal distribution of recruitment niches is likely to be key for many species, and it is likely to differ from one region to another. It is something about which little is known, but it is likely to provide new perspectives on the issue of why species may do better in an introduced range than in their native range.

SEE ALSO THE FOLLOWING ARTICLES

Biological Control, of Plants / Dispersal Ability, Animal / Dispersal Ability, Plant / Enemy Release Hypothesis / Life History Strategies / Range Modeling

FURTHER READING

Davis, A. S., D. A. Landis, V. Nuzz, B. Blossey, E. Gerber, and H. L. Hinz. 2006. Demographic models inform selection of biocontrol agents for garlic mustard (*Alliaria petiolata*). *Ecological Applications* 16: 2399–2410.

Govindarajulu, P., R. Altwegg, B. R. Anholt. 2005. Matrix model investigation of invasive species control: Bullfrogs on Vancouver Island. *Ecological Applications* 15: 2161–2170.

Jacquemyn, H., R. Brys, and M. G. Neubert. 2005. Fire increases invasive spread of *Molinia caerulea* mainly through changes in demographic parameters. *Ecological Applications* 15: 2097–2108.

Jongejans, E., O. Skarpaas, and K. Shea. 2008. Dispersal, demography and spatial population models for conservation and control management. *Perspectives in Plant Ecology, Evolution and Systematics* 9: 153–170.

Koop, A., and C. C. Horvitz. 2005. Projection matrix analysis of the demography of an invasive, non-native shrub (*Ardisia elliptica*). *Ecology* 86: 2661–2672.

Mendez, V., D. Campos, A. W. Sheppard. 2009. A model for plant invasions: The role of distributed generation times. *Bulletin of Mathematical Biology* 71: 1727–1744.

Neubert, M. G., and H. Caswell. 2000. Demography and dispersal: Calculation and sensitivity analysis of invasion speed for structured populations. *Ecology* 81: 1613–1628.

Parker, I. M. 2000. Invasion dynamics of *Cytisus scoparius*: A matrix model approach. *Ecological Applications* 10: 726–743.

Ramula, S., T. M. Knight, J. H. Burns, and Y. M. Buckley. 2008. General guidelines for invasive plant management based on comparative demography of invasive and native plant populations. *Journal of Applied Ecology* 45: 1124–1133.

Strayer, D. L., and H. M. Malcom. 2006. Long-term demography of a zebra mussel (*Dreissena polymorpha*) population. *Freshwater Biology* 51: 117–130.

DISEASE VECTORS, HUMAN

L. PHILIP LOUNIBOS

University of Florida, Vero Beach

In various circumstances, invasive bloodsucking insects have transmitted, in their expanded ranges, pathogens such as viruses, bacteria, protozoa, or parasitic worms to humans, thereby leading to human disease. In some cases, the pathogens were already present in the area of transmission before establishment of the invasive vector. For others, such as yellow fever and dengue viruses native to the Old World, invasive vectors and pathogens may have been transported together, or at about the same time, to the Americas. Mosquitoes (Diptera: Culicidae) are easily the most important group of invasive, pathogen-transmitting insects.

MOSQUITO-TRANSMITTED DISEASES

Yellow Fever

Aedes aegypti, known as the yellow fever mosquito, is believed to have migrated from West Africa to the New World in the fifteenth through seventeenth centuries aboard slave ships. Alternatively or additionally, *A. aegypti* may have first invaded Portugal and Spain before reaching the western hemisphere on European ships. In either case, the evolution of domestic traits in an originally feral species

TABLE 1

Mosquito-borne Diseases Transmitted by Invasive Species

Disease	Pathogen	Invasive Vectors	Transmission Locations	Timeframe
Yellow fever	Flavivirus	*Aedes aegypti*	New World	16th–20th centuries
Dengue fever	Flavivirus	*A. aegypti, Aedes albopictus*	New World, tropical Asia Pacific region	17th century–present
Chikungunya fever	Alphavirus	*A. aegypti* *A. albopictus*	Tropical Asia, India Indian Ocean, Italy	1950–present 2005–2008
Zika fever	Flavivirus	*A. aegypti*	Malaysia, Micronesia	1979, 2005
Ross River fever	Alphavirus	*A. aegypti*	Fiji, Samoa, Cook Islands	1979, 2003–2004
West Nile fever	Flavivirus	*Culex pipiens* complex	Europe, Americas	1990s
LaCrosse encephalitis	Bunyavirus	*A. albopictus*	eastern United States	1999
St. Louis encephalitis	Flavivirus	*C. pipiens* complex	eastern North America	19th century
Lymphatic filariasis	Nematode (*Wuchereria bancrofti*)	*Culex quinquefasciatus*	Tropical Asia, India Haiti, Brazil	19th century
Malaria	Protozoan (*Plasmodium* spp.)	*Anopheles gambiae* s.l. *Anopheles darlingi*	Brazil, Mauritius Amazonian Peru	1930–1941, 1866–1867 1990s

was crucial for enabling *A. aegypti* to occupy and flourish in water storage jars in the holds of these sailing vessels.

Yellow fever was absent from urban settlements in the New World until the arrival of *A. aegypti*, the only known vector of urban epidemics of this disease. The first clearly documented New World epidemic of yellow fever occurred in the Yucatan in 1648, although it has been suggested to have appeared in Haiti as early as 1495. Throughout the seventeenth through nineteenth centuries, yellow fever ravaged seaports on the Atlantic Coast, as far north as Philadelphia and New York. Presumably the yellow fever virus was being reintroduced by passengers, especially African slaves, on these ships.

In tropical and mild temperate regions of the Americas, *A. aegypti* became established and spread into the interior, while in the northeastern United States, where this species was unlikely to overwinter, new arrivals on ships may have vectored the yellow fever epidemics. Despite its abundance in tropical Asia for over 100 years (see below), *A. aegypti* has not transmitted yellow fever there.

Dengue Fever

Although dengue epidemics occurred centuries earlier in the Americas in association with the transport of *A. aegypti* from Africa on ships, the reemergence of dengue and appearance of dengue hemorrhagic fever followed the failed eradication scheme of the 1950s and 1960s against this species. In North America, where this species vectored epidemic dengue until early in the twentieth century, its distribution receded in the 1990s, concomitant with the spread of invasive *Aedes albopictus* in the southeastern United States.

In tropical Asia, *A. aegypti* is presumed to have arrived and established later, based on the absence of urban dengue in this region until late in the nineteenth century. The more recent dispersal of *A. aegypti* in the Oriental region is evidenced by comparatively low genetic diversity in tropical Asian populations of this species.

Although *A. aegypti* is responsible for most current mosquito-to-human transmissions of dengue, the presumed bridge vector of sylvatic dengue in its native range in tropical Asia was most likely *A. albopictus*, which is now regarded as the most important secondary vector of this same disease as well as the most invasive mosquito species of the twentieth century.

The native range of *A. albopictus* is centered in the Oriental region and India but extends west to the African island nations of Mauritius, the Seychelles, and Madagascar. In the hundred years preceding its most recent diaspora, *A. albopictus* had spread to Hawaii, Guam, and other Pacific islands. This species has become the primary vector of dengue in such localities as Hawaii, where *A. aegypti* has been successfully eradicated.

Chikungunya Fever

Chikungunya fever is a zoonotic arboviral disease native to Africa, which, until recently, received little attention, even though it was identified during the 1950s and 1960s from the Philippines, Thailand, and India, where invasive *A. aegypti* transmitted the virus to humans in the absence of evidence for a sylvatic cycle involving nonhuman primates, which is documented for Africa. However, in 2004 a strain of chikungunya originating from an outbreak in eastern Kenya was detected offshore in the Indian Ocean nation of Comoros and neighboring Mayotte, where both *A. aegypti* and the recently established *A. albopictus* were incriminated as vectors. As the virus was transported southward, Madagascar, Mauritius, and, especially, Reunion suffered epidemics. In the absence of *A. aegypti* on Reunion, *A. albopictus* was incriminated as the sole vector of the epidemic, which afflicted about 30 percent of the population and caused nearly 100 deaths. Subsequently, this viral wave moved on to India and tropical Asia, where both invasive *Aedes* species were involved in its epidemic transmission to humans. A relatively small epidemic in 2007 in northern Italy, vectored by *A. albopictus*, caught the attention of the European public because of the introduction of a tropical disease from an African virus transmitted by an invasive mosquito species, formerly regarded only as a pest.

Zika Fever

Zika virus, originally described as a zoonosis in Uganda, has been identified as a cause of mild disease in both Malaysia and Micronesia, where *A. aegypti* has been incriminated as the invasive vector.

Ross River Fever

On the Australian mainland, this arbovirus, which causes rashes and arthralgia in humans, is translated by native mosquitoes, but invasive *A. aegypti* are known to vector the virus on neighboring islands, such as Fiji and the Cook chain.

West Nile Fever

West Nile is another arbovirus originally from Africa—where it was relatively obscure and benign—that caused severe and widespread epidemics of human disease during its spread into Europe and North America late in the twentieth century. Because this arbovirus species seems to require birds for enzootic amplification, invasive species of the ornithophilic *Culex pipiens* complex have become either maintenance or epidemic vectors in its new range, particularly *Culex pipiens* s.s. in urban habitats of the northern United States and *Culex quinquefasciatus* in the South. Introgression between these closely related invasive species in the United States and elsewhere has led to local vector populations with geographically variable genetic backgrounds biased toward one species or the other.

Endemic Arboviral Encephalitides

The preceding sections have dealt primarily with mosquito-borne viruses that have been transported between continents, whereupon new transmission cycles involving invasive vectors have become established. However, when a competent vector species invades new sites where endemic arboviruses already circulate, the invasive vector species may insinuate itself into the virus–host life cycle to become a primary or secondary bridge vector. Such seems to have occurred in portions of the eastern United States, where LaCrosse virus is now suspected to be transmitted to humans by *A. albopictus*, which shares many habitat preferences with the native vector of this arbovirus, *Aedes triseriatus*. *Aedes albopictus* has also been found naturally infected with eastern equine encephalitis (EEE), which, like LaCrosse, may cause severe disease in humans, but to date there is no evidence of transmission of EEE to humans by this invasive vector species in nature.

A longer association between invasive vector species and an endemic arbovirus occurs in the eastern United States between members of the *C. pipiens* complex and St. Louis encephalitis virus. As with related West Nile virus, *C.*

pipiens s.s. is a more important vector species in the northern United States of St. Louis encephalitis, while *C. quinquefasciatus* is dominant in the southern United States.

Lymphatic Filariasis

So-called Bancroftian filariasis is caused by a parasitic nematode worm that is ordinarily transmitted by night-biting mosquitoes. Primarily a tropical affliction, it is most common in poor urban environments with inadequate sanitation where its cosmotropical vector, *C. quinquefasciatus*, thrives in polluted, eutrophic water and proliferates. As *C. quinquefasciatus* was spread widely by shipping in the nineteenth and twentieth centuries, fomenting multiple invasions in many locations, all transmission of human lymphatic filariasis by this species outside the vector's native Africa can be regarded as its invasive range (e.g., in India and Haiti), where this disease may be a significant public health burden.

Malaria

The arrival from West Africa in 1930 and establishment and spread into the northeast of Brazil of the African malaria vector *Anopheles gambiae* rivals the introduction of *A. aegypti* into the New World for epidemiological impact. Larvae or adults of this anopheline species are believed to have traveled by air or fast passenger ship from Dakar, Senegal, to Natal, Brazil, where the first malaria epidemic attributable to *A. gambiae* occurred in March to May of 1930. Although malaria was endemic in northeastern Brazil, the native anopheline vectors were inefficient transmitters compared to the highly anthropophilic and endophilic *A. gambiae*.

At the time of its invasion into Brazil (1930–1941), *A. gambiae* was not recognized as a species complex, which consists currently of seven closely related, named species. Of the two members of this complex known as efficient malaria vectors, *A. gambiae* s.s. and *Anopheles arabiensis*, the adaptation of the invader to the dry northeast of Brazil more closely resembled the habits of *A. arabiensis* in the African Sahel region. Recent analyses of DNA from preserved museum specimens from Brazil confirmed the Brazilian invader as *A. arabiensis*.

Although insecticide treatments eradicated *A. arabiensis* by 1932 from Natal, by this time the invading species had escaped and spread outside the city limits. In the so-called silent years, the invader spread first west, then south (Fig. 1), especially along water courses. Then a malaria epidemic said to rival the worst outbreaks described in the literature of this disease occurred in 1938, with mortalities ranging from 10 to 25 percent. Such high mortality

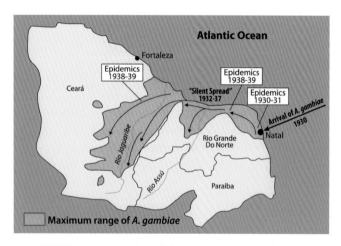

FIGURE 1 Invasion routes of *Anopheles gambiae* s.l. (subsequently resolved to be *Anopheles arabiensis*), in northeastern Brazil where it transmitted epidemic malaria in the 1930s. (Drawing based on information from Soper, E.F., and D.B. Wilson. 1943. *Anopheles Gambiae in Brazil, 1930 to 1940*. New York: Rockefeller Foundation.)

rates had also occurred on the island of Mauritius in 1866–1967, when accidentally introduced *A. gambiae* s.l. also fomented malaria epidemics. In both Mauritius and Brazil, the invasion of *A. gambiae* s.l. changed malaria transmission from endemic to epidemic.

Field observations indicated that the spread of *A. arabiensis* during 1930–1937 occurred primarily by infiltration (i.e., by movement from one habitat to another by natural dispersal). However, long distance dispersal in 1939–1940 may have been assisted by car, train, or boat. Concentrations of this species in discrete areas, especially during the dry season, facilitated the successful massive eradication scheme led by the Rockefeller Foundation, which treated larval habitats with the toxic insecticide Paris green. Coupled with house spraying with pyrethrum, the African invader was eliminated, without DDT, only 19 months after the eradication campaign began. As soon as *A. arabiensis* disappeared from northeastern Brazil, malaria incidence dropped precipitously.

Anopheles darlingi, the most efficient malaria vector native to the New World, is the principal transmitter of this disease in the Central Amazon region. Prior to the 1990s, this species was sparsely distributed in the Upper Amazon region of Peru and was absent from the Iquitos region as recently as 1991. However, sometime in the 1990s, this species appeared in abundance in the outskirts of Iquitos and became the primary vector of epidemic malaria, with more than 130,000 registered cases in 1997 among a population of less than 4 million persons. Although the mode of range expansion of *A. darlingi* into the Iquitos region is unknown, one speculation is that larvae were transported by boats with tropical fish, which are cultivated locally for export and sale in the aquarium trade.

DISEASES TRANSMITTED BY FLEAS, TSETSE FLIES, KISSING BUGS, AND LICE

Plague

The plague bacillus, responsible for the Black Death of the Middle Ages, is transmitted to humans from infected rodents, often rats, by flea bites, usually from the Oriental rat flea *Xenopsylla cheops*. Three pandemics of bubonic plague—in the sixth to eighth centuries AD, during medieval times, and from the late nineteenth into the early twentieth century—are estimated to have killed more than 200 million persons. The worldwide pattern of outbreaks in ports during the last pandemic suggests that the disease was probably introduced from infected Norway rats that disembarked from sailing ships or freight vessels accompanied by *X. cheops* and other flea species. During

that pandemic, plague began in inland China and reached Chinese port cities in 1894, from whence at the turn of the century it was disseminated by ships and erupted in epidemic form at major international ports such as San Francisco, Asuncion, Rio de Janeiro, Brisbane, Sydney, and Bombay. It is generally believed that the plague epidemic in San Francisco in 1899 led to the permanent establishment of plague as an endemic disease of wild rodents in the western United States.

African Trypanosomiasis ("Sleeping Sickness")

Tsetse flies are best known as disease vectors of domestic animals but remain locally important in Africa as transmitters of trypanosomes that cause human sleeping sickness. Some evidence for their invasiveness comes from reinfestations of areas from which tsetse flies were eradicated. The West African island of Principe was freed of *Glossina palpalis* for approximately 40 years before the species was reintroduced by ship or plane from Fernando Po, about 200 km distant. Human alterations of the landscape may promote range extensions of tsetse species, such as hedgerow plantings in the Nyanza District of Kenya that led to local expansions of the distribution of *G. palpalis*.

Chagas Disease ("American Trypanosomiasis")

Evidence supports recent, host-associated range expansions of species of the triatomine vectors of Chagas disease. The best documentation is for the human-aided spread of *Triatoma infestans*, the most important vector of *Trypanosoma cruzi* in South America. Domestication of this bug species is presumed to be pre-Columbian, whereafter dispersals followed pre-Inca tribal migrations from native Bolivia to southern Peru and northern Chile. Early in the 1900s, *T. infestans* accompanied migrant workers across Paraguay and Argentina to coffee farms in the vicinity of São Paulo, Brazil. Later, following the clearing of Brazilian coastal forests, this vector species colonized more southerly regions of the country and, independently, northeastern Brazil. These changes in distribution of *T. infestans* in the past century have been associated with outbreaks of acute Chagas disease in Brazil. Once *T. infestans* had been passively transported into an area, probably among domestic effects, it may have actively dispersed among houses by flights up to 200 m.

Rhodnius prolixus is an important Chagas disease vector with a disjunct distribution between Central American and Venezuelan and Colombian populations. Nymphs and eggs of this species were hypothesized to have been transported phoretically from South to Central America in the plumage of storks, which may harbor *R. prolixus* in nests and routinely migrate between these regions. However, some investigators favor a hypothesis that Central American *R. prolixus* are derived from an accidental release in El Salvador of laboratory-reared insects originally collected in Venezuela. These escapees and their descendants are presumed to have disseminated to other Central American countries by passive transport among immigrants. A recent establishment of this species west of Rio de Janeiro, Brazil, is also suspected of having resulted from a laboratory escape.

Typhus

Phoretic transport of the human body louse on migrating troops was responsible for the spread of typhus, which erupted in epidemic form during wars from the Middle Ages through World War II, when delousing with DDT interrupted transmission. Although human lice may have been present in an uninfected region before the arrival of troops, the massive influx of abundant and infected *Pediculus humanus* was the immediate cause of typhus epidemics during wartime.

SEE ALSO THE FOLLOWING ARTICLES

Ecoterrorism and Biosecurity / Epidemiology and Dispersal / Flaviviruses / Malaria Vectors / Mosquitoes / Pathogens, Human

FURTHER READING

Juliano, S. A., and L. P. Lounibos. 2005. Ecology of invasive mosquitoes: Effects on resident species and on human health. *Ecology Letters* 8: 558–574.

Laird, M., ed. 1984. *Commerce and the Spread of Pests and Disease Vectors.* New York: Praeger.

Lounibos, L. P. 2002. Invasions by insect vectors of human disease. *Annual Review of Entomology* 47: 233–266.

Marquardt, W. C., B. C. Kondratieff, C. G. Moore, et al., eds. 2005. *Biology of Disease Vectors*, 2nd ed. New York: Elsevier Academic.

Mullen, G., and L. Durden, eds. 2009. *Medical and Veterinary Entomology*, 2nd ed. New York: Academic Press.

DISPERSAL ABILITY, ANIMAL

R. M. McDOWALL

National Institute of Water and Atmospheric Research, Christchurch, New Zealand

Dispersal is simply the relocation of individuals of a species from a place of origin, where the species customarily occurs, to another place. The temporal or spatial scale of this relocation is not prescribed, and it varies hugely from taxon to taxon, from place to place, and from one time to another. At one extreme, dispersal is minor expansion at the periphery of a species' natural range, and at the other spatial extreme, it can involve displacement of individuals hundreds or even thousands of kilometers from a species' existing range.

MOVEMENT OF INDIVIDUALS

Dispersal can be deliberate or involuntary, and the number of dispersing individuals varies greatly. Furthermore, a dispersal event can involve the dispersal of an individual or individuals to novel locations or simply to other already occupied places. Dispersal is a natural process that goes on all the time, but it sometimes results in the presence of a species in novel places and becomes important to dispersal ecology. Humans do not always notice dispersal events, but when they do, they tend to regard them negatively. In many cases, dispersal has direct or indirect human involvement.

For some years during the later decades of the twentieth century, the idea that biotic dispersal was important was regarded by some, especially some biogeographers, as untestable. It was argued that the inability to test dispersal hypotheses rendered them unscientific. More recently, however, objections to biogeographic dispersal hypotheses have been substantially overcome with the advent of molecular technology. This has made it possible to estimate the dates of recent gene flow between disjunct populations, which has, in turn, resulted in a major change of perspective. Comparisons of date estimates with projections of when there might have been land (or sea) connections between locations where a species is known have enabled considerations of whether an observed distribution pattern could have developed when land connections existed, or whether the estimated gene flow was more recent. The result has been a substantial resurgence of interest in dispersal, to the point where molecular evidence is pointing to some dispersal events that few biogeographers would have predicted

MECHANISMS THAT DRIVE DISPERSAL

Many mechanisms drive dispersal. One issue to consider is the locomotory abilities of the species: do they walk, fly, or swim? Some animal species are able to travel vast distances, as is indicated by the huge migrations of some birds and fishes like godwits or salmon.

Sometimes dispersal is *assisted* by natural phenomena, such as prevailing or atypical winds, ocean currents, or the downstream flow of rivers; patterns of behavior may be adapted to make routine use of such assistance. Stylized posturing of ballooning spiders is an example.

Other times, dispersal may be entirely *driven* by such natural phenomena, may involve no active locomotion by individual animals, and so may be an entirely passive process. An example is transportation of usually flighted animal species across the Tasman Sea from eastern Australia to New Zealand, about 2,000 km to the east, by means of uplift currents caused by Australian bush fires. This can lift insects and birds high into the atmosphere, and they may be swept east by prevailing westerly winds to New Zealand. As a result of this process, new species arrive in New Zealand all the time, and a large proportion of the birds there are regarded as having dispersed relatively recently, such as species of swallow and plover in recent decades.

Modern human transport technology, especially across perceived barriers such as mountain ranges, deserts, or oceans (such as by aircraft or ships), is facilitating the spread of organisms from their natural ranges into new ecosystems, exemplified by the arrival of zebra mussels in the North American Great Lakes or of goby fish in the San Francisco Bay. The directions taken and distances involved are at a scale and frequency unprecedented and far exceeding what might occur naturally. This has led to the emergence of "invasion biology"—a lively and important field of study. A really interesting example is the human translocation of American shad to the Pacific Coast of North America, which was followed by natural dispersal across the Pacific Ocean to northeastern Asia.

Species are constantly being discovered in new locations. Sometimes it is not clear that they have been carried along with human activities or by some human-mediated influence, but in other cases, the only possible mechanism has been some form of human-associated transport activity. Examples include terrestrial insects carried in cabins of aircraft or among cargo of ships, and aquatic organisms in ballast water taken on board ships at one location and discharged at another, which is probably how zebra mussels reached the Great Lakes. In such instances, the cause may be obvious from the type of organism involved, the likely source, or the extreme oddity of the location of establishment, or there may be a clear correlation between the arrival of a species new to a location and some known human vector or transport systems.

EASE OF DISPERSAL

Some organisms disperse more easily than others. These differences in ability create a continuum, and where particular species fit along this continuum relates to many of the species' inherent characteristics. Our ability to predict where along such a continuum any species might be is less accurate than we think. Thus, we are often surprised at what novel species are turning up where, and how they got there. Nevertheless, we often assume that widespread species are good natural dispersers and may easily disperse under human influence.

Large species may disperse less easily, especially those taxa that actively disperse (though not necessarily); small species achieve passive dispersal more easily, though it cannot be assumed that all smaller species will do so more easily. Species with large natural ranges may disperse more easily, although some species' broad ranges may also mean little more than wide tolerances for different habitats. Other species disperse well because they produce huge numbers of offspring, often pelagic eggs and larvae, for example, that are discharged into the open ocean, where they can be transported in currents or are apt to be drawn into the ballast water of ships and are thereby carried long distances. This is especially true of some marine molluscs and fishes. Some animals, though highly sessile, attach to hulls of ocean-going vessels and discharge young along the route of passage, and so become widely invasive. There have been recent problems with oil rigs, which tend to spend long periods at one location, accumulate substantial marine epifaunal communities, and then be shifted long distances carrying that epifauna with them. Thus, sessile species may disperse as passengers.

Clearly, if a species cannot naturally cross environmental barriers, then it has no chance of becoming naturally established elsewhere, and in some cases, explicitly human-assisted arrival is crucial for establishment and eventual invasiveness; again, the zebra mussel is a good example. Once individuals of species have arrived at a new location, survival, reproduction, proliferation, and eventual invasiveness cannot be assumed, and it is probable that only a small minority of successful dispersers do become established. Some biologists have drawn an interesting distinction between "income breeders," which continue to produce offspring over a period and may be good at becoming established in a novel environment, and "capital breeders," which produce all their offspring in a single short reproductive burst, and this distinction has implications for dispersal ecology. Clearly, parthenogenetic species have an advantage, as a single individual can establish a population. Organisms that bear live young may disperse more successfully into new locations than egg bearers, partly because the females (usually) carry their progeny. This provides protection for their young and sometimes produces multiple offspring, serving as a basis for a breeding population.

Quite closely related animal taxa may have very different life histories and dispersal abilities. For instance, some have large numbers of small eggs and broadcast them into the pelagic zone of the ocean, where they can be swept passively long distances in ocean currents, while others may have large, yolky eggs that undergo more direct development; they have little or no pelagic stage, and so they do not disperse this way. This becomes evident when one compares tropical molluscs, with numerous small eggs and pelagic larvae, and sub-Antarctic molluscs, which tend to have fewer large, yolky eggs and no pelagic juvenile life stages. Some marine animals fasten their demersal eggs to substrates, and this limits dispersal to the extent that such substrates are immobile. However, some fish, especially gobies, fix their eggs to hard surfaces that can include floating vessels, or to seaweed substrates that may be carried long distances in ocean currents, enhancing the likelihood of successful dispersal. Some cryptic organisms may find refuge in floating rafts of seaweed or other vegetation or may seek the complex surface structures of ships or equipment, and their cryptic behavior aids their dispersal.

Endless attributes of animal species may interact with potential mechanisms that promote dispersal; while it can be predicted that some types of animals will be easily and widely dispersed, many give little or no hint of dispersal ability but turn up unexpectedly and become invasive. Aquatic organisms have inherent dispersal advantages, in that they are often naturally buoyant in a way that terrestrial organisms are not—very few animals are naturally buoyant in air, and so most must expend energy to achieve dispersal, apart from the extent to which winds sweep them into long-term suspension. Fluvial species, particularly, have the ability to move easily downstream in a river's flow, although dispersal distances may not be great and usually do not extend ranges.

POSTTRANSPORT ESTABLISHMENT

There is, of course, no simple connection between how easily a species may disperse and the extent to which that species either becomes established in a target locality or becomes invasive to the extent that it is regarded as a nuisance or that it causes ecological damage.

There are two aspects to understanding dispersal and postestablishment behavior of invasive species.

- One issue relates to how easily a species may disperse from its natural range to occupy new ranges whether under human influence or not: in such an instance, dispersal is a major driver of invasion.
- The other issue relates to how easily an invasive species, once it has been transported to a new site (whether deliberately by humans or by accident), may then spread from its new site of establishment, and that is an important aspect of dispersal ecology.

A species may differ greatly in these two abilities; some species that are not good at long distance dispersal, and are unlikely ever to reach some places where they are found without human assistance, may, having arrived at a new location, then spread widely, proliferate, and become seriously invasive and ecologically damaging (once again, the zebra mussel is a good example). Dispersal and establishment are very different processes and are influenced by different attributes of species.

Butterflies

A key element of post-transport establishment is the ability of a species to spread from its release site into optimal habitats for population expansion, a feature sometimes termed "outward scattering." This may be a matter of chance and may relate to the extent to which a species' habitat requirements are met at a destination. Sometimes the requirements are general and easily met, but in other instances, as when a dispersing species depends strongly on particular foods (such as insect caterpillars with a narrow host plant range), successful dispersal can be followed by a failure to establish. An interesting example is the monarch butterfly, which could not establish in New Zealand until acceptable food plants had become established there. Once these plants were present, the butterfly did become established, and in fact it is quite likely that this species had been dispersing for many years before this. Similarly, the Australian blue moon butterfly arrives regularly in New Zealand, sometimes in scores or more, but because it can find no suitable plants on which to oviposit, it has not become established. However, climate change, or perhaps the introduction of the appropriate food plants, could make possible the establishment of this butterfly.

Snails

A remarkable example of long-distance dispersal of a species that has become invasive is the New Zealand mudsnail, *Potamopyrgus antipodarum*, a very small, black snail, typically no more than 5 to 7 mm in height (Fig. 1). It is

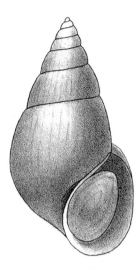

FIGURE 1 *Potamopyrgus*, the tiny New Zealand mud snail, has invaded aquatic ecosystems widely across North America, Western Europe, and Australia. (From Suter, H. 1915. *Manual of the New Zealand Mollusca.*)

primarily a freshwater species, although it spreads downstream into brackish river estuaries. It lives entirely on the substrate—which can include stones, mud, or aquatic macrophytes—where it grazes the periphyton. In many locations, it is parthenogenetic, with all individuals being female, and it is then a clonal species. It lives in very diverse conditions and tends to proliferate in waters of declining quality, often being seen as an indicator of anthropogenic water deterioration. It is endemic to New Zealand, where it is widespread. As a very abundant member of New Zealand aquatic ecosystems, it was found in the diets of introduced brown and rainbow trout there, and anglers became accustomed to catching trout whose stomachs appeared to be filled with fine gravel: in fact, they were full of *Potamopyrgus*. Studies of the calorific content of snails that had passed through the digestive tracts of trout showed that they had about as much energy as when first eaten, and from this it was realized that the trout were gaining little or no nutrients from the snails. This seems to be because *Potamopyrgus* has an operculum, and the snails were capable of "closing down" using their operculum, which effectively prevented their digestion by the trout. The observation that the snails can survive passage through the digestive tract of a trout with little damage hints at their success in spreading so early from New Zealand to Great Britain (see below) and beyond. An additional factor in their spread is their ability to reproduce parthenogenetically, because only one individual needs to be carried into a new habitat for the species to become established. Otherwise, not only must one of each sex be transported, but both sexes need to be present in a novel habitat, close enough together for mating.

In the middle of the nineteenth century, a species of *Potamopyrgus* was described from Great Britain, but it was not until the 1970s that it was realized that the British population was invasive and had originated in New Zealand. How it was transported to Great Britain is unknown, but this event was a harbinger of the future spread of this little snail into many parts of the world.

Probably deriving from its invasive British populations, *Potamopyrgus* began to spread widely throughout Western Europe and is now known in at least France, Switzerland, Spain, the Ukraine, Italy, Poland, Germany, Russia, Turkey, the Czech Republic, Greece, the former Yugoslavia, Finland, and countries around the Black Sea. It has even found its way to Iraq in the Middle East. It is also now present in Australia. In the United States and Canada, it appeared in western rivers and lakes in the 1980s and soon became known as an invasive species "of most concern" in North America. It is now widespread from the Great Lakes in the east to the Pacific coastal river systems and is a particular pest in Yellowstone National Park. The American populations are believed to have originated in Europe rather than New Zealand, although this is uncertain. What is of particular interest is the ability of this inconspicuous little snail to spread so widely and successfully. It is a disperser of the highest order. At least in some ecosystems, *Potamopyrgus* is becoming a dominant organism, leading to so much research that more is known about it in some introduced locations than in New Zealand. It is coming to dominate invertebrate communities in some streams, with alleged counts of up to 800,000 snails per square meter—that is, nearly one snail in every square millimeter.

Its impacts may not be limited to its ecological dominance. In New Zealand, *Potamopyrgus* hosts several parasites, some of which are present only at early larval stages but then go on to become reproductive adults in various fish species and also birds. It is not impossible that some fish species, such as brown and rainbow trout, or widespread aquatic birds such as cormorants, could end up infected by these parasites in ecosystems where *Potamopyrgus* has become invasive.

Potamopyrgus is thus a classic invader—it disperses readily, is parthenogenetic, and produces live young, so single individuals can establish new populations and proliferate rapidly. It has broad and generalized habitat tolerances and can thrive in degraded habitats. This little snail also gives the lie to the common notion, which originated with Charles Darwin, that island species are inherently maladapted for invasion and that it is always mainland species that ravage islands.

Fishes

The story of the spread of the European brown trout, *Salmo trutta*, is very different from that of the New Zealand mud snail. The brown trout is a favorite of recreational anglers. Its natural distribution is the cool-to-cold waters of Europe, south to the Atlas Mountains of Mediterranean Africa and east as far as the Black Sea. When European, especially British, settlers began to colonize other lands, especially Australia, New Zealand, Patagonian South America, and southern Africa, during the mid-nineteenth century, they went to great efforts to take the brown trout with them. It is also now widespread in North America.

Transporting stock into the southern hemisphere in the mid-nineteenth century, especially carrying trout across the tropics in slow-moving sailing ships, proved difficult, but by the 1860s, after much persistent effort (the ova were encased in large chests of ice), brown trout ova had arrived in Australia and were rapidly spread into other southern lands, especially New Zealand and Patagonian South America (Fig. 2). As a result, fabulous angling for brown trout developed. Brown trout translocated into the southern hemisphere were believed to be from stocks that were restricted to freshwater. However, in their European homelands, brown trout include some stocks that are sea migratory—known explicitly as anadromous sea trout. And although the translocated stock were ostensibly nonanadromous, it was soon found that, especially at higher southern latitudes, the introduced trout tended to become anadromous. Perhaps just a minority of individuals from the progeny went to sea. As a result, they

FIGURE 2 A nineteenth-century trout hatchery built for stocking New Zealand streams with brown trout. In many temperate countries, immense effort was invested in importing salmonid fishes from both Western Europe and North America to establish largely recreational angling fisheries. Hatcheries were established to rear stock for release. (Photograph courtesy of Nelson Acclimitisation Society.)

FIGURE 3 Releasing brown trout into a New Zealand stream for anglers to catch. Establishment of salmonids included the need for hatchery releases, at first to establish the species in natural habitats, and subsequently to augment natural reproduction to support the fisheries. (Photograph courtesy of D. Maindonald.)

grew faster and bigger than when restricted to freshwater. Brown trout tend to home—they can navigate back to the stream where they hatched in order to spawn. However, a few fish stray, and as a result, brown trout soon began to spread throughout coastal seas and into river systems where they had never been liberated.

One would think this would delight anglers, and in substantial measure it did: brown trout have become even more widespread and, in some instances, have grown bigger and faster than they would have if restricted to streams and lakes. Such sea-run brown trout, for example, form the basis of valued trout fisheries in the Falkland Islands and on Tierra del Fuego. However, there has been a serious downside to this spread. Brown trout have damaged the ecologies of the rivers and lakes where they became established by preying on both the stream insect communities and the indigenous fish populations. And with respect to dispersal, it has been the ability of brown trout to disperse through coastal seas (in addition to its popularity with anglers; Fig. 3) that has been a major driver of its spread. It is now ranked as one of the 100 most seriously invasive species globally (see Appendix). Thus, in this instance, while dispersal ability made no contribution to the initial establishment of an introduced species in distant locations, it has been hugely consequential after deliberate translocation by humans.

CONCLUSION

Dispersal is a fundamental global issue for invasive species' ecology: it is a key attribute for the spread of invasive species to new, often distant, habitats, and equally for their proliferation and spread once propagules can reach these new habitats. Dispersal is an outcome of the interplay of a range of phenomena: in part naturally occurring processes such as winds and ocean currents, in part a wide range of species attributes such as size, flying, or swimming ability, and in part various life history and ecological characteristics of the species. These processes and attributes are often mediated by human activities, especially relating to modern international travel and trade.

SEE ALSO THE FOLLOWING ARTICLES

Acclimatization Societies / Ballast / Fishes / Dispersal Ability, Plant / New Zealand: Invasions / Propagule Pressure / Snails and Slugs / Zebra Mussel

FURTHER READING

Clobert, J., E. Danchin, A. A. Dhondt, and J. D. Nichols. 2001. *Dispersal.* Oxford: Oxford University Press.

Floerl, O., G. J. Inglis, K. Dey, and A. Smith. 2009. The importance of transport hubs in stepping-stone invasions. *Journal of Applied Ecology* 46: 37–45.

Gammon, D. E., and B. A. Maurer. 2002. Evidence for non-uniform dispersal in the biological invasions of two naturalized North American bird species. *Global Ecology and Biogeography* 11: 1155–1161.

Holway, D. A., and A. V. Suarez. 1999. Animal behaviour: An essential component of invasion biology. *Trends in Ecology and Evolution* 14: 328–333.

Johnson, L. E., and J. T. Carlton. 1996. Post-establishment spread in large-scale invasions: Dispersal mechanisms of the zebra mussel, *Dreissena polymorpha. Ecology* 77: 1686–1690.

Lewis, P. N., D. M. Bergstron, and J. Whinam. 2006. Barging in: A temperate marine community travels to the subantarctic. *Biological Invasions* 8: 787–795.

Miglietta, M. P., and H. A. Lessios. 2009. A silent invasion. *Biological Invasions* 11: 825–834.

Schöpf Rehage, J., and A. Sih. 2004. Dispersal behaviour, boldness, and the link to invasiveness: A comparison of four species of *Gambusia. Biological Invasions* 6: 379–391.

Whinam, J., N. Chilcott, and D. M. Bergstrom. 2005. Subantarctic hitchhikers: Expeditioners as vectors for the introduction of alien organisms. *Biological Conservation* 121: 207–219.

Wonham, M. J., J. T. Carlton, G. M. Ruiz, and L. D. Smith. 2000. Fish and ships: Relating dispersal frequency to success in biological invasions. *Marine Biology* 136: 1111–1121.

DISPERSAL ABILITY, PLANT

EUGENE W. SCHUPP

Utah State University, Logan

Dispersal ability of plants is a key to understanding local population dynamics and population spread. A variety of biotic and abiotic dispersal modes have evolved in plants, some carrying propagules short distances and others contributing to long-distance movement, some dispersing propagules to sites suitable for plant establishment and others depositing seeds in inhospitable environments, some treating seeds gently and others destroying seeds. While dispersal distance is a major component of dispersal ability, a complete view also requires understanding dispersal effectiveness—the contribution of dispersal to plant population growth. Our present understanding of seed dispersal suggests that the species with the greatest potential for invasion are those that combine a relatively high proportion of longer-distance dispersal with highly effective local dispersal that promotes rapid population growth.

ADVANTAGES OF SEED DISPERSAL

There are three major hypothesized advantages of seed dispersal. The colonization hypothesis suggests that seed dispersal is advantageous because it allows parents to disperse seeds to unoccupied habitats—for example, dispersal of early successional species to recent disturbances. The escape hypothesis suggests that the advantage of dispersal lies in dispersing seeds away from the parent so that they escape distance- and density-dependent seed and seedling mortality around adults; the advantage lies in getting away from the parent, not in arriving in any particular place. Lastly, the directed dispersal hypothesis suggests that the advantage of seed dispersal is having seeds carried to predictably favorable locations, such as mistletoe seeds being deposited on suitable branches. In the context of invasive species, colonization encompasses the dispersal of seeds to new locations and the founding of new subpopulations—that is, the invasion process itself. In addition, all three hypothesized advantages can indirectly affect invasion by contributing to local population growth, and therefore to more long-distance dispersal events.

DISPERSAL MODES

Plants disperse their offspring in a variety of ways, including as a true seed, as an endocarp containing multiple seeds, and as a dry fruit such as an achene or a caryopsis. All of these are generally lumped as "seed dispersal" because functionally it is the dispersal of the seed that is the focus, even if it is being dispersed within a dry fruit. Some plants also can be dispersed as vegetative propagules, such as torn off stolons of *Carex* or bulbils of *Allium*, but the primary focus of this article will be on "seed" dispersal defined broadly.

Dispersal syndromes have been used to characterize dispersal systems based on linking the primary mode of

TABLE 1
A Sampling of Dispersal Modes, Actual Structures Involved in Promoting Dispersal, and Representative Invasive Species

Dispersal Mode	Structure	Species
Hydrochory	low-density fleshy aggregate fruit	*Annona glabra* (Annonaceae)
Hydrochory	plant fragments	*Potamogeton crispus* (Potamogetonaceae)
Anemochory	samara	*Ailanthus altissima* (Simaroubaceae)
Anemochory	winged seed	*Paulownia tomentosa* (Scrophulariaceae)
Anemochory	pappus	*Tragopogon dubius* (Asteraceae)
Anemochory	tufted hairs	*Tamarix* spp. (Tamaricaceae)
Anemochory	tumbling plant-releasing seeds	*Sisymbrium altissimum* (Brassicaceaae)
Autochory (ballistic)	explosive capsule	*Oxalis stricta* (Oxalidaceae)
Autochory (creeping)	hygroscopic pappus	*Centaurea cyanus* (Asteraceae)
Myrmechory	elaiosome	*Ulex europaeus* (Fabaceae)
Epizoochory (attached to fur)	recurved bract tips	*Centaurea virgata* (Asteraceae)
Epizoochory (attached to fur)	hooked prickles and curved spines	*Xanthium strumarium* (Asteraceae)
Epizoochory (attached to fur)	hooked hairs	*Galium aparine* (Rubiaceae)
Endozoochory	fleshy berry	*Lonicera maackii* (Capriifoliaceae)
Endozoochory	aggregate fruit	*Rubus armeniacus* (Rosaceae)
Endozoochory	drupe	*Prunus serotina* (Rosaceae)
Endozoochory	arillate seeds inside capsule	*Euonymus alatus* (Celastraceae)
Endozoochory	edible legume	*Prosopis juliflora* (Fabaceae)

dispersal with traits of the seed or fruit such as the presence of wings or hooks, color, odor, and so forth. For example, plants dispersed by frugivorous birds frequently have black, blue, or red odorless fruits, while those dispersed by bats often have pale yellow or green fruits with a musky odor. Thus, we frequently talk about bird, bat, or monkey fruits. As with plants in general, invasive species cover the spectrum of dispersal modes, being dispersed by wind, water, attachment to fur, frugivorous animals, and other vectors (Table 1; Fig. 1). In addition, many structures have evolved to carry out the same mode of dispersal. For example, wind dispersal (anemochory) can be promoted by a samara or other winged structure, a pappus, tufted hairs, or even by the dead plant breaking off, blowing in the wind, and releasing seeds as it tumbles (Table 1).

However, most plant species are dispersed by a variety of modes, either simultaneously or sequentially. In the western United States, the invasive riparian tree Russian olive (*Elaeagnus angustifolia*) is dispersed by birds, including both native species and the invasive European starling (*Sturnus vulgaris*); terrestrial mammals such as raccoons that eat fallen fruits; rodents that scatter hoard seeds; and as fruits or branches dispersed by water (Fig. 2). Similarly, the invasive annual cheatgrass (*Bromus tectorum*) is dispersed by wind and by sticking to the fur of mammals, tamarisk (*Tamarix* spp.) is dispersed by wind and by floating on water, and toadflaxes (*Linaria* spp.) appear to be dispersed by water, ants, birds, rodents, cattle, and deer.

In these examples, the alternative modes of dispersal can occur simultaneously; this can be viewed as simultaneous complementary dispersal. For example, dispersal by mammals complements dispersal by birds. Such complementary dispersal generally creates a spatial pattern of seed dispersal that differs from what any single dispersal mode would create because different dispersal modes carry seeds different distances and to different habitats and microhabitats. Alternatively, diplochory, or the sequential dispersal of seeds by two distinct modes of dispersal such as initial endozoochorous dispersal by birds and secondary dispersal by scatter-hoarding rodents, provides added benefits. This can be considered sequential complementary dispersal. In this light, Scotch broom (*Cytisus scoparius*) is dispersed from the parent ballistically by explosively dehiscing legumes that can toss seeds several meters, and then secondarily by ants that carry seeds away from the site of initial deposition, often to the nest, before removing the elaiosome to consume and then abandoning the seed. In such cases, it is assumed that the benefit of phase I dispersal is escape from mortality around the parent, while the added benefit of phase II dispersal is deposition of the seed in a suitable microsite (e.g., burial).

A key to successful plant invasion is the ability to "capture" local seed dispersal systems. This is obviously straightforward for some dispersal modes. A plume floats on the wind in the invaded range as easily as in the native range, and hooks attach to fur wherever fur

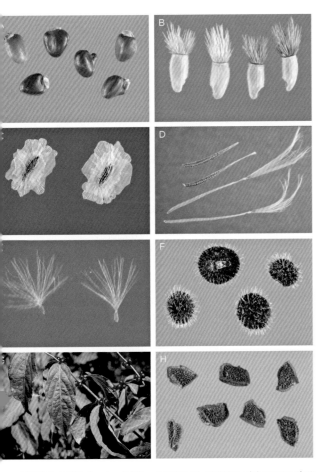

FIGURE 2 An American robin (*Turdus migratorius*), part of a large flock, feeding on the overwintering dried fruits of Russian olive (*Elaeagnus angustifolia*) in February 2009. In addition to being dispersed by a variety of birds, seeds are dispersed by frugivorous mammals, scatter-hoarding rodents, and water. (Photograph courtesy of the author.)

FIGURE 1 (A) Seeds of *Ulex europaeus* showing elaiosomes that attract ants for dispersal. (B) Hygroscopic pappus of *Centaurea cyanus* that help the seed creep across the surface. (C) Winged seeds of *Paulownia tomentosa* that assist in anemochory, or dispersal by wind. (D) *Tragopogon dubius* with pappus that carries the seed gently through even very light breezes. (E) Tufted hairs of *Tamarix chinensis* that aid in both anemochory and hydrochory. (F) *Galium aparine* seeds with hooked hairs that assist in epizoochory by attaching to fur, socks, and more. (G) Fleshy berry of *Lonicera maackii* which are fed on by birds. (H) Seeds of *Linaria dalmatica* that have no obvious adaptations for dispersal but appear to be dispersed by water, ants, rodents, livestock, wildlife, and farm equipment. (Photograph E is courtesy of Jose Hernandez, USDA-NRCS PLANTS Database. Photograph G is courtesy of USDA-NRCS PLANTS Database/Herman, D.E., et al., 1996, *North Dakota Tree Handbook*, USDA NRCS ND State Soil Conservation Committee; NDSU Extension and Western Area Power Administration, Bismarck. All other photographs are courtesy of Steve Hurst, USDA-NRCS PLANTS Database.)

is encountered. However, the assemblage of new dispersal systems is more complex in fleshy-fruited plants. Frugivore-plant seed dispersal systems tend to be generalist enough that fleshy-fruited plants find suitable endozoochorous dispersers virtually anywhere they arrive. Nonetheless, frugivores show preferences based on fruit traits that influence which native frugivores disperse how many seeds of any given exotic species.

Lonicera maackii, an invasive shrub in eastern North America, is dispersed by at least four native frugivorous birds: extensively by American robin (*Turdus migratorius*) and cedar waxwing (*Bombycilla cedrorum*), and only minimally by hermit thrush (*Catharus guttatus*) and northern mockingbird (*Mimus polyglottos*). As with Russian olive, this shrub has also incorporated the invasive European starling (*S. vulgaris*) into its dispersal assemblage. Because different species use the habitat differently, this disperser assemblage has resulted in wide dispersal of viable seeds to a range of habitats and in rapid population growth and spread.

POPULATION SPREAD

A primary contributor to population spread and invasion is the longer-distance seed dispersal making up the tail of the dispersal distribution—the minority of seeds carried unusually long distances from the parent. While shorter-distance dispersal drives local population dynamics, it is longer-distance events that found new populations and connect subpopulations in a metapopulation.

Seed rain is frequently expressed in terms of a dispersal kernel, a probability density function that represents the probability of seeds being dispersed to given distances from the source. Such curves are generally considered leptokurtic, with a greater peak density at the mean and a fatter tail than what would be expected from a normal distribution: that is, more seeds going very short distances and very long distances than expected. In particular, it has become widely accepted that a 2Dt distribution (bivariate Student's *t* distribution) might best reflect most natural dispersal. Such a highly leptokurtic curve produces a concave kernel at the source (i.e., it can represent high levels of seedfall under the entire crown, with a rapid falloff in density past the crown edge) and a fat tail (i.e., although few seeds are dispersed very far, long-distance dispersal, or LDD, is more common than expected from traditional models). Because LDD is infrequent, it is the hardest part of the distribution to quantify, and therefore the part we know least about. However, it is clear that the quantity of LDD is extremely sensitive to slight changes in the tail (Fig. 3A).

Longer-distance dispersal occurs biotically and abiotically. Larger mobile animals, such as carnivores, ungulates, large birds, and bats, can move over large distances and thus disperse at least some seeds over many kilometers. Dispersal distances can be especially long if dispersal occurs during migration, or even premigration, when many animals start to become more restless. Seeds can also be infrequently dispersed long distances by extreme events such as strong storms and floods.

Many species have more complex dispersal kernels than that captured by the 2Dt curve. Nut trees and stone pines frequently have complementary dispersal systems in which scatter-hoarding rodents disperse seeds short distances, often with median dispersal distances of less than 5 m, and scatter-hoarding jays disperse seeds longer distances, with median distances of hundreds of meters to kilometers. Similarly, seeds of species such as cheatgrass are dispersed centimeters to meters by wind and hundreds of meters to kilometers after being caught in the fur of a passing mammal. Such complementary dispersal can result in a total dispersal kernel with multiple peaks—a near peak and tail associated with local, within-population dispersal, and a further peak and tail associated with dispersal outside the population (Fig. 3B).

While dispersal distances are clearly important, population spread is also influenced by other characteristics of the spatial pattern of seed dispersal. Pond apple (*Annona glabra*) is invasive in the tropical rainforests of Queensland, Australia, where it is primarily dispersed by water but is also dispersed by southern cassowary (*Casuarius casuarius*; Fig. 4). While water disperses *A. glabra* seeds

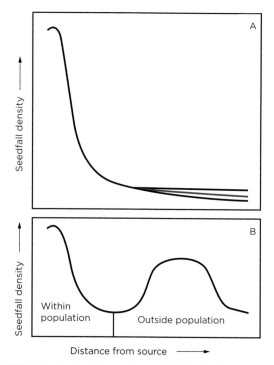

FIGURE 3 Hypothetical dispersal kernels. (A) A typical dispersal kernel has the majority of seeds moving only short distances from the source, with relatively few seeds being dispersed long distances. However, the shape of the tail, while having little impact on the overall pattern of dispersal, can have a very large impact on the actual number of long-distance dispersal events. (B) A number of plant species have multiple dispersal agents that have widely different patterns of movement and dispersal leading to a total dispersal kernel composed of two (or perhaps more) peaks and tails. For example, nut trees and stone pines are frequently dispersed by scatter-hoarding rodents that disperse seeds meters to tens of meters, and by scatter-hoarding birds that disperse seeds hundreds of meters to kilometers. Similarly, many herbaceous invasive species such as cheatgrass (*Bromus tectorum*) have seeds dispersed meters by wind and hundreds of meters to kilometers by sticking to fur of passing mammals.

FIGURE 4 The southern cassowary (*Casuarius casuarius*), an endangered terrestrial frugivore from Queensland, Australia. Due to its size and movement patterns, it can disperse large seeds relatively long distances, including those of *Annona glabra*, a rainforest invasive from the New World tropics. (Photograph courtesy of the author.)

downstream and along wetland edges, cassowaries can disperse seeds upstream and across divides to new drainages. This complementary dispersal results in a greatly altered spatial pattern of dispersal that accelerates the geographic spread of pond apple.

LOCAL POPULATION GROWTH: SDE

Local population growth depends on effective local seed dispersal. Seed disperser (or dispersal) effectiveness, or SDE, is the contribution of a dispersal agent to the growth of a plant population. SDE equals quantity times quality, where quantity is the number of seeds dispersed and quality is the probability that a dispersed seed survives to produce a new reproductive plant.

For animal seed dispersers, the quantity of seeds dispersed is determined by the number of visits the disperser makes to the plant and the number of seeds dispersed per visit. Thus, there are multiple ways to be a quantitatively effective seed disperser. Some species make frequent feeding visits while others visit only rarely, and some disperse many seeds per visit while others disperse few seeds per visit.

The number of visits is influenced by many factors, including local abundance, dependence on fruit as a resource, relative preference for the given fruit species, and availability of alternative resources. The number of seeds dispersed per visit is affected by disperser size, behavior, and physiology. Dispersers may consume pulp while dropping seeds beneath the parent, they may swallow fruits whole and remain at the plant to process the pulp and defecate or regurgitate seeds beneath the parent, or they may swallow fruits whole and quickly move elsewhere to process the pulp. The outcome is partly determined by gape size relative to fruit size, which affects the ability to swallow fruits, and partly by behavior, which affects the likelihood of depositing seeds beneath the parent as opposed to dispersing seeds away.

Although we do not have a complete understanding of patterns, a few generalizations are possible. First, it appears that there is no relationship between the number of visits a dispersal agent makes and the number of seeds dispersed per visit by that agent. Second, it appears typical for any given plant species to have only one or several species providing most feeding visits and seed dispersal, while there might be several to many more species visiting infrequently and dispersing few seeds. Third, it appears that the number of visits made to a plant is a slightly better predictor of the quantity of seeds dispersed than is the number of seeds dispersed per visit.

The quantity of seeds dispersed is important, but the quality of dispersal must also be considered. As with quantity, quality is determined by two major subcomponents: the quality of treatment in the mouth and gut and the quality of seed deposition. Some dispersal agents destroy many to most seeds, crushing, grinding, or digesting them, while others damage few to none. Gut treatment also can alter germination patterns of intact seeds, in some cases reducing germination while in others increasing the proportion of seeds germinating or the rate of germination. Although the consequences of altered germination patterns are not always clear, they likely vary among species depending on the relative advantages of rapid versus extended germination within a season and the relative advantage of having more seeds germinate this year versus having them remain in the seed pool.

The quality of seed deposition encompasses many processes affecting establishment. For example, seeds may be deposited singly or in clumps with many other seeds of the same or different species. The degree of clumping and neighbor identities both can influence seed predation, intra- and interspecific competition, disease transmission, and other factors. However, the primary influence on the quality of seed deposition is generally where seeds are deposited, in particular the distance from adult conspecifics and the habitat, microhabitat, or microsite of destination. Where a seed lands in the landscape affects the biotic and abiotic environment the potential recruit will face: seed predators, pathogens, browsers, competitors, mycorrhizae, solar radiation, temperature, soil structure, nutrient and water availability, and so forth. Because these environmental attributes vary widely, any given landscape is composed of potential seed deposition sites varying continuously from highly unsuitable to highly suitable. That is, sites form a continuously variable spectrum of degree of suitability for plant recruitment.

For animal-dispersed species, disperser behavior and physiology are probably the primary determinants of where seeds are deposited: postfeeding habitat and microhabitat selection, rate and directionality of movement, and gut passage rate. For wind and water-dispersed plants, the physical structure of the habitat, which affects movement through the landscape and eventual capture, appears to be most important. Because disperser species differ in behavior and physiology, they differ in the quality of dispersal they provide plants; some are more likely than others to deposit seeds in places where the probability of successful recruitment is especially high. Although these species can be ranked based on the quality of dispersal they provide, we do not understand well how disperser assemblages are arrayed along the quality axis. For example, is there a tendency to have a few high-quality species with most species providing low-quality dispersal?

Species dispersed by wind and water, of course, tend to have fewer dispersal agents: for example, wind, perhaps augmented with occasional water dispersal, sticking to mud in a hoof, catching on hair, or others. However, wind is not constant but varies in both sustained winds and in gusting, and it is likely that wind-dispersed seeds interact with habitat structure differently under different wind conditions, creating different patterns of seed fall and different qualities of dispersal at different times.

This is an example of the context dependence of SDE. Quantitative and qualitative components of SDE are not fixed attributes of plant or disperser species, but rather can vary greatly spatially and temporally. Disperser assemblages differ in space and time, with some species dispersing more seeds at some times and places and other species dispersing more at other times and places. The quality of any given pattern of dispersal can vary temporally and spatially as the components of the biotic and abiotic environment change. Thus, in many cases, plant species with a set of distinct complementary dispersers may more consistently recruit new individuals than may plant species with a suite of more similar dispersal agents.

In sum, both the quantity and quality of dispersal that a plant population receives affect SDE. In turn, SDE influences local population growth, and thus the success of invasive species populations once established. SDE also indirectly influences invasion of new habitats because rapid local population growth results in greater seed production and a greater number of seeds being dispersed long distances, increasing the likelihood of successful establishment of new populations.

DISPERSAL LIMITATION

Whether plants are limited more by dispersal or establishment has been much debated. Part of the difficulty is the variety of definitions of dispersal limitation. At one extreme, some refer to it only in the context of whether seeds move out of a population and arrive in unoccupied habitat. In this view, dispersal cannot be limiting within a population by definition: seeds are already there. Similarly, if seeds arrive in an unoccupied patch, then dispersal is not limiting; if the seed fails to recruit, then it is establishment limitation. This is a metapopulation perspective. On the other end of the continuum, some suggest that if greater seed arrival in a population, or in a given patch within a population, results in more successful

establishment, then dispersal limitation is occurring. This is a within-population perspective. Seed addition experiments have demonstrated both types of limitation. In many cases, sowing seeds into unoccupied habitats results in establishment, suggesting metapopulation-level dispersal limitation. Similarly, experiments sowing seeds into plots within existing populations frequently leads to greater establishment, suggesting within-population-level dispersal limitation.

Whether, and to what degree, seed dispersal limits plant invasions, then, depends to a large extent on one's perspective on dispersal limitation. From a metapopulation perspective, only colonization of unoccupied habitat is important. An absurd extreme view would be that as long as there is unoccupied habitat available in the world, dispersal limitation is occurring. More reasonably, we would be concerned with a relative measure of dispersal limitation. For example, how rapidly are new patches colonized, and does the rate change with the abundance of particular dispersal agents, with changes in habitat structure and connectivity, and degree of disturbance? In contrast, from a within-population perspective, dispersal limitation can also be an important attribute of already-occupied patches that affects the rate of local population growth.

ANTHROPOGENIC ALTERATIONS TO DISPERSAL PATTERNS

Humans have had huge impacts on seed dispersal patterns, both by acting as intentional or unintentional dispersal agents and by altering the environment in ways that affect the movement of seeds across landscapes. Many invasive species were intentionally dispersed to new regions for many reasons, including for use in erosion control, as fuel wood, in nitrogen fixation, and as ornamentals. Examples include the "dispersal" of Japanese honeysuckle (*Lonicera japonica*), Chinese tallow (*Triadica sebifera*), and kudzu (*Pueraria montana*) to the southeastern United States; black cherry (*Prunus serotina*) to Europe; mesquite (*Prosopis juliflora*) to eastern Africa; and prickly pear cactus (*Opuntia stricta*) to Australia. Unintentional intercontinental dispersal also occurs frequently: *B. tectorum* and many other weeds have been introduced to new continents as crop seed contaminants and are transported worldwide in the socks and boot laces of international travelers.

Human activities also contribute to dispersal following arrival in new continents. Toadflaxes (*Linaria* spp.) and many other species are dispersed by farm equipment, and aquatic invasives are transported among lakes and rivers attached to boats and trailers. Innumerable species are dispersed widely after becoming stuck in mud in tire treads or caught on the undercarriage of trucks; attached to pants, socks, and boot laces; caught in the fur of the pet dog at a quick stop; and so forth. Livestock management has also led to the spread of invasive species. In Ethiopia, *P. juliflora* is rapidly invading grasslands because cattle feed on the fleshy pods and defecate viable seeds when moved to open grassland. In western Utah, the distribution of squarrose knapweed (*Centaurea virgata*) reflects historic sheep trails from the west desert to sheep-shearing camps.

Anthropogenic influences on seed dispersal patterns can also be more indirect. Alteration of landscape structure can affect disperser movement, and thus seed dispersal. Habitat fragmentation, the creation of edges, and the availability of corridors affect which species use the landscape and how they move through it. Some species stick largely to edges while others are interior species that are more likely to require corridors, and yet others move easily across inhospitable terrain. Similarly, alterations to habitat structure can easily affect wind dispersal. Consequently, anthropogenic changes to the landscape presumably can increase or decrease the spread of an invasive species and can increase or decrease the rate of local population growth, depending on how different dispersal agents, which differ in SDE, respond to the changes.

Ultimately, human-assisted dispersal is so pervasive that it could be argued that metapopulation-level dispersal limitation has been largely eliminated. Seeds are moving long distances at an unprecedented rate and are, at least in principle, able to arrive anywhere on Earth. Dispersal limitation is still likely very important at the within-population level, however. Therefore, one critical key to managing plant invasions is to learn how to manage dispersal to increase both metapopulation and within-population dispersal limitation. However, we have much to learn before we can effectively do that.

SEE ALSO THE FOLLOWING ARTICLES

Demography / Dispersal Ability, Animal / Landscape Patterns of Plant Invasions / Propagule Pressure / Seed Ecology / Weeds

FURTHER READING

Buckley, Y. M., S. Anderson, C. P. Catterall, R. T. Corlett, T. Engel, C. R. Gosper, R. Nathan, D. M. Richardson, M. Setter, O. Spiegel, G. Vivian-Smith, F. A. Voigt, J. E. S. Weir, and D. A. Westcott. 2006. Management of plant invasions mediated by frugivore interactions. *Journal of Applied Ecology* 43: 848–857.

Chambers, J. C, and J. A. MacMahon. 1994. A day in the life of a seed: Movements and fates of seeds and their implications for natural and managed ecosystems. *Annual Review of Ecology and Systematics* 25: 263–292.

Clark, J., C. Fastie, G. Hurtt, S. Jackson, C. Johnson, G. King, M. Lewis, J. Lynch, S. Pacala, C. Prentice, E. W. Schupp, T. Webb III, and P. Wyckoff. 1998. Dispersal theory offers solutions to Reid's paradox of rapid plant migration. *BioScience* 48: 13–24.

Cousens, R., C. Dytham, and R. Law. 2008. *Dispersal in Plants: A Population Perspective.* Oxford, UK: Oxford University Press.

Dennis, A. J., E. W. Schupp, R. J. Green, and D. A. Westcott, eds. 2007. *Seed Dispersal: Theory and Its Application in a Changing World.* Wallingford, UK: CAB International.

Levey, D. J., W. R. Silva, and M. Galetti, eds. 2002. *Seed Dispersal and Frugivory: Ecology, Evolution and Conservation.* Wallingford, UK: CAB International.

Richardson, D. M., N. Allsopp, C. M. D'Antonio, S. J. Milton, and M. Rejmánek. 2000. Plant invasions: The role of mutualisms. *Biological Reviews of the Cambridge Philosophical Society* 75: 65–93.

Schupp, E. W. 1993. Quantity, quality and the effectiveness of seed dispersal by animals. *Vegetatio* 107/108: 15–29.

Vander Wall, S. B., and W. S. Longland. 2004. Diplochory: Are two seed dispersers better than one? *Trends in Ecology and Evolution* 19: 155–161.

Von der Lippe, M. 2007. Long-distance dispersal of plants by vehicles as a driver of plant invasion. *Conservation Biology* 21: 986–996.

DISTURBANCE

RICHARD J. HOBBS

University of Western Australia, Crawley

A disturbance is any event that produces a change in ecosystem structure and resource availability. A commonly used definition is "any relatively discrete event in time that disrupts ecosystem, community, or population structure and changes resources, substrate availability, or the physical environment." This encompasses events and processes at multiple spatial and temporal scales, ranging from small-scale soil disturbances by insects or mammals, through individual tree-falls in a forest, fires and floods of different extents, and hurricanes, volcanic eruptions, and the like. Disturbance is an important feature in many ecosystems and often drives ecosystem dynamics, provides opportunities for plant regeneration, and initiates habitat change.

DISTURBANCE REGIMES

The disturbance regime of an area describes the different types and characteristics of disturbance experienced there. Disturbance regimes affect ecosystem properties such as rates of soil erosion and formation, and pathways and temporal patterns of nutrient cycling and energy flow. Disturbance regimes can also act as a selective force affecting the life history traits of individual species and the composition, structure, and emergent properties of entire groups of organisms. Many ecosystems depend on some sort of disturbance for the provision of opportunities for release of limiting resources, and many species also depend on disturbance to provide regeneration opportunities. Thus, disturbance is also often important for the maintenance of species diversity.

Disturbances and disturbance regimes are described by a number of features, including intensity, frequency, extent, duration, and seasonality. Intensity refers to the severity of the disturbance, and frequency is the mean number of disturbances over a particular time (this can also be discussed in terms of mean return interval). Extent is the spatial spread of the disturbance, and duration refers to the temporal extent, while seasonality refers to the characteristic time of year at which disturbance occurs. The predictability of disturbance is another important feature referring to the regularity of expected occurrence of disturbance.

The term *disturbance* thus describes a wide range of phenomena. The relative discreteness of disturbances varies, with some, such as localized windstorms, lasting only a few minutes, and others, such as elevated grazing pressures, acting over much longer timeframes. Some commentators seek to differentiate between disturbances that are natural and those that are human induced and between endogenous and exogenous disturbances (those considered to originate from within the system, such as faunal disturbance, compared with those originating outside the system, such as hurricanes). Both of these dichotomies are, however, problematic, given an increasingly human-dominated world and one that is increasingly connected at multiple scales. An important human impact is the modification of disturbance regimes—for instance, changed fire frequencies or altered river flow regimes. Such modifications may be direct, as for instance in the deliberate suppression of wildfires, or they may be indirect, a result of other elements of global change such as land use and climate change.

DISTURBANCE AS A PRECURSOR FOR INVASIONS

Given that disturbance frequently acts to remove existing organisms from an area and to alter the availability of resources, it is generally accepted that disturbance can be an important precursor to invasions, particularly for plants. Thus, individual disturbances can provide entry points, or "invasion windows," for invasive species. In addition, alterations to historic disturbance regimes can also trigger invasions—for instance, more frequent fires, altered grazing regimes, or changed river flows can provide conditions that favor invasive species over native species.

The characteristics of the disturbance will determine their impact in terms of increasing the likelihood of invasions. Small-scale disturbances, such as animal diggings, can provide foci for the establishment of invasive plant species, which then can potentially spread into the surrounding area. At the other extreme, broad-scale climatic events such as hurricanes can produce dramatic changes to ecosystem structure over large areas, sometimes completely altering abiotic conditions and removing the pre-existing plant cover. This then increases the invasibility of the system and provides the potential for a shift to invasive dominance.

It is important to note, however, that disturbance does not always lead to invasion. The outcome of a particular disturbance depends on the local context and on whether propagules of invasive species are available to invade. Different species assemblages can display varying degrees of invasion resistance even following disturbance. Also, invasibility may increase due to the synergistic impacts of different types of disturbance, particularly where the combination of disturbances is a departure from the historic situation. Disturbance may also lead to increased invasion only when coupled with other environmental changes such as nutrient enrichment. Similarly, disturbances may behave differently in different landscape contexts: for instance, a fire in an intact landscape may have little impact with respect to invasions, whereas a similar fire in a fragmented landscape may result in significant invasion as a result of propagule pressure from the altered parts of the landscape. Hence, human modification and fragmentation of landscapes can change the rules significantly.

INVADING SPECIES AS A DISTURBANCE

In addition to benefiting from disturbance, invasive species can themselves constitute a disturbance or contribute to altering disturbance regimes. A contrast can be drawn between species that are "passengers" and those that are "drivers." The classic example of this is grass species that invade an ecosystem and alter the fire regime by producing a more uniform and more combustible fuel configuration. Grass invasion leads to more frequent fire,

FIGURE 1 Banksia woodland in Western Australia invaded by African veldt grass *Erharta calycina*. (Photograph courtesy of the author.)

and in turn, more frequent fire often promotes increased invasion, often leading to a transformation of the ecosystem from exhibiting woody dominance to exhibiting grass dominance (Fig. 1). This "grass–fire cycle" is seen in many parts of the world and can be exacerbated by increased nutrient inputs from, for instance, NOx pollution. Invasive plants can also, however, have the opposite effects on a fire regime when the invading species has a lower flammability than the native vegetation. In this case, the invasive species can prevent the spread of fire, and hence also reduce fire frequency and intensity in invaded locations.

Invasive plant species can also drive important ecosystem changes by altering biogeochemical processes, such as when a nitrogen-fixing species invades a nutrient-poor environment. Similarly, plant invasions of river margins can alter bank stability and influence the geomorphologic outcomes of flood events.

Other examples of invasive species that act as disturbance agents include introduced herbivores that alter the competitive interactions between plant species (either between native species or between native and nonnative species). Such animals may also produce significant physical disturbance in the form of diggings or wallows. These areas, in turn, may act as foci for plant invasions. Where the invading species acts as an ecosystem engineer, or a species with a pronounced influence on ecosystem structure or function, there is increased likelihood that the species' activities will have a follow-on effect on disturbance regimes.

Invasive predators can also alter disturbance regimes by impacting native herbivore populations. For instance, foxes and cats in Australia significantly reduced the populations of marsupial herbivores in many parts of the country, leading to altered grazing patterns and a reduction in soil turnover as the digging activities of these animals declined.

DISTURBANCE AS A MANAGEMENT STRATEGY FOR INVASIVE SPECIES

While many types of disturbance can increase invasibility, sometimes the same types of disturbance can be used as management tools to control or exclude invasive species. "Designed disturbance" is a technique increasingly used in conservation and restoration to drive particular processes and encourage particular species assemblages. For example, controlled water releases from dams on major rivers can simulate historic flood flows that scour river banks and remove invasive riparian plant species. Livestock grazing can be used to reduce the biomass of invasive grass species and encourage growth of native species in ecosystems, such as the serpentine grassland in California (Fig. 2).

Similarly, fire can be used as a means of temporarily reducing the biomass of invasive plant species and creating opportunities for the regeneration of native species. Also, altering fire frequency or intensity can be successful in controlling particular species if the life history characteristics of these species are sufficiently understood. Occasionally, a designed management regime involving fire and targeted grazing can be employed to drive the plant assemblage from being characterized by invasive dominance to being characterized by native dominance.

These approaches move away from the more traditional control methods that focus attention on the invasive organism to more ecosystem-based approaches

FIGURE 2 Serpentine grassland in Northern California, with different grazing regimes on either side of the fence. The left-hand portion is invaded by nonnative grasses such as *Lolium multiflorum* and *Bromus* spp., while the right-hand side is dominated by native forb species. (Photograph courtesy of the author.)

that focus on the system as a whole and attempt to alter conditions so that the invasive species cannot persist or reestablish.

SEE ALSO THE FOLLOWING ARTICLES

Fire Regimes / Invasibility, of Communities and Ecosystems / Land Use / Restoration / Succession / Transformers

FURTHER READING

Brooks, M. L., C. M. D'Antonio, D. M. Richardson, J. B. Grace, J. E. Keeley, J. M. DiTomaso, R. J. Hobbs, M. Pellant, and D. Pyke. 2004. Effects of invasive alien plants on fire regimes. *BioScience* 54: 677–688.

Davis, M. A., J. P. Grime, and K. Thompson. 2000. Fluctuating resources in plant communities: A general theory of invasibility. *Journal of Ecology* 88: 528–534.

Huston, M. A. 2004. Management strategies for plant invasions: Manipulating productivity, disturbance, and competition. *Diversity and Distributions* 10: 167–178.

Lockwood, J. L., M. F. Hoopes, and M. P. Marchetti. 2007. *Invasion Ecology*. Malden, MA: Blackwell.

Richardson, D. M., P. M. Holmes, K. M. Esler, S. M. Galatowitsch, J. C. Stromberg, S. P. Kirkman, P. Pyšek, and R. J. Hobbs. 2007. Riparian vegetation: Degradation, alien plant invasions, and restoration prospects. *Diversity and Distributions* 13: 126–139.

Sheley, R. L., J. M. Mangold, and J. L. Anderson. 2006. Potential for successional theory to guide restoration of invasive-plant-dominated rangeland. *Ecological Monographs* 76: 365–379.

EARLY DETECTION AND RAPID RESPONSE

RANDY G. WESTBROOKS
U.S. Geological Survey, Whiteville, North Carolina

ROBERT E. EPLEE
USDA APHIS PPQ, Whiteville, North Carolina

Prevention is the first line of defense against an introduced invasive species—it is always preferable to prevent the introduction of new invaders into a region or country. However, it is not always possible to detect all alien hitchhikers imported in cargo or to predict with any degree of certainty which introduced species will become invasive over time. Fortunately, the majority of introduced plants and animals do not become invasive. But, according to scientists at Cornell University, costs and losses due to species that do become invasive are now estimated to be over $137 billion a year in the United States. Early detection and rapid response (EDRR) is the second line of defense against introduced invasive species. EDRR is the preferred management strategy for preventing the establishment and spread of invasive species. Over the past 50 years, there has been a gradual shift away from large- and medium-scale federal and state single-agency-led weed eradication programs in the United States to smaller interagency-led projects involving impacted and potential stakeholders. The importance of volunteer weed spotters in detecting and reporting suspected new invasive species has also been recognized in recent years.

ADVANTAGES OF EDRR IN MANAGING INVASIVE SPECIES

EDRR provides certain advantages in managing new invasive species. One positive aspect of EDRR is that it does not restrict the commercial sale and trade of new exotic species that *might* become invasive—unlike restrictive introduction policies, which may result in lost economic opportunities. Indeed, many introduced species (e.g., wheat, rice, and soybeans) provide important benefits to society. By detecting and eliminating biological invaders early, public agencies and other groups will be able to focus more resources on a narrow range of widespread invaders that already pose a significant threat to the economy or the environment—without having to deal with more and more widespread invaders in the future. EDRR is a cost-effective strategy that aims to minimize those future costs and losses.

EDRR also provides other advantages over traditional pest control programs. It focuses exclusively on recently detected exotic plants that have established free living populations on managed lands or in natural areas. Also, unlike some long-term control efforts, EDRR has minimal and short-term impacts on the environment. The ultimate goal of EDRR is to restore an invaded habitat to a natural or seminatural state, as well as to provide improved habitat for native species. EDRR is not a panacea for the invasive species problem. It is one more tool that can help to address a problem that has been created through global trade and travel.

LARGE-SCALE FEDERAL–STATE NOXIOUS WEED ERADICATION PROGRAMS IN THE UNITED STATES

The concept of EDRR as a preferred strategy in managing invasive plants is not new. Before the introduction of organic pesticides after World War II, farmers scouted their fields constantly for new weeds. It was just common sense to eliminate a new weed before it became well established and spread. However, the concept of EDRR as the preferred management strategy in federal or state weed eradication projects in the United States first evolved as a guiding principle in the U.S. Department of Agriculture (USDA) Witchweed Eradication Program in the Carolinas in the early 1960s. Witchweed (*Striga asiatica*) (Fig. 1), which is native to Asia and Africa, parasitizes the roots of grasses such as large crabgrass (*Digitaria sanguinalis*) and grass crops such as corn, sorghum, millet, and sugarcane. Witchweed was officially confirmed to be in the United States by the North Carolina State University Plant Disease Clinic in July 1956 through analysis of corn plants and accompanying soil samples from Columbus County, North Carolina. Once it was confirmed, detection surveys (detection of incipient infestations) and delimiting surveys (surveys made to determine the total area infested) revealed that witchweed had infested over 432,000 acres (174,824 ha) of cropland from Florence, South Carolina, to Goldsboro, North Carolina.

FIGURE 1 Witchweed (*Striga asiatica*) parasitizing corn. (Photograph courtesy of R. Eplee.)

To protect the U.S. corn crop (12.2 billion bushels worth more than $48 billion in 2008) from permanent annual losses of 10 percent or more, a federal–state eradication program was established and first funded in the Carolinas in fiscal year 1958 (initially $3 million a year) under the Federal Organic Act (1944) and the Federal Plant Pest Act (1957). The long-term goal of the program was to detect, contain, and eradicate witchweed from the Carolinas (the only place it is known to occur in the Americas). To date, the infestation has been reduced from a high of 432,000 acres (174,824 ha) in 1970 to the current level of about 2,100 acres (850 ha) in a few counties of the North Carolina coastal plain. In 2009, the last remaining infested acres were released from quarantine in South Carolina—culminating a 51-year effort to eradicate witchweed from 80,000 acres in the Palmetto State.

At the height of the cooperative federal–state effort in the early 1990s, the USDA Animal and Plant Health Inspection Service (APHIS) employed 57 plant quarantine officers, numerous control technicians, summer survey crews, and contract sprayers to work in the Witchweed Eradication Program. However, in 1995, most of the work of the Witchweed Eradication Program was turned over to the Plant Protection Division of the North Carolina Department of Agriculture and Consumer Services (NCDA). NCDA currently employs several control technicians to continue the mop-up program in North Carolina. APHIS has one remaining plant quarantine officer working on the program in South Carolina. If current federal funding of approximately $1 million per year remains stable, witchweed will be eradicated from the Carolinas within the next few years.

The success of the Witchweed Eradication Program is largely due to the adoption of a number of important weed prevention and eradication principles that were identified by the APHIS Whiteville Methods Development Center in Whiteville, North Carolina, as outlined below:

1. Populations of suspected new invaders must be detected and reported to appropriate state and federal officials as early as possible.

2. Confirmed new species must be rapidly assessed to determine their potential for invasiveness and threat to natural and managed resources.

3. All incipient infestations in an affected area must be managed with biologically sound methods and procedures, to (a) prevent further spread and establishment (containment), and (b) deny further reproduction (control) until all viable propagules have been exhausted from the soil (eradication).

4. All impacted and potentially impacted stakeholders should have some role in the eradication effort.

With the exception of number 4, all these principles were put into practice and directly contributed to the eradication of witchweed from over 99 percent of infested lands to date. Principle 4 represents an important lesson learned through the Witchweed Eradication Program effort. Throughout the history of the program, impacted landowners have never been given any direct responsibility for eradication of witchweed on their properties.

With government contractors coming out to treat their witchweed infested crops (often on a farmer's timetable), there is little incentive for impacted landowners to take any responsibility for the problem—or for the program to ever end. There is no doubt that the USDA-Carolinas Witchweed Eradication Program is one of the most successful large-scale eradication efforts of all time. However, due to the high cost and the decades-long commitment required, it is unlikely that a federal–state weed eradication program of this magnitude and scope will ever be attempted again.

MEDIUM-SCALE FEDERAL–STATE NOXIOUS WEED ERADICATION PROGRAMS IN THE UNITED STATES

The USDA Witchweed Eradication Program demonstrated the need for broader federal authority to prevent the introduction and spread of foreign weeds in the United States. Thus, after more than a decade of debate and discussion by weed scientists nationwide, the Federal Noxious Weed Act (FNWA) was passed by the U.S. Congress in August 1974 and signed into law by President Gerald Ford in January 1975. The Plant Protection Act of 2000, which superseded the FNWA, provides authority to prohibit the introduction of listed species from abroad and to eradicate incipient infestations that become established within the country. Currently, the Federal Noxious Weed List includes 93 species of terrestrial and aquatic weeds: all species of the genera *Homeria*, *Aeginetia*, *Alectra*, and *Striga*, and all nonnative species of *Cuscuta* and *Orobanche*.

Beginning in 1981, USDA APHIS, the federal agency responsible for implementing the FNWA, started working with state and local agencies to eradicate several listed Federal Noxious Weeds (FNWs). The program included common crupina (*Crupina vulgaris*) in Idaho, Oregon, and Washington (about 30,000 acres, or 12,140 ha), goatsrue (*Galega officinalis*) in Cache County, Utah (39,000 acres, or 15,783 ha), wormleaf salsola (*Salsola vermiculata*) in California (1,500 acres, or 607 ha), branched broomrape (*Orobanche ramosa*) in south Texas (700 acres, or 283 ha), catclaw mimosa (*Mimosa pigra* var. *pigra*) in south Florida (500 acres, or 202 ha), and hydrilla (*Hydrilla verticillata*) in the Imperial Irrigation District of California (2,000 acres, or 809 ha).

FIGURE 2 Goatsrue (*Galega officinalis*) in Cache County, Utah. (Photograph courtesy of R. Eplee.)

The most noticeable thing about these single-agency-led projects is that they were much smaller in size than the Witchweed Eradication Program. Indeed, good progress was made toward eradication in each project. But, as with the Witchweed Program, impacted stakeholders had no direct involvement with the programs, and federal funding was eventually terminated after a period of time.

A good example is the federally funded Utah Goatsrue Eradication Program that was led by Utah State University from 1981 to 1996 (Fig. 2). Good progress was made in eradicating this poisonous legume, which is native to Eastern Europe, from Cache County, Utah, during the 1980s and early 1990s. However, when the Congressional earmark of $150,000 for the program was eliminated in 1996, federal support was withdrawn by USDA APHIS, and the program was terminated. Establishment of a goatsrue interagency task force would be the best approach for addressing goatsrue in Utah before it spreads throughout the American West.

A notable exception to the dearth of successful single-agency-led weed eradication projects is the California Department of Food and Agriculture's (CDFA) Weed Eradication Program. Over the past 100 years, CDFA has eradicated over 1,000 populations of 14 different introduced invasive plants across California. A few of its success stories include heartleaf nightshade (*Solanum cardiophyllum*), serrate spurge (*Euphorbia serrata*), wild marigold (*Tagetes minuta*), giant dodder (*Cuscuta reflexa*), and Austrian peaweed (*Sphaerophysa salsula*). EDRR is clearly a cornerstone of the CDFA Weed Eradication Program.

A notable success of the program is the eradication of the introduced aquatic weed hydrilla (*Hydrilla verticillata*) from irrigation canals in the Imperial Irrigation District (IID) of southern California over the past 25 years. Hydrilla (Fig. 3) was first discovered in the All American Canal of the IID in southern California in 1977. At one time, hydrilla heavily infested irrigation canals throughout the IID and reduced water flow by as much as 85 percent in some areas. Beginning in the early 1980s, a cooperative federal–state program led by CDFA was initiated to eradicate hydrilla from the IID. Since then, the program, which pioneered the use of sterilized (triploid) white amur grass carp (*Ctenopharyngodon idella*) from China, has reduced hydrilla in the IID by over 99 percent. CDFA has also eradicated hydrilla from nine other counties across the state and is working to eradicate it from nine additional counties. The largest effort involves eradication of an infestation in Clear Lake, in Lake County, California, which began in 1994. Clear Lake, which has a surface area of 43,785 acres (17,719 ha), is the largest lake that is entirely within the state of California.

Except in cases where there are influential stakeholders to promote the effort, or where the invasion is restricted to one land unit or owner, single-agency-led programs often lack the expertise, resources, and focus to sustain a prolonged eradication effort for new invasive plants. A long-term commitment to sustain the effort until eradication is achieved, as demonstrated by the CDFA Hydrilla Eradication Program, is paramount for success.

SMALL-SCALE FEDERAL–STATE NOXIOUS WEED ERADICATION PROJECTS IN THE UNITED STATES

In the late 1980s, with limited funding for the U.S. Federal Noxious Weed Program, there was a need to maximize available resources for weed eradication projects by detection of listed federal noxious weeds at the very earliest stage of invasion in the United States. To that end, a national herbarium survey was conducted in 1991 to help locate FNW infestations around the country.

FIGURE 3 Hydrilla (*Hydrilla verticillata*)–infested irrigation canal in the Imperial Irrigation District of Southern California. (Photograph courtesy of California Department of Food and Agriculture.)

FIGURE 4 Japanese dodder (*Cuscuta japonica*) parasitizing kudzu at the Clemson University Arboretum, Clemson, South Carolina, August 1992. (Photograph courtesy of R. Westbrooks.)

As a result of the survey, several new FNW infestations were identified, and eradication projects were initiated. Following the survey, several successful eradications in areas with small FNW infestations (generally less than 10 acres, or 4 ha) occurred by the mid-1990s. These included small broomrape (*Orobanche minor*) at several sites in Virginia, North Carolina, South Carolina, and Georgia; Japanese dodder (*Cuscuta japonica*, Fig. 4) at the Clemson University Horticultural Gardens in South Carolina; giant salvinia (*Salvinia molesta*) in Colleton County, South Carolina; tropical soda apple (*Solanum viarum*) in Georgia, South Carolina, North Carolina, and Pennsylvania; and wild red rice (*Oryza rufipogon*) in the Florida Everglades. Since that time, eradication of tropical soda apple in the Carolinas has been an ongoing effort. It is sometimes introduced with cattle from Florida that eat the fruit in infested pastures prior to shipment to other states around the Southeast.

During this same time period, efforts to initiate certain other FNW eradication projects were unsuccessful. These included wild sugarcane (*Saccharum spontaneum*) and wetland nightshade (*Solanum tampicense*) near Lake Okeechobee, in southern Florida. In these two cases, there was a lack of state or local interest in addressing the problem at an early stage of invasion. Stakeholder interest, support, and participation are clearly needed to make such programs work.

SMALL-SCALE WEED ERADICATION PROJECTS IN AUSTRALIA: KOCHIA IN WESTERN AUSTRALIA

EDRR is also being employed as the preferred management strategy for addressing new invasive species in other parts of the world, particularly in Australia and

FIGURE 5 Site in Western Australia infested with kochia (*Bassia scoparia*). (Photograph courtesy of Western Australia Department of Agriculture, Perth, Western Australia.)

New Zealand. The introduction and subsequent eradication of kochia (*Bassia scoparia*) in Western Australia is a good example.

Kochia, which is native to Eastern Europe and western Asia, is an annual weed of cereal crops and pastures in the warmer regions of the world (Fig. 5). Kochia was first imported into the wheat belt of Western Australia in 1990 as a salt-tolerant pasture forage plant. Although palatable to livestock, the plant is poisonous in large quantities. It also alters fire regimes and reduces the diversity and abundance of native plant species.

By 1992, the Western Australia Department of Agriculture recognized the weedy potential of kochia and targeted it for containment and eradication in 1993. Eradication strategies for kochia included burning, grazing, chemical control, and mechanical removal. By 1996, 46 of the 52 known infestations had been eliminated from the state. The last sightings of kochia in Western Australia were in March 2000.

The program's success was due in large part to the following:

- A detailed knowledge of where the plant had been planted
- The conspicuous nature of the plant
- Limited seed dormancy and a long vegetative period prior to seeding
- Fences around most of the planted sites that helped to contain it (mature plants break away from the root system and roll across the landscape in high winds, dropping seeds along the way)
- Rapid response within two years of introduction
- A well-organized and funded eradication program and cooperation from impacted farmers.

The total cost of the Kochia eradication effort in Western Australia was AUS $200,000 in combined federal and state funding, with an additional AUS $300,000 of in-kind control and monitoring activities provided by affected farmers. In retrospect, it is obvious that the introduction of kochia was a shortsighted idea that cost at least a half million dollars to correct. However, this pales in comparison to future kochia-induced losses and control costs that would have been suffered by Australian landowners without the eradication effort.

INTERAGENCY COOPERATION TO PREVENT THE ESTABLISHMENT AND SPREAD OF INVASIVE PLANTS: THE U.S. NATIONAL EDRR SYSTEM FOR INVASIVE PLANTS

By the mid-1990s, it was clear that funding would never be available to mount another large-scale government weed eradication project like that against witchweed in the United States—or even to sustain medium-sized projects such as those against goatsrue and common crupina in the West. By that time, it was also clear that building a separate coalition to address every new invader (even small, incipient infestations) was difficult and prevented a timely response. What was needed was a systematic approach for addressing new invaders—a *National EDRR System for Invasive Plants.* Early on, it was envisioned that the system would be coordinated nationally by member agencies of the Federal Interagency Committee for the Management of Noxious and Exotic Weeds (FICMNEW). However, from an operational standpoint, it would need to engage all federal, state, and local agencies and organizations that are responsible for land management and ecosystem protection across the country.

In 1997, under the auspices of the U.S. Departments of Agriculture and Interior, the U.S. Geological Survey, Biological Resources Discipline (USGS) began collaborating with USDA APHIS to provide technical and scientific support for interagency invasive species partnerships and to coordinate the development of the U.S. National EDRR System for Invasive Plants. Since then, numerous states and regional interagency partner groups have embraced EDRR as the preferred management strategy for dealing with new and emerging invaders. Conceptually, the National EDRR System includes a number of elements and processes. These are listed in Table 1.

NEW TRENDS IN VOLUNTEER DATA COLLECTION AND SYNTHESIS

Early detection and reporting is the foundation of the National EDRR System for Invasive Plants. The Invasive Plant Atlas of New England (IPANE) is a good example of

TABLE 1

U.S. National EDRR System for Invasive Plants

System Element	Functions
Federal Interagency Committee for the Management of Noxious and Exotic Weeds (FICMNEW)	National Interagency Coordination
Regional Invasive Plant Atlases (e.g., Invasive Plant Atlas of New England, Invasive Plant Atlas of the Midsouth)	Field Data Archiving and Volunteer Training
State Invasive Species Councils (e.g., Maryland Invasive Species Council)	State Interagency Coordination
State EDRR Coordinating Committees	State EDRR Capacity Building, Project Coordination, Volunteer Training, ID, Vouchering, Rapid Assessing, Technical Support
State Volunteer Detection and Reporting Networks (Trained Volunteers and Agency Field Personnel)	Detection and Reporting of New Invasive Plants to State and Federal Agencies
Invasive Plant Task Forces and Cooperative Weed Management Areas (e.g., Carolinas Beach Vitex Task Force)	Detection and Eradication of New Invasive Plants that Occur on Multiple Land Units
Land Management and Plant Protection Agencies	Detection and Eradication of New Invasive Plants that Occur on a Single Land Unit

this new trend in volunteer data collection and synthesis. IPANE is a cooperative effort between the University of Connecticut (UCONN), the New England Wild Flower Society, the U.S. Fish and Wildlife Service (USFWS), the USGS, and the USDA Cooperative States Research Education and Extension Service (CSREES). From 2000 to 2009, IPANE trained over 700 volunteers and agency field personnel to detect and report new infestations of 100 selected invasive plant species throughout New England. Once field reports are verified by project staff, the data are entered into the IPANE database. The database is maintained by the Center for International Earth Science Information Network (CIESIN) at Columbia University, with support from the USGS National Biological Information Infrastructure (NBII). Over the past several years, IPANE has become a model for volunteer field

data collection and synthesis and provides near-real-time information on the distribution of high-profile invasive plants such as oriental bittersweet (Fig. 6) throughout New England. Figure 7, which shows the distribution of oriental bittersweet (*Celastrus orbiculatus*) from the IPANE database, illustrates the real value of volunteer data collection and synthesis for research purposes, as well as for planning and executing "on-the-ground" management programs. To date, IPANE volunteers have submitted over 10,000 invasive plant occurrence records to the IPANE database.

FIGURE 7 Distribution of oriental bittersweet (*Celastrus orbiculatus*) in New England, United States. Red dots = herbarium records. Blue dots = field records submitted by IPANE volunteers from 2001–2008. Downloaded from the IPANE Web site on March 24, 2009. (Image courtesy of www.ipane.org.)

FIGURE 6 Oriental bittersweet (*Celastrus orbiculatus*) infestation in Connecticut. (Photograph courtesy of Leslie J. Mehrhoff, UCONN, Storrs, Connecticut.)

Currently, Mississippi State University is developing a second regional atlas, the Invasive Plant Atlas of the Mid-South, which is being patterned after IPANE. A long-term goal is to develop several more regional invasive plant atlases, and thus to collectively create a national invasive plant atlas for the United States.

THE BEACH VITEX TASK FORCE: AN INTERAGENCY TASK FORCE IN ACTION

Once a new invasive plant has been detected and confirmed in a state or region, an assessment is conducted to determine the proper course of action. If the infestation occurs on a single land unit, it is possible to initiate a single-agency-led weed eradication program. For new invaders that occur on multiple land units and in different jurisdictions, establishment of a geographically based cooperative weed management area or an invasive plant task force of impacted and potential stakeholders (species specific) may be warranted. The Beach Vitex Task Force is a good example of this new trend in interagency partnering.

Beach vitex (*Vitex rotundifolia*) is a woody vine that was imported from the beaches of Korea in the mid-1980s for use as an ornamental and as a dune stabilization plant in the southeastern United States (Fig. 8). By the mid- to late 1990s, dune restoration specialists with the U.S. Army Corps of Engineers began to notice beach vitex spreading from landscape plantings on beaches along the South Carolina coast, crowding out native species such as sea oats (*Uniola paniculata*) and sea beach amaranth (*Amaranthus pumilus*). Unlike native sea oats which have fibrous roots that help anchor sand dunes against storm waves, beach vitex has long tap roots that anchor the plant itself during

FIGURE 9 Beach vitex (*Vitex rotundifolia*) and eroded dunes at Debordieu Colony Beach, Georgetown, South Carolina, October 2004. (Photograph courtesy of R. Westbrooks.)

major storm events but do little to help protect the dunes against erosion (Fig. 9). The thick, deep root system also prevents sea turtles from nesting on primary dunes.

In November 2003, concerned sea turtle volunteers collaborated with the USGS to host the first Beach Vitex Symposium at the Belle W. Baruch Institute near Georgetown, South Carolina. At that meeting, the South Carolina Beach Vitex Task Force was organized to address the problem. In the spring of 2005, in recognition of the need to address beach vitex along the North Carolina coast, the effort was expanded into a bistate effort with the formation of the Carolinas Beach Vitex Task Force (www.beachvitex.org). Since that time, the task force has made significant progress over the past six years in eradicating beach vitex from beach communities along the Carolina coast. As part of the effort, volunteers work to remove beach vitex seedlings from public beaches and to detect and report large beach vitex plantings to task force coordinators. As of 2009, a total of 233 of the known 240 infested sites along the South Carolina coast had been treated or eradicated by a task force team. Similar efforts are under way to address infested sites along the North Carolina coast. Recently, the task force was expanded once again to include new efforts to address beach vitex infestations in beach communities around Norfolk, Virginia. There is no doubt that volunteer field data collection and assistance has been a major factor in the success of the Beach Vitex Task Force.

The Beach Vitex Task Force has shown the value of interagency partnering to address new invasive species and the importance of volunteers in early detection and reporting of incipient infestations. This type of collaborative, consensus-building process, which brings *principal partners* (lead agencies and organizations) and *cooperators*

FIGURE 8 Beach vitex (*Vitex rotundifolia*) at Debordieu Colony Beach, Georgetown, South Carolina, August 2004. (Photograph courtesy of R. Westbrooks.)

(impacted and potential stakeholders) together around a common cause, is a good approach for addressing new invasive species that threaten the managed and natural resources of the United States and other countries.

THE VICTORIA WEED ALERT PROGRAM: A STATE-LED EDRR PROGRAM IN ACTION

To enhance state efforts to address new weeds in southeastern Australia, the Victoria Department of Primary Industries (DPI) has established the Victoria Weed Alert Program. The overall goal of the program is to prevent the establishment and spread of new weeds in all parts of the state.

A critical element of the Victoria Weed Alert Program is trained volunteer weed spotters. Currently, over 2,000 trained weed spotters assist the DPI through early detection and reporting of unknown and suspected new invaders (Fig. 10). This includes a list of Victorian weed alert species, as well as state prohibited weeds. The list of target species does not include common weeds in Victoria such as Patterson's curse (*Echium plantagineum*), serrated tussock (*Nassella trichotoma*), and gorse (*Ulex europaeus*). Information gathered by the Victoria weed spotters helps the state detect and respond to new weed outbreaks before they become widespread and impractical to eradicate.

Weed spotter training is an important part of the Victoria Weed Alert Program. The primary focus of the training program includes weed identification, specimen collection, and protocols for reporting of new weeds to DPI. Registered weed spotters receive a weed alert handbook, a weed alert program newsletter, and a "WEEDeck" (a deck of pocket-sized cards with images and important information on weed alert target species).

FIGURE 10 Victoria Weed Alert Program: volunteer weed spotters in the field. (Photograph courtesy of Victoria Department of Primary Industries, copyright 2007. Used with permission.)

So far, WEEDeck cards for over 160 different species have been produced for use by the volunteers.

When a suspected new weed is observed, a weed spotter is asked to report the plant to a local DPI weed alert contact officer, and possibly to send a specimen of the plant to the National Plant Herbarium of Victoria for positive identification. If the national herbarium confirms that the plant is a species that is new to the state and not already regulated by the DPI, then a DPI specialist will conduct a weed risk assessment to determine its potential threat to natural and managed resources of the state and then develop an appropriate course of action—with input from impacted stakeholders. The goal is to reinforce the idea that weed spotters are community sentinels that provide a critical link between new weed outbreaks and DPI "first responders."

A good example of the Victoria Weed Alert Program in action involved the discovery of water hyacinth (*Eichhornia crassipes*) in Keysborough, a suburb of Melbourne, Victoria, in May 2008. Initially, a registered weed spotter reported an infestation of water hyacinth in a waterway within a new housing development in Keysborough. Further investigation by DPI officers revealed a second infestation in a nearby waterway. Two additional infestations were also detected in backyard water gardens during a survey of 200 homes in the area. As a result of the effort, over 400 kg of water hyacinth plants were removed from the four sites and destroyed by burial. As a precaution, follow-up detection surveys were conducted in the area over the following year to ensure all local populations were eradicated.

SUMMARY

The four most critical elements for success in preventing the establishment and spread of new invasive species are *early detection*, *reliable reporting*, *rapid assessment*, and *rapid response* through cooperative efforts by responsible agencies and impacted and potential stakeholders. Over the past 50 years, experience has shown that no single agency has adequate authority, resources, expertise, or ability to sustain the prolonged effort that is needed to effectively deal with most new invaders that are spreading across the landscape, especially invaders that pose a threat to natural areas. As a result, in the United States, there has been a gradual shift away from large, single-agency-led weed eradication programs, such as the USDA Witchweed Eradication Program, to small, interagency coalitions such as the Carolinas Beach Vitex Task Force.

Unlike most biological invasions around the world, the intentional introduction of kochia in Western Australia in 1990 was well documented. When the weedy potential of this plant was recognized by the Western Australia

Department of Agriculture in 1992, it was targeted for eradication in 1993 and was completely eradicated by 2000. The Kochia Eradication Program in Western Australia is a good example of how agencies and organizations can work with impacted and potential stakeholders to eradicate new invasive species at a modest cost before they spread out of control.

Currently, a national EDRR system is being developed in the United States to provide a systematic approach for early detection and rapid response to new invasive plants. Overall goals of the national EDRR effort are to promote EDRR as the preferred strategy for addressing new and emerging invasive species, through adoption of weed eradication principles and practices at all scales (country, state/province, county, park/forest/refuge/conservancy, farm/private property); to develop a coordinated framework for interagency partnering at the local, state, regional, and national levels; to continually develop and improve EDRR strategies and methods (volunteer detection and reporting networks, cooperative weed management areas and task forces, state EDRR coordinating committees, state invasive plant councils, regional invasive plant atlases); and to provide scientific and technical support to invasive species partner groups such as the Beach Vitex Task Force.

The Beach Vitex Task Force has shown the value of interagency partnering to address new invasive species and the importance of volunteers in detecting incipient infestations. The Victoria Weed Alert Program, which currently engages over 2,000 trained weed spotters, is a good example of a state-led EDRR program in action. Efforts by the program led to the detection and eradication of water hyacinth from waterways and home ornamental water gardens sites in Keysborough, Victoria, in May 2008. In evaluating such programs, it is quite evident that the costs associated with development and implementation of an effective EDRR state or provincial strategy such as this are far less than the increased costs that would be associated with more and more widespread invaders.

At the local level, EDRR is an effective tool that can help to address immediate problems that might otherwise go unnoticed at the national level. At the national level, EDRR is an important second line of defense against the establishment and spread of new invasive plants that complements ongoing federal efforts to prevent further introductions of foreign invaders at international ports of entry. The main idea is for people to detect and eliminate new invaders on the lands they manage, whether farms, parks, or county lands. With both foreign pest exclusion and EDRR systems in place, nations around the world will be better able to defend against future economic and environmental losses due to "plants and animals out of place."

SEE ALSO THE FOLLOWING ARTICLES

Eradication / Herbicides / Mechanical Control / Parasitic Plants / Risk Assessment and Prioritization / Weeds

FURTHER READING

Brooks, M., and R. C. Klinger. 2009. Practical considerations for early detection monitoring of plant invasions (9–33). In Inderjit, ed. *Management of Invasive Weeds*. Berlin: Springer.

FICMNEW. 2003. A National Early Detection and Rapid Response System for Invasive Plants in the United States: Conceptual Design. Washington, DC: Federal Interagency Committee for the Management of Noxious and Exotic Weeds. www.fws.gov/ficmnew/FICMNEW_EDRR_FINAL.pdf

Mehrhoff, L. J., J. A. Silander Jr., S. A. Leicht, E. S. Mosher, and N. M. Tabak. 2003. IPANE: Invasive Plant Atlas of New England. Department of Ecology and Evolutionary Biology, Storrs: University of Connecticut. www.ipane.org

Panetta, F. 2007. Evaluation of weed eradication programs: Containment and extirpation. *Diversity and Distributions* 13: 33–41.

Rejmánek, M., and M. Pitcairn. 2002. When is eradication of exotic pest plants a realistic goal? (249–253). In C. Veitch and M. Clout, eds. *Turning the Tide: The Eradication of Invasive Species*. Gland, Switzerland: IUCN.

Sand, P., R. Eplee, and R. Westbrooks, eds. 1990. Witchweed research and control in the United States. Weed Science Society of America Monograph 5.

U.S. Geological Survey, National Biological Information Infrastructure. National EDRR Framework. www.frogweb.gov/portal/community/Communities/Ecological_Topics/Invasive_Species/Early_Detection_Rapid_Response_(EDRR)/Early_Detection_Rapid_Response/

Victoria Weed Alert Program. www.dpi.vic.gov.au/DPI/nrenfa.nsf/LinkView/C52DE31C1BEEF7FBCA2573750020A2C60CC3718DB70C93ACA25737D001727E5

Westbrooks, R. 2004. New approaches for early detection and rapid response to invasive plants in the United States. *Weed Technology* 18(5): 1468–1471.

Westbrooks, R., L. Mehrhoff, and J. Madsen. 2009. Overview of the U.S. National Early Detection and Rapid Response System for Invasive Plants: Fact Sheet. www.nawma.org/documents/EDRR/U.S.%20National%20EDRR%20System%20-%20Fact%20Sheet%20-%20100308.pdf

EARTHWORMS

SAMUEL W. JAMES

University of Kansas, Lawrence

It will come as a surprise to most readers that invasive earthworms may have the greatest biomass per unit area of any invasive animal, and that most people have seen only invasive earthworms. Imagine that English sparrows and common pigeons are the only birds routinely seen. Out of several thousand earthworm species, about 1 percent account for the vast majority of ordinary human encounters with worms. The effects of invasive earthworms on plants, soils, and litter decomposition are well known because invasive species are the most common ones, but they are just beginning to be studied in the context of biological invasions.

TAXONOMY AND GEOGRAPHY OF INVASIVE EARTHWORMS

Prior to the recent growth of interest in biological invasions, the community of earthworm researchers referred to earthworms often found outside their native ranges as peregrine species. Peregrine literally means wandering, and the term is sometimes used in the context of pilgrimages, but when it is used to describe earthworm behavior, there is no implied purpose in the creatures' wanderings.

To say that an earthworm is a peregrine assumes some knowledge of where it originated. By 1900, the German specialist Michaelsen and others were formulating basic ideas about the natural distributions of earthworms. By that time collections in the research museums sufficed to indicate the approximate limits of the major earthworm families and to demonstrate that certain species were frequently found outside those boundaries. Esoteric hypotheses about earthworm dispersal via land bridges and transoceanic bird flights fell into disrepute. Then, Michaelsen's colleague Alfred Wegener proposed that continents move, and the pieces of the earthworm puzzle began to fall into place, at least in Germany. However, the term peregrine still ought to have some use, in order to differentiate among earthworms whose errant locations are uncommon and do not expand at a rapid pace. In contrast to the species mentioned in the next section, something like *Drawida japonica* (Moniligastridae) occurs in modest numbers in a very few locations outside temperate Asia. Why these merely peregrine species are not invasive is a subject for further thought and will be addressed in a later section.

For brevity, we will assume that deserts, most boreal forests, permafrost, and ice caps lack earthworms entirely, except for isolated areas and places modified by human activity. Of an estimated 120 species now domiciled outside their presumed native ranges (in some cases the native range can no longer be clearly determined), many have achieved global distributions within climate zones suitable to their existence. Some examples from the Lumbricidae (originally Holarctic) are the common invasive species *Lumbricus terrestris*, *Aporrectodea caliginosa*, *Aporrectodea trapezoides*, *Aporrectodea rosea*, *Eisenia fetida*, *Octolasion tyrtaeum*, and *Dendrobaena octaedra*. These

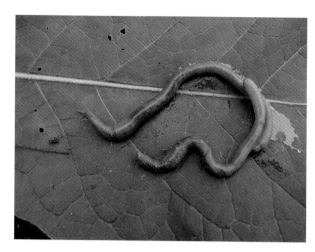

FIGURE 1 *Pontoscolex corethrurus*, originally from northeastern South America, is now found worldwide in tropical and subtropical climates and could be the world's most abundant earthworm. (Photograph courtesy of the author.)

plus other Lumbricidae, a total of about 30 species, are now present in the nontropical areas of all continents, as well as in cooler tropical montane habitats. Within these climatic constraints, essentially every place that European peoples or their plants have settled is home to invasive Lumbricidae, including glaciated parts of Europe itself where these are the only earthworms present.

A tropical invasive species, *Pontoscolex corethrurus* (Glossoscolecidae), by rough estimation, may be the most abundant earthworm in the world (Fig. 1). It probably originated from the Guyanan Shield region of South America, but it was described from specimens collected in the 1850s from Santa Catarina, Brazil, when it was already ubiquitous and extremely abundant. Now it is pantropical and occupies habitats ranging from seasonal dry forest to montane cloud forest, in addition to agricultural land.

Approximately 25 Asian species of *Amynthas*, *Metaphire*, *Perionyx*, *Pheretima*, and *Pithemera* (Megascolecidae) are collectively the next most abundant and widely distributed invasives, with some species (mainly *Amynthas*) in temperate zones and others in subtropical and tropical zones. *Pithemera bicincta* is thought to have traveled with Polynesians as they sailed throughout the tropical Pacific Ocean. Like *Pontoscolex*, many of these species were first discovered far from their native ranges. Taxonomy of this group is much complicated by many species having been described several times, including various morphs derived from parthenogenetic reduction of sexual organs. One species or complex of morphs is fast becoming a major concern in North American hardwood forests, because it rapidly devours forest-floor leaf litter.

Interestingly, several of the *Amynthas* species appear to have an annual life cycle, overwintering as egg capsules. They may retain their Asian habit of mass crawls and die-offs at the end of the growing (at home, the monsoon) season. This phenomenon is well known in tropical to subtropical Asia, but it has not been adequately studied. It may be an important factor in the management of this group of worms.

Several other families have important invasive species. There are about ten invasive Ocnerodrilidae (*Nematogenia, Ocnerodrilus, Gordiodrilus, Eukerria*), some of uncertain geographical origin (possibly Africa or South America), typically living in wet soil or saturated mud adjacent to fresh water. Taxonomy is difficult and often confused by parthenogenetic morphs, as in the Megascolecidae just mentioned. Five small sub-Saharan African species of *Dichogaster* (Acanthodrilidae) are now globally distributed, even turning up in bathhouse or shower drains in cold climates, indicating their ability to hitch rides in protected means of transport. More ordinarily, they are found in warm humid climates around the world, but seldom in large numbers. Two *Microscolex* (Acanthodrilidae) from southern South America are cosmopolitan; *M. dubius* has been found in remote southern California valleys and on the banks of the Tigris River in Turkey. *Eudrilus eugeniae* (Eudrilidae) is a tropical West African species with minor importance as an invasive, due mainly to its usefulness as a converter of plant debris to compost. Like many other composting worms, it can escape cultivation and establish populations in natural or artificial concentrations of organic matter, but this species is very limited by cold temperatures.

Recently the discovery of cryptic diversity within several invasive species such as *Allolobophora chlorotica* (Lumbricidae; five suspected lineages) offers both opportunities and challenges in understanding the history and ecology of earthworm invasions. The opportunities arise from the potential to track cryptic lineages to source populations in areas where the worms are indigenous, using molecular markers. This could shed some light on the history of earthworm movements. Challenges come from having to reexamine research done before the underlying diversity became known, because it is now not clear which species were actually studied.

CHARACTERISTICS OF INVASIVE EARTHWORMS

From the earliest days of realizing that some earthworms were living outside their natural ranges, scientists have speculated about the factors separating invasive from non-invasive earthworms. One may speak of tolerance to soil disturbance, but there does not seem to be any way to evaluate a particular species and predict its tolerance based on morphological or other data. To say that invasives are "broadly adapted" is not informative: invasives are, and noninvasives are not, simply because the former are found in many environmental circumstances and the latter are not.

Properties

A few basic features of earthworms are clearly relevant to the invasive way of life. Parthenogenesis or other uniparental reproduction enables a single individual to found a population. It is relatively straightforward to determine the functionality of male gonads in earthworms, but parthenogenesis is neither necessary nor sufficient for invasiveness. There are outcrossing invasive species and noninvasive parthenogenetic species.

A resistant stage of the life cycle, or the ability to adopt a resistant physiological condition, such as heat- or drought-induced diapause, would facilitate survival during transportation or in sites with frequent disturbance or highly variable soil climate. Again, there are numerous cases of this, typically diapause, in earthworms generally, so it is not sufficient. It may not be necessary; for example, *Lumbricus terrestris* does not enter diapause.

The ability to use diverse food resources, including those of poor quality, is another candidate for an essential quality of invasive earthworms. *Pontoscolex corethrurus* uses a muco-polysaccharide priming mechanism in its gut to increase bacterial decomposition of resistant soil organic matter. Therefore, it can survive in depleted agricultural soils and in a variety of low-carbon tropical soils, as well as in richer soils. Other invasive species may have comparable mechanisms, but little is known about how earthworms use different fractions of soil organic matter.

Other factors often mentioned are small body size and ability to spread rapidly once introduced to an area. Body size ranges widely in earthworms, from 15 mm to over 2 m, but in the present context, "small" means in the range of 20 to 150 mm. Most earthworms fall in this unremarkable size class, and so once again we have chosen a characteristic that is not diagnostic. On the other hand, there are no known invasive giant species, although a 400 mm parthenogenetic *Pheretima* in the Philippines is a notorious, but narrowly distributed, pest in montane rice terraces. Dispersal ability once on a site is really known only for invasive species, and so it is difficult to make comparisons with noninvasive species. Most of the data are derived from intentional introductions into agricultural land, with recent additions from studies of invasions in progress.

Transport

Human activities are the most important part of the invasive earthworm phenomenon. Earthworms are small and their egg capsules even smaller, so they travel without our knowledge. The most often cited means of transportation of earthworms are those involving the movement of soil and plants, often together. When people move to a new location, such as a different continent, they often like to have some familiar plants around them: the apple tree, the rosebush, or the rhubarb plant. European colonization of the Americas, parts of Africa, Australia, and elsewhere typically involved attempts to recreate the European agricultural base in the new location. Because the slow, long-distance transport of many plants required that they be in soil, the simultaneous arrival of earthworms is almost certain. Even in recent times, earthworms continue to arrive at international boundaries, as documented by border intercepts accumulated by the late earthworm specialist G. E. Gates.

Other movement of soils on large scales and over long distances has been mentioned in connection with the use of soils, sand, gravel, and the like in ships' ballast, a practice now replaced by the practice of using water, itself an agent of species invasion. It is impossible to verify that any of these actually transported an earthworm, but certainly it is conceivable. Topsoil is moved around by landscapers, and heavy equipment with bits of soil clinging to it may transport earthworms from one site to another. On shorter scales of time and distance, the modern nursery trade of potted or root-ball plants is a potential, and in some cases verified, agent of earthworm invasion. Temperate zone *Amynthas* in the eastern and central United States, including in the author's home city, Fairfield, Iowa, are spread from wholesalers to retail nurseries, and then to the homes of retail customers. Reforestation projects, generally hailed as a good thing, are often supplied by nurseries, in which the tree stock is grown in small plastic bags filled with soil. In tropical countries, where this silvicultural practice is almost universal, reforested lands are heavily infested with invasive earthworms, typically *Pontoscolex corethrurus*, *Polypheretima elongata*, and various *Amynthas*.

Earthworms are used all over the world as fish bait, and this is a very important factor in the establishment of new populations of invasive earthworms. In the Great Lakes region of the United States, the pattern of invasions is largely consistent with the hypothesis that discarded bait worms are the source of inoculum. *Amynthas hupeiensis*, a slightly malodorous green worm favored for catfish bait in central U.S. rivers, is easy to find at public river access sites and near road bridges. Other *Amynthas* are used as bait and probably have been spread by this mechanism.

Invasive vs. Peregrine

Some species have reached new territory through human intervention and yet have not become invasive, in the sense of undergoing rapid range expansion and population growth, whether or not at the expense of indigenous earthworms and other soil fauna. Why not? The noninvasive peregrines clearly have the ability to entrain in means of transport and to survive the voyage. The former would seem to involve a degree of tolerance of anthropogenic habitats, and the latter to imply the resistant qualities associated with invasive species. On arrival, the peregrines are able to live in new soils, something not universal to earthworms. In fact, several researchers have noted difficulty in establishing a (nonperegrine/invasive) species of interest in laboratory culture, even when they attempt to duplicate field conditions. Success, if achieved at all, is most likely met when the home soil is used. In these respects the peregrines are like invasives, but there must be some interesting differences. This will be the most productive approach to answering the question posed in the title of this section, because one has the opportunity to control for a variety of invasiveness factors across related species.

CONSEQUENCES OF EARTHWORM INVASIONS

Effects of earthworm invasions are diverse and highly variable in their intensity. In extreme cases, there can be complete alteration of soil profiles, drastic changes in nutrient cycling dynamics and in the distribution of organic carbon over the various classes found on and in soils, total overhaul of the soil-litter microflora, endangerment of forest-floor herbaceous plants, and probably significant changes in other soil fauna. All of the above can occur on the same site, an example of which is the situation in northern hardwood forests in the north central United States. These forests were devoid of earthworms prior to the invasion. Similar patterns occur in other glaciated forest sites across the northern United States and southern Canada. In the absence of earthworms, a thick, slowly decomposing layer of leaf litter develops, with anywhere from 4 to 7 years of accumulation present at any time. Below the litter layer lies mineral soil. Earthworms consume the litter capital along a wave front of invasion, simultaneously mixing the mineral soil with the finely divided organic matter of their fecal material and reducing the importance of fungi in the soil system. Nutrient mineralization is accelerated, and subsequently, a new state is reached in which earthworms persist on the annual input of leaf litter with rapid turnover. Any soil organisms dependent on the existence of a fungally dominated, structured leaf litter pack are probably severely reduced in numbers, although this situation has not been adequately studied. In some cases, the new soil environment is less cohesive, and deer feeding on small plants may uproot them in the act, rather than simply biting off an upper portion. This has led to reductions in forest-floor herb layers and to an influx of an invasive sedge that seems to tolerate the conditions.

Other habitats may have different responses. Where there is a preexisting native earthworm fauna, one might not expect such wholesale changes. For example, in Kansas tallgrass prairie, an indigenous earthworm fauna coexists with the invasive *Aporrectodea trapezoides*, but apparent impacts of the invader on soil and nutrient processes are small. To the extent that the invader partially replaces native earthworm biomass, the effects of the longer active season of the latter are reduced but not compensated for by the invader.

An Appalachian forest site invaded by *Amynthas agrestis* did not undergo profound changes in soil characteristics, except for reduction of the forest-floor litter layer. The invasion front advanced and retreated depending on moisture availability.

Invasive earthworms are most ecologically important in anthropogenic environments such as agricultural land and secondary forests, where they are often the only earthworms present. Their exclusive hold on these areas can be due to the extirpation of the native species during habitat conversion, followed by introduction of the invasives and by repeated disturbances. There has been considerable discussion of the sequence of events leading to domination of disturbed sites, regardless of whether they are disturbed chronically (e.g., cropland) or only at long intervals (e.g., pastures, tree plantations). There are some indications that undisturbed primary vegetation with a resident native earthworm community somehow resists invasion, even when potential invaders are residing only a few meters distant. Thus, the consensus is that some sort of disturbance is necessary to create conditions suitable for invasions, and then that propagules must be present, often in concert with the disturbance itself (e.g., use of earthmoving or farm machinery, reforestation, etc.). The frequency and intensity of disturbance will in part determine the degree to which invasive species dominate

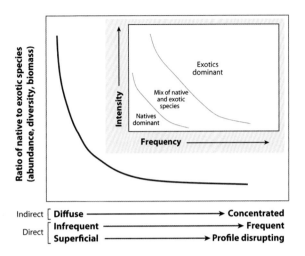

FIGURE 2 Hypothesized relationships between types of disturbance and expected proportions of native and exotic species. Direct disturbances are categorized by frequency of occurrence and by intensity or degree of physical perturbation of the soil profile. Frequency and intensity are related, but extreme disruption of the soil profile need not be frequent to effect change in earthworm communities (inset). Indirect disturbances include road density, human population density, degree of fragmentation, and diversity of land use types, among others.

the land unit (Fig. 2). The other determinants are the propagule pressure and the species content thereof.

Another fundamental question in the invasion of soils occupied, or perhaps occupied until recently, by indigenous earthworms is whether or not interspecific competition plays a role. There are anecdotal reports of native populations declining after the advent of invasive species, and some of these reports come from anthropogenic sites. These suggest that the natives can survive in at least some artificial habitats, such as suburban lawns and gardens, but that in some of these cases, an unknown factor intervened, be it the arrival of invasive species, a change in treatment of the site, a gradual post-conversion change in soil resources, or some other factor. On the other hand, recent research tests hypotheses of null models in the assembly of earthworm communities in northern France. The results reject the null models of no competitive interactions, leading to the conclusion that among the various invasive Lumbricidae in the region (there are no other kinds of worms there), competition does take place and limits the possible combinations of species that can make up a persistent community. An additional study of earthworm communities by the same French team reached similar conclusions in a Colombian tropical savannah, which in this case consisted of a mix of native and invasive species. In a southern Appalachian forest, there was a slight positive effect of invasives on native earthworm populations. The former were topsoil and litter feeders, and the latter were largely soil feeders, and so there would be a lack of niche overlap, and possibly even facilitation because the invasive species was transferring organic matter to the surface soil.

If competition occurs among invasive species themselves, it seems probable that there could also be competition within native communities and between native and invasive species. If one takes a precautionary approach to the question of competition among earthworms, there is reason to believe that invasions threaten endemic earthworm species, and for this reason, among others, there should be attempts to slow the spread of invasive earthworms. This concern can be tempered in cases where there is clearly little or no niche overlap between invading and indigenous species.

Invasional meltdowns involving earthworms are not numerous, or at least not frequently detected. The northern hardwood forest case discussed above may be one example with respect to forest-floor plants, and potentially to tree seedling recruitment. In central U.S. forests with the invasive shrub *Rhamnus cathartica*, the interaction of invasive earthworms with shrub leaf litter has altered soil and litter characteristics in a manner that promotes the growth of the shrub at the expense of native hardwoods. Removal of the shrubs caused significant reduction of invasive earthworms, indicating a positive feedback loop. A different situation occurs in New Jersey with two species of invasive shrubs, where the invasive shrubs create soil conditions (higher pH, greater nitrogen availability) more favorable to the worms. However, it is not certain that positive feedback exists in this system. In Hawaii the combination of an invasive lumbricid worm in montane forests, feral pigs foraging for worms, and the presence of a wind-dispersed invasive plant, *Morella (Myrica) faya*, on the periphery of the forest, is leading to rapid spread of the invasive plant at expense of native Hawaiian vegetation. Rooting pigs create disturbed bare soil, into which seeds fall and find a suitable site for germination and growth. Completing the positive feedback loop, higher nitrogen under the invasive plant leads to higher earthworm populations.

CONTROL

Currently, it seems that little can be done to reverse an earthworm invasion. Even a small earthworm population can number in the tens per square meter, and thus in the hundreds of thousands per hectare. Biocides specific for earthworms have not been invented, so nontarget organisms will also be affected if biocides are used. Predators are not very efficient, except in the case of the

New Zealand flatworm *Arthurdendyus triangulatus*, which is very effective against Lumbricidae in the United Kingdom. However, this flatworm introduction was not intentional and was considered a detrimental invasion itself. The earthworms appear to have no defense, and the flatworm almost totally eliminates earthworms and undergoes population crashes. As with other biological control agents, one would want to be extremely careful about this predator. In countries with indigenous earthworms (unlike the United Kingdom), a rampant flatworm predator could contribute to species extinctions.

The only viable control now available is to prevent further human-aided spread of invasive earthworms. Most countries now have strict border controls on the movements of plants and soils. The state of Hawaii is so strict that it does not even allow outbound shipment of live earthworms from the state, even though this would pose no imaginable danger to Hawaii, where the particular worm is already present. Education of sport fishermen not to release unused bait worms, restricting the availability of certain bait species, and alerting nursery businesses about their role in spreading undesirable earthworms could all be done for very little cost. Some of these efforts have begun, such as regulating bait in the Great Lakes region of the United States.

There are conflicting interests as well. In most agricultural land, if earthworms are present at all, they are invasive species. The agronomic benefits of their activities are many, and only in rare circumstances do earthworms exert net negative effects on soil structure, plant productivity, and other features of interest to humans. All of the known composting species can establish themselves outside the composting facility, provided the climate and microhabitat conditions are met. Very great benefits are to be gained from the use of vermicomposting in organic waste conversion, yet an increase in the scale of this process will almost certainly increase distributions of a few earthworm species. It is too late to close the door, so we have to decide how to live with the invasive earthworms we already have, and at the same time make sound judgments about when and how to reduce their expansion.

SEE ALSO THE FOLLOWING ARTICLES

Agriculture / Belowground Phenomena / Competition, Animal / Disturbance / Horticulture / Invasional Meltdown / Life History Strategies

FURTHER READING

Blakemore, R. J. 2006. *Cosmopolitan Earthworms: An Eco-Taxonomic Guide to the Peregrine Species of the World,* 2nd ed. Japan: VermEcology.
Bohlen, P. J., S. Scheu, C. M. Hale, M. A. McLean, and S. Migge. 2004. Non-native invasive earthworms as agents of change in northern temperate forests. *Frontiers in Ecology and the Environment* 2: 427–435.

Decaëns, T., P. Margerie, M. Aubert, M. Hedde, and F. Bureau. 2008. Assembly rules within earthworm communities in north-western France: A regional analysis. *Applied Soil Ecology* 39: 321–335.

Edwards, C.A., ed. 2004. *Earthworm Ecology*. Boca Raton, FL: CRC Press.

Gonzalez, G., C.-Y. Huang, X. Zou, and C. Rodriguez. 2006. Earthworm invasions in the tropics. *Biological Invasions* 8: 1247–1256.

Hendrix, P.F., ed. 2006. *Biological Invasions Belowground: Earthworms as Invasive Species*. Amsterdam: Springer-Verlag.

Hendrix, P.F., and P.J. Bohlen. 2002. Exotic earthworm invasions in North America: Ecological and policy implications. *BioScience* 52: 801–811.

Hendrix, P.F., M.A. Callaham Jr., J. Drake, C.-Y. Huang, S.W. James, B.A. Snyder, and W. Zhang. 2008. Pandora's box contained bait: The global problem of introduced earthworms. *Annual Reviews of Ecology, Evolution and Systematics* 39: 593–613.

Lee, K.E. 1985. *Earthworms: Their Ecology and Relationships with Soils and Land Use*. Sydney: Academic Press.

Tiunov, A.V., C.M. Hale, A.R. Holdsworth, and T.S. Vsevolodova-Perel. 2006. Invasion patterns of Lumbricidae into the previously earthworm-free areas of northeastern Europe and the western Great Lakes region of North America. *Biological Invasions* 8: 1223–1234.

ECONOMICS OF INVASIONS

SEE INVASION ECONOMICS

ECOSYSTEM ENGINEERS

SEE TRANSFORMERS

ECOTERRORISM AND BIOSECURITY

EMIEL ELFERINK AND WOUTER VAN DER WEIJDEN

Centre for Agriculture and Environment, Culemborg, The Netherlands

History is rich in cases of adverse outcomes of purposeful introduction of species. In most cases, such effects were not intended. Although biological warfare has happened several times in history, there are no documented cases of species introductions for this purpose. However, there have been several allegations, especially during the Cold War, of countries using insects as biological weapons (e.g., allegations of the United States spreading the Colorado beetle into eastern Europe). In 1996, in the United Nations, Cuba charged the United States with aerially releasing *Thrips palmi*, a pest of a wide variety of crops, on Cuban soil; the UN concluded in 1997, however, that due to technical complexity and passage of time, these charges could not be verified. Indonesia has been charged with introducing pigworms in western New Guinea, where Papuas and pigs live closely together. But those charges haven't been independently verified, either. In the future, both war and terrorism could be sources of intentional introduction of harmful species. Since the beginning of this century, over 100 terrorist attacks have been recorded, including the attacks in the United States on September 11, 2001, and in Madrid on March 11, 2004, and the many attacks with roadside bombs in Iraq and Afghanistan. In most cases, conventional weapons were used, but in one case—the anthrax letters in the United States in 2001—a biological weapon was applied. Because terrorists lack the ability to confront their adversaries directly, they often prefer an indirect approach. Introducing an invasive pathogen or parasite species to a new region, or reintroducing it in a region where it was eradicated earlier, is a potentially powerful weapon. Well-planned attacks can cause massive damage to humans, ecosystems, the economy, and society.

GLOBAL PATTERNS

Species introductions by countries into new areas for military purposes have been legally prohibited since April 1972, when biological warfare was banned under the Convention on Biological Weapons (BWC). However, there have been threats by other parties. In 2005, for example, the office of the prime minister of New Zealand received a letter claiming the deliberate release of the virus that causes foot-and-mouth disease. Such an attack could have a catastrophic impact on the country's huge dairy sector. In 2001, anthrax was disseminated by mail in the United States (where it is no alien), killing five people and sparking widespread panic even outside the United States (Fig. 1).

Invasive introduced species are, precisely by their invasive properties, interesting for a wide range of actors.

FIGURE 1 The anthrax letters of 2001, sent to U.S. media figures and politicians, were the first case of virulent anthrax spores being used for terrorist purposes. (Photograph courtesy of the U.S. Federal Bureau of Investigation.)

TABLE 1
Possible Means, Targets, Purposes, Motives, and Perpetrators for Terrorism

Means	Targets	Motives	Purposes	Perpetrators
Biological	Human	Political	Strategic damage	Lone wolf to global organizations
invasive species	Infrastructure	Ideological	Publicity for case	Radical religionist or nationalist
native species	Ecosystems/services	Personal	Disruption of society	Animal rights or green activist
Chemical	agriculture	Criminal	Undermining of authority	Mafia
Nuclear	animals	—	—	—
Radiological	plants			
Explosive	water			

NOTE: Highlighted areas are criteria for ecoterrorism.

Possible actors include terrorist organizations such as Al-Qaeda and Aum Shinrikyo, organized crime syndicates, such as mafia or drug cartels, radical political organizations, militia groups, and nationalist or religious organizations, as well as individual ("lone wolf") perpetrators.

Actors can have different motives for using invasive introduced species for purposes of terror. Traditionally, terrorists have been politically motivated, striving to challenge or overthrow a political order or to retaliate against an authority. "Idealistic" motives such as radical religious convictions or bioethics (anti-GMO, anti-abortion, pro–animal rights) are other well-known reasons for terror attacks. Other motives for terrorism include personal motives such as vengeance, or the perverted exercise of power, or even the thirst for sensation apart from any particular political or ideological conviction.

DEFINITION OF ECOTERRORISM

The term *ecoterrorism* is defined by various authorities in different ways, which causes confusion. For instance, the FBI defines ecoterrorism as

> the use or threatened use of violence of a criminal nature against innocent victims or property by an environmentally-oriented, subnational group for environmental-political reasons, or aimed at an audience beyond the target, often of a symbolic nature.

By contrast, the *Encyclopaedia Britannica* defines ecoterrorism as

> destruction, or the threat of destruction, of the environment by states, groups, or individuals in order to intimidate or to coerce governments or civilians.

The latter definition is adopted here. Table 1 summarizes means, possible targets, purposes, motives, and types of perpetrators relative to terrorism. Depending on the combination of these criteria, different types of terrorism can be defined. For instance, bioterrorism uses biological means,

while agrobioterrorism is a subset of bioterrorism using biological means to attack agriculture for whatever reason. Ecoterrorism is a combination of the highlighted criteria listed in Table 1. This article focuses on a specific subset of ecoterrorism, notably the use of invasive species that target ecosystems and ecosystem services. Figure 2 shows this delimitation (gray area). Five examples are presented below, with their numbers corresponding to those in Figure 2:

Group 1 refers to the use of biological agents of native origin used for a bioterrorist attack. One example is smallpox, caused by the viruses *Variola major* and *Variola minor*, which have been eradicated globally but are still present in some laboratories.

Group 2 refers to the use of biological agents that are native and can be used for ecoterrorism. One example is *Bacillus anthracis*, the agent used in the 2001 anthrax attacks in the United States. It can attack livestock in addition to humans.

Group 3 is ecoterrorism by means other than biological agents. Examples are forest fires and chemical agents such as the herbicide Agent Orange.

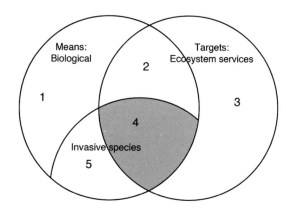

FIGURE 2 Schematic positioning of ecoterrorism. For further explanation, see the text.

Group 4 is the focus of this article: ecoterrorism by introduction of invasive species. One example is the South American leaf blight, a fungus that affects rubber trees of the genus *Hevea*. The causative fungus is native to tropical Latin America and the Caribbean, but if introduced in Southeast Asia, it could destroy the majority of global rubber production. Rubber is arguably the most important nonfood crop in the world, and a loss of rubber supplies could have a crippling effect on most economies.

Group 5 refers to bioterrorist use of invasive species for purposes other than environmental destruction: for instance, by introduction of the lethal malaria parasite *Plasmodium falciparum* from Africa into South America.

REQUIREMENTS FOR ECOTERROR AGENTS

Requirements for ecoterror agents largely depend on the tactical goal of the terror attack: Local damage, or widespread? Visible damage, or invisible? Acute damage, or chronic?

Most classic terror attacks have aimed at acute, visible, and local damage. This is true, for example, of the 1995 chemical attack by Aum Shinrikyo, in which sarin was released in the Tokyo subway. Some invasive species, too, can be used to cause such spectacular damage—for example, livestock pathogens—although they can be used to cause widespread structural damage as well. For a bioterrorist, the "ideal" invasive species weapon meets the following requirements:

- It is easily obtained.
- It is easily preserved and spread.
- It is easily introduced in a country.
- It carries low risk for perpetrators (unless they are suicidal).
- It is not quickly detected and identified.
- It is difficult to control (e.g., no appropriate vaccine, pesticide, or natural enemy is available).

No large numbers ("propagules") need to be introduced, just enough to start a population base. Introduction at multiple locations in numbers large enough to spark colonization would make an attack more effective by distributing the risks both of detection and of failure and rapidly exceeding the capacity for control. Compared to chemical and radiological weapons, biological weapons typically are more difficult to detect. In most countries, border inspectors are not trained to recognize all potential invasive species. Furthermore, it is extremely difficult to find out who has implanted an invasive species unless the perpetrator claims the attack. Once an invasive species has become established, it is often impossible to eradicate

it. History is full of examples, including *Phytophthora* in Europe (since 1845), the rabbit in Australia (since 1859), and the zebra mussel in North America (since 1986). Success stories of eradication are largely confined to islands.

POTENTIAL EFFECTS

One of the primary effects of ecoterrorism is economic damage. Indications of potential costs can be obtained from previous unintentional biological invasions and disease outbreaks:

- According to the World Bank, one single outbreak of flu costs $300 million a year in the United States alone.
- The cost of a pandemic of highly virulent and contagious flu may amount to $1,250 billion—3.1 percent of global GDP—mainly due to labor days lost.
- The World Health Organization estimates that $7 billion is required to fight HIV annually.
- The 2001 foot-and-mouth outbreak in the United Kingdom led to the destruction of 10 million animals, had a devastating impact on the livestock and dairy industries and on tourism, and cost more than €30 billion (Fig. 3).
- The Netherlands suffered outbreaks of the introduced livestock diseases BSE ("mad cow disease"; from 1997 onward), classic swine fever (1997), foot-and-mouth disease (2001), and bird flu (2003). The estimated costs were at least €550 million, €1,500 million, €650 million, and €850 million, respectively.
- Fighting the Asian long-horned beetle costs millions of dollars in the United States and threatens industries worth billions.

FIGURE 3 Burning of cows during the 2001 foot-and-mouth outbreak in England. Terrorists can cause similar damage by purposely spreading the virus. (Photograph courtesy of Murdophoto.)

• Invasive pathogens of soybeans in Brazil are responsible for an annual loss of $1 billion.

Bioterror attacks can cause even higher costs if the pathogen is simultaneously spread in multiple locations.

BIOSECURITY

Ecoterrorism can be counteracted by biosecurity measures. Biosecurity is a strategic and integrated approach that encompasses the policy and regulatory frameworks (including instruments and activities) that analyze and manage risks from the introduction of plant pests, animal pests and diseases, zoonoses, genetically modified organisms (GMOs), and invasive species and genotypes.

No specific policy on ecoterrorism using invasive introduced species is known. However, most countries under threat from terrorism have policies on terrorism in general. Furthermore, countries such as New Zealand, Australia, and the United States have specific policies on the introduction of invasive species (whether intentional or accidental). In 1999 President Clinton issued Executive Order 13112 to prevent the introduction of invasive species and provide for their control, and to minimize the economic, ecological, and human health impacts caused by invasive species. The order established the National Invasive Species Council with the charge of preparing and overseeing a national management plan and providing national leadership on invasive species.

The EU, too, is currently developing a strategy. After a study identified 10,822 nonnative species in the EU, the European Commission presented in 2008 a communication, "Towards an EU Strategy on Invasive Species," containing a series of policy options for developing a strategy to deal with invasive introduced species. The purpose of the document is to establish an EU strategy in 2010 with the aim of substantially reducing the impact of invasive species on European biodiversity.

While such policies discourage introduction of invasive species, they are certainly not sufficient to prevent introduction for terrorist purposes. Security measures can further reduce the risk. Frameworks for biosecurity typically focus on prevention and control of the threat. Elements include international treaties and standards, cooperative efforts, inspections in host countries and ports of entry, quarantine, intelligence, and treatment of shipments.

Small-scale preventive security measures are another important line of defense. Examples include farm management practices to protect animals and crops (Fig. 4); the eradication of introduced species; the use of disinfectants on people, clothing, equipment, and supplies; the

FIGURE 4 Agroterrorism is considered a serious threat in the United States, prompting laws to improve biosecurity. (Source: University Press of the Pacific.)

locking up of transports; and the screening of employees of laboratories.

Effective control primarily depends on early detection and rapid response to prevent establishment of the introduced species. Important aspects are

PREPAREDNESS AND COMMUNICATION: Capacity-building; the stockpiling of vaccines; the stockpiling of antiviral drugs; the presence of adequate supplies and equipment; effective coordination (scenarios, decision-trees, exercise); the enhancement of public awareness

SURVEILLANCE AND DETECTION: Early detection by detection technology and expertise; diagnostic facilities; specialist knowledge at research institutes; educated first contacts such as by border patrols, veterinarians, and rangers; cooperation and knowledge-sharing between local, national, and international authorities; surveillance of sensitive areas

RESPONSE AND CONTAINMENT: Confinement and elimination of introduced species

There can be little doubt that the world will face more bioterrorism in the future. And bioterrorism can come from many sides, including unexpected ones: the main bioterrorist of this era was an academic working in the U.S. Department of Defense. Because 100 percent prevention is impossible, society has to make sure its systems are resilient enough to eradicate or control invasive species that are accidentally as well as purposefully introduced.

SEE ALSO THE FOLLOWING ARTICLES

Agreements, International / Early Detection and Rapid Response / Eradication / Influenza / Invasion Economics / Pathogens, Animal / Pathogens, Human / Pathogens, Plant

FURTHER READING

Howard, R. D., J. J. F. Forest, and J. C. Moore. 2006. *Homeland Security and Terrorism: Readings and Interpretations.* New York: McGraw-Hill.

Lockwood, J. A. 2009. *Six-Legged Soldiers: Using Insects as Weapons of War.* Oxford: Oxford University Press.

Monke, J. 2006. *Agroterrorism: Threats and Preparedness.* Honolulu: University Press of the Pacific.

Pimentel, D., ed. 2002. *Biological Invasions.* New York: CRC Press.

Van der Weijden, W., R. Leewis, and P. Bol. 2007. *Biological Globalisation: Bio-invasions and Their Impacts on Nature, the Economy and Public Health.* Utrecht, The Netherlands: KNNV Publishing.

ELTON, CHARLES S.

DANIEL SIMBERLOFF

University of Tennessee, Knoxville

Charles S. Elton (1900–1991; Fig. 1) is widely viewed as the founder of modern invasion biology by virtue of his 1958 publishing of the monograph *The Ecology of Invasions by Animals and Plants*, the first synthetic treatment of the global scope and varied impacts of invasions by both animals and plants.

EARLY LIFE AND PUBLICATIONS

Elton's interest in invasions began much earlier. In his youth in Liverpool, he frequented a store displaying exotic animals, and in 1925 he wrote his mother a letter about the "war" waged by North American gray squirrels on native British red squirrels. On an early scientific expedition to Spitsbergen Island, he noted the presence of hover-flies and aphids that he deduced were blown there from the Kola Peninsula but that were clearly unable to sustain a population. Elton's first major work in 1927—*Animal Ecology*—became a highly regarded classic, one of the first textbooks of animal ecology; it referred to several devastating impacts of introduced animal species. He next treated introduced species in an article in the London *Times* on the need for a census of animals in Great Britain. The article was headlined "Animal Invaders" and began with a list of animals introduced to Britain (e.g., the muskrat), with a discussion of their impacts on native species. Two short introductory ecology texts followed in 1933 (one of which was based on a series of BBC radio broadcasts, as was his 1958 monograph). These featured many examples of impacts of introduced species, both animals and plants, and although Elton observed that some introductions benefit humans, he stressed cases in which they damage native species or human enterprises (e.g., agriculture).

Elton was founding editor of the *Journal of Animal Ecology* in 1932, and in this capacity he reviewed several books on introduced species. Two of these reveal what might be seen as a growing obsession with introduced species. In each, a review of a monograph on a specific introduction (a 1936 book on the Chinese mitten crab in Europe and a 1944 book on the eradication of the African malaria mosquito, *Anopheles gambiae*, in Brazil) became a platform for a broad-scale exposition of the international scope of invasions, the variety of their impacts, and some approaches to dealing with them.

Elton founded the Bureau of Animal Population at Oxford University in 1932 and headed it until its closure in 1967. In the 1930s, Bureau staff under Elton's direction undertook studies in Britain of the introduced North American muskrat (in support of a successful effort to eradicate it), the South American nutria, and the North American gray squirrel. During World War II, Elton shifted almost all of the Bureau's efforts toward studies of four introduced pests in Britain—the Norway rat, black rat, house mouse, and European rabbit—to aid the war

FIGURE 1 Charles S. Elton. (Photograph courtesy of Department of Zoology, University of Oxford.)

effort by identifying ways to lessen the tremendous losses they caused to agriculture and stored food. He quixotically suggested in 1944 that such research might well lead to the eradication of both rats in Great Britain. During the war, Elton also produced a short paper in an obscure journal published by Polish refugees in London (*Polish Science and Learning*) that set forth very briefly many of the themes about invasions that he later developed fully in his monograph.

THE MONOGRAPH ON INVASIONS

Elton did not publish on invasions after the war until his monograph appeared in 1958. This work was an expansion of the wartime paper noted above plus three 1957 BBC radio broadcasts. A remarkable piece of scholarship, it is vividly and wittily written for a lay audience but teems with astute natural history observations that build toward a scientifically compelling synthesis. Elton deals with animals, plants, and microorganisms from terrestrial, freshwater, and marine environments, and his examples are drawn from all over the globe. He begins by observing that widespread introduction of species is erasing the classical biogeographic distribution patterns that are crucial to understanding the history of life, but he quickly moves to the many and varied impacts of introduced species. He points to many of the phenomena that subsequently came to dominate the literature on the ecology of biological invasions. Perhaps the largest item on today's invasion research agenda that Elton did not treat was the evolution of invaders, but Elton was an ecologist, not an evolutionist.

The main scientific question Elton raised in the monograph was why some introductions lead to widespread invasions that produce massive impacts while in other cases the newcomer remains restricted geographically and is innocuous. Elton's proposed explanation for these differing outcomes has come to be known as the "biotic resistance hypothesis." In his view, the native resident community resists a newly arrived species by various combinations of predation, competition, parasitism, disease, and aggression, and areas in which this resistance is weak will be those that are subject to invasions and their impacts. Elton believed that the main determinant of the degree of biotic resistance was the number of species present in the resident community: communities with relatively few species would be more prone to damaging invasions. In support of this hypothesis, Elton pointed especially to famous invasions with great impacts on islands, with extended sections on Hawaii and New Zealand in particular. One well-known trait of islands

is, of course, a paucity of species. Elton also pointed to anthropogenically simplified or disturbed habitats, especially agricultural fields, as both species-poor and frequently heavily invaded, and to intact tropical rainforest as species-rich and seemingly impervious to invasions. Yet another piece of evidence he adduced in support of his biotic resistance hypothesis was that models of simple systems composed of one prey or host species and one enemy were inherently unstable mathematically, independently of external inputs such as climatic events. This idea of biotic resistance and its relationship to species diversity became part of an important research theme even beyond invasion biology: it was a key component of the debate among ecologists on the proposition that diversity confers stability.

Elton stressed that his monograph was largely a compendium of examples and lamented the dearth of general principles, weakness of patterns, and low level of predictability about the trajectory of introduced species. And he expressed the hope that invasion biology would rapidly advance on all these fronts. Yet many of his examples remain timely research subjects today—gypsy moths, nutria, woolly adelgids, sea lampreys, hybrid cordgrass. And some predictions in the monograph turned out to be uncannily accurate, such as that fouling organisms and those carried by ballast water would greatly increase in number; the zebra mussel is a now a prime example.

POST-MONOGRAPH PUBLICATIONS AND LIFE

After the monograph, Elton published the massive book he considered the culmination of his research, *The Pattern of Animal Communities* (1966), which focused especially on the plethora of research he, his students, and his colleagues had conducted for decades on the various species and communities of Wytham Wood, near Oxford. A chapter of this tome repeated the main themes of the monograph and detailed some of the invasions of Wytham Wood. This book did not excite much interest among contemporary ecologists and is rarely cited today, perhaps because his approach of attempted generalization from copious, detailed accounts of natural history was being superseded by a more quantitative and evolutionary approach at precisely the time the book was published.

Oxford University used the occasion of Elton's retirement to close the Bureau of Animal Population in 1967, over his strenuous objection. Elton lived for almost a quarter century after his retirement but wrote little. Although increasingly lionized for his contributions to animal ecology (especially the study of population cycles) and invasion biology, he was not an outgoing person and for the most part did not cooperate with or encourage the many scholars who sought him out to discuss his remarkable career. There is still no definitive comprehensive biography of Charles Elton.

IMPACT ON INVASION BIOLOGY

The view that Elton founded invasion biology is an exaggeration. Although a number of earlier workers had discussed introduced species, none had produced a synthesis like Elton's monograph, nor had any treated the scope of their impacts in nearly the breadth that Elton did. Thus, he did write the first "modern" monograph of invasion biology. Yet only two of his many students worked on invasions; so did only one of his long-term visitors at the Bureau. His monograph was not widely cited until the early 1990s, over 30 years after its original publication. The field of invasion biology really blossomed beginning in the late 1980s, which saw an international initiative on invasions by the Scientific Committee on Problems of the Environment (SCOPE) that engaged hundreds of ecologists and produced several widely read books. Many participants in that initiative had read Elton's monograph years earlier and cited him, although it is difficult to point to specific influences of Elton on their work. Nevertheless, these citations in the midst of an explosion of interest in invasions combined with a new edition of the monograph (2000) led a new audience to read Elton's monograph and become excited by it. An international conference on invasion biology at Stellenbosch University, South Africa, commemorated the 50th anniversary of the monograph.

Elton exerted an indirect and growing influence on conservationists and conservation biologists and their engagement with the issue of biological invasions through his extended interaction with Aldo Leopold, the most prominent conservation figure of the twentieth century. The two met in 1931 at the Matamek Conference on Biological Cycles in Labrador and became frequent correspondents on both scientific and personal matters. Elton became, in a sense, Leopold's tutor in matters of ecology, and Leopold's interest in introduced species, first expressed in 1918, expanded and evolved greatly under Elton's influence. Alone among prominent early conservationists, Leopold insistently and increasingly pointed to conservation problems caused by biological invasions, and it is easy to see Elton's ideas—and even some of his examples—in Leopold's copious writings on the subject, which culminated in several essays in the posthumously published *A Sand County Almanac* (1949).

SEE ALSO THE FOLLOWING ARTICLES

Evolution of Invasive Populations / Invasion Biology: Historical Precedents / Islands / SCOPE Project

FURTHER READING

Chew, M. K. 2006. Ending with Elton: Preludes to invasion biology. Ph.D. dissertation, Arizona State University, Tempe.
Crowcroft, P. 1991. *Elton's Ecologists: A History of the Bureau of Animal Population.* Chicago: University of Chicago Press.
Elton, C. 1927. *Animal Ecology.* New York: Macmillan (reprinted 2001, Chicago: University of Chicago Press).
Elton, C. S. 1958. *The Ecology of Invasions by Animals and Plants.* London: Methuen (reprinted 2000, Chicago: University of Chicago Press).
Ricciardi, A., and H. J. MacIsaac. 2008. The book that began invasion biology. *Nature* 452: 34.
Richardson, D. M., and P. Pyšek. 2008. Fifty years of invasion ecology: The legacy of Charles Elton. *Diversity and Distributions* 14: 161–168.
Simberloff, D. 2010. Charles Elton: Neither founder nor siren, but prophet. In D. M. Richardson, ed. *Fifty Years of Invasion Ecology.* New York: Wiley.
Southwood, R., and J. R. Clarke. 1999. Charles Sutherland Elton. *Biographical Memoirs of Fellows of the Royal Society of London* 45: 129–140.

ENDANGERED AND THREATENED SPECIES

RICHARD B. PRIMACK

Boston University, Massachusetts

Invasive species can have such a profound, negative impact on endangered species via competition, predation, and spread of disease that it would be reasonable to call them "endangering species." Invasive exotic species constitute one of the major threats to native species, negatively affecting approximately 30 percent of threatened birds and 10 percent of threatened amphibians (Fig. 1). Most threatened species and ecosystems face at least two or more threats to their continued existence, including habitat destruction, habitat fragmentation, pollution, climate change, overharvesting, and disease; many interact synergistically with the effects of invasive species to speed the way to extinction and hinder efforts at protecting biodiversity.

SPECIES THREATENED WITH EXTINCTION

Invasive species are considered to be one of the most serious threats facing the biota of many national parks and other protected areas. While the effects of habitat degradation, fragmentation, pollution, overharvesting, and disease potentially can be corrected and reversed in a matter of

TABLE 1

IUCN Red List Criteria for the Assignment of Conservation Categories

Red List Criteria	Quantification of Criteria for Red List Category "Critically Endangered"
A. Observable reduction in numbers of individuals	The population has declined by 80% or more over the last ten years or three generations (whichever is longer), either based on direct observation or inferred from factors such as levels of exploitation, threats from introduced species and disease, or habitat destruction or degradation
B. Total geographical area occupied by the species	The species has a restricted range (less than 100 km^2 at a single location) *and* there is observed or predicted habitat loss, fragmentation, ecological imbalance, or heavy commercial exploitation
C. Predicted decline in number of individuals	The total population size is less than 250 mature, breeding individuals and is expected to decline by 25% or more within three years or one generation
D. Number of mature individuals currently alive	The population size is less than 50 mature individuals
E. Probability of the species going extinct within a certain number of years or generations	Extinction probability is greater than 50% within ten years or three generations

NOTE: A species that meets the described quantities for *any one* of criteria A–E may be classified "Critically Endangered." Similar quantification for the Red List categories "Endangered" and "Vulnerable" can be found at www.iucnredlist.org/info/categories_criteria2001.

years or decades as long as the original species are present, well-established exotic species may be impossible to remove from communities. They may have built up such large numbers, become so widely dispersed, and integrated themselves so thoroughly into the community that eliminating them would be extraordinarily difficult and expensive. As climate change alters habitat in coming decades, however, the potential for new invasions by exotics grows. Likewise, opportunities for invaders to "hitch a ride" with human travelers is increased as human activities penetrate into the few remaining undisturbed areas of natural habitat.

The significance of invasive species to conservation measures is reflected in their status within global measures of threat to endangered species. On a global scale, the World Conservation Union (ICUN) has developed quantitative measures of threat for species based on their probability of extinction, using population trends and habitat condition. In practice, a species is most commonly assigned to an IUCN category based on the area it occupies, the number of mature individuals it has, or the rate of decline of its habitat or population (Table 1; Fig. 2). All of these criteria can be affected by pressure from invasive species. The IUCN has evaluated and described the threats to plant and animal species in its series of Red Data Books and Red Lists of threatened species. For mammals, 1,141 (21%) of 5,488 described species are listed as threatened; for birds, 1,222 (12%) of 9,990 species are listed as threatened; for amphibians, 1,905 (30%) of 6,347 species are listed as threatened; and for plants, around 8,000 (3%) of species are threatened out of 258,000. Invasive species are a major threat to many species; approximately 30 percent of bird species are threatened by invasive species. In addition, Red Lists have been developed for individual countries. The changes over time in the status of Red List species are also being used to evaluate the effectiveness of conservation policies. For all of these threatened and endangered species, invasive species are one of the most important threats to their continued existence.

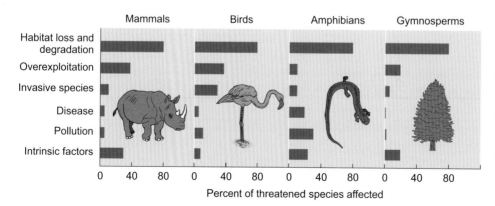

FIGURE 1 The loss and degradation of natural habitat is the greatest threat to the world's biodiversity, followed by over-exploitation. Invasive species are also a significant threat, particularly for birds. Because many species face multiple threats, the total threat adds up to more than 100 percent. (From Primack 2008, after IUCN 2004.)

THREATS TO ISLAND SPECIES

Island species are particularly vulnerable to invasive species. For birds, invasive species affect 67 percent of threatened species on oceanic islands, but only 8 percent of threatened birds in mainland areas.

The introduction of just one exotic species to an island may cause the local extinction (or extirpation) of many native species and threaten the existence of the ones that remain. For example, the Australian brown treesnake (*Boiga irregularis*) has been introduced onto a number of Pacific islands, where it devastates endemic bird populations. The snake eats eggs, nestlings, and adult birds. On Guam alone, the brown treesnake has driven 8 of the 11 forest bird species to extinction. Visitors have remarked on the absence of birdsong: "between the silence and the cobwebs, the rain forests of Guam have taken on the aura of a tomb." Perhaps in an attempt to locate new prey, brown treesnakes have been known to attack sleeping people. The government spends $4.6 million each year on attempts to control the brown treesnake population—so far without success. Brown treesnakes have appeared on other Pacific islands as well, presumably after stowing away on ships and planes. Such further invasions threaten additional groups of endemic species.

Similarly, the introduction of goats, rabbits, and other exotic grazing animals onto islands and mainland areas threatens the existence of large numbers of plant species. Aggressive invasive plants form pure stands in which native species can no longer grow and reproduce, leading to decline and local extinction of many endemic species and the specialist insects that feed on them. Invasive carnivores, such as predatory snails, mongooses, and weasels, threaten other groups of animals.

THREATS IN THE AQUATIC ENVIRONMENT

The introduction of exotic commercial and sport fish species into lakes has a long history. In most cases, these exotics have increased in numbers at the expense of local species; for example, following the introduction of exotic tilapia species and the Nile perch (*Lates niloticus*) into the East African great lakes, many endemic fish species have gone extinct, and those that remain have become scarce. Worldwide, more than 120 fish species have been introduced into marine and estuarine systems and inland seas. Although some of these introductions have been deliberate attempts to increase fisheries, most have been the unintentional result of canal-building and the transport of ballast water in ships. Often these exotic fish are larger and more aggressive than the native fish, and they may

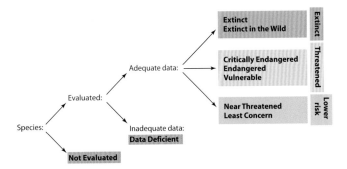

FIGURE 2 The IUCN categories of conservation status (colored boxes) depend on (1) whether species have been evaluated and (2) how much information is available for the species. If data are available, then the species can be categorized. (From Primack 2008, after IUCN 2008.)

push local fish toward extinction. Once these invasive species are removed from aquatic habitats, the native species are sometimes able to recover.

Aggressive aquatic exotics also include plants and invertebrates. One of the most alarming recent invasions in North America was the arrival in 1988 of the zebra mussel (*Dreissena polymorpha*) in the Great Lakes. This small, striped native of the Caspian Sea apparently was a stowaway in ballast tanks of a European tanker that discharged its ballast water into Lake Erie. Within two years, zebra mussels had reached densities of 700,000 individuals per square meter in parts of Lake Erie, encrusting every hard surface and choking out native mussel species in the process. Zebra mussels have now spread south throughout the entire Mississippi River drainage, as well as in all other directions via the Great Lakes and their tributaries. As it spreads (often via the boots or boat bottoms of unwitting local fishermen), this exotic species causes enormous economic damage to fisheries, dams, power plants, and water treatment facilities, and also threatens as many as 60 native mussel species with severe declines in population size and even extinction.

THREATS FROM BIOLOGICAL CONTROL AGENTS

When an exotic species becomes invasive, a common solution is to introduce an animal species from the exotic species' original range that will consume the pest and control its numbers. While biological control is often dramatically successful, there are many cases in which a biological control agent has itself become invasive, attacking native species along with, or instead of, its intended target species. For example, a parasitic fly species (*Compsilura concinnata*) introduced into North America to control invasive gypsy moths has been found to parasitize more than 200 native moth species, in many cases dramatically reducing population numbers. Many of the large moths that used to be abundant in the past are now rare or locally extinct because of the introduced fly—which has not succeeded in controlling gypsy moths.

THREATS FROM HYBRIDIZATION

Another special class of invasive species that threatens native species is introduced species that have close relatives in the native biota. When invasive species hybridize with native species and varieties, unique genotypes may be eliminated from local populations, and species boundaries may become obscured. This situation is sometimes called genetic swamping, and it appears to be the fate of native trout species when confronted by commercial species. In the American Southwest, the range of the Apache trout (*Oncorhynchus apache*) has been reduced by habitat destruction and by competition with introduced species. The species has also hybridized extensively with rainbow trout (*O. mykiss*), an introduced sport fish, blurring its identity as a distinct species.

THREATS FROM INFECTIOUS AGENTS

Human activities and species' interaction with humans often increase the transmission of disease, which is a major threat to species and biological communities. Disease-causing organisms such as bacteria, viruses, fungi, and protists can have major impacts on the structure of an entire biological community. These disease-causing species are sometimes introduced beyond their range and then spread rapidly, taking on the characteristics of invasive species. The increased mobility of people and the spread of exotic species have contributed to the spread of disease. For example, feral dogs and cats may be reservoirs of disease organisms that can spread to wild carnivore populations, endangering their existence. Furthermore, human activities, such as habitat fragmentation, may increase the incidence of both native and nonnative disease-carrying vectors, such as certain insects and ticks, leading to reduction in the size and density of vulnerable populations.

Disease is the single greatest threat to many rare species. The decline of many frog populations from pristine montane habitats in Australia, North America, and Central America is apparently due in part to the introduction and rapid spread of a disease caused by an exotic fungus (*Batrachochytrium dendrobatidis*). The last population of black-footed ferrets (*Mustela nigrepes*) known to exist in the wild was destroyed in 1987 by the canine distemper virus, a pathogen introduced to the area by domestic dogs. One of the main challenges of managing the captive breeding program for black-footed ferrets has been

protecting the population from canine distemper, human viruses, and other diseases; this is being achieved through rigorous quarantine measures and subdivision of the captive colony into geographically separate groups.

MOVING SPECIES TO AVOID THREATS FROM INVASIVE SPECIES

In many cases, invasive species have reached such high densities and are so difficult to eradicate that species may need to be brought into captivity to prevent their extinction in the wild. A goal of such programs is to establish new populations outside the range of these invasive species. These establishment programs are unlikely to be effective unless the factors that led to the decline of the original wild populations are clearly understood and eliminated, or at least controlled. For example, the kakapo (*Strigops habroptilus*), a flightless parrot, has been eliminated from the New Zealand mainland because of predation by introduced domestic cats, weasels, stoats, and ferrets. In order for an establishment program to be successful, these introduced predators would have to be removed from a large area, or the kakapo would have to be protected from predators in some way. Because neither of these options is possible at the present time, the birds have been established, via a deliberate introduction, on five small islands where there are no introduced predators. At these sites, management involving supplemental feeding is needed to increase clutch size. Other threatened bird and reptile species are also being maintained on small New Zealand islands, where invasive species can be eliminated or at least controlled.

CONCLUSION

Invasive species are a great threat to many endangered species. Preventing the introduction of new invasive species and controlling existing invasive species are important elements in the recovery plans for many endangered species and in the maintenance of healthy ecosystems.

SEE ALSO THE FOLLOWING ARTICLES

Biological Control, of Animals / Brown Treesnake / Grazers / Hybridization and Introgression / Islands / Nile Perch / Predators / Protected Areas / Zebra Mussel

FURTHER READING

IUCN. 2004. 2004 IUCN Red List of Threatened Species: A Global Species Assessment. IUCN Species Survival Commission. Gland, Switzerland: IUCN. For current status of species go to: www.iucnredlist.org/

Jaffe, M. 1994. *And No Birds Sing: A True Ecological Thriller Set in a Tropical Paradise.* New York: Simon and Schuster.

Mackay, R. 2006. *The Atlas of Endangered Species.* Berkeley: University of California Press.

Millennium Ecosystem Assessment (MEA). 2005. *Ecosystems and Human Well-Being.* 5 Volumes. Covelo, CA: Island Press.

Primack, R. B. 2008. *A Primer of Conservation Biology,* 4th ed. Sunderland, MA: Sinauer Associates.

World Wide Fund for Nature. 2004. Living Planet Report 2004. Gland, Switzerland: World Wide Fund for Nature.

ENEMY RELEASE HYPOTHESIS

BERND BLOSSEY

Cornell University, Ithaca, New York

Ecologists have long assigned natural enemies crucial roles in shaping successes and failures of biological invasions. Introduced species often arrive in new ranges without their suite of natural enemies and without herbivores, diseases, parasitoids, or predators and are thus thought to escape the regulating forces that would otherwise keep their populations at lower levels. This idea is called the enemy release hypothesis (ERH), and it assumes that lack of top-down control allows introduced species to flourish in their introduced range.

BASIC ASSUMPTIONS

The ERH is based on some fundamental assumptions about population regulation: (1) natural enemies are important abundance regulators of their food organisms; (2) native natural enemies have a greater negative impact on native than on introduced species; and (3) the lack of top-down control results in an ecological release that allows organisms that are freed of their regulating natural enemies to take advantage of this opportunity, resulting in increased population growth often associated with rapid range expansion and increased local abundance. Although, as detailed below, ERH is intuitively appealing and often referred to as a mechanism to explain the success of invasion by introduced organisms, we continue to lack a full understanding and sophisticated demonstration of its importance in biological invasions of plants and animals.

THE PATTERNS (LARGELY FOR PLANTS)

The first and most important step in assessing the importance of ERH is to examine how important top-down control of species is in their native ranges. Thousands of journal articles and many more books have examined the impact of various natural enemies on individual plants and occasionally on their populations. The outcome of these investigations shows that while nearly all green plants are eaten by at least some organisms, regulation of plant populations by natural enemies is not universal; some may even say it is the exception. And the record is biased toward large mammals and insect herbivores, while we know relatively little about pathogens or belowground organisms. What is clear is that natural enemies, competition, and abiotic conditions create complex patterns reducing the ability to predict outcomes or identify mechanisms. If regulation by natural enemies is not equivocal, are species that become invasive upon introduction into a novel environment somehow different than those without invasive tendencies? An interesting idea proposed by Blossey and Nötzold hypothesizes that plants with large natural enemy complexes that are under top-down control in their native ranges are more likely to become invasive in novel environments. These plants have the ability to redirect resources from defense to competitive ability (increased growth, seed output, etc.), thus leading them to be favored in their interactions with native plant species.

A similar lack of agreement exists for the top-down control of herbivores by predators and parasitoids. Biocontrol practitioners often assume that most herbivorous insects are under top-down control because usually less than 1 percent of native biota become pests. Disruption of top-down control, for example through pesticide use that gives rise to secondary pests (such as spider mites) that are freed of their natural enemies appears to support this notion. The famous HSS or "The World is Green" hypothesis by Hairston, Smith, and Slobodkin assumes that natural enemies protect green plants that otherwise would be devoured by herbivores. However, the patterns that emerge from detailed investigations are variable, at best, and are context-specific with few generalizations; top-down control is clearly not universal.

If the evidence for top-down control of species in their native ranges is lacking, then how about the evidence for lack of suppression of introduced species by natural enemies in the novel range? Similarly to the situation in the home range, sophisticated tests of ERH are lacking. Most of the evidence we have to date consists of studies on the presence or absence, and occasionally the abundance, of natural enemies in the introduced range. And even in the introduced range, not all species behave similarly; only 1 percent of all established species become seriously invasive (the tens rule). While all introduced plant and animal species accumulate natural enemies over time, we have evidence that the number of natural enemies in the introduced range is lower than in the native range. Moreover, it appears that plants that are highly invasive suffer less from herbivory in the introduced range than do plants considered less invasive. These plants may possess potent or novel anti-herbivore defense systems, allowing them to flourish in the presence of native generalists while benefiting from the absence of their specialized natural enemies (which would be adapted to their particular defense chemistry).

For insect herbivores, the spread and severe impact in the novel environments, for example by species such as the emerald ash borer (*Agrilus planipennis*), is often attributed to the absence of natural enemies. However, spread of fungal diseases (for example, chestnut blight, *Cryphonectria parasitica*) often shows that naive hosts lack any type of resistance or tolerance to the novel disease. Trying to tease apart contributions of enemy release and host resistance is a cornerstone of recent attempts at emergency management of agricultural and forest pests in North America, often resulting in classical biological control attempts paired with resistance breeding. For example, while the search for suitable biocontrol agents for emerald ash borer is ongoing, researchers try to identify resistant species or hybrids, and ultimately genetically based resistance traits, that may help in safeguarding North American ash species.

The third assumption of the ERH is that existing herbivores or predators and parasitoids (both specialists and generalists) in the introduced range have a larger negative impact on the demography of native species than on the demography of introduced species. Compared to the two previously examined conditions, this is the least supported hypothesis, owing to a lack of existing data. In fact, there is some evidence compiled by Parker and coworkers suggesting that native herbivores preferentially feed on naive introduced plant species, thus reducing their invasiveness somewhat, yet the very same species continue to spread, so the influence on demography by native herbivores is not sufficient for suppression below damage thresholds or to prevent range expansion. The general pattern of a small subset of introduced species that become invasive (plants or animals) from a much larger pool of established and naturalized species could be interpreted as evidence that most new arrivals are contained and are under control by native natural enemies. However, this assumption is as untested as the uncritical (but widely accepted) assumption that invasive species escape their natural enemies, and so it needs experimental confirmation.

TESTS OF ERH IN BIOLOGICAL CONTROL

The assumptions and predictions of the enemy release hypothesis have been most closely examined for introduced plants and their associated herbivores, and to a lesser extent for introduced herbivores and their associated invertebrate predators and parasitoids. This relatively extensive dataset and some large-scale experiments are the result of applications of the ERH in classical biological control of plants and herbivores. Classical biological control (the release of host-specific pathogens or invertebrate herbivores, predators, and parasitoids from the native range of a pest organism in the novel range) was a well-established practice long before a formal theoretical framework (and the name) was established (a search of ISI's Web of Science results in the first hit of the term "enemy release hypothesis" in 2000). The practice of biological control also provides an important, albeit somewhat problematic, test of the hypotheses and its predictions. A "true" test of the ERH would require that first population regulation of a species by top-down forces in the native range be established, and then second that lack of top-down control contribute to invasiveness in the introduced range, followed finally by suppression of the host after a release of natural enemies. Many of these tests would require extensive long-term manipulations in the native and introduced range that are logistically difficult and expensive, and thus have rarely been attempted.

Paul DeBach, in his classical book on biological control by natural enemies first published in 1974, chronicles the history of the practice of release of natural enemies from the native range for control of introduced pest plants and insects. The first deliberate release of an herbivore was the release of the cochineal insect *Dactylopius ceylonicus* for control of prickly pear, *Opuntia vulgaris*, to southern India from northern India, and subsequently to Sri Lanka, where it achieved successful control of its host plant. The scale had been introduced to India from Brazil in a case of mistaken identity, when it was believed to be *D. coccus*, which is the source of the crimson-colored dye carmine. The first deliberate international transport of herbivorous insects was the release of many different Mexican insects in Hawaii for control of *Lantana camara*; most of these failed to establish, but some are considered to contribute to successful control of the plant. The first internationally shipped insect parasitoid was *Apanteles glomeratus* (now *Cotesia glomeratus*), a parasitoid of cabbage whites, *Pieris* spp., shipped in 1883 from England to the United States, where it rapidly established and is now a common natural enemy of its host.

Two main success stories in the early days of biological control of insects and plants probably shaped the continued practice of biological control more than any program since. The first was the release of the vedalia beetle (*Vedalia cardinalis*), an Australian ladybird beetle credited with saving California's citrus industry from destruction by the introduced cotton-cushion scale (*Icerya purchasi*). The second was the spectacular success of the release of *Dactylopius* spp. and of the cactus moth *Cactoblastis cactorum* in Australia beginning in 1933, which cleared vast areas infested with *Opuntia* spp. and allowed use of these lands again for conservation, grazing, or agriculture. These natural enemies continue to provide suppression of their target hosts and alone are credited with saving entire industries and billions of dollars (ecosystem services in perpetuity). Biocontrol practitioners have attempted to repeat these successes across the globe, occasionally using the same natural enemies against the same pest species, sometimes with identical, and sometimes with very different outcomes. While many different factors (human, abiotic, and biotic) may influence the outcome of control programs, an important insight from these large-scale releases is that the success of top-down control is context-dependent. Overall, it appears that, of all natural enemies that were released, about a third establish (herbivores establish more successfully than parasitoids), and of those that establish, about a third contribute to suppression of their host. While theoretical and practical considerations may improve the science and practice of biological control, the outcome of releases of nearly 3,000 natural enemies suggests that top-down control is not universal. However, only a subset of natural enemies (the host-specific ones) are used in biological control, yet all natural enemies, specialists and generalists alike, may contribute to top-down control in the native range. Thus, success of biological control is a potentially powerful demonstration for ERH, although natural enemies are introduced without their own subset of parasitoids, hyperparasitoids, or other natural enemies, and thus their impact may be larger in the introduced range because they themselves benefit from enemy release. Failure of biological control and lack of pest suppression is not sufficient to reject ERH, while success cannot automatically be used to accept ERH either. The variable outcomes in programs working with the same pest and natural enemy complex in different areas provide evidence for intricate interactions involving more than host and natural enemy, and thus make predictions difficult. Moreover, the "players" in these systems, as well as their interactions, may evolve, and just because natural enemies control a target pest does not necessarily imply that a lack of natural enemies was the ultimate reason for invasiveness.

A FRAMEWORK FOR TESTING ERH

Given the significant financial and logistical obstacles of true and complete tests for the existence of enemy release in success (or failure) of invasions, how could sophisticated tests of ERH be conducted? Most of the evidence we have to date does document natural enemy "loads" (i.e., number of species) and occasionally their attack rates, but rarely do we have information on how these affect population growth rates of the host. Some of the main sources of evidence to collect would be the demographic response to removal of natural enemies. This is most easily done for herbivores, where through fencing, insecticide, molluscicide, or fungicide, major groups of enemies can be excluded (although beneficial microbes may be affected as well). In the introduced range, where natural enemies of native plants would be excluded, this should result in increased competitive ability of native plant species and reductions in abundance of introduced plant species. The situation is not quite as easy in the native range because it is rarely possible to exclude just natural enemies attacking a single plant species with the rather crude exclusion techniques at our disposal. However, we should see evidence for increased abundance of the plant if natural enemies are excluded in the native range. Such experiments need to be conducted using a range of abundances in the native and introduced ranges because the regulating ability of natural enemies may be affected by the abundance of its host. For example certain natural enemies in the native range may be able to control abundance of their host if the host is very abundant, but the same may not be true once the host becomes rare. In the introduced range, long-term site occupancy of the introduced species may have excluded individuals, seeds, or other propagules, and thus there are no native species that could numerically respond to enemy release.

While the experimental procedures to assess ERH for plants appear manageable, manipulating natural enemy complexes for herbivores (pathogens, predators, and parasitoids) or other animals appears daunting because there are no suitable exclusion mechanisms. The best we can probably attain is the development of detailed demographic models that show how organisms respond to different environments in the presence or absence of their different natural enemy complexes and then compare native and introduced ranges.

Although ERH is intuitively appealing and often referred to as a mechanism to explain the success of invasion by introduced organisms, we continue to lack a full understanding and sophisticated demonstration of its importance in biological invasions of plants and animals.

SEE ALSO THE FOLLOWING ARTICLES

Biological Control, of Animals / Biological Control, of Plants / Competition, Plant / Herbivory / Parasites, of Animals

FURTHER READING

Agrawal, A. A., P. M. Kotanen, C. E. Mitchell, A. G. Power, W. Godsoe, and J. Klironomos. 2005. Enemy release? An experiment with congeneric plant pairs and diverse above- and belowground enemies. *Ecology* 86: 2979–2989.

Blossey, B., and R. Nötzold. 1995. Evolution of increased competitive ability in invasive nonindigenous plants: A hypothesis. *Journal of Ecology* 83: 887–889.

Blumenthal, D. 2006. Interactions between resource availability and enemy release in plant invasion. *Ecology Letters* 9: 887–895.

Colautti, R. I., A. Ricciardi, I. A. Grigorovich, and H. J. MacIsaac. 2004. Is invasion success explained by the enemy release hypothesis? *Ecology Letters* 7: 721–733.

DeBach, P. P. 1974. *Biological Control by Natural Enemies*. London: Cambridge University Press.

Eppinga, M., M. Rietkerk, S. C. Dekker, and P. C. de Ruiter. 2006. Accumulation of local pathogens: A new hypothesis to explain exotic plant invasions. *Oikos* 114: 168–176.

Hawkes, C. V. 2007. Are invaders moving targets? The generality and persistence of advantages in size, reproduction, and enemy release in invasive plant species with time since introduction. *American Naturalist* 170: 832–843.

Keane, R. M., and M. J. Crawley. 2002. Exotic plant invasions and the enemy release hypothesis. *Trends in Ecology and Evolution* 17: 164–170.

Mitchell, C. E., and A. G. Powers. 2003. Release of invasive plants from fungal and viral pathogens. *Nature* 421: 625–627.

Parker, J. D., D. E. Burkepile, and M. E. Hay. 2006. Opposing effects of native and exotic herbivores on plant invasions. *Science* 311: 1459–1461.

EPIDEMIOLOGY AND DISPERSAL

ALAN HASTINGS

University of California, Davis

Introduced diseases can have a very large impact. They can be purposefully introduced for control of invasive species, or alternatively, they can be accidentally introduced and then can spread. At least some diseases are relatively easy to study, so there exist very good data and mathematical models that describe them. The spatiotemporal dynamics of diseases can be used as a way to understand basic underlying issues of invasive species. More recent advances in studying the role of networks to understand dispersal in an epidemiological context exemplify the way that the study of diseases will lead to insights into more general questions about invasive species.

BASIC MODELS OF EPIDEMIOLOGY

Diseases play a key role in the overall study of invasive species because our mechanistic understanding of the processes of invasion in an epidemiological context is particularly good. This work builds on the basic model of epidemiology, which goes back to Kermack and McKendrick in 1927. In the simplest description, the disease is assumed to spread quickly relative to the dynamics of the host population, and the dynamics of the disease within a host is assumed to be well described by assigning hosts to one of three classes: susceptible, infective, or removed. Susceptible individuals become infective through random contact with an infective individual. Infective individuals either die or recover and are immune at a fixed rate, in either case moving to the removed class. Interest is generally focused on the fate of the disease after the introduction of a small number of infective individuals. As shown by Kermack and McKendrick, the disease will spread, and there will be an epidemic if the reproductive number for the disease

$$R_0 = \beta N/\gamma$$

is greater than one, while the disease will die out relatively quickly if the reproductive number is less than one. Here, β is the rate of contacts between susceptibles and infectives that actually produce new infections, N is the population size, and γ is the rate of removal (by death or recovery) of infective individuals. Thus, R_0 is the mean number of infections produced by a new infective individual, calculated as the rate of infection when almost all individuals are susceptible times the mean time an individual is infective. The population size is relatively easily estimated, and the parameter γ can be estimated from the time an individual is infective. However, the parameter β, which is obviously key to understanding the spread of disease, is much more difficult to estimate, leading to uncertainty in the understanding of disease dynamics. This basic model applies essentially to metapopulations as well, so that the concept of reproductive number is a basic concept in invasion biology. Other kinds of diseases will lead to different descriptions with different dynamics. Diseases with free-living stages, or with multiple hosts, will produce very different dynamics.

DISPERSAL MECHANISMS AND DESCRIPTIONS

The basic model of epidemiology assumes a well-mixed population, and so it can be used to understand issues of establishment, but not issues of spread or the role of dispersal and spatial structure. Dispersal in epidemiology has been approached through descriptions of explicit movement of diseased individuals through space, randomly leading to a reaction–diffusion description, and also through the contacts of diseased individuals that are described by a kernel that specifies the distribution of contacts relative to the initial location where an individual becomes infective.

The reaction–diffusion approach has been applied to the spread of diseases in space since an early study of smallpox in London. In this kind of description, individuals move by a series of short steps in a random fashion. This model leads to a solution that is typically an expanding wave of disease that spreads at a constant rate through time, determined both by the movement rate of individuals and by the dynamics of the disease. This approach has proven effective in understanding a variety of diseases, including rabies in Europe, although it is important to recognize the limitations of this description. Because the model is solved as a deterministic system, local die-outs of the disease cannot be included, and long-range movement is ignored. In the descriptions of the spread of rabies, the difficulties implied by the effect on dynamics of very small levels of rabies have been explicitly considered. A reaction–diffusion description has also been used to describe spatial control measures.

More recent work has included other descriptions of movement, while still keeping some of the structure of the reaction–diffusion description. In descriptions that are continuous in space but discrete in time, movement is described by a kernel that gives the probability of a diseased or susceptible individual moving from one location to another. Other descriptions of movement given by Shigesada and Kawasaki and others allow for the possibility of long-range jumps that would produce new foci for disease, after which the disease would spread by more local mechanisms.

Alternatively, the role of restricted movement, which can make diseases die out because no local individuals are susceptible, can be considered using contact processes or other stochastic models. In a contact process, sites (which can be thought of as individuals) transition between different states in a probabilistic fashion depending on the states of neighboring sites. Thus, an infected individual surrounded only by other infected individuals cannot affect the spread of the disease. This description provides insights into how spatial structure affects the contact rate between individuals.

More recently, there has been an emphasis on the network structure describing contacts among individuals. This kind of description provides ways to understand how different dispersal structures affect disease spread and persistence. Disease spread can be much faster on small world networks than would be expected without the kind of structure that connects all individuals through relatively few steps.

It is clear from the discussion of different descriptions of movement that understanding the spread of diseases is very different for diseases that can be spread only by contacts between infected individuals than for diseases that have free-living stages. More complicated, and more poorly understood, are issues of spread in diseases that require multiple hosts.

IMPACTS AND DATA

There are classic examples of the impact and spread of diseases, such as the role of myxomytosis, which was introduced to Australia to curb rabbit populations, and diseases such as Dutch elm that have affected forests. Recent notable examples and studies of the role of movement in the spread or control of diseases are studies of the hoof-and-mouth epidemic in the United Kingdom in 2001 and retrospective studies of the very good spatiotemporal datasets available for childhood diseases such as measles or pertussis in the United Kingdom. In analyses of each of these systems, additional factors have been included. In particular, stochasticity has been included to allow the use of model-based statistical procedures that have allowed parameter estimates and forecasts that were used to guide the control of the hoof-and-mouth epidemic. Understanding the dynamics of childhood diseases required including seasonality and paying specific attention to the role of city size and connections between cities. The role of networks in studying diseases is highlighted by the role of air travel and also by the kinds of contacts that occur in the spread of sexually transmitted diseases. The lessons learned from these studies apply much more broadly than simply in the particular examples through which they were developed. Because at least some introduced diseases have been followed closely, the data and insights from studying introduced diseases has led not only to better understandings of disease dynamics, but also to fundamental insights into population biology.

SEE ALSO THE FOLLOWING ARTICLES

Biological Control, of Animals / Disease Vectors, Human / Flaviviruses / Parasites, of Animals / Pathogens, Animal / Rinderpest

FURTHER READING

Grenfell, B., and J. Harwood. 1997. (Meta) population dynamics of infectious diseases. *Trends in Ecology and Evolution* 12: 395–399.

Holmes, E. 1997. Basic epidemiological concepts in a spatial context (111–136). In D. Tilman and P. Kareiva, eds. *Spatial Ecology: The Role of Space in Population Dynamics and Interspecific Interactions*. Princeton, NJ: Princeton University Press.

Hudson, P., A. Rizzoli, B. Grenfell, H. Heesterbeek, and A. Dobson. 2002. *The Ecology of Wildlife Diseases*. Oxford: Oxford University Press.

Keeling, M., M. Woolhouse, D. Shaw, L. Matthews, M. Chase-Topping, D. Haydon, S. Cornell, J. Kappey, J. Wilesmith, and B. Grenfell. 2001.

Dynamics of the 2001 UK foot and mouth epidemic: Stochastic dispersal in a heterogeneous landscape. *Science* 294: 813–817.

Watts, D.J., and S.H. Strogatz. 1998. Collective dynamics of "small-world" networks. *Nature* 393: 440–442.

ERADICATION

PIERO GENOVESI

Institute for Environmental Protection and Research (ISPRA), Ozzano Emilia, Italy

Eradication—elimination from a site of all individuals of a species—is a management tool that allows us to prevent impacts caused by the undesired introduction of an alien organism. More precisely, eradication is the complete and permanent removal of all wild populations of an alien plant or animal species from a specific area by means of a time-limited campaign. Eradications are thus distinct from permanent control (reduction of population density) or containment (which is aimed at limiting the spread of a species by containing its presence within defined geographical boundaries). Eradications should also be distinguished from the control of a species to zero density, which aims at the complete removal of the target species, but through a continued removal effort.

HISTORY OF ERADICATIONS

The first eradications of alien species were carried on to prevent health risks to humans or livestock, rather than for conservation purposes. The first known eradication is probably the removal of the tsetse fly (*Glossina* spp.) from the 126 km^2 island of Principe in the Gulf of Guinea, carried out in 1911–1914, after sleeping sickness had dramatically reduced the local human population. There are also many examples of eradications carried out to protect humans from introduced vectors of pathogens (e.g., the eradication of mosquitoes to combat malaria from many regions). One remarkable example is the eradication of *Anopholes gambiae* from over 54,000 km^2 of northeastern Brazil, where this insect, inadvertently introduced from central Africa, caused an outbreak of malaria with over 14,000 deaths. The Rockefeller Foundation funded an eradication campaign, carried out through the spraying of larvae and adults, and it was successfully completed in 1940. Also, the eradication of the muskrat (*Ondatra zibethicus*) from Scotland in the 1930s was carried out mostly because of the destructive effects of this aquatic rodent, which feeds on crops and can severely damage watercourse embankments through burrowing, and not to protect local native species.

Only in the 1980s did eradication of alien species start to be carried out primarily to protect biological diversity, and in the 1990s eradication was formally identified as a priority conservation tool at the highest policy levels. In 1992 the Convention on Biological Diversity included an explicit reference to eradication as a response to biological invasions, and this principle was better detailed in the guiding principles adopted by the Convention in 2002 (decision VI/23, on alien species that threaten ecosystems, habitats, and species), which called on member state parties to adopt a hierarchical approach to address biological invasions, based primarily on prevention of unwanted introductions. However, once prevention has failed and an alien species has invaded a country, the best alternative is the rapid removal of all individuals, because this option prevents further impacts, and also because eradication can allow countries to avoid the costs of permanent control.

In recent years, eradications for conservation purposes have become a major management tool and arewidely applied, especially on islands. Over 1100 successful eradications have been carried out globally, with a success rate of 86 percent. Most programs have been realized on islands, targeting alien vertebrates. In Europe, eradications have in the past received less attention than in other regions of the world, but the situation is now changing, with over 100 removal campaigns completed so far, and about twice as many if we also consider eradications carried out in European overseas territories. Most eradications have been realized on islands, but there are several examples of successful campaigns completed on the mainland, such as those for the screw worm, which was extirpated from the southeastern United States, and the coypu (*Myocastor coypus*), which was successfully removed from East Anglia.

TARGET SPECIES

A large majority of eradications have targeted alien mammals, in particular rats, mice, rabbits, domestic goats, feral pigs and other ungulates, and domestic cats and other carnivores. Feral goats have been eradicated successfully from 107 islands, and rodents from 348, with a success rate of about 90 percent. There have been over 40 eradications of rabbits and more than 50 of domestic cats. In Europe, rat removals account for over 60 percent of all completed eradications of alien species. But although eradications usually target vertebrate species, there are also many examples of programs aimed at extirpating terrestrial invertebrates, such as the fruit fly, or the screw worm, which was extirpated from the southeastern United States, Central America, and North Africa.

Eradications in aquatic environments are in general much less frequent, because they are much more challenging than for terrestrial species, but there are many examples of successful eradications of aquatic or semiaquatic vertebrates (mink, coypu, muskrat, etc.), including freshwater fishes, as for example in the case of the removal of the European gudgeon (*Gobio gobio*) from the vicinity of Auckland, New Zealand. There are also examples of eradications of freshwater invertebrates, as in the case of the successful removal of the Australian marron crayfish (*Cherax tenuimanus*) from an area of New Zealand, but in general, eradication becomes very difficult—if not impossible—once species establish in open waters; in fact, although there have been successful eradications of marine species, these have always been realized at a very early stage of invasion, as in the case of the successful removal of a sabellid polychaete (*Terebrasabella heterouncinata*) from a mariculture facility in California. In the case of the alien mussel (*Mytilopsis sallei*) inadvertently introduced in Cullen Bay Marina in Darwin (Australia) in 1999, and detected a few months after the introduction, the removal required the immediate closure of the entire bay and the treatment of the area with large quantities of copper sulphate and sodium hypochlorite.

Plants can also be extirpated (although it is in general more challenging than for animal species), as confirmed by many successful eradications carried out on both islands and mainland areas. All known plant eradications have targeted invasions detected early, or plants with limited invaded range, and there is in fact no known case of successful eradication of a well-established alien plant (Fig. 1).

FIGURE 1 Spraying of herbicides to eradicate alien weeds in California. (Photograph courtesy of California Department of Food and Agriculture.)

An example of a plant eradication in a marine environment is the successful removal of the alien alga *Caulerpa taxifolia* from the lagoon of Agua Hedionda, in southern California. The eradication was started 17 days after the species was recorded and was carried out through containment of individual colonies with PVC frames covered with sheeting, and then treatment with liquid sodium hypocloride or with chlorine-releasing tablets. The program also required field monitoring for new growth, with the aim of removing any piece of alga, down to centimeter-sized pieces of fronds.

A particularly complex aspect of implementing an eradication is the fact that removal methods need to be effective even when the density of the target species decreases to a very low level. Predicting the effort required to complete an eradication can thus be difficult, as the removal of the very last individuals can require very significant and scarcely predictable efforts and resources. Many techniques have been developed to respond to this challenge. For example, the "Judas goat method" has been crucial to the success of many goat eradications, such as the program on San Clemente Island (California). A few individual goats are caught, fitted with radio-collars and released in the wild. Owing to their social nature, the radio-collared individuals tend to aggregate with other goats, allowing eradication teams to locate the residual animals and cull them.

METHODS

Eradications have been conducted using very different techniques, including trapping or shooting of vertebrates, poisoning of invertebrates, removing organisms by hand, using toxicants (for example, to eradicate *Mytilopsis sallei* from Darwin), using pesticides and herbicides for removing plants (as in the eradication of *Caulerpa taxifolia*), and draining to remove freshwater invaders from ponds. In many cases, integration of different methods has proved to be an effective approach. For example, several domestic cat eradications have been realized initially by using toxins and biological control, and when the population has been reduced, completing the process by hunting and trapping. Of course, the efficacy of removal techniques depends on the biology of the species; for example, the eradication of flies on Nauru Island was possible because the target species could be sterilized by irradiation.

Biological control agents and introduced pathogens have been used in many removal programs, but such methods should be viewed with caution and employed only after a careful assessment of associated risks. In fact, both of these methods can, in general, only reduce abundance of the target species; they are unlikely to permit complete eradication. Furthermore, biological control agents and pathogens can cause severe undesired impacts once established.

Another risk to consider is the possible impacts of the removal method on nontarget species. The undesired effects can be direct, such as those of toxicants or poison baits, or indirect, as when scavengers are poisoned by feeding on poisoned individuals. It should be noted that the impacts depend largely on the local contexts, which must be carefully assessed. For example, among rodent eradications, the many successful programs completed in New Zealand have been facilitated by the absence of any native rodent species in that country, thus allowing aerial poison baiting. The situation is different in other regions of the world, where islands are often inhabited by both native and alien rodents. In these cases, eradications must often employ selective baiting stations that prevent access to baits by nontarget species. For example, bait dispensers can be selective (Fig. 2), or the season of the program can restrict it to the target species, or the baits can attract only the target species. To eradicate house mice on Thevenard Island, in Western Australia, where a native mouse is also present, a specific bait-delivering station was developed; the size of the entrance allowed only the alien species to enter, and not the larger native individuals.

In some cases, it has not been possible to identify removal methods that ensure adequate selectivity, and alternative solutions have thus been adopted. In the eradication of rats from Channel Islands National Park (California), in order to avoid severe impacts on the local endemic subspecies of *Peromyscus maniculatus*, these were translocated before the treatment and then subsequently

FIGURE 2 The use of bait dispensers prevents impacts on nontarget species. (Photograph courtesy of Massimo Putzu.)

FIGURE 3 The use of aerial baiting allows eradication of rodents from islands larger than 10,000 ha. (Photograph courtesy of Paolo Sposimo.)

reintroduced to the island once the eradication was completed.

Impacts on nontarget species cannot always be avoided, as when eradications are conducted in freshwater or marine environments, and in these cases, it is crucial to predict the impacts on nontarget species, in order to assess if these effects are acceptable. For example, in the eradication of *Mytilopsis sallei* from Darwin, it was determined that native species affected by the toxicants could recolonize the bay from neighboring areas once the campaign had been completed. The risk assessment of removal methods is particularly crucial when biological control agents or genetically engineered viruses are considered, because release of these agents can be irrevocable.

It should be noted that the increased success rate of eradications is also due to significant advances in removal techniques, which can now allow the targeting of much more challenging species in much more challenging areas than in the past (Fig. 3).

BIOLOGICAL ASPECTS

Basic general rules of eradications are that they are biologically feasible when all reproductive individuals of the target population can be removed, and when risk of rapid reinvasion is zero (or close to zero). In this regard, the biological parameters of the target species—with particular reference to reproductive biology, life history, and dispersal ability—are crucial in assessing the feasibility of an eradication.

Removal is much more challenging when target species have a dormant life stage (e.g., a soil seed bank), a larval stage (e.g., most marine species), or high dispersal capacity and reproductive rates. This explains why marine species and plants are particularly difficult, if not impossible, to eradicate once they are established. However, there are many examples of successful removals of organisms in all taxonomic groups, and many plant eradications have been successful.

Another biological pattern that influences the feasibility of eradication is the lag phase observed in many introduction events. Invasions are often characterized by a phase of relative quiescence, followed by a rapid expansion. For this reason, eradications should be started at the earliest possible stage, when chances of successfully removing the infestation are greatest. If the removal methods do not allow removal of the very last individuals, then the eradication campaign can be at risk of failure, especially when it targets a species with a biology that enables a rapid recovery starting from few individuals. For example, because alien squirrels are generally able to establish in the wild even when only one or two pairs are released, any eradication of these rodents would require very effective techniques.

ERADICATION OF INTERACTING INVASIVE SPECIES

Analyzing the target species life history may not be enough, as ecological interactions with other species can play a major role in the final outcome of the eradication. In most cases, ecosystems are invaded by several alien species, which can be linked by complex interactions. Therefore, removal of an alien species may have unexpected effects on other invaders, which in some cases has led to new and severe conservation problems. One example is the so-called hyperpredation effect, where a population of predators is maintained at high densities by the presence of an abundant alien prey, causing higher predation pressures on native species. The removal of the alien prey may lead to a temporary increase in predation on native species, while the eradication of the predator may lead to an increase of the alien prey with other undesired effects. Another possible result of an eradication carried out in a multispecies system is the mesopredator effect, where the removal of a superpredator causes the demographic explosion of a mesopredator (such as the rat), with resulting increased pressure on native species. The removal of alien herbivores can permit the establishment of alien species or cause the explosion of invasive weeds, which can produce possibly severe effects in terms of habitat alterations. For example, the eradication of goats from Santiago Island (Galapagos) has caused the spread of introduced *Rubus niveus*, with significant impacts on several native species.

For these reasons, it is important that eradications in complex ecosystems be preceded by an assessment of the possible effects of the removal and that the simultaneous removal of several invaders be considered.

SOCIAL AND ECONOMIC ASPECTS

Eradications may require substantial economic resources: the attempted eradication of the medfly from California cost over €80 million, and the successful removal of the white-spotted tussock moth (*Orgyia thyellina*) from New Zealand cost $NZ 12 million. Even small projects can be costly; the removal of only 12 Himalayan porcupines from Devon required €230,000. However, the few studies that have compared the cost of eradication with the cost of inaction confirm that eradications are, in general, very cost beneficial in the long term, because they prevent economic losses due to impacts and to permanent control measures. For example, the successful removal of the coypu from East Anglia cost about €5 million, but this amount appears very cost effective if we consider that the economic losses caused by the species in Italy exceed €4 million a year and are expected to rise to over €12 million a year in the future.

Apart from economic resources, eradications also require support from the public, and social opposition has in several cases been a major constraint. For example, the strong opposition by a radical animal rights movement has been a major obstacle to eradicating the American gray squirrel from Italy. Also the eradication of the ruddy duck (*Oxyura jamaicensis*) from Britain has faced opposition from bird lovers, and even the rat eradication on Lundy Island (United Kingdom) was strongly opposed by animal right groups, which claimed that rats had lived on the island for a very long time, were an important part of the island's biodiversity, and deserved protection. But reactions to eradications may arise for different reasons than emotional ones; often people are concerned by the use of toxicants, especially when these are applied by aerial spraying, and the public can negatively react to cutting of trees and plants. Eradications can also be opposed by societal sectors interested in the use of the targeted alien species, as has often happened in the removal of introduced game species, which has often been vigorously resisted by hunters (e.g., Barbary sheep from the Canary Islands, tahr from New Zealand and South Africa).

In order to prevent such opposition, eradications should be preceded by adequate information and communication campaigns and should involve local communities when appropriate. For example, in the eradication of the American mink from the Outer Hebrides, an action

FIGURE 4 Nesting success of the Yelkouan shearwater (*Puffinus yelk-ouan*) increases dramatically in islands where rats have been eradicated. (Photograph courtesy of Massimo Putzu.)

that was widely supported by the inhabitants of the archipelago, a periodical newsletter (edited in both English and Gaelic) was published and widely circulated to inform the local communities, and the removal employed local staff.

OUTCOMES

The positive effects of eradications are often very evident, with rapid recovery of native species and habitats. For example, many eradications of rats from islands have caused the prompt recovery of bird populations (Fig. 4). And the removal of the American mink from small islands has led to an increase in the reproductive success of several vertebrates, including birds and amphibians. Eradications can be essential for recovery programs, as in the case of Campbell Island, where removal of rats was crucial for the recovery of the endemic Campbell Island teal (*Anas nesiotis*), or in the case of the cat eradication on Long Cay Island (Caicos Bank, British West Indies), which permitted recovery of the highly endangered iguana *Cyclura carinata*.

Eradications can also have beneficial effects on the local economy, as well as on health and human well-being. The eradication of fruit flies from Nauru Island permitted a significant increase in fruit production.

MONITORING

Monitoring is an essential element of eradication, and the design of any removal campaign should include a careful measure of the effects of techniques on nontarget species, as well as measures to detect reinvasions promptly. Verification of the completion of the eradication is another particularly important element, because a suspension of

the campaign when some individuals remain can undermine the entire campaign. Of course, verification of the success of an eradication also depends on the target species; for example, eradication of plant species that have long-lasting seed banks requires many years of monitoring before completion can be verified.

Given the importance of reliable verification of eradication outcomes, it has been suggested that eradication should be verified by independent bodies not involved in the eradication, as was done for the successful removal of the coypu from East Anglia.

KEY ELEMENTS FOR PLANNING ERADICATIONS

A key principle of biological invasions is that when a new species is detected at an early stage, before it establishes or spreads, it is crucial to extirpate it without delay. Prompt reaction to new incursions, in fact, significantly increases chances of success of the intervention and excludes most undesired effects of the removal campaign. Thus, when a new unauthorized incursion of an alien species is detected, the removal should be started immediately, without assessment of the related risks.

When, on the contrary, the aim of an eradication is to remove a well-established species, then the decision on whether to start the removal should be based on an assessment of its feasibility and on careful planning to take into account the most relevant biological, technical, organizational, sociological, and economic elements. A peer review of the eradication program should be obtained before a decision is made. Eradication techniques should be selected primarily on the basis of their efficacy, but also by taking into account risks to nontarget species and assessing the possibility of causing unnecessary stress and pain to target animal species. Ecological interactions of the target species with other alien species should be carefully assessed, in order to predict possible indirect effects of the removal at an ecosystem scale. When the public is sensitive to the eradication methods or effects, the program should include information efforts, a public awareness campaign, education programs, and the involvement of affected societal sectors (hunters, NGOs, foresters, landowners, fishermen, etc.) when possible and appropriate.

Particular attention should be paid to monitoring, which should be continued for an adequate time and should assess not only the results of the removal efforts and any undesired effects on nontarget species, but also the costs of the campaign and the positive effects in terms, for example, of native species recovery; it should also ensure prompt detection of reinvasion.

SEE ALSO THE FOLLOWING ARTICLES

Early Detection and Rapid Response / Endangered and Threatened Species / Herbicides / Invasion Economics / Islands / Lag Times / Risk Assessment and Prioritization

FURTHER READING

Clout, M. N., and P. A. Williams, eds. 2009. *Invasive Species Management: A Handbook of Techniques*. Oxford: Oxford University Press.

Genovesi, P. 2007. Limits and potentialities of eradication as a tool for addressing biological invasions (385–400). In W. Nentwig, ed. *Biological Invasions*. Ecological Studies, Vol. 193. Heidelberg: Springer-Verlag.

McNeely, J. A. 2001. *The Great Reshuffling: Human Dimensions of Invasive Alien Species*. Gland, Switzerland: IUCN.

Myers, J. H., D. Simberloff, A. M. Kuris, and J. R. Carey. 2000. Eradication revisited: Dealing with exotic species. *Trends in Ecology and Evolution* 15: 316–320.

Veitch, C. R., and M. N. Clout, eds. 2002. Turning the tide: The eradication of invasive species. Occasional Paper 26. Gland, Switzerland: IUCN.

EUCALYPTS

MARCEL REJMÁNEK

University of California, Davis

DAVID M. RICHARDSON

Stellenbosch University, Matieland, South Africa

Over 800 species of eucalypts (*Angophora, Corymbia*, and *Eucalyptus*) are native to Australia and a few Pacific islands. These genera include some of the most important solid timber and paper pulp forestry trees in the world. Besides pines, eucalypts are the most commonly and widely cultivated exotic trees. Almost 20 million ha (200,000 km^2) of eucalyptus plantations exist in tropical, subtropical, and temperate countries. In many countries, eucalypts are the most common and conspicuous nonnative trees. Over 70 species are naturalized (reproduce and maintain their populations) outside their native ranges. However, given the extent of cultivation, eucalypts are markedly less invasive than many other widely cultivated trees and shrubs. Reasons for this relatively low invasiveness are still not completely understood. Conclusions about positive or negative environmental and economic impacts of eucalypts are often anecdotal, highly controversial, and context-dependent.

TAXONOMY, GROWTH FORMS, REGENERATIVE STRATEGIES, AND HABITATS

Eucalypts (family Myrtaceae, subfamily Leptospermoideae) are currently classified into three genera: *Angophora* (14 species), *Corymbia* (113 species), and *Eucalyptus* (>740 species). *Angophora* and *Corymbia* are often treated as subgenera of *Eucalyptus*, sensu lato. The genus *Eucalyptus* (sensu stricto) is currently divided into ten subgenera, six of which are monotypic (having only one species). Naturalized eucalypts belong almost exclusively to the two largest subgenera, *Eucalyptus* (*Monocalyptus*, >140 species) and *Symphyomyrtus* (>360 species) (Table 1).

Most eucalypts are trees (10 to >50 m in height), some are "mallees" (multistemmed from ground level, usually <10 m in height), and a few are shrubs. Eucalypts are the tallest nonconiferous trees in the world. Several species (*E. regnans, E. grandis, E. deglupta*) reach more than 70 m in height. The tallest known specimen of *E. regnans* was 110 m. The tallest known living specimen of this taxon in Tasmania is 99.6 m tall. The maximum age of eucalypt species, estimated from dendrochronological (tree-ring) and radiocarbon measurements, is between 400 and 600 years. Interestingly, in spite of the otherwise enormous range of adaptations among eucalypts, shade-tolerant subcanopy species are not known.

The eucalypt breeding system is one of mixed mating with preferential outcrossing. A reduction in fruit (capsule) production, seed set, and seedling vigor has been demonstrated after self-pollination. However, eucalypts generally do not need any special pollinators. They are pollinated by many species of bees and wasps and, to a lesser extent, by birds, mammals, and wind.

Most eucalypts are well adapted to frequent fires. The most common adaptations are lignotubers and epicormic buds. The lignotuber is a woody swelling at the base of the stem; most eucalypts sprout from lignotubers. The epicormic buds (buds present in the outer bark) allow the sprouting of new branches from stems (after a fire or after a severe winter). Some species are able to sprout from both lignotubers and stems (combination sprouters). Another adaptation to fireprone environments is serotiny (most seeds are kept in fruits and released only after fire). Some successful invaders among eucalypts are obligate seeders, depending completely on seed production (*E. conferruminata, E. grandis*). Seeds can be shed in large numbers (up to 4,000 seeds per m^2). However, eucalypts produce very small seeds, usually 1–3 mm long and less than 2 mg. Some species have seeds even lighter than 0.5 mg (e.g., *E. grandis* and *deglupta*). Eucalypts produce seeds with no obvious endosperm (tissue that surrounds and nourishes the embryo). Therefore, the newly emerged seedlings are sustained by cotyledon photosynthesis, and their roots have to reach suitable substrate very soon.

Eucalypts are the dominant species of the wet coastal and near coastal parts of Australia. However, some species

TABLE 1

Naturalized Species of *Corymbia* and *Eucalyptus*

Corymbia

calophylla	C, LS	Australia (Western Australia), Hawaii?, New Zealand?
citriodora	B, LS	Australia (Victoria, Western Australia), California, Hawaii?, India, South Africa, Zimbabwe
ficifolia	C, LS	Hawaii, New Zealand, South Africa
maculata	B, LS	Australia (Victoria, Western Australia), South Africa?
torelliana	B, LS	Australia (Queensland), China, Florida

Eucalyptus

botryoides	S, CS	Australia (Norfolk Island, South Australia, Victoria, Western Australia), Hawaii, New Zealand
bridgesiana	S, LS	Hawaii, South Africa?
camaldulensis	S, CS/SS	Argentina?, Australia (Western Australia) Bangladesh, California,[a,b] Cyprus, France, Greece, Hawaii,[a] India, Israel, Italy, Pakistan, Portugal, Spain,[a,b] South Africa,[a,b] Zimbabwe
cinerea	S, LS	Hawaii, New Zealand, South Africa
cladocalyx	S, SS	Australia (Victoria, South Australia, Western Australia[b]), California, Hawaii?, South Africa[a]
conferruminata	S, OS, often confused with *E. lehmanii*	Australia (Victoria, Western Australia), California, South Africa[a,b]
crebra	S, LS	Hawaii, South Africa?
deanei	S, LS	Hawaii, South Africa?
deglupta	S, OS	Hawaii, Malaysia?
delegatensis	E, OS	New Zealand, South Africa?
elata	E, LS	New Zealand, South Africa?
fastigata	E, OS	California, New Zealand, South Africa
globulus	S, CS, incl. *E. maidenii*	Australia (Western Australia), Azores, California,[a,b] Canary Islands, Chile,[a] China,[a] France, Hawaii,[b] India, Italy?, Jamaica, New Zealand,[a,b] Peru,[b] Portugal,[a] Spain,[a] South Africa,[a] Zimbabwe
gomphocephala	S, SS	Australia (Victoria), Cyprus, Hawaii, South Africa, Spain
grandis	S, OS	Argentina?, California, Florida, New Zealand, Nigeria?, South Africa[a,b]
gunnii	S, LS	New Zealand, Portugal, Spain
leucoxylon	S, LS	Australia (Victoria), New Zealand?, South Africa
longifolia	S, LS	Australia (New South Wales), South Africa?
macarthurii	S, LS	California, New Zealand, South Africa?
marginata	E, LS	Hawaii, South Africa?
microcorys	S, LS	Australia (Western australia), Hawaii,[a] Sri Lanka, South Africa, Zimbabwe
muelleriana	E, LS	Australia (Western Australia), New Zealand?, South Africa?
nitens	S, SS	New Zealand, South Africa?[a]
obliqua	E, LS	New Zealand, South Africa
occidentalis	S, LS	Australia (Victoria), South Africa?
ovata	S, LS	California, New Zealand
paniculata	S, LS	Hawaii,[a] South Africa,[a] Zimbabwe
pilularis	E, OS	Hawaii,[a] New Zealand, South Africa
polyanthemos	S, LS	Australia (Western Australia), California, South Africa?
pulchella	E, LS	California, New Zealand
regnans	E, OS	New Zealand, South Africa?
resinifera	S, LS	Hawaii, Mexico?, New Zealand
robusta	S, OS	Brazil?, California, Florida, France, Hawaii,[a] La Réunion Island, Madagascar,[a] Malaysia?, New Zealand, Portugal, South Africa, Spain, Sri Lanka, Zimbabwe[a,b]
rudis	S, LS	California?, Hawaii
saligna	S, LS	Australia (Western Australia), Florida, Hawaii,[a] New Zealand, South Africa?, Sri Lanka,?, Uganda?
salubris	S, OS	Australia (Queensland), South Africa
sideroxylon	S, CS	Botswana?, California, Hawaii, New Zealand, Portugal, South Africa,[a] Spain
sieberi	E, SS	New Zealand, South Africa
tereticornis	S, LS	California, Cyprus, Hawaii, India,[a] Mexico (reported as *E. resinifera*), New Zealand?, South Africa, Zimbabwe
viminalis	S, LS	California, Hawaii, New Zealand, South Africa?

NOTE: Naturalized (reproducing and maintaining populations without human help) species of *Corymbia* and *Eucalyptus*. Based on published records, labels on herbarium specimens, information from experienced botanists and foresters, and personal observations. Question marks indicate lack of certainty as to whether the species is already naturalized, or is just a casual resident. Only species reported as naturalized from at least two countries are included. Subgenera: B – *Blakella*, C – *Corymbia*, E – *Eucalyptus* (*Monocalyptus*), S – *Symphyomyrtus*. Regenerative strategies: LS – lignotuber sprouter, SS – stem sprouter, CS – combination sprouter, OS – obligate seeder.

[a] Extensive planting.

[b] Species is classified as invasive in a particular state (spreading spontaneously away from points of introduction).

(*E. rameliana, E. pachyphylla*) extend to arid regions with annual precipitation below 350 mm. Within their native range, most eucalypt species grow in tall forests, woodlands, and savannas. Seasonal flooding is essential for successful regeneration of some species (*E. camaldulensis*). Over 20 species of eucalypts can grow on saline soils (e.g., *E. robusta* in swampy estuarine habitats or *E. camaldulensis* in valleys of old river systems). Some small trees or shrubs (e.g., *E. coccifera* or *E. pauciflora* subsp. *niphophila*) are adapted to subalpine areas (900–1,400 m) in Australia.

CULTIVATION

When Australia was settled in the eighteenth century, eucalypts were used for farm buildings, fencing, and firewood. The first eucalypt species to be cultivated from seeds outside its native range was *E. obliqua* in Royal Botanic Gardens at Kew, United Kingdom, in 1774. Other species were soon cultivated in botanical gardens and arboreta

TABLE 2
Naturalized *Corymbia* and *Eucalyptus* by Regenerative Strategies

Subgenus	Number of Species	Regenerative Strategy	Number of Species
Blakella	3	Lignotuber sprouters	28
Corymbia	2	Stem sprouters	4 (1 invasive)
Eucalyptus	10	Combination sprouters	4 (2 invasive)
Symphyomyrtus	30 (6 invasive)	Obligate seeders	9 (3 invasive)

NOTE: Numbers of naturalized *Corymbia* and *Eucalyptus* species by subgenera and by regenerative strategy. Only species reported as naturalized in at least two countries are included. Derived from Table 1.

in Europe as botanical curiosities and ornamentals. Once in cultivation, eucalypts appealed to foresters because of their fast growth (even on nutrient-poor soils) and because they yielded a variety of timber and nontimber products. Productivity of properly designed plantations is 20–70 m^3 ha^{-1} $year^{-1}$. Today, eucalypts provide saw timber, plywood, fiberboard, pulp for paper, poles, firewood, charcoal, essential oils, honey, and shelter. They are also considered to be suitable trees for biofuel production.

To date, the total area of eucalyptus plantations has been increasing exponentially. The global extent of eucalypt planting outside Australia reached 15.6 million ha in the 1990s—more than four times the global total in the 1970s. In 2008, the total area of eucalyptus plantations was estimated at 19.6 million ha. This is an area larger than the state of Washington, or the nations of Austria and Hungary together. Only one genus of trees is more extensively cultivated than eucalypts: *Pinus* (pines), with a global plantation

area of more than 52 million ha. The most commonly cultivated eucalypt is probably *E. globulus* (Tasmanian blue gum, with 2.3 million ha worldwide in 2008). *Eucalyptus globulus* is the primary source of global eucalyptus oil production, with China being the largest commercial producer. The other three most commonly cultivated eucalypts are *E. camaldulensis* (river red gum), *E. grandis* (flooded gum), and *E. tereticornis* (forest red gum). Current estimates of areas of eucalyptus plantations on individual continents and in Oceania are summarized in Table 3. The American statistic is dominated by Brazil, with over 3.5 million ha of plantations, while in Asia the leader is India, with over 5 million ha, including small cultivations on farms.

Eucalypt planting has accelerated recently in many tropical countries. However, extensive planting in the humid tropics has been inhibited by the incidence of pathogens and insect pests. Only a few species, such as *E. deglupta, E. pellita*, and *E. urophylla*, appear to be adapted to hot, humid environments. Eucalypts, because of their rapid growth and capacity for producing biomass, have been recently widely mentioned as feedstock for biofuels (e.g., *E. globulus, E. grandis, E. robusta, E. saligna, E. urophylla*). Widespread use for this purpose could substantially increase propagule pressure and increase the probability of local invasions.

ARE EUCALYPTS INVASIVE?

Eucalypts have enjoyed a history of widespread planting similar to that of pines. Given this, and the many species involved (with representatives of most of the major taxonomic and functional groups in the genus), we would expect to find the full range of outcomes: everything from taxa that are highly invasive and cover large areas as invaders to species that fail to recruit any offspring. Indeed, eucalypts, like pines, are prominent species on many national or regional weed lists in many parts of the world. However, they have been orders of magnitude less successful as

TABLE 3
Areas of Eucalyptus Plantations beyond Their Native Range in 2008

	Million ha	
Asia	8.3	(India: >5.0; China: >2.0; Vietnam: >1.0)
Americas	6.4	(Brazil: >3.6; Chile: >0.3; Argentina: >0.2; Peru: >0.2)
Africa	2.2	(South Africa: >0.4; Angola: >0.2; Morocco: >0.2)
Europe	1.3	(Portugal: >0.6; Spain: >0.5; Italy: >0.07)
Oceania	0.9	

NOTE: http://git-forestry-blog.blogspot.com/2008/09/eucalyptus-global-map-2008-cultivated.html and other sources.

FIGURE 1 A stand of *Eucalyptus globulus* on Santa Cruz Island, California. Saplings of eucalypts are usually found only close to plantations. (Photograph courtesy of M. Rejmánek.)

invaders than pines and several other widely planted trees, including many fleshy-fruiting trees (e.g., *Elaeagnus angustifolia*, *Ligustrum* spp., *Psidium* spp., *Morella (Myrica) faya*) and legumes (e.g., *Acacia* spp., *Prosopis* spp., *Leucaena leucocephala*). Where eucalypts have invaded, they have very seldom spread considerable distances from planting sites, and their regeneration is frequently sporadic (Fig. 1). Given that many eucalypts produce very large quantities of seeds, and in light of their diverse adaptations for dealing with disturbance (notably fire), their poor (or at best mediocre) performance as invaders is enigmatic. Many other Australian trees, including taxa that evolved under the same conditions as *Eucalyptus*, are much more invasive in other parts of the world (e.g., *Melaleuca quinquenervia*, *Hakea* spp., and many species of *Acacia*). What makes this difference? Are eucalypts inherently less invasive, or are they just a ticking time bomb? There seem to be three major reasons for the limited invasiveness of eucalypts: (1) relatively limited seed dispersal, (2) high mortality of seedlings, and (3) lack of compatible ectomycorrhizal fungi.

RELATIVELY LIMITED SEED DISPERSAL. Seeds of planted eucalypts are very small, but they have no adaptations for dispersal (wings or fleshy tissues) that would help them to proceed from local establishment (naturalization) to invasion. The passive release of seeds is undoubtedly aided by wind. However, all rigorous studies of eucalypt seed dispersal and seedling spatial distribution show that in general, seeds are dispersed over quite short distances. This is in agreement with measurements of terminal descent velocities of their seeds. While terminal velocities of seeds of invasive pine species are between 0.7 and 1.5 m s^{-1}, and for invasive maples (*Acer*) between 0.9 and 1.2 m s^{-1}, terminal velocities of seeds of all tested eucalypt species are between

2.0 and 5.5 m s^{-1}. Lower terminal velocity values mean that seeds can be carried by winds for longer distances.

HIGH MORTALITY OF SEEDLINGS. As noted above, eucalypts produce very small seeds (usually <2 mg) with no obvious endosperm. Therefore, the newly emerged seedlings are sustained by cotyledon photosynthesis, and their roots have to penetrate into suitable wet substrate very quickly. As a result, eucalypts can successfully regenerate from seeds only on wet, bare soil free of litter. However, seedlings in wet environments frequently die because of damping off caused mainly by parasitic fungi (*Botrytis*, *Colleotrichum*, *Cylindrocladium*, *Fusarium*) and water molds—oomycetes (*Phytophthora*, *Pythium*). Moreover, if there is any dense vegetation around, tiny eucalyptus seedlings are necessarily losers. Therefore, the window of opportunity for eucalyptus seedlings is rather narrow.

LACK OF COMPATIBLE ECTOMYCORRHIZAL FUNGI. It seems that the majority, if not all, of ectomycorrhizal fungi (EM) associated with eucalypts outside Australia are Australian species that have spread with their hosts. However, the importance of EM for establishment, growth, and spread of introduced eucalypts is not clear. There has been no example reported from exotic eucalypts that can compare with the dramatic response to mycorrhizal inoculation reported from exotic pine seedlings in Australia or South Africa in the early years of their introduction. With their more finely branched root systems and evolutionary adaptation to low phosphorus soils, eucalypts seem to be less dependent on EM than pines. The lack of EM may be not important for seedlings transplanted from nurseries, but, as some ecologists have suggested, it may be crucial for spontaneous establishment of seedlings away from plantations. However, colonization of eucalypt roots usually does not start with EM, but with ubiquitous nonspecific vesicular-arbuscular endomycorrhizae (AM) (*Gigaspora* spp., *Glomus* spp.). It is possible that AM play a more important role in initial eucalypt seedling establishment than EM.

Finally, we may ask whether some eucalypt species are inherently more invasive than others. Or is it only propagule pressure (the extent of planting) that makes some species more often naturalized, and some of them even somewhat invasive (spreading spontaneously over 100 m from points of introduction)? Tables 1 and 2 show that the majority of species naturalized in at least two countries (30) belong to the subgenus *Symphyomyrtus*. This seems to be just proportional to the size of this subgenus (>360 species). Still, all six species that can be classified as invasive in at least one

TABLE 4
Eucalyptus Species with No Conclusive Evidence of Naturalization

Species	Subgenus	Regenerative Strategy
acmenoides	*Eucalyptus (Monocalyptus)*	Lignotuber sprouter
bosistoana	*Symphyomyrtus*	Lignotuber sprouter
erythrocorys	*Eudesmia*	Lignotuber sprouter
fraxinoides	*Eucalyptus (Monocalyptus)*	Obligate seeder
jacksonii	*Eucalyptus (Monocalyptus)*	Lignotuber sprouter
macranda	*Symphyomyrtus*	Lignotuber sprouter
pauciflora	*Eucalyptus (Monocalyptus)*	Lignotuber sprouter
preissiana	*Eucalyptus (Monocalyptus)*	Lignotuber sprouter
pulverulenta	*Symphyomyrtus*	Lignotuber sprouter
punctata	*Symphyomyrtus*	Lignotuber sprouter
radiata	*Eucalyptus (Monocalyptus)*	Lignotuber sprouter
smithii	*Symphyomyrtus*	Lignotuber sprouter
spathulata	*Symphyomyrtus*	Obligate seeder
torquata	*Symphyomyrtus*	Lignotuber sprouter?

NOTE: Commonly cultivated *Eucalyptus* species with no conclusive evidence of naturalization. Species listed here are relatively commonly cultivated or tested, but the extent of their cultivation is still very limited when compared with many major plantation species listed in the Table 1. Therefore, "no conclusive evidence of naturalization" does not necessarily mean conclusive noninvasiveness.

country belong exclusively to the subgenus *Symphyomyrtus*. There seems also to be a somewhat larger proportion of subgenus *Eucalyptus* species among commonly planted eucalypts with no evidence of naturalization (Table 4). On average, seeds of species in the subgenus *Symphyomyrtus* are smaller (a notable exception, however, is *E. globulus*) than seeds of species in the second largest subgenus— *Eucalyptus* (>140 species). Therefore, terminal velocities of *Symphyomyrtus* seeds should be less, and they could be dispersed somewhat longer distances by wind. Also, none of the species classified as invasive in at least one country is a straight lignotuber sprouter (LS). This could also point to the primary importance of seed dispersal in invasiveness. Nevertheless, the more straightforward explanation is that species classified as invasive have been planted more extensively (see notes in Table 1). Then, however, reasons for their extensive planting may be correlated with their growth and timber characteristics that are associated with particular regenerative strategies.

ENVIRONMENTAL CONTROVERSIES AND CONTROL METHODS

Considering the amount of planting, eucalypts are relatively noninvasive species. If their potential spread is the only concern, then eucalypts should not be planted near rivers and streams. Temporarily flooded or eroded banks are suitable habitats for spontaneous establishment of their seedlings (Fig. 2). Moreover, their seeds can be dispersed for long distances by running water. However, there are other concerns.

Of all widely used plantation species, eucalypts have attracted by far the most criticism. There are four main concerns: (1) excessive water use and suppression of food crops growing nearby, (2) suppression of ground vegetation (possible allelopathic effects) and resulting soil erosion, (3) increased fire hazard, and (4) generally poor wildlife value. There is some substance to each of these concerns.

FIGURE 2 *Eucalyptus camaldulensis* is invasive along hundreds of kilometers of rivers in South Africa's Western Cape province. The images show (A) a general view of *E. camdulensis* established in the Berg River, (B) the extreme persistence of adult plants due to their ability to sprout from the roots, and (C) microsites for seedling establishment. (Photographs courtesy of D.M. Richardson.)

Nevertheless, it is important to realize that in many tropical countries, where eucalypts are grown on degraded soils unsuitable for continuing to support native trees (usually abandoned agricultural land), fuel and other products of resprouting eucalypts can greatly reduce the increasing human pressure on the remnants of natural forests. Even here, however, deleterious human practices associated with consecutive cutting cycles may eventually lead to yield decline and forest site degradation on a long-term basis. For long-term site quality and sustainability of biomass production, prolonging the cutting cycles and prohibiting or controlling litter raking appears to be imperative.

Eucalypts may be a major source of both nectar and pollen for honeybees. Because flowering of many eucalypts is abundant and lasts for long periods, some species (particularly *E. camaldulensis* and *E. cladocalyx*) are very valuable for the honey industry. When compared with native vegetation, usually significantly lower species diversity of arthropods, small mammals, and birds is reported from eucalyptus stands. For example, in Angel Island State Park, California, 41 species of birds were observed in native vegetation, but only 30 species in the eucalyptus forest. However, there may be also some other trends: approximately three times more California slender salamanders (*Batrachoseps attenuatus*) were found in eucalyptus vegetation than in native. Even more importantly, in California, eucalypts are major providers of shelter and nectar to the migrating monarch butterfly (*Danaus plexippus*) during winter months.

Allelopathic effects of eucalypts on native species are widely reported. Such reports are mostly based on laboratory bioassays. However, some field trials also point to decline of seed germination and increase of seedling mortality of some native species. If not chemical inhibition, then at least accumulation of dead material on the floor of eucalypt plantations retards regeneration of native species. Mixed-species plantations of eucalypts with native (mainly nitrogen-fixing) species have the potential to increase productivity while maintaining soil fertility and biodiversity.

Tasmanian blue gums (*E. globulus*) were planted in the San Francisco area of California as early as the second half of the nineteenth century. Having been in this landscape for such a long time, many old eucalypts are now treated as trees with "historical value" or as "heritage trees." Many people feel that eucalypts give California a "unique exotic flavor" lacking in other parts of the United States. This is the reason why removal of eucalypts on Angel Island in the San Francisco Bay (1990–1996) sparked a raging controversy. In a very balanced way, the history of this episode was described by Peter Coates.

FIGURE 3 It is undeniable that unmanaged stands of some eucalypt species can accumulate highly flammable dead material. To what extent Tasmanian blue gum (*Eucalyptus globulus*) groves contributed to the intensity of the tragic Berkeley–Oakland Hills fire in 1991 remains a subject of bitter discussions. This hazy but dramatic photograph of the 1991 fire was shot by former fire captain Wayne Drager through a plexiglass window, using a disposable camera on a bumpy helicopter flight. (Photograph courtesy of Wayne Drager.)

Because of accumulated litter, dense eucalypt stands are extremely flammable. The situation is exacerbated after winter freezes, when trees drop dead branches and foliage. During the last two weeks of 1990, a mass of frigid arctic air moved into California, and temperatures plunged to record lows along the Pacific Coast. It is very likely that fuel accumulation in unmanaged eucalyptus stands contributed to the intensity of the tragic fire in the Berkeley–Oakland Hills area in October 1991 (Fig. 3).

In arid and semiarid countries, where shortage of water is a big concern, benefits of eucalypt plantations may be outweighed by their negative environmental impacts: namely, their high water consumption. In South Africa, invasive eucalypts have been cleared over large areas as part of a national restoration program called Working for Water. In most areas, standing trees are felled and, where it is practically possible, their timber is harvested. Where recovery of the timber is impractical, felled plants are often stacked and burned. The most challenging management operations involve clearing river banks in large parts of the Western Cape province of dense stands of invasive *E. camaldulensis*. Clearing alone often causes destabilization of the river banks, and research is under way to determine the most effective ways of thinning the invasive stands gradually while simultaneously reintroducing key native plants to stabilize the sites and launch succession toward a sustainable community dominated by native plants.

The fact that eucalypt seeds do not have dormancy, with seed storage in the soil lasting less than a year, makes local eradication an achievable goal. However, resprouting of cut trees from stumps or lignotubers, which is advantageous in some situations, makes control of eucalypts difficult. Continuously cutting back the regrowth can eventually kill the tree, but this is a labor-intensive and expensive control method. Herbicide applications (triclopyr or glyphosate) to freshly cut stumps can greatly reduce resprouting. Because eucalypts are valued as timber and ornamental trees in many settings, biological control is very unlikely as an option.

SEE ALSO THE FOLLOWING ARTICLES

Allelopathy / Fire Regimes / Forestry and Agroforestry / Invasiveness / Mycorrhizae / Propagule Pressure / Trees and Shrubs

FURTHER READING

Coates, P. 2006. *American Perceptions of Immigrant and Invasive Species.* Berkeley: University of California Press.

Díez, J. 2005. Invasion biology of Australian ectomycorrhizal fungi introduced with eucalypt plantations into the Iberian Peninsula. *Biological Invasions* 7: 3–15.

Doughty, R. W. 2000. *The Eucalyptus: A Natural and Commercial History of the Gum Tree.* Baltimore: The Johns Hopkins University Press.

Keane, P. J., G. A. Kile, F. D. Podger, and B. N. Brown, eds. 2000. *Diseases and Pathogens of Eucalypts.* Collingwood, Australia: CSIRO Publishing.

Nicolle, D. 2006. A classification and census of regenerative strategies in the eucalypts (*Angophora, Corymbia* and *Eucalyptus*—Myrtaceae), with special reference to the obligate seeders. *Australian Journal of Botany* 54: 391–407.

Poore, M. E. D., and C. Fries. 1985. *The Ecological Effects of Eucalyptus.* FAO Forestry Paper 59: 1–87. Rome: FAO.

Rejmánek, M., D. M. Richardson, S. I. Higgins, M. J. Pitcairn, and E. Grotkopp. 2005. Ecology of invasive plants: State of the art (104–161). In H. A. Mooney, R. N. Mack, J. A. McNeely, L. E. Neville, P. J. Schei, and J. K. Waage, eds. *Invasive Alien Species: A New Synthesis.* Washington, DC: Island Press.

Ritter, M., and J. Yost. 2009. Diversity, reproduction, and potential for invasiveness of Eucalyptus in California. *Madroño* (in press).

Slee, A. V., M. I. H. Brooker, S. M. Duffy, and J. G. West. 2006. *Euclid: Eucalypts of Australia,* 3rd ed. Collingwood, Australia: CSIRO.

Williams, J., and J. Woinarski, eds. 1997. *Eucalypt Ecology: Individuals to Ecosystems.* Cambridge: Cambridge University Press.

EUTROPHICATION, AQUATIC

KATHARINA A. M. ENGELHARDT

University of Maryland Center for Environmental Science, Frostburg

Eutrophication is the natural or anthropogenic accumulation of nutrients in soil or water (from Greek *eu* = "well" and *trophe* = "nourished"). Oligotrophic (low-nutrient) waters

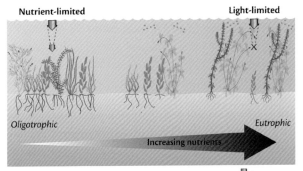

Increasing nutrients result in light limitation of native plants, lower diversity and root-to-shoot ratio of vascular plant communities, increase in phytoplankton, and greater risk of invasion.

FIGURE 1 The impacts of eutrophication on aquatic ecosystems. (Figure developed by the Integration and Application Network of the University of Maryland Center for Environmental Science.)

contain less than 5–10 µg L^{-1} phosphorus and less than 250–600 µg L^{-1} nitrogen. These nutrient concentrations are at least 2 to 10 times as high in eutrophic waters and can have major effects on biotic communities, including the loss of biodiversity and the invasion of nonnative species (Fig. 1).

NATURAL AND ANTHROPOGENIC EUTROPHICATION

Eutrophication is a slow natural process. The slow accumulation of nutrients is especially prevalent in depositional environments, such as lakes and wetlands, where nutrients and sediments derived from a watershed are collected in a basin and permanently or temporarily immobilized and stored. For example, estuaries are naturally eutrophic, and hence very productive, because they receive nutrients derived from watersheds and tidal flows. Lakes accumulate sediments and organic matter and over time convert into productive and nutrient-rich shallow lakes and emergent marshes. Other naturally eutrophic systems are areas along the coast where upwelling conveys nutrient-rich water to the surface. As an ecosystem's nutrient status changes over millennial time scales, so does its community structure, with local extinction and colonization of new species working in concert to produce species-rich and productive ecosystems.

However, the process of eutrophication can be greatly accelerated by human activities, such as runoff of excess fertilizer, sewage effluent, and stormwater runoff. In Australia, for example, sites affected by human activity have mean levels of 780 µg L^{-1} N and 95 µg L^{-1} P compared to 300 µg L^{-1} N and 21 µg L^{-1} P at less impacted sites. Because estuaries and lakes typically attract large human populations, many estuaries and lakes throughout the world are now highly eutrophic. Unlike the slow natural eutrophication process, anthropogenic eutrophication (or "cultural eutrophication") may be manifested in only a few years, leaving existing biota little time to adjust. Thus, anthropogenic eutrophication is typically associated with higher primary productivity (greater than 1 g carbon/m/day), higher oxygen demand, lower species richness, and changes in species abundances. Owing to their high-nutrient status and stressed existing biotic communities, eutrophic water bodies are also at higher risk of invasion by nonnative species, which can have their own positive or negative feedbacks on eutrophication.

MECHANISMS OF INVASION

Competition

Changes in community structure with eutrophication are predicted by plant competition theory. Competition for belowground nutrients is predicted to drive the outcome of species interactions under nutrient-limited conditions. In contrast, competition for light is the dominant mechanism when nutrients are abundant (Fig. 1). Thus, species with high belowground biomass are predicted to prevail under nutrient limitation, and species allocating biomass to light harvesting leaves will succeed when nutrients are in excess and light is more limiting. Not surprisingly, invasive species are often "weedy," with high aboveground growth rates under nutrient-rich conditions. These same species are often poor competitors when nutrients are limiting and the native community is intact. Invasive species may be initially very successful under nutrient-rich conditions, but without new nutrient additions, they may decline in population size as nutrients are immobilized and become more limiting.

Phosphorus is typically considered the main limiting nutrient in freshwater systems, while nitrogen is typically the more limiting nutrient in saltwater, although the nature of nutrient limitation varies greatly among ecosystems. Increasing the abundance of nutrients in systems that are nutrient-limited can dramatically change the competitive balance of plant communities (and entire food webs that depend on them) by releasing plants from competition and allowing some species to manifest their invasive potential. For example, the community structure of New England salt marshes has in some places shifted to monocultures of the invasive common reed, *Phragmites australis* (Fig. 2), owing to nitrogen eutrophication derived from shoreline development. *Myriophyllum spicatum*, an invasive submersed aquatic macrophyte of lakes and estuaries in the United States, gains competitive advantage over native species in eutrophic conditions.

FIGURE 2 Common reed, *Phragmites australis*, in a mid-Atlantic marsh of the United States. (Photograph courtesy of the author.)

Under nutrient-limited conditions, *M. spicatum* is a poor competitor, owing to its poor root system and the energetic demands of producing allelopathic compounds. But under nutrient-rich conditions, *M. spicatum* is a superior competitor because the species contains polyphenols that render it allelopathic towards epiphytic algae. Native species, on the other hand, accumulate algae on their leaves and are therefore not as efficient in competing for light. In the presence of algivorous snails, however, the abundance of *M. spicatum* and native species reverses because the snails decrease epiphytic biomass on the leaves of native species, and the native species become less light limited.

Disturbance

Since Charles Elton published his famous book *The Ecology of Invasions by Animals and Plants* on species invasions in 1958, ecologists have argued that the risk of invasion increases with declining species diversity. A possible explanation for this relationship is that a decline in native species richness decreases nutrient uptake by the community. Disturbance events may physically remove biomass or change environmental conditions enough to affect native species abundance and richness. Thus, disturbances, through their effects on the existing community, can decrease the ability of native species to respond to eutrophication, thereby providing opportunities for exotic species to invade and proliferate in an environment with ample nutrients and fewer competitors.

Enemy Release

Species that are adapted to high resource availability gain the most from being released from their native enemies ("enemy release"). This is because high resource-adapted plant species typically have high growth rates and do not produce energetically costly antiherbivory defenses. In their native habitat, therefore, these species are controlled by coevolved herbivores and do not become abundant. However, they become invasive when they are released from their enemies in the introduced habitat. Thus, enemy release and competitive release may act synergistically to allow invasive species adapted to high resource availability to establish and expand quickly under eutrophic conditions. Biocontrol—the introduction of enemies to combat invasive species—is a way of curbing the success of these invasive species.

INTERNAL EUTROPHICATION

Shallow lakes and wetlands are known for their ability to remove nutrients from water through direct uptake by plants, physical settlement of particles, and chemical immobilization of nutrients in sediments. However, long-term observations in the Netherlands have shown that wetlands can release nutrients bound in wetland sediments. The increase of nutrient concentrations without the external supply of nutrients is called "internal eutrophication" and is most prevalent when water levels are not allowed to fluctuate naturally. This changes the reduction-oxidation potential of the sediments. For example, ferric iron bonds with phosphorus to form an insoluble compound in aerobic conditions. But when sediments are constantly flooded, ferric iron is reduced to ferrous iron, and the iron complex releases the phosphorus. This process has led to changes in shoreline vegetation in Manitoba and Indiana. In Wisconsin, stable water levels and internal eutrophication have been associated with the expansion of invasive cattails (*Typha* sp.), which are phosphorus limited and can take immediate advantage of the released nutrients through clonal expansion (Fig. 3).

Internal eutrophication can also be caused by the invasion of novel species with different phenologies or ecophysiological traits. For example, purple loosestrife (*Lythrum salicaria*) is rapidly displacing native vegetation in some North American wetlands. The conversion of cattail wetlands to loosestrife wetlands alters the timing of litter input and releases phosphorus during a time when other plants cannot use it. This asynchrony in phenology increases downstream phosphorus loads, which effectively accelerates eutrophication in downstream water bodies.

The subtropical cyanobacterium *Cylindrospermopsis raciborskii* has invaded shallow areas of Lake Balaton, Hungary. It is a superior competitor for light and nutrients, and its invasion success is attributed, in part, to its ability to generate internal phosphorus. Two mechanisms can

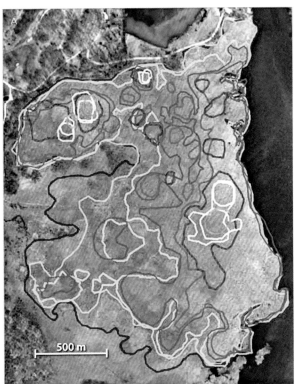

FIGURE 3 Expansion of nonnative cattail (*Typha × glauca* and *T. angustifolia*) in a Wisconsin wetland with stabilized water levels (A) and *T. angustifolia* and the native *T. latifolia* in a nearby wetland with fluctuating water levels (B) over a period of 37 years. Radii of the clones spread almost 4 m per year under stable water level conditions and 2.5 m under fluctuating conditions. (Figure courtesy of J. Zedler and A. Bairs.)

explain biologically generated internal eutrophication: (1) root oxidation or respiration can change the redox potential of the sediments, which can bind or release nutrients from the sediments (see above), and (2) plants can facilitate the release of nutrients to the water column when their leaves turn over rapidly throughout the growing season.

AMELIORATING EUTROPHICATION

The proliferation of invasive species under eutrophic conditions can have tremendous negative consequences for ecosystem services (e.g., decreased water supply, altered food web support through changes in fish and shellfish populations, lowered recreational opportunities, impeded navigation, damaged dams and support infrastructure). However, through their enormous productivity, invasive species have the capacity to remove nutrients from water, which ameliorates the effects of eutrophication. If additional anthropogenic eutrophication does not occur, invasive species may increase water quality and allow native species to return and outcompete invasive species.

Human-made lakes often experience an initial period of internal nutrient loading, which subsides as the ecosystem establishes. These lakes are often initially colonized by invasive species, such as water hyacinth in Lake Chivero, Zimbabwe, and water lettuce (*Pistia statiotes*) in Lake Volta, Ghana. The invasive plants are highly productive initially, covering a big portion of the lake surface, but then they often disappear or diminish in population size. These "boom and bust" periods of aquatic invasive plant species are typical and are often associated with changes in the external and internal nutrient loading of the water body, suggesting that the invasive species may ameliorate the effects of eutrophication. If anthropogenic eutrophication is minimized, therefore, invasive aquatic weed infestations may correct themselves with or without additional control measures.

The zebra mussel (*Dreissena polymorpha*), a freshwater mussel native to southeast Russia, has become a problematic invasive species in many countries and was first detected in the United States and Canada in 1988. Since then, the species has spread to many waterways with negative ecological and economic consequences. However, zebra mussels are filter feeders and can therefore ameliorate the effects of nutrient pollution. This has benefited native algae, which can grow to greater depths owing to higher water clarity, and smallmouth bass (*Micropterus dolomieu*) populations in Lake Erie.

Sometimes eutrophication may be ameliorated by the intentional introduction of a new species. For example, oysters are an important species in estuaries, where they filter water and ameliorate the effects of pollution. In the eastern United States, the native Eastern oyster (*Crassostrea virginica*) declined in the Chesapeake Bay, owing to many factors, including eutrophication of the estuary. The suminoe, or Asian oyster (*Crassostrea ariakensis*), was proposed for introduction in the Chesapeake Bay in the United States because it is more resistant to disease and

can grow more quickly. However, after extensive public comment, introduction of the Asian oyster was not recommended, owing to its invasion risk and potential competitive effects on the native oyster.

MANAGEMENT OF TWO INSEPARABLE ISSUES: EUTROPHICATION AND INVASION

Eutrophication and invasive species are typically assessed as isolated problems in lake management, yet they are often inseparably linked in aquatic ecosystems. The proliferation of invasive aquatic species is in most cases a result of anthropogenic eutrophication, often associated with other disturbances that decrease the integrity of existing communities. Managing eutrophication and invasive species needs to be done on a case-specific basis, and no management strategy will fit all cases because the aquatic community, the watershed properties, and the nature of limiting resources differ for each water body. Invasive species can contribute to eutrophication or ameliorate it. However, one thing is clear: to curb the success of invasive species and minimize eutrophication, external nutrient loading must be decreased, which requires a watershed approach that effectively and substantially decreases the multiple anthropogenic inputs of nutrients into water bodies. In addition, the native community may need to be actively restored to ensure that native species can replace introduced species in the food web.

SEE ALSO THE FOLLOWING ARTICLES

Competition, Plant / Enemy Release Hypothesis / Genotypes, Invasive / Invasibility, of Communities and Ecosystems / Lakes / Nitrogen Enrichment / Wetlands

FURTHER READING

Bertness, M. D., P. J. Ewanchuck, and B. R. Silliman. 2002. Anthropogenic modification of New England salt marsh landscapes. *Proceedings of the National Academy of Sciences USA* 99: 1395–1398.

Blumenthal, D. M. 2006. Interactions between resource availability and enemy release in plant invasion. *Ecology Letters* 9:887–895.

Chase, J. M., and T. M. Knight. 2006. Effects of eutrophication and snails on Eurasian watermilfoil (*Myriophyllum spicatum*) invasion. *Biological Invasions* 8: 1643–1649.

Hastwell, G. T., A. J. Daniel, and G. Vivian-Smith. 2008. Predicting invasiveness in exotic species: Do subtropical native and invasive exotic aquatic plants differ in their growth responses to macronutrients? *Diversity and Distributions* 14: 243–251.

Horne, A. J., and C. R. Goldman. 1994. *Limnology.* New York: McGraw-Hill, Inc.

Khan, F. A., and A. A. Ansari. Eutrophication: An ecological vision. *The Botanical Review* 71: 449–482.

Scheffer, M. 1998. *Ecology of Shallow Lakes.* Dordrecht, The Netherlands: Kluwer Academic Publishers.

Wetzel, R. G. 2001. *Limnology: Lake and River Ecosystems.* San Diego, CA: Academic Press.

Williams, A. E., and R. E. Hecky. 2005. Invasive aquatic weeds and eutrophication: The case of water hyacinth in Lake Victoria (187–225). In M. V. Reddy, ed. *Restoration and Management of Tropical Eutrophic Lakes.* Enfield: Science Publishers, Inc.

EVOLUTIONARY RESPONSE, OF NATIVES TO INVADERS

SHARON Y. STRAUSS

University of California, Davis

Introduced species may act as strong agents of selection on native species, either as a result of their direct interactions with natives or through *indirect interactions*. The ways in which introduced species can influence natives are as complex as the ways that any species can interact with any others in complex and diverse communities. Evolutionary changes of large magnitude can occur very rapidly, over decades or less, and are not relegated to geological timescales. Some of the best-documented cases of rapid evolution have come from human-caused impacts on natural populations, such as overharvesting or the effects of species' introductions.

THE DYNAMISM OF SPECIES AND THE DETECTION OF CHANGE

It is generally a mistake to think of either native or invasive species as static entities; the outcomes and impacts of invaders on native ecosystems will depend on the dynamic capacity of both introduced and native species to evolve. The ability of natives to evolve in response to invaders may lessen the negative impacts of some invaders or even allow natives to benefit from them. Alternatively, the inability of natives to adapt to antagonistic invaders can lead to extirpation or massive declines in native species.

While we expect that there have already been dramatic effects of invaders on the evolution of many native species, the ability to detect evolutionary changes in natives that one can ascribe specifically to effects of exotic species requires painstaking experimentation and correlative work. Often, we rely on time series data or historical collections to compare changes in traits as a result of history of a native with an invader. Moreover, the genetic basis of these changes must also be established in order to attribute changes in natives to evolution, as opposed to *plastic* responses of natives to invaders.

EVIDENCE FOR EVOLUTION IN NATIVES CAUSED BY INVASIONS

We have relatively few examples of evolutionary changes in native species caused by introduced species, despite the probable ubiquity of this phenomenon. This paucity of examples arises largely from a lack of documentation of native traits prior to introductions. To document rigorously that invaders are the causal agent of evolution in natives, one would ideally have samples of individuals from the same native populations both before and after invasion. One would also like to observe parallel responses in multiple native populations that have experienced independent invasions by the same invader. Breeding experiments would then be required to document that the observed changes in traits of native species are heritable. Together, such data would provide compelling circumstantial evidence that invaders are likely the source of differences in native populations before and after invasion.

We rarely have such a complete body of evidence. Some of the best cases in which there is documented evolution in response to an invader come from insect herbivores and their *host shifts* onto introduced host plants. Soapberry bugs (*Jadera*) and their close relatives (*Leptocoris*) have colonized several species of introduced ornamental trees and vines in the Sapindaceae that are related to the bugs' native host plants in the United States and Australia. Initially, performance of bugs on introduced hosts was lower than on native hosts, but bugs evolved behavioral, morphological (including longer beak size), physiological, and life history adaptations that increased the efficiency of host exploitation over 30 to 50 years (fewer than 100 generations). Similarly, some populations of *Euphydryas editha* butterflies have evolved to prefer an introduced weed *Plantago lanceolata* over their native host plants. In general, exotic plants are probably exerting selection on many native insect herbivore taxa and are likely causing evolutionary changes in these insects. For example, 82 of 236 butterfly species (34%) in California have been reported as ovipositing or feeding on introduced plant taxa. Not all of these host shifts require genetic changes, but the ability to use new hosts effectively likely does, as the cases above demonstrate. Few of these host shifts have been rigorously investigated with respect to their underlying genetic basis and their impacts on the evolution of native insects.

Another approach to documenting evolution in response to invaders takes advantage of the fact that, as an invader spreads, native populations experience different lengths of time in which they are exposed to the invader. One can then correlate changes in traits in natives with the length of time natives have coexisted with the invader. However, in the absence of multiple independent invasions, invaders usually spread in a spatially autocorrelated way (for example, from east to west). Thus, climatic and other environmental gradients may also coincide with invasion history. These and other gradients must be ruled out as the causes of evolutionary change in traits of natives.

Historical approaches provide good evidence that natives have evolved in response to both introduced predators and introduced prey. In Australia, the invasive cane toad *Rhinella marina (Bufo marinus)* is toxic to predators, and populations of naive native Australian predators, from snakes to frogs to crocodiles, have experienced dramatic declines as a result of consuming toxic toad prey. In response to the invasion of *Rhinella marina (Bufo marinus)*, two toad-eating snakes, *Pseudechis porphyriacus* and *Dendrelaphis punctulata*, have evolved longer bodies and reduced gape size (which limits the size, and hence dose of cane toad toxin) with time since exposure to toads (over 80 years). Two snake species that do not eat toads, but that cooccur with the other snakes and toads in these areas, showed no morphological changes over the same period, suggesting that no other environmental factor caused these changes. In a similar historical approach, there appears to have been rapid evolution in fence lizards (*Sceloporus undulatus*) in response to invasion by the red imported fire ant *(Solenopsis invicta)*. Hatchling fence lizards in the southeastern United States are often killed by introduced fire ants. Lizard populations with a longer history (around 70 years) of exposure to fire ant invasions have evolved longer hind legs and a greater twitch response to flick ants off bodies more effectively, as well as quicker escape behaviors when confronted with fire ant mounds, relative to naive lizard populations at the invasion front. Another way to determine the role of invaders in shaping traits of natives is through experimental removals of invaders and subsequent measurements of selection on traits in native species in the presence and absence of invaders. Using such an approach, Jennifer Lau experimentally manipulated all four combinations of the presence and absence of introduced herbivorous weevils and their introduced leguminous host plant *Medicago polymorpha*. Lau then measured patterns of *selection* on resistance and tolerance to weevil damage in a native legume, *Lotus wrangelianus*. In the presence of the invader *Medicago*, *Lotus* natives experienced both direct competition from *Medicago* and indirect effects of *Medicago* through greatly increased weevil densities. Lau showed altered patterns of selection on plant defense traits in *Lotus* in the four different treatments and corroborated these patterns in naturally invaded and uninvaded areas. Thus, invaders could be shown to have caused selection on traits of natives.

MULTIPLE INVASIONS

Multiple invasions may complicate adaptation of natives to exotic species when species invade either in sequence or simultaneously. If native populations are dramatically reduced in number after the first wave of an invasion, then they may be less resilient to other invasions. West Nile virus killed 72 percent of an Oklahoma population of American crows (*Corvus brachyrhynchos*) in a single year. White pine blister rust dramatically increased tree mortality, and pine populations had significantly reduced levels of genetic variation after outbreaks. While such mortality may result in rapid adaptive evolution favoring resistant genotypes in these populations, resulting *bottlenecks* may deplete the genetic variation and numbers of individuals available for responses to future episodes of selection from the next invader. Additionally, the nature of selective pressures can depend on the particular combination of species present in a community. In the *Lotus* example above, tolerance by *Lotus* to damage was favored when both exotics—weevil herbivores and *Medicago* competitors—were present, whereas resistance to damage by *Lotus* was favored when only the exotic weevil herbivore was present. Thus, the kind of plant defense traits favored in the native *Lotus* depended on the combination of exotic species that was present.

HYBRIDIZATION

Another important facet of the evolution of natives in response to invaders is the possibility of hybridization with introduced species. *Reproductive isolation* prevents closely related species from interbreeding, and many species have traits that maintain reproductive isolation from close relatives. When novel species that are closely related to natives are introduced to an area, mechanisms preventing genetic exchange between these taxa may be nonexistent, owing to a lack of prior evolutionary history. In one example, some populations of the federally endangered California tiger salamander (*Ambystoma californiense*) have hybridized with barred tiger salamanders (*Ambystoma tigrinum mavortium*) brought to California for fish bait.

In hybrid populations, genes from barred salamanders had the greatest representation in permanent bodies of water like cattle ponds and were less frequent in *vernal pools*, the natural habitat of California tiger salamanders. Human habitat modification (creation of numerous permanent bodies of water in grassland habitats) may be promoting the fitness of hybrids. To date, we do not know whether hybridization compromises the ability of native salamanders to persist in ephemeral vernal pools.

CAN EVOLUTION BY NATIVES PROVIDE RESILIENCE FOR INVADED NATIVE COMMUNITIES?

While many introduced species appear to have few dramatic effects, some pathological invasions clearly have huge impacts on ecosystem function and the persistence of natives. The ability of natives to evolve to lessen those impacts is important to conservation strategies. Adaptations such as the reduced gape size in snakes described above or resistance to introduced pathogens may allow native populations to persist and may even eventually compensate for the adverse effects of invaders. The native Amakihi bird in Hawai'i was forced to retreat to higher-altitude forests after the invasion of avian malaria; as resistance to malaria has evolved in Amakihi populations, they have been able to recolonize some lower-elevation forests.

Much of the ability to adapt will depend on extant genetic variation, population size, gene flow, and mutation rates. Conservation strategies that preserve genetic variation in native populations by conserving large populations and promoting gene flow in increasingly fragmented landscapes may provide genetic variation on which selection can act to ameliorate negative impacts of invaders.

SEE ALSO THE FOLLOWING ARTICLES

Climate Change / Evolution of Invasive Populations / Hybridization and Introgression / Lag Times

FURTHER READING

Carroll, S. P., J. E. Loye, H. Dingle, M. Mathieson, T. R. Famula, and M. P. Zalucki. 2005. And the beak shall inherit: Evolution in response to invasion. *Ecology Letters* 8: 944–951.
Fitzpatrick, B. M., and H. B. Shaffer. 2007. Hybrid vigor between native and introduced salamanders raises new challenges for conservation. *Proceedings of the National Academy of Sciences USA* 104: 15793–15798.
Langkilde, T. 2009. Invasive fire ants alter behavior and morphology of native lizards. *Ecology* 90: 208–217.
Lau, J. A. 2008. Beyond the ecological: Biological invasions alter natural selection on a native plant species. *Ecology* 89: 1023–1031.
Phillips, B. L., and R. Shine. 2006. An invasive species induces rapid adaptive change in a native predator: Cane toads and black snakes in Australia. *Proceedings of the Royal Society B—Biological Sciences* 273: 1545–1550.
Rhymer, J. M., and D. Simberloff. 1996. Extinction by hybridization and introgression. *Annual Review of Ecology and Systematics* 27: 83–109.
Singer, M. C., C. D. Thomas, and C. Parmesan. 1993. Rapid human-induced evolution of insect host associations. *Nature* 366: 681–683.

EVOLUTION OF INVASIVE POPULATIONS

CAROL EUNMI LEE

University of Wisconsin, Madison

Evolutionary mechanisms that operate at multiple stages of invasions (e.g., transport, introduction, establishment) could profoundly affect the invasive success of populations. Evolution within the native range shapes the intrinsic properties of populations (e.g., plasticity, evolvability) in a manner that would dictate their response to novel environments. Confronted with a sudden introduction into a novel range, a rapid evolutionary response is often essential for survival. Continued selection might be necessary following establishment, as traits that are beneficial during early stages of colonization would often differ from those that favor continued range expansion or long-term persistence. While much progress has been made in characterizing the process of evolution during invasions, more study is needed to understand the evolutionary forces that lead to populations with a greater propensity to invade.

BACKGROUND

Invasive populations experience abrupt and often dramatic ecological shifts on contemporary time scales, allowing observation of ongoing evolutionary changes in wild populations. In contrast to natural range expansions, which typically involve incremental geographic shifts, invasive populations are frequently transported across vast geographical distances that can span continents. These invasions often entail movement into novel habitats, at times with acute changes in abiotic (e.g., temperature, salinity) and biotic (e.g., competitors, predators, food) conditions. Thus, it is unsurprising that only a tiny fraction of introduced species become successful as invaders.

The evolution of invasive populations was first raised as a focal point of discussion by C. H. Waddington through a 1964 symposium, which resulted in the classic volume *The Genetics of Colonizing Species*, edited by Baker and Stebbins (1965). Waddington argued that recent and sudden transplantation of invasive species make them valuable models for studying constraints on adaptive evolution. He was interested in understanding limits to selection response and constraints on rapid phenotypic evolution, particularly in populations under acute stress, and the genetic consequences that would result. Waddington was also concerned with how phenotypic plasticity and natural selection might interact to produce an evolutionary response. For instance, environmental stress might induce a plastic response where an extreme phenotype is expressed. This extreme phenotype could then be subjected to selection and become fixed in the population. Waddington argued that the prevalence of this process, which he termed genetic assimilation, could be tested in the wild using invasive populations.

After a multi-decadal hiatus, interest in the evolution of invasive populations resurged in the 1990s. The advent of polymerase chain reaction (PCR) and advances in DNA sequencing allowed the proliferation of phylogeographic studies that reconstructed geographic pathways and evolutionary histories of invasions. Concurrently, application of quantitative genetic approaches uncovered patterns of phenotypic evolution associated with invasion events. At present, a few research programs are probing specific genetic mechanisms underlying adaptive evolution during invasions (see next sections). Epigenetic changes might play an important role in facilitating invasions, but explorations of such mechanisms are only in their infancy.

Although many questions raised by Waddington have remained largely unresolved and are current topics of active research, it has become widely recognized that evolutionary processes and genetic attributes of invasive populations can be important for their success in becoming established in a novel range. While extrinsic factors (e.g., transport vectors, dispersal barriers, resources, enemies) impose challenges and opportunities for invading populations, intrinsic properties of the organisms and populations dictate their response to the extrinsic factors, through mechanisms such as phenotypic plasticity or evolutionary adaptation.

Invasions are dynamic processes, with multiple stages of progression (i.e., transport, introduction, establishment, spread, stabilization, etc.) at which evolutionary mechanisms could impact invasive success. Prior to invasions, evolutionary forces within the native range could select for properties that make some populations more likely to invade than others, or more likely to invade specific habitat types. During the early stages of

colonization, selection could act on traits that promote dispersal or enable survival in the invaded range. In addition, demographic processes might result in population bottlenecks or admixture during invasions, affecting the nature of genetic variance available for selection in the invading population. Following establishment, selection might act on traits that influence ongoing range expansion and long-term persistence in the novel range. Alternatively, evolution might occur as a mere byproduct of relaxed selection in the invaded range, due to removal of selective forces that were formerly important in the native range, or from changes in evolutionary tradeoffs in the novel range. Some evolutionary mechanisms that are important at various stages of the invasion process are described briefly below, although these mechanisms might not operate exclusively at any given stage.

EVOLUTION WITHIN THE NATIVE RANGE

The evolutionary history of populations within their native ranges would influence their capacity and propensity to become invasive. Selection regimes in the native range would mold the response of populations to novel conditions, as selective forces would determine levels of plasticity, physiological tolerance, genetic variation, and other intrinsic properties of the population. For example, fluctuating selection might lead to the evolution of generalist strategies (e.g., broad tolerance or plasticity) or to enhanced evolvability, depending on the period of fluctuations relative to generation time. Nonselective evolutionary forces might also be important, as demographic processes in the native range, such as population bottlenecks and population expansions, could affect levels of genetic variation across the genome.

Invasive species commonly span a broad array of habitats within their native ranges, such that populations are exposed to a diversity of selection regimes. Such diverse selection regimes would lead to the evolution of populations that vary in their intrinsic capacities to invade. Phylogeographic studies are revealing a great deal of population genetic structure and large genetic divergences among invasive populations across their native ranges. Indeed, such studies are increasingly showing that certain populations might be more able to invade habitat boundaries than others, and that these populations tend to arise from particular lineages or from particular habitat types within their native ranges. For example, two genetically distinct clades of the copepod *Eurytemora affinis* cooccur in the St. Lawrence estuarine system, but populations from only one of the clades have invaded freshwater habitats, while the other clade has remained restricted to its native distribution.

Conditions in the native range would preadapt populations for similar habitat types in the invaded range, such that the invasion between similar habitats would essentially be that of habitat matching. In such cases where niche evolution is absent during invasions, the geographic distribution in the native range could be used to predict the potential range of an invasive species in its invaded range. However, accurate prediction requires information on the fundamental niche of an invasive species, which encompasses the full range of environmental conditions (biotic and abiotic) that the organism could inhabit. But most current methods of niche modeling capture the realized niche, which includes only the range where organisms are actually found. Environmental factors, such as competition and predation, could limit species distribution in the native range, such that the realized niche is smaller than the fundamental niche. Thus, species might have broader tolerances than those reflected in their native distribution, and the limits of such distribution data would hinder the predictive power of niche modeling.

Lineages that arose in certain environments would be preadapted to invading the conditions in which they evolved. For example, the plant family Poaceae (i.e., grasses) is overrepresented as both agricultural weeds and natural area invaders, relative to what would be expected based on estimated rates of introduction. Climate change of the mid-Miocene (approximately 6–8.5 million years ago) toward drier conditions led to the diversification of Poaceae and the prevalence of grass-dominated ecosystems. An evolutionary history in savannah environments is likely to have preadapted members of Poaceae to colonize human-impacted and deforested environments, as grasses tend to thrive under denuded, desiccated, and disturbed conditions. For instance, wind pollination by grasses (such that they do not depend on insect pollinators in the novel range) is a feature that aids in colonizing new and disturbed environments.

Populations that experience human-altered environments (e.g., agricultural fields, reservoirs, pavement) in their native range could become preadapted for invading similar habitats. For example, the corn root worm *Diabrotica virgifera virgifera* is adapted to infesting cultivated maize and has evolved resistance to a wide variety of insecticides and cultural control practices such as crop rotation. Some agricultural weeds have evolved crop mimicry to evade eradication. A striking example is the barnyard grass *Echinochloa crus-galli* var. *oryzicola*, which

resembles rice when hand-weeding (and morphological identification) is used for control. Reservoirs appear to provide havens for many aquatic invaders, as they possess properties that would tend to inhibit native freshwater species, such as high ionic load, eutrophic conditions, and high levels of disturbance. As the planet becomes increasingly blanketed by such human-impacted environments, effectively homogenizing habitats throughout the globe, invasive populations will increasingly adapt to such conditions.

More generally, it has been hypothesized by Di Castri (1989) and others that habitats characterized by disturbance or fluctuating conditions would tend to promote the evolution of populations that have a greater capacity to invade novel environments. Empirical observations appear to support this hypothesis. For example, brackish-water habitats that lie at the interface between marine water and freshwater typically exhibit fluctuating conditions across multiple spatial and temporal scales. Such brackish-water habitats have disproportionately given rise to invasive populations, including those that invade freshwater. In particular, the brackish Black and Caspian Sea basin has spawned many species invading freshwater habitats of Europe and North America. Such species include zebra mussels, quagga mussels, the fishhook waterflea *Cercopagis pengoi*, and many gammarid amphipods, such as *Corophium curvispinum*. An evolutionary history of fluctuating conditions in the Black and Caspian Sea basin has likely selected for lineages that have a tendency to become invasive. In another example, a variety of ant species invading North America from South America, such as the Argentine ant *Linepithema humile* and the fire ant *Solenopsis invicta*, show high levels of genetic differentiation across their native ranges. However, invasive populations appear to have originated from only a subset of their native range, from regions characterized by large-scale disturbances in the form of regular flooding. Whether the observed bias in sources of invasions is indeed due to the selection regime of the native habitat, or alternatively to the availability of transport vectors, warrants further investigation.

Habitats prone to disturbance or fluctuating conditions might promote the evolution of traits that confer invasive success. Theoretical studies indicate that fluctuating conditions might lead to the evolution of two distinct types of attributes that could increase the potential of populations to invade novel habitats. In environments where fluctuations are very rapid, occurring within an organism's lifespan, generalist strategies that allow the organisms to grow and persist under disturbed conditions are likely to

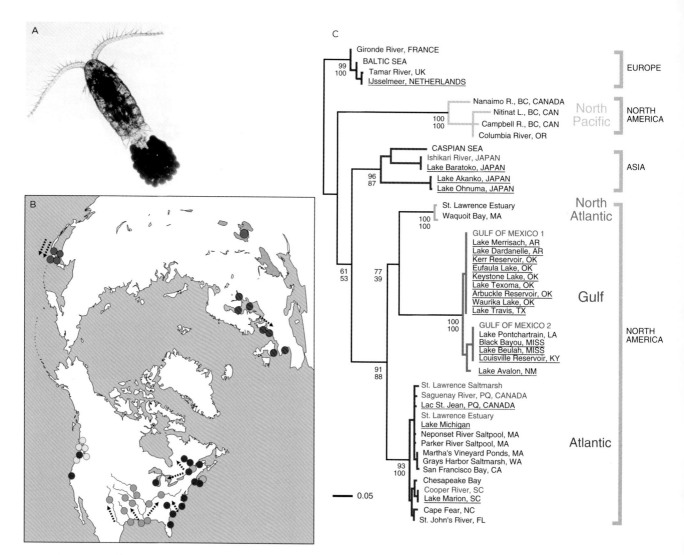

FIGURE 1 (A) The invasive copepod *Eurytemora affinis*. (Photograph by the author). (B) Geographic pathways of invasions (arrows) by the copepod *E. affinis* from saline sources into freshwater habitats, revealed using a phylogeographic analysis. (C) Phylogeny of *E. affinis* populations, showing saline source (colored) and freshwater invading populations (underlined). Many of the invaded freshwater sites are recently constructed reservoirs. Phylogeny used consensus sequences of haplotypes for each population. The invasions occurred primarily within the past 70 years or so, through human-mediated transport.

evolve. Generalist strategies might include life history and demographic traits that facilitate escape from harsh conditions or promote rapid population growth, broad tolerance, or phenotypic plasticity. Alternatively, environmental fluctuations over a slightly longer time scale, across a few generations, might select for increased evolvability, resulting in organisms with increased capacity to adapt rapidly to changing conditions. Theoretical models and empirical studies indicate that temporally fluctuating selection across generations could act to promote evolvability by increasing the propensity to produce adaptive variants (i.e., by increasing mutational variability or the recombination rate), or by promoting the maintenance of standing genetic variance for quantitative traits in a population.

Hypotheses on the role of disturbance or fluctuating conditions in promoting the evolution of invasive populations require further testing. In order to make definitive inferences, future studies would need to compare the invasive success of multiple populations from lineages that originate from fluctuating versus more constant habitats, preferably within a phylogenetic context. In general, studying the evolutionary history of invasive populations in their native ranges would provide insights into conditions that lead to the propensity to invade. The evolutionary history in the native range would dictate the types of habitats that populations could colonize without requiring niche evolution, while simultaneously influencing the capacity of these populations to extend their ranges into novel habitats.

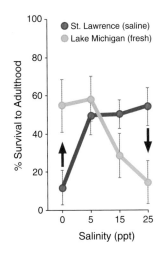

FIGURE 2 Evolutionary shifts in salinity tolerance following invasions. Common-garden experiments revealed heritable shifts in salinity tolerance between saline ancestral (St. Lawrence) and freshwater invading (Lake Michigan) populations (Fig. 1, Atlantic clade). Graph shows percentage (%) survival to adulthood ± SE for eight full-sib clutches split across four salinities.

EVOLUTION DURING INITIAL STAGES OF INVASIONS

When a population is confronted with a sudden introduction into a novel range, a rapid evolutionary response is often essential for survival. Empirical studies are showing that evolutionary changes during the process of invasions can be extremely rapid. Transport to a new location itself could select for survival during transit, while range expansion in the invaded range could select for dispersal capacity. Challenges imposed by biotic and abiotic factors in the novel range could induce rapid phenotypic evolution to enable survival in new niches in the invaded range.

A number of empirical studies demonstrate rapid phenotypic evolution in response to acute environmental shifts during invasions. For example, many species have invaded the North American Great Lakes from more saline habitats. In the case of the copepod *Eurytemora affinis* (Fig. 1), invasions from estuaries into the Great Lakes and other freshwater habitats have been accompanied by increases in freshwater tolerance and evolutionary shifts in ion regulatory functions (Fig. 2). In another case, the Eurasian barnyard grass *Echinochloa crus-galli* (Poaceae) recently extended its range into colder climates (Quebec, Canada) from southern regions of North America. This grass has a C_4 photosynthetic system, which confines this and many other species to warmer geographical regions. Relative to the southern population, the northern population has elevated specific activity for several oxygen-scavenging enzymes and greater catalytic efficiency for one enzyme tested. Also, plants often show latitudinal clines in flowering time to modulate the length of their growing season. For instance, evolution of reproductive timing, toward earlier flowering time, was critical for northern range expansion by the North American annual cocklebur (*Xanthium strumarium*, Asteraceae).

Adequate genetic variation within a population is necessary for adaptive evolution to occur. Given the rapid time scale of invasions, selection on standing genetic variation is likely to be more prevalent than selection from new mutations. Theoretical studies suggest that when populations move into a new niche, early stages of adaptation might rely more on selection on standing variation from the source habitat than on new mutations in the new "sink" habitat. Fluctuating environments in the native range could provide the conditions under which standing genetic variation would be available for selection in the source population (see previous section).

Invasions from multiple sources are resulting in the unprecedented mixing of genetically distinct populations, providing additional sources of genetic variation. For example, admixture among populations from multiple independent invasions of the brown Anole lizard (*Anolis sagrei*) in Florida has resulted in invasive populations with increased genetic variation. Hybridization is also occurring between invasive and native species, and between invasive and agricultural crop species, including genetically modified organisms (GMOs). Recombination among multiple invading populations or species could result in increased genetic variance, generation of novel phenotypes, transgressive segregation, and reduction in genetic load. Hybridization between different species, such as allopolyploid hybridization, leads to alterations in genomic architecture and also results in fixed heterosis in cases where populations can no longer reproduce sexually. Many of the most damaging invasive plant populations are the products of allopolyploid hybridization, and in many cases, such populations contain hybrid lines from multiple independent hybridization events. Thus, admixture or hybridization among invasive populations or species could profoundly affect the levels and types of genetic variation available for selection and impact the evolutionary responses of invading populations.

Aside from adaptive shifts, genetic drift during invasions could alter the nature of genetic variation within invading populations. For instance, reduction in population size during invasions could result in reduced allelic diversity and shifts in allele frequencies in the invading relative to source populations (i.e., a population bottleneck). The magnitude of this effect would depend on the size of the founding population, the extent and duration of the population bottleneck, the rate of population expansion following introduction, and the levels of ongoing migration from other populations. Literature surveys reveal a relatively modest loss of genetic variation following invasions. Recent surveys of animal, plant, and fungal taxa found a mean loss of heterozygosity of only around 18 percent. Rapid expansion in population size would mitigate effects of a population bottleneck and loss of genetic diversity. Conversely, loss of genetic diversity might potentially be more severe for populations that experience a long lag period prior to population expansion (see next section) as a result of a protracted population bottleneck.

Genetic drift acting on functional loci could also influence invasive success. For example, loss of genetic diversity in invading populations of ants could affect self-recognition systems, with profound consequences for invasive success. Genetic variation at loci affecting conspecific or colony mate recognition influences colony formation and size in many species of ants, and such colony structure affects the expansion of invasive ants into novel territory. For instance, the Argentine ant *Linepithema humile* typically reacts aggressively toward conspecifics that are genetically distinct in their native ranges. However, loss of genetic diversity during invasions has resulted in reduced aggression in the invaded range, as genetically proximate individuals tend to accept one another as colony mates. This loss of genetic diversity has led to the formation of massive supercolonies that have established hegemony over native ant species throughout their invaded ranges.

Invasive populations are regarded as useful models for studying rapid evolution, as invasions are events that occur on observable ecological time scales. The study of early phases of invasions could offer valuable insights into mechanisms of incipient evolution and limits to rapid evolution in response to acute stress. Waddington was particularly interested in adaptation under conditions of stress, and his questions regarding the prevalence of genetic assimilation remain unresolved. In addition, it would be worth exploring the extent to which invasive populations undergo selection at genomic sites with epigenetic modifications, as epigenetic mechanisms could provide a genetically labile source of heritable variation, enabling a rapid evolutionary response.

EVOLUTION FOLLOWING ESTABLISHMENT

Following the initial stages of introduction, evolutionary forces that continue acting on invading populations would influence the rate and extent of their spread and impact their long-term persistence. In many cases, continued selection might be necessary following establishment, as traits that are beneficial during early stages of colonization would often differ from those that favor continued range expansion or persistence in the long term. Also, compensatory evolution might occur in response

to initial evolutionary changes. On the other hand, evolutionary changes might occur simply as a byproduct of relaxed selection due to the absence of selective forces present in the native range. In addition, genetic correlations between traits could shift under new environmental conditions, in some cases reflecting changes in evolutionary constraints and tradeoffs.

Selective forces that act on invading populations are likely to change over progressive stages of invasions. Traits that confer an advantage during initial stages of an invasion would tend to be those that favor rapid individual and population growth. In such cases, selfing, asexual reproduction, and allocation of resources toward growth and reproduction would be favored. In contrast, once a population becomes established, selection might favor traits that are beneficial for responding to biotic or abiotic stressors in the novel environment. In such cases, asexual reproduction or selfing might become less favored, as recombination would increase genetic variance and enhance the efficiency of selection.

Several studies have documented the formation of clines following invasions, corresponding to environmental gradients within the invaded range. These clines appear to have arisen from selection on standing genetic variation in the invaded range. For several species of *Drosophila*, chromosomal inversions are associated with temperature adaptation. Latitudinal clines in chromosomal inversion frequencies have been recapitulated during independent colonizations to resemble those of the native range. A larger than expected number of genes showing differential expression in response to temperature and candidate genes for thermal tolerance were found localized within the inverted segments. For *Drosophila subobscura*, a cline of increased wing length with latitude formed in North America following introduction, replicating the pattern present in ancestral Europe. The cline in wing length that was not apparent after one decade appeared within two decades after introduction. Interestingly, these changes were achieved by disparate means for the different clines, through lengthening of different portions of the wings. In response to a thermal gradient in the invaded range, populations of the European wild rabbit *Oryctolagus cuniculus*, introduced into Australia in 1859, evolved leaner bodies and longer ears in the warmer climate. For the invasive plant St. John's wort (*Hypericum perforatum*), clines in the introduced range in North America tended to converge with those found in native Europe. The native and introduced populations show parallel latitudinal clinal variation in plant size, leaf area, and fecundity.

The need for adequate genetic variation might account for the lengthy lag period that often precedes the onset of population and range expansion, following the initial stages of invasions. The 150–170 year lag period that has been observed in some cases might result from the need for multiple invasions to facilitate a sufficient evolutionary response. Multiple or repeated invasions might be required to accumulate sufficient additive genetic variance upon which selection could act.

Rapid evolution following invasions could result from relaxed selection in the invaded range, due to release from a selective force that was important in the native range. A commonly invoked mechanism for invasive success of plants is the enemy release hypothesis (ERH), which postulates that invasive species become more vigorous in their invaded ranges due to escape from natural enemies from their native ranges (i.e., pathogens, parasites, and predators). Of the ERH hypotheses, the evolution of increased competitive ability (EICA) hypothesis, proposed by Blossey and Nötzold in 1995, has been particularly influential. According to the EICA hypothesis, there might be an evolutionary tradeoff between resource allocation toward defense against enemies and allocation toward fitness-related traits, such as increased growth or reproduction. Given such a tradeoff, relaxed selection for enemy defense in the introduced range should allow selection to favor increased growth or reproduction (termed "competitive traits"). A prediction of this hypothesis is that, when reared under identical conditions, introduced populations should display greater biomass or fitness than the source populations and should simultaneously show reduced resistance against specialist herbivores.

Support for the enemy release hypothesis, based on field comparisons and laboratory experiments, has been mixed. In general, introduced species do appear to lose their natural enemies at a biogeographical scale. Field comparisons generally show increased vigor and reduced herbivory in introduced plant populations, while some common-garden experiments appear to show rapid evolution of reproductive traits, consistent with the EICA hypothesis.

However, studies testing enemy release hypotheses have mostly been correlative, and few studies have rigorously tested whether a tradeoff actually exists between defense and fitness or growth, and whether enemy release is directly responsible for the phenotypic shifts observed. Other factors might account for phenotypic shifts in plant traits in the invaded range. For example, a meta-analysis of common-garden studies of many plant species found that differences in plant size between the native and invaded range depended heavily on latitudinal differences between the native and invading populations. Thus, observed differences in plant size between native and introduced populations might have resulted from clinal variation in traits (in both the native and invaded ranges) rather than from enemy release following invasions. Additionally, as many of the comparative studies did not include analyses of genetic structure, it is unclear whether they examined closely related populations from the native and invaded range, rather than highly divergent populations or those of different ploidy. A strong inference approach is required, where competing hypotheses are tested and all plausible mechanisms for phenotypic shifts are addressed.

Environmental shifts during invasions could also alter the structure of genetic correlations between traits, as genetic correlations are environment specific. Such effects are relevant for trait evolution in invasive populations, because genetic correlations dictate how selection on a particular trait would affect the evolution of other traits. For instance, negative genetic correlations between fitness-related traits might often be associated with fitness tradeoffs, such that shifts in genetic correlations with environmental change might be accompanied by altered fitness tradeoffs. A shift from resource-rich to resource-poor environments might shift genetic correlations between traits, as genetic correlations among traits might often be positive when resources are abundant, but negative when they are scarce. Such alterations in genetic correlations with environmental change would make evolutionary trajectories difficult to predict for populations undergoing environmental change.

In terms of long-term trajectories, do some invasive populations fail in the long run relative to others? It would be informative to determine how adaptations important for initial success differ from those that promote population persistence, and whether there are tradeoffs between early versus long-term success. It would be also be useful to investigate factors that are responsible for the observed lag periods that precede rapid growth and large-scale spread of invasive populations. Finally, knowledge on changes in genetic architecture of invasive populations over extended periods of time would provide key insights into evolutionary factors that affect long-term persistence.

FUTURE DIRECTIONS

Considerable progress has been made in reconstructing geographic pathways of invasions, characterizing patterns of phenotypic evolution, and documenting levels and effects of hybridization among invasive and between invasive and native populations. Some studies have also shed light on specific genetic changes that occur during adaptation in the novel range. However, the evolutionary mechanisms that make some populations or lineages more successful as invaders remain poorly understood. What evolutionary forces shape the intrinsic properties of invasive populations? The role of evolutionary history and selective forces in the native range has been hypothesized but inadequately tested.

SEE ALSO THE FOLLOWING ARTICLES

Disturbance / Enemy Release Hypothesis / Evolutionary Response, of Natives to Invaders / Genotypes, Invasive / Hybridization and Introgression / Lag Times / Ponto-Caspian: Invasions / Range Modeling / Taxonomic Patterns

FURTHER READING

Baker, H. G., and G. L. Stebbins. 1965. *The Genetics of Colonizing Species.* New York: Academic Press.

Blossey, B., and R. Notzold. 1995. Evolution of increased competitive ability in invasive nonindigenous plants: A hypothesis. *Journal of Ecology* 83: 887–889.

Colautti, R. I., J. L. Maron, and S. C. H. Barrett. 2009. Common garden comparisons of native and introduced plant populations: Latitudinal clines can obscure evolutionary inferences. *Evolutionary Applications* 2: 187–199.

Di Castri, F. 1989. History of biological invasions with special emphasis on the Old World (1–30). In J. A. Drake, H. A. Mooney, and F. Di Castri, et al., eds. *Biological Invasions: A Global Perspective.* Chichester, UK: John Wiley.

Dlugosch, K. M., and I. M. Parker. 2008. Founding events in species invasions: Genetic variation, adaptive evolution, and the role of multiple introductions. *Molecular Ecology* 17: 431–449.

Ellstrand, N. C., and K. A. Schierenbeck. 2000. Hybridization as a stimulus for the evolution of invasiveness in plants? *Proceedings of the National Academy of Sciences USA* 97: 7043–7050.

Havel, J. E., C. E. Lee, and M. J. Vander Zanden. 2005. Do reservoirs facilitate passive invasions into landscapes? *Bioscience* 55: 518–525.

Hoffmann, A. A., and L. H. Rieseberg. 2008. Revisiting the impact of inversions in evolution: From population genetic markers to drivers of adaptive shifts and speciation? *Annual Review of Ecology and Systematics* 39: 21–42.

Holt, R. D., M. Barfield, and R. Gomulkiewicz. 2005. Theories of niche conservatism and evolution: Could exotic species be potential tests? (259–290). In D. F. Sax, J. J. Stachowicz, and S. D. Gaines, eds. *Species Invasions: Insights into Ecology, Evolution and Biogeography.* Sunderland, MA: Sinauer Associates.

Huey, R. B., G. W. Gilchrist, and A. P. Hendry. 2005. Using invasive species to study evolution: Case studies with *Drosophila* and Salmon (139–164). In D. F. Sax, J. J. Stachowicz, and S. D. Gaines, eds. *Species Invasions: Insights to Ecology, Evolution and Biogeography.* Sunderland, MA: Sinauer Associates.

Lee, C. E. 2002. Evolutionary genetics of invasive species. *Trends in Ecology and Evolution* 17: 386–391.

Lee, C. E., and G. W. Gelembiuk. 2008. Evolutionary origins of invasive populations. *Evolutionary Applications* 1: 427–448.

Sgrò, C. M., and A. A. Hoffmann. 2004. Genetic correlations, tradeoffs and environmental variation. *Heredity* 93: 241–248.

Suarez, A. V., and N. D. Tsutsui. 2008. The evolutionary consequences of biological invasions. *Molecular Ecology* 17: 351–360.

Waddington, C. H. 1953. Genetic assimilation of an acquired character. *Evolution* 7: 118–126.

Wares, J. P., A. R. Hughes, and R. K. Grosberg. 2005. Mechanisms that drive evolutionary change: Insights from species introductions and invasions (229–257). In D. F. Sax, J. J. Stachowicz, and S. D. Gaines, eds. *Species Invasions: Insights into Ecology, Evolution and Biogeography.* Sunderland, MA: Sinauer Associates.

F

FIRE REGIMES

ROBERT C. KLINGER

U.S. Geological Survey–BRD, Bishop, California

A fire regime is the prevailing frequency, intensity, seasonality, and spatial extent (i.e., pattern) of fire in an area. Interactions between invasive plants and fire regimes have been recognized for at least 80 years, but interest in these relationships has increased dramatically over the last two to three decades, owing in large part to awareness of the worldwide scope of the situation. Strong interactions between fire regimes and invasive plants have been reported in deserts, temperate grasslands, tropical savannas and woodlands, and humid tropical forests. Regions with Mediterranean-type climates, high levels of biodiversity, and "fire-prone" ecosystems are of particular concern. These include fynbos and karoo in South Africa, heathlands and savannas in Australia, and chaparral in California. There is also great concern for the mattoral in Chile and tropical ecosystems in Hawaii and Australia, regions also rich in biodiversity but where, until recently, fire was a relatively infrequent event.

THE MAIN ISSUES REGARDING FIRE AND INVASIVE PLANTS

Studies on the relationship between fire regimes and invasive plants generally focus on three fundamental issues: (1) the degree to which invasions alter fuel characteristics and thereby alter fire behavior and fire regimes; (2) the degree to which invasions are facilitated by fire; and (3) the use of fire as a tool to manage nonnative species. Given the scope of the problem and the number of species involved, it is clear these issues have great practical importance. Not only do they offer insight on likely postfire succession patterns in areas where invasive plant species occur, but they can also provide guidance to land management agencies on the appropriate use of planned fires, whether for control of invasive species or, more generally, for manipulation of fire regimes to meet management goals. Moreover, they have helped improve our understanding of basic aspects of biological invasions, especially with regard to how invasions can influence community stability and pathways of succession leading to alternative ecological states.

ALTERATION OF FIRE BEHAVIOR AND FIRE REGIMES BY INVASIVE PLANTS

Single fires are events that can be characterized by fire behavior attributes such as rate of spread, flame length, and intensity. But when examining fire over longer periods and larger scales, the concept of the fire regime becomes a critical way to describe fire attributes. Fire behavior is influenced by local factors such as weather, topography, and fuel characteristics, while fire regimes are influenced by regional factors such as climate, landscape configuration, and emergent properties of previous fire events (for example, secondary succession patterns). Whereas fire behavior describes an individual fire event, the regime describes the historical pattern of many fire events in terms of their timing, frequency, intensity, severity, extent, patchiness, and fuel structure. It is a statistical concept where the attributes of the regime are typically expressed in terms of some kind of average (for example, a mean fire return interval of 30 years or a fire size distribution with a mode of 1,000 ha) and, critically, their variability.

Fire is a ubiquitous and natural disturbance in many ecosystems. Because fire is so pervasive, it is not surprising that many species have traits that allow them to persist or even thrive after a fire. Examples of these traits include

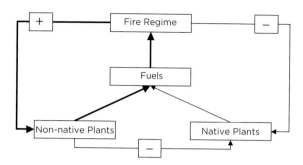

FIGURE 1 The generalized invasive plant–fire regime cycle, the feedback system through which invasive plants transform vegetation communities, fuel characteristics, and fire regimes. (Modified from Brooks et al., 2004, and Brooks, 2008.)

thick bark, resprouting, prolific seeding, serotiny, and long-lived seeds. The benefits of these traits tend to be constrained to a particular timing and pattern of burning, however. In other words, if a fire regime is moved outside of its historic range of variability, then rates of survival and reproduction of native species, which are adapted to the historic regime, may be vastly reduced.

Nonnative plants usually alter fire behavior and fire regimes through changes in fuel structure and continuity (Fig. 1). Changes in fire behavior can reduce survival rates of mature plants and seeds in the soil seed bank. If alterations to fuel structure and continuity persist, then the fire regime begins to change. By altering fire behavior and fire regimes, invasive species ultimately transform fundamental structural and functional components of entire ecosystems. Vegetation communities that once had relatively predictable postfire succession pathways are transformed to alternative states that bear little resemblance to historical ones. In ecosystems where burning was a recurrent natural event (for example, shrublands in Australia and South Africa, chaparral in the western United States, pine savannas in the southeastern United States) it typically takes a series of fires before major alterations to vegetation communities occur. However, transformations can be rapid in ecosystems where fire was historically infrequent or of low intensity (for example humid tropical forests).

The most widely reported process of how invasive species alter fire regimes is the "grass-fire cycle." This process is initiated when disturbance or land use practices lead to the introduction and establishment of invasive grasses in woody-dominated communities. As the invasive grasses increase in abundance, they form a layer of highly combustible fine fuel that previously did not occur in the community. These fuels have a much higher probability of igniting and promoting fire spread than does the woody vegetation, and so this change alters both fire behavior and fire regimes. Fire behavior is characterized by different

intensities and burn patterns than the historic norm, but the precise effect on the fire regime depends on whether the grasses are annual or perennial. When annual grasses drive the cycle, a positive feedback system is established between the grasses, short fire return intervals, and lower fire intensities. When perennial grasses drive the cycle, the feedback is between the grasses, short return intervals, and higher fire intensities. In either case, fire return intervals decrease due to the higher probability of ignition and greater postfire resiliency of grasses compared to woody species. Thus, landscapes once composed of woody-dominated communities are transformed to nonnative dominated grasslands.

With the annual grass-fire cycle, native woody species typically persist after initial fire events by resprouting or by experiencing high rates of establishment from the seed bank. But the grasses also persist and can fuel subsequent fires before the woody canopy develops enough to shade out the invaders. Re-sprouts and seedlings of the woody plants are vulnerable to fire, and so, as a result of the shortened fire return interval, their mortality rates in subsequent fire events are very high. Grasses become even more widespread and abundant, which results in fires of lower intensity. Survivorship of grass in the seed bank is greater in lower intensity burns, so the grasses have a positive response to the frequent and increasingly lower intensity fires. Thus, the layer of fine fuels is reestablished and becomes self-perpetuating (Fig. 2).

Perennial grasses also produce a dense layer of relatively continuous fine fuels, and thus increase the probability of ignition and fire frequency following invasion in woody-dominated ecosystems. However, with the perennial grass-fire cycle, fire intensity increases as well. In contrast with the annual driven cycle, the mortality of native woody species can be relatively high after an initial fire event. This allows the grasses to increase and the fire regime to shift from one of patchy, low- to moderate-intensity, infrequent fires to one of extensive, high-intensity, frequent fires.

The annual grass-fire cycle has been especially prevalent in the western United States. In the Great Basin, cheatgrass (*Bromus tectorum*), a highly invasive species from the steppes of Europe, has reduced fire return intervals in some communities from over a hundred years to less than a decade. This has resulted in the transformation of communities once dominated by perennial shrub and forb species into annual-dominated communities. Alteration of the vegetation communities has had higher-order effects as well, such as trophic cascades among vertebrate communities. The Mojave bioregion has begun to experience a similar transformation to that in the Great Basin, especially in lower-elevation communities. The species

FIGURE 2 In the Great Basin of North America, the annual grass-fire cycle has transformed communities dominated by sagebrush (*Artemisia tridentata*; A) into communities dominated by cheatgrass (*Bromus tectorum*; B). (Photographs courtesy of Steven Ostoja, U.S. Geological Survey.)

primarily responsible for altered fire regimes in the Mojave have been red brome (*Bromus rubens*) and Mediterranean grass (*Schismus arabicus* and *S. barbatus*). Annual grasses have also had a major transforming effect on fire regimes in chaparral and coastal scrub communities in southern California. Shrub density is high in intact stands of chaparral and coastal scrub, and the relatively continuous horizontal and vertical arrangement of vegetation creates low levels of light, making shade-intolerant annual grasses uncommon in intact stands. Fragmentation from urban development and other human land use activities has broken up many intact stands, however. This has resulted in a mosaic pattern of highly disturbed patches of invasive-dominated vegetation mixed among stands of chaparral and coastal scrub. These patches provide numerous source pools of invasive grass species with high propagule pressure into burned areas. In addition, most ignitions in chaparral and coastal scrub are human caused. The combination of numerous source pools, strong propagule pressure, and high rates of human-caused ignitions has resulted in fire return intervals decreasing from 50–80 years to less than 15.

Fire has been a major factor in the transformation of chaparral and coastal scrub stands to annual-dominated herbaceous communities, and destruction of coastal scrub stands has been so extensive that they are now widely recognized to be among the most imperiled in the world.

The perennial grass-fire cycle has been documented in many parts of the world, but a large number of the cases are from tropical regions in Central and South America, Australia, and Hawaii. Perennial grasses occurred naturally in these regions, but they were intolerant of intense grazing by livestock introduced by European settlers. Clearing of forests and woodlands followed by introduction of grazing-tolerant grasses became a common practice in many parts of the tropics. These grasses eventually spread from the pastures, especially along the edges adjacent to forests and woodlands. This has resulted in the alteration of fire regimes and the transformation of many vegetation communities throughout the tropics (Fig. 3). Fire frequency has increased

FIGURE 3 An example of the perennial grass-fire cycle from a lowland wet forest in Hawaii: (A) native tree fern (*Cibotium glaucum*) in an unburned area; (B) nonnative Hilo grass (*Paspalum conjugatum*) two years after a fire. (Photographs courtesy of Hawaii Volcanoes National Park.)

in seasonally dry forests in Costa Rica following invasion by jaragua grass (*Hyparrhenia rufa*). Riparian communities in northern Australia were historically a barrier to fire, but now readily burn because of invasion by buffelgrass (*Pennisetum ciliare*). In Hawaii, fire frequency and intensity have increased in a variety of ecosystems, including coastal woodlands and submontane dry forests and woodlands. Alterations to fire behavior and fire regimes in these ecosystems have been caused by a number of nonnative grass species, including *Andropogon virginicus*, *Hyparrhenia rufa*, *Melinis minutiflora*, *Pennisetum setaceum*, and *Schizachyrium condensatum*. As in southern California, human development and land use activities in Hawaii have contributed greatly to the increased levels of invasion and ignition. Some communities in Hawaii have been so drastically transformed by fire that management efforts are focused on creating novel communities with fire-tolerant natives rather than trying to restore communities with the original assemblage of fire-intolerant native species. Perennial grasses have altered fire regimes in many ecosystems in Australia, including woodlands, mesic forests, and savannas. A clear example of the development of the early stages of the cycle is in tropical savannas in northern Australia, where *Andropogon gayanus* was introduced into pasture areas as forage in the 1930s. It has since spread into savannas, where it forms dense stands that are 1–2 m taller than native understory species. This has resulted in a fourfold increase in fuel loads and a fire intensity eight times greater than that in uninvaded savanna.

Although the grass-fire cycle is the most widely reported culprit, invasive woody plants can also alter fire regimes. *Melaleuca quinquenervia*, a fire-adapted tree from Australia, has invaded a range of fireprone ecosystems in the southeastern United States. Fire regimes in invaded savanna and pine forest ecosystems have shifted from frequent, low-severity surface burns to less frequent, high-severity crown fires. In the same region of the United States, the climbing fern *Lygodium microphyllum* creates ladder fuels into canopies, which increases the likelihood of canopy fires in invaded savanna and woodland ecosystems. In Chile, the European shrub *Genista monspessulana* has increased fire intensity and extent in native mixed evergreen and deciduous forests.

Documentation of invasive plants suppressing rather than promoting fire is sparse, but anecdotal reports suggest this interaction may be more common than is realized. In Hawaii, the high moisture content of *Morella (Myrica) faya* may reduce rates of ignition, spread, and intensity in grasslands it has invaded. In the northeastern United States, a number of nonnative woody species, such as Japanese barberry (*Berberis thunbergii*), privet

(*Ligustrum* spp.), kudzu (*Pueraria montana*), and Japanese honeysuckle (*Lonicera japonica*), are thought to have reduced fire frequency in some ecosystems. In the southeastern United States, Chinese tallow (*Triadica sebifera*) has invaded extensive areas of the coastal plain. In heavily invaded sites, it forms dense thickets that suppress herbaceous species growth. This reduction in biomass and continuity of surface fuels makes the system less flammable and suggests the potential to reduce fire frequency.

FACILITATION OF INVASIONS BY FIRE

Many factors influence the introduction, establishment, spread, and abundance of invasive plants in burned areas. These include the proximity of pools of invasive plant species, propagule pressure, historic and contemporary land use, resource availability, competitors, plant–animal interactions, the types and attributes of other natural disturbances, and characteristics of the species themselves.

Fundamentally, the most important factors are species pools and propagule pressure. Fire alone cannot promote invasions unless there are a pool of species and pathways of introduction adequate to allow colonization of burned areas. In South Africa, a number of species of pines (*Pinus* spp.) introduced for commercial timber production have spread from plantations into highly diverse, fireprone communities such as fynbos and karoo. Other woody species intentionally or unintentionally introduced to South Africa have also spread into fynbos and karoo, including silky hakea (*Hakea sericea*), *Acacia* spp., and *Banksia* spp. Because of their strong positive responses to fire, these have now become the dominant species in the communities they have invaded. Similar processes have occurred elsewhere. In Australia and Central and South America, many invasive grasses have spread into burned areas from pastures where they were originally introduced as forage. In southern California, urban areas and land use activities provide source pools and vectors of spread into burned areas. Ironically, fire management and suppression activities can contribute to increased levels of invasion. Firebreaks and vehicles are pathways and vectors of spread, respectively, and seeding of burned areas with nonnative species for erosion control provides a large pulse of propagules.

Fire can facilitate establishment of invasive species by releasing resources such as light and nutrients. Fire consumes living and dead plant material, thereby creating space that invasive plants may otherwise not have been able to exploit. Many nonnative species are shade intolerant and increase in abundance only after the canopy in woody-dominated communities is fragmented or destroyed and light levels increase. Nutrients such as nitrogen, potassium, and phosphorous that were bound in unavailable forms in plant tissues and the soil are released after fires. Although a proportion of these nutrients are volatilized, especially in high-intensity fires, there is usually a pulse of increased nutrient availability in the first year after fire. Increased nitrogen availability may be particularly important in promoting invasions, because many nonnative annual species can exploit it more efficiently than native annual species. A key consideration in regard to postburn resource pulses is their transient nature. Within a few years of burning, species composition may often be determined more by factors related to physiographic features (e.g., elevation, aspect) and precipitation than space, light, and nutrients. Nevertheless, the initial resource pulse may be critical in shaping how species assemble, as well as the initial succession trajectory.

The overall increased availability of resources in burned areas can be exploited by both native and nonnative plants. However, by reducing abundance of native species, fire may facilitate invasions by reducing competition intensity. A great deal of indirect evidence points to the importance of reduced competition in burned areas, but studies directly demonstrating it are surprisingly lacking. One of the challenges in conducting competition experiments in burned areas is that, because so many resources increase in availability, it becomes problematic to identify which is most limiting. It is also important to distinguish a competitive effect from a change in resource availability. For now, probably the most appropriate assessment is that reduced competition likely does have an influence on invasion into burned areas, but its extent and importance relative to other factors cannot be accurately assessed.

Plant–animal interactions can promote invasions into burned areas. The relationship between herbivory and fire has been a topic of particular interest. Livestock grazing can alter vegetation structure that acted as a barrier to invasive plants, such as in shrubland and riparian communities, as well as reduce abundance of potential native competitors. Livestock are also vectors of spread for many nonnative plants species. Grazing by livestock is considered to have been a particularly important factor in invasions into burned (and unburned) areas in the western United States, Australia, and parts of Central and South America. But while livestock grazing often promotes invasions, it can also reduce the abundance of invasive plants in burned areas if the grazing regime is of low intensity and moderate duration. Besides herbivory, seed predation

and seed dispersal are two other important plant–animal interactions. However, their influence on invasion into burned areas has been far less studied than herbivory, and at this time, it is difficult to infer their importance in this regard.

Historical and contemporary land use and management activities can have significant influence on invasion into burned areas. As mentioned above, timber plantations, pastures and rangelands, and areas with intense human occupation or use often provide source pools of invasive plants and high levels of propagule pressure into burned areas. Long periods of fire exclusion may also be an important factor in plant invasions. Accumulation of heavy fuel loads during periods of fire exclusion can result in extremely large, intense fires with high rates of both aboveground and belowground (i.e., seed bank) mortality of native species.

A challenging issue in parts of North America is the "restoration" of fire after years of fire exclusion (rather than "restoration," this should more appropriately be considered a type of management manipulation of the fire regime). Extensive changes in many ecosystems have resulted from fire exclusion, so managing fire in a regime similar to the presuppression era is often seen as critical to the "proper" functioning of these systems. However, fire is a disturbance event, and, as just noted, this could increase levels of invasion in burned areas.

Invasive plants must have similar traits as native species to persist in ecosystems with recurrent fire. In addition, those species that transform fire regimes must not only be able to respond positively to fires but also have traits that alter fuel structure. Thus, invasive species that can dominate a community and even respond positively to fire may have little influence on fire regimes. For example, yellow star-thistle (*Centaurea soltitialis*) is an annual forb that dominates many California grasslands, including burned ones (but see below). However, fire behavior and fire regimes in these grasslands are determined by the assemblage of nonnative grasses, not star-thistle. Most invasive plant species that dominate areas after fire produce abundant seeds (e.g., annual grasses, some species of pines), sprout readily from roots and rhizomes (e.g., perennial grasses), establish dense and long-lived populations in the soil seed bank (e.g., shrubs such as *Acacia cyclops* and *Genista monspessulana*), or have a combination of the above traits. Phenology can also be a critical trait. One of the principal reasons *Bromus tectorum* has altered fire regimes is that it desiccates early in the year, thereby extending the length of the fire season.

MANAGEMENT OF INVASIVE PLANTS AND FIRE

Fire has often been recommended as a tool for managing invasive species. Strategies in control efforts targeted at single species have included increasing direct mortality of adult plants or seedlings, preventing flower and seed set, destroying seeds in the inflorescence or the seed bank, and depleting carbohydrate reserves. Options for manipulating fire behavior and attributes of the fire regime are limited, so tactics on most single species management burns have focused on the timing or frequency of burning. There are examples of short-term success in controlling some species, but few of these efforts have been unambiguously successful in the long term. Attempts to control *Centaurea soltitialis* in California grasslands provide one such example. Cover of *C. soltitialis* was reduced by 91 percent following three consecutive early summer fires; however, within four years of the cessation of annual prescribed burning, *C. soltitialis* seed bank density, seedling density, and plant cover had returned to near pretreatment levels. Additionally, while burning may reduce the target species abundance, it may not control its spread. There have also been instances where the abundance of the target species has increased after a prescribed fire, or where nontarget invasive species increased following fire.

Fire has also been used to manage multiple plant species or entire assemblages of invasive plants. The basic idea of these burns is that reduction in abundance of invasive plants would increase abundance and diversity of native species. Many of these studies have reported a decrease in abundance of invasive species in the first year after the burn, but studies that monitored the response beyond the first year have consistently shown that nonnative plants will return to pretreatment levels of abundance within two to four years. Dominance by any given species often shifts, but because there are multiple invasive species in the community, the overall dominance by nonnatives persists.

The unpredictable outcomes and short-term benefits of trying to control invasive plants with fire do not necessarily imply that its use for this purpose is inappropriate. However, it does make the determination of the ecological context of burning at least as critical a step as selecting where, when, and how fire is to be applied. For example, it is unlikely that burning is appropriate in an ecosystem composed of multiple invasive species if some or all of them have shown positive responses to fire. It would also be inappropriate to burn an area where plant invasions had altered fuel characteristics to such an extent that mortality of native species would be high. Fire alone is unlikely to be a consistently effective, long-term method of controlling invasive plant species. Burning as a sole management technique is likely to be most effective as a means of maintaining areas where fire-adapted native species already comprise a significant component of the community, but even then, great care must be taken to ensure that well-intentioned management burns do not promote invasions. In communities dominated by invasive species, fire may be most effective when integrated with other management techniques such as herbicide application, grazing, or mowing, followed by seeding and planting of native species. What may be the most crucial step in the planning process is setting realistic goals and expectations. Rather than expecting to have consistent, long-term control of invasive species by burning, using fire to create short-term windows of increased native species abundance would be an extremely important and achievable step in enabling them to persist in the community.

SEE ALSO THE FOLLOWING ARTICLES

Cheatgrass / Climate Change / Disturbance / Grasses and Forbs / Invasibility, of Communities and Ecosystems / Restoration / Succession

FURTHER READING

Bond, W. J., and B. W. van Wilgen. 1996. *Fire and Plants*. London: Chapman and Hall.

Brooks, M. L. 2008. Plant invasions and fire regimes (33–46). In K. Zouhar, J. Kapler Smith, S. Sutherland, and M. L. Brooks, eds. *Wildland Fire in Ecosystems: Fire and Nonnative Invasive Plants*. RMRS-GTR-42, Vol. 6. Ogden, UT: Department of Agriculture, Forest Service, Rocky Mountain Research Station.

Brooks, M. L., C. M. D'Antonio, D. M. Richardson, J. B. Grace, J. E. Keeley, J. M. DiTomaso, R. J. Hobbs, M. Pellant, and D. A. Pyke. 2004. Effects of invasive alien plants on fire regimes. *Bioscience* 54: 677–688.

D'Antonio, C. M. 2000. Fire, plant invasions, and global changes (65–93). In H. A. Mooney and R. J. Hobbs, eds. *Invasive Species in a Changing World*. Washington, DC: Island Press.

D'Antonio, C. M., and P. M. Vitousek. 1992. Biological invasions by exotic grasses, the grass/fire cycle and global change. *Annual Review of Ecology and Systematics* 23: 63–87.

Galley, K. E. M., and T. P. Wilson, eds. 2001. Proceedings of the Invasive Species Workshop: The Role of Fire in the Control and Spread of Invasive Species. Fire Conference 2000: The First National Congress on Fire Ecology, Prevention, and Management. Tall Timber Research Station Miscellaneous Publication 11. Tallahassee, FL: Tall Timbers Research Station.

Keeley, J. E. 2006. Fire management impacts on invasive plants in the western United States. *Conservation Biology* 20: 375–384.

Keeley, J. E., J. Franklin, and C. M. D'Antonio. In press. *Fire and Invasive Plants on California Landscapes*.

Klinger, R. C., M. L. Brooks, and J. A. Randall. 2006. Fire and invasive plants (726–755). In N. Sugihara, J. W. Van Wagtendonk, J. Fites, and A. Thode, eds. *Fire in California Ecosystems*. Berkeley: University of California Press.

Zouhar, K., J. Kapler-Smith, S. Sutherland, and M. L. Brooks, eds. 2008. *Wildland Fire in Ecosystems: Fire and Nonnative Invasive Plants*. General Technical Report RMRS-GTR-42. Vol. 6. Ogden, UT: U.S. Department of Agriculture, Forest Service, Rocky Mountain Research Station.

FISHES

PETER B. MOYLE
University of California, Davis

EMILI GARCÍA-BERTHOU
University of Girona, Spain

Alien fishes are a major cause of the loss of aquatic bio-diversity, especially in freshwater. They are present in waters worldwide, dominating environments as diverse as alpine streams, tropical lakes, and the Mediterranean Sea. A relatively small number of species make up most of the invaders, but representatives of 50 fish families have been moved to new areas, either purposefully or as the byproduct of other human activities, such as trade and canal building. Predicting the success of alien fish introduction is difficult, but humans are good at matching species to a receiving environment. The spread of alien fishes is continuing at a rapid pace, resulting in increased homogenization of fish faunas worldwide. Much of this spread is preventable.

BACKGROUND

The fishes are the most diverse vertebrate group, estimated at about 30,000 species: 58 percent of these species are marine, 41 percent live in freshwater, and the remaining 1 percent are diadromous. Much of this diversity is threatened by the current human-caused extinction crisis. An estimated 20–30 percent of freshwater fishes are likely to become extinct in the near future. While the major cause of extinction is habitat change (including pollution), it is synergistic with invasions of alien fishes. A few aggressive fish species now dominate many freshwater ecosystems worldwide, especially in temperate and arid areas. In addition, the movement of native fishes over natural barriers into nearby new waters is a common practice. The contagion has also spread to marine environments. Almost 20 percent of the fish fauna of the Mediterranean Sea now consists of alien species, and abundant aliens can be found in habitats as diverse as coral reefs, estuaries, and coastal oceans. The result is the growing global homogenization of fish faunas.

Virtually any fish species can be moved by humans and successfully invade an appropriate environment, but successful introductions are known from "only" about 50 (of 515) families of fishes. The families with the most

introduced species, many with worldwide distributions in freshwater, are the Cyprinidae (carps), Salmonidae (salmon and trout), Cichlidae (tilapias and cichlids), Gobiidae (gobies), Poeciliidae (livebearers), and Centrarchidae (sunfishes). These alien species have worldwide origins: Africa (most cichlids), eastern North America (sunfishes, livebearers), eastern Asia (most gobies, many cyprinids), and Europe (salmonids, carps). Among the most widespread species are common carp (*Cyprinus carpio*), Mozambique tilapia (*Oreochromis mossambicus*), rainbow trout (*Oncorhynchus mykiss*), brown trout (*Salmo trutta*), mosquitofish (*Gambusia affinis* and *G. holbrooki*), and largemouth bass (*Micropterus salmoides*).

The spread of such species has sometimes taken heroic efforts. Common carp, for example, were first carried by the Romans from the Danube River to Rome, from which they spread with Christianity across Europe. Independently, the Chinese moved carp around China at about the same time. From Europe, live common carp were carried by sailing ship, in barrels, to the United States, perhaps as early as the 1830s. By the 1870s, carp were held in such high esteem in the United States that they were raised in ponds near the capitol in Washington, D.C., and shipped by rail to most congressional districts in the country.

Rainbow trout (Fig. 1) and brown trout, in contrast with common carp, are sensitive, coldwater species that had to be sent by ship in the nineteenth century to distant ports as fertilized eggs. This meant that fish hatcheries had to be present at both the shipping and receiving ends of the route. The small juvenile fish were then transported considerable distances in milk cans or similar containers, often by horseback, to cold mountain waters. This was

FIGURE 1 The rainbow trout (*Oncorhynchus mykiss*), native to western North America and eastern Asia, is the most widely introduced salmonid, often escaping from fish farms. (Photograph courtesy of Eric Engbretson, U.S. Fish and Wildlife Service.)

once a major activity of angling clubs, and as a result, most coldwater streams and lakes around the world support at least one species. Today, however, many introductions are made either illegally or as a result of other human activity, mostly related to trade or the movement of water.

PATHWAYS

Humans move fish to new waters by a wide variety of means, but there are two basic pathways: *purposeful introductions* and *byproduct introductions*. Purposeful introductions are made in order to accomplish some goal, such as improving fisheries, while byproduct introductions are made as the secondary result of some other human activity such as aquaculture or trade. Purposeful introductions are made for five basic reasons: fisheries, biological control, ecosystem improvement, aesthetics, and conservation.

Fisheries improvement is the single biggest reason for the establishment of alien fishes worldwide. Some introductions, such as those of various Chinese carps and African tilapias, have been made to improve local food supplies. These fishes often support indigenous fisheries (e.g., tilapia in Asian reservoirs, pejerrey [*Basilichthyes bonariensis*] and rainbow trout in Lake Titicaca, Peru). Others were introduced to support both commercial and recreational fisheries, such as Chinook salmon (*Oncorhynchus tshawytscha*) into New Zealand and Chile, and American shad (*Alosa sapidissima*) into California. But much more common has been the spread of fish for angling. Thus, the worldwide distributions of eastern North American rainbow trout, largemouth bass, sunfish (*Lepomis* spp.), and bullhead catfish (*Ameiurus* spp.) are largely the result of the enthusiasm of anglers for them. Anglers also spread favored fish within regions; thus, the European catfish (*Silurus glanis*) is much more widespread in Europe than it was historically, because of its use to supplement reservoir fisheries. Many of these introductions were once promoted by governments but are now illegal.

Biological control of mosquitoes and gnats has resulted in the spread of a number of small, tolerant species of fish, especially guppies (*Poecilia reticulata*) and mosquitofish (Fig. 2), with mixed success in controlling pests. Likewise, grass carp (*Ctenopharyngodon idella*) and redbreast tilapia (*Tilapia zilli*) have been introduced to control aquatic weeds, with mixed success. In general, the more confined the environment into which fish are introduced, the more likely some degree of control will be achieved.

Ecosystem improvement for game fishes is the underlying reason for introducing "forage fishes," especially plankton-feeding fishes (usually herrings or shad, Clupeidae, or minnows, Cyprinidae), into bodies of water. Such

FIGURE 2 Mosquitofish such as the *Gambusia holbrooki* in the picture (pregnant female on top, male on bottom), as well as its close relative *Gambusia affinis,* have been introduced worldwide for mosquito control and have extirpated many native species with a similar ecological niche. (Photograph courtesy of Carles Alcaraz.)

introductions are usually made in order to improve the growth rates and sizes of game fishes such as largemouth bass, catfish, or trout.

Aesthetics is one reason goldfish and guppies are so widely distributed around the world. While many populations are the result of escapees, others were deliberately established, usually initially in ornamental ponds, from which they eventually moved into less confined situations.

Conservation is the goal of introductions made to protect a species threatened with extinction in its native range. Such translocations are typically made within fairly short distances of the species' historic habitats. Thus, several species of pupfish (*Cyprinodon*) in the American Southwest exist mainly in spring systems or ponds into which they were introduced after their native environments became uninhabitable.

Byproduct introductions are not made in as directed a fashion as purposeful introductions, but they still involve many species of fish and deliberate human actions (or lack of actions) that result in establishment in natural environments. These introductions are the byproducts of aquaculture, the pet trade, the use of fish as bait in fisheries, trade (ballast water), and canal and aqueduct use.

Aquaculture is a growing industry worldwide, and fish constantly escape into natural waterways from aquaculture operations, often in large numbers. Thus, Atlantic salmon (*Salmo salar*) kept in net pens in estuaries "leak" into the environment wherever they are raised, including Chile and Canada. In Chile, the numbers are sufficient to support small fisheries. In subtropical areas, a variety of tropical aquarium fishes are now established in the wild, having escaped from large commercial rearing operations. Thus, African walking catfish (*Clarias batrachus*) have spread widely over Florida, after escaping from fish farm ponds. Likewise, Siberian sturgeon (*Acipenser baerî*) are now breeding in the Danube River as the result of fish escaping from aquaculture ponds.

The *pet trade* creates introductions not only from its aquaculture operations but also from pet fish that are dumped into waterways when their owners tire of them. Most such fish die, of course, but the introductions are presumably so numerous that a few manage to survive when conditions are right. Thus, swordtails (*Xiphophorus* spp.) are common in streams and ditches in tropical areas worldwide, and oriental weatherfish (*Misgurnus anguillicaudatus*) are found in scattered localities in North America, Australia, Hawaii, and elsewhere.

Bait fish are most typically local fishes that are moved short distances with leftover fish released after a day of fishing. Nevertheless, commercial species in North America, such as the fathead minnow (*Pimephales promelas*) and red shiner (*Cyprinella lutrensis*), are established in many watersheds to which they are not native, as are some European cyprinids. Bait fish introductions are often cryptic because they often involve local species but they can eventually result in species becoming widespread in nonnative watersheds.

Ballast water of large transoceanic ships is carried in special tanks holding thousands of gallons, and it contains large numbers of organisms including fish. It is moved back and forth across the oceans on a regular basis. Many of the fish that have become successfully established through this method are gobies (Gobiidae), including three Asian species of *Tridentiger* in the San Francisco estuary and the round goby (*Neogobius melanostomus)* from the Black Sea in the Laurentian Great Lakes and in Europe.

Canals and aqueducts transport fish as well as water and can carry native fishes to new basins and increase the spread of alien species. The biggest single transfer of this kind has been through the Suez Canal, resulting in 65 species of Red Sea fishes becoming established in the Mediterranean Sea. However, any canal or aqueduct that connects watersheds can result in movement of fishes with the water. Thus, the 1150 km–long California Aqueduct has enabled the colonization of southern California by at least five fish species from northern California, including the abundant alien fishes shimofuri goby (*Tridenter bifasciatus*) and Mississippi silverside (*Menidia audens*). Similarly, most large rivers in Europe are interconnected by canals, facilitating the spread of many invaders.

CHARACTERISTICS OF SUCCESSFUL ALIEN FISHES

In the right circumstances, any species of fish can be successfully introduced into new habitats, in part because humans are remarkably good at matching species with habitats that will support them. Nevertheless, the limited taxonomic diversity of widespread alien fishes suggests that certain suites of characteristics favor success. Some of the best predictors of success have been (1) previous success, (2) desirability to humans, (3) physiological tolerance, and (4) large propagule sizes (>100 individuals) for the initial introduction. Of these characteristics, only physiological tolerance (ability to withstand extremes in temperature, salinity, and dissolved oxygen) is really a characteristic of the organism, per se. One reason for the lack of a universal list of characteristics of success is that different characteristics are needed at different stages of the invasion process. Thus, fishes with some degree of parental care (e.g., tilapias, livebearers) have an advantage early in the invasion because of improved survival when numbers are small, while species with high fecundities have an advantage during the spread stage of an invasion because of their ability to produce large numbers of young, provided they live long enough to persist through unfavorable periods. Moyle and Marchetti (2006) showed that "parental care is a good predictor of the ability of a fish to become established, long life span is a good predictor of its ability to spread following establishment, and moderate size is a good predictor of its ability to become abundant." Overall, fishes most likely to succeed as invaders have intermediate or mixed characters (i.e., they are not too specialized in any of their life history or ecological traits), although there are many exceptions, of course.

For the most part, humans overcome various limitations by closely matching potential introductions with conditions that are close to those in the species' native environments. Among the most remarkable cases of environmental matching has been the introduction of anadromous fishes, those that make extensive migrations between fresh- and saltwater, to many places in the world; examples are Chinook salmon and steelhead rainbow trout to New Zealand, and Chile and striped bass (*Morone saxatilis*) to California. The shad subsequently colonized rivers along the North Pacific Rim and now have populations in rivers in the Kamchatka Peninsula in Russia. In temperate regions, the sites of the greatest number of successful introductions have generally been those where native species richness is high, because of habitat diversity and abundant water, especially where combined with extensive human alteration of the waterways. Thus, the biological characteristics of an alien fish species must fit into the characteristics of the particular aquatic environment into which it has been introduced, making generalizations about the characteristics of successful invaders of limited usefulness.

GEOGRAPHY OF ALIEN FISHES

Alien fishes are found in most zoogeographic regions of the world, although they tend to be most pervasive in freshwaters at mid-latitudes, in temperate and arid countries. But they are also abundant in other areas as well, such as on remote islands from the tropics to the south Atlantic. The streams of the island of Oahu, in the Hawaiian Islands, for example, contain a shifting assemblage of tropical aquarium fishes, including mollies and swordtails (Poeciliidae), South American catfishes, and cichlids (part of the 35 or so alien fishes recorded from the islands). Alien species are much less common in marine environments, although temperate estuaries tend to support multiple alien species. The ability of diverse marine environments to support aliens is demonstrated nicely by the presence of a few alien species in the coral reefs of Hawaii (e.g., the snapper, *Lutjanus kasmiva*) and the massive invasion of the Mediterranean Sea by Red Sea fishes.

Not surprisingly, some of the most alien-dominated waters are found in industrialized countries with Mediterranean climates, such as California, Spain, and South Africa. These countries have limited water and intensive human land use, so they contain more of the highly altered habitats favored by alien fishes such as common carp, goldfish, largemouth bass, and similar species. Curiously, high mountain lakes and streams, seemingly among the most pristine places on Earth, are often dominated by alien trout species, put into historically fishless waters by anglers and fisheries agencies. Tropical countries tend to have fewer established aliens, although the presence of guppies and swordtails in urban ditches, tilapia species in reservoirs and lakes, and African catfishes (e.g., *Clarias gariepinus*) in some rivers in South America demonstrates that there is nothing intrinsic to tropical systems that prevents alien fishes from becoming established.

IMPACTS OF INTRODUCTIONS

As the carp and trout examples demonstrate, humans are very successful vectors for fish. For many fish species, their relationship to humans is a mutualistic one: they get to colonize new waters, while we humans benefit from their presence. Often, however, the relationship is more ambiguous or one-sided. For example, the invasion of the Laurentian Great Lakes by the sea lamprey (*Petromyzon marinus*) resulted in the collapse of commercial fisheries

as the result of lamprey predation on native salmonids. The invasion of Lake Victoria in Africa by the Nile perch (*Lates nilotica*) resulted in the destruction of native cichlid populations (and the subsistence fishery that depended on them), while providing a perch fishery that brought money into some parts of the local economy. Both of these invasions resulted in extinctions of native fish species, but dramatic extinctions following successful introductions are relatively uncommon. Much more common is reduction of native fish and invertebrate populations or restrictions of their distribution to relatively unaltered habitats. This results in the *biotic homogenization* of the more disturbed waters, especially large rivers, streams below dams, or heavily polluted streams. Rivers throughout the United States, Europe, South Africa, South Korea, and other places have largemouth bass, sunfish, bullhead catfish, common carp, mosquitofish, and a few other alien species as prominent parts of their fauna, with the distinctive natives often (but not always) having been reduced to minor status.

The effects of alien species on native fishes and their ecosystems are felt through (1) predation, (2) competition, (3) diseases and parasites, (4) hybridization, and (5) ecosystem engineering.

Predation is the mechanism of most rapid and conspicuous damage, especially if the alien predator presents a novel form of predation (as indicated by being different taxonomically from native predators, genus level or higher). Thus, Nile perch and sea lamprey (see above) represented styles of predation not present in the systems they invaded. The worldwide negative impact of largemouth bass is the result of their effectiveness in preying on soft-rayed fishes (e.g., small Cyprinidae) in lake and stream environments. In New Zealand, brown trout dramatically alter the structure of stream ecosystems by preying on native galaxiid fishes and invertebrates (Fig. 3).

Competition, in contrast to predation, tends to result in the reduction of native fish populations, but rarely in extinction. Thus, the invasion of vendace (*Coregonus albula*) into Scandinavian lakes depleted zooplankton populations and forced native whitefish (*C. lavaretus*) to switch to feed mainly on benthic and terrestrial insects (but it did not result in extinction). However, competition for multiple limited resources may result in the elimination of native species by an invader. Thus, in Lake Michigan, round goby successfully compete for food, space, and breeding sites with mottled sculpin (*Cottus bairdi*), resulting in drastic declines of sculpin populations.

Disease and parasites may be the least appreciated interaction among alien and native fishes. Whirling disease

FIGURE 3 The northern snakehead (*Channa argus*) was introduced illegally into the Potomac River in the eastern United States, where it now preys on native fishes, such as the pumpkinseed sunfish shown here. It is regarded as a major threat to native fish and fisheries. (Photograph courtesy of Ryan Saylor and Nick Lapointe.)

(caused by *Myxobolus cerebralis*), for example, is native to Europe, where brown trout are resistant to it; when picked up by native rainbow trout and other salmonids in North America, it had devastating effects on populations. Likewise, a variety of parasites, including tapeworms and flukes, can be spread by alien species to populations of endangered fishes, furthering their decline. The swimbladder nematode (*Anguillicola crassus*) was introduced to Europe with Asiatic eels and has added to the multiple factors causing the decline of European eel (*Anguilla anguilla*).

Hybridization occurs when an invading fish species breeds with a closely related nonnative species, resulting in hybrids that reduce the ability of the native population to sustain itself, especially if the native species is rare. For example, hybridization of endangered sterlet (*Acipenser ruthenus*) with alien Siberian sturgeon is regarded as a major threat to its survival. Likewise, hybridization of native cutthroat trout (*Oncorhynchus clarkii*) and other native trout species with alien rainbow trout in western North America has been a major factor leading to the extirpation of the former.

Ecosystem engineering is the process by which an invasive species dramatically alters the ecosystem it invades, making it much less suitable for multiple native species. The best-known example among fishes is the common carp, which can cause huge increases of suspended material in the water column by its benthic feeding activities. In shallow lakes, this results in dramatic decreases in water clarity, disappearance of aquatic macrophyte beds, loss of zooplankton, and the decline of visual-feeding fishes.

Equally dramatic are invasions of aquatic invertebrates such as the zebra mussel (*Dreissena polymorpha*), which cause shifts in the flow of energy in lake and river systems from pelagic invertebrates and fish to the benthos. The result can be a dramatic decline in pelagic fishes.

Of course, not every successful alien fish is harmful, although lack of harm is often difficult to demonstrate. The invasion of the abundant shimofuri goby in the San Francisco estuary was intensively studied using both field and laboratory studies of ecology, physiology, and behavior during the invasion process. No effects on native fishes could be detected. It is thought that the reason for this was that the goby was feeding largely on alien invertebrates not being used by any of the native fishes and was living in a highly disturbed, highly variable system in which physical and chemical change was overwhelming strongest biotic interactions.

For most alien invasion with no apparent negative effects, however, studies are lacking, and so problems are ignored. Thus, revelations that the Nile perch was causing cichlid extinctions came years after the introduction, with negative effects on local human nutrition and the local terrestrial environment. This is a reflection of the *law of unintended consequences* for species introductions (a.k.a. the *Frankenstein effect*): many, if not most, successful introductions will have unpredicted, often subtle, negative effects on the invaded ecosystems. These effects are usually unanticipated, reflecting our still-poor abilities to make predictions about invasion impacts.

MANAGEMENT

Given the poor record of predicting negative effects of fish introductions, the best strategy is to prevent more byproduct and purposeful introductions. Byproduct introductions need active management of sources to be prevented, including the creation of regulations to greatly restrict the ability of the pet and aquaculture industries to move live fish around. For pet and similar industries, each ecological region should have a list of species allowed to be legally sold—species that can be shown to be unable to survive if they escape or are released into the wild. The burden of proof would then be on the industry to prove that a new species would not be harmful should it escape. A mechanism to make such judgments has been developed through *risk analysis*, a systematic procedure that is fairly well developed for aquaculture and transgenic fishes. Risk analysis should be used for *any* proposed introduction or movement of fish to a new area. The costs of risk analyses and management of unintended introductions should be internalized by these businesses.

Once a species has become established, if its distribution is still fairly localized, then eradication measures may work, especially through the use of fish poisons such as rotenone and antimycin, although intensive use of nets and other gear can also work in some situations. An alternative is to dry up the habitat if the undesirable species is confined to a reservoir, pond, or similar area.

Once an invasive species has become widespread, local control and management become the main options. Techniques can vary widely, but in regulated rivers, restoring a more natural flow regime can often discourage undesirable aliens while favoring native fishes. Subsidizing fisheries (e.g., for Asian carps) also has the potential to at least reduce the environmental impacts of invasive fishes.

SEE ALSO THE FOLLOWING ARTICLES

Aquaculture / Canals / Carp, Common / Great Lakes: Invasions / Mediterranean Sea: Invasions / Nile Perch / Pet Trade / Rivers / Sea Lamprey

FURTHER READING

Baltz, D.M. 1991. Introduced fishes in marine systems and inland seas. *Biological Conservation* 56: 151–171.

García-Berthou, E., C. Alcaraz, Q. Pou-Rovira, L. Zamora, G. Coenders, and C. Feo. 2005. Introduction pathways and establishment rates of invasive aquatic species in Europe. *Canadian Journal of Fisheries and Aquatic Sciences* 62: 453–463.

Kapuscinski, A.R., K.R. Hayes, S. Li, and G. Dana. 2007. *Environmental Risk Assessment of Genetically Modified Organisms: Methodologies for Transgenic Fish*. Oxfordshire, UK: CABI.

Kolar, C.S., and D.M. Lodge. 2002. Ecological predictions and risk assessment for alien fishes in North America. *Science* 298: 1233–1236.

Lasram, F.B.R., and D. Mouillot. 2009. Increasing southern invasion enhances congruence between endemic and exotic Mediterranean fish fauna. *Biological Invasions* 11: 697–711.

Lever, C. 1996. *Naturalized Fishes of the World*. New York: Academic Press.

Lodge, D.M., S.L. Williams, H.J. MacIssaac, K.R. Hayes, B. Leung, S.H. Reichard, R.N. Mack, P.B. Moyle, M. Smith, D.A. Andow, J.T. Carlton, and A. McMichael. 2006. Biological invasions: Recommendations for U.S. policy and management. *Ecological Applications* 16: 2035–2054.

Moyle, P.B., and M.P. Marchetti. 2006. Predicting invasion success: Freshwater fishes in California as a model. *Bioscience* 56: 515–524.

Rahel, F.J. 2002. Homogenization of freshwater faunas. *Annual Review of Ecology and Systematics* 33: 291–315.

FLAVIVIRUSES

JOHN S. MACKENZIE AND DAVID T. WILLIAMS

Curtin University of Technology, Perth, Western Australia, Australia

Flaviviruses are a group of arthropod-borne viruses that have demonstrated a propensity to spread and establish in new geographic areas. This invasiveness of flaviviruses is due to the complex interplay of a number of different factors, some of which are properties of the viruses and their vectors, and others that are the consequences of human activities. The viral factors include the ability of some members of the genus to use a wide spectrum of vector mosquito species and vertebrate host reservoirs, especially when establishing in new ecosystems, their ability to survive and replicate in vectors and hosts under a variety of environmental conditions, and the ability of some members to be transported between regions and continents through migratory bird hosts. Human activities are a major component of the interplay and include the dispersal and spread of the viruses in infected mosquitoes or infected passengers on ships and planes, by commercial transport of animals and goods, and by providing new mosquito breeding sites through changes in land use such as in newly developed rice paddies or areas of irrigated agriculture, or in urban shantytowns. Also important has been the spread of major mosquito vector species and their subsequent colonization of new areas, as seen by the widespread infestation of *Aedes albopictus* on all continents, frequently through the international trade in old automobile tires, and of *Ae. aegypti* mosquitoes, which may move between regions and continents in containers on cargo ships or in aircraft holds. The latter is perhaps the most important vector mosquito species, and its international spread has probably been the single most influential factor in the spread of yellow fever and the dengue viruses.

THE BIOLOGY OF FLAVIVIRUSES

Flaviviruses are small-enveloped, positive, single-strand RNA viruses that comprise one of the three genera within the family *Flaviviridae*. There are approximately 55 species in the genus *Flavivirus* and 70 recognized virus strains. Most members are transmitted by arthropod vectors, either mosquitoes or ticks, although a few have no known vector. The members of the genus and the genetic relationships are depicted in Figure 1. The type-species of the genus is yellow fever virus, but the genus also contains a number of important human and animal pathogens, including dengue viruses, Japanese encephalitis, West Nile, St. Louis encephalitis, and tickborne encephalitis viruses. The other two genera in the family Flaviviridae are *Pestivirus* and *Hepacivirus*, with the former containing species such as classical swine fever virus (or hog cholera virus) and bovine viral diarrhea virus, and the latter containing a single species, hepatitis C virus. The members of these two genera, however, do not require or utilize arthropod vectors for their transmission.

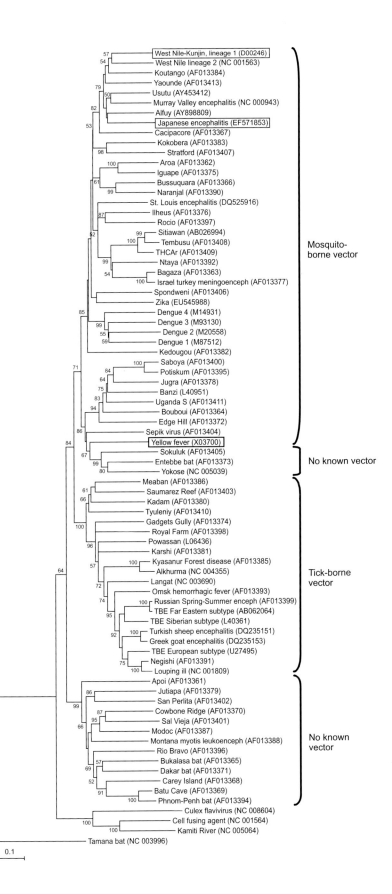

FIGURE 1 A phylogenetic tree of the *Flavivirus* genus based on genetic sequences from the NS5 gene, showing all major members of the genus except for Wesselsbron virus, a virus causing livestock disease and occasional human infections in Africa for which sequences in the NS5 gene were unavailable. Major vector clades are indicated. The three viruses described in the text, the West Nile, Japanese encephalitis, and yellow fever viruses, are shown boxed. The scale bar represents 0.1 nucleotide substitutions per site. Genbank accession numbers for the genomic sequences used in this analysis are bracketed next to the strain name. TBE = tickborne encephalitis. (This figure is based on the phylogenetic tree from Mackenzie, J.S., and D.T. Williams. 2009. Flaviviruses of southern, south-eastern and eastern Asia, and Australasia: The potential for emergent viruses. *Zoonoses and Public Health* 56: 338–356.)

Most members of the *Flavivirus* genus exist in natural transmission cycles between a vertebrate host and an arthropod vector. The viruses can replicate in the cells of both host and vector and are transmitted to new hosts through the "bite" of the arthropod vector. Recent examples of the invasiveness of flaviviruses have been mosquito-borne viruses, three of which are described in more detail below. The female mosquito becomes infected with a flavivirus when it takes a blood meal from an infected vertebrate host, a requirement in its life cycle to enable it to lay eggs. The virus then replicates in the cells of the mosquito, eventually reaching the cells of the salivary glands. The virus is then transmitted to the next vertebrate host in the saliva of the female mosquito when she bites the next host to acquire a further blood meal. There are two major groups of mosquito-borne flaviviruses, those with mammals as their primary vertebrate hosts and using *Aedes* spp. mosquitoes as their major vectors, and those with birds as their main vertebrate hosts and predominantly using *Culex* spp. mosquitoes as their vectors. The former include major pathogens such as yellow fever and the dengue viruses, both of which have jungle or forest mosquito–monkey transmission cycles, but which have also evolved human–mosquito urban transmission cycles, whereas the latter include Japanese encephalitis and West Nile viruses, two major encephalitic viruses, which use a variety of avian species. These viruses clearly demonstrate the invasive potential of flaviviruses and their ability to establish subsequently in new areas using local mosquito species and vertebrate hosts, but they also demonstrate the different ways in which they invade new geographic niches. As examples, yellow fever, West Nile, and Japanese encephalitis are described in more detail below.

EXAMPLES OF THE SPREAD OF MOSQUITO-BORNE FLAVIVIRUSES

Yellow Fever Virus: The Type Species of the Genus *Flavivirus*

Historically, the most important member of the genus *Flavivirus* has been yellow fever virus, which is believed to have been spread on sailing vessels from its original home in the forest canopies of West Africa to the Americas by the slave trade during the period between the late sixteenth century and the middle of the nineteenth century. It is hypothesized that the initial events may have been the transport of the anthropophilic mosquito vector, the *Ae. aegypti* mosquito, to the New World on sailing ships, where it probably rapidly became established. Subsequently, it is believed that the virus was taken to the New World through mosquito–human transmission cycles on the sailing ships, with mosquitoes breeding in the freshwater barrels and transmitting the virus between the sailors and slaves on board. Many explosive outbreaks of yellow fever, a hemorrhagic illness with significant mortality, were recorded when these vessels reached ports in the Americas and Europe. Thus, yellow fever has long demonstrated the propensity to spread and colonize new geographic areas, which has been a feature of other members of the genus. The natural sylvatic transmission cycle of the virus is in the forest canopy between monkeys as vertebrate hosts and *Aedes* spp. mosquitoes, usually referred to as jungle yellow fever, but the virus also occurs in urban mosquito–human transmission cycles, with *Ae. aegypti* being the vector mosquito species. These urban transmission cycles can result in explosive outbreaks with high mortality, especially in Africa, where annual cases can approach 200,000 with as many as 30,000 deaths. In South and Central America, the sylvatic cycle is maintained between *Haemagogus* spp. and *Sabethes* spp. mosquitoes and monkeys, but unlike the transmission cycles in Africa, American monkeys often die from the infection, probably because there has been insufficient time for adaptation to the virus since its relatively recent introduction. There has been little or no urban yellow fever for many decades in South and Central America, and almost all recent cases have been sporadic infections in jungle settings, but with the widespread infestation of *Ae. aegypti* mosquitoes in urban areas and around zones of human habitation throughout much of Central and South America north of latitude 35 °S, there is a continuing threat of a major urban yellow fever outbreak in the region, which could have catastrophic consequences. Thus, the invasiveness of yellow fever virus has been due largely to a dependence on human activities for the virus to move between continents and to the spread of *Ae. aegypti* mosquitoes, providing a very efficient vector for urban transmission. An excellent vaccine is readily available, the 17D live virus vaccine, which is relatively inexpensive, and there is continuing discussion about whether it should be included in the expanded program of immunization of affected countries.

West Nile Virus

West Nile virus, first isolated in Uganda in 1937, has long been a pathogen of humans and animals (some domestic bird species and horses) in Africa, parts of Europe, the Middle East, India, and Australia (where it was called Kunjin virus). It was responsible for outbreaks of relatively mild, febrile disease but with occasional cases of severe and sometimes fatal neurological disease. However, from 1994, it began to cause outbreaks of severe neuroinvasive disease in humans and horses in Europe, especially in Romania and eastern Russia, and in the Mediterranean basin including Israel, Italy, Morocco, and Algeria. In 1997 and 1998, West Nile virus was reported for the first time as the cause of a neurological illness and death in geese in Israel. Phylogenetic studies of a range of isolates has suggested that there are frequent reintroductions of virus from Africa into Europe, probably by way of the Middle East, through the movement of migratory birds such as white storks (*Ciconia ciconia*). West Nile virus was detected in the New World for the first time in New York in 1999, where it caused 59 cases of encephalitis in humans, seven of whom died, and the death of a number of different exotic bird species in the Bronx Zoo. Over the following years, it dispersed throughout the United States and contiguous provinces of Canada, causing significant morbidity and mortality in humans and horses, as well as an extensive and unprecedented avian mortality. It eventually became enzootic in all states except Alaska and Hawaii. Its rapid spread was due in part to the large number of avian species able to act as vertebrate hosts and to the wide range of mosquito species able to vector the virus; these conditions were probably exacerbated by the emergence of the virus in a novel ecosystem, thus providing a wealth of potential transmission cycles. West Nile virus has been responsible for over 30,000 human cases, with approximately 1,300 deaths, and for more than 27,500 equine infections since its arrival in North America. The equine infections would have been much greater if vaccines had not been developed and made available from 2003. More recently, the virus has spread southward into Central and South America and the Caribbean, but its virulence seems to have significantly diminished in these areas. How West Nile virus reached North America still remains to be determined. The most likely scenario, however, is through an infected mosquito hitching a lift on board an aircraft.

Japanese Encephalitis Virus

Japanese encephalitis virus, the most important cause of viral encephalitis in southern and eastern Asia, is believed to be responsible for over 40,000 cases of encephalitis and 10,000 deaths annually, but these figures are generally regarded as a significant underestimate, and the true figure may be closer to 175,000 cases of encephalitis annually. The natural transmission cycle of Japanese encephalitis virus alternates between marsh birds as the maintenance vertebrate hosts, particularly members of the Order *Ciconiiformes*, and Culicine mosquitoes, with domestic pigs being important amplifying hosts in both enzootic and epidemic situations. Recent phylogenetic analyses have indicated that Japanese encephalitis virus probably originated from an ancestral flavivirus in the Indonesia-Malaysia region, where it evolved into different genotypes that then spread out across Asia. Epidemics of encephalitis have been reported in Japan since the 1870s, and Japanese encephalitis virus was first isolated in 1935. Over recent decades, the spread of the virus may have been aided by deforestation and changes in land use, especially the development of rice paddy fields, which attract ardeid birds and provide conditions for culicine mosquito breeding, thus promoting the ideal ecological conditions for virus transmission. In addition, as rice paddies are usually close to villages where pigs are raised, the environment has been ideal for epidemic development. However, phylogenetic studies have indicated that virtually identical strains of virus can be isolated many miles apart, indicating that the virus is dispersed frequently and rapidly by migratory birds. The most recent expansion of Japanese encephalitis virus into new areas has been into the eastern Indonesian archipelago in the early 1980s, Papua New Guinea in the late 1980s, Pakistan in the early 1990s, and the Torres Strait of northern Australia in 1995. Interestingly, there has been one human case in mainland Australia that has been associated with windblown mosquitoes from Papua New Guinea during a cyclonic weather pattern, a meteorological feature that is predicted to increase under climate change. An inactivated cell culture–derived vaccine is now available, and a live attenuated vaccine has been used in China, and most recently in India. In addition, a chimeric vaccine based on the yellow fever 17D genome but with the Japanese encephalitis envelope protein is undergoing clinical trials.

FUTURE CONCERNS: HOW AND WHERE WILL FLAVIVIRUSES EMERGE NEXT?

These examples of the way mosquito-borne flaviviruses can spread and colonize new areas provide a number of lessons that may be useful in understanding how these and other viruses might spread in the future. They demonstrate the invasive characteristics of flaviviruses and the interplay of biological and environmental factors with human activities that potentiate and assist their spread and establishment in new geographic regions. In addition, the effects of global warming might provide the warmth and humidity in southern and central Europe for increased epidemic activity of West Nile and possibly dengue, and the conditions potentiating further spread into new regions, such as northern Europe, and for dengue and even yellow fever into North America. There is also the possibility of Japanese encephalitis virus spreading east across the Pacific into California.

SEE ALSO THE FOLLOWING ARTICLES

Mosquitoes / Pathogens, Animal / Pathogens, Human

FURTHER READING

Blitvich, B. J. 2008. Transmission dynamics and changing epidemiology of West Nile virus. *Animal Health Research Reviews* 9: 71–86.

Bryant, J. E., E. C. Holmes, and A. D. Barrett. 2007. Out of Africa: A molecular perspective on the introduction of yellow fever virus into the Americas. *PLoS Pathogens* 3(5): e75.

Gould, E. A., and T. Solomon. 2008. Pathogenic flaviviruses. *Lancet* 371: 500–509.

Gould, E. A., X. de Lambaillerie, P. M. Zanotto, and E. C. Holmes. 2001. Evolution, epidemiology, and dispersal of flaviviruses revealed by molecular phylogenies. *Advances in Virus Research* 57: 71–103.

Gubler, D. J., G. Kuno, and L. Markoff. 2007. Flaviviruses (1153–1252). In D. Knipe and P. M. Howley, eds. *Fields Virology*, Vol. 1, 5th ed. Philadelphia: Lippincott, Williams, and Wilkins.

Holmes, E. C., and S. S. Twiddy. 2003. The origin, emergence and evolutionary genetics of dengue virus. *Infection, Genetics and Evolution* 3: 19–28.

Kyle, J. L., and E. Harris. 2008. Global spread and persistence of dengue. *Annual Reviews of Microbiology* 62: 71–92.

Mackenzie, J. S., D. J. Gubler, and L. R. Petersen. 2004. Emerging flaviviruses: The spread and resurgence of dengue, Japanese encephalitis and West Nile viruses. *Nature Medicine* 10(12): S98–S109.

Wilder-Smith, A., and D. J. Gubler. 2008. Geographic expansion of dengue: The impact of international travel. *Medical Clinics of North America* 92: 1377–1390.

FORBS

SEE GRASSES AND FORBS

FOREST INSECTS

ANDREW M. LIEBHOLD

USDA Forest Service, Morgantown, West Virginia

DEBORAH G. McCULLOUGH

Michigan State University, East Lansing

Compared to other land uses, forests are relatively stable and diverse ecosystems that provide habitat for a variety of organisms. Unfortunately, forest ecosystems on virtually every continent are threatened by a variety of factors, including deforestation, climate change, and biological invasions. Among these factors, invasions by nonnative forest insects are a particularly serious problem with potentially severe ecological and economic consequences. This entry presents a brief overview of the pathways by which nonnative forest insects arrive, the impacts of invasive forest insects, and potential options for managing invaders.

BIOLOGY

Invasion by nonnative forest insects is not a new phenomenon. More than a dozen nonnative forest insects were damaging trees in the United States before 1850. However, subsequent to 1850, more than 400 species of nonnative forest insects, from virtually all insect orders and feeding groups, have become established in the United States and Canada (Table 1; Fig. 1). Among these species, sap-feeding insects such as aphids, adelgids, and scales, and foliage feeding insects such as sawflies and many species of Lepidoptera (moths and butterflies), are the most frequent invaders. Invasions by insects that feed under the bark on phloem or wood, however, have increased in recent years. This group includes bark beetles, ambrosia beetles, phloem-borers, and wood-borers. Several of these insects, such as the emerald ash borer *Agrilus planipennis*, the Asian longhorned beetle *Anoplophora glabripennis*, and the redbay ambrosia beetle *Xyleborus glabratus*, are capable of killing otherwise healthy trees and have been particularly destructive.

The biology of many forest insects is intimately intertwined with microorganisms such as fungi, and often it is necessary to consider such species complexes together in order to understand these problems fully. For example, considerable damage has occurred in pine plantations in the southern hemisphere following invasion by the woodwasp *Sirex noctilio*, along with its fungal symbiont *Amylostereum areolatum*. Female woodwasps infect trees with fungi as they lay eggs, and the woodwasp larvae feed on the fungi. The fungus invades the xylem of the tree, however, which disrupts water transport and causes the tree to die. In a very different example, beech scale *Cryptococcus fagisuga*, a sap-feeding insect native to Asia, was introduced into eastern Canada around 1890. At high densities, these tiny insects will cover the bark on the trunk and large branches of American beech trees. Feeding wounds of the scales enables native pathogenic fungi to enter and eventually kill the host tree. Over the past 120 years, beech scale has expanded across much of the range of American beech in Canada and the United States, killing millions of large beech trees and affecting the wildlife that feed on beech nuts or use beech trees for habitat.

In many parts of the world, large plantations of nonnative trees have been established to facilitate efficient

TABLE 1
Examples of Some Well-known Invasive Forest Insect Species

Species	Common Name	Native Range	Invaded Range	Host	Damage
Anoplophora glabripennis	Asian longhorned beetle	Eastern Asia	North America, Europe	*Acer* spp., *Aesculus* spp., *Populus* spp., *Betula* spp.	Dieback, breakage, mortality
Cameraria ohridella	Horse chestnut leaf miner	Southern Europe	Central and western Europe	*Aesculus hippocastanum*	Defoliation
Agrilus planipennis	Emerald ash borer	China, Korea, maybe other areas of Asia	North America, Russia (Moscow)	*Fraxinus* spp.	Dieback, mortality
Dendrocotnus valens	Red turpentine beetle	North America	China	*Pinus* spp.	Mortality
Sirex noctilio	Sirex woodwasp	Eurasia	Australasia, South America, Africa, North America	*Pinus* spp.	Mortality
Hyphantria cunea	Fall webworm	North America	Eurasia	Broadleaf trees	Defoliation
Adeleges tsugae	Hemlock woolly adelgid	East Asia, western North America	Eastern North America	*Tsuga* spp.	Defoliation, mortality
Gonipterus scutellatus	Eucalyptus weevil	Australia	Africa, North America, South America, Europe	*Eucalyptus* spp.	Defoliation
Lymantria dispar	Gypsy moth	Eurasia	North America	*Quercus* spp., *Populus* spp., *Larix* spp.	Defoliation
Xyleborus glabratus	Redbay ambrosia beetle	Asia	Southeastern United States	Redbay (*Persea borbonia*)	Mortality

FIGURE 1 Images of some important forest insect pests and the damage they cause. (A) The hemlock woolly adelgid, *Adelges tsugae*. (Photograph courtesy of USDA Forest Service, Region 8 Archive, Bugwood.org.) (B) Forest trees killed by the emerald ash borer, *Agrilus planipennis*. (Photograph courtesy of Daniel Herms, the Ohio State University, Bugwood.org.) (C) The Asian longhorned beetle, *Anoplophora glabripennis*, adult. (Photograph courtesy of Daniel Herms, the Ohio State University, Bugwood.org.) (D) Larva of the gypsy moth, *Lymantria dispar*. (Photograph courtesy of Daniel Herms, the Ohio State University, Bugwood.org.)

production of lumber and fiber. Many of these exotic plantation forestry efforts have been highly successful. However, much of the high productivity associated with growing exotic trees can be attributed to the "escape" of tree species from their associated herbivores. Unfortunately, this escape may not last forever, and in many parts of the world, herbivores are catching up with their hosts. For example, species in the genus *Eucalyptus* are native to Australia but have been widely planted elsewhere in the world. Many *Eucalyptus* plantations are highly productive, and several *Eucalyptus* species are used as landscape trees in urban forests. Recently, however, several Australian insects associated with *Eucalyptus* have invaded these regions and caused considerable damage.

INVASION PATHWAYS

The problem of forest insect invasions can be attributed largely to global trade. Many insects are inadvertently transported either on or within their host plant or because they hitchhike on commodities or items transported as cargo. Imported plants, destined for nurseries or landscapes, have historically been an important pathway by which plant-feeding insects have arrived in new habitats. Herbivorous insects that are transported with their host plant presumably have a high probability of becoming established because they need not search for hosts. Regulatory officials may require live plants to be treated or inspected either before shipping or upon arrival, to reduce the risk that nonnative insects or other organisms could become established. Despite these and other safeguards, ornamental trees for retail sale are increasingly propagated overseas, providing abundant opportunities for new introductions of potentially damaging insects.

Many insects have dormant life stages that may be accidentally transported on items that are completely unrelated to their biology. For example, thousands of used automobiles are shipped yearly from Japan to New Zealand, and these automobiles frequently arrive with gypsy moth (*Lymantria dispar*) egg masses.

Solid wood packing material has become another important pathway for invasive forest insects. Wood crating, dunnage, and pallets are often used as packing materials in cargo ships, airplanes, and trucks. Inexpensive and readily available wood is typically selected for this use. Unfortunately, such wood often contains bark- or wood-boring insects. International trade and containerized shipping has increased dramatically in the past few decades, and wood packing material now represents one of the most important pathways by which forest insects are accidentally transported to distant locations.

The International Plant Protection Convention recently enacted an international phytosanitary measure, ISPM 15, which limits the movement of raw wood. While this measure and other regulatory efforts should help, expanding global trade and travel will almost certainly result in more introductions of nonnative forest insects.

IMPACTS

Only about 15 percent of the nonnative forest insects known to be established in the United States have ever caused noticeable damage. The majority of nonnative insects remain at relatively low densities and cause no appreciable effect on trees or ecological processes. While the number of damaging species may be limited, a few of these species have severely affected the forests they have invaded. Emerald ash borer, for example, is an Asian insect that became established in Michigan at least ten years before it was discovered in 2002. To date, tens of millions of ash trees in Michigan and the upper Midwest have been killed by this phloem-feeding insect. At least 18 ash species native to North America appear to be threatened by this recent invader.

While most nonnative forest insects appear to be innocuous, some of these species may be altering ecosystems in subtle ways that are not easily detectable. Recent surveys have found that, in some regions of the world, most ambrosia beetles (Scolytinae) are not native. Most ambrosia beetles colonize dying or dead trees and play important roles in the decomposition of wood and nutrient cycling. It remains to be seen whether the replacement of native ambrosia beetles by exotics may be causing some alteration of ecosystem processes that to date has not been investigated.

MANAGEMENT

Regulatory officials and forest entomologists employ a variety of tactics to protect native forests from invasive insects. Preventing the arrival and potential establishment of nonnative forest insects is ideal. Numerous regulatory efforts including quarantines, mandatory pesticide treatments, and inspections are used to prevent nonnative insects from arriving on cargo or in baggage carried by travelers. Early detection of newly established nonnative insects is also critical. A localized, low-density population of an invader can sometimes be successfully eradicated, while eradication of a widely distributed or high-density population will be much more challenging.

The availability of sensitive detection methods remains a critical limitation of detection and eradication strategies.

Many species of Lepidoptera (moths and butterflies) produce unique sex pheromones that attract other members of the same species from long distances. Some bark beetles (Scolytinae) similarly produce aggregation pheromones or are strongly attracted to specific compounds emitted by their host trees. Pheromones can often be synthesized, produced efficiently, and used as lures in traps deployed across large geographic areas in order to detect newly founded populations. Many groups of insects, however, do not produce long-range pheromones, and detecting low-density populations of these insects is more difficult. Insects that spend most of their life beneath the bark, in roots or other hidden locations, are especially difficult to find and monitor until they reach high densities. Some of the most damaging forest insect pests, including the hemlock woolly adelgid, the Asian longhorned beetle, and the emerald ash borer, are very difficult to detect until damage becomes apparent.

Distinguishing nonnative species from native species can also be challenging for some groups of insects whose systematics are not well known. In one example, wood-boring beetles recovered from a pocket of dying spruce trees in an area of eastern Canada in the early 1990s were originally identified as a native species. Roughly ten years later, widespread spruce mortality in the province was determined to be caused by the brown spruce longhorned beetle, *Tetropium fuscum*. When the original specimens were reexamined, they were also found to be the nonnative beetles. The availability of taxonomic expertise, therefore, can be an important component of early detection.

Containment, or slowing the spread of established species, is a management approach increasingly considered for certain forest insect invasions. One of the best examples of this strategy is the Slow the Spread program currently implemented to reduce the spread of the gypsy moth in the United States. Under this program, a grid of pheromone traps is deployed annually along a 100 km band along the expanding invasion front in order to locate newly invaded, isolated populations. Once these populations are identified, their spatial extent is delimited using additional traps, and ultimately they are suppressed through the use of mating disruption or multiple applications of a microbial insecticide. Suppression of these nascent populations along the invasion front prevents their growth and coalescence, ultimately resulting in slower spread into new regions. Quarantines to restrict transport of potentially infested trees, logs, firewood, or related materials into uninfested areas can also serve to slow the spread of an invader.

CONCLUSIONS

Invasions by nonnative forest insects constitute a major threat to the sustainability of natural forest ecosystems. While most nonnative forest insects have had relatively minor effects, a small proportion of these species have had catastrophic impacts, altering ecosystem processes into the extended future. Over the last century, numbers of invasions have steadily increased, and presently, the majority of forest insect problems are the result of invasions. Given trends of increasing global trade, this problem is likely to expand. Increased resources and efforts to reduce arrival rates of nonnative insects, to intensify early detection and eradication, and to slow the spread of established invaders will be needed in the future.

SEE ALSO THE FOLLOWING ARTICLES

Eucalypts / Gypsy Moth / Hemlock Woolly Adelgid / Pathogens, Plant / Pesticides for Insect Eradication / *Phytophthora*

FURTHER READING

Brockerhoff, E. G., A. M. Liebhold, and H. Jactel. 2006. The ecology of forest insect invasions and advances in their management. *Canadian Journal of Forestry Research* 36: 263–268.

Haack, R. A. 2006. Exotic bark- and wood-boring Coleoptera in the United States: Recent establishments and interceptions. *Canadian Journal of Forestry Research* 36: 269–288.

Langor, D. W., L. J. DeHaas, and R. G. Foottit. 2009. Diversity of nonnative terrestrial arthropods on woody plants in Canada. *Biological Invasions* 11: 5–19.

Liebhold, A. M., W. L. Macdonald, D. Bergdahl, and V. C. Mastro. 1995. *Invasion by Exotic Forest Pests: A Threat to Forest Ecosystems*. Forest Science Monographs 30.

Niemelä, P., and W. J. Mattson. 1996. Invasion of North American forests by European phytophagous insects. *Bioscience* 46: 741–753.

Wingfield, M. J., B. Slippers, B. P. Hurley, T. A. Coutinho, B. D. Wingfield, and J. Roux. 2008. Eucalypt pests and diseases: Growing threats to plantation productivity. *Southern Forests* 70: 139–144.

FORESTRY AND AGROFORESTRY

DAVID M. RICHARDSON

Stellenbosch University, Matieland, South Africa

Afforestation has a long history in the northern hemisphere, but it was only in the twentieth century that many tree species began to be planted over large areas in environments far removed from their natural ranges. A small number of tree species now form the foundation of commercial forestry enterprises in many parts of the world. Hundreds of other tree species are widely planted for many purposes, including prevention of erosion and drift sand control and for the supply of fuelwood and other products. In the tropics and subtropics, the bulk of alien tree plantings date from the second half of the twentieth century. Agroforestry involves the integration of trees and shrubs with crops or animals on the same land management unit, either in a spatial mixture or in a temporal sequence, to derive the combined benefits of all components. This use of trees has a much longer history, stretching back to at least the Middle Ages. The widespread availability of thousands of species of nonnative trees for the last century or so has, however, revolutionized agroforestry and related nonconventional forestry activities and has exacerbated the problem of invasive alien trees.

DEFINING THE PROBLEM

The dimensions of planting of alien tree species are shaped by ecological, economic, cultural, and political factors that differ considerably in different parts of the world. These factors are totally different for commercial forestry and agroforestry. Unraveling the drivers of planting is crucial for understanding and managing pathways for the introduction and dissemination of alien trees around the world. These drivers, together with a range of ecological factors that determine levels of invasiveness and invasibility, define the extent and magnitude of the problem and delineate the options for intervention.

Alien trees and shrubs contribute significantly to the economies of many countries, but there are also important costs associated with their widespread use in forestry. Considerable attention has, for example, been given to assessing the effects of plantations of alien trees on water resources and biodiversity of afforested areas. The invasive spread of cultivated trees from planting sites into natural and seminatural habitats, where they have large impacts on a wide range of ecosystem properties and functions, has emerged as an important concern more recently. Such invasions are increasingly causing major conflicts between foresters, on one hand, and conservationists and watershed managers on the other.

A BRIEF HISTORY OF AFFORESTATION

There is evidence of large-scale afforestation in parts of the ancient Mediterranean basin, where timber- and crop-producing trees were planted as long ago as 255 BC. Commercial tree planting in the Orient was recorded before the sixth century AD. Conifers were planted to stabilize sand dunes in Portugal as early as the fourteenth century, and in subsequent centuries, large areas were afforested for this purpose on the Atlantic coast of France and along the shores of the Baltic. Afforestation for timber production and environmental management was carried out in Japan in the seventeenth and eighteenth centuries. Despite its long history, the scale of forestry remained fairly small until recently. Sustained, large-scale forestry was not widespread until the late nineteenth century in Europe, and it only expanded to other parts of the world in the twentieth century. There has been rapid growth, mainly in developing countries, in the use of alien trees in "nonconventional forestry" for fuelwood production and for restoration of badly eroded or exhausted lands. The 1980s saw the rapid expansion of tree planting, mainly with legumes, to meet the needs of rural communities and to support agricultural systems in agroforestry, to revegetate degraded lands, and to protect areas from soil erosion and desertification.

DRIVERS OF INTRODUCTIONS

The most important reasons for using alien trees instead of native trees in commercial forestry include the following:

- In many parts of the world, where none now exist or where the indigenous conifers grow poorly or do not respond well to intensive forestry management, there is a need for coniferous species and certain hardwoods to produce fibers and solid wood products. The wood of *Eucalyptus* and *Pinus* species are especially adaptable for a wide range of products.
- Alien trees are often preferred because they grow much faster than native species.
- Indigenous tree species are more difficult to manage than commercially available alien species.
- The biology of indigenous species is often poorly known, including how to collect, store, and germinate seed; how to produce seedlings in a nursery; and how to manage them in a forest. Consequently, foresters prefer to work with well-studied alien species.
- The availability of seed is a key to success in plantation forestry. Seeds of native species are often hard to obtain, whereas seeds from aliens (often genetically improved) are readily available.
- Alien tree species are better suited to planting in grasslands or scrublands (marginal forest lands), where most afforestation is required. Also, aliens are frequently particularly successful in degraded forest lands.

- It is frequently necessary to develop local forest industries to improve the balance of trade by reducing the need for imported wood products. Knowledge of markets and manufacturing technology currently favors the use of the wood of aliens such as pines or eucalypts.
- Occasionally, aliens are used to replace native species that are susceptible to diseases or insects and cannot be grown profitably.

The aforementioned reasons apply to some extent to agroforestry as well, but the specific use of the trees in different situations dictates many other criteria. For both forms of forestry, however, expectations in terms of required products and services effectively preclude the use of native tree species in many cases.

Alien trees selected for use in commercial forestry need to grow well in plantations, grow on bare ground without shelter, produce desired products, and be physiologically well adapted to take advantage of favorable growing periods in new environments. For agroforestry, these factors also apply to some extent, but criteria for selection depend on the particular application. Rapid growth, ability to grow in degraded habitats and to withstand droughts and browsing by animals, and capacity to deliver (often multiple) desired products and services are fundamental considerations. Only fairly recently have considerations relating to risks to the environment been added as criteria, especially in commercial forestry, but production issues frequently override such considerations, especially in developing countries. Weediness criteria are generally much less important in selecting trees for use in agroforestry; indeed, the capacities to reseed prolifically and spread are often seen as favorable characteristics for many agroforestry systems, especially in degraded areas in developing countries.

DOMINANT SPECIES BEING INTRODUCED

Pinus and *Eucalyptus* are by far the most important genera used in alien commercial forestry in the tropics and subtropics (Table 1). The main pine species are *P. caribaea*, *P. elliottii*, *P. kesiya*, *P. oocarpa*, *P. patula*, *P. pinaster*, *P. radiata*, and *P. taeda*; several other Central American and Mexican species are increasing in importance. Among the eucalypts, the most important species are *E. globulus*,

TABLE 1
Naturalized and Invasive Pine Species in the Southern Hemisphere

Species	Naturalized	Invasive
Pinus banksiana		New Zealand
P. brutia	Australia (WA)	
P. canariensis	Australia (WA, SA)	South Africa
P. caribaea[a]		Australia (WA, SA), New Caledonia
P. contorta	Argentina	Australia (NSW), Chile, New Zealand
P. elliottii[a]		Argentina, Australia (NSW), Brazil, New Zealand, South Africa
P. halepensis		Argentina, Australia (SA, Vic), New Zealand, South Africa
P. kesiya	Brazil, South Africa	
P. monticola	Argentina	
P. mugo		New Zealand
P. muricata		New Zealand
P. nigra		Australia (NSW, Vic, SA), New Zealand
P. patula[a]	Madagascar, New Zealand	Malawi, South Africa, Zimbabwe
P. pinaster[a]	Reunion	Australia (SA, Vic, NSW, TAS), Chile, New Zealand, South Africa, Uruguay
P. pinea	Australia (NSW)	South Africa
P. ponderosa[a]		Argentina, Australia (SA), Chile, New Zealand
P. radiata[a]		Australia (WA, SA, Qld, NSW, Vic, TAS), Chile, New Zealand, South Africa
P. roxburghii	South Africa	
P. strobus		New Zealand
P. sylvestris	Argentina	Chile, New Zealand
P. taeda[a]		Argentina, Australia (NSW, Qld), Brazil, New Zealand, South Africa
P. uncinata	New Zealand	

[a] Species widely planted for commercial forestry.

NOTE: *Pinus* and *Eucalyptus* species form the bulk of commercial forestry plantations of nonnative species worldwide. Whereas few species of eucalypts have become widely invasive, at least 22 pine species are naturalized and 17 pine species are invasive outside their natural ranges. The table lists pine species known to be "naturalized" or "invasive" in the southern hemisphere. Updated from Richardson and Higgins (1998), Richardson and Rejmánek (2004), and Richardson (2006).

FIGURE 1 Examples of invasions resulting from forestry and agroforestry. (A) *Acacia mearnsii* (Fabaceae; black wattle) is the most widespread invasive tree in riparian habitats over large parts of South Africa (seen here in the Western Cape near Paarl). The species is widely grown for its bark and for timber, especially in the eastern part of South Africa. (B) Widespread *Pinus ponderosa* invasions between Bariloche and El Bolson, Río Negro, Argentina. *Pinus ponderosa* is one of at least 17 pine species that are invasive outside the natural range of pines. Widespread invasions of alien conifers are a fairly recent phenomenon in South America, unlike the situation in Australia, New Zealand, and South Africa, all of which have much longer histories of intensive plantation forestry with alien pines. (Photographs courtesy of the author.)

E. grandis, E. camaldulensis, E. tereticornis, E. urophylla, and *E. deglupta*. Other important taxa used in forestry outside their natural ranges are *Acacia* spp., *Gmelina arborea, Tectona grandis,* and *Swietenia* spp. (Fig. 1).

Hundreds of tree species are fairly widely planted in nonconventional forestry. For example, the Agroforestree Database, administered by the World Agroforestry Centre (ICRAF; www.worldagroforestrycentre.org), in Nairobi, Kenya, has information on the management, use, and ecology of more than 500 tree species. Trees typically used in agroforestry may be divided into the following groups, corresponding to their typical functions in agroforestry systems.

Fast-Growing, Nitrogen-Fixing Legume Trees

The best-known example of this group is *Leucaena leucocephala*, which has been planted throughout the tropics for over 400 years and has been recorded as a weed throughout Africa, in India, and in Southeast Asia, Australia, and several Pacific, Indian Ocean, and Caribbean islands. Other legume trees used extensively outside their home ranges include *Calliandra calothyrsus, Gleditsia triacanthos, Gliricidia sepium, Parkinsonia aculeata,* and numerous *Senna* species; all are highly invasive in many localities. Australasian acacias, especially *Acacia auriculiformis* and *A. mangium,* are also very widely used and highly invasive. Numerous other acacias are used worldwide, and most are invasive; *Acacia cyclops, A. longifolia, A. melanoxylon, A. mearnsii, A. pycnantha,* and *A. saligna* are among the most widespread invasive alien trees in South Africa.

Trees for Dry Zones

Prosopis spp., *Acacia nilotica,* and *Azadirachta indica* (neem) have all been introduced into many countries outside their native home range, and are all invasive.

Non-Legume Service Trees

Cecropia peltata (which forms a complex with *C. concolor* and *C. pachystachya*) is used as a shade tree in coffee plantations in Africa and is highly invasive.

Fast-Growing Timber Trees

This group includes many tree species in demand in deforested regions for fuelwood, poles, and windbreaks. The group includes many species of eucalypts (some of which invade in some conditions), as well as *Casuarina equisetifolia* and *C. cunninghamiana,* which are popular for sand dune stabilization and shelterbelts and are highly invasive.

High-Value Timber Trees

Several high-value timber species, which are used in agroforestry in their native ranges, have become invasive in parts of the tropics to which they were introduced. Examples are *Cedrela odorata,* a valuable shade tree in coffee and cocoa plantations in Latin America, and *Cordia alliodora,* a common shade tree in coffee and cocoa plantations in Latin America.

Fruit Trees

These are important components of agroforestry systems such as home gardens, forest gardens, and mixed tree crop plantations, often providing an important basis for farmers' subsistence and income. *Psidium guajava* has been introduced throughout the tropics since colonial times and is highly invasive wherever it is planted. Introduced *Citrus* species (*C. aurantium*, *C. sinensis*, and intermediate types) readily spread into undisturbed forests and become a characteristic feature of their understory (e.g., in Paraguay).

EXAMPLES OF INVASIONS RESULTING FROM COMMERCIAL FORESTRY AND AGROFORESTRY

Two examples are presented to illustrate the scale and dimensions of the introduction pathways defined by commercial forestry and agroforestry.

Commercial Forestry with Pines in the Southern Hemisphere

The area afforested with pines and eucalypts in the southern hemisphere increased dramatically during the second half of the twentieth century. For pines, the most dramatic growth has been in Chile, where the first large-scale plantings (of *Pinus radiata*) were done only as recently as the early 1970s. The expansion of plantations of this species in Chile, Australia (since the early 1960s), and New Zealand (since the late 1960s) has been phenomenal: by 1996 roughly 4 million ha had been planted to *P. radiata*. Rapid expansions in pine afforestation have taken place in other South American countries, notably Brazil, Venezuela, and Argentina, and in other tropical and subtropical

regions. At least 17 *Pinus* species are now well established as invaders of natural ecosystems in the southern hemisphere, and eight species are weeds of major importance. Four of the most widespread invasive pine species have been widely planted (*P. halepensis*, *P. patula*, *P. pinaster*, and *P. radiata*). The remaining four *Pinus* species that are considered major weeds (*P. contorta*, *P. nigra*, *P. ponderosa*, and *P. sylvestris*) are not major forestry species in the southern hemisphere, although they have been widely planted for erosion control and other uses in New Zealand. Species with self-perpetuating invasive populations over large areas are *P. contorta* (mainly in New Zealand), *P. halepensis* (South Africa), *P. nigra* (New Zealand), *P. patula* (Madagascar, Malawi, South Africa), *P. pinaster* (South Africa, Australia, New Zealand, Uruguay), *P. ponderosa* (New Zealand), *P. radiata* (Australia, New Zealand, South Africa), and *P. sylvestris* (New Zealand). Pine invasions in the southern hemisphere can be explained by a model incorporating information on species attributes, residence time, the extent of planting, groundcover characteristics, locality (latitude), disturbance regime, and the resident biota in the receiving environment. Figure 2 summarizes information for pine introductions, naturalization, and invasion in the southern hemisphere.

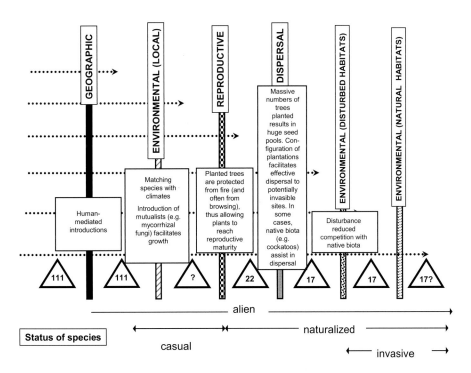

FIGURE 2 A conceptualization of the naturalization-invasion process, showing successive barriers (or invasion windows) that *Pinus* species introduced to the southern hemisphere must overcome to become naturalized or invasive (modified from Richardson, 2006). Numbers in triangles indicate the number of *Pinus* species: all 111 *Pinus* species have been moved by humans to nonnative habitats in the southern hemisphere, where at least 22 species are naturalized, and 17 species are classified as invasive (see Table 1). Boxes indicate key ways in which the pathway defined for forestry activities has assisted pine species in overcoming different barriers.

Invading pines have a wide range of impacts. These have been best studied in South African fynbos ecosystems, where dense forests of self-sown pines have radically reduced streamflow from watersheds, altered fire regimes, and reduced local-scale biodiversity. The impacts on water resources led to the instigation of the well-known Working for Water Programme which coordinates large-scale clearing operations. The extent of the problem is set to increase dramatically over the next few decades, both in areas where invasions are already evident and in other areas with recent plantings over large areas or where changes to land use practices are altering invasibility (see below).

The Use of Prosopis Species in Dry-Zone Agroforestry

Increasing the tree cover in degraded semi-arid areas has been the objective of many development projects. The Neotropical *Prosopis juliflora* and other *Prosopis* species have been widely used in the dry tropics to halt desertification, as they tolerate drought and poor soils and can stabilize sand dunes and enrich the soil through nitrogen fixation. They also provide fuelwood and charcoal, pods as fodder and food, and seed gum. *P. juliflora* has been widely introduced throughout the dry tropics and has spread over large tracts of Africa, Asia, and South America. Agroforesters often see the invasiveness of *P. juliflora* as a bonus, although perceptions are changing in regions such as Brazil, Ethiopia, Kenya, and India, due to its toxicity to livestock and impacts on biodiversity. In South Africa, several *Prosopis* species have been widely used as amenity trees, mainly for livestock fodder and shade, in the arid parts of the country. Following the spontaneous hybridization of several taxa (notably *P. glandulosa* var. *torreyana* and *P. velutina*), *Prosopis* spp. rapidly spread over huge areas, making large tracts of rangeland totally unproductive. There is an ongoing debate on whether *Prosopis* is "friend or foe" in South Africa, but opinions are converging on the latter view. The Working for Water Programme is spearheading the management of invasive *Prosopis* using mechanical and chemical control, the management of livestock, and biological control using seed-attacking insects.

Prosopis species are among the "weeds of national significance" in Australia and cause serious impacts over some 800,000 ha, mainly in northern Australia (Fig. 3). A strategic plan for managing these invasions aims to remove the current stands and to prevent impacts by (1) coordinating and maintaining management at a national level; (2) containing all core infestations and subjecting these to sustained management aimed at eventually

FIGURE 3 Dense invasions of mesquite (hybrids of *Prosopis* spp.; Fabaceae) in the Pilbara Region, northern Western Australia. (Photograph courtesy of R. Parr.)

eradicating these; (3) eradicating all isolated and scattered stands; and (4) preventing further spread.

ULTIMATE CAUSES OF PROBLEMS

The pathways created by the use of nonnative trees in commercial forestry and agroforestry have resulted in many serious problems with invasions. The fundamental reasons for this are as follows:

- The alien species are often planted in massive numbers, thus ensuring huge sources of propagules to initiate invasions.
- The configuration of plantings creates superb conditions for launching invasions. In the case of large plantations, there is often a large edge adjoining invasible habitat.
- Plantations are often adjacent to natural or seminatural vegetation; such areas are often managed for other uses, such as water production or biodiversity conservation, thus creating acute conflicts of interest when invasions occur.
- Establishment of the trees is often accompanied by disturbance (to reduce competition from native vegetation) and the intentional introduction of mutualists such as mycorrhizal fungi for pines and rhizobia for legume trees. All these factors favor growth and recruitment of the alien trees, not only in areas identified for silviculture but also in surrounding areas.

Table 2 summarizes the relevance of important advances in plant invasion ecology for understanding and managing invasions resulting from forestry and agroforestry with alien species.

TABLE 2
Generalizations Regarding Invasions of Tree Species Used in Forestry

1. Which species invade?	
Empirical, taxon-specific factors	• Some families are significantly overrepresented in invasive floras; among families used in different forms of forestry, Fabaceae and Pinaceae are especially notable. [Woody legumes and pines, the mainstays of agroforestry and conifer-plantation forestry, respectively, are inherently highly invasive.] • If a species is invasive anywhere in the world, then there is a good chance that it will invade similar habitats in other parts of the world. [Accurate global databases and lists are essential for assessing risks of invasiveness.]
Importance of biological characters	• Several characters linked to reproduction and dispersal are key indicators of invasiveness (e.g., simple or flexible breeding systems, small seed mass, short juvenile period, short intervals between large seed crops, long flowering and fruiting periods). [These properties are often exactly the ones that make species desirable for use in forestry, making problems with invasions inevitable.] • Seed dispersal by vertebrates is implicated in many plant invasions. Also, the ability to utilize generalist mutualists (seed dispersers, pollinators, mycorrhizal fungi, nitrogen-fixing bacteria) greatly improves an alien taxon's chances of becoming invasive [Generalist seed dispersers are almost always available to launch invasions, and intentional introductions (e.g., of rhizobia) are enhancing invasibility.] • Genetic change can facilitate invasions, but many species have sufficient phenotypic plasticity to exploit new environments. [Tree breeding and other forms of genetic manipulation can change invasiveness.]
Environmental compatibility	• Climate matching is a useful first step in screening alien species for invasiveness. [Climate matching for forestry was initially crude, but is becoming more sophisticated, thus ensuring better climate matching and facilitating more invasions.] Resource enrichment or release, often just intermittent (e.g., in exceptionally wet years, involving canopy opening through logging or fire), initiates many invasions. [Increasing time since the initiation of forestry plantings means increased opportunities to initiate invasions.] • Propagule pressure can override biotic or abiotic resistance of a community to invasion to some extent. [The larger the size of the founder population (i.e., plantation size or number of trees planted), the greater the chance of invasions.] • Determinants of invasibility (macro-scale climate factors, microclimatic factors, soils, various community or ecosystem properties) interact in complicated ways; evaluation of invasibility must therefore always be context specific. [There is much scope for transferring knowledge between regions, but provision must be made for unexpected outcomes.]
Stochastic factors	• The probability of invasion success increases with initial population size and the number of introduction attempts (propagule pressure) [See above.] • Long residence time improves the chances of invasion, and most plant invasions are preceded by a lag phase that may last many decades. [Increasing time since the initiation of forestry plantings means increased opportunities to initiate invasions.]
2. Control, contain, or eradicate?	
	• Early detection and initiation of management can make the difference between being able to employ feasible offensive strategies (eradication) and facing the necessity of retreating to a more expensive defensive strategy (mitigation, containment, etc.). [Proactive measures to reduce the chances of invasions and to deal with problems at an early stage must be incorporated in standard silvicultural practices.]

NOTE: Generalizations are given regarding alien plant invasions that relate strongly to invasions of tree species used in commercial forestry and agroforestry. Specific implications for interpreting current invasion patterns and implementing management are given in brackets. (Based on information in Richardson et al., 2004).

WHAT CAN BE DONE?

A wide range of management interventions are under way to deal with invasive trees spreading from planting sites in different parts of the world. These range from ad hoc interventions to remove invading plants from small areas, to systematic, integrated programs at the national scale. Options for intervention are very different for different forms of forestry and in different parts of the world.

Increasing awareness of problems associated with invasive forestry trees means that information on invasive species and ways of dealing with them is becoming more easily accessible, on the Internet, in scientific and popular publications, and via special interest groups. Ignorance is no longer an excuse for disseminating invasive trees. Afforestation policies need to include clearly stated objectives to reduce impacts outside areas set aside for forestry. Containment of alien trees to areas set aside for their cultivation must become an integral part of silviculture. Progress is being made toward this end, but coordination is required to optimize such measures. Protocols dealing with invasive forestry trees across the full range of conditions in which alien trees are used must be strived for. The

Food and Agriculture Organization of the United Nations (FAO; www.fao.org/), the Global Network for Forest Science Cooperation (IUFRO; www.iufro.org/), and the World Agroforestry Centre (ICRAF; www.worldagroforestry.org/) are the obvious institutions to drive such initiatives, and they have all made some contributions in this direction, but much more work remains to reduce the effectiveness of pathways created by the use of alien trees in generating and sustaining biological invasions. Solutions lie in the integrated application of legal instruments and pragmatic new approaches.

In commercial forestry, many companies currently implement the ISO 14001 Environmental Management System to ensure international standards on environmental performance. The Forestry Stewardship Council (FSC; www.fsc.org/) is an independent NGO that promotes responsible management of the world's forests. The FSC is promoting measures to reduce problems associated with invasive trees.

Some advances in addressing problems of invasive forestry trees have been made in different parts of the world, and these need to be considered when formulating principles and protocols for widespread adoption. First, biological control offers considerable potential for reducing conflicts of interest where indispensable crops cause problems as invaders. Excellent progress has been made in this regard in South Africa. The histories of control efforts on Australian *Acacia* spp. and *Prosopis* spp. are instructive, and good options exist for resolving conflicts. Problems exist for some highly invasive forestry trees. For example, pines are an obvious target for biocontrol action, but efforts to this end in South Africa were suspended because the best candidate agent for biocontrol, the seed-attacking weevil *Pissodes validirostris*, is possibly implicated as a vector of pitch canker. There is also scope for deploying genetically engineered sterile trees to reduce problems of invasiveness through reduced seed production, but one obstacle to this solution is that FSC regulations expressly forbid any use of genetically modified species. Several countries, notably South Africa, are considering "polluter pays" clauses, including possibly the imposition of a levy on timber products to fund management of invasive plants that originate in forestry plantations. In cases where invasions clearly impact the delivery of ecosystem services such as water, levies on the supply of services may be required to ensure sustainable control programs.

The situation is somewhat simpler in agroforestry systems, where decisions need to be made about whether any benefits derived from the invasive spread of an alien tree outweigh the reduced value of ecosystem services (e.g., the potential of grazing land in areas infested with

Prosopis). There is also more scope to use native species or less-invasive alien species as alternatives for highly invasive species. In many cases, options exist for managing invaded areas by manipulating disturbance regimes (e.g., fire cycles, grazing levels) to impede invasion. Improved solutions to problems caused by invasive alien trees lie in better integration of available control methods.

SEE ALSO THE FOLLOWING ARTICLES

Eucalypts / Fire Regimes / Forest Insects / Hemlock Woolly Adelgid / Horticulture / *Lantana camara* / Trees and Shrubs

FURTHER READING

Evans, J., and J. Turnbull. 2004. *Plantation Forestry in the Tropics*, 3rd ed. Oxford: Oxford University Press.

Haysom, K. A., and S. T. Murphy. 2003. The Status of Invasiveness of Forest Tree Species Outside Their Natural Habitat: A Global Review and Discussion Paper. Working Paper FBS/3E. Rome: Forestry Department, Food and Agriculture Organization of the United Nations, FAO.

Hughes, C. E.. and B. T. Styles. 1987. The benefits and potential risks of woody legume introductions. *The International Tree Crops Journal* 4: 209–248.

Richardson, D. M. 1998. Forestry trees as invasive aliens. *Conservation Biology* 12: 18–26.

Richardson, D. M. 2006. *Pinus*: A model group for unlocking the secrets of alien plant invasions? *Preslia* 78: 375–388.

Richardson, D. M., P. Binggeli, and G. Schroth. 2004. Invasive agroforestry trees: Problems and solutions (371–396). In G. Schroth, G. A. B. de Fonseca, C. A. Harvey, C. Gascon, H. Vasconcelos, and A.-M. N. Izac, eds. *Agroforestry and Biodiversity Conservation in Tropical Landscapes*. Washington, DC: Island Press.

Richardson, D. M., and S. I. Higgins. 1998. Pines as invaders in the southern hemisphere (450–473). In D. M. Richardson, ed. *Ecology and Biogeography of* Pinus. Cambridge: Cambridge University Press.

Richardson, D. M., and M. Rejmánek. 2004. Invasive conifers: A global survey and predictive framework. *Diversity and Distributions* 10: 321–331.

Richardson, D. M., B. W. van Wilgen, and M. Nunez. 2008. Alien conifer invasions in South America: Short fuse burning? *Biological Invasions* 10: 573–577.

Williams, M. C., and G. M. Wardle. 2007. *Pinus radiata* invasion in Australia: Identifying key knowledge gaps and research directions. *Austral Ecology* 32: 721–739.

FRESHWATER PLANTS AND SEAWEEDS

LARS W. J. ANDERSON

USDA–Agricultural Research Service, Davis, California

Introduced invasive aquatic plants and algae include flowering plants (angiosperms) and ferns, as well as freshwater and marine algae (e.g., "seaweeds") that have been

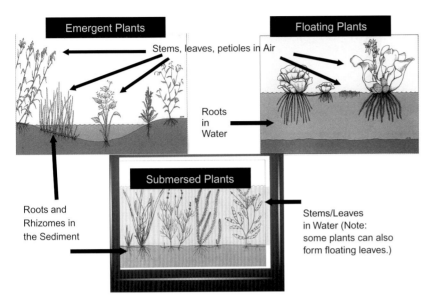

Emergent Plants

Stems, leaves, petioles in Air

Floating Plants

Roots
in
Water

Roots and
Rhizomes in
the Sediment

Submersed Plants

Stems/Leaves
in Water (Note:
some plants can also
form floating leaves.)

FIGURE 1 Diagram depicting a typical ecological classification of aquatic plants, showing growth habits of aquatic plants and their relationships to the aquatic habitat. Some submersed aquatic plants are able to produce "floating leaves," but they differ from true floating plants because they are anchored to the bottom with roots and rhizomes. Marine macroalgae ("kelp") fit into the "submersed plant" category with regard to their relationship to the habitat and access to nutrients and light. Marine macroalgae are anchored to the substratum via holdfasts and rhizoids, not true roots.

transported from areas where they are native to different continents or bioregions, for commercial purposes, for personal aesthetic uses (aquarium, landscape), or through inadvertent contamination of shipments of other plant or animal materials. Over 35 families of freshwater aquatic and riparian plants have invasive species representatives, and several hundred species are considered problematic. Over 270 species of marine macroalgae (seaweeds) have been introduced around the world, several of which cause serious ecological damage and economic losses, as well as facilitate the spread of diseases. Habitats invaded by these species include wetlands, lakes, ponds, rivers, bays, estuaries, coastal zones, highly developed irrigation storage and conveyance systems, hydroelectric power systems, harbors, marinas, and aquaculture facilities.

TYPES OF INTRODUCED AQUATIC PLANTS AND ALGAE

Plant biologists and aquatic ecologists categorize aquatic plants based on characteristics of their growth habit and by the way they obtain the resources needed for growth and reproduction (Fig. 1). This "ecological classification" approach is not based on genetic relationships (phylogeny), but it is very useful in understanding the important functions of features the groups possess, as well as in understanding why some species are more or less likely to be invasive than others (Fig. 2). Ecophysiological characteristics typical of the plant and macroalgae groups are provided in Table 1, and a typical classification scheme is summarized here:

Floating Plants

These float at the surface of the water. Their roots do not normally contact the bottom sediment. Examples include water hyacinth (*Eichhornia crassipes*), duckweeds (*Lemna* spp., *Landoltia* spp., *Spirodela* spp.), and waterferns (*Azolla* spp., *Salvina molesta*).

Emergent Plants

These are rooted in the sediment, and nearly all the leaves are aerial (supported on shoots). Examples include cattails (*Typha* spp.), the giant reed (*Arundo donax*), the common reed (*Phragmites australis* and potential hybrids), the yellow flag Iris (*Iris pseudacorus*), the flowering rush (*Butomus umbellatus*), invasive cordgrass (*Spartina alterniflora*, hybrids of *S. alterniflora* and *S. foliosa*, *S. densiflora*), and perennial pepperweed (*Lepidium latifolium*).

Submersed Plants and Marine Macroalgae

These are species that are mostly underwater and are anchored either by true roots (freshwater and marine flowering plants) or, in marine macroalgae (kelp or seaweeds), by "holdfasts" or "rhizoids." (It should be noted that, unlike floating and emergent plants, submersed plants are often subjected to very low light levels because light is quickly reduced (attenuated) as it penetrates the water column.) Examples of freshwater submersed plants are hydrilla (*Hydrilla vertcillata*), Brazilian egeria (*Egeria densa*), Eurasian watermilfoil (*Myriophyllum spicatum*), elodea (*Elodea canadensis*), curlyleaf pondweed

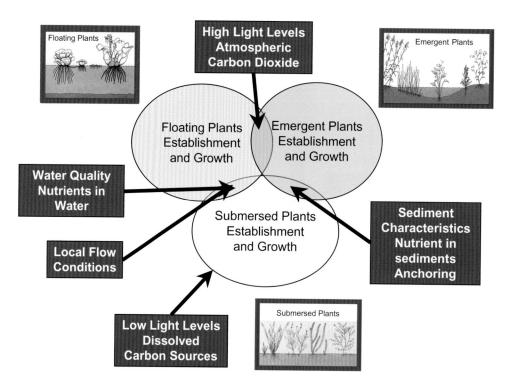

FIGURE 2 Diagram showing how the three main ecological categories of aquatic plants have access to needed resources for establishment, growth, and dispersal. Note that marine macroalgae fit into the "submersed plant" category with regard to their relationship to the habitat and access to nutrients and light. Marine macroalgae are anchored to the substratum via holdfasts and rhizoids, not true roots.

(*Potamogeton crispus*), and the marine plant Japanese eelgrass (*Zostera japonica*). Examples of marine macroalgae include undaria (*Undaria pinnatifida*), codium (*Codium fragile* ssp. *tomentosoide*), caulerpa (*Caulerpa taxifolia*), *Kappaphycus alvarezii*, *Gracillaria salicornia*, and *Sargassum muticum*.

Submersed Plants with Floating Leaves

These are species that typically sprout underwater from vegetative propagules or germinate from seeds, and subsequently form leaves that float on the surface. Examples include water chestnut (*Trappa natans*) and various water lilies such as *Nuphar luteum*, *Brasenia schreberi*, and *Nympoides aquatica*.

Unicellular and Filamentous Algae

These are capable of exhibiting very rapid growth rates, doubling in numbers in a few hours to a few days. They usually obtain their nutrients from the water column, although some are attached to sediments and may even have root-like structures for anchoring. They usually reproduce asexually by simple cell division, although some species do produce sexually differentiated cells under certain conditions. There are two major subcategories of

algae based upon their tendency to grow as either single cells or colonies of cells: (1) unicellular and (2) filamentous and colonial.

UNICELLULAR ALGAE Both green algae (having chloroplasts) and cyanobacteria ("blue-green algae") are microscopic free-floating (planktonic) cells that can become so prevalent that the water when placed in a glass container appears pea-soup green to nearly black or shades of light brown. Many normally form the base of the food chain in most ponds and lakes and are consumed by microscopic invertebrates (zooplankton). Typical invasive and introduced taxa are *Tetraspora* spp, diatoms such as "rock snot" or "didymo" (*Didymosphenia geminata*), and dinoflagellates such *as Gymnodinium catenatum* and *Alexandrium minutum*, which can produce paralytic shell poisons.

FILAMENTOUS AND COLONIAL ALGAE These form large colonies or mats and can be seen easily without a microscope or hand lens because the chains, or filaments, contain from thousand to millions of cells, although each single cell is microscopic. They may be composed of green algae or cyanobacteria (blue-green algae). Some

TABLE 1
Ecophysiological Characteristics of Introduced Aquatic Plants and Marine Macrophytes

Floating Plants	Emergent Plants	Submersed Plants and Marine Macroalgae
Full sunlight or partial shade (plants are high-light adapted)	Full sunlight (plants are high-light adapted)	Very low light is required (<1% of surface levels) (plants are shade adapted)
Carbon dioxide is obtained directly from air	Carbon dioxide is obtained directly from air	Carbon is from dissolved CO_2 in water, from bicarbonate in water (e.g., at high pH), or from sediments (some plants can use sediment-borne carbon); plants with floating leaves can obtain CO_2 from the air.
Nutrients are primarily from water	Nutrients are from sediment	Nutrients are primarily from sediments, but also water, especially in marine macroalgae
Rapid capture of water-borne nutrients ("spikes")	Nutrients must cycle via sediments	Some potential for rapid capture of nutrient spikes in the water
Subject to large daily changes in air temperature	Subject to large daily changes in air temperature	Generally subject to smaller daily changes in temperature
Less vulnerable to light competition by neighboring plant shading	Some vulnerability to shade from other plants	Highly vulnerable to shade from other submersed plants, and from floating and emergent plants
Canopy (foliage) structure is somewhat variable	Canopy (foliage) structure has little variability	Canopy (foliage) structure is highly variable, both seasonally and vertically (i.e., with depth)
Populations are mobile (e.g., wind, tide, or current-driven)	Populations are sedentary, stable due to rooting in sediments	Populations are generally rooted and sedentary, but are subject to flooding-related flows that result in fragmentation
Whole plant as dispersal units, as well as seed in some	Dispersal is primarily by seed	Dispersal is by fragments or specialized propagules (e.g., turions)
Generally produce one type of leaf	May produce two to three types of leaves ("heterophyllic")	Are often heterophyllic (can form submersed, floating, or aerial leaves)

species produce small gas vacuoles in each cell that enable the entire mass to float to the surface of the water. Many have the ability to anchor on the sides of cement-lined canals, where they impede the flow and capacity of the canal. They also tend to slough off pieces of the colony, which then are dispersed with the water flow and eventually block pumping systems, fish screens, and irrigation plumbing. Typical species include those in these genera: *Nostoc, Spirogrya, Cladophora, Pithophora, Rhizoclonium,* and *Hydrodictyon.*

HABITATS WHERE INTRODUCED AQUATIC PLANTS AND ALGAE BECOME ESTABLISHED

Aquatic sites are highly variable and often unique relative to physical features, ecology, and human use. They encompass relatively small, simple water storage tanks, or industrial holding ponds, to complex, multiuse reservoirs and natural river systems, where the primary concern is preservation of habitats for native plant and animal populations. Native aquatic plants and algae are essential components of aquatic ecosystems because they are the primary producers that form the base of complex food webs that include other native invertebrates, amphibians, and vertebrates (reptiles, fish, and waterfowl), and

because they provide physical structure for a variety of fauna. In addition, aquatic plants can interact among themselves and influence their own relative abundance and ability to spread. Figure 3 illustrates how the three ecological types of plants can affect each other. The general features of aquatic (freshwater and marine) ecosystems are described below.

Nonflowing Sites

These include lagoons, large lakes, ponds, and reservoirs, which vary tremendously in surface area, average depth, slope of shorelines, and exposure to wind and inflows. Although they are considered "nonflowing," in reality, water is always moving both horizontally and vertically, although often at a rate that is imperceptible without sensitive measuring equipment. Low-velocity flows (i.e., fractions of a centimeter per minute) are important in the overall distribution of nutrients and suspended sediments on seasonal time scales and can also affect localized establishment of introduced plants and algae. Finally, wind-driven movements can disperse fragments of noxious aquatic weeds and thereby worsen infestations in a lake or pond.

Nonflowing sites often have steep gradients (changes) in light, temperature, sediment type, and dissolved oxygen within a relatively short distance from the shoreline. During

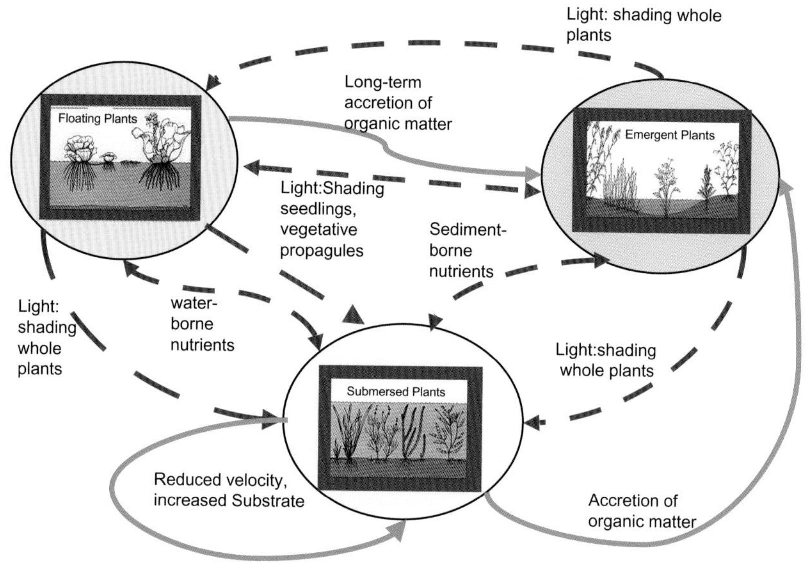

Red (dashed)=Negative effect
　　　　　　(causes reduced growth)
Green (solid)=Positive effect (more growth)
Double Arrows=Competitive Interactions

FIGURE 3 Diagram illustrating the typical aquatic macrophytes and how their interactions may affect access to resources needed for establishment and growth.

summer months, these gradients can be steep due to the slope and increases in depth within the zones where aquatic weeds thrive, and because warmer, less dense water creates a stable surface layer. The availability of light for bottom-dwelling (submersed) plants, for example, can be reduced by 90–95 percent at a distance of only 3–4 m from the edge of a lake.

Flowing Sites

Rivers, streams, and irrigation and flood-control systems have seasonally variable flows, whereas estuaries, bays, and coastal zones also have daily tidal-driven flows (and wave action), with rates ranging from fractions of meters per second to several meters per second. These flows are far greater than the water movements typical of the "non-flowing" sites described above. Invasive rooted plants and attached algae must be able to withstand very strong frictional forces (drag), or they may be uprooted and thus unable to obtain nutrients. Many have extensive, strong rhizomes, roots, or other anchoring systems and specialized foliage that can withstand the forces of flowing water. Establishment of "floating" plants is usually limited to protected areas where velocities are lower, such as at dams, bridges, or other flow control structures. However, slow, meandering rivers can become highly infested with floating plants such as water hyacinth and giant salvinia.

REPRODUCTION AND DISPERSAL OF INVASIVE INTRODUCED AQUATIC PLANTS AND ALGAE

Pathways and Vectors

The most problematic freshwater aquatic plants and marine macroalgae have been introduced via the aquarium and "backyard" aquascape trade over the past 50 years. A few have been introduced either as "contaminants" or because they were incorrectly identified and were thought to be native species. The commercial pathways have been augmented by hobbyists' exchanges of species and by the release of these species into natural sites such as lakes, ponds, rivers, marinas, and bays. Most commercial and private exchanges (both domestic and international) of aquatic plants and algae are still largely unregulated. Once a plant or algal species has become established, other pathways and vectors tend to spread the problem. Primary vectors are boats, boat trailers, construction equipment (e.g., for pier installations), waterfowl, and major storms. The success of these pathways and vectors is due to the viability and resilience of the various propagules that aquatic plants and algae can produce as part of their normal life cycles. The reproductive and dispersal strategies and mechanisms are summarized here and in Table 2.

Asexual (Vegetative) Reproduction

Aquatic plants and algae use a variety of asexual reproductive modes in addition to typical seed production (sexual reproduction). For example, ferns produce spores that eventually fall from the floating fronds and may be dispersed by water currents. Considering the mobility of the aquatic environment, and disturbances often experienced by aquatic plants such as floods, tidal movements, rapid river flows, and associated sediment erosion, it is not surprising that seed dispersal alone would not be the optimal mode of ensuring survival. In fact, the formation of

TABLE 2
Typical Reproductive and Dispersal Structures in Some Aquatic Plants, Algae, and Marine Macrophytes

Species	Structure	Location or Deposition Site
Floating Type		
Water hyacinth	Seed	Fruits; top of sediment
	Ramets (clones)	On water, floating
Giant salvinia	Ramets (clones)	On water, floating
Duckweeds	Ramets (clones)	On water, floating
	Turions (dormant buds)	Surface of sediment; water
Emergent Type		
Giant reed	Ramets, Rhizomes	Sediments, floating
Common reed (Phragmites)	Seed	Water surface, sediment surface
Cattails	Seed	Water surface, sediment surface
Perennial pepperweed	Seed	Surface of sediments
	Rhizomes	Surface of sediments
Primrose willows (water primrose) species	Seed, rhizomes	Surface of sediments
Submersed Type		
Hydrilla	Tuber (subterranean)	Within sediment
	Axillary turion	In leaf axils, water surface
	Stem turion	On tips of stem or shoots
	Root crowns	In and on sediment
Brazilian waterweed (*Egeria densa*)	Fragments; root crowns	Surface of sediments and water
Oxygen weed (*Lagarosiphon major*)	Fragments; root crowns	Surface of sediments; water
Pondweeds (*Potamogeton* spp.)	Seed	Fruits, floating on water, surface of sediments
Potamogeton nodosus	Winter buds	Within sediment
Potamogeton crispus	Turions	Stem (leaf axils)
Eurasian watermilfoil	Seed	Flowers, floating, surface of sediments
	Rhizomes	Within sediment
	Apical shoot "auto fragments"	Upper shoots, floating on water
Freshwater, marine microalgae	Cells, whole colonies, cysts	Surface of substratum; water; other aquatic plants
Marine macroalgae		
Undaria pinnafitida	Sporophytes, zoospores	Surface of substratum; water
Caulpera taxifolia	Fronds, fragments, rhizoids	Surface of substratum; water
Gracilliaria spp.	Sporophytes, fragments	Surface of substratum; water
Kappaphycus spp.	Fragments, zoospores	Surface of substratum; water

general and specialized "vegetative" structures is common among aquatic plants (Table 2). These structures have at least four important functions that facilitate establishment and dispersal:

· They provide stable, and anchored, propagules that can rapidly produce new plants in response to elevated temperatures and increased day length (spring conditions).
· They ensure that a high proportion of next year's plants emerge in suitable condition.
· They provide adequate carbohydrates and other nutrients so that plants can reach required light levels.
· They enable overwintering and dispersal.

Some of the world's worst aquatic weeds, such as hydrilla, exhibit all these features. Many submersed aquatic plants rely mostly upon asexual reproduction for year-to-year recruitment.

Sexual Reproduction

Almost all aquatic plants exhibit sexual reproduction. In emergent plants, flowers are usually borne on vertical shoots or horizontally branching shoots. These flowers may be wind or insect pollinated. Cattails produce a seed that has numerous fine protrusions that help support it in even light winds and which also enable it to bounce along the water surface until it lodges against the shoreline. In some submersed plants (e.g., *Potamogeton* spp.), pollination takes place underwater, and the pollen may be sticky and string-like instead of spherical.

Floating plants (except ferns) such as water hyacinth and false pickerelweed (*Monochoria vaginata*) form flowers on elevated petioles or shoots, but often, after pollination (e.g., by insects), the entire flower (inflorescence) drops down into the water. Subsequently, flowers or fruit break off the parent plant and may sink or float for some time. The ripened seed may not germinate for months, years, or decades. In many emergent floating and submersed plants, formation of flowers or vegetative reproductive structures occurs in response to day length (photoperiod). Hydrilla and American pondweed produce flowers in the summer under long days but produce tubers or winter buds in the sediment primarily in the fall under short days (<13 hours). In contrast, curlyleaf pondweed (*Potamogeton crispus*) produces turions under summer (long-day) conditions.

EXAMPLES OF IMPACTS FROM INTRODUCED AQUATIC PLANTS AND ALGAE

Table 3 summarizes the types of impacts caused by introduced aquatic plants and algae. The complexities and multiple-level ecological impacts and economic losses (e.g., loss of recreational revenue, decreased property values, health-related costs, negative impacts on fisheries), coupled with actual control costs, make accurate assessments extremely difficult. There have been no reliable studies on the overall global economic impacts (losses and costs of control) of aquatic plants and algae. Estimates range from $500 million to $2 billion per year in the United States.

The best economic information is from a few projects where narrowly targeted species have been managed or eradicated. For example, the biological control of *Azolla filliculoides* in South Africa cost $58 million. However the estimated "benefit-to-cost" was over 1,100:1. Control of the introduced riparian weed purple loosestrife in the United States is estimated to cost over $45 million per year, but that estimate is now nearly ten years old. The successful eradication of the marine alga *Caulerpa taxifolia*

TABLE 3

Examples of Impacts Caused by Introduced Aquatic Plants and Algae

Aquatic Site	Typical Use	Weed Impacts
Lakes, reservoirs (generally >10 acres)	Usually multiple uses: fisheries and wildlife, water storage, recreation, flood control, potable and irrigation source, hydroelectric power-production	Ecological impacts (reduced native plant habitat, altered food web relationships), clogged pumps, prevented access for recreation, fish-kills, hazards to swimmers, damage generator turbines
Ponds (<10 acres)	Recreation, fishing, small-scale irrigation, vineyard frost-control, aesthetics, fire control	Ecological impacts (fish kills, reduced native plant habitat, altered food wed interactions), odors, blocked pumps, unsightly aesthetics, mosquito habitat
Livestock watering tanks	Water for horses, cattle, sheep	Algae: odor, taste, can be toxic
Industrial and commercial ponds (generally <10 acres)	Cooling tower recycling, evaporative holding ponds, initial waste-holding	Algae and some higher plants: odor, toxicity, plugged pipes, clogged pumps, clogged spray emitters
Aquaculture ponds (generally <10 acres)	Fish production, shrimp production, horticultural aquatic plant production, algae production (e.g., *Spirulina*)	Algae and higher plants: taste, odor, lowered dissolved oxygen, harvest interference, mosquito habitats
Ornamental aquascapes	Aesthetics (Note: private and public sites, often with people-oriented activities, indirect recreation, and potential contact)	Algae and higher plants: odors, mosquito vector habitat, habitat for swimmer's itch parasite host (snails), clogged filters and aerators
Flood control canals	Seasonal conveyance of high flows (November–May)	Higher plants: interference with flows, damage related to overflowing banks and erosion, damage to bridges
Irrigation canals	Seasonal water conveyance (generally April–September in central and northern California; may be all year long in southern California)	Algae (mostly filamentous) and higher plants: reduced capacity, blocked flows, clogged pumps and control structures (gates, weirs), clogged drip emitters and subsoil irrigation systems
Rivers, streams, levees	Fisheries and waterfowl, plant habitats, recreational, commercial boating and shipping, potable, agricultural, and industrial water sources	Ecological impacts (reduced native plant populations, degraded spawning habitats, altered food web relationships), recreational activity prevention, interference with pumping and commercial ship movements, damage to levee banks, erosion
Marine environments: bays, lagoons, marinas	Boating, fishing, water skiing, swimming, native species, mariculture	Ecological impacts (reduced native plant habitat, altered food web relationships), recreational activity prevention, blocked cooling water systems

in the United States required $7 million over a six-year period.

MANAGEMENT: CONTROL AND ERADICATION

Due to the wide range of uses and functions of aquatic sites, management strategies must be tailored to each site. In all cases, any known and potential "nontarget" effects must be considered, along with regulatory limitations on use of the various control and eradication methods. In particular, the ability to alter water movement, water levels, and water uses, even if only temporarily, can greatly affect the success of management actions, as well as provide greater flexibility in protecting nontarget species and preserving water uses. "Eradication" means killing or physically removing all viable life stages (e.g., fruits, seeds, turions, tubers, rhizomes, root crowns, spores, cysts, and other vegetative growth) of the target plant or alga with the objective of preventing further spread and dispersal, preventing continued reproduction (sexual or asexual), and eliminating the infested site as a source of future offsite infestations. Eradication includes effective quarantines coupled with elimination of all pathways and vectors of the target species. "Control" means to reduce the impact of an introduced and invasive aquatic plant or alga to some acceptable level of cover, biomass, or other measure of detrimental impact. Opting for "control" usually means accepting very long-term actions that have no endpoint that will not eventuate in complete extirpation of the target species. Successful strategies and actions for control and eradication include the following.

Prevention

Actions that reduce the onset or promotion of aquatic weeds fall into two broad categories: (1) control of dispersal pathways and vectors and (2) limitation of required resources for growth. Because most aquatic plants require adequate levels of light, it is reasonable to assume that extensive shallow areas (e.g., less than 1.5 to 2 m) will facilitate growth of submersed aquatic plants. Likewise, extensive, shallow-sloping shorelines will provide ideal conditions for plants to spread from nearshore areas to deeper areas. Similarly, aquatic sites that act as "sinks" or low points for runoff from ornamental planting, such as turf, shrubs, or even orchards, will tend to receive nutrients from the adjacent areas through normal irrigation or rainfall. Unfortunately, nutrients tend to be applied to these areas in late winter to spring, precisely when the aquatic plants begin to grow. With this in mind, design and construction should minimize extensive shallow areas and should include physical barriers to reduce inflows of nutrients. Gutters or drains can be used around small lakes and ponds to carry runoff away from the site. Likewise, shorelines having steep slopes (e.g., a 2:1 ratio of depth to linear distance from shore) are useful in discouraging the creation of extensive areas with adequate light levels. In general, ponds and lakes that have most areas with average depths greater than 5 m tend to discourage establishment of submersed aquatic plants.

Physical or Mechanical Removal

For maintaining small areas, hand-operated, manual mechanical methods include small rakes, nets, floating booms (to gather floating plants), and grappling hooks. These methods usually result in the production of many fragments (pieces) of plants that cannot be contained and removed.

For large infestations, gasoline- and diesel-powered, hydraulically operated cutters and harvesters are effective and have improved dramatically over the past 20 years. They are often used with a "transport" barge that carries cut plant material from the harvester to conveyers that can offload the material onto trucks. These systems can cut and remove plants to a depth of 2 to 3 m and can operate in as little as 1 m of water, but they produce thousands of viable plant fragments. In addition, many thousands if not millions of invertebrates, fish, small reptiles, and amphibians may be removed along with the harvested plants. More aggressive methods employ suction devices of various sizes and designs, operated from boats or barges or manipulated by divers, who can direct the head of the system more accurately. Because roots, rhizomes, and other reproductive structures (e.g., tubers) can be removed along with the plants, control may be longer lasting than with cutters or harvesters. Most suction or dredging operations require various permits and usually necessitate settling basins and "silt" curtains to prevent disturbed materials from being distributed to the entire water body, or downstream from the dredge site. The dredge spoils (waste) must be transported and disposed in a proper permitted site (e.g., a landfill or other area). Thus, although dredging can be extremely effective in removing both submersed and emergent aquatic weeds, there are usually serious impacts on nontarget benthic organisms.

Plastic and fabric materials have also been used successfully to smother relatively small areas (e.g., <1 acre). These "benthic" barriers can contain the infestation and deprive the plants or algae of light and access to dissolved oxygen, and they usually can kill the vegetative portions of the plants within a few weeks to months. A similar system was used in conjunction with injection of sodium hypochlorite (bleach) to successfully eradicate the marine alga *Caulerpa taxifolia* in the United States between 2000 and 2007.

Herbicides and Algaecides (Chemical Control)

Aquatic herbicides and algaecides have been used for over 100 years (e.g., copper sulfate). However, compared to hundreds of herbicide active ingredients used in terrestrial habitats and cropland ecosystems, there are relatively few (up to 12, depending upon the country) active ingredients currently allowed ("registered" or "permitted") for use in and around aquatic sites. Due to the complexities and environmental sensitivity of aquatic sites, the extensive testing and review necessary for the approval of chemicals can take over ten years, and costs can exceed $100 million for a new aquatic herbicide label.

As with most pesticides, the "dose determines the toxicity" in aquatic herbicides and algaecides, and therefore, for each active ingredient, there are different "dosages," or concentrations, of the active ingredient required to control a given target plant or algal species. Contact herbicides affect only the plant parts (usually leaves and stems) that are in immediate contact with the herbicides. In contrast, systemic herbicides can translocate from the point of initial contact and uptake to other sensitive parts of the plants such as roots, newly forming leaves, stems, and rhizomes. And because plants translocate food (products of photosynthesis) to different parts of the plant at different times of the year (e.g., spring versus fall), systemic herbicides may be more effective in different seasons. Just as with terrestrial herbicides, the responses of target weeds often differ from both native aquatic plants and other introduced invasive species. This can lead to successful "selective" management of the target species while at the same time "releasing" the nontarget plants to thrive under reduced competition. However, it can also lead to simply a replacement of one invasive plant with another. Because the approval status of particular aquatic herbicides is different for each country, and because new ones are periodically approved in some nations, while the approval of others may be discontinued, it is essential to obtain the most current information.

Management of aquatic weeds in flowing water systems is difficult due to short contact time with herbicides placed in water and because mechanical methods easily distribute plant fragments downstream. Rapid transport of water precludes the use of many aquatic herbicides because downstream uses may be incompatible with these products. For this reason, it is critically important to know the full extent of downstream uses and the final endpoint of the water. It is not surprising, therefore, that few herbicides are registered (approved) by any country for flowing-water uses.

Tidal waters are a special case of "flowing water." Although most people assume tidal water means "seawater" or saline water, this is not always true. For example, the Sacramento–San Joaquin delta in California exhibits typical bimodal (twice-daily) tidal fluctuations, yet most of the water is fresh and, indeed, is vital for California's agriculture and commerce, as well as for providing drinking water for more than 20 million people. In some of the sloughs in the delta, tidal flows can exceed 6 km/hour.

Biological Control

The use of a living organism to control an aquatic weed is termed "biological control." Although this approach certainly has many advantages over other methods previously described, there are currently very few biological control agents that have been successfully used for the management of aquatic weeds. The most commonly employed biological control agents are those used in a "classical" release strategy, wherein an insect or pathogen that had coevolved with the target plant in its native environment is released into the infesting population. Demonstrated "host specificity" is the most important criterion for approving releases in most countries. The best examples of biological control of aquatic weeds are insects (a moth, a thrips, and a flea beetle) that were introduced to control alligatorweed in the southeastern United States and a weevil (*Cyrtobagous salviniae*), which was released for control of *Salvinia molesta* in South Africa, Southeast Asia, and Australia in the 1980s and in the United States in 2006. Research is continuing to determine the efficacy of flies (e.g., *Hydrellia pakistanae*) on hydrilla. Perhaps one of the most unusual approaches has been the augmentative (enhanced-population) release of a North American native weevil *Euhrychiopsis lecontei* for control of the introduced Eurasian watermilfoil (*Myriophyllum spicatum*) in the United States. In this case, the native weevil prefers the nonnative *Myriophyllum spicatum* to its normal native host, northern milfoil (*Myriophyllum sibericum*).

The most consistently successful biological control agent for submersed and floating freshwater aquatic weeds on a global scale has been the grass carp, or white amur (*Ctenopharyngodon idella*). Unlike other "classical"

TABLE 4

Examples of Biological Control Agents Released against Introduced Aquatic Plants

Target Weeds	Biological Control Agents
Alligatorweed (*Alternanthera philoxeroides*)	Flea beetle (*Agasicles hygrophila*)
	Thrips (*Amynothrips andersoni*)
	Moth (*Vogtia malloi*)
Hydrilla (*Hydrilla verticillata*)	Tuber weevil (*Bagous affinis*)
	Fly (*Hydrellia balciunasi; H. pakistanae*)
	Moth (*Paraponyx diminutalis*)
	Triploid grass carp (*Ctenopharyngodon idella*)
	Mycoleptodiscus terrestris ("MT") (native fungal pathogen)
Water hyacinth (*Eichhornia crassipes*)	Weevils (*Neochetina bruchi, N. eichhorniae*)
	Moth (*Sameodes albigutallis*)
Eurasian watermilfoil (*Myriophyllum spicatum*)	Weevil (*Euhrychiopsis lecontei*)
	Midge (*Cricotopus myriophylli*)
Giant salvinia (*Salvinia molesta*)	Weevil (*Cyrtobagous salviniae*)
Purple loosestrife (*Lythrum salicaria*)	Root-boring weevil (*Hylobius transversovittatus*)
	Leaf feeding beetle (*Galerucella calmariensis, G. pusilla*)
Various submersed and floating weeds	Triploid (sterile) grass carp (*Ctenopharyngodon idella*)

agents, the grass carp is a generalist herbivore that feeds on most submersed plants and many floating species (e.g., duckweed, giant waterfern). One of the most dramatic and successful uses of the grass carp was its introduction into 800 km of irrigation canals in the Imperial Irrigation District of California in the mid-1980s for control of hydrilla. Within two years after the introduction, biomass of hydrilla was reduced by 95 percent. However, because the grass carp can be so efficient and nonselective, there are concerns that the fish's voracious feeding habits may have significant impacts on aquatic habitats. To mitigate these concerns, many regulatory agencies require the use of only sterile (triploid) fish. The sterile fish cannot reproduce, yet they feed and grow at about the same rate as normal (diploid) grass carp. Fish are typically stocked at 20 to 30 cm lengths (about 0.1 kg to 0.2 kg), at rates of 20 to 50 fish per infested hectare. Table 4 summarizes the biological control agents that have either been released for aquatic weed control or that have the potential for release.

ENVIRONMENTALLY RESPONSIBLE PRACTICES AND USE OF BENEFICIAL PLANTS

Regardless of the use and characteristics of an aquatic site, the most important practice is to prevent introductions of nonnative aquatic plants. Use of beneficial native plants in aquatic sites has received increasing attention over the past ten years. Unfortunately, there are few suppliers of these plants, although there are some sources for native grasses that can be used in riparian and ditch bank settings. Due to the highly modified, disturbed nature of many aquatic sites (e.g., irrigation canals), successful establishment of native aquatic plants is difficult. For example, the potential use of the low-growing sedge slender spikerush (*Eleocharis acicularis*) in canals and lakes was investigated for several years. However, this plant grows quite slowly, requires relatively high light levels, and can be easily dislodged by physical wave action and foraging by waterfowl. Unfortunately, many of the "ideal" characteristics (open canopy, small stature) of beneficial plants tend to make them less competitive compared to introduced species.

The establishment of desirable shade trees near shorelines can significantly reduce light, and thereby lessen the opportunity for algae and rooted aquatic plants to grow. Trees planted around small ponds, or along the sides of canals, can reduce light by well over half of what is found in unshaded areas. Because the roots of some trees can extend into the pond or canal, these should not be used. Alternatively, some type of containment can be used to prevent roots from encroaching into the bank. For irrigation and drainage canals, tree plantings may be of limited practicality because vehicle access is usually required and canal-bank roads are usually quite narrow.

Aquascapes and other small ponds designed for aesthetic purposes are particularly prone to rapid invasion when nonnative plants are used. The desire for quick establishment of vegetation should be tempered with the knowledge that, in the long run, slower-growing, less aggressive native plants will usually provide more attractive ponds and will be less labor intensive to maintain.

FURTHER READING

Anderson, L. W. J. 2005. California's reaction to *Caulerpa taxifolia*: A model
for invasive species rapid response. *Biological Invasions* 7: 1003–1016.
Aquatic Plant Management Society (APMS). www.apms.org/
Ashton, P. J., and D. S. Mitchell, 1989. Aquatic plants: Patterns and modes
of invasion, attributes of invading species and assessment of control
programmes (111–154). In J. A. Drake et al., eds. *Biological Invasions: A
Global Perspective.* Chichester: Wiley.
Center for Invasive and Aquatic Plants (Florida). http://aquat1.ifas.ufl.edu/
Cohen, A., and J. Carlton. 1995. *Nonindigenous Aquatic Species in a United
States Estuary: A Case Study of the Biological Invasions of the San Fran-
cisco Bay and Delta.* U.S. Fish and Wildlife Service and National Sea
Grant (Connecticut Sea Grant). NOAA Grant # NA36RG0467.
Cook, C. D. K. 1990. *The Aquatic Plant Book.* The Hague: SPB Academic
Publishing.
Ecosystem Restoration Foundation (AERF). www.aquatics.org/index.html
Johnson, R. R. 2007. Seaweed invasions: A synthesis of ecological, eco-
nomic and legal imperatives. *Botanica Marina* 50: 1–144.
Pieterse, A. H., and K. J. Murphy, eds. 1990. *Aquatic Weeds: The Ecology and
Management of Nuisance Aquatic Vegetation.* Oxford: Oxford University
Press.
Sculthorpe, C. D. 1967. *The Biology of Aquatic Vascular Plants.* London:
Edward Arnold, Ltd. (Reprinted in 1985 by Koeltz Scientific Books.)
Williams, S. L., and J. E. Smith. 2007. A global review of the distribution,
taxonomy and impacts of introduced seaweeds. *Annual Review of Ecol-
ogy, Evolution and Systematics* 38: 327–359.

FUNGI

MARIE-LAURE DESPREZ-LOUSTAU

*Institute National de la Recherche Agronomique (INRA),
Villanave d'Ornon, France*

DAVID M. RIZZO

University of California, Davis

Fungi make up one of the major clades of the tree of life.
Compared with plants and animals, fungi are often poorly
represented on invasive species lists. However, over the
past several centuries, a number of fungal introductions
have had destructive impacts on agricultural and natural
ecosystems. Although most information on invasive fungi
concerns fungal parasites of plants or animals, fungi with
a wide range of ecological characteristics have been introduced into new ecosystems around the world. The long-
term impacts of most of these introductions are unknown.
Limited knowledge of fungal diversity and ecology may
explain the poor understanding of ecological impacts and
limited efficacy of prevention of fungal invasions.

GENERAL FEATURES OF FUNGI

Fungi are eukaryotic, heterotrophic organisms with
absorptive nutrition that are classified in the kingdom
Fungi. Some morphological and ecological characteristics of Fungi are shared by other organisms now demonstrated to belong to different evolutionary lineages
(sometimes called "pseudo-fungi"). The resemblance
between fungi and water molds and downy mildews,
belonging to Oomycetes (in Stramenopila), is an example of convergent evolution.

Most fungi are characterized by a vegetative thallus,
called mycelium, consisting of a network of microscopic
branched filaments, called hyphae. Fungi reproduce asexually or sexually by producing spores, which constitute
dispersal and survival propagules.

The kingdom Fungi encompasses a huge diversity of
taxa, of which only a small part has so far been recognized. While 75,000 to 120,000 species have been formally described, the total number of fungal species on
Earth is estimated to be greater than 1.5 million (fungi
and plants exist in a 6:1 ratio). Fungi may be found in
a wide range of habitats and niches. The majority of
fungi are saprobes (i.e., they decompose dead organic
matter). Fungi are especially important in cellulose and
lignin digestion, and therefore play an important role in
nutrient cycling in most terrestrial ecosystems. Fungi are
often symbiotic with other organisms (including plants,
animals, and other fungi), on which they depend for
their nutrition. These relationships range from mutualism (e.g., with lichens, mycorrhizae) to parasitism. As
symbionts of other species, fungi may affect the structure and dynamics of plant and animal populations and
communities.

Fungi are of considerable social and economic
importance for human societies. Their beneficial effects
include direct or indirect use for food (edible mushrooms; yeasts for baking or for beer and wine production) and medicinal purposes (antibiotics). But fungi
also have detrimental effects on human health (human
pathogens, poisonous species, mycotoxins, crop losses
due to fungal diseases).

WORLDWIDE MOVEMENT OF FUNGI
BY HUMANS

Fungi have probably always traveled with humans through
transportation on plants or other material. For example,
in 1991 a well-preserved 5,000 year old mummy was
found on a glacier in the Alps. Named Ötzi the Iceman,
he carried two species of shelf mushrooms collected from

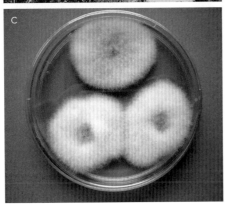

FIGURE 1 *Fusarium circinatum* (Fungi, Ascomycota Hypocreales; sexual stage = *Gibberella circinata*). (A) Tree death as a result of resinous cankers (B) caused by *Fusarium circinatum* (C). This fungus is the causal agent of pitch canker on pines, especially *Pinus radiata*, and can also infect Douglas fir. The fungus is believed to have originated in Mexico and has been reported in many regions: the eastern and western United States, Chile, South Africa, and recently Europe (Spain and Italy). It is a threat for native populations of Monterey pine in California and for the forest industry in several regions, such as New Zealand. (Photographs courtesy of Xavier Capdevielle and Renaud Ioos.)

decaying wood; one species was likely used for medicinal purposes, while the other was included in a fire-starting kit. Plant parasitic rust fungi (*Puccinia* spp.) have tracked their cereal hosts worldwide through domestication and cultivation from their center of origin. The devastating impact of rust fungi in Roman times is attested by the Robigalia, annual festivals to the god Robigus, who was supposed to protect the crops. Early European explorations in the fifteenth century were probably an introduction route for many fungi, as well as, presumably, the oomycete *Phytophthora cinnamomi*, the causal agent of ink disease of chestnut in Europe, first reported in Portugal.

With the exploding increase of travel and trade in recent decades, introductions of fungi to new regions have been rising exponentially (Figs. 1 and 2). This is explained by the fact that travelers and commodities carry many unintentional "contaminants." In particular, fungi present attributes (inconspicuous mycelium, microscopic propagules, dormant structures) that may favor their transport,

establishment, and spread. In Europe, levels of imported goods have been shown to be a much better predictor for the number of alien fungal pathogens within a country than have geographic variables such as country area, latitude, or longitude. Many fungal and similar pathogens have been introduced worldwide with their host plants. Examples include *Phytophthora infestans* in infected tubers from America to Europe, *Cryphonectria parasitica* on infected Japanese chestnut trees from Asia to North America, and *Diplodia pinea* in pine seeds imported in South Africa (Fig. 3). Exotic tree species have been a vector of mycorrhizal species, such as *Suillus amabilis* accompanying Douglas fir from North America to Europe.

In contrast to plant and animal invasions, deliberate introductions have played a minor role in fungal introductions. Desirable fungi introduced into new habitats mostly include pathogenic fungi used for classical biological control and species used for mycorrhization. Few, if any, examples of fungal invasions resulting from the

FIGURE 2 *Melampsora larici-populina* (Fungi, Basidiomycota, Pucciniales) is the Eurasian poplar rust fungus. In Europe it is widespread both in natural stands of the native *Populus nigra* and in cultivated stands of poplar (mostly Euro-American and inter-American hybrids). The impact is limited in natural stands, but the disease can cause severe damage in cultivated stands due to growth loss induced by defoliation. *M. larici-populina* was recently introduced to North America. (A) Resistant (right) and susceptible (left) clones defoliated after severe infection. (B) Typical orange sporulation (uredia producing urediniospores = asexual stage) of the fungus. (C) Differential reactions between poplar cultivars and rust races. (Photographs courtesy of Pascal Frey.)

escape of purposefully introduced pathogenic or mycorrhizal fungi have been reported. The low invasive success of these deliberate introductions may lie in the low genetic diversity of the introduced material and in the absence of close relatives of the target hosts to which fungi could extend their ranges. A few species of edible fungi have also been introduced outside their native range into various parts of the world. *Agaricus bisporus*, indigenous to southwestern Europe and California, has been cultivated worldwide. Escapes of cultivated strains have been suggested to occur in some regions, which may put at risk native diversity by introgression with wild strains. Conversely, *Lentinula edodes*, the Asian shiitake mushroom cultivated in Europe and America, has not been reported to occur in natural environments.

IMPACTS OF FUNGAL INVASIONS

The most obvious impacts of fungal invasions are epidemics caused by exotic pathogenic fungi. It has been estimated that 65 to 85 percent of plant pathogens worldwide are alien in the location where they were recorded. Introductions of fungal pathogens into novel areas result in changes in the coevolutionary process with their hosts. In the case of agricultural crops and forest plantations, the hosts themselves are exotic in most parts of the world where they are grown. During fungal invasions in these agroecosystems, host and pathogen populations are reunited under very different conditions (e.g., monocultures, reduced genetic variability in the host, high density, increased moisture regimes) from those that prevailed during natural coevolution. This often results in devastating epidemics, as exemplified by potato blight (*Phytophthora infestans*), coffee rust (*Hemileia vastatrix*) in Asia, soybean rust (*Phakopsora pachyrhizi*) in the Americas, grapevine powdery (*Erysiphe necator*) and downy mildew (*Plasmopara viticola*) in Europe, and wheat stripe rust (*Puccinia striiformis*) in Australasia. Such invasions can have significant ecological, economic, and social consequences. The Great Irish Famine in 1845–1849, triggered by the potato blight epidemics, is a striking example. In 2001, the cost of alien plant pathogens (mostly fungi) was estimated to be US $23.5 and $35.5 billion per year for the United States and India, respectively.

In contrast to agricultural situations, pathogens introduced into natural ecosystems form new interactions with species they have never encountered, by realizing host shifts. The new host species that do not share a coevolutionary history with the pathogens (under any environmental condition) may show limited genetic resistance to infection. There are many examples of exotic fungal plant pathogens in natural ecosystems, particularly forests,

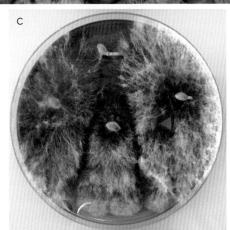

FIGURE 3 *Diplodia pinea* (syn. *Sphaeropsis sapinea*, Ascomycota, Incertae sedis, no known sexual stage) is a damaging pathogen to plantings of both exotic and native pine species on all continents (North and South America, South Africa, New Zealand, Europe). *Diplodia pinea* kills current-year shoots, and ultimately entire trees. The fungus can remain latent in trees without causing symptoms, turning pathogenic after hail or drought stress. Movements of infected seeds and latently infected plants probably assisted in the introduction of the fungus into many areas. (A) Dying trees, infected with *Diplodia pinea*. (B) Pycnidia (black spots = asexual fructifications of the fungus) on pine cones. (C) Outgrowth of *D. pinea* from infected seeds on selective medium. (Photographs courtesy of Dominique Piou and Thibaut Decourcelle.)

where naive hosts have shown very high susceptibility to infection and death. The chestnut blight pathogen, *Cryphonectria parasitica*, was introduced from Asia to North America in the late 1800s and had a rapid and devastating impact on eastern hardwood forests by removing the dominant American chestnut and inducing cascading effects in the whole forest community. The introduction of *Phytophthora cinnamomi* into southwestern Australia is another example of serious alteration of the native plant communities, resulting from the death not only of the dominant jarrah (*Eucalyptus marginata*) but also of many understory species (92% of Proteaceae are susceptible to this pathogen in Western Australia). The devastating epidemics of Dutch elm disease (*Ophiostoma ulmi*), white pine blister rust (*Cronartium ribicola*), and more recently sudden oak death in the United States (*Phytophthora ramorum*) are additional examples. Introduced fungi are also important causes of emerging diseases in animals. Chytridiomycosis, caused by *Batrachochytrium dendrobatidis*, has been associated with population declines and possible extinction of many amphibian species in several regions of the world (Fig. 4). In most of these cases of emerging diseases, the pathogens were completely unknown prior to their introduction, and in only a minority of cases has the geographic area of origin of the pathogen been determined.

Beyond plant and animal pathogens causing emerging diseases, the impacts of exotic mutualistic and saprotrophic fungi on ecosystem structure and function have not been extensively characterized. *Clathrus archeri*, the octopus stinkhorn, is a typical example (Fig. 5). After its first report in France in 1920, it spread throughout Europe, reaching high population levels in some locations, but its impact on indigenous fungal populations is unknown. The potential impact of introduced mycorrhizal fungi has long been assumed to be negligible or positive; however, this situation clearly deserves more attention. Interactions of exotic plants, animals, and microbes (including fungi) with native fungal communities are still poorly understood. Changes in saprobic and mycorrhizal fungal communities during plant community succession have been well documented, and it is probable that exotic species–mediated successional changes would result in changes in fungal communities.

FURTHER UNDERSTANDING AND PREVENTING FUNGAL INVASIONS

The impacts of introduced destructive plant pathogens in the nineteenth century (especially potato late blight) prompted efforts to regulate the movement of

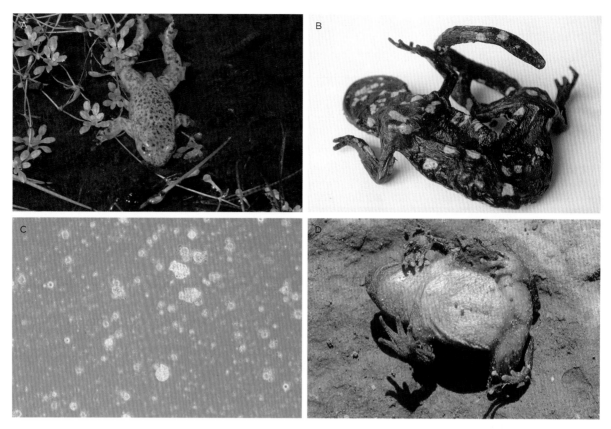

FIGURE 4 *Batrachochytrium dendrobatidis* is a chytrid fungus (Fungi, Chytridiomycota, Rhizophydiales) that was described in 1999 and recognized as a proximate driver of many amphibian declines. Over 350 amphibian species can be infected. Ecological and genetic studies suggest that the emergence of chytridiomycosis was caused by the human-mediated movement of this recently evolved pathogen from a yet unidentified source population. Distribution and impact of the disease are associated with environmental variables, especially temperature. (A), (B), and (D) Batrachochytrium-induced mortality in toads and salamander. (C) The chytrid fungus: sporangia in water. (Photographs courtesy of Matthew Fisher.)

agricultural commodities that may carry pests and diseases harmful to agriculture. The Plant Quarantine Act of 1912 was the first legal action taken in the United States to prevent the introduction of pests from foreign countries. The International Plant Protection Convention (IPPC) was developed in the 1920s. Phytosanitary regulations and quarantine policies have long been centered on individual species lists. However, several recent introductions, where pathogens were unknown prior to their destructive emergence, point to the limitations of such measures. Emphasis is currently shifting to a pathway approach, in which the focus is on the most likely pathways through which invasive organisms might be introduced. New regulations governing the import of wood-packing material (International Phytosanitary Standard ISPM15 of the IPPC) exemplify this pathway approach.

Fungi are an especially diverse group of organisms that harbor biological traits favorable to invasiveness (inconspicuousness, high reproductive and evolutionary potential, high plasticity). However, fungi are generally poorly represented on invasive species lists; for example, the IUCN global invasive species database includes 11 species of fungi and pseudo-fungi as compared to several hundred terrestrial plants. The low representation of fungi on invasive species lists and the limited efficacy of the prevention of introductions are thought to be mainly due to a lack of ecological and biogeographical information on this group.

Knowledge of the diversity of fungal species, either indigenous or exotic, that already occur in a given place is essential. However, baseline data on the diversity of resident fungal species is still limited even for well-studied environments such as temperate regions. Several recent initiatives have been made to develop databases of baseline fungal diversity. These include the Systematic Mycology and Microbiology Laboratory of the USDA database that includes over 94,000 species of plant fungal pathogens and the inventory of alien fungi in Europe associated with the DAISIE project. Several fungal barcoding initiatives

FIGURE 5 The octopus stinkhorn, *Clathrus archeri* (Fungi, Basidiomy-cota, Phallales), native from Australasia, was introduced to Europe in the early 1900s, maybe with Australian soldiers during World War I. Multiple introductions linked to the wool trade have been suggested. The fungus has been spreading in most of Europe, especially in areas with a temperate oceanic climate. It was also more recently introduced to California. Its impact on native fungal communities is unknown. (Photograph courtesy of Jacques Guinberteau.)

are rapidly progressing that will help with the detection of quarantine organisms and the description of fungal communities. The use of modern population genetic tools and analyses has increased insights into the biogeography of fungi. Many species previously thought to have cosmopolitan distributions have been shown to be actually a complex of cryptic species (morphologically similar but representing distinct evolutionary lineages) with distinct geographical ranges. When a new disease outbreak is caused by a previously unknown species, as has occurred in recent emerging epidemics, it is often difficult to determine between the endemic (with environmental or evolutionary changes) and the exotic pathogen hypotheses. In spite of intensive research efforts on *Batrachochytrium dendrobatidis* (Fig. 4) since its discovery, the question on its origin remains unanswered to date. Several potential sources (in Africa, America, and more recently Asia) have been proposed based on population genetic data (assuming increased levels of genetic diversity in the center of origin) or on histology on historical museum specimens (searching for long presence of the pathogen on endemic species, indicative of co-evolution). The reasons behind the recent emergence of chytridiomycosis thus remain unclear until more ecological and genetic data can help identify and rank the factors (genotype transfer, evolutionary change, environmental change).

Improving prevention of and management measures for fungal invasions will require a better knowledge of the drivers and consequences of these invasions. Predicting invasiveness of species based on biological information has been an important field of research in invasion ecology. Recent analyses of past fungal invasions have suggested that some taxonomic groups, such as Erysiphales and Peronosporales (powdery and downy mildews), or functional groups (foliar pathogens) are more likely to be successful invaders. Quantifying the impacts of exotic species is often impossible or difficult because of a lack of baseline ecological data on invaded ecosystems. An increased knowledge in the assemblage of fungal species in the resident community, and its disruption by invaders, should lead to a better understanding of invasibility.

SEE ALSO THE FOLLOWING ARTICLES

Biological Control, of Animals / Biological Control, of Plants / Disease Vectors, Human / Forestry and Agroforestry / Mycorrhizae / *Phytophthora* / Parasitic Plants / Pathogens, Plant

FURTHER READING

Desprez-Loustau, M.-L. 2009. The alien fungi of Europe (15–28). In *Handbook of Alien Species in Europe.* Dordrecht: Springer.

Desprez-Loustau, M.-L., C. Robin, M. Buée, R. Courtecuisse, J. Garbaye, F. Suffert, I. Sache, and D. M. Rizzo. 2007. The fungal dimension of biological invasions. *Trends in Ecology and Evolution* 22: 472–480.

Farr, D. F., and A. Y. Rossman. Fungal Databases, Systematic Mycology and Microbiology Laboratory, ARS, USDA. http://nt.ars-grin.gov/fungaldatabases/

Hawksworth, D. L. 2001. The magnitude of fungal diversity: The 1.5 million species estimate revisited. *Mycological Research* 105: 1422–1432.

Loo, J. 2009. Ecological impacts of non-indigenous invasive fungi as forest pathogens. *Biological Invasions* 11: 81–96.

Parker, I. M., and G. S. Gilbert. 2004. The evolutionary ecology of novel plant–pathogen interactions. *Annual Review of Ecology, Evolution and Systematics* 35: 675–700.

Pimentel, D., S. McNair, J. Janecka, J. Wightman, C. Simmonds, C. O'Connell, E. Wong, L. Russel, J. Zern, T. Aquino, and T. Tsomondo. 2001. Economic and environmental threats of alien plant, animal and microbe invasions. *Agriculture, Ecosystems, Environment* 84: 1–20.

Rizzo, D. M. 2005. Exotic species and fungi: Interactions with fungal, plant and animal communities (857–877). In J. Dinghton, P. Oudemans, and J. White, eds. *The Fungal Community,* 3rd ed. CRC Press.

Schrader, G., and J. G. Unger. 2003. Plant quarantine as a measure against invasive alien species: The framework of the International Plant Protection Convention and the plant health regulations in the European Union. *Biological Invasions* 5: 357–364.

van der Putten, W. H., J. N. Klironomos, D. A. Wardle. 2007. Microbial ecology of biological invasions. *ISME Journal* 1: 28–37.

G

GAME ANIMALS

LEONARD A. BRENNAN AND FRED C. BRYANT

Texas A&M University, Kingsville

People have been capturing game animals and introducing them to new locations for centuries. Marco Polo brought ring-necked pheasants (*Phasianus colchicus*) from Asia and introduced them to Europe. In another illustrious example, George Washington introduced fallow deer (*Dama dama*) to his Mount Vernon estate. The heyday of translocating exotic game animals to the United States spanned about four decades, from the 1890s to the 1930s. Although exotic game animals are still being introduced in the United States as well as many other countries, it is now usually done in relatively limited and regulated contexts, at least compared to the past. For hoofed mammals, there are presently about 230,000 individuals represented by over 80 species outside of zoos, in the United States. Comparable estimates for game birds are not known but are certainly significant because species such as ring-necked pheasants, gray partridge (*Perdix perdix*), and chukars (*Alectoris chukar*) are now widely established over hundreds of thousands of square kilometers of the United States.

MOTIVES FOR INTRODUCING GAME ANIMALS

The motives for introducing game species to locations outside of their native range were relatively simple. A perceived lack of native game, from whatever cause, was typically the rationale used to import exotic game species to the United States. State wildlife agencies also played a major role in the introduction of exotic species. In these organizations, the additional rationale of "providing hunting opportunities" was a source of motivation. The fact that budgets for these agencies rose and fell on the sales of hunting licenses was also a source of inspiration for translocating exotic species to new places. To complicate matters, documentation of introductions by wildlife agencies was typically scant, poor, or nonexistent. For example, many game bird species were introduced to the United States from the Old World, but only a few became established. Records documenting failed introductions have long faded into the mist of lost institutional memories. Also, translocation records in many states may be suspect for releases as frequent as 20 years ago.

Today, conservation of endangered game species is a new factor that is driving translocations and introductions. For example, the scimitar-horned oryx (*Oryx dammah*), which is hunted on Department of Defense lands in New Mexico, is considered highly endangered and possibly extinct in its native North Africa range. There are a number of other examples where populations of exotic game mammals in the United States now outnumber populations in their native habitats.

GAME BIRDS

Among birds, galliforms clearly constitute the bulk of introduced game species. A notable exception is the Eurasian collared dove (*Streptopelia decaocto*), a columbid originally from Asia that was introduced in the Bahamas in 1975 and is now spreading quickly across the United States and Canada.

Game birds have been introduced through three primary pathways: (1) species from other places that were brought to the United States, (2) species from the United States that were introduced to places overseas, and (3) species native to the United States that were introduced to places outside their native range in the United States. The section below summarizes what is known about introductions of nine species of galliforms that have been introduced, or attempted to be introduced, around the world. Although numerous galliforms such as pea fowl (*Pavo* spp.), guinea fowl (*Guttera* spp., *Numida* spp.), and jungle fowl (*Gallus* spp.) have been translocated far from their original ranges, these birds have been either domesticated or semi-domesticated and are not hunted here as wild game birds.

HIMALAYAN SNOWCOCK (*TETRAOGALLUS HIMALAYENSIS*) Native in mountains of central Asia. Introduced in the Ruby Mountains of northeastern Nevada.

CHUKAR (*ALECTORIS CHUKAR*) Native across broad expanse of southeastern Europe, Asia Minor, and central Asia. Introduced to the Great Basin region of the United States, where several hundred thousand are bagged yearly.

RED-LEGGED PARTRIDGE (*ALECTORIS RUFA*) Native to Western Europe, introduced in England. Also marginally established in New Zealand. Attempts to introduce this species to the United States failed.

BLACK FRANCOLIN (*FRANCOLINUS FRANCOLINUS*) Originally found in southern Asia and southern Europe. Successfully introduced to south Florida, Louisiana, Hawaii, and Guam. Reintroduced to Italy and Portugal.

GRAY PARTRIDGE (*PERDIX PERDIX*) Native to Europe and western Siberia. Established across a band of states and provinces from New York and southern Ontario, west through Montana, Alberta, and eastern British Columbia, and south through eastern Washington to northwestern Nevada. Also introduced to New Zealand but did not persist.

RING-NECKED PHEASANT (*PHASIANUS COLCHICUS*) Most widely distributed pheasant in the world, even prior to introductions. Native range covered most of temperate Asia. Introduced nearly worldwide, in places such as Chile, New Zealand, Tasmania, North America, Japan, Hawaii, and Europe. An extremely popular game bird, with millions bagged annually.

WILD TURKEY (*MELEAGRIS GALLOPAVO*) Native to the United States and Mexico. Range in the United States expanded greatly to include northern Vermont, southeastern Quebec, southern Ontario, and western states such

FIGURE 1 Northern bobwhite. Males have a white face. The female, in the middle of the covey, has a brownish, tawny-colored face. (Photograph courtesy of Larry Ditto.)

as California, Utah, Oregon, Wyoming, and Montana. From 1995 to 1998, at least 540 wild turkeys were trapped in Texas and relocated to undisclosed locations in Mexico. Also introduced widely around the world, including Hawaii, Australia, New Zealand, and still possibly persisting from introductions in Austria and Germany.

NORTHERN BOBWHITE (*COLINUS VIRGINIANUS*; FIG. 1) Original range in eastern and midwestern United States, south through the eastern third of Mexico and Guatemala. Introduced to the Palouse Prairie region of southeastern Washington. Worldwide introductions include the West Indies, France, and New Zealand. Attempted introduction to British Isles failed. Translocation of bobwhites from Mexico and Texas to replenish areas of depleted populations was a major, but poorly documented, activity of many state game agencies in the middle part of the twentieth century. For example, in the late 1930s, 20,000 bobwhites trapped in Mexico were introduced into East Texas. Banding returns indicated that less than 1 percent of these birds were bagged by hunters, and the program was abandoned.

CALIFORNIA QUAIL (*CALLIPEPLA CALIFORNICA*) Original range from southern British Columbia through parts of Washington, Idaho, Oregon, and Nevada, south to Baja California in Mexico. Introductions around the world include Chile, western Argentina, King Island Australia, New Zealand, and Hawaii. Attempts at introductions in Europe were unsuccessful. Because of its status as a highly popular and desirable game bird, many translocations from its original range to new locations in such places as Vancouver Island were attempted and apparently were successful.

HOOFED GAME ANIMALS

Introductions of exotic hoofed game animals literally span the alphabet—from axis deer (*Axis axis*) to zebras (*Equus* spp.)—as noted by E. C. Mungall in her *Exotic Animal Field Guide*. Compared to introductions of game birds, which have been relatively benign by most ecological standards (although introductions of any exotic species to islands such as Hawaii, Guam, or New Zealand are, by definition, problematic), introductions of hoofed game animals have caused serious wildlife management problems, as noted in some of the species accounts for this section. Pathways of introductions for hoofed game animals are largely the same as those for game birds. A primary difference, however, is that for hoofed game animals, the majority of the introductions have been into the United States from places such as Asia and northern Africa. Another difference is that, for a species such as bighorn sheep (*Ovis canidensis*), reintroductions of translocated animals to restock areas suffering local or regional extinctions have been a major focus of wildlife agencies for many decades, especially in the western United States.

Most exotic mammals require some form of fencing to attempt to contain their movements and minimize interactions with native wildlife. However, a common characteristic of fences is that they ultimately become porous, animals escape, some become naturalized, and a new set of wildlife conservation challenges ensues. Transmission of diseases and parasites by exotics is also a problem in some circumstances.

AXIS DEER (*AXIS AXIS*) Original range was Indian subcontinent. Naturalized throughout many areas of Texas, especially South Texas and Edwards Plateau. A gregarious deer that seems to thrive in suburban and rural habitats of southern and central Texas.

FALLOW DEER (*DAMA DAMA*) Original range was probably southern Europe and Asia Minor, but centuries of capture and translocation have obscured this. Known to hybridize with Persian fallow. Naturalized in great numbers on the Edwards Plateau of Texas. Also naturalized and now common in Patagonia. Attempts in the 1960s to 1980s at farming fallow deer in the United States for meat have largely been abandoned.

EUROPEAN RED DEER (*CERVUS ELAPHUS*) Considered conspecific with American elk. Originally found in Europe, but now widely distributed on deer farms throughout the United States for production of venison and velvet. Well established in New Zealand and Patagonia, both for sport hunting and farmed meat production. American elk have also been translocated to New Zealand.

SAMBAR DEER (*CERVUS UNICOLOR*) Original range was Southeast Asia. Widely introduced to many locations in Australia, New Caledonia, New Guinea, Mauritius, and New Zealand, as well as California, Texas, and St. Vincent Island in Florida. Known to hybridize with European red deer (*Cervus elaphus*) in British Isles and New Zealand.

SIKA DEER (*CERVUS NIPPON*) Found originally in China and Japan. Free-ranging in Maryland, Virginia, and elsewhere. Problematic as competitor with native white-tailed deer. Also problematic in Great Britain, where introduced from the 1860s to 1930s. Escaped and deliberately introduced animals are known to hybridize with European red deer; such hybrids may be a threat to native "pure" stocks of red deer in Great Britain.

SILK Hybrid of American elk and Sika deer. Originated in Missouri, but also were moved to Texas for venison ranching. Ability to hybridize with American elk is an example of risks exotics pose when brought into contact with native animals in new range.

WHITE-TAILED DEER (*ODOCOILEUS VIRGINIANUS*) Native to North and South America and introduced widely around the world, including to the British Isles, Bulgaria, the Czech Republic, Finland, the former Yugoslavia, New Zealand, Cuba, the Virgin Islands, and other Caribbean islands. Source populations of these translocations are poorly documented and largely unknown. This species is widely distributed across North and South America; thus, there could be any number of states, countries, or provinces that have been involved in the movement of these game animals. In Texas, records from 1961 to 1981 indicate that at least 565 deer of this species were trapped in 11 different counties and moved to Mexico. Additionally, from 1995 to 1999, another 596 white-tailed deer were moved into Mexico from Texas. The exact locations in Mexico where these animals were released were not documented, so far as we know.

MULE DEER (*ODOCOILEOUS HEMIONUS*) A widely distributed and highly important game species in western North America. Records indicate that from 1997 to 2009, a total of 477 mule deer were trapped in Texas and released at undisclosed locations in Mexico. The black-tailed deer (*O. h. columbianus*, or *O. h. sitkensis*) introduced to the Queen Charlotte Islands, has caused significant ecological

FIGURE 2 Male nilgai antelope. (Photograph courtesy of Larry Ditto.)

impacts, especially by limiting regeneration of western red cedar (*Thuja plicata*).

NILGAI ANTELOPE (*BOSELAPHUS TRAGOCAMELUS*; FIG. 2) Originally from the Indian subcontinent. Introduced to Texas in the 1930s. Widely distributed across deep South Texas, especially along the coastal region below Baffin Bay. Problematic competitor with domestic cattle. Does not respect fences. Harvested in the wild as meat for specialty restaurants. Has been transplanted to other parts of Texas for hunting opportunities, but subfreezing temperatures limit its northern distribution.

BLACKBUCK ANTELOPE (*ANTILOPE CERVICAPRA*) Originally from India. Widely distributed across the Edwards Plateau and also somewhat in South Texas. Also translocated to Argentina and Australia. Generally a benign introduction compared with larger introduced ungulates.

AOUDAD (*AMMOTRAGUS LERVIA*) Also known as Barbary sheep, from North Africa. Introduced to and flourished in New Mexico and northern and western Texas. Problematic competitor with desert bighorn sheep and desert mule deer. Nearly impossible to eradicate in rough terrain.

CATALINA GOAT (*CAPRA HIRCUS*) Named for feral goats on Catalina Island, California. Most likely, these first goats were from Spain and were left on Catalina by sailors. Now used as a generic term for any exotic goat offered for hunting. Eradication has been attempted on Catalina Island and in the Galapagos Islands because of damage to habitats.

MOUNTAIN GOAT (*OREAMNOS AMERICANUS*) Translocated from its historic range in Idaho, western Montana, the western Canadian Rockies, and southeastern Alaska into the central Rocky Mountains and the Olympic Peninsula; they have been considered destructive on the Olympic Peninsula.

REINDEER (*RANGIFER TARANDRUS*) Native along the coast in Greenland and in arctic habitats of Europe and Asia. Brought from Finland to Alaska in the 1890s to replace depleted caribou (now considered conspecific with reindeer; formerly *R. arcticus*, and *R. caribou*) populations, which have never fully recovered. Introduced population on St. Matthew Island in Bering Sea irrupted from 40 to over 1,300 animals, then crashed, over a 20 year period. Introduced in the early 1900s to St. Paul Island, Alaska, where it is still abundant. Brought to South Georgia Island in Antarctica by Norwegian whalers in 1911, where it is considered a pest and a threat to native plants.

MUSKOX (*OVIBOS MOSCHATUS*) Originally distributed from the North Slope of Alaska to the Hudson Bay. Currently found from the Canadian arctic mainland and islands to the northern coast of Greenland. Introduced to locations in Iceland, Spitsbergen, Nunivak Island (Alaska), and the Kerguelen Isles of Antarctica.

BIGHORN SHEEP (*OVIS CANIDENSIS*) Widely distributed throughout North American cordillera. Populations severely depleted by overhunting and diseases from domestic livestock. State wildlife agencies have been involved with translocations for reintroductions into areas of local and regional extinctions for more than 50 years.

MOUFLON (*OVIS MUSIMON*) Considered an early form of domestic sheep that became feral. Originally from Corsica and Sardinia. Abundant on the Edwards Plateau of Texas.

WILD BOAR OR FERAL HOG (*SUS SCROFA*) Two sources of stock were responsible for introduction of this species: (1) domestic pigs that escaped from barnyard life to become "feral," and (2) European wild boar brought to North Carolina from Germany in 1912 and transplanted to places such as Tennessee, Texas, and California. Geographic range and populations are continuing to expand in places such as Patagonia, the Galapagos, and elsewhere. Hogs are a huge menace to native vegetation and ground-dwelling animals on islands, as well as in mainland habitats. They are also capable of transmitting tuberculosis and pseudo-rabies to domestic livestock. Control of wild hog populations—eradication is probably out of the question except perhaps on certain islands—is one of the most challenging issues in wildlife conservation today.

FUR-BEARING MAMMALS

The term "fur bearers" refers to carnivorous, but also in some cases to herbivorous, wild animals that have historically been exploited for the value of their fur. Meat from fur-bearing species such as bears, although often of great volume and value to native peoples and frontier settlers, was usually of secondary consideration for most of these species. A common characteristic among most fur-bearing mammals is that they have historically been persecuted, either for the value of their pelts for all types of clothing or because of their threats to livestock and human safety.

This article addresses the topic of fur bearers because many of these species represent a special case of wild animal introductions, or more correctly, reintroductions, in the case of many species. The motive behind reintroductions of fur bearers back into their native ranges stems from what Aldo Leopold called an "atonement" approach to wildlife management. Today, it might be termed a guilt trip. Nevertheless, the case of using reintroductions of various species of fur-bearing mammals to conserve and sustain their populations as well as their legacy as wild animals is a major bright spot in the modern arena of wildlife science and conservation. Along with physical translocations of wild-trapped, or in some cases captive-bred, animals into their former range, the role of protective policy and regulations, the elimination of bounties, and cultural appreciation for the intrinsic value of knowing that these species roam the landscape have been key factors that have allowed some populations to recolonize areas. Conversely, purposeful introductions of various furbearing species from native ranges to overseas locations for economic development of fur industries have been problematic.

GRAY WOLF (*CANIS LUPUS*) Original range was most of North America, except for the southeastern United States and the Californias. Systematically exterminated from most of the lower 48 states, with the exception of parts of Michigan and Minnesota. Remains relatively abundant in Alaska and Canada. Reintroduction into Yellowstone National Park and surrounding ecosystem, although highly controversial, can be considered biologically successful in recent years. Other introductions into the northern Rocky Mountains have also been successful.

RED WOLF (*CANIS RUFUS*) Virtually eliminated from original range from Texas to Illinois to Florida. Heavily persecuted to the point of near extinction, then former range was usurped by the coyote. Reintroduction into the wild in parts of Carolinas seems to be met with initial, tentative success.

BLACK BEAR (*URSUS AMERICANUS*) Once found throughout all woodlands of North America and south into Mexico. Persecuted to the point of regional extinction in the eastern and midwestern United States. Populations are recovering, and in some places are becoming abundant to the point of being a nuisance. Louisiana and Arkansas have conducted major black bear reintroduction programs that lasted nearly a decade and were highly successful.

GRIZZLY BEAR (*URSUS HORRIBILIS*) Originally found in western North America, from Alaska to Mexico, through the Great Plains and Central Valley of California. Frequent encounters with what Lewis and Clark called "white bears" as they slogged along the Missouri River are a prominent element of their published journals. Exterminated in Mexico and most of North America, except far northern Montana and parts of Wyoming and perhaps Idaho. A remarkable proposal to reintroduce grizzlies into the more than 10,000 square miles of the Selway-Bitterroot and River of No Return Wilderness areas in Idaho was seriously considered and almost implemented in the late 1990s and early 2000s. Meanwhile, during this period of political posturing and hand-wringing, one or more grizzlies apparently found their own way into the Selway-Bitterroot from either Montana or Wyoming. A hunter killed one in the Selway-Bitterroot during September 2007, thinking it was a black bear. Perhaps this is an example of an indirect approach to reintroducing a species back into its native range?

PUMA (*FELIS CONCOLOR*) Also called mountain lion, cougar, panther, or catamount. Once found throughout the conterminous United States south through Mexico and into South America. Now largely extinct over the eastern two-thirds of United States, but thought to be expanding in mid-central states and known to be relatively abundant in western mountain states. Efforts to conserve and sustain the endangered Florida subspecies continue to be highly motivated and creative, even to the point of adding animals from Texas to thwart genetic problems.

FISHER (*MARTES PENNANTI*) Ranges from southern Canada to the western Rockies and northern New England. The highly prized fur of this relatively large weasel caused it to be exterminated over large parts of its range. Staunch regulation of trapping and purposeful reintroduction has allowed populations to recover. In New England, a key motivation of reintroducing fishers was their unique ability to prey on porcupines, and thus to exert control on porcupine populations that were becoming serious pests in farmland woodlots as well as in areas of commercial timber.

"NORTH AMERICAN" MINK (*NEOVISON VISON*) Original distribution in Alaska, lower two-thirds of Canada, and all of conterminous United States except for the arid Southwest. Widely introduced in northern Europe for development of the fur industry. Naturalized animals from escapes or intentional releases are considered pests and are a major problem for ground-nesting birds and native small mammals across many locations in Europe.

RIVER OTTER (*LUTRA CANADENSIS*) With the exception of the extreme west coast, river otters were found throughout nearly all of the United States and Canada, excluding the high Canadian Arctic, the Great Basin, and parts of extreme southern Arizona and New Mexico and the western two-thirds of Texas. Populations were reduced to the point of widespread regional extinctions by excessive trapping. Today, nearly a dozen states have some kind of active river otter reintroduction program, all of which appear to be highly successful.

SEA OTTER (*ENHYDRA LUTRIS*) Original range was nearly the entire northern half of the coastal Pacific Rim, in both North America and Asia. Fur was considered the finest in the world. Populations in the western United States were thought extinct until a population was rediscovered in California during the 1930s. Major reintroduction efforts were conducted along the coasts of Oregon, Washington, Alaska, and British Columbia and have been considered successful.

BEAVER (*CASTOR CANADENSIS*) Once found throughout the United States and Canada, except in tundra habitats. Nearly exterminated populations were once considered to number over 60 million. In the United States, regulations have largely been responsible for allowing populations to recover somewhat. In the United Kingdom, however, where native beaver populations (*Castor fiber*) were exterminated sometime in the 1600s, efforts are currently underway to restore populations using reintroduced animals from Norway. Fifty individuals of *C. canadensis* were introduced in the 1940s to Tierra del Fuego, Argentina, to start a fur industry, and have swelled to a population of over 100,000. They are now the focus of an aggressive eradication effort. Also imported to Finland to replace the European beaver (*C. fiber*), which was trapped to extirpation in 1868; today, about 1,500 native European and 12,000 "Canadian" beavers are present in Finland and are considered a cause for concern.

MUSKRAT (*ONDATRA ZIBETHICUS*) Distributed originally throughout most of central and northern North America except for arid southwest and extreme southeastern coastal plain. Widely introduced as a source of fur for economic development in Europe around 1905 and quickly became a naturalized pest and threat, especially to native plants.

NUTRIA (*MYOCASTOR COYPUS*) Original range in South America. One of the few fur bearers purposefully introduced from one continent to another when introduced into Louisiana in the 1930s for development of a fur industry, which never realized intended potential. Today, it is a problematic excavator in levees, and sometimes a crop pest, in sugar cane and rice plantations. Apparently successfully eradicated from Great Britain; introduced and still present in France, Israel, and elsewhere.

SUMMARY

The species accounts and background information in this article represent the tip of the iceberg with respect to the breadth and depth of information behind the legions of introduced game species. The need for more complete and better-documented coverage of where, when, and how many game animals were moved from one place to another is one of the glaring shortcomings of the modern wildlife science and conservation literature. A comprehensive treatise that details introductions and translocations of game animals around the world is sorely needed.

SEE ALSO THE FOLLOWING ARTICLES

Birds / Carnivores / Endangered and Threatened Species / Eradication / Grazers / Herbivory / Hybridization and Introgression / Mammals, Aquatic

FURTHER READING

Brennan, L.A. 2007. *Texas Quails: Ecology and Management.* College Station, TX: Texas A&M University Press.
Demarais, S., and P. Krausman. 2000. *Ecology and Management of Large Mammals in North America.* Upper Saddle River, NJ: Prentice Hall.
Leopold, A.S. 1977. *The California Quail.* Berkeley, CA: University of California Press.
Lewis, J.C., L.B. Flynn, R.L. Marchinton, S.M. Shea, and E.M. Marchinton. 1990. Ecology of Sambar Deer on St. Vincent Island National Wildlife Refuge, Florida. Bulletin 25, Tallahassee, Florida, Tall Timbers Research Station.
Madge, S., and P. McGowan. 2002. *Pheasants, Partridge and Grouse.* Princeton, NJ: Princeton University Press.

Mungall, E. C. 2007. *Exotic Animal Field Guide: Nonnative Hoofed Mammals in the United States*. College Station, TX: Texas A&M University Press.

Mungall, E. C., and W. J. Sheffield. 1994. *Exotics on the Range: The Texas Example*. College Station, TX: Texas A&M University Press.

GENOTYPES, INVASIVE

KRISTIN SALTONSTALL

Smithsonian Tropical Research Institute, Balboa, Panama

An invasive genotype is a genetic profile that links individuals within a species displaying invasive growth or spread; the term is typically used in the context of introduced species living outside their native ranges and is usually synonymous with "introduced genotype." The term is often used in experimental contexts to generically describe individuals from the introduced range of a species, particularly in growth studies where individuals from the introduced, and therefore invasive, range are being compared with those from the native range.

GENETICS OF INVASIONS

Genetic information can provide important insights into the process of introduction, colonization, and subsequent spread of invasive species. Introductions typically begin with a small number of individuals derived from a much larger source population. The degree to which genetic diversity is preserved in invasive populations versus those in the native range of a species can help us to develop an understanding of the potential for local adaptation and invasion success. This information can then inform risk assessments and help us to develop management strategies for invasive species.

Phylogeographic techniques provide a means of studying the historical processes that are responsible for the current distribution of organisms by linking their distribution to genetic data, often chloroplast (plants: cpDNA) and mitochondrial (animal: mtDNA) DNA sequences. Such data are typically well structured geographically and can be used to define invasive lineages within a species, to link them to locations within the native range of a species, and to reconstruct the timing and mechanisms of introduction. Other, more sensitive molecular markers, such as nuclear microsatellites, can provide additional information about reproduction and genetic structuring within populations of invasive species and can help to pinpoint both the location of introduction and the geographic origins of an invasive genotype. For example, European green crabs (*Carcinus maenas*) were first introduced to the east coast of North America in the early 1800s. Based on mtDNA, it has been shown this introduction originated from Europe, and, using microsatellites, a secondary introduction to the Pacific Coast of North America was identified as having resulted from spread of individuals from the east to the west coast.

GENETIC PATTERNS OF INTRODUCTION

Human-mediated transport of species into novel environments can substantially alter the genetic makeup of introduced populations relative to that seen in their native home range, both in the amount of genetic variation present and in the way the variation is distributed within and between populations. The standard expectation is that introduced species should experience a change or reduction, or both, in genetic diversity relative to their native source populations due to genetic drift, founder events, and postintroduction genetic bottlenecks. While this may be the case, recent studies have shown that genetic diversity can vary greatly in introduced species, depending on a variety of factors (Table 1), such as the

TABLE 1
Likely Patterns in Genetic Diversity Displayed by an Introduced Species Relative to Its Native Range

Genetic diversity	Likely introduction scenario	Example
Low	Single introduction, or founding event, of a few individuals; genetic bottleneck may also occur	Argentine ants (*Linepithema humile*) in North America
Similar or increased	Multiple introductions from different source populations in the native range successfully establish and interbreed	Reed canary grass (*Phalaris arundinacea*) in North America
Novel	Following introduction, hybridization with a different species creates a novel invasive genotype or a new species	Cordgrass in California (*Spartina alterniflora* × *foliosa* = hybrid that can interbreed with its parents) and cordgrass in England (*Spartina anglica* = novel species, no backcrossing)

number of founding individuals, whether or not multiple introductions occur, the extent of the species' native range that is represented in the introduced population(s), and the type of mating system.

In general, multiple introductions are probably the most common scenario, for both accidentally and intentionally introduced organisms. Successful introductions from different source populations can increase genetic diversity in the introduced range of a species, particularly if the native range of the species is large. Interbreeding between these differing individuals creates novel combinations of genetic variation from the native range and may increase the adaptive potential of the invader. For example, *Phalaris arundinacea* (reed canary grass) is a highly successful invader in North America that has been introduced from Europe multiple times. Invasive populations have been shown to be as genetically diverse as those in their native range and also to show different allele combinations, indicating that sexual reproduction has mixed genetic material from different parts of the native range of the species in ways that have never occurred previously in the evolutionary history of the species. Introduced plants also show higher levels of heritability of invasive traits and higher levels of phenotypic plasticity. Changes such as these in the genetic makeup of a species can allow invaders to evolve in response to novel environmental conditions in their introduced range, allowing them to invade new habitats and adapt to ongoing processes such as changing climate, ensuring that they will continue to impact native species and communities in the future.

THE IMPORTANCE OF GENETIC DIVERSITY

Genetic diversity arises through mutation and recombination and can be affected across generations by natural selection, gene flow, and random genetic drift. Although genetic diversity is important because it can determine the evolutionary potential and risk of extinction for a species, levels of genetic diversity in an invader, relative to the native range, may or may not be important in terms of determining whether or not a species will become invasive, because adaptive traits that make it aggressive may remain unaffected by low genetic diversity. Many invasive plants were first introduced as landscaping or horticultural plants that may have originally been clonally propagated from a limited number of individuals possessing desirable traits. This process tends to lower the genetic diversity and make it more likely that a single lineage will become invasive. An extreme example of this phenomenon is *Pennisetum setaceum* (fountain grass), an invasive grass that has been found to have only one genotype worldwide based on

multiple types of genetic markers. Similarly, many invasions are founded by only a few individuals that are able to colonize novel habitats and become aggressive invaders, despite low levels of genetic diversity in the invading population. In cases such as this, high fecundity and high survivorship of offspring rather than genetic diversity appear to be the adaptive factors facilitating invasion.

MATING SYSTEMS

A wide range of reproductive strategies are employed by different species. With invasive species, what is clear is that high reproductive capacity, high dispersal ability, and rapid maturation of offspring are important factors in the invasion process. Many organisms, particularly plants and some invertebrates, can reproduce asexually as well as sexually. When environmental factors are favorable, these organisms can reproduce and spread quickly via asexual reproduction. For plants this includes both production of seeds that are identical to the parent plant (apomixis) and sprouting of new individuals from fragments of roots or rhizome tissues (clonal propagation). While organisms that reproduce asexually typically have lower genetic diversity than sexually reproducing ones, this form of reproduction can produce the optimal "invasive genotype" that can then spread rapidly throughout a wide area. When the environment becomes less favorable, asexual organisms can often switch to sexual reproduction, which mixes up the gene pool and may create favorable genotypes that can better survive changes in the environment.

IMPLICATIONS FOR MANAGEMENT

Knowledge of the geographic origins and genetic diversity of an invader can help with management and control. For example, *Phragmites australis* is a widespread invader in North American wetlands. Although it was known to be a native plant, its distribution and abundance have dramatically increased over the past 150 years, prompting speculation that it was an introduced species. Recent work has found that a nonnative lineage of the species has indeed been introduced to North America and is now outcompeting native populations of the species across much of its historic range. Invasive populations are dominated by a single cpDNA haplotype, most likely originating from Europe, which has now spread from the east coast, where it was first introduced all across North America, and is also invading in places where *P. australis* did not grow historically (Fig. 1). Knowledge that both native and introduced populations of this species may be present at a site can help managers design control programs that avoid disturbing native populations while targeting invasive ones.

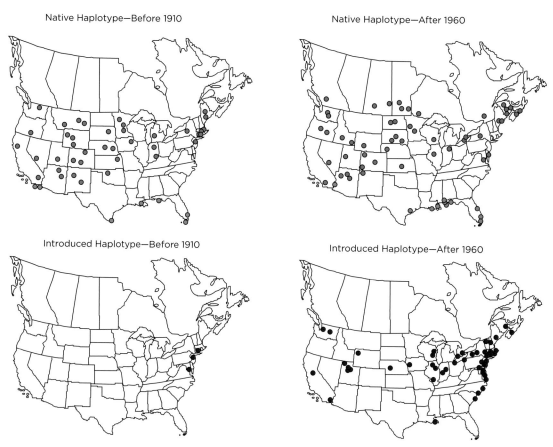

Native Haplotype—Before 1910

Native Haplotype—After 1960

Introduced Haplotype—Before 1910

Introduced Haplotype—After 1960

FIGURE 1 Historical and modern distributions of *P. australis* cpDNA haplotypes in North America. Red circles represent the invasive introduced genotype, green represent native *P. a. americanus*, and blue represent *P. a. berlandieri*, a third lineage found along the Gulf Coast. Invasive *P. australis* was likely accidentally introduced to North America in the 1800s along the mid-Atlantic coast and has spread west across the continent and south to locations that were historically outside the native range of *P. a. americanus*. (Figure reprinted from K. Saltonstall, 2002.)

It is generally accepted that increasing the number of propagules increases invasion risk. This can lead to the assumption that the introduction of only a few individuals poses little risk of a species becoming invasive. However, genetic studies of invasive species have shown that, in many cases, small numbers of founding individuals can lead to widespread invasions (e.g., the case of many frogs, Argentine ants). The implications of this fact are major for eradication programs, where species capable of becoming invasive despite low propagule pressure and genetic diversity represent a particular challenge. If only a few individuals are left behind, then reinvasions are likely, thus making previous management ineffective.

SEE ALSO THE FOLLOWING ARTICLES

Apomixis / Evolution of Invasive Populations / Geographic Origins and Introduction Dynamics / Hybridization and Introgression / Propagule Pressure

FURTHER READING

Duglosch, K. M., and I. M. Parker. 2008. Founding events in species invasions: Genetic variation, adaptive evolution, and the role of multiple introductions. *Molecular Ecology* 17: 431–449.

Gray, A. J. 1986. Do invasive species have definable genetic characteristics? *Philosophical Transactions of the Royal Society of London B* 314: 655–674.

Lavergne, S., and J. Molofsky. 2007. Increased genetic variation and evolutionary potential drive the success of an invasive grass. *Proceedings of the National Academy of Science* 104: 3883–3888.

Lee, C. E. 2002. Evolutionary genetics of invasive species. *Trends in Ecology and Evolution* 17: 386–391.

Novak, S. J., and R. N. Mack. 2005. Genetic bottlenecks in alien plant species (201–228). In D. F. Sax, J. J. Stachowicz, and S. D. Gaines, eds. *Species Invasions: Insights into Ecology, Evolution, and Biogeography.* Sunderland, MA: Sinauer Associates, Inc.

Saltonstall, K. 2002. Cryptic invasion by a non-native genotype of the common reed, *Phragmites australis*, into North America. *Proceedings of the National Academy of Science* 99: 2445–2449.

Wares, J. P., A. R. Hughes, and R. K. Grosberg. 2005. Mechanisms that drive evolutionary change (229–257). In D. F. Sax, J. J. Stachowicz, and S. D. Gaines, eds. *Species Invasions: Insights into Ecology, Evolution, and Biogeography.* Sunderland, MA: Sinauer Associates, Inc.

GEOGRAPHIC ORIGINS AND INTRODUCTION DYNAMICS

STEPHEN J. NOVAK

Boise State University, Idaho

One of the important areas of research on invasive species involves identifying an introduced species' region of origin. Identification of the geographic origins (source populations or regions) of an invasive species can provide researchers and land managers with (1) an assessment of the introduction dynamics of invasive species, (2) the relevant comparisons for determining the genetic consequences of introduction, including whether postimmigration evolution has occurred, (3) the ability to study the ecology of source populations in their native habitats, and (4) information that can be used to develop effective management or control strategies (e.g., that can guide the search for biological control agents) for reducing the impacts of these highly destructive species.

GEOGRAPHIC ORIGINS AND PROPAGULE PRESSURE

Long-distance dispersal and introduction events have genetic signatures. These signatures are influenced by several interacting factors: the amount and distribution of genetic diversity within and among source populations, the number of individuals sampled from the native range and established in the new range (founder population size), and the number of separate introduction events that occur (single versus multiple introductions). The amount and distribution of genetic diversity in native populations is determined by the biological characteristics and evolutionary history of a species. For instance, the genetic structure of plant species has been correlated with their reproductive and mating systems; plants with uniparental reproduction (i.e., self-pollination or clonal reproduction) typically exhibit higher structure than species that are predominantly outcrossing. However, populations of primarily outcrossing species often possess an isolation-by-distance structure across their entire native distribution. The genetic signature of introduction events is also influenced by the size of founder populations and the number of introductions that occur. Reconstructing such events and determining the geographic origins of invasive species is facilitated by sufficiently high genetic structure among native populations.

Propagule pressure has recently emerged as an important predictor of establishment success and invasion. The concept of high propagule pressure can also be described by the terms "multiple introductions" and "large founder populations." With multiple introductions and large founder populations, in comparison with one to a few introductions or small founder size, the overall genetic and phenotypic diversity of a species in its new range is likely higher, and the potential for severe genetic bottlenecks is likely decreased. Thus, propagule pressure not only holds demographic and ecological consequences for introduced species, but it also can have genetic consequences.

Multiple introductions may also lead to introduced populations that are admixtures (i.e., that contain the genetic information of several native populations). Finally, multiple introductions and large founder populations can also influence the likelihood of invasion through the introduction of preadapted genotypes and by increasing the potential for postimmigration evolution. Sufficient genetic diversity (especially additive genetic variance) within introduced populations is a prerequisite for adaptive evolution, and multiple introductions increase the likelihood that a high level of genetic and phenotypic (or morphological) diversity will occur within introduced populations. The combined analysis of native and invasive populations using molecular markers allows investigators to assess the geographic origins and introduction dynamics of an invasive species and is essential for determining whether phenotypic evolution has occurred due to stochastic events or natural selection.

USING MOLECULAR MARKERS TO IDENTIFY GEOGRAPHIC ORIGINS

Identification of the geographic origins of an invasive introduced species and reconstruction of the species' entry and spread in its new range can be accomplished more accurately if these investigations are begun soon after the species is initially detected, or even early in the invasion process, when the number and size of introduced populations is relatively small. In most cases, however, the urge, or need, to determine the geographic origins of an invasive species occurs after it has established, spread, and caused ecological and economic harm. Obviously, these studies require multiple sources of information: most notably the combined genetic analyses of native and introduced populations. The remainder of this section will outline the methods that can be used to determine the geographic origins (i.e., source populations or regions) of invasive introduced species.

Identifying the geographic origins and reconstructing the history of introduction and spread of invasive species requires (1) herbarium or museum specimens and historical information, (2) geographic distribution data for a species across its native and invasive ranges, (3) samples of the native and invasive populations, and (4) molecular genetic analyses of native and introduced populations. Herbarium or museum specimens are unambiguous records of a species' occurrence at a specific place and time and can provide information about early detection (introduction) sites and the pattern of spread of a species in its new range. Similarly, historical information such as the results of biological surveys, local floras, and published accounts can be used to identify early introduction sites. Clearly, early detection sites must be sampled and included in these analyses, but populations should likewise be sampled from across the known distribution of a species' invasive range.

Because many invasive species are deliberately introduced, commercial importation inventories and agricultural and horticultural records may provide information on human vectors and pathways of introduction. While this information may provide insights into specific localities or regions in the native range that should be sampled and included in these analyses, populations from across the geographic distribution of an invasive species' native range should also be sampled. This sampling strategy not only increases the likelihood of accurately pinpointing the geographic origins of an invasive species, but it also allows certain native populations or regions to be excluded as sources.

The geographic origins of invasive populations can be determined with molecular markers using two general methods: the phylogeographic approach and the multilocus genotype approach. While the choice of molecular markers will depend on multiple factors (e.g., cost, funding, availability of equipment, etc.), the marker chosen should mutate at a rate that is sufficient to result in a level of genetic diversity (polymorphisms) that is appropriate for this type of analysis. In the phylogeographic approach, the haplotype diversity of individuals from native and invasive range populations is usually determined by sequencing organelle DNA (mitochondrial DNA in animals and chloroplast DNA in plants). The geographic origins of invasive populations can be identified by tracing shared haplotypes back to certain native populations, but because they occur in historically ordered states, the phylogenetic relationships among different yet closely related haplotypes can also be determined. Genealogical relationships among haplotypes can be assessed using parsimony, maximum likelihood, Bayesian analysis, or nested clade analysis.

In the multilocus genotype approach, genotype data are typically generated over several to many nuclear DNA loci and compared among individuals from native and invasive populations. The type of nuclear molecular markers that can be used to generate multilocus genotypes can be classified into two broad categories: codominant or dominant markers. Codominantly expressed (or inherited) markers exhibit specific allelic states, allow individuals to be assigned to homozygous or heterozygous genotypes, and provide an accurate estimate of allele frequencies in a population. Codominant markers include allozymes and microsatellites. Dominant markers are scored as present or absent; thus, they do not allow specific genotypes to be assigned to individuals and do not provide an accurate estimate of allele frequencies. However, dominant markers usually generate more polymorphic loci than do codominant markers. Dominant markers include random amplified polymorphic DNA (RAPD), inter-simple sequence repeats (ISSR), and amplified fragment length polymorphisms (AFLP).

Following the generation of multilocus genotype data for native and invasive populations, these genotypes can be analyzed using assignment tests (two Bayesian-based assignment programs currently in use are BAPS and STRUCTURE). In these assignment methods, individuals from invasive populations are "assigned" to different source populations in the native range based on their overall genetic similarity across all of the loci examined. Clearly, the probability of correctly assigning an invasive genotype to its actual source population and excluding other native populations as possible sources increases with the number of polymorphic loci employed.

As indicated above, both of these approaches require sufficient genetic diversity, but the accuracy of both approaches is facilitated by sufficiently high genetic structure among native populations. However, because the multilocus genotype approach is based on the frequency of alleles at several to many loci, it can be employed even when genetic structure is weak. Investigators must always choose the molecular marker to be used in such analyses with care, and only after preliminary analyses of both native and invasive population samples has revealed ample genetic diversity to ensure success. Another technique that increases the overall accuracy of identifying the geographic origins of invasive populations is to sample and analyze as many source populations in the native range as possible. Historical information may aid in targeting certain populations for sampling, but as indicated above, comprehensive sampling across the entire native range is advisable. In summary, both methods described here have been used in a growing number of studies to provide a better understanding of biological invasions and to successfully identify the geographic origins of invasive populations of plants and animals.

CRYPTIC INVASIONS AND TAXONOMIC UNCERTAINTY

A cryptic invasion is defined as the occurrence of a species or genotype that was not previously recognized as alien in origin or not distinguished from other aliens. Such invasions can be detected through the combined analysis of native and introduced populations using molecular markers, and this approach was used to study the recent expansion of *Phragmites australis* populations in North America. Chloroplast DNA (cpDNA) haplotypes of *P. australis* were geographically structured among continents, with haplotype M being nearly worldwide in distribution. Haplotype M is now believed to be Eurasian in origin and was probably introduced into North America in the early nineteenth century. It is now the most common and widespread haplotype in North America. These findings suggest that the expansion of *P. australis* in the last 150 years likely occurred because of the introduction and invasion of an aggressive nonnative haplotype that until 2002 had gone undetected (see Fig. 1 in the article "Genotypes, Invasive").

Molecular markers have recently been used to investigate a cryptic invasion and disentangle the taxonomic uncertainty associated with the invasion of *Salsola* species (tumbleweeds) in California. *Salsola* is a taxonomically challenging genus because only a few morphological characters can be used to distinguish closely related species, and hybridization has apparently occurred in both the native and invasive ranges of these species. Although some taxonomic issues have not yet been resolved, it is believed that as many as six *Salsola* species have been introduced into California, with *S. tragus* L. (prickly Russian thistle) and *S. paulsenii* Litv. (barbwire Russian thistle) being widespread across the state. An earlier analysis of Russian thistle accessions from California using molecular markers clearly identified populations of *S. tragus* and *S. paulsenii* but also revealed the presence of a cryptic new species referred to as *Salsola* type B. Subsequent molecular analyses have revealed the following: (1) *Salsola* type B has now been identified as being *S. kali* subsp. *austroafricana* Aellen (Fig. 1), (2) source populations for the introduction of *S. kali* subsp. *austroafricana* into California appear to have been drawn from southern Africa (where it may also have been introduced) or from Australia, (3) the geographic origins of *S. tragus* in California appears to be the Ukraine (thus, the common name "Russian thistle" is apt), and (4) several *Salsola* hybrid taxa appear to have originated in California, including a fertile allopolyploid hybrid of *S. tragus* and *S. kali* subsp. *austroafricana*, which has been named *S. ryanii*. Thus, analyses of *Salsola* species from California and elsewhere using molecular markers have not only identified a source population, revealed

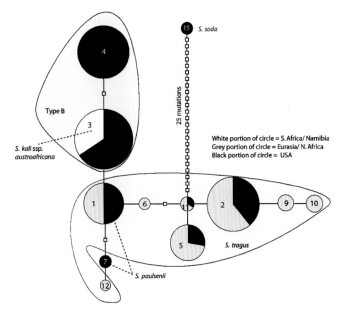

FIGURE 1 Haplotype network of DNA sequences for native and invasive populations of several *Salsola* species. Circles represent the haplotypes detected, and squares indicate haplotypes not detected. The size of each circle is proportional to that haplotype's frequency worldwide. The shadings indicate the percentage of haplotypes from the United States (black), Eurasia and North Africa (gray), and southern Africa (white). (Figure previously published in Gaskin et al., 2006, *Madroño* 53: 244–251, copyright 2006, California Botanical Society.)

a cryptic invasion, and resolved taxonomic uncertainties, they also have revealed contemporary speciation events occurring in the invasive range of these taxa.

MULTIPLE INTRODUCTIONS AND ADMIXTURES

The combined analysis of native and invasive populations was used to identify the geographic origins of *Bromus tectorum* L. (cheatgrass), and to reconstruct its introduction dynamics around the world. *Bromus tectorum* is an annual, predominantly selfing grass with a worldwide geographic distribution; it is broadly distributed across its native range in Eurasia, and the plant has been introduced into Australia, New Zealand, Japan, Hawaii, temperate South America, and much of North America. *Bromus tectorum* is now the dominant plant across at least 200,000 km^2 in the Intermountain West of North America, where it causes enormous ecological disruption and economic costs. To date, 51 Eurasian populations, 192 North American populations, and 38 populations from other naturalized ranges (Canary Islands, Argentina, Chile, Hawaii, New Zealand, Australia, and Japan) have been analyzed for their allozyme diversity.

Identifying the geographic origins and the introduction dynamics of *B. tectorum* is facilitated by the fact that multilocus genotypes in populations from the native

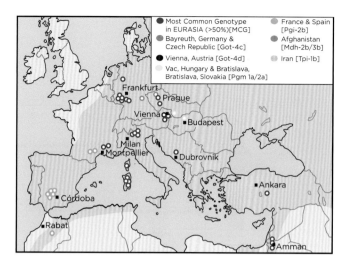

FIGURE 2 Distribution of multilocus genotypes among native range populations of *Bromus tectorum*. Note the high geographic structuring of multilocus genotypes: the *Pgi-2b* genotype occurs in populations from the western Mediterranean, and the *Got-4c*, *Got-4d*, *Pgm-1a*, and *Pgm-2a* genotypes occur in populations from central Europe. (Figure courtesy of Richard N. Mack.)

range, especially Europe, are geographically structured: approximately 75 percent of the genetic diversity in the native range is partitioned among populations (Fig. 2). Based on multilocus genotype distributions, introduced *B. tectorum* in North America and other regions around the world appears to have been drawn exclusively from either central Europe (Czech Republic, Germany, Hungary, and Slovakia; Fig. 3) or the western Mediterranean (France, Spain, or Morocco). The geographic pattern of multilocus genotypes in western North America supports historical evidence that this invasion stems from multiple introductions: a minimum of six to seven independent founder events. Multiple introductions may have also occurred in other naturalized ranges of *B. tectorum* (Argentina, Chile, and New Zealand), although the number of introductions varies. The level of genetic diversity within introduced populations in many regions is, on average, higher, and the level of genetic differentiation among introduced populations is lower than in the native range. This pattern suggests that populations in these regions (especially in North America) consist of admixtures of two or more native-range genotypes. Thus, the genetic diversity of introduced populations of *B. tectorum* appears to be strongly influenced by introduction dynamics. Clearly, multiple introductions have shaped the North American invasion of *B. tectorum* and the invasion of other plants, and also appear to have contributed to the invasion of some animal species.

Mitochondrial DNA (mtDNA) haplotype analyses of native and invasive populations of *Anolis sagrei*,

a small diurnal lizard, reveal patterns similar to those described above for *B. tectorum*. Invasive populations in the southeastern United States (Florida, Georgia, Louisiana, and Texas), the Caribbean (Grand Cayman, Grenada, and Jamaica), and the Pacific islands (Hawaii and Taiwan) stem from multiple introductions from several genetically distinct populations in the native range, especially different populations in Cuba. And many invasive populations of *A. sagrei* appear to be composed of admixtures of haplotypes from different native populations (Fig. 4). In addition, differential admixture from multiple source populations also correlates with morphological diversity: invasive populations derived from a larger number of source populations have the highest amount of diversity (regardless of the identity of particular source populations). Thus, morphological diversity of invasive populations of *A. sagrei* appears to be determined by stochastic events associated with the number of native populations sampled and introduced into the species' new range. A subsequent study documented multiple introductions and admixtures in seven of the eight *Anolis* lizard species examined, indicating that this pattern

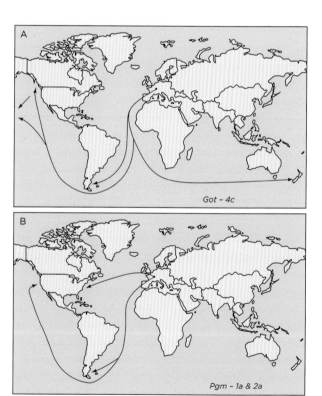

FIGURE 3 Distribution of populations of *Bromus tectorum* with the *Got-4c* (A) and *Pgm-1a* and *Pgm-2a* (B) multilocus genotypes. The geographic origins of both genotypes are in central Europe, and arrows indicate possible routes of dispersal to introduced populations in North America, Hawaii, South America, and New Zealand. (Figure previously published in Novak and Mack, 2001, *BioScience* 51: 114–122, copyright 2001, American Institute of Biological Sciences.)

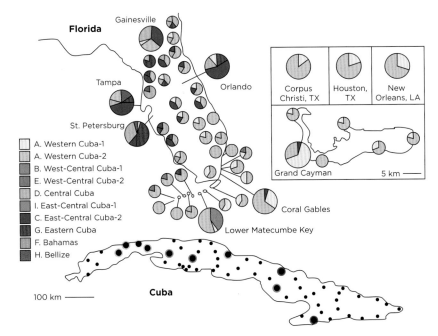

FIGURE 4 Mitochondrial DNA haplotype diversity in invasive populations of *Anolis sagrei*, and the distribution of this diversity in source populations from the native range. Black dots represent native Cuban populations sampled, and dots encircled with colors indicate Cuban populations that possessed haplotypes closely related to the haplotypes of introduced populations. Pie diagrams of introduced populations indicate the frequency of haplotypes from different source populations. (Figure previously published in Kolbe et al., 2007, *Molecular Ecology* 16: 1579–1591, copyright 2007, Blackwell Publishing Ltd.)

of introduction may be common among related taxa that share similar life history traits.

Multiple introductions, admixtures, and increased morphological (phenotypic) variability have also been documented for the recent invasion of the freshwater snail *Melanoides tuberculata* in Martinique. The native range of this predominantly parthenogenetic snail includes the intertropical belt of the Old World from Africa to southeastern Asia, and the species is now invasive in the New World from northern Argentina to Florida. It is capable of both clonal and sexual reproduction, but because sexual reproduction is so rare, a single genotype can dominate and persist within a population for many generations. Additionally, each clonally derived genotype exhibits a distinctive morphological variant (morph) based on the shape and color ornamentation of its shell. A worldwide phylogenetic survey of mtDNA diversity by Benoit Facon and colleagues revealed five separate introductions of *M. tuberculata* into Martinique over approximately 20 years, each characterized by a different shell morph (Fig. 5). Thus, the combined use of genetic markers and morphological traits allowed reconstruction of the history of introduction and invasion of *M. tuberculata* in Martinique.

The origins of the closely related FAL, MAD, and PAP morphs were traced to Indonesia and Malaysia; the source for the PDC morph was Okinawa, Japan; and the MA morph originated in Luzon, Philippines. Moreover, two morphs in Martinique (CPF and FDF) appear to have been locally produced by outcrossing events (sexual reproduction) between several of the introduced morphs:

CPF = FAL × PAP and FDF = FAL × PDC. The number of morphs detected, on average, within the populations from Martinique was much greater than that of native populations (Fig. 5). This finding indicates the presence of admixtures within most of the invasive populations of Martinique, likely a consequence of multiple introductions. In addition, using a quantitative genetic approach, Facon and colleagues detected high genetic variance for four of five life history traits in all seven morphs occurring on Martinique. The two sexually produced morphs, CPF and FDF, exhibited greater life history trait variability than their parental morphs, and they also appeared to have a selective advantage over (i.e., they outcompete) their parental morphs when they cooccurred. Results for *M. tuberculata* show that multiple introductions can increase the genetic diversity and evolutionary potential of alien species, and thereby increase the risk of invasion, even in a species that reproduces primarily by cloning. Such insights emerge with the accurate identification of the geographic origins of invasive species through the combined analysis of native and introduced populations.

ECOLOGY OF SOURCE POPULATIONS

Ecological studies with invasive species have focused almost exclusively on analyzing these species in their new range to either describe the negative ecological problems they cause or to determine the traits or ecological conditions that lead to invasion. However, two recent reviews have emphasized the importance of studying the ecology of invasive plants in both their native and introduced

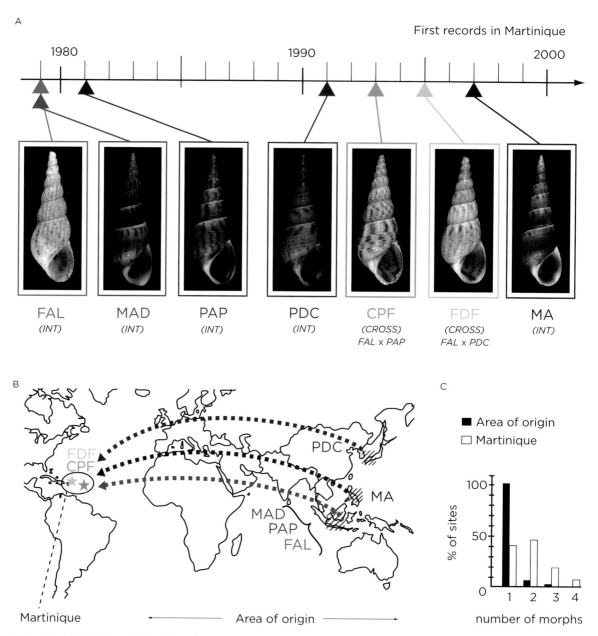

FIGURE 5 Introduction history of *Melanoides tuberculata* in Martinique based on shell morph variability (A) and the distribution of mitochondrial DNA haplotypes (B). FAL, MAD, PAP, PDC, and MA are classified as introduced (INT) morphs; CPF and FDF appear to have been locally produced by sexual reproduction (CROSS) between several of the introduced morphs. The geographic origins of the five introduced morphs are shown in (B). The number of morphs per local populations, given in (C), shows that more morphs were detected within sites from Martinique than in sites from the native range. (Figure adapted from Facon et al., 2008, *Current Biology* 18: 363–367.)

ranges. Such studies can determine whether an invasive species is more abundant, grows faster, accumulates more biomass (is larger), is more fecund, and experiences less herbivory and pathogen attacks in its new range, compared with its native range. This approach also allows investigators to examine and compare the expression of phenotypic (quantitative) traits in native and invasive populations within the same experimental design. Thus, all of these data can be used to assess several hypotheses

that have been established to explain the success of introduced species: the enemy release hypothesis, the empty niche hypothesis, the novel weapons hypothesis, and the evolution of increased competitive ability (EICA) hypothesis. Clearly, the first step in validly testing any or all of these hypotheses, through the simultaneous comparison of native and invasive populations, involves the accurate identification of the geographic origins of alien species.

Despite the wealth of ecological information that can be garnered using this approach, only a handful of studies have actually compared the growth and performance of native and introduced plant populations in the field. However, one of best examples of the ecological and evolutionary insights that can be gained through the direct comparison of native and introduced populations comes from the ongoing research of Lorne Wolfe and colleagues with the invasive plant *Silene latifolia* Poir (white campion). *Silene latifolia* is an herbaceous perennial that is native to Europe and was accidentally introduced into North America approximately 200 years ago. It is now invasive over much of North America and is listed as a noxious weed in Canada and the United States. Wolfe measured enemy attack in both the native and invasive ranges and found that floral herbivory and aphid attack by two generalists, fruit predation by a specialist, and attack by a fungal specialist were much higher in the native range. In addition, when plants from both the native and introduced ranges were grown in common-garden environments, the plants from North America exhibited life history traits associated with higher growth and performance, which likely contribute to invasion. These findings are in general agreement with the enemy release and EICA hypotheses and demonstrate the usefulness of studying the ecology of invasive plants in both their native and introduced ranges.

GEOGRAPHIC ORIGINS, INTRODUCTION DYNAMICS, AND BIOLOGICAL CONTROL

In its native range, an invasive introduced species can be attacked by an array of generalist or specialist natural enemies (e.g., predators, herbivores, seed predators, parasites, and pathogens). Because populations of an invasive species are often genetically structured across their native ranges, we would also expect populations of natural enemies in the native range to be structured. When the interaction between an enemy and its host is finely tuned, local adaptation (a coevolutionary relationship) occurs. This would be particularly true for specialist enemies. Classic biological control involves the deliberate release of nonnative organisms (biological control agents) with the goal of reducing the population density of an invasive introduced species in its new range. The classic biological control approach is based on the assumption described above: natural enemies often regulate populations of invasive species in their native ranges, and the most specific and effective control agents can be found in the localities from which the invasive species originates. Thus, the accurate delineation of an invasive species' geographic origins (i.e., source populations or regions) can guide and improve the search for biological control agents.

Employing molecular markers to reconstruct an alien species' introduction dynamics, pattern of spread, and genetic diversity in its new range can also provide valuable information to biological control programs. For instance, if one to a few introductions occurred, and alien populations are genetically depauperate due to population bottlenecks, then only one (or just a few) highly specialized control agents may be necessary. In such cases, control agents would probably experience fast population build-up and spread as they attacked the target species. With multiple introductions, genotypes from several to many native source populations would occur across the introduced range of an alien species, and several (or even many) agents from different portions of the native range would be required for control. Furthermore, if alien populations are genetic admixtures, several to many control agents would have to be established within the same locality in the introduced range. Thus, not only does the combined analysis of native and invasive populations using molecular markers provide a better understanding of the invasion process (i.e., introduction dynamics), but results from these analyses can also be used to study the ecology of invasive species in their native ranges and to develop and refine programs to control invasive species in their new ranges.

SEE ALSO THE FOLLOWING ARTICLES

Biological Control, of Plants / Cheatgrass / Enemy Release Hypothesis / Evolution of Invasive Populations / Genotypes, Invasive / Invasiveness / Propagule Pressure

FURTHER READING

Avice, J.C. 2004. *Molecular Markers, Natural History and Evolution*. Sunderland, MA: Sinauer Associates.

Ayres, D.R., F.J. Ryan, E. Grotkopp, J. Bailey, and J. Gaskin. 2009. Tumbleweed (*Salsola*, section *Kali*) species and speciation in California. *Biological Invasions* 11: 1175–1187.

Bossdorf, O., H. Auge, L. Lafuma, W.E. Rogers, E. Siemann, and D. Prati. 2005. Phenotypic and genetic differentiation between native and introduced plant populations. *Oecologia* 144: 1–11.

Cox, G.W. 2004. *Alien Species and Evolution*. Washington, DC: Island Press.

Facon, B., J-P. Pointier, M. Glaubrecht, C. Poux, P. Jarne, and P. David. 2003. A molecular phylogeography approach to biological invasions of the New World by parthenogenetic Thiarid snails. *Molecular Ecology* 12: 3027–3039.

Facon, B., J-P. Pointier, P. Jarne, V. Sarda, and P. David. 2008. High genetic variance in life-history strategies within invasive populations by way of multiple introductions. *Current Biology* 18: 363–367.

Gaskin, J.F., F.J. Ryan, G.F. Hrusa, and J.P. Londo. 2006. The genotype diversity of *Salsola tragus* and potential origins of a previously unidentified invasive *Salsola* from California and Arizona. *Madroño* 53: 244–251.

Hierro, J.L., J.L. Maron, and R.M. Callaway. 2005. A biogeographical approach to plant invasions: The importance of studying exotics in their introduced and native range. *Journal of Ecology* 93: 5–15.

Keller, S.R., and D.R. Taylor. 2008. History, chance, and adaptation during invasion: Separating stochastic phenotypic evolution from response to selection. *Ecology Letters* 11: 852–866.

Kolbe, J. J., R. E. Glor, L. R. Schettino, A. C. Lara, A. Larson, and J. B. Losos. 2007. Multiple sources, admixture, and genetic variation in introduced *Anolis* lizard populations. *Conservation Genetics* 21: 1612–1625.

Kolbe, J. J., A. Larson, and J. B. Losos. 2007. Differential admixture shapes morphological variation among invasive populations of the lizard *Anolis sagrei*. *Molecular Ecology* 16: 1579–1591.

Muller-Scharer, H., U. Schaffner, and T. Steinger. 2004. Evolution in invasive plants: Implications for biological control. *Trends in Ecology and Evolution* 19: 417–422.

Novak, S. J., and R. N. Mack. 2001. Tracing plant introduction and spread: Genetic evidence from *Bromus tectorum* (cheatgrass). *BioScience* 51: 114–122.

Novak, S. J., and R. N. Mack. 2005. Genetic bottlenecks in alien plant species: Influence of mating systems and introduction dynamics (210–228). In D. F. Sax, J. J. Stachowicz, and S. D. Gaines, eds. *Species Invasions: Insights into Ecology, Evolution and Biogeography*. Sunderland, MA: Sinauer Associates.

Roderick, G. K., and M. Navajas. 2003. Genes in new environments: Genetics and evolution in biological control. *Nature Reviews Genetics* 4: 889–899.

Ryan, F. J., and D. R. Ayers. 2000. Molecular markers indicate two cryptic, genetically divergent populations of Russian thistle (*Salsola tragus*) in California. *Canadian Journal of Botany* 78: 59–67.

Ryan, F. J., S. L. Mosyakin, and M. J. Pitcairn. 2007. Molecular comparisons of *Salsola tragus* from California and Ukraine. *Canadian Journal of Botany* 85: 224–229.

Saltonstall, K. 2002. Cryptic invasion by a non-native genotype of the common reed, *Phragmites australis*, into North America. *Proceedings of the National Academy of Sciences of the United States of America* 104: 3671–3672.

Wolfe, L. 2002. Why alien invaders succeed: Support for the escape-from-enemy hypothesis. *The American Naturalist* 160: 705–711.

GRASSES AND FORBS

CARLA D'ANTONIO, KAREN STAHLHEBER, AND NICOLE MOLINARI

University of California, Santa Barbara

Grasses and forbs are important components of "alien species" or "invasive plant" lists worldwide. A grass is any plant in the Poaceae lineage. These are herbaceous monocots with narrow, long leaves; jointed stems; and flowers in spikelets with glumes. Forbs are not taxonomically restricted. Here, they include any nongrass, herbaceous (nonwoody) plants, excluding those that are fully aquatic. They can be either monocots or dicots and include herbaceous plants emerging from bulbs (geophytes). As with grasses, forbs can be annual or perennial. For the purposes of this review, they do not include vines. Grasses and forbs make up a large fraction of the "invasive alien plants" on lists from throughout the world, including both natural and agricultural settings. They dominate agricultural

weed lists and continue to spread in cropland systems. In addition to their prevalence on invasive plant lists, they have been important as tools for ecological study: their more rapid lifecycle and ease of propagation have made them useful for investigations of the genetics, ecological interactions, and ecosystem consequences of invasion.

GENERAL FEATURES OF INVASIVE GRASSES AND FORBS

Representation on Weed Lists

The prominence of grasses and forbs relative to other types of plants varies from region to region and list to list (Table 1). Of regional lists reported here, forbs and grasses combined make up from 34 percent (Brazil) to 79 percent (southwestern Africa) of the invasive alien species. A worldwide list of the 100 worst invasive alien species (see Appendix) developed by the Invasive Species Specialist Group of the World Conservation Union contains 32 terrestrial plants, 9 percent of which are grasses and 16 percent of which are forbs. With the exception of New Zealand, forb species tend to outnumber grass species on invader lists.

Highly Represented Families

Forbs and grasses on invasive plant lists belong to a diversity of lineages, although there are a few standout genera with numerous species that have naturalized across the globe. In a consideration of the Global Invasive Species database maintained by the IUCN (www.issg.org/database/), for example, plants within Asteraceae (= Compositae) and Brassicaceae (= Cruciferae) make up 30 percent and 10 percent of the listed forbs, respectively. Grasses (Poaceae) have as many representatives as all of the forb families combined. A similar pattern appears in the flora of California, where ten families, including Asteraceae, Poaceae, and Brassicaceae, make up 75 percent of the naturalized nonnative species. In China, while the complete list of invasive plants includes 59 different families, the majority also belong to Asteraceae, Poaceae, and Brassicaceae. The largest families in Europe's nonnative flora likewise include Asteraceae, Poaceae, and Brassicaceae. These are, however, large families, irrespective of alien status. If the European alien flora is adjusted for diversity within the family, then Asteraceae are underrepresented, while Rosaceae, Fabaceae, and Onagraceae are overrepresented. Worldwide, such an analysis suggests that Brassicaceae, Leguminosae, Poaceae, Papaveraceae, and Chenopodiaceae are overrepresented families. These include many forb genera. When only "aggressive invaders" in these lists are considered, woody families tended to be overrepresented

TABLE 1
Frequency of Grasses and Forbs in Regional Lists of Invasive Species

List Name (Organization or Source)	Region	% Graminoids	% Forbs
100 of the World's Worst Invaders (IUCN)	worldwide	9	16
WeedsUS (www.invasive.org)	United States	15	39
100 of Europe's Worst Invaders (DAISIE)	Europe	11	33
Weeds Australia (www.weeds.org.au/)	Australia	8	30
Consolidated List of Environmental Weeds in New Zealand (New Zealand Department of Conservation)	New Zealand	32	13
Alien Plants of Korea (National Institute of Environmental Research)	South Korea	18	78
Invasive Alien Plants of Japan (http://invasive.m-fuukei.jp/jlist)	Japan	13	18
Invasive Alien Organisms in Western Southern Africa (Brown and Grubb, 1986)	southwestern Africa (semi-arid)	14	65
Exotic Species in Costa Rica (Red Interamericana de Información sobre Biodiversidad www.invasoras.acebio.org)	Costa Rica	35	38
Exotic Species in Brazil (http://i3n.institutohorus.org.br)	Brazil	16	18
Plant Threats to Pacific Island Ecosystems (USFS, PIER)	Pacific Islands	19	36

NOTE: Relative frequency of grasses and forbs within regional lists of invasive species. Graminoids include grasses, rushes, and sedges.

(Fig. 1), including Myrtaceae, Pinaceae, and Rosaceae. This analysis did not show grasses to be strongly represented as aggressive invaders. Likewise, a review of invaders declared "weeds of national significance" in Australia demonstrated that perennial forbs and woody species are overrepresented but Poaceae are not (Fig. 2). By contrast, Daehler surveyed aggressive invaders of "natural areas" worldwide and found Poaceae to be an overrepresented family.

Origins

The regions of origin of forb and grass species in the IUCN database are spread throughout the world (Table 2). The invasive grass origins are more equally distributed throughout Eurasia and the Mediterranean, Asia, and Africa. The forbs are skewed to representing Eurasia and the Mediterranean basin, as well as Central and South America. The primary geographic origin of invasive forbs and grasses differs among recipient regions. For example, the Americas are the major source of invasive alien plants in China, and most of these are forbs and grasses. By contrast, only 20 percent of alien species in Europe are of North American origin, and many are from within the region of Europe and Asia but are expanding their ranges due to human activities. Highly successful invaders of a region typically originate from regions with similar climates to the introduced location. For example, in semi-arid and arid Africa, although the majority of alien plants are from Europe, the species with the highest abundance and impacts are overwhelmingly from arid regions of Australia and the Americas.

Habitats Invaded

Invasive forbs and grasses have spread into almost every type of natural habitat. Many thrive in agricultural and other repeatedly disturbed areas (such as roadsides). Within "natural" areas, their abundance also tends to be closely tied to human modification of disturbance regimes. This includes places with increased frequency of disturbance, places with novel types of disturbances, and places where historic types of disturbances have been suppressed. Of the three grasses listed by the IUCN as among the world's worst aliens, two occur in wetland settings, and one occurs largely in disturbed pastures. They include *Spartina anglica*, a salt marsh invader; *Arundo donax*, a riparian and freshwater marsh invader; and *Imperata cylindrica*, an invader of tropical pastures and seasonal subtropical woodlands. Of the forbs on the IUCN list, most also occur in mesic environments. *Fallopia japonica* (Japanese knotweed, Polygonaceae) occurs in riparian habitat. *Hedychium*

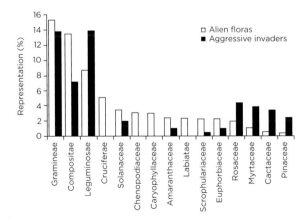

FIGURE 1 Representation of species by family in alien plant floras compared to "aggressive invaders," as determined by Cronk and Fuller (1995) to be species considered to have population, community, or ecosystem impacts. Alien floras represent lists from 26 regions throughout the world. (From Pyšek, P., 1998, Oikos; courtesy of P. Pyšek.)

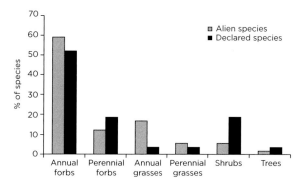

FIGURE 2 Distribution of established alien plants and "declared weeds" in central Australia and western New South Wales among flowering plants. "Declared weeds" are noxious weeds or weeds considered of national significance, in this case for "rangelands." (Figure courtesy of A. Grice, CSIRO, Australia.)

gardenerianum (kahili ginger, awapuhir, Zingerbaceae) occurs in wet tropical forest understories. *Lythrum salicaria* (purple loosestrife, Lythraceae) occurs in freshwater marshes, on lake edges, and in riparian zones. By contrast, *Euphorbia esula* (leafy spurge, Euphorbiaceae) occurs in rangelands and is not associated with high moisture. The fifth forb on the list, *Sphagneticola (Wedelia) trilobata* (wedelia, creeping ox-eye, Asteraceae), typifies many invasive species by occurring in an enormous range of tropical habitats from dry rangelands to wet forest understories. Examples of IUCN species are shown in Figures 3 and 4.

A group of ecologically significant invasive grasses and forbs are those that dominate in semi-arid and arid grasslands and rangelands in North America and parts of Australia. Some of the most notorious species include *Bromus tectorum* (cheatgrass) and *Bromus madritensis* (red brome), both Mediterranean basin annual species that have invaded extensively across the desert regions of the western United States, from hot deserts in Arizona to cold deserts in Idaho. Some of the most noxious forb invaders of rangelands are in the genus *Centaurea* (Asteraceae) (Fig. 5). *Centaurea solstitialis* (yellow star-thistle) is a winter annual that affects millions of acres of annual grassland and open shrublands

TABLE 2
Origins of Invasive Grasses and Forbs

Native Range	Grasses (%)	Forbs (%)
Asia	25	16
Central and South America	15	27
North America	5	4
Eurasia and the Mediterranean	35	51
Africa	20	2

NOTE: Origins of the invasive grasses and forbs listed within the Global Invasive Species Database. Species native to more than one region were included in both.

in California, Oregon, Washington, and Idaho. As of 2004, its relative *C. maculosa* (spotted knapweed) affected more than 2.8 million ha of montane grassland, mainly in the western United States and Canada. *Centaurea diffusa* (diffuse knapweed) was estimated to affect 0.8 million ha of grassland, successional old fields, prairies, and montane rangeland across the United States and Canada.

TRAITS OF INVADERS AND INVADED COMMUNITIES

A species must overcome numerous biotic and abiotic filters in order to establish successfully in a new region and become invasive. The process of a species becoming widespread and having a large impact can be divided into four stages: (1) transport to a new region, (2) initial establishment of a reproductive population ("naturalization"), (3) spread away from initial site ("invasiveness"), and (4) development of local and possibly regional impacts ("transformer species"). Very little work has focused on traits associated with transport and initial naturalization. Much more has focused on traits associated with spread and impact.

Transport and Initial Establishment

Grasses and forbs are transported beyond their native range for any number of reasons. All three of the grasses on the IUCN worst invaders list were purposefully moved to new areas. *Arundo donax* has been moved as an ornamental and for the establishment of plantations for woodwind reed production. *Spartina anglica* was moved for mudflat stabilization. *Imperata cylindrica* was moved as an ornamental. Many other grasses were transported for forage, although many ecologically detrimental annual species (*B. tectorum, B. madratensis*) were accidental imports. Of the IUCN's 100 worst aliens, all forbs were purposeful imports at least initially. For example, both *F. japonica* and *H. garderianum* were introduced via the nursery trade as ornamentals. Most of the forbs in the larger IUCN database were introduced intentionally: 54 percent for cultivation, food, or medicine, and 26 percent as escaped ornamental plants. Only 20 percent arrived accidentally, through contaminated seed, soil, or ballast water. The relative importance of each of these introduction methods varies by region. In China, for example, escaped ornamental herbs are prevalent, along with plants introduced for medicinal uses. In Hawaii, most introduced species were brought in as ornamentals or as pasture forage.

Are forbs and grasses more likely to be transported to other regions than other groups of plants? This is not well established. One of the few studies to quantify the traits associated with transport and initial establishment

FIGURE 3 Invasive grasses and forbs from IUCN, world's 100 worst alien species list (see Appendix). (A) *Imperata cylindrica*, (B) *Spartina anglica*, (C) *Lythrum salicaria*, and (D) *Hedychium gardnerianum*. (Photographs courtesy of A. Diamond, B. Kers, S. Buchan, and G. Carr, respectively.)

of nonnative species used a local Canadian flora to assess presence and absence data for European congeners. The authors found that, of the traits measured (life form, stem height, flowering period, and native range), the native range of a species was the best predictor of its ability to be transported and become established in a new region. Life form was not important in determining whether a species was transported and established, suggesting that grasses and forbs are not any more likely than woody species to be transported and to establish into a new range. Nonetheless,

it makes sense that, due to their generally smaller seeds and fruits compared to shrubs and trees, grasses and forbs may be more likely to be accidental introductions.

Traits Influencing Spread, Abundance, and Impact

Despite the frequently made assumption that life form is an important predictor of invasiveness (stage 3), few studies of invasive floras have found this to be the case. For example, only two of six reviewed multispecies studies that

FIGURE 4 More invasive grasses and forbs from IUCN, world's 100 worst alien species list. (A) *Euphorbia esula*, (B) *Fallopia japonica*, and (C) *Sphagneticola (Wedelia) trilobata*. (Photographs courtesy of sarcocenia .com, E. Haug, and S. and K. Starr, respectively.)

included life form as an explanatory variable found it to be significantly associated with invasiveness. One that used the North American flora found shrubs and trees to be more invasive than forbs and grasses. A study using Raunkiaer's life form scheme on the Czech alien flora found that life form success varied with habitat, with hemicryptophytes (plants that bud at or near soil surface) being successful in seminatural habitats and therophytes (annuals) being

most common in disturbed sites. In addition, a review by Lonsdale of grass and legume species purposefully introduced into Australia showed that grasses were more likely than expected to become important "weeds" (spreading species with perceived negative impact), with 17 percent of all introduced grasses becoming important weeds. Legumes were less likely than expected to become "weeds."

Numerous studies compare the performance of cooccurring, related, invasive, and noninvasive forb species, with a focus on traits that might define "invasiveness." Almost no studies have focused on traits of closely related grasses. One that did found that the highly invasive perennial grass *Cortaderia selloana* (pampas grass) exhibits sexual reproduction, whereas the less invasive *C. jubata* (jubata grass), is strictly asexual in California, where they are both invasive. Juveniles of the two species are similar morphologically, but the higher survival of *C. selloana* is likely due to physiological differences in water use efficiency and nutrient capture and may be related to greater genetic variability as a result of sexual reproduction. Also *C. selloana* has expanded the types of habitats it has invaded over time, whereas *C. jubata* has remained largely in ruderal (highly disturbed) habitats.

As with *Cortaderia*, reproductive traits are thought to be important influences on the successful spread of species. Clonality is an often considered factor allowing a species to become dominant rapidly. There are, however, no comparative studies that evaluate the contribution of clonality to invasiveness or level of impact exclusively, particularly within a grass and forb context. Nonetheless, some highly transformative invaders such as *A. donax*, *I. cylindrica*, *S. anglica*, and *L. salicaria* are clonal. Rhizome fragmentation and regeneration clearly contribute to spread. Clonality, once the plants are established, contributes to competitiveness and continued dominance. All forb species on the 100 worst invaders list are clonal. Likewise, *Cardaria draba* (whitetop, hoary cress, Brassicaceae), considered one of the worst weeds in England and also problematic in the western United States and Australia, has highly effective rhizomatous growth and is readily spread by root fragments. By contrast, none of the invasive species of *Centaurea* or *Cortaderia* are rhizomatous, and most *Bromus* are not rhizomatous, and so there are other ways for invasive forbs and grasses to become dominant. The production of large numbers of seeds and flexibility in seed production in response to the environment appear to be important.

Several studies provide interesting insights into invasiveness within a forb group. Researchers from the University of California, Davis, report that the most invasive species

FIGURE 5 Invasive *Centaurea* in rangelands of the United States. (A) *Centaurea maculosa*, (B) *Centaurea solstitialis*, (C) *Centaurea diffusa*. (Photographs courtesy of S. Thorsted, B. Bumgarner, and G. Blaich, respectively.)

in the genus *Centaurea* within California, *C. solstitialis*, is highly self-incompatible, while the less invasive *C. melitensis* and *C. sulphurea* are highly self-compatible. Perhaps more importantly, *C. solstitialis* showed a positive response in fecundity and size to gap-forming disturbances and simulated grazing plus a longer period of growth and flowering, thereby indicating that plasticity may be an important trait promoting invasiveness. Two other studies

have pinpointed reproductive and physiological traits contributing to the invasiveness of a forb species. They have found that higher individual growth rates, particularly speed to flowering, and higher seed production including sometimes longer flowering periods characterize invaders compared to their noninvasive congeners.

The performance of cooccurring *native* and *nonnative* or *invasive* and *noninvasive* species has been shown to vary with habitat. For instance, under low nutrient conditions, there is no difference in relative growth rates for noninvasive and invasive congeners in the herbaceous monocot family Commelinaceae, yet under high nutrient conditions, invasive species grow significantly faster. In addition, invaders had higher fecundity and vegetative reproduction, greater specific leaf area, and a more plastic root-to-shoot ratio than noninvaders. Differences between the invaders and noninvaders were maximized under the highest nutrient conditions.

Invasion Success and the Resident Community

Two related hypotheses have emerged that help explain the success of a given nonnative species in relation to the resident community, and most tests of both ideas have utilized forb or grass invaders. Charles Darwin proposed the "naturalization hypothesis," which posits that an introduced species is less likely to be successful if there is a closely related species already inhabiting the area. The hypothesis hinges upon related native and nonnative species sharing similar traits, and thus occupying similar niche space. Richard Mack added to this hypothesis by proposing that nonnative species with native congeners would also be subjected to biotic resistance by specialist herbivores and pathogens. The second hypothesis, "limiting similarity," predicts that successful invaders should differ functionally from native resident species without regard for phylogenetic relationships. Under the limiting similarity hypothesis, any species that shares a similar niche to the invader will contribute to invasion resistance. Currently, most of the research evaluating these hypotheses has involved forb and grass invaders in constructed or natural grassland type settings.

Multiple studies have evaluated the naturalization hypothesis, exploring whether nonnative grasses and forbs entering into a new community are more or less likely to become successful if a congener or a close relative is already present. An analysis by M. Rejmánek published in 1999 of nonnative Poaceae, Asteraceae, and Brassicaceae in California found that species within these families that lack native congeners were overrepresented when compared to nonnative species with native congeners. Using a phylogenetic approach, his colleagues Strauss et al. evaluated whether introduced invasive grass species in California were more closely related to natives than were introduced noninvasive taxa. They concluded that this was the case, supporting the naturalization hypothesis. The general consensus for nonnative grasses and forbs is that colonization of distantly related species is more likely to occur than colonization of species with a congener already present in the recipient community.

Support for "limiting similarity" is mixed. A study of invading herbaceous communities with a variety of invader species in Michigan did not find a consistent reduction in invasion success when sites were dominated by residents that were functionally similar to the invader. Other evidence that bears on the importance of limiting similarity for forb species includes reduced success of *Centaurea solstitialis* in communities with late-active forbs whose phenology is similar to *C. solstitialis* in California, and increased success (density and biomass) of planted *C. maculosa* in manipulated grassland communities with phenologically similar forbs removed (Montana).

Other factors besides competition from closely related resident species may influence the susceptibility of a system to invasion. A synthesis of the biotic resistance literature by J. Levine and colleagues revealed that resident herbivores, competitors, and species diversity all contribute to the ability of the native residents to reduce establishment and performance of invaders. Their meta-analysis incorporated all life forms, including woody species, yet 72 percent of the competition, 100 percent of the species diversity, 83 percent of the herbivory, and 78 percent of the fungal studies focused on grass and forb invaders. In addition, the performance of grass and forb invaders was more sensitive than that of woody species to competition from resident plant communities.

The ability of a community to resist grass and forb invasion may also be related to the degree of anthropogenic perturbation as it affects growing conditions. Wetland communities with increased nutrient supply and flooding regimes are typically invaded by nonnative grasses and forbs—for example, *L. salicaria* is known to take advantage of altered hydrologic regimes and enhanced nutrient runoff in wetlands, as is *Phalaris arundinaceae*, a potent grass invader of the upper midwestern United States.

IMPACTS

Impacts of Grasses on Fire Regimes and Subsequent Ecosystem Change

It is now widely recognized that invasive grasses frequently alter fire regimes in the invaded areas. Because a fire regime represents a temporal and spatial control over the cycling of nutrient, water, and energy flow, as well as over the immediate behavior and composition of plants and animals, alterations to the fire regime lead to changes in multiple aspects of ecosystem structure and functioning. Invasive grasses have led to increases in fire frequency in arid and seasonally arid habitats including shrublands in the Great Basin and Mojave deserts of the western United States, riparian corridors in coastal California (Fig. 6) and Texas, and woodlands and pastures in Hawaii, Indonesia, and other subtropical and tropical habitats. *Bromus tectorum* invasion in the Great Basin is perhaps the most notorious example: its invasion has increased fire frequency from a greater than 30 year return interval to less than 5 years. Likewise, *I. cylindrica* invasion into tropical pastures has greatly altered fire regimes in many regions. In northern Australia and Florida, invasive grasses have led to large increases in fire intensity in locations where fire is a regular structuring force, often with catastrophic consequences for native species. For example, recent research by Rossiter et al. in Australia has shown that fire intensities in the northern tropical Eucalypt savannas has increased eight times in sites invaded by the

FIGURE 6 Riparian invader *Arundo donax* in two California river systems. (A) *Arundo donax* in foreground, fire burning in *Arundo* stand in background. Santa Clara River, Ventura County. (B) Dead native willow tree in riparian area burned in *A. donax*-fueled fire. Ventura River, Ventura County. (Photographs courtesy of T. Dudley.)

FIGURE 7 (A) Sagebrush steppe shrubland with flowering native grasses (*Poa secunda*) but with some invasion by *Bromus tectorum* before fire. (B) Burned sagebrush steppe vegetation now dominated by *B. tectorum* with only scattered native shrubs. Fire fueled by *B. tectorum*. Nevada, United States. (Photographs courtesy of R. Blank.)

South American grass *Andropogon gayanus* (gamba grass) compared to native grass sites. This, in turn, has caused the loss of otherwise fire-resistant native understory species.

Impact of Grasses on Community Composition and Structure

Stands of particular invasive grasses are associated with decreased plant richness in many habitats including tropical forest edges, subtropical woodlands and shrublands, tall grass prairie stands, and grass-invaded sites in California. In some cases, this is a direct result of competition between the invaders and native species. For example, in California, *Cortaderia selloana* outcompetes native shrubs, resulting in reduced shrub diversity and density in invaded sites. This, in turn, reduces arthropod and rodent activity. In the case of Hawaii and the Great Basin, reduced diversity of associated plants is the result of grass-fueled fires selecting against fire-intolerant native species as well as competitive effects of grasses on native species in the postfire environment. Most of these studies are biased toward finding such an effect because they have been conducted on species that are considered among an area's worst invaders, and the species achieve this status because they are suspected of negatively affecting local scale richness or evenness.

Studies on two invasive grasses illustrate how changes in plant assemblage structure during grass invasions can influence other trophic levels. *Arundo donax* dominance has been shown to be associated with reduced abundance and diversity of emerging aquatic insects in riparian habitats of northern California. Its litter has been demonstrated to reduce aquatic insect growth compared to native litter. In southern California, *A. donax* invasion is associated with reduced diversity and abundance

of birds. *Bromus tectorum* has transformed much of the intermountain western United States via increased fire frequencies, and this transformation has been shown to reduce plant diversity (Fig. 7) and to decrease habitat for shrub dependent birds. It may also decrease forage and cover for ground squirrels and other prey for raptors.

Impacts of Grasses on Soils and Nutrient Cycling

Several studies have demonstrated that invasive grasses can alter nutrient cycling, but not enough studies have been done to be predictive across invader lineages or ecosystems. Studies in the eastern United States have demonstrated that an invasive understory grass slows down nitrogen cycling in a deciduous forest understory. By contrast, studies in Hawaii and in an eastern U.S. wetland have demonstrated that invasive grasses can speed up internal nitrogen cycling relative to uninvaded communities. In the Hawaiian example, this may lead to long-term leaching losses of nitrogen from invaded sites. *Agropyron cristatum* invasion of midwestern U.S. prairies has led to reduced carbon and nitrogen storage in invaded compared to uninvaded soils. Invasion of serpentine soils by *Aegilops triuncialis* in California retards the decomposition of aboveground plant litter and increases soil aggregate stability.

A small number of studies have evaluated the effects of invasive grasses on soil microbes. These have demonstrated differences in microbial community composition between invaded and "native" sites, including lower levels of mycorrhizal fungal diversity in invaded sites in Utah and California. In all cases, these studies involved annual European or Eurasian grasses, including *Aegilops triuncialis* (barb goatgrass) and *B. tectorum* (cheatgrass or downy brome), invading native-dominated communities in the western United States. Recent work by Batten et al. in California has demonstrated experimentally that these changes in microbial community structure negatively impact native plants from these sites.

Impact of Forbs on Other Plants

Similarly to those invaded by grasses, sites invaded by exotic forbs are also associated with decreased plant species richness. Thirteen selected forb invaders of the Czech Republic all reduced the diversity, richness, and evenness of native plants in a variety of invaded communities; however, the degree of these reductions differed between the species. The species with the largest impacts, *Fallopia* spp. (knotweeds, Polygonaceae) and *Heracleum mantegazzianum* (giant hogweed, Apiaceae), were those that grew taller and more densely than resident native vegetation, perhaps competing more effectively for light than the native plants. The smothering, perennial groundcover *Tradescantia fluminensis* (small-leaved spiderwort, Commelinaceae) also reduces the amount of light available to native plants. Invasive forbs also sometimes compete more readily for limiting resources; *Centaurea maculosa* displays significantly higher phosphorus efficiency than its native neighbors and gains more biomass in phosphorus-limited soils.

Recent work with various *Centaurea* species provides an interesting example of the negative impact of invaders on neighboring plants, as well as an exploration of the contributing mechanisms. *Centaurea maculosa* strongly reduces the abundance and diversity of native species. *Centaurea diffusa* (diffuse knapweed) greatly decreases the biomass and phosphorus uptake of several native North American bunchgrasses while having little effect on European grasses from its native range. This negative effect may be due in part to the release of racemic catechin from some species of *Centaurea*, according to recent research by Bais and coworkers from Colorado and Montana. The (+) enantiomer possesses antimicrobial properties, but the (−) enantiomer is phytotoxic and reduces the growth of North American grasses when applied to the soil. Studies reporting the presence and concentration of catechin in natural soils have presented conflicting conclusions, indicating that release of these chemicals may be seasonal, sporadic, or spatially uneven.

Impacts of Forbs on Other Trophic Levels

Invasive forbs can alter the abundance, diversity, and behavior of native species in other trophic levels. Insects are the most frequently studied group, due to their close association with the plant community. New Zealand forests invaded by *Tradescantia fluminensis*, for example,

possess a reduced diversity of beetles and fungus gnats, and have overall changes in the composition of epigaeic insects relative to uninvaded understory habitats. *Pieris napi oleraceae* butterflies will lay eggs on the invasive *Allaria petiolata* (garlic mustard, Brassicaceae), but larvae do not complete development; therefore, areas with abundant *A. petiolata* serve as population sinks for these rare butterflies. Exotic plants can also enable butterfly species to expand into new habitats by extending the breeding season and providing additional food sources.

Invasive forbs can also change the behavior of pollinators, drawing them away from native plants. *Lythrum salicaria* (purple loosestrife) has showier flowers and produces more nectar than the North American native *L. alatum*, which causes it to attract more pollinators and reduce seed set in the native by up to 34 percent. Loosestrife can also reduce seed set in native wetland plants outside the *Lythrum* genus. The presence of showy, dominant alien forbs can also reduce pollinator efficiency and quality by increasing the movement of pollen between incompatible species.

The influence of invasive herbaceous plants on vertebrate populations is poorly documented, and the scant evidence is mixed. As a case study of the confusion over this issue, a study in wetland communities near Lake Huron in North America showed that wetlands invaded by *L. salicaria* contained the highest *density* of breeding birds of all the habitats they surveyed. Yet responses by individual species were mixed and sometimes in opposite directions. Overall, the *diversity* of breeding birds was lowest in invaded sites.

Impacts of Forbs on Soils and Nutrient Cycling

Invasive forbs can have significant impacts on soil organisms such as mycorrhizal fungi. For example, *Allaria petiolata*, a significant invader of forest understories in eastern North America, has antifungal properties that inhibit ectomycorrhizae, which are important for native species. *Centaurea maculosa* reduces the abundance and diversity of arbuscular mycorrhizal fungi and has a different overall microbial community in its rhizosphere compared to natives. This may affect competitive interactions between it and its neighbors. *Carduus pycnocephalus* (Italian plumeless thistle, Asteraceae) also reduces the density of arbuscular mycorrhizal fungi in its rhizosphere and grows best in soils without these fungi. These types of plant–soil feedbacks have been hypothesized to favor continued dominance of the invasive species.

Where studied, soil invertebrate communities have been found to be altered by invasive forbs. Soil arthropods in the Rocky Mountains of the western United States show sensitivity to invasion by *Euphorbia esula* and *Cirsium arvense*;

they are more numerous and diverse in univaded communities. Likewise, soil invertebrates in a temperate lowland rainforest in New Zealand are less diverse where *Tradescantia* invasion of the understory has occurred.

When they have carbon-to-nitrogen ratios substantially different from most of the residents, forb invaders can increase or decrease the amount of decomposition and therefore the availability of soil nutrients. Baruch and Goldstein showed that, as a group, invasive plants including many herbs present in Hawaii possess greater leaf nitrogen concentrations than do residents. More recently, other researchers have demonstrated that one of these forbs, *Hedychium gardenerianum*, decomposes faster and loses nitrogen and phosphorus more rapidly from litter, compared to native species. Asner and Vitousek could even detect an effect of this species on ecosystem nitrogen stature using hyperspectral imagery. Likewise, litter beneath *Tradescantia fluminensis* mats decays much more swiftly than litter on the uninvaded forest floor, increasing available nitrogen in invaded temperate forests.

The timing of nutrient release can also be changed by invaders. For example, *Lythrum salicaria* leaves have two times the phosphorous concentration of similarly aged leaves of native *Typha* (cattail) species that they replace. These high-phosphorus leaves decompose very quickly in autumn, potentially increasing the amount of phosphorus available in wetlands during this period. In spring, however, *L. salicaria* stems are highly resistant to decomposition, compared to *Typha*, resulting in altered timing of release of nutrients from plant litter.

Invasive species alter the movement through and storage of water in soil, although quantitative data on this impact are relatively scarce. *Centaurea maculosa* plants possess a deep taproot and relatively little surface foliage and can increase surface runoff and stream sedimentation in invaded communities by 56 and 192 percent, respectively, when measured on similarly sloped hillsides in a Montana rangeland. Two different sources demonstrate that, in some settings, *Centaurea solstialis* reduces deep soil water storage by transpiring water later into the summer drought period in California.

Economic Impacts of Grasses and Forbs

Studies of economic impacts of invasive plants have focused largely on woody species and forbs and not grasses, despite the clear ecological importance of invasive grasses. Also, most economic studies involve noxious weeds in agricultural settings, and these are summarized in D. Pimentel's papers, although details of how costs were obtained are not provided. With regards to invasive

plants in noncrop systems, only a few species have been studied extensively. In the northern Great Plains of the United States, *Euphorbia esula* has been a focus of control programs in rangelands, where it infects over 2.5 million acres. Its milky latex is toxic to livestock, and it therefore greatly reduces forage capacity of rangelands. Reductions in grazing capacity, livestock sales, and grazing land values due to *E. esula* invasion have been estimated at approximately \$37.1 million per year. Additionally, the effects on ungrazed lands, such as losses of native species, reduced recreational use, and altered ecosystem services, add \$3.4 million in annual direct impacts. Including secondary impacts on the economy of the region, the total annual impact of *E. esula* invasion exceeds \$129 million. Using a similar model and approach, the economic effects of *Centaurea* species in Montana were estimated at \$42 million annually, including impacts on grazed rangeland and wildlands. Costs of control can be considerable: almost \$45 million is spent annually to control *Lythrum salicaria* in the continental United States. Despite the absence of data for many species, the spread of exotic forbs represents a serious economic concern in many settings worldwide.

Detailed evaluation of the economic impacts of invasive grasses has rarely been done. Individual itemized costs such as the annual cost of fighting fires fueled by *Bromus tectorum* (\$20 million per year) in the western United States can be found in published literature. Losses of rangeland quality for livestock grazing have not been estimated despite clear negative impacts and reduced palatability of many invasive grasses, in particular *Tanaeitherum caput-medusae* (Medusae head) and *Aegilops triuncialis* (barbed goatgrass), in rangeland settings in the western United States and Australia. Likewise, despite well-recognized costs of controlling *Arundo donax* in California's riparian areas, no economic impact assessments are available in the published literature. A dam removal project in Ventura County, California, spent over \$2 million removing *A. donax* from above the dam prior to undertaking deconstruction. In addition to direct costs associated with removal, costs associated with *A. donax* include losses of wildlife habitat, risk of and costs of fighting fire, costs of revegetation, costs of beach clean-up of *Arundo* debris after storms, and losses to fisheries due to massive influx of *Arundo* debris into the nearshore ocean environment after storms. Similarly, despite widely recognized ecological and health effects of fires fueled by *Imperata cylindrica*, and costs associated with lost timber and crop trees (such as rubber or coconut) when uncontrolled fire burns through an *I. cylindrica* understory, real-cost data of any sort are difficult to find for this species. The massive

Indonesian fires of 1997–1998 are estimated to have cost $2–$4 billion in timber and other crop losses, the direct loss of homes, health problems from extensive haze, and losses of wildlife, native forests, and tourism. The fires were not caused by *I. cylindrica*, but the prevalence of dry *I. cylindrica* fuel on previously forested landscapes likely contributed to the massive scale of these fires.

SEE ALSO THE FOLLOWING ARTICLES

Agriculture / Cheatgrass / Disturbance / Fire Regimes / Invasiveness / Transformers / Vegetative Propagation / Weeds / Wetlands

FURTHER READING

Allison, S. D., and P. M. Vitousek. 2004. Rapid nutrient cycling in leaf litter from invasive plants in Hawaii. *Oecologia* 141: 612–619.

Bais, H. P., R. Vepachedu, S. Gilroy, R. M. Callaway, and J. M. Vivanco. 2003. Allelopathy and exotic plant invasion: From molecules and genes to species interactions. *Science* 301: 1377–1380.

Callaway, R. M., G. C. Thelen, A. Rodriguez, and W. E. Holben. 2004. Soil biota and exotic plant invasion. *Nature* 427: 731–733.

D'Antonio, C. M., and P. M. Vitousek. 1992. Biological invasion by exotic grasses, the grass/fire cycle and global change. *Annual Review of Ecology and Systematics* 23: 63–87.

DiTomaso, J. M. 2000. Invasive weeds in rangelands: Species, impacts, and management. *Weed Science* 48: 255–265.

Duncan, C. A., J. J. Jachetta, M. L. Brown, V. F. Carrithers, J. K. Clark, J. M. DiTomaso, R. G. Lym, K. C. McDaniel, M. J. Renz, and P. M. Rice. 2004. Assessing the economic, environmental, and societal losses from invasive plants on rangeland and wildlands. *Weed Technology* 18: 1411–1416.

Ehrenfeld, J. G. 2003. Effects of exotic plant invasions on soil nutrient cycling processes. *Ecosystems* 6: 503–523.

Lambdon, P. W., P. Pyšek, C. Basnou, M. Hejda, M. Arianoutsou, F. Essl, V. Jarošík, J. Pergl, M. Winter, P. Anastasiu, P. Andriopoulos, I. Bazos, G. Brundu, L. Celesti-Grapow, P. Chassot, P. Delipetrou, M. Josefsson, S. Kark, S. Klotz, Y. Kokkoris, I. Kühn, H. Marchante, I. Perglová, J. Pino, M. Vila, A. Zikos, D. Roy, and P. Hulme. 2008. Alien flora of Europe: Species diversity, temporal trends, geographical patterns and research needs. *Preslia* 80: 101–149.

Levine, J. M., P. B. Adler, and S. G. Yelenik. 2004. A meta-analysis of biotic resistance to tic plant invasions. *Ecology Letters* 7: 975–989.

Liu, J., M. Dong, S. Miao, Z. Li, M. Song, and R. Wang. 2006. Invasive alien plants in China: Role of clonality and geographical origin. *Biological Invasions* 8: 1461–1470.

Lonsdale, W. M. 1994. Inviting trouble: Introduced pasture species in northern Australia. *Austral Ecology* 19: 345–354.

Lowe, S., M. Browne, S. Boudjelas, and M. De Poorter. 2000. *100 of the World's Worst Invasive Alien Species: A Selection From the Global Invasive Species Database.* Aliens. The Invasive Species Specialist Group (ISSG) of the Species Survival Commission (SSC) of the World Conservation Union (IUCN).

Milton, S. J. 2004. Grasses as invasive alien plants in South Africa. *South African Journal of Science* 100: 69–75.

Murray, B. R., C. R. Dickman, T. Robson, A. Haythornthwaite, A. J. Cantlay, N. Dowsett, and N. Hills. 2007. Effects of exotic plants in native vegetation on species richness and abundance of birds and mammals (216–221). In D. Lunney, P. Eby, P. Hutchings, and S. Burgin, eds. *Pest or Guest: The Zoology of Overabundance.* Mosman: Royal Society of New South Wales, Australia.

Pyšek, P., and D. M. Richardson. 2007. Traits associated with invasiveness in alien plants: Where do we stand? (97–127). In W. Nentwig, ed. *Biological Invasions.* Ecological Studies, Vol. 193. Berlin: Springer.

Rejmánek, M. 1999. Invasive plant species and invasible ecosystems (79–102). In O. T. Sandlunch, P. J. Schei, and K. Vilken, eds. *Invasive Species and Biodiversity Management.* Dordrecht: Kluwer.

Strauss, S. Y., C. O. Webb, and N. Salamin. 2006. Exotic taxa less related to native species are more invasive. *Proceedings of the National Academy of Sciences, USA* 103: 5841–5845.

GRAZERS

DAVID M. FORSYTH

Department of Sustainability and Environment, Heidelberg, Victoria, Australia

Terrestrial mammalian grazers (here including browsers, granivores, and true grazers) have been widely introduced to new locations in both hemispheres. Grazers can have major negative social, economic, and environmental impacts, but some populations (particularly of large grazers) are valued as subsistence, economic, and recreational (including aesthetic) resources. Management of invasive grazers can be controversial when the control method is perceived to be inhumane or when the population is considered an asset by some people but a pest by others.

GLOBAL PATTERNS

Grazers were often deliberately introduced to establish populations for food (e.g., goats [*Capra hircus*] to remote islands as food for shipwrecked mariners), for commercial exploitation (e.g., fur bearers to South America and New Zealand), and to create sporting opportunities (e.g., deer in many locations). Domestic livestock that escape or are abandoned often establish "feral" populations. Many commensal rodents (e.g., the house mouse [*Mus musculus*]) were accidentally introduced to new locations.

The success of grazer introductions depends on the number of individuals introduced and the similarity between the climate of the receiving location and native range. It is also likely that an absence of predators and disease increases the probability of establishment.

Ten grazers are listed among the IUCN's list of 100 of the world's worst invasive alien species (see Appendix): brushtail possum (*Trichosurus vulpecula*), European rabbit (*Oryctolagus cuniculus*), feral goat, feral pig (*Sus scrofa*), gray squirrel (*Sciurus carolinensis*), house mouse, macaque monkey (*Macaca fascicularis*), nutria (*Myocastor coypus*), red deer (*Cervus elaphus*), and ship rat (*Rattus rattus*).

POPULATION DYNAMICS

Many grazers undergo "irruptions," or rapid increases to peak abundances followed by rapid declines to much lower abundances. Classic examples of irruptions of large (reindeer [*Rangifer tarandus*]) and small (house mouse) grazers are shown in Fig. 1. A rapid increase occurs because the population is not limited by food; the crash occurs when food requirements outstrip supply. Few studies have measured how food changes during an irruption, but house mouse populations in New Zealand increased in winter following periodic autumnal seeding of indigenous tree species (Fig. 1B).

Disease can also be an important driver of the population dynamics of grazers. The abundance of rabbits in Australia (Fig. 2) is regulated by periodic outbreaks of two diseases, myxamotosis and rabbit hemorrhagic disease. Although small grazers (e.g., rabbits and rodents) are often important prey for invasive predators such as the red fox (*Vulpes vulpes*) and feral cat (*Felis catus*), there is debate about the extent to which these predators regulate the abundance of their primary prey.

Competition can also affect the population dynamics of invasive grazers. In New Zealand, the Alpine chamois (*Rupicapra rupicapra*; from Europe) has been displaced from previously occupied sites by another mountain ungulate, the Himalayan tahr (*Hemitragus jemlahicus*). The mechanism for competition appears to be behavioral avoidance of the more social tahr by chamois, although higher densities of tahr may also reduce the availability of food for chamois.

NEGATIVE IMPACTS

The negative impacts of grazers can be classified as social, economic, or environmental, and all vary with species, location, and time. The social impacts of small grazers can be particularly significant. House mice in the cereal zones of eastern Australia attain densities of more than 800 mice per ha^{-1} over thousands of kilometers every 3 to 9 years, and people living in these areas experience extreme stress from mice being in large numbers in and around their houses.

Major economic impacts are incurred by primary producers, particularly pastoralists, horticulturalists, and foresters. Competition with livestock for pasture by grazers is a widespread and important economic impact: for example, the annual economic cost of lost livestock production in Australia due to rabbits was estimated at 113 million Australian dollars in 2004. Another important economic impact occurs when grazers are vectors or end hosts of major livestock diseases (e.g., bovine tuberculosis in swamp buffalo [*Bubalus bubalis*] in Australia and

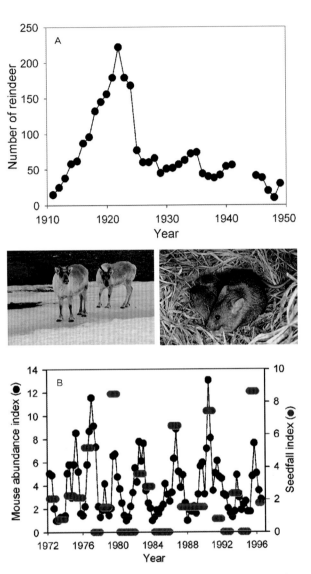

FIGURE 1 Irruptive dynamics of large and small grazers. (A) Number of reindeer counted annually on St. George Island, Pribilof Group, United States, following their deliberate introduction in 1911 (no counts were made during 1942–1945). (Photograph courtesy of Harald Steen, Norwegian Polar Institute.) (B) Number of house mice and seeds recorded quarterly in indigenous forest in the Orongorongo Valley, New Zealand, during 1972–1996. (Data courtesy of Wendy Ruscoe, Landcare Research New Zealand Ltd., photographs courtesy of Rod Morris, New Zealand Department of Conservation.)

brushtail possum in New Zealand). For example, nontariff trade barriers based on the prevalence of bovine tuberculosis could cost New Zealand up to 1.3 billion New Zealand dollars annually in lost export earnings. The costs of control can also be substantial: in New Zealand, at least 54 million New Zealand dollars have been spent annually since 2002 to control the brushtail possum.

The environmental impacts of grazers can be classified into those on plants (above and below ground) and those on other animals. Grazers typically reduce the abundance of

FIGURE 2 Small grazers such as rabbits can attain very high densities and cause substantial social, economic, and environmental impacts. (Photograph courtesy of National Archives of Australia A1200, L44186.)

preferred food plants and increase the abundance of some other plant species (through release from interspecific competition). Grazers sometimes cause the extinction of plant species. Rabbits introduced to Laysan Island (in the northwestern Hawaiian Islands) in 1903 apparently eliminated 22 plant species before they were eradicated in 1923. The amount of bare ground often increases when grazers are at high density. Grazers sometimes play an important role in suppressing the abundance of invasive plants if the invasive species are more palatable than the indigenous species.

Grazers may sometimes act as "ecosystem engineers." North American beaver (*Castor canadensis*) were released at Lago Fagnano in the Argentine part of Isla Grande in Tierra del Fuego in 1946 and have since spread to other smaller islands and to the mainland, over an area of about 7 million ha in which they occupy more than 27,000 km of waterway. Beavers modify the Tierra del Fuego ecosystem by changing the fluvial environment and killing riparian forest trees (Fig. 3).

There are three main ways that grazers might influence belowground processes. First, how grazers regulate the return of organic matter to the soil can influence biomass production and resource allocation. Second, grazers can alter the quality of resource inputs to decomposers through the return of feces and urine. Third, over long time scales, grazers may alter the quality and quantity of litter returned to the soil, and hence its decomposability.

There is interest in the impacts of grazers on carbon sequestration. The major carbon pools in natural ecosystems are live biomass, detritus and soil carbon, consumers, and decomposers. Grazers directly affect carbon by consuming foliage (i.e., they are consumers) and indirectly through plant–soil feedbacks. In New Zealand, soil carbon storage responded idiosyncratically to the exclusion of introduced browsing mammals at 30 sites. Because live biomass in forests is a large carbon pool, the most profound effect of grazers on carbon storage is likely to be where herbivory slows the succession of grassland to forest.

Invasive grazers can act as reservoir hosts for pathogens that can negatively affect native grazers. The gray squirrel (*Sciurus carolinensis*) has replaced the native red squirrel (*S. vulgaris*) throughout much of the United Kingdom following its introduction at the turn of the twentieth century. Although competition for food resources is important, experimental infections show that the parapoxvirus causes a deleterious disease in the red squirrel while having no detectable effect on gray squirrel health.

Grazers may also sustain native and introduced predators, which may have negative impacts on native species. Feral pigs enabled golden eagles (*Aquila chrysaetos*) to colonize the California Channel Islands. However, the eagles also preyed on the island fox (*Urocyon littoralis*), a competitor with the island spotted skunk (*Spilogale gracilis amphiala*): the reduced abundance of the island fox resulted in an increase in the abundance of the island skunk. The house mouse is prey for introduced stoats (*Mustela erminea*) in New Zealand: increased stoat abundance following mouse irruptions leads to higher mortality (via predation) in some native ground-dwelling and hole-nesting bird populations.

Hybridization between introduced and native grazers may pose a threat to species integrity. Red deer have been continuously present in Britain since the end of the last glaciation. The Japanese sika deer (*Cervus nippon*) has greatly increased its range in the United Kingdom

FIGURE 3 Riparian *Nothofagus* forest in Chile, killed by North American beaver. The forest has been replaced by grassland that is now used by feral horses. (Photograph courtesy of John Parkes, Landcare Research New Zealand Ltd.)

following introduction since 1860, and where they overlap with red deer, hybridization often occurs. Although a recent study in Scotland estimated that only 7 percent of 735 red and sika deer were "hybrids," the percentage varied with location up to a maximum of 43 percent.

HUMAN ATTITUDES AND MANAGEMENT

Attitudes to populations of grazers typically vary according to their social, economic, and environmental costs and benefits. Invasive grazers remain a commercial and recreational resource in many places, and subsistence harvesting may often be important for indigenous and rural peoples. However, the effects of invasive grazers on native biodiversity are now commonly (but not universally) considered undesirable, and there is often a desire to eradicate or at least control such populations. Many countries now have strict laws that aim to prevent invasive grazers from establishing in new locations, and the importation of new species for agriculture is often subject to a risk-assessment process.

Controversy about the management of feral horses (*Equus caballus*) in the United States, Australia, and New Zealand illustrates the conflicts that can arise about the management of some large grazers. In all three countries there has been intense debate about the ecological impacts of the horses, the size of populations, and the humaneness of management techniques. Management of many invasive grazer populations is controversial because the control methods are perceived to be inhumane.

Invasive populations may sometimes be discovered to be an important *ex situ* wild population for a species that has declined in its native range. For example, several species of wallaby (*Macropus* spp.) that established in New Zealand subsequently became rare or extinct in the wild in Australia, and animals captured from the invasive populations have been used to reestablish wild populations in their native ranges.

CONTROL

Commercial harvesting of invasive mammal species has been encouraged in many places because of its potential for economic and environmental benefits. Indeed, commercial harvesting is often viewed as a cost-effective means of reducing the abundance of invasive grazer populations. At least 1 million feral goats are harvested annually from the semiarid rangelands of Australia, and the size of the harvest is determined by the price paid per animal and by environmental conditions (Fig. 4). Harvesting and opportunity costs will also be important. Although commercial harvesting can have important benefits for conservation, when the harvest becomes unprofitable, the grazer

population will increase again. Moreover, harvesters are likely to ensure that populations are sustained and may even introduce them to hitherto uncolonized areas.

Bounties are financial incentives to harvest animals, and the hunter is required to present a body part for payment. However, bounty payments create a source of income that leads people to sustain the population as a resource. Bounty hunters often concentrate their efforts in areas where they most easily harvest the animal and not where the benefits from control are greatest. Bounty systems are also subject to fraud (e.g., animal parts from outside the bounty area are presented for payment) and do not guarantee a significant reduction in impacts.

Fences can be useful for excluding grazers from small areas but are expensive to construct and need regular checking and maintenance to minimize the risk of intrusion. A classic example of a failed fence is the 1,700 km rabbit-proof fence in Western Australia: the fence was started in 1901, but by the time it was finished in 1907, rabbits were already on both sides of the fence.

Biological control has been attempted for small grazers, but such efforts are unlikely to eradicate the population. Predators such as the red fox and stoat were introduced as biological control agents for rabbits, but where they have established, rabbits are still often a problem, and the predators themselves have sometimes precipitated extinctions of native mammals and birds. Myxomatosis and rabbit haemorrhagic disease virus (RHDV) were deliberately introduced to control rabbits in Australia: although both diseases were successful at reducing rabbit abundance and impacts, neither disease remains as effective now as

FIGURE 4 Feral goats living in the semiarid rangelands of Australia are commonly captured at watering points established for domestic livestock. Commercial harvesting of feral goats is an important source of income for many landholders. (Photograph courtesy of Tim Johnson, Department of Agriculture and Food, Western Australia.)

when first introduced due to the evolution of resistance in rabbits. Selection for virulence or attenuation in viruses such as myxomatosis and RHDV depends on the mode of transmission. There has been selection for attenuated strains of the myxomatosis virus (which is transmitted among live rabbits by biting insects) and for virulence in RHDV (which is transmitted from dead rabbits to live rabbits via blowflies).

Fertility control is often proposed as a humane method to reduce populations of grazers when translocation is impractical and culling is socially unacceptable. However, for long-lived large grazers (e.g., feral horses), it will take many years for this approach to achieve significant population reductions, and management costs can be very high. Fertility control is unlikely to be effective for most populations of small grazers due to the difficulty of treating a sufficiently large proportion of the population.

Although there are many examples of introduced grazers being eradicated from offshore islands, three prerequisite biological conditions must be met: (1) immigration must be prevented, (2) all animals must be placed at risk from the control method, and (3) the harvest must exceed births. An economic condition for eradication is that sufficient funds are available; however, the cost of removing individuals often increases exponentially as density is reduced. The eradication of nutria in southeastern England was successful because nearly all animals were trappable.

Control of grazers or their predators can sometimes have unintended consequences. When feral goats and feral pigs were eradicated from Sarigan Island, in the Commonwealth of the Northern Mariana Islands, a previously undetected invasive vine, *Operculina ventricosa*, increased greatly in abundance. Following the eradication of feral cats from sub-Antarctic Macquarie Island in 2001, the abundance of rabbits increased, and some plant communities were quickly transformed into grazed lawns and bare ground (Fig. 5).

SEE ALSO THE FOLLOWING ARTICLES

Acclimatization Societies / Australia: Invasions / Biological Control, of Animals / Game Animals / Herbivory / New Zealand: Invasions / Predators / Rats

FURTHER READING

Hone, J. 2007. *Wildlife Damage Control*. Collingwood: CSIRO.
Long, J. L. 2003. *Introduced Mammals of the World: Their History, Distribution and Influence*. Collingwood: CSIRO.
Lowe, S., M. Browne, S. Boudjelas, and M. De Poorter. 2000. *100 of the World's Worst Invasive Alien Species: A Selection from the Global Invasive Species Database*. Auckland: Invasive Species Specialist Group.
Pimentel, D., ed. 2002. *Biological Invasions: Economic and Environmental Costs of Alien Plant, Animal, and Microbe Species*. Boca Raton, FL: CRC Press.

FIGURE 5 Vegetation at two sites on sub-Antarctic Macquarie Island before (A, C) and after (B, D) the abundance of rabbits increased following the eradication of feral cats. (Source: Bergstrom, D.M., A. Lucieer, K. Kiefer, J. Wasley, L. Belbin, T.K. Pedersen, and S.L. Chown. 2009. Indirect effects of invasive species removal devastate World Heritage Island. *Journal of Applied Ecology* 46: 78. © The Authors. Reproduced with permission of Blackwell Publishing Ltd.)

Thompson, H. V., and C. M. King. 1994. *The European Rabbit: The History and Biology of a Successful Colonizer*. Oxford: Oxford University Press.
Zavaleta, E. S., R. J. Hobbs, and H. A. Mooney. 2001. Viewing invasive species removal in a whole-ecosystem context. *Trends in Ecology and Evolution* 16: 454–459.

GREAT LAKES: INVASIONS

EDWARD L. MILLS AND KRISTEN T. HOLECK

Cornell University, Ithaca, New York

The Great Lakes of North America (Fig. 1) have undergone accelerated ecological change since the arrival of European settlers some 250 years ago. These large, freshwater, inland sea ecosystems have been subjected to numerous stresses,

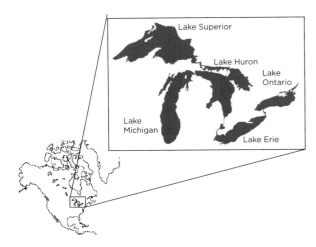

FIGURE 1 North America's Great Lakes.

including overfishing, cultural eutrophication, loss of habitat, toxic contamination, decline of native fish communities, and biological invasions by nonindigenous species (NIS). Nonindigenous species continue to be discovered in the Great Lakes, threatening ecosystem integrity. To date, 185 NIS have been discovered in the Great Lakes and have entered these waters through a wide variety of vectors. All Great Lakes have been invaded by NIS, and all have experienced significant ecological, economic, and social impacts as a result.

INVASION HISTORY

Exploration of the Great Lakes basin by Europeans commenced about four centuries ago and was followed by clearing of forests, settlement, and commercial development. The Great Lakes became a major transportation route for this region of the United States and Canada, and large cities grew around ports from which Midwestern grain, ore, lumber, furs, and other products were exported. The transport of these goods was enhanced through the construction of canals. By the late 1800s, a vast network of canals had been constructed in northeastern North America, thereby dissolving natural barriers to the dispersal of NIS into the Great Lakes. The Erie Canal (1825) and the Welland Canal (1829) were opened to facilitate trade to and from the basin. In April 1959, construction of the St. Lawrence Seaway and hydroelectric facility was completed, thereby allowing transoceanic ships to ply the largest freshwater lake ecosystem in North America from throughout the world. In fact, transoceanic shipping has been the primary entry mechanism, accounting for over one-third of NIS introductions to the Great Lakes (Fig. 2).

ORIGINS

Great Lakes NIS are native to a wide variety of geographic regions, including Eurasia, other basins within the

United States and the Atlantic and Pacific coasts of North America, Asia, Europe, and New Zealand (Fig. 3). Most NIS in the Great Lakes are native to Eurasia, including the Ponto-Caspian region (Black, Azov, and Caspian Seas), for two reasons. First, a high volume of goods is transported to the Great Lakes from this region; second, this area has a similar north temperate climate. Increasing world trade and travel will elevate the risk that additional species from around the globe will gain access to the Great Lakes. Existing connections between the Great Lakes watershed and neighboring basins such as the Chicago Sanitary and Ship Canal, as well as growth of industries such as aquaculture, live food markets, and aquarium retail stores, will also increase the risk of NIS introductions.

ENTRY VECTORS

The entry mechanisms that have acted singly or jointly in the movement of organisms into the Great Lakes basin include shipping, canals, unintentional release (for example, escape from cultivation), and deliberate release, related both to deliberate introduction of fish species to enhance fisheries and to private-sector activities (for example, release of aquarium and baitfish). The importance of ship ballast water (Fig. 4) as a vector for NIS introductions in the Great Lakes was recognized in the 1980s, prompting ballast management measures. In the wake of the establishment of both the Eurasian ruffe (*Gymnocephalus cernuus*) and zebra mussel (*Dreissena polymorpha*), Canada introduced voluntary ballast exchange guidelines in 1989 for ships declaring "ballast on board" (BOB) following transoceanic voyages, as recommended by the Great Lakes Fishery Commission and the International Joint Commission. In 1990, the United States Congress passed the Nonindigenous Aquatic Nuisance Prevention

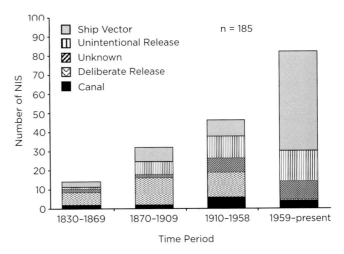

FIGURE 2 Vectors of introduction of NIS to the Great Lakes, 1830–2008.

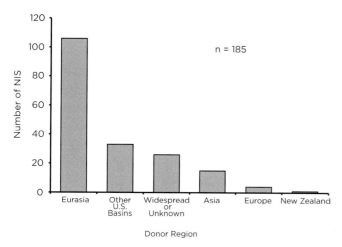

FIGURE 3 Donor regions for NIS introduced to the Great Lakes, 1830–2008.

and Control Act, producing the Great Lakes' first ballast exchange and management regulations in May of 1993. The National Invasive Species Act (NISA) followed in 1996, but this act expired in 2002. A stronger version of NISA entitled the Nonindigenous Aquatic Invasive Species Act has been drafted and awaits Congressional reauthorization.

Contrary to expectations, the reported invasion rate did not decline after initiation of voluntary ship ballast water exchange guidelines in 1989 and mandated regulations in 1993. More than 90 percent of transoceanic vessels that entered the Great Lakes during the 1990s declared "no ballast on board" (NOBOB) and were not required to exchange ballast, although their tanks contained residual sediments and water that would likely be discharged into Great Lakes waters. The residual waters and sediments of these ships have been found to contain species previously unrecorded in the basin. One likely introduction vector scenario is that such species could be discharged after a ship undergoes sequential ballasting operations as it travels between ports within the Great Lakes to offload and take on cargo. In 2006, Canada implemented new regulations for the management of residuals contained within NOBOB tanks, and requires the salinity of all incoming ballast water to be at least 30 ppt. NOBOB ships destined for U.S. ports were required to flush their tanks with saltwater beginning in 2008.

Recent studies suggest that the Great Lakes may differ among themselves in vulnerability to invasion. For example, Lake Superior receives a disproportionately high number of discharges by both ballasted and NOBOB ships, yet it has sustained surprisingly few initial invasions. Conversely, the waters connecting Lake Huron and Lake Erie are an invasion "hotspot" despite receiving disproportionately few ballast discharges. These hotspots appear to be ideal locations for the establishment of NIS because of their diverse lentic and lotic habitats, which provide more opportunities for species to establish as compared to open lake habitat.

Other vectors, including canals and the private sector, continue to deliver NIS to the Great Lakes and may increase in relative importance in the future. Silver (*Hypophthalmichthys molitrix*) and bighead carp (*Hypophthalmichthys nobilis*) escapees from southern U.S. fish farms have been sighted below an electric dispersal barrier in the Chicago Sanitary and Ship Canal, which connects the Mississippi River and Lake Michigan. Second only to shipping, unauthorized release, transfer, and escape have introduced NIS into the Great Lakes. Of particular concern are private-sector activities related to aquaria, garden ponds, baitfish, and live food fish markets. For example, nearly a million Asian carp are sold annually at fish markets within the Great Lakes basin. Until recently, most of these fish were sold live. All eight Great Lakes states and the province of Ontario now have some restriction on the sale of live Asian carp. Enforcement of regulations on many private transactions, however, remains a challenge and continues to pose a risk to Great Lakes waters.

IMPACTS

Damaging effects on Great Lakes ecosystems from unplanned species introductions are many, including habitat modifications, competition, predation, disease pathogens, and genetic effects. Several nonindigenous fish, including the sea lamprey, alewife, white perch, and introduced salmonids, have caused adverse predator–prey interactions on native species. The sea lamprey (*Petromyzon marinus*), a parasitic fish, has had a catastrophic impact on native lake trout (*Salvelinus namaycush*) in the Great Lakes (Fig. 5). In the late 1960s, alewife (*Alosa pseudoharengus*) populations accelerated the collapse of coregonid and bloater (*Coregonus hoyi*) populations and negatively

FIGURE 4 Ship deballasting. (Photograph courtesy of P. Le Calvez.)

FIGURE 5 Sea lamprey on lake trout. (Photograph courtesy of R. Greenough, Sure Strike Charters.)

impacted yellow perch (*Perca flavescens*) and other native fish. The white perch (*Morone americana*) gained access to the Great Lakes in the 1950s, where it has had substantial impacts on native fish species and community stability. The introduction of salmonids, either deliberately (to enhance the sport fishery) or unintentionally (as in the case of pink salmon), has had significant and permanent ecological (predation on native salmonids and other fishes and the introduction of parasites and diseases) and genetic impacts (caused by interbreeding with native fishes).

The poster children of invasive species in the Great Lakes are the zebra mussel and quagga mussel (*Dreissena rostriformis bugensis*) (Fig. 6). Both species are ecosystem engineers—they modify existing habitat by increasing water clarity and provide new habitat in that their shell clusters serve as refuge areas for other invertebrates. The zebra mussel was first detected in the Great Lakes in 1988. The quagga mussel, first observed in 1989, was observed coexisting with zebra mussels in 1991, and by 1995, quagga mussels had replaced zebra mussels in some areas at depths up to 85 m. As these dreissenid mussels became firmly established at shallow depths, the abundance of the amphipod *Diporeia* decreased by 68 percent at depths of 30–90 m. Historically, the burrowing amphipod *Diporeia* represented 60–80 percent of benthic biomass in the lakes and was a critical high-energy food source for fish like lake whitefish. *Dreissena* spp. (especially *D. r. bugensis*) appear to have played a major role in the demise of *Diporeia* in the lower Great Lakes, but to date the actual mechanism for the cause of the decline is unknown. The loss of *Diporeia* has major implications for the capacity of Great Lakes ecosystems to support fish that depend on these organisms, as well as for how benthic energy will effectively flow to fish. Lastly, both zebra and quagga mussels have generated enormous economic costs for management and control of fouling on water intake structures.

Colonization of Great Lakes waters by dreissenid mussels has greatly changed the physical and ecological character of these ecosystems. Infestation of the lakes by zebra and quagga mussels has led to the local extinction of at least 18 native clam species. Perhaps the greatest change in the lakes has been a systemwide increase in water clarity in response to intense removal of algal and particulate material through grazing by *Dreissena* spp. Increased water clarity and light penetration has resulted in reduced free-floating algae in offshore waters and increased macrophyte growth in shallow-water habitats. Consequently, dramatic ecological changes associated with *Dreissena* spp. have resulted in the coining of new terms like "benthification" and "nearshore shunt" to describe how energy and nutrients flow through Great Lakes ecosystems. Dreissenid mussels also play a role in the transmission of disease. The pathogen *Clostridium botulinum* has been implicated in recurring outbreaks of type E avian botulism in the lower Great Lakes. *Clostridium* bacteria thrive in the anaerobic conditions produced in the sediments beneath and within beds of dreissenid mussels. Round gobies (*Neogobius melanostomus*) ingest the botulinum toxin as they feed on the mussels and then transfer the toxin to higher trophic levels (e.g. waterbirds).

Viral hemorrhagic septicemia (VHS) is a deadly fish virus and an invasive species that is threatening Great Lakes fish. VHS was implicated for the first time ever in the Great Lakes as the cause of large fish kills in lakes Huron, St. Clair, Erie, and Ontario and in the St. Lawrence River in 2005 and 2006.

Invasive species can have a significant impact on the sustainability of Great Lakes food webs and native species. Native lake trout declined to extinction in Lake Ontario in the 1950s in response to an expanding population of sea lamprey. Control of sea lamprey began in 1971, and lake trout stocking was renewed in 1973 to restore a self-sustaining fish population. Unfortunately, stocked lake trout of hatchery-origin failed to reproduce and scientists have

FIGURE 6 Zebra mussel (left) and quagga mussel (right). (Photograph courtesy of NOAA, Great Lakes Environmental Research Laboratory.)

determined that the alewife contributed to reproductive failure. High mortality of lake trout fry (i.e., early mortality syndrome) is caused by thiamine deficiency resulting from a maternal lake trout diet of alewife.

THE FUTURE PROLIFERATION OF INVASIVES

NIS have invaded the Great Lakes basin from regions around the globe, and increasing world trade and travel will elevate the risk that additional species will gain access to the Great Lakes. Indeed, the recent arrival of the bloody red shrimp, *Hemimysis anomala,* was predicted, and scientists are now assessing its impact. Existing connections between the Great Lakes watershed and systems outside the watershed, such as the Chicago Sanitary and Ship Canal, and growth of industries such as aquaculture, live food markets, and aquarium retail stores will also increase the risk that NIS will be introduced.

Changes in water quality, global climate change, and previous NIS introductions may make the Great Lakes more hospitable for the establishment of new invaders. Evidence indicates that newly invading species may benefit from the presence of previously established invaders. That is, the presence of one NIS may facilitate the establishment or population growth of another ("invasional meltdown"). For example, the sea lamprey may have created enemy-free space (by preying on lake trout) that facilitated the alewife's invasion, and the round goby and *Echinogammarus* have thrived in the presence of previously established zebra and quagga mussels. Dreissenids have effectively set the stage to increase the number of successful invasions, particularly those of coevolved species in the Ponto-Caspian assemblage. Evidence also suggests that dreissenids have promoted the proliferation of other nuisance species, including native and exotic weeds and blue-green algae.

MANAGEMENT IMPLICATIONS

Researchers are seeking to better understand links between vectors and donor regions, the receptivity of the Great Lakes ecosystem, and the biology of new invaders in order to make recommendations to reduce the risk of future invasion. To protect the biological integrity of the Great Lakes, it is essential to closely monitor routes of entry for NIS, to introduce effective safeguards, and to adjust safeguards quickly as needed. The rate of invasion may increase if positive interactions involving established NIS or native species facilitate the establishment of new NIS. Moreover, each new invader can interact in unpredictable ways with previously established invaders, potentially creating synergistic impacts.

The identification of ship ballast water as a major vector transporting unwanted organisms into the Great Lakes has motivated control efforts. However, ships' hulls and both water and sediments in ballast tanks can all contain NIS. Ballast water exchange reduces the risk of invasion but does not totally eliminate unwanted invaders. While open ocean seawater may be sufficiently saline to kill freshwater organisms in the ballast tanks of transoceanic ships, it may not suffice to kill brackish or estuarine organisms.

Great Lakes managers will continue to wrestle with the invasive species issue, for NIS are stressors with predominantly negative impacts on ecosystem sustainability. The long history of invasions and the increasing rate of established introductions clearly indicate that the Great Lakes are highly vulnerable to invasion. Invasive species usually have characteristics such as high abundance, short generation time, broad physiological plasticity, high genetic variability, and tolerance for a wide variety of habitats. Management strategies aimed at prevention must consider the linkages between NIS and vectors. Without vector control (both known and emerging vectors), the Great Lakes will continue to be exposed both to ecological disruptions and to surprises linked to biological invasions.

SEE ALSO THE FOLLOWING ARTICLES

Ballast / Canals / Fishes / Invasional Meltdown / Ponto-Caspian: Invasions / Regulation (U.S.) / Sea Lamprey / Zebra Mussel

FURTHER READING

Holeck, K. T., E. L. Mills, H. J. MacIsaac, M. R. Dochoda, R. I. Colautti, and A. Ricciardi. 2004. Bridging troubled waters: Understanding links between biological invasions, transoceanic shipping, and other entry vectors in the Laurentian Great Lakes. *Bioscience* 54: 919–929.

Mills, E. L., J. H. Leach, J. T. Carlton, and C. L. Secor, 1994. Exotic species and the integrity of the Great Lakes: Lessons from the past. *Bioscience* 44(10): 666–676.

Mills, E. L., J. H. Leach, J. T. Carlton, and C. L. Secor. 1993. Exotic species in the Great Lakes: A history of biotic crises and anthropogenic introductions. *Journal of Great Lakes Research* 19(1): 1–54.

Ricciardi, A. 2006. Patterns of invasions in the Laurentian Great Lakes in relation to changes in vector activity. *Diversity and Distributions* 12: 425–433.

GYPSY MOTH

PATRICK C. TOBIN AND ANDREW M. LIEBHOLD

U.S. Department of Agriculture, Morgantown, West Virginia

The gypsy moth, *Lymantria dispar* (L.) (Lepidoptera: Lymantriidae), is a highly polyphagous foliage feeder whose larvae can feed on over 300 deciduous and coniferous host species, most notably oak, willow, aspen, larch, and birch. Its native range includes temperate Eurasia and the Mediterranean coast of North Africa, where it occasionally erupts in damaging outbreaks.

INTRODUCTION TO NORTH AMERICA

In 1869, gypsy moth egg masses were brought to Medford, Massachusetts, by an amateur entomologist, Étienne Léopold Trouvelot, after which life stages accidentally escaped from captivity. Trouvelot attempted to limit establishment success by aggressively searching for life stages and destroying them; however, the infestation persisted, increased slowly, and was first noticed by Medford residents about ten years later. By 1889, the gypsy moth became so abundant and destructive on fruit and shade trees that it attracted public attention.

In 1890, the State of Massachusetts appropriated $25,000 to eradicate the gypsy moth, whose infestation by then covered approximately 2,539 km². Control tactics included the use of copper acetoarsenite and "cyclone burners," an oil-fueled flame thrower used to destroy life stages. By 1899, little defoliation was detected, although complete eradication was not yet achieved; nevertheless, in 1900 the State of Massachusetts ordered the eradication effort discontinued because of the belief that the gypsy moth had been reduced to a minor pest.

Gypsy moth populations dramatically rebounded over the next several years, and in 1905 Massachusetts resumed control work, spending more than $25 million over the next 25 years. By this time, the gypsy moth had spread to Maine, New Hampshire, and Rhode Island. The gypsy moth has since continued to spread slowly and relentlessly to the north, south, east, and west and now occupies North America from Nova Scotia to Wisconsin, and Ontario to Virginia (Fig. 1). It currently occupies approximately one-third of the potential range of susceptible forests in North America.

LIFE HISTORY

The gypsy moth develops through a single generation each year. The life cycle of gypsy moth is presented in Fig. 2. Eggs overwinter and larvae hatch in the spring. Males and females pass through five and six instars, respectively, before pupating. The larval stage lasts approximately 4–6 weeks, and the pupal stage lasts about 2 weeks. Adults emerge in mid-summer and generally survive a week or less; males tend to emerge slightly before females. Adults do not feed. Females produce a sex pheromone that males use in mate location. Following mating, eggs are laid in a single mass that is buff-colored and that contains hairs originating from the female's abdomen. Like all poikilotherms, developmental rate is mostly driven by temperature, but host

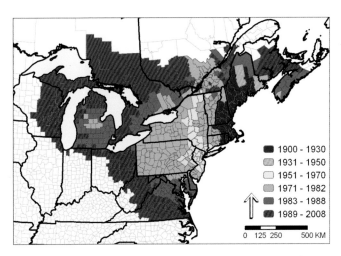

FIGURE 1 Distribution of the gypsy moth through time, as determined by U.S. and Canadian quarantine records. (Map courtesy of Laura Blackburn.)

plant quality can also play a role. Several robust phenology models are available for predicting the seasonal occurrence of gypsy moth life cycle events.

Across the native range of gypsy moth, populations exhibit considerable variation in life history traits. Perhaps the most variable characteristic is female flight capability. In populations located in various portions of Asia, at least some proportion of females is capable of directed flight. However, female flight is less common in Europe, and in many European populations females are completely incapable of flight. North American populations originated from western Europe, and female flight is completely absent.

Another characteristic that varies considerably among populations is body size, which is often higher in certain Asian strains than in European strains. This greater body size may reflect the increased musculature necessary in females to achieve guided flight. But greater body size also necessitates greater food consumption, and certain Asian strains are known to consume much larger quantities of foliage to complete development.

The most common primary host trees throughout much of the gypsy moth's range are those in the genera *Quercus, Populus, Larix, Salix,* and *Betula.* However, in some areas of the world, these hosts may not be present and populations instead exploit other species. For example, in Central Asia outbreaks are most common on *Pistacia, Juglans,* and *Malus.* With the exception of deciduous conifers (e.g., *Larix, Metasequoia,* and *Taxodium*), conifers are generally not primary gypsy moth hosts. Although late instars may feed on conifers, particularly when preferred species have been completely defoliated, early instars cannot complete development on these

FIGURE 2 Life cycle of the gypsy moth: (A) eggs, (B) larvae, (C) female (left) and male (right) pupae, and (D) male (left) and female (right) adults. (Photographs courtesy of Patrick Tobin [A, B] and USDA APHIS PPQ Archive, USDA APHIS PPQ, Bugwood.org [C, D].)

hosts. As a consequence, there are no cases in the world of gypsy moth outbreaks developing in stands dominated by nondeciduous conifers.

POPULATION DYNAMICS

In North America, gypsy moth populations are affected by a variety of trophic interactions and typically oscillate between low and high densities (Fig. 3). Although gypsy moth dynamics are not characterized by highly regular oscillations, most populations exhibit some degree of periodicity with the dominant period varying from 5 to 10 years. Low-density populations are most strongly affected by small mammal predators such as *Peromyscus* spp. These generalist predators feed on a variety of insects, fruits, and seeds; consequently, their populations are not numerically linked to gypsy moth cycles. Instead, their dynamics are affected strongly by the

availability of mast, upon which they depend as a primary food source during winter; thus, yearly variability in mast production indirectly influences the "release" of gypsy moth populations from low to high densities.

During outbreaks, populations reach very high levels but ultimately collapse after 1–3 yr, usually as a result of disease epizootics. In North America, the two common gypsy moth pathogens are the fungus *Entomophaga maimaiga* and the gypsy moth nucleopolyhedrosis virus (*Ld*MNPV), both of which infect only gypsy moths. The *Ld*MNPV is present in virtually every gypsy moth population around the world, and is thought to have been introduced to North America with the founding gypsy moth population. The range of *E. maimaiga* was previously limited to Japan. In 1989, it was discovered to be causing considerable mortality in North American gypsy moth populations, but it remains

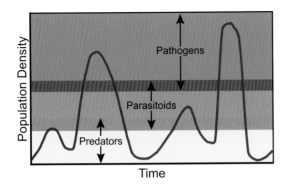

FIGURE 3 Generalized description of gypsy moth population dynamics and the primary sources of mortality at low, intermediate, and high densities.

unclear whether it arrived in North America through earlier attempts to introduce it in the early 1900s, or whether it was accidentally introduced later.

A plethora of parasitoids and predators were also introduced to North America in the early twentieth-century as part of classical biological control efforts against the gypsy moth, including the generalist larval parasitoid *Compsilura concinnata* (Meigen). Although *C. concinnata* can exert spatially density-dependent regulation of gypsy moth, it is also known to exert a detrimental effect on several native Lepidoptera. It is thus a classic example of the danger in introducing nonspecific biological control agents. Other important introduced enemies of the gypsy moth, which also tend to act in a spatially density-dependent manner, include the egg parasitoid *Ooencyrtus kuvanae* (Howard), the larval parasitoids *Cotesia melanoscela* (Ratzeberg) and *Parasetigena sylvestris* (Robineau-Desvoidy), the pupal parasitoid *Brachymeria intermedia* (Nees), and the predator *Calosoma sycophanta* (L.)

CONSEQUENCES OF OUTBREAKS

Outbreaks in North America have been both environmentally and economically costly. Since 1924, over 376,000 km^2 of U.S. forests have been defoliated, including over 50,000 km^2 in 1981 alone (Fig. 4). Gypsy moth outbreaks also tend to be spatially synchronized over areas spanning hundreds of kilometers, which can greatly exacerbate the ecological and socioeconomic impacts of high-density populations and overwhelm management resources allocated to mitigate impacts.

The critical factor in determining the susceptibility of stands to gypsy moth outbreaks is the relative abundance of preferred gypsy moth host species (e.g., oaks). During outbreaks, the gypsy moth can cause partial or total defoliation of canopies, often resulting in growth loss and severe physiological stress in trees. When defoliation occurs over consecutive years or in conjunction with other sources of

stress, such as drought or late-season frosts, trees will become weakened and die. Tree death is typically associated with secondary organisms such as the shoestring root rot fungus *Armillara* spp. or the two-lined chestnut borer, *Agrilus bilineatus* (Weber). In most cases, mortality is limited to forested areas consisting of previously weakened trees, but in occasional instances, outbreaks can kill almost all the trees in a stand. Average mortality across a forested landscape during outbreaks is roughly 25–30 percent of the trees.

Gypsy moth outbreaks can result in loss to timber and other traditional forestry values. Greater losses tend to occur to the ecosystem services that forests provide, such as wildlife habitat, carbon sequestration, and nutrient cycling. Outbreaks can also alter the composition of the community, including indirect changes to native herbivores that gypsy moth tends to outcompete and altering forest succession. The greatest economic impacts are considered to be in residential areas, and in addition to the loss of foliage and occasional mortality of trees in urban settings, gypsy moths often cause considerable nuisance in populated areas owing to large quantities of caterpillars and their frass. Although most people do not react strongly upon contact with the urticating hairs on gypsy moth larvae, some individuals develop severe allergic reactions.

RANGE EXPANSION

Females of the European strain of gypsy moth (currently established in North America) cannot fly and usually oviposit within 2 m of where they emerged as adults. Dispersal occurs mainly through early instar ballooning, which can be facilitated through atmospheric transport. Nevertheless, natural dispersal of larvae is believed to be relatively short-ranged, as most larvae do not disperse more than a few hundred meters. However, the characteristic behavior of late instars searching for cryptic resting sites often results in larvae pupating, and consequently emerging females ovipositing, on human-made objects, including recreational equipment and motor vehicles. Accidental movement of life stages on such objects, especially sessile egg masses that overwinter from late summer to the following spring, is a mechanism of dispersal that plays an important role in invasion spread. Both short-ranged (first instar ballooning) and anthropogenic dispersal mediate the spread of gypsy moth through a process known as stratified dispersal, in which local growth and dispersal is coupled with the long-range dispersal of propagules. The implications of stratified dispersal in the context of an invading species are profound. Long-distance dispersal can result in the establishment of colonies ahead of the invading population front, which can

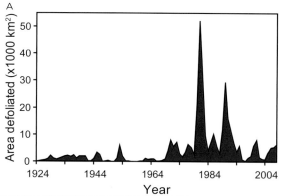

FIGURE 4 (A) History of defoliation in the United States, (B) widespread defoliation in Pennsylvania, and (C) nuisance impacts due to gypsy moth frass and body parts on a residential picnic table. (Photographs courtesy of Karl Mierzejewski, Centre County, Pennsylvania, and John Ghent, USDA Forest Service, Bugwood.org, respectively.)

grow in density, spatial extent, and eventually coalesce with the expanding population front (Fig. 5). The result is a much faster rate of range expansion than would be expected under simple local population growth and diffusive spread. Past work has highlighted this consequence in the gypsy moth: the use of a Skellam model parameterized from demographic data predicted gypsy moth spread owing to local growth and short-ranged dispersal to be 2.5 km/yr; however, historical distributional records of gypsy moth indicated a spread rate of 20.8 km/yr. This

greater rate of spread observed in nature is believed to result from stratified dispersal.

Long-distance dispersal of the gypsy moth in North America thus tends to contribute greatly to its range expansion and is typical of the stages of any biological invasion. However, once life stages arrive in a new area, they must successfully establish. Many biological and ecological factors influence the establishment success of newly arriving populations. Owing to Allee effects and stochastic forces that act upon low-density founder populations, not all newly arriving gypsy moth populations successfully establish, and establishment success is often positively related to the size of the initial arriving population. Because establishment rates also drive spread, the interaction between Allee effects and successful colony establishment can have landscape-scale consequences in the rate of gypsy moth spread.

The mechanisms responsible for Allee effects observed in isolated, low-density gypsy moth populations and their consequences to establishment and spread are probably better known than for any other insect species. A large amount of experimental and empirical evidence indicates that mate location failure is the primary cause of Allee effects in the gypsy moth. Despite the presence of a highly sensitive mechanism by which males can detect very low levels of sex pheromones released by females, this system is not always effective, and most females go unmated in

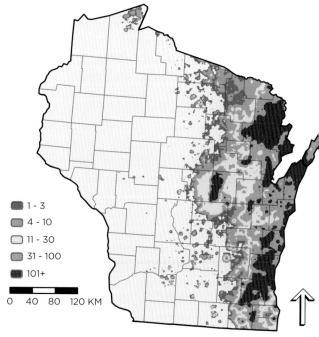

1 - 3
4 - 10
11 - 30
31 - 100
101+

0   40   80   120 KM

FIGURE 5 The 2001 gypsy moth invasion in Wisconsin, illustrating the importance of stratified dispersal in gypsy moth spread. (Map courtesy of Laura Blackburn.)

low-density, isolated populations. In particular, the isolated nature of colonizing populations leads to a phenomenon in which males "become lost" when searching for females that are both spatially and temporally disjunct. As a result of the strong Allee effect arising from mate location failure, the arrival of a single egg mass (from which several hundred larvae may hatch) is most likely not sufficient to establish a new population. This mechanism thus explains the observation that hundreds of egg masses are accidentally transported annually from established gypsy moth populations, but only a small fraction persist for one or more years.

MANAGEMENT

In North America, there are three primary types of gypsy moth management programs: (1) detection/eradication, which targets new colonies in areas uninfested by the gypsy moth (e.g., the west coast of North America), (2) the Slow-the-Spread program, which consists of a barrier zone along the invasion front in the United States, and (3) suppression of outbreaks in areas that are infested by the gypsy moth as a means to mitigate impacts. The first two programs rely on intensive monitoring of populations using pheromone-baited traps, and over 250,000 traps are deployed annually in support of these two programs. These traps are deployed with the synthetic sex pheromone, (+) disparlure, and are highly sensitive tools for detecting low-density populations. Upon the detection of male moths in pheromone-baited traps, more extensive networks of traps are deployed in the next year to determine if colonies have persisted, delimit the spatial extent of infestations, and locate the epicenter of the infestation. In this manner, management tactics can then be deployed in the most economically viable, site-specific manner while minimizing potential nontarget effects. Most eradication programs use aerial and ground applications of *Bacillus thuringiensis variety kurstaki* (*Btk*) to eliminate populations. In the Slow-the-Spread program, most treatments are carried out using mating disruption: synthetic pheromone is formulated in a slow-release material and aerially applied to foliage, which permeates the air with pheromone and interferes with the male moths' ability to locate females (Fig. 6). Mating disruption tactics are most effective in low-density gypsy moth populations.

Suppression of outbreaks is a different management approach. Management procedures vary among different jurisdictions but generally involve identifying areas of increased population densities based on the appearance of detectable levels of defoliation. In most cases, egg mass surveys are conducted to identify specific forested areas where outbreaks are present and likely to cause

FIGURE 6 Management of gypsy moth through aerial application of *Bacillus thuringiensis kurstaki.* (Photograph courtesy of Quentin Sayers.)

defoliation. Pheromone-baited traps are not used to monitor for outbreaks, because they become quickly replete with moths at population levels far below outbreaking densities. Most programs use aerial applications of *Btk* to suppress populations. In a few areas, populations are managed using Gypchek®,[1] a commercial formulation of *Ld*MNPV, or the insect growth regulator diflubenzuron. Since *Btk* potentially affects all Lepidoptera, Gypchek is an alternate tactic when there are concerns of nontarget effects to threatened and endangered species. Diflubenzuron, although highly effective, has broader nontarget effects and can potentially affect all molting arthropods; consequently, its use is heavily regulated, especially in areas near and around water sources. Although suppression of outbreaks may be effective in protecting specific areas from defoliation in a given year, these efforts generally do not alter the course of regional outbreaks.

CONCLUSION

Biological invasions continue worldwide because of trends in world trade and travel. The gypsy moth is an example of an "older" biological invasion that predates the current abundance of invasion pathways through which new species are arriving in new areas at alarming rates. Fortunately, much basic and applied research has focused historically on the gypsy moth, which reached its apex in the 1970s and 1980s. Consequently, gypsy moth managers can exploit this extensive research and have at their disposal some of the most effective tools of any management program targeting a biological invasion: (1) basic biological, ecological, and economic information that allows us to predict when

---

[1]Mention of a proprietary product does not constitute an endorsement or a recommendation for its use by USDA.

and where gypsy moth populations are likely to show up or reach outbreak levels, or have significant impacts, (2) a means by which to detect low-density, newly established populations, and (3) effective control tactics that can be used to suppress or eradicate populations. The other consequence of years of extensive research on the gypsy moth is that it provides an excellent system for understanding the population biology of invading species. Detailed information about mechanisms operating behind population establishment and spread is vast, and perhaps more extensive than for any other invading insect species.

SEE ALSO THE FOLLOWING ARTICLES

Forestry and Agroforestry / Life History Strategies / Pesticides for Insect Eradication

FURTHER READING

Doane, C. C., and M. L. McManus, eds. 1981. *The Gypsy Moth: Research toward Integrated Pest Management.* Technical Bulletin 1584. Washington, DC: U.S. Department of Agriculture.

Davidson, C., K. W. Gottschalk, and J. E. Johnson. 1999. Tree mortality following defoliation by the European gypsy moth (*Lymantria dispar* L.) in the United States: A review. *Forest Science* 45: 74–84.

Elkinton, J. S., and A. M. Liebhold. 1990. Population dynamics of gypsy moth in North America. *Annual Review of Entomology* 35: 571–596.

Johnson, D. M., A. M. Liebhold, P. C. Tobin, and O. N. Bjørnstad. 2006. Allee effects and pulsed invasion by the gypsy moth. *Nature* 444: 361–363.

Jones, C. G., R. S. Ostfeld, M. P. Richard, E. M. Schauber, and J. O. Wolff. 1998. Chain reactions linking acorns to gypsy moth outbreaks and Lyme disease risk. *Science* 279: 1023–1026.

Liebhold, A. M., J. A. Halverson, and G. A. Elmes. 1992. Gypsy moth invasion in North America: A quantitative analysis. *Journal of Biogeography* 19: 513–520.

Régnière, J., V. Nealis, and K. Porter. 2009. Climate suitability and management of the gypsy moth invasion into Canada. *Biological Invasions* 11: 135–148.

Sharov, A. A., and A. M. Liebhold. 1998. Bioeconomics of managing the spread of exotic pest species with barrier zones. *Ecological Applications* 8: 833–845.

Tobin, P. C., and L. M. Blackburn, eds. 2007. Slow the spread: A national program to manage the gypsy moth. General Technical Report NRS–6. Newtown Square, PA: U.S. Department of Agriculture, Forest Service.

Tobin, P. C., C. Robinet, D. M. Johnson, S. L. Whitmire, O. N. Bjørnstad, and A. M. Liebhold. 2009. The role of Allee effects in gypsy moth (*Lymantria dispar* L.) invasions. *Population Ecology.* In press.

HABITAT COMPATIBILITY

SCOTT STEINMAUS

California Polytechnic State University, San Luis Obispo

The habitat of an organism is the type of environment it inhabits, or can inhabit, and is characterized by abiotic (nonliving) and biotic (living) factors such as solar radiation and temperature regimes, nutrient concentrations, densities of competitors, prey sizes, and prey densities. However, not all habitat factors are relevant for persistence of a particular organism. An organism's niche represents the range of relevant environmental variables in which the organism can survive and reproduce. An organism's niche in the absence of other biota is called its *fundamental niche*. When other biota (competitors, predators, pathogens) are present, the organism's niche is usually restricted and is called its *realized niche*. The term niche is also used to describe the organism's role in an ecosystem. This includes its effects on the habitat. Habitat compatibility for an invasive species means that there is a match between its realized niche and at least one habitat in the area of its introduction.

GENERAL CHARACTERISTICS

Habitat characteristics can be used as predictors of invasion risk. Available data and models indicate that the significant predictors for compatibility are the suitability of microclimate, soil characteristics, frequency and magnitude of disturbance, and how the habitat is managed. Environmental niche models (ENM) were originally developed for biodiversity informatics but are also currently utilized to predict the potential distribution of an invader in a new location. These models use algorithms to summarize geographic information system (GIS) data such as GIS layers that describe climate, edaphic and biotic variables, and other features of a habitat to characterize the actual or potential distribution of individual species. They have also been used to identify significant components of an invasive species' niche that have provided it the opportunity to be successful in a new location. Examples of ENMs include maximum entropy (MAXENT) models, genetic algorithm for rule-set production (GARP), BIOCLIM, and relative environmental suitability (RES) models.

Human-associated activities are responsible for the majority of successful invasions because their disturbances result in substrate denudation, nutrient enrichment (eutrophication), destruction or reduction of biomass, and fragmentation of successionally advanced communities. These activities often reduce both abiotic and biotic resistance and provide the conduit by which propagules of invasive species have been introduced.

Open wetlands, riverbanks, coastal locations, lakes, and ponds are very compatible habitats for invasive species because water is not a limiting resource as it often is in terrestrial habitats. These habitats are often subject to frequent disturbance and nutrient pulses brought on by seasonal flooding. Additionally, aquatic habitats may have a buffering effect on environmental extremes that occur in adjacent terrestrial habitats.

Global climate change predictions indicate that as global average temperatures increase, habitats at higher latitudes or elevations will become more susceptible to invasions. Climatic variability will also increase, which could give rise to naturally occurring disturbances such as fire, wind, and flooding. Additionally, as precipitation patterns change, resources and conditions will likely change to those less favorable to natives, making their habitats more prone to invasion.

Environmental heterogeneity can be temporal at many scales, as when resources are seasonally available or periodically variable (e.g., sun flecks within plant canopies). An environment may also be spatially heterogeneous at many scales, such as where soil types or canopy openings are patchy at the square-centimeter to square-meter scale and beyond. Environmental homogeneity of some habitats may explain why exotic species introductions are unsuccessful—for example, cases where there is inadequate soil moisture in space or time to support seedling establishment. As environmental heterogeneity accentuates by increases in patch size, intensity of differences among patches, or time between resource flushes, habitats become more compatible to species that can complete their life cycles within the time span or spatial confines of favorable environmental conditions and resources. If recruitment limitations can be overcome, then another successful strategy may be that of large, long-lived species with wide dispersal capacity, which can better bridge favorable patches of growth-limiting resources and favorable conditions.

The following list provides *generalized* trends for habitat compatibility to invaders:

- Habitats with low net primary productivity such as those in extreme conditions or with low resource levels are not likely to support long trophic chains or invaders that are higher-trophic-level consumers.
- Habitats in close proximity to human habitation and activities are usually suitable for many invaders.
- Low-latitude temperate regions are invaded more than tropics, deserts, or savannas.
- Open wetlands may be more suitable for invasive species than mesic communities, and mesic communities are generally more susceptible than xeric communities.
- Islands can have several times higher proportions of exotic species than mainland sites, especially those along traditional trade routes, because of high propagule pressure of introduced species and other biogeographic features unique to islands.
- Habitats rich in native species tend to be less compatible to invasion than those with few native species at the square-centimeter to square-meter scale, but more susceptible to invasion at the landscape scale (km^2).

DISTURBANCE REGIME

Naturally occurring sources of above- and belowground disturbances to which natives have adapted include fire, windstorms, herbivory, and flooding. Human-associated disturbances include logging, overfishing, grazing, tillage, and grading. Any change, whether increase or decrease, in the frequency or magnitude of natural disturbance regimes will usually make a habitat more susceptible to invasion. The effects of any mode of disturbance can interact synergistically, antagonistically, or neutrally with one another. For example, increased fire frequency often results in increased soil erosion during subsequent precipitation. Disturbance usually increases with proximity to human habitation and visitation. Very few aliens can invade successionally advanced communities without a disturbance. However, an increasing number of shade-tolerant invasive plant species are being reported from undisturbed forests.

Fire

A majority of the world's terrestrial habitats have historically depended on fire for ecological sustainability because fire affects vegetation types and architecture, nutrient and carbon fluxes, and hydrologic properties of soils. These habitats may become more compatible to invasion when those frequencies are reduced as the result of fire suppression policies protecting human interests. This can happen because fire fuel loads increase under these policies, contributing to catastrophic fires that remove all aboveground vegetation, not just the understory as would normally occur. Invasive species themselves such as annual exotic grasses that invade shrubland (e.g., *Bromus tectorum*) or habitats where woody vegetation was removed can contribute to an increase in fire frequency. By growing and senescing earlier than natives, exotic grasses alter the distribution of fine fuel across space and time, thus extending the fireprone duration of a landscape and facilitating further invasion.

Soil Disturbance

Habitats experiencing major soil disturbance, such as tillage or erosion, are highly susceptible to invasion. Even if soil disturbance ceases, exotics can maintain dominance, excluding natives for long periods of time (e.g., decades) if native propagules have been depleted from the site. At the same time, natives in adjacent habitats can maintain their dominance, if the soil remains undisturbed. However, even a reduction in substrate disturbance can make a habitat more compatible for some invaders; for example, *Ammophila arenaria* (European beachgrass) reduces normal sand dune migration, because of rapid vertical growth, facilitating its own spread at the expense of natives.

Flooding

Habitat compatibility for invasive species may increase when flooding events are altered from normal levels. Draining swamps, regulating water flow from dams, stabilizing stream banks, and digging canals will alter normal flooding events. Invasive species themselves can transform aquatic habitats by reducing disturbance, such as how *Arundo donax* (giant reed) reduces water flow in waterways, facilitating its spread across a waterway. *Tamarisk ramossisima* invades sites where the flooding regime has been altered by dams and levees and may also contribute to increased fire frequencies in these habitats.

Herbivory and Predation

Habitat compatibility may increase when the normal intensity of herbivory is altered. Introduced nonnative grazers, such as cattle, have different feeding patterns from those of native browsers. When stocking rates are high, cows trample soils more than native browsers. Altered modes of grazing and fire contribute to an increased susceptibility to soil erosion. Under high grazing pressures, stream banks become altered while sedimentation rates and nutrient levels increase in the waterway. All of these impacts tend to increase habitat compatibility to invasion. Gypsy moths and unregulated deer are other examples of herbivores that have disturbed forest habitats, making them more prone to invaders. Introduction of exotic predators on islands disrupts food webs and may provide opportunities for new invaders.

ABIOTIC FACTORS

Resources

Because fundamental niches of many introduced species can overlap only with resource-rich habitats, intermittent resource enrichment and resource release following disturbance contribute to habitat compatibility. Mineral nutrient enrichment, especially nitrogen and phosphorus resulting from human activities such as a result of agricultural runoff, industrial contamination, or air pollution, may make a habitat more suitable for many fast-growing invaders. The larger the difference between resource supply and resource demand, the more susceptible the habitat is to invasion (Fig. 1). Habitat susceptibility increases as longer periods separate resource availability and subsequent uptake.

Invasive species themselves can alter resource levels spatially and temporally, thus transforming habitats and making them more susceptible to further invasion. This synergistic interaction has been called "invasional meltdown."

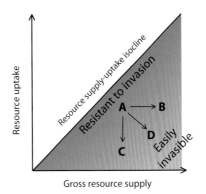

FIGURE 1 The theory of fluctuating resource availability holds that a community's susceptibility to invasion increases as resource availability (the difference between gross resource supply and resource uptake) increases. Resource availability can increase due to a pulse in resource supply (A→B), a decline in resource uptake (A→C), or both (A→D). In the plot shown, resource availability, and hence invasibility, increases as the trajectory moves further right or below the supply-uptake isocline (where resource uptake = gross resource supply). (From Davis et al., 2000.)

The following list provides some examples of transformation activities of invaders:

- Excessive resource users such as fast-growing annual exotic grasses (e.g., *Bromus tectorum*, *Centaurea solstitialis*) reduce resource availability at times and at soil depths critical for natives. Similarly, *Tamarix* spp. invasion in 25 southwestern U.S. states has depleted water for natives.
- Soil carbon storage is modified. In soils infested by exotic grasses, soil carbon-to-nitrogen ratios may increase in fire-prone habitats, and phosphorus cycles are altered in habitats that lack fire.
- Light suppression is achieved by excessive plant litter accumulation, such as in *Eucalyptus* forests and invaded grasslands or by the covering of the substrate with introduced benthic macroalgae, as occurs in the Mediterranean Sea.
- Some invaders can increase levels of limiting resources, for example by nitrogen fixation by rhizobia of invasive legumes such as broom species, *Acacia* spp., or Actinomycetes of the nonleguminous *Morella* (*Myrica*) *faya* and Australian pine (*Casuarina* spp.) in California, South Africa, Florida, and Hawaii.

Conditions

Habitats at the same latitude with a similar climate as the habitat from where an invasive species originated are often compatible to the invader. Similarity of climatic seasonality appears to be an important climatic determinant of habitat compatibility as revealed from ENMs discussed above. Factors such as duration of warm and cool or wet and dry periods and range of maximum and minimum daily and annual temperatures affect habitat susceptibility. However, even though predictions of habitat compatibility based on climate matching alone have been successfully implemented for some species, they have been inadequate for many potential invasive species, especially those with small native ranges.

As conditions become more extreme and resources become scarcer, abiotic resistance to invasion increases. Consequently, very few alien species are found in environmentally extreme habitats or nutrient-poor habitats (e.g., mires, heathlands, high mountain grasslands). Xeric communities have high abiotic resistance and therefore do not support the survival of exotics. Therefore, mesic communities are more susceptible than extreme environments.

As with resources, conditions can be altered from their original levels following the establishment of certain invasive species, thus contributing to further invasion:

- Accumulation or redistribution of salts as done by *Mesembryanthemum crystallinum* and *Tamarix* spp. can reduce native competitiveness.
- Compounds can accumulate that are allelopathic, such as those released by *Eucalyptus* spp. and *Juglans* spp. that suppress the growth of other plants or the compounds released by *Solidago canadensis* that suppress native soil pathogens.
- Soil chemistry can be altered, such as when *Rhododendron ponticum* acidifies soils.

BIOTIC FACTORS

Resident Biota

Habitats can be characterized by their biota, which can influence the potential for success of invaders (restricting their realized niches) through competition, facilitation, predation, disease, or parasitism. As mentioned previously, species-rich habitats are less compatible at the square-centimeter to square-meter scale but are more invasible at the landscape scale (10–1,000 km^2). It seems that competition through efficient resource preemption by residents (resource use complementarity) explains resistance at the smaller scales, while species-rich landscapes are usually environmentally heterogeneous and possess numerous potential niches. At the larger scales, a positive relationship may exist because of sufficient environmental heterogeneity for extrinsic factors that overwhelms the effect of competition for resources.

Human-made monocultures may be susceptible to invasion because of the lack of coevolutionary history among the resident species. This suggests that the history of species assembly may be more important than species diversity in determining compatibility. However, a recent review of the research literature on invasive species reveals that community assembly often occurs by ecological sorting and fitting of species and involves ecological interactions among all species, while long periods of coevolution may not always be necessary to explain final community composition. Essentially, communities with long histories of species assembly and exposure to larger species pools are more resistant to invaders.

Habitats with competitive natives may partially resist invasion, but usually other factors such as disturbance and impact of resident pathogens, herbivores, and predators determine compatibility (space for realized niches of invaders). Communities composed of plant species that regrow rapidly following disturbance (e.g., humid tropical forests) have high biotic resistance to potential invaders.

Resident Enemies and Mutualists

According to the enemy release hypothesis, alien plant and prey species will tend to have more advantages than native prey and plant species because the aliens will be less likely to have natural enemies in the new habitat. Indeed, in a review of 473 naturalized species in the United States, the species had 84 percent fewer fungal and viral pathogens than in their native habitats. Introduced populations of all taxonomic groups, including plants, insects, crustaceans, molluscs, reptiles, birds, and mammals, exhibit 29 to 86 percent parasite release compared with native populations.

Habitats can be characterized by the niches of the resident natural enemies such as predators, herbivores, and pathogens as well as the resident mutualists and facilitators such as pollinators, nitrogen fixers, mycorrhizae, and propagule dispersers. Theoretically, habitats are more susceptible to invaders when biotic resistance is low—for example, those occupied by natural enemies that are specialized for natives and by native mutualists or facilitators that are generalists (Fig. 2). This is because specialized enemies are less likely to successfully switch to an exotic species and because generalist mutualists and facilitators will aid all species, native and exotic. Conversely, habitats occupied by generalist predators or pathogens and by specialized facilitators or mutualists may be less likely to be invaded. In habitats with high host diversity, and thus low host density, generalist

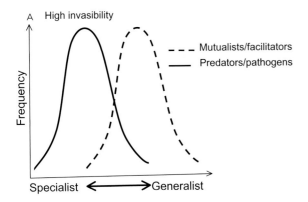

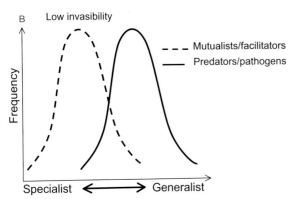

FIGURE 2 Specialization in species interactions and invasibility. Individual ecosystems can vary such that the frequency distribution of species along a continuum from absolute specialists to absolute generalists differs between predators and pathogens (solid lines) and mutualists and facilitators (dashed lines). (A) In a high-invasibility system, predators and pathogens are more frequently specialists, whereas mutualists and facilitators are more frequently generalists. Such systems are relatively easy to invade because few predators are able to prey upon exotic species (for which they are not specialized), whereas many mutualists are able to assist exotic species. (B) A low-invasibility system, with the opposite distribution and invasion outcome. The curves illustrated here are for heuristic purposes only; the actual shape of these curves is unknown empirically. Their impact on invasibility should operate as described here, however, as long as there is a difference in the mode of the two distributions, and as long as the frequency of interactions determines the average outcome of invasions. (From Sax et al., 2007.)

pathogens usually dominate, which may contribute to the high biotic resistance in tropical rainforests.

Agriculture, horticulture, and forestry provide several examples of endemic pathogens that have decimated introduced species such as crops. However, the outcome of this specialist-generalist theory depends on whether predators involved in the interaction have a foraging-switch behavior or whether the prey exhibits a defense-switch behavior. Defense-switch is less likely when species have no prior experience with similar threats as has occurred for bird species of most isolated Pacific islands prior to human arrival.

SEE ALSO THE FOLLOWING ARTICLES

Climate Change / Disturbance / Eutrophication, Aquatic / Invasibility, of Communities and Ecosystems / Land Use / Range Modeling / Transformers

FURTHER READING

Chytrý, M., L. C. Maskell, J. Pino, P. Pyšek, M. Vilà, X. Font, and S. M. Smart. 2008. Habitat invasions by alien species: A quantitative comparison among Mediterranean, subcontinental, and oceanic regions of Europe. *Journal of Applied Ecology* 45: 448–458.

Davis, M. A., J. Davis, J. P. Grime, and K. Thompson. 2000. Fluctuating resources in plant communities: A general theory of invasibility. *Journal of Ecology* 88: 528–534.

Kearney, M. 2006. Habitat, environment and niche: What are we modeling? *Oikos* 115: 186–191.

Lonsdale, W. M. 1999. Global patterns of plant invasions and the concept of invasibility. *Ecology* 80: 1522–1536.

Mack, M. C., and C. M. D'Antonio. 1998. Impacts of biological invasions on disturbance regimes. *Trends in Ecology and Evolution* 13: 195–198.

Mitchell, S. C. 2005. How useful is the concept of habitat? A critique. *Oikos* 110: 634–638.

Ohlemüller, R., S. Walker, and J. Bastow Wilson. 2006. Local vs. regional factors as determinants of the invasibility of indigenous forest fragments by alien plant species. *Oikos* 112: 493–501.

Sax, D. F., J. J. Stachowicz, J. H. Brown, J. F. Bruno, M. N. Dawson, S. D. Gaines, R. K. Grosberg, A. Hastings, R. D. Holt, M. M. Mayfield, and M. I. O'Conner. 2007. Ecological and evolutionary insights from species invasions. *Trends in Ecology and Evolution* 22: 465–471.

Schoener, T. W. 2009. Ecological niche (3–13). In S. A. Levin, S. R. Carpenter, H. C. J. Godfray, A. P. Kinzig, M. Loreau, J. B. Losos, B. Walker, and D. S. Wilcove, eds. *A Princeton Guide to Ecology*. Princeton: Princeton University Press.

Thuiller, W., D. M. Richardson, P. Pyšek, G. F. Midgley, G. O. Hughes, and M. Rouget. 2005. Niche-based modeling as a tool for predicting the risk of alien plant invasions at a global scale. *Global Change Biology* 11: 2234–2250.

HAWAIIAN ISLANDS: INVASIONS

LLOYD LOOPE

U.S. Geological Survey, Makawao, Hawaii

The Hawaiian Islands are justifiably famous for their biological uniqueness but have lost roughly half of their original native-dominated habitat. Dry and mesic forests have been reduced most drastically. Hundreds of species have been lost to extinction, including at least 106 of 1,300 plant taxa, three-fourths of approximately 1,000 endemic snail species, and at least 83 of 115 endemic land bird species. With only 0.4 percent of the land area of the United States, Hawaii harbors over 25 percent of the country's federally listed endangered species. While habitat destruction by humans has been a direct factor in Hawaii's ecological losses in the past, human-facilitated biological invaders are currently the primary agents of continuing degradation.

BACKGROUND: GEOLOGIC SETTING, ORIGINS, AND EVOLUTION OF HAWAII'S ENDEMIC BIOTA

The biological uniqueness of the Hawaiian Islands (Fig. 1) results from their great topographic and climatic diversity and their extreme isolation from continents and other islands. Shield volcanoes have been emerging over a stationary hot spot, then moving northwestward on the Pacific tectonic plate, for over 70 million years. After reaching their maximum height, the volcanoes subside and erode down to sea level. The state of Hawaii includes the low-lying northwestern Hawaiian Islands stretching northwest to 28 °N, but eight major high islands located at 19–22 °N make up more than 99 percent of the state's total land area. The youngest island, Hawaii, has an area of more than 10,000 km^2 (63% of the total area of the state) and has elevations reaching more than 4,000 m. The oldest rocks of the eight high islands range from 420,000 years old (Hawaii) to about 5 million years old (Kauai). Moisture-laden trade winds flow from the northeast over the complex terrain of the islands. Whereas rainfall has averaged 630–750 mm annually over the open ocean near Hawaii, the islands themselves have locally received on average as much as 10,000 mm of rain in windward sites and as little as 200 mm in leeward sites.

Colonization of the current islands by ancestors of the modern terrestrial animals and plants has taken place by long-distance dispersal—on the winds, by floating, or attached to storm-driven birds—mostly within the past 5 million years, although the island origins of several lineages date back much further. Evolution in isolation has led to the generation of species and higher lineages unique to one or more of the islands.

Some animal and plant groups reached the geologically developing Hawaiian Islands later than others, leaving the opportunity for some of the early immigrants to evolve into new roles and habitats. Beginning with a single colonizing species, certain animal or plant groups underwent a sequence of speciation events that produced large numbers of related species that live in a wide range of habitats and play a variety of ecological roles. Only 350–400 insect species, representing only 15 percent of the world's insect families, colonized the islands via long-distance dispersal, but these subsequently diversified to form an endemic insect fauna that may have exceeded 10,000 species. Some of the successful colonizing species became sources of spectacular evolutionary adaptive radiation. Among birds, over 50 species of Hawaiian honeycreepers (Fig. 2) evolved from a common cardueline finch-like ancestor, believed to have arrived from North America just over 5 million years ago. The silversword alliance in the sunflower family is a spectacular plant example.

VERTEBRATE ANIMAL INVASIONS IN HAWAII

With the exception of the endemic Hawaiian hoary bat (*Lasiurus cinereus semotus*), the Hawaiian monk seal (*Monachus schauinslandi*), and sea turtles that come on land occasionally, Hawaii's lack of terrestrial native mammals, amphibians, and reptiles is undoubtedly a significant contributor to the vulnerability of the islands to vertebrate invasions. Five species of amphidromous fish of freshwater habitats evolved from marine fish families. Among vertebrates, only birds were important prehistoric long-distance colonizers, resulting in remarkable evolutionary radiations, including an array of flightless goose-like birds ("moa-nalos"), the size of large swans, which were apparently derived from colonizing ducks. At least four species of moa-nalos are known from subfossil bones in lava tube caves or sand dunes; their extinctions occurred within the past 500 years. Evidence of their diets from associated coprolites strongly suggests their prehistoric role as major herbivores. Hawaii's vegetation and flora have apparently not evolved in the absence of vertebrate herbivory, but rather in the absence of mammalian herbivory.

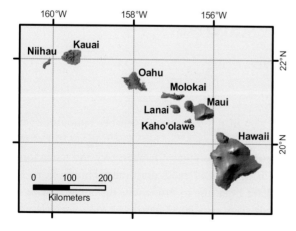

FIGURE 1 Map of the major Hawaiian Islands. (Map courtesy of Jenny Paduan and David Clague, Monterey Bay Aquarium Research Institution.)

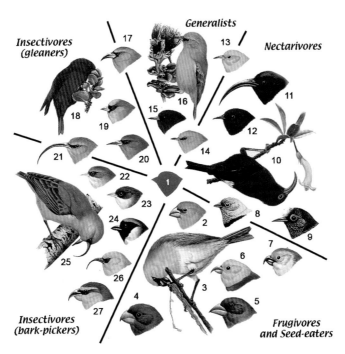

FIGURE 2 Hawaiian honeycreepers (Fringillidae: Drepanidinae) provide a textbook example of evolutionary adaptive radiation. Over 50 species evolved from a common cardueline finch-like ancestor, believed to have arrived from North America just over 5 million years ago. This diagram illustrates 26 of them. Key to adaptive radiation labels: 1. Cardueline ancestor; 2. Laysan finch, *Telespiza cantans*; 3. Palila, *Loxioides bailleui*; 4. Kona grosbeak, *Chloridops kona*; 5. Greater koa-finch, *Rhodacanthis palmeri*; 6. ʻOʻu, *Psittirostra psittacea*; 7. Lanaʻi hookbill, *Dysmorodrepanis munroi*; 8. ʻUla-ʻai-hawane, *Ciridops anna*; 9. ʻAkohekohe, *Palmeria dolei*; 10. ʻIʻiwi, *Drepanis coccineus*; 11. Hawaii Mamo, *Drepanis pacificus*; 12. ʻApapane, *Himatione sanguinea*; 13. ʻAnianiau, *Magumma parva*; 14. Maui ʻAlauahio, *Paroreomyza montana*; 15. Kakawahie, *Paroreomyza flammea*; 16. Hawaii ʻAmakihi, *Hemignathus (Chlorodrepanis) virens*; 17. Kauaʻi ʻAmakihi, *Hemignathus (Chlorodrepanis) kauaiensis*; 18. ʻAkepa, *Loxops coccineus*; 19. ʻAkekeʻe, *Loxops caeruleirostris*; 20. Greater ʻamakihi, *Hemignathus (Viridonia) sagittirostris*; 21. Lesser ʻakialoa, *Hemignathus (Akialoa) obscurus*; 22. Hawaii creeper, *Oreomystis mana*; 23. Akikiki, *Oreomystis bairdi*; 24. Poʻo-uli, *Melamprosops phaeosoma*; 25. ʻAkiapolaʻau, *Hemignathus (Hemignathus) munroi*; 26. Kauaʻi nukupuʻu, *Hemignathus (Hemignathus) hanapepe*; 27. Maui parrotbill, *Pseudonestor xanthophrys*. (Illustration © 2009 H. Douglas Pratt; modified from Pratt, 2005, *The Hawaiian Honeycreepers: Drepanidinae*, Oxford University Press [Plate 8].)

Rodents

Of four rodent species introduced to the Hawaiian Islands, the arboreal black rat (*Rattus rattus*) probably has the greatest effect currently on native fauna and flora. The other three species are the Polynesian rat (*R. exulans*), the Norway rat (*R. norvegicus*), and the house mouse (*Mus musculus*). Rodents feed on the fleshy fruits and flowers of Hawaiian plants and girdle and strip tender branches. Past and continuing effects of rat predation on birds, snails, and insects of Hawaiian rainforests have been enormous.

Polynesian rats arrived in Hawaii with colonizing Polynesians, most likely in the tenth or eleventh century AD. Recent paleoenvironmental and archaeological investigations from the lowlands of southwestern Oahu suggest surprisingly severe impacts of Polynesian rats slightly in advance of direct human impacts involving cultivation and fire. A dramatic piece of evidence has involved a pollen type that was once abundant in Oahu's lowland pollen record but had entirely disappeared in the pollen record by 1500 AD. It did not match the pollen of any known living plant. Serendipitously, two individuals of a plant whose pollen matched the once-abundant type were found on a cliff on the island of Kahoolawe and were described as a new leguminous genus and species (*Kanaloa kahoolawensis*).

Ungulates

Feral pigs are currently the primary modifiers of the remaining Hawaiian rainforest and have substantial effects on other ecosystems, through direct herbivory and trampling, as well as by dispersal of other invasive species, primarily flowering plants. Although pigs were brought to the Hawaiian Islands by Polynesians about 1,000 years ago, the severe environmental damage currently inflicted by pigs apparently began much more recently and is largely due to the release of domestic, non-Polynesian genotypes. Polynesian pigs were much smaller, more docile, and less prone to taking up a feral existence than those introduced in historical times. By the early 1900s, the

FIGURE 3 Typical chronic feral pig damage to the forest floor in koa/ohia forest at about 1,500 m elevation, in Maui, Hawaii. Many conservation areas on Maui have been fenced and pigs removed. (Photograph courtesy of Pat Bily, the Nature Conservancy of Hawaii.)

damage caused by feral European pigs in native rainforests was being recognized as a major threat to Hawaiian watersheds; it continues today (Fig. 3), except within pig-free fenced exclosures on conservation lands. There is at least anecdotal evidence that pigs were able to spread into wildlands only gradually, because a diet with adequate protein was lacking until earthworms were gradually dispersed into wildland areas.

The negative effect of goats on vegetation, especially on islands, is well known worldwide. Goats were introduced to the eight major Hawaiian islands before 1800, and within a few decades, they had reached remote areas. Goats have reduced or eliminated whole populations of native plants, have aided in plant invasions, and have hastened soil erosion. Sustained, locally high populations of goats have obliterated even the most unpalatable native plant species. Indeed, some plant species survive only on ledges and other sites that are inaccessible to goats, or they are older trees that had reached a size that made them less vulnerable by the time goats came to occupy their habitats. Elimination of feral goats within the fenced boundaries of Hawaii Volcanoes National Park in the 1970s and Haleakala National Park in the 1980s was a great achievement that provided an important impetus for further conservation efforts.

Other problematic ungulates in Hawaii include cattle, several species of deer, and mouflon sheep. Feral cattle persist in forests on Maui and other islands despite decades of effort at removal. Burgeoning populations of deliberately introduced (1959) axis deer (*Axis axis*) on Maui pose a notably severe threat to native ecosystems.

Predatory Mammalian Carnivores and Omnivores

Feral dogs (*Canis familiaris*) and especially feral cats (*Felis catus*) are very significant predators of seabirds and other bird species in Hawaii. The small Indian mongoose, *Herpestes auropunctatus*, is undoubtedly a serious predator of Hawaii's endemic birds, but little information is available on damaging the effects of mongooses versus cats or rats, for example. There are good data, however, on damage caused by mongooses to populations of Nene (*Branta sandwicensis*) and Hawaiian dark-rumped petrel (*Pterodroma phaeopygia sandwichensis*) populations.

The small Indian mongoose has been introduced intentionally into wild habitats more widely than any other mammal, because of its reputation as a control agent for rats and snakes. A Jamaican sugar planter introduced four male and five female small Indian mongooses to Jamaica from eastern India in 1872. From Jamaica, mongooses eventually were introduced to 29 islands throughout the Caribbean and to the northern coast of South America. In 1883, sugar planters in Hawaii imported the mongoose from Jamaica to four Hawaiian islands (Hawaii, Oahu, Maui, and Kauai). The mongoose failed to establish on Kauai but was later established on Molokai. Current conventional wisdom in Hawaii is that the introduction of mongooses failed to control rats in areas of introduction because the mongoose is diurnal and rats are primarily nocturnal, but there is scarce support and much contradiction for this anecdote based on literature from locations where the mongoose has been introduced.

Birds

Alien birds may threaten native birds directly through competition (Fig. 4) and transmission of diseases and parasites, and indirectly through aiding ecosystem alteration by dispersal of seeds of alien plants, ultimately affecting native birds dependent on native ecosystems. More than 170 species of birds, including parrots, owls, pheasants, egrets, finches, and geese, have been introduced to Hawaii, more than to any other place of comparable size; 54 species have established breeding populations in the wild, and there are at least 13 other potentially naturalized species. Introduced birds are now dominant, both in numbers of species and individuals, in nearly all terrestrial habitats in Hawaii.

Before 1945, nearly all releases were intentional, by government agencies, naturalization societies, or individuals motivated to fill a void created by the loss of native birds. After 1945, legislation started to restrict the release of birds, and new releases mainly consisted of escapees

FIGURE 4 The nonnative Japanese white-eye (*Zosterops japonicus*) was introduced to Hawaii in about 1929 and is now consistently common throughout Hawaii's high-elevation forests. A generalist feeder, it most commonly gleans for insects, but it also feeds on fruits and nectar. The white-eye is considered the most likely candidate as a harmful competitor with endemic honeycreepers. However, such competition is difficult to document, and it is far from clear how the birds could be effectively removed. (Photograph courtesy of Jack Jeffrey.)

from captivity and game bird releases. Thirteen of the 23 species naturalized since 1950 are common cage finches or parrots. Research on the success of Hawaii bird introductions suggests that interspecific competition among invaders is a primary force determining establishment, although other factors such as climate and introduction effort may be as important.

Owing to patterns of disease and habitat loss, most native songbirds in Hawaii are restricted primarily to upper elevation forests, while the majority of introduced birds reside in nonforested habitats in the lowlands, such as agricultural fields and urban areas. Thus far, 16 introduced bird species are reliably encountered in native forests, the primary forest-type of native birds, but only seven pose major direct threats to native bird species.

Amphibians and Reptiles

Hawaii lacks native amphibians and terrestrial reptiles; 27 species of amphibians and reptiles have established, but little is known about their ecological effects. The best-studied species is the coqui (*Eleutherodactylus coqui*), a small frog native to Puerto Rico that arrived in Hawaii through the plant trade. It first established in the late 1980s, and it currently occurs on four Hawaiian islands. Coqui have been shown to attain average densities of over 20,000 frogs per hectare and to remove 114,000 invertebrate prey items per night in Puerto Rico; coqui densities in Hawaii exceed the Puerto Rico density threefold, with the difference likely attributable to the apparent lack of native or alien predators in Hawaii

and the abundance of suitable retreat sites. Preliminary studies confirm sparse predation (<1 frog per night per animal) by alien mongooses (*Herpestes auropunctatus*) and no predation by rats and cane toads *Rhinella marina* (*Bufo marinus*). Coqui are still mostly at low elevations but can be expected to spread at least to an elevation of about 1,200 m, based on their upper limit in Puerto Rico. Coqui have been shown capable of producing substantial impacts at the ecosystem level (e.g., acceleration of leaf litter decomposition rates and new leaf production). Coqui are among the best-known invasive species in the state, because the loud calls of the males are a major annoyance to many people.

Invasion of Guam by the brown treesnake (*Boiga irregularis*) and its dramatic impacts there have sensitized Hawaii to the severe threat of transport and establishment of that snake. Several hundred of the world's snake species might have similar impacts; some that were repeatedly smuggled as pets into Hawaii could have similar colonization abilities and impacts. Although Hawaii has the most restrictive laws in the United States to exclude invasive animals, a large number of alien animals, many of them snakes, are smuggled into the state each year. The proliferation of small alien prey items such as the coqui is likely to increase the potential for establishment in Hawaii of snakes and other large reptiles by providing an abundant food base.

INVERTEBRATE ANIMAL INVASIONS IN HAWAII

Arthropod Invasions

HIGH NUMBERS OF NONNATIVE ARTHROPOD SPECIES Hawaii's susceptibility to invasions is dramatically illustrated by the fact that nearly as many nonnative arthropod species are established in Hawaii as in the other 49 states of the United States combined. How can this be possible, and why is it the case? A consultant's report to the U.S. Department of Agriculture (www.hear.org/articles/mcgregor1973) recognized the point that innate characteristics of Hawaii seem to make agricultural quarantine more difficult: "[For insects and mites] in the period 1942–72 the rate of colonization per thousand square miles was 40 species, 500 times the rate of continental United States," in spite of a larger quarantine force in relation to volume of commerce. He speculated on possible reasons: "Although there is much greater diversity of crops and habitats within the continental United States, these are dispersed over a vastly larger land area. In Hawaii . . . the various habitats are more readily accessible from the principal port of entry. The more moderate and stable climate is also more favorable for an invading species than is the climate over much of the United States."

INVASION OF SOCIAL INSECTS Hawaii's likely lack of native social insects, including ants, has important implications for its vulnerability to invasions. Ants are a major component of most ecosystems of the world, strongly influencing nutrient and energy flow through predation, scavenging, soil turning, and mutualisms. Absence of native ants in Hawaii likely left ecological roles to be filled by the evolution of insect groups that were successful long-distance colonizers. Invasive ants appear to be among the most potent forces threatening native arthropod species. Six ant species were documented as present in Hawaii by 1879, and an additional seven by 1899. Now that over 50 species of nonnative ants have established in the islands, major biotic rearrangements are taking place at the expense of Hawaii's endemic arthropods. For example, a refuge for 30–40 species of locally endemic arthropods, including essential pollinators of endemic plant species, in still mostly ant-free shrubland habitat at 2,000–3,000 m elevation in Haleakala National Park on the island of Maui is being gradually but steadily invaded by the Argentine ant (*Linepithema humile*). The endemic arthropod biota in most of Hawaii's lowland habitats was drastically reduced over a century ago by invasion of the big-headed ant (*Pheidole megacephala*), yet most remaining rainforest habitat is ant-free. Endemic *Tetragnatha* spiders have demonstrated the ability to coexist with some ant species, but they have been eliminated in areas dominated by big-headed and long-legged (*Anoplolepis longipes*) ants. These same endemic spider species were found to lack crucial behavioral and morphological defense mechanisms (compared with nonnative spiders) in laboratory trials involving exposure to ant predation.

APHIDS Aphids (Homoptera: Aphididae) may be considered representative of many groups lacking in the original Hawaii arthropod fauna, in that many species are established while almost nothing is known about their effects. No aphid species is native to Hawaii, but nearly 100 are established in Hawaii, all of them having been inadvertently introduced. Aphids, in general, cause both direct plant-feeding damage and the transmission of numerous pathogenic viruses. They feed on many agricultural crops but are also found on many native Hawaiian plants. A recent exploratory field survey found aphids feeding and reproducing on 64 native Hawaiian plant species (16 indigenous and 48 endemic) in 32 families; the majority of these plants are classified as endangered. Invasive aphids may have important impacts on the Hawaiian flora, but this remains to be documented.

MOSQUITOES (DIPTERA: CULICIDAE) Mosquitoes and the pathogens they transmit are ubiquitous throughout most of the tropical and temperate regions of the world. Yet the pre-European distribution and diversity of mosquitoes and mosquito-borne diseases throughout much of the Pacific region was sparse. The first invasive mosquitoes in the Pacific may have accompanied Polynesians on their voyaging canoes as they colonized the islands of the central and southern Pacific. However, the Hawaiian Islands lacked a single mosquito species until *Culex quinquefasciatus* was inadvertently introduced at Lahaina, Maui, in about 1826 by a whaling ship that had come from Mexico. Other mosquitoes established in Hawaii have been *Aedes aegypti* (1892), *Aedes albopictus* (1896), *Aedes vexans nocturnus* (1962), *Wyeomyia mitchelli* (1981), and *Aedes japonicus* (2004). By far the most likely route of entry, currently, would appear to be commercial air travel. Hawaii seems to have been fortunate in avoiding mosquito invasions, because more than 40 species of mosquitoes are said to have been detected in arriving aircraft.

Culex quinquefasciatus is the primary vector of avian pox and avian malaria (see Animal Pathogens, below). *Aedes aegypti* vectored dengue fever in humans in a severe epidemic in Hawaii in 1944, but it may have been entirely eliminated by military control efforts. Dengue fever resurfaced in 2001, primarily on Maui, but the vector in 2001 was *Aedes albopictus*. A current concern for both humans and native birds is West Nile virus, not yet present in the islands but potentially vectored efficiently both by *Culex quinquefasciatus* and by the recent arrival *Aedes japonicus*.

PARASITOIDS Alien parasitic wasps, including accidental introductions and purposefully released biological control agents, have been implicated in the decline of native Hawaiian insects, especially Lepidoptera, although causation is difficult to establish, and most of the evidence is ambiguous. More than 100 species of alien Ichneumonidae and Braconidae (Hymenoptera) have become established in the Hawaiian Islands. About 41 endemic species are found in these two families but represent only 5 of 64 subfamilies within the

Ichneumonidae and Braconidae. Results from several recent studies indicate that alien parasitoids have deeply penetrated native forest habitats and may have substantial impacts on Hawaiian ecosystems. Parasitoids are high-level consumers in food webs, and the domination of the Hawaiian parasitoid community by invasive species is likely to affect the structure and function of food webs and ecosystems.

SEVERE DAMAGE BY INSECTS TO NATIVE PLANT SPECIES Invasive insect herbivores have wrought substantial destruction on endemic plant populations in Hawaii. Particularly notable examples include the fern weevil (*Syagrius fulvitarsus*), established about 1900, which is especially damaging to species of the tree fern Sadleria; the black twig borer (*Xylosandrus compactus*), which established in the 1970s and is particularly damaging to Acacia koa and numerous endemic dry forest trees, such as mehamehame (*Flueggea neowawrea*); the two-spotted leafhopper (*Sophonia rufofascia*), which established in 1988; and the erythrina gall wasp (*Quadrastichus erythrinae*), which established in 2005.

Mollusc Invasions

Hawaiian land snails have been dramatically reduced by rats and by the predatory snail *Euglandina rosea*, introduced for biocontrol of another alien snail in 1958. No endemic Hawaiian land snail species can be regarded as secure. The entire genus *Achatinella*, with 41 species, was federally listed as endangered in the 1970s; all but two species are now either extinct or near extinction. The endemic Hawaiian family Amastridae, with more than 330 species, has been virtually eliminated from Hawaii.

Vulnerability of Hawaiian land snails to predation is to a large extent a result of life history patterns. Hawaiian snails in the genera *Achatinella* and *Partulina* mature slowly (six to seven years until reproduction), live to a maximum age of roughly 20 years, and produce one to seven offspring each year. In contrast, the invasive predator *Euglandina* takes less than a year to mature and produces more than 600 eggs per individual per year.

The Hawaiian Islands have no native slugs, but 38 species of slugs and snails have established in Hawaii, mostly due to the plant trade; over a dozen introduced slug species are established. Slug predation on endemic Hawaiian plant species is thought to be a highly significant factor in the severe decline of many species, but this is possibly only beginning to be investigated experimentally.

VASCULAR PLANT INVASIONS IN HAWAII

Plant invasions are currently rampant worldwide. Removed from the context of their native ecosystem through human commerce, some plant species thrive when transported to favorable new locations. One plausible explanation involves separation from herbivores or pathogens of the invader's native habitat. But Hawaii is exceptionally vulnerable. Although Hawaii still possesses extensive areas with relatively intact native vegetation, it has suffered more degradation of vegetation of wildland areas than most locations in the world, to the point of threatening the sustained survival of its unique biota. The primary reasons for this are a reduced competitiveness of native plant species in Hawaii, combined with historical and continuing high rates of plant introduction by humans.

Well over 100 of the more than 1,200 naturalized alien plant species in Hawaii pose significant threats to native ecosystems. Among the most destructive invading plant species are miconia (*Miconia calvescens*), strawberry guava (*Psidium cattleianum*) (Fig. 5), clidemia (*Clidemia hirta*), firetree (*Morella [Myrica] faya*), kahili ginger (*Hedychium gardnerianum*) (Fig. 6), Australian tree-fern (*Cyathea cooperi*) (Fig. 7), albizzia (*Falcataria moluccana*), and various grasses.

Some of the most disruptive weeds are such formidable competitors that they can gradually become community dominants, particularly when invasion is accompanied by disturbances such as a hurricane or ohia dieback. They can threaten entire native

FIGURE 5 More than any other plant species to date, strawberry guava (*Psidium cattleianum*) has invaded native Hawaiian forests on a massive scale. It overwhelmingly dominates this September 2007 aerial view of otherwise native forest at Wao Kele o Puna, on the island of Hawaii. (Photograph courtesy of the Carnegie Airborne Observatory, Carnegie Institution.)

FIGURE 6 *Hedychium gardnerianum*, from the eastern Himalayas, fondly known to some in Hawaii as kahili ginger, is a relatively recent introduction, having spread aggressively in Hawaii's middle to high-elevation rainforests only since the 1940s. It is a fierce competitor—native ohia stands have been shown to have reduced foliar nitrogen levels when this invader dominates the understory. Biological control is a much-needed tool to aid conservation efforts for this species and others among Hawaii's worst plant invaders. (Photograph courtesy of Jack Jeffrey.)

ecosystems and their accompanying biodiversity. Some can alter ecosystem-level processes such as nutrient cycling, sometimes aiding the invasion of other introduced species.

One way of explaining why native Hawaiian plant species may be relatively poor competitors when exposed to invasion by more efficiently exploitive species from elsewhere is to address what is being learned about ohia, Hawaii's dominant forest tree. In its evolutionary history

FIGURE 7 An invasive Australian tree fern (*Cyathea cooperi*) is aggressively invading the forests of Kauai. It has been shown to be extremely efficient at using soil nitrogen and is able to grow six times as rapidly in height (15 cm annually versus 2.5 cm) and to maintain four times more fronds than the endemic Hawaiian tree ferns, *Cibotium* spp. (Photograph courtesy of Trae Menard, the Nature Conservancy of Hawaii.)

in Hawaii prior to human contact, ohia was clearly a superior competitor to other plant species in many if not most habitats. Its adaptability to diverse habitats is remarkable, but its ability to specialize is limited by within-species gene flow. Ohia forests today are characterized by relatively open canopies, widely spaced crowns, and inefficient light absorption. Much light reaches the forest understory, where native and alien grasses, herbs, and shrubs are able to establish. Net CO_2 assimilation, leaf turnover, and growth rates of ohia are typically low, and ohia shows little response to increases in light or nutrient supply. Carbon fixation rates for other plant species found in ohia forests are typically lower in native than in alien species. Successful invasive species are better suited to capturing and using light resources, resulting in higher growth rates and more rapid spread than native Hawaiian species.

Aggressive alien plant invaders with nitrogen-fixing root nodules, such as firetree and albizzia, are able to quickly transform ohia-dominated ecosystems through dramatically increasing soil nitrogen levels by as much as one to two orders of magnitude. Direct competition has been demonstrated—native ohia stands have been shown to have reduced foliar nitrogen levels when kahili ginger has invaded the understory.

Alteration of the fire cycle in Hawaii by invasive introduced grasses merits special mention. In contrast to the influence of fire in many terrestrial environments of the world, it does not seem to have played an important evolutionary role in most native ecosystems of the Hawaiian Islands, and relatively few Hawaiian endemic plant species possess adaptations to fire. Lightning is uncommon on oceanic islands because small island land mass is not conducive to convective build-up of thunderheads. Many native Hawaiian ecosystems may have lacked adequate fuel to carry fires ignited by lightning or volcanism. Fires in modern Hawaii are mostly human-caused, being fueled primarily by invasive grasses, and are generally highly destructive to native plant species. The major grasses fueling fires in Hawaii are beardgrass (*Schizachyrium condensatum*), broomsedge (*Andropogon virginicus*), buffel grass (*Pennisetum ciliare* [*Cenchrus ciliaris*]), fountain grass (*Pennistum setaceum*), and molasses grass (*Melinis minutiflora*). Invasion by these grasses in otherwise undisturbed native ecosystems adds enough fine fuel to previously fire-free sites to carry fire. Most native species are eventually eliminated by successive fires, whereas invasive grasses recover rapidly after fire, increase the flammability of the site, and become increasingly dominant after repeated fires.

PATHOGEN INVASIONS

Animal Pathogens

AVIAN POX Avian pox is a viral infection of birds caused by a large, double-stranded DNA virus, *Avipoxvirus* spp. The geographic distribution of the 13 recognized species is essentially worldwide, and they are collectively able to infect all bird families, although some groups, such as pheasants and finches, are more susceptible than others. The disease in most birds is mild and rarely results in death. Introduction to the islands of the mosquito *Culex quinquefasciatus* in the 1820s provided an efficient vector for spread to native birds. Transmission is enhanced with increasing vector and host densities. Although avian pox was likely introduced into Hawaii during the 1800s with the importation of continental birds, the first definitive diagnosis of pox infection of native forest birds was not until 1902. At least two variant strains of the virus species in Hawaii occur currently in native forest birds.

Endemic Hawaiian birds are more likely to be infected with avian pox than are their introduced counterparts, and they are much more severely affected. The Hawaiian honeycreepers, for example, apparently evolved in the absence of the disease and lost much of their ancestral resistance. Avian pox may have contributed to the numerous bird extinctions in Hawaii in the 1800s, prior to the arrival of avian malaria (see below). Avian pox is only one factor that has contributed to native avian population declines of endemic Hawaiian birds, but it continues to influence present-day numbers and distribution patterns. In comparison, although avian extinctions are few to date in the Galapagos archipelago, there is a recent report of increasing prevalence and effect of avian pox virus in Darwin's finches.

AVIAN MALARIA Avian malaria is caused by a protozoan (*Plasmodium spp.*) that parasitizes erythrocytes and is transmitted by mosquitoes. More than 40 species of the parasite have been described worldwide from many bird families. Passerines are most commonly affected, but epidemic disease is rare outside Hawaii. A single lineage of *Plasmodium relictum*, believed to be of Old World origin, is currently known in Hawaii. The disease is believed to have established in the islands in the early 1900s; it was first documented in Hawaiian forest birds in the 1930s. The mosquito *Culex quinquefasciatus* provides an exceptionally efficient vector for spread within and among bird populations. Although *P. relictum* undoubtedly arrived in one or more introduced bird species, endemic forest birds are the main reservoirs of the disease today. The highest current incidence of malaria in Hawaii is in wet, mid-elevation (900–1,500 m) native forests, where mosquito populations overlap the ranges of the highly susceptible native birds.

A high frequency of concurrent malaria and pox infections complicates the identification of precise effects of either disease on endemic forest bird populations. Many of Hawaii's endemic birds commonly succumb to infection within 5–15 days of exposure to malaria, with more than 95 percent of erythrocytes parasitized. Some species are able to persist only above the upper elevational limit of mosquitoes, which is currently about 1,500 m. Other species are relatively resistant to malaria, and at least a few are in the process of evolving increasing resistance. The prognosis for the highly susceptible species is grim, given that human-induced climate change is predicted to result in 2–3 °C of atmospheric warming, enabling mosquitoes to thrive at higher elevations. Conservation managers are striving to provide conditions (e.g., reduced predation, reduced mosquito-breeding sites) favorable for maximizing opportunities for the evolution of avian resistance.

Plant Pathogens

Hawaii does not have a particularly dramatic history of plant pathogen introductions to date, but given the greatly increased movement of plant material with globalization and the tendency for a few endemic plant species to have dominance and broad elevational range, it is becoming increasingly recognized that prevention measures through rigorous pathway management are urgently needed. Hawaii's Department of Agriculture has succeeded in keeping important pathogens of coffee and coconut out of Hawaii for over a century using such a strategy.

PATHOGENS OF OHIA, *METROSIDEROS POLYMORPHA* Guava rust (*Puccinia psidii*), native on guava (*Psidium guajava*) in Brazil, has begun to spread to other areas of the world and has shown a dangerous tendency to infect and seriously damage a broad range of plant species in the myrtle family, including Brazilian eucalyptus plantations. A single genotype of *P. psidii* arrived in Hawaii in 2005, probably on commercial foliage of *Myrtus communis* from California. Hawaii's current *P. psidii* population does not infect guava but has decimated the nonnative tree rose apple (*Syzygium jambos*) and is mildly infecting

FIGURE 8 Guava rust on ohia, *Metrosideros polymorpha*. What is likely a single genotype of the notorious guava rust (*Puccinia psidii*), native to the neotropics, has been mildly infecting ohia, the most important forest tree in Hawaii, since its establishment in Hawaii in 2005. Keeping out new genotypes of this rust is a high priority for Hawaii's invasive species prevention efforts. (Photograph courtesy of Robert C. Anderson.)

ohia (Fig. 8), the most important forest tree in Hawaii (Fig. 9). Arrival of additional genotypes would likely threaten Hawaii's ohia forest. Hawaii is in the process of refining regulatory measures to prevent the arrival of additional genotypes of guava rust from South America, Florida, and California.

Primarily because of eucalyptus forestry, there is tremendous opportunity for the wholesale spread of pathogens (and insects) of Myrtaceae. For example, *Coniothyrium zuluense*, a serious fungal leaf pathogen of *Eucalyptus*, believed to be derived from a pathogen on native Myrtaceae in South Africa, has already arrived in Hawaii. Whether this pathogen can infect ohia is unknown, but its arrival further illustrates the need for careful management of the Myrtaceae pathway.

INVASIVE SPECIES OF MARINE AND FRESHWATER AQUATIC ENVIRONMENTS

Local or regional endemism is relatively rare in marine environments worldwide because most groups of marine species produce large numbers of offspring that can potentially travel long distances on ocean currents. Major barriers to gene flow in the sea are absent, reducing opportunities for small-scale differentiation. Taxonomic groups of Hawaii's marine ecosystems have endemism levels of roughly 25 percent at the species level, in contrast to terrestrial organism groups, where endemism is typically 50–100 percent. Although there is local adaptation, there has been no evolutionary radiation of marine species within the Hawaiian island chain. Nevertheless, Hawaii's marine endemism is higher than that of most areas of tropical marine waters, and invasions seem to be more numerous. Nonindigenous marine species in Hawaii include 287 invertebrates (of which, 86 are cryptogenic spp.), 5–7 percent of the 4,099 total number of invertebrate species. There are about 20 algae, 20 fish, and 12 flowering plant species.

Most invertebrates have been introduced by hull fouling and ballast, although a few were introduced intentionally. Many are cryptic or have remained in highly disturbed harbors and similar environments. It has been difficult to determine the impacts and interactions that the invertebrate invaders may be having on or with native marine flora and fauna. Notable exceptions for which substantial impacts are documented or are very likely include snowflake coral (*Carijoa riisei*), Caribbean barnacle (*Chthamalus proteus*), and Philippine mantis shrimp (*Gonodactylus falcatus*).

Invasive marine algae have attracted the most attention in recent years. At least five intentionally introduced (for the carageenan and agar industry) species of red algae (including *Gracilaria salicornia*, *Hypnea musciformis*, and *Kappaphycus* spp.), have established and dispersed around the Hawaiian Islands and have locally become ecologically dominant, forming extensive, destructive blooms in reef environments of some locations, where they appear to be outcompeting native benthic species. The destructive blooms and associated reef impoverishment are not entirely an invasive species problem, however, because several native algal species respond similarly to coastal nutrient enrichment.

Thirty-four species of marine fishes have been intentionally or accidentally introduced into Hawaiian waters, and at least 20 have become established. The Hawaiian government introduced 11 species of shallow-water snappers and groupers in 1955–1961 as potential food fish; three became established in the nearshore reef fisheries: blueline snapper or ta'ape (*Lutjanus kasmira*), peacock grouper or roi (*Cephalopholis argus*), and to'au (*Lutjanus fulvus*). Ta'ape and roi have become abundant and constitute a source of great (as yet unresolved) controversy regarding their impact on the abundance of native fish. Government-sanctioned introductions ceased after 1960, but introductions through shipping and the aquarium trade have since increased.

FIGURE 9 Hawaii's dominant tree ohia, *Metrosideros polymorpha*, is a remarkable generalist in Hawaii, with the ability to dominate a broad range of sites—from sea level to 2,500 m elevation, from the wettest to driest sites, and including the entire range of substrate age. The ohia forest in this photo is in Haleakala National Park's remote and near-pristine Kipahulu Valley, a moderately wet, middle-elevation site, of moderate age. The overwhelming dominance of ohia makes such forests particularly vulnerable to future pest invasions. (Photograph courtesy of Arthur C. Medeiros.)

More than 50 species of nonnative fishes, invertebrates, reptiles, amphibians, and plants are established in Hawaii's freshwater streams, reservoirs, and other inland waters. Hawaii's five remarkable native (four of them endemic) species of gobioid amphidromous fishes (o'opu) of freshwater streams are parasitized by six species of helminths that have been introduced with live-bearing aquarium fish.

CONCLUSIONS

The IUCN's Invasive Species Specialist Group created a list of the world's 100 worst invasive alien species (see Appendix); Hawaii has 50 of them. Almost as many nonnative arthropods have become established in Hawaii as in all of the other 49 states combined. The likely reason for Hawaii's high number of invasive species is high "propagule pressure" (from globalization) combined with a proximity of diverse habitats to ports of entry. Many invasive species, once established, appear to have much greater impacts on species in the endemic biota of the

Hawaiian Islands than they have on the biota of continents. There may be at least a tendency toward more ecosystem-level impacts of individual invaders and more synergy among biological invasions in Hawaii.

Hawaii provides fertile ground for those desiring documentation of biological invasions and the consequences of their interaction with an endemic island biota—"invasion biology meets conservation biology." Hawaii is by no means already saturated by invaders, and continuing new invasions can matter immensely (e.g., new genotypes or strains of bird diseases and of guava rust). Some feel that Hawaii could be used to greater advantage as a testing ground for how to improve "border biosecurity," given its limited ports of entry to a small and very discrete land area, to demonstrate that preventing the problem at the border is easier than dealing with it once it becomes established.

SEE ALSO THE FOLLOWING ARTICLES

Ants / Birds / Endangered and Threatened Species / Grasses and Forbs / Pathogens, Animal / Rats / Reptiles and Amphibians / Snails and Slugs

FURTHER READING

Atkinson, C.T., and D.A. LaPointe. 2009. Introduced avian diseases, climate change, and the future of Hawaiian honeycreepers. *Journal of Avian Medicine and Surgery* 23: 53–63.

Coutinho, T.A., M.J. Wingfield, A.C. Alfenas, and P.W. Crous. 1998. *Eucalyptus* rust: A disease with the potential for serious international implications. *Plant Disease* 82: 819–825.

Denslow, J.S. 2003. Weeds in paradise: Thoughts on the invasibility of tropical islands. *Annals of the Missouri Botanical Garden* 90: 119–127.

Krushelnycky, P.D., L.L. Loope, and N.J. Reimer. 2005. The ecology, policy and management of ants in Hawaii. *Proceedings of the Hawaiian Entomological Society* 37: 1–25.

Peck, R.W., P.C. Banko, M. Schwarzfeld, M. Euaparadorn, and K.W. Brinck. 2008. Alien dominance of the parasitoid wasp community along an elevation gradient on Hawai'i island. *Biological Invasions* 10: 1441–1455.

Pratt, T.K., C.T. Atkinson, P.C. Banko, J.D. Jacobi, and B.L. Woodworth, eds. 2009. *Conservation of Hawaiian Forest Birds: Implications for Island Birds.* New Haven, CT: Yale University Press.

Price, J.P., and D.A. Clague. 2002. How old is the Hawaiian biota? Geology and phylogeny suggest recent divergence. *Proceedings of the Royal Society of London B* 269: 2429–2435.

Shluker, A.D. 2003. State of Hawai'i, Aquatic Invasive Species (AIS) Management Plan. Honolulu, HI: Department of Land and Natural Resources, Division of Aquatic Resources. www.invasivespeciesinfo .gov/toolkit/hi.shtml

Sin, H., K.H. Beard, and W.C. Pitt. 2008. An invasive frog, *Eleutherodactylus coqui*, increases new leaf production and leaf litter decomposition rates through nutrient cycling in Hawaii. *Biological Invasions* 10: 335–345.

Vitousek, P.M. 1986. Biological invasions and ecosystem properties: Can species make a difference? (163–176). In H.A. Mooney and J.A. Drake, eds. *Ecology of Biological Invasions of North America and Hawaii.* Ecological Studies No. 58. New York: Springer-Verlag.

HEMLOCK WOOLLY ADELGID

DAVID A. ORWIG

Harvard Forest, Harvard University, Petersham, Massachusetts

The hemlock woolly adelgid (HWA), *Adelges tsugae*, is a small (~1.5-mm), aphidlike insect that was introduced into the eastern U.S. near Richmond, Virginia, from Japan during the early 1950s. Since its introduction, HWA has spread to 18 states from Georgia to southern Maine, threatening both eastern hemlock (*Tsuga canadensis*) and Carolina hemlock (*T. caroliniana*) currently growing on more than 19 million acres in North America (Fig. 1). Rapidly expanding since the 1980s, HWA has created extensive decline and mortality of eastern hemlock in the southern Appalachians, portions of the Mid-Atlantic region, and in southern New England. Hemlock is a long-lived foundation species, creating important structural diversity and habitat for a wide variety of wildlife species and modulating core ecosystem processes such as energy and nutrient flows and water balance. Because hemlock has no known resistance to HWA, continued infestation threatens a rangewide decline or elimination of this ecologically, culturally, and economically important tree species.

LIFE HISTORY

Hemlock woolly adelgid has a complex, polymorphic life cycle, with two parthenogenetic generations each year in the eastern United States. Adelgids settle on the underside of young hemlock twigs (Fig. 2) and feed on ray parenchyma cells via a long stylet, causing needle loss, bud mortality, and branch and tree mortality within 1–4 years in the southern United States and within 4–15+ years in the northern United States. Adult adelgids get their name from the woolly filaments they produce that protect them and their eggs (inset, Fig. 2). Adelgid eggs and crawlers can disperse rapidly via wind, birds, deer, and human activities such as logging and transfer of infested nursery plants. Rates of HWA spread have been estimated to be 8–20 km per year, with the higher rates in the southern portion of the hemlock range.

FACTORS CONTROLLING HWA POPULATIONS

Chemical

Population densities of HWA can be controlled on ornamental hemlocks by a variety of chemical insecticides. Foliar applications of horticultural oil or insecticidal soap are effective against HWA on accessible trees, but the procedure must be repeated often and is cost-prohibitive for forested settings. Systemic treatments of neonicotinoid insecticides such as imidacloprid and dinutefuran have proven most effective at controlling HWA. The chemicals can be injected directly into the tree stem or applied to the

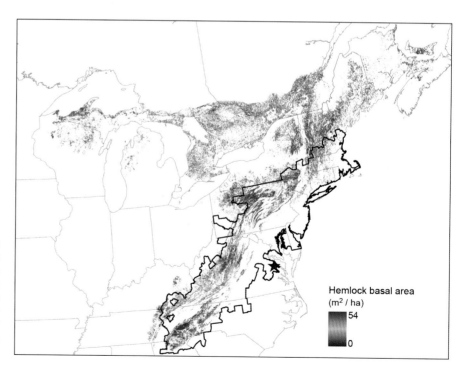

FIGURE 1 Abundance of eastern hemlock measured as basal area at locations with at least 100 ha of contiguous forest. Bold outline denotes county-level distribution of hemlock woolly adelgid in 2009, and star indicates first reported infestation of HWA, near Richmond, Virginia, in 1951. (Map courtesy of Matt Fitzpatrick.)

Hemlock basal area (m² / ha)

54

0

FIGURE 2 (A) Close-up image of an adult hemlock woolly adelgid. (Photograph courtesy of Tawny Virgilio.) (B) The woolly tufts covering adult adelgids on the underside of an eastern hemlock branch. (Photograph courtesy of Richard Hirth.)

soil beneath hemlock trees, where they will be taken up into the tree for several years of HWA protection. Soil chemical applications can cause reductions in species richness and abundance of detritivore and phytophaga insect guilds. A controlled-release tablet of imidacloprid is recommended for use in the soil when treating trees in sensitive aquatic habitats, minimizing the risk to nontarget species.

Biological

In the eastern United States, where hemlock has shown no resistance to HWA and where there are no effective native predators, control of this pest in forested situations is difficult and relies on a variety of biological control agents. Over 2 million coccinellid beetles from Japan (*Sasajiscymnus tsugae*), 25,000 beetles from China (*Scymnus* species), and over 50,000 derodontid beetles from western North America (*Laricobius nigrinus*) have been released in multiple eastern states but have not yet controlled this pest. Another potential line of attack being used against HWA is the use of entomopathogenic fungi. These naturally occurring fungi can penetrate the cuticle of HWA, deplete their nutrients, and kill them. The fungi then release spores from the dead host to infect and kill other HWA; major disease outbreaks can cause a pest population to collapse.

To suppress HWA populations in forested settings, a whey-based, fungal microfactory formulation has been developed with the fungus *Lecanicillium muscarium* and is being tested for use in large-scale foliar applications.

Climate

Naturally occurring populations of HWA in eastern North America can be drastically reduced by extreme winter temperatures or sharp decreases in temperature during mild winters. Laboratory studies have shown that HWA survival dropped significantly when insects were exposed to temperatures below −25 °C, a small percentage survived exposure to −30 °C, and none survived below −35 °C. Field studies corroborate these findings by documenting high (>90 percent) HWA mortality at sites with temperatures below −20 °C and a significant positive correlation of minimum winter temperatures and winter HWA survival. The northerly spread and ultimate range of HWA may therefore be controlled by the severity, duration, and timing of minimum winter temperatures. However, if recent climate projections suggesting warmer winter temperatures for North America come true in the future, HWA may eventually spread unimpeded throughout the range of hemlock.

DIRECT AND INDIRECT IMPACTS OF HWA

Structural and Compositional

HWA can feed on and kill all sizes and ages classes of hemlock, from the smallest seedling to the largest old-growth tree, and once HWA is introduced into a forest, the damage is chronic. Hemlock trees cannot sprout or refoliate after needle loss, and therefore hemlock can be eliminated from a site once infested. In forest settings, hemlock decline and mortality often lead to a complete change in forest cover from a cool, dark, conifer-dominated forest to a mix of deciduous species like black birch (*Betula lenta*), red maple (*Acer rubrum*), and oak (*Quercus*) species with species-rich understories (Fig. 3). Canopy gaps resulting from HWA can also lead to increases in invasive species like tree-of-heaven (*Ailanthus altissima*) and oriental bittersweet (*Celastrus orbiculatus*). In the southeastern United States, other species such as tulip poplar (*Liriodendron tulipifera*) and rhododendron (*Rhododendron maximum*) are likely to replace hemlock or increase in abundance following its loss from these forests.

Ecosystem

Overall, the largest ecosystem impacts come from the removal of hemlock, either directly by HWA, or indirectly from logging, a common management response

FIGURE 3 Dense black birch regeneration under dead hemlocks following a decade of HWA infestation in southern New England. (Photograph courtesy of Dave Orwig.)

to long-term HWA infestation. Continuous infestation by HWA leads to thinning canopies and eventual tree death, leading to microenvironmental changes such as increased light, soil surface temperatures, and subsurface soil moisture. These changes lead to cascading changes in ecosystem processes such as accelerated decomposition and nutrient cycling rates and increased nutrient availability. In addition, HWA can affect the flow of nutrients and energy from tree canopies to the soil. Its woolly covering provides energy for increases in foliar microbial populations that can lead to enriched nutrient content of throughfall as it passes through infested canopies and is deposited on the soil surface.

Wildlife

Eastern and Carolina hemlock play a unique role in North American forests. They are extremely long-lived, shade-tolerant conifers that grow in dense, monospecific stands and in combination with white pine (*Pinus strobus*) and deciduous species. Their deep crowns and associated understory conditions provide important habitat for a variety of wildlife species, including more than 120 vertebrate species and close to 300 insect species. Many different bird species spend at least part of their life cycle in hemlock forests, often feeding on insects and mites dispersed throughout the dense tree crowns. Species of particular concern that would be negatively impacted by the loss of hemlock include the black-throated green warbler (*Dendroica virens*), Acadian flycatcher (*Empidonax virescens*), Blackburnian warbler (*Dendroica fuscus*), Canadian warbler (*Wilsonia canadensis*), and hermit thrush (*Catharus guttatus*). Another animal strongly associated with hemlock is the white-tailed

deer (*Odocoileus virginianus*), which congregates under these evergreens in winter for food and cover. Red-backed salamanders (*Plethodon cinereus*) and red-spotted newts (*Notophthalmus viridescens viridescens*) thrive under fallen wood in these forests, as do several ant species.

Hemlock's demise across broad areas could dramatically affect both terrestrial and aquatic fauna. Owing to its affinity to grow along streams and in wetlands, hemlock modifies the conditions of aquatic ecosystems by maintaining cooler water temperatures. These conditions are extremely important for trout and the aquatic invertebrates they eat, which rely on the thermal modification and stable base flows provided by hemlock.

OUTLOOK FOR THE FUTURE

HWA continues to spread through the range of hemlock, causing widespread decline and mortality and altering forest dynamics and ecosystem function. Over the last 20 years, populations of another invasive insect from Japan, the elongate hemlock scale (EHS; *Fiorinia externa*) have increased dramatically, are distributed across at least 14 eastern states, and often cooccur with HWA. It is uncertain what impact two interacting invasive pests will have on hemlock ecosystems. The search continues for effective biological controls in a strategy of building a complex of natural enemies that contribute to HWA and EHS suppression. However, it is already too late for many forests in the southern Appalachians, where vast areas of hemlock forest now stand dead (Fig. 4). Cold winter temperatures in the northern U.S. appear to limit the rate of HWA spread and associated hemlock damage, but loss of this

FIGURE 4 Extensive HWA-induced hemlock mortality within Great Smoky Mountains National Park, North Carolina. (Photograph courtesy of Will Blozan.)

important foundation species nevertheless appears likely across broad portions of its range in North America, irrevocably altering forest communities and bringing attendant changes in terrestrial and aquatic ecosystems and the wildlife that inhabit them.

SEE ALSO THE FOLLOWING ARTICLES

Biological Control, of Animals / Climate Change / Disturbance / Forest Insects / Pesticides for Insect Eradication

FURTHER READING

Ellison, A. M., M. S. Bank, B. D. Clinton, E. A. Colburn, K. Elliott, C. R. Ford, D. R. Foster, B. D. Kloeppel, J. D. Knoepp, G. M. Lovett, J. Mohan, D. A. Orwig, N. L. Rodenhouse, W. V. Sobczak, K. A. Stinson, J. K. Stone, C. M. Swan, J. Thompson, B. Von Holle, and J. R. Webster. 2005. Loss of foundation species: Consequences for the structure and dynamics of forested ecosystems. *Frontiers in Ecology and the Environment* 3(9): 479–486.

McClure, M. S., and C. A. S.-J. Cheah. 1999. Reshaping the ecology of invading populations of hemlock woolly adelgid, *Adelges tsugae* (Homoptera: Adelgidae), in eastern North America. *Biological Invasions* 1: 247–254.

Orwig, D. A., and D. R. Foster. 1998. Forest response to the introduced hemlock woolly adelgid in southern New England, USA. *Journal of the Torrey Botanical Society* 125: 60–73.

Preisser, E., A. Lodge, D. Orwig, and J. Elkinton. 2008. Range expansion and population dynamics of co-occurring invasive herbivores. *Biological Invasions* 10: 201–213.

USDA Forest Health Protection Hemlock Woolly Adelgid. http://na.fs.fed.us/fhp/hwa/.

Ward, J. S., M. E. Montgomery, C. A. S.-J. Cheah, B. P. Onken, and R. S. Cowles. 2004. *Eastern Hemlock Forests: Guidelines to Minimize the Impacts of Hemlock Woolly Adelgid.* Morgantown, WV: USDA Forestry Service Northeastern Area State and Private Forestry, NA-TP-03-04.

HERBICIDES

JOSEPH M. DITOMASO

University of California, Davis

Herbicides are chemicals that kill plants by inhibiting growth or destroying tissues. They are an important method of invasive plant control. Understanding how and when they are used, as well as what they do in plants, is critical to the effective and safe use of herbicides. Like plants, insects, and other animals, herbicides share various properties that enable them to be classified into specific chemical families. In most cases, herbicides in the same chemical family inhibit plant growth by means of similar mechanisms (Table 1).

CLASSIFICATION OF HERBICIDES

In addition to their chemical structure, herbicides can be classified by a number of different methods, including method of application, selectivity, translocation pathway, and mode of action. The usefulness of each classification method depends upon the desired objectives of the user. For invasive plant management, the three most important aspects that influence an applicator's ability to use herbicides effectively include an understanding of the variety of methods of application, proper timing of treatment as it relates to herbicide movement in plants, and the factors that can influence selectivity of herbicides.

METHODS OF APPLICATION

Herbicides can be applied to the soil, directly into the water column, or on the foliage or stems of plants. Once applied, herbicides must enter the plant and move to the tissues where they exert their biochemical activity. Entry of an herbicide into a plant is generally through the roots or leaves, depending on where the herbicide is applied. The tissues where the herbicide is most easily absorbed, the pathway of translocation, and the site of action determine whether a compound is applied to the soil, stem, or foliage or directly into the water column (some herbicides).

The application of herbicides can be as broadcast treatments over large areas, as directed or spot applications to individual plants or small patches, or by means of specialized wicking techniques. Treatment can be made through aerial, boat, or ground equipment, or by individuals carrying backpack sprayers or wicks (Fig. 1).

Postemergence foliar treatment is one of the most common herbicide application techniques. This application is made to the leaves of newly emerged annual seedlings or to larger established herbaceous perennials, and even to some woody species. The waxy cuticle is the most important locus of herbicide absorption into the leaf. Although its primary function is to protect leaf surfaces from water and gas loss, it also acts as a significant barrier to the penetration of water-soluble herbicides. Most herbicides are polar (amine formulations) and thus are less likely to penetrate the waxy cuticle. In these cases, a surfactant is often added to, or formulated with, the herbicide, to increase uptake. Some postemergence herbicides are developed as nonpolar ester formulations that can easily penetrate the cuticular waxes. Herbicides such as 2,4-D, triclopyr, and imazapyr are available in both amine and ester formulations, each being used in different habitats, or with different timings and application techniques.

TABLE 1
Mode of Action, Chemical Class, and Herbicides Most Often Used in Invasive Plant Management

Mode of Action	Chemical Class	Herbicides	Susceptible groups	Symptoms
Growth regulator—synthetic auxin	Benzoic acid	Dicamba	Broadleaf species	Growth inhibition, twisting (epinasty), growth deformation
Growth regulators—synthetic auxins	Phenoxy carboxylic acids	2,4-D, MCPA	Broadleaf species	Growth inhibition, twisting (epinasty), growth deformation
Growth regulators—synthetic auxin	Pyridine carboxylic acids (picolinic acids)	Aminopyralid, Clopyralid, Picloram, Triclopyr	Broadleaf species	Growth inhibition, twisting (epinasty), growth deformation
Growth regulator—synthetic auxins	Pyrimidine carboxylic acid	Aminocyclopyrachlor	Broadleaf species	Growth inhibition, twisting (epinasty), growth deformation
Photosynthetic inhibitor—electron transport inhibitor	Triazinone	Hexazinone	Nonselective	Chlorosis (yellowing)
Photosynthetic inhibitors—electron transport inhibitors	Ureas	Diuron, Tebuthiuron	Nonselective	Chlorosis (yellowing)
Photosynthetic inhibitors—electron acceptor	Bipyridiliums	Diquat, Paraquat	Annuals	Rapid desiccation
Pigment synthesis inhibitor	None designated	Fluridone	Aquatic plants	Bleaching (whitening)
Lipid synthesis inhibitor—ACCase	Aryloxy phenoxy propionate	Fluazifop-P-butyl	Grasses	Growth inhibition, leaf chlorosis (yellowing)
Lipid synthesis inhibitors—ACCase	Cyclohexanediones	Clethodim, Sethoxydim	Grasses	Growth inhibition, leaf chlorosis (yellowing)
Amino acid inhibitors—acetolactate synthases (ALS) inhibitor	Sulfonylureas	Chlorsulfuron, Metsulfuron, Rimsulfuron, Sulfometuron	All control broadleaf species, some nonselective	Growth inhibition, deformation, leaf chlorosis (yellowing)
Amino acid anhibitors—acetolactate synthase (ALS) inhibitors	Imidazolinones	Imazapic, Imazapyr	Most grasses, many broadleaf species	Growth inhibition, deformation, leaf chlorosis (yellowing)
Amino acid inhibitor—EPSP synthase inhibitor	Glycine	Glyphosate	Nonselective	Growth inhibition, reddish to orange coloration, growth deformation
Unknown or poorly understood, considered to be membrane destruction	Inorganic	Copper	Algae, some aquatic plants	Rapid tissue destruction
Unknown or poorly understood, considered to be membrane destruction	None designated	Endothall	Aquatic annuals and algae	Rapid tissue destruction

Postemergence herbicides can be applied as broadcast treatments or directed (spot) applications to control early weed invasions or to prevent the spread of small infestations. Broadcast treatments are best in areas with large infestations and few if any sensitive nontarget species. They can be made aerially with fixed winged aircraft or helicopters, ground applicators using a variety of vehicles, or boats in aquatic settings. Directed or spot treatments have the advantage of selectively removing individuals or patches of targeted invasive species, and although they can be more labor intensive and expensive, they tend to use less total herbicide per area than do broadcast applications.

Although aquatic herbicides are generally applied by broadcast methods, including aerially or by boat, granular formulations can be used in a similar manner as a spot treatment, with the pellet slowly dissolving in the target zone where the weed is present. In terrestrial systems, directed treatments are generally applied with backpack sprayers. These foliar treatments are directed at the species of interest.

Directed applications can be applied using different techniques. Spray-to-wet treatments use high spray volumes but relatively low herbicide rates. They require better coverage than low volume treatments, which use higher concentrations of the herbicide. Low-volume treatments require more precision but can save labor, because they require fewer refills of the backpack sprayer.

Another recently developed application technique is the drizzle method, which ejects a fine spray stream. Unlike spray-to-wet applications with their small evenly spaced droplets, the drizzle technique is comprised of a few large, sparsely distributed droplets on the plant leaf surface. It is

FIGURE 1 Methods of herbicide application, including (A) aerial application by helicopter to a grassland, (B) ground application using ATV fixed with spray boom for control of invasives in a rangeland, (C) backpack sprayer with hand pump for spot applications, and (D) wick applicator with hollow handle for selective spot application. (Photographs courtesy of Jack Kelly Clark [A, B] and Guy Kyser [C, D].)

similar to the low-volume method, using high herbicide concentrations but reducing labor costs with fewer reloadings of the sprayer, allowing workers to cover more ground in less time. As another advantage, the spray stream reduces drift and allows the applicator to treat invasive plants on steep banks or in areas with limited access.

It is also possible to achieve selective control of a particular invasive plant with otherwise nonselective or relatively nonselective postemergence herbicides by employing a wick applicator. These can be either handheld or vehicle-mounted boom wipers that can be applied as a spot treatment or as a broadcast application. As a benefit, this application method reduces the potential for herbicide drift and injury to adjacent sensitive areas. With this technology, herbicides are applied at high concentrations directly to foliage. Handheld wicks can target individual plants, while selectivity using boom wipers relies on a height difference between the taller invasive plant and the lower-growing desirable vegetation.

Preemergence treatment is the application of an herbicide to the soil prior to the emergence of the seedling. Although a preemergence herbicide can be absorbed by emerging shoot tissues, the primary route of entry into the plant is through the roots. Preemergence treatments are typically broadcast applied using liquid formulations. On occasion, granular herbicide formulations can be applied aerially, particularly in forested or woody areas or in areas where herbicide drift can be a problem. Preemergence herbicides can also be used to make spot treatments to smaller incipient infestations. Preemergence herbicides require incorporation by rainfall to reach the root zone of the target plant.

Although many herbicides are only effective when applied postemergence or preemergence, there are a number that can be active in the soil or with foliar application. This is particularly true for herbicides that are used to control many invasive herbaceous plants. These compounds generally give postemergence control of seedlings and rosettes, as well as soil residual activity for at least a couple of months until spring rainfall is completed. Such compounds include aminopyralid, clopyralid, picloram, aminocyclopyrachlor, chlorsulfuron, sulfometuron, imazapic, imazapyr, and hexazinone.

While woody invasive species can be treated with postemergence applications, many can be more effectively managed using stem treatments. Stem treatments can be applied through techniques called cut stump, stem injection (hack-and-squirt), frilling, and basal bark. (These techniques are only used on resprouting species.) The first three of these techniques apply the herbicide directly into the vascular tissue. The use of these methods depends on the number and diameter of stems and the woodiness of the bark. The primary herbicides used in stem treatments include glyphosate, triclopyr, and imazapyr, all of which are systemic in the plant.

Cut stump treatments can be used for controlling plants from shrubs with multiple stems to very large trees. The stems are cut down with loppers or a chainsaw, and a concentrated herbicide solution is immediately applied to the outside cambial region of the cut stem. For vines, small-stemmed shrubs, and small trees, the entire stump can be treated. Stump treatments are most effective during periods of active growth, with the goal of eliminating resprouts from the base of the stem. In areas where aesthetics are important, cut stump treatments are desirable, because stems can be completely removed with no dead material left on the ground after treatment. The other advantage to this method is that no follow-up work is needed.

In stem injection (or "hack-and-squirt"), a hatchet or machete is used to make a series of downward cuts in the bark around the circumference of the tree trunk. Most species take one cut for every three inches of trunk diameter. About 1 milliliter of undiluted herbicide is added per incision, immediately after each cut. Application timing must be in summer or early fall, when sap flow is downward, towards the roots. This technique does not work on every species but has been shown to be effective on some important woody invasives. A similar frilling technique can also be employed. This requires that the trunk of the tree be completely girdled so that the cambium is exposed under the bark. Undiluted herbicide can be applied into the frill marks. Cordless drills can also substitute for a hatchet or machete, with the herbicide applied directly to the drill hole. Tools are commercially available that inject herbicide, or pellets containing herbicide, into the tree through a tube.

In shrubs with multiple small stems or in smaller trees (stem diameter generally <6 inches), concentrated chemical can be applied directly to the basal 12–18 inches of green stems using a basal bark application method. As with the stem injection technique, application timing must be later in the season, when carbohydrates are moving to the belowground reproductive structures. Because the herbicide has to penetrate the thin bark layer to reach the cambium, herbicides in a concentrated form must be applied in an oil carrier, and only lipophilic herbicides can be used. The most common herbicide used in basal bark treatments is the ester formulation of triclopyr. This technique leaves a dead standing plant. For wildland areas where snags are encouraged for wildlife habitat, this or the stem injection method can be a preferred control technique. In areas where snags could result in a fire or health hazard, removal of the tree should be about 1 year after treatment.

HERBICIDE MOVEMENT AND TIMING OF APPLICATION

Timing of herbicide applications can determine the effectiveness of the treatment. The best time to apply a specific herbicide will depend upon the pathway in which it is translocated and the target site where it acts. These same characteristics will, to a large degree, determine whether the herbicide is applied to the foliage, stem, or soil.

The rapid activity of contact herbicide precludes their ability to translocate much in plants. In contrast, systemic herbicides are slower-acting, giving them the opportunity to translocate in the xylem (apoplast) or phloem (symplast) to sites not directly in contact with the spray solution. Table 2 lists the primary pathway of movement for herbicides used for control of invasive plants.

A contact herbicide moves very little in plants and kills only plant parts in close proximity to the point of chemical contact. Parts that are not contacted (e.g., roots) may, however, die because they are deprived of essential compounds obtained from contacted parts (e.g., leaves). For contact herbicides to be effective, adequate distribution of the herbicide spray solution over the foliage is essential. Because contact herbicides are most useful for control of annuals, or perennials in the seedling stage, the timing of application has to be in early growth stages. In more mature plants and perennials, contact herbicides cause only temporary suppression, and plants resprout quickly from protected stem buds or underground reproductive structures. Nearly all organic herbicides have contact activity. In aquatic systems, many herbicides have contact activity, because they can come into close proximity with much of the plant surface without the need for translocation.

The apoplasm or nonliving portions of the plant is the primary pathway through which water moves from

TABLE 2

Herbicides Used in the Control of Invasive Plants, and Their Spectrums of Control and Application Methods

Herbicide	Spectrum of Control	Application Method	Primary pathway of Movement in Plant	Additional Comments
2,4-D	Most broadleaf species	Postemergence foliar and aquatic	Phloem	Often restricted use, formulated as ester and amine, no soil activity
Aminocyclopyrachlor	Many broadleaf species	Postemergence foliar	Phloem	Not registered as of 2009, moderate soil activity
Aminopyralid	Some broadleaf species	Activity both postemergence foliar and preemergence	Phloem	Good control of Asteraceae and Fabaceae, more active than clopyralid, moderate soil activity
Chlorsulfuron	Most broadleaf species, some grasses	Activity both postemergence foliar and preemergence	Phloem	Fairly long soil activity, good control of Brassicaceae, mixed selectivity favors perennial grasses
Clethodim	Grasses only	Postemergence foliar	Phloem	Fairly good on larger perennial grasses, no soil activity
Clopyralid	Some broadleaf species	Activity both postemergence foliar and preemergence	Phloem	Good control of Asteraceae and Fabaceae, being replaced by aminopyralid, moderate soil activity
Copper	Algae and some aquatic plants	Submerged aquatic only	Contact, little movement in xylem	Inorganic compound can accumulate in system, often restricted
Dicamba	Most broadleaf species	Postemergence foliar	Phloem	Often restricted use, usually tank-mixed, not as widely used as other growth regulators for invasive plants
Diquat	Aquatic annuals	Submerged or floating aquatic only	Contact, little movement in phloem	High toxicity to applicator, binds to soil quickly, not effective in turbid waters
Diuron	Most broadleaf and grass species	Preemergence and aquatic	Xylem	Long soil activity, not often used for invasive plant control
Endothall	Aquatic annuals and algae	Submerged aquatic only	Contact, little movement in xylem	Not as commonly used as other aquatic herbicides
Fluazifop-P-butyl	Grasses only	Postemergence foliar	Phloem	Fairly good on larger perennial grasses, not registered in many noncrop areas, no soil activity
Fluridone	Most aquatic plants	Submerged aquatic only	Phloem	Granular and liquid formulations, needs long contact time
Glyphosate	Most species, including woody	Postemergence foliar and stem, aquatic use	Phloem	Good control of woody species with cut stump treatment, no soil activity
Hexazinone	Many plants, including woody	Activity both postemergence foliar and preemergence	Xylem	Often restricted use, can control shrub species, mainly used in forestry, sometimes aerially applied as granular formulation
Imazapic	Mainly grasses, some broadleaf species	Activity both postemergence foliar and preemergence	Phloem	Good control of annual grasses, particular *Bromus* spp., long soil activity
Imazapyr	Most broadleaf and grass species	Activity both postemergence foliar or stem and preemergence, aquatic use	Phloem	Good control of woody species, long soil activity, formulated as ester and amine

(Continued)

TABLE 2 *(CONTINUED)*

Herbicides Used in the Control of Invasive Plants, and Their Spectrums of Control and Application Methods

Herbicide	Spectrum of Control	Application Method	Primary pathway of Movement in Plant	Additional Comments
MCPA	Most broadleaf species	Postemergence foliar	Phloem	Nearly always tank-mixed, not often used on invasives, no soil activity
Metsulfuron	Most broadleaf species	Primarily postemergence foliar	Phloem	Mainly used in rangeland, not registered in all states, moderate soil activity
Paraquat	Most annuals	Postemergence foliar	Contact, little movement in phloem	Restricted use, highly toxic to applicator, no soil activity, not often used for invasive plant control
Picloram	Most broadleaf species	Activity both postemergence foliar and preemergence	Phloem	Long soil activity, not registered in all states
Rimsulfuron	Many broadleaf and grass species	Primarily postemergence foliar, some preemergence	Phloem	Not yet registered for use in noncrop areas, fairly short soil activity
Sethoxydim	Grasses only	Postemergence foliar	Phloem	Not effective on some larger perennial grasses, not registered in many noncrop areas, no soil activity
Sulfometuron	Most broadleaf and grass species	Activity both postemergence foliar and preemergence	Phloem	Long soil activity
Tebuthiuron	Nearly all species, including woody	Preemergence	Xylem	Granular and pellets, long soil activity
Triclopyr	Most broadleaf species	Postemergence foliar and stem, also aquatic use	Phloem	Both amine and ester formulations, basal bark and cut stump treatments for woody plants, no soil activity

the soil to the foliage. This includes the cell walls and the xylem. The driving force for this upward movement is the removal of water from the leaves by transpiration. Transpiration of water from the leaves acts as a wick, bringing more water and herbicides up from the roots. Although an herbicide may be considered xylem-mobile, it must at some time enter the living tissues or symplasm to exert its phytotoxic effect. Xylem-mobile herbicides are usually soil-applied, accumulate in the leaves, and are photosynthetic inhibitors. Examples include diuron, hexazinone, and tebuthiuron (Table 2). Because these herbicides are primarily absorbed by the roots, application timing can depend on the soil mobility of the herbicide. For example, hexazinone is soil-mobile and can be applied before the rainy season, where it will then leach into the root zone providing shrub control in reforestation efforts. Other xylem-mobile herbicides are most effective on annuals when applied before germination. These compounds will be in the root zone as seedlings develop.

The symplasm consists of the living tissues of the plant. This encompasses the network of connecting cytoplasms throughout the plant, including the phloem. It is primarily within the plant cell symplasm that sugars move from photosynthetically active tissues, principally the leaves, to growing tissues including the apical meristems, expanding young leaves, rapidly elongating stems, developing flowers and fruits, and root tips. Sugars can also accumulate in storage tissues, particularly underground root crowns, taproots, tubers, rhizomes, and bulbs. Phloem-mobile herbicides primarily move along the same pathway as sugars, although they also move, to some degree, in the xylem. Because sugars are primarily produced in the leaves, most phloem-mobile herbicides are applied to the foliage and are active on both annuals and perennials, including woody species.

When a perennial plant with underground reproductive organs is treated with a postemergence, phloem-mobile herbicide, translocation of sugars and the herbicide to belowground parts is most rapid when large amounts of carbohydrates are moving toward the roots. This usually occurs after full leaf development. When perennial plants are beginning to emerge in the spring, postemergence application of a phloem-mobile herbicide will not provide effective control, as very little herbicide will translocate to the vegetative reproductive structures. Thus, timing of foliar herbicide application is important for the control of perennial weeds, and the most effective treatment time is in summer to early fall. Because these herbicides accumulate at the growing points and storage organs, they are used to control annuals and perennials, including woody plants. For this reason they are by far the most common group of herbicides used for the control of invasive plants (Table 2).

FACTORS THAT INFLUENCE HERBICIDE SELECTIVITY

Most herbicides can also kill desirable vegetation under a particular set of conditions. What determines selectivity is the total amount of herbicide that reaches a sensitive metabolic site. An herbicide is selective to a particular vegetation type or plant species only within certain limits, typically determined by a complex interaction between the plant, the herbicide, and the environment. Thus, control of weeds by selective herbicides is usually relative, not absolute. Because selectivity is typically rate-dependent, it is important to use herbicides at the recommended rate. Lower rates may not control the weeds, and higher rates may result in loss of selective control and injure nontarget plants. For general selectivity characteristics of herbicides most commonly used to manage invasive plants, see Tables 1 and 2.

Role of the Plant in Selectivity

The factors involved in plant susceptibility to herbicides are influenced by the growth stage and rate, morphology, and genetics of each species or particular ecotype of a species.

As a general rule, younger plants are easier to kill than older ones, and rapidly growing plants are more susceptible to herbicides than slower-growing ones. This is true for annuals and perennials. For this reason, many herbicides tend to be less effective when applied under environmental stress conditions, where growth rates are reduced. In addition, height difference between an invasive species and desirable vegetation can be the basis for selectivity using wick applicators or direct spray.

Plant morphology is also a common factor that can determine the susceptibility of a species to a particular herbicide. Morphological differences in root systems, location of growing points, and leaf properties most often account for variation in herbicide sensitivity. For example, when deep-rooted desirable species are present, shallow-rooting annual species can be selectively controlled with soil-applied herbicides having limited leaching capability.

The location of the growing points can play a major role in the activity of different herbicides. The growing points of grasses and broadleaf rosettes are located at the base of the plant and are protected from contact herbicides by the surrounding leaves. These same plants may be more susceptible to soil-applied herbicides. Perennial plants have protected buds below the soil surface that can escape contact herbicide injury, while many seedling dicot species have exposed susceptible growing points at the top of the young plant and in the leaf axils.

Leaf features, such as the shape, angle, arrangement, surface area, hairs, and wax thickness can influence the amount of herbicide that will contact the plant. Spray solution tends to bounce or run off when applied to a plant with narrow vertical leaves. A broadleaf plant with flat, wide, horizontal leaves can more easily retain the spray solution. Plants with fewer overlapping leaves will contact less herbicide than plants with numerous non-overlapping leaves. A dense layer of leaf hairs can hold the herbicide droplets away from the leaf surface, while a thin layer of leaf hairs causes the chemical to stay on the leaf surface longer than normal. Waxy surfaces adhere and absorb less of the herbicide solution.

Genetic differences among species can lead to differences in the rate of metabolic degradation of an herbicide, potential for a species to develop herbicide resistance, binding characteristics at the enzymatic site of action, and the ability of a plant to absorb and transport the herbicide to the site of action.

Role of the Herbicide in Selectivity

Among the factors influencing selectivity, the amount of herbicide applied to an area is perhaps most critical to obtaining proper selectivity. At high enough concentrations, most herbicides will kill all plants. At sufficiently low doses, herbicides will not kill any plants.

Slight variations in the molecular configuration of an herbicide will change its properties, which in turn will modify its effect on plants, its soil longevity, and its rate of metabolic detoxification. Similarly, altering the formulation of the product (e.g., granular, ester, amine) or adding a particular surfactant can narrow or broaden the spectrum of species an herbicide will control.

By placing the herbicide in contact with the invasive plant, it is possible to achieve selectivity with an otherwise nonselective compound. This can be accomplished with both preemergence and postemergence herbicides. With preemergence herbicides, for example, some compounds are not inherently selective but may be made to function selectively because of their location in the soil. Such selectivity depends upon a difference in seed germination depth,

seedling rooting habit between desirable and invasive plants, and herbicide mobility in the soil. With foliar applied herbicides, this can be accomplished by using shielded sprays, directed sprays, or wick applicators. Treatment timing can also achieve the effect of placement selectivity. Nonselective foliar herbicides are commonly used selectively to control emerged annual plants before emergence of more desirable species or bud break in woody species.

Role of the Environment in Selectivity

Environmental factors often account for much of the inconsistency in the performance of herbicides. There are many environmental conditions that influence herbicide selectivity and efficacy, including temperature, humidity, and precipitation. High temperatures can reduce the amount of time that an herbicide will remain in solution on the leaf surface, increase vaporization of some compounds, contribute to co-distillation where the herbicide evaporates with the water, and increase the rate of soil degradation through enhanced microbial activity. Cold temperatures can have the opposite effect.

Too much soil moisture can increase herbicide leaching, while too little can slow the movement of the herbicide into the root zone of germinating plants. With postemergence herbicides, rain during or soon after foliar applications may wash herbicides off the leaves and reduce uptake and effectiveness.

Any environmental condition that directly affects the cuticle has some effect on herbicide absorption of postemergence compounds. For the most part, maximum weed control occurs when a foliar-treated herbicide is applied under warm, humid conditions with adequate soil moisture. In contrast, minimum control generally occurs when plants are water-stressed at cool temperatures with low humidity. When plants are under drought stress, photosynthesis slows down and growth is reduced; this lowers translocation rates and reduces herbicide accumulation at the site of action. Other forms of stress, including nutrient, light, acidity, and temperature stress, can also reduce the efficacy of herbicides on plants.

INTEGRATED PEST MANAGEMENT

Although herbicides are effective for the control of small infestations, they generally do not provide long-term large-scale control of invasive plants when used alone. In the absence of a healthy plant community composed of desirable species, one invasive weed may be replaced by another equally undesirable species insensitive to the herbicide treatment. A successful long-term management program should be designed to include combinations of mechanical, cultural, biological, and chemical control techniques. There are many possible combinations that can achieve the desired objectives, but these choices must be tailored to the site, economics, and management goals.

SEE ALSO THE FOLLOWING ARTICLES

Biological Control, of Plants / Integrated Pest Management / Mechanical Control / Weeds

FURTHER READING

Bossard, C.C., J.M. Randall, and M.C. Hoshovsky, eds. 2000. *Invasive Plants of California's Wildlands*. Berkeley: University of California Press.

California Weed Science Society. 2002. *Principles of Weed Control*, 3rd ed. Fresno, CA: Thomson Publications.

DiTomaso, J.M. 2000. Invasive weeds in rangelands: Species, impacts, and management. *Weed Science* 48: 255–265.

Duncan, C.L., and J.K. Clark, eds. 2005. *Invasive Plants of Range and Wildlands and Their Environmental, Economic, and Societal Impacts*. Lawrence, KS: Weed Science Society of America.

Hance, R.J., and K. Holly, eds. 1990. *Weed Control Handbook: Principles*, 8th ed. Oxford: Blackwell Scientific Publications.

Radosevich, S.R., J.S. Holt, and C.M. Ghersa. 2007. Weed and invasive plant management approaches, methods, and tools (259–306). In *Ecology of Weeds and Invasive Plants*, 3rd ed. New York: John Wiley & Sons, Inc.

Ross, M.A., and C.A. Lembi. 2009. *Applied Weed Science: Including the Ecology and Management of Invasive Plants*, 3rd ed. Upper Saddle River, NJ: Pearson Prentice Hall.

Sheley, R., and J. Petroff, eds. 1999. *Biology and Management of Noxious Rangeland Weeds*. Corvallis: Oregon State University Press.

Vencill, W.K., ed. 2007. *Herbicide Handbook*, 9th ed. Lawrence, KS: Weed Science Society of America.

Weeds CRC. 2004. *Introductory Weed Management Manual*. Adelaide, Australia: CRC for Australian Weed Management.

HERBIVORY

MARA A. EVANS AND DONALD R. STRONG, JR.

University of California, Davis

Herbivores, organisms that primarily consume living plant tissue, range in size from microbes to the largest animals. Herbivory, both by invasive species and of invasive species, can significantly impact populations, communities and ecosystems. Foraging by herbivores varies spatially and temporally as herbivores discriminate in their consumption of plant species and specific parts of plants. Herbivory varies temporally with plant phenology and consumer life history. The multifarious interactions between herbivores and the autotrophs they consume can alter competitive interactions, push ecosystems to alternative steady states, and even facilitate the spread of nonnative species. Ultimately, the complexity of these interactions requires careful consideration when one attempts to manage invasive species and restore ecosystems. Management of introduced herbivores is part of a larger environmental restoration endeavor that combines functional human uses of the earth, such as agriculture and ecosystem services, with more intrinsic, ethical, and moral values of conserving biological diversity for its own sake.

DOMESTIC AND WILD HERBIVORY

Far and away the greatest nonnative herbivory is by ruminant livestock. Managed grazing is the most extensive human land use and covers more than 25 percent of terrestrial earth. It is a huge driver of global change, affecting ecosystem structure, biogeochemistry, trace gas emission, hydrology, biodiversity, and biosphere–atmosphere interactions. Plant community effects of grazing include prevention of forest regeneration, huge water demands, desertification, encroachment of weedy woody vegetation, and imperilment of endangered species. Livestock grazing also causes eutrophication and siltation of streams. It leads to destruction of native plants and animals that interfere, or are perceived to interfere, with livestock. While grazing was historically involved in large ecosystem changes, such as the diminishment of native grass and forb communities and concomitant invasion by Old World plant species in the North American West and the Neotropical savannas, the present rapid rate of increase in grazing presages even greater ecological influences for the future.

Beyond livestock, herbivory by introduced species extends to virtually all ecosystems. Whether an introduction is intentional or accidental, the trophic interactions between invasive and native species play into complex dynamics that have consequences for both the community and the ecosystem.

PLANT COMPETITION

Competition between plants is among the most common and powerful ecological forces. The first, and perhaps most pervasive, influence of introduced herbivores is to alter competition among plant species. Even native species have substantial ecological effects. For example, the reintroduced native Tule elk (*Cervus elaphus nannodes*) impacted the local plant community in California coastal grasslands by preventing the establishment of the native shrubs *Lupinus* and *Baccharis pilularis*. California grasslands are increasingly dominated by invasive grasses including the perennial velvet grass, *Holcus lanatus*. Grazing elk trampled vegetation, reduced accumulation of leaf litter, and created microsites that benefited grass over dicot species. Physical disturbance by the elk promoted both invasive and native grasses, yet this outcome varied spatially. In areas with established native *B. pilularis* shrubs, *H. lanatus* was protected from elk herbivory. Conversely, in open grasslands where the elk prevented the emergence of shrubs and consumed *H. lanatus,* the success of the invasive grass was limited, but other annual and perennial grasses flourished.

While direct effects of herbivores include weakening and killing autotrophs, herbivory also has indirect affects. For example, selective foraging by the European wild rabbit, *Oryctolagus cuniculus,* influences competition and dominance among plant species. Such apparent competition between plant species becomes evident once the herbivores are removed or excluded from a habitat. A native to Spain and Portugal, *O. cuniculus* has been introduced worldwide, including in England, New Zealand, and Hawaii, but the introduction has had particularly severe ecological effects in Australia. Biological control of rabbits in the form of introduced viruses promises direct effects of rabbit reduction, which reduces herbivory upon native flora. The indirect benefit of reducing the food subsidy that rabbits provide feral cats and foxes in Australia is another demonstration of the more complicated food web influences of introduced herbivores.

Introduced to the subantarctic Kerguelen archipelago, *O. cuniculus* foraged on a variety of plant species, including the native forb *Acaena magellanica.* Subsequent eradication of the rabbits did not benefit *A. magellanica,* but rather resulted in the success of invasive plant species such as the grass *Poa annua* and the low forb *Cerastium fontanum.* While the removal of grazing pressure largely contributed to this outcome, *A. magellanica* and other native plant species failed to reestablish because their seed banks were depleted by the prolonged period of foraging by *O. cuniculus.* In this case, one might infer that the native *A. magellanica* had lost in direct competition with the introduced grass and *C. fontanum.* However, the competition was but apparent, and the rabbit herbivory was the crucial factor in reducing *A. magellanica.*

ALTERNATIVE STEADY STATES

Introduced herbivores can drive an ecosystem that has maintained itself in one condition over a long term to a noticeably different state in which it remains indefinitely.

A prime example is the introduction of the herbivorous South American golden apple snail, *Pomacea canaliculata*, for aquaculture in Thailand. These generalist grazers quickly became threats to local wetland plant communities and rice crops. Even at low densities, the snails disturbed interactions between native species and transformed macrophyte-dominated wetlands to opaque, green open waters—dominated by dense phytoplankton. After consuming macrophytes, the snails excreted previously unavailable phosphorus nutrients into the water, making the environment eutrophic. The combined effect of increased nutrients and light, due to reduced shading by macrophytes, allowed plankton to become the dominant primary producer in the ecosystem. Thus the snails shifted the wetland ecosystem from one steady state to another (Fig. 1).

Such alterations in a food web have implications for the provision of services by the ecosystem. Wetland plants provide multiple resources that benefit other wetland species as well as local human populations. Macrophytes play an important role in nutrient cycling and water filtration. These large aquatic plants contribute to the physical structure of the wetland, slowing water flow, trapping sediment, and providing habitat for the fish, amphibians, mammals, invertebrates, and birds that feed on the plants. The animals use macrophytes for cover from predators and as habitat in which to rear their young. Humans, in turn, rely on wetlands for water, food, recreation, and flood control. By changing the food web structure from one dominated by aquatic plants to one dominated by plankton, *P. canaliculata* changed the native ecology. The unhealthy and unsightly

FIGURE 1 The golden apple snail (*Pomacea canaliculata*) has a shell measuring 40 mm by 45 mm. (Photograph courtesy of N.O.L. Carlsson.)

FIGURE 2 Golden apple snails severely reduce macrophytes in wetlands by consuming the aquatic plants. The snails lay their eggs out of the water in large, pink masses that each contain several hundred eggs. (A) Noninvaded wetlands. (B) Invaded wetlands. (Photographs courtesy of N.O.L. Carlsson.)

FIGURE 3 North American beavers (*C. canadensis*) introduced to Tierra del Fuego construct dams that impede the flow of water. Highly saturated soils are oxygen-deficient and result in the death of native trees. (Photograph courtesy of Sandra Chiang.)

eutrophication posed a threat to the local agricultural, fishing, and tourism industries, which were dependent on the wetlands (Fig. 2).

The introduction of the North American beaver, *Castor canadensis*, to South America is another example of an introduced herbivore that has shifted an ecosystem between states. Beavers dammed riparian corridors in Tierra del Fuego, converting river habitats into marshes. Prolonged periods of inundation shifted the native plant community from one dominated by trees to one predominantly composed of shrubs and forbs. Local *Nothofagus* trees were negatively affected by *C. canadensis* in three ways. First, having evolved in the absence of beaver herbivory, the native trees did not possess the natural defenses of low nutrient quality and low palatability that would make them undesirable to this invasive herbivore. Second, the trees that did avoid herbivory were not adapted to the waterlogged soil caused

by beaver dams. Saturated soils became anoxic and physically unstable, and both of these conditions resulted in tree death. Finally, the high foraging rates of beavers diminished contributions to the seed bank and negatively impacted *Nothofagus* reproduction (Fig. 3).

The introduction of *C. canadensis* reshaped hydrological regimes in invaded riparian communities. River ecosystems experience variable hydrology both temporally, with seasonal precipitation events, and spatially, with periods of inundation that spread laterally across flood plains. By building dams, *C. canadensis* altered stream hydrology and the seasonality and extent of flooding. Native riverine organisms were negatively affected when the disturbances changed in frequency and duration.

DISTURBANCE

Domestic livestock grazing, elk, and beaver introductions exemplify how introduced herbivory physically disturbs an environment. Similarly, the invasive aquatic rodent nutria, *Myocastor coypus*, created a patchy landscape of "eat-outs" in North American marshes. These rodents were introduced to the southeastern United States from South America and reared in captivity for their fur, yet escaped and spread in 1940 after a hurricane. In patches where *M. coypus* foraged, plant cover declined, soil temperatures and decomposition rates increased, and soil nutrient content declined. Higher soil temperatures also increased evaporation, which in turn increased soil salinity. The combination of these changes made it difficult for native plants, such as the emergent native macrophytes *Sagittaria latifolia* and *Sagittaria platyphylla*, to recolonize, even with subsequent *M. coypus* eradication. The highly disturbed patches provided opportunities for nonnative plants to invade wetlands.

The ongoing invasion of the woolly adelgid, *Adelges tsugae,* an aphidlike insect from Asia in eastern hemlock, *Tsuga canadensis,* forests in the northeastern United States is an example of disturbance driven by a nonnative herbivore changing a terrestrial ecosystem from one state to another. Unable to withstand the woolly adelgid attacks, the hemlock trees died within a couple years of infestation. Those that did not die were preemptively logged to slow the spread of the pests, and in the absence of *T. canadensis,* the native black birch, *Betula nigra,* and American rhododendron, *Rhododenron maximum,* became the dominant species at the northern and southern extremes of the hemlock range, respectively.

Characterized by damp, acidic soils, slow litter decomposition rates, and deep shade, evergreen hemlock forests transpired 50 percent less water than the deciduous trees during the summer; they moderated soil moisture and provided ample shade. Local stream flows and water temperatures were stabilized by hemlock, and streams shaded by *T. canadensis* had more diverse aquatic invertebrate communities than streams in the absence of *T. canadensis.* Hemlock served as the foundation species for these forests, and its loss to invasive herbivory has been detrimental for the associated native biota. Currently no successful control of the woolly adelgid exists to prevent the demise of the hemlock forests.

FACILITATION

Introduced herbivores can facilitate the spread of invasive plant species by dispersing their seeds. Movement of seeds gives invasive plants an advantage over native species that may be limited by dispersal. For example, invasive feral hogs, *Sus scrofa,* on the Hawaiian Islands consumed the sweet, ripe fruit of the invasive guava, *Psidium cattleianum.* The hogs covered large areas to forage and dispersed *P. cattleianum* seeds widely in their feces. The invasive hogs benefited from nutrients provided by *P. cattleianum* fruit, and in exchange the guava seeds were propagated and given the opportunity to invade widely. Of course, direct herbivory by cattle, sheep, goats, and hogs has been tremendously destructive to the endemic Hawaiian flora. The beautiful Halikala and Mona Kea Silverswords, *Argyroxiphium* spp., are cases in point, as several species have been driven virtually or actually to extinction by livestock grazing.

Endozoochorous seed dispersal, such as that by hogs, is a mutualistic relationship that aids the spread of invasive plants regardless of whether the consumer is native or invasive. But how does invasive herbivory impact the spread of native seeds? In England, native roe deer, *Capreolus capreolus,* were responsible for dispersing seeds for 34 percent of native vascular plant species. The invasive Chinese muntjac deer, *Muntiacus reevesi,* grazed on fewer species of plants than *C. carpreolus,* produced fewer fecal pellets, and as a result spread 15 percent fewer seeds overall. Even with a higher population density than the native deer species, *M. reevesi* was considered an inferior seed disperser. The combined effects of a narrow diet and reduced rates of seed dispersal suggested that *M. reevesi* could limit the extent to which native seeds were propagated.

The reverse situation, in which introduced plants facilitate the spread of invasive herbivores, is also likely. The two-spotted leafhopper, *Sophonia rufofascia,* a native of South East Asia, is currently found on the Hawaiian Islands, where it is a threat to numerous native plants and agricultural crops. The introduced faya tree, *Morella (Myrica) faya,* is also a threat to the Hawaiian flora as it fixes nitrogen in the soil, thereby increasing soil nitrogen and facilitating the spread of nonnative plants. As a nitrogen-fixing plant, *M. faya* tissue has high concentrations of nitrogen that make it a good source of nutrients for herbivores such as *S. rufofascia.* Stands of *M. faya* serve as reservoirs for *S. rufofascia* populations that then invade groves of native hirsute trees, *Metrosideros polymorpha.* The morphology of the *M. polymorpha* leaves vary from being hairy to smooth, and this affects the amount of damage a plant incurred due to *S. rufofascia* herbivory. However, regardless of its native defenses, *M. polymorpha* grown in the presence of *M. faya* were more heavily predated than those near areas where *M. faya* had been removed. The introduced tree served as a spatial and trophic refuge for the leafhoppers and appeared to sustain a continuous invasion front.

MANAGEMENT IMPLICATIONS

We have seen that herbivory by invasive species and of invasive plants altered competitive interactions, facilitated the spread of unwanted species, and disrupted ecosystem stability. These dynamics often result in complex management efforts. Removal of invasive herbivores is meant to restore ecosystems to a historical state, but this goal is complicated by ongoing climate change and the interactions between native and nonnative species, and it can result in unforeseen management outcomes.

In recent decades temperatures in the Kerguelen archipelago have warmed, and precipitation has declined.

Selective herbivory by invasive rabbits promoted the introduced dandelion, *Taraxacum officinale,* which was better suited to the recent Kerguelen climate than the native herb, *Acaena magellanica,* that it replaced. In an ecosystem with multiple invasive species, altered abiotic conditions are especially problematic when the invasive species is well adapted to the new environmental conditions while native species are not. As a result of continued anthropogenic climatic change, rabbit eradication in the Kerguelen islands did not fully restore the historical plant community.

In England, the European rabbit, *Oryctolagus cuniculus,* has filled a role once held by native herbivores. Heterogeneous landscapes generated by rabbit herbivory benefited the native bird community including the stone curlew, *Burhinus oedicnemus,* by promoting foraging and creating suitable nesting sites. Rabbits are also prey for the native foxes, but in absence of rabbits, foxes raided stone curlew nests instead. Managing these new interactions requires a great deal of ecological insight, since the removal of the rabbit populations has negatively impacted native birds. This suggests that the simultaneous removal of *both* the invasive herbivore *and* control of its predators must be considered among the possible effective management strategies.

Often continued control of invasive herbivores is preferred rather than total eradication. Despite the damage incurred by domestic livestock, complete cessation of grazing can be detrimental to native species. For instance, California vernal pool ecosystems often occur in rangelands with a long history of cattle grazing. While intense herbivory harms these sensitive seasonal wetlands, studies by Dr. Jaymee Marty demonstrated that limited exposure to cattle actually benefited native endemic plants in the degraded environments where they are now found. Abundant invasive grass species were superior competitors to native herbs such as *Lasthenia* spp. and *Veronica* spp. and had higher evapotranspiration rates that dried vernal pools to the detriment of both endemic plant and invertebrate communities. Vernal pools across California are important habitat for the endangered fairy shrimp, *Branchinecta* spp., that require sufficient periods of inundation to complete their life cycle. Limited cattle grazing suppressed invasive weeds, prolonged periods of inundation in the pools, and allowed vernal pool ecosystems to persist.

While limited grazing by cattle benefited native species in vernal pools, in other ecosystems complete removal of invasive plants is the preferred outcome. In some environments the most effective way to prevent the spread on an invasive plant is to simultaneously introduce its natural herbivore. The aquatic fern, *Salvinia molesta,* native to South America but highly invasive in wetland habitats, is combated with the introduction of its native herbivore, *Cyrtobagous salviniae,* a weevil that damages the floating plants by foraging selectively on delicate tissues and burrowing into the stems to reproduce. *Salvinia molesta* is a problematic aquatic weed, because it forms dense mats that block light, reduce dissolved oxygen content, and make waterways impenetrable to boats. Chemical control measures were ineffective, but in South Africa the introduction of *C. salviniae* successfully eliminated the weed. Introducing herbivores as a biological control measure is not a panacea. In parts of the United States, *S. molesta* cannot be controlled by *C. salviniae,* because winter temperatures kill the weevil population. Careful consideration must be made about when and where to introduce herbivores as a way to successfully manage weeds.

For these reasons, the management of invasive species calls for a keen understanding of their complex ecology. Successful outcomes are facilitated by well-defined goals, adaptive strategies, and collaborative efforts. In Hawai'i, where feral hogs have destroyed narrowly restricted endemic plants such as the precious silverswords, *Argyroxiphium* spp., as well as spread seeds of invasive plants, managers have designated habitats to be maintained free of hogs. It is a great achievement of sophisticated, contemporary management that this ambitious, quantifiable goal is met in many crucial areas. In other invaded habitats such as California and England, the complete eradication of cattle and rabbits was not feasible, so adaptive strategies informed by ecological knowledge were essential, and low levels of herbivory prevailed. In recent decades invasive species have been subject to intense ecological study that has been enhanced by collaboration between scientists and managers. Quantifying and monitoring eradication efforts are paramount to better understanding how populations, communities, and ecosystems are impacted by herbivory and invasive species.

SEE ALSO THE FOLLOWING ARTICLES

FURTHER READING

Asner, G. P., A. J. Elmore, et al. 2004. Grazing systems, ecosystem responses, and global change. *Annual Review of Environment and Resources* 29: 261–299.

Carlsson, N. O. L., C. Bronmark, and L. A. Hansson. 2004. Invading herbivory: The golden apple snail alters ecosystem functioning in Asian wetlands. *Ecology* 85(6): 1575–1580.

Chapuis, J. L., Y. Frenot, and M. Lebouvier. 2004. Recovery of native plant communities after eradication of rabbits from the subantarctic Kerguelen Islands, and influence of climate change. *Biological Conservation* 117: 167–179.

Crawley, M. J. 1983. *Herbivory: The Dynamics of Animal-Plant Interactions.* Berkeley: University of California Press.

Donlan, C. J., B. R. Tershy, and D. A. Croll. 2002. Islands and introduced herbivores: Conservation action as ecosystem experimentation. *Journal of Applied Ecology* 39: 235–246.

Drollette, D. 1997. Australia: Wide use of rabbit virus is good news for native species. *Science* 275(5297): 154.

Johnson, B. E., and J. Hall Cushman. 2007. Influence of a large herbivore reintroduction on plant invasions and community composition in a California grassland. *Conservation Biology* 21(2): 515–526.

Lees, A. C., and D. J. Bell. 2008. A conservation paradox for the 21st century: The European wild rabbit *Oryctolagus cuniculus*, an invasive alien and an endangered native species. *Mammal Review* 38: 304–320.

Wade, M. R., G. M. Gurr, et al. 2008. Ecological restoration of farmland: Progress and prospects. *Philosophical Transactions of the Royal Society B-Biological Sciences* 363(1492): 831–847.

Walker, L. R., and E. A. Powell. 1999. Regeneration of the Mauna Kea silversword *Argyroxiphium sandwicense* (Asteraceae) in Hawaii. *Biological Conservation* 89(1): 61–70.

HISTORY OF INVASION BIOLOGY

SEE INVASION BIOLOGY: HISTORICAL PRECEDENTS

HORTICULTURE

SARAH REICHARD

University of Washington, Seattle

Gardening is a popular leisure activity, with many people enjoying growing food and flowers. Horticulture, however, is responsible for more than half of wildland invasive plants overall and more than 80 percent of woody plant invaders in the United States. Horticulture, derived from the Latin words for garden (*hortus*) and growing (*culture*), is often defined as the science and art of growing plants. While vague, that definition also explains why it is so hard to deal with invasive plants introduced through this field—everything from small-scale food production to landscaping in residential areas is included. There is a long chain of professionals leading from the introduction and development of new cultivars for the market, to their production and sale, and finally to their use in our backyards. Each professional plays a role in ensuring that the introduced plants bring no harm to wildlands. In addition, amateur gardeners must

be informed and involved in preventing the movement of invasive plants.

THE HORTICULTURE VALUE CHAIN

Plant Introductions and Origins

Some of the most interesting and romantic figures in horticulture are the "plant explorers." For centuries, these hardy souls have braved mountaintops and leeches to discover species unknown in cultivation and bring them to use. In the 1700s, a farmer near Philadelphia named John Bartram stopped plowing his field to ponder the beauty of a flower, leading to a career collecting plants from the American colonies for introduction into Europe. David Douglas also explored parts of North America, introducing many species to Great Britain and Europe. Ernest Wilson was famous for introducing about 2,000 plant species from China in the early 1900s, before the borders were closed for decades for political reasons. More recently, explorers such as Daniel Hinkley from the United States have reexplored China, also combing other countries looking for plants of commercial value.

New plants are also introduced into trade by developing new cultivated varieties (cultivars). Cultivars are named variations of species and are usually the result of selection for various traits found in the wild populations, but not typical of them. Examples include such things as red-leaved forms of green plants or a blue flowered form of a species that naturally has white flowers. Modern cultivars are given "fancy names" that are not Latinized, such as *Cercis canadensis*, or Forest Pansy, a purple-leaved form of the familiar redbud. Older cultivars developed prior to 1959 may have Latinized names such as *Chamaecyparis lawsoniana*, or Aureomarginata. Cultivars may also be derived from hybridizing two different species.

Most of the time, the plants these explorers find or breeders develop are valuable additions to cultivation, whether for growing or for hybridizing with other species. Many of the species find their way into the horticultural industry, although few rival or replace top sellers such as rhododendron, petunias, and marigolds. Even fewer become problem invasive species.

Many factors determine whether a new species or cultivar has "garden merit" or is worth putting into production. This determination may be done by a company seeking exclusive distribution, a botanical garden that selected the forms, or organizations or groups of individuals interested in a taxonomic or functional group of plants, such as rose or rock garden societies. Factors such as growth form, flower or fruit color, temperature or water requirements, and ease of propagation are usually evaluated under strict conditions. However, few explorers or breeders consider invasive potential prior to introduction. Some are beginning to do so, but there are difficulties in determining invasive ability in species new to horticulture.

Commercial Horticulture

Production begins once a new species or cultivar is determined to be of garden merit, often by a wholesale or grower nursery company. Some wholesale companies are small-scale operations that sell only in their region, but, as with many modern industries, the market is increasingly dominated by a few large-scale companies. Wholesale operations generally grow large quantities of each species or cultivar to uniform standards, much as any agricultural business grows a crop. These large quantities mean that when a species in production is deemed invasive and no longer legal or desirable to grow, the companies may suffer a significant profit loss. Wholesalers propagate plants or purchase small plants (liners) to grow larger for resale. The plants are grown in-ground and potted before selling, or grown in increasingly larger pots until they are ready to sell. Some of the larger companies have numerous growing facilities located in different regions or even different countries. To keep costs down, most growing facilities are located in areas with mild climates so that plants can be grown out of doors. There is an increasing trend toward growing nursery plants in developing countries where land and labor are inexpensive and temperatures mild, but because strict laws regulate the importation of soil, which might contain harmful microorganisms, most are still grown in the countries in which they are sold. Some, especially plants to be sold by florists for indoor use, are grown under glass. This industry, along with the production of cut flowers, is called "floriculture" (flower growing) and is a subset of horticulture.

Commercial seed production is related to but separate from commercial plant production. Large agricultural-scale seed production operations for bulk or packaged seeds usually occur in Asian or Latin American countries, while seed trials and breeding may take place in Europe, Australia, or the United States. Even the seeds of native species used for restoration work are often grown overseas, increasing the potential for contamination with weed seeds local to the growing area. Most countries, however, have laws regulating the amount of contamination allowed in imported seeds, with stringent methods developed and overseen by the Association of Official Seed Certifying Agencies.

Retail operations may buy directly from wholesale nurseries, but they also often use "rewholesalers" or brokers who hold or purchase plants from numerous wholesale operations and deliver them to the garden centers. This allows retail nurseries to have a wide assortment of species available to the consumer without tracking down the individual offerings of wholesale businesses. Brokers generally buy from an array of regional and national companies.

The decentralization of the wholesale, seed, and broker operations points to one of the most difficult problems in preventing the sale of known invasive species: few species are universally invasive everywhere they are introduced. Some are acceptable for use in one region, but potentially very invasive in others. If businesses with wide product distributions are to prevent delivering invasive plants, they must know which species are invasive in each region in which they sell. In some cases the species may be legally quarantined by state/provincial or federal governments (although in general they are not); for instance, the Australian and U.S. governments both have lists of species that are prohibited from sale in the country, but some states in those countries also have lists that exclude additional species from commerce. Those same species may be available in other states where they are not invasive. Also, because species may begin to invade after a long period in a region, a species that was formerly considered acceptable may develop new status as an invader. Currently it is difficult for all the levels of the horticultural industry chain to keep up with emerging information about regional invaders, but it is especially difficult for those in the wholesale and broker industries, because they may be remote from the region in which the species invades. The environmental community could be of assistance in developing methods that use postal codes or other techniques of tracking species invasive in each region.

There has been a huge explosion of mail order and Internet sales in recent years. One North Carolina nursery, Plant Delights, saw Internet sales go from 1.5 percent of total sales in their first year of such sales to 40 percent in their sixth. This will only continue as people become more comfortable with electronic commerce. These types of sales suffer from the same problem of decentralization already discussed: a nursery in North Carolina has an

FIGURE 1 A retail nursery sign (bottom of photo) warning buyers that a species may become problematic in the garden. (Photograph courtesy of the author.)

uphill battle tracking species invading in all the states and, potentially, other countries. It is in the best interest of nurseries, legally and ethically, to not sell invasive plants, but consumers must be aware of what invades in their areas and not purchase them from any source until better methods of informing Internet companies are developed.

Aquatic gardening is very popular, and a study of Internet sales of aquatic plants had very alarming findings. Kristine Maki and Susan Galatowitsch placed 40 orders to aquatic plant nurseries throughout the United States and had them shipped to Minnesota. State or federal prohibited species were shipped 92 percent of the times they were ordered, indicating the nurseries were unaware or unconcerned about the prohibition. In addition, 93 percent had plants or animals in the shipments that appeared to be contaminations, and 10 percent of them were prohibited species. About 18 percent of the intended shipments were incorrectly identified by the shippers. While this study was specific to aquatic species, it is likely that many of the same problems may be found in other Internet sales shipments.

Gardeners are probably most familiar with retail nurseries and garden centers. These businesses can range from small operations run by a family or individual to huge chain stores. Retail nurseries usually carry a wide assortment of species, as well as other products such as tools, pesticides, garden pots, and furniture. They are the

unit of the horticulture industry chain that should be most familiar with regional invasive plants, and where the consumer should direct the most attention. In many cases, however, retail operators are even less aware of area invaders than their customer is. Retail units selling regional invasive plants should be politely informed of consumer concerns and directed to local agencies or universities that can provide more information for them. Many retail operations are responding to the information and pressure by not stocking invasive plants, letting people know why a species might be missing from the inventory (Fig. 1), or explicitly promoting alternative species (Fig. 2).

The rise of retail home "superstores," also called "big box" or "do-it-yourself (DIY)" stores, is more problematic. Unlike most retail nurseries, where decisions about inventory are made at a very local level, the superstores may have a centralized, or at best regionalized, decision-making process. Because of the local nature of invasions, buyers may be stocking plants appropriate for some areas but inappropriate for others. As with retail nurseries, stores selling regional invasive species should be informed at the store, ideally with a follow-up letter to the consumer relations department of the corporation. The lack of local buying decisions may require more persistence to achieve withdrawal of the species.

FIGURE 2 A garden center sign alerting customers that a popular garden plant they might expect to find is unavailable because it is locally invasive. (Photograph courtesy of the author.)

The "Taste-Makers"

The nursery and seed industries supply the plants, but other units of the horticulture chain may drive demand for the species. Public gardens, including parks, arboreta, and botanic gardens, display species in attractive settings that inspire gardeners to use the plants in their own gardens. The managers of the gardens have a special responsibility to ensure that the plants they display are not invasive in the area. In rare cases wherein the species is an important part of a curated collection (e.g., the garden has a complete collection of maples, and exclusion of an invasive species would affect their collection), the displayed plant should include interpretation that states it is invasive in the area, what impact it has, and that local home gardeners should not use it.

Gardeners may also be inspired by plants displayed in commercial or business settings. Most commercial landscapes are designed by landscape architects or designers. Landscape architects are usually highly educated and licensed professionals. Most are trained in aspects of engineering and their level of knowledge about plant materials varies by individual interests. They are members of a professional organization and receive additional career training through that group. This gives environmentalists and ecologists an opportunity to provide up-to-date information about current invasive plants. Landscape designers, on the other hand, may or may not have professional training or be members of a professional organization. They tend to be more focused on plants than on engineering and tend to have a wider plant palette in their designs. They may therefore be more in tune with the behavior of different species or cultivars, but they may also be more difficult to contact with updated information.

Garden writers and radio and television personalities make up the last, but certainly not least, of the taste-makers. Garden writers are often freelance authors writing for a number of publications, though some are staff with magazines or newspapers. The influence of even regional garden writers is much greater than previously, because many publications are now available over the Internet. Such writers and bloggers have a great responsibility to ensure that the species they promote are safe locally, as well as to mention other areas of the world where they invade. On-air personalities have less to worry about in this regard but still have a responsibility to promote noninvasive species in the regions in which their programs are broadcast.

Finally, amateur home gardeners are the last link in the horticultural value chain and have more power than they generally realize. All of the other links essentially serve the home gardener and will carefully listen to their well-articulated concerns. It is very important that home gardeners learn about invasive species in their areas, take appropriate actions to ensure that they do not use them in their own gardens, and encourage others to not grow or sell them.

In 2001, a small international group of horticulture professionals and amateurs came together at the Missouri Botanical Garden to develop the "St. Louis Codes of Conduct." Codes of conduct are voluntary statements of best practices. A summary of the codes developed there for nurseries, public gardens, landscape architects, and gardeners is in Table 1. Codes of nursery professionals developed in Great Britain (2004) and for the European Union (2008) are also included in Table 1.

WHY HORTICULTURE IS SUCH A SUCCESSFUL VECTOR

There should be little surprise that many invasive plants are introduced through horticultural channels. Appropriateness to temperature and precipitation extremes is considered first in evaluations. Species therefore are suited to growing conditions. In addition, many of the life history traits that make a species invasive are also traits that horticulturists prefer. For instance, fast vegetative growth in clonal species is desired for groundcovers or slope stabilizers, but also means a species could quickly grow from plantings and become a problem. Early attainment of flowering and fruiting allows growers to sell flowering plants sooner and recoup more profit, but may lead to species rapidly sowing seeds from cultivated plantings into wildlands. Similarly, long flowering and fruiting times may be desirable in ornamental plantings, but have been correlated with invasive plants. Many ornamental plants also have attractive red fruit, increasing their recognition by birds as a food source, and thus also the probability of dispersal by consumption. Ease of propagation is considered important for a species to enter wholesale production, but—again—may increase the likelihood of invasion.

There are other reasons that garden plants may be likely to invade. Many invasions have been traced to "propagule pressure." The concept is simple: if there are many plants grown in multiple gardens, potentially over a long period of time, the chances that some of the seeds will be able to be dispersed to suitable sites for germination increases as well. The length of time a species has consistently been available through the trade has also been identified in the UK and the United States as a factor in which species invade today.

TABLE 1
Voluntary Codes of Conduct for Best Practices for Horticulture Professionals and Amateurs

Conduct	Amateur Gardeners	Public Gardens	Landscape Architects	U.S. Nurseries	Great Britain Nurseries	European Union Nurseries
Learn which species invade your region	✓	✓	✓	✓	✓	✓
Discourage the sale and use of invasives, promoting safe alternatives	✓	✓	✓	✓	✓	✓
Verify the identity of plants you use					✓	✓
Do not exchange invasive plants or seeds	✓	✓				
Request that public gardens not display invasive plants	✓		✓			
Ask garden writers not to promote invasive plants	✓		✓	✓		
Work with others and help to educate them	✓	✓	✓	✓	✓	✓
Help remove invasive plants from private and public lands	✓	✓				
Report new invaders to relevant agencies	✓	✓				
Develop or use screening methods for new species		✓	✓	✓	✓	✓
Conduct an institutionwide review of policies		✓				
Know and follow all laws and international conventions		✓		✓	✓	✓
Help revise local ordinances			✓			✓
Phase out existing inventory of established invaders				✓	✓	✓
Be careful how you dispose of invasive plant material					✓	✓
Be aware of hitchhiking pests on plants and in soil					✓	
Know the best ways to control problem species					✓	
Take climate change into account						✓

NOTE: Lack of inclusion of a type of conduct by a group does not necessarily signify that the group considers it unimportant.

New introductions may undergo a flush of popularity as the new "in" plant and become widely grown in a short period of time but not have the longevity in cultivation to develop sufficient numbers to present propagule pressure.

It would be helpful if species immediately started to invade after introduction. If that were the case, growers could simply halt production at an early stage, minimizing both the loss of investment by wholesalers and the probability of invasion from multiple growing sites. Unfortunately, that rarely occurs. There are many possible reasons for the "lag phase" between the time of introduction and the time of spread. Regardless of the reason, it happens frequently and prevents using early spread as a determinant of invasive ability in plants.

SAFE ALTERNATIVES TO INVASIVE PLANTS

Cultivars are genetically different from the wild type of the species. The genetic differences result in the differences in growth form or flower, fruit, or leaf color that generally vary between the wild type and other cultivars. Usually, the genetic change is minor and has little to do with the traits that lead to invasiveness in the species. For instance, a difference in leaf color may have little to nothing to do with seed production or viability. A difference in flower color may have nothing to do with the desirability of the fruit to birds.

Many professional and amateur horticulturists accept that their favorite species is invasive, but hope that some cultivars are genetically different enough that they can use them instead. Whether this is true is unclear. One study found that selection of cultivars of *Ardisia crenata* actually increased invasiveness by increasing the number of fruits per plant, along with some other traits, ultimately increasing competitiveness.

Some cultivars are sterile, but even that is not a guarantee of noninvasiveness. There are many reasons for sterility and the cause must be understood before the cultivar can be verified as sterile. For instance, some species do

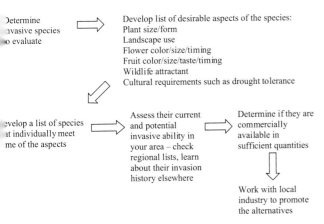

Determine invasive species to evaluate ⟹ Develop list of desirable aspects of the species:
Plant size/form
Landscape use
Flower color/size/timing
Fruit color/size/taste/timing
Wildlife attractant
Cultural requirements such as drought tolerance

Develop a list of species that individually meet some of the aspects ⟹ Assess their current and potential invasive ability in your area – check regional lists, learn about their invasion history elsewhere ⟹ Determine if they are commercially available in sufficient quantities ⟹ Work with local industry to promote the alternatives

FIGURE 3 Steps to determining noninvasive alternatives to known invasive plants.

not set fertile seed when the stigma of the flower receives pollen from the same flower, or other flowers from the same cultivar. When pollen from other cultivars, or from plants invading the area, reach the stigma, however, viable seeds can be produced. Studies on ornamental pear (*Pyrus calleryana*) have found that hybridization of cultivars after introduction leads to invasiveness. Cultivars of this species are unable to reproduce unless pollen from a different gene pool (cultivar) is used. It is therefore necessary to test the cultivar with pollen from a large pool of cultivars to verify sterility. Some cultivars may also revert to the wild type under some environmental conditions, a process that is not well understood.

Perhaps the most stable form of sterile cultivars arises from the development of triploid forms. Most animals and many plant species are diploid, which means that within the cell nucleus they have a set of chromosomes from the mother and another from the father. Plants, however, because of their more open developmental system may have polyploids, or multiples of diploids, such as tetraploids (four sets) or hexaploids (six sets). In animals, these mutations are usually aborted naturally, but in plants it is a common method by which new species arise. Triploids (three sets) generally occur when diploid and tetraploid forms of the same species, or closely related species, cross. Because triploid sets of chromosomes cannot pair up during meiosis, they are often sterile. However, it is possible for them to double and become hexaploids, so triploid cultivars may always not be the answer to invasive species.

The safest type of alternative to invasive plants is probably to find noninvasive species that share some of the desirable traits of an invasive species. No single plant may have all of the traits, and people may find a species to be a desirable garden plant for different reasons. One gardener might be attracted to a species for its flowers, another for its drought tolerance, and another for the timing of its flowering. A list of the important traits of each invasive species should be created, and then species with no invasion history and a lack of traits relating to invasion can be developed. Figure 3 explains the process. Alternative species could include those both native and nonnative to the region. This has been done in many areas around the world, and commercial and amateur horticulturists have responded positively.

THIRD-PARTY CERTIFICATION

Commercial horticulturists understand that the issue is confusing not only to them, but also to consumers. There is some discussion about developing protocols to have independent groups—perhaps government or nonprofit organizations—certify nurseries that do not sell known invaders in their regions. This is similar to programs such as dolphin-safe canned tuna or sustainably harvested lumber. Consumers would then know that plants at the identified nurseries are not invasive. Because there are many details to develop, it is unclear when or if such programs will be available. An early effort in New Zealand failed because only a few nurseries participated and consumers were not sufficiently informed to realize why their favorite plants were not available at participating nurseries. In Australia, a group of nurseries have set up a new trade organization called Sustainable Gardening Australia (SGA). Nurseries that join must train their staff in many aspects of sustainable practice, including invasive species. Although they are not true third-party certification (they do their own auditing), it is a positive step. Until such systems are expanded and true third-party monitoring is developed, consumers will have to be proactive and informed.

SEE ALSO THE FOLLOWING ARTICLES

Black, White, and Gray Lists / Hybridization and Introgression / Lag Times / Life History Strategies / Pollination / Propagule Pressure / Reproductive Systems, Plant

FURTHER READING

Codes of Conduct. www.centerforplantconservation.org/invasives/home .html.
Culley, T., and N. Hardiman. 2009. The role of intraspecific hybridization in the evolution of invasiveness: A case study of the ornamental pear tree *Pyrus calleryana*. *Biological Invasions* 11: 1107–1119.
Dehnen-Schmutz, K., J. Touza, C. Perrings, and M. Williamson. 2007. A century of the ornamental plant trade and its impact on invasion success. *Diversity and Distributions* 13: 527–534.
Kitajima, K., A. Fox, T. Sato, and D. Nagamatsu. 2006. Cultivar selection prior to introduction may increase invasiveness: Evidence from *Ardisia crenata*. *Biological Invasions* 8: 1471–1482.

Maki, K., and S. Galatowitsch. 2003. Movement of invasive aquatic plants into Minnesota (USA) through horticulture trade. *Biological Conservation* 118: 389–396.

Pemberton, R., and H. Liu. 2009. Marketing time predicts naturalization of horticultural plants. *Ecology* 90: 69–80.

Reichard, S. H., and P. White. 2001. Horticulture as a pathway of invasive plant introductions in the United States. *BioScience* 51: 103–113.

Taylor, J. M. 2009. *The Global Migrations of Ornamental Plants: How the World Got into Your Garden.* St. Louis: Missouri Botanical Garden Press.

Trusty, J., B. Lockaby, W. Zipperer, and L. Goertzen. 2007. Identity of naturalised exotic *Wisteria* (Fabaceae) in the south-eastern United States. *Weed Research* 47: 479–487.

HYBRIDIZATION AND INTROGRESSION

KRISTINA A. SCHIERENBECK

California State University, Chico

The movement of species across large distances such as continents, oceans, mountain ranges, deserts, and against ocean currents does not normally occur without human intervention but has become a common occurrence over the last 500 years. This transport can be intentional, as in the case of domesticated animals and horticultural specimens, or unintentional, as in the case of many weeds. The human-mediated dispersal of organisms outside of their home ranges increases the likelihood that phylogenetically distinct species, evolving separately over millions of years, will be brought into contact and hybridize. These hybrid events can result in the formation of new invasive species or populations, the loss of species diversity, and extinction.

HISTORY

Hybridization is defined as the interbreeding of reproductively isolated organisms; the term most commonly refers to reproduction between two defined species, but it is also used to refer to cross-breeding between divergent populations or morphological forms within a species. Hybridization is now recognized as an important mechanism of evolutionary adaptation and speciation in plants but is less accepted as an important source of variation in animals. Historically, the phenomenon has been controversial because it defies the pre-Darwinian belief that species were unchangeable entities placed on the Earth by a supreme being. Regardless, hybridization was recognized

as a source of novel variation long before the publication of Darwin's *On the Origin of Species* in 1859, or even before the rise of modern civilization. The emergence of domestic dogs (*Canis familiaris*) as a separate species from wolves (*C. lupus*) and dog–wolf hybridizations are well established by 10,000 years ago. Archaeological and genetic evidence indicates that over 10,000 years ago, Neolithic humans were in residence with hybridizing European cattle (*Bos taurus*), Near East cattle (*B. indicus*), and European aurochsen or wild ox (*B. primigenius*). It cannot be determined, however, whether these cooccurrences are reflective of intentional hybridization practices or are the passive result of human proximity and use. Multiple varieties of domesticated animals are depicted in a variety of cultures throughout the world by 5,000 years ago.

The first written reference in western literature to hybrids is thought to be Homer's mention of a mule (around 850 BC) (Eurasian horse [*Equus ferus caballus*]–African donkey [*Equus africanus asinus*] hybrid) in *The Odyssey*. The first modern, documented production of a hybrid plant was around 1717 by Thomas Fairchild, who produced a new variety of pink (*Dianthus*) through a carnation (*D. caryophyllus*) and sweet William (*D. barbatus*) cross. Hybridization and plant breeding became enormously popular in the latter half of the eighteenth century. Koelreuter conducted 136 artificial interspecific hybrid experiments from 1761 to 1766 but did not believe in hybridization as a speciation mechanism. Ironically, the father of biological classification and early species concepts, Linnaeus (1774), recognized speciation by hybridization. Darwin, following a number of interspecific hybridization experiments, initially concluded that as a speciation mechanism, hybridization had little importance; however, due to his later work with orchids, he acknowledged that hybridization may be an important avenue for speciation in plants. The idea that hybridization played a minor role in evolution was further emphasized during the neo-Darwinian movement, largely in the first half of the twentieth century, particularly by zoologists. The botanical neo-Darwinists, however, held a much different view and recognized the importance of hybridization to plant speciation.

Closely related species, whether plant, animal, fungal, or bacterial, are more likely to share enough chromosomal structure and allelic compatibilities to produce viable offspring. Rates of hybridization vary among different major phylogenetic groups, but estimates are that up to 25 percent of plant species are of hybrid origin. Hybridization within bird families has been documented to be as low as 0 percent in some families and up to 100 percent within other families. Estimates are that approximately 10 percent of mammal and 35 percent of lepidopteran species can hybridize.

Not all hybrid events produce viable offspring, as is often illustrated by the case of the mule. It is important to note, however, that there are rare cases of viable offspring in mules when they are mated with either of the parental species. Barriers to hybridization exist in nature, such as habitat isolation and behavioral patterns, which in plants could be incompatible flowering times and in animals could be differing mating behaviors. These barriers can be overcome when populations of previously separated taxa are repeatedly placed in close proximity, as illustrated by ongoing hybridization among wolves, which tend to occur in isolated wilderness areas, and coyotes (*Canis latrans*), which occur more frequently at the urban–wildland interface. Many commercially available nonnative plant cultivars in North America are capable of hybridizing with native taxa. These include pine (*Pinus* spp.), red cedar (*Juniperus* spp.), willow (*Salix* spp.), maple (*Acer* spp.), oak (*Quercus* spp.), rose (*Rosa* spp.), hibiscus (*Hibiscus* spp.), larkspur (*Delphinium* spp.), columbine (*Aquilegia* spp.), lupine (*Lupinus* spp.), clematis (*Clematis* spp.), azalea (*Rhododendron* spp.), lily (*Lilium* spp.), and peony (*Peonia* spp.); all of these have rare North American congeners. Plant cultivars are typically preselected or are developed for their ability to hybridize and are often bred to be drought-tolerant, to have large numbers of flowers, and to serve as pollinator attractants—traits that may act to accelerate the spread of nonnative genetic material into native populations.

INTROGRESSION

The phenomenon of the transfer of genetic material of one species into another was first described by Ostenfeld in 1927 and later termed "introgressive hybridization" in 1938 by Anderson and Hubricht. In contrast to hybridization, which technically refers to the mating and the subsequent production of offspring between two species, introgression requires repeated backcrossing of one parent species into another until the "foreign" alleles are present throughout the recipient species. Introgression can occur in the direction of one parent species or reciprocally between both parent species. If hybridization is extensive in the direction of one species, then the genetic material of the recipient species can be genetically "swamped" by the introgressing species. Alternatively, if introgression is extensive in both directions, a hybrid "swarm" may develop. Further complicating the scenario, any combination of hybridization, introgression, or swamping may

occur with more than two species and involve various levels of gene flow among these species. Genetic swamping or hybrid swarms may result in the lack of differentiation of one or more of the species involved. The first of these scenarios is described below as extinction by hybridization.

EXTINCTION BY HYBRIDIZATION

Hybridization can result in the extinction of at least one species in repeated hybridization events due to genetic swamping. As humans increase the movement of historically allopatric taxa throughout the globe, extinction by hybridization is preceded by bringing together species that have evolved separately. The scenarios that can result in extinction by hybridization include specialized species from divergent habitats that are brought together by habitat disturbance or human dispersal, species with narrow distributions that are brought into contact with a more common species, or an invasive species spreading and hybridizing with one or more native species.

Examples of extinction by hybridization are becoming commonplace. Hybridization of the Australian common blue butterfly (*Zizinia labradus*), introduced into New Zealand, with the native butterfly (*Zizinia oxleyi*) has resulted in the loss of many populations of the New Zealand native. The intentionally introduced barred tiger salamander (*Ambystoma tigrinum mavortium*) is hybridizing with the native, threatened California tiger salamander (*Ambystoma californiense*) and is having dramatic genetic and ecological effects on the native species. It is estimated that 38 percent of all fish extinctions in the United States are due to hybridization with nonnative species. Rainbow trout (*Oncorhynchus mykiss*) have been planted in many western waterways as sport fish and have hybridized with a number of endemic species and subspecies. Examples of congeners threatened through a combination of hybridization and habitat loss include the western cutthroat trout (*Onchorynchus clarki lewisii*) throughout the western slope of the Rocky Mountains, the gila trout (*O. gilae gilae)* in Arizona and New Mexico, the apache trout (*O. gilae apache)* in Arizona, and the golden trout (*O. mykiss aguabonita*) in California. The sheepshead minnow (*Cyprinodon variegatus*), introduced in Texas, is implicated in the extinction by hybridization of the endemic Pecos pupfish (*C. pecosensis*). The Amistad gambusia (*Gambusia amistadensis*) was extirpated in the wild in 1968 following the construction of the Amistad Reservoir over its warm spring habitat; it was later declared extinct, as the last captive populations were lost to hybridization with the western mosquito fish (*G. affinis*).

Endemic species, particularly those with limited geographic ranges or those recently evolved, are susceptible to hybridization with their nonnative congeners. These taxa are often found on islands, in mountain streams, or in other geographically complex landscapes. These limited areas are easily disturbed, which can result in the production of hybrid habitat. The flora and fauna of large or small archipelagos such as Hawaii, the Galapagos, and New Zealand have evolved in isolation from their mainland congeners, and some of their species are particularly vulnerable to hybridization from more widespread taxa. Domesticated ducks are descended from the wild mallard (*Anas platyrhynchos*) and have been widely dispersed. These descendents, which are now widespread and abundant, are hybridizing with their native congeners in many areas throughout the world. For example, close relatives of the mallard are threatening the endemic Hawaiian duck (*A. wyvilliana*), which has hybridized extensively with mallards on the island of Oahu; the New Zealand gray duck (*A. superciliosa*); and Meller's duck (*A. melleri*) of Madagascar.

HYBRID REPLACEMENT

Hybrid replacement is the phenomenon of a hybrid entity replacing one or both of its parental taxa. This is usually the result of heterosis (hybrid vigor) of the hybrid offspring, pressure on the habitat of one or both of the parental species, or human-mediated spread of a hybrid genotype that may or may not backcross with its native congeners but is reproductively and competitively superior to one or both of its parents.

Increasing examples of hybrid replacement abound. Pineland lantana (*Lantana depressa*), an endemic species found on limited substrates in Florida, is hybridizing with the popular ornamental lantana (*Lantana camara*) from South America. The ornamental was introduced in the late eighteenth century to North America and is producing triploid offspring with the endemic that are competitively replacing the rare native. The rusty crayfish (*Orconectes rusticus*), endemic to the Ohio River Valley, has been widely dispersed throughout eastern North America for sport fishing and has established in areas where other native *Orconectes* reside. The rusty crayfish is a particular threat to the northern clearwater crayfish (*O. propinquus*), native to northeastern North America, with which it is hybridizing extensively. The hybrid offspring of the rusty and northern clearwater crayfish are better able to compete than either parent for food in a variety of habitats, including the parental habitats. Additionally, there is evidence that hybridization is also resulting in the genetic swamping of the northern clearwater crayfish.

The story can be even more complicated. Dutch elm disease, a fungal species (*Ophiostoma novo-ulmi*) that may itself be a hybrid, was introduced to North America in the 1930s and resulted in extensive losses of the native American elm (*Ulmus americana*). Because of the devastation by Dutch elm disease of native American elms, other elm species have been introduced as replacements. However, the Siberian elm (*U. pumila*), introduced into midwestern North America as a replacement for the American elm, is hybridizing with another native elm, the North American red elm (*U. rubra*). Genetic data now support extensive hybridization between Siberian and red elm. Introgression is also occurring with the replacement of red elm alleles by those of Siberian elm.

HYBRIDS AS NEW SPECIES OR INVASIVE ENTITIES

In addition to hybrids replacing one or both of their parental species in an ecosystem, there are increasing examples of new hybrid species or populations created as a result of historically allopatric species or different genotypes being brought together via human dispersal. These hybrids can result in the formation of a new species, as demonstrated with cordgrass (*Spartina anglica*; Fig. 1), formed as a result of a single hybrid, allopolyploid event in Europe between hexaploid parents (*S. alterniflora*, native to the Atlantic coast of North America, with *S. maritima*, native to Europe). The new cordgrass species occurs in a broader range of habitats than either of the parental species and has serious impacts on salt marshes throughout the world. Alternatively, the repeated formation of hybrids can occur with multiple introductions of two nonnative parental

FIGURE 1 *Spartina anglica* off the coast of Brittany, France. (Photograph courtesy of Malika L. Ainouche, University of Rennes.)

taxa as illustrated by hybrid populations of two Eurasian salt cedar species (*Tamarix ramosissima* with *T. chinensis*; Fig. 2) that are invasive throughout riparian areas in western North America.

HYBRID SWARMS

Some hybridization events will result in the development of a hybrid swarm in which there is extensive hybridization among individuals across generations. These hybrid swarms often do well in disturbed habitat or in "hybrid habitat" that is unsuitable to either parental species. Human disturbances occurring at the interface between urban areas and wildlands often create hybrid habitats and thus provide a conduit for the movement of domesticated and other nonnative plant and animal taxa to interface with their native congeners. In the early twentieth century, the white sucker fish (*Catostomus commersoni*) was introduced into the Colorado River, where the flannelmouth sucker (*C. latipinnis*) and the bluehead sucker (*C. discobolus*) are native. Although the flannelmouth and bluehead suckers previously occurred in the same area, they did not interbreed. Each species is now hybridizing with the white sucker fish, resulting in a hybrid swarm between all three species and introgression between the two native species. Mammal species, particularly those associated with human habitation, are also subject to hybridization. Eurasian wildcats (*Felis silvestris silvestris*) are hybridizing with domestic cats (*F. catus*) throughout Europe, with 8 to 31 percent of individuals within the wildcat populations being of hybrid origin. In Scotland, Sika deer (*Cervus nippon*) introduced from Japan in 1860, are hybridizing widely with the native red deer (*Cervus elaphus*), producing up to 43 percent

FIGURE 3 Large, male native red deer (*Cervus elaphus*) with smaller, introduced male sika deer (*C. nippon*), and ahead of them a hybrid female. (Photograph courtesy of Josephine Pemberton, University of Edinburgh.)

hybrid progeny in some areas (Fig. 3). Hybridization with London plane tree (*Platanus* x *acerifolia*), a widely planted horticultural cultivar, has led to the genetic erosion of native populations of the native California sycamore (*P. racemosa*), of the western sycamore (*P. occidentalis*) in eastern North America, and of the Oriental sycamore (*P. orientalis*), in their respective ranges.

BIOTIC HOMOGENIZATION AND REVERSAL OF SPECIATION

Biotic homogenization can be described as the increased presence of a few, widespread species that thrive with humans and replace native species and communities or as the loss of native, locally adapted genotypes through the hybridization and introgression of dominant taxa. Seehausen has termed the latter phenomenon the "reversal of speciation." The loss of both ecological and genetic diversity can result in the increased susceptibility of ecosystems to invasion. The loss of reproductive isolation due to increased migration among separately evolved taxa will continue to lead to the loss of biological diversity worldwide. If continued at current rates, human population growth, human-caused disturbance, the movement of propagules, hybridization, introgression, and the reversal of speciation will lead to the dominance of human-adapted species worldwide.

SEE ALSO THE FOLLOWING ARTICLES

Evolution of Invasive Populations / Genotypes, Invasive

FURTHER READING

Ainouche, M., P. M. Fortune, A. Salmon, C. Parisod, M.-A. Grandbastien, K. Fukunaga, M. Ricou, and M.-T. Misset. 2009. Hybridization,

FIGURE 2 Salt cedar hybrids, *Tamarix ramosissima* x *T. chinensis*. (Photograph courtesy of John Gaskin, USDA, Agricultural Research Service.)

polyploidy, and invasion: Lessons from *Spartina* (Poaceae). *Biological Invasions* 11: 1159–1173.

Hertwig, S. T., M. Schweizer, S. Stepanow, A. Jungnickel, U.-R. Böhle, and M. S. Fischer. 2009. Regionally high rates of hybridization and introgression in German wildcat populations (*Felis silvestris*, Carnivora, Felidae). *Journal of Zoological Systematics and Evolutionary Research* 47: 283–297.

Koblmuller, S., M. Nord, R. K. Wayne, and J. Leonard. 2009. Origin and status of the Great Lakes wolf. *Molecular Ecology* 18: 2313–2326.

McDonald, D. B., T. L. Parchman, M. R. Bower, W. A. Hubert, and F. J. Rahel. 2008. An introduced and a native vertebrate hybridize to form a genetic bridge to a second native species. *Proceedings of the National Academy of Sciences, USA* 105: 10837–10842.

Muhlfeld, C. C., S. T. Kalinowski, T. E. McMahon, M. L. Taper, S. Painter, R. F. Leary, and F. W. Allendorf. 2009. Hybridization rapidly reduces fitness of native trout in the wild. *Biology Letters* 5: 328–331.

Ryan, M. E., J. R. Johnson, and B. M. Fitzpatrick. 2009. Invasive hybrid tiger salamander genotypes impact native amphibians. *Proceedings of the National Academy of Sciences, USA* 106: 11166–11171.

Schierenbeck, K. A., and N. C. Ellstrand. 2009. Hybridization and the evolution of invasiveness of plants and other organisms. *Biological Invasions* 11: 1093–1105.

Seehausen, O., G. Takimoto, D. Roy, and J. Jokela. 2008. Speciation reversal and biodiversity dynamics with hybridization in changing environments. *Molecular Ecology* 17: 30–44.

Senn, H. V., and J. M. Pemberton. 2009. Variable extent of hybridization between invasive sika (*Cervus nippon*) and native red deer (*C. elaphus*) in a small geographical area. *Molecular Ecology* 18: 862–876.

Zalapa, J. E., J. Brunet, and R. P. Guries. 2009. Patterns of hybridization and introgression between invasive *Ulmus pumila* (Ulmaceae) and native *U. rubra*. *American Journal of Botany* 96: 1116–1128.

HYDROLOGY

DAVID LE MAITRE

Council for Scientific and Industrial Research (CSIR), Stellenbosch, South Africa

Hydrology is the science concerned with the movement, distribution, and quality of water on and in the Earth. Hydrology includes the study of both the hydrological cycle and water resources. The hydrological cycle describes the movement of water on the Earth and is dominated by the flow of water from the atmosphere to the surface as precipitation (mainly rainfall) and the return process through evaporation. Water resource studies focus on the quantity or quality of water available for human use but have expanded to address both water as an ecosystem service that provides multiple benefits to society and the role of water in sustaining wetlands, rivers, and estuaries. Invasions by introduced plant species have impacts on hydrology that affect ecosystems and

the water available for use by society. The nature and degree of these impacts varies with species and the environments they invade.

VEGETATION AND WATER RESOURCES

On the land, vegetation plays a key role in the movement of water by regulating both the processes that control the transfer of water from rainfall through to wetlands, rivers, lakes, and groundwater, and the process of transpiration through leaves back to the atmosphere. Some of the published literature has created the impression that all introduced plant species can have exceptionally large impacts on water resources. This is not correct because all plants use water in the same way and are affected by the same factors. The critical issue is how the attributes of the invading species differ from those of the indigenous species they replace and to what extent they replace them. Measurements of the impacts of introduced plant species on water resources have been made for only a few species. It is precisely this similarity with plants in general, either as species or in communities, that enables scientists to have confidence in the estimates they derive and in generalizing from these limited observations based on a broader understanding of the general hydrological effects of vegetation. Obviously, there are exceptions to these general patterns, but it is not possible to deal with those here.

The generalizations given here are based on the vegetation being in a reasonably natural and well-managed state. Land degradation can have significant impacts on both water quantity and quality because it alters key processes, but a discussion of these is beyond the scope of this entry.

IMPACTS ON WATER QUANTITY

Syntheses of more than 250 watershed studies have shown that the influence of vegetation on the balance between rainfall and evaporation increases as rainfall increases, but tends toward an asymptote (Fig. 1). These patterns are due to a variety of limiting factors that can be grouped into four major types, three of which are physical and one of which is based on plant attributes:

1. Energy available from solar radiation or, in certain situations, advected energy

2. Soil moisture availability, especially in strongly seasonal climates

3. Precipitation droplet size and its effect on interception

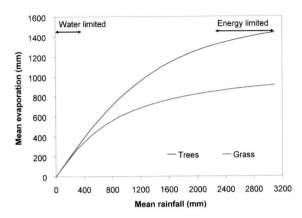

FIGURE 1 Global trends in the relationship between mean annual rainfall and mean annual evaporation for tree- or tall shrub- and grass-dominated vegetation.

4. Biological factors:

   (a) Plant physiology—for example, whether vegetation is evergreen or deciduous and what its moisture stress tolerance is

   (b) Plant size (height, stem diameter, leaf area) and depth of the root system

In general, no more than two of the four factors act as the primary controls in a particular situation.

Physical Factors

The asymptotic trend in Figure 1 emerges because evaporation is driven by the water deficit in the atmosphere—the difference between its vapor content and the potential vapor content at that air temperature. In tropical forests and many high-rainfall mountain areas, the high percentage of cloud cover reduces the amount of solar energy available to heat the air, vegetation, and ground, and thus limits evaporation. The same applies in arctic and alpine environments, where the low temperatures limit evaporation. At low rainfall, the evaporation in dryland environments is limited by the availability of water both to evaporate from the soil and for the vegetation to transpire. However, if there is additional water available in a particular area, then evaporation from that area can be much greater. Examples include springs; wetlands; areas with groundwater within the plant's root depth, such as floodplains or oases; and vegetation in and adjacent to river channels. Evaporation in oasis-type situations can be increased when dry, hot air from the desert moves through an oasis—a phenomenon called advection of energy. Evaporation losses through interception are strongly influenced by the duration and intensity of the rainfall, the area of plant surface and litter available, and

the energy available to evaporate the intercepted water. If these factors are held essentially constant, then drop size becomes important because small drops (e.g., mist, drizzle) result in greater interception losses than do large drops (e.g., tropical downpours).

Biological Factors

Where the annual rainfall is less than about 400 mm, the vegetation structure (grasses or trees) makes almost no difference, but as the rainfall increases, the type of vegetation has an increasing influence. The two kinds of vegetation in Figure 1 represent the extremes: (a) seasonally dormant grasslands with shallow root systems (generally <1.5 m deep) and (b) evergreen, deep-rooted (10–20 m or more deep), tree- or tall shrub–dominated woodlands and forests (Fig. 2). The lack of a difference at low rainfall (Fig. 1) is primarily due to water shortage, which limits evaporation regardless of the vegetation structure.

Plant physiology is important where it determines the ability of the plant to continue transpiring despite low soil moisture, high air temperatures, and a high atmospheric water deficit. For example, moisture stress–resistant plants such as the mesquite (*Prosopis* spp.) will continue transpiring even with very low moisture availability in the soil, but others such as cottonwoods (*Populus* spp.) are very sensitive and reduce transpiration rapidly in response to even moderate soil moisture shortages. Whether or not the plant loses its leaves or goes dormant is important, especially where this happens in the dry season.

Impacts of Invasions

Some of the greatest contrasts due to biological attributes are found where seasonal grasslands have been changed to evergreen woody vegetation, or vice versa, as

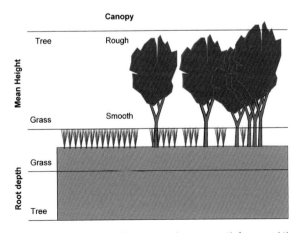

FIGURE 2 Characteristics of the tree and grass growth forms and the vegetation types they form, highlighting some key contrasts that influence how much water they use.

has happened in South Africa, parts of South America, and Australia. In these examples, two more biological factors are important (in addition to structure and physiology): the smooth canopies of grasses limit air circulation by building up a boundary layer of moist air that limits transpiration, and the relatively shallow root systems of grasses also limit the soil moisture they can access. The trees have rough canopies (which increases air circulation), a larger area of leaves, and deeper root systems. The greater leaf area of trees may also increase rainwater interception losses. In South Africa, the replacement of fynbos shrublands and seasonal grasslands with evergreen pine plantations (*Pinus* spp.) increased evaporation from around 900 mm per year to 1,200 mm, a difference of 33 percent. It is likely that dense invasions by pines have a similar impact. In South Africa extrapolations from studies of the hydrological impacts of plantations have been used to estimate that invasions by woody plant species account for about 5–7 percent of the total surface water resource and reduce the yields of water supply systems by about 6 percent.

Measurements of the water use by invasions of woody species in situations where water is not limiting and energy is readily available indicate that water use is high. Woodlands of invading wattle (*Acacia mearnsii*) growing in riparian zones in South Africa have been found to have an annual evaporation of 1300–1500 mm, which is more than plantations of this species. Measurements of the water use by introduced tamarisk (*Tamarix* spp.) on floodplains in the southwestern United States reported very high water use (>1,200 mm/year) and significant economic impacts. But more recent work suggests that the early work significantly overestimated the impacts and that the annual water use is about 980 mm and is comparable with that of the native cottonwoods (*Populus* spp.) and willows (*Salix* spp.). However, the total water use could still be greater if *Tamarisk* is able to invade a greater proportion of the flood plain or form denser stands than the native species. Mesquite (*Prosopis* spp.) invades floodplains that are the key source of groundwater for settlements and farming in the arid parts of South Africa and many other parts of Africa. Measurements of the water use by mesquite woodlands in the southwestern United States give an annual evaporation of 350–500 mm in areas where groundwater is less than about 10 m below the surface. The areas of South Africa that they invade have an annual rainfall of about 150–450 mm. If the available estimates apply in South Africa, then these invasions could be significantly affecting groundwater

availability. The extensive invasion of *Prosopis* species in similar habitats in Ethiopia, Kenya, and Tanzania is likely to be having the same kinds of impacts on water resources, particularly groundwater.

These impacts are not limited to woody species. Annual pasture grasses (*Bromus* spp.) from Europe have replaced desert shrublands in parts of California. These shallow-rooted grasses are now being replaced by a deeper-rooted, annual thistle (*Centaurea solstitialis*), also from Europe. The initial invasions replaced a mixture of seasonal forbs and grasses and may have decreased vegetation water use. The replacement of the invading annual grass by the thistle has resulted in an increase of 15–25 percent in the amount of water used and is likely to have a similar impact on water resource availability.

The key to understanding and assessing the hydrological impacts of an introduced species is to understand (1) where the area is in terms of the relationship between rainfall and evaporation (Fig. 1), (2) how dominant it is, and (3) how its invasions may have changed vegetation structure (Fig. 2) and seasonality.

IMPACTS ON WATER QUALITY

Relatively little research has been done on the impacts of invasions on water quality. Invasions by introduced plant species can alter water quality in two main ways. One is that by decreasing the amount of water reaching rivers, they can reduce the amount of water available to dilute pollutants discharged directly into rivers or indirectly via return flows of polluted groundwater from irrigated lands. The other is directly by their physical effects or by chemicals they produce or sequester. Tamarisks are able to tolerate and use saline groundwater by exuding salt through their leaves. The salt is then deposited or washed into the topsoil, where it may affect less tolerant species, but the true extent of this impact is not known.

A number of introduced, nitrogen-fixing *Acacia* species were used to stabilize windblown coastal sands in South Africa and have become major invaders of these habitats. These dune fields store large quantities of groundwater that are used to supply many of the towns and farms. When these trees are cleared, there is a flush of nitrate into the groundwater that can reach levels that may affect human health. Babies are particularly sensitive to nitrates, which can result in the "blue baby" syndrome and impact brain development.

Aquatic invaders, both submerged and floating, can affect water quality and alter aquatic ecosystems by shading the water surface, limiting oxygenation of the water, competing for nutrients, and displacing indigenous species. They can obstruct waterways and otherwise interfere with human activities. They also provide a habitat for the breeding of the vectors of waterborne diseases and for disease organisms.

SEE ALSO THE FOLLOWING ARTICLES

Climate Change / Eutrophication, Aquatic / Freshwater Plants and Seaweeds / Invasion Biology / Invasion Economics / South Africa: Invasions / Trees and Shrubs / Wetlands

FURTHER READING

Calder, I. R. 1999. *The Blue Revolution*. London: Earthscan.

Dye, P., and C. Jarmain. 2004. Water use by black wattle (*Acacia mearnsii*): Implications for the link between removal of invading trees and catchment streamflow response. *South African Journal of Science* 100: 40–44.

Gerlach, J. D., Jr. 2004. The impacts of serial land-use changes and biological invasions on soil water resources in California, USA. *Journal of Arid Environments* 57: 365–379.

Jarvis, P. G., and K. G. McNaughton. 1986. Stomatal control of transpiration: Scaling up from leaf to region. *Advances in Ecological Research* 15: 1–49.

Le Maitre, D. C., D. B. Versfeld, and R. A. Chapman. 2000. The impact of invading alien plants on surface water resources in South Africa: A preliminary assessment. *Water SA* 26: 397–408.

Levine, J. M., M. Vilà, C. M. D'Antonio, J. S. Dukes, K. Grigulis, and S. Lavorel. 2003. Mechanisms underlying the impacts of exotic plant invasions. *Proceedings of the Royal Society of London B* 270: 775–781.

Millennium Ecosystem Assessment. 2005. *Ecosystems and Human Well-Being: Wetlands and Water Synthesis*. Washington, DC: World Resources Institute.

Stromberg, J. C., M. K. Chew, P. L. Nagler, and E. P. Glenn. 2009. Changing perceptions of change: The role of scientists in *Tamarix* and river management. *Restoration Ecology* 17: 177–186.

Zavaleta, E. 2000. Valuing ecosystem services lost to *Tamarix* invasion in the United States (261–300). In H. A. Mooney and R. J. Hobbs, eds. *Invasive Species in a Changing World*. Washington, DC: Island Press.

Zhang, L., W. R. Dawes, and G. R. Walker. 2001. Response of mean annual evapotranspiration to vegetation changes at catchment scale. *Water Resources Research* 37: 701–708.

INFLUENZA

C. W. POTTER

University of Sheffield, United Kingdom

Influenza is a severe respiratory infection that occurs as epidemics in most countries in some years, and in some countries in most years. Epidemics can affect 5–50 percent of a population. Death rates are approximately 0.10 percent of those infected, but since the number infected is large, the number of human deaths worldwide in each year can reach 100,000 and more. History records severe outbreaks of influenza from AD 1510 to the present, and this will probably continue into the future. This underlines the importance of influenza as an invasive infection, constantly changing to evade past immune response and constantly erupting to threaten public health. Two epidemiological patterns are known.

HUMAN INFLUENZA

Influenza spreads chiefly as an inhaled, droplet infection. The incubation period is approximately 24–48 hrs, and symptoms start abruptly. The temperature rises rapidly; there is sore throat, cough, running nose, and headache; muscles ache; there is severe prostration which forces patients to bed. Complications include viral pneumonia and secondary bacterial pneumonia, particularly associated with *Staphylococcus aureus* and others. For the elderly, those with chronic heart or lung problems or metabolic diseases, and young children, and even those pregnant, there is a disproportionate increase in the number of fatalities.

Epidemiology

ANTIGENIC DRIFT Probably the most important component of influenza virus structure is a rod-shaped protein termed the haemagglutinin (HA), multiple copies of which radiate from the virus surface; this is the protein that binds to the cell receptor to initiate virus entry, replication, and cell death. Antibody to the HA constitutes the most important immune mechanism to infection. There are sixteen subtypes of influenza A virus based on major differences in the HA molecule, termed HA_1, HA_2, HA_3, etc.: these different HAs do not cross-react serologically.

Replication of influenza virus A is error-prone, since there is no repair mechanism, and mutant virus strains of the same subtype regularly occur. Although influenza infection is followed by immunity, this does not necessarily provide immunity to the mutant virus, which may cause a new epidemic. These mutations accumulate over time, a phenomenon termed antigenic drift, allowing annual epidemics to occur.

ANTIGENIC SHIFT Periodically, at intervals that are historically calculated at 10–40 yrs, influenza occurs as a pandemic. This is defined as an outbreak arising and spreading from a specific geographical area, causing higher infection rates and fatalities. Secondly, a pandemic is caused by a new influenza A virus subtype, the HA of which is not related to that of influenza viruses circulating immediately before the outbreak. Thus, the majority of the world's population has no effective immunity acquired from past infection.

Pandemics are so dramatic that they are recorded from the sixteenth century; however, accurate records begin only in the eighteenth century (Fig. 1). No information of the causative virus can be obtained for the pandemics prior to 1889, but data are available from this date. The viruses

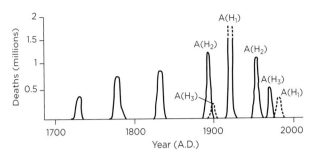

FIGURE 1 History of influenza pandemics.

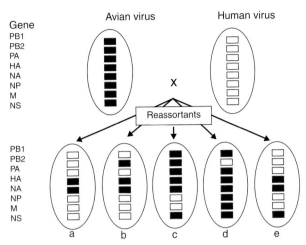

FIGURE 2 Reassortment of influenza A virus. Eight RNA gene fragments coding for proteins (PBI-NS) from an avian and a human virus infect a single cell. Assembly of a new virus can be derived from either parent.

of the pandemics of 1957 and 1968 are known, and were caused by influenza virus A(H$_2$) and A(H$_3$), respectively. The virus responsible for the pandemic of 1918, responsible for some 50 million deaths, has been resurrected from surviving tissues and is an influenza virus A(H$_1$). Antibody tests on serum samples from persons alive during the pandemics of 1889, 1900, and 1918 indicate that these were due to influenza viruses A(H$_2$), A(H$_3$), and A(H$_1$), respectively. A pattern emerges in Fig. 1 that indicates that the next pandemic may be caused by an influenza virus A(H$_2$); it is noted that this virus disappeared in 1968 after causing the 1957 pandemic, and no one under the age of 40 yrs has antibodies to this subtype. The pandemic of 1977 is controversial, since many discount this as a pandemic as it affected only persons under the age of 23 yrs: accepting this view suggests the next pandemic could be by an influenza virus A(H$_1$). However, nothing can be deduced with certainty from this short history of the past.

ORIGIN OF PANDEMIC INFLUENZA VIRUSES The most persuasive theory for the origin of pandemic influenza viruses suggests that reassortment generates these new strains. All the influenza A virus HA subtypes are present in birds, while humans are infected only by A(H$_1$), A(H$_2$), and A(H$_3$) subtypes, with few exceptions (but usually only one subtype at a time). Human and bird influenza viruses are largely segregated; however, some species can be readily infected by both human and avian types—notably pigs. The influenza A virus RNA is present as eight distinct genetic units: should a pig cell be concurrently infected with both a human and an avian influenza virus, then two sets of eight genes are present. Providing that on assembly a new virus contains one of each of eight RNA fragments, whether from the avian or human parent, a viable reassorted virus can be produced (Fig. 2). This new virus may contain the HA gene from the avian parent that has not been seen by most humans previously, and against which they have no immunity.

An alternate mechanism could be a major mutation in the virus HA gene, producing an HA against which there is no immunity. Since by this mechanism the new virus would be related to the original virus, it would fail to qualify as being a new pandemic strain; however, it would satisfy part of the definition of a pandemic strain in causing widespread infection.

Avian Influenza

Avian influenza is an intestinal disease; the viruses are excreted into water and ingested by contact birds; thus, spread is most common in water birds (ducks, etc.) and in domesticated species (chickens, turkeys, etc.). Since it is thought that human pandemic influenza viruses are commonly derived from avian sources by reassortment—and all 16 HA subtypes occur in birds—great attention has been paid to avian influenza virus epidemiology made possible by international cooperation and using new, rapid, and sophisticated technology. The incidence of infection among domestic birds is high, but clinical evidence of infection is not always present.

Concern was highlighted in 1997 when a boy aged 3 yrs contracted and died of complications of an influenza virus A(H$_5$) in Hong Kong, contracted from imported chickens from China, where this infection was rampant. In the following 6 months a further 17 cases were diagnosed, of whom 5 died; all the infections were from birds, and there was no evidence of person-to-person spread. To limit infections, and to reduce the chances of this virus mutating to allow a person-to-person spread or to participate in a reassortment to produce a pandemic virus strain, a ban was placed on chicken imports from China, and

1.5 million birds in Hong Kong were slaughtered and the carcasses destroyed; no further human cases occurred.

Since the above incident, over 200 publications have reported outbreaks of influenza virus A(H_5) infection in chickens, other domestic bird species, and wild birds, and these reports have come from all over the world; the policy of containment and mass slaughter has been widely applied. Surveillance has also identified outbreaks of influenza in birds by serotypes A(H_7), A(H_9), and others, but few human infections have occurred; in contrast, serotype A(H_5) exists as a low pathogenic strain and a highly pathogenic avian influenza virus (HPAIV); HPAIV can be lethal in both birds and humans.

AVIAN INFLUENZA A(H5) IN BIRDS Comparative studies of HPAIV and less virulent strains of influenza A(H5) have identified reasons for their difference. Although the virulence of influenza viruses is multigenetic, much attention has been focused on the NS gene product. One function of this protein is to block cell defense mechanisms against virus infection, in particular the interferon response; however, HPAIV contains a mutation in this gene that permits enhanced virus replication, allowing increased pathogenicity.

Influenza virus A(H_5) infection of birds is widespread in many species and is a relatively mild intestinal infection, and often asymptomatic. In contrast, HPAIV infection is not limited to the intestinal tract, but is a multiorgan infection with virus present in lungs, kidneys, and brain. Lungs show cell destruction and adhesions and initiate an exaggerated immune response with fluid accumulation and macrophage and lymphocyte infiltration. Death follows in the majority of affected birds. Experimental studies in mice and other animal species have confirmed the lethality of HPAIV virus; thus, in mice similar findings have been reported with multiorgan infection, severe weight loss, and death.

HPAIV Infections in Humans

Although HPAIV is highly infectious among avian species, only some 300 humans have been infected, and only from infected birds. This is despite the millions of persons who have been exposed, particularly in eastern countries, where intense farming of chickens and other birds is traditional; thus, this virus is unlikely to cause a pandemic. Indeed, this situation has been a feature of HPAIV infection for the past 12 years, and the continued inability to mutate or reassort to form a pandemic strain has led some authorities to question whether HPAIV can initiate a pandemic. Others question why this human infection has occurred in such a small number of exposed humans; one suggestion is that some genetic factor links the infected.

Of the 300 infected persons, most have been in the eastern countries of China, Vietnam, Thailand, and Indonesia; more than 50 percent of these cases were fatal. The incubation period is longer than for other influenza cases, at 2–7 days; patients develop high temperatures with lower respiratory tract symptoms, diarrhea, bleeding from the nose and gums, and encephalitis. Lower respiratory symptoms occur early and progress; multiorgan disease of intestine, kidney, brain, or heart may occur as disease progresses. Elevated levels of cytokines, which indicate immune responses, may cause toxic shock and fluid accumulation in the lungs, to add to the other pathologies. For fatal cases, death occurs 1–4 weeks after onset.

PORCINE INFLUENZA

Influenza virus A(H_1) has been a continuing infection of pigs in the Americas throughout the known history of influenza viruses. The original A(H_1) virus caused the great pandemic of 1918, and this subtype mutated to cause the widespread epidemics of 1933 and 1947; in addition, in 1976 it caused an alarming outbreak at Fort Dietrich that failed to spread into the civilian population. At the time of writing (May 2009), an outbreak of influenza virus A(H_1) infection has been reported in Mexico, with some deaths; vacationers returning from Mexico could carry the virus to other countries. However, there is no evidence that this virus can spread rapidly from person to person outside Mexico. The outcome could be that the outbreak will peter out, but at worst further mutation or reassortment could generate a pandemic strain. All are anxiously awaiting future events.

SEE ALSO THE FOLLOWING ARTICLES

Disease Vectors, Human / Epidemiology and Dispersal / Pathogens, Animal / Pathogens, Human

FURTHER READING

Crosby, A. W. 1989. *America's Forgotten Pandemic: The Influenza of 1918.* London: Cambridge University Press.

Li, X. S., Y. Guan, and J. Wang. 2004. Genesis of highly pathogenic and potentially pandemic H_5N_1 influenza virus in Easter Asia. *Nature* 430: 209–213.

Potter, C. W. 1998. Chronicle of influenza pandemics (3–15). In K. G. Nicholson, R. G. Webster, and A. J. Hay, eds. *Textbook of Influenza.* Oxford: Blackwell Science Ltd.

The writing committee of WHO consultation on human influenza A H_5. 2005. *New England Journal of Medicine* 353: 1370–1385.

World Health Organisation Global Influenza Program Surveillance Network. 2005. Evolution of H_5N_1 avian influenza in Asia. *Emerging Infectious Diseases* 11: 1575–1521.

INTEGRATED PEST MANAGEMENT

ROBERT F. NORRIS

University of California, Davis

Integrated pest management, known widely as IPM, originated with the need to decrease insecticide use in crops. The earliest concept was proposed by Stern et al. in 1959 and involved monitoring insect populations and assessing damage potential prior to applying an insecticide; it was referred to as integrated control. There are now in excess of 70 definitions of IPM.

COMPONENTS OF THE IPM APPROACH

The concepts can be most easily understood based on components that compose the IPM approach (Fig. 1), including the following:

1. Pest populations should be monitored, and to the extent feasible, control measures should be initiated based on the pest population level that is anticipated to result in unacceptable loss or injury.

2. Control should be based on an ecologically sound mix of all available tactics—genetic, cultural, physical, biological, behavioral, and chemical—as appropriate for the organism to be controlled.

3. Control tactics should be chosen to minimize detrimental impacts on nontarget organisms and ecosystems.

4. Control tactics must be economically sustainable for the manager.

The concept of a pest organism is anthropocentric and is defined as any organism that is perceived by humans to interfere with their activities. Ecologically, there are no such organisms as pests. Organisms in several different phyla are considered to be pests, including weeds, arthropods, nematodes (eelworms), molluscs, pathogens, and vertebrates. Weeds are ecologically different from all other pests, which are consumers, as the former are producers. Pathogens and weeds do not possess a nervous system and therefore cannot modify behavior based on external stimuli. The optimum IPM program necessarily differs between management of these different classes of pests.

IPM STRATEGIES

There are five overall IPM strategies. Prevention is aimed at stopping the pest organism from becoming established. This is the preferred IPM approach if it is feasible, and

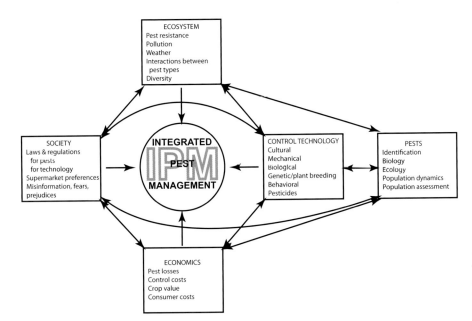

FIGURE 1 Diagram depicting relationships between components of an integrated pest management program. (Redrawn after Norris et al., 2003.)

it is especially relevant to management of nonindigenous species. Temporary alleviation uses specific control tactics on an emergency basis to limit localized pest outbreaks. Ongoing management of within-field pest populations is the third strategy and accepts the presence of the pest organism with the understanding that continuing actions are necessary to keep the pest population at acceptable levels. This approach to IPM acknowledges that a species, either native or nonindigenous, has established populations that require action but are not feasible to eradicate. The fourth approach is area-wide pest management for pests whose populations extend beyond the boundaries of a single managed unit and is especially applicable to mobile arthropod pests and many pathogens. The fifth IPM strategy is eradication, which uses all available control tactics to eliminate the entire pest population from an area or region. This approach can usually only be considered when populations are still at limited size or are restricted to relatively small geographic areas, and it is typically not considered to be part of an ongoing IPM program. It can often be appropriate for limited infestations of nonindigenous species.

The use of economic thresholds, often shortened to "ETs," is central to the original IPM concept for management of multivoltine arthropod pests. Economic thresholds are less ecologically sound for management of all other categories of pests, including for univoltine arthropods. The use of the economic injury level or economic threshold approach to pest management implies that there is a pest density that can be tolerated and that control action is required only when the pest population exceeds that tolerable density. Pest establishment has therefore occurred prior to initiation of control action. For this reason, adoption of economic thresholds should probably not be a component of IPM for invasive species, as there is, arguably, no pest density that can be tolerated before control action needs to be taken.

IPM STEPS

IPM programs require that several steps be performed before a control action is initiated. Correct identification of the pest is the first step. This step is critical as all subsequent decisions are based on correct knowledge of the pest involved. The second step is surveillance to determine the extent of the pest population. Once this is established, the potential for damage or loss must be determined. The next step entails reviewing all available feasible control tactics. A step that is often omitted involves assessment of possible interactions between the selected control tactics and pests in other categories.

Completing this step can avoid complicating the control of other pests. Once the control tactics have been chosen, it is necessary to determine if there are any legal or environmental constraints. Assuming that all the preceding steps have been followed and deemed satisfactory, it is then possible to make an action decision and to implement a control program.

IPM TACTICS

There are three categories of control tactics: those intended to manipulate the desired host plants (crop), those intended to manipulate the pest organism, and those aimed at manipulating the environment.

Host Plant Manipulation

ALTERATION OF CULTURAL PRACTICES This tactic includes such practices as changing rotations, altering sowing dates, modifying crop density, changing irrigation timing, altering fertilizer use, and changing harvesting schedules. This type of control usually provides only partial pest reduction and is typically relatively slow-acting.

ALTERATION OF HOST PLANT RESISTANCE (GENETIC) This approach involves modifying the desirable plant (usually crop) using plant breeding techniques in order to increase tolerance, or resistance, to the pest. The technique probably has little relevance to management of invasive species.

Manipulation of the Pest Organism

PHYSICAL CONTROLS These can include physical barriers for exclusion of pests, hand removal, and various forms of mechanical tillage. Many of these tactics form the cornerstone of IPM for weeds and several species of vertebrate pests and are a major component of management for invasive species.

BIOLOGICAL CONTROL This tactic uses a higher-trophic-level organism to feed on a pest organism to the point that the latter is rendered less fit. This type of control is central to most arthropod and many rangeland weed IPM programs. Biological control is potentially relevant to management of invasive species, but it has the disadvantage that it typically does not provide 100 percent suppression. Also, there is usually a lag phase before satisfactory control is attained, and population establishment can occur during this period.

BEHAVIOR MODIFICATION This tactic applies only to organisms with a nervous system and relies on visual, auditory, or olfactory signals. The tactic is not applicable to plants and pathogens and is probably not relevant to managing invasive species.

CHEMICAL APPROACHES The use of pesticides is the tactic employed when all other tactics do not provide adequate pest population suppression, and it is a component of many IPM programs. Pesticides are a major part of a control program for many nonindigenous species. Employed in the absence of other control tactics, pesticides alone may, however, not provide satisfactory long-term control.

Environment Manipulation

This category includes the following:

- The use of heat and cold, altered irrigation patterns, flooding, and similar practices to favor the desired vegetation.
- The use of various types of refugia to harbor beneficial organisms.

IPM IMPLEMENTATION SCALES AND LEVEL OF INTEGRATION

IPM can be implemented at three ecosystem scales. The simplest scale involves only a single crop field or defined managed ecosystem. A more complex scale involves IPM at the level of the whole farm or managed unit and requires a much more complicated assessment of pest biology and ecosystem interactions. The most complex IPM involves landscape or regional programs that include numerous different kinds of managed units (e.g., farms, woodlands, towns, riparian areas, etc.). Implementation of a regional IPM program is often difficult because of the multiple managers and agencies involved.

IPM is considered at three levels of integration. Level I IPM entails only monitoring of pest populations and scheduling of pesticide applications as required. Calendar pesticide spraying programs that do not require population assessment are not considered to be IPM. Level II IPM uses all pest management tactics in a coordinated approach for the integrated management of a single pest species. Level III IPM integrates pest management tactics in relation to all categories of pests and the regional ecosystems. Most current IPM programs are at level II. One of the earlier examples of level II IPM was insect management in cotton in Peru, where adoption of the IPM approach saved the crop. More recent examples include spider mite management on strawberries in

Honduras, aid to tomato production in several regions of the world, leafy spurge control in the north central United States, and management of orange-striped oakworm in urban oak trees in eastern North America; in each case, insecticide use was drastically reduced. IPM programs for cotton in several production regions approach level III; likewise, IPM programs for tree fruit production approach level III in some regions. It is doubtful, however, that any current program achieves true level III integration.

IPM AND INTEGRATED CROP MANAGEMENT

IPM is a component of the broader concept of integrated crop management. Management of pests cannot be divorced from such factors as regional crop choice, soil types, weather, irrigation, fertilizer management, crop rotation, harvesting schedules, and other management aspects not directly related to pest management. IPM programs should be initiated prior to planting a crop (or other managed ecosystem), should take place during crop growth, and should be continued after crop harvest. To this end, development of a crop timeline shows which actions should be undertaken at which stage in the crop.

Successful management of invasive species requires using the IPM approach if long-lasting ecologically sound solutions are to be found. It is universally recognized that one IPM tactic alone is unlikely to be completely satisfactory for long-term prevention, management, or eradication of any species deemed to cause losses to humans.

SEE ALSO THE FOLLOWING ARTICLES

Agriculture / Biological Control, of Animals / Biological Control, of Plants / Herbicides / Mechanical Control / Pesticides (Fish and Mollusc) / Pesticides for Insect Eradication / Pesticides (Mammal)

FURTHER READING

Bajwa, W. I., and M. Kogan. 2002. *Compendium of IPM Definitions (CID): What Is IPM and How Is It Defined in the Worldwide Literature?* IPPC Publication No. 998, Integrated Plant Protection Center (IPPC). Corvallis: Oregon State University Press.

CAST. 2003. *Integrated Pest Management: Current and Future Strategies.* Ames, IA: Council for Agricultural Science and Technology.

Flint, M. L., and P. Gouveia. 2001. *IPM in Practice: Principles and Methods of Integrated Pest Management.* Publication #3418. Oakland: University of California Press, Division of Agriculture and Natural Resources.

Norris, R. F., E. P. Caswell-Chen, and M. Kogan. 2003. *Concepts in Integrated Pest Management.* Upper Saddle River, NJ: Prentice-Hall.

Stern, V. M., R. F. Smith, R. van den Bosch, and K. S. Hagen 1959. The integrated control concept. *Hilgardia* 29: 81–99.

INVASIBILITY, OF COMMUNITIES AND ECOSYSTEMS

JASON D. FRIDLEY

Syracuse University, New York

Invasibility describes the susceptibility of biological communities to colonization and dominance by introduced organisms. Biologists have long observed that some ecosystems contain few introduced invaders, while others contain many; such patterns include both differences in invader richness (number of introduced or invasive species) and abundance (total number of introduced or invasive individuals). Identifying the causes of this variation has proven a difficult endeavor and is the subject of considerable research in invasion biology. This research is important for conservation efforts that seek to prioritize areas for invasive species control and prevention, for ecological restoration efforts that seek to create invasion-resistant communities, and for understanding the processes that govern community assembly across a variety of ecosystems.

HOW BIOLOGISTS STUDY COMMUNITY INVASIBILITY

Since its inception, the field of invasion biology has called upon a diverse array of scientific hypotheses and methods to examine the nature of community invasibility. For example, the animal ecologist Charles Elton, in his 1958 book *The Ecology of Invasions by Animals and Plants*, argued that naturally more complex ecosystems are less prone to invasion by introduced organisms. Elton supported his argument with theoretical and laboratory studies of simple predator–prey dynamics and decades of observational data on the occurrence of invasive species in natural and anthropogenic ecosystems. In the decades since Elton's work, the dialogue between theoreticians, experimentalists, and field biologists has expanded to include other forms of biological data, particularly from evolutionary biologists (e.g., population genetics, animal behavior). However, effective integration of these approaches has proven to be one of the greatest challenges of modern invasion studies, in that predictions from one type of analysis (e.g., theoretical models of community assembly) are often in conflict with those based on empirical studies (e.g., patterns of native and invader richness in surveys of plant and animal communities). Although clever experiments can sometimes reconcile empirical patterns and theory, such experiments are rare, due to the expense and logistical difficulty of controlling natural processes across a wide range of environmental conditions. Like any truly predictive science, progress in understanding the basis of community invasibility will ultimately depend on the extent to which a relatively simple theoretical framework can be effectively applied to the many circumstances that describe the full variety of invasion events in real communities.

PATTERNS OF APPARENT INVASIBILITY IN NATURE, FROM REGIONAL TO LOCAL SCALES

Variation in invasion levels across ecosystems is distributed hierarchically in space: there are large regions of the globe that contain relatively few invasive species (e.g., tropical forests) while others contain many across a variety of habitat types (e.g., warm-temperate regions). Within continents, there are habitats of highly invaded vegetation (e.g., shrublands and grasslands in Mediterranean-type climates in the Americas, Australia, and South Africa) and habitats that remain relatively invader-free (e.g., remote forestlands). Within habitats, there are certain landscape positions that are prone to invasion (e.g., streamsides) and others that appear more resistant (e.g., uplands).

Regional Invasibility

Biomes that represent climatic extremes, including deserts, boreal forests, and tundra, tend to contain relatively few invasive introduced organisms. Many tropical ecosystems also harbor few introduced organisms, although adequate data on invasions are lacking for many large tropical regions. Within these extremes, invasion patterns in temperate latitudes are habitat-dependent. For example, the Mediterranean-type climates of the world are natively dominated by grasslands and shrublands,

FIGURE 1 Old-growth hardwood forest in the Upper Peninsula of Michigan, near the shore of Lake Superior. To date, there are virtually no introduced plants, birds, or mammals in this forest. (Photograph courtesy of Kerry Woods.)

FIGURE 2 Mature hardwood forest (Hutcheson Memorial Forest, Rutgers University) in central New Jersey. Although most of the canopy trees are native, the understory is dominated by introduced plant species. (Photograph courtesy of the author.)

many of which have been transformed over the past 200 years by introduced plants. In some instances, these invasions have taken place in the absence of obvious human disturbance (e.g., pine invasions in South African fynbos), while in others it is difficult to determine how concomitant land use changes (introduced grazers, plowing, fire promotion or suppression) have contributed to invasibility. Other temperate vegetation types contain few invaders. Some old-growth mixed hardwood forests in the interior of North America, for example, contain almost no introduced plant or animal species (Fig. 1).

Fine-Scale Invasibility Patterns

Environmental factors that vary over relatively fine spatial scales, such as disturbance rates and soil resource availability, influence the frequency and extent of species invasions in natural plant communities. For example, floodplain and scour bar communities along streams and rivers in the southern Appalachian Mountains of the eastern United States contain many introduced plant species, and yet adjacent upland communities contain few. The riparian communities are frequently disturbed, have high rates of nutrient enrichment from flooding, and are less shaded than adjacent uplands. They also receive greater rates of propagule supply from introduced species that disperse as floating seeds or vegetative material. Indeed, the rate of propagule influx can exert strong control over the invasibility of natural communities. For example, even in a protected, physically undisturbed forest in central New Jersey, introduced plant species such as multiflora rose dominate the understory, partly because the forest

is relatively small and is surrounded by a well-populated suburban landscape (Fig. 2).

Oceanic Islands

Biogeographers have long noted the unusually high rates of invasion by introduced organisms on oceanic islands, and the extraordinary degree to which these invaders transform island ecosystems. For example, the vascular flora of the contiguous United States has increased by about 25 percent on average as a result of post-Columbian plant introductions; in contrast, the flora of Hawaii has increased nearly 60 percent. Of the 108 recorded global bird extinctions since 1600, 97 have been on islands, largely due to human introduction of new predators and diseases. Many invasion biologists attribute a significant component of island invasion susceptibility to their historical biological isolation, in terms both of their lower overall species diversity compared to mainland areas of similar size, and of the unfamiliarity of many insular organisms to efficient predators (e.g., mammalian carnivores). For example, introduced herbivores such as rabbits and goats tend to decimate native island plant species and thus indirectly promote dominance by introduced plants that are well defended against herbivory.

THEORIES OF INVASIBILITY AND SPECIES COEXISTENCE MODELS

Modern invasion biology has its theoretical roots in classical species coexistence theory. Coexistence in a simple two-species system is stable at equilibrium if each species can successfully invade a stable population of the other. For example, if a potential invader can access resources that remain unused by a resident species that has reached its carrying capacity (maximum local population size), then theory predicts that a species may successfully invade and stably coexist with the resident, depending on the similarity of their resource requirements. Although precise conditions for invasibility can be established given assumptions about resource use requirements for a given species assemblage, it has been difficult to apply this theory with any precision to real communities. Another way theory predicts that native communities could allow invasions is if resident species abundances are kept below their carrying capacities due to recurrent disturbances. Although resident species may competitively displace introduced species given enough time, disturbance rates may be sufficient across a variety of ecosystems to allow periods of open space or other unused resources to be exploited by novel species. Even in the absence of disturbance, natural mortality events temporarily make new resources available for use by new propagules; if introduced species can more quickly disperse to such sites than resident species, then invasibility patterns could be driven by such dispersal differences, even if introduced species are competitively inferior (a "competition–colonization tradeoff"). Finally, although not a criterion for coexistence, potential invaders that are more efficient at converting resources into offspring under competition in a given environment are predicted to competitively displace residents, thus facilitating their own invasion to the detriment of certain native populations.

SPECIES DIVERSITY AS A BARRIER TO INVASION

In his 1958 book, Charles Elton argued that "the balance of relatively simple communities of plants and animals is more easily upset than that of richer ones; that is, more subject to destructive oscillations in populations, especially of animals, and more vulnerable to invasions." Although many of Elton's examples involved outbreaks of invasive pest species in ecosystems greatly modified by humans, one component of Elton's hypothesis—the diversity–invasibility hypothesis, or the suggestion that communities containing more species are less invasible—has come to dominate much theoretical and empirical research on community invasibility. Although a variety of recent theory and experimental data have supported Elton's claims, other theoretical work, and in particular new empirical data, has called into question the basis of the diversity–invasibility hypothesis. For example, the theoretical ecologist Robert May demonstrated in the early 1970s that population abundances in more complex food webs were not necessarily more stable than those of less diverse communities, undermining Elton's notion that ecological complexity was the key to understanding species invasions. But other theoretical work based on partially overlapping resource requirements among species generally supports the contention that more diverse ecosystems are less prone to invasion.

There has also been a steady accumulation of field survey data to quantify empirical relationships of native and alien (introduced) species richness across a wide variety of ecosystems worldwide. In contrast to Elton's theory, these data typically indicate strong positive correlations between native and alien species richness across many spatial scales, from large regions (e.g., the Untied States) to small vegetation plots. Conversely, many experiments have shown that species richness can reduce both the species richness of successful invaders and total invader abundance. Most of these examples come from experimental plant communities of relatively small and homogeneous experimental units, where resident diversity also acts to increase the probability of having a particularly invasion-resistant species in residence (the "sampling effect"). Extrapolating these studies to natural systems has proven difficult, because experiments can rarely include all the pertinent complexities of natural ecosystems (e.g., abiotic heterogeneity and realistic microbial, herbivore, and predator assemblages). Most invasion biologists today would agree that there are circumstances under which native species richness can play a role in reducing the frequency and extent of invasions, but considerable disagreement remains as to whether such "diversity effects" are strong enough to have applied significance against a background of natural variation in other factors (e.g., disturbance rates, ambient resource levels, supply of invader propagules).

ECOSYSTEMS MADE VULNERABLE VIA DISTURBANCE AND FREE RESOURCES

Most invasive plants and animals that have spread widely as a consequence of human assistance are characteristic of disturbed or human-modified habitats, from house mice to urban Eurasian sparrows to annual bluegrass in golf courses. This has led many biologists to seek invasibility causes related to disturbance processes themselves, whether via the coevolution of certain plant and animal species to human habitation and agricultural practices or through disturbance as a means to alter the ecological resistance of native plant and animal communities. Generally, observational and empirical data support a strong role for disturbance intensity or frequency in the spread of many invasive introduced species. For example, the prevalence of "bare ground" or open space is often the best predictor of invasibility in vegetation studies, and the prevalence of introduced species in vegetation tends to decrease over the course of succession. There are many invaders, however, that thrive in relatively undisturbed ecosystems. For example, many shade-tolerant understory shrubs from Asia and Europe have spread into mature forests in eastern North America. Thus, although disturbance processes are unquestionably an important dimension of community invasibility when viewed across all ecosystems, disturbance alone cannot be the sole driver of variation in community invasion rates.

In addition to disturbance rates, gross rates of resource availability vary among ecosystems, and low resource levels or harsh environmental conditions can reduce community invasibility if potential invaders cannot endure certain environmental stresses. Such "abiotic resistance" factors can include low temperatures (e.g., most tropical invaders are frost-sensitive), drought severity, and high soil acidity or salinity. Indeed, any mismatch between a potential invader's environmental tolerances (fundamental niche) and the prevailing environmental conditions of a particular ecosystem leads to abiotic resistance, regardless of the biotic composition of the native community. However, because natural selection has allowed species to adapt to a large array of different environmental stresses, it remains unclear whether some ecosystems can be generally more abiotically resistant than others.

ECOSYSTEMS EXPOSED: THE ROLE OF INVADER PROPAGULE SUPPLY

One view of community assembly in ecology supposes that species inhabit particular niches, and new species are only able to invade a community if a suitable niche is available. The opposing view of this strict "niche-driven" assembly process supposes that community membership simply reflects those species that are able to disperse there, and that communities differ greatly in the overall number and species richness of new immigrants that arrive in any given year ("propagule pressure"). This latter dispersal-driven view of community assembly was first popularized in the book *The Theory of Island Biogeography* by Robert McArthur and E. O. Wilson in 1967. MacArthur and Wilson argued that, in addition to local processes related to habitat availability and species interactions, certain ecosystems were more isolated from new immigrants, and thus should contain fewer species than communities closer to propagule sources.

The principles of island biogeography also apply to the susceptibility of communities to species invasions. Generally, those communities that have a greater influx of propagules from introduced species exhibit higher invasion rates. This is true across many types of communities, marine to terrestrial. Global shipping ports, for example, are among the most highly invaded ecosystems worldwide, due to the transport of marine organisms in ship ballast water. One study estimated that a new introduced estuarine or marine species establishes in major ports once every one or two years. Historically important centers of the global shipping trade also have interesting alien floras due to propagules contained in shipping or ballast materials. For example, the city of Philadelphia has among the richest floras of any North American city, but a significant proportion of these are rare or nonsustaining (waif) populations of introduced species from coastal areas worldwide. In terrestrial landscapes in general, there are significantly more invasive plant species in the old fields and forests surrounding urban and suburban areas, near where invaders were introduced for horticultural or economic purposes. Even in landscapes historically isolated from major sources of propagule input from introduced species, such as the interior of large national parks, there are always a few habitats that, due primarily to their proximity to human visitation or other sources of propagule supply, contain some invasive introduced species (e.g., roadsides, hiking or horse trails, riparian areas).

EVOLUTIONARY DISEQUILIBRIA: EMPTY NICHES AND POORLY ADAPTED NATIVES

Charles Darwin was among those biologists who view species invasions as a natural consequence of mixing plant and animal communities that have been historically isolated. In *On the Origin of Species* (1859), Darwin observed that because "natural selection acts by competition, it adapts the inhabitants of each country only in relation to the degree of perfection of their associates," such that, "we need feel no surprise at the inhabitants of any one country, although on the ordinary view supposed to have been specially created and adapted for that country, being beaten and supplanted by the naturalised productions from another land." From this perspective, native species are not necessarily the best competitors or the most effective exploiters of their environment, but instead may leave "empty niches" available for colonization by novel species or be outcompeted by foreign species that are able to convert resources into offspring more efficiently. For example, European beachgrass (*Ammophila arenaria*), native to the coasts of Europe and the Mediterranean, has established extensively on previously unvegetated dunes from plantings across the North American West Coast in the late nineteenth century; some biologists interpret this as a case of the beachgrass occupying a habitat that the native vegetation could not colonize. There are many potential examples of this in the literature of global plant invasions. Timberline species in New Zealand, for example, are less cold-tolerant than conifers from the northern hemisphere; when introduced, these conifer species have increased the elevation of timberline and established reproductive individuals in areas formerly uninhabited by trees.

SEE ALSO THE FOLLOWING ARTICLES

Competition, Animal / Competition, Plant / Disturbance / Elton, Charles S. / Islands / Landscape Patterns of Plant Invasions / Propagule Pressure / Succession

FURTHER READING

Davis, M. A., J. P. Grime, and K. Thompson. 2000. Fluctuating resources in plant communities: A general theory of invasibility. *Journal of Ecology* 88: 528–534.

Elton, C. S. 1958. *The Ecology of Invasions by Animals and Plants*. London: Methuen (reprinted 2000, Chicago: University of Chicago Press).

Fridley, J. D., J. J. Stachowicz, S. Naeem, D. F. Sax, E. W. Seabloom, M. D. Smith, T. J. Stohlgren, D. Tilman, and B. Von Holle. 2007. The invasion paradox: Reconciling pattern and process in species invasions. *Ecology* 88: 3–17.

Levine, J. M., and C. M. D'Antonio. 1999. Elton revisited: A review of evidence linking diversity and invasibility. *Oikos* 87: 15–26.

Lockwood, J. L., P. Cassey, and T. Blackburn. 2005. The role of propagule pressure in explaining species invasions. *Trends in Ecology and Evolution* 5: 223–228.

Rejmánek, M. 1989. Invasibility of plant communities (369–388). In J. A. Drake et al., ed. *Biological Invasions: A Global Perspective*. New York: Wiley and Sons.

INVASIONAL MELTDOWN

BETSY VON HOLLE

University of Central Florida, Orlando

Invasional meltdown is the process by which two or more nonnative species facilitate each other's establishment or exacerbate each other's impact on native species. These interactions need not be between species with a shared evolutionary history. Interactions between two or more nonnative species have the potential to lead to synergistic impacts on the recipient ecosystem, wherein the impacts of a group of nonnative species are greater than the summed impacts of the individual species. The outcome of species introductions could result in accelerating rates of invasion, driving ecosystems to be invaded by greater numbers of species, each invasive species having the potential to facilitate further invasions or enhance impacts of other nonnative species.

BACKGROUND

In 1958, in the seminal book *The Ecology of Invasions by Animals and Plants,* Charles Elton popularized the notion of "biotic resistance" to invasion, whereby negative interactions such as predation, parasitism, and competition with native species act to resist the invasion of nonnative species into communities. This hypothesis was in line with the reigning paradigm at the time that negative interactions were the primary forces structuring ecological communities. Thus, the primary research focus of nonnative species impacts traditionally centered on negative interactions between nonnative invaders and native resident species. Daniel Simberloff and Betsy Von Holle introduced the term invasional meltdown in a paper published in the first issue of the journal *Biological Invasions* in 1999 and suggested that there are many reports of mutual facilitation between nonindigenous species in the literature. However, many of these reports of facilitation were incidental, and usually not the focus of the research. In this paper, as well as a paper published in 2000 in Biological Review by David Richardson and colleagues, it was posited that positive interactions between nonnative species had largely been ignored by the research community. Simberloff and Von Holle suggested that community-level invasional meltdown should be investigated as a potential phenomenon that could lead to an accelerating accumulation of introduced species, contrary to the deceleration of species accumulation predicted by the biotic resistance model. This term was met with great favor in the scientific community and garnered media attention as well.

In 2006, six years after the original invasional meltdown paper, Daniel Simberloff reviewed invasional meltdown studies conducted thus far and found that most studies reporting invasional meltdown had demonstrated the weakest degree of facilitation (one species aiding another), and very few had clearly demonstrated population-level impacts using demographic studies. Additionally, only a handful of studies had clearly quantified a mutualism between populations of two or more species. The community-level phenomenon of invasional meltdown, whereby facilitations between nonnatives lead to increasing rates of establishment or increasing impacts, has rarely been demonstrated. Jessica Gurevitch commented on the review paper and suggested that positive feedbacks should be the focus of future research on the topic, rather than facilitation between two species. Gurevitch (2006) distinguishes between simple facilitation, whereby one or more species has a positive effect on another, and positive feedbacks, whereby "[r]unaway positive feedbacks in a system create snowball effects in which a phenomenon builds on itself in an accelerating fashion, becoming unstoppable." She suggested that to assess the relative magnitude and/or importance of the effect, quantitative approaches such as metaanalysis are necessary. In a rejoinder, Simberloff responded that ecology is a highly idiosyncratic science, and one best served by amassing a catalogue of case studies to understand species- and community-level interactions and impacts on specific regions and habitats. Both authors agreed that more empirical research on the positive interactions between nonnative species should be undertaken.

EMPIRICAL EVIDENCE FOR THE HYPOTHESIS

Indirect Effects

There are many examples of the weakest form of facilitation, with one species facilitating another, though indirect effects. The invasion of bullfrogs in Oregon is facilitated by the presence of a coevolved nonnative sunfish, which increases tadpole survival by reducing predatory native macroinvertebrate densities.

Pollinator visitation rates and seed output of the invasive annual species Italian thistle (*Carduus pycnocephalus*) in Argentina are increased when it grows in association with shrubs of the invasive, nitrogen-fixing yellow bush lupine (*Lupinus arboreus*), with the yellow bush lupine acting as a magnet species for the thistle. There was no difference in soil nitrogen content with and without lupine; thus, the increased seed set of the thistle is attributed to the increased pollinator visitation rate associated with lupine. The magnet effect of facilitated pollination and reproduction of *C. pycnocephalus* by *L. arboreus* could promote its naturalization in the community.

On Isla Victoria, Argentina, two nonnative species of deer, the red and fallow deer (*Cervus elaphus* and *Dama dama*, respectively) preferentially browse seedlings of the two native dominant species of tree: a conifer, *Austrocedrus chilensis,* and a broadleaf, *Nothofagus dombeyi*. The deer avoid browsing the introduced species of conifer, the Douglas fir (*Pseudotsuga menziesii*) and the ponderosa pine (*Pinus ponderosa*), which could potentially facilitate the invasion of these nonnative pines.

In a metaanalysis of the effects of native and nonnative herbivory on more than 100 nonindigenous plant species,
it was found that native herbivores suppressed nonnative plants, while nonindigenous herbivores facilitated both the abundance and species richness of nonnative plants. Furthermore, the replacement of native with nonnative herbivores eliminates the intrinsic biotic resistance of native herbivores to nonnative plant invaders, facilitating plant invasions.

Population-Level Effects

AQUATIC INVASIONS A historically benign invasive species, the gem clam (*Gemma gemma*), introduced from the east coast of the United States to the west coast, became much more abundant in Bodega Bay, California, after the introduction of the European green crab (*Carcinus maenas*) which was introduced 30 years after the introduction of the gem clam. This supports the idea that lag times of nonnative species—the length of time from introduction to spread—may be due to successive introductions of nonnative species having the potential to facilitate one another. Evidence for facilitation of the gem clam by the nonnative green crab was found, but there was no evidence of the gem clam's facilitating the green crab. Further tests need to be conducted to understand whether the interaction between these two species is a mutualism.

A literature review of the interactions between nonnative species of the Great Lakes revealed 3 cases of mutualisms as well as 14 cases of commensalisms, more than the number of documented negative interactions between invasive species in the Great Lakes. These population-level impacts were, for the most part, not tested empirically.

TERRESTRIAL INVASIONS South African succulent plant species of the genus *Carpobrotus* are considered major pests across the Mediterranean. Two species of *Carpobrotus* were introduced into southeastern France in the early 1800s: *C. edulis* and *C.* aff. *acinaciformis,* which became naturalized on Mediterranean coastlines in the early 1900s. These species are invading the littoral ecosystems of southeastern France, and their rapid vegetation growth threatens rare plant communities. On coastal islands, *Carpobrotus edulis* and *C.* aff. *acinaciformis* are readily dispersed by two species of nonnative mammals: rabbits (*Oryctolagus cuniculus*) and the ship rat (*Rattus rattus*), both of which were introduced into this region prior to 1751 (with the potential for rats invading the coastal islands as early as 2400 BP). Low levels of dispersal of *Carpobrotus* occur on the mainland, and, when dispersal does occur, it is by native mammal species. Seed digestion by rats and rabbits on the islands significantly enhances the speed and seed germination of *Carpobrotus* as compared to dried fruits removed from the same site, with seeds from rat feces having significantly greater germination than seeds from rabbit feces, which have greater germination than dried fruits from the plants. Feral cats (*Felis catus*) on the islands also occasionally eat *Carpobrotus* seeds. The maximum seed dispersal distances for cats is far greater than that of rats and rabbits, so they serve as a potential long-distance vector of *Carpobrotus*. The prolific *Carpobrotus* fruits are twice as large as native fruits and provide nonnative island mammals a source of water and energy-rich food during the summer drought season. In this mutualism, *Carpobrotus* gains seed dispersers which enhance germination success, and the nonnative mammals obtain an energy and nutrient-rich food source. The continued spread of *Carpobrotus* threatens the natural vegetative communities of these islands, and other impacts of the nonnative mammals may not yet have been quantified. Thus, the facilitation between these species may result in synergistic impacts on the native biota of these islands.

Community-Level Effects

AQUATIC INVASIONS The introduced Asian hornsnail, *Batillaria attramentaria*, occurs in very high densities in the mudflat ecosystem of northern Puget Trough, in Washington state. The Asian hornsnail facilitates two other nonnative species that use the hard substrate provided by the shells of this species, with no comparable native species filling this role. The nonnative Atlantic slipper shell, *Crepidula convexa,* and the introduced Asian anemone, *Diadumene lineata,* use the hard substrate provided by the shells of *B. attramentaria* almost exclusively. *Batillaria* indirectly facilitates the nonnative eelgrass, *Zostera japonica,* by modifying oxygen and nutrient levels. Selective grazing by *Batillaria* may indirectly augment the preferred diatom resources of the nonnative mudsnail, *Nassarius fraterculus,* which may be the cause for the higher densities of this species associated with *Batillaria*. Additionally, two native hermit crabs, *Pagurus hirsutiusculus* and *P. granosimanus,* habitually use the shells of *Batillaria*. The invasibility of this mudflat ecosystem is enhanced by the Asian hornsnail, as this species increases the densities of four nonnative species, which may enhance their population sizes and the probability (and impacts) of persistence. However, empirical evidence of increased impacts on native fauna or positive effects on *Batillaria* by the four associated nonnative species has not been determined.

TERRESTRIAL INVASIONS The yellow crazy ant, *Anoplolepis gracilipes,* is one of the six most widespread and damaging invasive ant species in the world. This species invaded Christmas Island, located in the northeastern Indian Ocean, in the early twentieth century, persisting at low population densities and having negligible impacts on the native biota for the first 70 years of its invasion. Starting in 1989, supercolonies became more widespread and by 2001 had covered one-quarter of the rain forest on this island, attaining densities of up to 2,254 foraging ants per m^2 in invaded supercolony sites. *Anoplolepis gracilipes* kills the dominant native omnivore, the red land crab (*Gecarcoidea natalis*), by spraying formic acid over the eyes and mouthparts of the crab. The expansion of the yellow crazy ant supercolonies on the island results in the occupation of red crab burrows by the ants, which kill and consume the native crab and occupy the burrows as nest sites. Red land crab abundance is 42 times lower in ant-invaded sites. As a result of the severe reduction of the red crab, the dominant consumer of understory vegetation in this system, there is a doubling of litter cover, a thirtyfold increase in understory seedling density, and 3.5 times greater seedling species richness. Population outbreaks of two species of scale insects, the lac insect of unknown origin, *Tachardina aurantiaca,* and the nonnative *Coccus celatus* are associated with sites invaded by the yellow crazy ant. The ants feed on the carbohydrate-rich honeydew of the scale insects and are assumed to defend the scales from predators and parasites. Furthermore, outbreaks of the scale insects and their associated honeydew spur the growth of a sooty mold in the canopy, causing widespread tree dieback and altering canopy and understory species composition. In areas with high densities of the yellow crazy ant the abundance of the native ground foraging emerald dove, *Chalcophaps indica,* has been significantly reduced. In sum, the yellow crazy ant has a direct negative effect on the red land crab, and the reduction of populations of the dominant herbivore of this system results in dramatic changes in understory community composition and litter cover. Nonnative scale insects are used by the yellow crazy ant as an energy source and spur the growth of sooty mold, resulting in widespread canopy dieback. This ant–scale insect mutualism results in an invasional meltdown whereby the impacts of the mutualism on this ecosystem are greater than the summed impacts of each individual species (Fig. 1).

Introductions of nonnative mammalian predators to islands have additive impacts on island avifauna, with the probability of bird extinction related to predator species richness as a positive logistic function. If competition occurred between island bird predators, or if there was functional redundancy of these species, the impact on native bird fauna would be expected to saturate with

FIGURE 1 Impacts of invasion of the Christmas Island rain forest by the yellow crazy ant, *Anoplolepis gracilipes,* and scale insects. (A) Uninvaded site with open understory maintained largely by the foraging activities of the native red land crab, *Gecarcoidea natalis.* (B) Invaded site 1–2 years after invasion, with a dense and diverse seedling cover and thick litter layer. (Photographs courtesy of Peter Green.)

increasing numbers of introduced predators. Empirical evidence suggests that every additional nonnative mammal predator added to islands results in an increased likelihood of island bird extinctions, suggesting that the impact of each additional predator may be facilitated by those present, rather than the impacts of predators being dampened via competition or functional redundancy.

Mutual facilitations between nonnative earthworms and nonindigenous nitrogen-fixing plant species have been found in the continental United States and Hawai'i. Earthworm growth rates and carrying capacities are increased in the field through the provision of higher-quality food that nitrogen-fixing plant species provide in the form of nitrogen-rich litter. Additionally, populations of earthworms are thought to be enhanced through the modification of soil properties by nitrogen-fixers, including increased nitrogen, soil moisture, and pH. Earthworms act to increase the rate of nitrogen burial and litter breakdown with their feeding activities, rapidly increasing soil biogeochemical impacts. Thus, a mutually reinforcing positive feedback cycle results

from the interactions between earthworms and nitrogen-fixing plant species. Cascading community-level effects from the changes in soil biogeochemistry are prone to occur in areas with naturally low levels of soil nutrients. These impacts include increased colonization and persistence of other nonnative plants in the elevated nutrient areas, as well as the attraction of native and nonnative bird species that serve as vectors for the nonnative plants that thrive in these elevated nutrient environments. Additionally, in the highly invaded system of Hawaii, nonnative earthworms are a protein source for invasive feral pigs, which may have been a causal factor for the increased densities and expansion of feral pigs in the twentieth century. Feral pigs are estimated to disturb one-third of the diggable area of the mountain rain forest of Hawaii every year, increasing nutrient levels through fecal deposition as well as spreading propagules of nonnative plants such as the nitrogen-fixing nonnative firetree (*Morella (Myrica) faya*), which is associated with high levels of earthworms across this landscape. The runaway effects of changes in soil biochemical properties, as well as concomitant community-level impacts caused by the positive feedback loop between nonnative earthworms and nonindigenous nitrogen-fixing plants, is an example of invasional meltdown.

Accelerating Rates of Invasion

Observational studies of historical invasions have documented accelerating rates of invasion for a wide variety of taxa over time in the aquatic ecosystems of the Chesapeake Bay, the Great Lakes, San Francisco Bay, and the Baltic Sea; the marine systems of coastal North America, the Mediterranean Sea, and the Northeast Pacific Ocean; and terrestrial island ecosystems. Accelerating rates of invasion could be due to invasional meltdown, increased anthropogenic vectoring of species, increased detection rates, greater sampling efforts, or changes in the resistance of the recipient ecosystem to invasion. A model has been built using constant introduction rates and invader success that generated the exponential distribution of introduction times that has been observed in a variety of systems. Thus, without careful demonstration of direct and indirect interactions at the population and community levels, observational studies of systems demonstrating accelerating rates of invasion cannot be used as examples of invasional meltdown.

CONSEQUENCES OF INVASION MELTDOWN AND MANAGEMENT

Facilitations between a group of nonnative species can lead to greater impacts on native species than their individual impacts summed. This has been demonstrated empirically with the devastating ant-scale mutualism on Christmas Island and the interaction between nonnative earthworms and nitrogen-fixing plant species. In order to understand the relative frequencies of facilitative versus detrimental interactions between nonnative species, there is a pressing need for additional empirical studies that quantify interactions between nonnative species. It is possible that invasional meltdown may occur infrequently, but with devastating impacts. Thus, ecosystem managers and scientists should be aware of its existence in order to record and predict the species and ecosystems most likely to experience invasional meltdown. The potential for positive interactions with invaders counters the criticism that prevention efforts are wasted if some future invasions are inevitable. It has been clearly demonstrated that there is the potential for a new species introduced into an ecosystem to have facilitative interactions with future and present nonnative resident species of that ecosystem. Thus, this warrants increased efforts and policies for preventing new invasions by dramatically reducing the frequency of new introductions to reduce the successful establishment of invaders in the future, as well as overall ecosystem susceptibility to invasion. Some of the greatest challenges of invasion biology have been the prediction of nonnative species likely to invade a given habitat, impacts of those species, as well as highly invasible habitats. Likewise, identifying high-impact species likely to form facilitations with established nonnative species, as well as the habitats and ecosystems most susceptible to invasional meltdown, may become the focus for future research on this topic.

SEE ALSO THE FOLLOWING ARTICLES

Dispersal Ability, Plant / Earthworms / Invasibility, of Communities and Ecosystems / Mutualism / Mycorrhizae / Pollination

FURTHER READING

Bourgeois, K., C. M. Suehs, E. Vidal, and F. Medail. 2005. Invasional meltdown potential: Facilitation between introduced plants and mammals on French Mediterranean islands. *Ecoscience* 12: 248–256.

Gurevitch, J. 2006. Commentary on Simberloff (2006): Meltdowns, snowballs and positive feedbacks. *Ecology Letters* 9: 919–921.

Kourtev, P. S., J. G. Ehrenfeld, and W. Z. Huang. 1998. Effects of exotic plant species on soil properties in hardwood forests of New Jersey. *Water, Air and Soil Pollution* 105: 493–501.

O'Dowd, D. J., P. T. Green, and P. S. Lake. 2003. Invasional "meltdown" on an oceanic island. *Ecology Letters* 6: 812–817.

Richardson, D. M., N. Allsopp, C. M. D'Antonio, S. J. Milton, and M. Rejmánek. 2000. Plant invasions: The role of mutualisms. *Biological Reviews* 75: 65–93.

Simberloff, D. 2006. Invasional meltdown 6 years later: Important phenomenon, unfortunate metaphor, or both? *Ecology Letters* 9: 912–919.

Simberloff, D., and B. Von Holle. 1999. Positive interactions of nonindigenous species: Invasional meltdown? *Biological Invasions* 1: 21–32.

Wonham, M. J., and E. Pachepsky. 2006. A null model of temporal trends in biological invasion records. *Ecology Letters* 9: 663–672.

INVASION BIOLOGY

MARK A. DAVIS

Macalester College, St. Paul, Minnesota

Invasion biology is a scientific discipline that studies the human transport and introduction of species throughout the world, as well as the subsequent spread of these species and their health, economic, and environmental impacts. Although some scientists and naturalists observed and commented on the introductions of new species as long as several centuries ago, and even though Charles Elton published his famous book on invasions (*The Ecology of Invasions by Animals and Plants*) in 1958, a formally defined field of invasion biology did not emerge until the early 1980s. Since then, the field has grown enormously. While only a few dozen articles were published annually during most of the 1980s, this number exceeded 1,000 by the early years of the twenty-first century. Today, thousands of biologists around the world are studying introduced species and contributing to the field of invasion biology.

THE HISTORY OF INVASION BIOLOGY

One would imagine that as far back as several thousand years ago, careful observers would have noticed the establishment and spread of species brought into their region by travelers. The first known documented accounts of such observations appeared in Western writings in the 1700s, when European naturalists traveled to North America and described some of the European plants and insects they observed there. During the 1800s and into the twentieth century, as global travel became more common, biologists and geographers often reported on nonnative species that had become established in regions far from their native environments. However, despite the observations and accounts, and even though some of the species were causing problems in their new regions, few scientists focused on introduced species as a specific research topic during this time. Somewhat surprisingly, even in the immediate decades following the 1958 publication of Elton's famous book on animal and plant invasions, there was little widespread interest in studying nonnative and invasive species.

By the early 1980s, concerns over the undesirable impacts of some introduced species had grown, and in 1983, the Scientific Committee on Problems of the Environment (SCOPE), an international network of scientists and scientific institutions, created a special advisory committee and charged it with encouraging and facilitating research and understanding of species introductions and their impacts, and with applying this knowledge to management of such species. During the remainder of the 1980s, a small but committed group of scientists focused their studies on invasive species. Although the number of publications during the 1980s was modest, the efforts by the founding group of scientists had attracted the attention of a large number of ecologists throughout the world. As a result, the field of invasion biology experienced exponential growth during the 1990s (Fig. 1). In the last few years of the twentieth century, the Global Invasive Species Programme (GISP), a new international collaborative effort, was initiated to minimize the spread and impacts of nonnative invasive species. At the same time, two journals with a primary focus on species introductions were founded: *Diversity and Distributions,* in 1998, and *Biological Invasions,* in 1999. The intense and widespread interest in species introductions and their effects continued throughout the first decade of the twenty-first century. During this time, invasion biologists began to work more with scientists in other disciplines, including geographers, molecular biologists, climatologists, and computer scientists, collaborations that have increased our understanding of the invasive process and heightened our ability to manage invasive species.

PRIMARY RESEARCH AREAS WITHIN THE FIELD OF INVASION BIOLOGY

Species Transport and Dispersal

Before a species can spread in a new region and cause problems, it must first be transported to the new environment. While transport does not assure establishment and spread of a species, it is a necessary first step. Some species are able to disperse very long distances by themselves (e.g., aerial-borne microbes and plant seeds), but the vast majority of species introduced into new regions of the world during the last few centuries have been transported by humans. In some cases, these species have been introduced intentionally (e.g., plants for horticulture, animals for pets, and both plants and animals for agriculture). Some of the species introduced for these purposes

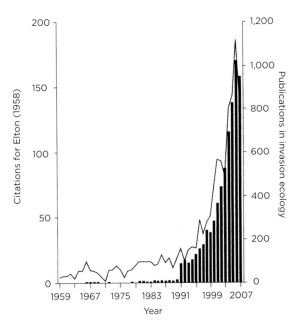

FIGURE 1 The number of biological invasion publications (columns) since Charles Elton published *The Ecology of Invasions by Animals and Plants* in 1958. Also shown are the number of publications that cited Elton's 1958 book during this time period (line). (Reprinted, with permission, from Anthony Ricciardi and Hugh MacIsaac, 2008, copyright Nature Publishing Group.)

escaped and have been able to establish and persist in surrounding environments without any further assistance from humans. Other introductions are not intentional. For example, some marine organisms attach themselves to the hulls of oceangoing ships and then become disengaged or disperse their eggs in waters distant from their native range. In other instances, plants and animals intentionally introduced often serve as hosts for parasites and pathogens, which are then transported and introduced along with their hosts. Some human diseases, once confined to particular regions of the world, are becoming more widespread as a result of the pathogens' being transported when their human hosts travel from one region of the world to another. When crop seeds are transported from one region of the world to another, seeds of other plant species are often inadvertently transported as well. When the crop seeds are planted, so are the seeds of the other species.

Many studies have shown that the likelihood that an introduced species will become established in a new region is a function of the number of propagules (e.g., seeds, eggs, larvae) in a transport or dispersal event, as well as of the number of dispersal events. The combination of these two variables, number of propagules in a dispersal event and number of dispersal events, is often referred to as *propagule pressure*. Because humans are responsible for the long-distance transport of most species, as the number of human travelers has increased, so has propagule pressure, and hence the number of introductions. In recent decades, owing to the dramatic increase in international commerce, most regions of the world are experiencing an increase in propagule pressure for many types of organisms.

Establishment

Establishment of an individual in a new region can be defined as propagules' persisting long enough in the new region to be able to reproduce. To do this, they need to accomplish four tasks. First, they need to find an abiotic environment (e.g., temperature, salinity, moisture) they can tolerate. Second, they need to be able to access resources (e.g., water, soil nutrients, space) needed for growth, maintenance, and reproduction. Third, unless the initial propagule is a fertilized female, or capable of reproducing asexually, propagules need to find one another to mate. And, fourth, founding propagules must avoid prereproductive mortality—e.g., getting eaten by a predator or herbivore. The ability of the founding propagules to accomplish these tasks is influenced by their own traits as well as by the characteristics of the new environment.

The susceptibility of an environment to the colonization and establishment of a species is often referred to as its *invasibility*. Many factors can influence an environment's invasibility, including the availability of resources, abiotic factors, and the presence or absence of enemies (e.g., predators and pathogens) or mutualists, which help the new species to establish (e.g., pollinators and species that may modify the environment to the advantage of the new species) (Fig. 2). In any case, invasibility is not a static or permanent characteristic of an environment. Because invasibility is influenced by an array of biotic and abiotic

events and processes, as these change, so will the environment's invasibility. It is also important to remember that invasibility varies from species to species. The same conditions that may make an environment quite invasible to one species may make it quite resistant to another.

Charles Darwin suggested that a species should be better able to establish in a new region if it is not closely related (in an evolutionary sense) to the resident species in the new environment, his reasoning being that the new species would not experience a high level of competition from the resident species if they are not close relatives. Now referred to as Darwin's Naturalization Hypothesis, this prediction has been tested in many studies in recent years. The results have been mixed, some supporting the hypothesis and some not. It may be that Darwin's hypothesis can be supported in situations in which invasibility is primarily being determined by competition for resources or by the presence of enemies shared by the new species and its resident relatives. In cases in which invasibility is determined more by the presence of shared mutualists or by the abiotic conditions, species more closely related to resident species may actually enjoy greater establishment success.

Persistence and Spread

Ultimately, the field of invasion biology is not interested in all the species that have been transported to and become established in new regions of the world. The subject of the field is primarily the subset of these species that spread widely once established, often producing undesirable consequences (health, economic, or environmental) while doing so. Once a species has been transported to a new region, its subsequent dispersal and spread may still be primarily facilitated by humans. For example, many introduced freshwater organisms are accidentally

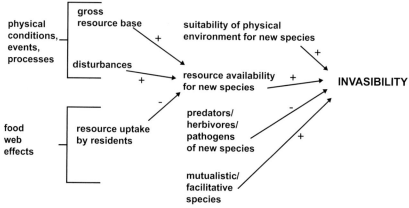

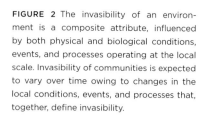

FIGURE 2 The invisibility of an environment is a composite attribute, influenced by both physical and biological conditions, events, and processes operating at the local scale. Invasibility of communities is expected to vary over time owing to changes in the local conditions, events, and processes that, together, define invasibility.

dispersed from one lake or river to another by recreational boaters. And many introduced plants are distributed into national parks accidentally by tourists when seeds of the plants are brought into the parks on the tourists' shoes and clothing. However, in many cases, after the human-mediated initial transport, secondary dispersal within the new region takes place mostly independently of direct human activity using traditional dispersal vectors of the species (e.g., wind, water, and animals).

Many studies have been conducted to try to determine whether certain species traits are associated with invasiveness (the ability to spread rapidly in the new environment). If such traits can be identified, they can be used to evaluate the invasive potential of species not yet introduced into a region, a process known as risk analysis. While some traits do seem to be associated with invasiveness in some organisms (e.g., rapid growth rate and phenotypic plasticity, or the ability of an organism to change some of its traits, such as behavior or amount of energy allocated to reproduction), the best predictor of invasiveness seems to be whether the species has been invasive elsewhere.

Evolution after Introductions

It was originally thought that founding populations would exhibit low genetic diversity, since they likely would have been established by only a few individuals, and that this low genetic diversity would impede the ability of the species to adapt to its new environment. However, a number of studies have shown that founding populations often are not lacking in genetic diversity at all. It is believed the high genetic diversity in many established populations of nonnative species is due to the fact that introduced individuals are often transported from a variety of different gene pools in their native range. This genetic diversity means that many introduced species are able to adapt to new biotic and abiotic conditions. For example, some introduced species have been found to adapt genetically to new climatic conditions. The new species can be similarly affected by resident predators and pathogens. Over time, and as a result of its adaptations with the residents, a recently arrived species may be able to evolve adaptations enabling it to persist in and exploit its new environment better. At the same time, the new species begin to affect the evolution of the residents. If the new species is a predator or pathogen, it will represent a new selection pressure that would favor better predator avoidance behavior among resident prey, and pathogen-resistant strains among resident hosts.

Although rapid spread of nonnative species has received the most attention from invasion biologists,

rapid declines of an introduced species after a period of rapid spread have commonly been documented as well. Researchers believe that this boom–bust phenomenon may be partly the result of the evolution within resident species that enable them to adapt better to the new species and to impose some regulation over its growth and spread. It is important to remember that genetic adaptation takes place only over generations. If an introduced predator is too successful, or an introduced pathogen too virulent, the resident prey and hosts may not have sufficient time to adapt before being extirpated by the new species.

One important type of evolutionary process that can occur following the introduction of a new species to a region is hybridization. In this case, the new species mates with a related resident species, or another recently introduced related species, thereby bringing together two sets of gene pools, which previously had been separate. Sometimes biologists view this blending of two gene pools as undesirable, since it can lead to a loss of species uniqueness and diversity. However, the new combination of genes resulting from hybridization also creates new genotypes and thereby new evolutionary opportunities.

Impacts of Nonnative Invasive Species

Like native species, nonnative species can affect human health, local and national economies, and ecosystems. Although most nonnative species do not have large impacts, a small proportion of introduced species are considered harmful or undesirable owing to their effects. The greatest threats involve human pathogens that are being transported throughout the world spreading disease, including avian influenza, AIDS, severe acute respiratory syndrome (SARS), West Nile encephalitis, Ebola hemorrhagic fever, dengue hemorrhagic fever, and avian influenza. Many of these and other diseases are zoonotic, which means that the pathogens can also use other animals as hosts. As a result, these diseases can be spread by the global transport of animals (e.g., for food production, and through the pet trade). Some nonnative species are causing great economic harm, costing countries billions of dollars. Examples are species that substantially reduce the productivity of crops, timber, and fisheries, contaminate or otherwise disrupt water supplies, and kill farm animals through disease.

Many invasion biologists study how introduced species affect the species and ecosystem processes in their new environment. In some instances an introduced species reduces the abundance of some of the native species (e.g., through predation, disease, or competition). Although there are few instances of introduced species causing extinctions in marine systems or in terrestrial environments on continents, many cases are documented of introduced predators and pathogens driving native species to extinction on islands and in freshwater systems (e.g., lakes and rivers). Other introduced species have been found to alter dramatically some ecosystem processes, such as the rate at which nutrients are cycled and the availability of certain resources, such as nutrients and water. Still other introduced species are known to alter the frequency and intensity of disturbances, such as fire.

Although the undesirable impacts of introduced species are usually emphasized, it must be remembered that some new species produce desirable effects. Some species have been found to increase the diversity of native species in an environment. For example, many sessile invertebrates require a hard surface to which to attach themselves, and in some aquatic environments, hard surfaces are uncommon and ultimately limit species diversity. In some of these environments, the shells of introduced mussels have served as attachment sites for these other invertebrates, thereby increasing the number of resident native species. Some nonnative trees have been found to alter the soil conditions in previously deforested environments, thereby making it possible for native trees to reestablish on these sites. The ecological impacts of introduced species are many, and they vary from species to species and environment to environment. Invasion biologists have only begun to understand the array and extent of these effects.

Management of Nonnative Invasive Species

In addition to learning how nonnative species may impact their new environments, and sometimes cause economic harm and health threats as well, invasion biologists also try to learn how to manage these species in order to reduce their undesirable effects. One way is to reduce the likelihood that undesirable species are transported from one region of the world to another. Education, regulations and laws, the use of inspectors, and screening efforts to identify which species should be targeted during inspection procedures are all strategies currently employed to try to reduce the initial introductions. Even if introductions of undesirable species can be reduced, prevention will never be one hundred percent successful for all species. Some invasion biologists are investigating the use of early detection or early warning programs to identify the arrival of undesirable species before they have had time to spread. Experience has shown that efforts to eradicate unwanted species are very expensive and usually fail once the species has become widespread. In some cases, biological control

efforts have succeeded in dramatically reducing the abundance of undesirable nonnative species. However, because many biological control efforts involve introducing new nonnative species (e.g., parasites, pathogens, or predators of the target species), considerable testing needs to be conducted prior to the release of the control organism to make sure that it will not adversely affect other species.

A fundamental challenge of most invasive species management is that regardless of success in removing the undesirable species, management efforts may produce other undesirable consequences. Because all species in an environment exist within a web of interactions with other species and the physical environment, removing one species, particularly an abundant species, inevitably produces an array of effects, some predictable and some not. In some instances, the nonnative species may have been providing some unknown desirable effects that will be eliminated upon its removal. In many cases, a nonnative invasive species is more a symptom than the cause of a problem. For example, nonnative plant species are often believed to be the cause for the decline of native plant species, but studies have shown that in many instances, the decline of the native species is due mainly to other factors, such as changes in disturbance regimes and resource availability, which then favored the establishment of the nonnative species. In such cases, removing the nonnative species will likely not substantially benefit the native species unless the underlying causes are also addressed.

CHALLENGES FACED BY THE FIELD OF INVASION BIOLOGY

The field of invasion biology is a very young discipline. Like any growing enterprise, the field has experienced some growing pains. In recent years, scholars within and outside the field have raised questions regarding value-laden terminology that has often been used within the field of invasion biology—e.g., referring to species introductions as "invasions" and to nonnative species as "biological pollution." Some common claims of invasion biologists have also been recently challenged—e.g., claims that nonnative species threaten native species with extinction through competition, or that introduced species cause declines in biodiversity. In fact, much evidence suggests the opposite is true. At regional and local scales (e.g., state and county scales), introductions often far exceed the number of extinctions that they cause, thereby resulting in an increase in species diversity. Some scientists have not been happy with the dichotomous nature of invasion biology, in which species are declared to be either native or nonnative, because they believe this distinction has no sound ecological or evolutionary basis. Other invasion biologists disagree. These sorts of disagreements are common within science, particularly when a field is just developing.

The field of invasion biology also faces fundamental challenges from its very subject, species introductions. Theory has suggested, and empirical data have shown, that species introductions are usually very difficult to predict ahead of time. The reason seems to be that very small changes in environmental conditions or in the number of arriving propagules can sometimes dramatically change the likelihood of successful establishment. The apparent presence of a tipping point (a point at which small changes can produce very large consequences) in the invasion process likely explains many well-known aspects of invasion dynamics, including why invasions are often episodic, why there is often a lag phase prior to invasion spread, why some invasions experience a rapid collapse following a period of irruption, and, in general, why invasions typically have been so difficult to predict (and, unfortunately, why invasions are likely always going to be difficult to predict). This finding has important management implications. For example, managers working to prevent initial establishment of a species could be deluded into thinking that their prevention efforts were very successful since no establishment had yet occurred, when, in fact, the system they were managing was moving closer and closer to the tipping point at which successful establishment becomes very likely.

THE FUTURE OF INVASION BIOLOGY

Current developments in the field indicate that interdisciplinary research will become more common. Geographers, computer scientists, molecular biologists, and even social scientists (economists, anthropologists, sociologists) and humanists (philosophers and ethicists) have much to offer to the field. Within the field of ecology, invasion biology is reconnecting to other subdisciplines that also study species movements. For example, climate change is causing ranges of many species to shift, creating new combinations of species in much the same way that introduced species create new combinations. Whether produced by climate change or species introductions, these new communities are being referred to as "novel communities" or "novel environments" by some ecologists, who are calling for the field of ecology to focus on this general phenomenon of ecological novelty.

The field of invasion biology has exhibited remarkable growth since the creation of the SCOPE advisory committee in 1983. In recent years there has been a remarkable influx of new investigators, many bringing with them new ideas, and some challenging old ones. This bodes very well for the field's future. Disciplines begin to stagnate in the absence of new participants and new ideas. The influx of new minds and perspectives into the field is what will ensure the field's vitality in upcoming years. The new investigators and ideas represent opportunities, perhaps not unlike the ecological and evolutionary opportunities created when new species are introduced into an environment.

SEE ALSO THE FOLLOWING ARTICLES

Elton, Charles S. / Evolution of Invasive Populations / Invasibility, of Communities and Ecosystems / Invasion Biology: Historical Precedents / Propagule Pressure / SCOPE Project

FURTHER READING

Cadotte, M. C., S. M. McMahon, and T. Fukami, eds. 2006. *Conceptual Ecology and Invasion Biology: Reciprocal Approaches to Nature.* Dordrecht, the Netherlands: Springer.

Davis, M. A. 2009. *Invasion Biology.* Oxford: Oxford University Press.

Lockwood, J. L., M. F. Hoopes, and M. P. Marchetti. 2007. *Invasion Ecology.* Malden, MA: Blackwell.

Sax, D. F., J. J. Stachowicz, and S. D. Gaines, eds. 2005. *Species Invasions: Insights into Ecology, Evolution and Biogeography.* Sunderland, MA: Sinauer Associates.

INVASION BIOLOGY: HISTORICAL PRECEDENTS

MATTHEW K. CHEW

Arizona State University, Tempe

Invasion biologists traditionally cite Charles Elton's 1958 *The Ecology of Invasions by Animals and Plants* as their foundational document. That book was written to warn about introduced species, not to acknowledge those who had studied them, leaving Elton's followers with few clues to the existence of earlier work. Had he attempted a comprehensive literature review, Elton might have discovered that introduced species interested Englishmen and their colonial colleagues by the early 1600s and that natural historians, botanists, zoologists, and (eventually) ecologists from many countries had recorded thoughts and observations regarding introductions. Like Elton, most of his predecessors studied introductions briefly or intermittently, as sidebars to major research programs. Whether and how previous accounts might have affected Elton's conceptions is unclear. We now know that invasion biology's intellectual genealogy is centuries deep and globally broad, and that it included ideas unfamiliar to Elton.

WHAT'S NEW? HOW DO WE KNOW?

Studying biotic *re*-distribution requires noticing unfamiliar organisms or finding evidence of a different, bygone distribution. Fifty centuries ago, Sumerians recorded different kinds of plants and animals living in different places. Ever since, travelers' accounts have listed species characterizing various regions. Aristotle and Theophrastus laid tentative foundations for biogeography, while other ancients such as Pliny the Elder mixed hearsay with natural history, leaving a messy datum. For another 15 centuries, natural history was the bailiwick of adventurers, noblemen, clerics, apothecaries, and physicians. Wide-ranging taxa acquired many local names. Similar endemics from various locations were easily conflated. Rampant complications resulted from competing authorities and schools of practice often motivated as much by chauvinism as accuracy.

During European exploration and colonization, the number of named species ballooned. Problems of identifying and sorting became acute. Studying *where* different species occurred, and attempting to explain *why*, was a rich man's pursuit, far more difficult and expensive than collecting and describing specimens. Alexander von Humboldt, a true pioneer of biogeography, spent enough for a lifetime's comfortable living in Europe underwriting his expeditions to the Americas. Wealthy physician Robert Darwin financed his son Charles' five year H.M.S *Beagle* adventure. Less prosperous natural historian-explorers sometimes doubled as ships' surgeons on commercial or military expeditions. They sought the unusual, exotic, and potentially profitable rather than news of familiar species abroad. But nostalgic wonder at finding familiar species in an unfamiliar setting sometimes elicited comments, providing glimpses of both biogeographical dynamics and human motivations.

Practicalities dictated which introduced taxa were noticed and studied. Terrestrial, temperate organisms were more familiar than tropical, alpine, polar, or aquatic ones and were available to be seen by more trained eyes. Sessile or sluggish taxa such as plants, barnacles, and molluscs that could be scrutinized at leisure were most easily identified or declared unfamiliar. To paraphrase an early Elton publication, motile taxa complicated study by fleeing. Biota redistributed for economic or subsistence purposes are not a core concern of modern invasion biology, but earlier authors saw crops and livestock as among the world's most significant introductions. By the mid-nineteenth century, their origins, like those of their accidental counterparts, became objects of study.

Neither Elton's forerunners nor his potential colleagues organized, developed standard methods, or agreed on particular ways of understanding introduced species. New knowledge requires new categories and terminology. Competing proposals were based on objective criteria such as precision and utility, but also on nationalism, institutional pride, and expediency. Terms we take for granted changed meanings over time. "Native" once meant anything not produced directly by artifice, including feral animals and naturalized plants. An early translator of Linnaeus explained that "exotic" plants survived only with intervention (e.g., in hothouses). By 1835 some English botanists found the existing terminology insufficient, provoking discussions that produced "native" and "alien" as they are applied today. Conceived by early Victorian minds in a world whose dynamic nature was barely imagined, these basic categories have persisted since 1847, a dozen years before Darwin proposed evolution by natural selection, 20 before Haeckel proposed "œcologie," and 55 before Wegener argued that continents "drift." Natives and aliens have been subdivided by uncertainty of origin and degree of naturalization. The familiar "invasive species" may have first appeared in the colonial journal *The Indian Forester* in 1891. Over the ensuing century, many other categories and terms were proposed, but most were rarely used or forgotten. Classifying species by geographical origin has been called anekeitaxonomy, from a Greek word for "belonging." Anekeitaxonomies have often included categories for species "grandfathered" into local acceptability by long tenure, usefulness, or aesthetic appeal.

EARLIEST REPORTS AND BASELINE ISSUES

It is difficult to identify the "first scientific" discussion of an *intentionally* introduced species. Ancient accounts involve little science and evidence no concern, simply documenting the advent of a curious or useful organism. Explorers wrote of introducing plants and animals to and from places they visited, but not with scientific aims or methods. The likeliest candidate in this respect is Peter Kalm, whose observations (1750) are discussed below. A very early mention of *accidentally* introduced species comes from Francis Bacon's *Sylva Sylvarum* (1627): "Earth that was brought out of the Indies and other remote countries for ballast for ships, cast upon some grounds in Italy, did put forth foreign herbs, to us in Europe not known."

Here the mode of introduction becomes important. Before watertight metal hulls were constructed in the mid-

1800s, ships' ballast consisted of rocks and soil, deposited ashore to preserve the quays. By the Middle Ages, solid ballast had been liberally exchanged among Baltic, northeastern Atlantic, Mediterranean, and accessible upriver ports. Only by the 1600s had botany advanced enough that skilled practitioners might notice unfamiliar plants growing from discarded foreign soil and report them to others. Contemporaneous publishing advances meant reports could increasingly be exchanged among interested parties. Aquatic species redistribution began long before the advent of marine biology in the 1870s. Meanwhile, overland dispersal of biota proceeded incrementally between established settlements, with few sudden advents occurring far from known populations. Invasion biology's biogeographical baselines must therefore be inferred wherever commerce preceded science.

PRACTITIONERS

This section briefly identifies major contributors to the scientific discussion of species introductions, organized by publication era.

1750s–1858

PETER KALM was sent by Linnaeus to acquire useful American plants for acclimatization in Sweden. In (mostly) Pennsylvania and southern New England, Kalm recognized at least 15 European plant species. He found honeybees, called "English flies" by Native Americans; encountered bedbugs, houseflies, cockroaches, mice, and rats; and reported a case where one mosquito species apparently replaced another. Unfortunately, most of his original journals were destroyed, unpublished.

GEORGE LOUIS LE CLERC, COMTE DE BUFFON'S extensive *Histoire Naturelle* attempted to inventory the Earth's diversity. He mentioned introductions of rats, goats, sheep, mice, ferrets, and dogs. Buffon was one of the first respected intellectuals to realize species found in the New and Old Worlds were often similar but rarely really identical. This idea helped set the stage for later concerns about keeping species from different regions separated, restricting them to places where they belonged.

KARL WILLDENOW was named Director of the Berlin Botanical Garden in 1801. His ideas conflicted with Buffon's, as he expected that plants in similar situations on every continent would be identical but that a species found in multiple latitudes must have been redistributed. His best-remembered student was Alexander von Humboldt, who conducted the first large-scale, quantitative studies of plant geography, describing the relationships of vegetation to factors such as climate, soil, latitude, and elevation. He noted that wind, currents, birds, and humans all aided the migration of plants, including the spread of American cacti in the Mediterranean region.

AUGUSTIN PYRAMUS DE CANDOLLE made foundational contributions to plant geography in two major works in the early nineteenth century. He identified at least 21 species of foreign plants around Montpellier, France. Candolle, nominally a Biblical creationist, nevertheless believed vegetation expressed local conditions. His authority bolstered the idea that plants belong to places.

CHARLES LYELL'S *Principles of Geology* (vol. 2, 1832) addressed many biogeographical issues, commented on Candolle's work, discussed modes of plant dispersal, and included humans among the important (and unwitting) dispersal agents.

REV. JOHN HENSLOW is usually recalled for putting Darwin aboard the H.M.S. *Beagle.* In 1835, he called for a category of plants of uncertain origin. This led to efforts to standardize ideas about degrees of geographical belonging based on time since introduction and establishment of self-sustaining populations.

HEWETT COTTRELL WATSON apprenticed as a lawyer and studied medicine but practiced neither. Watson suspected that competition to find new plants, or known plants in new locations, led unscrupulous botanists to "discover" populations by establishing them. In response (1847, 1868) he offered the categories *native, alien, denizen, colonist,* and *casual,* borrowing the first three from English common law and the others from general use. Watson distinguished between *geographical botany,* which focused on the distribution of taxa, and *botanical geography,* which involved inventorying taxa of bounded areas.

ALPHONSE DE CANDOLLE succeeded his father at the Geneva Botanical Garden. His 1855 alternative to Watson's categories was a hierarchical system first dividing plants into *cultivated* or *spontaneous* types, and then splitting cultivated plants into crops and weeds, and spontaneous plants into five types: *adventitious,* or temporary; *naturalized,* or established but introduced; *probably foreign,* with the more doubtful being labeled *perhaps foreign;* and finally, *indigenes, aboriginals,* or *natives.* In 1885 Candolle postulated that naturalized plants spread rapidly but that indigenes declined more gradually.

GEORGE BENTHAM (1856) reviewed de Candolle's 1855 book in English. He believed phytogeography informed debates about the origin of life but felt the field was too speculative. He commented on the effects of farming and accidental transportation, and the rarity of redistribution without human agency. He redrew the categories, dividing plants first among districts, and then into cultivated and wild species, further subdividing cultivated plants into crops and weeds, and wild plants by origin but not intention.

PHILIP SCLATER (1858) declared: "An important problem in Natural History . . . too little agitated, is that of ascertaining the most natural primary divisions of the earth's surface, taking the amount of similarity or dissimilarity of organized life solely as our guide." Botanists had been addressing that problem for 50 years, but ornithologist Sclater was typically unaware of outside work. He proposed six ornithological realms, thus underpinning much later zoogeography. Sclater attributed changes in bird distribution to "some external cause, generally referable to the agency of man."

Darwin and the Years 1859–1898

CHARLES DARWIN'S *Beagle* voyage journal (1837) included perhaps the earliest application of *invasion* to a stand of European thistles and cardoon (wild artichokes) in Argentina. Raised in a militaristic age, and visiting a continent in conflict under military escort, perhaps Darwin was bound to imagine prickly European weeds as armies. During the *Pax Britannica*, Darwin reframed the scene for the *Origin of Species* (1859), postulating that circumstances favored the introduced species. To a Victorian Briton abroad, "native" connoted inferiority.

SIDNEY I. SMITH (1867) was a comparative anatomist of arthropods who speculated about rapid diffusion and the problem of recognizing baseline conditions. Smith believed humans disturbed otherwise harmonious forces of natural distribution. He highlighted the speed with which a European butterfly had become naturalized around Quebec City, wondering whether others had previously done likewise, unnoticed.

JOSEPH DALTON HOOKER (1867, 1884) explicitly deferred to Watson but cautioned that his categories were vague. Hooker, too, employed imprecise terminology, using "cultivated," "semi-naturalised," "naturalized," and "quasi-indigenous" to suggest a continuum. He traveled widely and was dismayed by the replacement of remote island floras with European plants because it destroyed scientifically valuable information, but his tone anticipated later conservationists.

ALFRED RUSSELL WALLACE (1876, 1881, 1889) revised Sclater's regions, making them "Wallace's Realms." In *Island Life*, he considered cases of small vertebrates recently redistributed among different island groups, discussed the advent of terrestrial mammals in New Zealand, and (echoing Hooker) deplored replacement of St. Helena's biota. None of this inspired him to predict similar events elsewhere. In *Darwinism*, Wallace used introductions to elucidate natural selection, but he neither endorsed nor indicted redistribution.

HARRY WALLIS KEW contributed *The Dispersal of Shells* (prefaced by Wallace) to the respected "International Scientific Series" (1893). John Carrington reviewed the book for *Science-Gossip* magazine, saying it showed "the influence of the civilisation of mankind in unintentionally contributing to the equalization of the fauna of various regions, if climatically suitable. This agency must necessarily increase in force now that travel and transit has become so rapid and certain with the aid of steam."

H. G. WELLS, who coauthored a major biology textbook, speculated on plant introductions in his 1898 novel *The War of the Worlds*, basing the progress of the "red weed" that accompanied his Martians on the advent of *Elodea canadensis* in British waterways. Meanwhile, Leland O. Howard (1898) and T. S. Palmer (1897) were gathering cautionary tales of introduced insect pests for the U.S. Department of Agriculture.

1900–1924

STEPHEN T. DUNN attempted to comprehensively inventory the *Alien Flora of Britain* in 1905 by collecting accounts from then-active botanists. He briefly reviewed the terms used by Watson and Alphonse de Candolle, pausing to note that plants could not be "native" to roadsides or hedgerows. Alfred J. Ewart followed suit in Australia with his *Weeds, Poisoned Plants and Naturalized Aliens of Victoria* (1909) but never explicitly defined his "alien" category.

ALBERT THELLUNG worked within the Zurich-Montpellier school of botany. After collaborating with established botanists, he developed a monograph (1912) on introduced species around Montpellier, a textile-processing center where fleece-washing liberated seeds. Thellung showed how riparian "wool gardens" reflected recent imports. He sought to replace adopted terms like *native* and *alien* with Greek and Latin neologisms, hoping that categories like "anthropochores" (benefitting from human activities), "neophytes" (introduced, persisting independently), and "epœkophytes" (weeds of plowed or cleared ground) would put the study of introductions on better footing. He died young without leaving students to promote his system, but his terminology persists in some modern European accounts. Thellung assisted Ida Hayward and George Druce with their Scottish wool garden study (1916). Morten Porsild used Thellung's categories in a 1932 booklet on introduced plants in Greenland.

ALDO LEOPOLD produced the first of his several introduction-related papers and essays on "Mixing Trout in Western Waters" (1918) for an American Fisheries Society meeting. He followed with "Birds and Cats" (1918), a scathing, "op-ed" indictment of housecats and their owners for the Albuquerque, New Mexico, newspaper. He also wrote two pieces disparaging introduced game birds and their proponents: "Chukaremia" (1938) and an unpublished fragment "The Ring-Necked Parvenu" (1944). Leopold's "Cheat Takes Over" (1941) spares little sympathy for introduced grasses, but a 1943 book review ("What is a Weed") seems to take exception to reflexive anti-alien sentiment. All are polemical and make little pretense at scientific study.

JAMES RITCHIE wrote *The Influence of Man on Animal Life in Scotland* (1920) using Scottish issues as points of departure for discussing worldwide phenomena. A quarter of the book dealt with introductions: "The Deliberate Introduction of New Animals"; "Effects Upon the Spread of Animals of Canals, of Roads and Bridges, and of Railways"; "Camp-Followers of Commerce, or Animals Introduced Unawares" including "hangers-on" (parasites), "stowaways on ships," "skulkers in dry [stored] food materials," "foundlings amongst fruit," "creatures conveyed by plants and vegetables," and "timber transportees." Ritchie summarized a 1916 California report listing nearly 50 species of insects discovered by quarantine inspectors. His 1931 *Birds and Beasts as Farm Pests* discussed rats, muskrats, and gray squirrels. In 1932 he delivered a six lecture BBC radio series, *The Changing Face of Nature*. Episode four, "The War upon Wild Animals," called the advent of domestic grazing in Britain a "terrific invasion of aliens," terminology Charles Elton echoed a year later. Episode six, "Gains and Losses in Animal Life," examined introductions (to Britain) of little owls from Holland and gray squirrels from North America, the Australian rabbit debacle, and English sparrows and European starlings in North America, and warned about American muskrats in continental Europe and Britain. Ritchie was wary of species introductions well before Elton but appreciated and deplored natives and aliens alike as cases warranted.

GEORGE M. THOMSON'S *The Naturalisation of Animals and Plants in New Zealand* (1922) focused on fauna. He concluded that New Zealand's history of introductions was uniquely well documented. Many were undertaken by "acclimatisation societies," that maintained records. Thomson wrote some 1,200 species accounts, closing with chapters including "Interaction of Endemic and Introduced Faunas," "Alteration in Flora since European Occupation of New Zealand," and "Inter-relation of Native and Introduced Flora." His ecological speculations anticipated modern discussions of multispecies interactions. Thomson believed that poor colonial land management facilitated introductions. He rejected the idea that introduced plants, mostly weeds likely to survive mainly in modified habitats, would inevitably displace indigenous types. He argued that the disappearance of pesky native insects likely signaled the demise of other, less noticeable species.

Elton and the Years 1925–1957

CHARLES ELTON'S earliest paper about introductions was "The Dispersal of Insects to Spitsbergen" (1925), which attributed the advent of blowflies on an Arctic island to Roman Catholic Europeans' Lenten appetite for herring. Elton's several books (published between 1927 and 1967) addressed the topic, as did his 1932 and 1957 BBC radio talks and various book reviews. Elton's 1943 "The Changing Realms of Animal Life," published in a wartime journal for Polish expatriate scholars, provided a partial outline for his 1957 talks and his 1958 book, *The Ecology of Invasions*, both of which were written for general audiences. Elton's only substantial published research on introduced species concerned rodent control, but staff and students of his Oxford Bureau of Animal Population worked on several introduced species. Notable among these was Dennis Middleton's 1931 study of gray squirrels.

DOUGLAS H. CAMPBELL was a respected student of mosses, but his 1926 *An Outline of Plant Geography* was poorly received. He was an early proponent of an anti-modern outlook, idealizing precolonial America. He deplored introductions and dismissed imported crops as little better than weeds, but he was also concerned about the advent of new agricultural pests.

HENRY N. RIDLEY directed the Singapore Botanical Gardens. His 1930 *Dispersal of Plants throughout the World* was turgid but informative, including a chapter on "Dispersal by Human Agency." Ridley believed a complete accounting of plant introductions was impossible. He noted that humans were causing rare plants to become common, and common ones rare.

EVGENIĬ VLADIMIRICH WULFF curated the Herbarium of Cultivated Plants at the Soviet Institute of Plant Industry. He translated Humboldt's works into Russian, believing one must know the history of botany in order to understand it as science. Wulff recognized that transportation changes were accelerating the rate of plant redistribution. His *Introduction to Historical Plant Geography* (1933; translated in 1943) mentioned many botanists touched on here; he distinguished between intentional and accidental introductions but also showed weeds and crops undergoing parallel artificial selection.

FRANK EGLER studied introduced plants in lowland Hawaii (1942) and later at his Connecticut research farm (1961). He was puzzled by some botanists' antipathy to introduced species, concluding (much as Thomson had in New Zealand) that poor land management practices rendered conditions sometimes more favorable to imports than to indigenes.

RUDOLF BIGALKE, Director of the South African National Zoological Gardens, took Europeans to task in 1943 for their "adulteration" of African floras and faunas and their "strong tendency to introduce wild animals" from homelands to colonies. Referencing the preservation mission of national parks, he decried introductions of foreign plants or animals to such areas and quoted a 1921 resolution of the American Association for the Advancement of Science to that effect. Bigalke also quoted George Thomson and a 1944 British Ecological Society statement authored by Elton about immoderate interest in "new and bizarre" wildlife.

KAZIMIERZ WODZICKI focused *Introduced Mammals of New Zealand* (1950) almost as narrowly as its title. Wodzicki demonstrated an awareness of previous work reminiscent of E. V. Wulff's. He cited Elton's *Animal Ecology* and *Voles, Mice and Lemmings*; both books by Ritchie; several by Thomson; and Aldo Leopold's *Game Management*. In review, Elton wrote that Wodzicki's book "gives an extremely clear and interesting account of their present status, and is informed by an ecological outlook," and reproduced three of its figures for *The Ecology of Invasions*.

Wodzicki favored demonstrably beneficial species, but his book was a government document; politics may have affected its content.

MARSTON BATES prepared "Man as an Agent in the Spread of Organisms" for the 1955 symposium Man's Role in Changing the Face of the Earth after Frank Egler withdrew. Working from scratch, entomologist and "human ecologist" Bates reviewed 57 introduced plants, animals, and microorganisms, noting that such cases produced "experiment-like conditions" but "the possibilities for study [have not been] fully realized." Bates contacted Elton while preparing the paper but received little information. In his own ensuing book, *The Ecology of Invasions*, Elton mentioned "Man as an Agent" only as a source of citations, but Bates may have inspired Elton to revisit the topic.

ANDREW H. CLARK examined immigration of all kinds to New Zealand (1949). The Canadian geographer later participated in the same Man's Role symposium as Marston Bates with his paper "The Impact of Exotic Invasion on the Remaining New World Mid-latitude Grasslands."

CARL LINDROTH examined *The Faunal Connections between Europe and North America* in 1957, concentrating on entomological evidence but including some plants in his chapter "The Human Transport of Animals across the Northern Atlantic." Lindroth's book reported more basic research than most of those coming before but was not concerned with conservation issues.

TEXTBOOKS

Between 1895 and 1954, introduced species were discussed in various textbooks. Eugenius Warming's 1895 *Plantesamfund* included cases from Argentina and New Zealand. Coulter, Barnes, and Cowles's 1911 *Textbook of Botany* briefly mentioned "the influence of man upon vegetation." Victor Shelford's 1913 *Animal Communities in Temperate America* offered scattered references to introduced biota but ignored already notorious house sparrows and starlings. Richard Hesse's 1924 *Tiergeographie auf œkologischer Grundlage* devoted two pages to introductions and a paragraph to quarantines. Charles Elton's 1927 *Animal Ecology* mentioned several cases of animal and plant introductions. Allee, Emerson, Park, Park, and Schmidt's 1949 *Principles of Animal Ecology* focused on "preadaptation," "biotic barriers," human-modified habitats, island biotas, and competition between congeners. Eugene Odum's 1953 *Fundamentals of Ecology* concluded that managing natives was preferable to introducing alien substitutes.

Andrewartha and Birch's 1954 *The Distribution and Abundance of Animals* used examples of introductions mainly to make or support more general contentions. Henry J. Oosting's 1956 *The Study of Plant Communities* (2nd ed.) employed anecdotes about introduced plants and animals as cautionary tales to discourage tampering with "the balance among the species of a community."

CONCLUSION

Introduced species have attracted interest since the beginnings of modern science. Empire builders exploited the fact that some plants and animals could be moved from place to place and established anew. Pioneering biogeographers and taxonomists had to reconcile new learning with biblical accounts or show that their empiricism superseded such traditions. The fact that some plants and animals could survive and thrive in new lands challenged notions of an orderly creation; that they might replace indigenous island biota added fuel to that fire.

Botanical anekeitaxonomies began appearing in the second quarter of the nineteenth century. The concept of permanently attaching taxa to places survived the Darwinian revolution and remains contentious as twenty-first century ecologists address notions of wildness, balance, propriety, nostalgia, and fears of irreversible change.

SEE ALSO THE FOLLOWING ARTICLES

Acclimatization Societies / Agriculture / Ballast / Darwin, Charles / Elton, Charles S. / Mediterranean Sea: Invasions / Xenophobia

FURTHER READING

Bates, M. 1956. Man as an agent in the spread of organisms. In W. L. Thomas, C. O. Sauer, M. Bates, and L. Mumford, eds. *Man's Role in Changing the Face of the Earth.* Chicago: University of Chicago Press.

Chew, M. K. 2006. Ending with Elton: Preludes to Invasion Biology. Doctoral Dissertation. Arizona State University, Tempe, AZ.

Clark, A. H. 1949. *The Invasion of New Zealand by People, Plants and Animals: The South Island.* Westport, CT: Greenwood Press.

Davis, M. A. 2009. *Invasion Biology.* Oxford: Oxford University Press.

INVASION ECONOMICS

CHARLES PERRINGS

Arizona State University, Tempe

Invasive species are defined in the Convention of Biological Diversity as ". . . those alien species which threaten ecosystems, habitats or species." They are introduced species that establish and spread, and that cause harm. Historically, the most damaging introduced species have been human pathogens. Introduced pathogens have been the source of a long line of epidemics from the bubonic plague in the fourteenth century to HIV AIDS today. But the introduction of alien species into new environments has caused widespread disruption to agriculture, forestry, fisheries, and conservation landscapes, all of which matters to the people who are affected. The economics of invasive species is ultimately concerned with the harm that such species do. It deals with the allocation of resources either to reduce the risks of future introductions or to control the damage inflicted by harmful species that have already been introduced. It answers three main questions: What is the harm done by invasive species? What should be done to regulate the dispersal of potentially invasive species? What should be done to control or contain the damage done by proven invaders?

THE HARM DONE BY INVASIVE SPECIES

The majority of introduced species cause little or no harm, either because they fail to establish or because they have few effects on the diversity and functioning of native communities, and thus few implications for the production of ecosystem services (the things that people care about). However, a small proportion of introduced species does have a significant impact. These species are covered by international agreements to limit damage—such as the International Health Regulations (IHR), the Convention on Biological Diversity (CBD), the International Plant Protection Convention (IPPC), and the Sanitary and Phytosanitary (SPS) Agreement of the General Agreement on Tariffs and Trade (GATT). The damage costs of invasive species include both direct and indirect effects. Not surprisingly, human health impacts are perceived to be the most important of the direct effects. We do not have good estimates of the cost of most introduced diseases, but rough estimates do exist of the cost of particular outbreaks—SARS, for example, is estimated to have cost in order of $50 billion. However, there is no good way to test whether the damage avoided by most disease control programs outweighs their cost.

Aside from human health impacts, direct effects are those associated with predation by introduced pests on valued local species, whether native or exotic. Invasive pests are of primary concern in agriculture, forestry, and fisheries, wherein direct effects of this kind are most obvious and rather easier to calculate. For this reason, invasive species are thought to have greater impacts on livelihoods in parts of the world where people depend most

TABLE 1
Estimated Annual Cost of Invasive Species (USD billion)

	U.S.	UK	Australia	South Africa	India	Brazil	Total
Plants	0.148			0.095			0.243
Mammals							0.000
Rats	19.000	4.100	1.200	2.700	25.000	4.400	56.400
Other	18.106	1.200	4.655				23.961
Birds	1.100	0.270					1.370
Reptiles/Amphibians	0.006						0.006
Fishes	1.000						1.000
Arthropods	2.137		0.228				2.365
Molluscs	1.305						1.305
Diseases							0.000
Livestock	9.000		0.249	0.100			9.349
Human	6.500	1.000	0.534	0.118		2.333	10.485
Total	58.299	6.570	6.866	3.013	25.000	6.733	106.484

SOURCE: Pimentel, D., S. McNair, S. Janecka, J. Wightman, C. Simmonds, C. O'Connell, E. Wong, L. Russel, J. Zern, T. Aquino, and T. Tsomondo. 2001. Economic and environmental threats of alien plant, animal and microbe invasions. *Agriculture, Ecosystems and Environment* 84: 1–20. These estimates represent a subset of the costs of invasions only, and are based on readily available data. The data for countries outside the United States cover fewer categories and are likely to be less accurate. The total costs are indicative only, and should be thought of as a lower bound.

heavily on these activities. One set of damage estimates by David Pimentel and colleagues concluded that in the first years of this century, invasive species caused annual damage costs equal to about half the value of agricultural gross domestic product (GDP) in the United States and Australia, but greater than agricultural GDP in South Africa and Brazil (Table 1). In light of the importance of agriculture in those countries, this indicates a disproportionate impact.

The indirect effects of invasive species are of greatest concern to ecologists. Introduced pathogens, predators, or competitors have, for example, been implicated in the loss of native species over a wide range of ecosystems. This has affected the capacity of ecosystems to deliver the services that underpin the production of the things that people care about, and to absorb anthropogenic and environmental stresses and shocks without losing resilience. Maintenance of functional diversity, in particular, has an insurance value—in the sense that it supports the provision of ecosystem services over a range of environmental conditions. There are no reasonable estimates of the damage done by the many invasive species whose effects involve subtle changes in ecological functioning, ecosystem processes, or ecosystem resilience. Unless these effects translate into the loss of marketed products, the damage involved is difficult to calculate. It requires an understanding of the many ways in which functioning ecosystems

support well-being, and while this is well known for some managed ecosystems, for others it is not.

The current lack of information on the expected damage cost of many invasive species means that it is hard to assess whether current control measures are too much or too little. The estimates of Pimentel and colleagues suggested that invasive species might be imposing annual damage costs of around 5 percent of global GDP. If so, there are currently insufficient resources committed to both the mitigation of future invasive species risks and costs of current control.

REGULATING THE DISPERSAL OF POTENTIALLY INVASIVE SPECIES

The risks associated with biological invasions depend both on the probability that they will occur and on the damage that will be caused if they do. Thus, mitigation of those risks means reducing either the likelihood or the cost of future invasions. The likelihood of future invasions is growing with the widening and deepening of international trade. This is because every container of goods moved from one country to another includes a sample of the species in the exporting country. There is clear evidence for the role of trade in the spread of a number of human diseases such as HIV AIDS, West Nile virus, SARS, avian flu (H_5N_1), swine flu (H_1N_1), and a series of livestock and plant diseases that includes sudden oak death (*Phytophthora ramorum*) and foot and mouth disease (*Aphtae epizooticae*). The same is true for many insects, plants, and animals. The opening of new markets or trade routes is linked to the introduction of new species, and the growth in trade volumes increases the frequency with which the species from the exporting country are introduced (propagule pressure)—and, with it, the probability that they will establish and spread. Indeed, the volume and direction of trade, along with environmental similarities between trading partners, are good empirical predictors of which introduced species are likely to become invasive. A good example is western flower thrips (*Frankliniella occidentalis*), a highly polyphagous insect that damages horticultural and ornamental crops. Until the 1960s, it was restricted to the western part of North America, but since then has been spread through trade in plants to Europe, North and East Africa, Australia, New Zealand, Japan, South Korea, Malaysia, and Central and South America. In other cases, the problem has lain with the mode of transport rather than the goods traded. Introduction of most of the nonindigenous species discovered in ports—as with the poster children of invasive aquatic species, the zebra mussel (*Dreissena polymorpha*) and the Asian clam (*Corbicula fluminea*)—depended on ballast water exchange in ships.

Invasive species and the damage they cause are said to be an externality of trade. An externality in this context is an effect of some market transaction that is not taken into account by those engaged in the transaction. If the importers or exporters of commodities that harbor invasive species are unaware either of the existence of those species or of the damage they might do, for example, they will not take that damage into account. The same tends to be true if they are aware of the potential for harm but are not legally liable for any damage. The economic problem in this case is to internalize the externality—i.e., to make sure that the transactions between people take all relevant effects into account. One difficulty at the international level is that there are no markets for many sanitary and phytosanitary services, and thus no means of compensating countries for costs incurred in protecting the rest of the world from potentially invasive pests and pathogens (Fig. 1). The recent development of systems of payments for ecosystem services (PES) is designed to address the problem, but while PES schemes exist for some global services, such as carbon sequestration or habitat protection, they do not yet exist for SPS measures. The SPS Agreement authorizes countries to undertake defensive measures. This is largely limited to screening and exclusion programs, such as port inspections, interceptions, quarantine, and similar measures, but it allows individual countries to pay trading partners for enhanced SPS measures.

A second issue is that the control of invasive species is a public good—that it confers benefits to all, and

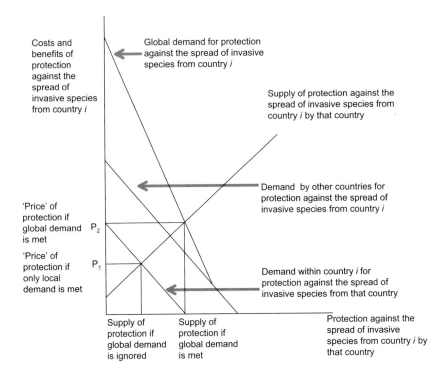

FIGURE 1 This figure describes demand for protection against invasive species in a particular country *i* by the citizens of that country and by the rest of the world. If exports from country *i* threaten other countries, the global community may be expected to wish higher levels of protection than the citizens of that country. It follows that in the absence of markets for global protection against invasive pests or pathogens, country *i* will undertake only enough protection to satisfy local demand and will equate the marginal local cost of protection with the marginal local benefits it offers. However, if the global community can express its willingness to pay for protection through direct payments, such as payments for ecosystem services, the level of protection can be increased to the point that the marginal local cost of protection is equal to the marginal global benefits. This is the socially optimal outcome.

that none can be excluded from those benefits. More importantly, it is a "weakest link" public good, in that the benefits to all are limited to the benefits offered by the least effective provider. Consider an infectious disease whose spread is determined by public health efforts at the national level. Since any country where the disease is not contained is a potential source of infection to others, the benefits to all cannot be better than the benefits offered by the least effective provider. International problems of this kind are best addressed by collective action to strengthen the weakest link. In contrast to the International Health Regulations, however, the SPS Agreement precludes collective action to address global risk in this way.

CONTROLLING AND CONTAINING THE DAMAGE DONE BY PROVEN INVADERS

At the national level, there is reason to believe that prevention is generally better than cure—that the costs of mitigating risk through port and pathway interventions are lower than the costs of eradication or control of established pests. Nevertheless, given the growing risks associated with globalization and trade growth and the limited set of options open under the GATT, countries increasingly have to identify options for the control of established alien species. Prevention (primarily screening to prevent introductions) and the control of established invasive species (eradication or containment) are substitute strategies. Reducing the cost or raising the effectiveness of one will increase its use relative to the other. It follows that the weakness of most defensive efforts to screen imports makes the eradication or containment of established invaders the default strategy.

In every case, efficiency requires balance between the marginal costs and benefits of the strategy. A well-known invasive species control program in South Africa, the "Working for Water" program, bases control of invasive *Pinus, Hakea,* and *Acacia* species on their impact on the water available both to other species in the fynbos system and to human users. Such measures of benefit provide a benchmark against which to assess the costs of the program. In the absence of information about the damage avoided through adoption of some control strategies, there is no way to judge whether the strategy is warranted. Indeed, this remains a general problem in invasive species control.

At present, the commitment of resources to the problem of invasive species appears to be increasing in most countries. During the 1990s, for example, annual spending

on emergency eradication programs to avoid the costs of introduced pathogens in the United States increased from $10.4 million to $232 million. More recently, annual expenditure on invasive species in the United States has exceeded $1 billion. While this implies that expectations about the potential damage cost of invasive species are increasing in the United States, not all countries are raising expenditures on invasive species in the same way, and even within the United States, the signals are mixed. At least one major conservation organization, the Nature Conservancy, has largely abandoned its invasive species program. In India, the emphasis on many invasive species is switching from control to exploitation in an attempt to reduce the net cost of ineradicable species such as *Lantana camara*.

INTERNATIONAL COOPERATION

While individual countries have a responsibility to address the costs of invasive species that are currently being ignored in trading decisions, the solution to the invasive species problem requires global information flows and global cooperation. As an externality of international trade, the invasive species problem requires actions that force trading partners to take account of the risks they impose on the rest of the world. While the IHR is currently one of the strongest international cooperative agreements in existence, the other instruments governing invasive species—the SPS Agreement, the IPPC, and the CBD—are weak in this regard. Since renegotiation of such agreements is notoriously difficult, it may be expected that national defense against harmful invaders will be the dominant approach for some time to come.

SEE ALSO THE FOLLOWING ARTICLES

Agreements, International / Pathogens, Human / Risk Assessment and Prioritization

FURTHER READING

Leung, B., D. M. Lodge, D. Finnoff, J. F. Shogren, M. A. Lewis, and G. Lamberti. 2002. An ounce of prevention or a pound of cure? Bioeconomic risk analysis of invasive species. Proceedings of the Royal Society of London. *Biological Sciences* 269(1508): 2407–2413.

Levine, J. M., and C. M. D'Antonio. 2003. Forecasting biological invasions with increasing international trade. *Conservation Biology* 17: 322–326.

Millennium Ecosystem Assessment. 2005. *Ecosystems and Human Well-Being: Synthesis*. Washington, DC: Island Press.

Perrings, C., H. Mooney, and M. Williamson, eds. 2010. *Bioinvasions and Globalization: Ecology, Economics, Management and Policy.* Oxford: Oxford University Press.

Perrings, C., M. Williamson, and S. Dalmazzone, eds. 2000. *The Economics of Biological Invasions*. Cheltenham, UK: Edward Elgar.

INVASIVENESS

MARCEL REJMÁNEK

University of California, Davis

The degree to which a species is able to reproduce, spread from its place of introduction, and establish in new locations is called invasiveness. One of the basic questions of invasion biology is whether some species are more invasive than others, and if so, which of their biological attributes are responsible for that difference. The current consensus is that species are not equal in their invasiveness; however, different biological attributes may be important in different groups of organisms and in different environments. Many other factors, namely propagule pressure (introduction effort) and residence time, often mask differences in invasiveness that are due to biological attributes. Over the last two decades, substantial progress has been made in our understanding of attributes responsible for invasiveness of seed plants, molluscs, fishes, birds, and some other groups of organisms. The ultimate goal is to use biological attributes responsible for, or at least correlated with, invasiveness for screening of nonnative species and predictions of potential invasions.

THE NATURALIZATION-INVASION CONTINUUM

Several prerequisites and stages of biological invasions are usually recognized: (1) selection of species and genotypes → (2) transport → (3) introduction → (4) establishment (consistent reproduction) = naturalization → (5) spread (invasion *sensu stricto*) → (6) environmental or economic impact. The first three steps entail intentional or unintentional human assistance. The remaining steps are spontaneous but may still be assisted by human activities. The first three steps determine the species pool of potential invaders. Species that are invasive may be introduced due to different selection processes operating during these stages. These steps are necessary conditions for any invasion and are not the focus of this article. Step 6 (whether a species becomes a pest or not) is also only marginally covered here. This entry focuses on step 5 (spread), which implicitly includes step 4 (reproduction). It is conceptually useful to distinguish between steps 4 and 5, but they

are tightly interconnected. For a species to be invasive, it has to reproduce (establish), successfully disperse (spread), and reproduce (establish) again in new locations, and so on. After this clarification, one may ask what differs between invasive and non- or less-invasive species.

EXTRAPOLATIONS BASED ON PREVIOUSLY DOCUMENTED INVASIONS

Extrapolations based on previously documented invasions are fundamental for predictions in invasion ecology. With the development of relevant databases, this approach should lead to immediate rejection of imports of many taxa known to be invasive in similar habitats elsewhere (prevention) and to prioritized control of those that are already established. Such transregional, taxon-specific extrapolations are very useful in many situations, but our lack of mechanistic understanding makes them intellectually unsatisfying. Understanding how and why certain biological characters promote invasiveness is extremely important, since even an ideal whole-Earth database will not cover all (or even most) potentially invasive taxa. In New Zealand, for example, Peter Williams and his colleagues reported that 20 percent of alien weedy species collected for the first time in the second half of the twentieth century had never been reported as invasive outside New Zealand.

Basic taxonomic units used in plant invasion ecology are usually species or subspecific taxa. However, genera are certainly worth considering. Plant species belonging to genera notoriously known for their invasiveness or "weediness" (e.g., *Amaranthus, Cuscuta, Echinochloa, Ehrharta, Myriophyllum*) should be treated as highly suspicious. However, a continuum from invasive to noninvasive species is also common in many genera (*Acacia, Acer, Centaurea, Pinus*). Rigorous testing should be used to determine which pattern is more typical. Recently, attention has been paid to taxonomic patterns of invasive plants. In terms of relative numbers of invasive species, some plant families seem consistently overrepresented: Amaranthaceae, Brassicaceae, Chenopodiaceae, Fabaceae, Gramineae, Hydrocharitaceae, Papaveraceae, Pinaceae, and Polygonaceae. Among large families, the only conclusively underrepresented one is Orchidaceae. Crustacea (crustaceans), Bivalvia (bivalve molluscs), and Gastropoda (gastropods or univalve molluscs) are overrepresented among invasive freshwater macroinvertebrates in Europe and North America. It is an imperative that, among animals, some functional groups such as carnivorous mammals, amphibians, and reptiles should be prevented from introductions not only because of their invasiveness, but also because of damaging consequences to native faunas.

ATTRIBUTES OF INVASIVE SPECIES

Assuming abiotic environment compatibility, five biological attributes are, to different degrees, responsible for invasiveness of all kinds of organisms: population fitness homeostasis, population fitness, minimum generation time, rate of population expansion, and organismal competitiveness or self-suitable modification of the environment (Fig. 1).

The relative importance of these attributes varies depending on the amount of critical resources, on disturbance regimes, and on the spatial heterogeneity of the environment. Their components are not necessarily compatible and may be important under different circumstances. For example, the ability to use available resources quickly is important in disturbed habitats, while the ability to reduce the amount of critical resources (lower R^*) is important when a species is invading successionally advanced communities. Also, short minimum generation time (positively influencing fitness) is usually associated with short longevity (negatively influencing fitness).

Population Fitness Homeostasis

Population fitness homeostasis (PFH) means consistent fitness on a population level over a broad range of environments. PFH is determined by individual fitness homeostasis and genetic polymorphism. Individual fitness homeostasis (IFH, Herbert Baker's *general purpose genotype*) is the ability of an individual to maintain consistent fitness across a range of conditions through phenotypic plasticity. Phenotypic plasticity is responsible for both IFH and PFH of many plant invaders with little or no genetic diversity (e.g., *Alternathera philoxeroides* in Asia, *Arundo donax* and *Hieracium aurantiacum* in North America, *Clidemia hirta* and *Pennisetum setaceum* in Hawaii). Additionally, rapid evolution of plasticity can play an important role in species invasiveness. On the other hand, there is also abundant evidence for local adaptations through selection acting on population genetic diversity of introduced plant species (e.g., *Escholzia californica* in Chile, *Hypericum perforatum* in North America, *Phyla canescens* in Australia) and animal species (*Potamopyrgus antipodarum*, a freshwater snail). One important source of genetic diversity within invasive species is their repeated introduction from multiple sources. Multiple introductions often transform among-population variation in native ranges to within-population variation in introduced areas. One of the best examples is *Anolis sagrei*, a Cuban lizard introduced at least eight times from different geographic source populations to Florida. Genetically highly variable Florida populations with high PFH now represent potent sources for introductions and invasions elsewhere.

High PFH of a species translates into its broader ecological niche. Successful invasions of ants and earthworms seem to be supported by their omnivory and dietary flexibility. The most recent analysis by Tim Blackburn and colleagues of traits responsible for successful invasions of exotic birds concluded that the breadth of habitats a species uses has the strongest effect of all variables considered. Bird species with large brains appear to be better invaders because of their higher cognitive skills. For all kinds of organisms, it is reasonable to expect that wide native habitat range of a species is a good indicator of its high PFH, and therefore of its high invasiveness.

Population Fitness

Actual levels of population fitness in particular environments are the key component of all invasions. Unfortunately, fitness quantified as finite rate of population increase (λ) is only rarely properly measured, and comparisons of fitness between invasive and noninvasive species are almost nonexistent. In a thus far completely exceptional study, Jean Burns found that invasive plant species in the family Commelinaceae had significantly

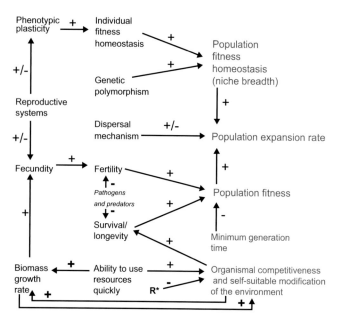

FIGURE 1 Positive (+) and negative (–) causal relationships among biological attributes responsible for species invasiveness. R^* is the level to which the amount of the available form of the limiting resource is reduced by a monoculture of a species once that monoculture has reached equilibrium (i.e., once it has attained its carrying capacity).

larger λ values than noninvasive ones, but only under high nutrient conditions. More often, fitness is just estimated on the basis of its components, fertility or fecundity. For example, in one such study, Robert Mason and his colleagues found significantly higher seed production per plant for herbaceous and woody invasive species when compared with the same groups of native species. Similarly, higher fecundity was recognized as the best predictor of invasive pests among freshwater molluscs (Fig. 2).

Fecundity is usually positively dependent on biomass growth rates. However, there are many exceptions among animals. Positive correlation between individual plant biomass and seed production per plant is one of the most robust generalizations of plant ecology. Therefore, higher values of relative growth rates (RGR) in plants may often indicate higher fitness. The recent meta-analysis of all available studies by Mark van Kleunen revealed that both growth rates and fitness-related attributes are significantly higher for invasive plant species when compared with either noninvasive or native plant species. High relative biomass growth rate was also recognized as one of the most important attributes of successfully established fish species in the Great Lakes of North America (Fig. 3). However, there are tradeoffs between biomass growth rate and survival—another component of fitness. There are both benefits and costs to fast living. For example, because RGR of plants is usually negatively related to

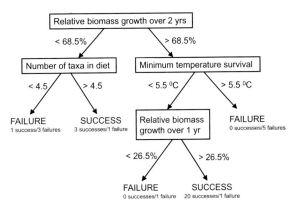

FIGURE 3 Decision tree of successful and failed introduced fishes in the Great Lakes of North America. The numbers of known successful and failed nonnative species categorized into each terminal point are given, illustrating that 3 of 45 species were misclassified. Fast relative biomass growth was the most important attribute of successful invaders at this stage. Among established fishes, however, quickly spreading species had slower relative growth rates. (After Kolar and Lodge, 2002, *Science* 298: 1333–1336, simplified.)

water use efficiency, fast growth is not the best strategy for perennial plant invaders in arid environments. Based on their studies in resource-poor habitats in Hawaii, Jennifer Funk and Peter Vitousek showed that invasive plant species were generally more efficient than native species at using limited light, water, and nitrogen. Michelle Leishman and her colleagues recently questioned the generality of this conclusion.

Last but not least, fecundity depends on reproductive systems. Consistent offspring production in new environments is usually associated with rather simple or flexible breeding systems. For example, rare and endangered taxa in the plant genus *Amsinckia* (e.g., *A. furcata, A. grandiflora*) are heterostylic, while derived invasive taxa (*A. menziesii, A. lycopsoides*) are homostylic and self-compatible. Self-pollination has been consistently identified as a mating strategy in colonizing species. Nevertheless, not all sexually reproducing successful invaders are selfers. John Pannel and Spencer Barrett examined the benefits of reproductive assurance in selfers versus outcrossers in model metapopulations. Not surprisingly, their results suggest that an optimal mating system for a sexually reproducing invader in a heterogeneous landscape should include the ability to modify selfing rates according to local conditions. In early stages of invasions, when populations are small, plants should self to maximize fertility. However, later, when populations are large and pollinators or mates are not limiting, outcrossing will be more beneficial, mainly due to increasing genetic polymorphism.

Vegetative reproduction can compensate more than sufficiently for sexual reproduction in some invasive plant

The Laurentian Great Lakes basin

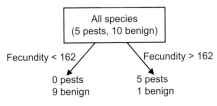

The 48 contiguous states of the US

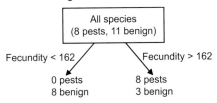

FIGURE 2 Decision trees for determining whether a freshwater mollusc species will be benign or a pest (i.e., cause economic or environmental harm). For the Great Lakes, fecundity was one of eight predictor variables but the only one chosen by the categorical and regression tree analysis (CART). For the U.S. analysis, fecundity was the only predictor used. Fecundity was defined as the annual number of eggs or live offspring released per one female. (After Keller et al., 2007, *Conservation Biology* 21: 191–200.)

species. Water hyacinth (*Eichhornia crassipes*) and infertile hybrid giant salvinia (*Salvinia molesta*) are well-known examples. The ability to allocate energy to different modes of reproduction depending on environmental conditions is one type of phenotypic plasticity and increases IFH and PFH. Apomictic plants (such as dandelions) and parthenogenetic animals (such as the New Zealand mudsnail, *Potamopyrgus antipodarum*) have an advantage, at least initially, as a single individual can establish a population. Consistently, ovovivipary was detected as one of the most important attributes discriminating invasive from native European stream macroinvertebrates.

Minimum Generation Time

Short minimum generation time, also called juvenile period, is an obvious advantage for invasive species. Not surprisingly, substantial proportions of nonnative floras in temperate zones are annual species. Short minimum generation time is usually a prominent attribute used for identification of (potentially) invasive woody species (Fig. 4).

Invasiveness of woody taxa in disturbed landscapes is associated with short juvenile period (<10 years), small seed mass (<50 mg), and short intervals between large seed crops (1 to 4 years). These three attributes contribute,

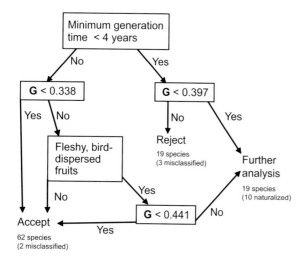

FIGURE 4 Decision tree for prediction of the risk of naturalization of nonnative woody seed plants in Iowa, produced by categorical and regression tree analysis. The numbers of species (out of 100) correctly classified as acceptable, nonacceptable, and requiring further analysis are indicated. G is the geographic-risk value. $G = (\Sigma P_i)/n$, where $P_i = N_{natur}/N$, N_{natur} is the number of species native to geographical area i known to naturalize in Iowa, N is the total number of species native to that area in the examined species set, and n is the number of areas in the species' native range. (Derived from M.P. Widrlechner et al, 2004, *Journal of Environmental Horticulture* 22: 23–31.)

directly or indirectly, to higher values of three parameters critical for population expansion: net reproduction rate, reciprocal of mean age of reproduction, and variance of the marginal dispersal density. For wind-dispersed seeds, the last parameter is negatively related to terminal velocity of seeds, which is positively related to √ (seed mass). Because of the tradeoff between seed number and mean seed mass, small-seeded taxa usually produce more seeds per unit biomass. Invasions of woody species with very small seeds (<3 mg), however, are limited to wet and preferably mineral substrates (Table 1). Based on invasibility experiments with herbaceous species, it seems that somewhat larger seeds (3 to 10 mg) extend species–habitat compatibility. As seed mass seems to be positively correlated with habitat shade, large-seeded aliens may be more successful in undisturbed, successionally more mature plant communities.

Rate of Population Expansion

Fast dispersal of propagules or offspring is another crucial component of invasiveness. Rate of dispersal always depends on two species-specific characteristics: fertility and efficiency of dispersal mechanism. This is also the substance of Fisher-Kolmogorov's classic formulation of population rate of expansion of the population front (how many meters a constant population density can propagate in one dimension in one year) in a homogeneous environment: $2\sqrt{(rD)}$, where r is the intrinsic rate of population increase (fertility minus mortality, i.e., individual/individual/year), and D is the diffusion coefficient (m^2yr^{-1}). The first term is directly related to population fitness: $r = \ln \lambda$.

A trivial but very important difference between plants and animals is that animals, particularly in terrestrial environments, do spread actively, but long-distance plant dispersal is always passive. The most important long-distance dispersal agents are people, other vertebrates (mostly birds), water, and wind. Plants have many different adaptations or preadaptations for dispersal by these vectors. For example, wind-dispersed fruits may have pappi (as in the case of many Asteraceae) or wings (as in the case of many Pinaceae and some Sapindaceae). Fleshy-fruiting plants (e.g., *Chrysanthemoides molinifera*, *Rubus* spp.) and plants producing seeds with large elaiosomes (*Acacia cyclops*) may be dispersed over long distances by birds; plants producing seeds with small elaiosomes (*Cytisus scoparius*) are usually dispersed over short distances by ants. Corky "wings" of *Rumex* and *Sagittaria* fruits aid dispersal by water. Plant species with seeds without any dispersal-promoting appendages are usually less invasive

TABLE 1

		Opportunities for Vertebrate Dispersal	
		Absent	**Present**
$Z < 0$		Noninvasive unless dispersed by water[a]	Possibly invasive[b]
$Z > 0$	Dry fruits and seed mass > 3 mg	Likely invasive[c]	Very likely invasive[d]
	Dry fruits and seed mass < 3 mg	Likely invasive on wet mineral substrates[e]	
	Fleshy fruits	Unlikely invasive[f]	Very likely invasive[g]
Any Z value	Vegetative propagation	Likely invasive, usually short-distance spread[h]	

NOTE: General rules for detection of invasive woody seed plants based on values of the discriminant function Z*, seed mass values, presence or absence of opportunities for vertebrate dispersal, and vegetative propagation. (Modified after Rejmánek et al., 2005.) *$Z = 23.39 - 0.63\sqrt{M} - 3.88\sqrt{J} - 1.09S$, where M = mean seed mass (in milligrams), J = minimum juvenile period (in years), and S = mean interval between large seed crops (in years). This discriminant function (Z) was derived on the basis of differences between invasive and noninvasive pine (*Pinus*) species. Positive Z indicates invasive species; negative Z indicates noninvasive species. The function was later successfully applied to other gymnosperms and, as a component of this table, to woody angiosperms.

[a]Examples of noninvasive species are *Aesculus hippocastanum* (horse chestnut), *Araucaria araucana* (monkey puzzle), *Bertholletia excelsa* (Brazil nut), *Camellia* spp. (camellias), *Fagus* spp. (beech trees), *Pinus lambertiana* (sugar pine), *Swietenia macrophylla* (Honduras mahogany), and *Tilia* spp. (lime trees). However, some large-seeded species may be dispersed by water: *Nypa fruticans* (nipa palm) is spreading along tidal streams in Nigeria and Panama, and *Thevetia peruviana* (yellow oleander) can be dispersed over short distances by surface runoff in Africa.

[b]Examples of invasive species in this group are *Pinus pinea* (Italian stone pine) and *Melia azedarach* (Persian lilac) in South Africa, *Olea europaea* (olive) in Australia, *Quercus rubra* (American red oak) in Europe, *Mangifera indica* (mango) in the Neotropics, and *Persea americana* (avocado) in the Galapagos.

[c]Examples of mainly wind- and ant-dispersed species are *Acer platanoides* (Norway maple), *Clematis vitalba* (old man's beard), *Cryptomeria japonica* (Japanese cedar), *Cryptostegia grandiflora* (rubber vine), *Cytisus scoparius* (Scotch broom), *Pinus radiata* (Monterey pine), *Pseudotsuga menziesii* (Douglas fir), *Robinia pseudoacacia* (black locust), and *Tecoma stans* (yellow trumpetbush).

[d]Species with seeds possessing large arils (e.g., several species of *Acacia*) or with seeds coated with a wax (*Triadica sebifera*, Chinese tallow) are dispersed by birds.

[e]Examples of these are *Alnus glutinosa* (black alder) and *Salix* spp. (willows) in New Zealand, *Eucalyptus camaldulensis* (river red gum) in South Africa, *Melaleuca quinquenervia* (melaleuca) in southern Florida, *Tamarix* spp. (salt cedar) in the southwestern United States, and *Baccharis halimifolia* (eastern baccharis) in Australia.

[f]*Acca sellowiana* (pineapple guava), *Rhaphiolepis indica* (Indian hawthorn), and *Nandina domestica* (heavenly bamboo) are frequently cultivated but noninvasive species in California because very few vertebrates eat their fruits. *Nandina domestica*, however, is dispersed by birds and water in the southeastern United States.

[g]Examples of these are *Berberis* spp. (barberries), *Clidemia hirta* (soapbush), *Crataegus monogyna* (common hawthorn), *Lantana camara* (lantana), *Ligustrum* spp. (privets), *Lonicera* spp. (honeysuckles), *Passiflora* spp. (passion flowers), *Pittosporum undulatum* (sweet pittosporum), *Psidium guajava* (guava), *Rosa multiflora* (multiflora rose), *Rubus* spp. (blackberries), *Schinus terebinthifolius* (Brazilian pepper), and *Solanum* spp. (nightshades).

[h]*Pueraria montana* (kudzu) can spread from rooting stems up to 30 m in one growing season. Viable branches of *Salix* spp. (willows), *Populus* spp. (cottonwoods), and tubers of *Thunbergia grandiflora* (Bengal trumpet) can be dispersed by water in streams and rivers over long distances.

(*Eucalyptus* spp.). However, because increasing volumes of soil are moved around by people (in topsoil, in mud on cars, with horticultural stock), plant species with numerous, dormant, soil-stored seeds are preadapted for this kind of dispersal. Similarly, as Jonathan Jeschke and David Strayer showed, human affiliation of vertebrates (domesticates, pets, human commensals) is one of the strongest determinants of their invasion success.

Seed dispersal by vertebrates is responsible for the success of many invaders in disturbed as well as "undisturbed" habitats. Even some very large-seeded alien species such as mango (*Mangifera indica*) can be dispersed by large mammals. The proportion of naturalized plant species dispersed by vertebrates seems to be particularly high in Australia: over 50 percent. Assessment of whether there is an opportunity for vertebrate dispersal is an important component of the screening procedure for woody plants (Table 1).

Organismal Competitiveness and Self-Suitable Modification of the Environment

Undisturbed (natural and seminatural) plant communities in mesic environments are more likely invaded by tall plant species. The most prominent examples are new, taller life forms (*Pinus* spp. and *Acacia* spp. in South African fynbos, *Cinchona pubescens* in shrub and fern or grassland communities of the Galapagos highlands). Undisturbed plant communities in *semiarid* habitats seem to be invasible, especially by environmentally compatible species that rapidly develop deep root systems (e.g., *Bromus tectorum* or *Centaurea solstitialis*). In short, in undisturbed plant communities, efficient competitors for limiting resources will very likely be successful invaders and the worst environmental weeds. Theoretically, given a set of R_i^* values (R_i^* is a level of resource below which an ith species cannot survive), for a pool of potential invaders, it should be possible to predict the average likely success of each invading species in undisturbed communities. However, if seasonality, senescence, or even very low levels of natural disturbance allow establishment of shade-intolerant taxa that are taller than resident vegetation at maturity, then such taxa can still be highly successful and influential invaders in spite of their high R^* for light.

Ability to use available resources quickly is an attribute of many successful plant invaders in disturbed habitats. Obviously, there is a tradeoff between this kind of strategy and possession of low R^*. Whether some species can quickly use resources and also reduce their levels below those tolerable by resident species remains to be seen. Such species would be the most successful invaders. Something similar, however, has been concluded about successful ant invaders. Most successful exotic ant species are good at quickly discovering food resources, but, at the same time, they are able to displace native competitors.

Recently, there has been a renewal of interest in the role of allelopathy in plant invasions. It seems that some chemical substances released from living or decaying biomass of nonnative species can inhibit growth of native plants or soil microorganisms. This can increase the invasiveness of such species. However, with the exception of some consistent effects (e.g., juglone released by walnuts, *Juglans* spp.), results are highly inconsistent, depending on climate and soil properties. Allelopathic substances are potentially more influential in soils with low organic content and in habitats with low precipitation.

In general, reduction of the amount of critical resources below the level needed by resident species or release of chemicals inhibiting growth of residents by nonnative species are examples of "niche constructions" accelerating plant invasions, particularly in undisturbed environments. Some invasive grasses (e.g., *Andropogon gayanus, Bromus tectorum, Hyparrhenia rufa*) can initiate and maintain a positive grass-fire feedback and can transform whole ecosystems to their benefit. Among animals, introduced beavers (*Castor canadensis*) and social insect species are the most visible self-suitable niche constructors.

Long-term population invasiveness, however, does not depend only on organismal anatomical, physiological, or behavioral properties treated above, but on relationships between population fitness values of invaders and residents and the degree of niche overlap between invaders and residents (Fig. 5). As Peter Chesson and more recently Andrew MacDougall and colleagues showed, there are essentially three possible invasion outcomes for all possible combinations of niche and fitness differences: (1) when fitness of residents is greater than fitness of the

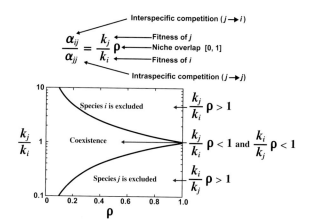

FIGURE 5 According to Robert MacArthur's consumer-resource model, the ratio of interspecific (between species, α_{ij}) and intraspecific (within species, α_{jj}) competitive effects can be expressed in terms of population fitness (k_j and k_i) of species j and i and resource-use (niche) overlap (ρ) between those species. Species j competitively excludes species i if the ratio is greater than one. In general, when $\rho = 1$ (niche overlap is complete), the species with the larger population fitness excludes the other. However, niche overlap smaller than 1 constrains the fitness differences compatible with coexistence. This model provides a theoretical framework for potential outcomes of interactions between invading and resident species. (Derived from Chesson, P., 1990, *Theoretical Population Biology* 37: 26–38, and Chesson, P., and J. J. Kuang, 2008, *Nature* 456: 235–238.)

invader and niche overlap is large, residents will repel the potential invader; (2) when either there is no difference in fitness or niche overlap is small, the invader and residents can coexist; (3) when fitness of the invader is greater than fitness of the residents and niche overlap is large, the invader can exclude residents. High PFH may contribute to the third outcome. In general, successful invasion can result from either fitness differences that favor the dominance of the invader or niche differences that allow the invader to establish despite lower population fitness. However, the outcomes of invasion will differ. Only the former leads to displacement of resident species. The latter leads to coexistence and not to local extinctions of residents. This model explicitly connects two major topics of invasion biology that are often treated independently: species invasiveness and invasibility of biotic communities. Despite the fact that quantifications of both fitness and niche overlap are far from simple measurements, this model provides a useful theoretical framework that will very likely guide research on biological invasions in years to come.

SEE ALSO THE FOLLOWING ARTICLES

Demography / Dispersal Ability, Plant / Genotypes, Invasive / Habitat Compatibility / Life History Strategies / Reproductive Systems, Plant / Taxonomic Patterns / Tolerance Limits, Plant

FURTHER READING

Baker, H.G. 1965. Characteristics and modes of origin of weeds (147–172). In H.G. Baker and G.L. Stebbins, eds. *The Genetics of Colonizing Species*. New York: Academic Press.

Blackburn, T.M., P. Cassey, and J.L. Lockwood. 2009. The role of species traits in the establishment success of exotic birds. *Glogal Change Biology* 15: 2852–2860.

Boivin, T., G. Rouault, A. Chalon, and J.-N. Candau. 2008. Differences in life history strategies between an invasive and competing resident seed predator. *Biological Invasions* 10: 1013–1025.

Burns, J.H. 2008. Demographic performance predicts invasiveness of species in the Commelinaceae under high-nutrient conditions. *Ecological Applications* 18: 335–346.

Dawson, W., D.F.R.P. Burslem, and P.E. Hulme. 2009. Factors explaining alien plant invasion success in a tropical ecosystem differ at each stage of invasion. *Journal of Ecology* 97: 657–665.

Funk, J.L., and P.M. Vitousek. 2007. Resource-use efficiency and plant invasion in low-resource systems. *Nature* 446: 1079–1081.

Garcia-Berthou, E. 2007. The characteristics of invasive fishes: What has been learned so far? *Journal of Fish Biology* 71D: 33–55.

Grotkopp, E., M. Rejmánek, and T.L. Rost. 2002. Toward a causal explanation of plant invasiveness: Seedling growth and life-history strategies of 29 pine (*Pinus*) species. *American Naturalist* 159: 396–419.

Jeschke, J.M., and D.L. Strayer. 2006. Determinants of vertebrate invasion success in Europe and North America. *Global Change Biology* 12L: 1608–1619.

Keller, R.P., and J.M. Drake. 2009. Trait-based risk assessment for invasive species (44–62). In R.P. Keller, D.M. Lodge, M.A. Lewis, and J.F. Shogen, eds. *Bioeconomics of Invasive Species*. Oxford: Oxford University Press.

Leishman, M.R., V.P. Thomson, and J. Cooke. 2010. Native and exotic invasive plants have fundamentally similar carbon capture strategies. *Journal of Ecology* 98: 28–42.

MacDougall, A.S., B. Gilbert, and J.M. Levine. 2009. Plant invasions and niche. *Journal of Ecology* 97: 609–615.

Miller, A.W., G.M. Ruiz, M.S. Minton, and R.F. Ambrose. 2007. Differentiating successful and failed molluscan invaders in estuarine ecosystems. *Marine Ecology Progress Series* 332: 41–51.

Pyšek, P., V. Jarošík, J. Pergl, R. Randall, M. Chytrý, I. Kühn, L. Tichý, J. Danihelka, J. Chrtek Jr., and J. Sádlo. 2009. The global invasion success of Central European plants is related to distribution characteristics in their native range and species traits. *Diversity and Distributions* 15: 891–903.

Rejmánek, M., D.M. Richardson, S.I. Higgins, M.J. Pitcairn, and E. Grotkopp. 2005. Ecology of invasive plants: State of the art (104–161). In H.A. Mooney, R.N. Mack, J.A. McNeely, L.E. Neville, P.J. Schei, and J.K. Waage, eds. *Invasive Alien Species: A New Synthesis*. Washington, DC: Island Press.

Rejmánek, M., D.M. Richardson, and P. Pyšek. 2005. Plant invasions and invasibility of plant communities (332–355). In E. van der Maarel, ed. *Vegetation Ecology*. Oxford: Blackwell Publishing.

Richards, C.L., O. Bossdorf, N.Z. Muth, J. Gurevitch, and M. Pigliucci. 2006. Jack of all trades, master of some? On the role of phenotypic plasticity in plant invasions. *Ecology Letters* 9: 981–993.

Statzner, B., N. Bonada, and S. Dolédec. 2008. Biological attributes discriminating invasive from native European stream macroinvertebrates. *Biological Invasions* 10: 517–530.

Tecco, P.A., S. Diaz, M. Cabido, and C. Urcelay. 2010. Functional traits of alien plants across contrasting climatic and land-use regimes: Do aliens join the locals or try harder than them? *Journal of Ecology* 98: 17–27.

van Kleunen, M., E. Weber, and M. Fischer. 2010. A meta-analysis of trait difference between invasive and non-invasive plant species. *Ecology Letters* 13: 235–245.

INVERTEBRATES, MARINE

JAMES T. CARLTON

Williams College, Mystic, Connecticut

Human activity has dramatically altered the global distributions of thousands of species of marine invertebrates and protists in only the past few millennia. The movement of peoples and their goods created, and continues to maintain, novel corridors that serve to connect coastal habitats around the world in a manner unprecedented in Earth's history. By bridging the natural barriers of oceans and continents, marine invertebrates were and continue to be instantaneously moved across or between oceans to regions where they did not evolve. Human-mediated movements do not enhance the rate or volume of natural species movements; rather, synanthropic vectors move marine invertebrates to new areas of the world (such as from Europe to Australia) that they would never have been able to reach otherwise. A challenge for evolutionary biologists, geneticists, biogeographers, and ecologists is that many of these human-mediated invasions commenced long before scientists evolved.

VECTORS

No coastline in the world has been immune to the potential for invasions by exotic species, as there is no shore that a vessel, bearing a rich biota from a distant shore or ocean, has not landed upon or passed (experimental studies on a replica of Sir Francis Drake's vessel *Golden Hinde* demonstrated that ships passing along a coastline in the open ocean can shed species in transit). Although Charles Elton heralded the global movement of edible oysters for commercial purposes as "the greatest agency of all that spreads marine animals to new quarters of the world," that honor belongs to ships and boats. Humans first went to sea tens of thousands of years ago, often moving in directions opposite to those of ocean currents. Such is the antiquity of human movement on the seas that we do yet understand the depth and breadth to which many marine invertebrates now owe their distribution to ancient vessel movements, whether it be by early humans colonizing South Pacific islands or, later, by the translocation of species on and in the hulls of new generations of oceangoing vessels in the days of the Lapita, Polynesian, Chinese, Phoenician, and other empires. The result is that many common marine and estuarine invertebrates around the world are cryptogenic.

FIGURE 1 The Asian seasquirts *Botrylloides violaceus* (orange) and *Styela clava* (brown), and the Asian green alga *Codium fragile fragile* (green) in New England fouling communities. (Photograph courtesy of the author.)

That humans were creating novel marine corridors long ago is signaled by the appearance of the American clam *Mya arenaria* in Europe circa AD 1200, in close association with Norse movements across the North Atlantic at the same time. By the 1500s, European vessels were bringing allochthonous species from the Pacific theater to the Atlantic; 200 years later, these species would then be described by European biologists as native taxa. Striking among these is the Japanese oyster *Crassostrea angulata*, long embedded in Euroculture as the iconic "native" Portuguese oyster. A wide range of Pacific species, ranging from shipworms to isopods to seasquirts, bear European type localities (and are often still interpreted as native to the Atlantic Ocean).

Ships carried or carry marine invertebrates on their hulls and sea chests as *fouling* organisms (such as sponges, hydroids, sea anemones, barnacles, amphipods, mussels, bryozoans, and ascidians); in their hulls, when there were wooden ships, as *boring* organisms (such as shipworms, which are not worms, and gribbles, which are isopods); as *crevicolous* organisms (in burrows and openings in wooden or steel hulls); with *rock and sand ballast* (into the twentieth century); and in *water ballast* (since the nineteenth century), the latter two mechanisms capable of transporting almost every phylum of marine life. Ocean-going commercial vessels, barges, yachts, dry docks, drilling platforms, and amphibious planes and seaplanes all serve as effective vectors for thousands of species.

Numerous other vectors move marine invertebrates. These include fisheries activities, such as the aquaculture (mariculture) industries; the long-distance translocation of commercial shellfish (such as oysters, mussels, and clams), seaweeds, and fish; the movement of live worm bait dunnaged in seaweed; and the movement of fisheries gear,

along with the many invertebrates inadvertently associated with these enterprises. The live seafood and aquarium pet industries, facilitated by the even more widespread availability of the same species through Internet trade in the twenty-first century (the "bioweb"); the marine biological supply industry and the associated research and education communities; salt marsh and eelgrass restoration activities; and the increased amount of floating anthropogenic debris all add to the plethora of vectors, and often to the lack of vector awareness, which continue the profound global bioflow of marine invertebrates. The number of human-mediated vectors that disperse marine organisms doubles approximately every 100 years.

In addition, sea level and lock canals have played a significant role in creating corridors between biogeographic provinces. Examples include the Suez Canal (1869) and the associated "Lessepsian migration" from the Red Sea to the Mediterranean; the Panama Canal (1914), which despite its freshwater lakes and lochs permitted the passage of many species; the Cape Cod Canal (1914), which opened a corridor between colder-water and warmer-water biogeographic provinces, and the Kiel Canal (1895), which links the Baltic and North seas.

THE DIVERSITY AND GENETICS OF MARINE INVERTEBRATE AND PROTIST INVASIONS

Table 1 shows the diversity of global marine protistan and invertebrate invasions in tropical, subtropical, and temperate seas, with log scale estimates of the minimum number of probable introductions. Conservatively, more than 4,000 species of marine invertebrates may have been moved around the world over the past thousand and more years; the actual number may be much larger. As noted above, many of these species remain cryptogenic. There are extensive gaps in our knowledge relative to invasions in many groups, and for most groups, we can only approximate the number of invasions. Invasions of smaller and planktonic invertebrates around the world are especially poorly known and remain rich arenas for exploration.

Application of molecular genetic techniques to marine invertebrate invasions has been critical in elucidating the biogeographic tracks (sources and routes) of many introductions and in revealing cryptic genospecies clades and invasions among so-called cosmopolitan species. For example, genetic studies revealed that the widespread mussel *Mytilus edulis* consisted of three species in the northern hemisphere, one of which, *M. galloprovincialis*, has been introduced by ships from its Mediterranean home to widespread locations throughout the Pacific theater, including California, Japan, Chile, and Australia.

TABLE 1
Global Marine Protist and Invertebrate Invasions

Taxon	Tropical Subtropical	Temperate	Probable Minimum Scale of Invasions, as Relative Numbers of Species (Magnitude)	Taxon	Tropical Subtropical	Temperate	Probable Minimum Scale of Invasions, as Relative Numbers of Species (Magnitude)
Protista				Copepoda: parasitic	+	+	100s
Foraminiferida	+	+	100s	Branchiura	NS	NS	1s
free-living	NS	NS	100s	Cirripedia	+	+	10s
parasitic/ endocommensal "protozoa"	NS	+	100s	Leptostraca	NS	NS	1s
				Mysidacea	+	+	10s
symbiotic/ ectocommensal "protozoa"	+	+	100s	Cumacea	+	+	10s
				Isopoda	+	+	10s
				Tanaidacea	+	+	10s
Placozoa	NS	NS	1s	Amphipoda	+	+	100s
Porifera	+	+	100s	Stomatopoda	+	NS	1s
Cnidaria				Euphausiacea	NS	NS	1s
Hydrozoa	+	+	100s	Decapoda	+	+	100s
Scyphozoa	+	+	10s	**Pycnogonida**	+	+	10s
Anthozoa	+	+	10s	**Arachnida**			
Ctenophora	+	+	1s	Acari	NS	NS	10s
Platyhelminthes				Pseudoscorpiones	NS	NS	1s
free-living	+	+	100s	**Insecta**	+	+	100s
parasitic	+	+	100s	**Mollusca**			
Nemertea	NS	+	10s	Scaphopoda	NS	NS	1s
Nematoda	+	NS	100s	Cephalopoda	+	NS	1s
Kinorhyncha	NS	NS	1s	Polyplacophora	+	NS	1s
Nematomorpha	NS	NS	1s	Gastropoda	+	+	100s
Acanthocephala	NS	NS	1s	Bivalvia	+	+	100s
Gnathostomulida	NS	NS	10s	**Phoronida**	+	NS	1s
Rotifera	NS	+	10s	**Brachiopoda**	NS	NS	1s
Kamptozoa	+	+	1s	**Bryozoa**	+	+	100s
Sipuncula	+	NS	1s	**Chaetognatha**	NS	NS	1s
Echiura	NS	NS	1s	**Hemichordata**	NS	+	1s
Tardigrada	NS	NS	1s	**Echinodermata**			
Annelida				Crinoidea	NS	NS	1s
"Oligochaeta"	+	+	10s	Echinoidea	+	+	1s
Hirudinida	NS	NS	1s	Asteroidea	+	+	1s
Polychaeta	+	+	100s	Ophiuroidea	+	+	10s
Crustacea				Holothuroidea	+	NS	1s
Mystacocarida	NS	NS	1s	**Chordata**			
Cephalocarida	NS	NS	1s	Ascidiacea	+	+	100s
Branchiopoda	NS	+	10s	Larvacea	NS	NS	1s
Ostracoda	NS	+	10s	Thaliacea	NS	NS	1s
Copepoda: free-living	+	+	100s				
				Number of Species			>4,000

NOTE: Column 3 total based on minimum of 20 percent of log scale range for each taxon. + = invasions known; NS = no studies.

The "native" clam *Macoma balthica* in San Francisco Bay proved to be the morphologically identical *Macoma petalum*, introduced from New England with commercial oysters in the nineteenth century. After the American Pacific coast barnacle *Balanus glandula* invaded Japan in the 1990s, genetic studies demonstrated that the source was Alaska or Puget Sound, in concert with the patterns of shipping traffic to Hokkaido Island. Genetic studies demonstrate that the European snail *Littorina littorea* is an eighteenth or nineteenth century invader of North America, rather than being native to the western Atlantic Ocean, as has been occasionally proposed.

ASPECT DOMINANCE AND HABITAT DIVERSITY OF INVERTEBRATE INVASIONS IN THE OCEANS

Along many coasts of the world, many of the most visible (aspect-dominant) species have arrived and become established only since the mid-nineteenth century. This knowledge is based upon historical evidence from the first "baseline" surveys by scientists in a given region. This fact suggests that many invasions, which may also now be prominent if not regulatory (engineering) species, occurred in earlier centuries prior to baseline surveys but are misinterpreted as being members of naturally occurring and naturally evolved communities.

Thus, in the Port of Le Havre, France, the retaining walls of the inner quay are coated with the Asian seasquirt *Styela clava*, the Australian tubeworm *Ficopomatus enigmaticus*, and the Japanese sea anemone *Diadumene lineata*. On the shores of the northwest Atlantic Ocean from Canada to Long Island Sound, billions of the omnivorous European periwinkle snail *Littorina littorea* blanket intertidal shores. In Cape Town, South Africa, the harbor floats support waving beds of millions of the European seasquirt *Ciona intestinalis*, while not far away, the Mediterranean mussel *Mytilus galloprovincialis* now dominates parts of the open wave-swept coast. In the Hawaiian Islands, harbor communities and open coral reefs support hundreds of exotic species, many of which have arrived, or were imported, only since World War II. In San Francisco Bay, California, there are nearly 300 established nonnative marine and estuarine animals and plants, most of which have arrived since 1850 (Fig. 2). Similar pictures can be painted for most coastal regions of the world.

Many bays, estuaries, and lagoons—now converted to ports, harbors, and marinas—support vast populations of species imported with ships and shellfish, to the extent that the same species (many of whose original homes are not yet known) may now be found in harbors around the world—the marine equivalent of urban "weedy lots." Marine invertebrate invasions, however, are not restricted to ports and harbors: invasions occur in a wide range of habitats, as suggested above.

Pelagic environments have sustained invasions by jellyfish (as well as by diatoms and fish) in the Gulf of Mexico, the Hawaiian Islands, the eastern Mediterranean, and elsewhere.

Open ocean sublittoral benthic habitats support invasions in many parts of the world of a wide range of taxa, including the New Zealand marine snail *Maoricolpus roseus* in Australia, the tiny European flatworm *Convoluta convoluta* in the Gulf of Maine, and the American razor clam *Ensis directus* in western Europe.

In the deep sea, the North Pacific king crab *Paralithodes camtschatica* is now abundant in the Barents Sea and is proceeding south along the coast to mainland Europe. Kelp beds in the northwestern Atlantic have been invaded by the European bryozoan *Membranipora membranacea* (Table 2). Exposed rocky and coral intertidal shores support nonnative species on the Atlantic coast of North America and in Europe, South Africa, Argentina, the Hawai'ian Islands, Micronesia, Polynesia, and elsewhere. Mangroves, salt marshes, and supralittoral shores have been invaded around the world. Although most studies on exotic species have occurred in estuaries, it is clear that most marine habitats are not immune to invasions, and many of these communities also remain rich arenas for further study of the anthropogenic influences on the origins of their late Holocene biodiversity.

THE ECOLOGY OF MARINE INVERTEBRATE INVASIONS

Impacts of a few (of the thousands of) marine invertebrate invasions have been studied quantitatively or experimentally. These studies demonstrate the remarkable shifts that can take place in community and ecosystem structure following the arrival of new species (Table 2). These impacts include all of the classic drivers of community structure (such as predation, competition, and disturbance), as well as geological impacts by introduced bioeroders. As there are no quantitative or experimental studies on perhaps 95 percent or more of all marine invertebrate invasions, no statistical conclusion can be drawn on the extent or scale of their impact. Although a common statement is that "most invasions do no harm,"

FIGURE 2 "Sphaeroma topography": Eroded shores along the margins of San Francisco Bay, California, caused by the burrowing activities of the New Zealand isopod (pillbug) *Sphaeroma quoianum*. (Photograph courtesy of the author.)

TABLE 2

Examples of Ecological Impacts of Marine Invertebrate Invasions

Species (Native Region, Vector)	History and Impact
Northwest Atlantic Ocean	
Periwinkle snail *Littorina littorea* (Europe; with ballast rocks or intentional release as food, or both), and the shore crab *Carcinus maenas* (Europe; ships)	Reorganizing Intertidal Shores of the Northwest Atlantic Ocean (Part I): Established no later than the 1830s in the Gulf of St. Lawrence, Canada, this European snail spread from Nova Scotia in the 1850s to New Jersey by 1880. A habitat and trophic generalist, *L. littorea* became the dominant invertebrate macrograzer on rocky intertidal shores but also invaded marshes and mudflats. Experimental studies have demonstrated that this invasion has led to fundamental alterations in plant diversity, density, and distribution, in native gastropod abundance, and in hermit crab shell utilization patterns, among many other impacts. Rounding Cape Cod in the 1870s, *L. littorea* met the European shore crab *Carcinus maenas*, present in Long Island Sound by the early 1800s, a large and aggressive multihabitat omnivore that subsequently began a northward march in the 1890s. The biology and ecology of the rocky shores on the northwest Atlantic Ocean coasts are poorly known prior to the arrival of *Littorina* and *Carcinus*.
Crab *Hemigrapsus sanguineus* (Japan; ballast water)	Reorganizing Intertidal Shores of the Northwest Atlantic Ocean (Part II): First detected in 1988 in New Jersey, by 1998 this crab was the most abundant intertidal crab from Massachusetts to New York; by the 2000s, it had spread into Maine. This Asian shore crab displaces and replaces both native crabs and the previously introduced *Carcinus*, eliminates mussel populations, and significantly decreases other prey species, among its other impacts.
Ascidian guild: *Styela clava* (Japan), *Styela canopus* (Japan), *Ascidiella aspersa* (Europe), *Botryllus schlosseri* (Europe, but perhaps originally Pacific), *Botrylloides violaceus* (Japan), *Didemnum vexillum* (Japan), and *Diplosoma listerianum* (Indo-Pacific?); all with hull fouling.	Restructuring Hard-Bottom Sublittoral Communities: Commencing in the late 1700s and early 1800s, but particularly since the 1970s, foreign ascidians (sea squirts) have invaded the Northwest Atlantic Ocean. By mid-summer in Long Island Sound, a dense, multilayered, polychromatic array of largely nonnative solitary and compound ascidians coats nearly every available submerged hard surface in higher salinities (Fig. 1), replacing the former native fouling communities that may have been composed of tubeworms, bryozoans, and mussels.
Bryozoan *Membranipora membranacea* (Europe; ballast water)	Unpredictable Cascades: This beautiful and well-known gray bryozoan forms huge colonies on the blades of the kelp *Laminaria*. The blades become brittle and are more easily defoliated during storms, and the bryozoan colonies may also block nutrient uptake and photosynthesis. *Membranipora* appeared off the Maine coast in the 1980s; kelp beds declined, and the open space so created was then occupied by another introduced species, the Asian green alga *Codium fragile fragile*; once *Codium* is established, kelp bed reestablishment is prevented.
Northeastern Pacific Ocean	
Isopod *Sphaeroma quoianum* (New Zealand; vessel hull fouling and boring)	Bioerosion of Marshes and Soft Sediment Shores: Introduced in the mid or late 1800s to the California coast, this New Zealand isopod was soon recognized as a potentially major bioeroder of intertidal marsh banks, sandstone shores, and other friable habitats (Fig. 2). Having been previously restricted by the formerly colder climates of the Pacific Northwest to California, this isopod invaded the now-warmer southern Oregon in the 1990s and within a decade had caused extensive erosion to the margins of Coos Bay.
Mud snail *Ilyanassa obsoleta* (New England) and mud snail *Batillaria attramentaria* (Japan) (both with commercial oyster imports)	One-Two Punch and Demise of Native Species: In the early 1900s, the Atlantic mud snail was introduced to San Francisco Bay, driving the once-abundant native mud snail *Cerithidea californica* (whose populations were rapidly being destroyed by the filling of the Bay) off the mudflats and into salt marsh refugia. In the 1940s, the Japanese mud snail was introduced to neighboring central California bays; *Batillaria* aggressively competes with *Cerithidea* and may lead to its estuary-wide extinction.
Clam *Corbula amurensis* (China, ballast water)	Benthic-Pelagic (De)coupling: Arriving in the 1980s in San Francisco Bay, this Chinese estuarine clam soon occurred by the billions throughout the estuary, leading to the disappearance of phytoplankton stocks, declines in zooplankton (upon which the clams also feed), and declines in the abundance of other species, such as mysid crustaceans (an important member of fish diets), that relied on phytoplankton. *Corbula* accumulates selenium in high concentrations, and this bioaccumulation is propagated up the food chain in San Francisco Bay, leading to teratogenesis and reproductive failure in clam predators.
Hawaiian Islands	
Barnacle *Chthamalus proteus* (Caribbean; hull fouling)	Invader Replaces Former Invaders: This Atlantic barnacle was first collected in the Hawaiian Islands in the early 1990s and is now the most abundant intertidal barnacle on Oahu. It has reduced the abundance in certain habitats of the Indo-Pacific *Amphibalanus reticulatus*, a 1920s invader of the islands, and it appears to have led to niche compression of the Indo-Pacific *Amphibalanus amphitrite*, an even earlier invader.

(Continued)

TABLE 2 *(CONTINUED)*

Examples of Ecological Impacts of Marine Invertebrate Invasions

Species (Native Region, Vector)	History and Impact
Multiple Oceans	
Blue mussel *Mytilus galloprovincialis* (Mediterranean; ships)	Displacement of Native Species: A superb competitor for space on middle and upper intertidal shores, this mussel replaced the native mussel *Mytilus trossulus* in southern California in the mid-20th century and the native mussel *Aulacomya ater* in South Africa in the late 20th century. The native mussel *Perna perna* remains secure on low intertidal shores of South Africa: with its much stronger byssal threads, it endures wave exposure better while also outcompeting *Mytilus* on the lower shores.
Mussel *Musculista senhousia* (Japan; ships, mariculture)	Carpeting Bay Floors: This mat-forming Asian mussel now forms dense byssally intertwined beds of many thousands of individuals per square meter in bays and estuaries on the North American Pacific coast and in New Zealand, Australia, and the Mediterranean. These surface mats smother and eliminate infaunal clams, eelgrass, and other prior soft-bottom residents. At the same time, as is typically the case with engineering species that serve to form imported three-dimensional habitats, numerous invertebrates find enhanced shelter in the mussel communities.
Comb jelly *Mnemiopsis leidyi* (North America; ballast water)	Pelagic Consumer: The American ctenophore *Mnemiopsis* arrived in the Black Sea in the 1980s, in the Caspian Sea in the 1990s, and in northwestern Europe in the early 2000s. A consummate consumer of zooplankton, including ichthyoplankton, it caused significant decline of fish stocks in the Black Sea. Populations of *Mnemiopsis* have since been reduced by the invasion in the late 1990s of the ctenophore-eating ctenophore *Beroe ovata* from the Mediterranean, but *Beroe* has not yet entered the Caspian Sea.

most invasions have not been studied to determine what changes they may or may not have wrought.

THE FUTURE OF MARINE INVERTEBRATE INVASIONS

Invasions of marine invertebrates into new areas of the world will continue. On the one hand, increasing management measures globally—focused on vectors such as ballast water, mariculture, or the transport of live seafood—will no doubt lead to the prevention of certain invasions. On the other hand, the plethora of vectors remain a challenge to address, such as the diversity of species transported on ship hulls (despite antifouling treatments) coupled with the increased survival of species as ships cross ever faster through the inhospitable waters of the open ocean.

Warming waters are leading to new invasions. "Caribbean creep" consists of the invasion of the U.S. southern and mid-Atlantic coast by a host of species previously restricted to southern Florida and the Bahamas, including the porcelain crab *Petrolisthes armatus*, the barnacle *Megabalanus coccopoma*, and the marsh snail *Creedonia succinea*. Species—both native and introduced—long restricted to southern or central California are now inexorably moving north, including the southern California barnacle *Tetraclita rubescens* colonizing northern California and southern Oregon, the New Zealand isopod *Sphaeroma quoianum* moving north into Oregon, and a host of nonnative ascidians moving from the south into the Pacific Northwest. On all continental shores where historical distribution records are available, species are known to be moving north in the northern hemisphere and south in the southern hemisphere. In addition, as shores warm, they will become susceptible to invasions by species previously excluded by too-cold temperatures.

SEE ALSO THE FOLLOWING ARTICLES

Algae / Ballast / Canals / Crustaceans / Geographic Origins and Introduction Dynamics / Ostriculture / Seas and Oceans

FURTHER READING

Carlton, J. T. 1999. The scale and ecological consequences of biological invasions in the world's oceans (195–212). In O. T. Sandlund, P. J. Schei, and Å. Viken, eds. *Invasive Species and Biodiversity Management.* Dordrecht: Kluwer Academic Publishers.

Carlton, J. T. 2001. *Introduced Species in U.S. Coastal Waters: Environmental Impacts and Management Priorities.* Arlington, VA: Pew Oceans Commission.

Carlton, J. T. 2003. Community assembly and historical biogeography in the North Atlantic Ocean: The potential role of human-mediated dispersal vectors. *Hydrobiologia* 503: 1–8.

CIESM Atlas of Exotic Species in the Mediterranean. 2002. Monaco: CIESM Publishers.

Leppäkoski, E., S. Gollasch, and S. Olenin, eds. 2002. *Invasive Aquatic Species of Europe: Distribution, Impacts, and Management.* Dordrecht: Kluwer Academic Publishers.

Rilov, G., and J. A. Crooks, eds. 2009. *Biological Invasions in Marine Ecosystems: Ecological, Management, and Geographic Perspectives.* Berlin: Springer.

Ruiz, G. M., P. W. Fofonoff, J. T. Carlton, M. J. Wonham, and A. H. Hines. 2000. Invasion of coastal marine communities in North America: Apparent patterns, processes, and biases. *Annual Review of Ecology and Systematics* 31: 481–531 + Appendix 1 (online).

ISLANDS

WILLIAM G. LEE

Landcare Research, Dunedin, New Zealand

Islands have always been at the forefront of issues surrounding the impacts and control of introduced invasive species and, more recently, of major attempts at eradicating introductions and facilitating ecological restoration. Studies on islands have been crucial for improving our understanding of the factors causing invasion success, the range of species and ecosystem-level impacts, and the challenges involved in restoring ecosystems once introduced species have been removed. Compared with most continental regions, islands have several properties that make them ideal for ecological studies: relatively small area with clear boundaries, simplified systems in distinct environments, low species richness, the presence of endemic species, and the absence of some functional groups.

HUMAN INFLUENCES

Introduced species are human-assisted biotic immigrants, and consequently, they bypass a major driver important in the evolution of island biotas: namely, the effects of insularity (distance from potential source biota, colonization rate). Humans have made contact with nearly all the islands around the globe, and our activities have vastly increased dispersal distances of thousands of plant, animal, fungal, and protozoan species. The mixing of species and functional groups that have never been associated throughout their evolutionary history increases the trend toward the global homogenization of biota. Humans also increase arrival rate, population propagule pressure, and the availability of modified habitats, all of which facilitate the establishment and spread of introduced biota.

The rapid demise of distinctive endemic birds and mammals was an early indication of the negative effects of invasive introduced species distributed either deliberately or accidentally by humans during the settlement of islands. Darwin was among the first to record these extraordinary impacts and the general success of introduced species on islands, from observations while voyaging on the H.M.S. *Beagle* during the 1830s. He was probably responsible for initiating the widespread perception that island biotas and ecosystems were more vulnerable to nonnative invasions, compared with those of continental landmasses. This was a difference that he attributed to the more benign climate and absence of competitors and strong predators on

many island systems. However, while there is abundant evidence for catastrophic impacts of introduced invasive animals and plants on islands, generalizations about the abiotic and biotic factors responsible for these strong effects continue to be widely debated. Today many argue that islands are in fact so distinctive and idiosyncratic that comparisons with mainland regions are meaningless.

CHARACTERISTICS OF ISLAND BIOTAS

There are two fundamental sorts of islands: oceanic, that is those formed de novo usually by volcanic eruption far from any mainland, and continental or nearshore islands, which were formed by relative sea level rise. Nearshore islands generally maintain a resident biota closely related to that of adjoining mainland areas because of fragmentation of the original biota and continuing biotic contact over narrow sea gaps. The biota of oceanic islands was, until human intervention, exclusively derived from immigrants or evolution from founder species. Age, isolation, size, and the characteristics of potential source biotas are major factors influencing communities on islands. The small size of many islands coupled with steep environmental gradients has led to small population sizes and ranges. Abundant marine mammals and birds often form extensive colonies in the absence of strong terrestrial predation and, coupled with marine aerosol, provide large quantities of extra nutrients in coastal regions. Natural disturbances (cyclonic storms, volcanic eruptions) can also disproportionately impact terrestrial systems on islands.

Recent studies have identified clear differences between island and mainland areas in the relative importance of competition and predation in structuring communities. For example, a worldwide study of lizard populations from islands and mainland areas showed that island population densities, under similar resource availability, were ten times higher than those on the mainland, reflecting lower numbers of competitor and predator species. However, lizard densities declined significantly where competitor and predator species richness was higher on islands but not on the mainland, suggesting greater susceptibility of island lizard populations to both competition and predation. A similar sensitivity to the presence of top predators (lizards) also controls the densities of spiders on islands. These results are some of the few indicating that insular biotas may be inherently more vulnerable than mainland biotas to additional competition and predation.

Isolated island biotas are generally less regularly affected by disease, compared with nearshore islands and continental regions, because infectious agents that fail to persist (i.e., that burn out) lack a metapopulation structure to generate recolonization. Therefore, over time, island biota are generally challenged by a more limited range of disease-causing agents, leading to their inability to mount effective responses to newly introduced ones.

Globally, islands occur in virtually all bioclimatic regions and come in diverse shapes and sizes, making it difficult to generalize about their biotas. However, as an oversimplification, the following attributes are common to most islands compared with continental regions: smaller populations with reduced genetic diversity; fewer species and, as a consequence, lower representation of different functional types (especially at higher trophic levels); high levels of endemism (often through local differentiation); presence of novel forms (e.g., flightlessness, gigantism, asociality in invertebrates); and the dominance of birds and reptiles rather than mammals. Indeed, mammals are often absent, even from many large islands.

Individually or in combination, these environmental, biotic, and ecosystem features have all been suggested to be significant in affecting the vulnerability of islands to invasion by introduced species. However, unequivocal evidence of the relative importance of the different features is difficult to obtain because of many confounding factors.

INTRODUCED SPECIES

Escalating human migrations, commercial interchange, and new settlements over the last 200 years have led to unprecedented numbers of plant, animal, fungi, and microbe translocations around the globe for food production, horticulture, agriculture, and forestry, as well as incidentally as contaminants. Human importation for food, shelter, trade, and gardens has produced extremely high introduction rates for some groups, usually commensurate with growth of human populations. On islands the origin of the first settlers (European, Polynesian), duration of human settlement, extent of native habitat loss, and type of agricultural intensification (shifting cultivation, crops, pasture, forestry) determine the number, range, and local abundance of introduced species. Increasingly, the accidental introductions of cryptic organisms (small invertebrates, microbes), either on or in larger animals, are becoming obvious, often through their disease and other impacts on native biota.

Plant species far outnumber most other introductions on islands and are the best documented; they show the proportion of introductions that were able to develop self-sustaining wild populations (i.e., naturalized populations) independently of human plantings. Naturalized introduced plant species generally result in a net increase (approximately twofold) in species richness on islands, because the gain of new species is usually vastly in excess of losses through extinction of native plant species. In New Zealand, of the 20,000 introduced plant species present, at least 10 percent have become naturalized, and these now outnumber the native vascular flora. On the Galapagos Islands, total introductions (438 species) remain fewer than natives (550 species), but the naturalization proportion (42.5%) is very high. The success of particular plant species on islands appears to depend more on human and biogeographic factors and less on the biological attributes of the species. In a comparative study of *Trifolium* species in New Zealand, human activities dominated the main success factors explaining introductions (economically important species introduced early), naturalizations (large-scale planting), and the relative rate of spread (frequent contamination of other agricultural crops). A large native range was a significant factor at all three invasion stages, but biological and ecological attributes were of secondary importance.

For birds, gains of naturalized species on islands often approximate the number of native species lost by extinction, irrespective of island area. The reason for the relative stability of the total avian fauna is unclear because processes regulating introduction success (size and number of introduction attempts) and native bird extinctions (mammalian predation) are ecologically unconnected. It is possible that species interactions constrain bird species numbers on islands, while for plants, numbers of species reflect the habitat and disturbance heterogeneity enhanced on islands during human occupancy.

Introductions of mammals, fish, amphibians, and invertebrates have also caused a net increase in species on many islands, with importations greatly exceeding extinctions of native species. However, lizards are less frequently transported and naturalized, and therefore often show declines in species richness following human settlement. Extremely small animals are rarely detected among introductions but are invariably present as contaminants on plants, on animals, and in soils taken to islands, but they are poorly documented.

IMPACTS OF INTRODUCED SPECIES

Introduced species often enable humans to settle and survive on islands, providing food and an economic basis for trade. However, introduced species can also negatively affect many activities on islands, including the economics of agriculture, horticulture, and plantation forestry, but the focus here is on the effects on native biodiversity and natural ecosystems.

Plants

The impacts of introduced plants on islands unfold over long periods, but the most consequential introduced plants are those that readily capture resources (nutrients, light, or water) and change local disturbance regimes (e.g., fire, flooding, sedimentation). Although they rarely cause native plant extinctions, many introduced plants create major ecosystem disruptions. Smothering lianes and shade-tolerant species, fire-promoting grasses, nitrogen-fixing legumes, herbivore-resistant shrubs and succulents, and clonal aquatic macrophytes invariably feature on lists of the most devastating plants on islands, and their impacts are generally proportional to the area of the island occupied. They particularly threaten endemic plants in lacustrine, sand dune, and coastal shrubland and forest environments.

Vigorous vines are perhaps the most serious invasive weeds on islands, collapsing native vegetation (including tall forest) and preventing reestablishment. They are widespread on subtropical South Pacific islands, and several of the most problematic species are derived from garden escapees native to South America (*Passiflora*), Eurasia (*Lonicera*), and Asia and Africa (*Dioscorea*). Many shade-tolerant, bird-dispersed trees are also infamous invaders on islands, transforming native forests and reducing ecosystem services. A tropical South American tree widely planted by gardeners because of attractive foliage, *Miconia calvescens*, has been spread by fruit-eating birds on Tahiti and now occurs over 50 percent of the island. This canopy dominant threatens several endemic forest species, and its superficial roots fail to stabilize slope soils, increasing the frequency of landslides and sediment loadings in streams.

Nitrogen-fixing legumes (Fabaceae) are also notorious invaders on islands, irrespective of whether they are from islands (e.g., *Falcataria moluccana*) or continents (e.g., *Lupinus arboreus, Ulex europaeus*). These species often establish in abandoned pastures, on disturbed surfaces such as volcanic debris or sand dunes, or beneath canopy gaps in native shrublands or forests. They raise soil nitrogen and phosphorus levels, accumulate thick litter, and have prodigious, persistent seed banks. These features enable them to consolidate their occupancy and eventually facilitate establishment of other nutrient-demanding nonnative species, which often totally displace native successional species. Invasive plant species may also exhaust resources: for example, Australian *Eucalyptus* species deplete soil moisture and lower water tables on Mediterranean islands while exacerbating the fire cycle, further adversely affecting native species.

Animals

Although invasive plants are problematic, and omnivores and herbivores may cause substantial changes in island communities, the primary agents of extinction on islands are introduced carnivores (e.g., rodents, cats [*Felix catus*], mustelids, mongooses). Rodents (*Rattus exulans, R. norvegicus, R. rattus*) have eliminated at least 50 bird species on approximately 40 islands and are implicated in ongoing declines of many island endemic species. The small Indian mongoose (*Herpestes auropunctatus*), introduced on many islands to control rats, toads, and snakes, is implicated in the extinction of two species of rails, a petrel species, and two species of skinks on Fijian islands; three shrew species on Haitian islands; and a snake on Antigua and St. Lucia. Introduced carnivorous reptiles have also been responsible for multiple extinctions on islands. The most infamous example is the New Guinea brown tree snake (*Boiga irregularis*), an unintended introduction on the Pacific Island of Guam around 1950. The snake rapidly became a strong top predator, exterminating at least 12 bird, two bat, and three reptile species and transforming and simplifying the vertebrate food web, which currently comprises predominantly introduced prey species (rats, birds, and skinks). Nowadays, only three native vertebrates, all lizards, remain widespread in forests in Guam.

Community-Level Interactions

The impact of nonnative species on islands has revealed complex indirect interactions that often have community-level consequences. One of the most remarkable examples is the shift in competitive balance between native carnivores on the Californian Channel Islands following the introduction of pigs (*Sus scrofa*). The island fox (*Urocyon littoralis*) and spotted skunk (*Spilogale gracilis amphiala*) were historically the top native predators on the islands, with the foxes competitively and numerically dominant. Following the arrival of pigs, the relative abundance of these native predators changed dramatically, primarily in response to the presence of the golden eagle (*Aquila chrysaetos*), which is native on the neighboring mainland. Sustained by juvenile pigs as a new food source, the eagle colonized and spread rapidly on the islands and began to destabilize the competitive balance between fox and skunk. The eagle had little impact on the fecund pig population, but its increased abundance significantly reduced the foxes (hyper-predation), which facilitated an increase in the skunk population by reducing competition with the fox (competitive release).

Trophic Cascades

Introductions on islands can cause profound indirect changes across several trophic levels that change the major ecosystem processes, in addition to the composition of above- and belowground assemblages. Introduced social insects, particularly ants, can have rapid, broad impacts, drastically modifying community structure. The yellow crazy ant (*Anoplolepis gracilipes*) from Africa established supercolonies on Christmas Island in the northeastern Indian Ocean late last century and now occupies 25 percent of the forested area. The island supports millions of red land crabs (*Geocarcoidea natalis*), which selectively eat fruit, seedlings, and leaves and extensively disturb soil by forming burrows and depositing droppings. The crabs are the primary consumer on the ground, and they control the structure and composition of the rainforest. However, the crazy ants also occupy these burrows for nest sites, and when they are disturbed by the crabs, they respond defensively, spraying corrosive formic acid. An estimated 10–15 million crabs have been removed by the ants, initiating major changes including the proliferation of seedlings of diverse forest species and reduced rates of litter breakdown. Ant invasion in this system has also increased canopy dieback due to increasing populations of the ant-protected honeydew-secreting scale insects and sooty mold on canopy stems. As the forest canopy declines, there has also been infiltration by introduced weeds such as the stinging tree (*Dendrocnide peltata*).

Equally spectacular and pervasive indirect effects have followed from introduced carnivores eradicating colonies of nesting seabirds from islands. Foxes became top predators after being introduced in the early twentieth century on many islands in the Aleutian archipelago, devastating resident seabird populations. Subsequently, the 60-fold reduction in nutrient subsidies on the invaded islands reduced ecosystem productivity and initiated a transition from vegetation dominated by grasses, sedges, and large forbs toward dwarf-shrub tundra vegetation. The disturbances associated with nesting, combined with cessation of nutrient transfer, may drastically alter ecosystem processes following the elimination of seabird colonies on islands by introduced predators. On islands in northern New Zealand, the removal of seabirds by rats has lowered litter depth and decomposition rates, the abundance of primary consumers (predatory nematodes) and secondary consumers (e.g.,

microbe-feeding nematodes) in soils, the species richness of spiders, and plant productivity and foliar nitrogen levels. Without disturbance from seabird activity, there is also a greater density of tree seedlings and plant biomass.

Invasional Meltdown

The negative impacts of introduced species on native biodiversity can be compounded by the presence of other introduced species, a phenomenon termed "invasional meltdown." The spectacular endemic radiation of honeycreepers in Hawaii was nearly eliminated by introduced microparasites. Avian malaria (*Plasmodium relictum*) and avian pox (*Poxvirus avium*) probably arrived on the islands in chickens in the eighteenth century but became virulent and widespread only when a vector, the exotic mosquito (*Culex quinquefasciatus*), was introduced in the 1820s. The honeycreepers were highly susceptible to the new infectious diseases, and only upland species remain because the mosquitoes are limited by cold temperatures.

ERADICATING INTRODUCTIONS

Islands are at the forefront of invasive species eradication attempts, which have traditionally been driven by the need to create safe sites for threatened species and more recently as part of wider ecosystem restoration attempts. Invasive plants on islands are generally managed rather than eradicated, because of the practical difficulties in removing species with large seed banks, or those that are valued as garden plants. Raoul Island in the South Pacific is a notable exception, with systematic eradication of approximately 30 introduced plant species being attempted, but with only seven having been eliminated in the last 30 years. Instead, the main focus has been on eradication of introduced animals. The larger invasive species (e.g., cattle, sheep, goats, foxes, pigs) have been removed from many islands, but a greater challenge has been to eradicate small predatory mammals (e.g., rodents, mustelids) and invertebrates (e.g., ants). Invasive ant control is being undertaken on islands in Hawaii, New Zealand, the Galapagos, and Australia; although populations have been reduced, eradication has proved elusive. Rodents are a primary candidate for eradication because of their ubiquity and severe impact on islands, and techniques have been developed using rodenticides that have successfully removed them from 280 islands worldwide, the largest of these being 11,300 ha in area.

Island eradications often target one introduced species and leave others that appear to have less impact or that are more difficult to eliminate. However, this approach is increasingly being questioned, as major ecosystem adjustments following removal of key introduced top predators, in particular, are producing unexpected outcomes detrimental to important native species and communities. As others have noted, on many islands, the ecological context of eradication is increasingly complex, with many introduced species existing across different trophic levels and food webs and vegetation types that have established over hundreds of years in the presence of the introduced biota.

The unanticipated consequences of management actions to eradicate invasive species singly is particularly apparent on the subantarctic Macquarie Island where cats, rabbits (*Oryctolagus cuniculus*), and rodents have all established since introduction during the nineteenth century. Concern about rabbits destroying the maritime tall grassland and mega-forb associations has led to 30 years of biological control starting in 1968, based on regular introduction of the *Myxoma* virus, which has depressed the rabbit population tenfold. As rabbits declined, cats increasingly shifted to feeding on the dwindling seabird population, countering conservation efforts to protect the birds. A cat eradication program was eventually implemented until the last animal was removed in 2001. Subsequently, with the removal of the top predator and the waning effectiveness of the virus, rabbit numbers rapidly increased, transforming the vegetation to short turf or bare ground and threatening the island's conservation values.

After the expense of major pest eradication, intentional and accidental reintroduction of pest species must be avoided. This has resulted in islands developing rigorous biosecurity strategies that focus on preventing new introductions, providing early detection and eradication of new incursions, and sustaining a framework for risk assessment and management should an introduced species become established.

CONSERVATION GOALS

Removal of introduced species challenges the formulation of modern goals for conservation on islands. Extinctions of irreplaceable endemic taxa, inadequate repopulation sources, habitat modification, and persistence of mixed native and nonnative communities all preclude the return of island ecosystems to anything approaching their original state. Islands have frequently been seen as biodiversity sanctuaries, naturally sheltered from harmful invasive species through their geography, and therefore potentially useful for protecting threatened mainland species, irrespective of the ecological effects on, or evolutionary association with, the island's biota. However, the designation of islands solely as conservation safety nets for miscellaneous taxa overlooks both their intrinsic biodiversity values and the ecological challenge of sustaining communities with no natural analogs. The removal of critical introduced species is but the first step in island restoration. Reintroductions of native consumers and predators, and other keystone species (e.g., seabirds) that influence disturbance and nutrient processes, will be required. Replacement of extinct species with functional surrogates is also being considered to enhance the viability of islands and their contribution to global biodiversity.

SEE ALSO THE FOLLOWING ARTICLES

Acclimatization Societies / Brown Treesnake / Hawaiian Islands: Invasions / Invasional Meltdown / New Zealand: Invasions / Small Indian Mongoose

FURTHER READING

Buckley, L. B., and W. Jetz. 2007. Insularity and the determinants of lizard population density. *Ecology Letters* 10: 481–489.

D'Antonio, C. M., and T. L. Dudley. 1995. Biological invasions as agents of change on islands versus mainlands (103–121). In P. M. Vitousek, L. L. Loope, and H. Adsersen, eds. *Islands: Biological Diversity and Ecosystem Function*. Berlin: Springer-Verlag.

Howald, G., C. J. Donlan, J. P. Galván, J. C. Russell, J. Parkes, A. Samaniego, Y. Wang, D. Veitch, P. Genovesi, M. Pascal, A. Saunders, and B. Tershy. 2007. Invasive rodent eradication on islands. *Conservation Biology* 21: 1258–1268.

Sax, D. F., and S. D. Gaines. 2008. Species invasions and extinction: The future of native biodiversity on islands. *Proceedings of the National Academy of Sciences, USA* 105: 11490–11497.

Simberloff, D. 1995. Why do introduced species appear to devastate islands more than mainland areas? *Pacific Science* 49: 87–97.

Zavaleta, E. S., R. J. Hobbs, and H. A. Mooney. 2001. Viewing invasive species removal in a whole-ecosystem context. *Trends in Ecology and Evolution* 16: 454–459.

KUDZU

DANIEL SIMBERLOFF

University of Tennessee, Knoxville

Kudzu (*Pueraria montana* variety *lobata*), a large, woody, perennial vine in the family Fabaceae (legumes), was introduced from Japan to the United States in 1876 and was subsequently planted widely in the South as animal forage and to prevent soil erosion. Its popularity peaked in the 1930s, but by the 1950s it was recognized as highly invasive; it was listed by the U.S. Department of Agriculture as a weed by the 1970s. Today it is iconic: it is probably the best known invasive plant in North America, a widely used general metaphor for rampant, uncontrolled growth; perhaps no other invasive environmental feature is as closely identified with the South in the public mind.

HISTORY OF THE INTRODUCTION AND SPREAD

Kudzu was introduced to the United States in 1876 at the Philadelphia Centennial Exposition, and again at the 1883 New Orleans Exposition. It was initially marketed as an ornamental plant, especially good for providing shade on southern porches. Seeds were widely available, including through mail-order catalogs. By the early 1900s, the U.S. Department of Agriculture promoted kudzu as a forage crop for cattle, and a Florida couple, Charles and Lillie Pleas, developed a thriving business selling root crowns and seedlings for this purpose through the mails. By the 1930s and 1940s, the Soil Erosion Service, later renamed the Soil Conservation Service (SCS), was recommending kudzu to stem soil erosion, which was an increasingly well-publicized problem. Channing Cope in Georgia became known as the "Kudzu King" and founded the Kudzu Club of the Americas during this period, aggressively promoting kudzu on his daily radio show and through other media as a way to stop erosion and replenish depleted soils. The SCS grew kudzu seedlings in nurseries and distributed over 85 million of them to southern landowners; the U.S. government paid landowners $19.75 per hectare to plant kudzu, while the Civilian Conservation Corps was assigned to plant kudzu in parks and other public lands. These plantings, combined with those for forage, led to widespread plantings—over 1.2 million ha by 1946.

Social and agricultural changes in the South greatly aided the spread of kudzu. Cotton farming failed widely because of poor and depleted soils and the invasion of the boll weevil, and the first half of the twentieth century also saw a major migration from the rural South to cities. These factors led to the conversion of much farmland to secondary forest, which, in turn, left vast areas of kudzu to uncontrolled growth. The vine spread rapidly, and the U.S. Department of Agriculture removed kudzu from its list of permissible cover plants in 1953 and listed it as a weed in 1979. In 1997, kudzu achieved the status of Federal Noxious Weed by an act of the U.S. Congress. In 1999, an end-of-the-millennium issue of *Time* magazine cited the introduction of kudzu as one of the 100 worst ideas of the century. It has been recorded in at least 28 states, is currently spreading by about 50,000 ha per year, and occupies around 3,000,000 ha in the United States.

BIOLOGY

Kudzu requires the support of other plants (or other structures) to reach high light levels near the top of the forest canopy, and it is superbly adapted to climb by high

FIGURE 1 Kudzu overgrowing native vegetation in Tallahassee, Florida, with the Florida state capitol building in background. (Photograph courtesy of Don C. Schmitz.)

allocation of fixed carbon to stems, branches, and leaves instead of support structures (Fig. 1). In particular, its high allocation to photosynthetically active leaves rather than woody material fosters its famed rapid growth; stems can grow up to 30 m in a growing season, or about a foot per day. It is also characterized by frequent rooting of stems where they contact the soil, high leaf area and photosynthetic rates, large hydraulic capacitance, and (as a legume) the ability to fix atmospheric nitrogen through symbiotic *Rhizobium* bacteria in root nodules. This combination of traits renders kudzu a ferocious competitor in many southeastern forests. Its high productivity results in kudzu layers up to 2.5 m thick in high light environments, but kudzu maintains reduced photosynthesis and growth rates even in substantial shade. By fixing nitrogen and building up soil nitrogen beneath its canopy, kudzu can thrive in areas of nutrient-deficient soil that inhibit many other plants.

Because kudzu does not allocate carbon to woody support structures, it is able to allocate it to root growth, with primary roots reaching over 180 kg and penetrating the soil at a rate of 0.03 m per day to depths of over 3 m. These roots are succulent, storing starch, nitrogen, and water, and are used as a root crop in Asia. Vines grow out in all directions from a rooted node, and stems can reach over 60 m in length. Stems from a single node become disconnected from one another within three years, so individual ramets quickly become physiologically independent of one another, and rooted nodes from primary, secondary, and even higher-order branches lead to densities up to tens of thousands of ramets per hectare. This makes control difficult, because all rooted nodes must be treated chemically or physically to eliminate a population.

Kudzu quickly orients its leaves to stay parallel to solar rays, which reduces excess radiation at midday that cannot be used in photosynthesis, raises leaf temperature, and causes water loss. This ability tends to keep kudzu leaves near the optimum irradiance and temperature for photosynthesis. Furthermore, light not intercepted by upper layers of a kudzu canopy can pass through to lower layers, where it may be used for additional photosynthesis. This photosynthetic efficiency helps kudzu overtop even high trees within a few years of colonization. There has been little research on the ability of kudzu to withstand drought, but the large volume of water in its root structures suggests that anecdotal reports of good drought resistance are accurate.

Although vegetative reproduction is believed to be mainly responsible for kudzu's astounding spread in the United States, kudzu does flower as well. However, seed set is low at the northern edge of its American range because kudzu is pollinator-limited; seed set is substantially higher in the Deep South, where several native and nonnative insects are known to pollinate kudzu. Kudzu is believed to be primarily cross-pollinated, but some selfing may occur. Without scarification, seeds germinate at a low rate, and scarification rates (by fire or other means) may be low in nature. Most seeds are lost to fungal pathogens and insects. Kudzu has a great amount of genetic variation in the United States with little regional patterning, probably reflecting the multiple introductions, and perhaps gene exchange through sexual reproduction between individuals from different populations.

IMPACTS

Kudzu is legendary as a high-impact invader, partly because it can so easily be seen blanketing vegetation, billboards, and even buildings and abandoned cars along highway and railroad rights-of-way. However, surprisingly little quantitative research has been published on its ecological impact. There are many reports of kudzu's shading and ultimately overgrowing other plants, girdling saplings, and replacing native vegetation in an ever-spreading area that looks like a monoculture but that may contain a number of plant species that leaf out in the spring before kudzu does. Importantly, there is no evidence that such stands undergo succession, at least in the short term—that is, the replacement is long-term in the absence of intervention. Kudzu is also suspected of exacerbating storm damage by linking trees so that they fall together, and of retarding recovery by quickly dominating treefall areas. Kudzu's fixing of nitrogen and occupancy of such a large area of the South suggest that it may contribute to several problems associated with increased nitrogen, including local extinction of native species adapted to low-nitrogen soils and eutrophication of water bodies. There is also concern that massive kudzu stands affect regional air quality through the release of isoprene, a photochemically reactive hydrocarbon.

The northern range limit of kudzu is believed to be set by cold temperature, as leaves are killed by the first hard frost, and leafing out in the spring follows that of many native trees. However, kudzu responds rapidly and very positively to increases in carbon dioxide. The fact that carbon dioxide and temperature are increasing implies that kudzu will spread northward, a movement that may be exacerbated by increased forest fragmentation that will produce the kind of high-light conditions favoring kudzu.

Kudzu has a substantial economic impact. The cost in lost forest productivity is estimated at up to $500 million per year, while power companies spend about $1.5 million annually to manage kudzu, which nevertheless causes occasional power outages, primarily by bringing down power lines. In addition, railroads and national and state parks incur costs controlling kudzu. A growing major cost arose when Asian soybean rust (*Phakopsora pachyrhizi*), a fungus, first appeared in the continental United States in 2004, probably blown by one of several hurricanes that year, as in 2002 it had reached South America. The rust is now found in 14 states, mostly in the South, and kudzu serves as a host, as do soybeans and several other legumes. That kudzu is so widely distributed, and that it serves as a reservoir for a damaging pest of soybeans, adds another cost to those it already imposes, although the exact amount is difficult to estimate. It includes fungicide that must now be applied to many soybean plantings, plus lost production. Kudzu genotypes vary greatly in susceptibility to the rust, complicating efforts to predict the impact of kudzu on soy production via this route.

MANAGEMENT

Controlling kudzu on a large scale in the United States has proven difficult, although small stands have been successfully cut back and kept from proliferating by manually cutting the stems and ripping out as much of the root as possible. Herbicides can also be used on small stands, but such herbicides are expensive and must be used repeatedly for many years. The fact that connections between rooted nodes are quickly severed means that every ramet must be treated independently.

The impossibility of controlling large areas of kudzu physically or chemically has led to great interest in biological control. In fact, in the United States, many insects attack kudzu, and it is prone to bacterial blights. None of these have substantially stemmed its spread, however. Several insects, both native and introduced, that attack kudzu also attack soy, so their development and rearing as biological controls would be problematic, if not impossible. Hope springs eternal, however, and explorations in the native range of kudzu continue for a natural enemy, or a group of them, that might help to limit kudzu in the United States. In its native Asian range, kudzu does not achieve the dramatic dominance that it has in the South. Many natural enemies, including insects and pathogens, have been identified in Asia, but there is no evidence that they limit wild kudzu populations there. Probably at least part of the reason they are not limiting is that few of them attack the massive root. However, Asians use the root for food and medicine, which may aid in control. In the United States, frequent advocacy of employing kudzu in cooking (there are even kudzu cookbooks) and occasional exploration of its herbal properties have not led to its widespread use, much less served as a measure of its control in nature. However, biofuel enthusiasts in the United States have suggested that kudzu could be a superb biofuel feedstock because of its rapid growth, although no current technology can yet effect this transformation on a commercial scale.

SYMBOL AND METAPHOR

Although kudzu is found in the Middle Atlantic states and the Midwest, its growth is most exuberant and evident in the South, where it has embedded itself in the culture in many ways. By 2001, there were at least 19 streets named "Kudzu," as well as innumerable businesses (Fig. 2). In any state in the Deep South, postcards abound showing

FIGURE 2 Kudzu Grill in Sandersville, Georgia. (Photograph courtesy of Derek H. Alderman.)

kudzu blanketing trees, houses, and other objects. Reports of corpses being found in kudzu are frequent—and not all apocryphal. A popular southern-themed comic strip is "Kudzu," by Doug Marlette, and several bands have included "Kudzu" in their names.

Even beyond the South, kudzu has come to serve as an exploded metaphor, representing not only the South itself but also other invasive species, horribly grasping and smothering interlopers of all kinds, disastrous unintended consequences, and damaging unconstrained development. Thus, federal legislation changing telecommunication practices was seen as "eliminating the regulatory kudzu that has been strangling our industry," while a judicial decision on arbitration clauses in contracts analogized the practice to kudzu vegetation creeping across the South, and a lawyer protesting prosecutorial aggressiveness by the Justice Department attacked the False Claims Act as having "grown like kudzu," with its "tendrils" having been spotted in a variety of types of actions. A newspaper termed nutria the "kudzu of the marsh," while another termed hydrilla "kudzu's aquatic cousin."

SEE ALSO THE FOLLOWING ARTICLES

Biological Control, of Plants / Competition, Plant / Life History Strategies / Nitrogen Enrichment / Vegetative Propagation / Vines and Lianas

FURTHER READING

Alderman, D. H. 1998. A vine for postmodern times: An update on kudzu at the close of the twentieth century. *Southeastern Geographer* 38: 167–179.

Alderman, D. H. 2005. Channing Cope and the making of a miracle vine. *Geographical Review* 94: 157–177.

Alderman, D. H., and D. G. Alderman. 2001. Kudzu: A tale of two vines. *Southern Cultures* 7(3): 49–64.

Blaustein, R. J. 2001. Kudzu's invasion into southern United States life and culture (55–62). In J. A. McNeeley, ed. *The Great Reshuffling: Human Dimensions of Invasive Species*. Gland, Switzerland: IUCN.

Forseth, I. N., Jr., and A. F. Innis. 2004. Kudzu (*Pueraria montana*): History, physiology, and ecology combine to make a major ecosystem threat. *Critical Reviews in Plant Sciences* 23: 401–413.

Miller, J. H. 1996. Kudzu eradication and management (137–149). In D. Hoots and J. Baldwin, eds. *Kudzu: The Vine to Love or Hate*. Kodak, TN: Suntop Press.

Miller, J. H., and B. Edwards. 1983. Kudzu, *Pueraria lobata*: Where did it come from and how can we stop it? *Southern Journal of Applied Forestry* 7: 165–169.

Pappert, R. A., J. L. Hamrick, and L. A. Donovan. 2000. Genetic variation in *Pueraria lobata* (Fabaceae), an introduced, clonal, invasive plant of the southeastern United States. *American Journal of Botany* 87: 1240–1245.

Sun, J.-H., Z.-C. Li, D. K. Jewett, K. O. Britton, W. H. Ye, and X.-J. Ge. 2005. Genetic diversity of *Pueraria lobata* (kudzu) and closely related taxa as revealed by inter-simple sequence repeat analysis. *Weed Research* 45: 255–260.

Sun, J.-H., Z.-D. Liu, K. O. Britton, P. Cai, D. Orr, and J. Hough-Goldstein. 2006. Survey of phytophagous insects and foliar pathogens in China for a biocontrol perspective on kudzu, *Pueraria montana* var. *lobata* (Willd.) Maesen and S. Almeida (Fabaceae). *Biological Control* 36: 22–31.

L

LADYBUGS

EDWARD W. EVANS
Utah State University, Logan

WILLIAM E. SNYDER
Washington State University, Pullman

Until recently, it might have been surprising to think of ladybugs as invasive. Called ladybugs by Americans, ladybirds by Europeans, and more formally, lady beetles or ladybird beetles by professional entomologists, these insects constitute the beetle family Coccinellidae (insect order Coleoptera). Their bright colors, diverse feeding habits, and widespread abundance have always attracted attention. They were first named ladybirds centuries ago by European Christians, in honor of Our Lady, the Virgin Mary. Most are predators of other insects or mites, but some solely consume plants or Ascomycete fungi (sooty molds). With inadvertent or intentional human assistance, many ladybug species have spread to new parts of the world. A few have also become invasive.

ESTABLISHMENT IN NEW GEOGRAPHIC REGIONS

Among the most destructive introduced species of ladybugs throughout the world are pests of cultivated crops. The Mexican bean beetle (*Epilachna varivestris*) is a well-known example. Host plant ranges are fairly narrow for these pest ladybugs, which are members of the plant-feeding beetle subfamily Epilachninae. Consequently, these introduced species largely confine themselves to crop fields. Ladybugs from other subfamilies that attack and consume aphids, scale insects, and other pest insects or mites have been released intentionally and repeatedly throughout the world to enhance natural pest control ("biological control"). Upon release, these ladybugs in some cases have spread from agricultural settings and have attained very high numbers in other habitats. In so doing, they have joined the ranks of invasive introduced species. It is often unclear whether these ladybugs' establishment resulted directly from purposeful introductions, or whether such efforts failed and establishment instead arose from inadvertent introduction. Either way, these invasive introduced ladybugs bring a complex, worrying mix of potentially strong negative as well as positive impacts to new regions of the world.

INTENTIONAL INTRODUCTIONS FOR BIOLOGICAL CONTROL AND INVASIONS

Enthusiasm among biological control practitioners for introducing ladybugs, although now often greatly tempered, was previously strong and widespread. It was fueled by stunning, early success. In the late nineteenth century, the nascent citrus industry of southern California verged on sudden collapse, under life-threatening attack by an introduced scale insect, the cottony cushion scale (*Icerya purchasi*). Upon being dispatched to Australia, USDA entomologist Albert Koebele in 1888–1889 sent back to Los Angeles 129 adults of the ladybug *Rodolia cardinalis* (the vedalia beetle). These took well, and their many descendants soon rid the citrus groves of serious scale infestations. Shipments of oranges resumed their rapid expansion. The vedalia beetle continues to this day to be an effective biological control agent and is well behaved in restricting its activities largely to targeted crops. The spectacular success of the vedalia introduction inspired many additional efforts worldwide to introduce ladybugs. These have included species that feed primarily on aphids but which can be quite broad in the range of aphid species (and other prey) that they attack. In North America alone, at least 179 species of ladybugs have been introduced for biological control following the success of the vedalia beetle.

In North America, recent experiences, especially with two introduced ladybug species that primarily consume aphids, have proved alarming to both the general public and the scientific community. This alarm now has spread worldwide along with the second of these two invasive ladybugs. This has resulted in a fundamental reassessment of the biological control practice of introducing ladybugs that are not strict specialists in what they eat. Although the source of establishment (intentional or inadvertent release) is unclear, these species have invaded habitats other than the agricultural areas for which they were targeted, and they may well be having large adverse effects that go far beyond the intended suppression of insect pests.

SPREAD OF INVASIVE LADYBUGS AND PUBLIC REACTION

The first of these two species to become invasive in North America was *Coccinella septempunctata*, the seven-spot ladybug (Fig. 1). It is among the most common, familiar ladybirds of Europe. Efforts to introduce the seven-spot ladybug to both eastern and western North America occurred from the 1950s through the early 1970s. These efforts seemed unsuccessful; no beetles were subsequently recovered near release sites. However, feral populations of the seven-spot ladybug were later found in New Jersey and Quebec in 1973; these may have established from accidental introductions associated with transoceanic shipping. Over the next five years, more than 500,000 individuals

FIGURE 1 *Coccinella septempunctata* on bark, Carisbrooke Castle, Isle of Wight, England. (Photograph courtesy of P. M. J. Brown, Anglia Ruskin University.)

FIGURE 2 An aggregation of *Harmonia axyridis* overwintering indoors. (Photograph courtesy of M. Potter, University of Kentucky.)

were collected in New Jersey and released throughout the eastern United States (and as far west as New Mexico and Utah). Additional release efforts in western states were made in the 1980s for biological control of the Russian wheat aphid (*Diuraphis noxia*). The seven-spot ladybug also dispersed strongly on its own. By the early 1990s, it had become established essentially throughout all of North America, and it had penetrated many habitats and ecological communities.

As in Europe, the seven-spot ladybug is most abundant in herbaceous habitats in North America. But these include many natural and semi-natural habitats (e.g., former agricultural fields undergoing succession to natural vegetation) as well as targeted agricultural crops (e.g., alfalfa, wheat, and soybean). Furthermore, the seven-spot ladybug occurs in arboreal habitats, and these also include woodlots and shelterbelts as well as orchards. The high numbers that the seven-spot ladybug frequently attains are widely thought to reduce pest aphid populations in North America. It is unclear, however, whether such impact is simply replacing that of outcompeted, native ladybugs now in decline (see discussion below).

On the whole, the lay public expressed only modest concern over the possible loss of native ladybug diversity, and other potential adverse, nontarget effects, as the invasive seven-spot ladybug established itself as one of North America's most abundant ladybugs. Far more strongly negative public reaction has occurred following the more recent increase in abundance and spread of another invasive ladybug, first in North America and subsequently elsewhere (especially in Europe). This is the harlequin ladybug (*Harmonia axyridis*, often referred to also as the multicolored Asian ladybug or Halloween ladybug Fig. 2), a species that lives at least as much on trees (arboreal habitats) as in herbaceous habitats. Numerous introductions of this species to North America were made from 1915 on, but, as with the seven-spot ladybug, these appeared unsuccessful. Then, in 1988, a population was discovered in southern Louisiana, 300–400 km from the nearest known release locations; again, accidental introduction associated with transoceanic shipping may be implicated. The harlequin ladybug spread rapidly thereafter (with how much human assistance is unclear), and, like the seven-spot ladybug before it, within 10 to 15 years, it had established itself throughout North America. Its rapid colonization of the entire continent likely occurred in part through spread also from a probable second accidental introduction in the Pacific Northwest. Since the mid-1990s, the harlequin ladybug has also been intentionally introduced for biological control to at least 12 European countries. It has now spread widely throughout continental Europe, and it crossed the English Channel to arrive in Britain in 2004. Transoceanic shipping from the southeastern United States is apparently responsible for the spread of this ladybug to other parts of the world as well, including South America and South Africa.

As an introduced predator, the harlequin ladybug does appear important in biological control; for example, it is the most numerous and voracious of ladybugs consuming the soybean aphid (*Aphis glycines*), a major pest of soybeans in the American Midwest. Interestingly, the soybean aphid is itself an invasive species also native to Asia, such that the two invasions have restored a natural predator–prey interaction between beetle and aphid. But, as with the seven-spot ladybug, and as discussed below, serious ecological concerns over adverse nontarget effects surround introductions of the harlequin ladybug.

Public concern in North America over ladybug introductions has grown, especially because of widespread, negative personal experiences with the harlequin ladybug. This species invades homes to spend the winter in mass aggregations of individuals that often number in the thousands. Homeowners' headaches are compounded because the ladybug oozes body fluid from its joints when disturbed ("reflex bleeding" that results in yellow-orange stains) and because it bites people and can trigger allergies. The harlequin ladybug also becomes an agricultural pest by feeding on fruits in the late summer and autumn. When accidentally crushed with grapes, it taints the flavor of wine. Much of the American public no longer views the introduction of aphid-eating ladybugs favorably. A similar change of heart is rapidly transforming many other parts of the world as well, as the harlequin ladybug continues to spread to new regions.

WHY SO SUCCESSFUL?

Both the seven-spot and harlequin ladybugs increased quickly in a variety of invaded habitats to become the dominant ladybugs present. Why are these invasive ladybugs so successful? Success is often suggested to arise from their generalist habits (e.g., broad habitat preferences and ranges of prey attacked), flexible life histories (e.g., ability to produce more than a single generation per year), and high reproductive potentials (e.g., ability to produce large numbers of relatively small eggs). The large expanses in North America of crops such as alfalfa and soybean and (for the harlequin ladybug especially) the large numbers of shade trees in urban settings often support very high numbers of aphids. These habitats may provide ideal places where these species can multiply. Both the seven-spot and harlequin ladybugs are large members of their family; their large body sizes promote high rates of reproduction to exploit these and other habitats. It is presently unclear whether agricultural and urban habitats strongly subsidize the large populations of these species found also in other habitats.

NONTARGET EFFECTS

As the seven-spot and harlequin ladybugs have spread as invaders, concerns have arisen over unintended, adverse effects on nontarget species. Even as these predators attack targeted aphid pests, ironically they might undermine existing biological control by interfering with other natural enemies; they can do so, for example, by consuming aphids parasitized by wasps and by adversely affecting the wasps' foraging behavior (adult females of wasps that are parasitoids [i.e., parasitic insects that seek out and lay their eggs in or on other insects such as aphids] have been found to avoid areas with chemical traces of the seven-spot ladybug's former presence). Furthermore, these predators may interfere with parasitism of other pest insects (e.g., larval beetles and caterpillars) if they consume sufficient numbers of aphids; such consumption denies parasitoids access to honeydew, the sugar-rich fluid excreted by aphids that is often an important source of nutrition for foraging parasitoid females. In North American wetlands, the harlequin ladybug feeds on larvae of beetles introduced as biological control agents of purple loosestrife.

Introduced ladybugs feed widely on aphid species, including many that are not pests. Predation by the seven-spot ladybug (a generalist predator) on lepidopteran eggs and larvae may also further endanger rare butterflies such as the Karner blue butterfly (*Lycaeides melissa samuelis*), a dwindling species associated with specialized and increasingly restricted habitats (e.g., greatly fragmented oak savannahs of the American Midwest). Similar concerns have been raised for monarch butterflies (*Danaus plexippus*) as preyed upon by the harlequin ladybug.

Potential adverse effects on native ladybugs have received the most attention. Numbers of some native species of ladybugs have declined markedly in various settings following the introductions of the seven-spot and harlequin ladybugs. In North America, for example, population declines have been reported in such crops as alfalfa, corn, and wheat as the seven-spot ladybug has established. The spread of the harlequin ladybug in Europe has led to dramatic declines especially of the formerly widespread and abundant two-spot ladybug (*Adalia bipunctata*) in a large number of arboreal habitats; a similar decline had been observed previously in North America.

Several general mechanisms to account for these declines have been proposed. First, multispecies aggregations of ladybugs typically develop rapidly as local populations of prey (e.g., aphids) reach high numbers. Invasive ladybugs may prevail in asymmetric, exploitative competition for prey among larvae. Second, ladybugs frequently prey upon each other (both intra- and interspecifically; adults and larvae eat eggs and other larvae, especially as aphids or other primary prey become scarce). Invasive ladybugs may also prevail in asymmetric interspecific (intraguild) predation. The harlequin ladybug is an especially aggressive and successful intraguild predator; it is well adapted chemically, morphologically, and behaviorally for both overcoming the defenses of other ladybugs and defending itself against them. (It may similarly be especially effective as an intraguild predator of other species such as lacewings and predatory flies that also accumulate at aphid outbreaks.) New molecular techniques to detect prey DNA within the digestive tract of their predators should illuminate how frequently invasive lady beetles eat their native relatives. Third, in another outcome of exploitative competition, adults of native ladybugs may abandon habitats by dispersing elsewhere if invasive ladybugs drive prey to very low levels or otherwise undermine habitat suitability. Such habitat compression or shift may explain the marked declines in abundances of native ladybugs in alfalfa fields of western North America, with reduction in the standing crop of aphids in these fields following the arrival of the seven-spot ladybug. A fourth possibility, hybridization (and possible sexual transmission of disease), may arise when invasive ladybugs mate with closely related native species, but evidence for this is lacking so far. Each of these four general mechanisms remains to be evaluated further for its possible role in the apparent population declines of native ladybugs when they interact and share habitats with invasive ladybugs.

THE FUTURE OF LADYBUG INVASIONS

It is striking that, for decades following the introduction of the vedalia beetle to California, subsequent worldwide introductions of other ladybugs did not result in alarming invasions, and yet in the late twentieth century, two of these ladybugs suddenly became widely and very alarmingly invasive. Is the world on the cusp of many additional such invasions? It is difficult to answer this question. On one hand, biological control practitioners are becoming increasingly cautious and likely will not introduce additional generalist predators such as the seven-spot and harlequin ladybugs. On the other hand, it may well be that neither of these species became invasive in North America directly as the result of biocontrol efforts. Instead, transoceanic shipping may be the culprit. If so, likely we can anticipate additional ladybug invasions in the future, associated with increasing global trade and travel (seven-spot ladybugs recently sailed with The Splendour of the Seas, as the cruise ship traveled for two weeks from Casablanca to Venice, with many ports of call along the way; see Further Reading below). In addition, other ladybug invasions are currently under way, but as yet, they are little studied, including those of such species as *Propylea quatuordecimpunctata* in northeastern North America and *Hippodamia* spp. in Chile. These points underscore the importance of continuing to study intensively the effects and impacts of the seven-spot ladybug, the harlequin ladybug, and other ladybugs as invasive species.

SEE ALSO THE FOLLOWING ARTICLES

Agriculture / Biological Control, of Animals / Competition, Animal

FURTHER READING

Brown, P. M. J., H. E. Roy, P. Rotheray, D. B. Roy, R. L. Ware, and M. E. N. Majerus. 2008. *Harmonia axyridis* in Great Britain: Analysis of the spread and distribution of a non-native coccinellid. *BioControl* 53: 55–67.

Elliott, N., R. Kieckhefer, and W. Kauffman. 1996. Effects of an invading coccinellid on native coccinellids in an agricultural landscape. *Oecologia* 105: 537–544.

Gordon, R. D. 1985. The Coccinellidae (Coleoptera) of America north of Mexico. *Journal of the New York Entomological Society* 93: 1–912.

Harmon, J. P., E. Stephens, and J. Losey. 2007. The decline of native coccinellids (Coleoptera: Coccinellidae) in the United States and Canada. *Journal of Insect Conservation* 11: 85–94.

Koch, R. L., and T. L. Galvan. 2008. Bad side of a good beetle: The North American experience with *Harmonia axyridis*. *BioControl* 53: 23–35.

Minchin, D. 2010. A swarm of the seven-spot ladybird *Coccinella septempunctata* (Coleoptera: Coccinellidae) carried on a cruise ship. *European Journal of Entomology* 107: 127–128.

Obrycki, J. J., N. C. Elliott, and K. L. Giles. 2000. Coccinellid introductions: Potential for and evaluation of non-target effects (127–145). In P. A. Follett and J. J. Duan, eds. *Nontarget Effects of Biological Control*. Boston: Kluwer Academic Publishers.

Snyder, W. E., and E. W. Evans. 2006. Ecological effects of invasive arthropod generalist predators. *Annual Review of Ecology, Evolution, and Systematics* 37: 95–122.

Weber, D. C., and J. G. Lundgren. 2009. Assessing the trophic ecology of the Coccinellidae: Their roles as predators and prey. *Biological Control* 51: 199–214.

LAG TIMES

JEFFREY A. CROOKS

Tijuana River National Estuarine Research Reserve, Imperial Beach, California

Lag times in invasion biology typically refer to either the delayed onset or relatively slow rate of an invasion event or process. These temporal lags are often associated with population-level dynamics such as increases in the number of invaders or their geographic range, but lags occur throughout all aspects of invasions, including human responses to them. Although lags provide an opportunity to examine basic ecological and evolutionary dynamics, a primary interest in them relates to their tendency to cause "ecological surprises," often resulting in profound changes within invaded ecosystems. An important principle regarding lags is that they, in and of themselves, do not necessarily decrease predictive ability. Some lags are inherent and expected, but a startling array of decades- or centuries-long lags has also been documented. Because scientific and management activities typically take place over far more compressed timescales, it is necessary to incorporate longer-term perspectives to understand and manage invasions effectively.

TEMPORAL ASPECTS OF INVASION AND THE LAG EFFECT

One of the fundamental goals of invasion biology is to understand the temporal nature of invasions—to answer the question "when?" In some instances, the initial invasion event can represent an ecological release whereby the invader, having left behind natural enemies, rapidly comes to dominate new territory. Notorious examples include the brown tree snake (*Boiga irregularis*) on Guam and the zebra and quagga mussels (*Dreissena spp.*) in North America. However, in many cases an ecological explosion appears to come only after the burning of a long fuse. For example, studies of tree naturalization in Germany have found that the average interval between first planting and first appearance in the wild is well over one hundred years, with some lags greater than three centuries. Similarly, ships have been moving organisms around on their hulls for centuries, yet new species are still routinely transported by this vector (even on long-established routes). Even more broadly, positive correlations between time since introduction and geographic range emerge on timescales of millennia, indicating that the dynamics of invasion have yet to play themselves out over long time scales.

The recent growth of the field of invasion biology has refined our understanding of the temporal development of invasions. Studies of invaders across all ecosystem types, as well as conceptual and mathematical models, have better defined the suite of factors that can produce lags. In some cases it is possible to assign specific causes to observed cases, but for many invasions, identifying the factors driving lags, or their cessation, remains problematic. Nonetheless, even the recognition of this lag effect can lead to better long-term invasion prevention and control strategies.

TYPES OF LAGS

Usage in Invasion Biology

In invasion biology, the term lag implicitly conveys the comparison of events or rates. It is typically used in relation to three different types of comparisons (Fig. 1), although this is rarely explicitly stated. In the first two types of lags, different phenomena are compared. In the first case, a lag describes a temporal offset between discrete events (Fig. 1A). For example, there is often some delay between the onset of vector activity and the actual arrival of new species in recipient areas. The second usage of lag relates to rates of processes when one rate is slower than the other (Fig. 1B). This is typically most evident when the processes initiate at the same time—an example being when two species introduced simultaneously spread at different rates.

The third type of lag involves comparing the same process at different times (Fig. 1C). A lag in this instance occurs when initial rates are slower than later rates, thus creating a concave curve. Common examples are plots of exotic species accumulation in an area over time, which often show increasing invasion rates in recent years

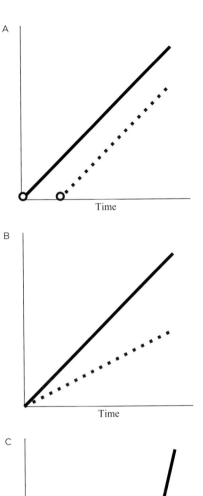

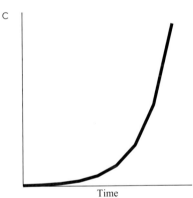

FIGURE 1 Three different types of lags, including (A) a temporal off-set between events or processes, (B) the relatively slow rate of one process compared to another, and (C) initial rates that are slower than later rates within the same process. (Adapted from Crooks, J. 2005, *Ecoscience* 12: 316–329.)

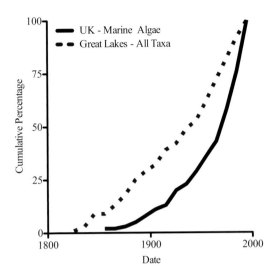

FIGURE 2 Increasing number of invaders appearing in different locations. Note the concave curves and the initial, slow rates of species accumulation. (Adapted from Crooks, J. 2005, *Ecoscience* 12: 316–329; Ribera Siguan, M.A. 2003, pp. 183–226 in G.M. Ruiz and J.T. Carlton, eds., *Invasive Species: Vectors and Management Strategies,* Washington, DC: Island Press; Ricciardi, 2001, *Canadian Journal of Fisheries and Aquatic Sciences* 58: 2513–2525.)

(Fig. 2). An important feature of these sorts of lags is that they often are not recognized as such until they are over.

Intrinsic versus Prolonged Lags

Although changes in invasion dynamics, especially abrupt ones, can challenge our understanding of invasions and impede effective management, some types of invasion-related lags are not surprising and should be expected. It is therefore worth distinguishing between those lags that are predicted (or predictable) and those that are prolonged beyond expectations. The most effective predictions typically come from quantitative models of well-documented invasion dynamics. The numbers of a given invader and the geographic area it occupies are among the most common invasion metrics, and both have lags that emerge from simple (but often robust) dynamics.

For numerical increase in an invasive population, an apparent lag arises from the familiar pattern of exponential growth. Here, populations grow relatively slowly early in an invasion, although the intrinsic rate of population increase (r) actually remains constant. Exponential growth is easily visualized by plotting population size on a logarithmic scale, which produces a linear relationship. For example, the population growth of invasive collared doves in the Netherlands on an arithmetic scale displays this characteristic early lag, but on a log scale the pattern is a straight line, indicating a lag due only to the dynamics of exponential growth (Fig. 3A). For a lag to be prolonged, then, plots on a log scale would have to reveal curves that inflect upward (reflecting an increase in r-values over time). Such a pattern is seen for the growth of the human population over the last 10,000 years (Fig. 3B).

A similar phenomenon causes lags in the geographic spread of an invasive population over time. If a population is expanding in all directions at a constant rate, it will form a series of equally spaced concentric circles, and the area

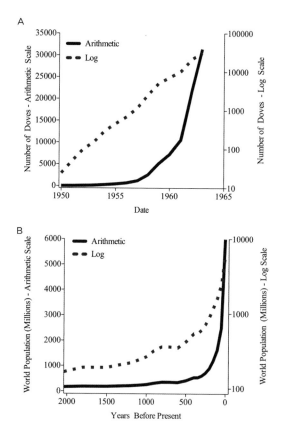

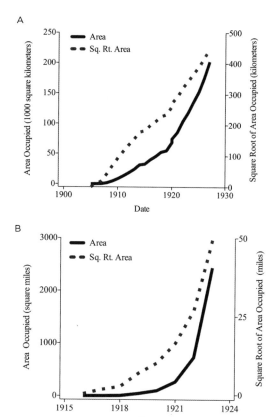

FIGURE 3 Intrinsic and prolonged lags related to numerical increase of invasive populations. (A) On an arithmetic scale, exponential population growth gives rise to a long lag before rapid population increase, as evidenced by collared doves in the Netherlands. On a logarithmic scale, however, the relationship is the straight line expected for exponential growth, thus demonstrating an intrinsic lag. (B) The pattern of growth of the world population over the last 10,000 years reveals a longer-than-expected lag, as the curve is nonlinear on a log scale. (Adapted from Hengeveld, R. 1989, *Dynamics of Biological Invasions*; Crooks, J.A. 2005, *Ecoscience* 12: 316–329; U.S. Census Bureau Historical Estimates of World Population.)

FIGURE 4 Intrinsic and prolonged lags related to the geographic spread of invasive populations. (A) The total area occupied by an invader spreading at a constant rate gives rise to an apparent lag, as seen with introduced muskrats in Europe. When plotted as square root of area, however, this intrinsic lag will be linearized. (B) Prolonged lags are revealed by nonlinear trends on such plots, as seen with the early spread of Japanese beetles in the United States. (Adapted from Hengeveld, R. 1989, *Dynamics of Biological Invasions*; Crooks, J.A. 2005, *Ecoscience* 12: 316–329.)

occupied will be a squared function of time (because the area of a circle is proportional to the square of its radius). When plotted as total area occupied over time, an early lag thus appears. As with numerical growth, this relationship is again easily transformed to determine whether this is an intrinsic or prolonged lag, in this case by plotting the square root of area against time. This is exemplified by the spread of muskrats introduced into Europe (Fig. 4A), which reveals that the initial lag was an expected, intrinsic one. The early spread of Japanese beetles in the eastern United States, however, reveals a curve that inflects upward, and thus a lag that is longer than expected (Fig. 4B).

It should be noted that a basic assumption for assessments of intrinsic versus prolonged lags is that invasions are documented when they initiate. For many

unintentional introductions, however, we tend to notice invaders only after they become abundant and conspicuous. This detection lag will be particularly important for invaders that are cryptic, in difficult-to-study habitats, or established before scientific study of the area. A term, "minimum residence time," has been developed to reflect the length of time since detection, explicitly recognizing that this might be an underestimate. Lags in detection will generally decrease the ability to detect prolonged lags.

LAGS ACROSS INVASION PHASES

Although population-level lags during invader establishment are best known, and it is relatively straightforward to distinguish conceptually between intrinsic and prolonged lags in these cases, there is potential for lags across all phases

TABLE 1
Examples of Lags and Their Potential Causes, Categorized by Phase of Invasion

Species	Invasion Characteristics	Lag (yrs)	Suggested Cause of Lag or Its Cessation
Arrival to New Location			
Zebra and quagga mussels (*Dreissena* spp.)	Ballast-borne invaders into Great Lakes, which began receiving ballast water in 1959	30	Unknown, but assumes mussels were found soon after invasion (i.e., no detection lag)
Jellyfish (*Rhopilema nomadica*)	Invaded Mediterranean through the Suez canal in the 1970s	100	More favorable conditions and increased traffic in the canal
Establishment (Numerical Increase and Geographic Spread)			
English cordgrass (*Spartina anglica*)	Coastal invader in the UK	40	Hybridization with native (*S. maritima*) and subsequent chromosome doubling
Figs (*Ficus* spp.)	Ornamental tree invaders in Florida	45	Introduction of pollinating wasp species
Brown anole (*Anolis sagrei*)	Caribbean lizard invading Florida	50–80	Repeated introduction increasing genetic variability
Cut-leaved teasel (*Dipsacus laciniatus*)	Weed invasion in the eastern United States	70	Spread along interstate highway system
Gypsy moth fungus (*Entomophaga maimaga*)	Fungal parasite of invasive gypsy moth in United States	80	Unknown, but suggestions include resting spores or serial introductions
Brazilian pepper (*Schinus terebinthifolius*)	Tree invader in Florida everglades	100	Anthropogenic changes such as lowering of water table
Smooth cordgrass (*Spartina alterniflora*)	Mudflat invader in Washington state	100	Pollen limitation in low-density patches (Allee effect)
Red sea mussel (*Brachidontes pharaonis*)	Early invader through the Suez Canal, but remained in low abundances in the Mediterranean	120	Increased space availability due to loss of native reef-forming molluscs
Collared dove (*Streptopelia decaocto*)	Spread into Europe and Africa	200	Increasing urbanization and changing climate
Oxford ragwort (*Senecio squalidus*)	Weed that initially escaped containment from an Oxford botanical garden in the 1700s	200	Anthropogenic spread and disturbance (including effects associated with World War II)
Establishment and Ecological Effects			
Milfoil weevil (*Euhrychiopsis lecontei*)	Native beetle in the United States that switched from native to introduced watermilfoil (*Myriophyllum* spp.)	30	Increasing abundance of exotic milfoil and time needed for weevils to switch species
Velvetleaf (*Abutilon theophrasti*)	Invasive weed in North America	200	Evolution of improved competitive abilities and anthropogenic spread

of invasion. However, for many other types of lags, it is difficult to generate all but the most coarse of predictive models with which to guide our expectation of how invasion dynamics should change over time. In such cases, we often revert to intuition and relative comparisons, although increased study should improve our ability to understand the temporal aspects of invasions and increase predictive power. A useful framework for assessing lags, as well as their causes, is to consider three distinct invasion phases: arrival, establishment, and ecological integration (Table 1).

Arrival

The successful arrival of a new invader in a region consists of that organism associating itself with a vector of transport (e.g., ballast water or nursery stock) and surviving transit and release into the receiving environment. Lags in

species arriving in a new area most commonly arise when there is a gap between initiation of the vector and species utilization of it. For a single species, this could be viewed as a temporal offset between two discrete events—vector initiation and initial invasion. Construction of new corridors of transport, such as roads or canals, provides a good opportunity to examine species utilization of newly available invasion vectors. In the case of the Suez Canal, there has been a steady increase in the rate of invasion, with lags in arrival of over a century for some invaders. This lag is due in part to changing conditions within the canal itself and highlights that characteristics of the vector will be important in shaping temporal responses of invading species.

Lags in arrival can also be viewed as changes in the cumulative number of species arriving into an area over time. At present it is difficult to generate expectations of how long species should take to use vectors, and how the cumulative curve of invasions should change over time. In general, we would expect that vector operation and invasion rate should be correlated—an increase in vector activity should lead to an increase in invasion rate. Many patterns of invader accumulation over time display early lags that have given way to more rapid rates of species accumulation, such as invasions into British shorelines and the Great Lakes.

The recent hastening of invasive species arrivals into new systems, with anthropogenic invasion rates now outpacing natural rates by many orders of magnitude, is no doubt fueled in large part by increased vector activity associated with globalization. Changes in the invader or the receiving environment, independently of vector characteristics, also can lead to changes in temporal patterns of species arrival. For example, shifts in the distribution or abundance of an invader in its native range might make it more amenable to transport. Similarly, changing abiotic or biotic characteristics of the receiving environment might allow a previously unsuccessful invader to suddenly find success. For arrival, this mostly relates to the ability of transported individuals to survive initial release, with the expectation that the more similar the receiving environment is to the source environment, the more likely it is that the transported individual will survive to form a founder population.

Establishment

As discussed previously, the temporal nature of the establishment phase of invasion has received the most scrutiny, and population models demonstrate that lags are apparent even with constant rates of intrinsic increase and radial expansion. However, lags that far exceed these intrinsic lags can also exist. One striking example is that of the Oxford ragwort, which is known to have escaped containment from a British botanical garden in the 1700s but failed to spread extensively until the mid-1900s. This was likely due in part to increased transport and bombing-induced disturbance associated with World War II. Decades- or century-long lags in naturalization may even be the norm, especially for longer-lived woody perennial species such as the trees in Germany. The sobering implication of these long lags is that even if all new introductions were to cease today, we would continue to see the continued naturalization, population increases, and spread of already introduced invaders well into the future (Fig. 5).

There are many possible causes of these prolonged lags. For geographic spread of species, some of the

same processes affecting initial arrival can also influence secondary spread, such as the opening of new corridors of transport. Demographic factors such as "jump dispersal," whereby occasional long-distance migration establishes invasion foci beyond the more slowly spreading invasion front, can lead to initially slow rates of spread followed by more rapid ones. Intraspecific interactions at low population-densities, or "Allee effects"—such as decreased probabilities of finding mates—might also cause lags until population densities are high enough to overcome undercrowding. For example, an Allee effect related to pollen limitation in low-density patches was suggested as the cause of the lag in the spread of exotic cordgrass (*Spartina alterniflora*) in Washington state.

Genetics have long been considered to be a potential cause of prolonged lags, given that founder effects and lack of genetic diversity typically (but not always) negatively influence incipient invasions with small population sizes. Recently, a host of evidence has demonstrated the critical role that genetics play. Even very small genetic changes can lead to dramatically different dynamics of invasive populations, such as a one-nucleotide mutation in the cucumber mosaic virus that allowed the exploitation of a new host. It has also been demonstrated that repeated, or serial, introductions can overcome founder effects and allow invasive populations to increase rates of population increase or geographic spread. For example, populations of the exotic lizard, *Anolis sagrei*, in Florida now have more genetic variability than populations within

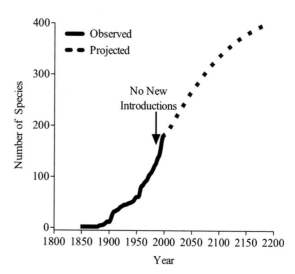

FIGURE 5 The rate of naturalization of woody plants in South Australia, including observed numbers (blue line) and predicted naturalizations (red dotted line), assuming no new introductions after 1985. (Adapted from Caley P., et al., 2008, *Diversity and Distributions* 14: 196–203.)

its source range, suggesting serial introductions that have allowed spread of the species after a decades-long lag.

Changing interactions with biotic and abiotic elements of ecosystems can also cause increases in invader populations. These ecological releases might be caused by the disappearance of species previously keeping invasive populations in check, such as consumers or competitors. In a number of cases (especially on islands), management efforts aimed at eradicating exotic consumers have allowed explosions of other invaders formerly held in check. Addition of species that facilitate invaders can also allow delayed population explosions of exotics. For example, the introduction of pollinating wasps into Florida and New Zealand have allowed invasive figs (*Ficus* spp.) to become much more invasive. Changing habitat factors will also play a role in invader dynamics. In particular, given observed relationships between degraded habitat quality and invader success, increasing disturbance may facilitate invasives at the expense of natives. Global climate change will also likely affect the distributions of invaders, and some invaders currently surviving in marginal environments may find conditions more hospitable in coming years.

Integration

Ecological integration refers to the emergence of invader impacts on community, ecosystem, and evolutionary dynamics. The total impact of an invader can be considered a function of the invader's density and range (both population-level dynamics discussed above), as well as of its per capita impact. Although predicting the time course of the development of impacts is fundamentally difficult, given that it involves both the invader and other biotic and abiotic elements of the ecosystem, it is useful to consider integration-related lags from two perspectives. First, some invader effects will take longer to manifest themselves than others, owing to the actions of the invader, the response of the recipient biota, or both. For example, the direct effects of an introduced predator consuming native prey might be apparent relatively quickly, as observed with the decimation of native bird species on Guam by the brown tree snake. If predation also leads to fragmentation of prey populations within refugia, however, the genetic and evolutionary effects associated with small, isolated populations would undoubtedly take longer to become apparent (but could be no less important). Although some types of impacts appear to be occurring more quickly than we might have predicted just a short while ago (e.g., evolution of new species in response to anthropogenic invasions), in general it is safe to assume

that the interactions we see today are the result of invasions that happened years ago.

Probably even more problematic than the slow development of ecological impacts are sudden changes in the types or intensity of impacts. These changes in per-capita impacts can arise from thresholds in invader dynamics, such as overcoming Allee effects related to predation efficiency or habitat alteration (e.g., critical densities needed for molluscs to form reefs). There also can be shifts in the species impacted by invaders. The biological control literature provides examples of this "host-switching," such as that of an introduced weevil (*Rhinocyllus conicus*) gaining the ability to attack native as well as exotic thistles. This highlights that over long time scales, invaders will likely be involved in evolutionary arms races with resident biota. A further example of this is a phenomenon termed "evolution of improved competitive abilities" (EICA), in which, once liberated from natural enemies, invaders are able to allocate more resources to increased competitive abilities at the expense of defense mechanisms. The long-term increase in the ability of species to outcompete natives has been suggested to be a potential contributor to the end of a 200-year lag in the velvetleaf (*Abutilon theophrasti*) in North America.

MANAGEMENT IMPLICATIONS

Lags have the potential to interact strongly with management action (or inaction) related to invasive species. Typically, potential management decisions are based on current conditions and perhaps some historical information. Recognition of lags emphasizes that decades or even centuries might pass before species undergo ecological explosions. Although it is typically difficult to determine whether a species is in a lag phase while the lag phase is still occurring, it is likely safest to employ the precautionary principle and assume that any invader has the potential to become an abundant nuisance. There are many examples of species that did nothing for decades or even centuries, only then to become problems. Thus the lag effect can lead to a false sense of security; it also highlights the highly idiosyncratic nature of invasions and emphasizes that even low-probability events may happen over the long term.

How to incorporate such considerations into management action is not always clear cut, however. Given the limited resources typically allocated to invader control, decisions must be made on how to balance effort devoted to a currently uncommon invader, which may or may not become a problem, versus known nuisances with demonstrable effects. Although achieving this balance will likely happen on a case-specific basis, it does argue that invasion control should be considered in the context of restoration—wherein the ultimate goal is not to get rid of target invaders but rather to restore natural integrity. It is also important to recognize that there also might be lags related to assessing effectiveness of control efforts caused by persistent legacies of removed invaders. Most fundamentally, however, it further reinforces the need to prevent unwanted invasions rather than to deal with them later.

Although the inability to recognize and predict lags poses serious management problems, there can be benefits of both time lags and understanding the processes that produce them. Lags imply that in many instances there might be relatively long windows of opportunity when invasive populations are small and confined. Somewhat paradoxically, however, action during this phase must occur at just the time when we often know least about the invasion. Working to understand the factors causing lags suggests that we might exploit invader vulnerabilities in order to control biological invasions better.

SEE ALSO THE FOLLOWING ARTICLES

Demography / Dispersal Ability, Animal / Dispersal Ability, Plant / Disturbance / Evolutionary Response, of Natives to Invaders / Evolution of Invasive Populations / Habitat Compatibility / Invasibility, of Communities and Ecosystems

FURTHER READING

Caley, P., R. H. Groves, and R. Barker. 2008. Estimating the invasion success of introduced plants. *Diversity and Distributions* 14: 196–203.

Crooks, J. A., and M. E. Soulé. 1999. Lag times in population explosions of invasive species: Causes and implications (103–125). In O. T. Sandlund, P. J. Schei, and A. Viken, eds. *Invasive Species and Biodiversity Management*. Dordrecht, the Netherlands: Kluwer Academic Publishers.

Crooks, J. A. 2005. Lag times and exotic species: The ecology and management of biological invasions in slow-motion. *Ecoscience* 12(3): 3116–3329.

Doak, D. L., J. A. Estes, B. S. Halpern, U. Jacob, D. R. Lindberg, J. Lovvorn, D. H. Monson, M. T. Tinker, T. M. Williams, J. T. Wooton, I. Carroll, M. Emmerson, F. Micheli, and M. Novak. 2008. Understanding and predicting ecological dynamics: Are major surprises inevitable? *Ecology* 89(4): 952–961.

Hengeveld, R. 1989. *Dynamics of Biological Invasions*. New York: Chapman and Hall.

Kowarik, I. 1995. Time lags in biological invasions with regard to the success and failure of alien species (15–38). In P. Pyšek, K. Prach, M. Rejmánek, and M. Wade, eds. *Plant Invasions: General Aspects and Special Problems*. Amsterdam: SPB Academic Publishing.

Liebhold, A., and J. Bascompte. 2003. The Allee effect, stochastic dynamics and the eradication of alien species. *Ecology Letters* 6: 133–140.

Pyšek, P., and V. Jarošík. 2005. Residence time determines the distribution of alien plants (77–96). In S. Inderjit, ed. *Invasive Plants: Ecological and Agricultural Aspects*. Basel: Birkhauser Verlag.

Sakai, A. K., F. W. Allendorf, J. S. Holt, D. M. Lodge, J. Molofsky, K. A. With, S. Baughman, R. J. Cabin, J. E. Cohen, N. C. Ellstrand, D. E.

McCauley, P. O'Neil, I. M. Parker, J. N. Thompson, and S. G. Weller. 2001. The population biology of invasive species. *Annual Review of Ecology and Systematics* 32: 305–332.

Strayer, D. L., C. T. Eviner, J. M. Jeschke, and M. L. Pace. 2006. Understanding the long-term effects of species invasions. *Trends in Ecology and Evolution* 21: 645–651.

LAKES

HUGH J. MACISAAC

University of Windsor, Ontario, Canada

Lakes are essential resources for and are extensively exploited by human populations, which has resulted in a number of perturbations including introduction of nonindigenous species (NIS). Once established, some NIS spread in a series of secondary introductions to new lakes by either natural or human-mediated mechanisms. Nonindigenous species are associated with a wide array of physical, chemical, and biological impacts. Management efforts for invertebrates and fishes have focused principally on preventing new invasions and curtailing the spread of established NIS to new lakes, while programs for macrophytes are directed at both prevention and in-lake control.

SPECIES DISPERSAL TO LAKES

Species may enter lakes by natural or human-mediated dispersal. Because lakes provide drinking and industrial water, food, recreation, hydroelectricity, and transportation, their condition is of paramount importance. One consequence of this extensive use has been the introduction of many NIS to lakes, particularly macrophytes, fishes, and invertebrates (Fig. 1). The human-mediated introduction of species can cause severe industrial impairment and disrupt ecological and aesthetic qualities of lakes. For this reason, the comparative rate of introduction of NIS by humans versus natural dispersal is of interest. Ecologists dating to Darwin have assumed that natural dispersal was widespread and reflected by the semi-cosmopolitan distribution of zooplankton species in lakes. Recent analyses have indicated that the rate of human-mediated species introduction may exceed the background rate by 10,000 times for some species. This exceptional difference has focused attention on the need to prevent new invasions. Before this goal can be achieved, however, it is necessary to understand how species can be transported to lakes.

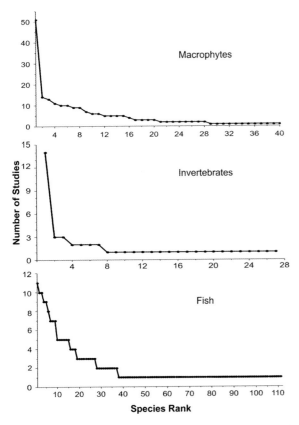

FIGURE 1 Relative dominance in published studies on lakes of nonindigenous macrophytes, invertebrates, and fishes based upon ISI Web of Science searches. In each case, searches consisted of combining searches for six synonyms for *nonindigenous* ("introduced," "exotic," "alien," "colonizing," "nonnative") with "lake" and with a taxon identifier. A seventh term, "invasive," was used in addition to the six terms for *nonindigenous*, and results were used only if the study considered the species as nonindigenous. Taxon identifiers were "macrophytes or aquatic plants," "invertebrate," or "fish or Pisces." Searches were conducted from 1965 through August 12, 2009, for macrophytes and invertebrates, and for 2006 and 2007 for fish, because the latter are much better studied than either of the other groups. The best-studied nonindigenous macrophytes in lakes include *Myriophyllum spicatum* (51 studies), *Hydrilla verticillata* (14), *Elodea canadensis* (13), *Potamogeton crispus* (11), and *Lagarosiphon major* and *Eichhornia crassipes* (10 each). The best-studied nonindigenous invertebrates in lakes include *Bythotrephes longimanus* and *B. cederstroemi* (18 studies), *Dreissena polymorpha* and *D. rostriformis bugensis* (12 studies), and *Cercopagis pengoi* (3 studies). The best-studied nonindigenous fishes in lakes include *Cyprinus carpio* (11 studies), *Oncorhynchus mykiss* and *Salmo trutta* (10 studies each), and *Lepomis gibbosus* and *Neogobius melanostomus* (9 studies each).

Natural Dispersal Mechanisms

Unlike Darwin, we are now aware that the distribution of many lake species is limited to specific continents, and often to specific regions within continents. These large-scale patterns suggest that constraints on dispersal have strongly limited the distribution of life in lakes across the globe. A variety of natural mechanisms are available for

NIS to enter lakes. In the simplest case, species extend their ranges by streams or rivers that connect colonized and uncolonized lakes. This form of advective introduction is notable in that it typically occurs in a single direction only, downstream from the source population. This mechanism can introduce virtually all species present—ranging from viruses and bacteria to fishes—in bulk transfers of water. This mechanism was undoubtedly important to initial dispersal of lake species following retreat of continental glaciers. For example, the opossum shrimp *Mysis diluviana* (formerly *relicta*) extended its range to deep lakes around the Great Lakes following glacier retreat at the end of the Pleistocene Epoch. The importance of advective dispersal diminishes strongly over time, as all lakes that can be colonized through the rivers connecting them eventually become so. In cases where a new NIS is introduced by humans, however, advection can be a major mechanism for subsequent spread of the species (see below). Advection influences only regional spread and cannot account for long distance transfers, including between continents. Advection has played an important role in the secondary spread of species including zebra mussels *Dreissena polymorpha*, quagga mussels *D. rostriformis bugensis*, fanwort *Cabomba caroliniana* in North America, and many Ponto-Caspian species in the Rhine River (see below).

A second mode of natural dispersal is wind. Large windstorms blowing from arid regions—such as the Sahara-Sahel region in Africa—contribute billions of tons of aerosolized dust to the atmosphere each year and can be important agents of long-range dispersal of bacteria, viruses, fungi, and plant pollen. Unlike the spatially constrained dispersal that may occur using stream connections, dispersal of species by wind (and rain) can occur over a broad region to lakes unconnected to one another. This mechanism also is strongly unidirectional. The importance of wind is inversely dependent on the size of the species, with small organisms such as bacteria and fungi being moved readily and larger species such as fishes being moved rarely and only by water spouts or gale-force winds. Over time, the importance of wind dispersal is likely to diminish, as species able to disperse will have already colonized downwind areas. However, if climate change alters wind patterns, then species that have not yet spread may have an opportunity to do so.

Two attributes favor dispersal of many aquatic taxa: asexual reproduction and production of resting stages. Asexual reproduction is advantageous because colonization by a single parthenogenetic female can be followed by repetitive cycles of asexual reproduction, resulting in rapid population growth, high population density, and a reduced probability of local extinction. Aquatic plants like the highly invasive Eurasian watermilfoil *Myriophyllum spicatum* exhibit extreme flexibility with respect to reproduction. This and many other invasive macrophyte species can reproduce sexually, resulting in seed production, or asexually by production of stolons or fragmentation of parental tissues. The latter vegetative capability poses a severe challenge for mechanical control of introduced plant populations (see below). A second reproductive attribute that favors colonization by many NIS is production of small, light, diapausing stages (e.g., cysts, gemmules, statoblasts, resting eggs, ephippia) that resist adverse conditions including desiccation. If these stages are carried aloft to aquatic habitats with favorable conditions, they may ex-cyst or hatch into larger, heavier free-living adult stages that are far less likely to be transported by wind. As with adults, the larger and heavier the resting stage, the less likely it is to be transported by wind.

Experimental studies show that large numbers of lake-dwelling species are transported by wind and rain. Two groups of taxa tend to dominate these colonization studies: crustacean zooplankton and rotifers. A study in South Africa identified 17 taxa of invertebrates that were collected in windsocks. Most of these specimens were resting eggs of cladoceran crustaceans, although anostracans, turbellarians, copepods, and branchiopods were also collected. Another study revealed dispersal by wind or rain of 13 rotifer species, two copepods, and seven cladoceran species to experimental mesocosms from which animal vectors were excluded. Other groups that colonized mesocosms included water mites, flatworms, and ostracods. Colonization can occur very rapidly, but in some cases, it can be delayed by up to a year. Early colonists tend to be asexual rotifer species. A suite of other zooplankton species may never colonize, highlighting the stochastic nature of colonization. Other studies on experimental ponds have had similar results, with colonization of some zooplankton species occurring within days or weeks, particularly those over short distances. As distance from the source increases, colonization tends to decline.

Colonization of a pond by one species may positively or negatively alter establishment by subsequent invaders. These so-called "priority effects" have been demonstrated in both laboratory cultures of lake phytoplankton and with zooplankton in experimentally created ponds. Priority effects are potentially important in terms of community composition of lakes and ponds, because different community assembly trajectories and species assemblages

may result depending on the order in which species invade. Recent work suggests that priority effects may extend to genotypes as well, with early colonizing genotypes producing sufficiently large propagule banks as to saturate the environment and repel future invasions by other genotypes of the same species.

The application of molecular markers has helped to identify sources and frequencies by which lake species such as crustaceans and bryozoans disperse to lakes. These studies provide information that is otherwise unobtainable from studies that employ experimental pools, mesocosms, or newly created ponds. Initial studies of genetic composition of colonizing species were limited to allozyme differences between source and destination populations, although these markers have largely been supplanted by mitochondrial and microsatellite markers that provide greater resolution. For example, spread of the fishhook waterflea *Cercopagis pengoi* to the Great Lakes was traced using a mitochondrial gene to a population living in the Baltic Sea, which itself was introduced from a Ukrainian port on the Black Sea. Mitochondrial markers have also been used to examine the spread of many other NIS that have spread to lakes in Europe and North America, including mysid shrimp *Heimysis anomala* and *Limnomysis benedeni,* round goby *Neogobius melanostomus,* tubenose goby *Proterorhinus semilunaris,* spiny waterflea *Bythotrephes longimanus,* and amphipod *Echinogammarus ischnus.* Microsatellite markers provide the highest resolution and have recently been used to survey introduced populations of spiny waterfleas, zebra and quagga mussels, round goby, sockeye salmon *Oncorhynchus nerka,* lake trout *Salvelinus namaycush,* northern pike *Esox lucius,* sculpin *Cottus* spp., whitefish *Coregonus,* and common reed *Phragmites australis* in lakes across Europe and North America.

ANIMAL-MEDIATED DISPERSAL Interactions between species may play a pivotal role in the dispersal of lake species. Animals transfer other species either externally (ectozoochory) or internally (endozoochory). While vectors for spread could potentially involve any species that uses shoreline or lake habitat and thereafter moves to another lake (e.g., moose, beaver), most attention has focused on waterbirds and fish vectors. Nonindigenous species could be transferred live or as resting stages. Local movements by waterbirds could disperse NIS regionally, while long-distance migrations have the potential to effect long-distance dispersal. Studies dating back to Darwin demonstrated that some species (e.g., molluscs) disperse between lakes on the external surfaces of waterfowl, notably their legs, feet, bills, and plumage. More recently, live-captured sanderling *Calidris alba*, dunlin *Calidris alpina*, and curlew sandpiper *Calidris ferruginea* have been observed with cockles *Cerastoderma edule* attached to their digits, confirming Darwin's hypothesis.

Plumage may offer an even more inviting opportunity for dispersal of aquatic NIS. It is not clear how many lake-dwelling species can be moved alive in waterfowl plumage. Some bryozoans produce statocysts with hooks that could easily attach to waterfowl plumage. Resting eggs or plant seeds that float also would be most likely to get tangled in waterfowl plumage. Fishhook and spiny waterfleas are two invasive NIS that have colonized the Great Lakes from Europe over the past 30 years (Fig. 2). One study has confirmed fishhook waterfleas in plumage of a dead lesser scaup *Aythya affinis* that was dipped into Lake Ontario during the summer when the waterflea was present. In addition, "foam" consisting of millions of spiny waterflea carcasses—some with resting eggs—has been observed on the leeward shoreline of Lake Huron on windy days. These eggs could easily attach to the plumage of foraging dabbling ducks, with the possibility of secondary transfer to other lakes. It is not clear how important this mechanism has been in spread of spiny waterfleas around the Great Lakes, as patterns of spread are more consistent with human than waterfowl vectors (see below).

Dispersal of invertebrate resting stages and plant seeds by their predators is a growing avenue of interest. Several studies have demonstrated that resting stages and or seeds may pass through the digestive tracts of predators—mainly ducks and shorebirds—and remain viable. Experimental studies have identified a range of invertebrate species (e.g., gastropods, crustaceans), strictly aquatic plants (*Chara, Riella americana*, diatoms, cyanophytes, *Najas marina*), and wetland plants (*Potamogeton natans, Scirpus paladosus*, and 23 other species) that may be ingested and survive passage through duck digestive tracts. Germination success is higher for seeds that are hard or otherwise difficult to digest, and following passage through ducks with smaller gizzards. Many resting eggs of the spiny waterflea are destroyed by passage through ducks, although even a small proportion (<4%) surviving can have important consequences for dispersal in light of the enormous populations of migrating waterfowl that may carry the species. The distance these eggs are carried will vary according to the flight speed of the bird, the composition of other foods ingested, and the size of the eggs. Available studies suggest that endozoochory could contribute to dispersal of plants and animals over short (<300 km) distances.

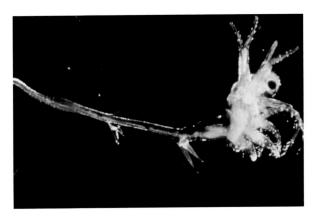

FIGURE 2 Native to northern Europe, the spiny waterflea (*Bythotrephes longimanus*) was first discovered in 1982 and had spread to all of the Great Lakes by 1987. Scientists think international cargo ships first carried the spiny waterflea to North America in their ballast water. The species has changed the food webs of the Great Lakes by causing declines in native zooplankton through direct predation, thus impacting sport and commercial fisheries. (Photograph courtesy of United States Environmental Protection Agency, Great Lakes National Program Office.)

The importance of endozoochory to long-distance dispersal is likely quite low because migrating waterbirds will completely evacuate their digestive tracts long before they reach their destination.

Evidence of endozoochorous dispersal by fishes in lakes is more limited. The waterflea *Daphnia lumholtzi* is native to parts of Australia, Southeast Asia, and Africa, and was first reported in North America in a reservoir in Missouri in 1991. The most plausible hypothesis for its introduction is via ephippia carried in the digestive tract of Nile perch *Lates niloticus* introduced to the same reservoir in 1983.

It is also possible that secondary dispersal of the spiny waterflea has been facilitated by fish predators. Many spiny waterflea resting eggs consumed by Great Lakes' fish remain viable following passage through the digestive tract. Migrations of these fishes between connected lakes could disperse the species. Some of the fish species that ingest spiny waterfleas are collected and sold commercially as bait. If these were used on lakes that have not yet been invaded by the waterflea, then viable eggs released in feces could potentially establish a new population.

Human-mediated Introductions to Lakes

Natural dispersal has resulted in a slow but consistent spread of NIS to lakes within continents for millennia. This background process of natural species colonization of lakes by fishes, invertebrates, and macrophytes has been joined by the much faster process of human-mediated introductions. For example, human introduction of fishes increased steadily after 1860 and dramatically between 1950 and 1980. Human-mediated introductions to lakes can be classified into two categories: intentional and inadvertent. Intentional releases include species stocked as food for humans or other species, for sport fishing, for biocontrol, and for pet, aquarium, or ornamental purposes (Table 1).

INTENTIONAL RELEASES Intentional stocking of lakes with fishes or with species intended as fish food is widespread. Aquaculture, sport fishing, and fishery improvement have been responsible for more introductions than any other mechanisms. A Web of Science search revealed that the top three fish species introduced to lakes were common carp *Cyprinus carpio*, rainbow trout *Oncorhynchus mykiss*, and brown trout *Salmo trutta*, all of which were introduced for human food. Common carp has been introduced worldwide to more than 117 countries as either a food or sport fish. While originally stocked as a human food source, the species is now widely considered an ecological menace because it destroys macrophyte beds, thereby adversely affecting species associated with these plants as well as increasing water turbidity. Rainbow trout have been introduced globally as a sport or human food fish to at least 80 countries, while brown trout were introduced and have established in at least 28 countries. Both trout species are used in aquaculture (e.g., Chile, Argentina), owing to their high commercial value, and both are associated with strong, adverse ecological effects on native communities. Other fishes introduced commonly to lakes for human food or sport include largemouth bass (*Micropterus salmoides*; successfully introduced to 58 countries), brook trout (*Salvelinus fontinalis*; 40 countries), Nile tilapia (*Oreochromis*

TABLE 1
Possible Dispersal Mechanisms for Species Introduced to Lakes.

Mechanism	Mode	Examples
currents	N	macrophytes, zebra mussels
birds	N	invertebrate resting stages, plant seeds
other animals (e.g., fish)	N	zebra mussels
canals	A	sea lamprey, blueback herring, zebra mussels
ballast water	A	zebra and quagga mussels, phytoplankton, round and tube nose gobies, New Zealand mud snail, and many others
vessel hulls	A	*Enteromorpha flexuosa, Bangia atropurpurea*
vessel interiors	A	zebra mussels
fishing boat live wells	A	zebra mussels
recreational boat bilge water	A	zebra mussel larvae
recreational boat engine water	A	zebra mussel larvae
recreational boat trailers	A	macrophytes and associated fauna (e.g., zebra and quagga mussels)
navigation buoys	A	zebra mussels
marina/boatyard equipment	A	zebra mussels
fishing equipment/cages	A	zebra mussels
fish stocking water	A	zebra mussels
bait and bait bucket water	A	zebra mussels
commercial products	A	zebra mussels
marker buoys and floats	A	zebra mussels
fire truck water	A	zebra mussels
float planes	A	zebra mussels
recreational equipment	A	zebra mussels, New Zealand mud snails
litter (garbage)	A	zebra mussels
scientific research	A	zebra mussels
pet/aquarium releases	I	water hyacinth, cabomba, hydrilla, Canadian waterweed, Egeria waterweed, catfish, mollies, bullfrog, salvinia, and many more
human food	I	salmonids, bass, walleye, Nile perch, Nile tilapia, common, bighead, grass and silver carp, catfish, crayfish, and many more
food for other species	I	mysids, amphipods
ornamental	I	macrophytes
sport fishing	I	salmonids
biocontrol	I	grass, silver, bighead and black carp, mosquitofish, and many more

NOTE: Mode refers to what the mechanism is: N = natural colonization, I = intentional human transfer, A = inadvertent human transfer. Modified from Carlton's (1993) list for zebra mussels.

niloticus niloticus; 75 countries), and bighead carp (*Hypophthalmichthys nobilis*; 43 countries).

A number of species have been introduced to facilitate growth of other species, usually fish, that are exploited for human benefit. Some of these introductions have occurred with spectacular unintended consequences. For example, both mysid and amphipod crustacean species have been stocked into lakes to enhance fisheries in lakes in Soviet Russia, Scandinavia, and western North America. The opossum shrimp *M. diluviana* was introduced to Flathead Lake, Montana, and Kootenay Lake, British Columbia, to enhance kokanee salmon *Oncorhynchus nerka* production. While opossum shrimp can be eaten by adult fishes, the mysids outcompete kokanee fry for zooplankton prey, resulting in a "bottleneck" in survival of young fish. Adult populations of kokanee subsequently collapsed in both systems.

The amphipod *Gmelinoides fasciatus* is native to the Lake Baikal region of Siberia but has been intentionally stocked into a series of lakes and reservoirs in order to enhance production of fish predators. The species was stocked into Gorkovskoye Reservoir on the Volga River cascade in the 1960s, after which it spread upstream and downstream to other major reservoirs on the river. It was also stocked into lakes on the Karelian Isthmus in northwestern Russia, which resulted in its eventual colonization of lakes Ladoga and Onega, which are among the largest lakes in Europe and are 5,000 km west of Lake Baikal.

Other amphipods have also been transferred between continents. The North American estuarine amphipod *Gammarus tigrinus* was stocked as fish food into the salt-contaminated Werra and Weser rivers in Germany. The species was also introduced to Europe via ballast water. Its introduced distribution in Europe and North America now includes a number of freshwater lakes, including Lough Neagh in Northern Ireland and the Laurentian Great Lakes.

Other crustaceans, notably crayfishes, have been stocked as human food. European stocks of crayfish were devastated prior to 1900 by the introduction of crayfish carrying a fungal disease (*Aphanomyces astaci*). Disease-resistant stocks of other species were subsequently introduced, including a number of North American species (*Pacifasticus leniusculus, Orconectes limosus, Procambarus clarkii, Procambarus zonangulus*). Transfer of crayfish within continental Europe has also occurred fairly commonly (e.g., *Astacus leptodactylus, Astacus astacus*). Within North America, rusty crayfish *Orconectes rusticus* has been moved between lakes through its use as a sport-fishing bait, often with adverse ecological effects.

It has recently become increasingly popular to introduce species to lakes for biocontrol purposes. Western mosquitofish (*Gambusia affinis*; 73 countries) and eastern mosquitofish (*Gambusia holbrooki*; 29 countries) have been widely introduced globally for mosquito control, although only the former species has demonstrated effectiveness, and both species have adverse ecological impacts on invaded lakes. Black carp (*Mylopharyngodon piceus*; 16 countries) have been introduced to aquaculture ponds for control of snails, which are intermediate hosts of trematode parasites of cultivated fishes. Silver carp (*Hypophthalmichthys molitrix*; 58 countries) have been introduced to aquaculture ponds and lakes for control of algal blooms, while grass carp (*Ctenopharyngodon idella*; 67 countries) have been introduced to lakes and reservoirs both to control aquatic vegetation and to serve as a food source. Black, silver, and bighead carp raised in aquaculture facilities in the southern United States escaped confinement and have migrated up the Mississippi River, with the last two posing a current invasion risk to the Great Lakes via the Chicago Ship and Sanitary Canal.

Other biocontrol agents have been introduced to eliminate undesirable vegetation. For example, water hyacinth *Eichhornia crassipes* was stocked into many waterways as an ornamental species despite its reputation as a highly invasive NIS. The species was reported in Lake Kyoga, Uganda, in 1988 and in Lake Victoria, the world's second largest lake, in 1989. Water hyacinth severely impedes nearshore boat travel and fishing in Lake Victoria, adversely affects water intakes for municipalities around the lake, dramatically increases evapotranspiration, and greatly transforms the ecology of colonized areas. The weevils *Neochetina bruchi* and *Neochetina eichhorniae* were successfully introduced from their native range in South America to Lake Victoria to feed on and control water hyacinth. Hyacinth aerial coverage has declined from its maximum of 178,374 ha in 1998–1999 to less than 2,000 ha by 2001, at least in part due to weevil herbivory.

Giant salvinia *Salvinia molesta* is a highly invasive aquatic fern sold as an ornamental for use in aquariums and garden ponds. Like water hyacinth, this species has been released into lakes around the world, where it can completely cover the surface, impeding boaters and adversely affecting lake communities. Giant salvinia has been successfully controlled using another herbivorous weevil, *Cyrtobagous salviniae*, from the plant's South American native range.

Eurasian watermilfoil *Myriophyllum spicatum* is an introduced invasive macrophyte in the United States and

Canada. Work on biological control of watermilfoil is much less well established than for the previously mentioned macrophytes, although a native North American weevil, *Euhrychiopsis lecontei,* has some potential. Use of native species in biological control of NIS holds great appeal because no additional NIS would be introduced—with their attendant risks—to the system. Other methods currently employed to control invasive macrophytes include mechanical harvesting and herbicide application. Mechanical harvesting can clear lakes of standing biomass, although fragments created or left behind can allow recolonization or dispersal. Some jurisdictions have implemented laws banning sale, possession, or shipment of invasive macrophytes.

Aquarium trade, water garden, and human food species also are introduced intentionally to lakes. Fishes are by far the most commonly established introduced aquarium species, followed by gastropods and aquatic plants. Enhancing establishment success of these species is the fact that they are often released as large, healthy, mature individuals. Indeed, aquarium fishes are typically discarded because their owners tire of them, or because they become too large or prolific. For example, the red-bellied piranha *Pygocentrus nattereri* is a South American fish very popular in the aquarium trade. Two large individuals were captured from a reservoir in Oahu, Hawaii, and otolith analysis indicated that one fish originated from aquarium stock before being released. In some cases, aquarium and water garden pathways can be the dominant mechanism by which species are introduced to lakes. For example, in Australia, 22 of 34 nonindigenous fish species originated from the aquarium or ornamental trades. Fishes that are more commonly sold in stores are more likely to be introduced to and established in lakes than are rare species. Analyses of aquarium stores around the Laurentian Great Lakes revealed six families commonly for sale: Cichlidae (e.g., cichlids), Poeciliidae (e.g., guppies, mollies, and swordtails), Characidae (e.g., tetras), Cyprinidae (e.g., goldfish, koi), Beloontiidae (e.g., Siamese fighting fish, gourami), and Loricaridae (e.g., armored catfish). One of the fishes most likely to colonize this region successfully is the white cloud mountain minnow (*Tanichthys albonubes*), which has moderate "propagule pressure" from aquarists and can survive environmental conditions around the Great Lakes.

A number of invasive aquatic plants, including Eurasian watermilfoil, Canadian waterweed *Elodea canadensis,* and Egeria waterweed *Egeria densa,* are available in aquarium shops or by Internet sale. Surveys of aquatic plants purchased from pet stores and biological supply houses have revealed that contamination with epiphytic animals is very common. Such "fellow travelers" pose a significant introduction risk if the water in which the desired species is shipped is discarded directly into local waterways.

A number of other taxa are also available in aquarium shops. For example, a parthenogenetic crayfish (*Procambarus* "Marmorkrebs") is sold in human food markets in Madagascar and aquarium shops in Germany. This species recently colonized aquatic ecosystems in Madagascar, where it is expected to have adverse ecological effects on the country's highly endemic flora and fauna. Unlike most of the animal species sold in the aquarium trade that are introduced to lakes in low numbers, and which can suffer from Allee effects owing to a scarcity of potential mates, this asexual crayfish has a very high reproductive potential.

INADVERTENT HUMAN INTRODUCTIONS Many mechanisms exist by which humans introduce or spread NIS inadvertently (Table 1). Principal among these are the release of contaminated ballast water, the creation of canals, and the movement of recreational boats on trailers. Not all species in an area have the same likelihood of dispersal via these vectors. For example, planktonic organisms may have a higher probability of being loaded into ballast water than strictly benthic ones, and motile species such as sea lamprey *Petromyzon marinus* and blueback herring *Alosa aestivalis* may use canals to colonize systems like the Great Lakes while less vagile species are unable to do so.

Ballast water is a potent mechanism for the introduction of NIS to coastal marine habitats because of the tremendous volume of water moved and because nearly entire communities may be transferred at once. Ballast water plays a much smaller role in introducing NIS to freshwater systems because transoceanic vessels usually do not load and discharge ballast into freshwater habitats on the same trip. One exception to this pattern is provided by the North American Great Lakes, where up to 70 percent of species invasions over the past 50 years are attributable to ships. Many of these species originated in western Europe or the Ponto-Caspian basin. Virtually all life forms have been successfully introduced to the Great Lakes by contaminated ballast water, including notorious species such as zebra and quagga mussels, round gobies, fishhook waterfleas, and bloody red shrimp, all of which are Ponto-Caspian natives. Many of the NIS introduced by ballast water are first reported in the corridor between Lake Huron and Lake Erie or in areas surrounding major ports. Many of these species then spread through a combination of natural and human-mediated mechanisms to other Great Lakes and to inland lakes. The dominance of ballast water as a vector of NIS introductions resulted in policy changes by the U.S. (2008) and Canadian (2006) governments that require all transoceanic vessels to exchange filled ballast water tanks or flush empty ones while operating on open ocean prior to entering the lakes. This policy change is expected to dramatically reduce propagule pressure associated with freshwater species in the ballast vector.

Canals have been instrumental in spreading NIS around the world. This importance is highlighted by Europe, where a network of an estimated 28,000 km of rivers, lakes, and canals has been established over the past few centuries. This system has facilitated north–south and east–west spread of many species. One of the most dramatic of these invasions was the spread of the ctenophore *Mnemiopsis leidyi* from the Black and Azov seas to the Caspian Sea. This invasion was facilitated by discharge of contaminated ballast water by a ship that traveled through the Volga-Don canal, which was opened in 1952 to link the Black-Azov and Caspian seas.

Once canal connections have been established, many species will spread downstream in currents. An entire guild of Ponto-Caspian invertebrates including amphipods (*Echinogammarus trichiatus, Chelicorophium robustum, Dikerogammarus villosus, Dikerogammarus haemobaphes*), mysids (*Limnomysis benedeni, Hemimysis anomala*), ispods (*Jaera istri*), water mites (*Caspihalacarus hyrcanus danubialis*), flatworms (*Dendrocoelum romanodanubiale*), oligochaetes (*Potamothrix vejdovskyi*), polychaetes (*Hypania invalida*), leeches (*Caspiobdella fadejewi*), and bivalves (*Dreissena rostriformis bugensis*) spread from the Danube River into the Rhine River via the Main-Danube canal after it was reopened in 1992. This canal is fed by Danube River water, allowing species to spread downstream. Once these species establish beachheads in major freshwater ports like Rotterdam and Antwerp, ships may carry them all over the world. In this manner, these ports can serve as hubs for invasions far removed from western Europe.

While zebra mussels have spread across much of continental Europe over the past 200 years, the quagga mussel has only recently begun to spread outside of its native range in the Black Sea basin using the Danube-Rhine and Volga-Don canal systems to invade the lower Rhine and upper Volga rivers, respectively. Canals and rivers have dispersed zebra and quagga mussels in both Europe and North America.

Two dramatic invasions illustrate the importance of trailered recreational boats to the spread of NIS. Zebra mussels were first found in Ireland in 1997 in Lough Derg. DNA fingerprinting and field surveys suggest that the species was introduced to Ireland, fouled on the hull of a recreational boat ferried from England, owing to a highly unusual combination of events: enhanced licensing requirements for secondhand vessels implemented in England, elimination of value-added taxes on secondhand vessels in Ireland, and favorable exchange rates. Since the initial invasion, the mussel has spread widely across Ireland through a combination of natural dispersal in rivers and canals as well as by pleasure craft fouled with mussels.

An even more dramatic example of jump dispersal occurred with the discovery of the quagga mussel in Lake Mead, Nevada, in 2007, some 2,700 km from the closest known source population in Lake Michigan. Adult mussels fouling an exterior surface of a trailered pleasure boat were likely responsible for this invasion, although many other boating mechanisms can transport veliger larvae or adult individuals (Table 1). This invasion reveals the highly unpredictable nature of long-range dispersal, as successful colonization required mussel survival while the boat was trailered across semi-desert environments. Only slightly less impressive was the dispersal of zebra mussels to reservoirs in Colorado and California from eastern North America, illustrating that functional vectors exist for transcontinental movement of both mussel species.

Movement of trailered recreational boats between lakes is a potent mechanism for the spread of invasive macrophyte species. Surveys of trailered boats departing Lake St. Clair, Michigan, revealed that 33–36 percent were fouled by stranded macrophyte species, many of which are nonindigenous to the lake. The human role in spreading nonindigenous macrophyte species was highlighted by a New Zealand survey that found NIS in 83 percent of lakes used for boating or fishing but 0 percent in lakes without these activities. Trailers appear to be the predominant vector for moving nonindigenous macrophytes, rather than the boats themselves, likely because submerged trailers are highly conducive to fouling by floating plants. Inspections at lakes in British Columbia found Eurasian watermilfoil on 4.3 and 0.06 percent of trailers and pleasure boats, respectively. Other NIS likely to be transported by recreational boats include *Hydrilla verticillata* and *Cabomba caroliniana*, although many other macrophyte species could be transported by this mechanism.

Boaters and anglers may congregate on particular lakes, and if these lakes become invaded, they can serve as hubs for secondary spread to other systems. As an example, *Hyrdilla verticillata* recently colonized Pongolapoort Dam, a large reservoir in South Africa. This reservoir hosts the largest fishing competition in the southern hemisphere. Trailered boats departing this lake pose a clear risk of introducing *Hydrilla* to other systems in South Africa unless preventative measures are implemented to ensure the trailers are cleaned before they depart from the reservoir. Trailered boats pose an additional risk of spreading "fellow traveler" invertebrates that colonize nonindigenous plants in lakes.

Reservoirs are formed when rivers are dammed. In some parts of the world, reservoirs are the dominant standing water bodies. Reservoirs have played a central role in spreading some NIS. For example, after the waterflea *Daphnia lumholtzi* first colonized a reservoir in Texas in 1990, it spread across the southeastern, midwestern, and southwestern United States, with most populations becoming established in reservoirs. Dispersal of this species most likely involved inadvertent transport in recreational boats (e.g., live well or bilge water) or on fishing gear of anglers associated with major fishing competitions. Reservoirs have also been implicated in serial invasions by the copepod *Eurytemora affinis*.

Reservoirs are good candidates for invasion because they have higher water flow than do nonconnected lakes; they also have large surface areas and high visitation rates by recreational boaters. Each of these factors may increase propagule pressure of NIS entering reservoirs. More controversially, reservoirs have high disturbance rates, high but more variable productivity, and human-constructed food webs, all factors linked to invasibility. Whatever the mechanism, the examples provided—*Daphnia lumholtzi*, *Eurytemora affinis*, *Dreissena polymorpha*, *D. rostriformis bugensis*, and *Hydrilla verticillata*—demonstrate that reservoirs may serve as stepping stones in serial invasions of landscapes once an invader first arrives.

Other mechanisms associated with human activities also spread NIS among lakes. The spiny waterflea first invaded North America (Lake Ontario) in 1982 and has since spread to more than 144 lakes in Ontario plus about a quarter as many in Great Lakes states. Natural dispersal may have played a role in some of these invasions (see above), as the species is found in and around a river and canal system that runs through the invaded range in Ontario. However, invaded lakes are

significantly closer to roads than their nearest neighbor lakes that are not invaded, which suggests a role for human activity. These waterfleas can foul fishing lines that are trolled through invaded lakes, resulting in the formation of "knots" on fishing lines. Adult animals retrieved on fishing lines undoubtedly die shortly after immersion, but their resting eggs remain viable for an extended period. Mathematical models that use information on travel patterns of boaters leaving invaded lakes has allowed reconstruction of the invasion sequence. Lake Muskoka, an extremely popular recreational inland lake, was invaded relatively early (1989). Up to thirty-nine other lakes were either directly or indirectly invaded from Lake Muskoka, with a stepping-stone chain spanning at least five links. The number of lakes that may become directly invaded from Lake Muskoka has declined because many of the systems linked to this lake are already invaded.

Other lake systems with other NIS may behave similarly. For example, Lake Mead in California and Nevada will be a source of future dispersal of quagga mussels, because it has the highest boater visitation rate in the western United States. Lake Mead was identified as a strong candidate for invasion and hub status, and the "100th Meridian Initiative" was specifically designed to prevent *Dreissena* mussels from colonizing the western United States, yet the invasion still occurred. Efforts to manage invasions must initially focus on prevention, followed thereafter by preventing propagule transfer out of the lake if the system becomes invaded.

IMPACTS OF INVASIONS ON LAKES

Nonindigenous species may produce an array of direct and indirect changes in invaded lakes. These include altered physical and chemical characteristics, as well as changes in biological composition. In many cases, it is difficult to separate effects because physical changes may drive chemical or biological ones, or vice versa.

While impacts of introduced aquatic NIS are often negative, many instances exist in which species are intentionally stocked into lakes (see above). Lakes with high water clarity are desirable both because they have high aesthetic value and because they command high cottage prices. Impaired water clarity often can be improved by reducing loading of limiting nutrients, especially phosphorus. Aquatic ecologists have used two strategies that may involve introduction of NIS to achieve this objective. First, some lakes may be "biomanipulated" by stocking high densities of piscivorous fishes to reduce planktivorous fish abundance and thus enhance algal grazing by large zooplankton. Favoring large zooplankton at the expense of planktivorous fishes can enhance overall grazing of phytoplankton and improve water clarity. An alternate strategy involves eliminating the entire fish community through netting or application of poisons, followed by stocking of piscivorous fishes. Problems with either method include attenuation of beneficial effects with each trophic level and the ephemeral nature of some benefits. Biomanipulation efforts ideally involve stocking native piscivores (e.g., northern pike, or pikeperch *Stizostedion lucioperca*), although concern about introducing nonindigenous predators (e.g., largemouth bass) is usually secondary to improving water quality.

A second example of biomanipulation involves stocking mussels into systems with poor water quality. Introduced zebra and quagga mussels—and a South American analog, the golden mussel *Limnoperna fortunei*—filter silt, sediment, edible and inedible phytoplankton, and small zooplankton from water, improving light transmission. Inedible particles are rejected as pseudofeces and tend to accumulate with feces around mussel colonies on the lake bottom. Dutch researchers have exploited this characteristic of mussel filtering to improve water clarity. The addition of zebra mussels to lakes and ponds suffering from eutrophication has generally improved water clarity and reduced algal biomass. In fact, there may be no stronger effect of zebra mussels on invaded lakes than improved water clarity. Despite this fact, in most parts of the northern hemisphere, the perceived drawbacks exceed the potential benefits, and zebra mussels are not intentionally stocked in lakes as a result. A number of jurisdictions in North America have banned live transport of the species, including Vermont, Indiana, New York, Minnesota, Kansas, Florida, California, and Manitoba, to reduce invasion risk.

Many introduced fishes have the opposite effect on water quality, particularly if the fish is benthivorous and adversely impacts macrophytes or resuspends sediments. For example, before-and-after comparisons of lakes in New Zealand in which fishes including rudd *Scardinius erythrophthalmus*, tench *Tinca tinca*, European perch *Perca fluviatilis*, brown bullhead *Ameiurus nebulosus*, goldfish *Carassius auratus*, and koi *Cyprinus carpio* were introduced demonstrated profound reductions in water clarity following the introductions. Key to changes in water clarity is the relative abundance of macrophytes and phytoplankton. If macrophytes are consumed or uprooted by benthivorous fishes, then the physical structure and habitat formerly associated with the plants is lost, as is the ability to stabilize

lake sediments and reduce resuspension of sediments. Once macrophytes are lost, the lake may shift rapidly to an "alternative stable state" characterized by perpetual turbidity and enhanced phytoplankton production.

Introduced fishes are not the only species capable of inducing dramatic shifts in ecosystem state. Introduction of red swamp crayfish *Procambarus clarkii* into shallow Chozas Lake, Spain, switched the system from one rich with macrophytes, invertebrates, amphibians, and birds to a turbid system in which abundances of these groups were greatly diminished. Similarly, in North America, the rusty crayfish *Orconectes rusticus* is currently spreading in lakes around the Great Lakes. In some but not all invaded lakes, rusty crayfish directly or indirectly reduce macrophyte abundance and species richness through herbivory, destruction, or increased turbidity, and they reduce abundances of many invertebrate taxa as well. These lakes are also characterized by very low abundances of *Lepomis* sunfish, whose macrophyte habitat crayfish destroy. On the other hand, invaded lakes where macrophytes continue to flourish are often characterized by abundant sunfish, which regulate crayfish abundance through intense predation on juveniles.

Introduced common carp and red swamp crayfish change lake chemistry in addition to or in conjunction with physical and biological changes. Both species cause significant increases in organic and inorganic suspended solids and significant declines in macrophyte abundance. Carp also negatively affect ammonia concentration and chironomid abundance, while chlorophyll *a*, total phosphorus concentration, and rotifer and copepod abundances increase significantly. Crayfish increase total and ammonia nitrogen and total and orthophosphorus concentrations, but decrease populations of both chironomids and oligochaetes.

Intense colonization of nearshore areas of lakes by introduced macrophytes may partially or completely obstruct navigation by ferries and recreational boats, reduce lake mixing, and result in dramatic light attenuation. In turn, reduced light or shading affects processes including phytoplankton productivity and sediment chemistry. For example, the macrophyte swollen bladderwort *Utricularia inflata* has invaded lakes in the Adirondack Mountains in upper New York State. Growth of this plant just above the lake bottom decreases growth of native macrophytes and indirectly reduces sediment oxygen and redox potential while increasing pH, carbon dioxide, ammonia, total phosphorus, and iron.

As with many invaders, the magnitude and direction of change following macrophyte invasions depends on species' functional roles. Prolific growth of introduced Eurasian watermilfoil can either benefit or harm fish populations. The plant can obstruct swimming by pelagic and piscivorous species, can reduce piscivory, and can reduce the availability of invertebrate prey by replacing native plants upon which these species live. In other systems, watermilfoil encourages invertebrate colonization and provides refuge from piscivores for young-of-year fishes. *Hydrilla verticillata* appears to perform a similar function in Lake Izabal, Guatemala, as biomass of the most important subsistence fish species is higher in the hydrilla habitat than in the five habitats with native plants. Beds of introduced African elodea *Lagarosiphon major* and Canadian waterweed in a New Zealand lake had higher standing stocks and production of epiphytes, higher diversity and abundance of invertebrates, and higher native fish abundance than did native plant beds. Consequently, certain functional groups appear to benefit from colonization of lakes by nonindigenous macrophyte species, although native plants may suffer from competition, and human access may be adversely affected.

Zebra and quagga mussels are prime examples of "ecosystem engineers." In addition to their effects on water clarity, they can cause a wide range of direct and indirect physical, chemical, and biological changes in the lakes they invade. For example, extensive accumulations of mussel shells—both living and dead—may be found in many lakes. These shells enhance the structural complexity of the lake bed, particularly where the substrate is mud or sand. Although not all species benefit, invertebrate species diversity and total nonmussel abundance is typically much higher in these shell beds than in adjacent soft sediment. Many invertebrates also benefit from the accumulation of rich organic wastes produced by mussel feces and pseudofeces. A number of groups clearly have been adversely affected by zebra and quagga mussel colonization of the Great Lakes. These species include native unionid mussels, which through both exploitative and interference competition, become very rare or absent in areas with large *Dreissena* populations. Lake whitefish *Coregonus clupeaformis* populations in lakes Huron and Michigan appear to have declined sharply in response to large quagga mussel populations in both lakes. Whitefish body condition has deteriorated concomitantly, reflecting a near collapse of its primary food source, *Diporeia* spp., in both lakes. Other fishes suffering steep population declines in Lake Michigan since 1999 include alewife *Alosa pseudoharengus*, bloater *Coregonus hoyi*, deepwater sculpin *Myoxocephalus thompsoni*, and slimy sculpin *Cottus cognatus*. Crustacean zooplankton abundance has also declined sharply in Lake Huron, possibly reflecting the focusing of predators on remaining plankton following collapse of the *Diaporeia* populations. Although the precise mechanisms by which these changes are occurring is not understood, they are linked to *Dreissena* expansion. Earlier workers had speculated that introduced *Dreissena* mussels cause a nearshore shunt of phosphorus through prolific population suspension feeding, locking this essential nutrient in nearshore sediments and away from open waters. Available evidence from some of the large deep Great Lakes is consistent with this hypothesis, as large-scale changes are no longer limited to the shallow lakes (Erie, St. Clair) for which the hypothesis was developed.

Adverse effects of *Dreissena* mussels are not limited to fish and plankton. In recent years, large numbers of waterbirds, including common loons *Gavia immer*, red-breasted mergansers *Mergus serrator*, and red-necked grebes *Podiceps grisegena*, among others, have died of botulism E on lakes Erie, Ontario, and Huron. The exact mechanism responsible for the growth of the bacteria *Clostridium botulinum* that produce the toxin, and the manner by which the birds acquire the toxin, is not yet clear, although it appears to be associated with consuming contaminated *Dreissena* by introduced round gobies. The gobies are, in turn, consumed by the waterbirds, which get sick and drown.

Introduced Predators

Perhaps the strongest adverse effects of NIS in lakes are associated with introduced predatory fish. Introduced species are the second leading cause of extinction for North American fishes (27 of 40 species) and for fishes worldwide (11 of 23 species). Nile perch is native to lakes in northern Africa but was introduced to Lake Victoria—the second-largest lake in the world—around 1960 to establish a fishery based upon biomass derived from the lake's many small-bodied species of cichlid fishes. The introduction has proven remarkably successful, and today Nile perch constitutes one of the major commercial fishes in the lake. However, the introduction has also caused extinction of hundreds of the cichlid fish species exploited by Nile perch in the lake.

Some of the best examples of species extinction following introduction have occurred in Central and South American lakes. In Lake Atitlán, Guatemala, introduction of largemouth bass was followed by extinction of two-thirds of the native fish species, and bass possibly contributed to loss of the endemic Atitlán grebe *Podilymbus gigus* by preying on native crabs, one of the grebe's principal foods. Similarly, introduction of the peacock bass *Cichla ocellaris* to Gatun Lake from South America resulted in myriad changes, including the loss of six of eight previously common native fish species and a highly altered food web structure. Introduction of lake trout, sea trout *Salmo trutta*, and pejerey *Basilichthys bonariensis* into Lake Titicaca caused either direct (predation) or indirect (competition) suppression, and likely extinction, of a native endemic fish *Orestias cuvieri*, as well as causing declines in at least three other species. Introduction of a suite of 12 fishes into Lake Banyoles, Spain, was associated with an apparent loss of the fishes *Gasterosteus aculeatus* and *Tinca tinca*, and a decline of three others.

Fishless alpine lakes are extremely susceptible to species losses from fish introductions. The introduction of Arctic charr *Salvelinus alpinus* into a mountainous lake in Slovenia caused the elimination of the copepod *Arctodiaptomus alpinus* and the virtual elimination of *Cyclops abyssorum taticus*. A similar pattern unfolded in the mountains of Austria where high abundances of alpine charr *Salvelinus umbla* were stocked into a fishless lake. Within nine years, two copepods had virtually disappeared, and *Daphnia rosea* had become very rare. In a fishless mountain lake in western Canada, the introduction of brook trout resulted in the loss of two large crustaceans, the copepod *Hesperodiaptomus arcticus* and the cladoceran *Daphnia middendorffiana*. Following removal of the fish population, *Daphnia* recolonized the lake via ephippial eggs, while the copepod, which does not produce resting stages, did not recover. These differences highlight the importance of life histories in determining possible responses by impacted species once a stressor has been removed.

Adverse effects of introduced fishes in mountain lakes also extend to amphibians. Introduced trout species now inhabit 90 percent of total lentic surface area in a mountainous region of Idaho. Water bodies inhabited by trout have significantly lower amphibian abundances than those that lack fish. Although it might appear that amphibians can persist in these remaining fishless areas, these habitats tend to be too shallow to support reproduction or overwintering.

Introduced Diseases

Introduced diseases can profoundly affect native populations in lakes. As many as 40 different game and sport fish species in the Great Lakes have been infected since 2003 by a strain of viral hemorrhagic septicemia (VHS) virus, a disease common in coastal marine waters and in farmed freshwater salmonids in Europe but never seen previously in the Great Lakes. Populations of muskellunge *Esox masquinongy*, smallmouth bass *Micropterus dolomieu*, northern pike, freshwater drum *Aplodinotus grunniens*, and round goby have substantially declined coincident with viral infection.

Parasites may be carried as "fellow travelers" with species that are either intentionally stocked or inadvertently introduced. For example, introduced round and tubenose *Proterorhinus marmoratus* gobies, introduced via ballast water to the Great Lakes, carried at least two and possibly up to four parasites from the Black Sea. Whirling disease is a myxosporean parasite *Myxobolus cerebralis* that infects salmonids, causing both skeletal and neurological problems and resulting in characteristic spiraling swimming behavior. It was first reported in introduced rainbow trout in Europe, but the disease is now common in salmonids in many countries. It arrived to North America in frozen rainbow trout imported from Europe. In Europe, introduced Asian topmouth gudgeon *Pseudorasbora parva* carries an infectious pathogen *Spaerothecum destruens* that increases mortality and causes reproductive failure in foy *Leucaspius delineatus*, an endangered native cyprinid species. Such "spillover" effects, in which an introduced species carries a nonindigenous pathogen or parasite that is even more harmful to native species, are not uncommon. Spillback effects, in which NIS serve as competent hosts for native parasites that can later infect other native hosts, can also occur. For example, introduced rainbow trout and brook trout in Lake Moreno, Argentina, acquire four of their five helminth parasites from native fish and account for almost 25 percent of total helminth egg production. The result is a larger reservoir of parasites available to colonize the native species. Another example is provided by introduced African cichlids (*Oreochromis*) in Lake Chichancanab, Mexico. Following cichlid introduction, six of seven native *Cyprinodon* species declined dramatically, and one was extirpated. The cichlids served as an intermediate host for transmission of the native parasite *Crassiphiala cf. bulboglossa*, and because these fishes were heavily preyed upon, they increased the transmission of infective stages to the bird host and, indirectly, to the definitive *Cyprinodon* hosts. Of course, many species other than fishes are vulnerable to introduced diseases. The crayfishes of Europe long ago sustained massive population declines and species extirpations resulting from the co-introduction of North American crayfishes and their diseases.

Hybridization with Native Species

Introduction of nonindigenous species may result in hybridization with native species. Hybridization is a concern because it may adversely affect persistence of endangered species, jeopardize the legal status of these endangered species, or alter the ecological, behavioral, or genetic specialization of native species. Hybridization has contributed to 38 percent of native fish extinctions in North America over the past century. As introductions of species to lakes continue unabated, opportunities for hybridization should increase in the future.

In many cases, more than one factor is implicated in species extinctions. Hybridization can become an especially important factor when very small native populations interact with much larger populations of introduced species. For example, stocks of longjaw cisco *Coregonus alpenae*, deepwater cisco *C. johannae*, and blackfin cisco *C. nigripinnis* in lakes Michigan and Huron were severely depleted by overfishing, and, to a lesser extent, sea lamprey predation, following which remaining individuals are thought to have hybridized with common ciscoes. The last recorded finds of native individuals of these species were in 1967, 1951, and 1923, respectively.

While hybridization between introduced and native species poses the greatest conservation concern, hybridization involving multiple lineages within a species or between multiple nonindigenous species can also be a problem. For example, mitochondrial DNA analyses of common carp from Lake Biwa, Japan, uncovered an infusion of five separate European strains into the native population, jeopardizing the integrity of this ancestral population. In the Murray-Darling river system in Australia, hybrid carp became far more widely distributed than either of the introduced stock species from which it was formed. Two and possibly three feral tilapia (*Oreochromis*) species also form hybrids in parts of Australia.

SEE ALSO THE FOLLOWING ARTICLES

Canals / Carp, Common / Crayfish / Great Lakes: Invasions / Hybridization and Introgression / Nile Perch / Ponto-Caspian: Invasions / Rivers / Water Hyacinth / Zebra Mussel

FURTHER READING

Carlton, J. 1993. Dispersal mechanisms of the zebra mussel (*Dreissena polymorpha*) (677–697). In T.F. Nalepa and D.W. Schloesser, eds. *Zebra Mussels: Biology, Impacts, and Control.* Boca Raton, FL: Lewis Publishers.

Havel, J.E., C.E. Lee, and M.J. Vander Zanden. 2005. Do reservoirs facilitate invasions into landscapes? *BioScience* 55: 518–525.

Johnson, P.T.J., J.D. Olden, and M.J. Vander Zanden. 2008. Dam invaders: Impoundments facilitate biological invasions into freshwaters. *Frontiers in Ecology and the Environment* 7: 359–365.

Leung, B., J.M. Bossenbroek, and D.M. Lodge. 2006. Boats, pathways, and aquatic biological invasions: Estimating dispersal potential with gravity models. *Biological Invasions* 8: 241–254.

MacIsaac, H.J., J.V.M. Borbely, J.R. Muirhead, and P.A. Graniero. 2004. Backcasting and forecasting biological invasions of inland lakes. *Ecological Applications* 14: 773–783.

Padilla, D.K., and S.L. Williams. 2004. Beyond ballast water: Aquarium and ornamental trades as sources of invasive species in aquatic ecosystems. *Frontiers in Ecology and Environment* 2: 131–138.

Rahel, F.J. 2007. Biogeographic barriers, connectivity and homogenization of freshwater faunas: It's a small world after all. *Freshwater Biology* 52: 696–710.

Ricciardi, A., and R. Kipp. 2008. Predicting the number of ecologically harmful species in an aquatic system. *Diversity and Distributions* 14: 374–380.

Vander Zanden, M.J., and J.D. Olden. 2008. A management framework for preventing the secondary spread of aquatic invasive species. *Canadian Journal of Fisheries and Aquatic Sciences* 65: 1512–1522.

Ward, J.M., and A. Ricciardi. 2007. Impacts of *Dreissena* invasions on benthic macroinvertebrate communities: A meta-analysis. *Diversity and Distributions* 13: 155–165.

LAMPREY

SEE SEA LAMPREY

LANDSCAPE PATTERNS OF PLANT INVASIONS

THOMAS J. STOHLGREN

U.S. Geological Survey, Fort Collins, Colorado

Natural landscapes can be viewed as microcosms of the global environment. At global scales, plant diversity is highest in warm-wet areas, with noticeable decreases in plant diversity in extreme environments: near the north and south poles, on cold mountaintops, and in deserts. Plant invasions generally follow the same predictable patterns at landscape scales. Many low-elevation meadows, riparian zones, and canopy gaps in zones of moderate climates contain the greatest native species richness and the greatest nonnative species richness. This "rich get richer" pattern of invasion often scales from landscapes to continents, as most species track favorable environmental conditions (e.g., high light, warm climates, ample precipitation, and high soil nutrients). Disturbances, such as fire, flooding, and insect outbreaks, often facilitate plant invasions at landscape scales. Many exceptions occur in abiotically or biotically special environments: places where there is too much water or too much canopy cover, or where plant biomass is controlled by a few dominant native or nonnative species (e.g., *Arundo donax*).

NATIVE SPECIES RICHNESS PATTERNS

There are generalities apparent in landscape patterns of native species richness and cover (abundance), although there are exceptions:

· High-productivity sites with high light, warm temperatures, ample precipitation, and adequate soil nutrients generally have higher native species richness than low-productivity (high-stress) sites. Likewise, low-elevation zones generally have higher species richness and cover than do high-elevation sites. Exceptions occur where a few species attain dominance on high-productivity sites (e.g., full-canopy tree species, sod-forming grasses, and monoculture sites).

· Heterogeneous areas generally have higher species richness than homogeneous areas. Heterogeneous areas are maintained by underlying patterns of soil, geology, grazing, disturbance, pathogens, diseases, and other factors.

· Areas of high environmental stress (e.g., severe drought conditions or anoxic conditions from standing water) usually contain fewer plant species.

· Areas with limiting resources such as low soil nutrients, water stress, or low light from shading by dominant species generally result in lower plant diversity and cover.

· Moderate disturbance areas and mid-successional areas generally support more species than immediate post-fire areas or late-successional forests. Frequent massive disturbances such as huge floods or frequent intense fires can decrease local plant diversity and cover.

· Later-age geological substrates (i.e., older surfaces) generally support more species than younger substrates.

The general patterns for native plant diversity often apply to other organisms. High-productivity sites attract or respond to areas of high numbers of soil organisms, pollinators, birds, small mammals, grazers, and all manner of pathogens and diseases. It is this palette of native species diversity that paves the way for invasive plant species.

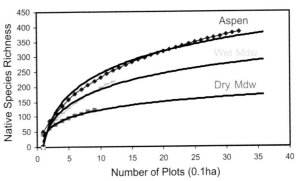

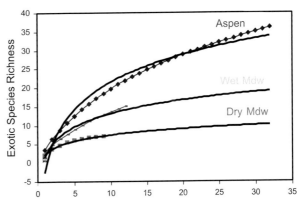

FIGURE 1 Species accumulation curves of native and nonnative species in Rocky Mountain National Park, Colorado. Curves were extrapolated from different numbers of sample plots. (Photograph courtesy of the author.)

INVASIVE SPECIES RICHNESS PATTERNS

Nonnative plant species often follow the same patterns as native species in many landscapes, assuming propagules are not limiting. Since every county in the United States contains invasive species, and since many invasive plant species are dispersed by wind, birds, small mammals, and insects, we can safely assume that some propagules have arrived (or will arrive) in most landscapes. Once the propagules arrive, they face the same suite of difficulties as native species propagules. They must find suitable habitat, including microsites with reduced competition, predators (herbivores), pathogens, and diseases.

Most field studies based on detailed vegetation plots strongly suggest that nonnative plant species are successfully establishing, growing, and reproducing in many landscapes. In fact, at landscape scales, hotspots of native species richness are often hotspots of invasion. Consider two examples of plant diversity patterns in mountains and deserts to understand typical invasion patterns. Species-accumulation curves (i.e., counting novel species encountered when adding more vegetation plots) illustrate the patterns.

In the mid-elevations of Rocky Mountain National Park (1800 m to 3200 m), native plant diversity is greatest in habitats with high light (little or no conifer canopy), high soil moisture, and warm temperatures (Fig. 1). Native plant species richness declines with increasing elevation and canopy cover. Native plant species accumulate quickly in stands of aspen (*Populus tremuloides*) along the edges of wet meadows. Such areas have high soil nitrogen and attract many species of birds, small mammals, and herbivores. Nonnative plant species have successfully invaded these same habitats. The dry meadow habitats have less soil moisture and nitrogen during the growing season, and thus have correspondingly fewer native and nonnative plant species.

Similar patterns can be found in the Grand Staircase-Escalante National Monument in southern Utah (Fig. 2). Riparian zones, with ample soil moisture, generally contain more native and nonnative species than xeric upland habitats. *Tamarix* sp. (salt cedar) has successfully invaded many riparian zones throughout the southwestern United States. However, fires in upland habitats can promote invasion of nonnative annual grasses (e.g., *Bromus tectorum* [cheatgrass]).

The prevailing pattern of invasion is "the rich get richer"—sites with high native species richness and cover are prone to invasion. However, there may be exceptions.

ROLE OF COMPETITION AND OTHER FACTORS

Many additional factors are superimposed on these general patterns. Competition can play a role locally, where large, tall, high-biomass native species, such as large trees, preempt resources and inhibit invasion. Likewise, dominating species such as sod-forming grasses (*Poa pratensis* [Kentucky bluegrass]) and monoculture-producing species (*Arundo donax*, *Tamarix* sp.) can reduce the ability of native and other nonnative species to coexist locally.

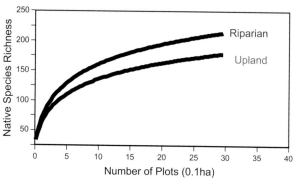

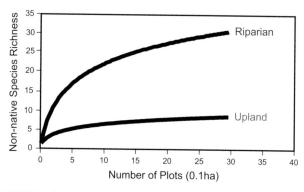

FIGURE 2 Species accumulation curves of native and nonnative species in Grand Staircase-Escalante National Monument, Utah. (Photograph courtesy of Paul Evangelista.)

The life history of species also plays a role. Early-succession habitats with highly variable precipitation and frequent disturbances such as fire, conducive to native annual plant species, are often conducive to nonnative annual plant species (e.g., *Avena fatua* grasslands in California). Likewise, nonnative perennial trees such as *Tamarix* and Norway maple (*Acer platanoides*) tend to invade habitats occupied by native perennial trees such as cottonwoods and eastern deciduous forest species, respectively.

Invading nonnative plant species may be one step ahead of their native pathogens and diseases. They may be exploiting unused resources or underused resources. Invaders might benefit from native pollinators and soil organisms. Thus, there are many potential factors influencing the establishment, growth, spread, and persistence of nonnative species in natural landscapes.

THE FUTURE LANDSCAPE

Landscape patterns of plant invasions are still unfolding. The invasion may be just under way in many areas. There is little sign of "saturation." Many microsites within landscapes may be filled by current invaders, or by new invaders assisted by trade and transportation. Current invaders, like native plant species, keep adapting to new environments. Large-scale effects of succession, land use change, climate change, fires, pollution, fertilization, insect outbreaks, floods, and hurricanes, and small-scale effects of grazing, diseases, and competition, provide for ever-changing patterns of plant invasions.

SEE ALSO THE FOLLOWING ARTICLES

Competition, Plant / Dispersal Ability, Plant / Invasibility, of Communities and Ecosystems / Protected Areas / Range Modeling / South Africa: Invasions / Tolerance Limits, Plant

FURTHER READING

Chytrý, M., V. Jarošík, P. Pyšek, O. Hájek, I. Knollová, L. Tichý, and J. Danihelka. 2008. Separating habitat invasibility by alien plants from the actual level of invasion. *Ecology* 89: 1541–1553.

Hobbs, R. J., and L. F. Huenneke. 1992. Disturbance, diversity, and invasion: Implications for conservation. *Conservation Biology* 6: 324–337.

Huston, M. A. 1994. *Biological Diversity: The Coexistence of Species in Changing Landscapes.* New York: Cambridge University Press.

Kumar, S., T. J. Stohlgren, and G. W. Chong. 2006. Spatial heterogeneity influences native and nonnative plant species richness. *Ecology* 87: 3186–3199.

Rejmánek, M., D. M., Richardson, P. Pyšek. 2005. Plant invasions and invasibility of plant communities (332–355). In E. van der Maarel, ed. *Vegetation Ecology.* Oxford: Blackwell Science.

Rouget, M., and D. M. Richardson. 2003. Understanding actual and potential patterns of plant invasions at different spatial scales: Quantifying the roles of environment and propagule pressure (3–15). In L. E. Child et al., eds. *Plant Invasions: Ecology and Management Solutions.* Backhuys.

Stohlgren, T. J. 2007. *Measuring Plant Diversity: Lessons from the Field.* New York: Oxford University Press.

Stohlgren, T. J., D. Barnett, C. Flather, J. Kartesz, and B. Peterjohn. 2005. Plant species invasions along the latitudinal gradient in the United States. *Ecology* 86: 2298–2309.

Stohlgren, T. J., D. T. Barnett, C. S. Jarnevich, C. Flather, and J. Kartesz. 2008. The myth of plant species saturation. *Ecology Letters* 11: 313–326.

LAND USE

RICHARD J. HOBBS

University of Western Australia, Crawley

Humans use ecosystems in many different ways and with differing degrees of transformation. Land uses range from conservation-related use, with relatively little intentional transformation, to the purposeful utilization of particular ecosystem components, as, for instance, in grazing systems, to the replacement of the original ecosystem with crops or plantations, and to the complete removal of virtually the entire ecosystem, as is the case in extractive uses such as mining. Possible changes include both ecosystem decline and recovery. A particular human activity may result in a sudden or a more gradual change in ecosystem properties. Such changes often result in conditions that lead to increased invasibility.

HUMAN LAND USE AND LAND USE CHANGE

The recent history of the world has been one of a dramatic increase in the incidence of human-induced disturbances, as humans use an increasing proportion of the Earth's surface in some way or another and appropriate an increasing amount of the Earth's productive capacity and natural resources. Increasing human modification of ecosystems has been noted for some time but has been reemphasized lately as a result of global assessments of the state of the world's ecosystems. Direct (e.g., land conversion) and indirect (e.g., long-range transport of pollutants) human influence takes many forms and is difficult to quantify in simple metrics, but global assessments using several available datasets have resulted in analyses that indicate that large proportions of the global land surface are significantly impacted by human activities. Prevailing trends include increasing urbanization, deforestation and ecosystem fragmentation, agricultural intensification in some areas, and abandonment of agricultural land in others. The areas involved in transformations such as deforestation are truly massive and are arising with considerable rapidity.

Land uses vary in their degree of modification of the existing ecosystem. Conservation-oriented use will tend to limit the extent of human-induced disturbance, either by design or by default (for instance in remote or harsh environments). Utilization of native ecosystems is a prevalent land use over large parts of the world and includes rangelands and other pastoral areas, forestry in native forest ecosystems, other extractive uses such as animal or plant harvesting and apiculture, and a growing use of lands for recreation and tourism. Replacement, on the other hand, involves the removal of the native ecosystem and its replacement with a simpler system geared toward the production of particular crop or livestock species. Finally, complete removal results when a native ecosystem is destroyed, such as occurs during surface mining or with urban and industrial development. Each land use entails a suite of deliberate and inadvertent impacts of varying severity on the natural ecosystem.

Any particular area can have a mixture of these land use categories, with, for instance, remnants of native ecosystems remaining (in a more or less altered state) within urban areas. Thus, the interactions between different land uses, either on the same or adjacent areas, are also important.

Human modification has often caused increasing degradation of ecosystem components, resulting in a decline in the value of the ecosystem for either production or conservation purposes. This has been met with an increasing recognition that measures need to be taken to halt or reverse this degradation; hence, the importance of restoration or repair of damaged ecosystems is increasingly recognized.

While it is possible to obtain some estimates of the more obvious changes caused by deforestation and land clearing, it is more difficult to assess the totality of land use changes. Predicting future trends is also rendered difficult by the multifaceted nature of land use and land use change. Likely interactions between land use change and climate change further render predictions of future trends difficult. Most trends in land use change are also driven primarily by economic, political, and social, rather than biophysical, factors.

Recent studies have indicated a close connection between the intensity of human activity and an array of ecosystem changes, including the incidence of invasive species. Human land use and spatial and temporal changes in this are therefore likely to be important determinants of the invasibility of ecosystems and the availability and success of invasive species. A two-way interaction between ecosystem transformation and invasions can be envisioned: land use change can increase the potential for invasions, and conversely, invasive species can drive or enhance land use change.

THE ROLE OF LAND USE CHANGE IN ENHANCING INVASIONS

Ecosystems are dynamic, species compositions are variable, and species can appear and disappear in any given system, depending on the current abiotic conditions,

levels and types of disturbance, and composition of the regional species pool. Increasing levels of human transformation of ecosystems coupled with the dramatic increase in the deliberate and inadvertent transport of biota across the globe provide increased opportunities for invasion and radical alteration of ecosystem dynamics.

In general, any change in ecosystem properties will provide opportunities for species colonization or population expansion. This is equally the case with land use change. Changes in ecosystem state are often accompanied by changes in amounts and flows of resources such as water, nutrients, and so on. For instance, removal of the native vegetation of a region and its replacement with agriculture or urban structures can result in a much more "leaky" system, with greater flows of water and nutrients out of the system and less internal recycling. These resources are then transported and become available elsewhere. Alternatively, activities such as building dams on rivers reduce environmental flows by trapping sediment and reducing flood levels. Disturbances and land transformations afford invasive species opportunities to colonize and spread, and they are often able to do so as well as or better than the species native to the area.

Land transformation also results in fragmentation of the native ecosystem, leading to the creation of edges between original and transformed areas: such edges are frequently the places at which nonnative species initially come in contact with the native system, and they are also places where resource availabilities (for instance, of light or nutrients) change most dramatically.

Land use changes sometimes involve an increased use of introduced species—for instance, new forage species, plantation trees, ornamental species in cities, and so on. Such species are transported around the globe with little attention being paid to their potential to spread and become problems elsewhere. For instance, the extensive use of northern hemisphere pine species in the southern hemisphere results in major problems when these pines invade adjacent systems.

Land transformation thus acts to encourage biotic change, first by causing system changes that provide the opportunity for biological invasion, and second by bringing new species from different biogeographic regions into contact with these altered systems. Entirely novel species assemblages are possible. The extent to which the biotic changes in any given area present a problem depends on the goals of management in that area. For instance, radical biotic changes have taken place in urban areas throughout the world, both through the removal of large proportions of preexisting ecosystems and through the introduction of nonnative plants and animals. Many of these species show no signs of becoming invasive, but a subset can cause significant problems if they spread rapidly and threaten conservation values or ecosystem services.

THE ROLE OF INVASIONS IN ENHANCING LAND USE CHANGE

As well as being a result of land transformation, invasion by nonnative species can itself act as a driver of land transformation. This will occur when the invading species cause ecosystem change (i.e., when they become transformer species). Such changes are possible when, for instance, an invading plant species becomes dominant and changes the type of vegetation present. Invading trees can transform a grassland or shrubland into a forest, and invading grasses can, via land clearance or an altered fire regime, change a woody perennial system into an open grassland (Fig. 1).

Less extreme changes caused by species that invade and alter system structure or function can also result in sometimes severe reductions in the value of land for current land uses. This then leads to the requirement to either develop the land for alternative uses or to increase the level of management required to maintain the existing land use. Examples include the invasion of rangeland systems by woody perennials and the invasion of forest and shrubland by disease organisms. Nonnative species can also pose problems for restoration projects, especially where nonnative species prevent the establishment of other desired species. On the other hand, invasive species often possess traits that make them potentially useful in restoration, such as rapid growth and spread and ability to persist in conditions unsuitable for native species. Hence, there may be sensible arguments for the use of potentially invasive species in restoration, if the attendant risks are adequately assessed and planned for. There are also many invasions that have little or no impact on land use, and some invasions even prove beneficial from some perspectives. The net costs and benefits of particular invasions may often have to be assessed across a number of conflicting land uses.

Often, the causal link between land transformation and invasions is convoluted. A change in land use, or the continuation of an inappropriate land use or of inappropriate levels of use, can provide the conditions necessary for an invading species to become established. For instance, in rangelands, inappropriate grazing or fire regimes can result in invasion by shrubs or grasses. Thereafter, however, the invading species initiates further system change, which precipitates the need for a change in land use or for increased management to maintain the existing land use. Hence, the change in system state initiated by the disturbance or management regime is enhanced or sped up by the invasion of nonnative species.

SEE ALSO THE FOLLOWING ARTICLES

Agriculture / Disturbance / Fire Regimes / Forestry and Agroforestry / Restoration / Succession / Transformers

FURTHER READING

Domènech, R., M. Vilà, J. Pino, and J. Gesti. 2005. Historical land-use legacy and *Cortaderia selloana* invasion in the Mediterranean region. *Global Change Biology* 11: 1054–1064.

Ellis, E. C., and N. Ramankutty. 2008. Putting people in the map: Anthropogenic biomes of the world. *Frontiers in Ecology and the Environment* 6: 439–447.

Hobbs, R. J. 2000. Land use changes and invasions (55–64). In H. A. Mooney and R. J. Hobbs, eds. *Invasive Species in a Changing World*. Washington, DC: Island Press.

Marini, L., K. J. Gaston, F. Prosser, and P. E. Hulme. 2009. Contrasting response of native and alien plant species richness to environmental energy and human impact along alpine elevation gradients. *Global Ecology and Biogeography* 18: 652–661.

Richardson, D. M. 1998. Forestry trees as invasive aliens. *Conservation Biology* 12: 18–26.

Robinson, R. 2008. Human activity, not ecosystem characters, drives potential species invasions. *PLoS Biology* 6: 196.

Vitousek, P. M., C. M. D'antonio, L. L. Loope, M. Rejmánek, and R. Westbrooks. 1997. Introduced species: A significant component of human-caused global change. *New Zealand Journal of Ecology* 21: 1–16.

Von Holle, B., and G. Motzkin. 2007. Historical land use and environmental determinants of nonnative plant distribution in coastal southern New England. *Biological Conservation* 136: 33–43.

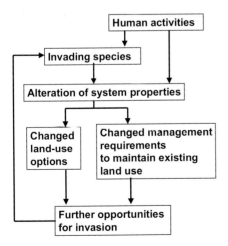

FIGURE 1 Interrelationships between human activities (management of ecosystems) and invasions: land-use change can enhance the potential for invasion, while invasive species can also drive or enhance land-use change.

LANTANA CAMARA

STEPHEN JOHNSON

Industry and Investment New South Wales, Orange, Australia

Lantana camara (lantana) is one of the world's worst invasive species, found in at least 75 countries and island groups throughout tropical, subtropical, and temperate areas of the world. Invasive populations belonging to this species aggregate negatively impact a range of environmental and primary production areas. Conversely, the species is widely planted as an ornamental. A persistent integrated weed management strategy is needed to control this weed.

AN AGGREGATE SPECIES

Lantana camara (lantana) is an aggregate species. It is highly variable and is composed of a number of interbreeding taxa. These taxa, in many cases called varieties, vary in terms of their ploidy, plant morphology, response to environmental conditions and natural enemies, chemical composition and toxicity, and response to herbicides. These varieties arise from horticultural and natural hybridization, intentional selection, and somatic mutation.

The origin of many of these varieties is unclear. This situation was recognized by Roger Sanders in his 2006 study, which determined the "current usage of the name *Lantana camara* includes a widely cultivated and naturalized hybrid species that is taxonomically distinct from native *L. camara*." The weedy material he examined was determined to be a new species, *L.* x *strigocamara*. This research is continuing. In lieu of this, this article uses the aggregate species name *L. camara*.

PLANT DESCRIPTION AND BIOLOGY

Lantana camara is a multibranched perennial shrub that generally grows 1–4 m in height (Fig. 1). The branches are square in cross-section and 2–4 mm in diameter when young, becoming more rounded, gray/brown, and up to 150 mm in diameter when mature. The young stems of the weedy varieties are hairy and have short recurved prickles, while those on the nonweedy varieties are often rounder and more slender and do not have prickles.

Plants have oval-shaped leaves that are borne opposite each other, 20–120 mm long and 15–80 mm wide. Leaf size, shape, texture, and color vary depending on the variety and environmental conditions.

The flat-topped to dome-shaped flower heads, composed of 20–40 tubular flowers, are produced in pairs between the young leaves and the stem. The heads are 10–30 mm in diameter. The flowers are often colored in combinations of white, cream, yellow, orange, red, purple, and pink, and change color with age. The fruit (a drupe) is 4–8 mm in diameter, green and hard when immature and a shiny purple/black when ripe. Fruit are borne in clusters of up to 20, with each fruit containing one "seed," pear-shaped and pale straw in color, 1.5–4 mm long and wide. This seed is in fact two fused pyrenes, each with a single embryo. Nonweedy varieties produce few fruit—mostly failing to set seed after pollination—but do have fertile pollen.

The root system of *L. camara* is brown and woody and has a short taproot with many shallow lateral branches forming a dense root mat. Plants regrow vegetatively from the base of the stem, from the plant crown if defoliated, from lateral root fragments when broken, and slowly from horizontal or broken stems that come into contact with moist soil.

Although germination of *L. camara* occurs throughout the year, seedlings are more likely to establish under high soil moisture, temperature, and light conditions, particularly after disturbance. Seedling growth rates are slow, and plants generally do not flower in their first year of growth. Plants flower and set fruit all year round in tropical and subtropical areas and throughout spring to autumn in more temperate areas, particularly in response

FIGURE 1 A flowering and fruiting branch of the common pink variety of *Lantana camara* in Australia. (Photograph courtesy of the author.)

to rainfall. In excess of 12,000 fruit may be produced on large *L. camara* plants.

A large number of bird species spread seed, as do a range of native and feral animals, and livestock. Aside from deliberate spread via the horticultural trade, spread may also occur in water, in soil, on machinery, on people, and vegetatively, in discarded garden waste.

NEGATIVE IMPACTS

Lantana camara is a weed in at least 75 countries or island groups in tropical, subtropical and temperate areas, outside what is presumed to be its native range in Central and South America. The species is weedy in the southern United States, in the western Mediterranean, throughout Africa, central and southeast Asia, and in various countries throughout the Pacific and Indian ocean regions, including New Zealand and Australia. *Lantana camara* has wide altitudinal and soil type tolerances.

A weed of disturbance, varieties of *L. camara* form thickets and invade environmental and primary production areas (Fig. 2). For example, 302 native plant and animal species and 153 ecological communities were recently considered at high risk because of *L. camara* invasion in eastern Australia. This includes three World Heritage–listed areas. More generally, *L. camara* impacts coastal dune, sedge and heath land, riparian, woodland, and many forest communities (including rainforest), reducing biodiversity and negatively impacting ecosystem functioning.

Lantana camara is a major weed of soft- and hardwood timber plantations, particularly during tree establishment. It is also a weed in natural areas used for timber harvest. The species is a weed of plantations, particularly banana, citrus, coconut, coffee, copra, oil palm, pineapple, rub-

ber, and tea crops, but also in areas planted to tropical fruits, nuts, pome, and stone fruits.

One of the most significant negative impacts caused by *L. camara* is on pasture production and animal health. For example, several million hectares of pasture in eastern Australia have been affected. The weed shades out more desirable pasture species, with infestations often becoming inaccessible. Generally unpalatable, most varieties are also toxic to cattle and sheep, as well as to other animals.

Lantana camara has a wide range of other negative impacts, including affecting corridors used for railways, roads, and electricity infrastructure; hosting pest animals, insects, and plant pathogens; reducing visual and recreational amenity; and occasionally causing poisoning to humans if ingested. Research continues into the allelopathic interactions of *L. camara* on other plant species.

BENEFITS

Lantana camara has been used as an ornamental/horticultural species since the mid-1600s. Extensive hybridization has resulted in widespread use of varieties and hybrids derived from the species in horticulture throughout the world, not only in areas that have suitable terrestrial habitats.

Not only is *L. camara* used as a container plant, it is used extensively in landscape design and function when planted as individual plants and hedges. The species is used in public and private gardens, in parks, and beside roads and footpaths because of its colorful flowers, low maintenance requirements, and drought tolerance.

The essential oils present in *L. camara* flowers and leaves can be extracted for use in perfumes. The species is also used as a herbal medicine, with extracts showing antimicrobial, fungicidal, insecticidal, and nematicidal activity.

Lantana camara is widely used as firewood and as an alternate food and habitat source for insects, birds, and other animals where none other is present.

CONTROL

An integrated weed management strategy is needed to manage *L. camara*. This will include the prevention of the movement of plants into clean areas via various hygiene practices, and considering restrictions on sale. There are a number of effective herbicides that can be applied to actively growing foliage (including regrowth) as cut stump or intact plant basal bark applications. There are three general factors to consider with

FIGURE 2 The common pink variety of *Lantana camara* invading pasture in eastern Australia. (Photograph courtesy of the author.)

herbicide applications: (1) climatic and varietal differences may affect herbicide efficacy; (2) repeat herbicide applications on regrowth material are needed; and (3) the off-target effects of some herbicides need consideration.

Fire is another useful management tool, particularly for clearing thickets of *L. camara*. This is particularly the case when fire is used prior to, or as a follow-up treatment to, chemical or mechanical clearing. Mechanical clearing using a range of machinery is effective, although regrowth needs to be controlled via herbicide application or hand pulling. Cultivation, followed by the planting of competitive pastures, is also useful in some situations. Both revegetation and grazing management are important tools in preventing reinvasion.

At least 40 biological control agents have been released against *Lantana camara* varieties around the world. The variable nature inherent in the species aggregate has meant that, at best, only partial or seasonal control has been achieved in many cases.

Overall, vigilance and repeated control of regrowth are the keys to success in managing *L. camara*.

SEE ALSO THE FOLLOWING ARTICLES

Australia: Invasions / Herbicides / Horticulture / Hybridization and Introgression / Trees and Shrubs / Weeds

FURTHER READING

Day, M. D., C. J. Wiley, J. Playford, and M. P. Zalucki. 2003. *Lantana: Current Management Status and Future Prospects*. Canberra: Australian Centre for International Research. www.aciar.gov.au/publication/MN102

Gooden, B., K. French, P. J. Turner, and P. O. Downey. 2009. Impact threshold for an alien plant invader, *Lantana camara* L., on native plant communities. *Biological Conservation* 142: 2631–2641.

Holm, L. G., D. L. Plucknett, J. V. Pancho, and J. P. Herberger. 1977. *The World's Worst Weeds: Distribution and Biology*. Honolulu: East-West Centre.

The National Lantana Management Group. 2009. Draft plan to protect environmental assets from Lantana. Yeerongpilly, Queensland: Department of Employment, Economic Development, and Innovation. www.environment.nsw.gov.au/LantanaPlan/index.htm

Sanders, R. W. 2006. Taxonomy of *Lantana* sect. *Lantana* (Verbenaceae): 1. Correct application of *Lantana camara* and associated names. *Sida* 22: 381–421.

Sharma, G. P., A. S. Raghubanshi, and J. S. Singh. 2005. *Lantana* invasion: An overview. *Weed Biology and Management* 5: 157–165.

Swarbrick, J. T., B. W. Wilson, and M. A. Hannan-Jones. 1998. *Lantana camara* (119–140). In F. D. Panetta, R. H. Groves, and R. C. H. Shepherd, eds. *The Biology of Australian Weeds*, Vol. 2. Meredith, Australia: Richardson.

van Oosterhout, E. 2004. Lantana Control Manual. *Current Management and Control Options for Lantana* (Lantana camara) *in Australia*. Brisbane: Queensland Department of Natural Resources, Mines and Energy. www.weeds.org.au/WoNS/lantana/

Walton, C. 2006. *Lantana camara*. Global Invasive Species database. www.issg.org/database/

LAWS, FEDERAL AND STATE

MARC L. MILLER

University of Arizona, Tucson

There is no clear answer to the seemingly simple question "What laws govern private and government behavior with regard to harmful invasive species?" There is little statutory law, and for important dimensions of the nonindigenous species (NIS) problem, including identifying new NIS invasions, tracking the impact of known harmful invasive species, and responding to emerging threats, there is none. Most fundamentally, there is no *general* law of invasive species either at the federal level or in any U.S. state. Some federal laws have responded to threats from particular invasive species, or threats from particular pathways for alien species, in particular ballast water as a source of aquatic NIS. But no federal law has ever responded to the general problem of prohibiting, preventing, screening, identifying, removing, and understanding NIS.

RECENT TRENDS IN NIS LEGISLATION

Federal and state legislative activity since 2005 has trended toward increasing recognition of the importance of responding to harmful NIS. This uptick in legislative action and references to invasive species focuses on particular invaders, especially those with significant economic impacts, and even more specifically, on aquatic invasives in general and the threat from nonnative mussels in particular.

With a few interesting exceptions, including legislation in Hawaii and Minnesota, most U.S. states at best weakly echo the general aspects of federal statutory law. As with federal legislative activity, since 2005 there has been increasing attention to invasive species in state laws and appropriations. The state activity varies by the kinds of invaders and pathways relevant to each jurisdiction and is highest in Hawaii and states with coastal areas and high levels of commerce. There is a bias in state legal responses toward marine invasives similar to that seen in federal law.

Legislation is not the only federal law that relates to harmful NIS. Two presidential executive orders have addressed NIS issues directly. The first was a 1977 executive order issued by President Jimmy Carter, and the second was a 1999 executive order issued by President William J. Clinton. The Clinton executive order created a National Invasive Species Council staffed by cabinet-level officers, and the council promulgated a National Invasive Species Management Plan in 2001 and issued a new management plan for 2008–2012.

FEDERAL LAW

The 1993 Office of Technology Assessment (OTA) report on the problem of harmful nonindigenous species concluded that "the current Federal effort is largely a patchwork of laws, regulations, policies, and programs. Many only peripherally address NIS, while others address the more narrowly drawn problems of the past, not the broader emerging issues." An April 1999 Congressional Research Service (CRS) report, titled *Harmful Non-Native Species: Issues for Congress*, summarized U.S. federal law this way:

> In the century or so of congressional responses to harmful, non-native species, the usual approach has been an ad hoc attack on the particular problem, from impure seed stocks, to brown tree snakes on Guam. A few attempts have been made to address specific pathways, e.g., contaminated ballast water, but no current law addresses the general concern over non-native species and the variety of paths by which they enter this country.

The general observations of the OTA and CRS reports remain true today.

"Black List" and "Exclusion" Acts

There is a compensation in the distribution of plants, birds, and animals by the God of nature. Man's attempt to change and interfere often leads to serious results.

REP. JOHN LACEY (R-IND.)
33 CONG. REC. 4871 (1900)

The recognition that NIS might cause harm has been evident in U.S. federal law at least since the Lacey Act, first enacted in 1900 and substantially revised in recent years. It was originally enacted to protect native wildlife, especially birds that were being commercially harvested for their feathers in the Florida Everglades. The sponsors focused on the lack of controls on commerce in wild species among the states as well as between nations, but they also recognized that alien species could harm native species and ecosystems. The legislative history illustrated these points through harm from sparrows and starlings that had been introduced in the latter half of the nineteenth century.

The Lacey Act currently provides the federal government with authority to ban the import, export, or transportation of "any fish or wildlife" or "any plant" that is made illegal by "any law, treaty[,] or regulation" of the United States or of any individual state. The Act provides for both civil and criminal penalties of a modest nature. For example, knowingly or negligently violating the Act may result in a penalty of "not more than $10,000 for each such violation . . . and criminal penalties, up to five years in prison and a $20,000 fine" for each violation. The Lacey Act explicitly leaves U.S. states free to make or enforce laws "not inconsistent" with the federal provisions.

The major limitation on the Lacey Act's general authority to ban animal and plant species is that these powers apply only to animals and plants that are made illegal under federal or state law. The Lacey Act is also built around the idea of a "black list," or forbidden list: it does not authorize the exclusion of animals whose threat is unknown.

While the Lacey Act provides the Secretary of the Interior the power to exclude several species of particular concern, it does not provide for the exclusion of seeds, plant pests, or most plants. This gap in the Lacey Act has been partially closed by a host of federal statutes that, together, provide federal officials with power to exclude many kinds of harmful plant pests, seeds, and noxious weeds. These acts include the Plant Pest Act, the Plant Quarantine Act, the Federal Noxious Weed Act of 1974, and the Federal Seed Act.

In May 2000, Congress passed the Plant Protection Act, which consolidated and revised the Plant Quarantine Act, the Plant Pest Act, the Federal Noxious Weed Act, aspects of the Department of Agriculture Organic Act, and several less prominent acts.

The 2000 Plant Protection Act expands the regulatory and enforcement powers over plant pests and noxious weeds and includes new civil penalty structures. The Act also encourages the steady use of science, the wide involvement of experts and stakeholders in policymaking, the consideration of "systems approaches," the development of integrated management plans on the basis of geographic and ecological regions, and the authorization of new types of classification systems.

Several provisions, however, plant their own substantial seeds for mischief. For example, the Act encourages the use of biological pest controls, finding that "biological control is often a desirable, low-risk means of ridding crops and other plants of plant pests and noxious weeds." Biological controls are themselves invasive species—additional biological pollution.

The Plant Pest Act preempts state efforts to regulate plant pests and noxious weeds. The judgement about whether these preemption provisions interfere with valuable state laws will depend on how courts and agencies interpret the provision that allows regulation of interstate commerce when regulations "are consistent with" federal regulations and on how sympathetic and wise the Secretary of Commerce will be in responding to state requests for waivers based on "special need."

National Invasive Species Act

The National Invasive Species Act (NISA) of 1996 reauthorized a previous federal statute with a less encompassing title: the Non-Indigenous Aquatic Nuisance Prevention and Control Act (NANPCA) of 1990. NANPCA focused on one place (the Great Lakes) and on one pathway (ballast water), and was driven by concerns about one harmful NIS (zebra mussel). It was a statute designed to organize state and federal forces against the zebra mussel and other NIS that had been, and might be, introduced through ballast water.

In 1996, NISA expanded the focus of the NANPCA to mandate regulations to prevent the introduction and spread of aquatic nuisance species. In NISA, Congress encouraged the federal government to negotiate with foreign governments to develop an international program for preventing NIS introductions through ballast water. The geographical scope of the Act was expanded as well, to include funding authorization for research on aquatic NIS in the Chesapeake Bay, San Francisco Bay, Honolulu Harbor, and Columbia River system.

A series of bills to reauthorize NISA has been introduced in both houses of Congress over the past decade.

Over the past five years, members of Congress have shown increasing awareness of threats specific to their jurisdictions, and especially to agricultural, commercial, and industrial interests, with strong concerns about harm from invasive species. Several new laws have been enacted—and other laws have been proposed—which, for the most part, focus on specific threats from harmful NIS in specific regions of the country. Federal laws dealing with harmful NIS and passed since 2003 include the following:

- The Nutria Eradication and Control Act of 2003.
- The Brown Tree Snake Control and Eradication Act of 2004, which provides for the control and eradication

of the brown tree snake on the island of Guam and the prevention of the introduction of the brown tree snake to other areas of the United States.
- The Water Resources Development Act of 2007, which focuses on the Asian carp dispersal barrier demonstration project in the upper Mississippi River. The act authorizes the Secretary of the Army to study, design, and carry out the project to delay, deter, impede, or restrict the dispersal of aquatic nuisance species into the northern reaches of the Upper Mississippi River system.
- The Great Lakes Fish and Wildlife Restoration Act of 2006, which amends the Great Lakes Fish and Wildlife Restoration Act of 1990. The bill reauthorizes provisions concerning the Great Lakes Fishery Commission's retention of authority and responsibility for the formulation and implementation of a comprehensive program for eradicating or minimizing sea lamprey populations in the Great Lakes basin.
- The Salt Cedar and Russian Olive Control Demonstration Act, which directs the Secretary of the Interior to carry out an assessment and demonstration program to control salt cedar and Russian olive species.
- Section 6006 of the Safe, Accountable, Flexible, and Efficient Transportation Equity Act, which includes a provision that makes activities for the control of noxious weeds and the establishment of native species eligible for federal aid funds under the National Highway System (NHS) and the Surface Transportation System (STP).
- The National Plan for Control and Management of Sudden Oak Death, which directs the Secretary of Agriculture, acting through the Animal Plant and Health Inspection Service, to develop a national plan for the control and management of Sudden Oak Death, a forest disease caused by the fungus-like pathogen *Phytophthora ramorum*.
- The Consolidated Natural Resources Act of 2008, which authorizes the Secretary of the Interior to enter into cooperative agreements with state, local, or tribal governments and other federal and public agencies for the purpose of protecting natural resources in the National Park System, including prevention and control of NIS within the National Park System.

These new laws—and a much longer list of proposed legislation—suggest that harmful invasive species, or at least specific harmful invasive species, are on the Congressional radar. But it is also safe to say that Congress has not yet addressed the general problem of harmful invasive species; the laws that are being enacted and most of the proposed legislation do not provide an overall regulatory framework for dealing with NIS threats.

General Environmental Policy Acts

Harmful NIS have played a surprisingly small role in the courts, either through traditional common law claims (such as nuisance or negligence) or through litigation based on general environmental laws.

There are several major federal environmental policy statutes and a set of public lands statutes that might apply to harmful NIS in some situations. The National Environmental Policy Act (NEPA) requires federal government agencies to assess the environmental impact of their actions through the promulgation of environmental assessments (EA) and environmental impact statements (EIS).

Claimants have argued that the federal government has failed to take account of the impact of invasive species under NEPA. NEPA is primarily directed at the actions of federal agencies. NEPA assumes the possibility of expertise in recognizing and assessing future environmental harms from present actions. In the case of potentially harmful NIS, this kind of information and expertise may not be present; even when this information is available, the Environmental Protection Agency (EPA) has been reluctant to act. Environmental policy decisions in many areas have trouble incorporating limited present information and great future uncertainty. Even when NEPA applies, it only requires analysis of environmental impacts but does not itself impose substantive barriers, preferences, or limits on government action.

The Endangered Species Act (ESA) might apply to NIS whenever a government or private action threatens an endangered species. The ESA might also lead to direct actions against harmful NIS in the development of recovery plans for listed species. Because harmful NIS have been identified as a significant source of ecosystem change (which may lead to pressures on rare or endangered species), and in some contexts as a direct extinction threat through predation, competition, or displacement, the ESA might bar some introductions or lead to some efforts at removal. The situations where the powerful effects of the ESA apply, however, are likely to be few. If the ESA applies at all in terms of introductions, it will most likely apply only to intentional introductions of NIS, and only to those introductions where a nexus can be found between the NIS and a listed species.

In at least one prominent case, the federal courts have several times upheld an order to the Hawaiian Department of Land and Natural Resources to remove nonindigenous goats and sheep that threatened the endangered palila bird. The legal oddity of the litigation over efforts to protect the palila bird confirms the small likelihood of the ESA and recovery plans becoming a major mechanism for control of harmful NIS.

Other general environmental policy acts might also allow for regulation of some harmful NIS or particular pathways. In 2005, in the case of *Northwest Environmental Advocates v. EPA*, a federal district court in California determined that the Clean Water Act (CWA) applies to ballast water. Ballast water discharges from boats have historically been exempt from CWA permitting requirements. The suit was brought against the EPA by an environmental nongovernmental organization, but the suit was joined by the attorneys general from six Great Lakes states. The trial court concluded that the EPA had exceeded its statutory authority by exempting ballast water, and this decision was affirmed on appeal to the United States Court of Appeals for the Ninth Circuit. In response, Congress passed the Clean Boating Act of 2008, which restored the permit exemption that the district court vacated.

Another broad class of federal laws that provide authority to federal agencies are the federal public lands laws, especially the Multiple Use Sustained Yield Act of 1960, the National Forest Management Act of 1976, and the Federal Land Policy and Management Act of 1976. These acts, and related historical and contemporary legislation governing grazing, timber, and other uses, provide a broad array of authorities and responsibilities with respect to federal public lands. Similar legislation aimed at the governance of smaller federal land units includes the National Wildlife Refuge System Administration Act. In addition, the powers granted under these sweeping public land laws may be magnified, and even extended to some activities on state and private lands, under the expansive interpretation of the U.S. Constitution's "property clause," which provides that "the Congress shall have power to . . . make all needful rules and regulations respecting the Territory or other property belonging to the United States."

In addition to these major federal environmental statutes, a host of more focused environmental and nonenvironmental laws also have some relevance to harmful NIS. For example, the Wild Bird Conservation Act of 1992 regulates the importation of some wild birds, and thus might limit the introduction of birds that pose a special risk of becoming harmful NIS should they escape; it might also reduce the chance of accidental introductions of bird diseases through the careless importation of wild birds.

The most important federal agencies for dealing with NIS, including the Animal and Plant Health Inspection Service (APHIS), the Agricultural Research Service (ARS), the U.S. Forest Service (USFS), the U.S. Fish and Wildlife Service (FWS), and the National Park Service (NPS), all fall under the authority of two departments—the U.S. Department of Agriculture (USDA) and the U.S. Department of the Interior (DOI). Another federal agency with responsibility for a surprisingly large amount of federal lands—and therefore for the administration and protection of those lands consistent with other federal laws—is the Department of Defense. DOD lands have distinctive qualities, partially because they are often protected from regular uses of public lands and partially because they are subject to uses that can have detrimental environmental impacts.

Government authority to respond to harmful NIS arises from international law reflected in treaties signed by the United States. Perhaps the best example of such legal authority is the Convention on International Trade in Endangered Species of Wild Fauna and Flora (CITES), which provides additional authority for border inspections and creates an independent basis (indeed, an independent obligation), even in the absence of listing a species under one of the "black list" acts, for exclusion. The OTA report lists seven treaties with direct effects on harmful NIS and seven treaties with indirect effects on harmful NIS, including CITES.

Executive Orders Addressing Harmful NIS

Two executive orders, one issued by President Carter in 1977 (E.O. 11987) and the other issued by President Clinton in 1999 (E.O. 13112), directly address the problem of harmful NIS. Executive orders are an odd species of law, issued on occasion by the president. They direct one or more federal agencies to act in a particular policy direction specified by the president.

While executive orders cannot create new legal authority, the executive orders on invasive species assert the maximum available authority in support of federal NIS efforts.

While now part of federal executive legal history, President Carter's E.O. 11987 is worth remembering. Executive Order No. 11987 is an astounding document, as striking and unexpected, although not nearly as profound, as Charles Elton's classic 1958 book *The Ecology of Invasions by Animals and Plants.* Some aspects of harmful NIS were, of course, part of public policy and debate by 1977, but

NIS as a general issue had yet to strike public and political consciousness.

Executive Order No. 11987 is not only unexpected because of its topic, but also because of its brevity, its clarity, and its political timing. Executive Order No. 11987 is one page long. Discussions about it began within the White House only weeks after Carter took office in January 1977, and the order itself was issued as part of the first public policy statement on the environment by the Carter Administration. The heart of the order provides the following policy directives:

(a) Executive agencies shall, to the extent permitted by law, restrict the introduction of exotic species into natural ecosystems on lands and waters which they own, lease, or hold for purposes of administration; and, shall encourage the States, local governments, and private citizens to prevent the introduction of exotic species into natural ecosystems of the United States.

(b) Executive agencies, to the extent they have been authorized by statute to restrict the importation of exotic species, shall restrict the introduction of exotic species into any natural ecosystem of the United States.

The short executive order included at least one other visionary aspect: it directed executive agencies to prevent the export of native (U.S.) species "for the purpose of introducing such species into ecosystems outside the United States where they do not naturally occur." President Carter was not concerned just with U.S. ecosystems; he was concerned with the threat of NIS to the naturalness of all ecosystems.

Executive Order No. 11987 included no complete procedure for implementing its policy directive and disappeared from federal policy as dramatically as it first appeared.

Twenty-two years later, on February 3, 1999—in the last two years of his presidency—President Clinton issued Executive Order No. 13112, a longer and more complex document, substantively and procedurally, than Executive Order No. 11987. Executive Order No. 13112 states its goal as preventing "the introduction of invasive species and provide for their control and to minimize the economic, ecological, and human health impacts that invasive species cause."

In some ways, its policy goals are more sweeping than those of Executive Order No. 11987. Executive Order No.

13112 includes control of existing invasive species as one of its primary goals. "Alien species" are defined in ecological, not political terms, as "with respect to a particular ecosystem, any species, including its seeds, eggs, spores, or other biological material capable of propagating that species, that is not native to that ecosystem." "Introduction" is defined to include "intentional and unintentional escape, release, dissemination, or placement of a species into an ecosystem as a result of human activity." "Invasive" is defined to include economic harms and harms to human health, both elements that went beyond the traditional focus of ecologists on ecological harms. Section 2 of the new executive order directs "each federal agency . . . to the extent practicable and permitted by law to use its programs and authority, subject to available funds, to pursue the following objectives:

(i) to prevent the introduction of invasive species;

(ii) to detect and respond rapidly to and control populations of such species in a cost-effective and environmentally sound manner;

(iii) to monitor invasive species populations accurately and reliably;

(iv) to provide for restoration of native species and habitat conditions in ecosystems that have been invaded;

(v) to conduct research on invasive species and develop technologies to prevent introduction and provide for environmentally sound control of invasive species; and

(vi) to promote public education on invasive species and the means to address them."

Executive Order No. 13112 also created an Invasive Species Council, made up of all cabinet officers with significant responsibility for NIS (Secretary of State, Secretary of the Treasury, Secretary of Defense, Secretary of the Interior, Secretary of Agriculture, Secretary of Commerce, Secretary of Transportation, and Administrator of the Environmental Protection Agency). The council was required to issue an invasive species management plan within 18 months. The council is advised by an advisory committee, whose responsibility is to "recommend plans and actions at local, tribal, State, regional, and ecosystem-based levels to achieve the goals and objectives" of the management plan.

Executive Order No. 13112 uses many contemporary federal management tools. The interagency council made up of cabinet officers places responsibility as high as it

can go. Involving a wide range of cabinet-level officers increases the likelihood of a full airing of views and revelation of conflicts, and perhaps increases the chance of reasonable policy consistency, efficiency, and success of enforcement. Requiring a plan provides a device for action and commentary. Creating an advisory committee increases the chance of expert input and invests a number of people and organizations outside the government in the details of the council's work.

The 2001 National Invasive Species Management Plan was approved by the National Invasive Species Council on January 18, 2001. The 80 page plan was full of bureaucratic but relatively nonspecific steps; it was even less clear about its systematic aims. Despite their generality, most and perhaps all of these goals have not been met.

Two general problems with the Invasive Species Management Plan stand out. The first is the extent to which the plan continues to define the invasive species problem largely in terms of current federal agency jurisdiction and authority, rather than as a crosscutting issue for the federal government (and of immense relevance to states, localities, Indian tribes, and private actors). Second, the plan does not include or require any measures of current collective harm, and therefore offers no basis other than expenditure of energy and money for determining whether the policies proposed are effective or as efficient as possible. The U.S. General Accounting Office (GAO) issued a report in October 2002 that reached similar conclusions:

While the National Invasive Species Council's 2001 management plan, Meeting the Invasive Species Challenge, calls for actions that are likely to help control invasive species, it lacks a clear long-term outcome and quantifiable performance criteria against which to evaluate the overall success of the plan . . . . [T]he only available performance measure that can be used to assess overall progress is the percentage of planned actions that have been completed by the due dates set in the plan. By this measure, implementation has been slow. Specifically, the council departments have completed less than 20% of the planned action items that were called for by September 2000 . . . . [W]hile the national management plan calls for many actions that would likely contribute to preventing and controlling invasive species, even if the actions in the plan were more fully implemented their effect would be uncertain because they typically do not call for quantifiable improvements in invasive species management or control.

A revised National Invasive Species Management Plan was issued in August 2008 and by its title claims to govern the period 2008–2012. The revised National Invasive Species Management Plan is relatively short and is organized around five basic strategic goals: prevention, early detection and rapid response, control and management, restoration, and organizational collaboration.

At the federal level, the effort to organize federal efforts through the National Invasive Species Council does not appear to have achieved what the optimistic executive order that led to its creation had sought.

STATE LAWS

The story of harmful NIS laws in the states is similar to the federal story. Generally, states with more commerce, and especially more foreign commerce, such as California and Florida, have been subject to greater harm from NIS, and therefore have more developed (but still incomplete) laws. The state with the most substantial and aggressive invasive species legal and policy regime is Hawaii.

Most state laws respond to particular invaders and pathways; as with federal law, there is far greater attention to threats from aquatic invaders, and especially to those from invasive mussels and clams. The special state legislative attention to aquatic invaders appears even in states such as New Mexico that have not yet experienced direct harm.

California, for instance, has enacted two statutes specifically oriented toward the control and eradication of invasive mussels. Several states have enacted aquatic invasive species acts that could in theory be applied to the control of any aquatic invasive species, but these acts are generally oriented toward the control of invasive mussels and other aquatic organisms that attach themselves to watercraft.

The Idaho Invasive Species Act stands out among these newly passed aquatic invasive species statutes for the breadth of its language. The statute defines "invasive species" as any "species not native to Idaho, including their seeds, eggs, spores, larvae or other biological material capable of propagation, that cause economic or environmental harm and are capable of spreading in the state." Again, the legislative history of the act, as well as its application post-passage, shows an orientation toward invasive mussels. Proposed legislation in Massachusetts would implement a comprehensive program aimed primarily at invasive plants, to prevent and curb the spread of harmful nonindigenous species, to require annual reports, and to require state agencies to avoid doing activities that could spread harmful invasives.

The increasing attention to threats from harmful NIS, even in states that have seen little harm from invasive species, is illustrated by the passage of the New Mexico Aquatic Invasive Species Act. New Mexico is a landlocked desert state with few large water bodies or major rivers. New Mexico passed its Aquatic Invasive Species Act before any known infestations of aquatic mussels or other invasives had occurred. The impetus for this legislation was the fact that reservoirs in neighboring Colorado had been infested by invasive mussels. The bill received a good deal of support from NGOs and passed unanimously in 2008.

Hawaii has made its Invasive Species Council a permanent council, has passed a prohibition on the importation of portions of invasive plants in addition to a previous prohibition on the importation of whole plants, and has imposed fees on every air or sea freight shipment for invasive species inspection, quarantine, and eradication (even overriding a gubernatorial veto to impose the fee on air freight shipments). Pending legislation would impose severe penalties (at least $100,000 per violation) on "any person or organization who intentionally imports, possesses, harbors, transfers, or transports . . . any prohibited or restricted plant, animal, or microorganism without a permit, with the intent to propagate, sell, or release that plant, animal, or microorganism." Puerto Rico has what appears on the books to be a relatively strong "white list" invasive species law—a policy framework that creates a presumption against importation of species unless they are affirmatively approved as noninvasive.

CONCLUSION

A more coherent legal framework is necessary at both the federal level and in the states to adequately and systematically address the complex range of issues raised by invasive species.

SEE ALSO THE FOLLOWING ARTICLES

Ballast / Black, White, and Gray Lists / Endangered and Threatened Species / Great Lakes: Invasions / Hawaiian Islands: Invasions / Regulation (U.S.) / Zebra Mussel

FURTHER READING

Environmental Law Institute. 2002. *Halting the Invasion: State Tools for Invasive Species Managers.*

Fowler, A. J., D. M. Lodge, and J. F. Hsia. 2007. Failure of the Lacey Act to protect U.S. ecosystems against animal invasions. *Frontiers in Ecology and the Environment* 5(7): 353–359.

General Accounting Office, Invasive Species. 2001. *Obstacles Hinder Federal Rapid Response to Growing Threat* [July, GAO-01-724].

General Accounting Office, Invasive Species. 2002. *Clearer Focus and Greater Commitment Needed to Effectively Manage the Problem* [October, GAO-03-1].

General Accounting Office, Invasive Species. 2003. *Federal Efforts and State Perspectives on Challenges and National Leadership* [June 17, GAO-03-916T].

General Accounting Office, Invasive Species. 2005. *Progress and Challenges in Preventing Introduction into U.S. Waters via the Ballast Water in Ships* [September 9, GAO-05-1026T].

Miller, M. L. 2004. The paradox of U.S. alien species law. In M. L. Miller and R. Fabian, eds. *Harmful Invasive Species: Legal Responses.* Washington, DC: Environmental Law Institute.

Miller, M. L., and G. H. Aplet. 1993. Biological control: A little knowledge is a dangerous thing. *Rutgers Law Review* 45: 285.

Miller, M. L., and G. H. Aplet. 2005. Applying legal sunshine to the hidden regulation of biological control. *Biological Control* 35: 368.

Miller, M. L., and R. Fabian, eds. 2004. *Harmful Invasive Species: Legal Responses.* Washington, DC: Environmental Law Institute.

Miller, M. L., and L. H. Gunderson. 2004. Biological and cultural camouflage: The challenges of seeing the harmful invasive species problem and doing something about it. In M. L. Miller and R. Fabian, eds. *Harmful Invasive Species: Legal Responses.* Washington, DC: Environmental Law Institute.

LIANAS

SEE VINES AND LIANAS

LICHENS

SEE BRYOPHYTES AND LICHENS

LIFE HISTORY STRATEGIES

JENNIFER L. BUFFORD AND
CURTIS C. DAEHLER

University of Hawaii, Honolulu

In 1965, Herbert G. Baker proposed that agricultural weeds and colonizers of disturbed areas were successful because of a suite of life history traits that allowed them to quickly reproduce and spread. Baker's ideas sparked a new avenue of research, searching for life history strategies associated with or indicative of invasive species. A life history strategy describes the growth habits, timing and magnitude of reproductive effort, and mortality within a population. Successful life history strategies maximize the lifetime number of surviving offspring produced under the prevailing environmental conditions. There is still much debate about the general predictive value of life histories when applied to invasiveness. Nevertheless, studying life history strategies common among invaders may help us understand the patterns and impacts of invasions and may direct and improve detection of potential invaders and control efforts.

LIFE HISTORY STRATEGIES: THE CONCEPT

A life history describes an organism's lifetime allocation of resources towards growth, reproduction, and maintenance. The concept of a life history "strategy" originates with the premise that environmental and physiological limitations force organisms to allocate resources to one area at the expense of another, with a particular focus on tradeoffs between growth and reproduction. Adaptive life history strategies maximize reproductive success under specific conditions. Many organisms have been classified as having either r- or K-selected life history strategies (Fig. 1), where r and K refer to parameters of the logistic population growth model. Organisms are classified as r-selected by their high reproductive output and early maturation, and they are often colonizing species. K-selected organisms are usually slow-growing, long-lived, and produce well-provisioned young late in life.

A question that has occupied both researchers and managers is whether some life history strategies or traits predispose an organism to invasiveness. Baker specified a list of traits *a priori* that ought to predispose a plant to weediness. Some, though not all, of these claims have been supported through quantitative studies. To complicate matters, the terms "weeds" and "invaders" were sometimes used interchangeably. Weed is a subjective term used to describe an unwanted plant, with no distinction between native and introduced species. Weeds have historically been associated with agricultural fields or other highly modified anthropogenic habitats and thus are often ruderals.

In contrast, the term "invasive" is usually used to describe a species, plant or animal, that has spread extensively

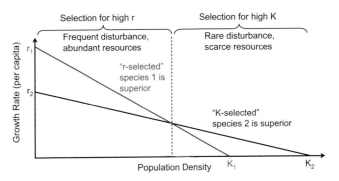

FIGURE 1 Where resources are abundant and population densities are low, often because of frequent disturbances, species with a higher intrinsic growth rate (r) are most successful (r-selected species). Where resources are scarce and population densities are high, species with lower intrinsic growth rates, but more efficient use of scarce resources (higher carrying capacity, K) are more successful (K-selected species). (Adapted from MacArthur, R. H. 1972. *Geographical Ecology: Patterns in the Distribution of Species.* New York: Harper and Row.)

beyond its native range; however, it may also refer specifically to species that have negative impacts. The distinction is crucial, as differences in the way "invasive" is defined may also lead to differences in associated life history traits, since invaders with a serious impact on the native environment are a distinct subset of all invaders. Although many invaders become established in disturbed habitats, some may spread into relatively undisturbed communities, and thus a diversity of life history strategies is represented in invaders.

COMMON INVADER LIFE HISTORY TRAITS

An invasion begins with the introduction of a nonnative species, which then establishes a reproducing population, spreads through the introduced range, and sometimes has major impacts on native ecosystems. Life history strategies can influence success at all stages of invasion (Table 1), and optimal strategies may be different for different stages, and may even be contradictory. For example, birds are more likely to be introduced if they have a small body size and are migratory but are more likely to establish if they have a large body size and are nonmigratory. Mode of arrival is also relevant: deliberate introductions are often biased toward different life history strategies than unintentional introductions. Finally, life history strategies vary by habitat, particularly between anthropogenically disturbed and natural environments. The majority of invasions occur at disturbed sites, and most research has focused on these areas.

Arrival

Unintentional introductions tend to be small, fast-reproducing, and short-lived (r-selected). They may have strategies to survive in a dormant state or otherwise facilitate unwitting anthropogenic transport. By contrast, deliberate introductions for horticulture, agriculture, silviculture, aquaculture, hunting, pets, and biocontrol are often selected for high establishment success (survival and growth) under diverse and sometimes adverse conditions, and for high productivity. These introductions may include large and long-lived species. These biases in the life history traits of the introduced species pool could explain some of the life history trends reported among invaders. Studies that compare the life histories of invasive and noninvasive introductions may provide the most insight, but the introduced species pool may not be representative of all potential life histories, thus narrowing the scope of inference.

Establishment

Once propagules are introduced, a successful invader must establish a reproducing population. For successful establishment, reproduction in the new environment must sometimes occur independently of mutualists and conspecifics. The ability of a single female to establish a population is an important characteristic of many invasive fish and other vertebrates. In plants, pollination through selfing or through abiotic or generalist pollinators facilitates establishment and subsequent invasion.

Asexual reproduction can also establish a population, and vegetative reproduction is a key trait of many plant invaders, facilitating both establishment and short-distance dispersal. Vegetative reproduction may be advantageous, because it avoids sensitive seed and seedling stages and reduces the need for pollinators. Vegetative reproduction is also effective in disturbed environments, where physical disruption can lead to frequent fragmentation. Humans frequently transport species that have high reproductive potential in disturbed anthropogenic environments, creating a link between life history and propagule pressure, the latter being one of the most important factors associated with establishment and naturalization.

Of course, for a plant or animal to successfully establish, its propagules must acquire resources and grow in the new environment. Growth characteristics and their tradeoffs have been studied extensively, especially in plants. Plant growth is usually measured as a relative growth rate (RGR), which is related to photosynthetic efficiency and leaf characteristics, including the specific leaf area (SLA = leaf area/leaf mass). SLA represents the potential for photosynthesis and indicates nitrogen content. It describes a tradeoff between building thin leaves quickly and building fewer, denser leaves that live longer and may be better protected against herbivores. A high SLA with high RGR is common in invasive species of disturbed environments, especially in herbs and seedlings, perhaps because it indicates an ability to quickly take advantage of available nutrients.

In animals, the trends are not as clear. A high relative growth rate was important in fish establishment in the Great Lakes, but a slow RGR was correlated with faster spread, perhaps because slow growth was correlated with other confounding factors, such as broad environmental tolerance. Body size is also important, since it indicates the practical results of growth. In insects, a small body size has been correlated with a high reproductive rate and a higher rate of invasiveness. In vertebrates, a relatively large body size may increase invasiveness, especially when competitors are present.

Plasticity, the ability of an organism to adapt to a wide variety of habitats and conditions, is also an important component of invasiveness. Organisms that exhibit

TABLE 1
Key Life History Traits by Stage of Invasion

	Introduction	Establishment	Spread	Impact
Growth	High productivity (deliberate introductions)	High growth rate Environmental tolerance Phenotypic plasticity Allocation to resource acquisition in productive environments (good competitor)	Lateral growth	Good competitor (often high RGR) Generalist predator Ecosystem-level changes
Reproduction	Small propagules (stowaways) Propagation under artificial conditions (deliberate introductions)	Abiotic/generalist pollination or selfing Asexual reproduction	Short MGT High fecundity Long-distance dispersal	"Swamps" native species (high fecundity)
Defense	Relatively weak (survival and productivity in anthropogenic environments with few enemies)	Plasticity	Plasticity	Few natural enemies (resource reallocation)

high flexibility in their allocation of resources to growth, reproduction, and defense may be better able to establish in novel environments (Fig. 2). Furthermore, phenotypic plasticity and a broad environmental tolerance allow

FIGURE 2 *Verbascum thapsus*, a Eurasian invader, photographed on Mauna Kea, Hawaii, showing normal and fasciated morphologies. The fasciated inflorescences produce more seeds per plant but apparently begin flowering at a later age. These life history changes appear to be an example of phenotypic plasticity and may be advantageous when the juvenile death rates are low in the invaded environment. (Photograph courtesy of J. Bufford).

invasive species, including plants, fish, and other vertebrates, to occupy a greater range of environments. Genetic diversity and the formation of ecotypes may also be important, allowing invaders to adapt to their new environment over the course of the invasion. For example, within 100 generations, introduced house sparrows in North America adapted to environmental clines by differentiating in coloration and body size, both important in local adaptation.

Spread

Invasive introduced species, by definition, disperse and increase their range. In this spreading phase, an organism's patterns of reproduction and dispersal are extremely important.

Age of reproductive maturation can dramatically affect an organism's population dynamics. A short minimum generation time (MGT) results in faster production of offspring, is associated with a high RGR, and can promote invasiveness. In plants, these traits are especially beneficial in a disturbed environment, allowing reproduction even when permissive conditions are temporary. In some vertebrates, a short MGT also correlates with invasiveness.

The frequency of reproduction also influences opportunities for spread. A life history that consistently includes multiple seed crops produced in rapid succession increases invasiveness, and some invaders are successful through mass reproduction and germination, effectively swamping native species and increasing the probability of successfully spreading at new sites. Invasive agricultural weeds are often annuals, which flower early and produce many seeds in one event. Perennial life histories may be more successful in cold climates and less disturbed habitats. Both clutch frequency and size are indicators of bird invasiveness, and invasive fish tend to have high fecundity.

In plants, low seed weight is correlated with higher seed production and greater dispersal and thus is an indication of spread potential and invasiveness. Small seeds tend to have high germination rates and lower rates of dormancy, which can result in rapid spread. Similarly, a small genome (the amount of DNA in one haploid set of chromosomes) is associated with small seeds and is correlated with faster mitosis, a short MGT and high growth rate in disturbed habitats. The total amount of DNA in the nucleus, however, depends on the organism's ploidy level. When phylogenetically related species are compared, polyploids are usually more likely to be invasive than diploids.

Seed size is often correlated with dispersal, including long distance dispersal, which is a crucial part of the invasion process. Abiotic seed dispersal promotes invasiveness in agricultural weeds and woody species. Small seeds are often abiotically dispersed, facilitating long-distance dispersal events. Small seeds packaged in small fruits may be dispersed by vertebrates, facilitating long-distance and directed dispersal into appropriate microhabitats. Heavy seeds and seeds requiring specialized biotic dispersal are usually a disadvantage that can restrict invasion.

Impact

The impact of a species on its environment may also be influenced by its life history strategy. Plants that allocate more resources to growth and competition, for example through unusually efficient nutrient uptake, nitrogen fixation, or a tall and spreading growth form, often have the greatest ecosystem and community impacts.

Competitive ability is a well-established factor in invasion success and is important in diverse taxa, including birds, amphibians, fish, and plants. Competitive ability is a reflection of resource allocation toward growth and aggression at the expense of other areas, such as defense. In animals, competition includes the struggle for food and territory. In plants, competition is often for light, and some plants seem to be effective invaders because of their ability to exploit canopy gaps through quick growth, outcompeting slower natives. Other invasive species are shade-tolerant and grow underneath the canopy. Allelopathy occurs when a plant prevents others from growing near it by releasing toxic compounds, thus reducing competition. Allelopathy may be more effective in an invaded environment, because native species often have low resistance to a novel toxin.

Growth in both animals and plants is driven by nutrient acquisition. Symbiotic nitrogen fixation in plants can increase the likelihood and impact of invasions. Similarly, efficient phosphate acquisition explained some of the difference in success between aquatic plant invaders. Increased efficiency in nutrient acquisition could improve an organism's competitive ability and its impact on the native community.

Normally, organisms experience a tradeoff between growth (and competitive ability) and defense, but this tradeoff does not seem to be realized in some invasions. This observation is linked to the hypothesis that invasive plants experience reduced herbivory because they have escaped their natural enemies (the enemy release hypothesis), allowing them to reallocate resources to growth and reproduction that would otherwise be invested in defense and giving them an edge over native species (the evolution of increased competitive ability, EICA, hypothesis). A recent metaanalysis supported the EICA hypothesis but emphasized that further empirical testing is needed, as most evidence to date is correlative rather than causative.

THE IMPORTANCE OF HABITAT

Invasions are complex processes in which many factors interact. Thus generalizations about life history strategies are difficult to make, especially as effective life history strategies vary dramatically by habitat type. One important distinction is between anthropogenically disturbed habitats and natural areas with minimal human disturbance. The majority of studies and the trends discussed here are based on research done in disturbed habitats. Disturbance is associated with high resource availability and creates a unique environment with plentiful opportunities for the establishment of new species, favoring *r*-selected life history strategies. For example, small seeds and high RGR may be especially beneficial in a disturbed, high nutrient environment. Relatively larger seeds with lower RGR may be better in low-resource environments. Human activities such as the application of salt to winter roads in temperate regions or the use of herbicides on crop lands and roadsides have promoted invaders with adaptations to human disturbances. In aquatic environments, dams and water enrichment by fertilizers have promoted invasion by aquatic organisms adapted to thrive in slow-moving water and more productive (high turnover) ecosystems.

In contrast, natural areas often have lower rates of anthropogenic disturbance and invasion, and the range of invasive life history strategies is expected to mirror the range of strategies exhibited by successful native species, within the limits of the pool of human introductions. However, the degree of anthropogenic disturbance, even in remote natural habitats, is likely to increase (e.g., through ongoing deposition of atmospheric nitrogen). Such environmental modification could promote invasion by organisms with the types of life history strategy currently associated with anthropogenic habitats. Thus, life history trends currently observed in disturbed environments may reflect future trends in invaders of increasingly modified natural habitats.

PREDICTIVE VALUE

The study of life history strategies in invasive species has been sparked largely by a desire to use life history traits to predict invasiveness. This question is extremely important to managers, as life history traits would be a cheap, easy way to screen for potential invaders. This is especially desirable given the huge cost of controlling noxious invaders.

An understanding of the life history strategies of invasive species can also aid in management and control efforts. A knowledge of an invader's population biology can help management efforts target sensitive life stages and can optimize the frequency and timing of control efforts. Understanding the life history strategy of a species can also allow us to predict the efficacy of biological control and can lead to more effective types of control.

Unfortunately, the potential predictive value of life history traits has not been fully realized, because of the complications described above. Studies that have focused on large-scale patterns have found trends in life history strategies, but studies comparing invasive and noninvasive species pairs have often been unable to explain their differences based on life history traits alone. Other factors, such as date of introduction or the frequency and degree to which humans have promoted establishment and spread of an invader, may be important.

More successful predictive models have been habitat- and taxon-specific and include other factors, such as home range, which may indicate inherent dispersal ability and environmental tolerance (Fig. 3). Less biologically informative but nonetheless useful characteristics, such as a history of invasiveness, are also important components of such models. Models that incorporate life history traits make more accurate predictions than those that use only a history of invasiveness. Thus, while life history traits are not exclusively predictive or

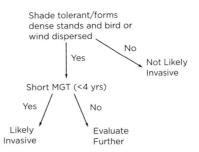

FIGURE 3 Sample decision tree for evaluating the potential invasiveness of new introductions to Hawaii, used when a more complete risk assessment is indecisive. Life history traits, including environmental tolerance, dispersal, and minimum generation time (MGT), are important in assessing the likely impact of a species. (Adapted from Daehler et al. 2004.)

highly consistent across invaders and invaded habitats, they are an important consideration, lending power to predictive models and helping to direct management efforts.

SEE ALSO THE FOLLOWING ARTICLES

CART and Related Methods / Demography / Dispersal Ability, Plant / Disturbance / Invasiveness / Reproductive Systems, Plant / Risk Assessment and Prioritization

FURTHER READING

Caswell, H. 1989. Life-history strategies (285–307). In J.M. Cherrett, ed. *Ecological Concepts: The Contribution of Ecology to an Understanding of the Natural World*. Boston: British Society/Blackwell Scientific Publications.

Craine, J.M. 2009. *Resource Strategies of Wild Plants*. Princeton, NJ: Princeton University Press.

Daehler, C.C. 2003. Performance comparisons of co-occurring native and alien plants: Implications for conservation and restoration. *Annual Review of Ecology, Evolution and Systematics* 34: 183–211.

Daehler, C.C., J.S. Denslow, S. Ansari, and H-C. Kuo. 2004. A risk-assessment system for screening out invasive pest plants from Hawaii and other Pacific islands. *Conservation Biology* 18(2): 360–368.

Grime, J.P. 2001. *Plant Strategies, Vegetation Processes, and Ecosystem Properties*, 2nd ed. Chichester, UK: John Wiley and Sons.

Grotkopp, E., M. Rejmánek, and T.L. Rost. 2002. Toward a causal explanation of plant invasiveness: Seedling growth and life-history strategies of 29 pine (*Pinus*) species. *The American Naturalist* 159(4): 396–419.

Kolar, C.S., and D.M. Lodge. 2001. Progress in invasion biology: Predicting invaders. *Trends in Ecology and Evolution* 16(4): 199–204.

Lockwood, J.L., M.F. Hoopes, and M.P. Marchetti. 2007. *Invasion Ecology*. Singapore: Blackwell Publishing.

Rejmánek, M. 2000. Invasive plants: Approaches and predictions. *Austral Ecology* 25: 497–506.

Sakai, A.K., F.W. Allendorf, J.S. Holt, D.M. Lodge, J. Molofsky, K.A. With, S. Baughman, R.J. Cabin, J.E. Cohen, N.C. Ellstrand, et al. 2001. The population biology of invasive species. *Annual Review of Ecology and Systematics* 32: 305–332.

Stearns, S.C. 1977. The evolution of life history traits: A critique of the theory and a review of the data. *Annual Review of Ecology and Systematics* 8: 145–171.

MALARIA VECTORS

LOUIS LAMBRECHTS

Institut Pasteur, Paris, France

ANNA COHUET AND VINCENT ROBERT

Institut de Recherche pour le Développement, Montpellier,
France

Invasion by insect vectors can significantly contribute to outbreaks of human and animal disease. Dispersion of insects can occur either naturally (by active flight or passive dispersal) or mediated by human activities (by land, air, or water transportation). Malaria is a mosquito-borne disease caused by blood parasites of the genus *Plasmodium* that infect a wide range of reptiles, birds, and mammals. Following successful introduction into a nonnative area, establishment and spread of mosquito vectors of disease can profoundly alter the transmission dynamics of the pathogens they transmit.

INVASIVE MALARIA VECTORS

With the exception of a single species infecting reptiles (transmitted by phlebotomine sandflies), all known *Plasmodium* parasites are mosquito-borne. Over 170 *Plasmodium* species have been described in the classic literature, but several new species have been identified in the last few years, and certainly more remain to be discovered. Malaria parasites of mammals, including humans, are exclusively transmitted by mosquitoes of the genus *Anopheles* (subgenus *Anopheles*) of the subfamily *Anophelinae*. By contrast, *Plasmodium* species infecting other vertebrates can be transmitted by vectors belonging to different genera of the subfamily *Culicinae* (e.g., avian malaria is transmitted by *Culex, Aedes, Aedomyia*, and *Coquillettidia* mosquitoes). Distributions of *Anopheles* species range from highly local to subcontinental. For example, *An. ovengensis* is found only in forests of southwestern Cameroon, and *An. bwambae* occurs exclusively within a 10-km radius of geothermal springs located in Bwamba County, Uganda. Conversely, *An. gambiae* and *An. funestus* occur in most of sub-Saharan Africa, and *An. darlingi* is found across all of tropical South America east of the Andes.

This article reviews several documented cases of invasive malaria vectors that altered transmission cycles of native or nonnative *Plasmodium* parasites. Although invasive malaria vectors may have other effects on resident species and ecosystems, the focus of this entry is their impact on human and animal health. The impact of malaria on nonimmune host populations can be disastrous, as exemplified by the estimated 130,000 human deaths in Egypt in 1942 that followed the introduction of an efficient malaria vector species from Sudan. Introduced malaria vectors can affect human and animal health by (1) independent introductions of a novel vector and a novel *Plasmodium* species, (2) acquisition of a native *Plasmodium* species by an introduced vector, and (3) simultaneous introduction of a novel vector and a novel *Plasmodium* species.

INVASION BIOLOGY OF MALARIA VECTORS

Introduction of a nonnative species relies strongly on the frequency of successful transport from its native area to a new area (i.e., propagule pressure). Although over relatively short distances (up to a few hundred kilometers) adult mosquitoes can be transported by wind, most introductions of mosquitoes over long distances are thought to have occurred via human-aided transportation,

FIGURE 1 *Anopheles gambiae s.s.* female, a major vector of human malaria parasites in tropical Africa. (Photograph courtesy of IRD/ Nil Rahola.)

especially by hitchhiking on ships and aircraft. Introductions have therefore been considerably facilitated by the establishment of transcontinental steamer ship lines in the nineteenth century and international airlines in the twentieth century. Extreme globalization of human and goods transportation during the last half-century undoubtedly enhanced propagule pressure and has been implicated in multiple events of mosquito introductions. Among factors that can influence the frequency of mosquito introductions, production of dessication-resistant eggs, preference for urban habitats, autogeny, use of human-made containers as larval habitats, and diapause favor human-aided transportation over large distances. In this regard, vectors of human malaria are much less likely to be transported and introduced into new areas than, for example, arbovirus vectors of the genus *Aedes* (especially the invasive species *Ae. aegypti* and *Ae. albopictus*). Indeed, *Anopheles* species that are efficient vectors of human malaria generally do not produce dessication-resistant eggs, are not autogenous, breed in natural habitats, and are not known to diapause. Because *Anopheles* species usually cannot breed on ships or aircraft, their successful introduction requires that at least one inseminated adult female survive the entire duration of the trip. Reduction of long-distance travel duration to a few days or a few hours in the last two centuries allowed introduction of *Anopheles* mosquitoes to remote islands, distant parts of the same continent, and even new continents. Following successful introduction, factors that influence the subsequent steps of invasion—establishment and spread—include adaptation to the new environment and ability to overcome negative interactions with predators and competitors. International travel and trade generally connect major cities, making nonnative species that are

adapted to urban environments more likely to succeed in the early phase of establishment. Thus, adaptation of most *Anopheles* vectors of human malaria to rural environments does not particularly favor the first step of their establishment upon arrival in a very urban environment. However, no universal rule can be formulated on the general ability of malaria vectors to establish and spread in a new area, because it depends on the specific location and circumstances of the introduction.

INDEPENDENT INTRODUCTIONS OF MALARIA VECTORS AND *PLASMODIUM* SPECIES

The opening of the steamer ship line between Tamatave, Madagascar, and Port-Louis, Mauritius, in 1864 coincided with a malaria outbreak in 1865. This outbreak was attributed to the introduction of mosquitoes of the *An. gambiae* species complex (*An. gambiae s.l.*) and *An. funestus*, the most important African malaria vectors. Malaria was almost certainly present among railway workers from India prior to the introduction of *An. gambiae s.l.* and *An. funestus*. It is believed that soon after, a cyclone was responsible for the introduction of *An. gambiae s.l.* from Mauritius into La Réunion Island (about 200 km away), where a malaria outbreak occurred in 1868. As in Mauritius, malaria parasites had probably been imported by human migration prior to the introduction of *An. gambiae s.l.* Thus, independent introductions of parasite and vectors resulted in the emergence of epidemic malaria on Mauritius and La Réunion Islands.

The Seychelles Archipelago had been free of *Anopheles* mosquitoes and therefore of endemic human malaria until 1908, when a malaria outbreak occurred on the Aldabra Islands. No *Anopheles* specimens could be found that year, so that the mosquito species that vectored the

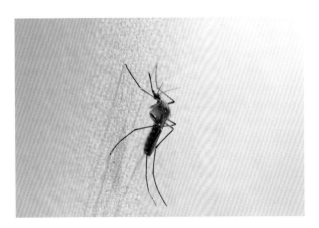

FIGURE 2 *Culex pipiens* female, vector of avian malaria but incompetent to transmit human malaria parasites. (Photograph courtesy of IRD/ Nil Rahola.)

outbreak remained a mystery. Following another malaria outbreak at Aldabra in 1930, larvae and adults of *An. gambiae s.l.* breeding on the islands were discovered. Malaria parasites had probably been present among workers from East Africa and Madagascar, so that subsequent introduction of the vector sufficed to trigger an outbreak. That 22 years elapsed between the two outbreaks suggests that *An. gambiae s.l.* may have been introduced a first time in 1908 but then disappeared before it was introduced a second time in 1930. If this hypothesis is true, it indicates that natural elimination of the invader mosquito does not preclude the occurrence of epidemic malaria.

Culex quinquefasciatus became established in the Hawaiian Islands in the early nineteenth century, several decades before the introduction of avian malaria with introduced birds from Asia in the early twentieth century. It is believed that *Cx. quinquefasciatus* was initially introduced into Hawaii as early as 1826 by a ship from Mexico. Analysis of genetic markers indicates that current Hawaiian *Cx. quinquefasciatus* populations probably resulted from multiple introductions. Avian malaria has devastated many native Hawaiian bird populations, especially at lower elevations, where *Cx. quinquefasciatus* occurs. The disease probably contributed greatly to the extinction of many Hawaiian bird species between 1910 and 1930 and the subsequent range reduction of other species. As in Mauritius, La Réunion, and the Seychelles Islands, malaria emergence was due to independent introductions of the vector and the parasite.

INTRODUCED MALARIA VECTORS TRANSMITTING NATIVE *PLASMODIUM* SPECIES

The African vector *An. gambiae s.l.* was introduced in Natal, Brazil, in 1930, causing malaria outbreaks in the vicinity in 1930–1931. Larvae or adult mosquitoes are believed to have traveled by air or steamer ship from Dakar, Senegal. Although insecticide treatments allowed elimination of *An. gambiae s.l.* from the Natal area by 1932, some mosquitoes escaped out of the city limits. They subsequently spread to a vast portion of northeastern Brazil, especially along river courses, causing devastating malaria epidemics in 1938–1939. At the time, *An. gambiae s.l.* was not recognized as a species complex, but molecular identification of museum specimens confirmed later that the invader was *An. arabiensis*, an arid-adapted species typical of the Sahel region of Africa. Human malaria was endemic in Brazil, but local *Anopheles* species had much lower vectorial capacity than *An. arabiensis*. Thus,

although *An. arabiensis* had a long history with the same *Plasmodium* species in its native range, it caused malaria outbreaks by transmitting native parasites more efficiently than native vectors. A massive eradication campaign conducted by the Rockefeller Foundation, mainly based on larval control, rapidly led to complete elimination of the invader in 1941, immediately followed by a drop in malaria incidence. Successful elimination of invasive *An. arabiensis* in Brazil remains a strong argument for proponents of larval control of malaria vectors.

Among *Anopheles* species native to the New World, *An. darlingi* is the most efficient malaria vector and contributes to malaria transmission in the Central Amazon Region. Before the 1990s, it was totally absent from the Peruvian region of Iquitos, in the Upper Amazon Region. However, invasive populations of *An. darlingi* were implicated in a large malaria outbreak in Iquitos in 1997. The observation that artificial ponds used for tropical fish culture around Iquitos were used by *An. darlingi* as larval habitats suggested that *An. darlingi* larvae might have been transported by boat with tropical fish. Again, introduction of a malaria vector more efficient than native vector species resulted in epidemic transmission of native *Plasmodium* species.

SIMULTANEOUS INTRODUCTION OF MALARIA VECTOR AND *PLASMODIUM* SPECIES

In 1959, a malaria outbreak occurred along the Mediterranean Israeli coast, a region that had been considered malaria-free. It was hypothesized that the outbreak had been caused by introduction of infected *An. pharoensis* from the Nile delta region (about 280 km away) that would have been transported by a strong southwesterly wind. Massive migration of *An. pharoensis* in the direction of the prevailing wind had been previously reported at locations in the Egyptian desert, several tens of kilometers away from the nearest breeding places. Wind-aided transportation of mosquitoes has been described as a form of long-distance migration resulting from "active" behavioral changes. If 89 confirmed malaria cases that occurred in 1959 in Israel were due to parasites introduced by *An. pharoensis*, a large number of mosquitoes (in the order of a million) must have been transported by wind.

Many long-distance flights connect malaria-ridden countries with malaria-free regions of the world. Among malaria vectors that hitchhike aboard aircraft and are released in malaria-free regions, some are infected with malaria parasites. Insect import by aircraft can be limited by treatments with insecticide either in flight or on the ground, but this measure is not systematically applied or not totally efficient. Introductions of infected vectors are held responsible for cases of "airport malaria," when people with no recent history of travel to tropical countries but living or working near international airports become infected with malaria parasites. Diagnosis of these cases is often delayed because of their unexpected nature. Between 1969 and 1999, 89 confirmed or probable airport malaria cases were reported, mostly in Europe. The highest risk is at airports in Belgium, France, and the Netherlands, where numerous flights from sub-Saharan Africa arrive daily. To date, introduction of infected malaria vectors in malaria-free countries has never caused more than a few sporadic malaria cases. Airport malaria does not require establishment and spread of an invasive vector but provides direct evidence of frequent introductions of malaria vectors (the vast majority uninfected) into malaria-free regions. That multiple "silent" introductions of malaria vectors occur every year without successful invasion, even temporary, demonstrates the existence of efficient barriers to invasion. Global spread of African malaria vectors may have been limited so far by climatic constraints because most flights from Africa have a European destination.

CONCLUSIONS

Although they are frequently introduced into nonnative areas, malaria vectors rarely establish populations. Historical records indicate that independent introductions of a novel vector and a novel malaria species and acquisition of a native *Plasmodium* species by an introduced vector occurred more often than simultaneous introduction of a novel vector and a novel parasite. Biological characteristics of *Anopheles* vectors of human malaria (eggs susceptible to desiccation, lack of autogeny, natural breeding sites, absence of diapause) make them relatively inefficient invaders. Most often, introduced malaria vectors fail to establish permanently or are quickly eliminated. The overall invasion success of *Anopheles* vectors of human malaria remains much lower than that of *Culex* vectors of avian malaria (and several other pathogens of humans and animals) or some *Aedes* vectors of numerous human and animal arboviruses. For example, *Cx. pipiens* and *Ae. albopictus* have become nearly cosmopolitan in the last half-century, which is mainly attributed to their ecological plasticity and adaptation to urban environments. Despite their lower overall invasive potential, the considerable impact that invasive malaria vectors can have on human or animal health makes it essential to maintain prevention measures and surveillance programs to reduce the risk of their introduction and establishment.

SEE ALSO THE FOLLOWING ARTICLES

Disease Vectors, Human / Mosquitoes / Pathogens, Human / Pesticides for Insect Eradication / Propagule Pressure

FURTHER READING

Garrett-Jones, C. 1962. The possibility of active long-distance migrations by *Anopheles pharoensis* Theobald. *Bulletin of the World Health Organization* 27: 299–302.

Gratz, N. G., R. Steffen, and W. Cocksedge. 2000. Why aircraft disinfection? *Bulletin of the World Health Organization* 78: 995–1004.

Hermitte, L. C. D. 1931. Occurrence of *Anopheles gambiae* (*costalis*) in Aldabra Islands (Seychelles). *Records of the Malaria Survey of India* 2: 643–657.

Juliano, S. A., and L. P. Lounibos. 2005. Ecology of invasive mosquitoes: Effects on resident species and on human health. *Ecology Letters* 8: 558–574.

Killeen, G. F., U. Fillinger, I. Kiche, L. C. Gouagna, and B. G. Knols. 2002. Eradication of *Anopheles gambiae* from Brazil: Lessons for malaria control in Africa? *Lancet Infectious Disease* 2: 618–627.

Lounibos, L. P. 2002. Invasions by insect vectors of human disease. *Annual Review of Entomology* 47: 233–266.

Mouchet, J., T. Giacomini, and J. Julvez. 1995. Human diffusion of arthropod disease vectors throughout the world. *Cahiers d'études et de racherches francophones/Santé* 5: 293–298.

Tatem, A. J., S. I. Hay, and D. J. Rogers. 2006a. Global traffic and disease vector dispersal. *Proceedings of the National Acadamy of Sciences of the United States* 103: 6242–6247.

Tatem, A. J., D. J. Rogers, and S. I. Hay. 2006b. Estimating the malaria risk of African mosquito movement by air travel. *Malaria Journal* 5: 57.

van Riper, C., S. G. van Riper, M. L. Goff, and M. Laird. 1986. The epizootiology and ecological significance of malaria in Hawaiian land birds. *Ecological Monographs* 56: 327–344.

MAMMALS, AQUATIC

CHRISTOPHER B. ANDERSON

University of Magallanes, Punta Arenas, Chile

ALEJANDRO E. J. VALENZUELA

Centro Austral de Investigaciones Científicas (CONICET), Ushuia, Terra del Fuego, Argentina

Aquatic mammals include species that live their entire lives in water, such as whales, dolphins or manatees, and those that use both terrestrial and aquatic environments, like seals, otters, and hippopotamuses. Aquatic mammals introduced by humans outside of their native range constitute only a small fraction of the more than 100 introduced invasive mammals found worldwide, but at least four species merit special attention, having attained extensive distributions as exotics and producing large ecological and economic impacts.

REASONS FOR AND CONSEQUENCES OF AQUATIC MAMMAL INTRODUCTIONS

North American Beavers

North American beavers (*Castor canadensis*) are the largest (11–35 kg) rodents native to North America, inhabiting both flowing and lentic freshwater environments from Canada to Mexico. Intensive trapping for their pelts caused the near extermination of beavers by the late 1800s. Subsequently, conservationists began reintroductions of *C. canadensis* not only throughout its native range, but also in parts of Europe (Finland, Poland, France, and Austria) and Russia, where *C. fiber*, the only other extant species in the genus, had also been driven to near extinction. Today in Eurasia, the North American beaver is found only in Finland and Russia with potential expansion into China. Known as "ecosystem engineers," beavers create and modify riparian and in-stream habitats by damming and foraging activities. Upon introduction, therefore, beavers help ensure their own protection and food supply and in the process alter the habitats where they become established. Unlike Eurasia, southern South American forests did not evolve with any species similar to the beaver. Consequently, introducing *C. canadensis* to Argentina in 1946 and its later expansion into Chile has caused extensive riparian deforestation and modified watershed biogeochemical cycles, thereby constituting the largest alteration of the subantarctic forest biome in the Holocene (Fig. 1).

FIGURE 1 Introduced North American beavers (*Castor canadensis*) in southern South America have caused extensive clearing of riparian subantarctic forests, constituting the largest landscape alteration since the retreat of the last glaciation 10,000 years ago. (Photograph courtesy of C.B. Anderson).

American Minks

American minks (*Neovison vison*; formerly *Mustela vison*) are members of the weasel family native to North America with semiaquatic adaptations that allow them to inhabit wetlands, stream banks, coastal shorelines, and lakes. Their generalist carnivorous diet encompasses a broad prey selection ranging in size from small amphibians, crustaceans, fish, and insects to larger items, including muskrats and geese. Mink's valuable pelts and easy domestication led to farm raising them in North and South America, Europe, and Asia (including Japan). Farm escapes and deliberate releases introduced populations into as many as 33 countries, where as semiaquatic carnivores they affect both terrestrial and aquatic prey. For example, mink are known to reduce populations of small mammals, like water voles (*Arvicola terrestris*), and ground-nesting birds through direct predation, but they also significantly influence other native predators, including the European mink (*Mustela lutreola*) and otters, due to niche overlap, competition for space and resources, and, in some cases, hybridization.

Muskrats

Muskrats (*Ondatra zibethicus*) are omnivorous rodents native to North America. They consume vegetation and small aquatic prey (freshwater mussels, crustaceans, fish, amphibians, and reptiles), and their burrowing activities, while not as extensive as the ecosystem engineering of beavers, constitute an important cause of stream bank erosion and can be particularly acute in marsh habitats. Around the world, farm raising and direct introductions into the wild of *O. zibethicus* have brought this species to as many as 42 countries in Europe, Asia, and South America. While many of these introductions originated from direct human efforts, the muskrat also bears the distinction of having invaded the greatest number of countries indirectly, due to expansion of human-established populations from one country to another.

Coypus

Coypus (*Myocastor coypus*; known as nutrias in North America) are medium-sized rodents (6–7 kg) native to South America. They are morphologically and ecologically similar to muskrats but tend to have a strictly vegetarian diet. Inhabiting marshes and bogs, coypus feed particularly on plant's basal meristems, which, rather than cropping vegetation, instead causes plants to die. Furthermore, their burrowing along banks and in marshes significantly impacts water control devices, such as dikes and irrigation channels, as well as the integrity of marsh ecosystems. For example, studies show that the foraging and burrowing of coypus degrade the structure of coastal wetlands. These impacts then allow greater penetration of sea waves, thus converting large areas of marsh into open water in places like the Chesapeake Bay. As with minks and muskrats, coypus have been extensively farmed for their fur. In fact, this species has been directly introduced into more countries than any of the other taxa considered here, which, added to accidental escapes and invasion from adjacent areas, means that *M. coypus* has been introduced to 50 countries in North America, Europe, Asia (including Japan), and Africa. While generally coypus are introduced for fur, in Florida they were deliberately established as a potential biological control of invasive aquatic plants.

GENERAL CONSIDERATIONS OF INTRODUCED AQUATIC MAMMALS

These four aquatic mammals have been introduced to at least 68 countries on all continents except Oceania and Antarctica, with the greatest effort to establish nonnative populations concentrated between the 1920s and 1950s (Fig. 2). While not all initiatives succeeded in establishing naturalized populations, the invasion success of these four mammals is in part the result of their intrinsic abilities to found and expand viable populations owing to high fecundity rates, constituting a new top predator, or being able to modify their own habitat. At the same time, being part of repeated, and

often government-sponsored introduction programs to "enhance" ecosystems has made propagule pressure high as well. Escapes from fur farms, especially in the case of mink, represent another cause of introductions, and some deliberate releases from farms by animal rights activists have occurred.

Clearly, these aquatic mammal introductions have been motivated principally by their potential economic value, but overall, the contribution of these introduced species to development is questionable when their deleterious impacts are considered. On the one hand, several countries in northern Europe, where cold climates produce high-quality pelts, maintain lucrative mink farming operations, and Russia still harvests both muskrats and minks from the wild. Yet, governments frequently have been forced to finance programs to deal with the dramatic and large impacts of these invasive species on recipient biotic communities and ecosystems.

CONTROL AND ERADICATION EFFORTS

Until recently eradication of invasive species was viewed as impractical or prohibitively expensive, but it is increasingly apparent that efficient techniques can be developed to eradicate problematic nonnative species completely. Furthermore, a single large investment in a successful eradication campaign is nearly always cheaper than an endless control effort. Beavers, minks, muskrats, and coypus have all been the subject of various such mitigation programs, and analyzing these experiences provides lessons about dealing with invasion impacts in general, and the subsequent restoration of native ecosystems.

Pioneering Efforts

Two of the world's earliest large-scale control and eradication efforts for introduced mammals were carried out on muskrats and coypus in the United Kingdom. Both species were brought to Britain during the late 1920s in failed fur farming enterprises. Muskrats were believed to be harmful initially, based on experiences in other parts of Europe, but coypus were not expected to present an ecological problem, harsh winters being thought likely to control their population. Therefore, early mitigation efforts focused on *O. zibethicus*, and a rapid, concerted trapping effort advised by scientists succeeded in exterminating this rodent, with the last captures occurring in Scotland in 1937. Later, emphasis shifted to the coypu as its population expanded to cover about 28,000 km^2 in East Anglia, England, and attained up to 200,000 individuals in the 1960s. Political and scientific attention

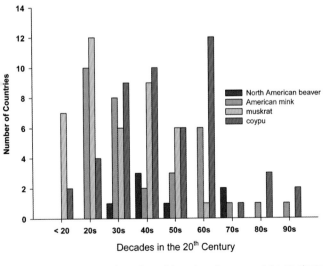

FIGURE 2 The number of countries where beavers, minks, muskrats and coypus have been introduced during the twentieth century by deliberate releases, accidental escapes, and range expansions of exotic populations into adjacent jurisdictions.

were brought to bear on this species owing in large part to the economic consequences involved in undermining water control devices. Consequently, an array of government and local actors worked to control coypus in the 1960s, which led to a later extermination campaign that achieved their eradication from England between 1981 and 1989. Key innovations demonstrated in these two successful eradication programs were that (1) research and management were solidly linked with formalized alliances between academic and governmental organizations, (2) novel financing schemes were designed whereby multiple agencies contributed to the program and trappers received a bonus equivalent to a three-year salary for fulfilling the goal of eradication in ten years, and (3) pelts were sold only to offset operation costs and not to provide hunter incentives.

New Top Predators on Islands

Scotland's Western Isles are home to an internationally important ground-nesting avian assemblage. To protect these breeding populations from introduced American minks, a five-year program was created to exterminate this nonnative predator from priority islands and reduce the population in adjacent areas to minimized recolonization. Set up in 2001 and funded by the European Union's Life Program with a consortium of local government agencies and conservation organizations, the project is the largest successful mink eradication program to date. Using a combination of trapping methods (e.g., line trapping, dog searching) and the constant refinement of strategy based on research, the project demonstrated the feasibility of eradicating minks from an archipelago. Important lessons about the ecology and trapping of mink emerged, including that (1) research generated improvements in censuses by using DNA fingerprinting from scat, thin-layer chromatography to identify bile acids, trap cameras, hair tubes, and clay-based footprint plates, and (2) capture methods were made more effective by studying the appropriate bait, the use of scent gland lures, the best time of the year to capture, and the best trap type (e.g., line trapping, dogs to find mink dens, floating traps). Additionally, the vital role of local involvement was evidenced as citizens contributed to monitoring efforts in the absence of researchers and trappers. Likewise, effective education and outreach activities ensured a general acceptance on the part of the general public of these conservation efforts. Ultimately, ecological and social benefits accrued from mink removal, allowing increases in the threatened bird populations and improving local economic activities such as aquaculture, small farming, and wildlife tourism.

New Challenges

Southern South America's experience with exotic North American beavers presents a new scenario and challenge for invasive species eradication efforts. From the 25 pairs introduced to Tierra del Fuego Island in 1946, beavers expanded their range to portions of the mainland, with more than 100,000 individuals by the 1990s. This ecoregion is one of the world's last great wilderness areas. Yet while the pervasive impacts of beavers caused concern, little early action was taken, in part due to the archipelago's remoteness. For example, Argentina did not authorize hunting beavers until 1981, and Chile classified them as "harmful" only in 1993. Furthermore, national efforts were uncoordinated until 2006, when the first binational meeting was held. Early efforts, as happened in England with the coypu, considered only control (rather than eradication) the only realistic goal. However, preliminary programs provided a testing ground for the local technical and scientific personnel and also, on a political and administrative level, permitted the initiative to gain experience and credibility. Currently, binational plans for beaver eradication remain in their infancy, but in 2008 agreements to deal with this problem were signed between both countries' ministries of foreign relations. Furthermore, an Argentine–Chilean working group, advised by international experts, has been established to determine the necessary strategies that would enable such an effort to be successful. The scale and dimensions, and consequently the cost, of this potential program are without precedent. Nonetheless, given the global track record of increasingly large and complex eradication initiatives, it is feasible to now believe in the possibility of eliminating invasive beavers from the Fuegian Archipelago. Yet, it also must be acknowledged that the southern tip of South America presents particular challenges due to its remoteness, extreme terrain, and rigorous climate, all of which make operations and success more difficult.

SEE ALSO THE FOLLOWING ARTICLES

Eradication / Islands / Predators / Rodents / Wetlands

FURTHER READING

Anderson, C. B., G. Martínez Pastur, M. V. Lencinas, P. K. Wallem, M. C. Moorman, and A. D. Rosemond. 2009. Do introduced North American beavers engineer differently in southern South America? An overview with implications for restoration. *Mammal Review* 39: 33–52.

Bremner, A., and K. Park. 2007. Public attitudes to the management of invasive non-native species in Scotland. *Biological Conservation* 139: 306–314.

Clout, M. N., and J. C. Russell. 2007. The invasion ecology of mammals: A global perspective. *Wildlife Research* 35: 180–184.

Gherardi, F., and C. Angiolini. 2004. Eradication and control of invasive species. In F. Gherardi, M. Gualtieri, and C. Corti, eds. *Biodiversity*

Conservation and Habitat Management, in *Encyclopedia of Life Support Systems (EOLSS).* Oxford: UNESCO, Eolss Publishers. www.eolss.net

Invasive species—Bay pressures—Nutria. Chesapeake Bay Program. www .chesapeakebay.net/nutria.aspx?menuitem=19616

Long, J.L. 2003. *Introduced Mammals of the World: Their History, Distribution and Influence.* Collingwood, Australia: CSIRO Publishers.

Macdonald, D., and R. Strachan. 1999. *The Mink and the Water Vole: Analyses for Conservation.* Stafford, UK: George Street Press.

MECHANICAL CONTROL

ANDREA J. PICKART

U.S. Fish and Wildlife Service, Arcata, California

Mechanical control of invasive species is broadly defined to include all physical manipulation or removal of plants or animals, whether or not machinery is involved. Mechanical control is frequently a component of an integrated pest management (IPM) strategy that also includes chemical or biological control agents. For invasive plants, mechanical methods occur along a spectrum of complexity from simple hand-pulling to the use of specialized heavy equipment. Mechanical control of invasive animals is usually restricted to vertebrates, although an IPM strategy for invertebrates may include a mechanical component such as trapping. Vertebrates are controlled or eradicated using direct methods such as trapping, hunting, and poisoning, or indirectly through habitat modification or the use of physical barriers. Mechanical control of plants and animals is often labor-intensive and costly, but it can have advantages in certain situations such as in rapid response to small early detections; in situations where the use of pesticides or biological controls have proven ineffective or controversial, have resulted in resistance, or have been determined to have unacceptable risk levels; and when the physical disturbance associated with mechanical methods promotes restoration of a disturbance-adapted or early successional community of native species.

MECHANICAL CONTROL OF INVASIVE PLANTS

Hand-Pulling and Hand Tools

The simplest form of mechanical plant control is uprooting by hand. Although labor-intensive, this method is commonly used when volunteer labor is available, or to achieve selectivity when working in sensitive habitats or among rare plants. The addition of simple tools such as shovels can increase efficiency. Digging combined with pulling has been used to successfully eradicate European beachgrass (*Ammophila arenaria* [L.] Link) on upwards of 40 ha of protected dunes under both private and public ownership in California. Initial removal is timed when stored carbohydrates are at their lowest level, a strategy that has been used in other types of mechanical treatments as well. In Oregon, the volunteer, nonprofit No Ivy League has removed over 100 ha of invasive English ivy (*Hedera helix* L.) by hand. Ground-colonizing ivy is pulled or raked up, and aerial vines are "girdled" (the vines are severed at the base of trees that have been colonized). Invasive shrubs such as brooms (*Genista monspessulana* [L.] L.A.S. Johnson, *Cytisus scoparius* [L.] Link, and *Spartium junceum* L.) can be manually chopped at ground level using tools such as Swedish axes and pulaskis, or can be completely uprooted with a specialized tool such as the Weed Wrench. Designed by a member of a weed crew, this steel tool is equipped with jaws that, in combination with the handle, provide powerful leverage. There are many other tools available for manual weed removal. Macleods are a type of combination rake/scraper that can be used for invasives like ivy and for removing duff and litter, as is sometimes desirable. Herbaceous plants can be cut with a machete, and woody plants with pruners, loppers, and a variety of hand saws and blades.

Although manual removal is often prohibitively expensive, the availability of inexpensive prison crews, conservation corps, youth corps, and volunteers make these methods feasible in some cases, especially for limited or new infestations. The use of volunteers, which can include school children, has an added educational and community-building benefit. Manual removal is also useful as a follow-up treatment to herbicides or mechanical removal, which frequently do not result in 100 percent mortality.

Portable Mechanized Tools

Invasive weed control often occurs in remote, roadless areas, making portable mechanized tools such as weedwhips, brushcutters, and chainsaws an important resource. These propane-, diesel-, or gas-fueled, handheld machines can greatly improve efficiency compared with hand tools. Invasive trees such as Russian olive (*Elaeagnus angustifolia* L.), Tasmanian blue gum (*Eucalyptus globulus* Labill.), and edible fig (*Ficus carica* L.) can be felled or sometimes girdled with chainsaws, but this is rarely sufficient to kill the tree and requires follow-up stump treatment by grinding or herbicide application. Brushcutters (similar to weedwhips but with attached handles and metal blades) have been used to kill invasive denseflowered cordgrass (*Spartina densiflora* Brongn.) over 10 ha in

a northern California salt marsh (Fig. 1). By angling the blade into the ground at the base of the plant and grinding the root crown and shallow rhizomes, one kills the plant in as few as one to two treatments. Another portable, mechanized approach is the use of backpack propane torches to "wilt" invasive seedlings or annual grasses and forbs (Fig. 2). This technique has been used successfully on early successional vegetation where there is not a deep litter layer. Although plants may initially catch, the fire doesn't carry. Alternatively, there are portable radiant heaters that commonly use a ceramic heating element to generate infrared radiation of up to 538 °C. Held close to the plant, the heat ruptures cells. Disadvantages to

portable mechanized tools include their labor intensiveness, bulkiness, need for refueling, and lack of selectivity.

Solarization, Weedmat, and Bottom Barriers

Covering of plants with landscaping material, sometimes referred to as solarization, can cause mortality of weed seed banks in addition to aboveground plants. Although clear polyethylene plastic is usually recommended to achieve lethal temperatures, opaque geotextile fabric used in a cool, foggy climate in northern California caused mortality of hard-coated legume (lupine) seeds. Fabric is placed over herbaceous vegetation and stapled to the ground with landscaping staples. Plants are kept covered anywhere from six weeks to two years. This method has been shown to work even on rhizomatous species such as iceplant (*Carpobrotus edulis* (L.) L. Bolus) and smooth cordgrass (*Spartina alterniflora* Loisel.). In aquatic systems, this method is referred to as a bottom barrier. Smothering of *Spartina anglica* swards after cutting in a Northern Ireland estuary was found to be as effective as herbicides for eradication. Many other examples of the effectiveness of this treatment have been documented, but the method is expensive and logistically demanding. Landscaping fabric is also commonly used to cover piled material after hand or mechanical removal to prevent resprouting, when burning or removal from the site is not feasible (Fig. 3).

Shading/Light Attenuation

Shading to discourage light-dependent species can be achieved in a number of ways. Shade cloth is expensive and is generally used over limited or new infestations. In aquatic systems, natural and synthetic dyes have been used. Probably the most common method of light attenuation is simply to promote growth of native shade-producing trees. Reed canarygrass (*Phalaris arundinacea*), a rhizomatous grass that is a prolific invader of wetlands in the Pacific Northwest of the United States, is intolerant of deep shade. Managers have found that introducing shade trees, especially conifers such as Sitka spruce (*Picea sitchensis* [Bong.] Carrière), leads to reduced dominance. This latter technique overlaps with a broader invasive plant strategy known as successional management, which, although not widely tested, calls for modifying ecological processes to direct plant community composition toward desired assemblages.

Vehicular Equipment

Both standard and modified vehicular equipment have been used extensively for the control of invasive plants.

FIGURE 3 Landscaping fabric covers piles of pulled English ivy (*Hedera helix*) to prevent resprouting. Fabric is removed after six months, and remains are allowed to decompose. Burning or hauling of piles is not practical in this isolated, forested natural area. (Photograph courtesy of the author.)

Bulldozers and backhoes, ranging from compact Bobcats to large Caterpillars, have been used to clear herbaceous to woody vegetation. In some cases, buckets have been customized for a particular job, such as adding tines to extract vegetation from soil (Fig. 4). Simply scraping surface vegetation does not prevent regrowth by rhizomatous species, which must be post-treated by hand-pulling or herbicides. An alternative technique used by dune managers in California employs an excavator to create a trench in an adjacent uninvaded area, excavate to the depth of viable roots or rhizomes, and backfill the trench, capping it with sufficient clean sand to prevent rhizomes from emerging. This method requires virtually no follow-up. Heavy equipment can be especially useful in removing an invader that has caused subsequent soil enrichment, such as Sydney goldentwattle (*Acacia longifolia* [Andrews] Willd.), Japanese knotweed (*Fallopia japonica* [Hourruyn] Ronse Decraene) and yellow bush lupine (*Lupinus arboreus* Sims). The duff or litter layer can be removed and composted. This entails the additional use of dump trucks and loaders. Plowing, rototilling, and disking are sometimes used to destroy root systems of species such as tamarisk (*Tamarix* spp.), although this treatment is generally followed by herbicides to prevent regrowth. In a Texas prairie restoration, Chinese tallow trees (*Triadica sebifera* [L.] Small) were treated with shredding mowers after being killed with herbicides, and the mulch was then used to damp diurnal temperature fluctuations needed for regeneration of the invasive tree.

Machines used for aquatic weed control have a history going back 100 years. In Florida, the notorious water hyacinth (*Eichhornia crassipes* [Mart.] Solms) was being controlled using "elevator" and "crusher boats" until the 1940s. These boats used machines to remove or crush plants, for disposal on land or water. "Hiballers," floating machines that collected and crushed plants, used a jet of water to deliver the slurry to land. Other machines used on aquatic invasives such as hydrilla (*Hydrilla verticillata* [L.f.] Royle) and Eurasian watermilfoil (*Myriophyllum spicatum* L.) have included harvesters (plants are removed and disposed of elsewhere), aquatic vegetation cutters or "cookie cutters" (a boat with hydraulically driven blades on the front of the hull that provide propulsion as well as cutting and clearing invasive vegetation), rotovators (underwater rototillers), flail choppers (a device mounted on an extending arm with blades that rotate rapidly inside a hood), dredges, pumps, and draglines. Many of these machines are used primarily in reservoirs and canals rather than natural areas.

Another type of machine developed for weed control delivers steam or hot water directly onto plants. Portable systems are available, but as they require a supply of water, machines are usually vehicle mounted. The Waipuna system uses water at temperatures above the boiling point delivered through a hose and wand. Water tanks and the heating mechanism are mounted on a truck or trailer. A biodegradable foam mixture of corn and coconut sugars produces a film that prevents immediate dissipation of heat. This type of equipment has been shown to be

FIGURE 4 A customized backhoe bucket with tines is used to clear a European beachgrass (*Ammophila arenaria*)-infested area in British Columbia. The material is subsequently buried. (Photograph courtesy of Sybilla Helms.)

effective in some cases and not in others, and it is limited (as is all mechanized equipment) to areas of suitable accessibility.

Hydrologic Manipulation

Mechanical control of plants can also take the form of modifying hydroperiods. Lakes and diked wetlands can be drawn down to promote extreme temperatures or desiccation, or flooded up to smother terrestrial species. While this is usually done in areas where water control structures already exist, devices such as sand bags and gabions (caged riprap) have been used to exclude tidal influence or to prolong inundation.

MECHANICAL CONTROL OF INVASIVE ANIMALS

Mechanical control of invasive animals commonly employs hunting or trapping of larger vertebrate species such as feral swine (*Sus scrofa*), horses (*Equus caballus*), donkeys (*E. asinus*), and goats (*Capra hircus*), and trapping of smaller vertebrate species such as black rats (*Rattus rattus*) and Norway rats (*R. norvegicus*). Trapping may be combined with lethal toxicants, as it was in the successful 30-year eradication program that removed 18,000 wild swine from Santiago Island in the Galápagos Archipelago, Ecuador. Traps may consist of snares, large cages, or pen traps. A well-known amphibian invader of Guam, the brown treesnake (*Boiga irregularis*) has been the subject of testing of over 49 different trap designs, resulting in a highly customized and effective design consisting of a mesh cylinder with inward-pointing funnel ends. Traps are baited with lure animals such as mice that are in a separate secure inner chamber and are provided with food. Other control and containment methods for this species include canine detection teams, vegetation removal (to reduce favorable habitat), and physical barriers constructed from materials such as concrete. As with game mammals, fish such as Asian carp (*Mylopharyngodon piceus, Hypophthalmichthys nobilis, H. molitrix*, and *Ctenopharyngodon idella*) are subjected to recreational and commercial harvests to reduce numbers. Mechanical barriers such as screens are used to prevent further spread, and hydrologic manipulations have been used to reduce habitat suitability. Netting, electrocuting, and explosives have been used with varying degrees of controversy and success to locally eradicate northern pike (*Esox lucius*) in Australia and the United States. Although mechanical control of invertebrates is not common, techniques such as water drawdown in reservoirs and mechanical cleaning of recreational boat hulls and trailers are used

to prevent spread and for localized control of zebra mussels (*Dreissena polymorpha*) and quagga mussels (*Dreissena rostriformis bugensis*). Targeted trapping efforts have been used to control small infestations of green crab (*Carcinus maenas*). Control of invasive insects such as the Africanized honeybee (*Apis mellifera scutellata*) may rely in part on mechanical traps used to lure and capture swarms.

SEE ALSO THE FOLLOWING ARTICLES

Early Detection and Rapid Response / Eradication / Freshwater Plants and Seaweeds / Integrated Pest Management / Rats / Tolerance Limits, Plant / Weeds

REFERENCES

Aquatic Plant Information System. http://el.erdc.usace.army.mil/aqua/apis/mergedProjects/Mechanical/html/apis_mechanical_and_physical_control_information.htm

Hammond, M. E. R., and A. Cooper. 2002. *Spartina anglica* eradication and inter-tidal recovery in Northern Ireland estuaries (124–132). In C. R. Veitch and M. N. Clout, eds. *Turning the Tide: The Eradication of Invasive Species.* Gland, Switzerland: IUCN Species Survival Commission.

Tu, M., C. Hurd, and J. M. Randall. 2001. *Weed Control Methods Handbook: Tools and Techniques for Use in Natural Areas.* The Nature Conservancy Wildland Invasive Species Team. Accessible at www.invasive.org/gist/handbook.html

Ver Steeg, B., ed. 2002. *Invasive and Exotic Species Compendium* [CD ROM]. Bend, OR: Natural Areas Association. Available from www.naturalarea.org/books.asp

The Watershed Project and California Invasive Plant Council. 2004. *The Weed Worker's Handbook: A Guide to Techniques for Removing Bay Area Invasive Plants.* Available from www.cal-ipc.org/ip/management/wwh/pdf/18601.pdf

Wittenberg, R., and M. J. W. Cock. 2005. Best practices for the prevention and management of invasive alien species (209–232). In H. A. Mooney, R. N. Mack, J. A. McNeeley, L. E. Neville, P. J. Schei, and J. K. Waage, eds. *Invasive Alien Species: A New Synthesis.* Scientific Committee on the Problems of the Environment (SCOPE) Series no. 63.

MEDITERRANEAN SEA: INVASIONS

BELLA S. GALIL

National Institute of Oceanography, Haifa, Israel

More than 600 introduced marine multicellular species have been recorded in the Mediterranean Sea. The majority of these species have been introduced through the Suez Canal, but shipping and mariculture have also contributed to the influx of species, especially in the western basin. Climate change is a significant factor assisting the spread and establishment of introduced species. Some introduced species have outcompeted or replaced native species locally, some are considered pests or nuisances, whereas others are of commercial value. Marine bioinvasions in the Mediterranean have far-reaching ecological and economic impacts.

INVENTORY: WHICH, WHEN, WHERE?

More than 600 introduced marine multicellular species have been recorded in the Mediterranean Sea. Nearly all are littoral and sublittoral benthic or demersal species. Because the shallow coastal zone, and especially the benthos, has been extensively studied and is more accessible, the chances that new arrivals will be encountered and identified are higher. Also, the species most likely to be introduced by the predominant means of introduction in the Mediterranean (Suez Canal, vessels, mariculture) are shallow-water species.

A taxonomic classification of the introduced species shows that the alien phyla most frequently recorded are Mollusca (33%), Arthropoda (17%), Chordata (17%), Rhodophyta (10%), and Annelida (8%). The data are presumably most accurate for large and conspicuous species that are easily distinguished from the native biota and for those occurring along a frequently sampled or fished coast and for which taxonomic expertise is readily available. Data are entirely absent for many of the small-sized invertebrate phyla.

The native range of introduced species in the Mediterranean is most commonly the Indo-Pacific Ocean (41%), the Indian Ocean (16%), the Red Sea (12%), and pantropical areas (9%). However, the actual origin of the Mediterranean populations of a species widely distributed in the Indo-Pacific Ocean may be its populations in the Red Sea or the Indian or Pacific oceans, or it may be through secondary introduction from populations already established in the Mediterranean itself. With few notable exceptions, the source populations of introduced species in the Mediterranean have not been ascertained by molecular means. Even taking into account these caveats, it is clear that most of the introduced species in the Mediterranean are thermophilic, originating in tropical seas.

With the exception of documented intentional introductions (i.e., aquaculture), rarely are the means and route of introduction of an alien species known from direct evidence. Mostly they are deduced from the biology and ecology of the species, the habitats and locales it occupies in both the native and introduced ranges, and its mode of dispersal (i.e., for a fouling species frequently recorded from ports, shipping is assumed to be the most probable vector). Inference from one case of introduction of a species to another may be fraught with

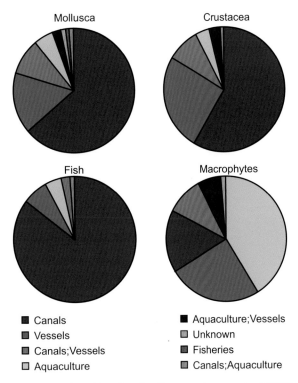

FIGURE 1 Number of introduced molluscs, crustaceans, macrophytes, and fish recorded in the Mediterranean Sea, and their means of introduction.

uncertainty, as pathways may differ between regions and between primary and secondary introductions. As far as can be deduced, the majority of introduced species in the Mediterranean entered through the Suez Canal (53%), followed by vessels (21%), and mariculture (10%). The means of introduction differ greatly among the phyla: whereas of the 92 introduced macrophytes, 41 percent and 25 percent were introduced with mariculture and vessels, respectively, the majority of alien crustaceans, molluscs, and fish are introduced Red Sea species (59%, 64%, and 86%, respectively), and mariculture introductions are few (4%, 5%, and 4%, respectively) (Fig. 1A–D).

The numbers of introduced species that have been recorded in the Mediterranean each decade over the past century have increased, especially recently (Fig. 2). The data may reflect political crises, economic development, and scientific interest in recording marine alien species. The few introduced species recorded prior to the twentieth century probably reflect ignorance of the phenomenon coupled with lack of detailed marine biological surveys. The gap in the 1910s indicates the First World War, whereas the dip in the 1930s and 1940s may be ascribed to the economic recession and devastation of the Second World War. The rather sharp decline in the number of

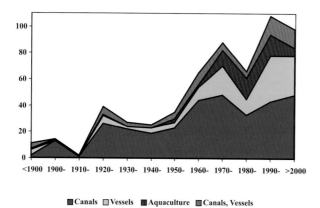

FIGURE 2 Number of introduced species recorded in the Mediterranean Sea, 1900–2009, and their means of introduction.

records in the 1980s may be due to the closure of the Suez Canal and the impact of the Arab oil embargo on oil shipping and international trade that limited the number of vessels transiting the Mediterranean. The increase in vessel-transported species in the past two decades may be attributed to the increase in shipping volume throughout the region, owing to the development of Middle Eastern oil fields, and later to the increasing trade with southeastern Asian nations. The increasing role of the Mediterranean as a hub of international commercial shipping, a surge in development of marine shellfish farming over the last 25 years, and the continuous enlargement of the Suez Canal all contribute to the resurgence of introductions

since the 1950s. A spate of records in the 1920s and 1970s reflects the publication of the results of two research programs focused on the biota of the Levantine basin. There seem to be as many introductions recorded in the first years of the twenty-first century (106 species) as in the 1990s (115 species). Many introduced species have established durable populations and have extended their range: 214 species have been recorded from three or more peri-Mediterranean countries, and 132 have been recorded from four or more countries.

A comparison of introduced species recorded along the Mediterranean coasts of Spain and France, and an equivalent length of coast in the Levant (Port Said, Egypt, to Marmaris, Turkey), shows marked differences in numbers, origin, and means of introduction (Fig. 3A and B). There are over four times as many introduced species along the Levantine coast (456) as in the westernmost Mediterranean (111). The majority of introduced species in the easternmost Mediterranean entered through the Suez Canal (68%, with 14% vessel transported and 2% through mariculture), whereas mariculture (42%) and vessels (38%) are the main means of introduction in the western Mediterranean.

ECOLOGICAL IMPACTS OF INTRODUCED SPECIES

With few exceptions, the ecological impacts of introduced species on the native Mediterranean biota are poorly known, although it is believed that invasive species may

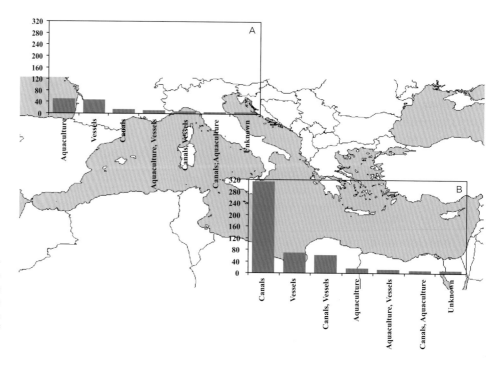

FIGURE 3 Number of introduced species in (A) the westernmost (Spain and France) Mediterranean Sea and (B) the easternmost (Egypt to Turkey), and their means of introduction.

cause major shifts in community composition. Several cases of sudden decline in abundance concurrent with proliferation of introduced species have been recorded, as is a single local extirpation of a native species. But even when populations of native Mediterranean species appear to have been wholly outcompeted or partially displaced from their habitat space by an introduced species, the causes cannot be disentangled from other potential factors such as the profound anthropogenic alteration of the marine ecosystem through habitat destruction, pollution, and rising Mediterranean seawater temperature. Thus, it remains difficult to extricate potential confounding factors in evaluating impacts of many (if not most) invasions.

Still, a handful of Mediterranean introduced species have drawn the attention of scientists, managers, and the media for having conspicuous impacts on the native biota. Perhaps the most notorious and best-studied introduced species in the Mediterranean are a pair of algae, the "killer alga" *Caulerpa taxifolia* and the sea grape seaweed *C. racemosa* var. *cylindracea*. A few of the discernible and sometimes dramatic changes in native biota associated with introduced species in the Mediterranean Sea are related below.

Caulerpa Seaweeds

An invasive strain of a tropical green alga, *Caulerpa taxifolia*, widely available through the aquarium trade, was unintentionally introduced into the Mediterranean with aquaria outflow in 1984, by the Musée Océanographique de Monaco. Within 15 years, the introduced alga had spread to Spain, France, Italy, Croatia, and Tunisia. The alga's rapid spread, high growth rate, and ability to form dense meadows (up to 14,000 blades per m²) on various bottom types, especially in areas with high nutrient loads, led to its replacement of native algal and seagrass species. Its presence is associated with a substantial reduction of species richness of native hard-substrate algae, and under certain conditions, it may outcompete the native seagrasses. Because the seagrass meadows are immensely important ecologically, any changes in their architecture, density, and quality would be reflected in the food web they support and in the highly diverse biota for which they serve as spawning ground and nursery. The alga's dense clumps obstruct fish feeding on benthic invertebrates, and density and biomass of fish assemblages are significantly lower in *C. taxifolia*–invaded seagrass beds. Moreover, the most potent of its suite of repellent endotoxins, caulerpenyne, is toxic for molluscs, sea urchins, and herbivorous fish, at least during summer and autumn

when metabolite production peaks, and it can devastate microscopic organisms. The diminution in the structural complexity of the invaded habitat, together with the replacement of the rich native biota with a species-poor community, results in a dramatic reduction in the richness and diversity of the affected littoral and constitutes a threat to the balance of the marine coastal biodiversity.

Another extremely invasive variety, *C. racemosa* var. *cylindracea*, endemic to southwestern Australia, was discovered in the Mediterranean in 1990 and has since spread from Cyprus to Spain. The alga may attain total coverage within six months, overgrowing other algae and curtailing species number, cover, and diversity of the algal community. Off Cyprus, within six years, it replaced the native seagrass community with a concomitant change in the biota.

Rabbitfish and Goatfish

Two species of rabbitfish, *Siganus rivulatus* and *S. luridus*, entered the Mediterranean from the Red Sea through the Suez Canal. The schooling, herbivorous fish form thriving populations as far west as the southern Adriatic Sea, Sicily, and Tunisia. The rabbit fish comprise 80 percent of the abundance of the herbivorous fish in shallow coastal sites in the Levant and have replaced native herbivorous fish. Their diet has a significant impact on the structure of the local intertidal rocky algal community, and their grazing pressure may have aided the proliferation of the introduced pharaonic mussel (*Brachidontes pharaonis*) by providing suitable substrate for its settlement.

The Red Sea goldband goatfish, *Upeneus moluccensis*, was first recorded in the Levant in the 1930s and has since established populations from Rhodes to Libya. Following the exceptionally warm winter of 1954–1955, it increased to 83 percent of the catch, replacing the native red mullet, *Mullus barbatus*, in the commercial fisheries. Both mullid species have a similar diet and occupy muddy bottoms shallower than 75 m, but whereas the red mullet spawns from April to June with a peak in May, the goldband goatfish spawns from June to September. In previous years, the young red mullets would settle to the bottom from July through September, where they have had a distinct size advantage over the later-spawned goldband goatfish. The failure of the 1955 red mullet year class may have left their niche only partly occupied, to the advantage of the Red Sea species, and the unusually warm waters enhanced the latter species' survival rate. The native mullet has since then been displaced into deeper, cooler waters: 87 percent of the mullid catch off the Israeli coast consists of introduced mullids at a depth of 20 m, and 50 percent at 55 m, but only 20 percent in waters deeper than 70 m.

Mussels, Limpets, Oysters, Snails

A small Red Sea mytilid mussel, *Brachidontes pharaonis*, was considered in the early 1970s to be 250 times rarer than the native mytilid along the Levant coast. However, by the late 1990s, dominance had shifted, with the introduced mussel forming dense populations on rocks, piers, and debris of up to 300 ind/100 cm², whereas the native species was only rarely encountered. The rabbit fish may have triggered the population increase by clearing the intertidal platform of algae. The pharaonic mussel has spread as far west as Malta and Sicily, probably through ship fouling, and it is common in both polluted harbors and hypersaline lagoons, where its densities reach an 10,000 ind/m².

A Red Sea limpet, *Cellana rota*, first collected in the Mediterranean in 1961, dominates the upper rocky littoral along the southeastern Levantine coast, where it has been replacing the native limpet. Similarly, the Red Sea spiny oyster, *Spondylus spinosus*, and the Red Sea jewel box oyster, *Chama pacifica*, have supplanted their native congeners. The native Mediterranean cerithiid gastropods, common in the southern Levant Sea until the 1970s, were supplanted by the Red Sea gastropods *Cerithium scabridum* and *Rhinoclavis kochi*.

Penaeid Prawns

Ten introduced penaeid prawns are recorded in the Mediterranean. The Red Sea prawns, in particular *Marsupenaeus japonicus*, *Metapenaeus monoceros*, and *Penaeus semisulcatus*, are highly prized and are considered a boon to the Levantine fisheries (see below). However, that boon came at the expense of the native prawn, *Melicertus kerathurus*. Formerly abundant on sandy or mud bottoms along the Levantine coast and supporting a commercial fishery throughout the 1960s, it has since nearly disappeared, and its habitat has been overrun by the Red Sea penaeid prawns.

Killifish

The Red Sea killifish, *Aphanius dispar*, is markedly euryhaline, whereas its peri-Mediterranean endemic congener, *A. fasciatus*, occurs mostly in brackish lagoons. Naturally occurring hybrids of the two killifish species were found in the hypersaline Bardawil Lagoon on the Mediterranean coast of Sinai, Egypt, where 45 percent of the killifish were identified as *A. dispar*, 17 percent as *A. fasciatus*, and 38 percent as hybrids. The first Red Sea killifish was collected in Israel in 1943; the last specimens of the native killifish were collected in 1976. Within a generation, an endemic genotype appears to have been locally lost through hybridization, and *A. dispar* and its hybrids replaced *A. fasciatus* along the Mediterranean coast of Israel.

ECONOMIC IMPACTS OF INTRODUCED SPECIES

Market-driven demands for introduced fish and shellfish are on the rise with the increasing affluence of Mediterranean countries. Coupled with the crisis in wild fisheries, this demand has created a surge in development of mariculture farming along the shores of the Mediterranean in the last 30 years. Two commercially important shellfish, the Pacific cupped oyster, *Crassostrea gigas*, and the Pacific carpet clam, *Ruditapes philippinarum*, were intentionally introduced to the Mediterranean in the 1960s and 1970s, respectively. Acclaimed as a replacement for the dwindling crop of the native carpet clam, the Pacific clam established breeding populations in the northern Adriatic lagoons by the mid-1980s and colonized all suitable areas. In 2006 the commercial yield from its cultured and feral beds reached 56,731 tons.

Some unintentionally introduced species are economically important as well and were exploited commercially almost as soon as they entered the Mediterranean. The blue swimmer crab, *Portunus segnis*, was already on sale in the fish markets of the Levant by the early 1900s. By mid-century, introduced Red Sea fishes and prawns were an important part of the Levantine fisheries. The dominant fishes in the inshore fisheries (trammel-netting and hook-and-lining) are the rabbitfish *Siganus rivulatus* and *S. luridus*, the obtuse barracuda *Sphyraena chrysotaenia*, and the Red Sea jack *Alepes djedaba*. The above species, together with *Sillago sihama* and *S. commerson*, two species that underwent population explosions in the early 1980s, are common in purse-seine landings. Nearly half of the trawl catches along the Levantine coast consist of Red Sea–introduced fish. The Red Sea lizardfish, *Saurida macrolepsis*, was first recorded from the Mediterranean in 1952 and has since spread in the entire eastern Mediterranean, from Albania to Libya. In 1955–1956, the lizardfish became commercially important, constituting an important staple of the coastal fishery in the area stretching from Port Said, Egypt, to Mersin, Turkey.

The Red Sea prawns, in particular *Marsupenaeus japonicus*, *Metapenaeus monoceros*, and *Penaeus semisulcatus*, are highly prized, and beginning in the 1970s, a shrimp fishery developed in the Levant using a small

FIGURE 4 *Rhopilema nomadica*, Haifa Bay, Israel. (Photograph courtesy of the National Institute of Oceanography, Israel.)

fleet of coastal "mini-trawlers" that specialize in shrimping. The increasing exploitation of Red Sea aliens meant the shifting of the trawling grounds nearshore because their densest populations occur at depths up to 50 m. The shoreward displacement of the fishing grounds raises the ratio of introduced to native taxa in the Levantine trawl landings.

Together with commercially exploitable species, the invasion of the Mediterranean has swept in venomous and poisonous species. Each summer since the mid-1980s, swarms, some 100 km long, of the Red Sea jellyfish *Rhopilema nomadica* have appeared along the Levantine coast (Fig. 4). When these draw nearer to shore, they affect tourism, fisheries, and coastal installations. As early as the summer of 1987, severe jellyfish envenomations requiring hospitalization had been reported in the medical literature. The annual swarming brings reports of envenomation victims suffering burning sensations, eurythema, and papulovesicular and urticaria-like eruptions that may last weeks and even months. Coastal municipalities report a decrease in tourists frequenting the beaches because of the public's concern over the painful stings inflicted by the jellyfish. Coastal trawling and purse-seine fishing are disrupted for the duration of the swarming owing to net clogging and the inability to sort yield. Fishermen, especially purse seiners, discard entire hauls because of the overwhelming presence of jellyfish in their nets. Jellyfish-blocked water intake pipes pose a threat to cooling systems of port-bound vessels, coastal power plants, and desalination plants.

FIGURE 5 *Lagocephalus scleratus*, captured off the Mediterranean coast of Israel in 2005. (Photograph courtesy of Limor Shoval.)

The recent introduction and spread of the silver stripe blaasop *Lagocephalus sceleratus* (Fig. 5) and the striped catfish *Plotosus lineatus* (Fig. 6) poses severe health hazards. The blaasop's internal organs, in particular the gonads during the spawning season, contain a strong paralytic neurotoxin. Several cases of tetrodoxin poisoning associated with consuming the fish have been reported from the Levantine basin, including a recent fatality. Injuries caused by the barbed and venomous first dorsal spine and pectoral spines of the striped catfish produce pain levels requiring hospitalization—injuries have been reported by local professional and amateur fishermen.

RISING TEMPERATURES AND INTRODUCED SPECIES

Evidence is accumulating worldwide that, in addition to habitat destruction, pollution, and overexploitation, changes in biodiversity patterns in freshwater and

FIGURE 6 *Plotosus lineatus*, captured off the Mediterranean coast of Israel in 2008. (Photograph courtesy of Kfir Gayer.)

marine systems are linked to climate change and invasive species. By the middle of the century, climate change and invasive species may be the dominant direct drivers of biodiversity loss and increased risk of extinction for many species, especially those already at risk due to low population numbers, restricted or patchy habitats, or limited climatic ranges. Climate change plays a role in influencing the rate at which new species are added to communities: rising temperatures may facilitate the establishment and spread of introduced thermophilic species.

The last decades of the twentieth century saw pronounced thermal fluctuations and a significant increase in the average temperature of the waters in the Mediterranean and the spread of Red Sea biota beyond the southeastern Levantine basin. In fact, the appearance of six Red Sea fish species in the Adriatic Sea was concurrent with a rise in the sea surface temperatures and the timing of a significant increase in the number of Red Sea species along the southwestern Anatolian and southern Aegean coasts, with a more extensive inflow of the warmwater Asia Minor Current. We are mostly ignorant of the mechanisms through which climate translates into ecologically important changes in populations. But it is likely that rising seawater temperatures may change the pool of species that could establish themselves in the Mediterranean, enable the thermophilic species (native and alien) to expand beyond their present distributions, and impact a suite of population characteristics (reproduction, survival) that determine interspecific interactions, therefore affecting the dominance and prevalence patterns of both native and introduced species and providing the thermophilic aliens with a distinct advantage over the native Mediterranean biota.

SEE ALSO THE FOLLOWING ARTICLES

Algae / Canals / Climate Change / Fishes / Invasion Economics

FURTHER READING

Galil, B.S. 2007. Loss or gain? Invasive aliens and biodiversity in the Mediterranean Sea. *Marine Pollution Bulletin* 55(7–9): 314–322.

Galil, B.S. 2009. Taking stock: Inventory of alien species in the Mediterranean Sea. *Biological Invasions* 11(2): 359–372.

Mineur, F., M.P. Johnson, C.A. Maggs, and H. Stegenga. 2007. Hull fouling on commercial ships as a vector of macroalgal introduction. *Marine Biology* 151(4): 1299–1307.

Verlaque, M. 2001. Checklist of the macroalgae of Thau lagoon (Hérault, France), a hot spot of marine species introduction in Europe. *Oceanologia Acta* 24(1): 29–49.

Verlaque, M., J. Afonso-Carrillo, M. Candelaria Gil-Rodríguez, C. Durand, C.F. Boudouresque, and Y. Le Parco. 2004. Blitzkrieg in a marine invasion: *Caulerpa racemosa* var. *cylindracea* (Bryopsidales, Chlorophyta) reaches the Canary Islands. *Biological Invasions* 6: 269–281.

MELASTOMES

JEAN-YVES MEYER

Délégation à la Recherche, Tahiti

ARTHUR C. MEDEIROS

U.S. Geological Survey-BRD, Maui, Hawaii

The melastomes, family Melastomataceae, comprise an important and species-rich family of tropical regions, both in the Old and New Worlds. Despite their high number of species, relatively few have been introduced and cultivated outside their native range, mainly as garden ornamentals, compared to other representatives of "weedy" families (Asteraceae, Fabaceae, Rosaceae, Acanthaceae). A relatively high proportion but low numbers of species have escaped cultivation and have become broadly naturalized and serious disrupters and modifiers of native forests worldwide. Much of the invasive reputation of the family is based on the track record of two melastome species, among the most notorious of plant invaders documented globally, *Miconia calvescens* and *Clidemia hirta*. Despite the fact that the relatively low number of cultivated melastome species has resulted in several extremely serious forest weeds, some horticultural literature continues to tout the family as an untapped source of unexploited species for the green industry.

NATURALIZED AND INVASIVE MELASTOMES

Derived from the Greek *melas* (black) and *stoma* (mouth), because of the purple-black color of the juice of the fleshy berries of most taxa, the family Melastomataceae (sometimes spelled as Melastomaceae) comprises about 180 genera and over 5,150 species that are found mainly in tropical and subtropical regions of the Old World and the neotropics. At least 25 species of melastomes are known to be naturalized (Table 1), including some of the most invasive plants in the tropical island rainforests of the Indo-Pacific. The worst melastome invaders are *Miconia calvescens* (called the "purple plague" or "*le cancer vert*") in Oceania (Australia, New Caledonia, French Polynesia, and Hawaii) and *Clidemia hirta* ("Köster's curse") in Africa, Asia, and Oceania. Lesser but also serious invaders are *Melastoma malabathricum* ("Indian rhododendron") in the Seychelles, and *Tibouchina herbacea* ("cane Tibouchina") and *Oxyspora paniculata* in Hawaii. Other widely naturalized melastomes include *Dissotis rotundifolia*, *Heterocentron subtriplinervum*, and several other *Melastoma*, *Miconia*, and *Tibouchina* species.

TABLE 1
List of the Documented Naturalized (nat.) and Invasive (inv.) Melastomes

Scientific Name	Life Form	Native Range	Areas of Introduction
Arthrostema ciliatum Pav. Ex D. Don	scandent perennial herb (1/4 m long)	tropical America	Hawaii (nat.)
Clidemia hirta (L.) D. Don	subshrub to shrub (0.5–3 m tall)	Central and South America, West Indies	Australia (inv.), Fiji (inv.), Hawaii (inv.), La Réunion, Mauritus (inv.), Madagascar (nat.), Malaysia (nat.), Singapore (inv.), Taiwan (nat.), Guam (inv.), Samoa (inv.), Tonga (inv.), Palau (inv.), Solomon (inv.), Vanuatu (inv.), Wallis and Futuna (inv.)
Dissotis rotundifolia (Sm.) Triana (syn. *Heterotis rotundifolia* (Sm.) Jacq.-Fél.)	decumbent herb	tropical Africa	Hawaii (nat.), Fiji (cult.), Samoa (nat.), Tahiti (nat.), Seychelles (cult.), Singapore (inv.)
Heterocentron subtriplinervium (Link & Otto) A. Braun & C. Bouché	erect suffrutescent herb or subshrub (1–3 m long)	Central America	Tahiti (nat.), Hawaii (nat.), Mauritius (nat.)
Heterotis decumbens (P. Beauv.) Jacq.-Fél. (syn. *Dissotis decumbens* (P. Beauv.) Triana)	prostrate herb (0.2 m long)	tropical Africa	La Réunion (nat.)
Medinilla venosa (Blume) Blume	erect shrub (2–3 m tall)	Celebes, Moluccas, Philippines	Hawaii (nat.)
Medinilla cumingii Naudin	scandent epiphytic shrub (1–3 m tall)	Philippines	Hawaii (nat.)
Melastoma candidum D. Don	erect shrub to small tree (1.5–5 m tall)	Vietnam, China, Philippines, Taiwan, Ryukyu, Japan	Hawaii (nat.)
Melastoma malabathricum L.	shrub (1–3 m)	tropical Asia	Seychelles (nat.), Mauritius (cult.)
Melastoma sanguineum Sims	shrub to small tree (2–8 m tall)	Malay Peninsula, Java, Sumatra, Vietnam, China	Hawaii (nat.)
Melastoma septemnervium Lourceiro	shrub	Japan, China, Taiwan, Vietnam	Hawaii (nat.)
Memecylon caeruleum Jack (*Memecylon floribundum* sensu auct.)	shrub (1–4 m tall)	Indonesia, China, Cambodia, Vietnam	Seychelles (nat.)
Miconia calvescens DC. (syn. *M. magnifica* Triana)	small tree (4–16 m tall)	Central and South America	French Polynesia (inv.), Hawaii (nat.), New Caledonia (inv.), Australia (nat.)
Miconia nervosa (Sm.) Triana	shrub (3 m tall)	Central and South America	Australia (nat.)
Miconia racemosa (Aubl.) DC.	shrub (2–3 m tall)	Central and South America	Australia (nat.)
Osbeckia octandra (L.) DC.	erect subshrub (1 m tall)	Sri Lanka	Mauritius (nat.)
Ossaea marginata (Desr.) Triana	suffrutescent herb (2.5 m long)	Brazil	Mauritus (inv.)
Oxyspora paniculata (D. Don) DC	shrub (2–4 m tall)	Himalayas, Nepal, Bhutan, India, China	Hawaii (nat.)
Pterolepis glomerata (Rottb.) Miq.	erect suffrutescent herb or subshrub (0.5 m tall)	South America, Lesser Antilles	Hawaii (nat.)
Tetrazygia bicolor (Mill.) Cogn.	shrub to small tree (up to 6 m tall)	Cuba, Bahamas, southern Florida	Hawaii (nat.)
Tibouchina herbacea (DC) Cogn.	herb to subshrub (1 m tall)	South America	Hawaii (nat.)
Tibouchina longifolia (Vahl) Baill. Ex Cogn.	subshrub (0.5–2 m tall)	South and Central America	Hawaii (nat.)
Tibouchina urvilleana (DC) Cogn	shrub to small tree (1–4 m tall)	South America	Hawaii (nat.)
Tibouchina viminea (D. Don) Cogn.	shrub (3–4 m tall)	South America	Mauritius (nat.), La Réunion (inv.)
Trembleya phlogiformis DC var. *parvifolia* Cogn.	subshrub to shrub	Brazil	Hawaii (nat.)

NOTE: Data according to Wagner, et al., 1990, and C. Imada in litt. for Hawaii; Smith, 1985, for Fiji; Friedman, 1994, for the Seychelles; Bosser, et al., 1990, for La Réunion and Mauritius.

DIVERSITY OF LIFE FORMS AND HABITATS

Most of the naturalized melastomes are shrubs to small trees (e.g., *Clidemia, Melastoma, Memecylon, Miconia, Oxyspora, Tetrazygia,* and *Tibouchina*), some of them are erect suffrutescent herbs or subshrubs (e.g., *Heterocentron, Osbeckia, Ossaea*), and a few are decumbent or creeping herbs (e.g., *Dissotis, Heterotis*) or large woody epiphytes (*Medinilla*). Of all melastome weeds, *Miconia calvescens* stands alone for its dramatic modification of native forests of Tahiti (French Polynesia), resulting in seemingly permanent, landscape-level loss of biodiversity. A small tree up to 18 m tall in its native range of South and Central America, but usually between 6 and 12 m tall in the introduced range and characterized by large leaves (up to 80 cm long and 30 cm wide) with purple undersides (Fig. 1), it invades semi-open areas and the understory of native rainforests from sea level up to 1,400 m in Tahiti, covering more than 70,000 hectares. *Clidemia hirta* (Fig. 2) is a pantropical weed, found in disturbed open areas such as agrosystems (forestry plantations, pasturelands, agroforestry, fields, roadsides, trailsides) as well as treefall gaps, forest edges, and river banks up to 1,500 m elevation in Hawaii and Fiji, where it develops dense, almost impenetrable thickets up to 2–3 m tall. In many locales, *C. hirta* has developed into a serious forest invader and modifier. It is unusual, in that long-distance human-assisted dispersal in its worldwide spread is apparently largely inadvertent, apparently as seed stored in soil.

FIGURE 2 Ripe berries of *Clidemia hirta* on the island of 'Uvea, Wallis and Futuna. (Photograph courtesy of Jean-Yves Meyer Photography.)

REASONS FOR SUCCESS

Many invasive melastomes were introduced as showy garden ornamentals (*Miconia* spp., *Medinilla* spp., *Tibouchina* spp.). Though many species are popular in horticulture and landscaping, numerous other species are found in botanical gardens in the tropics. Some species were deliberately introduced to control soil erosion on trail and roadsides (*C. hirta, D. rotundifolia*). In their native ranges, many are colonizers of landslides, small treefall gaps, and riverbanks and are found in secondary forests, pastures, and forest edges. As such, their life history strategies include adaptations such as rapid growth and high levels of production of fleshy fruits containing tiny seeds, typically bird-dispersed, and with high germination rates and efficient seed dispersal. A number of cultivated and even naturalized melastomes (e.g., some *Tibouchina, Dissotis,* and *Heterotis*) do not produce viable seeds in their regions of introduction and spread only vegetatively. Despite this, prolific seed production is a key characteristic of nearly all of the most highly invasive melastome species, with annual seed production in the millions per plant for *M. calvescens* and *C. hirta*. The reproductive system of the most invasive melastomes is self-compatible allowing a single plant to quickly form spreading populations and colonize new areas of introduction. The fleshy berries are attractive to and the small seeds easily dispersed over long distances by generalist frugivorous birds, both native and alien species (notably the silvereye *Zosterops lateralis*, the white-eye *Zosterops japonicus*, the red-vented bulbul *Pycnonotus cafer*, and the common myna *Acridotheres tristis* on Indo-Pacific islands). The tiny seeds (often 1 mm or less) are also easily spread with infested soil associated with humans, agriculture,

FIGURE 1 Monotypic forest of *Miconia calvescens* in Tahiti, French Polynesia. (Photograph courtesy of Jean-Yves Meyer Photography.)

vehicles, bulldozers, and hoofed animals such as feral pigs and goats. Melastome seeds, although small, have hard testa and are often long lived, forming dense and persistent soil seed banks; *M. calvescens* seeds in French Polynesia have persisted in excess of 15 years. The highly invasive species at least are characterized by high rates of germination (90 percent or more) and rapid germination (15–20 days) in favorable conditions, by relatively fast growth (1 m/year height increase for *M. calvescens*), and by early reproduction or sexual maturity (2–3 years for *C. hirta*; 4–5 years for *M. calvescens*). Life history traits associated with their role in their native range as pioneer species of rainforests have functioned in their invaded range as key elements of their spread and perturbations. Freedom from coevolved predators and pathogens in their native ranges has heightened their fitness in invaded ranges. Both *C. hirta* and M. *calvescens* are extremely successful invaders in spite of their relatively low genetic diversity.

IMPACTS AND CONTROL METHODS

Though a number of melastomes have the potential to become invasive, *Miconia calvescens* and *Clidemia hirta* have become case examples of the potential of tropical weeds to invade and degrade natural ecosystems. Both species are capable of invading primary forests and forming dense monotypic forest for *M. calvescens*, and thickets for *C. hirta*, progressively replacing and suppressing native biota and being presumed to increase soil erosion and mass-transfer landslides, ultimately damaging watershed function.

Manual uprooting and chemical treatment of cut-stumps can be effective in killing individual plants, but established populations of highly invasive species are often so dense and extensive that mechanical and chemical control is considered in these areas to be largely a holding action. Many melastomes appear excellent candidates for biological control as their success in naturalized conditions in part is apparently derived from their freedom from coevolved pathogens and insect herbivores that suppress them in their native forests.

The thrips *Liothrips urichi* from Trinidad has served as a partially effective biocontrol agent for *Clidemia hirta* in Hawaii and Fiji. There has also been some level of success in the use of the fungal pathogens *Colletotrichum gloeosporioides* f. sp. *clidemiae* in Hawaii to control *C. hirta* and *Colletotrichum gloeosporioides* f. sp. *miconiae* to control *M. calvescens* in Hawaii and French Polynesia, where it was released in 1997 and 2000, respectively. In Tahiti, the pathogen has successfully established and spread efficiently to all the *M. calvescens* invaded forests from sea level to the cloud forest and has caused partial defoliation of *M. calvescens* canopy trees from 10 to 47 percent according to elevation, thus allowing the recovery of some threatened endemic plants.

CONCLUSIONS

Melastomes in the genera *Melastoma*, *Miconia*, and *Tibouchina* and the species *Clidemia hirta*, *Medinilla venosa*, and *Oxyspora paniculata* have been legally declared "noxious weeds" in the State of Hawaii. All *Miconia* species are declared "noxious" in the Queensland and New South Wales regions of Australia, and *Miconia calvescens* is declared a "threat to biodiversity" in French Polynesia and an "invasive plant" in New Caledonia. Their introduction, cultivation, propagation, and dispersal are banned in these countries. Despite this, melastomes are still popular cultivated and garden plants worldwide, especially *Tibouchina* and *Medinilla*. Recently, the highly invasive *Miconia calvescens* was discovered as a potted plant on La Réunion Island (Mascarenes, Indian Ocean), with the seeds having been bought in Europe, and the species received an award as a prize-winning potted plant at a horticultural show in Manila, Philippines, in 2008!

Because of their life history traits as rainforest pioneers, the importation of additional species of melastomes will likely result in additional naturalizations. The absence of existing worldwide regulation, even for a highly invasive but attractive species such as *Miconia calvescens*, is likely to result in greater scope of worldwide naturalization of this serious forest pest. At least for *Miconia calvescens* and *Clidemia hirta*, once established, the development of safe (i.e., host-specific) biological control from their native ranges appears the best option for reducing their impacts on regional native biota and ecosystem functioning.

SEE ALSO THE FOLLOWING ARTICLES

Biological Control, of Plants / Dispersal Ability, Plant / Horticulture / Landscape Patterns of Plant Invasions / Risk Assessment and Prioritization / Trees and Shrubs

FURTHER READING

Brooks, S. J., F. D. Panetta, and T. A. Sydes. 2009. Progress towards the eradication of three melastome shrub species from northern Australian rainforests. *Plant Protection Quarterly* 24: 71–78.

Conant, P. 2009. *Clidemia hirta* (L) D. Don (Melastomataceae) (163–174). In R. Muniappan, G. V. P. Reddy, and A. Raman, eds. *Biological Control of Tropical Weeds Using Arthropods*. Cambridge: Cambridge University Press.

DeWalt, S. J., J. S. Denslow, and K. Ickes. 2004. Natural-enemy release facilitates habitat expansion of the invasive tropical shrub *Clidemia hirta*. *Ecology* 85(2): 471–483.

Global Invasive Species Database. www.issg.org/database/welcome/

Le Roux, J.J., A.M. Wieczorek, and J.-Y. Meyer. 2008. Genetic diversity and structure of the invasive tree *Miconia calvescens* in Pacific islands. *Diversity and Distribution* 14: 935–948.

Medeiros, A.C., L.L. Loope, P. Conant, and S. McElvaney. 1997. Status, ecology, and management of the invasive tree *Miconia calvescens* DC (Melastomataceae) in the Hawaiian Islands (23–35). In N.L. Evenhuis and S.E. Miller, eds. *Records of the Hawaii Biological Survey for 1996.* Honolulu, HI: Bishop Museum Occasional Papers No. 48.

Meyer, J.-Y., and J. Florence. 1996. Tahiti's native flora endangered by the invasion of *Miconia calvescens* DC (Melastomataceae). *Journal of Biogeography* 23(6): 775–783.

Meyer, J.-Y., R. Taputuarai, and E.M. Killgore. 2008. Dissemination and impact of the fungal pathogen *Colletotrichum gloeosporioides* f. sp. *miconiae* on the invasive alien tree *Miconia calvescens* (Melastomataceae) in the rainforests of Tahiti (French Polynesia, South Pacific) (594–599). In M.H. Julien, R. Sforza, M.C. Bon, H.C. Evans, P.E. Hatcher, H.L. Hinz, and B.G. Rector, eds. *Proceedings of the XII International Symposium on Biological Control of Weeds.* Wallingford: CAB International.

Murphy, H.T., B.D. Hardesty, C.S. Fletcher, D.J. Metcalfe, D.A. Westcott, and S.J. Brooks. 2008. Predicting dispersal and recruitment of *Miconia calvescens* (Melastomataceae) in Australian tropical rainforests. *Biological Invasions* 10: 925–936.

Pacific Island Ecosystems at Risk. www.hear.org/pier/index.html

MELTDOWN

SEE INVASIONAL MELTDOWN

MONGOOSE

SEE SMALL INDIAN MONGOOSE

MOSQUITOES

L. PHILIP LOUNIBOS
University of Florida, Vero Beach

Owing to their blood-questing habits, mosquitoes (order Diptera, family Culicidae) are familiar to nearly everyone. Invasive mosquitoes can be distinguished from nonnative, noninvasive mosquitoes based on their effects on (1) resident species and ecosystems or (2) human or animal health. Although these two categories are composed of a heterogeneous mix of mosquito genera and species, there are a few biological characteristics that help separate them from mosquito species not known to be established outside their native ranges. In particular, the evolution of a drought-resistant egg stage and adaptations to human environments are better represented in nonnative and invasive mosquito species.

NONNATIVE, NONINVASIVE SPECIES

Twenty-two species are currently recognized in this category, which belong to the genera *Aedes* (9), *Aedeomyia* (1), *Mansonia* (1), *Culex* (4), *Anopheles* (6), and *Wyeomyia* (1). Most known establishments of these nonnatives are confined to islands, including many records from the South Pacific, where at least some introductions were associated with maritime military movements during and after World War II. Nonnative species in this category that have become established in the continental United States, such as *Aedes togoi* in Washington State and *Aedes bahamensis* in Florida, are restricted to local coastal areas.

NONNATIVE INVASIVE SPECIES

A further nine nonnative species are considered invasive because of their potential for transmitting disease or disrupting resident species or ecosystems. A previous analysis showed that desiccation resistance of the egg stage was significantly associated with the probability of a mosquito species becoming nonnative (i.e., established outside its endemic range). It was also shown that association with human-dominated habitats was significantly more likely among the nine invasive species than the nonnative, noninvasive species.

The nine invasive species are members of the three most familiar mosquito genera: *Aedes*, *Culex*, and *Anopheles*. Shipping is the most probable means of intercontinental transport for their long-distance establishments. Although precise dates cannot be affixed to mosquito invasions before the twentieth century, the most broadly distributed invasive mosquitoes are believed to have spread in three temporally separate "waves": *Aedes aegypti* in the fifteenth to seventeenth centuries; the *Culex pipiens* complex in the nineteenth to twentieth centuries; and *Aedes albopictus* in the late twentieth century. In all cases, intercontinental "jump" dispersals were followed by regional spread, usually assisted by human transport in the cases of these three species adapted to anthropogenic environments.

SPECIES ACCOUNTS

Aedes (Stegomyia) aegypti

Commonly known as the Yellow Fever Mosquito, this species has been most important in recent decades as the primary epidemic vector of urban dengue. A domesticated subspecies of *A. aegypti* is believed to have evolved in Africa and to have arrived in the New World on ships, many of which were carrying slaves, crossing the Atlantic in the fifteenth through seventeenth centuries. This

TABLE 1
Invasive Mosquito Species and Their Origins, Invaded Ranges, and Preferred Habitats

Species	Origin	Invaded Range	Larval Habitats	Macrohabitats
Aedes aegypti	Africa	cosmotropical, warm	human-made containers	urban, domestic[a]
Aedes albopictus	Asia	cosmopolitan, temperate	human-made containers, phytotelmata	suburban, rural
Aedes atropalpus	eastern N. America	midwestern N. America Europe[b]	rock pools, human-made containers	riparian
Aedes japonicus	Asia	N. America	rock pools, human-made containers	suburban, rural
Aedes notoscriptus	Australia	New Zealand	phytotelmata, human-made containers	urban-rural
Culex pipiens	Africa	temperate N. Amer, Asia, Europe	pools, ponds, containers	urban-suburban
Culex quinquefasciatus	Africa	cosmotropical (warm)	eutrophic ponds and pools, containers	urban-suburban
Anopheles darlingi	Neotropics	Amazonian Peru	river margins and lagoons	rural
Anopheles gambiae[c]	tropical Africa	Mauritius, Brazil	groundwater pools, catchments	domestic

[a] Domestic status refers only to the *A. aegypti aegypti* subspecies.

[b] Apparently eradicated from local establishments in France and Italy.

[c] Complex of seven named species, of which only *A. gambiae* s.s. and *A. arabiensis* are probable invasives.

species invaded the Asian subcontinent much later, probably in the middle of the nineteenth century. Because of its role in urban yellow fever transmission, a massive eradication campaign was conducted against this species in the Americas during the 1950s and 1960s, but it ended in overall failure, despite temporary regional successes.

The domestic subspecies of *A. aegypti* is exceptionally well adapted to domiciliary and urban environments. Preferred larval habitats are artificial containers, indoors or outdoors depending on locality, and oviposition outdoors is positively associated with paved or asphalted landscapes. Where this subspecies is an important vector of arboviruses, females preferentially consume human blood, may take several bloodmeals in one gonotrophic cycle, and have reduced or eliminated sugar consumption, which is used by most mosquito species to power flight. The egg stage of domesticated *A. aegypti* is exceptionally resistant to drought, which not only facilitates survivorship during extended dry periods but also promotes transport of this stage between regions and continents.

Aedes (Stegomyia) albopictus

Owing to its major range expansions in the last three decades, the Asian tiger mosquito is considered currently to be the world's most invasive mosquito species. Because its recent diasporas occurred in the era of heightened interest in biological invasions, more is known about its means of introduction and establishment than for any other invasive mosquito species.

The native range of *A. albopictus* is the Asian subcontinent, from which source it is believed to have colonized island nations, including Madagascar, in the southwestern Indian Ocean, by accompanying human trade and migration in the second millennium. Hawai'i was invaded by this species late in the nineteenth century, and many other Pacific islands were colonized in the twentieth century, especially in association with maritime travel from the Asian mainland during wartimes.

Late in the twentieth century, this species became established and spread rapidly in the Americas, continental Africa, and Europe. Colonization of the temperate United States was facilitated by the introduction in tire shipments of a diapausing population of *A. albopictus* from Japan in the early 1980s, and at about this same time, a tropical population of this species invaded southeastern Brazil. The mosquito was first detected in the port cities of Houston and Rio de Janeiro, and coastal introductions preceded widespread invasions inland in both countries. Subsequently, *A. albopictus* also became established in selected countries of continental Europe and sub-Sahelian Africa and, most recently, in northern Australia. All major land masses, except the polar regions, have now been colonized by *A. albopictus*, although limited cold tolerance of this species has restricted its spread to south of the 40° N latitude isotherm.

Biotic and abiotic effects of, and resistance to, the *A. albopictus* invasion have been examined in the United States. The highly visible displacements in range and reductions in abundance of *A. aegypti* in the southeastern

United States have been attributed to the competitive superiority of *A. albopictus* larvae in artificial container habitats where accumulated leaf litter is an important organic nutrient base. The two species continue to coexist in certain areas of the Southeast, where drier and hotter local climates inflict proportionally greater egg mortality on *A. albopictus* than *A. aegypti*. Although *A. albopictus* larvae are competitively superior to other species of resident *Aedes* mosquitoes, evidence for displacements of these other species is weak. On the other hand, the presence of *Wyeomyia* spp. larvae in the axils of bromeliads reduces the suitability of these phytotelmata for growth and development of *A. albopictus* in Florida.

The survivorship of one inferior competitor, the native eastern treehole mosquito *Aedes triseriatus*, improves in the presence of *A. albopictus* with dipterous larval predators, which prey selectively on the invasive Asian tiger in preference to the native mosquito species. As for climatic adaptations since its colonization of the United States, the descendents of the original invaders have reduced the incidence of diapause among establishments in subtropical climates, such as in peninsular Florida.

Aedes albopictus is an important vector of two widespread arboviruses that cause human disease, dengue and chikungunya. Although *A. aegypti* has been recognized previously as the more important epidemic vector of these two emergent pathogens, recent outbreaks of chikungunya in the southwestern Indian Ocean and in Italy were vectored primarily by *A. albopictus*.

Aedes (Ochlerotatus) atropalpus

This rock-hole specialist vastly expanded its native distribution in the United States, probably sometime in the 1970s, by colonization of discarded, water-containing automobile tires, which facilitated range expansions into midwestern states, such as Ohio, Illinois, Indiana, and Nebraska. An introduction of this species through tires into Italy has apparently been followed by eradication. Because *A. atropalpus* is autogenous during its first gonotrophic cycle and is a weak competitor in its larval stages, its invasions are less likely to be an ecological or public health concern compared to those of other invasive mosquitoes.

Aedes (Finlaya) japonicus

This newest invasive mosquito species in the United States first arrived in the Northeast around 1998, apparently in tire shipments, and possibly in multiple introductions, from Japan. This species shares artificial container habitats with *A. albopictus* in zones of current sympatry in

Known U.S. distribution of *Aedes japonicus*

FIGURE 1 The current distribution of *Aedes japonicus* in the United States, showing temporal patterns of invasion in the East and disjunct establishments in the West.

the United States. However, *A. japonicus* is more prone to occupy rock holes and tolerates colder temperatures than *A. albopictus*, leading to a broader latitudinal range of the former species, extending into Canada. Thus, although *A. japonicus* larvae have been shown to be competitively inferior to *A. albopictus* in containers, the incomplete niche overlap of these two species mitigates interspecific interactions. Independent introductions, most likely via shipping, have led to local establishments of this species in Hawaii, in coastal Washington State, and in two countries of central Europe.

Aedes (Finlaya) notoscriptus

This native Australian species became established early in the twentieth century in New Zealand, where it has since become the most common mosquito species occupying container habitats in urban and peridomestic settings. Although no mosquito-borne diseases of humans are known in New Zealand, *A. notoscriptus* is recognized as the most important vector of dog heartworm in that country.

Culex pipiens Complex

In addition to several locally distributed species, this important Complex includes the cosmopolitan northern house mosquito *Culex pipiens* s.s. and the pantropical *Culex quinquefasciatus*, the southern house mosquito. Older literature often refers to *C. quinquefasciatus* as *Culex fatigans* or *Culex pipiens fatigans*. Although both *C. pipiens* and *C. quinquefasciatus* are invasive and of probable Old World origin, their native ranges have proven difficult to decipher owing to multiple transcontinental introductions, introgression between the two species in their invasive ranges, and recent range expansions within their presumed native continent, Africa.

CULEX (CULEX) QUINQUEFASCIATUS Major range expansions of this species, an important vector of nematode worms that cause human lymphatic filariasis, occurred via shipping in the nineteenth century to the New World, Australia, and the Pacific region. Within Africa, population expansions of this species were recognized in the mid-twentieth century associated with urbanization. Immatures of this species have adapted to human wastewater and runoff of high organic nutrient contents, where few larval competitors or predators occur. Female *C. quinquefasciatus* are known to take blood from avian or mammalian hosts and, as such, may vector to humans arboviruses ordinarily confined to enzootic cycles involving birds. As an invasive vector of avian malaria, establishments of *C. quinquefasciatus* in Hawaii may have led to the extinctions of some endemic species of birds.

CULEX (CULEX) PIPIENS S.S. The northern house mosquito, as its common name implies, occupies more temperate latitudes than its sibling *C. quinquefasciatus*. In their invasive ranges, such as in the Americas and the Orient, both morphological and molecular evidence points to geographic clines in characters of the two species, indicating introgression. Also like *C. quinquefasciatus*, *C. pipiens* s.s. is urban-adapted and common, for example, in catch basins and sewage drains in north-temperate cities, where it may be an important carrier of West Nile virus. Because it is more ornithophilic than *C. quinquefasciatus*, in temperate North America *C. pipiens* is more likely than the former species to maintain West Nile virus in enzootic cycles.

Anopheles gambiae Complex

Anopheles gambiae also is composed of a species complex, currently consisting of seven named species, all native to Africa. As the constituent species were not known when the two major invasions of its members occurred, post-hoc species identifications are not straightforward. However, modern methods of DNA extraction and sequencing applied to museum specimens have, in one instance, corroborated a suspected species identification.

Two species in this Complex, *A. gambiae* s.s. and *Anopheles arabiensis*, are the most important vectors of human malaria in Africa. Owing to the fact that the immature stages of anopheline mosquitoes occupy container habitats less frequently than the invasive aedine and culicine mosquito species described previously, their successful intercontinental transport by sea and subsequent invasion and establishment in new suitable habitat seems to have been less likely. However, owing to the efficiency of *A. gambiae* s.s. and *A. arabiensis* as malaria vectors, the impacts of two particular invasions are noteworthy.

The first recorded outbreak of malaria transmitted by Gambiae Complex mosquitoes outside of continental Africa occurred on the island of Mauritius in 1886–1887, as described in a 1910 book by Sir Ronald Ross, who was knighted for his discovery that malaria was transmitted by mosquitoes. It is believed that the invasive mosquitoes arrived on Mauritius by boat from the East African mainland.

The arrival and spread of Gambiae Complex mosquitoes in northeastern Brazil, and their ultimate eradication, were the subject of a book and the only example of successful extirpation of an invasive mosquito species from a broadly established range. As suspected in light of its establishment in the arid Northeast of Brazil, the relatively drought-tolerant *A. arabiensis* has recently been shown to be the Gambiae Complex species that was introduced from Senegal to the port city of Natal in 1930. Although control measures were applied after *A. arabiensis* caused a malaria epidemic in Natal, descendents of the colonizing invaders escaped detection and spread northward and inland, largely following river courses, where the immature stages developed in backwaters and in nearby shallow wells. After a few years of silent spread, severe malaria epidemics vectored by *A. arabiensis* again erupted in 1938–1939, leading to many deaths and to high emigration from this area of Brazil. Concentrations of the immature stages of this invasive species during the dry season assisted an extensive larviciding program that led to the eventual eradication of *A. arabiensis* from Brazil.

Anopheles darlingi

Although less efficient as a malaria vector than either *A. gambiae* s.s. or *A. arabiensis*, *A. darlingi* is recognized as the most anthropophilic and important native transmitter of this human pathogen in the New World. Previously unknown or rare in the Peruvian Amazon region, it was commonly detected early in the 1990s in mosquito collections in the port city of Iquitos. A few years later, this species was incriminated as the primary vector of epidemic malaria in the Iquitos region.

Subsequent population genetic analyses of *A. darlingi* specimens from the Iquitos region are consistent with a recent range expansion, with the Peruvian specimens showing some genetic affinities with *A. darlingi* individuals collected near the mouth of the Amazon River, 3,000 km downriver in Brazil. These genetic results support a hypothesis that *A. darlingi* were transported among river cargo from Brazil to Peru. One speculation is that immatures of this species may have accompanied shipments of freshwater ornamental fish, which are cultivated commercially in human-made ponds in the Iquitos region. In undisturbed habitats of its native range, *A. darlingi* larvae frequent margins of slow-moving rivers, but they also adapt to backwaters and ponds, including those used for fish farming.

SEE ALSO THE FOLLOWING ARTICLES

Disease Vectors, Human / Habitat Compatibility / Malaria Vectors / Pesticides for Insect Eradication

FURTHER READING

Benedict, M. Q., R. S. Levine, W. A. Hawley, and L. P. Lounibos. 2007. Spread of the tiger: Global risk of invasion by the mosquito *Aedes albopictus*. *Vector-Borne and Zoonotic Diseases* 7: 76–85.

Juliano, S. A., and L. P. Lounibos. 2005. Ecology of invasive mosquitoes: Effects on resident species and on human health. *Ecology Letters* 8: 558–574.

Laird, M., ed. 1984. *Commerce and the Spread of Pests and Disease Vectors.* New York: Praeger.

Lounibos, L. P. 2002. Invasions by insect vectors of human disease. *Annual Review of Entomology* 47: 233–266.

Soper, E. F., and D. B. Wilson. 1943. *Anopheles Gambiae in Brazil, 1930 to 1940.* New York: Rockefeller Foundation.

Tabachnick, W. J. 1991. Evolutionary genetics and arthropod-borne disease: The yellow fever mosquito. *American Entomologist* 37: 14–24.

MUSSEL

SEE ZEBRA MUSSEL

MUTUALISM

JOHN J. STACHOWICZ

University of California, Davis

Mutualism is an interaction between two species that benefits both and harms neither. These benefits are varied and include directly provided benefits such as food, pollination, or living space, or indirect benefits that are mediated through the presence of another species such as protection from natural enemies. Mutualisms between introduced species or between introduced and native species can foster the establishment or spread of introduced species, while mutualisms between natives can reduce the establishment of introduced species.

GENERAL IMPORTANCE

Mutualisms are ubiquitous in the natural world and are at the root of such key phenomena as the evolution of green plants (a mutualism between a eukaryotic cell and a photosynthetic bacterium) and the accretion of coral reefs (a mutualism between the coral and microscopic algae). While these "obligate" mutualisms capture our immediate attention, facultative mutualisms in which both partners benefit but are capable of living independently may be much more common in the natural world. In these interactions, species are able to expand their distribution into new habitats or grow faster and become more abundant in old habitats than they could without the mutualism. Given the diversity of mechanisms by which species interact mutualistically, they have the potential to influence the invasion processs at every stage: from colonization to establishment and spread.

However, invasion biology, like most of ecology, has historically focused on the importance of negative interspecific interactions like competition, predation, and parasitism in determining invasion success. Many of invasion biology's most cherished paradigms (such as the enemy release hypothesis, see related article) are based on the ideas that antagonistic interactions among species are of great importance in determining species' distributions and abundances. More recently ecologists have "rediscovered" the importance of beneficial interactions, including mutualisms, and modern treatments give these interactions equal weight with negative interactions. Invasion biology has begun to follow suit in this regard and now considers the importance of mutualism and other facilitative interactions as important potential determinants of invasion success.

MUTUALISMS AID IN THE ESTABLISHMENT AND SPREAD OF INVADERS

In the same way that the "enemy release hypothesis" predicts that lack of natural enemies is one reason for the success of invaders, leaving mutualists behind might be a reason for the failure of invaders. For example, the establishment of pine trees in parts of the southern hemisphere was limited by the presence of mutualistic fungi that help the trees acquire nutrients and water from soils. It was only after appropriate fungi were introduced in soil transplants, and sufficient populations of the fungi accumulated, that pines became invasive in these areas. Similarly, few fig tree species that have been introduced have become invasive, and those have done so only after the subsequent introduction of their specialist wasp pollinator.

However, most mutualisms involving introduced species are not obligate, and in most cases, pollinators, root symbionts, and seed dispersers can be recruited from assemblages in the invaded range to perform mutualistic functions. Sometimes, these "novel" mutualists may

FIGURE 1 Many introduced plants rely on native insects for pollination. (A) The invasive introduced plant purple loosestrife (*Lythrum salicaria*) being pollinated by the native broad-winged skipper (*Poanes viator*) in the Hackensack Meadowlands of New Jersey. (Photograph courtesy of Erik Kiviat/Hudsonia, Ltd.) (B) The introduced plant Himalayan Balsam (*Impatiens glandulifera*) being pollinated by a native bumblebee in Coventry, Connecticut. (Photograph courtesy of Nava Tabak.)

be natives; other times, they may be species introduced from different regions. Native insects are often generalist pollinators and will pollinate both introduced and native plants (Fig. 1). Similarly, native vertebrates and ants will spread seeds of many introduced plant species. In the Great Lakes, zebra mussels are involved in mutualistic interactions with other invasive molluscs that use the interstices between their shells as shelter; the snails prevent overgrowth of the mussel's shell by other attached organisms.

In general, it seems that the absence of specific mutualistic partners from an invader's native range only rarely limits the establishment and spread of introduced species. However, it has been proposed by some researchers that mutualists are more species-specific in the tropics; this could contribute to the relatively low number of species that have successfully invaded most tropical regions.

MUTUALISM BETWEEN NATIVES PROMOTES RESISTANCE AGAINST INVADERS

Mutualisms between natives can also promote the resistance of native communities to invasion. In one example, marine snails native to New England benefit a native seaweed by preventing the overgrowth of the seaweed by invasive marine invertebrates. In the absence of these native snails, the seaweeds become overgrown and die, leaving the area dominated by invertebrates, including several invasive species. There are relatively few examples of native mutualisms' repelling invasions by such active means, though it is not clear whether such interactions are rare or just poorly studied. However, mutualisms between natives do commonly affect the invasion process indirectly. For example, mutualism between a native plant and its bacterial or fungal symbionts can give natives an advantage in competition for nutrients over invasive species that benefit less from the presence of the mutualist.

INTRODUCED SPECIES CAN ALSO BENEFIT NATIVES

Although introduced species often harm natives, there is also the potential for native and introduced species to enter mutually beneficial relationships. Given that mutualisms are common, and that many are not species-specific, these might be more widespread than previously appreciated. Introduced species are known to benefit natives in a number of ways, including (1) creating novel habitat types or modifying existing habitat, (2) providing limiting resources, including serving as novel host plants or providing pollination services, and (3) releasing natives from their own enemies, such as competitors or predators.

Habitat Modification

Habitat modification is probably the most commonly reported means by which introduced species facilitate natives. Clams and mussels (like zebra mussels) that have invaded soft-bottom systems in rivers, lakes, and estuaries provide a prime example: the complex, three-dimensional structure provided by the bivalve shells on an otherwise featureless bottom provides refuge and attachment space for many species of sessile and mobile invertebrates, often greatly increasing local diversity of native species. Nonetheless, these habitat modifications can have detrimental effects on particular native species and can also benefit introduced species. Invasive plants can also alter the habitat by changing nutrient regimes in the soils, such as occurs when nitrogen-fixing plants invade areas without such plants, increasing total nutrient content in the soil (which can benefit some native plant species).

Novel Hosts

In addition to providing habitat, many invasive plants serve as a food source for native species. A survey of California butterflies found that over one-third of native species use introduced plants, and many native butterflies currently have no native host plants. In some cases, this has been the result of evolution of native species, such as the checkerspot butterfly, to take advantage of abundant invasive weeds as hosts. Similarly, invasive pollinators now serve as primary pollinators for native plants in areas where native pollinators have declined in abundance.

Release from Enemies

When introduced species preferentially consume particular native species, they potentially indirectly benefit the competitors or prey of the native species they consume. For example, island endemic skunks increased when golden eagles were introduced, because the eagles preferentially consumed foxes, the main competitors of skunks on the island.

The reciprocal effects of natives on invasive species are rarely measured, so it is often unknown whether these interactions are mutualisms. In some cases it is clear that the introduced species is unaffected, or at least does not benefit, so these are not true mutualisms. In other cases, the interactions may be mutually beneficial (as in the example of the zebra mussel and the snail).

Mutualism between natives, between introduced species, and between natives and introduced species are all common and important interactions in determining invasion success. The lack of an appropriate mutualist only rarely seems to limit invasion, because most interactions are facultative, and native species can be recruited to perform mutualistic services for introduced species.

SEE ALSO THE FOLLOWING ARTICLES

Disturbance / Enemy Release Hypothesis / Invasional Meltdown / Mycorrhizae / Pollination / Transformers

FURTHER READING

Bruno, J. F., J. D. Fridley, K. F. Bromberg, and M. D. Bertness. 2005. Insights into biotic interactions from studies of species invasions (13–40). In D. F. Sax, J. J. Stachowicz, and S. D. Gaines, eds. *Species Invasions: Insights into Ecology, Evolution and Biogeography.* Sunderland, MA: Sinauer Press.

Olyarnik S. V., M. E. S. Bracken, J. E. Byrnes, A. R. Hughes, K. M. Hultgren, and J. J. Stachowicz. 2008. Ecological factors affecting community invasibility (215–238). In G. Rilov and J. A. Crooks, eds. *Biological Invasions of Marine Ecosystems: Ecological, Management, and Geographic Perspectives.* Heidelberg: Springer.

Ricciardi, A. 2005. Facilitation and synergistic interactions among introduced aquatic species (162–178). In H. A. Mooney, R. N. Mack, J.

McNeely, L. E. Neville, P. J. Schei, and J. K. Waage, eds. *Invasive Alien Species: A New Synthesis.* Washington, DC: Island Press.

Richardson, D. M., N. Alsopp, C. M. D'Antonio, S. J. Milton, and M. Rejmánek. 2000. Plant invasions: The role of mutualisms. *Biological Reviews* 75: 65–93.

Rodriguez, L. F. 2006. Can invasive species facilitate native species? Evidence of how, when and why these impacts occur. *Biological Invasions* 8: 927–939.

MYCORRHIZAE

ANNE PRINGLE AND BENJAMIN WOLFE

Harvard University, Cambridge, Massachusetts

ELSE VELLINGA

University of California, Berkeley

Symbioses between mycorrhizal fungi and plant roots are referred to as mycorrhizae. Mycorrhizal fungi play three critical roles in biological invasions (Fig. 1). First, the mycorrhizal fungi can themselves invade novel habitats, either with introduced plants, or after association with native plants. Second, introduced mycorrhizal fungi can facilitate the invasion of introduced plants. Third, native mycorrhizal fungi will respond to invasions by other exotic species, for example by associating and spreading with introduced plants, or by declining after introduced insects or pathogens attack native host plants.

THE BIOLOGY OF MYCORRHIZAE

Mycorrhizal symbioses are ubiquitous. In these symbioses, the fungal mycelia scavenge through soil for resources (often phosphorus or nitrogen) and give these resources to plants in exchange for carbon. The associations are mutualisms but can sometimes function as parasitisms. Mycorrhizal associations may involve any of four different fungal phyla and a broad range of plants including mosses and liverworts, ferns, and seed plants. The mycorrhizal status of many plants is unknown, but for the 6,507 species that have been examined, only 18 percent do not form mycorrhizal associations. The symbioses are often classed as either arbuscular mycorrhizal (AM) or ectomycorrhizal (EM), and the different types are defined by both the taxonomy of the fungi and the structures formed in or around plant roots. In addition to AM and EM symbioses, mycorrhizal associations include arbutoid, monotropoid, ericoid, and orchid forms. This entry focuses on AM and EM symbioses, because there is more information about these mycorrhizal types and because it is increasingly clear that other forms involve the same fungal species as associate in EM symbioses.

MYCORRHIZAL SPECIES INVADE

Mycorrhizal symbioses are obligate; although many fungi and plants can be grown alone in the greenhouse or laboratory, in nature most species require the symbiosis. For this reason, mycorrhizal fungi have been introduced to novel ranges when plants are managed for commercial purposes; species are moved to facilitate agriculture, forestry, and horticulture. For example, in the southern hemisphere, pine forests did not grow until soils with mycorrhizal fungi were imported and mixed with the native soils around planted pine seedlings. Moreover, the potted plants sold by nurseries are often associated with mycorrhizal fungi, and fungi are introduced to novel ranges when potted plants are moved and replanted in local soils. Although specificity is a feature of some associations, and, for example, the fungal genus *Suillus* is specific to the plant genus *Pinus*, other mycorrhizal fungi are generalists, and these species may jump to new hosts in novel habitats.

Introductions of AM fungi are difficult to track because the diagnostic features of different species appear in soil and may be complicated to isolate or identify. However, all of the world's major crop plants associate with AM fungi, and because commercial mixes of the fungi are sold as alternatives to phosphorus fertilizers, species have probably been carried by humans to new ranges. But an understanding of how often fungal species are introduced, whether introduced species typically establish, if they establish whether they associate with native hosts, and if and how introduced AM fungi impact the local biodiversity of plants or fungi is prevented by an almost total lack of basic knowledge about species numbers and native ranges. Although the global diversity of AM fungi appears to be low (fewer than 200 described species), estimates of species richness are hindered by the potential for many species to remain undescribed. For example, intensive sampling in one North Carolina field demonstrated that more than one-third of the species described from this single 1 ha site were novel. Nor do biologists have a clear understanding of the native ranges of AM fungi. For example, the morphological species *Glomus mosseae* is assumed to have a global distribution, but recent data suggest a cryptic genetic diversity that may translate to distinct genetic species. If a morphological species concept is used, an introduction of *G. mosseae* from Europe

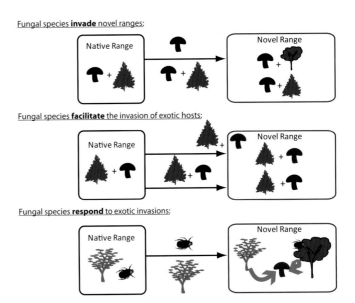

FIGURE 1 The roles of mycorrhizal fungi in biological invasions.

to North America is the movement of a species within its native range. Using a genetic species concept, this kind of introduction would be the movement of a species outside of its native range. Even apparently endemic species, for example the novel species of the North Carolina field, are considered as endemic only because of limited sampling from other sites. Until the biology, taxonomy, and biogeography of AM fungi are understood, introductions of AM fungi will be difficult to define or identify.

More is known about EM fungal introductions because many of these fungi form obvious, macroscopic reproductive structures: mushrooms (Fig. 2). At least 200 species from a diverse group of over 50 genera have been moved among different continents and islands (Fig. 3). As discussed above, EM fungi are likely to have been moved in the soil associated with plants used for forestry or horticulture. More is known about the biogeography of EM fungal species, although historical records of species presence or absence are sometimes incorrect. Inaccuracies are caused by the use of European names for native species; for example, early sources often list European species as found in North America, although in many cases the species being listed is a North American species misidentified with a European name. Most introduced EM fungal species appear not to spread and seem constrained to grow with the tree species on which they were introduced—for example, the species found with Australian eucalyptus in Spain. Exceptions include *Amanita phalloides*, invasive on the West Coast of North America and associated with the endemic *Quercus agrifolia*, and *Amanita muscaria*, invasive in Australia and

New Zealand and associated with native *Nothofagus* spp. Mechanisms facilitating the spread of *A. phalloides* and *A. muscaria* are poorly understood but may involve an ability to establish symbioses with local trees or competitive dominance over local mycorrhizal species. A limited diversity of native *Amanita* species may also provide niche space for *A. phalloides* in California, but this hypothesis remains untested.

MYCORRHIZAL FUNGI INFLUENCE PLANT INVASIONS

Mycorrhizal associations may also facilitate or limit the spread of introduced plants, although symbiotic constraints on the spread of introduced species are poorly understood. As discussed above, it is clear that *Pinus* species introduced to the southern hemisphere did not grow until associated mycorrhizal fungi were deliberately introduced. *Pinus* trees require introduced EM fungi either because the native trees don't associate with EM fungi or because they associate with a different set of EM fungal species that cannot associate with *Pinus*. Other trees may also rely on introduced mycorrhizal associates to spread, for example, eucalypts introduced to Spain. Mycorrhizal fungi also play a role in natural range expansions, and native birch expanding into the lowland heathlands of England grow better at sites where mycorrhizal fungi have also dispersed. And native fungi may also facilitate the

FIGURE 2 Mycorrhizal fungi are rarely discussed as introduced or invasive species. Fungal individuals are typically hidden within soil or other substrates, but when mushrooms appear, they can be obvious and charismatic features of the landscape. Pictured in A–C are the introduced species *Hydnangium carneum, Amanita muscaria, and Suillus luteus*. These are examples of commonly introduced ectomycorrhizal species (Vellinga et al., 2009). (A) *Hydnangium carneum* is a *Eucalyptus* associate from Australia, first described when it was found in Europe growing with potted plants. It forms subterranean fruit bodies. The species has also been recorded from North Africa, North and Central America, New Zealand, and more recently China. In New Zealand, it can form ectomycorrhizae with the native *Nothofagus* species, although reports of the species occurring in native forests are rare. (Photograph courtesy of Celestino Gelpi Pena.) (B) *Amanita muscaria* s.l. is a very conspicuous and easily recognizable mushroom, the classic red-and-white spotted mushroom of fairy tales. Originally from the northern hemisphere, members of this species complex have traveled to Australia, New Zealand, Africa, Hawaii, and South America. In Tanzania the species is a health hazard, as the local people confuse it with edible, equally orange- to red-colored *Amanita* species from the native miombo vegetation. As of this writing, the species appears to be confined to plantations within Tanzania, but in Australia and New Zealand, the species appears to be invading native *Nothofagus* forests. It also associates with introduced *Eucalyptus* in Uruguay. (Photograph courtesy of Tom May.) Two *Suillus* species, *S. luteus* and *S. granulatus*, have also been moved across the planet. (C) *Suillus luteus* is a Eurasian native, and *S. granulatus* is native to northern temperate *Pinus* forests. *Suillus luteus* was taken to North America with *Pinus sylvestris*. It is now grown with the California endemic *P. radiata* all over the world, even though it does not associate with *P. radiata* in its native habitat. Thus, the species is being taken to novel places in association with a host it would not normally find, and the two species travel together. The fungus now grows in Africa, South America, Australia, New Zealand, and in other places where pines have been introduced. *Suillus granulatus* has been recorded from Africa, South America and the Falkland Islands, Australia, New Zealand, and Hawaii. (Photograph courtesy of Dimitar Bojantchev.)

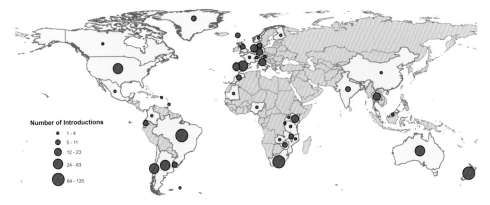

FIGURE 3 The global distribution of ectomycorrhizal introductions. Numbers of introductions are strongly correlated with the number of publications from any given country; for many countries there are no data. Colors and circles are proportional to the number of species that have been reported as introduced. (Reprinted from Vellinga et al., 2009.)

Number of Introductions
- 1 - 4
- 5 - 11
- 12 - 23
- 24 - 63
- 64 - 120

spread of introduced plants, as AM fungi appear to have done for *Centaurea maculosa* in North America.

NATIVE MYCORRHIZAL FUNGI ARE AFFECTED BY INTRODUCTIONS OF OTHER SPECIES

Exotic plants, pathogens, and insects may cause direct or cascading effects on native mycorrhizal communities, although data on these phenomena are rare. Introductions can clearly cause damage; for example, the introduced plant *Alliaria petiolata* (garlic mustard) kills local communities of both AM and EM fungi in North America. But introduced plants that associate with local fungal species may also facilitate the growth of these native species in new habitats. Introduced pathogens or insects impact native fungal communities by attacking the trees that host mycorrhizal fungi; *Phytophthora ramorum* (sudden oak death) is one example of an introduced pathogen that kills a diversity of tree species. As it kills the trees that host mycorrhizal communities, the mycorrhizal fungi are also likely to die.

SEE ALSO THE FOLLOWING ARTICLES

Agriculture / Belowground Phenomena / Eucalypts / Forestry and Agroforestry / Fungi / Hemlock Woolly Adelgid / Horticulture / Mutualism

FURTHER READING

Bever, J. D., P. A. Schultz, A. Pringle, and J. B. Morton. 2001. Arbuscular mycorrhizal fungi: More diverse than meets the eye, and the ecological tale of why. *BioScience* 51: 923–931.

Collier, F. A., and M. I. Bidartondo. 2009. Waiting for fungi: The ectomycorrhizal invasion of lowland heathlands. *Journal of Ecology* doi 10.1111/j.1365-2745.2009.01544.x.

Díez, J. 2005. Invasion biology of Australian ectomycorrhizal fungi introduced with eucalypt plantations into the Iberian Peninsula. *Biological Invasions* 7: 3–15.

Marler, M. J., C. A. Zabinski, and R. M. Callaway. 1999. Mycorrhizae indirectly enhance competitive effects of an invasive forb on a native bunchgrass. *Ecology* 80: 1180–1186.

Mikola, P. 1970. Mycorrhizal inoculation in afforestation. *International Review of Forestry Research* 3: 123–196.

Nuñez, M. A., T. R. Horton, and D. Simberloff. 2009. Lack of belowground mutualisms hinders Pinaceae invasions. *Ecology* 90: 2352–2359.

Pringle, A., R. I. Adams, H. B. Cross, and T. D. Bruns. 2009. The ectomycorrhizal fungus *Amanita phalloides* was introduced and is expanding its range on the West Coast of North America. *Molecular Ecology* 18: 817–833.

Pringle, A., J. D. Bever, M. Gardes, J. L. Parrent, M. C. Rillig, and J. N. Klironomos. 2009. Mycorrhizal symbioses and plant invasions. *Annual Review of Ecology, Evolution and Systematics* 40: 699–715.

Pringle, A., and E. C. Vellinga. 2006. Last chance to know? Using literature to explore the biogeography and invasion biology of the death cap mushroom *Amanita phalloides* (Vaill. Ex Fr.: Fr.) Link. *Biological Invasions* 8: 1131–1144.

Richardson, D. M., N. Allsopp, C. M. D'Antonio, S. J. Milton, and M. Rejmánek. 2000. Plant invasions: The role of mutualisms. *Biological Reviews* 75: 65–93.

Schwartz, M., J. Hoeksema, C. Gehring, N. Johnson, J. Klironomos, L. Abbott, and A. Pringle. 2006. The promise and potential consequences of the global transport of mycorrhizal fungal inoculum. *Ecology Letters* 9: 601–616.

Smith, S. E., and D. J. Read. 2008. *Mycorrhizal Symbiosis*, 3rd ed. San Diego, CA: Academic Press.

Stinson, K. A., S. A. Campbell, J. R. Powell, B. E. Wolfe, R. M. Callaway, G. C. Thelen, S. G. Hallett, D. Prati, and J. Klironomos. 2006. Invasive plant suppresses the growth of native tree seedlings by disrupting belowground mutualisms. *PLoS Biology* 4: 727–731.

Vellinga, E. C., B. E. Wolfe, and A. Pringle. 2009. Global patterns of ectomycorrhizal introductions. *New Phytologist* 181: 960–973.

Wolfe B. E., V. L. Rodgers, K. A. Stinson, and A. Pringle. 2008. The invasive plant *Alliaria petiolata* (garlic mustard) inhibits ectomycorrhizal fungi in its introduced range. *Journal of Ecology* 96: 777–783.

N

"NATIVE INVADERS"

DANIEL SIMBERLOFF

University of Tennessee, Knoxville

It is widely assumed that a species in its native range is unlikely to become "invasive" (to expand its population and spread into more or less natural habitats) or problematic, because it will be controlled by various species of predators, competitors, pathogens, parasites, and herbivores that have coevolved with it. This notion accords with the ancient and persistent idea of an inherent "balance of nature," in which tremendous fluctuations of population size or range will not happen unless humans do something (such as introducing a species) that upsets the balance. However, a number of native species have been observed to exhibit behavior that, if seen in an introduced species, would qualify it as "invasive": that is, their population sizes grow more or less suddenly, and their geographic or habitat range expands dramatically. Whether or not to term such species "invasive" is a controversial issue among invasion biologists and ecologists.

A CONTROVERSY REGARDING TERMINOLOGY

The term "biological invasion" is strongly associated with species introduced, deliberately or inadvertently, to new areas. One hypothesis often proposed by scientists to explain particular invasions by introduced species is termed the "enemy release hypothesis" and is grounded in the belief that, in its native range, a species is kept in check by an array of natural enemies, and it is the absence of these that allows populations of introduced species to grow explosively and spread. This hypothesis is the basis of classical biological control, in which natural enemies from the native range of an invasive introduced pest are deliberately introduced in an attempt to restore the balance by bringing the pest under control.

However, "invasion" has occasionally been used in the literature for more or less rapid spread of a native species into a new habitat, or for rapid expansion of its native range. Even Charles Elton, often considered the father of invasion biology, used "invasion" to refer to such spread of native species in his classic 1927 monograph, *Animal Ecology*. Others have used different terms for the same phenomenon. For example, "encroachment" is widely used in the plant ecology and range science literature for this purpose. It is important to distinguish the concept of "native invasions" from the simple dramatic vicissitudes that certain native populations typically undergo, which are sometimes termed "invasions" in both the popular and scientific literature. Red tides, for example, are caused by population explosions of native planktonic microscopic dinoflagellates. As these microorganisms are toxic to many vertebrates, these algal blooms often attract great attention. Because they are caused by the confluence of particular environmental conditions, they recur. Similarly, locust plagues in both the Old World and (in the past) North America arise occasionally when a particular range of environmental circumstances interacts with the population biology of certain native locust species. However, such dramatic outbreaks as those of the dinoflagellates or locusts, even aside from the fact that these occur where the organisms are native, do not accord with the definition of biological invasion adopted in this work and used by the majority of scientists working on invasions. That is, we restrict the definition of "invasion" to singular rather than recurrent events.

Most but not all invasion biologists prefer also to restrict the term to nonnative species. A minority, however, view this restriction as artificial and believe that the ecological features and impact of a population in nature, and not its geographic origin, should determine whether it is termed "invasive." Thus, several species in their native ranges have been called "native invaders," and several others can be viewed by some as falling into this category.

"INVASIONS" TRIGGERED BY ANTHROPOGENIC CHANGE IN THE ENVIRONMENT

The great American conservationist Aldo Leopold remarked in 1924 on the "invasion" of various grasslands in the American Southwest by native juniper and other woody species, and he surmised that such "invasions" were caused by increased grazing or fire suppression, or both. Substantial later research on western grassland "invasions" by western juniper (*Juniperus occidentalis*), one-seed juniper (*J. monosperma*), white fir (*Abies concolor*), and Douglas fir (*Pseudotsuga menziesii*) have shown that he was correct. Later, Leopold ascribed the "invasion" by native oaks of Wisconsin prairie to sufficient fragmentation and modification of the prairie that it would no longer support the spread of fire. In Maryland and Ohio, Virginia pine (*Pinus virginiana*) "invades" native serpentine grasslands when fire is suppressed while grazing is excluded. In addition to the direct effects of the pine on the grassland plants, this pine "invasion" is affecting mutualistic mycorrhizal fungi of some of these plant species, hastening the change in the plant community.

A particularly striking example of a native species' becoming "invasive" occurs in coastal regions of western Europe, where the native clonal grass *Elymus athericus* has, in many areas, spread dramatically seaward from a narrow band in the high intertidal, in some areas for hundreds of meters, replacing such marsh plants as *Atriplex portulacoides* and dominating the entire intertidal (Fig. 1). It is believed that the physiology of *E. athericus* is such that increased nitrogen from both aerial deposition and fertilizer and other runoff has allowed it to survive much longer and more frequent bouts of immersion in seawater, thus triggering the "invasion." The *Elymus* "invasion" drastically affects carbon and nitrogen dynamics as well as community composition and physical structure. Although the response of *E. athericus* to intensification of the nitrogen cycle is particularly pronounced, similar but less dramatic spread that has been observed for several other European grasses has been attributed to the same cause. On the west coast of the United States, invasion by

FIGURE 1 (A) Native *Elymus athericus* spreading over intertidal at Mont St. Michel, France. (Photograph courtesy of André Mauxion.) (B) Native *Elymus athericus* making up almost the entire mass of vegetation in the foreground spreading toward Mont St. Michel. (Photograph courtesy of Loïc Valéry.)

introduced smooth cordgrass (*Spartina alterniflora*) from the U.S. east coast has been exacerbated by increased environmental nitrogen concentrations.

In understories of many conifer and hardwood forests throughout eastern North American, hay-scented fern (*Dennstaedtia punctilobula*) has been termed a "native invader" because it has proliferated greatly over a short period in response to both overstory thinning, which increases light intensity, and heavy browsing of seedlings of competing plants by white-tailed deer (*Odocoileus virginianus*). The deer populations have grown spectacularly since European colonization because of habitat modification and the removal of almost all natural predators, such as mountain lions and wolves. Deer eat almost all woody seedlings and saplings but avoid the fern because of the high concentrations of phenolic compounds in its fronds. Once dense mats of the fern are established, they can prevent establishment of tree seedlings indefinitely. In the southern Yucatan Peninsula, coverage by native lacy bracken fern (*Pteridium caudatum*) has increased enormously over the last two decades because of changing patterns of land ownership and management, and a consequent lessening of the practice of attacking bracken quickly once it begins to establish in an area cleared mechanically or by fire. This increase has been called an "invasion." Once bracken is well established, it resists succession to other vegetation communities. In many other parts of the world, native bracken species have spread dramatically in the wake of modifications of land use practices, with the exact nature of the land use changes and details of the bracken "invasion" varying from site to site.

The common feature in all these cases is that humans have changed some aspect of the environment—fire cycles, grazing pressure, the nitrogen cycle, predator populations, mechanical clearing—as a result of releasing a native species from a degree of previous control. The outcome in each case is completely analogous to that observed in some invasions by nonnative species.

"INVASIONS" TRIGGERED BY CHANGED GENOTYPES

A major thrust of current invasion biology is the use of genetic techniques to determine the sources of particular populations of introduced species. Some research of this sort, combined with historical and other research to establish the native or nonnative status of a species in a region, has shown that the introduction of nonnative genotypes into native species has fostered "invasions." Common reed (*Phragmites australis*), once believed to be a nineteenth century introduction to North America, is now known to have been present in the southwestern U.S. for at least 40,000 years, and on the Atlantic and Pacific coasts for at least several thousand years. However, it has greatly increased its North American range and abundance in the last 150 years and is now viewed as a highly "invasive" weed, especially in disturbed wetlands but even in some undisturbed sites, including inland sites. This explosion resulted from the introduction of a nonnative Old World genotype that probably arrived in ship soil ballast early in the nineteenth century, then later spread during road and railway construction.

Similarly, reed canarygrass (*Phalaris arundinacea*), formerly considered introduced to North America, is in fact native there, and it was a relatively minor component of various grasslands, especially those on wet soils that are either poorly drained or subjected to repeated flooding. Beginning in the mid-nineteenth century, European genotypes were deliberately introduced as a forage crop (and subsequently for shoreline revegetation and phytoextraction of soil and water contaminants). By the late nineteenth century, reed canarygrass had begun to spread widely to many habitats, and it has now become so common and vigorous in some regions that it threatens wetlands and renders wetland restoration difficult. This "invasion" has been demonstrated to have been caused by recombinants involving several European genotypes.

In both of these cases, the introduction of the nonnative genotypes that led to the invasion was "deliberate." Whether accidental introduction of nonnative genotypes has led to such invasions is unknown.

INTERMEDIATE CASES

The relatively sudden, great contiguous geographic range expansion of certain native species has sometimes been termed "invasion" with little objection. For example, until about 1930, the Eurasian collared dove (*Streptopelia decaocto*) was native only as far west as Turkey and the Balkans. Then, over a 30-year period, it spread rapidly westward, ultimately reaching Great Britain and Iceland. This spread is often called an "invasion," and the species is usually not viewed as native in western Europe, but the case is ambiguous, as the cause of the sudden spread is mysterious. Many have argued that anthropogenic land use change in central and western Europe allowed this range expansion, but this hypothesis is far from proven. Others have suggested (but never proven) that a mutation was the trigger, in which case the nonnative designation in western Europe would seem arguable, which in turn would cause some to question use of the term "invasion" for this spread. The same species was kept in captivity in the Bahamas, escaped in 1974, established a wild population there, reached Florida apparently unaided by humans by the early 1980s, and has now established populations in sites at least as distant as California, the Great Lakes, British Columbia, and Veracruz in Mexico. No one would contest the designation of "invasive" for these New World populations.

A similar case is that of eastern groundsel (*Senecio vernalis*), native to the steppes of central and southern Russia until the late eighteenth century, when it began to spread northward and westward, primarily over the course of the nineteenth century, ultimately reaching France in the twentieth century. Again, it has not been conclusively shown that human activities caused this spread, although this is strongly suggested. The species is found largely on rubble, on embankments, and in other disturbed habitats often called "wasteland," and it has light, plumed, wind-dispersed seeds. The hypothesis, then, is that the amount of suitable habitat increased greatly and simultaneously with the spread.

RELATIVE NUMBER AND IMPACT OF "NATIVE INVASIONS"

In 1945, Leopold termed biological invasions "runaway populations," and he remarked that "[u]sually these runaways have been foreigners (like the carp, Norway rat, Canada thistle, chestnut blight, and white pine blister rust)," but added that native species are capable of such behavior. There has not been a systematic examination of the relative frequency and importance of events that, if one accepted the possibility of a "native invader" as described above, would qualify as biological invasions by native as opposed to introduced species, but Leopold was surely correct in saying that the majority of problematic invasions are of introduced species. This is not to minimize the real impact of some of the "invasions" by native species noted above. The spread of native bracken is a substantial global problem, while the various native tree and shrub "invasions" in prairies and grasslands of the American West are of conservation as well as economic concern.

SEE ALSO THE FOLLOWING ARTICLES

Biological Control, of Animals / Biological Control, of Plants / Genotypes, Invasive / Hybridization and Introgression / Land Use / Nitrogen Enrichment

FURTHER READING

Berendse, F., R. Aerts, and R. Bobbink. 1993. Atmospheric nitrogen deposition and its impact on terrestrial ecosystems (104–121). In C.C. Vos and P. Opdam, eds. *Landscape Ecology in a Stressed Environment.* London: Chapman and Hall.

Burkhardt, J.W., and E.W. Tisdale. 1976. Causes of juniper invasion in southwestern Idaho. *Ecology* 57: 472–484.

Cumming, J.R., and C.N. Kelly. 2007. *Pinus virginiana* invasion influences soils and arbuscular mycorrhizae of a serpentine grassland. *Journal of the Torrey Botanical Society* 134: 63–73.

De la Cretaz, A.L., and M.J. Kelty. 1999. Establishment and control of hay-scented fern: A native invasive species. *Biological Invasions* 1: 223–236.

Lavergne, S., and J. Molofsky. 2007. Increased genetic variation and evolutionary potential drive the success of an invasive grass. *Proceedings of the National Academy of Sciences, USA* 104: 3883–3888.

Richardson, D.M., P. Pyšek, M. Rejmánek, M.G. Barbour, F.D. Panetta, and C.J. West. 2000. Naturalization and invasion of alien plants: Concepts and definitions. *Diversity and Distributions* 6: 93–107.

Saltonstall, K. 2002. Cryptic invasion by a non-native genotype of the common reed, *Phragmites australis*, into North America. *Proceedings of the National Academy of Sciences, USA* 99: 2445–2449.

Schneider, L.C. 2006. Invasive species and land-use: The effect of land management practices on bracken fern invasion in the region of Calakmul, Mexico. *Journal of Latin American Geography* 5(2): 91–107.

Valéry, L., V. Bouchard, and J.-C. Lefeuvre. 2004. Impact of the invasive native species *Elymus athericus* on carbon pools in a salt marsh. *Wetlands* 24: 268–276.

Valéry, L., H. Fritz, J.-C. Lefeuvre, and D. Simberloff. 2008. In search of a real definition of the biological invasion phenomenon itself. *Biological Invasions* 10: 1345–1351.

NEW ZEALAND: INVASIONS

DAVID R. TOWNS AND PETER DE LANGE
New Zealand Department of Conservation, Auckland

MICK N. CLOUT
University of Auckland, New Zealand

Although settled only 1,000 years ago, New Zealand has been deluged with exotic species, many of which have become invasive. The flora has been hugely increased by the addition of naturalized species. The list of invasive plants equals the entire flora of endemic species. Likewise, introductions of mammals have massively increased the total fauna. Few extinctions are attributable to introduced plants, but the arrival of mammals has unleashed a cascade of extinctions among native species, especially birds. Invasive mammals are common in archipelagos, but the introduction of the arboreal brushtail possum was unusual and has proved disastrous for forest ecosystems. Further introductions are now strictly controlled to protect agriculture and biodiversity.

POLYNESIAN INTRODUCTIONS

New Zealand is distinctive as the last major archipelago to be settled by people, who then rapidly transformed it into one of the most heavily invaded archipelagos on Earth. Seeds and landsnails gnawed by Pacific rats or kiore (*Rattus exulans*) transported by Polynesians indicate that New Zealand was first settled in about 1280. These first people found a land inhabited by vast numbers of birds. Dense populations of burrowing seabirds (Procelariiformes) occupied the coast and many mountainous areas. In wetlands and forests were flightless ducks, geese, and moa—an endemic family (Dinornithidae) with species ranging from the size of chickens to others larger than ostriches. The importance of moa as food for this new Maori culture is illustrated by huge collections of bones at early settlement sites.

The Polynesian settlers brought four species of edible plants and one for making clothing, plus two species of mammals. Kiore were probably stowaways, but the Polynesian dog, or kuri, was highly valued as a companion animal and as a source of food and fur. Kuri were usually confined and tethered although some probably escaped. Dogs are well known to devastate populations of ground-dwelling birds and even large reptiles such as Galapagos iguanas. The combined effects of kiore, kuri, burning, and harvesting were devastating for the birds. By about 1500, all nine species of moa and their predator, the giant Haast's eagle (*Harpagornis moorei*), were extinct, along with flightless geese and the predatory adzebills (*Aptornis* spp.). In the following 200 years, a further 15 species died out, including waterfowl, wrens, raptors, ravens, and burrowing seabirds. The first, brief contact with Europeans was in 1642. But it was the arrival of James Cook and his scientific exploration of New Zealand beginning in 1769 that highlighted the uniqueness of the New Zealand flora and fauna. These explorers precipitated new invasions by alien species and the next wave of extinctions.

EUROPEAN INTRODUCTIONS

Cook's ships were like floating farmyards complete with farm pests. The first ship *Endeavour* was only 33 m long but somehow housed a crew of 85 and pens of cattle, sheep, goats, chickens, and ducks, while dogs and cats roamed under foot. There were also swarms of Norway rats (*Rattus norvegicus*) and cockroaches. When vessels were careened or tied to the shoreline, they provided a path for invasion by rats. By 1777, Cook's ship *Resolution* was effectively a floating menagerie of cattle, sheep, goats, pigs, rabbits, turkeys, geese, ducks, and peafowl, some of which were destined for deliberate release to supplement the diet of Maori in New Zealand. Cook's parties also planted gardens that were then left for subsequent voyagers—the beginnings of a wave of plant introductions from all over the planet.

The final and continuing wave of extinctions began with the arrival of permanent settlers from Europe early in the nineteenth century. Between about 1300 and 1840, forest cover was reduced by burning from 78 percent of the land area to 53 percent. Since 1840, this has been reduced to 23 percent. Feral cats may have appeared on some islands by 1806 and were followed by mice and ship (black) rats between 1858 and the 1890s. While the forested area of New Zealand declined, the number of introduced species rapidly increased. Between 1855 and 1988, introduced plants naturalized (spread without human assistance) at an average rate of about 11 species per year. By the early 1990s, there were almost as many species of naturalized plants (1,623) as those endemic to New Zealand (2,066). Over the same period, attempts were made to introduce 54 species of land mammals, which greatly exceeded the native fauna of three species of bats. From these introductions, 15 species became widespread, and a further 26 formed local populations. Likewise, 20 species of freshwater fish were introduced

into habitats occupied by only 23 endemic species. There were also attempts to introduce at least four species of frogs, equaling the endemic fauna of four species.

FROM INTRODUCTIONS TO INVASIONS

Many introductions, particularly of mammals, were from Europe. Some plants and animals were introduced to soften the new environment with familiar species. Others reflected a desire to establish agriculture, develop a fur industry, and provide additional game for the table. Most early colonists were escaping the strictures of a class system in Britain and wished to hunt and fish, something denied them by the owners of estates at home. The introductions were supported by acclimatization societies that identified and fostered the spread of desirable species; half of the mammal species were introduced between 1860 and 1880. These early introductions included species such as rabbits, hares (*Lepus europaeus*), and pheasants (*Phasianus colchicus*), some of which were protected by legislation passed in 1861 to ensure introduced game were not overexploited. They were further protected by additional laws in 1867. One of these set up hunting rules and prohibited the import of foxes, venomous reptiles, and birds of prey that might threaten introduced game. The other was for the preservation and propagation of salmon and trout.

The vertebrates introduced to New Zealand include 23 species of mammals that are invasive or would become so if they were not confined or controlled. Among these are 8 of the 14 most invasive mammals on the Global Invasive Species Data Base (GISD; Table 1). In contrast, three of the species of *Litoria* frogs introduced from Australia have established and are now widespread. The wetland habitats they use do not appear to overlap with those of native frogs and there are no documented detrimental effects of the frogs on native species or ecosystems. They might be viewed as neutral introductions, but even these have unexpected indirect effects (see below).

There may now be up to 50,000 species of plants in cultivation in New Zealand, 3,000 of which have naturalized and about 2,000 of which have become invasive weeds. Most of the early invasive species were escaped agricultural plants such as species of pines and gorse (see below). However, about 80 percent of newly naturalized plants are garden escapees, some of which (such as figs; see below) had been in cultivation for many years.

Many problem species are rapid invaders of disturbed sites. For example, in the past, open sites might have been colonized by native species of *Cortaderia* grasses, but these areas are now more commonly dominated by the aggressive introduced pampas grasses (*C. jubata* and *C. selloana*) from

TABLE 1
Proportion of the World's 100 Worst Invasive Alien Species Present in New Zealand

Groups and Total Listed	Present in New Zealand	Status in New Zealand
Micro-organisms (3)	avian malaria (*Plasmodium relictum*)	possibly widespread; vectored by native mosquitoes
Macro-fungi (5)	Dutch elm disease (*Ophiostoma ulmi*)*	localized
	frog chytrid fungus (*Batrachochytrium dendrobatidis*)	possibly widespread
	phytophora root rot (*Phytophthora cinnamomi*)	widespread
Aquatic plants (4)	common cord grass (*Spartina anglica*)	fully naturalized
	wakame seaweed (*Undaria pinnatifida*)	spreading
	water hyacinth (*Eichhornia crassipes*)*	local infestations eradicated
Land plants (32)	black wattle (*Acacia mearnsii*)	fully naturalized
	Brazilian pepper tree (*Schinus terebinthifolius*)	non-invasive
	cogon grass (*Imperata cylindrica*)	fully naturalized
	cluster pine (*Pinus pinaster*)	fully naturalized
	giant reed (*Arundo donax*)	fully naturalized
	gorse (*Ulex europaeus*)	fully naturalized
	kahili ginger (*Hedychium gardnerianum*)	fully naturalized
	kudzu (*Pueraria montana* var *lobata*)	non-invasive
	lantana (*Lantana camara*)	fully naturalized
	purple loosestrife (*Lythrum salicaria*)	fully naturalized
	strawberry guava (*Psidium cattleianum*)	fully naturalized
Aquatic invertebrates (9)	Mediterranean mussel (*Mytilus galloprovincialis*)	possibly native?
Land invertebrates (17)	Argentine ant (*Linepithema humile*)	established and spreading
	Asian tiger mosquito (*Aedes albopictus*)	incursions intercepted
	big-headed ant (*Pheiodole megacephala*)	confined to Kermadec Islands
	common wasp (*Vespula vulgaris*)	widespread
	crazy ant (*Anoplolepis gracilipes*)*	incursions intercepted
	khapra beetle (*Trogoderma granarium*)	incursions intercepted
	red imported fire ant (*Solenopsis invicta*)*	incursions intercepted
	sweet potato whitefly (*Bemisia tabaci*)	localized
Amphibians (3)	none of the most invasive species	
Fish (8)	brown trout (*Salmo trutta*)	widespread
	carp (*Cyprinus carpio*)	widespread
	rainbow trout (*Oncorhynchus mykiss*)	widespread
	western mosquito fish (*Gambusia affinis*)	widespread
Birds (3)	Indian myna bird (*Acridotheres tristis*)	widespread in north
	red-vented bulbul (*Pycnonotus cafer*)	eradicated and incursions intercepted
	starling (*Sturnus vulgaris*)	widespread
Reptiles (3)	none of the most invasive species	
Mammals (14)	brushtail possum (*Trichosurus vulpecula*)	widespread
	domestic cat (*Felis catus*)	widespread
	goat (*Capra hircus*)	widespread
	mouse (*Mus musculus*)	widespread
	rabbit (*Orcyctolagus cuniculus*)	widespread
	red deer (*Cervus elaphus*)	widespread
	ship rat (*Rattus rattus*)	widespread
	stoat (*Mustela erminea*)	widespread

NOTE: From GISD. When applicable, invasive plants are listed as "fully naturalized" and notifiable organisms with mandatory control are identified. All listed mammals have been eradicated from New Zealand islands. * control identified

South America. Furthermore, their spread is assisted by road embankments that provide invasion corridors and copious windblown seed, which has enabled colonization of important island reserves, such as Little Barrier Island. Similarly, clear felling in plantation forests, especially of Monterey pine (*Pinus radiata*), provides ideal ground for the spread of these grasses. Additionally, the pines themselves can be invasive into native grasslands at high altitudes, although the more aggressive species in these areas is the lodgepole pine (*P. contorta*). Pampas grasses and the plantation pine

FIGURE 1 Escaped (wilding) pine trees in high-altitude grassland, South Island, New Zealand. (Photograph courtesy of New Zealand Department of Conservation, Neville Peat.)

escapees can modify entire landscapes and are extremely expensive to control in difficult terrain (Fig. 1).

Introductions to Control Introductions

Serious problems from introduced game arose in the 1870s, when native grasslands burned and heavily stocked with sheep were invaded by newly released rabbits. Many farmers faced ruin, and some turned to the rabbits as a resource, with the 1 million skins that were exported in 1877 rising to 18 million by 1884. Despite opposition from the scientific community, sheep farmers successfully lobbied the government for the introduction of three species of mustelids to control the rabbits. The introductions began with ferrets (*Mustela furo*) in 1879, followed by weasels (*M. nivalis vulgaris*) and stoats in 1885. The limited impact these species had on the rabbits is revealed by exports of rabbit skins, which between 1920 and 1929 still averaged almost 17 million per annum. However, the mustelids, especially stoats, were devastating for native birdlife. As mustelids extended their range, a government caretaker, Richard Henry, transferred hundreds of vulnerable native birds to a sanctuary on Resolution Island. Unfortunately, the island was reached by stoats, which are strong swimmers, and the conservation effort failed. This depressing event was an object lesson: the invasions of rabbits indicated that a benign ideal can have disastrous downstream effects. The introduction of mustelids as a political response added another dimension to the disaster, and conservation efforts for native species were too little too late.

Fortunately, some early introductions failed for species that have become pests elsewhere, including mongoose, raccoons, and squirrels. However, there is a component of New Zealand introductions, especially of mammals, that is unique: the release of marsupials from Australia.

Disastrous Marsupials

Five species of wallabies and the brushtail possum (Fig. 2) were introduced before the end of the nineteenth century. Motives for the release of wallabies are unclear, although most likely they were intended largely for sport hunting. However, perhaps as souvenirs of time spent in Australia, four species of wallaby became established on Kawau Island after releases in about 1870 around his residence by the then Governor, Sir George Grey. These and other wallabies remain confined in their distribution, but the possum is now a widespread pest with serious economic consequences.

The brushtail possum is a cat-sized 2–4 kg, solitary, nocturnal, arboreal marsupial that was a presumed herbivore of foliage in Australia. Releases of possums from southeastern Australia began in the late 1850s, with the intention of establishing a fur industry. Further introductions were made from the 1860s, but colonization was also assisted by 127 liberations within New Zealand by acclimatization societies and government agencies between 1865 and 1926. Furthermore, acclimatization societies concerned that the releases would suffer from excessive hunting and trapping successfully lobbied for licensed harvest, with income from levies on the sale of skins to be shared by the societies and government. Orchardists, farmers, and horticulturalists also lobbied the government, but because of the damage possums caused to crops. Finally, in the 1940s, the first scientific evidence of the impacts of possums on native forests was released, and in 1947, all restrictions on harvest were removed, penalties were instituted if possums were liberated or harbored, and poisons were licensed for their control. From 1951 to 1961, there was also a bounty system to encourage control. This may have further increased spread from possums deliberately released as a source of income from bounties.

FIGURE 2 Australian brushtail possum. (Photograph courtesy of New Zealand Department of Conservation, Rod Morris.)

In the 1960s, possums were found to be a wild reservoir and vectors of bovine tuberculosis (BTb), a serious threat to New Zealand's largest export industries—dairy products and beef—and to domesticated deer herds. Ironically, the disease was also one of the few natural controls on wild possum populations and must have originally entered the possum populations from cattle. Furthermore, video surveillance of native nesting birds indicated that possums were also predators of eggs and chicks. Possums were thus transformed from fur-producing herbivores to a pest that damages crops, affects forest structure, impedes regeneration, spreads agricultural diseases, and preys on native birds. The threats to agriculture led to the formation of the Animal Health Board, which used epidemiological models to predict maximum population densities of possums where BTb is prevalent and determine the frequency of control required. Wide-scale control is by the aerial spread of carrot or cereal laced with compound 1080, an anti-browse toxin found in plants. By 2000, possum control and associated research cost the government and private agencies US $59 million per year. Probably no other country has tackled an invasive vertebrate on such a scale. The results for livestock have been encouraging, with 53 percent decreases in BTb infected herds of cattle and 58 percent in domesticated herds of deer after five years of nationally coordinated possum control. The Animal Health Board now intends to completely eliminate BTb from domestic stock. However, it is less clear whether possum control alone is of benefit to forests and their wildlife because other nontarget introduced species such as rats, mustelids, and deer are all affected by poison spread against possums. Deer may be the key to success or failure of this program. The rapid spread of BTb in some areas is likely to have been helped by transmission through wild deer populations. Reductions in the density of deer as a byproduct of possum control may thus be beneficial. However, there is vocal opposition to aerial spread of 1080 from deer-hunting enthusiasts. Such groups strongly advocate for ground-based shooting and trapping of possums, which is prohibitively expensive, or possum-specific controls such as specific poisons, which at present are unavailable.

Invaders with Direct and Indirect Effects

A conflict between commerce and conservation has repeatedly developed. This is particularly well illustrated over the recreational importance of two more invasive species: rainbow and brown trout. Brown trout were introduced in 1867, and acclimatization societies made sustained efforts to spread them into almost every water body accessible to anglers. The fishery has been managed by the acclimatization societies with assistance from government agencies for almost 150 years and is funded by license fees. The recreational salmonid fishery is now conservatively valued at more than US $300 million per year. There is no direct evidence of global extinctions caused by introduced trout, but comparisons between colonized and trout-free streams indicate that they have been responsible for local extinctions of native fishes and invertebrates. These losses include economically important native freshwater fisheries such as eels and whitebait, which are the marine sourced juveniles of species in *Galaxias*. Beyond these losses, however, are indirect changes to freshwater ecosystems that are only now being understood.

Studies of interactions between brown trout and native invertebrates revealed that, unlike native fish, trout suppress the movement of grazing mayflies, with lowered foraging on algae and lower propensity to enter the water column as drift. At the population level, the trout influence distribution of native fish, crayfish, and large invertebrates. They have particularly strong effects on nonmigratory *Galaxias*, through predation and competition, thus excluding the native species from their preferred habitats. The trout also reduce populations of crayfish, which are unable to detect the introduced fish using the chemical cues used to detect native predatory eels. Large, slow-moving invertebrates are also eaten by trout, including species that are algal grazers and invertebrate predators. The cumulative effects of trout on native fish and invertebrates result in modified communities, with significantly higher algal biomass in invaded streams, an effect termed a trophic cascade. The ecosystem effects of this cascade are to increase primary production by up to sixfold in streams with trout. In streams with native fish, invertebrates consume 75 percent of the primary production, but in invaded streams trout eat almost all of the annual production of invertebrates, which consume only 21 percent of primary production. Furthermore, enhanced algal production has resulted in more rapid uptake of ammonium, nitrate, and phosphate from stream water. In sum, the invaded streams have a higher biomass of fish, a lower biomass of invertebrates, and a far higher primary production than do uninvaded streams.

Invaders with Devious Life Cycles

Although invasive species can have dramatic effects on native species that ripple through entire ecosystems, others require intermediate species before they become invasive. Two examples from New Zealand involve species with complex life cycles and unusual relationships with native species.

For at least the last 100 years, a tropical Australian strangler fig, the Moreton Bay fig (*Ficus macrophylla*), has been grown as an ornamental tree in the warmer parts of northern New Zealand. The trees reach up to 60 m in height and with impressive root buttressing can attain an equivalent width. Each species of fig has specific wasps (Agaonidae) to fertilize the flowers, and the wasps also require the flowers in order to reproduce. In the absence of fig wasps, New Zealand trees were sterile, but they nonetheless produced copious quantities of infertile fruit. The fruit are attractive to the native fruit pigeon or kereru (*Hemiphaga novaseelandiae*), which congregate around the fig trees. Because kereru are declining on the mainland, such congregations in urban areas are welcomed. However, in 1996, the Moreton Bay fig wasp *Pleistodontes frogatti* was recorded, presumably having naturally dispersed to New Zealand in strong westerly winds from Australia. The first naturalized fig plants were then discovered. With kereru as long-distance dispersers of the now fertile fruit, a previously noninvasive plant has suddenly become a potential invader of native forests with the ability to establish in the crown of large forest trees. Furthermore, the Moreton Bay fig wasp appears to have unexpectedly entered cultivated figs such as *F. elastica* and *F. obliqua*, with potential for a new series of invasions.

The second example is a particularly insidious invasive species of frog skin fungus, chytridiomycete. The fungus causes the disease chytridiomycosis ("chytrid"), which is fatal to some frogs and has been implicated in declines and extinctions of frogs in the northern hemisphere and Australia. Chytrid appears to have originated in the African clawed frog *Xenopus laevis*, widely used in early pregnancy testing and still used in laboratory culture for genetic research. The fungus has subsequently become widely distributed in waterways throughout the world and was discovered in introduced *Litoria* frogs in New Zealand in 2000. A year later, chytrid was found in the rare endemic Archey's frog *Leiopelma archeyi*, which, in part of its range, was undergoing an unforeseen decline. This discovery suggested that the seemingly noninvasive *Litoria* frogs were in fact a reservoir for the highly invasive chytridiomycosis, which in turn threatened the native frog fauna. However, subsequent studies of virulence of chytrid in native frogs indicated that they are far more resistant to chytrid than are the introduced species. The incidence of chytrid in wild populations of native frogs has remained low. The sudden declines recorded for Archey's frog may have been in response to other unidentified environmental changes, but declines of introduced frogs may indeed reflect the effects of chytrid.

Asexual Invaders

Rapid invasions of new environments are often assisted by high reproductive output. For plants, this is usually by producing numerous easily dispersed seeds. However, some species are serious weeds despite being sterile. Hornwort (*Ceratophyllum demersum*) and three species of oxygen weeds (Hydrocharitaceae), *Egeria densa*, *Elodea canadensis*, and *Lagarosiphon major*, have escaped from aquariums and ornamental ponds to smother the beds of lakes and slow-flowing rivers. None of the species has sexual stages in New Zealand, but their vegetative spread has been assisted by boat users, hydroelectric dam construction, and misguided plantings for waterfowl. Hornwort is still spreading, but the three species of oxygen weeds have already irreversibly altered most of the larger freshwater lakes. Manual control is estimated to cost over US $5,000 per hectare, but the clogging of water intakes and turbines costs millions of dollars in lost power generation.

On land, similar smothering effects are typical of *Tradescantia*. The species present in New Zealand is assumed to be *T. fulminensis* and is sterile. Nonetheless, the species aggressively covers open ground, including sites that are partially shaded. The *Tradescantia* mats, which effectively suppress regeneration of many native shrubs and ground-cover plants, are extremely difficult to remove because of vegetative spread from residual plant fragments.

Beneficial Invasions?

One of the most ubiquitous woody weeds in New Zealand is gorse, which covers thousands of hectares of farmland and costs millions of dollars each year to control. Gorse has long-lived seeds. On hilly land and in large river beds this persistent seed bank enables gorse to permanently alter the successional pathways of native species. In some areas, however, the same persistence benefits native species of plants because the gorse provides protective cover against drying winds and grazing mammals. Over time, native species are then able to grow through the gorse, which dies out under a closed forest canopy. At one site, gorse appears also to have prevented the extinction of a giant (10–17 gm) flightless cricket, the Mahoenui weta (*Deinacrida mahoenui*). The weta were discovered deep in the gorse bushes, especially those that are older and opened periodically by goat activity. It is possible to conclude that the gorse may be sheltering weta from predation by species such as rats and mustelids because the weta are absent from adjacent native forests. The abundance of weta in this area of gorse prompted the Department of Conservation to create a 240 ha reserve to protect the two invasive species (gorse and goats) for the benefit of weta.

Another invasive plant that shelters native species is willow (*Salix* spp.). These were planted to stabilize river banks but have since invaded extensive wetland areas that have lost their native forests to logging. However, the willows now provide a habitat for threatened liverworts, mosses, and orchids that were once a feature of the destroyed swamp forest ecosystems. Unfortunately, such benefits from invasive species seem rare. More commonly, suites of invasive species interact to overwhelm native species and ecosystems.

Invasives Helping Invasives

In the simplest systems, such as on offshore islands, effects of an invasive top predator such as cats may be assisted by mesopredators such as rodents. For example, on sub-Antarctic Campbell Island, cats introduced during early attempts at settlement fed on native birds and Norway rats. The rats were essential to the cats as an alternative food source when seabirds were unavailable. Eventually, these food resources became so depleted that the cats died out, although the rats survived. On Moutohora Island, a population of Norway rats coexisted for many years with large numbers of a native burrowing seabird, the gray-faced petrel (*Pterodroma macroptera*). However, when rabbits were released onto the island, rat numbers greatly increased owing to the increased food supply provided by young and dead rabbits. The rats then turned on petrels, and the petrel population rapidly declined. Indirectly, the rabbits had raised predation intensity on the birds, a phenomenon sometimes called hyperpredation. On the main islands, the suite of invasive species is much larger, disturbance effects much greater, and the relationships between top predators and mesopredators far more complex. For example, on the southern part of South Island, declines of large endemic grand and Otago skinks (*Oligosoma grande* and *O. otagense*) were attributed to predation by feral cats and ferrets, which often had lizard remains in droppings. Cats and ferrets are also top predators among a complexly interacting group of introduced mammals including two additional species of mustelids, possums, two species of rats, mice, hedgehogs, and rabbits. Extensive control of the cats and ferrets failed to benefit the lizards, possibly because mesopredators released from control by top predators were able to attack lizards at greater intensity. Subsequently, two mammal-proof exclosures, capable of stopping all mammals including mice, were constructed and compared with the simultaneous control of all predatory mammals. Exclusion and control both resulted in increases in the abundance of lizards. In this system, the predator guild is largely fueled by rabbits, and food web models indicate that rabbit control may be the cheapest indirect method of protecting the lizards.

UNDOING THE DAMAGE

Responses to the effects of invasive species in New Zealand have come in two forms. The first has been to eradicate problem organisms wherever possible and politically acceptable. The second has been to develop legislation that would provide for rapid responses should new invasive species arrive.

Pest Control and Eradications in Real and Virtual Islands

The list of extinctions in New Zealand would have been much longer but for numerous offshore islands beyond the reach of invasive mammals. These islands provided a glimpse of largely unmodified ecosystems and also indicated what to aim for if pests and weeds could be eradicated or controlled. The invasion of Big South Cape Island by ship rats in 1962 also demonstrated the fate of island ecosystems if the further spread of such species could not be contained or reversed. This single invasion was followed by the global extinction of two species of birds and one species of bat, extinction in the wild of one species of bird (saved in captivity), and local extirpation of a giant weevil, three additional species of bird, and one of bat.

Eradications of released populations of cats, goats, and pigs on islands were achieved on a limited scale before the Big South Cape event. However, it spurred a much more systematic approach to the development and implementation of island eradications, especially of rats. Once effectiveness of the aerial spread of baits was demonstrated in the early 1990s, a way was opened to routinely clear many islands of all introduced mammals (Fig. 3). By 2008, over

FIGURE 3 Helicopter spreading bait for Norway rats, Campbell Island, New Zealand. (Photograph courtesy of New Zealand Department of Conservation, Peter Tyree.)

70 islands with a total land area exceeding 30,000 ha were free of introduced mammals, with benefits for hundreds of species of plants and invertebrates and at least 70 species of native vertebrates. These successes also prompted the development of numerous sites throughout New Zealand where pests and weeds are intensively managed as projects run by the Department of Conservation, regional councils, and privately initiated community groups. Some of these latter groups receive government grants or other forms of assistance. By June 2007, privately initiated fencing, pest control, and restoration projects were operating on more than 438,000 ha. Six showcase sites are operated by the Department of Conservation as mainland islands, where, in areas ranging from 250 to 50,000 ha, all introduced mammals are controlled to the lowest possible levels. There has also been a proliferation of privately initiated fenced sites where attempts are made to exclude all introduced mammals (Fig. 4). The first of these was the 252 ha Karori Sanctuary, which has a 500 year plan to restore a hardwood forest ecosystem in the center of the city of Wellington. This sanctuary houses the first restored mainland populations of tuatara (*Sphenodon punctatus*) and birds such as saddleback (*Philesturnus carunculatus*) and hihi (*Notiomystis cincta*), as well as a burgeoning population of forest parrots (*Nestor meridionalis*). The example has been followed by other community groups. The largest fenced area to exclude introduced mammals is operated by the Maungatautari Ecological Forest Trust, which raised sufficient funds to build a 47 km fence around 3,400 ha of regenerating forest on an extinct volcano. The fence was completed in 2006, and by December 2008, 11 species of pests had been eradicated from inside, with only small pockets of

FIGURE 4 Privately funded mammal-proof fence, Stewart Island, New Zealand. (Photograph courtesy of David R. Towns.)

mice, rabbits, hares, and goats remaining. The fenced area has already received four species of threatened birds.

Many forest areas are threatened by weeds. The level of control achieved usually does not match that attained for mammals, partly because of the persistence of many species as seeds. However, there have been heroic attempts. One of the longest running is a 30 year program of weed control on Raoul Island, 1,000 km northeast of New Zealand's North Island. At least 29 species of problem plants escaped from abandoned attempts at farming and gardening. Seven of these may have been eradicated. One of the remaining species, Mysore thorn (*Caesalpinia decapetala*), seems largely eradicated as reproducing plants, but seeds may continue to germinate for another 50 years. Others, such as olives (*Olea europaea* subsp. *cuspidata*) and guava (*Psidium cattleianum* and *P. guajava*), are so prone to dispersal by birds that restoration of seed-dispersing birds is on hold until effective control of these weeds is achieved.

Responses to New Arrivals

Recent legislative responses to proliferating pests and weeds have been two Acts of Parliament. First, the Biosecurity Act 1993 provides a legal basis for excluding, eradicating, and effectively managing pests and unwanted organisms. This empowers Biosecurity New Zealand (part of the Ministry of Agriculture and Forestry), other government agencies, regional councils, and pest management agencies to develop national and regional pest management strategies that target problematic pests and weeds. There are also enforcement provisions against the sale, propagation, breeding, release, or display of defined unwanted organisms or pests. Second, the Hazardous Substances and New Organisms Act 1996 regulates the import of organisms and products that could affect the environment. The use of new products and organisms is possible only if approved by the Environmental Risk Management Authority.

Using these powers, Biosecurity New Zealand has reacted swiftly to several recent arrivals. One of these, the white-spotted tussock moth (*Orgyia thyellina*), is a relative of the gypsy moth native to Asia and the Russian Far East and was discovered in eastern Auckland in 1996. Laboratory trials showed the moths to be strongly attracted to fruit trees, other introduced deciduous trees, and native beech. Given the risks posed by this plant defoliator, eradication was attempted using *Bacillus thuringiensis* var. *kurstaki* (Btk), which is a soil bacterium effective against gypsy moths in North America. Initially, 40 km² containing 80,000 residences were sprayed from aircraft, with this then being reduced to a 3 km² core area within 12 months. Final mopping up was achieved using a synthetic sex pheromone, and in 1998 eradication of the moth was formally announced. The final cost was US $6 million.

In 1999, another defoliating moth species, the painted apple moth (*Teia anartoides*) was discovered to be well established in western Auckland. This native of Australia posed a risk to native and plantation forests, crops, and amenity plantings, with potential economic costs of up to NZ $356 million over 20 years. A similar approach was applied to that used in eastern Auckland, but with more emphasis on ground control owing to concerns from residents about aerial spraying of Btk. Ground control proved ineffective, and after aerial spraying 69 times between 1999 and 2003, and the use of pheromone traps, eradication was announced in 2006. The cost was US $31 million. Other insect species including ants (Table 1) have similarly been detected and eradicated, but with lesser effects on the public and at much lower costs.

Despite strong legislation and willingness to apply it at considerable cost, two other notable pest organisms have arrived and were unable to be eradicated. The first of these was the varroa bee mite (*Varroa jacobsoni*), which was discovered in 2000, when infected areas were already too large to control. The mite most likely entered the country on illegally imported bees and is now found throughout the North Island. So far, the South Island has remained free of varroa. The mite poses a threat only to introduced colonial bees because native bees are solitary, with little means of mite transmission. The second escape was didymo (*Didymosphenia geminata*), an unsightly encrusting alga that is now widespread in rivers of the South Island and for which no control agent was known. This species probably arrived in imported contaminated drinking water or machinery. Most concern has centered on the potential effects on trout fisheries. The likely impact on native species is unclear.

LEARNING FROM INVASIONS

Additions and Deletions

The long history of well-documented introductions in the nineteenth century provided useful lessons on why invasions succeed or fail and has helped with design of translocations of native species. A study of 79 species of birds involved in 496 introductions showed that success was highest for species that were originally nonmigratory, when release populations were large, and when releases were repeated. Successful invaders were those with high birth-to-mortality ratios and which had been liberated at multiple sites.

Similarly, studies of the responses of communities of plants and animals to the removal or intensive control of pest species have often revealed the effects of invasive species while they were present. For example, kiore exclosures and island removals demonstrated negative effects on at least 17 species of native plants, potential effects on forest structure, detrimental effects on recruitment of juvenile tuatara, and suppressing effects on coastal skinks and geckos. The sequential removal of pests on islands has also revealed how the invasive species interacted and how these interactions affected native species. For example, cats were removed from Hauturu/Little Barrier Island in 1980 after they had largely destroyed seabird populations. One surviving species, Cook's petrel (*Pterodroma cookii*), failed to recover, with reduced fledging rates in the absence of cats. One mechanism for this failure was mesopredator release; burgeoning populations of kiore, previously kept in check by the cats, subsequently began to prey on Cook's petrel chicks. Increased chick survival was achieved only after kiore were eradicated in 2004.

In forested mainland sites, incremental control of ship rats, stoats, and possums has revealed target densities suitable for management of native species such as kokako (*Callaeas cinerea*). Fully fenced exclosures on the mainland are an innovation that should reveal much about both the vulnerability of these forest systems to weeds when mammals are excluded and the economics of exclusion versus control.

Ecosystem Effects of Invaders

Intensive studies of the ecosystem effects of invaders inform island restoration plans when mammals are removed. A series of studies comparing islands inhabited by rats with those never invaded showed that rats reduced forest soil fertility by disrupting the sea-to-land transport of nutrients by seabirds. This led to wide-ranging cascading effects on belowground organisms and the processes they drive. For example, invaded islands had deeper forest litter, and 8 of the 19 orders of invertebrates inhabiting the litter were more abundant on uninvaded islands. Furthermore, because of the indirect effects on soil composition, soil nematodes were less abundant on invaded than uninvaded islands. Plant growth rates on uninvaded islands were also higher than on invaded islands. The studies concluded that even when the rats have been removed, invertebrate and plant communities are unlikely to respond until seabird populations recover through a process that may take many decades. These results are reminiscent of the subtle ecosystem effects recorded for trout.

Public Attitudes toward Control of Invasive Species

Public attitudes toward introduced species and methods used to control them vary. Current support for the removal of problem organisms is reflected in the effort of community groups working on private and public lands to remove pests and weeds and to restore and replant native species. However, this more aggressive stand toward invasive species, especially mammals, can lead to polarized attitudes within local communities. For example, the control of possums and associated by-kill of deer spawned a coordinated campaign against compound 1080 and the agencies that use it; this has even included threats to sabotage conservation sites through the deliberate release of pests. Furthermore, within Auckland, attitudes to the spread of Btk against introduced moths differed between suburbs, with orchestrated campaigns of resistance to its use in western Auckland and claims of serious health effects and allergic reactions not encountered when the same product was used in eastern suburbs. The government has been willing to react massively with campaigns and funds against potential economic threats to agriculture, but less so with the same threats to biodiversity. For example, in 2006, the Animal Health Board had sufficient funds to treat 5.3 million ha against possums to contain BTb and protect beef and deer herds. However, the Department of Conservation could commit sufficient funds to treat only 302,000 ha against the effects of possums (and other vertebrate pests) on biodiversity. Once there is sufficient control of BTb, large-scale control of possums will cease, with subsequent increases in possum numbers. Unless other methods of large-scale control of possums are found, any gains to biodiversity as a spinoff of BTb control will then be lost.

Nonetheless, island eradications continue on an increasing scale, with proposals to rid Auckland Island (51,000 ha) of all pigs and the Rangitoto-Motutapu Islands (3,800 ha) of seven species of introduced mammals including mice. The technology developed in New Zealand to enable these eradications is now exported globally. Much can also be learned about the effects and effectiveness of fenced sites and intensive pest control on the mainland. Well-researched accounts of these enterprises are also needed if they are to maintain public support. However, this understanding will also require patience because the outcomes of the island eradications and mainland pest control may take decades to be fully understood.

SEE ALSO THE FOLLOWING ARTICLES

Acclimatization Societies / Databases / Eradication / Game Animals / Invasibility, of Communities and Ecosystems / Predators / Restoration

FURTHER READING

Allen, R. B., and W. G. Lee, eds. 2006. *Biological Invasions in New Zealand.* Berlin: Springer.

Atkinson, I. A. E., and E. K. Cameron. 1993. Human influence on the terrestrial biota and biotic communities of New Zealand. *Trends in Ecology and Evolution* 12: 447–451.

Fukami, T., D. A. Wardle, P. J. Bellingham, C. P. H. Mulder, D. R. Towns, G. W. Yeates, K. I. Bonner, M. S. Durrett, M. N. Grant-Hoffman, and W. M. Williamson. 2006. Above- and below-ground impacts of introduced predators in seabird-dominated island ecosystems. *Ecology Letters* 9: 1299–1307.

Hosking, G., J. Clearwater, J. Handisides, M. Kay, J. Ray, and N. Simmons. 2003. Tussock moth eradication: A success story from New Zealand. *International Journal of Pest Management* 49: 17–24.

King, C. M., ed. 2005. *The Handbook of New Zealand Mammals*, 2nd ed. Melbourne: Oxford University Press.

McDowall, R. M. 1994. *Gamekeepers for the Nation: The Story of New Zealand's Acclimatisation Societies, 1861–1990.* Christchurch: Canterbury University Press.

Montague, T. L., ed. 2000. *The Brushtail Possum: Biology, Impact and Management of an Introduced Marsupial.* Lincoln: Manaaki Whenua Press.

Parkes, J., and E. Murphy. 2003. Management of introduced mammals in New Zealand. *New Zealand Journal of Zoology* 30: 335–359.

Towns, D. R., and K. G. Broome. 2003. From small Maria to massive Campbell: Forty years of rat eradications from New Zealand Islands. *New Zealand Journal of Zoology* 30: 377–398.

Townsend, C. R. 2002. Individual, population, community, and ecosystem consequences of a fish invader in New Zealand streams. *Conservation Biology* 17: 38–47.

NILE PERCH

ROBERT M. PRINGLE

Harvard University, Cambridge, Massachusetts

The Nile perch (*Lates niloticus*) is a large (over 2 m in length—and over 200 kg in weight) piscivorous fish native to East, Central, and West Africa, including the Congo, Niger, and Nile river systems. In the 1950s and 1960s, this species was introduced into multiple lakes and dams in the Lake Victoria region—where it did not occur naturally. Over the subsequent 40 years, the Nile perch has become one of the most famous invasive species in history, inspiring hundreds of technical publications, a popular book (*Darwin's Dreampond*), and an Oscar-nominated documentary film (*Darwin's Nightmare*). The fish did two things to deserve such celebrity. First, it played a major role in the extinction of 200 or more of the approximately 500-species radiation of endemic cichlid fishes (*Haplochromis* spp.) in the Lake Victoria region. Second, it transformed a local artisanal fishery into a half-billion-dollar global industry, transfiguring millions of human lives in the process.

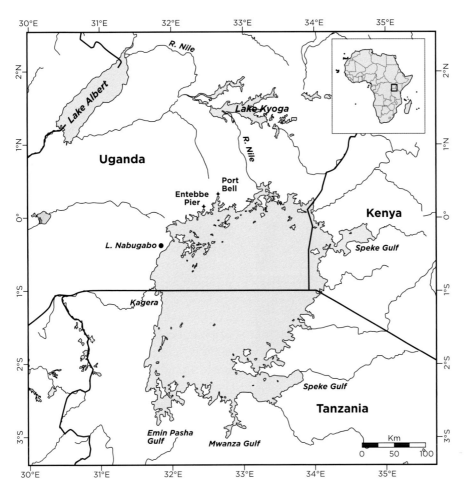

FIGURE 1 The Lake Victoria region. Stars mark the probable original sites of Nile perch introduction in the mid-1950s at Entebbe Pier and Port Bell; labels mark the locations of catch statistics presented in Fig. 3.

ORIGINS

The transfer of Nile perch into Lake Victoria was discussed for decades within the British colonial administration of East Africa. Researchers opposed the proposition on precautionary grounds, while some management officials—most vocally members of the Uganda Game and Fisheries Department—supported it as a means of enhancing production and creating a sport fishing industry. No formal decision was ever reached, as Nile perch started appearing in Lake Victoria in 1960. Written records and interviews with former colonial officials suggest that Nile perch from Lake Albert were introduced deliberately (but secretly) at Port Bell and Entebbe Pier in 1954 (Figs. 1, 2). Throughout the early 1960s, officials performed additional sanctioned introductions into Lake Victoria, as well as into Lakes Kyoga, Nabugabo, and others.

A NEW ECOLOGY

At first, little happened. In the mid-1970s, several ecologists deemed the introduction a success: the catch of native fishes appeared stable, while Nile perch were supplementing yields and drawing tourism revenue from anglers.

Ten years later, a commentary in *Nature* magazine declared that Lake Victoria's fisheries had been "not merely damaged but destroyed" by Nile perch. For reasons that remain incompletely understood, Nile perch densities had increased roughly 100-fold between the mid-1970s and mid-1980s, with a corresponding crash in the density and diversity of haplochromine cichlids (Fig. 3). One hypothesis proposed to explain this "Nile perch boom" is that intensifying human exploitation of haplochromine stocks released juvenile Nile perch from predation by and competition with cichlids; the recruitment of these juveniles to sizes at which they could consume cichlids

FIGURE 2 Freshly caught Nile perch (A) and the native minnow *R. argentea* (B). (Photographs courtesy of the author.)

further depleted cichlid populations and enhanced Nile perch recruitment in a positive-feedback loop. Migratory subadult Nile perch then colonized new parts of the lake, where the process repeated. This hypothesis is consistent with both the asynchronous upsurge of Nile perch in different parts of the lake and the astonishing rapidity of the faunal shift.

Nile perch was not the only factor contributing to haplochromine extinctions. A century of human population growth and agricultural expansion in the basin increased nutrient inputs to the lake, causing a doubling of primary production, a nearly order-of-magnitude increase in algal biomass, and the development of anoxic conditions in deeper waters. The corresponding decrease in water clarity reduced the ability of cichlids

to mate assortatively based on color morph, slackening reproductive isolation and likely leading to widespread hybridization.

It is currently impossible to determine the relative importance of top-down (Nile perch predation) and bottom-up (eutrophication) forces—and their interactions—in driving the cichlid extinctions in Lake Victoria. Haplochromine populations also plummeted in invaded lakes, such as Uganda's Nabugabo, where eutrophication was not so dramatic, confirming the importance of Nile perch predation. In any case, these two forces were linked: the Nile perch boom spurred migration toward the lakeshore, prompted expansion and development of lakeshore villages and cities, and hastened landscape conversion and tree clearing (for wood to smoke fish, among other things), all of which accelerated nutrient loading of the lake.

The winners of the Nile perch boom included Nile tilapia (*Oreochromis niloticus*, itself introduced), the native minnow *Rastrineobola argentea* (Fig. 1), and the freshwater shrimp *Caridina nilotica*, all of which expanded dramatically in the absence of competition and predation from haplochromines. These species also provided a prey base for Nile perch after haplochromines became scarce. By 2000, Nile perch, Nile tilapia, and *R. argentea* constituted more than 99 percent of the fishery. One scientist compared this shift to clear-cutting a rainforest and replacing it with a monoculture of fast-growing plantation trees. This contraction of the fish community essentially created a brand new ecology, one with lower functional diversity, shorter food chains, less oxygen, and greater numbers of the true apex predator in the system, *Homo sapiens*.

A NEW ECONOMY

The changes for the roughly 30 million people living in the lake basin were no less dramatic. As the estimated annual catch from Lake Victoria mushroomed from about 30,000 tons in the late 1970s to about 500,000 tons at the turn of the century, export earnings increased from approximately $1 million to around $260 million—90 percent of it from Nile perch. The estimated number of fishermen and fishing vessels operating on Lake Victoria also rose sharply; available statistics suggest that both have increased more than 300 percent from the early 1980s and continue to rise. With such tremendous profit potential came development aid, private investment, and class stratification. Fish-processing factories appeared around the lake, and the number of roads and refrigeration facilities at landing beaches increased. A new class of middlemen arose, who purchased fish at beaches and delivered them to factories for processing. Wealthier fishermen invested in multiple boats and nets, employing crews who did the actual fishing. Others borrowed money to purchase new boats and gears, creating enduring debts to entrepreneurs and processing facilities.

One can draw different conclusions about the local responses to Nile perch depending on where in the literature one looks. Many have argued that the Nile perch has been a disaster for small-scale fishermen, who dislike the fish. Others have highlighted positive effects of economic growth and favorable reactions by fishermen. In reality, local perceptions of the Nile perch and its effects on life are nuanced, variable, and temporally dynamic. Many early reactions were negative, because the fish was too big to be fished or preserved in traditional ways. But as the Nile perch boom created a new, globally integrated, cash-based economy, people adapted. Walking into a lakeside fishing community today and asking people how they feel about Nile perch is roughly analogous to walking into a General Motors plant and asking workers and job applicants how they feel about cars. Many respondents will tell you about their own financial troubles and the vagaries of the industry as a whole, and about the difficulties and indignities of life on the lower rungs of a multinational, multimillion dollar industry. They may or may not personally use the product that they produce, but their livelihoods depend on the global demand for it, and most are too young to remember the horse-and-buggy days. Many fishermen desire the restoration of a more diverse fish community, but few would opt to return to the quieter economy of the pre–Nile perch era. Indeed, many are attracted to the increasingly urban culture that has developed at the more successful landings, where a profusion of microbusinesses offer food, drink, clothing, entertainment, and sex to fishermen with cash in hand.

Ironically, given the increases in fish yields and value, childhood malnutrition rates remain comparable to national averages in Kenya, Uganda, and Tanzania. Household gender dynamics may explain this apparent paradox. Men control the fishing, and hence the cash, and they tend to spend money profligately outside the household, often on alcohol and affairs; women, who are active in onshore trading, consistently identified children as their first- or second-greatest expense. To bring income back into the household, many women market alcohol and sex to fishermen. Transactional sex and migrancy have both facilitated the spread of HIV/AIDS, which has had major (albeit poorly characterized)

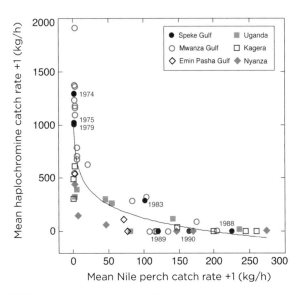

FIGURE 3 The Nile perch boom. Catch rates of haplochromine cichlids decreased precipitously as catch rates of Nile perch increased at six sites across Lake Victoria. The temporal dynamics (exemplified by the years of data points from Speke Gulf, shown) suggest that intensifying local exploitation of haplochromines may have facilitated the Nile perch boom (1980–1990), reinforcing haplochromine declines, followed by intensive exploitation and declines of Nile perch. Together, location and Nile perch catch rate explain 77 percent of the variance in haplochromine catch rate. (Data from Goudswaard, et al., 2007.)

impacts on the demographic and social structures of the region.

RECENT DEVELOPMENTS AND FUTURE OF THE FISHERY

In the early 1990s, Nile perch catches peaked and began to decline, as did the average size of individuals and the catch per unit effort. Because fishing effort continues to increase, the total annual catch hovers around 500,000 tons, but biologists question the sustainability of the industry. In 2004, Ugandan fish-processing facilities operated at less than 50 percent of capacity.

Nile perch declines enabled a partial recovery of native fishes, including some *Haplochromis* species feared extinct. By 2005, densities of zooplanktivorous haplochromines in Mwanza Gulf actually exceeded pre–Nile perch levels, but diversity had declined from 12 species to 8. Five (of an original 13) detritivorous species appeared regularly in 2005, but at a tenth of their previous density. Piscivorous and insectivorous species were hardest hit: only one of an original 30 was found in Mwanza Gulf from 2001–2005. Physical and physiological refugia—habitats too structurally complex or low in oxygen to support Nile perch—enabled some native species to withstand the onslaught and to repatriate subsequently.

And the haplochromines, models of explosive radiation, continue to evolve rapidly: many remaining species now display adaptations (such as increased gill surface area for enhanced oxygen uptake) that facilitate their coexistence with Nile perch.

The immediate future of the fisheries and remaining biodiversity of Lake Victoria hinges largely on fishing pressure. Recent models suggest that a 20 percent increase in fishing effort could drive Nile perch biomass down further—to extinction in some areas, stabilizing at low levels in others—enhancing the resurgence of the remnant native fauna. But eradicating the Nile perch would have complex and unpredictable consequences on biodiversity and human welfare, many of them undesirable. The seemingly optimal strategy is maintenance of fishing effort at levels high enough to allow the persistence of native fishes yet not high enough to extinguish Nile perch, while attempting to reverse eutrophication—a delicate balancing act. At present, management organizations and the lake's three governments are incapable of regulating fishing effort so precisely; Lake Victoria remains, for all practical purposes, an open-access resource.

SEE ALSO THE FOLLOWING ARTICLES

Eutrophication, Aquatic / Fishes / Invasion Economics / Lakes

FURTHER READING

Balirwa, J. S., C. A. Chapman, L. J. Chapman, I. G. Cowx, K. Geheb, L. Kaufman, R. H. Lowe-McConnell, O. Seehausen, J. H. Wanink, R. L. Welcomme, and F. Witte. 2003. Biodiversity and fishery sustainability in the Lake Victoria basin: An unexpected marriage? *BioScience* 53: 703–715.

Chapman, L. J., C. A. Chapman, L. Kaufman, F. Witte, and J. Balirwa. 2008. Biodiversity conservation in African inland waters: Lessons of the Lake Victoria region. *Verhandlungen Internationale Vereinigung Limnologie* 30: 16–34.

Geheb, K., S. Kalloch, M. Medard, A. Nyapendi, C. Lwenya, and M. Kyangwa. 2008. Nile perch and the hungry of Lake Victoria: Gender, status and food in an East African fishery. *Food Policy* 33: 85–98.

Goldschmidt, T. 1996. *Darwin's Dreampond: Drama in Lake Victoria.* Cambridge, MA: MIT Press.

Goudswaard, P. C., F. Witte, and E. F. B. Katunzi. 2007. The invasion of an introduced predator, Nile perch (*Lates niloticus*, L.) in Lake Victoria (East Africa): Chronology and causes. *Environmental Biology of Fishes* 81: 127–139.

Kaufman, L. 1992. Catastrophic change in species-rich freshwater ecosystems. *BioScience* 42: 846–858.

Pringle, R. M. 2005. The Nile perch in Lake Victoria: Local responses and adaptations. *Africa* 75: 510–538.

Pringle, R. M. 2005. The origins of the Nile perch in Lake Victoria. *BioScience* 55: 780–787.

Witte, F. M., J. H. Wanink, and M. Kishe-Machumu. 2007. Species distinction and the biodiversity crisis in Lake Victoria. *Transactions of the American Fisheries Society* 136: 1146–1159.

NITROGEN ENRICHMENT

JAE R. PASARI, PAUL C. SELMANTS, HILLARY YOUNG, JENNIFER O'LEARY, AND ERIKA S. ZAVALETA

University of California, Santa Cruz

Nitrogen (N) is abundant on Earth in the form of atmospheric dinitrogen (N_2), but its availability to organisms is strongly limited by processes that convert N_2 to biologically available (reactive) forms. Historically, fixation by N-fixing microbes and (to a lesser extent) lightning provided all reactive nitrogen. Through the production of synthetic nitrogen fertilizers, land management changes, and fossil fuel combustion, humans now add approximately 150 Tg/year more reactive nitrogen (N) to the surface of the Earth, more than doubling the nitrogen available to biotic organisms over the last 200 years. Nitrogen enrichment is particularly acute in most of the world's biodiversity hotspots, and it is expected to double again in the next 50 to 100 years. Nitrogen enrichment favors nitrophilic organisms (often exotic invaders) over many native species, resulting in biodiversity changes and losses.

NITROGEN ENRICHMENT IN TERRESTRIAL SYSTEMS

Nitrogen Enrichment Effects on Plant Communities

Most terrestrial ecosystems are N-limited, and 70 percent of global N enrichment occurs on land. The majority of this enrichment is a result of the production of synthetic nitrogen fertilizers, land management changes, and fossil fuel combustion. Fertilization of N-limited environments typically results in increased productivity, reduction in plant density and diversity, and increases in the size and abundance of nitrophilic species, particularly grasses. In addition to anthropogenic N enrichment, increases in abundance or extent of plants with N-fixing symbionts, high N litter, or susceptibility to periodic mass insect herbivory can also substantially enrich localized areas, advantaging nitrophilic invaders. These phenomena have increased the abundance of invasive grasses in coastal California grasslands after N enrichment from leguminous shrubs and have led to the dominance of grasses across many northern European heathlands.

Communities most at risk from N enrichment include those in historically nutrient-poor environments, those containing grasses with a high potential for competitive dominance after N enrichment, and those containing rare species that are likely to go locally extinct under competition with expanding nitrophilic species. The response of plant species to N enrichment varies according to plant traits, community traits, abiotic factors such as climate and precipitation, the form of N deposited (ammonia, nitrate, or gaseous NO_x), and the amount of N deposited, with high deposition leading to acidification in addition to fertilization (Fig. 1).

Effects of Nitrogen Enrichment on Invasions

Examples of N enrichment facilitating plant invasions are most pronounced in low-nutrient, forb- and shrub-dominated ecosystems invaded by nitrophilic grasses. These systems include mesotrophic fens, ombrotrophic bogs, calcareous grasslands, neutral-acid grasslands, montane subalpine grasslands, lowland dry heathlands, lowland wet heathlands, arctic and alpine heaths, serpentine grasslands, and some deserts. Loss of plant species diversity and the accompanying increase in grass dominance occur at deposition levels as low as 5 kg/ha/year—levels currently experienced in and around many urbanized regions worldwide. In addition, ecosystems naturally dominated by C4 grasses are at risk of invasion by nitrophilic C3 grasses under conditions of long-term, low-level N fertilization, as has occurred in Minnesota tallgrass prairies.

Other more nutrient-rich systems are also susceptible to invasion under N enrichment, including forest understories and cold deserts. In the cold deserts of the Colorado plateau, an N-fixing biological soil crust historically provided N to plants. Destruction of this crust in combination with aerial nitrogen deposition has altered the type and timing (although not the amount) of N deposition, resulting in increased success of the invasive thistle *Salsola iberica*.

Finally, there are numerous cases in which nitrogen enrichment, though expected to advantage certain exotics, has not been found to influence invader impact, including grass invaders in Dutch dune grasslands and California sage scrublands, and diffuse knapweed (*Centaurea diffusa*) in grasslands of western North America.

Mechanisms of Invasion

There are several N use and acquisition stages that determine a plant's growth and competitiveness under N enrichment. These stages include photosynthetic tissue allocation, photosynthetic nitrogen use efficiency, nitrogen fixation, nitrogen-leaching losses, gross nitrogen mineralization, and plant nitrogen residence time. In many cases, invasiveness in plants can be traced back to

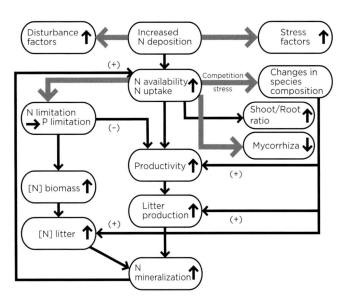

FIGURE 1 Schematic representation of the main effects of increased atmospheric N deposition on vegetation processes in terrestrial ecosystems. (From Aerts, R., and R. Bobbink. 1999. The impact of atmospheric nitrogen deposition on vegetation processes in terrestrial, non-forest ecosystems. In S. J. Langan, ed. *The Impact of Nitrogen Deposition on Natural and Semi-natural Ecosystems*. Kluwer Academic Publishers. With kind permission of Springer Science and Business Media.)

strategies at one of these stages relative to the strategies of native competitors. For example, exotic C3 plants are generally observed to outcompete native C4 grasses in N-enriched systems because of the lower photosynthetic N use efficiency of C3 plants, a phenomenon which is magnified as atmospheric carbon dioxide concentrations increase. Similarly, one mechanism explaining the success of alien Eurasian annual grasses in formerly perennial-dominated, nutrient-poor California grasslands is the difference between their photosynthetic tissue allocation strategies under N enrichment. While native California grasses and forbs tend to store added N in existing tissues, Eurasian annuals use added N to produce new shoot tissue, a strategy they evolved in their more fertile region of origin. Likewise, species that facilitate a high soil nitrification ratio (nitrification rate/ammonification rate × 100) tend to grow more under N enrichment. Invasive species in some regions also have higher nitrate reductase activity than natives, giving the former an advantage.

Even though these traits help explain and predict plant invasiveness under N enrichment, meta-analyses of N enrichment experiments suggest that complex interactions among invader traits, community traits, and abiotic environmental factors determine which plants will win and which will lose under N enrichment. While the ability of a plant to increase production following enrichment appears to be a good predictor of its invasibility, the impact of N enrichment and invasion on community richness is also a function of soil cation exchange capacity (CEC) and regional temperatures (with lower CEC and temperatures resulting in greater impact). Finally, N enrichment (particularly at high ammonia concentrations) can decrease frost, drought, and disease tolerance, and thereby reduce the competitive ability of natives compared to invaders that can tolerate high enrichment levels and environmental stress.

Cessation of Nitrogen Enrichment

The effects of N enrichment on plant communities can persist well beyond the period of enrichment. While species richness (but not abundances) returned to preenrichment levels after cessation of N enrichment in Minnesota tallgrass prairies, periodic cycles of invasion and depressed richness continued for decades after cessation of N enrichment in Colorado shortgrass steppe, suggesting that N enrichment can initiate time-lagged biotic inertias that preclude return to previous conditions.

NITROGEN ENRICHMENT IN COASTAL MARINE SYSTEMS

Coastal waters are being enriched with N and phosphorus (P) through both point-source and non-point-source pollution. However, because most coastal systems are N-limited, N generally has a higher impact than P in these systems. On a global scale, human activity has doubled the N flux from land to ocean, and most of this increase has occurred in the last several decades.

Nitrogen additions to coastal waters can facilitate invasions of alien marine algae (seaweeds) by increasing their competitiveness relative to native seaweeds. Siphonous, unicellular seaweeds are some of the most common invaders and include the genera *Codium*, *Caulerpa*, and *Bryopsis*. These seaweeds grow rapidly, quickly heal wounds, and can reproduce both sexually and asexually.

One third of all seaweed invasions have occurred in the Mediterranean Sea. For example, invasive seaweeds *Caulerpa taxifolia* and *C. racemosa* outcompete the endemic Mediterranean seagrasses *Posidonia oceanica* and *Cymodocea nodosa* and reduce overall species richness. *Caulerpa* spp. especially outcompete natives in areas of high nutrient input from urbanization. While native seagrasses (such as *P. oceanica* and *C. nodosa*) obtain much of their nutrient demand via roots, seaweeds with rhizomes (such as *Caulerpa* spp.) can utilize nutrients in both sediments and the water column. Furthermore, *Caulerpa* reduces the sediment nutrient supply to native seagrasses, leading to decreased growth and increased mortality.

Another member of the genus *Caulerpa* responds similarly to nutrient loading associated with urbanization in coastal waters along the eastern coast of the United States. In southeastern Florida's coral reef habitats, N enrichment contributes to an explosive invasion of *Caulerpa branchypus* var. *parvifola*, which outcompetes native algae because of lower N limitation.

These examples illustrate how anthropogenically derived N can enhance the success of invasive seaweeds, particularly in the genus *Caulerpa*. However, direct testing of the nutrient enhancement hypothesis has been limited, and experimental results are mixed. For example, nutrient additions do not stimulate growth rates for the red seaweeds *Acanthophora spicifera* or *Hypnea musciformis*, which were introduced from the Caribbean to Hawaii.

NITROGEN ENRICHMENT ON OCEANIC ISLANDS

Island ecosystems are disproportionately impacted by invasive species. There are many potential drivers for the apparently higher vulnerability of island ecosystems to invasions, and the importance of a particular driver varies by system. However, the relatively high availability of nutrients, including N, on some islands appears to be important among these invasion drivers. The high availability of N (as well as other resources) on many oceanic islands is in part a result of the limited plant communities present in these systems. The composition of indigenous plant communities on oceanic islands is usually limited by both space and propagule availability. These communities therefore often possess some combination of low species richness, few specialist species, and missing functional groups. As a result, island plant communities often use N incompletely in space and time, creating high net N availability compared to continental systems. Alien species, which often have stronger growth responses to nutrient enrichment than native island species, as well as higher plasticity to varying resource environments, can take better advantage of underutilized N. Invasive plants may also possess functional traits not represented in native island flora, allowing acquisition of unexploited N resources and sometimes triggering further N enrichment.

In Hawaii, the establishment of *Morella (Myrica) faya* provides an excellent example of a positive feedback loop between exotic invasion and N enrichment. *Morella (Myrica) faya* is an N-fixing tree introduced to Hawaii, where it colonizes recent lava flows, an environment where no native nitrogen fixing native plants can establish. *Morella (Myrica) faya* vastly increases N availability, which changes the course of ecosystem development to further facilitate other invasive organisms. Invasive grasses, including *Andropogon virginicus* and *Schizachyrium condensatum*, and invasive earthworms flourish in areas enriched by *Morella (Myrica) faya*. These invaders alter ecosystem processes, including fire regimes and nutrient cycling, further facilitating invasions.

Islands are also notable for the large quantity of their N that arrives as subsidies from the marine environment via marine wash, terrestrial animals foraging in the intertidal, and marine animals that forage at sea but rest or breed on land (primarily seabirds but also marine mammals and sea turtles). Changes in any of these subsidy patterns can have large impacts on the entire nutrient budget of the island and can thus trigger invasions. Nitrogen enrichment by ring-billed gulls and king penguins has facilitated plant invasions on islands in the Georgian Bay and Great Lakes region, and in the Crozet Archipelago, respectively.

MANAGEMENT, RESTORATION, AND POLICY

Prevention, management, and restoration of invaded and N-enriched systems can be attempted at several scales. At the local scale, managers can apply "ecological filters" to reduce competitive advantages conferred on nitrophilic invaders by N enrichment. The majority of research aimed at managing excess N through the application of ecological filters has been done in low-nutrient, forb- and shrub-dominated ecosystems of North America and western Europe. Amending soil with labile carbon such as sucrose or sawdust can increase microbial N immobilization, thus reducing N available to introduced nitrophilic plant species. This approach has been shown to successfully reduce plant-available N and exotic plant biomass in Minnesota wetland sedge meadows, Colorado shortgrass steppe, Manitoba and Minnesota tallgrass prairies, and California coastal grasslands. However, the benefits to native species are inconsistent, calculating the amount of carbon necessary to stimulate increased N immobilization is difficult, the effect is short-lived, and repeatedly adding labile carbon across large areas is expensive. In addition, enhanced N immobilization by soil microorganisms can reduce N losses, leading to a greater accumulation of N within the ecosystem. Repeated prescribed burns have been used with some success in California chaparral and tallgrass prairie ecosystems of the central United States to control invasive annual grasses over the short term, while volatilizing excess N contained in biomass at the same time. Prescribed burns are most effective in fireprone ecosystems; they can be counterproductive in the case of some invasives such as *Bromus tectorum* and *Taeniatherum*

caput-medusae, both of which spread more rapidly with frequent fires. Light to moderate grazing and mowing combined with biomass removal before seed set have also been used to control exotic invasion of N-enriched ecosystems in California serpentine grasslands, European calcareous grasslands, and semiarid grasslands in Utah and Colorado. Both prescribed fire and grazing result in a net export of N from the ecosystem, but grazing may be the more targeted approach because cattle often preferentially consume exotic, nitrophilic grasses over native forbs. If grazing is preferred and feasible, then careful monitoring of stocking levels is necessary because overstocking may exacerbate invasion. A relatively untested approach to restoring N-enriched ecosystems that are already heavily invaded involves the use of early seral "bridge species" with traits similar to existing or potential invaders. This approach may facilitate establishment of natives that can compete with exotic invaders and can be used in combination with one or more of the N-reducing management techniques described above.

At regional scales, mitigation and the establishment of critical loads are the primary policy tools used to address the spread of invasions under N enrichment. Mitigation requires that parties that cause N enrichment must purchase and manage sensitive lands to reduce the impacts of invasive species and preserve native communities. In addition, some regions in Europe with high levels of N enrichment have set critical loads that have been calculated to estimate the highest amount of N enrichment that an ecosystem can tolerate and still maintain its native plant communities.

SEE ALSO THE FOLLOWING ARTICLES

Eutrophication, Aquatic / Freshwater Plants and Seaweeds / Islands / Land Use / Mediterranean Sea: Invasions / Transformers

FURTHER READING

Adams, M. B. 2003. Ecological issues related to N deposition to natural ecosystems: Research needs. *Environment International* 29: 189–199.

Aerts, R., and R. Bobbink. 1999. The impact of atmospheric nitrogen deposition on vegetation processes in terrestrial, non-forest ecosystems (85–122). In S. J. Langan, ed. *The Impact of Nitrogen Deposition on Natural and Semi-natural Ecosystems.* Kluwer Academic Publishers.

Fenn, M. E., J. S. Baron, E. B. Allen, H. M. Rueth, K. R. Nydick, L. Geiser, W. D. Bowman, J. O. Sickman, T. Meixner, D. W. Johnson, and P. Neitlich. 2003. Ecological effects of nitrogen deposition in the western United States. *Bioscience* 53: 404–420.

Galloway, J. N., A. R. Townsend, J. W. Erisman, M. Bekunda, Z. Cai, J. R. Freney, L. A. Martinelli, S. P. Seitzinger, and M. A. Sutton. 2008. Transformation of the nitrogen cycle: Recent trends, questions, and potential solutions. *Science* 320: 889–892.

Gilliam, F. S. 2006. Response of the herbaceous layer of forest ecosystems to excess nitrogen deposition. *Journal of Ecology* 94: 1176–1191.

Howarth, R. W., A. Sharpley, and D. Walker. 2002. Sources of nutrient pollution to coastal waters in the United States: Implication for achieving coastal water quality goals. *Estuaries* 25(4b): 656–676.

Inderjit, D. Chapman, M. Ranelletti, and S. Kaushik. 2006. Invasive marine algae: An ecological perspective. *The Botanical Review* 72(2): 153–178.

Scherer-Lorenzen, M., H. O. Venterink, and H. Buschmann. 2007. Nitrogen enrichment and plant invasions: The importance of nitrogen-fixing plants and anthropogenic eutrophication. In W. Netwig, ed. *Biological Invasions*. Dordrecht, the Netherlands: Springer.

Vasquez, E., R. Sheley, and T. Svejcar. 2008. Creating invasion resistant soils via nitrogen management. *Invasive Plant Science and Management* 1: 304–314.

Williams, S. L., and J. E. Smith. 2007. A global review of the distribution, taxonomy, and impacts of introduced seaweeds. *Annual Review of Ecology, Evolution, and Systematics* 38: 327–359.

NOVEL WEAPONS HYPOTHESIS

RAGAN M. CALLAWAY

University of Montana, Missoula

The novel weapons hypothesis (NWH) is the idea that some exotic plant species may become invasive because they produce biologically active secondary metabolites that are not produced by species in the communities that are invaded, and that this novelty provides exotics with advantages against native competitors, consumers, or microbes that are not adapted to tolerate the chemical. Put another way, the NWH posits that some invasive exotic plants are spreading through and destroying native plant communities because they produce and release harmful chemicals that the naive native inhabitants have never experienced. So far, evidence for such novel weapons has focused on phytotoxic interactions among plants, novel defense chemicals, and the biochemical suppression of mutualistic fungi that are crucial for the growth of native species.

HISTORY

T. A. Rabotnov, an ecologist at Moscow State University, argued that plants could evolve in response to the chemicals exuded from the roots or washed from the leaves of their neighbors, stating that the "resistance of plant species to the vital secretions [i.e., secondary metabolites] of other components of biocenoses [communities] . . . has been created through the acquisition of properties [adaptation] by plants preventing the harmful action of

secretions of other organisms." Rabotnov stated that in natural conditions "allelopathically neutral" or "allelopathically homeostatic" biotic systems form in which allelopathic interactions are relatively weak because plants and microbes adapt to the chemicals produced by their neighbors (much like they rapidly adapt to herbicides and other chemicals). With this background, he suggested that "disturbed homeostasis" occurs when interactions take place among species without an evolutionary history—such as occurs with exotic invasions.

BIOGEOGRAPHIC EXPERIMENTS

Centaurea diffusa (diffuse knapweed) is native to Europe and Asia but invasive in western North America. In experiments with pairs of grass species in the same genera, *C. diffusa* suppressed the growth of North American species by 70 percent more than it suppressed the growth of Eurasian congeners. Activated carbon, which adsorbs to and deactivates some organic molecules, reduced the inhibitory effects of *C. diffusa* on the North American plants, but not the effects of the invader on the Eurasian plants. Similarly, experimental communities built from North American grass species were far more successfully invaded by *C. diffusa* than were communities built from Eurasian species. The root exudate 8-hydroxyquinoline may play a role in these biogeographic differences, as the chemical applied to plants in field soils suppressed the growth of North American species about 30 percent more than it did the growth of Eurasian species. These results raise the possibility that plant species from the native communities of *C. diffusa* have evolved tolerance to its root exudates, while North American plant species have not.

Phytotoxic, or allelopathic, effects have also been widely reported for a congener of *C. diffusa*, *C. maculosa* (*C. stoebe micranthos*, spotted knapweed), also native to Eurasia and invasive in North America. A number of different chemicals produced by the plant are biologically active, but a great deal of research has focused on the root exudate (±)-catechin, an isomeric phenolic compound exuded from the roots of *C. maculosa*. The ecological relevance of (±)-catechin phytotoxicity has been controversial, in part because of widely varying measurements of soil concentrations, inconsistency of detecting the chemical in root exudates, and variable results for phytotoxic effects among species and substrates, but the allelopathic effects of (±)-catechin, or the isolated (+) or (−) forms, have been demonstrated in vitro, in sand culture, in controlled experiments with field soils, and in the field at reasonably natural applied concentrations. Some of the variation in the effects of

(±)-catechin may be due to the effects of metal-catechin complexes, which can enhance or mitigate its effects on other plants. In competition experiments, *C. maculosa* has been shown to be a much better competitor against plant species native to North America than against species in the same genera but native to Europe and collected in communities with *C. maculosa*. In similar experiments with (±)-catechin and large numbers of North American and European species, the root exudate had stronger effects on native North American species than on native European species. In field experiments where (±)-catechin was applied to the rhizospheres of native species in Montana and Romania, the effects of the exudate were stronger in the invaded range of Montana. It has also been found that (±)-catechin has a more inhibitory effect on soil bacteria isolated from soils in the native range than on that in soils in the invaded range.

Alliaria petiolata (garlic mustard) is one of North America's most aggressive invaders of undisturbed forests in the Midwest and Northeast. Part of the impact of *A. petiolata* on native species appears to be through its effects on arbuscular mycorrhizal fungi and ectomycorrhizae, fungal mutualists of many North American native plants. Furthermore, *Alliaria petiolata* has far stronger inhibitory effects on arbuscular mycorrhiza fungi in soils from invaded regions of North America than on these fungi in soils from Europe, where *A. petiolata* is native. This biogeographically explicit antifungal effect appears to be caused by flavonoid chemicals produced by *A. petiolata*. The suppression of North American mycorrhizal fungi by *A. petiolata* corresponds with strong inhibition of many North American plant species that depend on the fungi, but European plants are weakly affected.

LITERATURE SURVEYS

A recent literature survey of the primary and secondary metabolites produced by plants divided the list into two categories of plant species that are exotic in North America: exotics that are highly invasive, and exotics that are widespread but not invasive. A comparison of these two lists of primary and secondary metabolites to lists of metabolites produced by plants native to North America found that invasive exotic plants were almost five times more likely than noninvasive exotics to produce novel biologically active secondary compounds (those never reported for plants native to North America). Many of these novel compounds possess allelopathic, herbivore defense, and antifungal properties.

IMPLICATIONS FOR COMMUNITY THEORY

Evidence for the NWH suggests a new, but not mutually exclusive, hypothesis for successful plant invasion. Also, the diffuse evolutionary relationships among organisms native to a region that are central to the NWH suggest that the organization of communities may be based, to a larger degree than previously thought, on regionally specific evolutionary trajectories with communities and interdependent evolutionary relationships among species within a trophic level. Furthermore, evidence consistent with the NWH suggests that disruption of evolutionary trajectories has the potential to destroy native communities.

SEE ALSO THE FOLLOWING ARTICLES

Allelopathy / Belowground Phenomena / Fungi / Mycorrhizae

FURTHER READING

Callaway, R. M., and E. T. Aschehoug. 2000. Invasive plants versus their new and old neighbors: A mechanism for exotic invasion. *Science* 290: 521–523.

Callaway, R. M., D. Cipollini, K. Barto, G. C. Thelen, S. G. Hallett, D. Prati, K. Stinson, and J. Klironomos. 2008. Novel weapons: Invasive plant suppresses fungal mutualists in America but not in its native Europe. *Ecology* 89: 1043–1055.

Callaway, R. M., and W. M. Ridenour. 2004. Novel weapons: A biochemically based hypothesis for invasive success and the evolution of increased competitive ability. *Frontiers in Ecology and the Environment* 2: 436–433.

Cappuccino, N., and J. T. Arnason. 2006. Novel chemistry of invasive exotic plants. *Biology Letters* 2: 189–193.

He, W., Y. Feng, W. M. Ridenour, G. C. Thelen, J. L. Pollock, A. Diaconu, and R. M. Callaway. 2009. Novel weapons and invasion: Biogeographic differences in the competitive effects of *Centaurea maculosa* and its root exudate (±)-catechin. *Oecologia* 159: 803–815.

Inderjit, J. L. Pollock, R. M. Callaway, and W. Holben. 2008. Phytotoxic effects of (±)-catechin in vitro, in soil, and in the field. *PLoS ONE* 3: e2536. doi:10.1371/journal.pone.0002536

Metlen, K., E. T. Aschehoug, and R. M. Callaway. 2009. Plant behavioral ecology: Dynamic plasticity in secondary metabolites. *Plant, Cell and Environment* 32: 641–653.

Tharayil, N., and D. J. Triebwasser. 2010. Elucidation of a diurnal pattern of catechin exudation by *Centaurea stoebe*. *Journal of Chemical Ecology* 36: 200–204.

Thorpe, A. S., G. C. Thelen, A. Diaconu, and R. M. Callaway. In press. Root exudate is allelopathic in invaded community but not in native community: Field evidence for the novel weapons hypothesis. *Journal of Ecology*.

Vivanco, J. M., H. P. Bais, F. R. Stermitz, G. C. Thelen, and R. M. Callaway. 2004. Biogeographical variation in community response to root allelochemistry: Novel weapons and exotic invasion. *Ecology Letters* 7: 285–292.

OSTRICULTURE

JENNIFER RUESINK

University of Washington, Seattle

Oyster culture (ostriculture) can be traced back thousands of years to practices, in both China and Europe, of collecting small wild oysters and moving them to more controlled areas to improve their growth, survival, and access. Complete control over their life cycle did not emerge until the 1970s, with the advent of successful hatchery production of oyster larvae. In 2005, more than 4.3 million tons of oysters (shucked weight) worth $3 billion were produced worldwide, of which 96 percent came from cultivated stocks (UN Food and Agriculture Organization). Production derives primarily from native oyster species (e.g., 80 percent from China alone). Nevertheless, oysters have played two major roles in marine species invasions (deliberate introduction and vector for other species), and by 1958, Charles Elton wrote that "[t]he greatest agency of all that spreads marine animals to new quarters of the world must be the business of oyster culture." In recent years, practices have changed to reduce opportunities for invasion associated with oyster culture.

OYSTER BIOLOGY

Morphology and Taxonomy

Edible oysters are bivalve molluscs (phylum Mollusca, class Bivalvia, order Ostreoida, family Ostreidae), in which a soft body is contained within two hinged shells (valves) of roughly equal but irregular size (Fig. 1). The left valve tends to be slightly cupped and attached to a hard surface, while the right valve is flatter. Inside the shell, major morphological parts include the mantle (outer tissue layer involved in secreting shell), gills (for gas exchange and filtration of small food particles via water motion from beating cilia), crystalline style within the digestive system (protein rod grinds food particles), one adductor muscle (which closes the shell), and gonad. Once shucked, oysters are eaten cooked or raw in their entirety.

Because oyster shells have notoriously variable shapes depending on where they are growing, taxonomic relationships have been markedly improved by recent genetic comparisons. Of the more than 50 living species now recognized within the family Ostreidae, only those in three genera are used in aquaculture (*Crassostrea*, *Ostrea*, and *Saccostrea*). In contrast to other members of Ostreidae, these live primarily in temperate areas and can form abundant wild populations. From these three

FIGURE 1 Two oyster species used in aquaculture. (A) *Crassostrea gigas* (Pacific oyster), native to Asia and established on all continents except Antarctica. (B) *Ostrea edulis* (European flat oyster), native to Europe and introduced widely but only reported to have established in one area (eastern United States) outside its native range. (Photograph courtesy of the author.)

TABLE 1

Introductions of Oyster Species

Species	Native Region	Introductions[a]	Successful Establishment[b]	Failed Establishment[b]
Crassostrea gigas Pacific cupped oyster	Asia	90 (62)	38 (20)	29 (20)
C. virginica American cupped oyster	Northwest Atlantic	18 (14)	3 (2)	10 (7)
C. angulata (*C. gigas*) Portuguese oyster	Asia	7	2	3
C. ariakensis	Asia	4 (3)	0 (0)	4 (3)
C. rhizophorae Mangrove cupped oyster	West Atlantic	3	0	2
C. sikamea Kumamoto oyster	Asia	3	0	2
C. iredalei Slipper cupped oyster	Asia	2	0	1
C. belcheri	Asia	1	0	1
C. corteziensis Cortez oyster	Tropical East Pacific (Mexico)	1	0	0
Saccostrea (*Crassostrea*) *echinata*	Asia	6 (5)	0 (0)	2 (1)
S. glomerata (*commercialis*) Sydney rock oyster	Australia	8 (6)	0 (0)[c]	5 (3)
S. cuccullata Hooded oyster	Asia/Africa	2	0	1
Ostrea (*Crassostrea*) *densalamellosa*	Asia	1	0	0
O. edulis European flat oyster	East Atlantic	12 (11)	1 (1)	5 (4)
O. puelchana	West Atlantic	1	0	0
O. lurida/conchaphila Olympia flat oyster	Northeast Pacific	1	0	0
O. (Tiostrea) chilensis (*lutaria*) Chilean flat oyster	South Pacific (Chile, New Zealand)	4 (3)	2 (1)	1 (1)
Total		164 (126)	46 (26)	66 (49)

NOTE: Includes variations on scientific name and common name, when available.

[a]Introduction data are reported in terms of all source–recipient combinations (and, if different numbers are provided within parentheses, this indicates that some introductions came from multiple sources to the same recipient location; the numbers in parentheses reflect recipient locations only).

[b]Known successes and failures in establishing naturalized, self-sustaining populations do not sum to the number of recipient locations because the fates of many introductions are not clearly reported.

[c]*Saccostrea glomerata* was introduced to Venice Lagoon in 1985 and was recently found on the coast of Turkey. This natural range expansion indicates successful establishment in two countries, although it has not been reported as such.

genera, 17 oyster species have been introduced outside of their native range (Table 1).

Reproduction

The life cycle of oysters is detailed in Figure 2. "Conditioning" is the development of gonad, which occurs when food is adequately available and temperatures are sufficiently warm. Oysters are sequential hermaphrodites, often maturing first as males, then alternating between producing eggs and sperm in successive reproductive bouts. "Spawning" is the release of gametes (eggs and sperm), often triggered by temperature extremes and occurring en masse in oyster populations. After fertilization, "larval development" involves several stages

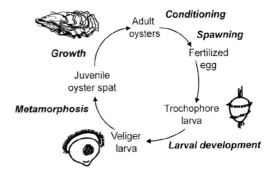

FIGURE 2 Life cycle of oysters.

(e.g., trochophore, veliger) of increasing cell number and size. Free-floating oyster larvae feed on phytoplankton to fuel their growth and development, and growth rates are positively related to water temperature. After two to four weeks, "metamorphosis" occurs at a body size of less than 0.5 mm, when oyster larvae select a hard surface on which to change into the adult body plan. This hard surface is often an adult oyster, enabling the construction of oyster reefs of cemented individuals in some species, but it may also be other shells, wood, or human constructions. Growth and maturation rates of oysters vary by species and by location, but many species involved in aquaculture are sexually mature in one year and reach market size in one to three years.

From the perspective of invasion biology, an important point is that a female oyster may produce a million eggs each year, which may travel as larvae for up to a month in the ocean before attaching to the bottom. In the subfamily Crassostreinae (cupped oysters), which includes *Crassostrea* and *Saccostrea*, fertilization is entirely external to the parents, and the larvae develop while swimming in the water. In the subfamily Ostreinae (flat oysters, including *Ostrea*), the eggs are held by female oysters within the mantle cavity, where fertilization occurs when water containing sperm from nearby males is pumped into the interior of the shell. The larvae are brooded by the female until they reach the veliger stage, when they are released to complete development in the water. The admonition to avoid eating oysters in months without an "R" is because gonad development occurs in these warm months and makes the oysters soft; additionally, brooding oysters have a "grainy" texture because the shells are present on the larvae.

METHODS OF CULTURE

For much of the history of oyster culture, production relied on the availability of young oysters (spat) through natural reproduction. These spat were collected on hard substrates such as oyster shell (cultch) and sticks, which could then be moved among growing areas or even to distant countries. For instance, from 1924 to 1977, over 1.6 million cases (36 × 18 × 12 inches) of cultch were shipped from Japan to the west coast of North America to support the culture of nonnative Pacific oysters. By the late 1970s, shellfish hatcheries had developed to enable the controlled reproduction of oysters, as well as the production of the massive amounts of phytoplankton necessary for growing larvae. Oyster growers may now purchase settling-competent larvae, which are packed into a cooled ball of several million individuals, then released

into a tank full of settling substrate and heated seawater to metamorphose, or they may purchase spat that have already spent several months growing as juveniles. Both products can be shipped around the world as overnight airfreight. There are few constraints to importing larvae or spat to countries where a particular oyster species is already cultivated, although some inspections and disease-free certification are generally required.

In addition to the historical practice of growing oysters on cultch, which results in clumps of adult oysters, a new technique of cultchless oysters results in single individuals suitable for the high-margin half-shell market. Cultchless oysters are produced by exposing larvae to small shell fragments or sand, on which they settle and metamorphose. Oysters are grown to market size in a wide variety of ways, including directly on tideflats, in clumps attached to lines, and in hanging baskets or suspended mesh trays. The lines, baskets, and trays may be suspended from floats in deep water or held on frames or stakes built on tideflats.

Triploids

Oysters, like most animals, are naturally diploid with two copies of each chromosome. Nevertheless, it is possible to create individuals artificially with three copies of each chromosome; these individuals are termed "triploid." They are created either by adding a chemical (e.g., cytochalasin B) during the first cell division that results (80% of cases) in an extra copy of maternal chromosomes remaining in offspring, or by mating diploid females with patented tetraploid (four-copy) males. Because each parent contributes half of its chromosomes to an offspring, essentially 100 percent contain three chromosome copies. The rationale for triploid oysters is that they do not develop gonads and instead devote this energy to growth. This strategy has the twin benefits of improving product quality during summer months and reducing time to market size. Triploids are also being considered for introduction to new areas where establishment of wild self-sustaining populations is not desired. Reversion to diploidy occurs rarely.

INTRODUCTIONS AND INVASIONS OF OYSTERS

Introduction

The earliest oyster introduction happened in the seventeenth century, when *Crassostrea angulata* was transported by ship from Asia to Europe with explorers and traders. The common name of *C. angulata* is Portuguese oyster, which reflects the fact that, over time, people forgot its nonnative origin and began to think of the species as a natural part of the coastal environment. Recent genetic data indicates that *C. angulata* may not actually be a separate species at all, but instead simply the earliest instance of the most widely introduced oyster, *Crassostrea gigas*. Over hundreds of years, 17 species of oysters (or 16, if *C. angulata* is synonymized with *C. gigas*) have been deliberately introduced outside of their native ranges. Because such introductions are generally tracked across national boundaries, most of the summary data reflect the movement of oysters from one country to another. However, some records reflect better geographic specificity, for instance, distinguishing different coasts of Australia. In total, there have been at least 164 separate introductions in which a nonnative oyster species has arrived in a new place, covering 73 recipient locations. The number of introductions is higher than the number of recipient locations because many locations have received multiple oyster species or have received the same species from several sources (but multiple introductions from a single source country are counted only once).

Oyster transfers among locations occurred most rapidly in the 1970s (Fig. 3). From 1970 to 1980, at least 72 introductions occurred, of which 44 percent were to South Pacific islands, with another 14 percent going to other tropical locations. This pattern reflects an international initiative to develop aquaculture and improve food security in impoverished nations. The other main reason for oyster introduction has been commercial failure of production by a native oyster, decimated by overfishing or disease. On the west coast of North America, the native oyster *Ostrea lurida* (= *O. conchaphila*) was commercially extinct in most of its range by 1900, and first eastern (*C. virginica*) and then Pacific oysters (*C. gigas*) were

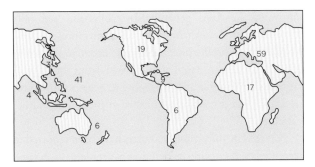

FIGURE 4 Number of oyster introductions to locations (generally countries) within regions (generally continents).

introduced as faster-growing replacements for aquaculture. In France, an unsustainable fishery was successfully replaced by aquaculture of the native European flat oyster (*Ostrea edulis*), but production was severely reduced by a disease (caused by *Bonamia ostreae*) beginning around 1980. High mortality in *O. edulis* shifted aquaculture to focus on *C. gigas*, which had been introduced to Europe in the mid-1960s but was considered an inferior oyster by European consumers until the native species became largely unavailable. Scientific research is responsible for 12 transfers, and larval dispersal across the boundary of a neighboring country for four; in three cases, a new oyster is known to have been transferred because it was mixed in with a shipment of a different oyster species.

Asia has been the source of roughly half of the introduced oyster species, a logical outcome of a long Asian history of aquaculture and presence of up to 30 native oyster species in China alone. Interestingly, the Atlantic and Caribbean coasts of South America, which also have high oyster diversity (ten species), have contributed just two of the introduced species, and Pakistan (with nine native oysters) none. The geographic scope of introductions reflects the main purposes for which oysters have been introduced (Fig. 4). To replace commercial ventures when native species failed, oysters of several species and from multiple source populations were introduced to European countries and to the west and east coasts of North America, each of which has just a single native oyster species. To enhance economic opportunities and food security, many introductions occurred to Pacific islands, which are outside the native range of any commercial oyster species.

Despite the large total number of introduced oyster species, one-half of all transfers involve a single species, *Crassostrea gigas* (Pacific oyster). *Crassostrea gigas* is native to Japan, Korea, and China, where it tends to occur intertidally on rocks or hard artificial surfaces. It is commonly

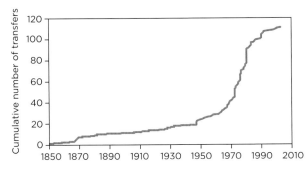

FIGURE 3 Time series of oyster introductions, showing the first reported occasion when a species was transported from one country to another. Introductions without information on the year of first transport are not included. Transfers within or back to the native range are not included.

cultivated in Asia, generally by collecting naturally recruited spat. *Crassostrea gigas* now contributes to commercial oyster production worldwide, occurring on every continent except Antarctica. It is grown from the tropical latitudes of Namibia and Singapore to near-polar regions in Alaska, but it has not established at these temperature extremes.

Few new introductions of oysters have been reported in recent years. A growing recognition of problems associated with nonnative species, along with new introduction protocols by the International Council for Exploration of the Seas (ICES), have prompted more explicit consideration of potential costs and benefits of oyster introduction. For example, oyster introductions to replace the production and ecosystem function of Eastern oysters (*Crassostrea virginica*), decimated by overexploitation and disease, were widely discussed in the scientific literature by the 1990s. In 2009, after millions of dollars had been invested in research, the U.S. Environmental Protection Agency determined in a final environmental impact statement to prohibit efforts to establish *C. ariakensis* in the Chesapeake Bay. In this case, *C. ariakensis* was tabulated as "introduced" to the eastern United States because of its use in laboratories and small-scale field trials, but no established population has resulted to date.

Establishment

The combined total of all oyster species introduced to different locations is 126. The number of recorded transfers is higher (164) because of imports of the same species from multiple locations. Out of 126 introductions, nonnative oyster populations have established self-sustaining populations in 21 percent and failed to establish in 39 percent, with the remainder (40%) being unreported. *Crassostrea gigas* has particularly high rates of establishment, with 32 percent establishment across all locations where it has been introduced and 74 percent establishment when tropical locations are excluded. Naturalized populations of *C. gigas* occur in western North America, eastern South America, European countries from Scandinavia to the Black Sea, southern Africa, Australia, and New Zealand. While it may be that *C. gigas* is simply the "best" invader of the bunch, it seems more likely that the pattern is driven by propagule pressure, because the availability of larvae and spat internationally has been much higher for *C. gigas* than for any other oyster.

Crassostrea gigas provides some remarkable examples of time lags between introduction and apparent establishment. First cultured in South Africa in 1950, it appeared in natural habitats 50 years later. In Europe, *C. gigas* was introduced to France and the Netherlands in the mid-1960s. Spatfall in the Oosterschelde (Netherlands) first occurred during warm summers of 1975 and 1976, and then again in 1982 and 1989 (even though imports had stopped in 1977). In Germany, Pacific oysters were first imported in 1971 but did not recruit widely until 1991 and 1994. In the early 2000s, *C. gigas* suddenly appeared in vast numbers on tideflats along the Wadden Sea coast. From these examples, *C. gigas* might be expected soon to establish in Chile, where the oyster was first introduced in 1983, and where local hatcheries now generate significant output of spat for both domestic and international markets. The mechanisms underlying these time lags are not entirely known, but chance events are a plausible explanation, given that small improvements in larval survival dramatically increase the number of offspring recruiting into a population. An additional likely candidate involves climate change and variability. Because larval development in oysters is temperature sensitive, unusually warm years at temperate latitudes may enable larger spawning events, faster larval development, and reduced larval mortality. Indeed, years of particularly large spatfall in Europe coincide with warm years.

Impacts

Oysters are ecosystem engineers that alter habitat, the physical environment, and the flow of energy and materials. They create complex hard surfaces with their shells, and *C. gigas*, in particular, forms novel reef structures in soft-sediment environments in Europe and western North America. Through filtration, oysters remove particles from the water and deposit materials in the sediment via feces and pseudofeces. These engineering capabilities have been studied extensively in native *C. virginica*, and to a lesser extent in introduced *C. gigas* grown in aquaculture, but the ecological effects of feral nonnative oysters are currently poorly studied. On the west coast of North America, *C. gigas* competes for space with seagrasses and serves as an intertidal recruitment substrate that may actually interfere with native oyster recovery. In eastern Australia, it outcompetes the native *Saccostrea glomerata* on rocks in the mid- (but not upper) intertidal zone. In the Wadden Sea, many mussel beds have been transformed into oysters, reflecting differential recruitment. The Nature Conservancy considers *C. gigas* as a category 3 species (of 4) in ecological impact, disrupting "multiple species, some wider ecosystem function, and/or keystone species or species of high conservation value."

OYSTERS AS VECTORS

The role of oysters as habitat has contributed to vectoring many other species during aquaculture transfers. Organisms that live on or among oyster shells have been packaged with oysters shipped worldwide. These associated species tend to be intertidal and adapted to periods of exposure to air, which improves their chance to survive in shaded and moist conditions during transport. This problem was historically acute. In the northeastern Pacific, the arrival of nonnative species due to oyster culture was most rapid during the period of spat imports from Japan. Between 1920 and 1970, about 16 species were transferred with oysters, but just five more arrived in the following 30 years. Across the globe, locations with more oyster introductions show a higher fraction of introduced marine species vectored by oysters (Fig. 5). In two regions (western North America, North Sea), oyster introductions, as a sole pathway or as one of several pathways, have contributed 40–50 percent of all nonnative marine species. Across eight regions where sufficient data are available to determine invasion owing to oyster culture, 3.5 nonnative species have established for each oyster species introduced (6.5 per oyster species if nonnatives vectored by multiple pathways are included).

At least 18 serious diseases affect adult oysters of cultured species, caused by agents as different as viruses, bacteria, fungi, and protists. Of these, two of the most problematic are apparently nonnative and are introduced with oysters. *Bonamia ostreae*, a haplosporidian parasite, causes 70–80 percent mortality of mature European flat oysters (*Ostrea edulis*). It infected *O. edulis* that were introduced to the United States and then traveled with individuals that were moved back to the native range in the late 1970s. The disease has since been the primary cause of failed *O. edulis* aquaculture in Europe. *Haplosporidium nelsoni*, which causes MSX disease, particularly in Eastern oysters (*C. virginica*), has sublethal effects in other oyster species. This trait apparently allowed it to be transferred with oysters from Asia to the west coast of the United States by 1900 and to the east coast of the United States by 1957. MSX persists today in the Chesapeake Bay, even though Asian oyster imports occurred in such low numbers that the original host itself never established. For an additional three diseases, transfers of oysters within their native ranges have contributed to outbreaks of disease in naive populations of the same oyster species.

Oysters today play a reduced role in the transport of nonnative species among marine ecosystems. They are now rarely moved long distances as clusters on cultch, but instead are generally transported as larvae or single (cultchless) spat. Additionally, hatcheries exist in most of the major locations of nonnative oyster culture, which reduces the need for transfer from distant locations. Nevertheless, spat are a global commodity, and shortfalls in hatchery production in one location may be remedied through imports from another. Given the devastating effects of oyster diseases for aquaculture, and the fact that disease agents are cryptic hitchhikers, disease transfer represents an ongoing risk of globalized oyster aquaculture. In addition, the speed of airfreight spat transport may continue to allow hitchhiking by invertebrates, protists, and algae.

SEE ALSO THE FOLLOWING ARTICLES

Aquaculture / Lag Times / Pesticides (Fish and Mollusc) / Transformers

FURTHER READING

Elton, C. S. 1958. *The Ecology of Invasions by Animals and Plants.* Chicago: University of Chicago Press.

Jacobsen, R. 2007. *A Geography of Oysters: The Connoisseur's Guide to Oyster Eating in North America.* New York: Bloomsbury.

McKindsey, C. W., T. Landry, F. X. O'Beirn, and I. M. Davies. 2007. Bivalve aquaculture and exotic species: A review of ecological considerations and management issues. *Journal of Shellfish Research* 26: 281–294.

Molnar, J. L., R. L. Gamboa, C. Revenga, and M. D. Spalding. 2008. Assessing the global threat of invasive species to marine biodiversity. *Frontiers in Ecology and the Environment* 6: 485–492.

National Research Council. 2004. *Nonnative Oysters in the Chesapeake Bay.* Washington, DC: National Academies Press.

Ruesink, J. L., H. S. Lenihan, A. C. Trimble, K. W. Heiman, F. Micheli, J. E. Byers, and M. Kay. 2005. Introduction of non-native oysters: Ecosystem effects and restoration implications. *Annual Review of Ecology, Evolution and Systematics* 36: 643–689.

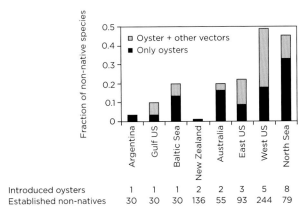

	Argentina	Gulf US	Baltic Sea	New Zealand	Australia	East US	West US	North Sea
Introduced oysters	1	1	1	2	2	3	5	8
Established non-natives	30	30	30	136	55	93	244	79

FIGURE 5 Established nonnative marine species brought in with transfers of oysters for aquaculture. Data are divided according to whether oyster aquaculture is the only likely pathway or whether it could have contributed along with other vectors. Established oyster species are included in the results. (Reprinted, with permission, from the *Annual Review of Ecology, Evolution, and Systematics* 36. ©2005 by Annual Reviews www.annualreviews.org.)

P

PARASITES, OF ANIMALS

BERND SURES

University of Duisburg-Essen, Germany

Although awareness about species introduction is increasing during recent years, most research efforts concentrate on free-living organisms. However, parasites may also be transported into new areas where they can establish. Analysis of the published literature reveals that less attention is paid to non-endemic organisms exhibiting a parasitic life style than to free-living organisms.

PARASITES AND INVASIONS

Apart from their role as nonindigenous species, parasites can also be an important factor determining the success of the invasion of free-living organisms into new regions. The "enemy release hypothesis" suggests that reduced control by enemies such as pathogens, parasites, and predators constitutes a fitness advantage for nonindigenous species. Additionally, endemic parasites might profit from the introduction of new free-living host species.

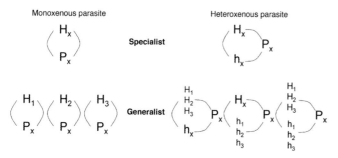

FIGURE 1 Life cycles of parasites. H = final host; h = intermediate host.

Examples from the literature suggest that the invasion of new susceptible host species allows indigenous parasites to colonize new habitats within their home range, provided the habitat requirements of the introduced species differ from those of the endemic host(s).

INTRODUCTIONS OF PARASITES

Parasites comprise species of various taxa. It has been suggested that at least 40 percent of all known species exhibit a parasitic lifestyle, at least during certain phases of their life. Compared to free-living nonindigenous species (NIS), less information is available for nonindigenous parasites (NIP). In the literature, rather few examples are available that show that parasites can successfully colonize new regions. It can be assumed that the colonization process of parasites is more complicated and tenuous than that of NIS, because parasites are usually unable to colonize new areas without their hosts, and their survival depends on a variety of prerequisites. As for free-living organisms that mainly rely on the availability of appropriate ecological conditions (e.g., a free niche) in the new area, parasites need at least to meet susceptible hosts. Depending on their life cycle (Fig. 1), parasites are either monoxenous (single-host life cycle) or heteroxenous (multiple-host life cycle). In either case, host specificity adds another variable to the invasion process of a parasite.

While specialists are specific parasites of one host species (final or intermediate host[s]) only, generalists can infect a range of host species. Accordingly, parasites appear to be less effective invaders, as their successful establishment relies not only on the presence of appropriate intermediate and final hosts but also on the availability of free niches for their hosts. In several cases, the introduction of free-living organisms offers excellent opportunities for their parasites to spread into new areas. Following their successful arrival and establishment, NIP can remain with their introduced host or they can infect indigenous species of the new territory as hosts.

NONINDIGENOUS PARASITES REMAINING WITH THEIR INTRODUCED HOSTS

A well-known example of an NIP that remains with its introduced host is the raccoon roundworm *Baylisascaris procyonis*. An endemic species in North America, the raccoon (*Procyon lotor*) was introduced by humans to areas outside the United States. Its introduction to Germany, for example, is well documented. In 1934, two pairs of pet raccoons were released close to the Edersee in the north of Hesse; additionally, 25 raccoons escaped in 1945 from a fur farm in Brandenburg close to Berlin. The estimated number of raccoons was 285 animals in the Hessian region in 1956, over 20,000 animals in the Hessian region in 1970, and between 200,000 and 400,000 animals in all Germany in 2008. Together with the raccoons in Hesse, their intestinal nematode *Baylisascaris procyonis* was also released to the field, while this roundworm was not detected in the Brandenburgian population. The two populations can still be distinguished parasitologically: 70 percent of the raccoons of the Hessian population are infected with the nematode, but none of the Brandenburgian population is. As is known for other ascarid species, the life cycle of *B. procyonis* needs only one host and is thus very easily completed. Eggs are excreted with the feces and must develop externally, typically in the soil, to become infectious again for raccoons. After ingestion of eggs containing infective larvae by a raccoon, the larvae hatch, enter the wall of the small intestine, and subsequently develop to adult worms in the small intestine. Accidental ingestion of eggs by other host animals such as small mammals leads to extraintestinal migration of larvae (larva migrans), with an estimated 5–7 percent of larvae invading the brain. These accidental hosts serve as so-called paratenic hosts, as no further development of the parasite occurs. However, raccoons may become infected when they eat larvae that have become encapsulated in the tissues of small mammals such as rodents. More than 90 species of wild and domesticated animals have been found to be infected with *B. procyonis* larvae. Not only small mammals but also humans can be confronted with a larva migrans after ingestion of infective eggs. Human infections with *B. procyonis* are typically fatal. Because eggs are transmitted by the fecal-oral route, young children are especially prone to infection with eggs of this nematode due to their intimate contact with soil (soil-eating behavior). However, infection of humans with eggs

of *B. procyonis* seems to be a rare event; only 11 recognized human cases, four of them fatal, have been reported in the United States. A much worse situation would have arisen if *B. procyonis* had infected other carnivores such as dogs, cats, or foxes as appropriate final hosts. In this case, contamination of the environment with infective eggs as well as cases of human infection would have increased dramatically.

NONINDIGENOUS PARASITES SPREADING TO ENDEMIC HOSTS OF THE NEW AREA

Instead of remaining with its introduced host, a parasite can also spread to other (new) host species of the new habitat, with detrimental effects on the respective host populations. Such negative effects of introduced parasites result from a missing host–parasite coevolution, during which the host adapts its immune response and the parasite counteracts immune effector mechanisms. The outcome of such an evolutionary "arms race" is a balance that allows the parasite and host populations to survive. In the case of introduced parasites that spread to new hosts, this balance between host and parasite species does not exist. A well-known example is the accidental introduction of the hematophagous eel swim bladder nematodes of the genus *Anguillicoloides* (originally described as *Anguillicola*) into populations of European eel (*Anguilla anguilla*) and American eel (*Anguilla rostrata*, Fig. 2).

Two different species of swim bladder nematodes were introduced during the last 33 years to Europe, from which at least one species can now be considered as successfully established. The best-studied and most

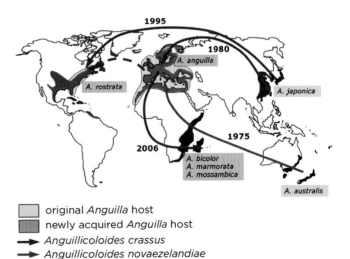

FIGURE 2 Worldwide translocation of eel species infected with *Anguillicoloides* species to new areas and subsequent establishment of the worms. (Figure courtesy of Kerstin Geiß.)

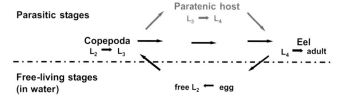

Parasitic stages

Paratenic host
$L_3 \rightarrow L_4$

Copepoda
$L_2 \rightarrow L_3$

Eel
$L_4 \rightarrow$ adult

**Free-living stages
(in water)**

free $L_2 \leftarrow$ egg

FIGURE 3 Life cycle of species of the genus *Anguillicoloides*, with L1–L4: larval stages. Black = obligate hosts; gray = hosts potentially used for transmission to final host.

abundant species is *Anguillicoloides crassus*, introduced to Europe in the early 1980s in imported Japanese eels. This nematode had coevolved with its endemic final host, the Japanese eel, *Anguilla japonica*, in the Far East, resulting in a balanced host–parasite interaction. After its arrival in Europe, *A. crassus* infected the European eel, *Anguilla anguilla*, and spread rapidly throughout the continent. Soon after its first report from Europe, *A. crassus* was described in European eels in North Africa and was also able to infect the American eel (*Anguilla rostrata*) in North America. At the moment, prevalences and mean intensities of *A. crassus* in wild European eels are much higher compared to wild Japanese eels in the Far East. Furthermore, pathological alterations of the swim bladder tissue owing to infection with *A. crassus* are much more severe in European eels than in Japanese eels. It is thus assumed that this parasite is a threat to the European eel population and might have contributed to the recently observed decline of European eel recruitment by 99 percent since 1980. Apart from higher infection intensities in European eels, field studies suggest that "European" *A. crassus* in their newly acquired host grow larger than in their original host and that morphological features differ clearly between European and Asian populations of *A. crassus*. Laboratory infection experiments revealed that fewer nematodes were able to establish in *A. japonica* than in *A. anguilla*. Obviously, the natural host of *A. crassus* is able to mount efficient protective immune responses against its parasite, while the newly acquired host seems to lack this ability.

In addition to *A. crassus*, another species of *Anguillicoloides* has been introduced into Europe. With the importation of *Anguilla australis* from New Zealand in 1975, its endemic parasite *Anguillicoloides novaezelandiae* was introduced into Lake Bracciano in Italy, where it infected European eels throughout the lake population with a prevalence of up to 80 percent. However, it appears that *A. novaezelandiae* has been unable to escape from Lake Bracciano and, in contrast to *A. crassus*, has thus failed to disperse over the European continent. Moreover,

data on eel parasites in Lake Bracciano suggest a decrease in infection parameters of *A. novaezelandiae* after *A. crassus* was introduced into the same lake where both NIP occur in sympatry. This is a striking example of how virulence between closely related parasite species of the same genus, which both use the same set of hosts, may differ.

In contrast to *Baylisascaris procyonis*, the eel swim bladder nematodes have an obligatory two-host life cycle (Fig. 3), which—on first view—would make a successful invasion less likely.

Although both *Anguillicoloides* species are able to infect other species of the genus *Anguilla* as adequate final hosts, they cannot switch to other host genera of anguilliformes. The host specificity concerning intermediate and paratenic hosts is unclear for the endemic regions of the nematodes, but it is expected that the intermediate host range is much narrower than in the introduced region. In Europe a broad spectrum of invertebrate species can be used as first intermediate hosts, and many different taxa may serve as paratenic hosts of *A. crassus*.

A number of other examples are known in which NIP were able to spread to new hosts of the introduced areas (as described in the resources listed under Further Reading).

THE ROLE OF PARASITES DURING INVASION PROCESSES OF INTRODUCED SPECIES

The "enemy release hypothesis" suggests that among other enemies the loss of parasites during an invasion process gives NIS a fitness advantage. Organisms that are not infected by parasites do not have to invest in a costly immune response but can use their energy reserves for growth or can allocate them to other life history traits. This reallocation often results in a more efficient reproduction and thus a bigger population size, which is an important advantage in competition with other species such as endemic ones. The reasons for the lower percentage of parasitized NIS are manifold. Introduced propagules are a small subset of the native population, which reduces the probability of introducing parasites along with a host species. Complex life cycles might further limit the invasion success of introduced parasites. If more than one host is involved in parasite development (heteroxenous parasites), and not all are present in the new area, then it is less likely that an NIP will be able to establish. Furthermore, populations of NIS may result from uninfected life history stages, such as larvae, that were transferred to the new area. Apart from the possible loss of parasites during the invasion process, a lower susceptibility to endemic parasites of the colonized area also gives a fitness advantage compared to endemic species, which are confronted with endemic parasites. As mentioned above, some parasites might exhibit strong host specificity and thus infect only one host species.

Irrespective of the reasons for a possible reduced parasitic infection, Torchin et al. (2003) compared parasites from 26 NIS comprising molluscs, crustaceans, fishes, birds, mammals, amphibians, and reptiles between their native and introduced ranges. On average, only 3 of 16 native parasites successfully accompanied an invader to its introduced range. In addition, an average of four parasites from the new area was able to infect the introduced host. As a consequence, introduced populations had roughly half the number of parasite species compared to native host populations. Not only was the diversity of parasites of NIS reduced in the colonized areas, but also the prevalence (percentage of infected hosts) was considerably lower.

The European green crab, *Carcinus maenas*, is a well-known successful worldwide invader and is listed among the 100 world's worst invasive species (see Appendix). It is a common littoral crab, native to European and North African coasts, and it has been introduced to the United States, Australia, and South Africa. As a food generalist, it has caused the decline of other crab and bivalve species in its introduced range. Studies carried out on European populations of green crabs indicate that body size and biomass are negatively associated with the prevalence of parasitic castrators such as the rhizocephalan barnacle, *Sacculina carcini*. A comparison of introduced and native populations of *C. maenas* gave evidence that the introduced populations, which are bigger, suffer less from parasites than do the native populations. This even leads to the idea of using endemic parasites of the home range, for example the parasitic barnacle *S. carcini*, as a possible biological control agent for green crabs.

ENDEMIC PARASITES AND NIS

Although several examples show a reduced parasite infection of NIS, endemic parasite species of the colonized area may also spread to NIS and accept them as additional susceptible hosts. Infection of the raccoon dog, *Nyctereutes procyonoides*, with adults of the fox tapeworm, *Echinococcus multilocularis*, is such an example. The raccoon dog is indigenous to East Asia but has been introduced to the eastern part of Europe and has started to spread to Western Europe. It is now abundant throughout Finland, Estonia, Latvia, and Lithuania, and has been also reported in Germany, France, Romania, Italy, and Switzerland. It turned out that the raccoon dog is a suitable final host

for *E. multilocularis*, which matures and reproduces in it. The fox tapeworm causes a serious disease (alveolar echinococcosis) in humans if eggs of *E. multilocularis* are ingested. After oral infection with eggs, the larval stages (metacestodes) develop primarily, almost exclusively, in the liver with lesions ranging from small foci of a few millimeters in size to large areas of infiltration (15–20 cm in diameter). Alveolar echinococcosis is characterized by a tumor-like, infiltrative, and destructive growth. From the liver, the metacestode tends to spread to both adjacent and distant organs by infiltration or metastasis formation. After an initial asymptomatic incubation period of 5 to 15 years, a chronic course follows with high fatality rates in untreated or inadequately treated patients. The main final host for *E. multilocularis* in Europe is the red fox, *Vulpes vulpes*. Although the diet of the red fox and raccoon dog overlap to some extent, as both are carnivores and feed on small mammals, invertebrates, and plants, the preferred habitats differ to some degree. In contrast to red foxes, raccoon dogs are often found close to watercourses. It is suggested that water systems are used as migratory corridors by raccoon dogs. Interestingly, other introduced species that show a high affinity for surface waters, such as the muskrat, *Ondatra zibethicus*, and the coypu, *Myocaster coypus*, can be used as new intermediate hosts, in which the tumor-like metacestodes can develop. By using these NIS, the fox tapeworm can now successfully complete its life cycle close to water bodies, which thus creates a new habitat for *E. multilocularis*. Accordingly, changes of the parasite transmission cycle might also be associated with altered infection risks for humans.

SEE ALSO THE FOLLOWING ARTICLES

Biological Control, of Animals / Crabs / Enemy Release Hypothesis / Habitat Compatibility / Life History Strategies / Mosquitoes / Pathogens, Animal

FURTHER READING

Anderson, R. M., and R. M. May. 1986. The invasion, persistence and spread of infectious diseases within animal and plant communities. *Philosophical Transactions of the Royal Society B* 314: 533–570.

Colautti, R. I., A. Ricciardi, I. A. Grigorovich, and H. J. MacIsaac. 2004. Is invasion success explained by the enemy release hypothesis? *Ecology Letters* 7: 721–733.

Dunn, A. M. 2009. Parasites and biological invasions. *Advances in Parasitology* 68: 161–184.

Keane, R. M., and M. J. Crawley. 2002. Exotic plant invasions and the enemy release hypothesis. *Trends Ecology and Evolution* 17: 164–170.

Lowe, S., M. Browne, S. Boudjelas, and M. De Poorter. 2000. 100 *of the World's Worst Invasive Alien Species: A selection from the Global Invasive Species Database.* Auckland: New Zealand Invasive Species Specialist Group.

Prenter, J., C. MacNeil, J. T. A. Dick, and A. M. Dunn. 2004. Roles of parasites in animal invasions. *Trends Ecology and Evolution* 19: 385–390.

Taraschewski, H. 2006. Hosts and parasites as aliens. *Journal of Helminthology* 80: 99–128.

Torchin, M. E., K. D. Lafferty, A. P. Dobson, V. J. McKenzie, and A. M. Kuris. 2003. Introduced species and their missing parasites. *Nature* 421: 628–630.

Torchin, M. E., K. D. Lafferty, and A. M. Kuris. 2002. Parasites and marine invasions. *Parasitology* 124: S137–S151.

PARASITES, OF PLANTS

SEE PARASITIC PLANTS; PATHOGENS, PLANT

PARASITIC PLANTS

HENNING S. HEIDE-JØRGENSEN

University of Copenhagen, Denmark

Parasitic flowering plants exploit other flowering plants for water and nutrients with the help of one or more haustoria. Part of the haustorium, the intrusive organ, penetrates host tissue to establish contact with the conductive tissue of the host. Introduced parasitic plants occur throughout the world, and some are considered invasive. They often cause considerable economic damage when attacking monocultures in agriculture, orchards, and forestry, and much effort is spent to avoid and control invasive, harmful parasitic plants.

HISTORY

Parasitic flowering plants have been known and described since the days of Theophrastus. However, for a long time even esteemed botanists were doubtful about the nature of parasitic plants, and the class Sarcophytae was established for monstrous excrescences such as members of Rafflesiaceae and some Balanophoraceae; other plants were classified as fungi. In 1969 Job Kuijt published his *The Biology of Parasitic Flowering Plants*, for close to 40 years the only comprehensive book on parasitic plants. Intensive research on the physiology and control of harmful parasites began in the 1950s, when a witchweed, *Striga asiatica*, was introduced into the United States and threatened the cultivation of maize.

NUMBER AND TYPES OF PARASITIC PLANTS

With one possible exception among gymnosperms (*Parasitaxus usta* in New Caledonia), parasitic plants are limited to eudicotyledons with about 4,500 species in about 280 genera belonging to 20 families. The majority, about 4,100 species, are hemiparasites (i.e., they are green plants meeting most or all of their needs for carbon through their own photosynthesis). Hemiparasites may be attached to roots and called root parasites. In this case, water and other nutrients are achieved partly from the soil through the roots and partly from the host through haustoria. If hemiparasites are attached to stems (and consequently called stem parasites), then they obtain all water and inorganic nutrients from the host. A minority of about 390 species are holoparasites (i.e., they lack chlorophyll and photosynthesis); hence, carbon must be obtained along with water and other nutrients from the host. Holoparasites may also be either root parasites or stem parasites. A few holoparasitic root parasites develop a reduced root system that may contribute to water and nutrient absorption, but this is not well established.

Parasitic plants may be either facultative or obligate parasites. The latter cannot survive without a host, while the former may survive for a longer period and even produce some seeds, but productivity is better when water and organic and inorganic nutrients are supplied from one or more hosts. Only hemiparasitic root parasites can be facultative parasites. However, there are no records from nature of a parasitic flowering plant that has completed at least part of its life cycle without haustorial connections to host plants. Competition from other species in the plant community will sooner or later eliminate a potential facultative parasite. Therefore, the terms facultative and obligate should be avoided until facultative parasites have been demonstrated to occur in nature. They may be used under laboratory conditions where it is possible to grow some hemiparasitic Orobanchaceae throughout the reproductive phase without a host.

There are parasite "lookalikes." These may be green orchids, bromelias, or ferns sitting on tree branches, but they neither develop haustoria nor obtain nutrients or water from the branches supporting them. Such plants are called epiphytes. Other "lookalikes" have lost all or nearly all chlorophyll and therefore look like holoparasites, but they have a three-part relationship wherein a mycorrhizal fungus interconnects the chlorophyll-free plant with a normal green plant having photosynthesis. Such plants used to be called saprophytes but are now called mycoheterotrophic plants (mycotrophic) plants. Examples are *Monotropa*, *Sarcodes*, some *Pyrola*, and orchids such as *Neottia nidus-avis* and *Corallorhiza trifida*.

HAUSTORIA

The development, structure, and function of the haustorium are essential subjects—as Job Kuijt has put it, "the haustorium is the *defining* part, the *essence* of parasitism."

FIGURE 1 *Notanthera heterophylla* (Loranthaceae), Chile. (A) shows the adhesive disc of a primary haustorium (center) and several secondary haustoria on two cortical roots running parallel with the host branch. (B) shows several cortical roots with young leafy shoots emerging above secondary haustoria. (Photographs courtesy of Job Kuijt. Reproduced from H. S. Heide-Jørgensen, 2008.)

Available space does not allow for many details. From a developmental point of view, there are two types of haustoria, primary and secondary. The primary haustorium (Fig. 1) develops directly from the primary root apex; in the more advanced parasites, it is the only haustorium, and functions throughout the lifetime of the parasite. Evolution of the primary haustorium made holoparasitism possible, because the generally small-seeded holoparasites need water and nutrients from a host immediately after germination. Secondary haustoria (Fig. 1) develop on lateral and adventitious roots, and they may be short-lived, sometimes living only a few months. They may occur in numbers of up to several hundred per plant. Regarding nutrient absorption, it may be an advantage to have the secondary haustoria placed on roots from different hosts, because different hosts absorb various nutrient ions in varying amounts.

A haustorium may consist of an outer part called a holdfast with an adhesive surface used for preliminary attachment to the host. Within the holdfast, an intrusive organ develops, which penetrates the outer layers of the holdfast and then penetrates the host by a combination of enzymatic dissolution of cell walls and mechanical compression of cells. When the intrusive organ reaches the conductive tissue of the host, a bridge of xylem cells differentiates and connects host xylem with parental xylem of the parasite. The parasite always maintains a lower water potential than the host. The main route for water and nutrients from host to parasite is through the xylem bridge, although the complete interface between the two partners also plays an important role in nutrient uptake. In the most advanced holoparasites, the intrusive organ comprises all vegetative tissue of the parasite. It splits into cellular strands, which penetrate large parts of the host, although they rarely reach the shoot tips. This internal tissue is called the endophyte, as opposed to the exophyte for external parts such as shoots and flowers.

SYSTEMATIC AFFINITY OF INVASIVE PARASITES

Introduced invasive parasites are known from 5 or 6 of 20 families, including parasites. Most important are Orobanchaceae, a family that now also includes parasites earlier placed in Scrophulariaceae, and Convolvulaceae (*Cuscuta*); there are further examples in Loranthaceae, Viscaceae, and Santalaceae. The majority of invasive species are hemiparasitic, mostly annual root or stem parasites. Invasive holoparasites are known only from Orobanchaceae, and they are annual root parasites.

HOST RANGE

Parasitic flowering plants only occur as introduced species when acceptable hosts are available. Parasites with one or few acceptable hosts have no possibility of becoming introduced or invasive outside the natural distribution of their hosts unless the hosts also become introduced. To predict the possibility of a species becoming introduced, it is necessary to know the range of its acceptable and preferred hosts, a factor that is often underestimated. If a parasite is not found on a certain species, this species may still be an acceptable host. The reason for the absence may be ecological, such as the lack of a suitable dispersal agent (e.g., birds), or may have to do with other environmental conditions (e.g., the light conditions may be insufficient for the parasite). In bird dispersal, the proper bird species must be available. When present, their flying behavior is important (e.g., many birds prefer to search for food, rest, and seek nesting possibilities in hedges, solitary trees, or wood edges, while the inside of the forest may be avoided). For root parasites, it may be physically impossible to follow a host root with attached haustoria back to the mother plant. In herbarium collections, the host species is rarely identified and noted.

Host range varies from one acceptable host (e.g., the dwarf mistletoe *Arceuthobium minutissimum* on *Pinus griffithii*) to at least 343 different host species for the loranth *Dendrophthoe falcata*. In general, holoparasites have fewer hosts than hemiparasites. To be counted as a host, the species must be able to support the parasite throughout its life cycle. Genetics and biochemical tissue incompatibility determine the maximum number of acceptable hosts, but in practice parasite range is mainly influenced by geographical (host distribution) and ecological (dispersal biology and environmental factors) relationships.

BIOLOGY OF INVASIVE PARASITES

Generally, perennial parasites reduce the vigor of the host but do not kill the host, because to do so would destroy the possibility for survival of the parasite. A weakened host produces fewer flowers, fruits, and viable seeds and is more susceptible to fungal diseases and harmful insects. However, annual parasites can allow themselves to kill the host, provided seed set is completed before the host dies. Therefore, some of the most harmful invasive parasites are annuals. As mentioned, dispersal biology and host distribution limit the possibility of a parasite's becoming introduced and invasive. When a parasite becomes introduced, the dispersal agent has often been humans, whether deliberately (as for *Viscum album*, see below) or accidentally.

Loranthaceae

This family contains more than 900 species of hemiparasitic stem parasites and three root parasites, mainly from tropical and subtropical regions. Flowers are generally showy, and birds pollinate the flowers and disperse the fruits. Host range is generally high. Only a few species are considered introduced on some tropical islands in Southeast Asia (e.g., the aforementioned *Dendrophthoe falcata*). However, several loranths act like invasive species in orchards and plantations of monocultures. In India, *D. falcata* causes enormous damage in plantations of teak (*Tectona grandis*), and the parasite may lead to death of entire trees. One reason for the success of *D. falcata* on teak may be that it is more shade-tolerant than most other loranths. On average, the parasite receives only 40 percent of the light received by the host, and it will survive even when the host leaves block 70 percent of the incident solar radiation.

In West Africa, some of the larger loranths have become real pests. *Tapinanthus bangwensis* uses a wide variety of hosts but has become invasive in plantations since cocoa was introduced as a crop in the 1870s. It has been shown that germination of seeds and establishment of seedlings of this light-dependent parasite are up to three times more likely in unshaded compared to shaded cocoa trees. Therefore, the problem increased with deforestation and the practice of growing cocoa without shade trees. *Phragmanthera capitata* has invaded plantations of teak and rubber, and the presence of other large species of *Tapinanthus*, *Agelanthus*, and *Globimetula* only worsens the situation.

For many years, the only method to control attacks by members of Loranthaceae was cutting down these stem parasites. Some of the host branches must also be cut due to the spreading endophyte inside the branches and to prevent regeneration from secondary haustoria placed on so-called epicortical roots (Fig. 1). Otherwise, new shoots may arise from the endophyte or the adhesive disk. In recent years, herbicides have been tried, but very few herbicides are available for a system where both host and pest are dicotyledons. The substance 2,4-D dichlorophenoxyacetic acid has been sprayed onto the leaves of various members of Loranthaceae and Viscaceae or injected into the trunk of the host, but with inconsistent results. Herbicides may be used to control *Dendrophthoe falcata* on teak if used during the deciduous stage of the host tree.

Viscaceae

All members of Viscaceae are hemiparasitic stem parasites. The distribution is similar to Loranthaceae but with more species in the northern temperate zone. Only a primary haustorium is present, and the most advanced genera have a widely distributed endophyte. The flowers are small and inconspicuous, and the fruits are dispersed by birds except in the case of *Arceuthobium*, which relies on self-dispersal by explosive fruits. *Arceuthobium* species are the most harmful parasites on conifers in North America, but the maximum dispersal distance is 20 m from the mother plant, and long-distance dispersal rarely occurs. Although present in Washington State and British Columbia, no *Arceuthobium* has spread to any of the minor west coast islands. A population on Mt. Constitution, Orcas Island, is interpreted as an Ice Age relict. There are seven genera, but only *Viscum album* occurs as introduced.

Around the year 1900, the European *Viscum album* ssp. *album* (Fig. 2) appeared in Sonoma County north of San Francisco, California, not spread by birds but introduced by the highly respected plant breeder Luther Burbank. By 1984, the parasite had spread by birds to about 114 km². The average distance of spread from the point of introduction was 5.8 km. In 1991 the corresponding figures were 184 km² and 8 km. *Viscum album* ssp. *album* occurs on more than a hundred different hosts of broad-leaved trees,

FIGURE 2 *Viscum album* ssp. *album* on apple tree, March. The European mistletoe is introduced and invasive in California. Female plant with ripe fruits, seven years after sowing. (Photograph courtesy of the author. Reproduced from www.viscum.dk/abstracts/text/snylteplanter.pdf.)

and in California it has at least 22 hosts. Many of those are introduced species from Europe, but native North American species are also attacked, such as *Acer saccharinum*, *Robinia pseudoacacia*, *Alnus rubra*, *Populus fremontii*, and *Salix lasiandra*. Because mainly ornamental trees in urban areas are attacked, damage is considered moderate. Further spread is expected to be limited due to the presence of few acceptable hosts in the surrounding area. However, if spread by humans to gardens at long distances from Sonoma County, the parasite could be a real pest. *Viscum album* was recently also reported in Victoria, Canada. As for loranths, the control method is cutting off host branches. The cut must be at least 30 cm below the haustorium to ensure removal of all endophyte tissue.

Santalaceae

The sandalwood family, with 35 genera, consists of both woody and herbaceous species, which, with a few exceptions, are hemiparasitic root parasites. Here, too, the fruits are dispersed by birds. The family is represented in all climatic zones except the arctic zone. The mainly African *Thesium* is by far the largest genus, with about 250 species. The small, white-flowered, Eurasian *T. arvense* is reported near Calgary in Canada and in Montana and North Dakota in the United States. It most likely arrived with seeds of agricultural plants. The root parasite is mainly a grassland species that can attack vegetables, but due to its sporadic occurrence, it is not a threatening invasive species.

Cuscuta (Convolvulaceae)

Cuscuta (dodder) is the only parasitic genus in Convolvulaceae. It has a worldwide distribution and is absent only in the most northern parts of the northern hemisphere. There

are at least 150 species (and possibly more), but there are many unsolved taxonomic problems. All species are herbaceous, winding, stem parasites with only secondary haustoria. Host range is high for most species but often difficult to determine, because *Cuscuta* haustoria attach to any subject within reach. However, many haustoria develop only a holdfast and no intrusive organ or endophyte. In such cases, the supportive species is not counted as a host.

Cuscuta species are fast growing (Fig. 3). This may in part be explained by faster nutrient translocation because the xylem bridge is accompanied by phloem. The presence of both xylem and phloem continuity is a unique feature in *Cuscuta*, and only one species of *Orobanche* is reported to have a similar advanced haustorium. *Cuscuta* species are annuals, and this life form, along with the fast growth, makes several species serious invasive weeds in agriculture, where crops such as tomato, potato, carrot, sugar beet, alfalfa, clover, avocado, coffee, and citrus species are attacked. The seeds are less than a millimeter in size. Very little is known about seed dispersal, but both birds and wind may be dispersal vectors. It is known that seeds survive the passage of the digestive canal of sheep. However, introduced invasive *Cuscuta* species probably always originate from contaminated seeds of crop plants. The invasive species causing most problems in many countries is the North American *C. campestris* (Fig. 3). In Asian countries, yield loss in sugar beet crops has been on the order of 3,500 to 4,000 kg/ha. In addition, *Cuscuta* may also be toxic to some domestic animals. No fully effective control method seems available. Mechanical methods such as flaming, harrowing, and hand-pulling

FIGURE 3 *Cuscuta veatchii* on *Bursera* sp., Baja California. The species is native, but its habit looks like the American *C. campestris*, which is invasive in many countries. The fast development of *Cuscuta* is illustrated by the fact that germination of the pictured species took place less than three weeks before this photograph was taken. (Reproduced from H. S. Heide-Jørgensen, 2008.)

have been used, and selective herbicides are also available but do not give full seasonal control. In 2004 the Asian *C. japonica* was discovered in California, and by 2007 it had appeared in 14 counties, indicating very fast dispersal. Furthermore, the growth rate is about 15 cm/day, and the host range is very wide. This indicates that *C. japonica* may soon be a troublesome invasive species.

Orobanchaceae

The broomrape family now includes witchweeds and other parasitic figworts. Of the about 90 genera, 75, representing 1,700 species, are hemiparasitic root parasites transferred from Scrophulariaceae. Furthermore, the family includes 17 genera of holoparasitic root parasites. The family is represented in all climatic zones and on all continents except Antarctica. Orobanchaceae contains the most troublesome introduced invasive parasites.

Parentucellia viscosa (Fig. 4) and *P. latifolia* from the Mediterranean region are annual root parasites in moist pastures and on heath land. Like other hemiparasitic root parasites, they have a wide host range, which includes native species in countries where they are introduced. They spread by tiny seeds carried by wind and water. *P. viscosa* occurs as introduced around the world in places such as Hawaii, the west coast of North America, Texas, Kansas, Denmark (where it is not a problem species), Japan, and Western Australia, and it is spreading further into Australia. Both species have recently been observed

FIGURE 5 *Striga asiatica* on partly wilted sorghum. This is an invasive species in the United States and Australia causing serious losses in crops from the grass family. (Photograph courtesy of Arne Larsen. Reproduced from H. S. Heide-Jørgensen, 2008.)

in the South Gippslands east of Melbourne. They can be fairly invasive and can degrade pastures if left unattended, but they may be controlled by use of selective herbicides.

Striga is another annual root parasitic genus. Seven of the 40 species are considered to be among the most damaging weeds within their mainly tropical African–Asian distribution. *Striga* is most common in semidry vegetation, where most species use grasses or sedges as hosts. In crop plants, *Striga* and *Orobanche* have found well-nourished, abundant hosts, allowing the parasites to develop extremely well and set lots of seeds. Therefore, these parasites become real pests, whether occurring as natural or introduced species. Two harmful *Striga* species, *S. asiatica* and *S. gesnerioides*, are known to be invasive in several countries. Long-distance dispersal is by wind or by insufficiently rinsed seed corn. Short-distance dispersal also occurs through water and by seeds sticking to claws, hoofs, footwear, wheels, and machinery.

S. asiatica (Fig. 5) was introduced into North and South Carolina, where it appeared in the 1950s, and into Southeast Australia. It is a serious threat in fields of maize, sorghum, and sugar cane. *S. gesneroides* was introduced into Florida. It mostly uses dicotyledons as its host—in particular, legumes. The seed set of *Striga* is on the order of 100,000 per plant, and the primary haustorium is so effective that by the time the parasite is visible above ground, it is too late to save the crop. In the most severe attacks, the yield loss may be up to 100 percent.

The large number of tiny seeds and a viability of more than 20 years are major problems for effective control of *Striga*. The most effective control is the development of resistant crop strains, and some success has been achieved in

FIGURE 4 *Parentucellia viscosa* introduced to Hawaii, the mainland United States, and many other countries. (Photograph courtesy of Forest and Kim Starr. Reproduced from H. S. Heide-Jørgensen, 2008.)

several crop plants. However, there are a number of methods, both mechanical and chemical, to avoid seed set, seed dispersal, and germination. These includes deep plowing to bury parasite seeds, hand-pulling, burning, cleaning tools and shoes, covering the soil with polyethylene to increase temperature, fallowing, fertilizing the soil, crop rotation, intercropping with catch crops, sowing early ripening strains late, practicing biological control using fungi and herbivorous insects, using chemical germination stimulants before sowing, fumigating soil with methyl bromide (for example), and using herbicides. None of these methods are effective or practical when used alone; it is necessary to use several of the methods simultaneously or successively as an integrated control system. It may also be noted that the biochemical and biological control methods are so expensive that they are not feasible in developing countries.

Orobanche is the largest genus in the family, with about 150 species (including *Phelipanche*). These are holoparasitic root parasites and mostly annuals. The root system is highly reduced, and several species have only a primary haustorium (Fig. 6). Seed production is enormous (up to 350,000 per plant). Dispersal biology is similar to that of *Striga*. At least six species are as problematic in agriculture as the harmful *Striga* species. They attack only dicotyledonous crops: mainly legumes (Fabaceae), but also others such as tomato, carrot, tobacco, hemp, and sunflower. The Mediterranean *O. minor* and *Phelipanche ramosa* (*O. ramosa*) have been introduced into several countries—*O. minor* into the United States, Chile, southern Africa, Australia, and New Zealand, and *P. ramosa* into Mexico, Cuba, Australia, New Zealand, and several U.S.

FIGURE 7 *Orobanche flava* on *Petasites hybridus* (large leaves). Both species are introduced to Denmark. (Photograph courtesy of the author. Reproduced from www.viscum.dk/abstracts/text/snylteplanter.pdf.)

states (it arrived in Texas as recently as 2000). The control methods and problems are the same as mentioned for *Striga*. In addition, soil application of an inhibitor of gibberellin synthesis prevents seed germination. *Orobanche amethystea* has recently been introduced to Israel, where it is invasive in vetch fields. Interestingly, introduced *Orobanche* species may be used to control other introduced species. *Orobanche flava* (Fig. 7) is introduced into Denmark, where it locally takes a heavy toll on the introduced and invasive *Petasites hybridus*.

OTHER TAXA

More introduced parasitic species than mentioned above are known, mainly from Australia and New Zealand. Most of the species are annual hemiparasitic root parasites that are not yet considered invasive. *Cassytha filiformis* (Lauraceae) is a still-spreading stem parasite with a similar potential to be invasive as *Cuscuta*. As a curiosity, it may be mentioned that *Euphrasia frigida* (Orobanchaceae) arrived in 2000 as the first parasitic plant on the volcanic island Surtsey, formed 33 km south of Iceland after an eruption in 1963.

SEE ALSO THE FOLLOWING ARTICLES

Dispersal Ability, Plant / Forestry and Agroforestry / Horticulture / Invasion Economics / Seed Ecology / Weeds

FIGURE 6 *Orobanche hederae* on roots ("white") of ivy, *Hedera helix*. From the primary haustorium (center), a tubercle develops. Three inflorescences with chlorophyll-free, scaly leaves rise from inside the tubercle, along with a number of very short adventitious roots. Note that the host root has wilted beyond the primary haustorium, indicating very effective water and nutrient absorption by the parasite. (Reproduced from H.S. Heide-Jørgensen, 2008.)

FURTHER READING

Heide-Jørgensen, H. S. 2008. *Parasitic Flowering Plants.* Leiden: Brill.

Joel, D. M., J. Hershenhorn, H. Eizenberg, R. Aly, G. Ejeta, P. Rich, J. Ransom, J. Sauerborn, and D. Rubiales. 2007. Biology and management of weedy root parasites. *Horticultural Reviews* 33: 207–349.

Kuijt, J. 1969. *The Biology of Parasitic Flowering Plants.* Berkeley: University of California Press.

Mathiasen, R. L., D. L. Nickrent, D. C. Shaw, and D. M. Watson. 2008. Mistletoes: Pathology, systematics, ecology, and management. *Plant Disease* 92: 988–1006.

Press, M. C., and J. D. Graves. 1995. *Parasitic Plants.* London: Chapman and Hall.

Press, M. C., and G. K. Phoenix. 2005. Impacts of parasitic plants on natural communities. *New Phytologist* 166: 737–751.

Sand, P. F., R. E. Eplee, and R. G. Westbrooks, eds. 1990. Witchweed research and control in the United States. *Weed Science Society of America Monograph* 5: 1–154.

PATHOGENS, ANIMAL

GRAHAM J. HICKLING

University of Tennessee, Knoxville

Animal pathogens are disease-causing agents of wild and domestic animal species, at times including humans. In the context of invasion biology, the term usually refers to infectious microorganisms such as bacteria and viruses and excludes nonliving agents such as toxins and toxicants. These infectious organisms are sometimes termed *microparasites* to distinguish them from *macroparasites*. Introduction of new pathogens into areas occupied by susceptible animal host species threatens native wildlife, disrupts animal-based food production systems, and puts human and companion animal health at risk.

INVASIVE ANIMAL PATHOGENS AND EMERGING INFECTIOUS DISEASES

Pathogens are a natural component of all ecosystems. Long-term association with their vertebrate hosts results in coevolutionary responses that reduce the virulence of the pathogens or boost the ability of the host to resist or recover from infection. Human-induced changes to the environment disrupt these natural pathogen–host relationships, often with adverse consequences. A disease agent transported to a new area may trigger outbreaks of disease among hosts previously naive to that pathogen—the introduction of rinderpest virus to Africa in the 1880s provides a grim example. Changes in transmission pathways of endemic pathogens can trigger unanticipated epizootics, such as when the 1827 introduction of *Culex quinquefaesciatus* mosquitoes triggered outbreaks of avian pox among birds on the Hawaiian Islands; the pox virus was already present on the islands but had not previously been causing significant disease. Change in habitat or climate can alter the biogeographic distribution of the vectors and hosts of animal pathogens, leading to disease invasion (or reinvasion of areas previous cleared of that disease). Collectively, these kinds of outbreak are termed emerging infectious diseases (EIDs), defined as infections that have newly appeared in a population or that are rapidly increasing in incidence or geographic range (see Table 1 for examples of EIDs affecting animals). The majority of EIDs affecting humans originate from pathogens originally carried by other animal species; these diseases are termed zoonoses. Examples of zoonoses include some strains of avian influenza, and flaviviruses such as West Nile virus.

FACTORS RESPONSIBLE FOR ANIMAL PATHOGEN INTRODUCTION AND INVASION

Increased frequency and speed of local and international travel, increased human-assisted movement of animals and animal products, and changing agricultural practices have all favored the introduction of animal pathogens to new areas. Genetic and environmental changes also facilitate animal pathogen invasion.

Changes in the Genetic Make-up of Pathogens

Animal pathogens sometimes become invasive as a consequence of natural changes in their genetic make-up, producing new strains with increased transmission rates or pathogenicity. For example, a new calicivirus closely related to the virus responsible for European brown hare syndrome emerged in rabbits in China in 1984 and spread to other countries via trade in farmed rabbits. The resulting outbreaks of rabbit hemorrhagic disease were highly lethal to unvaccinated European rabbits.

Humans, animals, and environmental sites are all reservoirs of bacterial communities that include some bacteria resistant to common antimicrobial agents. Our agricultural practices are increasingly providing environments in which these resistant bacteria can amplify and spread, so there is growing concern that enhanced microbial resistance will lead to future pathogen outbreaks. Nevertheless, most animal pathogen introductions are triggered by the movement of humans and other animals, or are a consequence of human-induced environmental change.

TABLE 1
Infectious Diseases with High Potential for Local Invasion

Disease	NIAID List	OIE List
African horse sickness		A
African swine fever		A
Anaplasmosis		B
Anthrax	Y	B
Atrophic rhinitis		B
Avian infectious bronchitis		B
Avian infectious laryngotracheitis		B
Avian influenza		A
Avian tuberculosis		B
Babesiosis	Y	B
Bacterial kidney disease (aquatic)		
Bluetongue		A
Borreliosis	Y	
Bovine spongiform encephalopathy		B
Bovine tuberculosis	Y	B
Brucellosis		B
Caprine arthritis/encephalitis		B
Chlamydiosis/psittacosis	Y	B
Chronic wasting disease and other prion diseases		
Contagious agalactia of sheep and goats		B
Chytridiomycosis		
Contagious bovine pleuropneumonia		A
Contagious caprine pleuropneumonia		B
Contagious equine metritis		B
Cysticercosis		B
Dermatophilosis		B
Dourine		B
Duck virus enteritis		B
Duck virus hepatitis		B
Echinococcosis/hydatidosis		B
Ehrlichiosis	Y	
Enterovirus encephalomyelitis		B
Enzootic abortion of ewes		B
Enzootic bovine leukosis		B
Epizootic hematopoietic necrosis (aquatic)		B
Epizootic hemorrhagic disease		
Epizootic lymphangitis		B
Equine encephalomyelitis (eastern and western)	Y	B
Equine infectious anemia		B
Equine influenza (virus type A)		B
Equine piroplasmosis		B
Equine rhinopneumonitis (EHV-1 and EHV-4)		B
Equine viral arteritis		B
Exotic Newcastle disease		A
Foot-and-mouth disease		A
Fowl cholera		B
Fowl pox		B
Fowl typhoid		B
Genital campylobacteriosis		B
Glanders	Y	B
Goat and sheep pox		A
Hantavirus infections	Y	
Heartwater		B
Hemorrhagic septicemia		B
Highly pathogenic avian influenza (fowl plague)		A
Hog cholera		A
Horse mange		B
Horse pox		B
Infectious bovine rhinotracheitis		B
Infectious bursal disease		B
Infectious hematopoietic necrosis (aquatic)		B
Japanese encephalitis		B
Leptospirosis		B
Listeriosis		
Lumpy skin disease		A
Lyssa virus diseases	Y	
Maedi-visna/ovine progressive pneumonia		B
Malignant catarrhal fever		B
Marek's disease		B
Monkeypox and other pox viruses	Y	
Mycoplasmosis		B
Nairobi sheep disease		B
Newcastle disease		A
Onchorhynchus masou virus disease (aquatic)		B
Ovine epididymitis		B
Ovine pulmonary adenomatosis		B
Paratuberculosis		B
Peste des petits ruminants		A
Plague	Y	
Porcine reproductive and respiratory syndrome		B
Pseudorabies		B
Pullorum disease		B
Q-fever	Y	
Rabies		
Rift Valley fever	Y	A
Rinderpest		A
Salmonellosis	Y	
Scabies		
Scrapie		
Screwworm		
Spring viremia of carp (aquatic)		
Surra		
Swine vesicular disease		A
Theileriosis		
Toxoplasmosis	Y	
Trichomoniasis		
Tularemia	Y	
Tuberculosis, human and bovine		
Vesicular exanthema		
Vesicular stomatitis		A
Viral hemorrhagic fever	Y	
Viral hemorrhagic septicemia (aquatic)		
West Nile virus and other viral encephalitides	Y	

NOTE: Selected infectious diseases in domestic and wild animals with high potential for regional or local invasion of the corresponding pathogen. Diseases that appear on the National Institute for Allergy and Infectious Diseases (NIAID) list of emerging and reemerging human diseases or on Office International des Epizooties (OIE) List A or B are indicated. OIE List A diseases are considered "capable of very serious and rapid spread." List B diseases are considered "of significant concern in the international trade of animals and animal products."

Much of the world's natural animal diversity historically has been maintained by separation of species by geographic barriers. For vertebrate animals, these barriers are often seen as providing protection from deleterious competitors or predators; protection from harmful pathogens is a further key benefit. A consequence of rapid expansion of the human population, and of globalization of trade, has been increased movement of animal pathogens across these historical barriers.

Invasive spread of animal disease associated with trade is not a recent development: the bubonic plague that devastated Europe in the 1300s was introduced by rats and fleas arriving along trade routes originating in the Gobi Desert. European colonization of Australasia, southern Africa, and South America was in each case accompanied by the settlers' domesticated animals and associated pathogens; brucellosis is an example of a disease transported worldwide by European settlers' cattle. The colonial era has passed, but global transport and trade continue to increase exponentially. Consequently, recent decades have seen frequent outbreaks in developed nations of unexpected pathogens, such as the 2003 introduction of monkeypox into the United States (via Gambian pouched rats brought in from East Africa for the pet trade).

Pathogens associated with animal production systems are among the most likely to be moved globally, often reinvading regions previously cleared of infection. Classical swine fever was once widespread, but many countries have now eradicated this disease from domesticated swine. Reintroduction of the virus can be devastating: in 1997–1998, an outbreak in the Netherlands cost $2.3 billion to eradicate, with approximately 12 million pigs being killed in order to contain the outbreak. North America is at risk for the introduction of this disease, which remains endemic in much of South and Central America.

Mycobacterium bovis, the causal agent of bovine tuberculosis and arguably the most important zoonotic agent in human history, was spread worldwide by European settlers transporting their cattle to new colonies. Intensive eradication efforts have been implemented to eradicate the disease from animal production systems in many counties, with a key feature of such campaigns being restrictions on moving cattle from test-positive herds. Movement control, when combined with test-and-slaughter or depopulation of cattle, is a proven effective eradication strategy—except in certain countries including England, Ireland, and New Zealand, where

the pathogen has crossed into wildlife species whose subsequent movements cannot easily be controlled. All three countries now face the problem of ongoing reinfection of domestic herds from these adjacent wildlife populations.

The bovine tuberculosis example illustrates how a chain of species interactions—in this case, from cattle to badgers or possums and then back to cattle again—can allow a pathogen to invade (or reinvade) an area. Another example is the plague that was introduced to California in 1900 by an infected human immigrant. The resulting epidemic among the human population infected commensal rodents, which in turn led to a rapidly spreading wave of infection among wild rodents that eventually reached human settlements as far north as Seattle.

In addition to the movement of infected animals, introduction of pathogens that cause direct-transmitted diseases such as classic swine fever and foot-and-mouth disease can also occur through trade in products made from such animals, or through the movement of contaminated fomites. In the case of classic swine fever, for example, the United States must guard against a wide range of potential transmission routes that include movement of infected domestic swine, contact with feral pigs, and importation of contaminated pork, pork products, vaccines, and artificial insemination products.

Vector-borne diseases add a further dimension to the invasion process, in that the pathogen can now potentially be introduced through the arrival of either the vector or the host. Furthermore, the vector may travel on the host or may be independently mobile. For example, heartwater disease of cattle is readily introduced into new regions by host animals or by ticks infected with *Ehrlichia ruminantium*. Vector-competent ticks are widespread and can be found on a variety of animals including reptiles; leopard tortoises and African spurred tortoises imported into Florida have both been found carrying infected ticks. One such tick, *Amblyomma variegatum*, was introduced into the Caribbean early in the nineteenth century. During the 1970s and early 1980s, this tick spread rapidly from island to island, carried in part by cattle egrets. The egrets' movement range is known to include the U.S. mainland, so the potential exists for future introduction of the heartwater pathogen to the mainland through natural wildlife movements.

The arrival of West Nile virus into New York City in 1999 was likely through a mosquito brought into the United States from overseas. Having invaded successfully, West Nile virus then spread progressively across the United States in subsequent years, with the rate and direction of its spread being closely tied to the seasonal migration of avian hosts. Similarly, migratory waterfowl are considered one of the most likely routes by which highly pathogenic avian influenza might invade the United States.

Fishers, hunters, and other recreationalists can introduce new pathogens into wildland areas. There is concern that inter-lake movement of bait fish by anglers has contributed to the spread of viral hemorrhagic septicemia and bacterial kidney disease in the Great Lakes system. Public access to many caves in the eastern United States was curtailed in 2009, given fears that such access might be contributing to the spread of the aetiological agent of white nose syndrome in bats. Recreationalists and researchers may have contributed to the spread of chytridiomycosis (caused by a fungus) and *Ranavirus* infections among wild amphibian populations that have declined in recent decades.

There has been concern that hunters transporting deer and elk carcasses across state lines could contribute to the spread of prions responsible for chronic wasting disease (CWD) in several North American ungulates. CWD has indeed spread rapidly in recent decades; however, this has most likely resulted from the movement of infected cervids among captive husbandry facilities in different states. Hunters and wildlife viewers also feed and bait deer in some situations; such feeding has been associated with persistence and spread of several diseases, including bovine tuberculosis and brucellosis.

In the late 1970s, an epizootic of raccoon rabies arose on the Virginia–West Virginia border that was attributed to the translocation of raccoons by hunters from the southeastern United States. This epizootic has since spread throughout the eastern United States, triggering one of the most expensive wildlife disease mitigation programs to date. A major component of the program is an oral rabies vaccine (ORV) "buffer" designed to curtail westward spread of the disease; the efficacy of this buffer is presently threatened, however, by illegal movement across the buffer of wild coyotes and foxes to supply hound-coursing pens in nearby states.

Environmental Change

Changes in habitat, land use, and climate can create favorable ecological conditions for animal pathogens that increase their chances of establishment after introduction. Such changes may also alter the abundance and movement of host and vector species in ways that increase the chances of invasion. These effects of habitat and climate on pathogen–vector–host population dynamics are complex, and so the consequences of changes in these variables are not easily forecast. Mammalian and avian hosts, being warm-blooded, are to some extent buffered against variation in climate and weather variables, while arthropod vectors such as ticks and mosquitoes are tightly coupled to climatic and other environmental factors. Vector-borne pathogens (*Rickettsias*, *Ehrlichias*, *Borrelias*, arboviruses, and others) often have distinct biogeographic distributions strongly determined by abiotic factors.

A 2008 study group concluded that the six high-risk animal pathogens in Europe most likely to be affected by climate change were bluetongue, Rift Valley fever, West Nile fever, visceral leishmaniasis, leptospirosis, and African horse sickness. Indeed bluetongue already seems to be responding to climate change, having emerged in the Mediterranean basin at the end of the 1990s, and then in August 2006 suddenly emerging as a severe epizootic in northern Europe. Persistence of the epizootic for several years suggests that the climatic and vector requirements for bluetongue virus establishment in regions north of its traditional range are now fulfilled.

Lyme borreliosis—vectored by blacklegged ticks—has emerged since the late 1970s to become the most prevalent vector-borne disease in the eastern United States. Models of future distribution under likely climate scenarios suggest that these ticks will expand northward into southern Ontario, putting humans and companion animals in Toronto and surrounding areas at increased risk. Such modeling studies have faced criticism, however, from those who argue that local agricultural practices and socioeconomic patterns will overwhelm the proposed effects of climate change.

In less temperate regions, such as parts of Africa and Australia, climate change–associated droughts could promote the spread of contagious diseases through increased contact between animals around limited patches of water and forage. Anthrax transmission, for example, is affected by the distribution of waterholes; vultures that feed on carcasses of animals killed by anthrax often return to waterholes and defecate anthrax spores nearby, where they are later encountered by herbivores coming to drink. Drought increases interaction between wildlife and domestic species at waterholes, thereby increasing pathogen transmission rates.

Climate change is similarly predicted to have important effects on pathogens in freshwater and marine ecosystems. Numerous disease outbreaks, especially in marine organisms, have been associated with climatic events such as the El Niño–Southern Oscillation, which will be altered by future climate change.

PREVENTION AND MANAGEMENT OF ANIMAL PATHOGEN INTRODUCTIONS

Most developed countries implement import controls and surveillance procedures relating to animal pathogens and have contingency plans in place for key animal diseases of public heath or economic concern. Computer modeling is increasingly used in developing contingency plans for animal pathogen introductions. Such models may be used to investigate the factors behind such outbreaks; to design efficient surveillance systems; to predict patterns and rates of disease spread; to evaluate the consequences of disease outbreaks; and to predict the efficacy of alternative mitigation methods. Many developing countries have flaws in their surveillance systems, however, which weakens overall global surveillance capability. Most developing countries also have only limited capabilities for outbreak response.

A particular concern surrounding animal pathogens is their potential for use in an act of bioterrorism. Governments face extreme difficulties in managing this risk, as illustrated in 1997 when a group of New Zealand farmers, frustrated by governmental and official responses to their problems of rabbit control, clandestinely introduced and spread rabbit hemorrhagic disease virus over a sizeable area on the central South Island. By the time the disease was detected by government officials, eradication was neither technically nor economically feasible; the disease was consequently reclassified as being endemic. This incident highlights the importance of surveillance and response capabilities, because preventing entry of biological agents may be difficult to achieve in the face of a determined adversary.

When disease outbreaks occur in aquatic environments, there is often little scope for successful intervention (the situation is different in aquaculture, where some degree of prevention and mitigation of disease is possible). Thus, the introduction of pathogens into aquatic systems will, in most cases, be irreversible events that the fishery manager must aim to avoid.

The continued occurrence of outbreaks of new animal disease has highlighted the need for closer integration of veterinary and medical communities and for improved education of the general public and policymakers on the risks of moving animal pathogens globally or even locally. Strategies to control emerging diseases will likely be more effective if public and animal health objectives are combined within a single strategy. Recent multidisciplinary initiatives such as "One Health" and "Conservation Medicine" and the coordinating role of international organizations such the OIE will be crucial for such strategies.

SEE ALSO THE FOLLOWING ARTICLES

Disease Vectors, Human / Ecoterrorism and Biosecurity / Flaviviruses / Influenza / Hawaiian Islands: Invasions / Parasites, of Animals / Pathogens, Human / Rinderpest

FURTHER READING

Brown, C., and C. Bolin, eds. 2000. *Emerging Diseases of Animals.* Washington, DC: ASM Press.

de la Rocque, S., S. Morand, and G. Hendrickx, eds. 2008. Climate change: Impact on the epidemiology and control of animal diseases. *OIE Scientific and Technical Review* 27(2): 1–613.

Friend, M. 2006. *Disease Emergence and Resurgence: The Wildlife-Human Connection.* Reston, VA: U.S. Geological Survey, Circular 1285.

King, L. J., ed. 2004. Emerging zoonoses and pathogens of public health concern. *OIE Scientific and Technical Review* 23(2): 1–726.

Ostfeld, R. S., F. Keesing, and V. T. Eviner, eds. 2008. *Infectious Disease Ecology: Effects of Ecosystems on Disease and of Disease on Ecosystems.* Princeton, NJ: Princeton University Press.

Roth, J. A., and A. R. Spickler. 2008. *Emerging and Exotic Diseases of Animals,* 3rd ed. Ames: Iowa State University Press.

PATHOGENS, HUMAN

PIETER BOL

Delft University of Technology, The Netherlands

Certain plagues that infect humans have the potential to spread rapidly over the globe. Although these diseases are said to be caused by *human* pathogens, their effects are generally *inhumane*, and their origins are often found in the animal kingdom.

Biologist Jared Diamond has stated that many factors contribute to the leap of microorganisms from beasts to humans. One of these factors was the start of husbandry after the last Ice Age, leading eventually to the modern crowded conditions of the bioindustry. The still-growing agglomerations of humans in concentrations of millions are another. The increasingly rapid transport of animals and people over the globe is a third.

DEFINING THE FIELD

Because the hominids *Homo erectus* and, later, *Homo sapiens* are descended from animals, one could postulate that virtually every infectious disease has derived from the animal realm. But it is not that simple. There has always been an ample supply of pathogens from the soil, such as worms, and germs from the water, such as typhoid bacteria. Moreover, bacterial meningitis, which until recently afflicted at least one out of ten nomadic children, seems

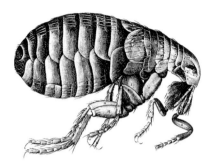

FIGURE 1 The primary culprits in transmitting plague were fleas carried in the fur of black rats. (Flea drawn by Robert Hook for his *Micrographia,* September, 1665, the first major publication of the Royal Society, London.)

to be restricted to human beings. But such diseases are confined to limited areas, and if they spread throughout the world, they do so slowly.

Recent pandemics (worldwide epidemics) have all had a connection with husbandry. In 1997 an outbreak of "chicken flu" was stopped by killing all the poultry in Hong Kong. But an outbreak of bird flu did occur in 2003, killing a few hundred people globally. In 2002 the severe acute respiratory syndrome (SARS) pandemic was evidently caused by close contact between animals and humans. During the writing of this article, the Mexican "swine flu" was declared a pandemic. And before all of these, mad cow disease (the onset of which occurred at the end of the 1980s), which leads to a variant of Kreuzfeldt-Jakob disease in humans, mainly affected Europe, yet another example of the adverse effects of the manipulation of animals by humans.

This article will focus on a few major pests: plague (the Black Death), smallpox, cholera, and HIV/AIDS (the most severe pandemic today).

PLAGUE

In the year 1346, Europe counted a population of about 100 million people. By 1352 there were 25 to 33 million fewer Europeans on what had become a defeated continent. What had happened? Tiny bacteria (a thousandth of a millimeter in diameter) had arrived from the east: the bacteria of the plague. They arrived not on their own, but aboard fleas that transmitted them to animals and humans when biting them (Fig. 1).

How did these insects reach Europe with their deadly load? There are two versions of the story of this invasion; the more likely one is that of the siege of Kaffa, a Genoese colony on the Crimean peninsula, during which parts of corpses of plague victims were flung into the city by means of catapults. But even this leaves us with the question of where this material came from. A more romantic version has the bacterium *Yersinia pestis* arriving from one of the Silk Routes; this might have happened in 1346 in

Odessa, on the Black Sea, when crates from Mongolia were opened that contained furs of giant marmots to be used to make the large wheel hats that can still be seen in the paintings of Bosch.

In a flash, the skin of the openers was black with fleas, which eagerly bit them, craving their blood. Contaminated fleas hosted the bacteria in the proximal stomach, where the bacteria formed clusters and prevented the meals of the flea from reaching the digestive system. The fleas could then leave the rodents on which they dwelt, especially if they were dying in large numbers, and jump onto humans. Biting, they vomited the bacteria into the humans' blood. During its course, the original bubonic form occasionally erupted into lung pestilence: direct contamination from the lungs of a victim to the lungs of other humans.

Plague is a horror that has been unforgettably described by Boccaccio in his introduction to the *Decameron* (1355). Boccaccio was in Ravenna during the plague outbreak in his home town of Florence in 1348. The vividness of his description was largely inspired by the record of general Thucydides of the Athens flu of 430 BC, from which Boccaccio heavily borrowed (Fig. 2).

Boccaccio describes the utter disruption of life, and the accompanying despair, agony, and moral decay. The loss of 25 to 33 percent of the European population had an enormous impact on the economy. Labor became precious, property and capital changed hands (often falling to young people), and towns had to reorganize their economies. The expansion of commerce overseas, along with the exploration of other continents, was greatly endorsed by shipowners, who directed their new capital to adventurous endeavors. This produced a strong stimulus for the impending capitalist revolution, which created the modern world.

FIGURE 2 In 1348 seven young women and three young men fled the plague-infested city of Florence. Their pastime was telling the 100 stories that Boccaccio composed in his *Decameron*. (John William Waterhouse, 1916, *A Tale from the Decameron,* oil on canvas, Liverpool, UK.)

FIGURE 3 A plague doctor in the sixteenth century. His costume is completely closed, and there are even glasses covering his eyes. Without knowing the cause of bubonic plague, he was well protected against flea bites. (Painting by unknown artist after the copper engraving *Dr. Schnabel von Rom* by Paul Furst (1608–1666), Nuremberg, 1656.)

In several European countries, there are still barren landscapes that have never recovered from this blow in the mid-fourteenth century. In Germany, these are called "Wüstung." When many key persons such as priests, tutors, midwives, bankers, blacksmiths, and merchants are taken from a region, the society can collapse entirely (Fig. 3). The situation can be beyond repair; the remnant population wanders off, and the effort to repopulate the region occasionally can be without success. If the AIDS epidemic continues its present ominous course, many a "Wüstung" may result in some African rural areas.

At the end of the nineteenth century, the plague bacterium was introduced to California from China. The authorities concealed the epidemic for a while, considering publicity to be bad for the image of their state, which wanted to attract people and industry. By the time they changed their minds, the spread of the disease was already out of control.

Plague is still with us, for there is a sylvatic reservoir in many countries from which humans can be infected. Mongolia, Vietnam, Madagascar, Tanzania, and Congo are among such countries. The World Health Organization (WHO) reports every year 1,500 to 2,000 victims (presumably an underreporting), of whom 10 percent die (not 50 percent, thanks to antibiotics). In about 13 states in the southwestern United States, there are still 10 to 15 victims a year—mainly hunters, trappers, and campers. A

vaccine against plague exists, but it is not generally distributed, and is given only to people who go into risky areas.

SMALLPOX

Smallpox has been humanity's companion since sedentary life began and large population concentrations came into existence. This viral infection has to be transmitted in an endless chain from person to person, because survivors become immune and no longer carry the smallpox virus. This implies that after an epidemic has raged in a certain region, the virus is not circulating there any more; only after several years can reintroduction lead to another epidemic, striking victims mainly among fresh cohorts of children. Smallpox needs a continent, or at least a large subcontinent, such as India, for its survival.

In the last millennium, smallpox was an enormous problem, at times surpassing plague as a demographic scourge. During a Chinese epidemic in the thirteenth century, half the population seems to have succumbed, a score even worse than that a century later in Europe, when the Black Death struck. In India children were counted during censuses only if they had had smallpox already, a rule that was applied well into the last century.

It has been calculated that smallpox may have killed more than a billion Europeans in the period from AD 1000 to 2000. This tally may seem unlikely, but one must realize that until the end of the eighteenth century, there was a high turnover in the population; half of all children died before their fifth birthday, and smallpox was among the main killers. In the Americas, smallpox was an even more ruthless killer reducing the population of Mexico, for example, by almost 90 percent in the sixteenth century (Fig. 4).

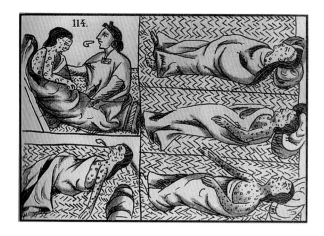

FIGURE 4 Aztec Indians dying from smallpox, unintentionally imported by the Spanish conquistadores in 1520. (Woodcut accompanying the text in Book XII of the sixteenth-century *Florentine Codex* [12 books compiled in 1540–1585 under the supervision of Bernardino de Sahagun], showing Nahua Indians of conquest-era central Mexico.)

Variolation is the intended introduction of smallpox virus into a person who has not yet had the disease. This practice started in China about one millennium ago. Dried, ground pox crusts were blown high into the nose by means of a blowpipe. This approach is surprisingly modern: at the moment, vaccines against several diseases are being developed that can be sprayed onto the nasal mucosa. The Chinese doctors were already aware of the fact that the powder should not be too old, but also not too fresh, for severe complications or death would then be impending. Of course, some of the treated persons died, but the toll so exacted was by no means comparable to that exacted by natural infections.

About the end of the seventeenth century, messages about variolation reached Europe. Lady Mary Wortley Montagu, "femme de lettres," became familiar with variolation in Constantinople as the wife of the British ambassador. The crusts or pus from patients were applied in small slits in the lower arm; the oculation spot was covered with a walnut shell and carefully bandaged in order to prevent spread of the disease to other body parts (mucal membranes, eyes) and to other persons.

Lady Montagu published about variolation with success, for in 1722, Amalia and Carolina, daughters of the Princess of Wales, were (in)oculated. The terms "(in)oculation" and "transplantation" well suited that century of horticulture. In the same year, Cotton Mather introduced variolation in Boston, having been taught the technique by African slaves.

But variolation did not become a great success. The deaths among the variolated and the contagion of people around them stood in the way—and this in a century when at least 60 million Europeans *alone* died from smallpox, and many millions of survivors became blind from it. However, Edward Jenner's 1798 introduction of inoculation by the cowpox virus was an immediate success. Cowpox renders cross-immunity against smallpox; Jenner predicted that the combination of vaccination with the tracking down and isolating of smallpox patients would eventually result in the extinction of the virus.

Indeed, in the 1970s, the entire world became too small for the virus. Encircled, it was denied effective continuation of its spread: the Earth was free of smallpox! Now, officially, there are only two virus stocks: one in the United States, and one in Russia. In 1999 it was believed that planned destruction of these last virus strains would definitively eliminate one of the most dangerous germs that has ever existed.

But the final blow was not executed at the end of June 1999. Some scientists now doubt whether the virus is present in only two places on Earth. We live in an era of terror, and smallpox (like plague) is one of the candidates for biological warfare, and particularly for terrorist attacks. But it is reassuring to know that there are still large stocks of the cowpox vaccine in many countries.

CHOLERA

The first reliable descriptions of cholera date from India in about 300 BC. The disease was caused by the bacterium *Vibrio cholerae*, and *Vibrio El Tor* became dominant starting at the beginning of the twentieth century. The toxins of the bacterium cause severe diarrhea, leading to a vast loss of water and minerals, which, without medical intervention, often leads to death. The natural habitat of *V. cholerae* is saline river estuaries, such as those of the Indus and Ganges rivers. The bacterium's toxins impair the host's bowel function, and the leaking of salt fluid provides the bacterium with its desired environment. Furthermore, the host's death is of no evolutionary disadvantage to the bacterium, because the victim will have contaminated much of its surroundings before dying, thus producing more victims.

The massive loss of water and minerals caused by cholera can kill a person in a day. Antibiotics help, but an even simpler treatment that has saved tens of millions of lives from the twentieth century until now is oral rehydration therapy (ORT). Salts dissolved in clean water can work wonders in cholera patients. Applying just water is dangerous: it dilutes the minerals that have been left.

Cholera was presumably restricted to South Asia and eastern Africa for a long time. It reached Europe around 1830 by two routes, through the Balkan region and through Russia, closing the final distance on war ships visiting harbors such as those of Hamburg and London. This may have been the route of the first invasion, but more important was the frequent reintroduction by (tea) clippers that sailed fast enough from the estuaries of India to those of Elbe, Thames, and Maas to provide *V. cholerae* from their ballast water (Fig. 5). Previously, the bacterium would have succumbed in the water before arrival, and patients would have either died or recovered before they reached their destination. From 1830 on, the many reintroductions of *V. cholerae* by windjammers caused epidemics in cities like London, Hamburg, and Paris (Fig. 6). Eventually this became the main stimulus for extensive sanitation projects later in the nineteenth century.

In 1854 the London physician John Snow published about the cause of cholera. His was one of the thousands

FIGURE 5 Clippers, the maritime triumph of the nineteenth century, were carriers not only of tea and other products to Europe, but also of the cholera bacterium (in their ballast water) that the swift ships delivered alive to their home ports. (*The Cutty Sark*, painting by John Alcot (1888–1973).)

of such publications on the matter in his time, but his was right. After studying more than 500 fatal cases occurring in ten days in London, he wrote, "We must conclude from this outbreak that the quantity of morbid matter which is sufficient to produce cholera is inconceivably small." Nearly 30 years later, Robert Koch demonstrated the tiny bacterium in Egypt and India. In the meantime, everywhere in Europe, enormous sanitation investments had been realized: clean water was supplied and sewages created, thanks to *Vibrio*.

Cholera is still a significant global problem. The vaccine is extremely ineffective, and most countries provide their traveling citizens with a vaccination certificate

FIGURE 6 In 1830 cholera reached Paris, where people died in the streets. (Lithograph by Honoré Daumier, 1840.)

without giving them the cholera vaccine. The only effective measure is sanitation, which is still a large challenge in many developing countries. At the beginning of the 1990s, there was a cholera epidemic in South America, caused by the fecal pollution of coastal waters and the consumption of raw fish dishes. Cholera was a menace in the refugee camps in Kenya and Tanzania in recent years, and it is an increasing health risk in the northern areas of South Africa.

The American microbiologist Rita Colwell showed that cholera is, to a large degree, an environmental problem. The bacterium can survive in a quiescent state in certain crustaceans found in ocean plankton; each tiny creature can harbor up to 10,000 bacteria. She proved that local warming of oceans, and the resulting expansion of seawater further into the estuaries of the large Indian and Bangladesh rivers, delivers the crustaceans deeper inland, where *V. cholerae* can be released, causing an epidemic. This effect is not to be underestimated, because the slope of the river bed is only a few meters over hundreds of kilometers.

Moreover, strong currents such as those of El Niño or La Niña can transport the bacterium over vast distances to foreign shores. However, remote sensing of oceans (especially warming patterns) and local microbiological surveillance in littoral and estuarine areas could contribute to an early warning system, preventing epidemics.

Colwell notes that the simple preventive measure of boiling water is too expensive for most people in the Indian and Bangladesh deltas. Firewood is scarce, and food preparation consumes it all. She has introduced the practice of filtering the water through the women's saris (folded into four or more layers). This reduces the cholera bacteria by 98–99 percent—scientists can be very imaginative!

HIV INFECTIONS AND AIDS

In 1981 what was to become a pandemic began in California. Two doctors described five young men, all sexually active homosexuals, who had been treated for biopsy-confirmed *Pneumocystis carinii* pneumonia (PCP) at three different hospitals in Los Angeles during 1980–1981. This feat shows that effective exchange of information is vital for early detection in infectious disease epidemiology.

The disease was soon named gay-related immune deficiency (GRID). Only two years later, the name acquired immune deficiency syndrome (AIDS) was coined. Later that decade, it became clear that—mainly through sexual intercourse—women could acquire the causal virus as well; at the moment, the sex ratio of HIV infections in the world is about 1. At first, the authorities in California were not overly alarmed, thinking that the disease was rare and confined to the homosexual population. Appropriate measures were not taken in due time, much in the same way that, a century earlier, an outbreak of plague in California had not been met with a timely response.

The virus was first described by the Frenchman Luc Montaignier in 1983, after which a sharp conflict arose with the American virologist Robert Gallo about its discovery. Then, the virus was called human T-lymphotropic virus (HTLV); only a year later, the present name, human immunodeficiency virus (HIV), came into use.

In 1984 suspicions were raised about the possible transmission of the virus by donated blood. In the same year, a test for HIV-antibodies in serum became available and pooled blood was shown to be frequently infected. Over the course of 1985, blood banks all over the world started testing for HIV in serum, but by that time many hemophiliacs had been infected via blood products.

In Europe especially, blood and blood products (from pooled serum) from the United States had proven fatal. In most European countries, blood was donated as a gift, whereas most donors in the United States were paid; this attracted people from some high-risk groups. But blood transfusions were not the only factors spreading the disease. The explosion of transport (air flights) around 1980, along with "free sex," contributed greatly to the expansion of AIDS. Among the first 200 diagnosed cases in the United States, some 100 had been caused by one homosexual man (Patient Zero), who was an attendant on domestic flights and was very sexually active. Intravenous drug use with contaminated needles and syringes is still a large source of HIV cases.

There is evidence that the origin of the HIV pandemic was the virus's crossing over from primates to humans somewhere in Africa, possibly hundreds of years ago. Once it had escaped from its niche, it started spreading globally. AIDS is the final stage of HIV infection, in which the so-called AIDS-defining afflictions occur. There are more than 100 of these, which, in combination with an existing (perhaps still undetected) HIV infection, can kill the patient. Kaposi sarcoma, the previously mentioned PCP, candidiasis, cervical cancer, and cytomegalovirus infections are among these afflictions. For 15 years, there was no adequate treatment for HIV infection, and the case fatality rate was near 100 percent.

During 1996, combination therapy became available, preventing HIV infection from becoming AIDS and saving the lives of many patients diagnosed with AIDS. This means that in many countries, the reservoir of people who have HIV is increasing, demanding a growing budget for care and a rise in the risk of HIV transmission. Simultaneously, the term "AIDS" has progressively fallen into disuse and is being replaced by "HIV infections."

The impact of the HIV pandemic is such that demographers have corrected the population prospects for the third quarter of this century by subtracting half a billion if the disease cannot be abated. Essential for the fight against AIDS are the use of condoms, the affordability of combination therapy in Third World countries, and the development of a vaccine. The last solution still seems far away, as does a malaria vaccine. Both the malaria parasites and HIV have a high mutation frequency; moreover, they can switch varying parts of their genome on and off, thus fooling an immune system prepared for the former and not the present war.

The World Health Organization reports that at the beginning of 2008, the number of people living with HIV/AIDS was 33 million. Two-thirds of them live(d) in sub-Saharan Africa (22 million). Of the 33 million, about 15.5 million were women; 2 million were children. Two million victims died in 2007; 25 million had preceded them in death between 1981 and 2006. In Africa alone, there were 11.6 million AIDS orphans. In Third World and transition countries, only 30 percent of the patients were treated adequately.

CONCLUSION

These four examples provide a brief impression of the impacts of invasions of human pathogens on individual life, on society, and on history. Other invasive diseases have had devastating effects on humans. Tuberculosis ravaged the isolated populations of Inuit and the people of Tierra del Fuego. German measles devastated the population of the Fiji Islands in the nineteenth century. The Spanish flu of 1917–1919 claimed between 30 and 60 million lives worldwide—far more than World War I. Syphilis was the analog of HIV infection in the late fifteenth and sixteenth centuries. And yellow fever, introduced from Africa via Barbados into the Americas in 1647 by the Dutch West Indies Company, rampaged and led to the first defeat of Napoleon, on Haiti, and also prevented the first French attempt to build a Panama Canal; both events occurred in the nineteenth century.

SEE ALSO THE FOLLOWING ARTICLES

Disease Vectors, Human / Early Detection and Rapid Response / Ecoterrorism and Biosecurity / Epidemiology and Dispersal / Influenza

FURTHER READING

Boccaccio, G. 1355/2003. *Decameron.* D. Wallace and J. P. Stern, eds. Cambridge: Cambridge University Press.
Bol, P., J. G. M. Scheirs, and L. Spanjaard. 1997. Meningitis and the evolution of dominance of righthandedness. *Cortex* 33: 723–732.

Colwell, R. R. 2002. A voyage of discovery: Cholera, climate and complexity. *Environmental Microbiology* 4(2): 67–69.

Diamond, J. 1997. *Guns, Germs, and Steel: The Fates of Human Societies.* W. W. Norton and Company.

Fenner, F., D. A. Henderson, I. Arita, Z. Jezek, and I. D. Ladnyi. 1988. *Smallpox and Its Eradication.* Geneva: World Health Organization.

Gottlieb, M. S., and J. D. Weisman. 1981. *Pneumocystis* pneumonia: Los Angeles. *MMWR Weekly* 30(21): 250–252.

McNeill, W. H. 1976. *Plagues and Peoples.* New York: Anchor Books.

Pollitzer, R. 1959. *Cholera.* Geneva: World Health Organization.

Van der Weijden, W., R. Leewis, and P. Bol. 2007. *Biological Globalisation.* Utrecht: KNNV.

Wills, C. 1996. *Plagues: Their Origin, History and Future.* London: Harper Collins Publishers.

PATHOGENS, PLANT

MEGAN A. RÚA AND CHARLES E. MITCHELL

University of North Carolina, Chapel Hill

Pathogens are biological organisms that infect other living organisms (hosts) and potentially cause disease (negative impacts on the host). Plant pathogens include fungi, viruses, bacteria, and nematodes. Pathogens can play several roles in biological invasions (Fig. 1). They may infect native host species or introduced host species, pathogens may themselves be introduced species, and they

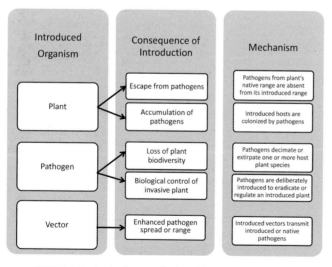

FIGURE 1 A simple framework describing three major introduced organisms and their relationship with pathogens. Introduced plants may either be released from pathogen pressure or accumulate pathogens. Introduced pathogens can decrease biodiversity or serve as biological control agents for introduced plants. Introduced vectors enhance the spread of both native and introduced pathogens.

may be transmitted by introduced vector species (chiefly arthropods).

INTRODUCED PLANTS

When a host species is introduced to a novel geographic region, many of the natural enemies from its native range may be absent in its introduced range. This may result in enhanced demographic success of the host species in its introduced range. Thus, introduced populations may "escape" or be "released" from the negative effects of natural enemies. Introduced species may also accumulate natural enemies. Some of these natural enemies may be co-introduced with the host species, while others will have previously been supported by native species. When introduced and native species share pathogens or other natural enemies, which species becomes dominant can be determined by the impacts of the shared natural enemy.

Release from Pathogens

When plant introductions occur, usually only a small number of individuals, at a given point in time, from a limited geographic area, and of a limited number of life history stages (most commonly, seeds), are introduced. Thus, this limited sample of host plants is unlikely to harbor the full diversity of natural enemies from the plant's native range. Furthermore, the small number of plants that typically start an invasion will lower the probability that any pathogen that is introduced with the plant will actually become established. While the factors that regulate native plant populations are poorly known, several studies have indicated that pathogens can play an important role. Thus, plants that are able to escape their enemies may demographically benefit from enemy release. Additionally, if plants are introduced to an environment with less virulent pathogens or where environmental conditions do not favor disease development, then plants may also demographically benefit from enemy release.

Comparisons of introduced plant populations with native populations of the same species generally suggest that the introduced populations are exposed to fewer species of natural enemies. A study considering 473 plants introduced to the United States from Europe showed that among harmful plant invaders, plants that were released from a larger species richness of fungal and viral pathogens were more widely classified as highly damaging (Fig. 2). On average, while introduced plants escaped over 90 percent of their native fungal and viral pathogens, plants only accumulated an average of 13 percent as many new fungal and viral pathogen species as they escaped. On the whole, plants were reported to be infected by less

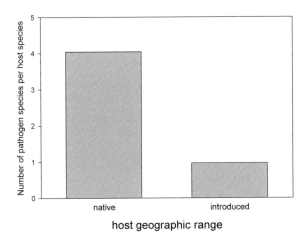

FIGURE 2 In published reports of fungal and viral pathogens infecting 473 plant species introduced to North America from Europe, each plant species was reported to be infected by over four times as many pathogens in its native range as in its introduced range. (Data from Mitchell and Power, 2003.)

than one-quarter as many pathogens in their introduced range as in their native range. Further study of 243 of these introduced plant species found that plant species native to habitats with moderate to high availabilities of water and nitrogen escaped a greater proportion of pathogens than those from habitats with lower availability of water and nitrogen. In addition to these broad surveys of pathogen species richness, experimental manipulations of natural enemy attack on introduced and native populations of the same plant species also generally suggest that introduced populations experience less negative impacts from enemies. Thus, release of introduced plants from pathogens may contribute to their invasive ability.

In contrast, comparisons of introduced plant populations to native plant populations of different species occurring within the invaded community have not generally found that the introduced plants are less attacked by natural enemies. However, most of these studies have relied on visual estimates of foliar damage. When compared across species, this does not necessarily indicate actual impacts on plant fitness. Additionally, these studies commonly use introduced plants that have native plants in the community which are closely related to the introduced plant (for example, in the same genus). Because more closely related plants are more likely to share natural enemies, these introduced species are less likely to experience enemy release than introduced plants lacking a closely related native species.

A well-studied example of an introduced plant species whose invasive success may be explained by pathogen escape is the successful North American invasion of white campion (*Silene latifolia*) from Europe. In large-scale

geographic surveys, plants were 17 times more likely to be damaged in Europe than in North America from enemies. Furthermore, when exotic genotypes were grown in their native range with native genotypes, exotics had higher susceptibility to fungal infection, fruit predation, and aphid infestation. This increased susceptibility may indicate that introduced populations of the plant are evolving increased competitive ability. These results suggest that pathogen release is one of the factors that has allowed *S. latifolia* to successfully invade North America.

Introduced plants can also experience release from belowground pathogens infecting their roots. Specifically, some studies have found that plants grown with soil organisms from their introduced range have greater biomass than when grown with soil organisms from their native range. This suggests that introduced plant populations may be less inhibited by soil pathogens and may benefit from belowground pathogen release. For example, in native European soils, *Centaurea maculosa* interacts with soil pathogens that have negative effects on plant growth. But in nonnative soils from North America, *C. maculosa* experiences positive feedbacks, which may contribute to its success. The introduced species *Carpobrotus edulis* experiences a similar release. When grown with rhizosphere soil from its native range (in South Africa), *C. edulis* exhibited a 32 percent reduction in biomass production, while plants grown with rhizosphere soil from the introduced range (the Mediterranean region) did not have a significant effect on plant biomass. In another example, in the native range in the United States, soil taken from areas close to black cherry (*Prunus serotina*) inhibited the establishment of neighboring conspecifics and reduced seedling performance in the greenhouse. In contrast, in the introduced range, black cherry readily established in close proximity to conspecifics, and the nearby soil community enhanced the growth of its seedlings. These experiments together suggest that release from soil-borne enemies may enhance the demographic success of introduced plants in a variety of ecosystems. Theoretical models have begun to build on these experimental results, and they suggest that the next step is to put these feedbacks from soil organisms into a community context by examining the degree to which they depend on the frequency (e.g., percent) of the community comprised by the introduced species.

Specific species or groups of natural enemies may be key to enemy release. For instance, in one experiment, European beachgrass (*Ammophila arenaria*) in California was not demographically released from pathogens, even though it was infected by 33 percent fewer fungal pathogen species and 90 percent fewer pathogenic nematode species in its introduced range than in its native range. Further studies with *A. arenaria* synthesizing data from around the world have shown that invasiveness correlates with escape from feeding specialist nematodes, rather than total number of pathogen and nematode species. All in all, evidence is strong that introduced plants are attacked by fewer pathogens and other natural enemies in their introduced ranges than in their native ranges. However, there is not yet sufficient data to conclude to what degree this enemy release contributes to the demographic success of invasive species.

Accumulation of Pathogens

Enemy release may decrease over time. The longer an introduced plant is established in a region, the more it will be exposed to pathogens. Over time, the introduced plant may thus accumulate pathogens both from its native range and from its introduced range. A recent meta-analysis comparing across published studies concluded that release from herbivores and pathogens faded during the 50 to 200 years after host introduction. Thus, it is important to consider time since introduction when one examines the impacts of pathogens on introduced plants.

There are two potential consequences of pathogen accumulation for introduced plant populations. The simplest expectation is that the pathogens will decrease plant population growth and density, reducing invasiveness. This has been called "biotic resistance" to invasion. In a field experiment in the Netherlands, communities with a greater abundance of root nematodes (resulting from the presence of the native oxeye daisy *Leucanthemum vulgare*) were less invaded. Biotic resistance may be strong when plants are introduced to a new locale with novel, virulent pathogens or where environmental conditions favor disease development. Furthermore, a lack of evolutionary history between introduced plants and new enemies may mean new enemies have stronger negative effects than enemies with which the plant has coevolved because there has been no selection for greater resistance or reduced virulence.

Alternatively, pathogens accumulated by an introduced host species may have negative demographic impacts not only on the introduced host, but also on native hosts. Introduced host populations that can sustain high rates of pathogen infection may increase the rate at which the pathogen infects native hosts. For example, a field experiment in a heavily invaded California grassland found that the invasive wild oat species, *Avena fatua*, increased transmission of barley yellow dwarf viruses to the native species *Elymus glaucus* (Fig. 3). When pathogen

FIGURE 3 Barley yellow dwarf virus species PAV infects hundreds of grass species, including both native and introduced grasses in many communities. (Photograph courtesy of Megan A. Rúa.)

infection rates in one host population are increased by transmission from another host population, this is called "pathogen spillover." Negative impacts of pathogen spillover are a form of "apparent competition." In some cases, apparent competition from an introduced species may decrease population growth and density of native species. This could contribute to dominance of native species by invasive species. A mathematical model has predicted that this may be the case for barley yellow dwarf viruses in California grasslands, but this prediction has yet to be experimentally tested.

Thus, both accumulation of pathogens and release from pathogens can be important components of the invasion process for introduced plants. Much more research is needed to be able to conclude how commonly pathogens inhibit and facilitate introduced plant species. Additionally, greater study of pathogens could facilitate prediction of successful introductions and their outcomes.

INTRODUCED PATHOGENS

Introduced pathogens can be just as destructive to plant communities as introduced plants. Invasions by pathogens are usually characterized by an absence of coevolution between the pathogen and its new hosts, referred to as "new encounter" or "novel interaction." Due to the lack of coevolutionary history, novel hosts may lack resistance to infection and tolerance of disease impacts. This can lead to detrimental impacts of introduced pathogens in their new range. Approximately 65–85 percent of plant pathogens worldwide are nonnative to the location where they were recorded. It is estimated that introduced plant pathogens are responsible for a large portion of crop losses, causing $23.5 billion in crop damage and approximately $2.1 billion in forest products to be lost each year in the

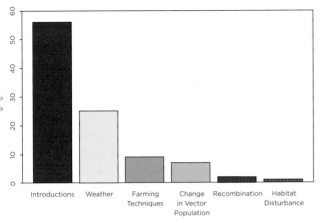

FIGURE 4 In a compilation of reports from the ProMED database, the majority of outbreaks of emerging infectious diseases of plants were reported to be caused by pathogen introductions. (Data from Anderson et al., 2004.)

United States. Furthermore, pathogen introductions are expected to increase in number, due to international trade, land use change, and severe weather events. Pathogen introduction generally occurs via two mechanisms: as accidental introductions or deliberately as a means of biological control.

Accidental Introductions

Most introductions of plant pathogens are accidental, and most of these occur as contaminants of a commodity. For example, around 1844, the pathogen causing potato late blight (*Phytophthora infestans*) was introduced from America to Europe via infected tubers. This introduction resulted in the Irish potato famine, spurring global human migration. Today, accidental introduction of pathogens is cited as the most important driver of emerging infectious diseases of plants globally (Fig. 4).

Such accidental introductions can have serious impacts on biodiversity. Around 1904, the fungus that causes chestnut blight (*Cryphonectria parasitica*) was introduced from Asia to North America via infected Japanese chestnut trees. Prior to this introduction, the American chestnut (*Castanea dentata*) was a dominant plant species of the eastern deciduous forest of North America. The fungal pathogen effectively eliminated this host species. This has had negative repercussions for humans, wildlife, and some insects, all of which depended on the chestnut for food resources.

In the 1920s, the pathogen *Phytophthora cinnamomi* was introduced to western Australia in contaminated nursery stock. It infects over 2,000 Australian tree, shrub, and wildflower species, often causing high mortality. Its introduction has changed thousands of square kilometers of sclerophyll shrubland and woodland to open woodland

dominated by sedges (close relatives of grasses) that are resistant to infection.

Another well-known introduction is Dutch elm disease, caused by three fungi: *Ophiostoma ulmi, O. himal-ulmi,* and *O. novo-ulmi.* In fact, multiple introductions of these pathogens have resulted in two pandemics of the disease. In the first pandemic, *O. ulmi* was introduced to Britain and North America around 1927, killing millions of elm trees in both regions over the next several decades. The second pandemic was caused by *O. novo-ulmi.* Around 1960, a strain of *O. novo-ulmi* that was distinct from the European strain of that species was introduced from Toronto to Britain. *O. novo-ulmi* has much worse impacts than *O. ulmi,* causing the second pandemic to be much more severe than the first. This pandemic killed 30 million elm trees in the United Kingdom alone and hundreds of millions of elm trees in North America. As well as killing valuable urban shade trees, the disease has also impacted forests. By removing elm trees from forests, it has even impacted different bird species that depended on the elms.

In many cases, the geographic origins of plant pathogens are not known, but a combination of three factors strongly suggests that the pathogen was recently introduced: a sudden local appearance followed by geographic spread, strong negative impacts on hosts, and low genetic diversity in the pathogen population. Based on these three criteria, the fungal pathogen that causes butternut canker (*Sirococcus clavigignenti-juglandacearum*) is regarded as introduced in North America. It was discovered in Wisconsin in 1967. The pathogen has since spread throughout the host species range, the eastern United States and Canada. The butternut tree (*Juglans cinerea*) is a close relative of the walnut tree, and its nuts help support wildlife populations. This valuable tree has been listed as threatened in both the United States and Canada.

The pathogen causing the disease sudden oak death (*Phytophthora ramorum*) is similarly regarded as introduced to North America. It was first observed in California in the mid-1990s. It has since spread across coastal California and Oregon, causing large-scale mortality of certain oak species (*Quercus* spp.) and tanoak (*Lithocarpus densiflorus*; not closely related to oaks). While its long-term effects are as yet unclear, it is expected to fundamentally alter these forests, indirectly impacting the entire community of species.

Despite these devastating examples, introduced pathogens do not represent a large portion of the invasion literature. For instance, fungi represent only 5 percent of the species listed in the Global Invasive Species Database. This may result from limited knowledge of fungal communities and biogeography, rather than from a low invasion success in fungi. It is often difficult to assess invasion status for fungi because the assignment of such status takes into account systematics, geographical range, and time of introduction, which are usually unknown for microbes. Understanding and predicting microbial invasions would benefit from the analysis of more comprehensive lists of alien microbes, supplemented with characteristics relevant to invasion biology and history (such as phylogeny, ecology, reproductive biology, region of origin, pathways of introduction, and past invasiveness).

Agents of Biological Control

Some plant pathogens have been deliberately introduced as biocontrol agents to eradicate or control introduced plant populations. Classical biocontrol is based on the assumptions that (1) the pathogen(s) regulate populations of the host plant in its native range, and (2) the absence of the pathogen contributes to the invasiveness of the plant in its introduced range. These assumptions have been supported for some plant species. However, the factors that regulate plant populations are poorly known in general, and the role of pathogens has received less attention than other factors.

Perhaps the best-known example of the introduction of a single pathogen to control a specific weed is the introduction of the rust fungus *Puccinia chondrillina* as a biocontrol agent for the invasive weed *Chondrilla juncea* in Australia. Later, the same pathogen was introduced to California, Oregon, Washington, and Idaho, where it has also reduced populations of *C. juncea*. Other examples of pathogens used for classical biological control include the introduction of the Acacia gall rust fungus (*Uromycladium tepperianum*) into South Africa from Australia to control *Acacia saligna* and the introduction of blackberry rust (*Phragmidium violaceum*) to control European blackberry (*Rubus vestitus*) in Australia.

While pathogens can play an important role in classical biological control, there is also the potential for biocontrol agents to have negative impacts on nontarget plants. Some studies have concluded that nontarget impacts of plant pathogens introduced as biocontrol agents are predictable, and therefore avoidable. However, nontarget impacts of pathogens have received less study than those of herbivorous insects and thus require further scrutiny.

INTRODUCED VECTORS

A vector is an organism that does not cause or experience disease itself but instead serves as a means of transmission, spreading pathogen infection from one host to another.

Three-quarters of known plant viruses are transmitted by vectors, as are many bacteria and fungi. Most vectors of plant pathogens are hemipteran insects (aphids, whiteflies, leafhoppers, planthoppers, and others). While some vector-transmitted pathogens also have other means of transmission, other pathogens depend obligately on vectors.

Without vectors, some introduced pathogens would not become invasive. For example, the pathogen causing Dutch elm disease is transmitted by bark beetles. In North America, it is spread by both a native and an introduced species of beetle. Citrus tristeza virus (CTV) likely originated in China and was probably introduced into South America between 1927 and 1930. However, it was not until the introduction of an efficient aphid vector, *Toxoptera citricidus*, from Asia, that CTV was able to emerge as a destructive pathogen. Thus, not only pathogen introduction but also the introduction or presence of vectors is often critical for the establishment of an introduced pathogen.

The introduction of a nonnative vector can also increase the spread and impacts of a native pathogen. The bacterium *Xylella fastidiosa* infects scores of plant species and is transmitted by insects that suck plant xylem. It causes various diseases on different host species, including Pierce's disease (PD) of grapevines. The pathogen has long been present in California, but in the 1990s, a more efficient vector species, the glassy-winged sharpshooter (*Homalodisca vitripennis*), was introduced into the region from the southeastern United States. The introduced vector species has greatly increased spread and impacts of the bacterium on commercial vineyards.

Perhaps the greatest concern about introduced vectors worldwide is the invasion and spread of whiteflies (particularly *Bemisia tabaci*). *Bemisia tabaci* has multiple biotypes (genetic variants) that transmit different viruses (Fig. 5). The B biotype of *B. tabaci* was introduced from the Old World to Florida in 1986 and has since spread throughout the tropical and subtropical Americas. As a group, whiteflies transmit over 114 virus species. The most economically significant viruses spread by whiteflies are the geminiviruses (family *Geminiviridae*) in the genus *Begomovirus*. These viruses have caused crop losses and consequential income loss throughout the world. However, reliable estimates of the true economic impact of *B. tabaci* are difficult to calculate because of the far-reaching areas affected, the number of crops and ornamentals affected, and variation in the nature and extent of injury with plant species, season, and localities. Consequently, the full extent of possible impacts of whitefly-transmitted viruses for natural communities is largely unknown.

FIGURE 5 The whitefly *Bemesia tabaci* feeding on a watermelon leaf. Whiteflies have been introduced to many regions and transmit over 114 viruses. (Photograph courtesy of Stephen Ausmus.)

Although critically important for the spread of some pathogens, introduced vectors have been given little consideration in theoretical and conceptual models of biological invasions. Incorporation of introduced vectors into these frameworks could help to predict the consequences of introduced plant pathogens for both agricultural and unmanaged communities.

CONCLUSION

Predicting the impact of invasions and viable threats of invasions is a growing field; however, it is one that often neglects some of the necessary basic components of terrestrial ecosystems: microbes. As discussed here, the impact of pathogens on plants in their introduced range, the impact of introduced pathogens, and the impact of introduced vectors all play a pivotal role in community and ecosystem processes for both natural and agricultural communities. Furthermore, anthropogenic change, particularly changes that impact the distribution of pathogens and plants, are predicted to become increasingly prevalent. Thus, in order to further understand these processes and make predictions about future invasions, pathogens and their vectors will need to be incorporated into invasion models.

SEE ALSO THE FOLLOWING ARTICLES

Belowground Phenomena / Biological Control, of Plants / Enemy Release Hypothesis / Fungi / Horticulture / Novel Weapons Hypothesis / *Phytophthora*

FURTHER READING

Anderson, P. K., A. A. Cunningham, N. G. Patel, F. J. Morales, P. R. Epstein, and P. Daszak. 2004. Emerging infectious diseases of plants: Pathogen pollution, climate change and agrotechnology drivers. *Trends in Ecology and Evolution* 19: 535–544.

Blumenthal, D. M. 2006. Interactions between resource availability and enemy release in plant invasion. *Ecology Letters* 9: 887–895.

Burdon, J. J. 1987. *Diseases and Plant Population Biology*. Cambridge: Cambridge University Press.

Desprez-Loustau, M.-L., C. Robin, M. Buée, R. Courtecuisse, J. Garbaye, F. Suffert, I. Sache, and D. M. Rizzo. 2007. The fungal dimension of biological invasions. *Trends in Ecology and Evolution* 22: 472–480.

Hallett, S. G. 2006. Dislocation from coevolved relationships: A unifying theory for plant invasion and naturalization? *Weed Science* 54: 282–290.

Hierro, J. L., J. L. Maron, and R. M. Callaway. 2005. A biogeographical approach to plant invasions: The importance of studying exotics in their introduced and native range. *Journal of Ecology* 93: 5–15.

Loo, J. 2009. Ecological impacts of non-indigenous invasive fungi as forest pathogens. *Biological Invasions* 11: 81–96.

Mitchell, C. E., A. A. Agrawal, J. D. Bever, G. S. Gilbert, R. A. Hufbauer, J. N. Klironomos, J. L. Maron, W. F. Morris, I. M. Parker, A. G. Power, E. W. Seabloom, M. E. Torchin, and D. P. Vazquez. 2006. Biotic interactions and plant invasions. *Ecology Letters* 9: 726–740.

Parker, I. M., and G. S. Gilbert. 2004. The evolutionary ecology of novel plant-pathogen interactions. *Annual Review of Ecology, Evolution, and Systematics* 35: 675–700.

Torchin, M. E., and C. E. Mitchell. 2004. Parasites, pathogens, and invasions by plants and animals. *Frontiers in Ecology and the Environment* 2: 183–190.

PERCH

SEE NILE PERCH

PESTICIDES (FISH AND MOLLUSC)

RAM PRATAP YADAV AND AJAY SINGH

Gorakhpur University, India

The term "pesticide" embraces a wide range of toxic chemicals used to control or eradicate undesirable forms of life. Compounds specially designed for the control of insects and other arthropods (i.e., insecticides) make up the bulk of these. Another group of pesticides is designed to deal with undesirable fish (both predatory and competitive), while still others were developed for use against aquatic snails that harbor intermediate stages of human parasites.

PESTICIDE CONTAMINATION OF WATER

Pesticide contamination of water can occur in many different ways and from many different sources; it may be of short duration or it may be prolonged. Presence of pesticides in water is the direct consequence of control operations carried out against undesirable fauna or flora. In some cases, the application of pesticide is made directly to the water, as, for example, in controlling aquatic insect larvae, predatory fish, or predatory fish larvae, or in removing undesirable aquatic weeds by application of herbicides. Static water bodies such as lakes, ponds, pools, marshes, and rice fields are also subject to direct pesticide application for control of mosquitoes, midge larvae, rice pests, undesirable fish, aquatic weeds, and snails.

FASCIOLA AND SNAIL INTERMEDIATE HOSTS

A neglected foodborne disease in the international public health arena is fascioliasis, a serious infectious parasitic disease of humans and animals worldwide and, from this standpoint, the worst of all the zoonotic helminths. Human cases are increasingly reported from Europe, the Americas, Oceania, Africa, and Asia; hence human fascioliasis is now considered a zoonosis of major global and regional importance. It is caused by liver flukes (trematodes) belonging to the genus *Fasciola*. Many animals, such as sheep, goats, cattle, buffalo, horses, donkeys, camels, and rabbits, show infection rates that may reach 90 percent in some areas.

Two major species, *F. hepatica* and *F. gigantica*, are the causative agents of fascioliasis in ruminants and in humans worldwide. *Fasciola hepatica* has a wider range than its tropical counterpart, *F. gigantica*, but their geographical distributions overlap in many African and Asian countries, although in such cases the ecological requirements of the flukes and their snail hosts are distinct. *Fasciola* develops in molluscan intermediate hosts. Species of the Lymnaeidae family (Fig. 1) are well known as intermediate hosts in the life cycle of *F. gigantica* or *F. hepatica,* but an increasing number of molluscan intermediate hosts of *F. hepatica* and *F. gigantica* have been reported.

CHEMICALS USED AS SNAIL CONTROL

The development of new, specifically molluscicidal chemicals was promoted by the World Health Organization, interested from the 1960s in controlling fresh-

FIGURE 1 Snails (*Lymnaea acuminata*) floating on the surfaces of leaves in a pond. (Photograph courtesy of Dr. Ram Pratap Yadav.)

water gastropods that act as intermediate hosts for trematodes causing human schistosomiasis. Trifenmorph (N-tritylmorpholine) was developed for this market and sold as Frescon, but it also proved effective against the amphibious species *Lymnaea truncatula* (Lymnaeidae) in pastures and was employed to reduce transmission of the liver fluke *Fasciola hepatica* to domesticated ungulates. Its use, however, was never extended to terrestrial gastropod pests. Another synthetic molecule developed around this time for schistosomiasis control was niclosamide (2,5-dichloro-4-nitrosalicylanilides), subsequently marketed as Bayluscide. The mode of action of molluscicidal chemicals depends on their properties and is best understood for carbamates such as methiocarb and cloethocarb. These act primarily as cholinesterase inhibitors. Perhaps the simplest mode of action of molluscicides is the desiccation caused by metaldehyde. Two other aspects of the actions of molluscicidal chemicals are their influence on metabolic enzymes and energy supply, and the detoxification processes that may limit their effectiveness.

Dichlorodiphenyltrichloroethane (DDT) began the era of synthetic organic pesticides, and many were subsequently evaluated for molluscicidal activity; often the results of tests for molluscicidal activity were conflicting. Getzin and Cole (1964) tested a number of pesticides against gastropods in the laboratory and concluded that chlorinated hydrocarbons and organophosphate were poor molluscicides, except for Zinophos (o,o-diethyl-o-pyrazinyl phosphorodithioate) and phorate (o,o-diethyl s-ethylthiomethyl phosphorodithioate).

MOLLUSC TOXICANTS

Dichlorodiphenyltrichloroethane (DDT) directly affects the freshwater habitat by disturbing the distribution and reproduction of the freshwater snail *Lymnaea stagnalis*. Concentrations of Nuvan (organophosphorus pesticides) ranging from 0.20 mg/L to 0.35 mg/L killed the snails *Lymnaea acuminata* and *Bellamya bengalensis* in pond water. Some organophosphorus pesticides including malathion and diazinon persisted in the organs of the bivalves *Anodonta woodiana* and *Corbicula leana*. The freshwater snail *Lymnaea acuminata* exposed to carbamate pesticide suffered inhibition of acetylcholinesterase activity in its nervous tissue. The pyrethroids permethrin, cypermethrin, and fenvalerate are highly toxic to the freshwater snail *L. acuminata*. For three different groups of pesticides (Rogohit [dimethoate, organophosphorus], Sevin [carbaryl, carbamate], and Stop [alphamethrin, synthetic pyrethroid]) for the freshwater snail *Lymnaea acuminata* at different time intervals at different exposure periods, the LC_{50} values of Rogohit were 19.65 mg/L to 10.81 mg/L, of Sevin were 20.05 mg/L to 14.19 mg/L, and of Stop were 0.020 mg/L to 0.005 mg/L.

Streams and stagnant water bodies such as ponds are the preferable habitats for *Physa* and *Lymnaea*; both are hosts of sheep liver fluke, *Fasciola gigantica*, a parasite infecting local sheep fed on vegetation growing beside such water bodies. This fluke can infect humans. The use of synthetic molluscicides in areas where both *Physa acuta* and *Lymnaea auricularia* occur can have negative effects on the control of *Fasciola gigantica* by upsetting a competitive balance between *Physa* and *Lymnaea* with increased net transmission of the fluke. The alcoholic extract of *Melia azadirchta* fruit used as molluscicide would potentially affect the *Physa acuta* population, causing fluctuations in both *Physa* and *Lymnaea* populations; field collection revealed that fluctuations were greater in *Physa* than in *Lymnaea,* and that the *Physa* population competes with the *Lymnaea* population. The use of molluscicides from *Melia azadirchta* fruit in areas where both *Biomphalaria pfeittei* and *Melanopsis tuberculata* occur could have negative effects on the control of schistosomiasis for the same reason. The molluscicide can potentially affect the *M. tuberculata* population, diminishing competition with the cooccurring *B. pfeittei* population.

FISH TOXICANTS

Off-target movement of chemicals used in industry and agriculture is usually unavoidable. When these chemicals get into natural water bodies, they may significantly damage fishes. The degree of increase in the activity of cellular enzymes in sera depends primarily on the magnitude and severity of cell damage. Furthermore, pesticides may enter water in combination with each other, causing additive harmful effects on fish. Recent studies by the U.S. Geological Survey detected many organochlorine pesticides in stream sediment and aquatic biota, of which 44 percent of the pesticides targeted were detected in sediments and 64 percent in aquatic biota, even though their agricultural uses in the United States were discontinued during the 1970s.

Malathion and carbaryl are esterase inhibitor neurotoxicants, with acute effects preceded by inhibition of acetylcholinesterase. As neurotoxicants, they interfere with many vital physiological functions and consequently alter the levels of various body components in fishes. Inhibition of AChE resulted in accumulation of acetylcholinesterase, which causes muscle twitching, leading to tetanus and eventual paralysis (paralysis of respiratory muscle may lead to death). Neurotoxicants can affect the physiology of several systems, resulting in metabolic disorders.

The fat-solubility of pesticides allows them to concentrate in tissues of their hosts as they move up the food chain. High concentrations of organochlorine pesticides in ovaries of the European eel *Anguilla anguilla* adversely affect its reproduction. Carbamate, more than the other two major pesticide classes (organophosphorus and organochlorine pesticides), is injurious to the liver of fishes. The pyrethroid fenvalerate accumulates in body tissues of the fishes *Labeo rohita, Catla catla,* and *Cirrhinus mrigala.*

RISKS VS. BENEFITS

The extensive benefits that humans accrue from these chemicals often appear to overweigh the costs in classical risk–benefit models. The economic impact of pesticides on nontarget species (including humans) has been estimated at approximately $8 billion annually in developing countries. However, what is required is the weighing of all risks against all benefits. The total cost–benefit picture from pesticide use differs appreciably between developed and developing countries. For developing countries, it is imperative to use pesticides; as no one would prefer famine, hunger, and communicable diseases at the cost of reasonable risks, it may be expedient to accept as reasonable a degree of risk that would seem unacceptable in a developed country.

CONCLUSION

The improved formulation of existing molluscicides and piscicides has probably achieved as much as it can, and innovation is now required to make any further significant advances in chemical control of snails and fishes. The recent research interest in mode of action of molluscicides and piscicides provides a more rational approach to the search for better materials, while the extension of screening to naturally occurring compounds has produced a chemically diverse range of active materials. It is hard to escape the conclusion that what is required is a number of new, better, active ingredients. If the impetus of current research can be maintained, these may well be found.

SEE ALSO THE FOLLOWING ARTICLES

Fishes / Herbicides / Snails and Slugs

FURTHER READING

Environews Forum. 1998. Pesticide use in developing countries. *Environmental Health Perspective* 106: A372.

Farag, H. F. 1998. Human fascioliasis in some countries of the eastern Mediterranean region. *East Mediterranean Health Journal* 4(1): 156–160.

Getzin, L.W., and S.G. Cole. 1964. *The Evaluation of Potential Molluscicides for Slug Control.* Technical Bulletin of the Washington Agricultural Experiment Station 653.

Nowell, L.H., P.D. Capel, and E.D. Dileanis. 1999. *Pesticides in Stream Sediments and Aquatic Biota: Distribution, Trends and Governing Factor.* Pesticides in the Hydrologic Systems Series Vol. 4. Boca Raton, FL: CRC Press.

Panigrahi, A. 1998. Evaluation of molluscicidal effect of the pesticide nuvan on the disease transmitting snails *Lymnaea acuminata* and *Bellamya bengalensis. Environmental Ecology* 16: 257–259.

Sancho, E., M.D. Ferrando, E. Andreu, and M. Gamon. 1992. Acute toxicity, uptake and clearance of diazinon by the European eel *Anguilla anguilla. Journal of Environmental Science and Health B* 27: 209–221.

Souslby, E.J. 1982. *Helminths, Arthropods and Domesticated Animals*, 7th ed. London: Bailliere, Tindall and Cassell.

Susan, T.A., K. Veeraiah, and K.S. Tilak. 1999. Biochemical and enzymatic changes in the tissues of *Catla catla*, exposed to the pyrethroid fenvalerate. *Journal of Ecobiology* 11: 109–116.

Zakaria, S.J. 1979. Studies on some factors affecting the development and viability of *Fasciola gigantica* and the life span of miracidia. MS Thesis. University of Mosul-Iraq.

PESTICIDES FOR INSECT ERADICATION

DANIEL A. HERMS

Ohio State University, Columbus

DEBORAH G. McCULLOUGH

Michigan State University, East Lansing

Pesticides remain a critical tool in efforts to eradicate nonnative species that can potentially affect forests, agriculture, livestock, and human health. This entry presents an overview of selected eradication programs in which pesticides played a significant role. Although pesticides have been used occasionally in eradication efforts geared at mammals, reptiles, or other organisms, this article focuses on eradication programs for nonnative insects affecting human health, agriculture, or forest ecosystems.

GOAL AND SCALE OF ERADICATION EFFORTS

Historically, the goal of insect eradication programs was the elimination of the entire population of the target pest. Substantial costs were often incurred in efforts to locate and control the last few individuals. More recently, population ecologists have demonstrated that inverse density dependence can lead to eradication of an invader population without requiring direct control of every individual. At low densities, for example, invaders are much less likely to find mates or overcome host defenses through group attack behavior than at high densities. Over time, these difficulties lead to negative population growth and eventual eradication. As we gain a better understanding of how inverse density dependent factors act on specific invaders, successful eradication at reduced costs may become feasible.

Both the scale and the success of insect pest eradication programs have varied widely, ranging from programs for the rapid elimination of a newly established, localized infestation to expansive programs implemented over years across vast geographic areas. In some cases, insecticides were the primary tool used to extirpate the invader, while in others, insecticides were integrated with additional strategies such as sterile male release, trap crops, cultural controls, or regulatory actions. This summary illustrates the often valiant efforts made to eradicate important insect pests, documents the evolution of insecticides and their role in eradication programs, and identifies factors contributing to the success or failure of these programs, some of which are currently under way.

GYPSY MOTH

The gypsy moth (*Lymantria dispar*), one of the most infamous invasive pests in North America, has been the focus of intensive eradication, management, and research efforts for more than 150 years. The long history of gypsy moth eradication efforts in the United States illustrates how insecticide products, technology, and philosophies have evolved.

In mid-summer, male gypsy moths locate nonflying female moths using a sex pheromone produced by the females. After mating, each female moth lays a single egg mass containing 200 to 1,200 eggs. Eggs overwinter and hatch the following spring. Egg masses are often laid in dark, protected places such as beneath bark flaps on trees or firewood or on the underside of lawn furniture, vehicles, or other items. Dozens of localized gypsy moth infestations in western, southern, and midwestern regions of the United States have been initiated by people who inadvertently transported items with egg masses.

Larvae feed on leaves of oaks, aspens, and other trees for approximately six weeks, then pupate, and adult moths emerge about two weeks later. Biological traits, including a high reproductive capacity and a host range encompassing at least 300 species of trees and shrubs, enable populations to reach high densities in urban settings as well as in hardwood forests. During outbreaks, an abundance of large, hairy caterpillars, their frass (insect feces), and severe defoliation of landscape and forest trees

FIGURE 1 Widespread defoliation caused by the gypsy moth, *Lymantria dispar.* (Photograph courtesy of Daniel A. Herms.)

create unpleasant conditions for people living or recreating in affected areas (Fig. 1).

The gypsy moth, native to much of Europe, Asia, and northern Africa, was introduced into North America in 1869. An amateur naturalist, E. L. Trouvelot, imported gypsy moths from Europe with the well-meaning but misguided intent of breeding disease-resistant silk moths. Some of the insects escaped from Trouvelot's home in Medford, Massachusetts, and 20 years later, the first major outbreak occurred. Residents demanded that officials take action. The first gypsy moth eradication program in the United States began in 1890 and targeted a 200 square mile area in Massachusetts. Over ten years, the program cost approximately $1 million, an amount equivalent to nearly $25 million today.

The insecticide known as Paris green (copper acetoarsenite) was sprayed on trees and shrubs to kill gypsy moth larvae. Paris green, however, was often not effective at controlling gypsy moth, and it was certainly not popular with residents. It could be toxic to plants and trees, and reports indicate that property owners used garden hoses to wash the insecticide off once the spray truck passed. In 1893, lead arsenate replaced Paris green in the eradication program. Lead arsenate was somewhat more efficacious, but problems with timing sprays and adequately covering the canopy of large trees persisted. Moreover, the arsenate was toxic to animals that grazed on vegetation beneath sprayed trees. Lead arsenate continued to be used on gypsy moth populations, however, until the mid-1940s, when DDT became available.

Cultural controls were also a part of eradication efforts, first on a local scale and then on a larger scale, as officials attempted to establish a barrier zone along the Hudson River Valley to prevent further expansion of this pest. Men were hired to climb into trees to destroy or treat egg masses with creosote. Sticky bands or hiding bands were wrapped around the trunks of thousands of trees to capture the caterpillars as they moved up and down the trees. In some areas, sections of infested forests were felled and burned. The density of gypsy moth populations dropped, and some localized infestations to the west and south of the barrier zone were eradicated. Once gypsy moth populations dropped to low levels, however, people largely forgot about them, and funding for the eradication program was cut. Gypsy moth populations began to build and spread, mostly as a result of people moving egg masses. By the early 1920s, much of New England had become infested, and eradication in these areas was largely abandoned. Officials began the first introductions of European parasitoids into New England for biological control of gypsy moths around this time.

Eradication continued to be pursued in other areas, however. A large infestation in New Jersey was the target of an eradication effort from 1920 to 1932, which was apparently successful. For ten years, officials attempted to eradicate a gypsy moth infestation discovered in Pennsylvania in 1932. Activities included egg mass treatments with creosote, lead arsenate sprays, and the burning of 10,000 acres of infested forest. In 1942, however, surveys showed that at least 1,000 square miles remained infested. Eradication was abandoned in Pennsylvania as well as in most of the rest of the northeastern United States, where efforts switched to suppressing populations during outbreaks.

Aerial spray technology advanced considerably following World War II, and aerial applications of DDT replaced lead arsenate ground sprays in gypsy moth programs. The relatively long persistence of DDT resolved problems related to timing sprays in years when egg hatch extended over two to three weeks. Studies indicated that a single spray applied at a rate of 1 lb per acre resulted in virtually 100 percent mortality of gypsy moth larvae. That level of efficacy led one U.S. Department of Agriculture study group to suggest that five years of DDT treatments across 30 million acres could be used to reestablish a barrier zone. An effort of that magnitude was not attempted, but from 1945 to 1958, DDT was applied to more than 10 million acres of forested land and communities in the northeastern United States. This included Michigan, where DDT sprays were applied from 1954 to 1963 to eradicate a gypsy moth infestation discovered in 1954. Eradication appeared to be successful in Michigan until improved pheromone traps deployed in 1963 showed that the infestation not only persisted but was more widespread than was originally thought. In addition, by the late 1950s, public concern about nontarget and environmental effects of DDT were building. Concerns intensified when DDT residues were found in the milk of dairy cows that grazed in areas in New York that had been aerially sprayed. Use of DDT by agencies involved in gypsy moth programs was phased out between 1958 and 1964.

Carbaryl, a broad-spectrum insecticide sold under the trade name of Sevin, replaced DDT in gypsy moth eradication programs in the United States from 1962 to 1977. An organophosphate insecticide, trichlorfon (Dylox), was also used in aerial applications for gypsy moth eradication in the 1970s. From 1967 to 1988, approximately 110,000 acres were treated with aerial applications of either Sevin or Dylox for gypsy moth eradication. Although Sevin and Dylox do not have the same environmental impacts as DDT, these broad-spectrum insecticides act on the nervous system of virtually all insects as well as other organisms. Organic farmers, beekeepers, and environmentalists challenged the use of these insecticides in gypsy moth programs, and program managers looked for alternatives.

A different kind of product, diflubenzuron, sold as Dimilin, was developed in the 1970s. Dimilin is an insect growth regulator that disrupts the processes by which insects and other arthropods shed and regrow their exoskeletons. It is somewhat more selective than the broad-spectrum insecticides because it does not affect adult insects or organisms that lack an exoskeleton. It is also persistent, which simplified spray timing and ensured that a single application remained effective for the season. Dimilin was aerially applied to 273,895 acres for gypsy moth eradication from 1977 to 1993. Its long persistence, however, also generated concern about effects on nontarget insects and arthropods. Several studies showed that insects and other arthropods, including aquatic species, continued to be affected months after an application.

In the 1930s, French scientists reported that a bacterium, *Bacillus thuringiensis* var. *kurstaki*, which occurs naturally in the environment, could be used to kill caterpillars. Scientists in the United States began developing formulations of this bacterium that could be produced commercially and applied aerially like a conventional insecticide in the early 1960s. Referred to as Bt or Btk, this microbial insecticide has been the most commonly used product for gypsy moth eradication in North America since the early 1980s. Btk is much more selective than the insecticides used previously. Unlike broad-spectrum insecticides that act on the nervous system of insects and other animals, Bkt affects the gut of the caterpillars. Caterpillars

that consume treated leaves die within a few days from the combined effects of starvation, damage to their digestive system, and bacterial growth within their bodies. Btk can affect the larvae of many nontarget Lepidoptera, but only if the caterpillars consume treated leaves. Additionally, Btk products are characterized by relatively short persistence. This can be an advantage because it reduces potential effects on nontarget Lepidoptera. On the other hand, short persistence, combined with the fact that young gypsy moth caterpillars are more sensitive than older caterpillars, means that Btk applications must be carefully timed to be effective. To ensure efficacy, gypsy moth eradication projects typically involve two or three sprays during a single season.

Between 1980 and 2008, at least 4,154,870 acres in 26 states were treated with Btk to eradicate localized gypsy moth infestations established when people accidentally transported egg masses from infested areas in eastern states. More than 826,800 additional acres were also treated with Btk between 1993 and 2008 as part of the national Gypsy Moth Slow-the-Spread program, which uses intense grids of pheromone traps to locate newly established populations just beyond the generally infested area. Although Btk is much more specific than broad-spectrum insecticides or diflubenzuron, it can affect populations of nontarget Lepidoptera, especially when multiple sprays are applied in consecutive years. Studies have shown that populations of these insects generally rebound after spraying ceases, but monitoring nontarget species of concern can be an important aspect of eradication programs.

One other product, Gypchek, was used on six occasions between 1979 and 1999 in gypsy moth eradication programs. The active ingredient in Gypchek is a virus disease that affects only gypsy moth larvae. Although Gypchek has virtually no impacts on any nontarget species, it is expensive and difficult to produce. Production issues limit its availability and have caused efficacy to vary. Gypchek has been used primarily in sensitive areas where threatened or endangered species of Lepidoptera preclude application of Btk.

WHITE-SPOTTED TUSSOCK MOTH

Many of the tools used in North America for gypsy moth eradication were incorporated into a New Zealand eradication program for the white-spotted tussock moth, *Orgyia thyellina*. This moth, a native of Asia, is a close relative of the gypsy moth with a similarly broad host range. An intensive eradication program was launched after a white-spotted tussock moth infestation was discovered in suburban Auckland, New Zealand, in 1996. Action included multiple aerial applications of Btk over

4,000 ha of suburban Auckland, combined with ground spraying in sites known to be infested from 1996 to 1998. Intensive monitoring efforts using both pheromone traps and live, female sentinel moths were also employed along with a significant public outreach program. This tussock moth was declared eradicated in 1998 and has not been detected in New Zealand since.

GAMBIAE MOSQUITO

In 1930, *Anopheles gambiae*, an important vector of malaria in Africa, was discovered along the northeast coast of Brazil. This mosquito proved to be a more efficient vector of malaria than the endemic *Anopheles* species, and severe malaria epidemics were soon reported in the region. Malaria seemed poised to spread rapidly throughout tropical South and Central America.

In 1932, the Yellow Fever Service, which was focused on eradication of *Aedes aegypti*, the vector of yellow fever, initiated a limited but successful program to eradicate *An. gambiae* from the region of Natal, which was the site of the first malaria epidemics. However, the mosquito continued to spread in the rural areas to the north and west, where malaria outbreaks occurred over the next several years. Initial control measures in these regions were largely limited to environmental modifications such as draining and filling larval habitat to decrease mosquito populations.

In 1939, The Malaria Service of the Northeast was formed, and a much more aggressive and eradication program was initiated, with financial and logistical support provided by the Rockefeller Foundation and the Yellow Fever Service. Cultural controls were largely replaced by insecticides, and all potential breeding sites were treated several times per month. Control was aided by easily defined larval habitat that was spatially limited during the dry season, as well as by the high affinity of adults for human dwellings. All water sources, including drinking supplies, were treated with Paris green to control larvae, and the interior surfaces of homes and other structures were treated on a weekly basis with pyrethrum. To reduce the spread of adults to noninfested areas, all cars, planes, trains, and boats were treated with pyrethrum at disinsectization posts prior to exiting infested zones. The insecticide-intensive program proved successful, and by 1941, within just two years, *An. gambiae* had been successfully eradicated from the South American continent.

IMPORTED FIRE ANTS

Imported fire ants in the genus *Solenopsis* are feared throughout the southeastern United States for their propensity to swarm relentlessly and sting any intruder that disturbs their nest, even inadvertently. Millions of people suffer painful stings each year. The imported fire ant complex consists of two species (and their hybrid): the red imported fire ant, *S. invicta*, which is the most notorious and widely distributed, and the black imported fire ant, *S. richteri*, which has a much more restricted distribution. Large fire ant mounds can exist in high enough densities to disrupt agricultural practices and damage equipment. Feeding on crops by the omnivorous fire ants can have an economic impact, and imported fire ants can also disrupt wildlife habit and can invade structures in urban environments.

Both species of imported fire ant were accidentally introduced from South America to the southeastern United States at Mobile, Alabama. The black imported fire ant was introduced around 1918, and the red imported fire ant became established in the 1930s or 1940s. The red imported fire ant proved to be a much more aggressive colonizer and spread quickly, ultimately displacing two native fire ant species as well as the black imported fire ant, which now occupies a very limited geographic distribution in northeastern Mississippi and neighboring Alabama and Tennessee. By 1952, the red fire ant was present in ten states. Artificial spread in infested turf sod and nursery stock played an important role in its rapid dispersal. Despite one of the most ambitious eradication programs ever attempted, imported fire ants continued to expand their range in the southeastern United States, although their rate of spread has been limited by quarantines and their intolerance of cold winters.

The program to eradicate imported fire ants was based almost exclusively on insecticidal control, and the value of alternative tactics such as biocontrol was discounted. Although insecticidal treatment of the black imported fire ant began in 1937 using calcium cyanide dust, widespread eradication of imported fire ants was first attempted in Mississippi in 1944, where mounds were treated with chlordane dust. Imported fire ants continued to spread throughout the state, and the program was discontinued in 1951. The U.S. Department of Agriculture initiated an expansive eradication program in 1957, treating millions of acres in Alabama, Georgia, Louisiana, and Florida with dieldrin and heptachlor. Treatments were initiated in Texas the following year.

From the onset of the program, there were substantial concerns and controversy surrounding the environmental and health effects associated with widespread use of persistent, broad-spectrum chlorinated hydrocarbon insecticides. The use of heptachlor in the fire ant eradication program was terminated in 1962, but only after

2.5 million acres had been treated with little effect as the range of imported fire ants continued to increase.

Widespread application of mirex bait replaced heptachlor in the fire ant eradication program in 1962, but this also proved largely ineffective at eliminating fire ant colonies. A 1967 U.S. National Academy of Sciences report concluded that fire ant eradication was biologically and technologically impossible. Nevertheless, several million acres were treated with mirex that same year after Congress requested that the U.S. Department of Agriculture evaluate the efficacy of mirex for fire ant eradication. Controversy surrounding the politically charged eradication program intensified as it became clear that mirex was biomagnifying in the food chain and that it posed a serious threat to human health. In 1970, the newly formed U.S. Environmental Protection Agency announced that it intended to ban the use of mirex (although its use against fire ants continued until 1979), and the U.S. Department of Agriculture announced that it would abandon eradication in favor of more localized suppression programs.

Imported fire ant nests proved quite resilient to insecticide applications due to the structure of their nests as well as the social and behavioral organization of their colonies. Furthermore, imported fire ants proved to be superior colonizers and competitors, and they quickly reinvaded and often increased their dominance of previously treated areas. Ultimately, insecticides had little effect on the distribution of imported fire ants and in fact facilitated their invasion. Conversely, native biodiversity was decreased by widespread use of broad-spectrum insecticides. In spite of what is perhaps the most insecticide-intensive eradication program ever attempted, the imported fire ant continued to spread, slowing only when its distribution reached the limits of its environmental tolerance to cold winter temperatures to the north and dry soils to the west.

BOLL WEEVIL

One of the most successful examples of the use of insecticides in an expansive insect eradication program is that of the boll weevil, *Anthonomus grandis grandis*, which spread rapidly across the cotton belt of the southeastern United States following its migration into southern Texas in 1892 from its native Mexico. By the mid-1950s, the boll weevil, having evolved resistance to chlorinated hydrocarbon insecticides, had devastated the U.S. cotton industry.

Buoyed by the initial successes of the screwworm eradication program, scientists began to contemplate a strategy to eliminate the boll weevil from the cotton belt, and a formal committee to study eradication was established by the National Cotton Council in 1969. Experiments and pilot projects were conducted in the 1970s to evaluate the feasibility of boll weevil eradication, and the full-scale Beltwide Boll Weevil Eradication Program was initiated in 1983.

The boll weevil eradication program strategy integrated a combination of tactics of varying effectiveness, including cultural controls, sterile male technique, trap crops, and pheromones. Insecticide applications, particularly malathion, timed at critical junctures in the insect life cycle, played a prominent role. Applications of insecticides targeted the emerged overwintering boll weevil adults just before flower buds became suitable for oviposition and larval development, and also were used late in the season to control diapausing adult weevils. Despite adamant contentions of some prominent scientists who argued that boll weevil eradication was impossible, this pest has been eliminated from the vast majority of the territory it has invaded, and final eradication from the United States seems imminent.

ASIAN LONGHORNED BEETLE

Systemic neonicotinoid insecticides are a key component of the eradication programs under way in the United States for the Asian longhorned beetle (*Anoplophora glabripennis*). The Asian longhorned beetle (ALB) was discovered in New York City in 1996 and in Chicago in 1998. Additional populations were subsequently found in two locations in New Jersey in 2002 and 2004; in Toronto, Ontario, in 2006; and in 2008, in Worchester, Massachusetts.

In the United States, federal, state, and local officials implemented aggressive eradication programs. Larvae feed under the bark on phloem tissue for a few weeks, then bore into the sapwood. As larval density builds, branches begin to decline and then die, and eventually entire trees can be infested and killed. Limbs or entire trees may break in strong winds. The extensive host range of the ALB includes maple, willow, birch, and other hardwood species. Unlike some insects such as gypsy moths, the ALB does not use long-distance pheromones. Detecting low-density populations and identifying infested trees is challenging. Eradication programs for the ALB therefore employ intensive visual surveys, often with tree-climbers and bucket trucks, to find the dime-sized exit holes left by emerging adults, oviposition pits or larval galleries on infested but nonsymptomatic trees.

Trees found to be infested are removed and destroyed to prevent insects from completing development and emerging. Healthy, uninfested trees within one-eighth of a mile of an infested tree are treated with a soil drench or trunk injection of imidacloprid, a systemic neonicotinoid insecticide. These application methods eliminate issues of drift associated with insecticide sprays, and the insecticide affects only those insects that actually feed on a treated tree. Imidacloprid products generally control beetle larvae and affect some other groups of insects such as sap feeders, while other insects such as Lepidoptera are rarely affected.

Successful eradication of the ALB was declared in 2007 for Chicago and in 2008 for one of the New Jersey infestations. Successful eradication was credited to several factors including relatively early detection of the infestations, intensive survey efforts, and aggressive treatment of infested and surrounding trees. Extensive public awareness efforts and cooperation among federal, state, and local agencies helped to build support for the project among residents, including affected property owners. Biological attributes of the ALB including its relatively slow natural dispersal and the fact that ALB infestations did not become established in extensive areas of forest also likely contributed to effective eradication. Other eradication programs in Massachusetts, New Jersey, New York, and Toronto continue. Eradication in Massachusetts and New York is complicated by the spatial extent of infestations and the ongoing discovery of additional infestations aided by human transport of infested wood.

EMERALD ASH BORER: OHIO

Officials implemented a similar approach when the first emerald ash borer (*Agrilus planipennis*) population in Ohio was discovered in January 2003 near Toledo. This beetle, native to Asia, had been identified as the cause of widespread ash mortality in southeastern Michigan in July 2002. Larvae feed under the bark on phloem in summer, overwinter, and then pupate, and then adults emerge the following summer. The S-shaped feeding galleries of the larvae disrupt transport of nutrients and water within the tree. In its native range, the emerald ash borer (EAB) is a secondary pest, colonizing only ash trees that are severely stressed or dying. North American ash trees have little resistance to this pest, however, and tens of millions of ash trees have now been killed. Little information about the biology of this phloem-feeding insect was available at the time of its discovery, and nothing was known about its control or management.

Once the EAB was confirmed, visual surveys were immediately initiated in the area to define the extent of the infestation. Surveyors looked for trees with declining canopies, epicormic shoots, holes made by woodpeckers feeding on EAB larvae, or the small, D-shaped exit holes in the bark left by emerging adult beetles. Officials delimited the infestation and then established a 0.50 mile radius around trees known to be infested. All ash trees within 0.25 miles of a symptomatic tree were felled and destroyed. Ash trees in the next 0.25 radius were treated with a soil drench of imidacloprid.

Unfortunately, as entomologists learned more about the EAB, it became apparent that trees with low densities of EAB exhibit no external symptoms and that the actual extent of the infestation likely extended well beyond the 0.50 mile radius. In addition, while imidacloprid has effectively protected some landscape ash trees from EAB, larval mortality in treated trees rarely approaches 100 percent, and at least some beetles likely emerged from the treated trees. Scientists also eventually determined that the EAB had been established in southeastern Michigan for at least ten years before it was identified. During that time, infested ash nursery trees, logs, and firewood were transported into many areas, initiating infestations in many locations in Ohio and other states. Quarantines now regulate transport of ash trees, logs, and untreated wood, but restricting movement of infested ash firewood remains challenging. Research has also shown that the flight ability, rate of spread, and reproductive rate of the EAB are substantially greater than those of the ALB. Other eradication efforts have been attempted for apparently localized EAB infestations in Michigan, Maryland, and other states, but few have been successful, largely because of the difficulty of detecting newly infested trees and the dispersal ability of adult beetles.

JAPANESE BEETLE

The Japanese beetle (*Popillia japonica*) has been the target of intensive eradication and management efforts since it was discovered in the United States in 1916 at a nursery in New Jersey. Native to Japan, the Japanese beetle found the eastern United States to its liking and quickly achieved status as a key pest. Larvae feed on the roots of many grasses, while the highly polyphagous adults feed on the foliage, flowers, and fruit of a wide range of ornamental plants, fruit trees, and vegetables (Fig. 2).

The first Japanese beetle eradication program was initiated in 1918. The strategy was to construct a 0.5 to 1 mile barrier zone around the infestation by destroying host plants or treating them with lead arsenate and lime. This strategy proved ineffective, as Japanese beetle adults easily breached the barrier zone, and the infested area in the United States had increased from 2.5 to 6.7 square miles by the end of the 1918. In the fall of 1918, 17 acres of pasture were treated with sodium cyanide, but the infestation expanded to 48 square miles, and an additional 50 acres were treated in the fall of 1919.

FIGURE 2 Japanese beetle adults, *Popillia japonica*. (Photograph courtesy of Daniel A. Herms.)

A final attempt to eradicate the Japanese beetle from the United States began in 1920. This program included a combination of using herbicides to kill host plants, poisoning adult beetles by treating host plants with lead arsenate, and collecting beetles by hand. Again, the eradication program proved ineffective, and the Japanese beetle continued to spread throughout the eastern seaboard states and eventually throughout the Midwest. As of 1998, the Japanese beetle was known to exist in all states east of the Mississippi River except for Florida.

The Japanese beetle is a superb hitchhiker, and isolated outlier infestations have been detected periodically beyond its contiguous range. Insecticides have been used successfully to eradicate such infestations. In 1936, an aggressive, and ultimately successful, program was initiated to eradicate an infestation in St. Louis, Missouri, that had been discovered two years earlier. Over several years, hundreds of acres were treated with lead arsenate and carbon disulfide emission.

DDT was first employed in the battle against the Japanese beetle in a pilot program in 1945 in North Carolina, which was designed to test its efficacy as an eradication tool. Beginning in 1954, aerial applications of DDT and dieldrin were applied to thousands of acres over a five year period in Indiana and Illinois, where the Japanese beetle was considered a major threat to field crops. Carbaryl and chlordane were first used in an attempt to eradicate the Japanese beetle in Sacramento, California, when an infestation was discovered there in 1961. Host plants were treated with carbaryl to kill adult beetles, using hydraulic and ultra-low-volume mist applicators, as well as dust

formulations. Larvae were targeted with soil applications of chlordane. The isolated nature of the population, coupled with very limited larval habitat (turfgrass), facilitated the success of the intensive insecticide treatment program, and eradication was deemed successful in 1964.

A combination of chlordane and carbaryl were also used successfully to eradicate the Japanese beetle from San Diego in 1974. However, unintended consequences were documented in fruit orchards, as the insecticide applications disrupted what had been a successful biological control program of woolly whitefly (*Aleurothrixus floccosus*) and triggered secondary outbreaks of citrus red mite (*Panonychus citri*). An isolated infestation that had been detected in the vicinity of Sacramento, California, in 1983 was eradicated in 1985. Other infestations also have been successfully eradicated from Nevada, Idaho, and Oregon.

A localized infestation of Japanese beetle was detected in Palisade, Colorado, in 2002, and an eradication program there was initiated that combined mass trapping and cultural controls (withholding irrigation from lawns) with soil applications of imidacloprid that targeted larvae. As humans continue to aid the dispersal of the Japanese beetle, isolated outlier infestations will continue to be detected, and insecticidal approaches used in eradication programs will continue to evolve.

CONCLUSION

This entry demonstrates the important role of insecticides in eradication programs that targeted a wide range of insect pests affecting natural and agricultural systems as well as human health. These programs have often engendered controversy, in part because of concerns about health or environmental effects associated with the insecticides or the potential evolution of insecticide resistance within the invader population.

Despite the short generation times and high reproductive rates of many insect species, insecticide resistance has rarely been an issue in either successful or unsuccessful eradication programs. Of the eight cases presented in this entry, insecticide resistance evolved only in the boll weevil, but this occurred in the 1950s, prior to the initiation of eradication efforts, when populations developed resistance to chlorinated hydrocarbons. Emergence of insecticide resistance in populations of nonnative insects that experienced genetic bottlenecks may be deferred for some time if the founding population does not include the rare mutations that confer resistance.

Over time, the availability of more selective insecticides, improved application technology, and integrated strategies that combine insecticides with other forms of control have helped to reduce some negative impacts of large eradication projects. These programs also illustrate the importance of human behavior for the success of eradication efforts. The willingness of growers, government agencies, and the general public to support eradication programs once a pest population has been suppressed below economic thresholds has strongly influenced the outcome of many projects. Success of future pest eradication efforts will likely depend on technological improvements, advances in understanding the population dynamics of target pests, and outreach education programs that engender public support.

SEE ALSO THE FOLLOWING ARTICLES

Agriculture / Ants / Eradication / Forest Insects / Gypsy Moth / Integrated Pest Management / Mosquitoes

FURTHER READING

Buhs, J. B. 2004. *The Fire Ant Wars*. Chicago: University of Chicago Press.

Dahlsten, D. L., and R. Garcia, eds. *Eradication of Exotic Pests: Analysis with Case Histories*. New Haven, CT: Yale University Press.

Doane, C. C., and M. L. McManus. 1981. *The Gypsy Moth: Research Toward Integrated Management*. USDA Forest Service Technical Bulletin 1584.

Graham, O. H., and J. L. Hourrigan. 1977. Eradication programs for the arthropod parasites of livestock. *Journal of Medical Entomology* 6: 629–657.

Hardee, D. D., and F. A. Harris. 2003. Eradicating the boll weevil. *American Entomologist* 49: 82–97.

Klassen, W. 1989. Eradication of introduced arthropod pests: Theory and historical practice. *Miscellaneous Publications of the Entomological Society of America* 73: 1–29.

Liebhold, A. M., and J. Bascompte. 2003. The Allee effect, stochastic dynamics and the eradication of alien species. *Ecological Letters* 6: 133–140.

Myers, J. H., A. Savoie, and E. van Randen. 1998. Eradication and pest management. *Annual Review of Entomology* 43: 471–491.

Popham, W. L., and D. G. Hall. 1958. Insect eradication programs. *Annual Review of Entomology* 3: 335–354.

Spear, R. J. 2005. *The Great Gypsy Moth War*. Amherst: University of Massachusetts Press.

PESTICIDES (MAMMAL)

DESLEY A. WHISSON

Deakin University, Melbourne, Victoria, Australia

A "pesticide" is defined here as any toxic material that kills a pest. For mammals, this includes materials that are inhaled (i.e., fumigants) or consumed (i.e., toxic baits). These materials are often referred to as "vertebrate pesticides," a category that also includes materials used to control birds, reptiles, fish, and amphibians; "rodenticides" (for rodents); or "predacides" (for predators). Pesticides have been used in the management or eradication of a diversity of invasive mammals including rodents, possums, rabbits, cats, canids such as the European red fox (*Vulpes vulpes*), mustelids (ferrets and stoats in New Zealand), and feral pigs (*Sus scrofa*). Vertebrate pesticides have a long history of use in urban and agricultural situations; however, they are increasingly being used in natural environments (especially islands) to mitigate impacts of invasive species. In most countries, vertebrate pesticides must be approved for sale and use by a government agency. Regulatory toxicology studies are usually conducted before a vertebrate pesticide is registered for use and are used proactively to assess the risk of the compound to humans, pets, livestock, wildlife, and the environment. They may also be conducted on older products to provide additional toxicology data required to meet new registration standards.

BURROW FUMIGANTS

Burrow fumigants include carbon monoxide, aluminium phosphide, hydrogen cyanide, carbon disulfide, methyl bromide, acrolein, and chloropicrin. Many of these are no longer used due to animal welfare concerns. Depending on the fumigant and target species, gases may be allowed to disperse passively or are mechanically propelled throughout burrows, warrens, or dens. Because burrow fumigation is labor intensive and costly, it is generally used only as a follow-up to other methods.

TOXIC BAITS

Toxic baits generally fall into two categories: anticoagulants (compounds that inhibit the synthesis of vitamin K–dependent clotting factors in the liver) and nonanticoagulants (all other toxicants).

Anticoagulant pesticides have predominantly been used for commensal rodent control but have also played a major role in the eradication and management of rodents in natural environments. Anticoagulants are also used for the management of common brushtail possums (*Trichosurus vulpecula*) in New Zealand. Anticoagulants were developed as pesticides in the 1940s following their use in human medicine. They are chemically separated into two general groups: the hydroxycoumarins (e.g., warfarin) and the indandiones (e.g., diphacinone), and they act by inhibiting synthesis of vitamin K–dependent blood-clotting factors in the liver. Animals poisoned with anticoagulants typically die within 3 to 10 days from internal haemorrhaging as a result of a loss of the blood's clotting ability and increased permeability of capillaries throughout the body. The lengthened clotting time (prothrombin time, or PT) from a toxic dose of anticoagulant may be evident within 24 hours but usually reaches a maximum in 36–72 hours. Prior to death, the animal may exhibit increasing weakness due to blood loss. Because of the slow action of anticoagulants (due to the long half-life of blood-clotting factors), the target animal does not associate poisoning symptoms with the bait eaten and does not become "bait shy." This is an advantage when one is dealing with neophobic species that may hesitate to feed on a novel food. The animal can accumulate a lethal dose after multiple small feeds on the bait. The slow action of anticoagulants also has a safety advantage because it provides time to administer the antidote (vitamin K_1) to nontargets (humans, pets, other wildlife) that may have ingested bait. A disadvantage of anticoagulants is that toxic residues accumulate in tissues and in the liver of the animal consuming the bait. This presents a risk to predators and scavengers that may feed on a poisoned animal (i.e., secondary poisoning).

Warfarin was the first anticoagulant pesticide developed and is one of a group of compounds known as "first-generation" anticoagulants. Other first-generation anticoagulants include pindone, diphacinone, chlorophacinone, and coumatetralyl. With these anticoagulants, animals must consume multiple doses of the bait over a period of up to two weeks to elicit a toxic effect. The development of resistance to first-generation anticoagulants in commensal rodent populations has been a major issue affecting use of these compounds. Resistance of rats to warfarin was first observed in Scotland in 1958 following several years of continued use of this compound. Soon afterward, anticoagulant resistance was identified in both rats and house mice in other European countries, and later in the United States. Rats and mice that are resistant to warfarin are cross-resistant to all first-generation anticoagulants. Warfarin resistance stimulated developmental research on new rodenticides (both anticoagulant and nonanticoagulant) and resulted in the "second-generation" anticoagulants bromadiolone, brodifacoum, difenacoum, flocoumafen, and difethialone.

Second-generation anticoagulants have higher toxicity (lower LD_{50}), and longer persistence than the first-generation anticoagulants, and they require only a single feed of sufficient bait to elicit a toxic response. The effects of these compounds are also cumulative. As with the first-generation anticoagulants, death is delayed for several days following ingestion of a lethal dose. The greater persistence and toxicity of second-generation anticoagulants also increases the risk of poisoning of nontarget animals. Residues can remain in body tissues for long periods (months), because they are not readily metabolized. Secondary poisoning with anticoagulants has been well documented in a wide range of native birds and mammals. Resistance to second-generation anticoagulants has been observed, primarily in European countries.

For control of invasive mammals in natural environments, diphacinone and brodifacoum have had the most widespread use. Brodifacoum (3-[3-(40-bromo-[1,10-biphenyl]-4-yl)-1,2,3,4-tetrahydro-1-naphthalenyl]-4-hydroxy-2H-1-benzopyran-2-one), a second-generation anticoagulant, has been successfully used to eradicate invasive rats (*Rattus rattus*, *R. exulans*, *R. norvegicus*) on many islands worldwide. The greater persistence and potency of brodifacoum makes it ideal for use in rat eradications. Although there is a high risk of nontarget poisoning associated with this compound, the risks are generally considered to be short term and to be outweighed by the long-term benefits of rat removal. Rapid recovery of native species' populations following invasive rat eradication with brodifacoum is commonly reported. Brodifacoum has also been used in New Zealand for control of common brushtail possums.

When invasive rodents must be managed in areas where the risk of nontarget poisoning is unacceptably high, less persistent or less toxic anticoagulants are often used. Diphacinone (2-(diphenylacetyl)-1,3-indandione), a first-generation anticoagulant, has been successfully used to eradicate rats from islands including Buck Island (Virgin Islands of the United States) and the South Island of the San Jorge Islands (Mexico). Diphacinone also has been used for controlling invasive rat populations in forests in Hawaii and Puerto Rico.

Nonanticoagulant pesticides (organic and inorganic compounds) include strychnine, sodium cyanide, zinc phosphide, sodium monofluoroacetate (1080), cholecalciferol, calciferol, bromethalin, alpha-chlorohydrin, arsenic, red squill, flupropadine, and para-aminopropiophenone. They have different modes of action that may be either acute (i.e., with a single feed required) or chronic (i.e., with multiple feeds required). Many of the older pesticides, formally referred to as the acute toxicants (e.g., arsenic and red squill), either are no longer registered or are rarely used due to their ineffectiveness or high risk relative to newer pesticides. Ineffectiveness of nonanticoagulants has often been attributed to the rapid onset of poisoning symptoms resulting in bait shyness. Newer nonanticoagulant pesticides (e.g., cholecalciferol and bromethalin) have a slower action so that bait shyness rarely occurs.

Nonanticoagulants are commonly used for commensal rodent control, although some (e.g., zinc phosphide, cyanide, cholecalciferol, and sodium monofluoroacetate) are used in field baiting programs. Of these, 1080 (sodium monofluoroacetate) has had the most widespread use and application for control of a diversity of invasive mammals. It is well known as a predacide but has also been used to manage common brushtail possums in New Zealand, feral pigs (*Sus scrofa*) in Australia, and European rabbits (*Oryctolagus cuniculus*). Its use in some countries has been discontinued due to concerns over its risk to nontarget species, persistence in the environment, and humaneness. A naturally occurring secondary plant compound, 1080 has evolved at high concentrations in some plant species as a defense mechanism against browsing invertebrates and vertebrates. Once ingested, monofluoracetate is converted within the animal to fluorocitrate, which inhibits the tricarboxylic acid cycle. This results in an accumulation of citrate in tissues and plasma, energy deprivation, and death as a result of cardiac or respiratory failure. Clinical signs of 1080 poisoning in mammals occur between 0.5 and 3 hours following ingestion and may include drowsiness, tremors, convulsions, nausea, and vomiting. Although 1080 is rapidly eliminated from living animals, it can persist in carcasses for periods of up to several months and therefore generate high secondary poisoning risks.

Sensitivity of mammals to 1080 varies widely. Dogs are extremely susceptible, and most other carnivores are highly sensitive. In some areas, native animals that forage in areas where fluoracetate-producing plants (e.g., plants of the genus *Gastrolobium*) are common have evolved a tolerance to the pesticide. This tolerance therefore reduces the nontarget hazards of baiting with 1080. In Western Australia where this occurs, 1080 has been an important component of a program known as "Western Shield," which was initiated in 1996 and aims to recover native fauna that have been adversely impacted by invasive predators (foxes and cats). The program, has involved aerial application of 1080 baits to around 3.5 million hectares of land several times each year.

Controversy over the use of 1080 has led to research into other predacides. Para-aminopropiophenone (PAPP) has been identified as an effective predacide that may be more target specific and humane than 1080. PAPP induces methaemoglobinaemia, which prevents oxygen from binding to red blood cells. This reduces the oxygen supply to the brain, and animals become lethargic and then unconscious prior to death in one to two hours.

BAITING STRATEGIES

Choice of a pesticide and how it is applied is influenced by many factors, including the target species, pesticide type and efficacy, desired outcome (i.e., eradication or control), location, potential environmental and nontarget hazards, resources available, regulations, and socio-political issues. As with other control methods, timing and the area treated are important considerations in developing an effective program using pesticides.

Vertebrate pesticides may be applied to a variety of baits including grains, vegetables, meats (fresh or dried), offal, and eggs, and there are commercially manufactured baits such as pellets, blocks, pastes, and gels that aim to improve target specificity. Mold inhibitors, attractants (olfactory or visual lures), insect repellents, and dyes may be added to improve the attractiveness, target specificity, or shelf life of baits. Concerns over the humaneness of some vertebrate pesticides have prompted research into the addition of analgesics into baits to reduce possible pain and distress associated with poisoning symptoms.

Bait application rates vary depending on the target species (population density, home range size, and habitat use) as well as the pesticide and the method of bait presentation. Bait must be applied at a rate that allows each target animal to obtain a lethal dose while minimizing the risk of excessive bait being available to nontargets. The pattern of bait placement is also an important consideration, as this can affect the frequency with which baits are encountered by both target and nontarget animals. In predator control programs, placement of baits along roads or tracks can increase the bait encounter rate of dogs and foxes that use these paths.

Bait may be applied in bait stations or other delivery devices, or by hand or aerial broadcasting. Bait may also be buried (e.g., for control of European red foxes in Australia) to reduce the potential for nontarget poisoning. In many cases, multiple delivery methods are used. Bait stations are commonly used to deliver multiple-feed anticoagulant pesticides. They can be designed to be accessible only to the target species, so they are often useful in areas where the risk of nontarget poisoning is high. The spacing of bait stations must consider the home range and habitat use by the target species so that all target animals have access to the bait. The

M-44 ejector is a bait-delivery device used to deliver predacides. This mechanical device delivers a dose of toxicant (in powder form) into the mouth of an animal biting the trigger mechanism (Fig. 1). Activation of the ejector requires significant upward force such that only relatively large animals are likely to be capable of releasing the trigger. Because the ejector is anchored in position, the risk of bait caching (common with some predators) is eliminated. Sodium cyanide is commonly used in these units. The powder reacts with the moisture in the animal's mouth, releasing hydrogen cyanide gas. Death occurs from ten seconds to two minutes after the device is triggered (Fig. 2).

In addition to minimizing bait exposure to nontarget species, bait stations allow bait uptake to be monitored and can be used in combination with nontoxic baits or tracking boards or pads to monitor the effectiveness of a control program. However, the approach is labor intensive and potentially expensive at large scales and may be impractical in rugged terrain with inaccessible areas. Regular visits to monitor bait stations can also result in disturbance of sensitive species (e.g., breeding seabirds).

Aerial broadcast is a common delivery method for vertebrate pesticides and is often used where concerns about nontarget poisoning are low. It is more cost-effective than bait stations, and bait can be applied to large or inaccessible areas. Broadcasting bait also increases the potential

FIGURE 2 A discharged M-44 ejector (foreground) with the carcass of the targeted fox nearby. (Photograph courtesy of Rob Hunt, NSW DECCW.)

for all individuals in the population to access bait. It has been used as the primary method of delivering poison bait in rodent eradication programs on islands, for predator control (e.g., fox control in Western Australia), and for possum control in New Zealand.

SEE ALSO THE FOLLOWING ARTICLES

Eradication / Islands / Mammals, Aquatic / Rats / Rodents

FURTHER READING

Brakes, C. R., and R. H. Smith. 2005. Exposure of non-target small mammals to rodenticides: Short-term effects, recovery and implications for secondary poisoning. *Journal of Applied Ecology* 42: 118–128.

Donlan, C. J., G. R. Howald, B. R. Tershy, and D. A. Croll. 2003. Evaluating alternative rodenticides for island conservation: Roof rat eradication from the San Jorge Islands, Mexico. *Biological Conservation* 114: 29–34.

Greaves, J. H., A. P. Buckle, and R. H. Smith, eds. 1994. *Rodent Pests and Their Control*. Wallingford: CAB International.

Howald, G., C. J. Donlan, J. P. Galvan, J. C. Russell, J. Parkes, A. Samaniego, Y. Wang, D. Veitch, P. Genovesi, M. Pascal, A. Saunders, and B. Tershy. 2007. Invasive rodent eradication on islands. *Conservation Biology* 21: 1258–1268.

Hyngstrom, S. E., R. M. Timm, and G. E. Larson, eds. 1994. Prevention and Control of Wildlife Damage. Nebraska Cooperative Extension Service, University of Nebraska, Lincoln, USDA-APHIS-Animal Damage Control, and Great Plains Agricultural Council.

Marsh, R. E. 2001. Vertebrate pest control chemicals and their use in urban and rural environments (251–262). In R. I. Krieger, ed. *Handbook of Pesticide Toxicology: Principles and Agents*, 2nd ed. San Diego, CA: Academic Press.

Nelson, J. T., B. L. Woodworth, S. G. Fancy, G. D. Lindsey, and E. J. Tweed. 2002. Effectiveness of rodent control and monitoring techniques for a montane rainforest. *Wildlife Society Bulletin* 30: 82–92.

FIGURE 1 A set M-44 ejector. When the target animal pulls on the baited ejector head, a spring-loaded plunger propels through a capsule containing the toxicant (center of the head), discharging the contents into the animal's mouth. (Photograph courtesy of Rob Hunt, NSW DECCW.)

PEST MANAGEMENT

SEE BIOLOGICAL CONTROL; INTEGRATED PEST
MANAGEMENT; HERBICIDES; MECHANICAL CONTROL;
PESTICIDES

PET TRADE

PETER T. JENKINS

Defenders of Wildlife, Washington, DC

The United States has a long history of importing myriad nonnative animals for the pet and home aquarium trades. Animals representing hundreds of different species have been released into the wild intentionally by their owners or inadvertently allowed to escape from confinement. A significant portion of these have invaded and caused serious environmental and economic problems, as well as transmitted diseases to other animals (both domesticated and wild) and to people. This article does not analyze either the quantitative or qualitative benefits provided by the pet and home aquarium trade; however, these are multibillion dollar sectors of the economy.[1] Rather, the focus is the biological "externalities" of this trade (i.e., unintended harms it can cause if regulated too liberally). This entry covers only the United States, a world "leader" in nonnative animal imports; however, the lessons learned may apply elsewhere.

THE PET AND HOME AQUARIUM PATHWAY

It is broadly accepted that the pet and home aquarium pathway has increased in importance as a source of nonnative animal invasions. Formerly, many such invasions (e.g., of cane toads, nutria, and mongoose) resulted as side effects of intentional stocking for sport, agriculture, and other utilitarian purposes, where the introduction went awry, and harmful effects were unanticipated. Many of those past introductions were done by federal and state agencies, which are now doing far fewer of them. However, the importation of novel pets and tropical fish for private enjoyment continues a steady upward trend—and the inevitable escapes, releases, and disease outbreaks that accompany it have continued unabated. Examples of harmful pet- and aquarium-trade invasions and disease

---

[1] It should be noted that the pet and home aquarium sector sells what are basically "luxury goods," not food, fiber, or other essential goods.

outbreaks resulting from these increasing imports include the following:

- Gambian giant rats imported in the pet trade caused the 2003 (first ever) U.S. outbreak of monkeypox virus, resulting in 72 cases nationwide. In humans, the symptoms of monkeypox are similar to those of smallpox but are typically milder. In Africa, it is fatal in as many as 10 percent of cases. The Centers for Disease Control and Prevention and the Food and Drug Administration banned importation of all African rodents as pets. Nevertheless, released Gambian giant rats themselves invaded in the Florida Keys, requiring an expensive, multiagency eradication campaign.
- The suckermouth catfish is a very common home aquarium fish that has invaded in several southern states from Florida to Texas, where it grows to large body size and burrows into waterway shores to build nests in massive numbers, degrading levees, dikes, and natural shoreline ecosystems and threatening populations of native birds that consume its sharp spines and then choke to death.
- Several pet tortoise species in the Geochelone and Kinixys genera were banned from further importation by the U.S. Department of Agriculture in 2000 because they carry a tick that transmits heartwater, a disease that sickens livestock and native deer.

Harmful escapes, releases, and disease outbreaks associated with nonnative pets and aquarium fish are largely avoidable through a science-based risk-screening approach. Indeed, at least Australia, Israel, and New Zealand already have adopted screening approaches, and they have been effective in stopping further harmful introductions. Comprehensive new legislation has been proposed to achieve these protections in the United States as well, but its status remains unclear as of this writing.

SIZE AND NATURE OF THE NONNATIVE WILD ANIMAL IMPORT TRADE

Quantity and Diversity of Imports

The U.S. pet and home aquarium import trade has grown significantly since the 1990s, driven in part by the popularity of novel, exotic species. The ease of international travel, shipping, and financial transactions has made importing animals easier than ever. Importers have access to the latest communications technology and overnight air cargo services from almost anywhere in the world. The Internet allows online sellers of nonnative animals to deal more directly with buyers, which circumvents pet stores and often avoids licensing and inspection systems.

TABLE 1
Quantities of Live Imported Vertebrate Animals

Vertebrates	Number of Individual Animals Imported			Weight of Imports, for Imports Not Counted Individually, in Kilograms (kg)		
	Identified Species	Unidentified Species	Proportion Identified	Identified Species	Unidentified Species	Proportion Identified
Amphibians	23,780,548	2,577,619	90%	1,288,908	9,839	99%
Birds	1,470,703	693,647	68%	350	0	100%
Fish	15,218,584	876,792,759	2%	1,760,016	980,830	64%
Mammals	214,871	23,450	90%	11,833	915	93%
Reptiles	6,733,326	2,364,015	74%	0	0	—
All Taxa (five years)	47,418,032	882,451,490	5%	3,061,107	991,584	76%

NOTE: From United States Fish and Wildlife Service (FWS), Law Enforcement Management Information System (LEMIS). Includes those identified and unidentified to the species level over five years (2000–2004). "Unidentified species" typically were minimally identified in the publicly available data to class, order, family, or genus, or were labeled broadly, such as "tropical fish" or "non-CITES fish." Due to time and personnel constraints, the FWS does not enter every species name in the LEMIS data that the agency compiles.

Table 1 indicates the scope of the animal import trade for vertebrates. It also shows the significant proportion of shipments for which the available import data do not include identification to the species level ("unidentified species") in the publicly available summary data for these imports. Data presented herein on animal imports are from federal compilations for which the ultimate end use of the animal is not documented consistently. Thus, not all quantities presented here are entirely of imports for the pet and home aquarium trades—some may be for other sectors, such as the live seafood, live bait, zoo display, and scientific markets. Nevertheless, several studies have shown imports for the pet and home aquarium sector to dominate the other sectors quantitatively.

In sum, more than 950 million individually counted vertebrates were imported during the five year period 2000 to 2004, and more than 4 million kg of animals by weight were imported—for example, large numbers of fish or live frogs weighed together in a shipment, rather than counted individually. This high level of "propagule pressure" illustrates the scope of the risk from pet escapes and releases.

According to detailed accounting by Romagosa (unpublished), import data show that, since 1998, the United States has added, on average, more than 185 *new* species of mammals, birds, amphibians, turtles, lizards, and snakes to the vertebrate import pool annually. The accumulated number of different identified species in these classes (which excludes the largest single category of imports, fish for home aquaria) ever imported totaled approximately 2,470 in 2006 and likely now exceeds 2,800 in 2009. In short, the diversity of identified imported species of mammals, birds, amphibian, turtles, lizards, and snakes vastly exceeds the cumulative total of native U.S. species in those groups (approximately 1,800 native species). In short, native species are being "swamped" by the import trade. While not all of these are for the pet or home aquarium industries, various studies indicate that the bulk of them are. And, again, large numbers of imported species are not fully identified to the species level.

Nature of the Pet and Home Aquarium Import Trade

Other than specified mammals, birds, and a few fish species, no other imported species for pet and home aquarium use need to undergo a quarantine period or have proof of prior veterinary clearance from their country of origin. Shipments of a vast variety of nonnative species that cross the border may quickly disseminate throughout the 50 states, none of which, besides insular Hawaii, possess the ability to exercise effective state control over interstate commerce and travel. Thus, by way of illustration, a tropical snake, fish, or bird caught in the wild in Africa could arrive in a home in, say, Florida as a pet and then be moved by its owner to, say, Louisiana, all in a matter of a few days. If it somehow escapes from captivity, or is released on purpose by its owner, and it manages to survive, then this animal could become a new wild resident of North America in short order. This scenario in fact has occurred repeatedly.

Not all live nonnative animals shipped and sold in the United States are imported. Many species are now captive bred from imported ancestor stock in commercial enterprises (e.g., aquaculture facilities, pet breeders, or hobbyists); thus, they would not appear in the Fish and Wildlife Service import data. No comprehensive lists exist of all the nonnative pet species being captive bred in the United States. Nor do any federal regulatory requirements exist that mandate or facilitate collection of these data. Many pet breeders are small-scale or "backyard" family enterprises that may stay in business just a few years.

PET INVADERS

For a variety of reasons, people regularly "free" their nonnative pets and aquarium fish without permission, and often in violation of state laws. Their motivation may be that the owner cannot or will not keep an animal further for any number of reasons, but he or she is unable or unwilling to find a new home for, or euthanize, the animal. And, of course, escapes and accidental releases occur. Becoming established successfully after release or escape is a process, not a single event, and it can take many years or decades. Released or escaped nonnative animals often will die off in their new location, but a proportion, estimated at up to 50 percent for some species, may establish a self-sustaining population. At the next stage of invasion, many such species remain localized, and most probably are not even detected by humans. Some populations remain localized for many years. However, a proportion of such established species, again up to 50 percent for some animals, may spread further and become abundant in many new locales.

Several studies have addressed invasive animals, but none as comprehensively as the 1993 U.S. Congress Office of Technology Assessment (OTA) report, *Harmful Non-Indigenous Species in the United States*. The OTA found that, of the 212 total vertebrate species of non-U.S. origin that had established wild populations in the United States, 48 of them (or 23%) were originally imported as cage birds or other pets. It also found the aquarium trade to be the source of at least 27 established fish species. In just an 11 year period (from 1980 to 1991), releases from aquariums led to successful invasions by at least seven new nonnative fish species. The OTA further stated: "Given the high U.S. rates of pet imports—estimated to be hundreds of thousands to millions of wild birds, aquarium fish, and reptiles annually—the potential for pet escapes and releases is great." Since completion of the OTA report in 1992, there has been no indication of a slackening of the risk—indeed, the risk has become markedly greater as the numbers and variety of pet imports have increased.

Infectious pathogens and harmful parasites have accompanied recent pet animal imports. Exotic Newcastle's disease (END), heartwater, and the deadly amphibian disease chytridiomycosis are examples. Not all of these are exclusive to the pet pathway; they can also arrive via poultry imports (END) and the live frog food trade (chytridiomycosis), for example.

Examples of Recent Invasions through the Pet Trade

BURMESE PYTHON Thousands of giant Burmese pythons (*Python molurus bivittatus*; Fig. 1) are competing with and attacking native wildlife in Everglades National Park. First detected in the mid-1990s, the population likely started as pets released by owners who did not want to keep them. After several years the invasion became established, and it is now spreading rapidly. The pythons prey on native animals including rare species protected under the Endangered Species Act, they are harming the ability of Everglades restoration projects to meet their benchmarks by removing large numbers of native indicator species, and these large constrictors present a long-term human safety risk. Pet pythons in captivity have killed more than ten humans (mostly their owners or their owners' children) in this country over the last two decades. A multimillion dollar control program is needed to stop their further spread, but there are not nearly enough funds available to carry it out.

FIGURE 1 A National Park Service biologist examines a 10 foot, 3 inch Burmese python captured on an access road to Everglades National Park. The pythons—some of the world's largest snakes and common in the pet trade—have established a huge population in southern Florida, likely due to being released by disenchanted owners. (Photograph courtesy of Lori Oberhoffer, National Park Service.)

MONK PARAKEET Monk parakeets (*Myiopsitta monarchus*) were sold broadly in the pet trade and, by the 1970s, had established themselves outside captivity. By 2001, they were reported in 12 states, with the largest population in Florida (50,000 to 150,000 birds), and with Texas a close second. Monk parakeets nest in utility poles, transmission towers, and electric substations in urban areas. Their sturdy, long-lasting nests can be up to a meter in diameter, can weigh up to 220 kg, can house multiple pairs, and allow the birds to survive in harsh winters. Nests have short-circuited and damaged utility transmission lines, caused power outages, and sometimes set nesting poles ablaze, causing hundreds of thousands of dollars in damage. The U.S. monk parakeet population doubles approximately every five years and will continue to grow dramatically unless control efforts are stepped up, but controlling these charismatic birds frequently draws opposition from animal rights activists.

RED LIONFISH The red lionfish (*Pterois volitans*; Fig. 2) is a carnivorous tropical pet fish originating from Indo-Pacific areas. It was introduced in the 1990s, likely via aquarium release(s), and has formed fast-growing wild populations in the western Atlantic. This is the first documented establishment of a nonnative *marine* fish thought to have originated via escape or release from private aquaria. The numerous barbs of the red lionfish pose a safety hazard to divers and fishers because of their very painful venom. Negative ecosystem effects on reef life have been observed. Red lionfish are aggressive predators that feed on shrimp and other fish, including the young of important commercial species such as snapper and grouper. They continue to spread around the Florida

FIGURE 2 Poison-spined red lionfish now ply the western Atlantic and Caribbean—the first documented occurrence of a tropical, nonnative, marine fish established in the wild, likely due to accidental or intentional releases from home aquaria. (Photograph courtesy of John Randall, Hawaii Bishop Museum, U.S. Geological Survey.)

Keys and the Caribbean islands, much to the dismay of the affected recreational and commercial sectors.

REGULATION OF THE WILD ANIMAL IMPORT TRADE

Existing (2009) federal laws regulating the pet import trade leave no practical legal avenue to keep several classes of potentially harmful species out of the country. For nonnative species that are likely to be invasive or injurious in the United States, but that are *not* also known threats to human, livestock, or plant health, the Lacey Act "injurious species" process is the only available regulatory law. Yet this process requires notice and comment rulemaking that takes three to four years on average to complete for each taxon listing. No authority exists under the Lacey Act to keep imports of a species out of the country before or during this lengthy rulemaking. This over-100-year-old "backstop" law, written in 1900 and requiring cumbersome procedures to regulate just one species, is a classic example of a reactive approach.

The Ecological Society of America (ESA) issued a detailed "Position Paper on Biological Invasions," which states (emphasis added):

> [ESA] recommends that the federal government, in cooperation with state and local governments, take the following . . . actions: (1) Use new information and practices to better manage commercial and other pathways to reduce the transport and release of potentially harmful species. (2) *Adopt new, more quantitative procedures for risk analysis and apply them to every species proposed for importation into the country.*

In short, ESA recommends that the federal approach should change from primarily allowing species to enter the country until they are shown to be harmful to the proactive approach of prohibiting their entry until they are shown to be acceptably safe. Many other experts and advocacy organizations have adopted this position.

Keller et al., in the first national cost–benefit analysis of such an import risk-screening program, analyzed the existing Australian plant screening program and showed that it is generating large net long-term economic benefits for that country. Risk-based prescreening protocols are likely to generate even greater net benefits for animal imports because of two main factors: (1) animals generally have higher rates of successful invasion than do plants, and (2) animals tend to have a shorter lag time between introduction and invasion. Thus, the costs of invasive species of animals generally will accrue earlier than those for plants. The economic net-present value from a risk-screening system that prevents those costs would therefore be greater.

New laws have been proposed that would regulate the pet and home aquarium trade's unwanted externalities. The IIIth Congress's House Resolution 669, to enact the *Non-Native Wildlife Invasion Prevention Act*, would require the U.S. Secretary of the Interior, acting through the Fish and Wildlife Service, to assess the risk of nonnative wildlife species (both aquatic and terrestrial animals) proposed for importation and, with public input, decide whether the species should be allowed or prohibited. By providing an expeditious screening process for wild animals, H.R. 669 would address inadequacies of the existing law—the slow, reactive Lacey Act. It would address nondomesticated animals imported for any use (e.g., for the pet and home aquaria sectors, but also for aquaculture, scientific products, live bait, and seafood). As of this writing, it has not passed, but advocates for a better-regulated import trade for live wild animals are hopeful.

In addressing risks of the pet and home aquarium trade, it is vital to recognize that there are about 3,000 identified species already in the "import pool" (i.e., they have already been imported into the United States, some long ago, but many more in recent decades). These species already in the import pool need risk-screening too. Risk assessments that looked only at new (novel) species proposed to be added to this huge import pool would miss the vast bulk of the potential invasion risk: the species that are already here.

The small number of domesticated animals that make up the vast majority of U.S. pets are not the main challenge. The challenge is undomesticated, wild animals such as rare snakes, birds, and tropical fish. New species additions to the import pool actually likely do not add significant "companionship" value for society as a whole that could not be met by the thousands of pet species already imported. The current U.S. system of unregulated imports of thousands of species solely for their novelty, curiosity, or collection value is unsustainable in view of the environmental, health, and economic risks this system poses.

SEE ALSO THE FOLLOWING ARTICLES

Aquaria / Birds / Burmese Python and Other Giant Constrictors / Fishes / Laws, Federal and State / Pathogens, Animal / Reptiles and Amphibians

FURTHER READING

Fowler, A., D. Lodge, and J. Hsia. 2007. Failure of the Lacey Act to protect U.S. ecosystems against animal invasions. *Frontiers in Ecology and the Environment* 5: 353–359

Franke, J., and T. Telecky. 2001. *Reptiles as Pets. An Examination of the Trade in Live Reptiles in the United States.* Washington, DC: The Humane Society of the United States.

Jenkins, P., K. Genovese, and H. Ruffler. 2007. *Broken Screens: The Regulation of Live Animal Imports in the United States.* Washington, DC: Defenders of Wildlife. Available at www.defenders.org/animalimports

Jeschke, J., and D. Strayer. 2005. Invasion success of vertebrates in Europe and North America. *Proceedings of the National Academy of Sciences, USA* 102: 7198–7202.

Karesh, W., R. Cook, E. Bennett, and J. Newcomb. 2005. Wildlife trade and global disease emergence. *Emerging Infectious Diseases* 11: 1000–1002.

Keller, R., D. Lodge, and D. Finnoff. 2007. Risk assessment for invasive species produces net economic benefits. *Proceedings of the National Academy of Sciences, USA* 104: 203–207.

Kroeger, T. 2007. *Economic Impacts of Live Wild Animal Imports in the United States.* Washington, DC: Defenders of Wildlife. Available at www.defenders.org/animalimports

Lodge, D., S. Williams, H. MacIsaac, K. Hayes, B. Leung, S. Reichard, R. Mack, P. Moyle, M. Smith, D. Andow, J. Carlton, and A. McMichael. 2006. Biological invasions: Recommendations for U.S. policy and management. *Ecological Applications* 16: 2035–2054.

Padilla, D., and S. Williams. 2006. Beyond ballast water: Aquarium and ornamental trades as sources of invasive species in aquatic ecosystems. *Frontiers in Ecology and the Environment* 2: 131–138.

Simons, S. A., and M. De Poorter, eds. 2009. Best practices in pre-import risk screening for species of live animals in international trade: Proceedings of an expert workshop on preventing biological invasions, University of Notre Dame, Indiana, USA, 9–11 April 2008. Global Invasive Species Programme, Nairobi, Kenya. Available online at www.gisp.org/publications/policy/workshop-riskscreening-pettrade.pdf

PHYTOPHTHORA

DAVID M. RIZZO

University of California, Davis

Phytophthora is a genus of approximately 100 species of fungal-like, plant pathogenic organisms classified in the kingdom Stramenopila. Often known as "water molds," *Phytophthora* species have a swimming spore stage (when they are known as zoospores). *Phytophthora* species are well known as pathogens of agricultural, ornamental, and forest plants. Across the genus, individual *Phytophthora* species may infect roots, stems, leaves, flowers, or fruits of susceptible plants and cause dieback, decline, or death. Some *Phytophthora* species are specialists (infecting only one or a few plant species), and others are generalists (infecting many plant species across several or many plant families). The diversity of *Phytophthora* is quite astounding, and many new species have been described in the past ten years. In addition, *Phytophthora* species have become one of most important causes of emerging diseases in native plant communities.

GEOGRAPHIC ORIGINS AND SPREAD

Phytophthora species are commonly found in soil and water, and they often cause plant disease in managed and natural ecosystems throughout the world. Many individual species are currently considered to have a cosmopolitan distribution because they are found in many locations around the world. Much of the current distribution of these *Phytophthora* species is thought to have occurred through the movement of plants and soil during the spread of agriculture over hundreds of years. A number of *Phytophthora* species have been introduced into natural ecosystems around the world, sometimes with important ecological consequences.

Currently, the center of origin is unknown for most of the ecologically important invasive species of *Phytophthora* (Table 1). Population genetic (e.g., very reduced genetic variability in the pathogen population) and ecological (e.g., high susceptibility of some hosts) research supports the hypothesis that many of these species have

probably been recently introduced to new locations. Many *Phytophthora* species were originally known as important pathogens of cultivated plants before being detected in native plant communities. In recent years, a number of species (e.g., *P. ramorum, P. austrocedrae*) were first recognized as causing damage in natural ecosystems before they were identified in agricultural or horticultural settings. Recent research has pointed to the importance of the horticultural industry in the continued spread of *Phytophthora* species. Examples of potential dispersal pathways into native plant communities include the planting of infected ornamental plants at the urban–wildland interface and the use of infected plants in restoration efforts. Once established in a plant community, *Phytophthora* species may disperse by rainsplash, wind, movement of water through soil, stream water, or movement of soil by animals or humans.

Another potential source of invasive *Phytophthora* species is through interspecific hybridization events. The evolution of new species through mating events of two known *Phytophthora* species has been documented several times within the genus. For example, in Europe an

TABLE 1

Examples of Emerging Diseases Caused by *Phytophthora*

Phytophthora Species	Plant Host(s)	Geographic Area of Origin	Area of Invasion	Impact
P. cinnamomi	At least 266 genera in 90 families including both woody and herbaceous plants	Possibly Southeast Asia	Australia, Europe, North America (southeastern United States, California, Mexico)	Mortality associated with many genera
P. lateralis	Port Orford cedar (*Chamaecyparis lawsoniana*), Pacific yew (*Taxus brevifolia*)	Unknown, possibly Southeast Asia	California, Oregon	Extensive mortality in *Chamaecyparis*
P. ramorum	At least 26 genera in 17 families including both woody and herbaceous plants	Unknown	California, Oregon, United Kingdom	Mortality restricted to select species of Fagaceae (*Quercus, Lithocarpus densiflorus*)
P. alni	Alder (*Alnus* spp.)	Unknown, possibly of hybrid origin	Europe, Alaska	Extensive mortality of *Alnus* throughout Europe; not associated with mortality in Alaska
P. austrocedrae	*Austrocedrus chilensis*	Unknown	Argentina	Extensive mortality in *Austrocedrus*
P. kernoviae	Over 10 genera	Unknown	United Kingdom	Mortality noted in select species of Fagaceae (especially *Fagus*) and Ericaceae (*Rhododendron* and *Vaccinium*) found in woodlands and heathlands

NOTE: Recent examples of emerging diseases caused by *Phytophthora* species apparently introduced into natural ecosystems.

emerging pathogen of alder, *P. alni*, is thought to have recently evolved as a hybrid between two *Phytophthora* species that may not have been pathogens on alder. *Phytophthora alni* has now spread throughout Europe via water courses and nursery stock.

IMPACTS OF INVASIVE *PHYTOPHTHORA* SPECIES

Phytophthora species have caused, and continue to cause, devastating epidemics in both agricultural and native ecosystems. The ecology of *Phytophthora* invasions, however, often differs between the two types of ecosystems. During pathogen invasion of agroecosystems, plants are often reunited with pathogens with which they coevolved. Epidemics may result if the host–pathogen system is brought back together in a different geographic setting under very different conditions (e.g., monoculture, reduced genetic variability in the host, and increased moisture regimes). Perhaps the most famous example of this situation was the epidemic of late blight of potatoes, caused by *P. infestans*, associated with the Irish potato famine of the 1840s. The center of origin of *P. infestans* is thought to be southern Mexico, where it infects native species of potatoes without causing major epidemics. Potatoes were grown for many years in exotic locations (e.g., Europe, the United States) prior to the introduction of *P. infestans* from Mexico. In addition to *P. infestans*, there are many other very important *Phytophthora* species of agricultural crops. These include *P. sojae* (soybeans), *P. capsici* (many crops), *P. fragarie* (strawberries), and *P. parasitica* (many crops). The social and economic costs of these agricultural epidemics have been very high. Billions of dollars are spent each year to combat *Phytophthora* diseases through the use of fungicides, resistant plants, and other horticultural practices.

In contrast, *Phytophthora* species introduced into nonagricultural plant communities usually do not share a coevolutionary history (under any environmental condition) with potential hosts. In this situation, hosts have often (but not always) shown very high susceptibility to infection and death. Losses of plants may occur at multiple spatial scales, ranging from individual trees to entire landscapes. In some cases, millions of trees have been lost to infection leading to significant changes in the composition of plant communities. Two examples are discussed below.

Phytophthora cinnamomi is a soil-borne species that is known to cause root or crown rot of hundreds of plant species worldwide; it is especially important on woody plants. This pathogen has been introduced into native plant communities in Australia, North America, and Europe. *Phytophthora cinnamomi* appears to have been introduced in

the early 1900s into Australian forests and has been associated with dieback and death of native plants in the jarrah (*Eucalyptus marginata*) forest of Western Australia and in the Brisbane Ranges National Park, in Victoria (Fig. 1A). The pathogen has virtually eliminated tree species over hundreds of thousands of hectares of the jarrah forests, often converting them to grassland or shrubland. A number of rare species in the genera *Banksia*, *Darwinia*, *Grevillea*, and *Verticordia*, as well as the very rare Wollemi pine (*Wollemia nobilis*), have been threatened by the pathogen.

The emerging plant disease sudden oak death has caused extensive mortality of oaks (*Quercus* spp.) and tanoak (*Notholithocarpus densiflorus*) in coastal forests of California and Oregon since the mid-1990s (Fig. 1B). The disease is caused by *P. ramorum*, a generalist pathogen that also infects over 40 plant genera including

FIGURE 1 Examples of tree mortality caused by introduced species of *Phytophthora*. (A) *Phytophthora cinnamomi* in jarrah (*Eucalyptus marginata*) forest of Western Australia. (Photograph courtesy of Chris Dunne.) (B) Sudden oak death, caused by *P. ramorum* in a tanoak forest (*Lithocarpus densiflorus*) in coastal California. (Photograph courtesy of Kerri Frangioso.)

ferns, conifers, and herbaceous and perennial angio-sperms. However, extensive mortality appears to be restricted to oak and tanoak, leading to shifts in plant community structure. Cascading ecosystem impacts are beginning to be documented on birds, mammals, insects, and mycorrhizal fungi. In areas where *P. ramorum*–associated overstory mortality has significantly impacted the composition of forests, changes in forest floor inputs, organic matter, decomposition rates, and nutrient and water dynamics are also expected. High levels of tree mortality may lead to increased risks and impacts of wildfire due to excessive fuel loads. Loss of trees in urban and wildland situations due to *P. ramorum* has also led to increased costs of removal of hazardous dead and dying trees, loss of timber trees, and changes in ecosystem aesthetics.

MANAGEMENT

Management of *Phytophthora*-associated diseases can be very difficult once the pathogen is established, especially in native plant communities. Disease prevention and mitigation at the individual plant or field levels have focused on chemical control or other programs designed to maintain health of plants. Some fungicides have been developed that act as protectants (e.g., phosphites) against infection, but few chemicals have been developed that work once the plant is infected. A few breeding programs have developed plant varieties that are resistant to infection by specific *Phytophthora* species. For example, several *P. lateralis*–resistant varieties of Port Orford cedar have recently been developed for planting in landscapes.

In native plant communities, landscape management strategies for *P. ramorum* have included prevention, eradication, treatment, and restoration. Eradication has been attempted in some cases, most notably with *P. ramorum* in Oregon, but it has met with mixed success. Various other sanitation programs (i.e., infected tree removal, barrier fences, vehicle cleaning) have been implemented on a more limited scale to control the soil-borne pathogens *P. cinnamomi* in Australia and *P. lateralis* in Oregon and California. Fungicides have been applied widely in some forests in Australia to reduce the spread of *P. cinnamomi*. The effectiveness of these programs has been variable; important successes have been balanced by continuing tree mortality in many areas.

Ultimately, the best way to manage this genus is to prevent introduction in the first place. The broadest scale for management, regional to international, is driven by regulations and management practices designed to prevent further spread of *Phytophthora*. In recent years, broadening of national and international quarantines designed to prevent pathogen movement has led to a renewed impetus to manage *Phytophthora* diseases in nursery settings.

SEE ALSO THE FOLLOWING ARTICLES

Agriculture / Australia: Invasions / Forestry and Agroforestry / Fungi / Parasitic Plants / Pathogens, Plant

FURTHER READING

Cahill, D. M., J. E. Rookes, B. A. Wilson, L. Gibson, and K. L. McDougall. 2008. *Phytophthora cinnamomi* and Australia's biodiversity: Impacts, predictions and progress towards control. *Australian Journal of Botany* 56: 279–310.

Cline, E. T., D. F. Farr, and A. Y. Rossman. 2008. A synopsis of *Phytophthora* with accurate scientific names, host range, and geographic distribution. *Plant Health Progress* doi:10.1094/PHP-2008-0318-01-RS.

Erwin, D. C., and O. K. Ribeiro. 1996. Phytophthora *Diseases Worldwide*. St. Paul, MN: APS Press.

Hansen, E. M. 2008. Alien forest pathogens: *Phytophthora* species are changing world forests. *Boreal Environment Research* 13: 33–41.

Hansen, E. M., D. J. Goheen, E. Jules, and B. Ullian. 2000. Managing Port-Orford cedar and the introduced pathogen *Phytophthora lateralis*. *Plant Disease* 84: 4–14.

Hardham, A. 2005. Pathogen profile: *Phytophthora cinnamomi*. *Molecular Plant Pathology* 6: 589–604.

Rizzo, D. M., M. Garbelotto, and E. M. Hansen. 2005. *Phytophthora ramorum*: Integrative research and management of an emerging pathogen in California and Oregon forests. *Annual Review of Phytopathology* 43: 309–335.

PLANTS, AQUATIC

SEE FRESHWATER PLANTS AND SEAWEEDS

POLLINATION

DIEGO P. VÁZQUEZ

Instituto Argentino de Investigaciones des las Zonas Áridas, CONICET, Mendoza, Argentina

CAROLINA L. MORALES

Instituto de Investigaciones en Biodiversidad y Medio Ambiente, CONICET, Comahue, Argentina

Pollination is the transfer of pollen grains, which contain male gametes, to the receptive part of the ovule-bearing organ of seed plants (the micropyle in gymnosperms and the stigma in angiosperms). Pollination, a crucial step in sexual reproduction of seed plants, can occur through abiotic vectors (wind, water) and biotic vectors (insects, birds, other vertebrates). Introduced species may influence pollination either directly, when the alien is a plant or a pollinator, or indirectly, when the alien is another kind of species that modifies the pollination process by affecting the plant, the pollinator, or both. In addition, pollination can modulate the establishment and spread of introduced plant species.

PLANT INVADERS

Impact of Alien Plants on the Pollination of Native Plants

Alien plants can influence the pollination of native plants through competition for pollinators and interspecific pollen transfer. Competition for pollinators occurs when an alien plant draws pollinators away from the native plant, particularly when pollinators are scarce. The native plant may then suffer from decreased male (pollen dispersal) and female (seed production) reproductive success. Traits that may make alien plants likely to compete with native plants for pollinator visits include rich floral rewards, showy floral colors, high population size and density, high degree of generalization in the interactions with pollinators, and high phenotypic similarity in floral traits with native plants.

There are several documented cases of reduced reproduction of native species owing to pollinator-mediated competition with plant invaders. A well-known example is the Himalayan balsam (*Impatiens glandulifera*, Balsaminaceae), which has invaded many river banks in Europe and North America. Floral nectaries of this species produce sugar at a rate higher than any other plant in the invaded communities, which attracts pollinators away from coflowering species, reducing their seed set.

Although evidence suggests that most animal-pollinated alien plants have negative effects on the pollination of native coflowering species, effects are highly contingent on species identities and the ecological context. For instance, the dandelion (*Taraxacum officinale*), a weed of European origin, reduces the pollination and reproductive success of native *Taraxacum* species in Japan. However, in central Chile, it affects the native high-Andean species *Hypochaeris trinchioides* and *Perezzia carthamoides* only when growing at high densities. Likewise, in mountain meadows in Colorado, removal of *T. officinale* has not influenced the visitation and reproductive success of native *Delphinium nuttallianum*.

In addition to competition for pollinators, alien animal-pollinated plants can influence the pollination of native plants through interspecific pollen transfer—the transport of pollen from anthers of one species to the stigma of another species. Interspecific pollen transfer occurs when pollinators switch between flowers of different species while foraging. Thus, even if alien plants do not compete for pollinators, they can still affect native species either by depositing alien pollen on stigmas of native species or by wasting native pollen on alien flowers. For instance, the alien thistle *Carduus nutans* frequently invades highly disturbed forests in the southern Argentine Andes, where it co-occurs with the endemic herb *Alstroemeria aurea*. Bumblebees visit flowers of both species (Fig. 1), depositing alien pollen on stigmas of *A. aurea* and reducing the deposition and germination of conspecific pollen.

Although the amounts of heterospecific pollen deposited in nature are often too low to affect reproduction, when native and alien species are closely related, even small amounts of heterospecific pollen may have devastating reproductive consequences for native species

FIGURE 1 Plant–pollinator interactions involving native and alien species in the southern Argentine Andes. (A) Native bumblebee *Bombus dahlbomii* visiting the alien thistle *Carduus nutans*, a frequent invader of highly disturbed forests in the southern Argentine Andes. (Photograph courtesy of Néstor Vidal.) (B) Introduced bumblebee *Bombus ruderatus* visiting a flower of the endemic herb *Alstroemeria aurea*. (Photograph courtesy of Carolina L. Morales.)

through hybridization. This is what happens in Canadian red mulberry (*Morus rubra*) forests invaded by the alien white mulberry (*Morus alba*). In highly invaded areas, the availability of alien or hybrid pollen may be an order of magnitude greater than that of native pollen, negatively affecting both male and female reproductive success of the native mulberry.

The Role of Pollination in the Establishment and Spread of Alien Plants

One can also ask whether pollination limitation can be a barrier to the establishment of an alien plant. The relatively low degree of specialization of most plant–pollinator interactions makes this possibility unlikely, because most alien species find suitable pollinators in the invaded community. There are, however, a few cases of extreme specialization, such as figs and yuccas, for which pollination may indeed limit spread. For example, of the 60 species of *Ficus* introduced to Florida, only three have become invasive, and they have done so only after the accidental introduction of their specific pollinating wasp.

Even if pollination does not limit establishment, it may limit the rate of spread of alien plants when their reproduction is highly pollen-limited. For example, in Washington State, pollinators limit the population growth rate of Scotch broom (*Cytisus scoparius*) in recently invaded habitats, which in turn limits the broom's rate of spread into new habitats.

POLLINATOR INVADERS

Alien pollinators can influence pollination, particularly when their morphology, behavior, and phenology or activity period is different from those of the native pollinators of the plants they visit. For example, an exotic bumblebee introduced to Australia, *Bombus terrestris*, is a much less effective pollinator of the native *Eucalyptus globules* than the main native pollinator, the swift parrot *Lathamus discolour*, mainly because the bumblebee moves less frequently between plants and therefore transfers less outcross pollen than the native parrot does.

However, as for plant invaders, pollinator invaders do not necessarily disrupt pollination of native plants. The exotic bumblebee *Bombus ruderatus* is less effective than its native congener *B. dahlbomii* as a pollinator of *Alstroemeria aurea*, an herb endemic to the southern Andes of Argentina and Chile (Fig. 2). Yet, in many areas, *B. ruderatus* is replacing the native bumblebee, which makes it nowadays the most reliable pollinator of *A. aurea*. In some cases, the alien pollinator may be even more effective than native pollinators. For example, the introduced

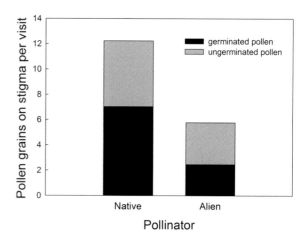

FIGURE 2 Differences between the native bumblebee *Bombus dahlbomii* and the alien *B. ruderatus* in their per-visit effectiveness as pollinators of the endemic herb *Alstroemeria aurea* in Nahuel Huapi National Park, Argentina. The alien *B. ruderatus* is less effective both in terms of total pollen grains and of the proportion of viable (germinated) pollen deposited in flower stigmas. (Adapted from Madjidian et al., 2008, *Oecologia* 156: 835–845.)

Africanized honeybee (*Apis mellifera scutellata*) is more effective than native pollinators in pollinating isolated individuals of *Dinizia excelsa*, a tree native to the Amazonian forest; the greater effectiveness of the honeybee is due to its greater flight distances, which increase gene flow among isolated *D. excelsa* individuals.

RECIPROCAL FACILITATION BETWEEN ALIEN PLANTS AND ALIEN POLLINATORS

Alien species interact not only with native species via pollination but also among themselves, sometimes facilitating each other's establishment and spread (i.e., an invasional meltdown). The establishment of *Ficus* and its specialist wasp pollinators in Florida discussed above is one striking example. But extreme reciprocal specialization is not a precondition for reciprocal facilitation between alien plants and pollinators. In California, introduced *Apis mellifera* strongly increases the reproductive output of the weedy thistle *Centaurea solstitialis*. Likewise, visitation by the European *Bombus terrestris* enhances reproduction of *Lupinus arboreus*, a North American species introduced to Tasmania. Although *L. arboreus* is currently a minor weed in Tasmania, the recent introduction of *B. terrestris* is facilitating its spread, making it likely to become as problematic as in New Zealand now that it has an effective pollinator.

OTHER TYPES OF INVADERS

Other types of alien species besides seed plants and animal pollinators can affect pollination. Introduced predators, parasites, and diseases of pollinators are one possibility. A dramatic example is the introduction of avian malaria to the Hawaiian Islands, where it contributed to the extinction of a substantial fraction of the native birds, some of which were pollinators of native plants. Some plant species, most notably those in the Campanulaceae, are believed to have gone extinct owing to pollination failure.

Plant consumers can interfere with pollination by directly consuming flowers, or indirectly, by causing changes in flower attractiveness. For example, the eradication of introduced possums *Trichosurus vulpecula* and wallabies *Petrogale penicillata* from some islands in New Zealand has resulted in significant increases in the flowering of two plant species, pohutukawa (*Metrosideros excelsa*) and rewarewa (*Knightia excelsa*).

More subtly, decreased plant population density caused by introduced herbivores can also affect pollinator visitation frequency and the extent of interspecific pollen transfer. For example, trampling by introduced cattle in Nahuel Huapi National Park, Argentina, drastically decreases the population density of the herb *Alstroemeria aurea*; this effect on population density in turn results in increased rates of interspecific pollen transfer, because its generalist pollinators (mainly bumblebees) visit other species more frequently than in denser, cattle-free *A. aurea* populations.

SEE ALSO THE FOLLOWING ARTICLES

Bees / Competition, Plant / Dispersal Ability, Plant / Horticulture / Hybridization and Introgression / Invasional Meltdown / Mutualism / Reproductive Systems, Plant

FURTHER READING

Barthell, J. F. R., J. M. Randall, R. W. Thorp, and A. M. Wenner. 2001. Promotion of seed set in yellow star thistle by honey bees: Evidence of an invasive mutualism. *Ecological Applications* 11: 1870–1883.

Bjerknes, A., Ø. Totland, S. Hegland, and A. Nielsen. 2007. Do alien plant invasions really affect pollination success in native plant species? *Biological Conservation* 138: 1–12.

Brown, B. J., R. J. Mitchell, and S. A. Graham. 2002. Competition for pollination between an invasive species (purple loosestrife) and a native congener. *Ecology* 83: 2328–2336.

Burgess, K. S., M. Morgan, and B. C. Husband. 2008. Interspecific seed discounting and the fertility cost of hybridization in an endangered species. *New Phytologist* 177: 276–284.

Madjidian, J., C. L. Morales, and H. Smith. 2008. Displacement of a native by an alien bumblebee: Lower pollinator efficiency overcome by overwhelmingly higher visitation frequency. *Oecologia* 156: 835–845.

Parker, I. M. 1997. Pollinator limitation of *Cytisius scoparius* (Scotch broom), an invasive exotic shrub. *Ecology* 78: 1457–1470.

Richardson, D. M., N. Allsopp, C. M. D'Antonio, S. J. Milton, and M. Rejmánek. 2000. Plant invasions: The role of mutualisms. *Biological Reviews* 75: 65–93.

PONTO-CASPIAN: INVASIONS

TAMARA SHIGANOVA

Russian Academy of Sciences, Moscow

The Black, Azov, and Caspian seas (Ponto-Caspian) were a united basin several times in the past, most recently in the Pliocene, when they were connected in the almost freshwater Pontian Lake-Sea. The marine biota was eliminated, and a brackish-water biota was formed then. Its representatives still dwell in the Caspian Sea, in the Sea of Azov, and in desalinated regions of the northwestern Black Sea, and they are referred to as Ponto-Caspian species. All three seas were reconnected by the Volga–Don Canal in 1952. The Black Sea is also a part of the Mediterranean basin and is connected via the Bosporus Strait with the Sea of Marmara, and further by the Dardanelles Strait with the Mediterranean Sea. Owing to accelerating human activities such as shipping, deliberate stocking, unintentional releases, and canal constructions, many nonnative species have arrived and established in these seas. After the construction of ballast water tanks, this process became global. In addition, since the 1980s, warming of the upper water layer of the Black Sea has led to increased populations of thermophilic species and to the northward expansion of their ranges. The Black Sea has thus become the main recipient for nonnative marine and brackish-water species and acts as a donor to the other Ponto-Caspian seas. An important detrimental role of the whole Ponto-Caspian basin is dispersal of its native brackish-water Ponto-Caspian species outside their geographical range, mainly because of canal constructions connecting previously separated waters.

COMMON ORIGIN OF THE PONTO-CASPIAN: ENVIRONMENT AND BIOTA

The three seas (Fig. 1) are all temperate basins characterized by lower salinity than standard ocean values. All basins are isolated from the World Ocean, and all seas have low biodiversity but high productivity. The physical evolution of the marine environments on the southern flanks of Europe and Asia during the Tertiary and Quaternary affected their biotas in many ways. The paramount variable causing biotic change was salinity, which fluctuated strongly. This turbulent history caused the heterogeneity of the regional fauna and flora.

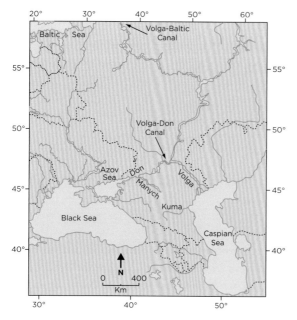

FIGURE 1 The Black, Azov, and Caspian seas (Ponto-Caspian), with the Volga–Don Canal.

The present biotas of the seas are largely relicts, the remains of an ancient Tethys fauna (including Sarmatian, Pontian, and Caspian species) after it was exposed to repeated alternation of desalination and salination phases. This biota is supplemented by comparatively recent invaders of freshwater origin and by immigrants (pseudorelics) from the Arctic. The latter worked their way south via the Volga River and are therefore mainly restricted to the Caspian Sea. In addition, Atlantic-Mediterranean elements advanced from the west through the Mediterranean with decreasing forms from one sea to another in an eastward direction.

The biotas of the basins differ in the proportions of the groups considered. The flora and fauna of the Caspian Sea best preserved the remarkable relict Ponto-Caspian fauna, while the Sea of Azov and the Black Sea are inhabited by a mixed Ponto-Caspian, marine Mediterranean, and fresh water biota in brackish-water areas.

THE BLACK SEA

Environment, Native Biota, and Disturbance

The Black Sea is a nontidal, meromictic basin with a thin (60–200 m) oxygenated surface layer. Beyond this layer, the water column is a virtually lifeless hydrogen sulfide environment. A pycnohalocline blocks vertical mixing. The narrowness of the active layer renders the ecosystem of the Black Sea extremely vulnerable to climatic changes and anthropogenic impacts. The present-day flora and fauna, which were formed under conditions of relatively

low salinity (17–22‰) and the existence of an anoxic zone beneath the upper oxygen-containing layer, are distinguished by low species diversity of most taxonomic groups and complete absence of many of them. However, the biota is highly productive, especially in nearshore regions, because of a high abundance of key planktonic and benthic species and large populations of commercial fish species.

The Black Sea biota is 80 percent of Atlantic-Mediterranean origin, 10.4 percent of freshwater origin, and 9.6 percent of Ponto-Caspian origin. Another component is a small Arctic assemblage that contains mainly flora. Biota of Atlantic-Mediterranean origin comprise species of the Lusitanian province and of the boreal zone of the Atlantic Ocean. The Lusitanian origin species are warmwater species that inhabit the upper layer of the Black Sea. The Atlantic boreal origin species are found in moderately cold water and have features of coldwater relicts. They are mainly benthic, demersal, and pelagic species that inhabit the cold intermediate layer and below. In addition to salinity, impoverishment of the Black Sea biota is due to the absence of deepwater species at depths below the oxygenated layer. The number of species in the Black Sea is presently relatively small: 3,786. Of these, 1,619 are fungi, algae, and higher plants; 1,983 are invertebrates; 180 are fish; and 4 are mammals.

Since the 1970s, under the influence of climatic and anthropogenic factors, the biota has changed greatly. Among the most pronounced anthropogenic drivers are regulation of river runoff and an increased supply of dissolved phosphates and nitrates accompanied by reduced silicate from great rivers. Changes include a switch in phytoplankton domination from diatoms to dinoflagellates, a significant increase in phytoplankton biomass, and massive development of harmful algae with subsequent eutrophication, a corresponding twofold increase in primary production, and subsequent explosions of native gelatinous species such as *Aurelia aurita* and *Noctiluca scintillans*. In addition, fishing pressure has resulted in decreasing stocks of top predators, large pelagic fishes—migrants from the Aegean Sea and dolphins. The Black Sea has thus become mesotrophic or eutrophic (in its northwestern parts), which has facilitated invasion by nonnative species.

Vectors, Pathways, and Composition of Invaders

Among the Ponto-Caspian seas, only in the Black Sea is intensive shipping accelerating, with routes to different regions of the World Ocean. Since the late twentieth century, enormous numbers of marine and brackish-

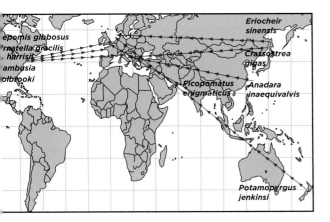

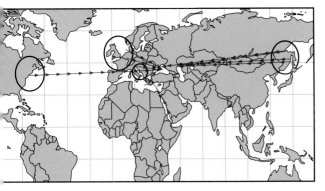

FIGURE 2 Scheme of the donor areas of invasion into the Black Sea. (A) Primary pathways of selected species. (B) Donor areas and scheme of pathways of invasion.

water species have been arriving in its harbors in ballast water and on fouled hulls. Disturbance has facilitated their establishment. Biotic changes are also caused by other human activities: release by aquarists, deliberate stocking of commercial species and release of species that accompany them, and penetration via canal systems connecting previously isolated basins. During recent decades, a new vector has accelerated introductions: expansion of warmwater species from the Sea of Marmara and the Mediterranean, owing to rising upper layer temperatures.

Pathways of species penetration to the Black Sea based on an analysis of established nonnative species and their donor regions are shown in Figures 2 and 3. The greatest number of species (36%) arrived from the Mediterranean through natural range expansion or in ballast water. This proportion keeps increasing with rising upper layer water temperatures. Only a few of these species have become abundant, and few of them have reached the other Ponto-Caspian seas. Among these species are representatives of micro-, phyto-, and zooplankton; demersal plankton; macrophytes; zoobenthos; and fishes. These species,

as a rule, are of subtropical or even tropical origin; most still occur only in the southern parts of the Black Sea and near the Bosporus Strait, where salinity is higher. Certain species have penetrated to the nearshore regions off Bulgaria, Romania, and the Ukraine (Odessa, Crimea) in the course of their migration or when they have been carried in currents or ballast water.

By contrast, some species of Adriatic origin (2%) now have self-sustaining populations and are abundant. Chief among them is a bivalve, *Anadara inaequivalvis*, which established in most coastal areas and spread to the Sea of Azov. The successful establishment of propagules from the Adriatic Sea is explained by the lower salinity of some Adriatic regions compared to other parts of the Mediterranean Sea and by the fact that Adriatic water temperatures are similar to those of the Black Sea. In addition, shipping traffic between ports of the Adriatic and Black seas favors introductions.

Many established invaders from North American Atlantic regions were introduced in the 1980s and 1990s (11%). All species of this group are rather eurythermal and, importantly, euryhaline and widely distributed in coastal waters of the World Ocean. Invaders from this region especially affected the Black Sea. Among these are the ctenophores *Mnemiopsis leidyi* and *Beroe ovata* and the copepod *Acartia tonsa*. The barnacles *Balanus eburneus* and *B. improvisus* were introduced from the same area much earlier. This pathway also characterizes accidentally introduced brackish-water species represented by inhabitants of brackish bays and estuaries. The most successful of these are the hydromedusas *Blackfordia virginica* and *Bougainvillia (Perigonimus) megas*.

One more source of nonnatives is nearshore European Atlantic waters (8%); most of these species were brought

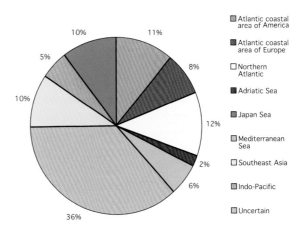

FIGURE 3 Donor areas of the nonnative species and their proportions (in %) in the Black Sea.

by ships. The most successful is the crab *Rhithropanopeus harrisii*. Some of these species dispersed from coastal estuaries of Europe via rivers and canals to deltas and brackish areas of the Black Sea: the Chinese mitten crab *Eriocheir sinensis* and the pumpkinseed fish *Lepomis gibbosus*. Others arrived from the northern Atlantic (12%); most notable are the diatom *Pseudosolenia calcar-avis* and the bivalve *Mya arenaria*. The Japan Sea is another source of invaders (6%), including the rapa whelk *Rapana venosa* and the intentionally introduced mullet *Liza haematochilus (Mugil soiuy)*. Together with this mullet came three species of fish parasites. Aquarists occasionally released *Oryzias latipes*, originally from Japanese freshwater. Except for the rapa whelk, no spontaneous invaders from the Japan Sea have established permanent populations.

Of interest are Indo-Pacific species that might arrive with ballast water or as Lessepsian migrants (5%). In recent years, this group has expanded substantially. Other species from the nearshore waters of the Pacific or Indian oceans first established in coastal Europe and the Adriatic Sea and next in the Black Sea, including the polychaetes *Capitellethus dispar* and *Glycera capitata*, the gastropod *Potamopyrgus jenkinsi*, and the Chinese mitten crab *Eriocheir sinensis*. None of these became abundant and widely distributed. The attempted introduction of five fish species from estuaries of the Japan Sea, the Amur River, and other rivers of northeastern Asia did not succeed. However, two fern species and strains of *Vibrio cholerae* were brought from these regions.

As seen from the pathways listed above, some species are nonnative to the regions from which they reached the Black Sea. Acclimation to the Black Sea followed adaptation to conditions of a primary recipient area. The crab *Rhithropanopeus harrisii* moved to Europe from coastal north America; the Chinese mitten crab came from eastern Asia to the North Sea; the aquarium fish pumpkinseed was brought to Europe from North America; the bivalve *Anadara inaequivalvis* from the coastal Philippines was first released into the Adriatic Sea estuaries; the mosquitofish *Gambusia holbrooki* from Central America was first transferred to the Adriatic Sea; the polychaete *Ficopomatus enigmaticus (Mercierella enigmatica)* came from coastal India to coastal Europe; and the gastropod *Potamopyrgus jenkinsi* from coastal New Zealand first invaded Europe.

In sum, 156 nonnative species have established, including Mediterranean and freshwater species (171 species if some doubtful cases are counted). In addition, 97 zoobenthic species occur continuously only off the Straits of Bosporus and some are also in the southern part of

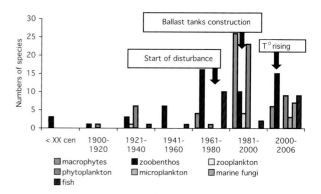

FIGURE 4 Chronology of species invasions into the Black Sea.

the Black Sea; 117 Mediterranean species occur as isolated individuals in the Bosporus region, and a few individuals of 64 more mainly Mediterranean zooplankton and phytoplankton and Indo-Pacific species are found in the western and northern Black Sea. These species are not considered as established. Invasion rates keep increasing for a variety of ecological, systematic, and functional groups (Fig. 4). The main factors facilitating invasion were disturbance in the 1970s, construction of ballast tanks in ships in the early 1980s, and rising upper layer temperatures most recently.

Ecosystem Impacts

Most negative effects are caused by invaders that form massive populations. These include benthic species (e.g., the rapa whelk, which consumes oysters, mussels, and other bivalves), pelagic species (e.g., *Mnemiopsis leidyi*, which consumes zooplankton, fish eggs, and larvae and affects all trophic levels), and fresh- and brackish-water species (e.g., the pumpkinseed, which consumes zooplankton, fish eggs, larvae, fry, and small adult fish). At the same time, however, the rapa whelk has become a valuable commercial species. The most pronounced event was the invasion by *M. leidyi*, which caused cascading effects at higher trophic levels, from a decrease in zooplankton to collapsing planktivorous fish to large pelagic fish and dolphins. Similar effects occurred at lower trophic levels, with a decrease in zooplankton leading to an increase in phytoplankton released from zooplankton grazing pressure, and increases in bacterioplankton, zooflagellates, and ciliates. Ten years later, another warmwater ctenophore, *Beroe ovata*, an obligate predator on *M. leidyi*, was introduced with ballast water from North America. Its invasion led to "invasional meltdown," in which one introduced species facilitates invasion by its predator. The *B. ovata* invasion reduced *M. leidyi* populations, and the ecosystem began to recover at all trophic levels.

The cholera bacterium *Vibrio cholerae* may provoke epidemics from time to time in coastal populated areas. The common shipworm *Teredo navalis* attacks wooden structures. The *Pseudosolenia calcar-avis* diatom is an additional food item for large zooplankton species, but when its populations explode, it supplants more valuable native phytoplankton species. The role of the fouling species *Balanus improvisus* is negative, but its larvae are consumed by small pelagic fishes.

The introduction of other organisms may be regarded as positive. The bivalves *Anadara inaequivalvis* and *Mya arenaria*, although replacing native species, constitute valuable food for zoobenthic species and benthophagous fishes, while their larvae are consumed by small pelagic fishes. The crab *Rhithropanopeus harrisii* has become an additional food source for benthophagous fishes. The mullet *Liza haematochilus* competes with native mullets but is itself a valuable commercial species.

Most established nonnative species are temperate, euryhaline, and eurythermal. Species with all of these traits not only establish but also become explosive and expanding, first reaching the Sea of Azov and the Caspian Sea; some then invade the Sea of Marmara and the Aegean Sea. Thus, the Black Sea has become a secondary donor for the expansion of nonnative species.

THE SEA OF AZOV

Environment, Native Biota, and Disturbance

The Sea of Azov is a remnant of an ancient system of straits that formerly occupied the Caucasian foredeep. It is the smallest and shallowest among the southern Eurasian seas, has a salinity of 0.1–14‰, and is connected with the Black Sea by the narrow, shallow Kerch Strait. The history of its origin and evolution is closely related to that of the Black Sea.

The biota of the Sea of Azov was formed of representatives of freshwater, brackish-water, and marine assemblages. The freshwater and brackish-water assemblages are mainly represented by Ponto-Caspian relics that populate rivers and lagoons of the Sea of Azov and consist of 54 species in 32 genera. The bulk of the relics are concentrated near river mouths, where the water is virtually fresh. The small marine assemblage is represented by Mediterranean flora and fauna. The biota includes 350 species of free-living invertebrates (without turbellaria and nematodes), 605 species of phytoplankton, 30 species of macrophytes, and 103 fish species. For well-studied groups, the Black Sea biota is 3.4 times as rich as that of the Sea of Azov. Not only does the number of species decrease from the Black Sea to the Sea of Azov, but the

proportions of groups change as well. Many species with a Mediterranean origin, which presently dwell in the North Atlantic, disappear.

The regulation of the Don (1952) and Kuban (1973) rivers and the withdrawal of river runoff to fill reservoirs reduced flooded and spawning areas. In the sea proper, one observes a growth in the vertical temperature and salinity gradients and an increase in the oxygen-deficient zones near the bottom. Increased salinity spurred dispersal of Black Sea species into the Sea of Azov; most of them disappeared after salinity decreased to the original level, but a few persisted.

Vectors, Pathways, and Composition of Invaders

All nonnative species entered from the Black Sea via the Kerch Strait with currents or ships. All except the mussel *Mytilus galloprovincialis*, goby *Gobius niger*, and probably the macrophyte *Ectocarpus caspicus*, which established during a period of high salinity, are nonnative in the Black Sea. Species nonnative in the Black Sea that penetrated into the Sea of Azov derive ultimately from the North Atlantic (13 species), coastal North America (5 species), coastal northern Europe (6 species), undetermined parts of the northern Atlantic (2 species), the Adriatic Sea or its basin (4 species), the Japan Sea and its estuaries (1 species), and the Mediterranean (2 species). Five fish species were intentionally brought from the freshwaters of northeastern Asia, as were three fish species from the Atlantic coast of America. Freshwater species arrived from the Volga basin and settled in deltas and the very brackish Taganrog Bay (Fig. 5A). Thus, in fact, all species arrived from the Black Sea, except for intentionally introduced fish and freshwater species that spread from the Volga basin (Fig. 5B).

The distribution of nonnative species is similar to that of seasonal migrants from the Black Sea, depending on their salinity tolerance. Some of them (rapa whelk, blue crab *Callinectes sapidus*) live only in the southern area and the Kerch Strait, where salinity is higher. Others (the comb jellies *M. leidyi* and *B. ovata*) can survive only during warm seasons and are reintroduced every spring (*M. leidyi*) or later summer (*B. ovata*). The copepod *Acartia tonsa* and diatom *Pseudosolenia calcar-avis* can survive in the Sea of Azov but develop only in warm seasons.

The main events that favored nonnative species were increased salinity, which allowed many Black Sea species and Black Sea invaders to invade the Sea of Azov (some of which remained after salinity declined toward original levels), and ballast tanks on ships (Fig. 6). It is important to note that the species that established are euryhaline,

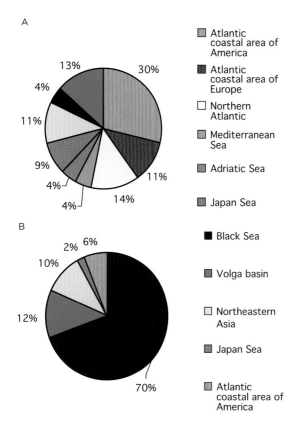

A

B

- ◻ Atlantic coastal area of America
- ◼ Atlantic coastal area of Europe
- ◻ Northern Atlantic
- ◻ Mediterranean Sea
- ◻ Adriatic Sea
- ◻ Japan Sea
- ◼ Black Sea
- ◻ Volga basin
- ◻ Northeastern Asia
- ◻ Japan Sea
- ◻ Atlantic coastal area of America

FIGURE 5 Donor areas of the nonnative species and their proportions (in %) in the Sea of Azov. (A) Primary areas. (B) Secondary (present) areas.

eurythermal, have wide oxygen level tolerance, and thrive in shallow water. Some freshwater species established in bays and deltas. There are 47 alien species total.

Ecosystem Impacts

Some nonnative species have harmed the ecosystem, while others have few apparent negative effects. All aspects of the ecosystem, fish resources included, were greatly affected by the predatory ctenophore *Mnemiopsis leidyi*.

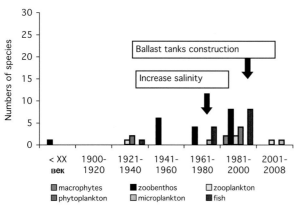

FIGURE 6 Chronology of species invasions into the Sea of Azov.

Negative effects of other harmful invasive species that spread from the Black Sea were mentioned in the Black Sea section: the cholera bacterium, common shipworm *Teredo navalis*, barnacle *Balanus improvisus*, and diatom *Pseudosolenia calcar-avis*.

The introduction of other organisms may be positive. Benthic species such as the bivalves *Anadara inaequivalvis* and *Mya arenaria* spread over regions with low oxygen unfavorable for other benthic representatives; they provide food for benthic-feeding fishes, while their larvae are consumed by small pelagic fishes. The crab *Rhithropanopeus harrisii* also became a food item for benthophagous fishes. The mullet *Liza haematochilus* became a valuable commercial fish in the Sea of Azov, where native mullets are not abundant and do not reproduce.

The ctenophore *Beroe ovata* is a useful invader; unfortunately, because of its seasonal dynamics, it appears in the Sea of Azov too late, when *M. leidyi* has already reproduced, spread, and undermined the stocks of trophic zooplankton. The development in the Black Sea of *B. ovata* influences the size of the *M. leidyi* population; therefore, after *B. ovata* appeared, *M. leidyi* entered the Sea of Azov later, and its abundance was lower.

THE CASPIAN SEA

Environment, Native Biota, and Disturbance

The Caspian Sea is the largest inland water body; its shelf zone (<100 m deep) occupies 62 percent of its surface area. Physical geography and bottom topography divide the Caspian into northern, middle and southern regions. Sea-level oscillation is one of the main factors that determine the status of its ecosystems. During the twentieth century, environmental conditions deteriorated significantly, mainly owing to sea-level changes, river runoff, and pollution from multiple sources including petroleum hydrocarbons and phenols.

Inhabitants belong to four groups. The most ancient and abundant are autochthonous (Ponto-Caspian) species (84%). Arctic species (3%) arrived during the last glaciations. Atlantic-Mediterranean species (1%) penetrated about 13,000 years ago. They have become full members of Caspian communities, have evolved considerably, and have generated new species and subspecies. Freshwater species (13%) have entered on several occasions.

The present-day Caspian is relatively species-poor. Species richness is lower than that of the Black Sea by a factor of 2.5, although the biota contains 733 species and subspecies of plants and 1,814 species and subspecies of animals, of which 1,069 are free-living invertebrates, 325 are parasites, and 415 are vertebrates (the last are mainly

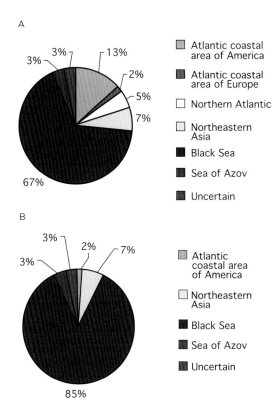

A

- ■ Atlantic coastal area of America — 13%
- ■ Atlantic coastal area of Europe — 2%
- ☐ Northern Atlantic — 5%
- ☐ Northeastern Asia — 7%
- ■ Black Sea — 67%
- ■ Sea of Azov — 3%
- ■ Uncertain — 3%

B

- ■ Atlantic coastal area of America — 7%
- ☐ Northeastern Asia — 2%
- ■ Black Sea — 85%
- ■ Sea of Azov — 3%
- ■ Uncertain — 3%

FIGURE 7 Donor areas of the nonnative species and their proportions (in %) in the Caspian Sea. (A) Primary areas. (B) Secondary (present) areas.

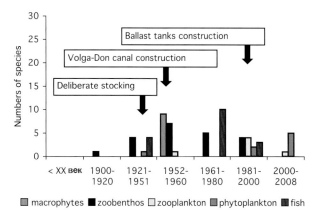

FIGURE 8 Chronology of species invasions into the Caspian Sea.

represented by freshwater species, and this list is still growing). The principal causes of the high degree of faunal endemism lies in the long-term isolation of the basin and its salinity regime. The low salinity (0.1–11‰ in the northern part, 12.6–13‰ in the other parts) and its native biota restricted penetration by many marine species and, at the same time, constrained access by freshwater species. In spite of low biodiversity, the Caspian Sea has high productivity, particularly in the northern Caspian, and rich fish stocks.

Vectors, Pathways, and Composition of Invaders

The appearance of nonnative species and changes in biodiversity may be divided into three phases (Figs. 7 and 8). The first comprised deliberate large-scale introductions beginning in the 1930s within the framework of the Soviet Union program for enriching commercial stocks, aimed at enlarging the resources of either commercial fishes themselves or their food organisms. However, among these introductions, only two finfish (the mullets *Liza saliens* and *L. aurata*) from the Black Sea and two benthic species (the polychaete *Hediste* (*Nereis*) *diversicolor* and the bivalve *Abra ovata*) from the Sea of Azov achieved significance. Two Black Sea shrimps, *Palaemon adspersus* and

P. elegans, were also released and became a valuable food source for benthic-feeding fishes. Eight other fish and fish parasites were introduced during these actions.

The second introduction phase started when the Volga–Don Canal opened in 1952. Most of these species were carried from the Black Sea by ships as fouling organisms. The third phase began in the early 1980s, when mainly phyto- and zooplanktonic species began to arrive in ballast water. All established nonnative species carried by ships were brought from the Black Sea. A first group includes 21 widely distributed and often abundant euryhaline species. These are of Atlantic-Mediterranean origin but have lived for 1,500 to 2,000 years in the Black Sea and have adapted to its low salinity. They are Cladocera (*Pleopis polyphemoides*, *Penilia avirostris*, *Podon intermedius*), the amphipod *Corophium volutator*, the bivalve *Mytilaster lineatus*, phytoplankton, and macrophytes.

Another group consists of nine Black Sea brackishwater species that were adapted to life in low-salinity areas before invading similar areas of the Caspian: the hydromedusa *Moerisia maeotica*; the isopods *G. aequicauda* and *Iphigenella shablensis*; the molluscs *Monodacna colorata*, *Dreissena bugensis*, *Lithoglyphus naticoides*, and *Tenellia adspersa*; the kamptozoan *Barentsia benedeni*; and the pearlwort *Conopeum seurati* (= *Membranipora crustulenta*).

In addition to long-established Black Sea species, some Black Sea invaders were also introduced. These include seven species from Atlantic inshore regions of North America, among them *Mnemiopsis leidyi*. Also arriving via this route were the North American hydromedusae *Blackfordia virginica* and *Bougainvillia megas*. From the European Atlantic coast came the polychaete *Ficopomatus enigmaticus* and two diatom species: *Pseudosolenia calcaravis* and *Cerataulina pelagica*. The origin of other invaders is uncertain, including the pearlwort *Lophopodella carteri*, the three-spined stickleback *Gasterosteus aculeatus aculeatus*, and the Chinese mitten crab.

About 60 species are established, although some fishes are known only from single individuals. Some introduced species also failed to establish. Except for a few freshwater invertebrates and deliberately introduced freshwater fish, plus two species deliberately introduced from the Sea of Azov, all established invaders were introduced from the Black Sea (Fig. 7). Most established species are euryhaline; many are widely distributed in coastal waters and therefore have wide ecological tolerances. Fewer established invaders are from brackish water, and the only freshwater species are deliberately introduced fish.

Marine nonnative species settled in the middle and southern Caspian, often replacing native species. Brackish and freshwater species settled in the northern Caspian, although the most euryhaline of them might penetrate the middle and southern Caspian.

Ecosystem Impacts

The Caspian Sea ecosystem was the most vulnerable to invaders because of its long isolation and high level of endemism. Most Atlantic invaders had major impacts; some, such as *Mnemiopsis leidyi*, affected all trophic levels and, finally, ecosystem functioning. Although few in number, these species occupy dominant community positions. They include the phytoplankter *Pseudosolenia calcar-avis*, the cladoceran *Pleopis polyphemoides*, and the copepod *Acartia tonsa*. The biomass of *M. lineatus*, *A. ovata*, *N. diversicolor*, and *Balanus improvisus* makes up more than 60 percent of the total benthos. Fouling communities consist almost wholly of nonnative species. Native species dominate only among the fishes.

Some introductions may be positive, but many are harmful, especially that of *M. leidyi*. *H. (N.) diversicolor* and *A. ovata* became the favorite food of stellate and Russian sturgeons. The crab *Rhithropanopeus harrisii*, competing with sturgeon for food, is itself one of their food items. The mussel *Mytilaster lineatus*, which forms the bulk of the benthic biomass, replaced native species and is scarcely used by benthophagous fish and sturgeons. The fishery for mullets achieved only limited importance. The diatom *Pseudosolenia calcar-avis*, having increased phytoplankton biomass, was of limited nutritive value for zooplankters and pelagic phytophagous fish.

After the invasion of *M. leidyi*, the functioning of the Caspian ecosystem changed, as in the Black Sea previously. In addition, after invasion by *M. leidyi*, only nonnative zooplankton, meroplankton, and zoobenthic species survived. Among the factors that permitted the outbreak of *M. leidyi* was the absence of predators. There is much optimism, therefore, about the accidental arrival in the Black Sea of *Beroe ovata*, which preys exclusively on zooplanktivorous ctenophores. If intentially introduced, *Beroe ovata* would be able to live in salinities of the middle and southern Caspian and could perhaps control *Mnemiopsis* abundance if it were introduced.

CONCLUSION

In the latter half of the twentieth century, increased shipping and construction of canals caused the Black Sea to become both a recipient and a donor area for temperate marine and brackish-water species. It serves as a hub for species that then spread further to the Sea of Azov and the Caspian Sea. All three seas have low diversity but high productivity and are inland semi-closed or closed seas with limited water exchange with the ocean (in the case of the Black Sea) or with no such exchange (in the case of the Caspian Sea), all features that make them vulnerable to invasion.

The location of the Black Sea is crucial. Situated between the species-rich Mediterranean Sea and the species-poor Azov and Caspian seas, the Black Sea is also intermediate between these seas in abiotic conditions: it has much lower salinity than the Mediterranean but higher salinity than the Sea of Azov and the Caspian Sea. Anthropogenic disturbance of the Black Sea has also increased vulnerability; eutrophication and overharvesting of top fish predators and dolphins are especially important.

All the introduced species mentioned here are widely distributed in the coastal areas of the world ocean. They have abundant genetic variation and are generally dominant in their native habitats. They often became dominant in the Black Sea and in some instances in the other seas. Native biodiversity has declined, and invaders now dominate. The main driver in all these ecosystems has become the most aggressive gelatinous invader *Mnemiopsis leidyi*. Since *Beroe ovata* appeared in the Black Sea, these two ctenophores have largely determined ecosystem status there.

A new trend has recently appeared in the Black Sea. Mediterranean species increase their dispersal through currents and ballast water, aided by rising temperatures. These organisms include phyto- and zooplankton, macrophytes, benthic or demersal organisms, and fishes. They are of subtropical or even tropical origin. In addition, some species of Indo-Pacific origin have also invaded the Black Sea; some are Lessepsian migrants, and others arrived in ballast water. Only two of these species have reached the Sea of Azov, and none have reached the Caspian Sea.

SEE ALSO THE FOLLOWING ARTICLES

Ballast / Canals / Eutrophication, Aquatic / Great Lakes: Invasions /
Invertebrates, Marine / Lakes / Mediterranean Sea: Invasions /
Seas and Oceans

FURTHER READING

Dumont, H., T. Shiganova, and U. Niermann, eds. 2004. *The Ctenophore*
    Mnemiopsis leidyi *in the Black, Caspian and Mediterranean Seas and*
    Other Aquatic Invasions. NATO ASI Series, 2. Environment. Kluwer.
Kasymov, A. G. 1987. *Biota of the Caspian Sea.* Baku: ELM. [In Russian.]
Mordukhai-Boltovskoi, F. D. 1969. *Guide to Fauna in the Black and Azov*
    Seas. Kiev: Naukova Dumka. Ed. Vodyanitsky V. [In Russian.]
Shiganova, T., and H. Dumont. 2010. *Non-native Species in the Inland*
    Southern Seas of Eurasia. Berlin: Springer.
Zaitzev, Y. P., and B. G. Alexandrov, eds. 1998. *Biological Diversity, Ukraine.*
    New York: United Nations Publication.

PREDATORS

MICK N. CLOUT

University of Auckland, New Zealand

JAMES C. RUSSELL

University of California, Berkeley

Predators are animals that kill other animals for food. Examples of invasive introduced predators extend across the animal kingdom and include many species of insects, spiders, molluscs, fish, amphibians, reptiles, birds, and mammals. Predatory species have been deliberately introduced outside their natural range for a variety of reasons. These include companion animals (e.g., dogs, cats); species introduced as fur bearers, for sport or for aquaculture (e.g., mink, foxes, salmonid fish); those introduced for biological control purposes (e.g., predatory insects, cane toads, mustelids); and many that were simply introduced as novelties or for "faunal enrichment" (e.g., ornamental fish, frogs, birds). Many invasive predators, especially arthropods and rodents, but also snakes and marine molluscs, have also been introduced accidentally with cargoes or by "hitchhiking" in or on vessels.

IMPACTS

The impacts of invasive predators on native species and natural ecosystem processes are arguably more severe than those of other types of introduced species. This is because, by definition, a predator kills its prey. Predation therefore has a more severe and direct effect than does a herbivore eating part of a plant, or any species (plant or animal)

competing with others for resources. An analysis of the IUCN database of known extinctions of species over the past 500 years reveals that most have occurred on islands and that terrestrial vertebrates have disproportionately gone extinct, compared with plants. This analysis also reveals that predation (by itself or in combination with other factors) has been a cause of extinction for approximately 80 percent of all terrestrial vertebrates. Predation by humans has, of course, been a cause of such extinctions in several cases.

The mechanisms by which predation occurs can vary greatly between predators. Some "super-predators," such as large carnivores, are at the top of the food chain. These predators usually eat entire organisms at any life stage (e.g., young through adult). Other "mesopredators," such as small omnivores, may rest lower down on a food chain, above their prey but below other higher-order predators. These predators may prey upon only specific life stages of their prey, such as vulnerable young. These relationships between predators and prey in an ecosystem can be represented by simple food web diagrams (Fig. 1). Predators can also switch between different prey, based upon seasonal availability or relative abundance of prey species.

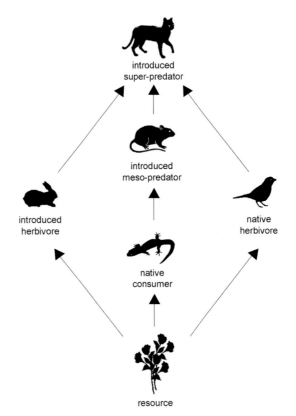

FIGURE 1 Simple food web with two introduced predators. Arrows represent the flow of energy.

Because of these intricacies of predation, an individual-level impact (e.g., a predator eating a prey) may not lead to a population-level impact (e.g., a species being affected by predation). The overall population-level effect of predation will depend not just on prey survival rates but also on their relative birth rates (i.e., their ability to replenish their population size).

CASE STUDIES

Dogs (*Canis domesticus*) were among the first predators to be introduced by people. For example, dingoes were introduced to Australia at least 3,500 years ago and contributed to the extinction of several endemic Australian mammals. These included the thylacine, an endemic predatory marsupial that apparently suffered from competition from dingoes. This makes the point that introduced predators can adversely affect other predator species in addition to their prey. Another domestic predator that has been widely introduced by people is the cat (*Felis catus*). The impact of feral cats has been devastating in many ecosystems to which they have been introduced, especially on oceanic islands where many species have not evolved defenses to the tactics of predatory mammals, such as hunting nocturnally and by scent.

The introduction of predators to control pest species has a recent record of many successes (especially with arthropod introductions), when proper risk analysis precedes the introduction. Unfortunately, the history of biological control is also littered with some very bad mistakes. One of the most notoriously disastrous attempts at biological control was the introduction of predatory snails and a predatory flatworm to several tropical islands beginning in the 1950s to control the giant African snail (*Lissachatina fulica* [*Achatina fulica*]), itself an introduction. Control of the target species remained elusive, but the introduced predatory snails caused the extinction of several species of endemic snails on Tahiti, Hawaii, and other islands. Another biological control disaster was the attempt to control introduced rabbits (*Oryctolagus cuniculus*) in New Zealand in the 1880s by releasing predatory mustelids. Control of rabbits was not achieved, but the mustelids (especially stoats, *Mustela erminea*) preyed extensively on native wildlife and contributed to the extinction or decline of several endemic bird species. Similarly disastrous effects have resulted from the introduction of another mustelid, the small Indian mongoose (*Herpestes auropunctatus*), to various tropical islands as a potential biological control agent for rats

FIGURE 2 Black rats (*Rattus rattus*) are among the most widespread invasive rodent species and have damaged native biodiversity on many islands around the world. (Photograph courtesy of James Russell.)

in sugarcane fields. Control of rats was generally not achieved, but losses of native wildlife often occurred due to mongoose predation.

Introduced rodents (including *Rattus* species and house mice) are now present on every continent except Antarctica and in over 80 percent of the world's archipelagoes (Fig. 2). Invasive rodents have been implicated in over 50 percent of vertebrate extinctions on islands. Extinctions of vulnerable species typically occur rapidly after the arrival of predatory rodents and can have flow-on effects to the entire ecosystem. For example, the extinction of seabirds on islands alters the flow of resources from the marine to the terrestrial systems, an important source of nutrients for plants and insects. Even after introduced rodents have been removed, it may be a very long time, if ever, before ecosystem processes can be restored to their status prior to invasion.

Several species of predatory social invertebrates (ants and wasps) have invaded many parts of the world through accidental introduction and have had severe impacts on native species and ecosystem processes. Notorious examples include red imported fire ants (*Solenopsis invicta*), yellow crazy ants (*Anoplolepis gracilipes*), and the common wasp (*Vespula vulgaris*). Each of these species is omnivorous (scavenging, and then harvesting sugary excretions), but also predatory.

Red imported fire ants are generalist predators that have now been accidentally introduced to many parts of the world and prey on a wide range of invertebrates and some vertebrates, including reptiles and nesting birds. They are capable of transforming the structure and composition of entire communities. Yellow crazy ants are similarly invasive, generalist predators that form super-colonies, recruit rapidly to food resources, and aggressively disrupt invertebrate communities. A specific example of their impacts is on Christmas Island, in the Indian Ocean, where they formed associations with honeydew-secreting Hemiptera, increased rapidly in abundance, and killed vast numbers of native land crabs (*Gecarcoidea natalis*) that are key herbivores in the local rainforest ecosystem.

In the honeydew beech (*Nothofagus*) forests of New Zealand, invasive common wasps reach very high densities in summer and fall, fueled by abundant honeydew excreted by native scale insects. These wasps deplete the honeydew and compete with native wildlife for this resource. They also prey on a range of native invertebrates, including spiders, moths, and stick insects. Probable elimination of some species of spiders from honeydew beech forests by these invasive wasps has been demonstrated. Those species of native moths with a caterpillar stage that coincides with the peak wasp season are also eliminated from wasp-invaded forests.

One of the best-documented recent invasions by an accidentally introduced vertebrate predator is that of the brown treesnake (*Boiga irregularis*) on the island of Guam. This species arrived in Guam shortly after World War II, probably in the undercarriage of military aircraft. Over the next 40 years, 17 of 18 native bird species on Guam were severely affected by snake predation. Twelve of these bird species became extinct as breeding residents and others suffered declines of more than 90 percent. The snake is also a major predator on young of the endangered Marianas flying fox (*Pteropus mariannus*). Overall, it has transformed the natural ecosystems of Guam. By the time it became clear that the brown treesnake was responsible for these extinctions and declines, it was too late to react. In addition to its effects on biodiversity, the brown treesnake also has many nonbiological impacts, such as regularly causing power outages through short-circuiting electricity transformers and crawling into the bedrooms of young children, where its nonpoisonous bite can still be damaging.

Invasive predators also have severe impacts in marine and freshwater ecosystems. For example, the introduction of Nile perch (*Lates niloticus*) to Lake Victoria in Africa has apparently caused the extinction of hundreds of species of endemic haplochromine cichlid fish and has transformed the lake ecosystem and its environs. Nile perch are now used extensively by local people as a food resource, and preservation of these large oily fish is commonly done by smoking. This has raised the local demand for firewood and has led to increased levels of deforestation around the lake. However, the availability of an abundant food source has also led to increased economic demand for Nile Perch, both from within Africa and overseas.

INDIRECT EFFECTS

Introduced predators can also have complex indirect effects on ecosystems by changing how other species interact with one another. For example, by driving a prey species to extinction, an invasive predator may also affect other species that depend on this prey. The presence of other introduced species can also create complex interactions. For example, one invasive predator may prey on another, or an invasive predator may depend on an invasive herbivore as its main prey.

Because invasive predators often prey on other introduced species, their introduction to (or removal from) ecosystems where such interactions occur can have unexpected consequences. On sub-Antarctic Macquarie Island, the introduction of European rabbits in 1879 was followed shortly afterward by extinction of the endemic parakeet (*Cyanoramphus novaezelandiae erythrotis*). It is now thought that the introduction of rabbits supported a substantial increase in the numbers of previously introduced feral cats, which then drove the vulnerable parakeet to extinction. In the 1970s the control of rabbits, in order to protect the island's native vegetation, was achieved through the release of a *Myxoma* virus. This led to a diet switch by the cats to seabirds. Control and then eradication of cats from Macquarie Island was therefore undertaken. Rabbits have once again increased in numbers, although it is unclear whether this release is attributable to the removal of cats or a decrease in efficiency of viral control. Vegetation cover on Macquarie has also changed with rabbit density, although this effect has been confounded by climate warming over the past 50 years.

On Little Barrier Island (New Zealand), feral cats were impacting adult survival of threatened Cook's petrels (*Pterodroma cookii*). However, the nesting success of Cook's petrels paradoxically decreased after the eradication of feral cats. The probable explanation is that the impact of invasive Pacific rats (*Rattus exulans*), which are nest predators, increased after cat eradication, owing to a "mesopredator release" effect. When the rats were also eradicated from Little Barrier Island, the nesting success of Cook's petrels improved dramatically, in the absence of any invasive predators. Such examples highlight the importance of understanding the interactions between predators and their prey when planning for action against invasive species for conservation purposes.

Invasive predators can sometimes have ecological effects that extend beyond those on their prey: their influence can potentially extend to the very foundation of food webs, through "trophic cascades." Well-documented examples of this effect come from the Aleutian Islands, where introduced Arctic foxes (*Alopex lagopus*) and brown rats (*Rattus norvegicus*) prey on seabirds and greatly reduce their abundance. On islands invaded by foxes, the loss of seabirds disrupted their transport of marine nutrients to the land. This had flow-on effects to soil fertility and ultimately resulted in radical changes to the vegetation, changing grassland to communities dominated by shrubs and forbs. On islands invaded by rats, the loss of seabirds caused an increase in intertidal invertebrates which seabirds preyed upon, and a decrease in algal cover that the invertebrates fed upon.

Predators can also cause shifts in the biology of their prey. When predators selectively target only some individuals in a population, traits in the prey population can evolve over time. Introductions of the predatory ground-dwelling curly-tailed lizard (*Leiocephalus carinatus*) on Caribbean islands caused brown anoles (*Anolis sagrei*) to increase the height at which they perched in trees. Introduced salmonids in Sierra Nevada alpine lakes preferentially target large *Daphnia*, causing subsequent generations to be selected for smaller body size. Such ongoing selection processes at the population level will ultimately change the evolutionary pathway of the prey populations.

The dynamics of predator–prey populations are commonly studied using predator–prey (also known as Lotka–Volterra) population models. These mathematical models couple the population size, birth rates, and death rates of predators to their prey. The simplest models relate only one predator and prey to each other (Fig. 3), but many more complex models can be examined, incorporating such factors as the carrying capacity of each population in the environment, the presence of alternative predators or prey, or the effect of seasons or human harvesting on populations. Such models allow the prediction of long-term outcomes for populations, such as population size fluctuations, coexistence of the predator and prey, or extinction of the prey species.

CONTROL/ERADICATION

Because invasive predators can have such severe effects, their control or eradication usually benefits native species and ecosystems, provided it does not result in unanticipated indirect effects. On geographically isolated areas, such as on islands or within fenced sanctuaries, it is possible to eradicate predatory invasive vertebrates, provided that the removal rate is faster than the population growth rate. However, in most situations, eradication is impossible, and control is the only option. Well-established invasive predators occupying large areas of habitat on continents, large landmasses, or large water bodies usually present insurmountable eradication challenges and the eradication of invertebrates is generally more difficult than that of vertebrates. For these reasons, prevention is always better than cure. It is generally far more cost-effective to prevent biological invasions, or (failing that) to detect and eradicate newly established populations, than to manage the consequences of widespread, well-established invasive predators. Reinvasion following eradication is also possible, whether intentional or accidental. It is therefore important to have measures in place to detect invasive predators in the early stages of invasion, allowing for eradication before their severe effects on other species become irreversible.

SUMMARY

Predators are a natural part of all environments, and their presence is vital for maintaining a balance in nature, although population sizes of predators and prey will always fluctuate over time. Introduced predators, however, can drastically change ecosystems, as shown by examples from different ecosystems across the world. Deliberate introductions of invasive predators for biological control purposes, or for other reasons, therefore need to be subject to rigorous controls. Accidental introductions, especially of invertebrates and commensal rodents, will continue, and no area can ever be considered completely safe from them. We are only now beginning to understand fully the consequences of predator introductions, as mathematical models and field studies suggest that the introduction of invasive predators can have long-term effects on ecosystems that they invade.

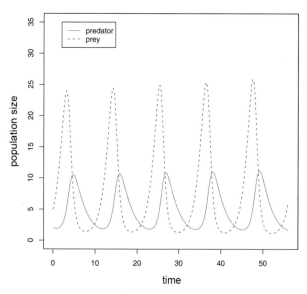

FIGURE 3 Standard predator-prey model with cyclic population dynamics.

FURTHER READING

Courchamp, F., J.-L. Chapuis, and M. Pascal. 2003. Mammal invaders on
    islands: Impact, control and control impact. *Biological Reviews* 78(3):
    347–383.
Croll, D. A., J. L. Maron, J. A. Estes, E. M. Danner, and G. V. Byrd. 2005.
    Introduced predators transform subarctic islands from grassland to
    tundra. *Science* 307(5717): 1959–1961.
Goldschmidt, T., F. Witte, and J. Wanink. 1993. Cascading effects of
    the introduced Nile perch on the detritivorous/phytoplanktivorous
    species in sublittoral areas of Lake Victoria. *Conservation Biology* 7(3):
    686–700.
Wiles, G. J., J. Bart, R. E. Beck, and C. F. Aguon. 2003. Impacts of the
    brown tree snake: Patterns of decline and species persistence in Guam's
    avifauna. *Conservation Biology* 17(5): 1350–1360.

PRIORITIZATION

SEE RISK ASSESSMENT AND PRIORITIZATION

PROPAGULE PRESSURE

RICHARD P. DUNCAN

Lincoln University, New Zealand

Propagule pressure is a measure of the number of individuals of a species arriving in a region to which they are not native. It has been identified as a key factor determining whether species introduced to a new region are likely to establish wild populations; put simply, the more individuals that arrive, the greater the chances of successful establishment. It is also a key factor in determining the pattern and rate of subsequent spread of species throughout a newly invaded region.

MEASURING PROPAGULE PRESSURE

Propagule pressure (also termed introduction effort) is a term describing the number or rate at which propagules enter a region. For species with good records of their introduction and release, it is often possible to distinguish two components of propagule pressure: the number of individuals included in any one release event and the number or rate of discrete release events. Birds that were introduced to New Zealand in the nineteenth century under the auspices of acclimatization societies, for example, arrived in shipments for subsequent release into the wild.

The greater the number of shipments (discrete release events) or the greater number of individuals included in each shipment, the greater the propagule pressure.

In many situations, it is difficult to determine the number of individual propagules of a species that arrive in a region, and hence to measure propagule pressure directly. This is especially true for accidental introductions, which typically go unrecorded. It is also true for plants introduced initially into cultivation, where propagule pressure is often a function of the amount of seed escaping into the wild, which is determined by the number of individuals planted and the rate of seed production and dispersal, all of which can be difficult to quantify. As a consequence, studies often use surrogates for propagule pressure—some measure known or considered to be highly correlated with it. Boats, for example, are the primary vector by which invasive zebra mussels are transported between lakes in North America, and estimates of boat traffic between invaded and noninvaded lakes thus provide a useful surrogate for propagule pressure.

THEORETICAL UNDERPINNINGS

There is a strong theoretical justification for the link between propagule pressure and the probability that a species will establish a population in the wild. The link stems from the theory and observation that large populations are less likely to go extinct than small populations for several reasons. First, large founding populations are less likely to decline to extinction than are small populations, due to stochastic fluctuations owing to demographic or environmental stochasticity, or to natural catastrophes. Second, large populations are less prone to negative effects, such as inbreeding or difficulty in finding mates, which can occur at low population densities (Allee effects). Third, greater propagule pressure associated with a greater number of release events may improve establishment by increasing the chance that at least one release population encounters favorable conditions for population growth.

The combined effect of these factors can be summarized as a curve relating the probability of establishment to the number of propagules arriving, termed the establishment curve (or dose–response curve). The theoretical shape of this curve is determined by the relative role that various processes play in affecting establishment, but Leung et al. (2004, *Ecology* 85: 1651–1660) have derived a simple yet general form for the relationship. If we define a population as establishing at a location to which it is released if at least one propagule from the release event establishes, then the probability of establishment is one minus the probability that all propagules failed to establish:

$$E = 1 - (1 - P)^N$$

where E is the overall probability of establishment, N is propagule pressure expressed as the number of individual propagules released, and P is the probability of a single propagule establishing (assuming that the chance of establishment for each propagule is independent—e.g., that there are no Allee effects). The probability of a single propagule establishing (P) will be determined by the attributes of the introduced species and the conditions encountered at the introduction site. Where conditions at a site are more favorable for a species, the chance of a single propagule establishing will be higher. Figure 1 shows the expected form of the relationship between overall probability of establishment and the number of propagules released (the establishment curve) for various values of P, with propagule pressure shown on both untransformed and $\log_{10}$ transformed axes (a common transformation in studies of propagule pressure).

While Figure 1 implies that greater propagule pressure will enhance establishment success, there will be situations where species fail to establish regardless of the level of propagule pressure because site conditions are completely unfavorable (i.e., $P = 0$, as we might expect, for example, when tropical species are introduced to very high latitudes).

When populations are subject to Allee effects, the probability of a single propagule establishing will be further reduced at low propagule pressure, and the shape of the curve will shift to something like a logistic form on the untransformed scale. At low propagule pressure, the overall probability of establishment will be small due to the inverse density–dependent Allee effects. As propagule pressure increases, the probability of establishment accelerates, eventually reaching a plateau where the addition of further propagules has little influence.

IMPORTANCE TO UNDERSTANDING INVASION PATTERNS

Establishment

Observational and experimental studies confirm that greater propagule pressure is a key factor increasing the chances of an introduced species establishing. To illustrate, 120 bird species are recorded as being introduced to New Zealand, for which data on propagule pressure (here, the number of individuals released) are available for 112 species. The probability of establishment increases progressively with increasing propagule pressure: of the 62 species for which fewer than 20 individuals were released, only one species established; of the 23 species with 21–100 individuals released, 7 established; and of the 27 species with more than 100 individuals released, 20 established.

The importance of propagule pressure is emphasized by the finding that even relatively crude surrogates are often useful predictors of establishment. For example, while it is difficult to quantify directly the propagule pressure associated with introduced plants in cultivation, we would expect propagule pressure to be strongly linked to the total number of individuals planted. Even indirect measures of planting effort, such as their historical frequency in nursery catalogs, can explain differences among species in their probability of having established wild populations.

Indeed, where data are available, studies of the factors associated with the successful establishment of introduced species have repeatedly highlighted the key role that propagule pressure plays: it is consistently the most reliable predictor of variation in establishment success. Moreover, its importance in explaining introduction outcomes most likely accounts for two general features of invasions.

A

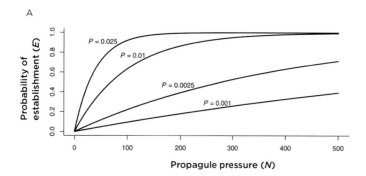

B

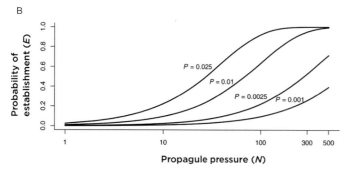

FIGURE 1 (A) The relationship between overall probability of establishment, E, and propagule pressure, N (the number of propagules released or arriving at a site), for different values of P (the probability of a single propagule establishing), derived from the equation. (B) The same relationship as in (A), but with propagule pressure shown on a $\log_{10}$-transformed axis.

First, the large majority of species introduced to new regions fail to establish wild populations. Given that introductions typically involve small founding populations (low propagule pressure), it is not surprising that most fail. Second, the importance of propagule pressure can explain the apparently idiosyncratic nature of invasions. Despite intensive searching, it has proven difficult to identify species' traits, or features of the locations to which species are introduced, that consistently explain introduction outcomes. However, studies reveal that variation in propagule pressure can vary markedly: a single species introduced to several locations can have widely varying propagule sizes, which can explain why, in the absence of data on propagule pressure, outcomes appear idiosyncratic.

Variation in propagule pressure may further confound the interpretation of invasion patterns in other ways. Specifically, propagule pressure is often found to be correlated with other traits linked to establishment success. For example, large native range size is an attribute of species that has been shown to correlate with establishment success. This has been given an ecological interpretation, in that species with larger range sizes may exhibit wider environmental tolerance and thus be more likely to find a suitable niche in a new environment. However, a large native range size also means that species are more likely to be encountered, picked up, and transported by humans, leading to greater propagule pressure, and where data on both range size and propagule pressure are available, propagule pressure is invariably the stronger predictor. Hence, while some traits may be correlated with establishment outcomes, such correlations may frequently arise only indirectly through a link with propagule pressure. As Williamson (1996) stated: "Looking for real differences in invasibility requires looking at the residuals from the relationship between invasion success and propagule pressure."

Spread

The subsequent spread of an invasive species from its initial point of establishment can be thought of as a series of further establishment events that involve populations colonizing new territory. Here again, the likelihood of spread into new locations is a function of propagule pressure: locations with higher rates of propagule input because, for example, they are located close to the initial establishment site are likely to be more rapidly invaded. Indeed, the intensity of propagule pressure in different parts of the landscape resulting from the initial patchy distribution of individuals is a key factor determining the subsequent pattern and rate of spread of invasive species.

The increase in the size of an invading population means also that the rate of propagule production will increase. As a consequence, habitats that pose an initial barrier to invasion (those with a low probability of establishment at low rates of propagule input) may eventually succumb because they are swamped with sufficient propagules to overcome those barriers. Thus, the question of whether some habitats are more invasible than others makes sense only with reference to the rate of propagule input. It is becoming increasingly clear that a full understanding of the invasion processes requires an understanding of the interactions between propagule pressure, species attributes, and the characteristics of invaded locations.

SEE ALSO THE FOLLOWING ARTICLES

Acclimatization Societies / Demography / Invasibility, of Communities and Ecosystems / Range Modeling / Risk Assessment and Prioritization

FURTHER READING

Beirne, B. P. 1975. Biological control attempts by introductions against pest insects in the field in Canada. *Canadian Entomologist* 107: 225–236.
Cassey, P., T. M. Blackburn, D. Sol, R. P. Duncan, and J. L. Lockwood. 2004. Global patterns of introduction effort and establishment success in birds. *Proceedings of the Royal Society, London, Series B (Supplement)* 271: S405–S408.
Colautti, R. I., I. A. Grigorovich, and H. J. MacIsaac. 2006. Propagule pressure: A null model for biological invasions. *Biological Invasions* 8: 1023–1037.
Davis, M. A. 2009. *Invasion Biology.* Oxford: Oxford University Press.
Drake, J. M., and D. M. Lodge. 2006. Allee effects, propagule pressure and the probability of establishment: Risk analysis for biological invasions. *Biological Invasions* 8: 365–375.
Lockwood, J. L., P. Cassey, and T. Blackburn. 2005. The role of propagule pressure in explaining species invasions. *Trends in Ecology and Evolution* 20: 223–228.
Rouget, M., and D. M. Richardson. 2003. Inferring process from pattern in plant invasions: A semimechanistic model incorporating propagule pressure and environmental factors. *American Naturalist* 162: 713–724.
Von Holle, B., and D. Simberloff. 2005. Ecological resistance to biological invasion overwhelmed by propagule pressure. *Ecology* 86: 3212–3218.
Williamson, M. 1996. *Biological Invasions.* London: Chapman and Hall.

PROTECTED AREAS

JOHN RANDALL

The Nature Conservancy, Davis, California

The World Conservation Union (IUCN) defines a protected area as an area of land or sea especially dedicated to the protection and maintenance of biological diversity, and of natural and associated cultural resources, and managed through legal or other effective means. Nonnative species that establish and spread in protected areas can cause damaging and undesirable change and can block the achievement of management objectives. Protected area managers and planners sometimes take actions to prevent invasions, to control or eliminate established invaders, or to mitigate their effects. In many protected areas around the world, however, the numbers of different invasive species that are already established plus those that have the potential to invade, along with the sheer abundance of particular species, may be so large that there can be no realistic expectation of managing them all, much less eliminating them.

PROTECTED AREA MANAGEMENT CATEGORIES

The IUCN also recognizes that the specific objectives for which different protected areas are managed differ widely, and so has defined six protected area management categories as follows:

Category Ia: Strict Nature Reserve: protected area managed mainly for science.

Category Ib: Wilderness Area: protected area managed mainly for wilderness protection.

Category II: National Park: protected area managed mainly for ecosystem protection and recreation.

Category III: Natural Monument: protected area managed mainly for conservation of specific natural features.

Category IV: Habitat/Species Management Area: protected area managed mainly for conservation through management intervention.

Category V: Protected Landscape/Seascape: protected area managed mainly for landscape/seascape conservation and recreation.

Category VI: Managed Resource Protected Area: protected area managed mainly for the sustainable use of natural ecosystems.

THE EXTENT OF INVASIONS IN PROTECTED AREAS

Invasions by nonnative species were recognized as a threat to nature conservation and native species at least as early as the mid-1800s. For example, within a year of the 1864 establishment of Yosemite Valley as a park, concerns about weeds of European origin choking out native vegetation were raised by the commission appointed to manage the valley. Today, invasive species are found in

terrestrial, freshwater, and nearshore marine systems around the world, and even high-elevation sites with very low densities of humans are susceptible to invasions. A 2007 IUCN study found that they have been reported from a total of 487 protected areas in 106 countries and from 277 wetland sites designated as internationally important under the Ramsar Convention on Wetlands (17% of the 1,641 Ramsar sites dedicated worldwide as of February 2007) in 84 countries. Nonetheless, there is evidence that protected areas are less invaded than areas managed for other purposes. A 1999 study found that protected areas contain about half the number of nonnative plant species contained by sites outside of them. A 2003 analysis of lists of vascular plants from 302 nature reserves established between 1838 and 1996 in what is now the Czech Republic determined that reserves established earlier harbor a significantly lower number of nonnative plant species than those established later. It also suggested that native vegetation in these reserves helps to slow the establishment of nonnative plant species. On the other hand, a 2008 study found that U.S. national parks with high native plant species richness, and parks with many threatened and endangered plants, had higher numbers of nonnative plant species.

A variety of factors influence the number and abundance of invasive species in protected areas. Several studies have found that numbers of invasive species in protected areas are positively correlated with numbers of human visitors and with length of roads, trails, and rivers. Cars and other vehicles can carry nonnative plants and other invasive species long distances into protected areas. Domesticated animals may also have a role in introducing invasive species to protected areas. A 2009 study found that nonnative species richness and the ratio of nonnative to native species were greater in plots located along active horse trails than in plots located along old roads or in intact native communities. Thus, where designation of a site as a protected area draws more visitors and greater traffic by domesticated pack animals, it may actually increase the risk of invasions. Climate and geographic isolation also play a role. Temperate protected areas tend to have higher numbers of nonnative species than tropical and arid subtropical protected areas, and islands generally have more nonnative invaders than do mainland sites in the same climatic zone.

Invasive species are frequently cited as second only to habitat destruction as a threat to biological diversity worldwide. Quantitative information that would allow comparisons with other threats to protected areas is hard to come by, but some data are available that support the contention that invasive species are among the most severe threats to biological diversity in protected areas. Invasive species were the most frequently cited threat in a 2009 summary of 974 Nature Conservancy projects around the world, listed by 587 (60%) of the projects (http://conpro.tnc.org/reportThreatCount; unpublished data, The Nature Conservancy), far more common than the number two threat category of residential and commercial development, with "just" 488 (50%). In a 1994 poll of U.S. national park superintendents, 61 percent of 246 respondents indicated that invasions of nonnative plants alone were moderate or major problems in their parks. According to a 1998 analysis, invasive species contributed to the imperilment of nearly half of the plants, birds, and fish now considered rare, threatened, endangered, or extinct in the United States at that time, second only to direct habitat loss and degradation.

A quick survey of papers in the journal *Biological Invasions* and other leading ecological and conservation journals reveals that invasive species come from nearly every major taxonomic group including every class of vertebrate animals; many phyla of terrestrial and aquatic invertebrates and particularly arthropods, molluscs, annelids, and platyhelminthes; algae; nonvascular plants; gymnosperms; flowering plants; many types of fungi; and a wide variety of bacteria and other microorganisms.

INVASIVE SPECIES EFFECTS IN PROTECTED AREAS

Nonnative species can undermine or even eliminate the species, communities, and ecosystems that protected areas were established to protect. Their impacts may range from significant alterations of ecosystem processes to changes in community composition and structure to suppression or elimination of native species populations to genetic effects. Some nonnative species, however, may have positive effects in protected areas, for example by controlling populations of another harmful invasive species or serving as nurse plants for native species in sites undergoing restoration. Biological control agents for nonnative invasive species are almost always nonnative themselves but are released with the expectation that they will do more good than damage. Careful follow-up monitoring has revealed this to be the case in some protected areas.

The invasive species that cause the most pervasive damage to biological diversity at a given site are those that alter ecosystem processes and factors. The best-known examples are invasive plants such as cheatgrass (*Bromus tectorum*) that have invaded millions of acres in North America's intermountain west, including protected areas like Dinosaur National Monument on the Colorado–Utah border and the Snake River Birds of Prey National Conservation Area in Southwestern Idaho. Cheatgrass has been responsible for increases in fire frequency from means of once every 60–110 years to once every 3–5 years that have converted large areas of shrublands to grasslands. Recent research concluded that the expansion of cheatgrass in the Great Basin alone has released an estimated 8 ± 3 Tg (teragram = 10^{12} grams) of carbon to the atmosphere, and that it will likely release another 50 ± 20 Tg C in the coming decades. It has changed portions of the western United States from a carbon sink to a carbon source.

The nitrogen-fixing invasive plants *Morella (Myrica) faya* and *Morella cerifera* have increased soil nitrogen in Hawaii Volcanoes National Park's soils, which resulted in changes in plant succession and promoted populations of nonnative earthworms that in turn increased rates of nitrogen burial, further altering soil nutrient cycling in these systems.

Recent studies have revealed that invasive animals can drive ecosystem-level changes in protected areas, too. Predation by arctic foxes (*Alopex lagopus*), introduced to Aleutian Islands within the Alaska Maritime National Wildlife Refuge, extirpated or greatly suppressed some of the large seabird colonies the refuge was established to protect. This led to sharp decreases in deposition of guano rich in nitrogen and other plant nutrients on these islands and ultimately to changes in the vegetation from lush beach rye grasslands to far less productive dwarf shrub– and forb-dominated communities.

One of the many invaders that alter community structure and composition is the Australian paperbark, *Meleleuca quinquenervia*, which forms large stands of swamp-forest with little or no herbaceous understory that replace marshlands dominated by sedges, grasses, and herbs in Florida's Everglades. Australian paperbark may also alter fire regimes, promoting crown fires that favor it at the expense of native plants in the Florida plant communities it invades, which have all evolved with higher fire frequency, but lower-intensity surface fires. Selective predation or herbivory can also cause great changes in community composition and suppression or even the extirpation of native species populations. For example, some argue that the factor most responsible for driving nearly 20 Australian mammals to extinction in the past 200 years has been predation by introduced European red foxes (*Vulpes vulpes*) and cats (*Felis catus*). Invasive diseases and insect pests can also alter community composition. In the first half of the twentieth century, the introduced chestnut blight fungus, *Cryphonectria parasitica*, killed more than 3 billion American chestnut trees (*Castanea dentata*) in eastern North America, virtually eliminating this species, which had dominated large areas of deciduous forest, including large portions of Great Smoky Mountains National Park and other national and state parks, forests, and preserves in the Appalachians and surrounding regions.

INVASIVE SPECIES MANAGEMENT IN PROTECTED AREAS

Where protected area management objectives are to protect biological diversity or ecological integrity, the ultimate purpose of preventing or controlling invasives is to further the goal of preserving valued populations of native species, natural communities, or ecosystem processes and states that sustain them. To ensure that time and money are used most effectively, site managers must identify the introduced species that truly harm biodiversity or have the potential to do so if allowed to invade and increase in abundance, and distinguish them from introduced species that do not pose such a threat. Some U.S. public agencies with land management responsibilities have processes for prioritizing the effects of different invaders or even different populations of a given invasive species within a protected area such as the Alien Plant Ranking System developed for use by the National Park Service. Note that this system focuses exclusively on plants and does not pay attention to invasive species not present on the site but that may be potentially poised to invade and that should therefore be given highest priority in some cases.

Protected area managers have been taking actions to control invasive species in protected areas since at least the 1930s. For example, Civilian Conservation Corps workers pulled tens of thousands of nonnative thistles from Yosemite Valley's meadows in those years. Attention to invasive species and actions to address them in protected areas have grown rapidly, particularly since the 1980s, following a renewed burst of scientific interest in the issue exemplified by the publication of the SCOPE series on invasive species in the late 1980s and early 1990s. The past decade has also seen sharp increases in activities to address invasive species by the federal agencies that manage public lands and waters in the United States, with total expenditures of $63,506,000 in fiscal year 2007, up over 36 percent from $47,235,000 five years earlier, in fiscal year 2002. Efforts to control invasive plants and terrestrial vertebrate animals have been most common and prominent worldwide, but efforts to control other taxa, including particularly troublesome insects and plant pathogens as well as some fish and shellfish, have also been attempted.

The best strategies and methods to control invasive species will vary widely according to the conservation targets being protected; the identity, behavior, and distribution of the invaders; the control methods available; and the funds and other resources available to carry out the control. A great deal of information on control strategies and methods, particularly for invasive plants, is now available in published literature and on the Internet. Nearly every weed and pest animal control method employed by farmers, ranchers, foresters, and commercial pest controllers has also been employed in protected areas. Nonetheless, certain methods that may be used successfully to achieve other management goals such as providing abundant, nutritious vegetation for domestic grazing animals, may not be appropriate or legal for use in nearby protected areas. Control methods should always be selected and implemented in a manner designed to yield the greatest net benefit in terms of the conservation goals for the site. In all cases, the effects of control activities on the abundance and impacts of the invasive species, and more importantly on the native species, natural communities, and ecosystems whose protection is the ultimate goal of the protected area, should be monitored and assessed. Without this, a great deal of time and money may be spent on strategies and methods that provide no net conservation benefit or which lead to net losses of biodiversity. For example, plots in leafy spurge–infested areas of a Nature Conservancy preserve that had last been treated with herbicide at least nine years earlier actually had lower diversities of native plants, particularly broadleaved species, and densities of leafy spurge that were not significantly lower than plots in untreated areas: the worst of both worlds. Observations made the first year or so after herbicide treatment had suggested that herbicide treatment significantly reduced leafy spurge abundance, and it was hoped that native grasses and forbs would move in to take advantage of this. While this was unfortunately not the long-term outcome, it is fortunate that follow-up monitoring of the impacts on targeted native plant diversity was conducted and that it revealed this fact so that this method will not be repeated.

PREVENTION AND EARLY DETECTION OF INVASIONS IN PROTECTED AREAS

A comprehensive plan for addressing invasive species threats to a protected area may include strategies for preventing new invasions, quickly detecting and eliminating or containing new invasions that do occur, controlling established invaders to contain or reduce their harmful impacts, and taking other actions to mitigate the harmful impacts of invaders on biodiversity. The goal should be to minimize the overall negative impacts of invasive species on the site's conservation objectives. Strategies for preventing new invasions often involve actions taken to close important pathways for invasions. For example, emerald ash borer (*Agrilus planipennis*) and other invasive forest pests and diseases are believed to have been carried to many new sites on firewood carried from distant infested areas by campers. Some states and parks are banning or discouraging the movement or import of firewood in an effort to close this pathway. Similarly, the highly invasive zebra and quagga mussels (*Dreissena polymorpha* and *D. bugensis*) and other invasive freshwater plants and animals have been carried long distances and introduced to new watersheds on recreational boats. Managers of some uninfested waterways now require boats to be inspected or at least owner-certified to be clean and free of invasive species. These strategies, of course, require managers to identify potential invaders that are not yet present and pathways by which they would be most likely to invade. And they are likely to be effective only when the pathway is readily subject to control, for example when they rely largely or completely on human activities. Unfortunately, prevention strategies of this sort are not always effective over the long term. The 100th Meridian Initiative, which sought to prevent the westward spread of zebra mussels and other freshwater invaders, was unable to prevent the introduction of these mussels to the lower Colorado River and several other waterways in California and Colorado.

Effective early detection and rapid response programs likewise require conservation managers to identify in advance those species likely to invade; to carry out regular, systematic searches for them; and to ensure appropriate actions are quickly taken when a new invasion is discovered. The design of efficient search protocols usually requires identifying and focusing more attention on portions of the site most likely to be invaded first, such as along roads and trails or around firewood piles and boat ramps. Unfortunately, to date, rapid and effective responses have proven to be far less common than new detections.

SEE ALSO THE FOLLOWING ARTICLES

Early Detection and Rapid Response / Regulation (U.S.) / Restoration / Risk Assessment and Prioritization / SCOPE Project / Transformers

FURTHER READING

Brockie, R. E., L. L. Loope, M. B. Usher, and O. Hamann. 1988. Biological invasions of island nature reserves. *Biological Conservation* 44: 9–36.

De Poorter, M., S. Pagad, and M. I. Ullah. 2007. Invasive alien species and protected areas: A scoping report. Part I: Scoping the scale and nature of invasive alien species threats to protected areas, impediments to IAS management and means to address those impediments. Washington, DC: World Bank, as a contribution to the Global Invasive Species Programme (GISP). *www.gisp.org/publications/reports/IAS_ProtectedAreas_Scoping_I.pdf*

Dudley, N., ed. 2008. *Guidelines for Applying Protected Area Management Categories.* Gland, Switzerland: IUCN.

IUCN. 1994. *Guidelines for Protected Area Management Categories.* CNPPA with the assistance of WCMC. Gland, Switzerland: IUCN.

Kurten, E. L., C. P. Snyder, T. Iwata, and P. M. Vitousek. 2008. *Morella cerifera* invasion and nitrogen cycling on a lowland Hawaiian lava flow. *Biological Invasions* 10: 19–24.

Lonsdale, W. M. 1999. Global patterns of plant invasions and the concept of invasibility. *Ecology* 80: 1522–1536.

MacDonald, I. A. W., D. M. Graber, S. DeBenedetti, R. H. Groves, and E. R. Fuentes. 1988. Introduced species in nature reserves in Mediterranean type climatic regions of the world. *Biological Conservation* 44: 37–66.

Pyšek, P., V. Jarošik, and T. Kufera. 2002. Patterns of invasion in temperate nature reserves. *Biological Conservation* 104: 13–24.

Usher, M. B. 1988. Biological invasions of nature reserves: A search for generalizations. *Biological Conservation* 44: 119–135.

Wilcove, D. S., D. Rothstein, J. Dubow, A. Phillips, and E. Losos. 1998. Quantifying threats to imperiled species in the United States. *BioScience* 48: 607–615.

Zavaleta, E. S., R. J. Hobbs, and H. A. Mooney. 2001. Viewing invasive species removal in a whole-ecosystem context. *Trends in Ecology and Evolution* 16: 454–459.

R

RANGE MODELING

JONATHAN M. JESCHKE

Ludwig-Maximilians-University Munich, Planegg-Martinsried, Germany

Range modeling means modeling the geographic ranges of species. Within the context of invasive introduced species, range models are used to predict how far an invasive species will spread in the region to which it was introduced. Different range models exist: for example, bioclimatic models, envelope models, ecological niche models, and species distribution models. These models are very popular and potentially powerful, yet their accuracy is currently unclear.

APPLICATIONS

Range models are applied in many contexts: to predict future or present geographic ranges of species, or to estimate past climatic conditions on the basis of past geographic ranges of species.

Predicting Future Ranges of Species

Predicting the future ranges of species is the most popular application. It includes predicting eventual ranges of invasive introduced species and predicting range shifts due to climate change.

PREDICTING EVENTUAL RANGES OF INVASIVE SPECIES Range models are often used to predict how far an invasive introduced species will spread. For example, the zebra mussel (*Dreissena polymorpha*) was introduced to North America in the late 1980s and started to spread from the Great Lakes. A number of range models have been constructed to predict its eventual North American range. It has been of great interest where in North America this invasive species will find suitable establishment conditions. If policymakers know well in advance that an invasive species will probably be able to establish itself in a given region, then they may be able to arrange to reduce the risk of the species reaching this region and to minimize the damage it will cause.

PREDICTING RANGE SHIFTS DUE TO CLIMATE CHANGE Environmental conditions, especially climatic conditions, are important determinants for geographic ranges of species. It is thus no surprise that changing climatic conditions cause changes in geographic ranges. But how exactly does the range of a species shift due to climate change? Range models are used to address this question. This application includes predicting range shifts of diseases such as malaria. Range models can also help estimate how many species will go extinct due to climate change.

Predicting Present Ranges of Species

Range models are further used to predict present-day ranges of species. Data on the distribution of many species are sparse, especially for those that are cryptic or are living in poorly investigated regions. Some locations where such a species is present will be known, but its actual range may be unknown. Range models can help estimate its range. Other applications of range models with respect to present-day ranges of species include the identification of suitable sites for the stocking or culture of valuable species, the clarification of systematic relationships, and the investigation of mechanisms underlying geographic ranges of species.

Estimating Past Climatic Conditions

Another application of range models is to estimate past climatic conditions on the basis of known past geographic ranges of species. Here, a range model can be constructed by comparing a species' current geographic range with current climatic conditions. The resulting model will tell under which climatic conditions the species occurs. The model can then be used to predict past climatic conditions if the species' past geographic range is known.

MODELING APPROACHES

Two different approaches exist to construct range models. First, mechanistic models follow a bottom-up approach. Here, physiological tolerances of the focal species are measured in the laboratory (e.g., minimum and maximum temperature). The measured tolerances are then compared to field conditions, so that locations are identified that match the species' tolerances. With this mechanistic method, the fundamental ecological niche of the species is modeled. A mechanistic range model has, for example, been constructed to predict the eventual range of invasive introduced itchgrass (*Rottboellia cochinchinensis*) in the United States.

The second group of range models is the larger one and is formed by empirical models. These correlative models follow a top-down approach. They try to capture the realized ecological niche of the focal species. Basic empirical models are restricted to climatic variables (e.g., temperature, precipitation, or evaporation), while other models include additional variables (e.g., geographic variables). If the goal is to predict the eventual range of a currently spreading invasive species, then the model is typically constructed by dividing the native range of the species into grids. For each grid, the values of the chosen climatic, and possibly other, variables are then compared to the presence or absence of the species. A variety of statistical methods have been applied for this comparison, for example logistic regression analysis, generalized linear modeling (GLM), generalized additive modeling (GAM), climate envelope techniques such as BIOCLIM, classification and regression tree analysis (CART), neural networks, and genetic algorithms such as GARP. Independently of the applied method, the constructed model is then used to predict the eventual range of the species in the region to which it was introduced. Again, this region is divided into grids. If we know the values of the chosen variables (the same variables as for the native range), then the model will predict in which grids the invasive species will eventually occur. Empirical range models have been constructed for many invasive species, including fireweed

(*Senecio madagascarensis*) and the red imported fire ant (*Solenopsis invicta*).

ASSUMPTIONS

Range models make the following critical assumptions: (1) Species interactions do not affect geographic ranges, (2) the genetic and phenotypic composition of species is spatially and temporally constant, and (3) species are not limited in their dispersal.

Species Interactions Do Not Affect Geographic Ranges

Mechanistic range models assume that species interactions (e.g., competition, mutualism, predation) do not affect geographic ranges. They measure the physiological tolerances of a species in the laboratory and assume that a species will occur at all locations where the environmental conditions are within these tolerances. They assume that the realized ecological niche of a species is equal to its fundamental ecological niche, an assumption that is, of course, in stark contrast to ecological knowledge and not met in nature.

Empirical range models assume that the effects of species interactions on geographic ranges are constant over space and time. They do not usually explicitly consider species interactions, but because they are constructed by comparing climatic and other variables to *realized* geographic ranges, they implicitly capture the influence of species interactions for the space and time they are constructed. When one applies them to predict geographic ranges for another space or time, however, it is assumed that this influence of species interactions remains the same. For example, if a model for an invasive species is constructed with data from its native range, then the model's predictions for the exotic range are based on the assumption that the native and exotic ranges are identically affected by species interactions. Again, this assumption cannot hold in nature because the species is confronted with very different species in the exotic range as compared to the native one.

The Genetic and Phenotypic Composition of Species is Spatially and Temporally Constant

Both mechanistic and empirical range models assume that the genetic and phenotypic composition of species is spatially and temporally constant. Again, the genetic and phenotypic composition is not explicitly considered by range models, but they are applied as if this composition were constant. For example, if a model for an invasive species is constructed with data from its native range and applied to predict its eventual exotic range, then it is assumed that the individuals of this species do not differ genetically and phenotypically between the native and exotic range. Of course, this second crucial assumption of range models is not met in nature either. It is, for instance, frequently observed that individuals of invasive species reach larger body sizes in their exotic than in their native ranges. To mention just one example, the green crab (*Carcinus maenas*) was reported to have a 30 percent larger carapace width in its exotic than in its native range. These differences between individuals of the native and exotic range cause differences in the ecological niche between the native and exotic range. They challenge the usefulness of models that are constructed in the native range and applied in the exotic range.

Species are Not Limited in Their Dispersal

Finally, both mechanistic and empirical range models assume that species are not limited in their dispersal. Species are assumed to be at all locations where conditions are suitable. This third assumption is in stark contrast to reality as well. It is known that barriers to species dispersal exist and that, even in the absence of such barriers, species often need a substantial amount of time to reach locations with suitable conditions. The topic of this encyclopedia is a good example: invasive species are species that were introduced by humans to locations where they did not occur before and where they have found suitable conditions. They did not reach these regions before (e.g., due to a dispersal barrier), and so they needed humans to transport them there.

USEFULNESS AND ACCURACY OF RANGE MODELS

As outlined in the last section, range models make assumptions that do not hold in nature. Yet this is true for all models. We know that all models are wrong. The question is whether range models are useful. As range models are mainly used for predictions (see above), their usefulness mainly depends on the accuracy of these predictions.

Unfortunately, it is currently unclear how accurate the predictions of range models really are. Although predicted ranges have been compared to observed ones, these comparisons have typically been done for spatially and temporally dependent datasets. The box that follows outlines three different strategies to compare model predictions with observations. The preferable strategy depends on what a model is used for. As outlined above, range models are typically used for predicting species ranges that are temporally or spatially independent. For

---

DIFFERENT STRATEGIES OF MODEL VALIDATION

resubstitution After fitting the model to a given dataset, its predictions are compared to the same dataset.

data splitting After splitting a dataset into two parts, the model is fitted to the first part of the dataset, and its predictions are then compared to the second part; examples for this strategy are jackknifing, boot-strapping, and cross-validation.

independent validation Two temporally or spatially independent datasets *A* and *B* are needed for this strategy: after fitting the model to dataset *A*, its predictions are compared to dataset *B*; an example for this strategy is to fit the model to data from a species' native range (dataset *A*) and compare it to data from the species' exotic range (dataset *B*).

---

example, predicting the eventual *future exotic* range of an introduced invasive species based on its *current native* range means predicting a temporally and spatially independent range. Model predictions should be evaluated accordingly; hence, the strategy of independent validation is recommended. Yet predictions of range models have typically been evaluated by means of resubstitution or data splitting (cross-validation), and only rarely by means of independent validation. The studies that did apply independent validation suggest that current range models have an intermediate accuracy.

OUTLOOK

The accuracy of range models can be improved by considering their underlying assumptions. Specifically, researchers have started to develop extended range models that consider species interactions (e.g., for owls interacting with woodpeckers) and dispersal limitations (e.g., for the lizard orchid, *Himantoglossum hircinum*). It will be vital to test such extended models by means of independent validation. If their predictions turn out to be accurate, then extended range models will be powerful tools in the hands of invasion biologists.

SEE ALSO THE FOLLOWING ARTICLES

CART and Related Methods / Climate Change / Dispersal Ability, Animal / Dispersal Ability, Plant / Evolution of Invasive Populations / Habitat Compatibility / Tolerance Limits, Animal / Tolerance Limits, Plant

FURTHER READING

Araújo, M. B., and A. Guisan. 2006. Five (or so) challenges for species distribution modelling. *Journal of Biogeography* 33: 1677–1688.

Gaston, K. J. 2003. *The Structure and Dynamics of Geographic Ranges.* Oxford: Oxford University Press.

Guisan, A., and W. Thuiller. 2005. Predicting species distribution: Offering more than simple habitat models. *Ecology Letters* 8: 993–1009.

Heikkinen, R. K., M. Luoto, M. B. Araújo, R. Virkkala, W. Thuiller, and M. T. Sykes. 2006. Methods and uncertainties in bioclimatic envelope modelling under climate change. *Progress in Physical Geography* 30: 751–777.

Jeschke, J. M., and D. L. Strayer. 2008. Usefulness of bioclimatic models for studying climate change and invasive species (1-24). In R. S. Ostfeld and W. H. Schlesinger, eds. *The Year in Ecology and Conservation Biology. Annals of the New York Academy of Sciences* 1134.

Kearney, M., and W. P. Porter. 2009. Mechanistic niche modelling: Combining physiological and spatial data to predict species' ranges. *Ecology Letters* 12: 334–350.

Pearson, R. G., and T. P. Dawson. 2003. Predicting the impacts of climate change on the distribution of species: are bioclimate envelope models useful? *Global Ecology and Biogeography* 12: 361–371.

Peterson, A. T. 2003. Predicting the geography of species' invasions via ecological niche modeling. *Quarterly Review of Biology* 78: 419–433.

Sax, D. F., J. J. Stachowicz, J. H. Brown, J. F. Bruno, M. N. Dawson, S. D. Gaines, R. K. Grosberg, A. Hastings, R. D. Holt, M. M. Mayfield, M. I. O'Connor, and W. R. Rice. 2007. Ecological and evolutionary insights from species invasions. *Trends in Ecology and Evolution* 22: 465–471.

Williams, J. W., and S. T. Jackson. 2007. Novel climates, no-analog communities, and ecological surprises. *Frontiers in Ecology and the Environment* 5: 475–482.

RAPID RESPONSE

SEE EARLY DETECTION AND RAPID RESPONSE

RATS

MICHEL PASCAL

Institut National de la Recherche Agronomique (INRA), Rennes, France

Among the 65 described extant species in the genus *Rattus*, humans have introduced five to various regions: the large spiny rat (*Rattus praetor*), the ship rat or black rat (*R. rattus*), the Oriental house rat (*R. tanezumi*), the Pacific rat (*R. exulans*), and the Norway rat or brown rat (*R. norvegicus*).

HISTORIC INTRODUCTIONS AND THEIR IMPACTS

Various combinations of the rat species listed above have been introduced to more than 80 percent of the world's archipelagos, as well as to all continents except for Antarctica. Some of these introductions date to around 10,000 years ago, and others to less than 500 years ago or less. Finally, some of these species have been reservoirs of pathogens that have caused dreadful epidemics or epizootics and huge losses to agriculture and stored products, as well as to human infrastructure. For instance, in 2005 the estimated annual loss to the U.S. economy caused by the ship rat, the Norway rat, and the house mouse was $19 billion, a figure that does not account for environmental and public health costs. As another example, although most people know about the plague epidemics of the Middle Ages caused by the introduction to Europe of the ship rat, a flea that lives on the rat, and the bacterium *Yersinia pestis* that the flea carries, few realize that there were plague outbreaks in Europe in the first half of the twentieth century and that plague remains a persistent problem in several regions, including Madagascar. Finally, although the public knows about the role of rodents in plague epidemics, few are aware of the many other parasitic, viral, or bacterial diseases for which introduced rats are reservoirs and vectors. A good example is leptospirosis, which kills at a rate 20–50 times higher on tropical islands than in temperate countries. These various detrimental economic and public health effects explain why the ship rat, Norway rat, and house mouse are the three rodent species most studied historically and why they are still studied actively.

A plethora of recent zooarchaeological, ecological, and genetic research has elucidated several aspects of the history of rat invasions and the role that their commensalism with humans has played in their successful colonization of a vast portion of the globe. This research has also revealed the nature and scope of the impacts of some rat species on the functioning of various ecosystems, especially on islands.

On islands, the ship rat, Norway rat, and Pacific rat have all been shown to have substantially disrupted ecosystem functions. They have negatively affected at least 170 animal and plant species on more than 40 islands or archipelagos and have directly contributed to at least 50 species extinctions. Species eliminated by introduced rats span a wide range of taxa, including mammals, birds, reptiles, insects, crustaceans, and snails. Three comparative methods have been used to trace rat impact on island ecosystems. The first entails comparing ecosystem status before and after an invasion, the second involves comparing ecosystem status on both invaded and rat-free islands in the same archipelago, and the third involves comparing ecosystem status before and after rat eradication.

A recent survey found 344 attempts to eradicate island populations of the ship rat, Norway rat, and Pacific rat, of which 318 were successful. The frequency of such eradication projects demonstrates just how alarmed resource managers and conservationists are about introduced rats.

LARGE SPINY RAT (*RATTUS PRAETOR*)

The large spiny rat is native to the western part of Papua New Guinea. Archaeologists have shown that it was introduced to the Bismarck Archipelago around 8,000 BC. This rat probably contributed to the extinction of 12 bird species in the Bismarck Archipelago, although its exact role in these extinctions is uncertain because six other mammals were also introduced to these islands, including the Pacific rat, the dog, and later the pig. More recently, humans have introduced the large spiny rat to the Solomon Islands, Vanuatu, and Fiji, and possibly to Palau. However, the only introduced populations that have persisted are those in the Bismarck Archipelago and the Solomon Islands, and the impact of these introduced populations on ecosystems of these islands is unknown.

NORWAY RAT OR BROWN RAT (*RATTUS NORVEGICUS*)

Contrary to what is implied by its common and scientific names, the Norway rat (Fig. 1) is not originally from Norway (the provenance of the specimens that Linnaeus used to describe the species); it is native to southeastern Siberia, northern China, and the northern islands of Japan. Although a fourteenth century archaeological site in Tuscany, Italy, contained many Norway rat remains, it was not until the eighteenth century that a massive invasion of western Europe occurred, as documented in reports and archaeological remains. For instance, the oldest archaeological remains of the Norway rat in France date from 1750 and were found in Paris, which at that time was a hub of river traffic. From European ports, the Norway rat began to colonize all continents, save Antarctica, as

FIGURE 1 Norway rat, *Rattus norvegicus*, in the rainforest of Basse-Terre on Guadeloupe, French West Indies. (Photograph courtesy of the author.)

well as most island groups. The speed and breadth of this conquest can be attributed to the fact that this rodent thrives in anthropogenic habitats and to the explosion of commerce, voyages of discovery, and advances in marine technology in the eighteenth and nineteenth centuries. The economic and health consequences of these introductions are outlined above.

Its geographic origin may lead one to believe that the Norway rat is better adapted colder climates than are the ship rat and the Pacific rat. However, and paradoxically, although widespread invasions in the northern hemisphere support this hypothesis, the situation in the southern hemisphere does not: many sub-Antarctic islands have been invaded by the ship rat and not by the Norway rat. Moreover, the Norway rat, unlike the ship rat and the Pacific rat, is a poor climber. However, it is well adapted to water and is frequently found along water courses, on the banks of river mouths, and at the rear of mangrove swamps in tropical and subtropical regions. The differences in habitat preferences and behavior of the Norway rat, the ship rat, and the Pacific rat lead to differential access to various resources and help to explain differences in the impact of these species on island ecosystems.

A recent survey showed that, of 109 attempts to eradicate island populations of the Norway rat, 104 succeeded. The largest island so far cleared of Norway rats is 11,300 ha Campbell Island (New Zealand). These eradication projects have greatly aided the identification of ecosystem impacts of the Norway rat. A 2006 review showed that eradicating the Norway rat from islands in the English Channel and along the European Atlantic coast led to a spectacular increase in the reproductive success of several seabird populations, the number of nesting pairs of land birds, and the abundances of two shrew species. An eradication on a small island near Mauritius led to a significant increase in native skink and gecko populations.

In addition, a comparison between sub-Antarctic islands (in the South Georgia and Falkland Islands) invaded by the Norway rat and those without rats shows that this species preys heavily on marine and land birds. Finally, following the introduction of the Norway rat in 1995 to Fregate Island in the Seychelles, populations of a tenebrionid beetle plummeted, and mortality greatly increased for chicks of the native magpie-robin (*Copsychus sechellarum*).

It is noteworthy that the province of Alberta, Canada, which is surrounded by rat-infested areas, has remained nearly free of Norway rats because of the highly efficient and aggressive Alberta Rat Patrol, which works to minimize habitats favorable to this rat and quickly springs into action when a rat is observed.

ORIENTAL HOUSE RAT (*RATTUS TANEZUMI*)

Only recently has consensus been reached on the identification of *R. rattus* and *R. tanezumi*, which are morphologically very similar, and they may often have been confounded in the past. Moreover, it is possible that future systematic revisions including *R. tanezumi* will show it to comprise several taxa that are, in fact, separate species. This fact explains existing uncertainty about its native distribution. It is currently believed that the native range of *R. tanezumi* extends in southern Asia from eastern Afghanistan and northern and eastern India, where there is a possible zone of coexistence with *R. rattus*, through central and southern China, Korea, and continental Indo-China plus its continental islands. Whether this species is introduced or native in Taiwan and Japan is currently debated. It is very likely that *R. tanezumi* is introduced in the Malay Peninsula and the islands of the Sunda Shelf. Finally, *R. tanezumi* is definitely introduced in the Philippines, Sulawesi, and many islands stretching from the Moluccas and Nusa Tenggara east of New Guinea to the islands of the Eniwetok Atoll and Fiji.

When and by whom this species was introduced to the various islands is currently very poorly understood. The ecological and other consequences are also poorly known, but it is likely, given the nearness of their systematic positions and the similarity of their life histories, that impacts of introduced Oriental house rats resemble those of introduced ship rats.

PACIFIC RAT (*RATTUS EXULANS*)

The Pacific rat (Fig. 2) originated in Southeast Asia. At least 10,000 years ago, it was introduced to islands in the Bismarck Archipelago. Subsequently, waves of human colonization by the Lapita people (3,600–3,000 years ago), and then by Polynesians (beginning around 2,500 years ago, but probably in great numbers beginning only about 1,000 years ago), introduced this rodent to almost all Pacific islands, including the most isolated ones to which Polynesians traveled, such as Easter Island, and also to the most recently colonized ones, such as New Zealand.

Whether or not the Pacific rat was deliberately introduced by the Polynesians, and if so, why, are outstanding questions. Nevertheless, this rat, whose remains have been recovered from archaeological sites (even the earliest) of Pacific islands, is an excellent marker for the spread of humans in this part of the world. It is not easy to determine the relative importance of the roles played by the

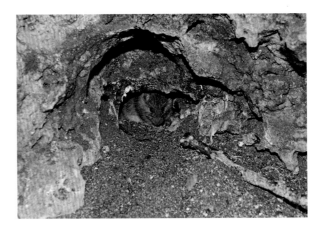

FIGURE 2 Pacific rat, *Rattus exulans*, in the Limestone plateau of Teuaua islet, Ua Huka Island, Marquesas Islands. (Photograph courtesy of the author.)

Pacific rat and by humans in the great changes these island ecosystems have undergone since the earliest human colonization. In fact, these changes have had many causes: anthropogenic habitat destruction and change, but also impacts of other species, often introduced at the same time as the Pacific rat, such as dogs and pigs. During the last 50 years, 61 eradication projects have been mounted against the Pacific rat, of which 55 succeeded.

The key role of predation by the Pacific rat on chicks of endemic land birds was unequivocally established following rat eradication from areas of reproduction of the Rarotonga flycatcher (*Pomarea dimidiata*) in the Cook Islands. It was also demonstrated for endemic seabirds such as the Henderson petrel (*Pterodroma atrata*), for which two independent monitorings of 60 nests each showed predation of all nestlings by rats within a week of hatching. Furthermore, in New Zealand, comparison of islands invaded by Pacific rats and those lacking rats, and of islands before and after Pacific rat eradication, has shown that this rat causes drastic declines in populations of seabirds, land birds, reptiles, amphibians, insects (e.g., endemic wetas and tenebrionid beetles), and terrestrial snails, to the extent of causing local extirpations. Observations of the same islands show that the Pacific rat hinders forest regeneration by removing seeds and seedlings. This activity often leads to local disappearance of several plant species.

SHIP RAT OR BLACK RAT (*RATTUS RATTUS*)

The ship rat (Fig. 3) is native to the Indian subcontinent. Recent archaeological and molecular genetic research shows that, over the last five millennia, humans have accidentally introduced this species to all continents except Antarctica, as well as to 80 percent of all island groups. In

FIGURE 3 Ship rat, *Rattus rattus*, in the coastal dry forest of Grande-Terre on Guadeloupe, French West Indies. (Photograph courtesy of the author.)

fact, so far, the oldest archaeological remains of *R. rattus* that have been discovered outside the Indian subcontinent come from the Middle East and date to less than 5,000 years ago; earlier published dates from this region are probably in error. In any event, the oldest reliable records of the ship rat outside of Asia come from Egypt in the second millennium BC. The species then apparently remained restricted to the banks of the eastern Mediterranean for nearly a thousand years before spreading to the western Mediterranean, where it suddenly appeared between the fourth and second centuries BC in Corsica, the Balearic Islands, and Italy. It was only during the first century AD that the ship rat appeared in western Europe outside the Mediterranean, and it was restricted to seaports and river ports. In France, it was not until the middle of the Middle Ages that ship rats increased greatly, as evidenced by the presence of their remains in archaeological sites of this era. They are present in 90 percent of archaeological sites from the eleventh century and in 100 percent of sites from the beginning of the fourteenth century. By then, the ship rat was present in sites that were not located at ports or along major transportation routes. Owing to voyages of discovery and the growth of maritime traffic that began in the fifteenth century, the ship rat was introduced from western Europe to the majority of regions where it occurs today.

A second ancient introduction route allowed the ship rat to colonize the east coast of Africa and islands in the Indian Ocean. Recent research in molecular phylogeography indicates that the founders of ship rat populations on Madagascar and Grand Comoro arrived from Oman about 3,000 years ago, at approximately the time of the first human settlements on Madagascar (archaeologists currently date these to 2,300 years ago).

The fact that the ship rat is commensal with humans greatly contributed to the extraordinary rapidity and scope of its spread. Did ship rats associate with humans before they became commensal, and did they become commensal before the advent of agriculture? Or was it the beginning of agriculture that led to the evolution of commensalism in this species? These controversial questions cannot presently be answered. In any event, it is this commensal behavior that makes this species particularly invasive and especially likely to reach those few islands it has not yet colonized. This is why environmental managers and authorities that regulate agricultural activity and public health are extremely vigilant for early signs of ship rat introduction, in order to attempt to eradicate the rats before they cause their customary and often irreversible damage.

Over the past 50 years, eradication projects have targeted ship rats on 174 islands and have succeeded on 159. The largest such cleared island is 1,022 ha Hermite Island (Western Australia). Elimination of the ship rat from islands of the western Mediterranean, the Caribbean, and the eastern Pacific has led to spectacular increases in nesting pairs of seabirds and seabird reproduction, as well as to reestablishment of several locally extirpated bird populations and to pronounced population growth of terrestrial crabs and two mammal species. Moreover, comparison of invaded and uninvaded islands in the Balearic Archipelago shows that the ship rat has caused the extinction of several endemic species of tenebrionid beetles and the local extirpation of at least six plant species. Finally, the recent invasion of Lord Howe Island in the Tasman Sea caused the local extirpation of the stick insect *Dryococelus australis*. This list of inimical effects of the ship rat on island ecosystems may be greatly extended in the future, if one takes into account the many observations of predation on a large spectrum of animal and plant species. These observations testify to the remarkable ability of the ship rat to take advantage of very diverse resources. In any event, impacts of the ship rat are seen as being so severe that the species was included on the list of 100 worst invaders published by the International Union for the Conservation of Nature (see Appendix).

SEE ALSO THE FOLLOWING ARTICLES

Agriculture / Eradication / Islands / Pathogens, Human / Predators / Rodents (Other)

FURTHER READING

Howald, G., C.J. Donlan, J.P. Galván, J.C. Russell, J. Parkes, A. Samaniego, Y. Wang, D. Veitch, P. Genovesi, M. Pascal, A. Saunders, and B. Tershy. 2007. Invasive rodent eradication on islands. *Conservation Biology* 21(5): 1258–1268.

Matisoo-Smith, E., and J.H. Robins. 2004. Origin and dispersal of Pacific peoples: Evidence from mtDNA phylogenies of the Pacific rat. *Proceedings of the National Academy of Sciences, USA* 101(24): 9167–9172.

Pascal, M., O. Lorvelec, and J.-D. Vigne. 2006. *Invasions Biologiques et Extinctions: 11 000 ans d'histoire des Vertébrés en France.* Paris: Coédition Belin-Quæ.

Steadman, D.W. 2006. *Extinction and Biogeography of Tropical Pacific Birds.* Chicago: University of Chicago Press.

Towns, D.R., I.A.E. Atkinson, and C.H. Daugherty. 2006. Have the harmful effects of introduced rats on islands been exaggerated? *Biological Invasions* 8: 863–891.

REGULATION (U.S.)

PHYLLIS N. WINDLE

Union of Concerned Scientists, Washington, DC

Regulations, or legal rules, lay out the specific means by which agencies interpret and implement the laws passed by legislatures. Regulators in the executive branches of federal, state, and local governments create these rules. The steps by which laws are passed often receive far more attention from the press and the public than does the process of how regulations come about. Yet the hundreds of significant regulations that federal agencies issue each year ultimately determine how well laws function. These regulations affect the health, safety, and environment of everyone. They affect every stage of biological invasions, ranging from which nonnative species may be imported to which pesticides may be used in their eradication. Many of the worst invasive species problems are directly attributable to lax or outmoded regulation, along with the sluggishness of the regulatory process.

THE RELATIONSHIP BETWEEN LEGISLATION AND REGULATION

U.S. law is created by (1) legislative bodies, which draft, pass, and amend two separate kinds of bills, those to authorize action and those to fund it; (2) executive branch offices and agencies, which issue regulations, procedures, executive orders, and formal guidance; and (3) decisions made by courts. Thus, like the legislation or statutes, passed by Congress and signed by the president, regulations are law. There may be penalties for failing to comply, and under certain circumstances, courts can step in.

Executive branch regulators, and the presidents, governors, and other government executives for whom they act, differ on *whether* to regulate, based on their philosophies of government. Both presidents and congresses have directed federal agencies, boards, and other bodies *how* to regulate as well, via presidential executive order or in specific legislation. On a given topic, such as invasive species, the underlying statutes themselves also provide more or less guidance on *whether, how,* and *what* to regulate. This gives the relevant agencies varying degrees of regulatory discretion. A statute that says "must" or "shall" makes it the duty of the president to execute its provisions. If, however, the nature and scope of regulations are not addressed, then a different set of issues or conflicts will arise.

In the United States, executive branch agencies at all levels of government enact regulations. International treaties also play a role. For example, the 2004 International Convention for the Control and Management of Ships' Ballast Water and Sediments, which is not in force yet, set a minimum standard, or acceptable level of pollution, for ships' ballast water. Thus, countries signing the treaty would be expected to issue national pollution regulations at least this strong for ships discharging ballast water in their waters. Congress could set this standard by statute or the U.S. Coast Guard could do the same via agency regulations. So, on a given issue, statutes and regulations can overlap, providing different ways to accomplish the same goal.

There is significant interaction between federal and state regulations. For example, state agencies implement much of the federal Clean Water Act. States often model their overall regulatory approaches on federal ones—for example, by issuing regulations to ban lists of animals and plants from import, called dirty, black, or red. State plant health agencies work closely with the U.S. Department of Agriculture (USDA) when it comes to issuing quarantines for specific invasive pests and diseases, for example, the emerald ash borer (*Agrilus planipennis*), sudden oak death (*Phytophthora ramorum*), and viral hemorrhagic septicemia, a disease of fish.

Where there are gaps in law, an agency cannot regulate regardless of need. Gaps are subject to interpretation. In the 1970s, the U.S. Fish and Wildlife Service (FWS) claimed it had legal authority to reverse U.S. policy on the import of nonnative animals. It proposed to create a regulatory clean (or white or green) list of species approved for import. Opponents contended that the FWS lacked legal authority for this change and forced FWS to end its rulemaking. The FWS is now in general agreement that added authority, provided by new legislation, would be needed for this change.

In the opposite case, where Congress provides regulatory authority but the pertinent executive branch agency has failed to act, groups can sue *if* the specific authorizing law allows. A coalition of groups petitioned the U.S. Environmental Protection Agency (EPA) in 1999, then sued in 2001, claiming that the Clean Water Act requires the EPA to regulate ships' ballast water. The courts agreed and, after EPA lost its appeal, ordered the agency to begin a new regulatory process for ballast water in 2008.

HOW AGENCIES MAKE RULES

The Administrative Procedure Act of 1946 defined the modern regulatory process. It furnished a way citizens could learn about proposed regulations, provided for citizen participation at certain stages via public comments, and required a minimum level of agency accountability. These basics have not changed. Federal agencies publish proposed regulations in the *Federal Register* and accept public comments before issuing final rules. Final rules also are published in the *Federal Register* and appear as formal legal language in the *Code of Federal Regulations* (*CFR*).

A number of pieces of legislation, plus several presidential executive orders, have altered or supplemented the basic requirements set in 1946. These include the National Environmental Policy Act, the Paperwork Reduction Act, the Regulatory Flexibility Act, the Congressional Review Act, and executive orders on economic impacts, energy supply, and property rights. The result is that federal rulemaking is a long and complex, multistep process that can be inscrutable to anyone not regularly and deeply involved (Fig. 1). Some agencies take ten years to prepare what are called major rules. How regulations are regarded depends on one's point of view. Descriptions of the U.S. federal regulatory system run the gamut from dysfunctional, largely unsupervised, politicized, pervasive, overreaching, burdensome, complicated, time-consuming, expensive, and inscrutable to inevitable, versatile, indispensable, and little appreciated.

The most significant recent change to the rulemaking process has been the increasing role played by the president's Office of Management and Budget (OMB) in reviewing federal agencies' proposals. This change began under President Reagan and is a continuing source of controversy. The primacy of OMB versus that of the rulemaking agencies and their scientific experts; when and how cost–benefit analysis is used in regulatory decision-

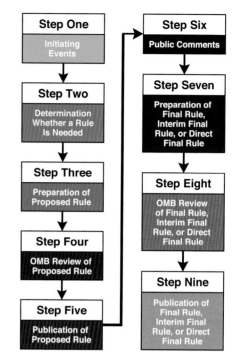

FIGURE 1 Major stages in developing federal regulations. (Modified from ICF Incorporated "The Reg Map" ©2007. All rights reserved. This document may not be reproduced in any form without permission. Online at www.reginfo.gov/public/reginfo/Regmap/index.jsp.)

making; and the transparency of the regulatory process to the public are all significant ongoing issues.

Rulemaking has become more influential in the last half century. People outside of the public eye preside, and too often, agencies regulate with less-than-desirable accountability (Fig. 2). While agencies are required to

FIGURE 2 Dorothy looks behind the curtain. (Cartoon courtesy of Keith Bendis.)

accept the public's comments, they are not required to use them. For instance, USDA's Animal and Plant Health Inspection Service (APHIS) was required by the 2000 Plant Protection Act to set up a system by which states could strengthen their plant pest regulations beyond the federal level if they demonstrated a "special need." In 2006, APHIS proposed a way to implement this. It received public comments from 17 state departments of agriculture, environmental groups, industry organizations, and private citizens, many of which asked for significant changes to APHIS's proposed rules. When final rules were issued in 2008, APHIS noted that "only minor" changes were made, a common occurrence.

In a longstanding practice, outgoing presidents push through multiple regulatory changes just before their administrations end. There are few ways to eliminate or change such "midnight regulations." However, a new administration will reverse the regulations it finds most offensive if it has the tools to do so. In April 2009, for example, President Obama reversed President Bush's last-minute regulations that weakened the Endangered Species Act.

SIGNIFICANT AREAS OF FEDERAL REGULATION OF INVASIVE SPECIES

A host of federal and state regulations apply to different sorts of already or potentially invasive species. These include restrictions on plant and animal imports into and within the country; quarantines of shipments, facilities, or geographic areas to prevent the spread of pests and diseases; fines for violating labeling or inspection requirements; veterinary aspects for owning and possessing animals; bonding and licensing rules for retailers and importers; and so on. The following sections do not describe these comprehensively. Instead, they highlight the major federal regulatory players and cases in which important changes have been made or are under way.

The USDA, the FWS, and the U.S. Coast Guard (USCG) play the most significant roles in writing, implementing, and revising invasive species regulations. The EPA is poised to play a new and larger role for ships' ballast water. A few other federal agencies enact rules on specific topics. For example, the Centers for Disease Control and Prevention regulate the import of fewer than a dozen animal groups to protect public health.

The Dirty Lists: Regulating "Injurious Wildlife" and "Noxious Weeds"

The most critical invasive species rules have not been updated significantly in decades. For example, the major

rules governing the import of nursery plants—the source of over half of the weeds in natural areas—are more than 50 years old. The original law on importing wildlife, the Lacey Act, was passed in 1900.

The Lacey Act and the weed, or invasive plant, section of the Plant Protection Act of 2000, first passed as the Federal Noxious Weed Act of 1974, are the two chief laws used to prevent the intentional import of invasive species. These laws provide the models on which numerous state laws are based. Each, however, was originally intended to attain a much narrower goal. Just one portion of the Lacey Act gave authority to the FWS to designate animals as "injurious wildlife," which cannot be imported or moved between states without a permit. The Federal Noxious Weed Act likewise authorized USDA to ban the import of designated "noxious weeds." Organisms are named to these two lists by federal regulation. At present, about 20 groups of animals and some 95 plant groups are on these two lists. Usually, the entries are species, although taxonomic families, genera, and at least one subspecific variety are included, too. Organisms are added to these lists for various reasons, not just invasiveness. They may be poisonous to people or livestock or they may carry diseases.

Neither the two federal lists, nor the regulatory process which created them, reflects today's understanding of the geographic distribution of invasive species, their costs, or biology. For plants, the list is strongly biased toward crop weeds, plant parasites, and species that are problems in the eastern United States. Invasive plants of natural areas get short shrift. A few taxa have been added to or removed from these federal lists in the last decade, but neither the USDA nor the FWS has overhauled them in light of the world's known or potential invasive species.

The list-making processes themselves are fundamentally flawed: they take too long, they are more complex than necessary, they are neither transparent nor repeatable, and they are susceptible to political interference. Lately, FWS and USDA have focused their decisions more on possible economic hardships for importers and retailers than on the cost to U.S. taxpayers from failing to carry out the broadest intent of the law. Although scientists have developed new and more sophisticated ways of evaluating potential invasiveness and identifying organisms, neither agency routinely incorporates these in its regulatory processes. Data that the USDA and FWS collect have multiple shortcomings, and often regulators are not well versed in the latest quantitative methods.

The time it takes for the federal listing process to run its course is especially worrisome. Often, years pass before a final decision is made, especially if there is opposition from affected industries or trade groups. By the time a species is prohibited, it is likely to have already become established and to be spreading. The FWS spent more than seven years considering whether to regulate black carp (*Mylopharyngodon piceus*), for instance. When the process began, these carp were contained in aquaculture ponds, were not yet found in open U.S. waters, and could be shipped legally between states. By 2005, evidence suggested that the species had escaped from containment and was firmly established in the Mississippi River basin. The FWS added black carp to the regulated list in 2007, which stopped interstate shipping and acknowledged the impact the fish could have on rare native mussels, part of the U.S. freshwater fauna thought to have an extinction rate as high as that of tropical forests.

Marine scientists and environmental groups petitioned USDA's APHIS in 2003 to further regulate the marine alga *Caulerpa*, popularly known as the "killer alga" for its impacts in the Mediterranean Sea. At the time, only the Mediterranean, or aquarium, strain of the species *Caulerpa taxifolia* had been named a federal noxious weed and was thus banned from import. The petitioners asked that all species in the genus *Caulerpa*—or alternately, the entire species *C. taxifolia*—be banned. Because the single regulated strain is indistinguishable from close relatives, the petitioners claimed that the narrow ban was not scientifically justifiable. The USDA has yet to provide a final response to this petition, seven years later. Meanwhile, federal, state, and local officials, plus a variety of private groups, spent seven years and about $7 million to eradicate two small infestations of the aquarium strain, according to the Southern California Caulerpa Action Team. *Caulerpa racemosa* has become an even worse problem than *C. taxifolia* in the Mediterranean. In contrast to the federal rules, California strictly regulates nine species of *Caulerpa*, including *C. racemosa*.

Regulating Ballast Water

From 2002 to 2007, Congress considered new legislation that would have significantly expanded the underlying 1990 and 1996 laws on aquatic invasive species. In its various forms, the proposed National Aquatic Invasive Species Act would have authorized new programs to screen deliberate imports of aquatic species and to detect new invaders early. Also, these bills would have provided substantial new funding for research and more resources to states for their aquatic nuisance species plans. The largest part of the bills addressed the elimination of live organisms in the ballast water of ships. Along with closing loopholes in current law, ships would have been required to treat their ballast water. However, efforts to pass this comprehensive legislation slowed, and momentum shifted to a simpler legislative approach, dealing only with ballast water. A number of Ballast Water Management Acts were also proposed, including ones that specified a ballast water pollution standard, rather than leaving standard-setting to agency regulators. However, none of these bills passed both the House and the Senate.

Which federal agency would write, review, and enforce regulations was among the issues holding up legislation. Under 1990 and 1996 laws, the U.S. Coast Guard regulates ballast water. Many of its regulations apply to the records ships must maintain regarding their ballast water management and to whom these records must be available.

While new federal legislation was being considered, an alternate system of regulation was being addressed by the U.S. District Court of Northern California. After the EPA lost a Clean Water Act lawsuit, the agency offered a relatively weak proposal for new ballast water regulations. Environmental groups sued again, in 2009, and the issue remains unresolved.

Quarantine 37: Regulating Invasive Plants and Their Pests and Diseases

Quarantine 37, or Q37, is the group of federal regulations that govern the import of nursery materials or, as the USDA terms them, "plants for planting." Preventing entry of insects, fungal diseases, plant viruses, and other organisms that are introduced unintentionally and then cause harm to the nursery trade has always been a major objective of Q37. However, the regulations have never addressed the potential of nursery plants themselves to be invasive. The sale of horticultural plants and their routine escape from gardens and yards constitute important and unregulated pathways by which plants invade. Aquatic plants, especially, are often unidentified, misidentified, or available for sale despite being banned by federal or state regulation.

In 2001, a large group of stakeholders met and agreed to the St. Louis Accord, a set of voluntary protocols designed to reduce the import and spread of invasive plants via many industry- and research-related pathways. Industry groups hoped these voluntary measures would head off federal regulation. However, disappointment in these and similar efforts has contributed to proposals for formal regulatory change at the USDA.

In 2005, APHIS officials announced a plan to substantially revise and update Q37 in a process expected to take ten years. Chief among the changes would be new regulations for a category of plants that are "Not Authorized for

Importation Pending Pest Risk Assessment" (NAPPRA) as well as the addition of nonvascular plants to Q37. A notice of proposed rulemaking was published by APHIS in the Federal Register on the NAPPRA category in December 2004. The related proposed regulations were published in July 2009. This is the first of several regulatory updates planned for Q37. Final rules will be issued after APHIS receives and responds to public comments.

One goal of these changes is to bring Q37 more in line with the regulations for the import of fruits and vegetables, known as Quarantine 56. The Q56 rules are much more restrictive, although APHIS admits that nursery plants pose a significantly higher risk: larger volumes are imported; the number of taxa imported is tenfold greater (several thousand versus about 200 fruit and vegetable genera), the country of origin is difficult to detect, far fewer pest risk assessments are performed before import, and imported weeds and pests are more likely to find matching U.S. hosts and environments.

Regulating Wood Imports and Packing Material

In the early 1990s, the USDA's Forest Service and APHIS began formally assessing the risk that imported logs and lumber would carry pests and diseases into the country. At the same time, individual scientists were looking at risks posed by the solid wood in which imports, like marble and machinery, were packed. The results of this work led to new regulations in 1995. These focused on requirements that logs, lumber, and other unmanufactured wood materials be treated before import, but, for the first time, wood packing material was also addressed.

In 1998, APHIS issued additional rules intended to limit the arrival of wood-boring insects such as the Asian longhorned beetle (*Anoplophora glabripennis*) and the emerald ash borer. These required that solid wood packing material arriving from China be treated with heat or methyl bromide. The regulations were enacted quickly, after infestations of Asian longhorned beetles were found, first in New York and then in Chicago. APHIS amended and updated these rules throughout the 2000s. By 2007, *all* imported wooden packing material was required to be treated before entry—and labeled as such.

For both logs and packing materials, the agencies moved relatively quickly once the importance of the pathways was determined. For packing material, though, experts remain concerned that treatments will not reach pests deep inside wood and that the requirements for labeling do not account for the frequency with which packing materials are reused nor for the ease with which labels can be forged. The USDA did not include in its

environmental impact assessment a fair and unbiased evaluation of replacing wood packing material with something else, such as plastic, an option that could eliminate pests.

ENFORCEMENT

Studies that evaluate the effectiveness of specific federal (or state) regulations on invasive species are relatively recent. Generally, they document what critics have long claimed: that current regulations are too out of date, lax, and inadequately enforced to prevent invasive species problems.

For ballast water, the National Invasive Species Act of 1996 required that the Coast Guard assess compliance with voluntary provisions for exchanging ballast water at sea and, if it was unsatisfactory, proceed with mandatory rules. These were finalized in 2004.

The Environmental Law Institute surveyed state weed managers in the Great Lakes regarding enforcement of their state dirty lists for weeds. Michigan tested compliance with quarantines to limit the movement of firewood, which can carry forest pests. Results of both were disappointing.

There are longstanding regulations in place requiring that animals and plants be accurately identified and labeled during shipment and sale. Yet a large proportion of shipments are unidentified, misidentified, mislabeled, or inadequately recorded. Scientific names are the only unambiguous source of identification. However, many commercial shipments are not labeled with these, and federal databases that record border inspections often do not include this information. Also, shipments arrive too often contaminated with other live organisms, such as snails, other plants, parasites, or pathogens, which also lack identification.

Better addressing invasive species problems depends on strong statutes as well as effective regulations to implement them. The pace of rulemaking on invasive species seems to have accelerated in the past decade. So far, though, U.S. policy remains a poor fit for the problems at hand.

SEE ALSO THE FOLLOWING ARTICLES

Black, White, and Gray Lists / Endangered and Threatened Species / Horticulture / Laws, Federal and State / Risk Assessment and Prioritization

FURTHER READING

Bass, G. D., M. Bird, C. S. DeWaal, et al. 2008. Advancing the public interest through regulatory reform: Recommendations for President-elect Obama and the 111th Congress. Washington, DC: OMB Watch. www.ombwatch.org/taxonomy/term/204

Cohen, A. N., and B. Foster. 2000. The regulation of biological pollution: Preventing exotic species invasions from ballast water discharges into California coastal water. *Golden Gate University Law Review* 30: 787–883.

Fowler, A. J., D. M. Lodge, and J. F. Hsia. 2007. Failure of the Lacey Act to protect US ecosystems against animal invasion. *Frontiers in Ecology and the Environment* 5: 353–359.

Jenkins, P. T., K. Genovese, and H. Ruffler. 2007. *Broken Screens: The Regulation of Live Animal Imports in the United States.* Washington, DC: Defenders of Wildlife.

Kerwin, C. M. 2003. *Rulemaking: How Government Agencies Write Law and Make Policy*, 3rd ed. Washington, DC: CG Press.

Lodge, D. M., S. Williams, H. J. MacIsaac, et al. 2006. Biological invasions: Recommendations for U.S. policy and management. *Ecological Applications* 16: 2035–2054.

Skrzycki, C. 2003. *The Regulators: Anonymous Power Brokers in American Politics.* Lanham, MD: Rowman and Littlefield Publishers.

Sumner, D. A., ed. 2003. *Exotic Pests and Diseases: Biology and Economics for Biosecurity.* Ames: Iowa State Press and Wiley-Blackwell.

REMOTE SENSING

GREGORY P. ASNER

Carnegie Institution for Science, Stanford, California

CHO-YING HUANG

National Taiwan University, Taipei

Remote sensing has become a key tool for mapping ecosystems, but it is relatively new in studies of alien invasive plants. With the advent of high spatial, temporal, and spectral resolution sensors since 1990, the role of remote sensing in invasive species research and management has gradually increased. Different types of sensors offer varying information on invasive species or the habits they invade. This entry highlights the major types of sensors currently available, as well as new systems, and how they are being used for alien invasive species work.

REMOTE SENSING TECHNOLOGIES AND APPROACHES

There are many types of field, airborne, and satellite remote sensing technologies, but a few have proven useful for alien invasive species detection and mapping. The most intuitive and straightforward remote sensing approach for alien plant detection is to use high spatial resolution images (<10 m pixel size; Table 1) to visually inspect the spatial distribution of nonnative species. High spatial resolution imagery is usually acquired from aircraft, but a few satellite sensors also offer this capability. Independently of the image source, the goal with high spatial

resolution imagery is to pinpoint species based on their unique spatial patterns or phenological characteristics. For example, in 1995 Everitt and colleagues used aerial photographs taken during the flowering seasons of Eurasian *Euphorbia esula* and Asian *Tamarix chinensis*, finding the visible reflectance (400–700 nm) of the invaders to be significantly higher than surrounding vegetation due to their bright inflorescences. Color infrared (CIR) aerial photographs are also often used, with resolutions ranging from a few centimeters to about 2 m. In 2005 Müllerová and colleagues used CIR aerial photographs acquired during the onset of *Heracleum mantegazzianum* invasion (the largest central European forb native to the western Caucasus) in the Czech Republic. The brightness of *H. mantegazzianum* was much greater than that of the neighboring vegetation, not only during the flowering, but also in the early fruiting season, due to its distinct structure of fruiting umbels.

At times, the spatial configuration or structure of the invader can be delineated from that of its native neighbors. In 2005 Fuller used multispectral IKONOS satellite imagery with a 4 m spatial resolution to map an Australian tree, *Melaleuca quinquenervia*, in southern Florida. He found that the spatial pattern of *Melaleuca* was highly aggregated and regularly shaped, compared to the nearby patches of woody plants, which were likely associated with infrastructural development. However, Fuller's analysis pointed out that the 4 m spatial resolution of IKONOS was insufficient to identify the alien tree when canopy cover dropped below 50 percent. A variety of other high spatial resolution satellite sensors have been used to estimate the location and extent of invasive plants, but most (if not all) fall short in detecting single individuals or clusters prior to reaching about 50 percent cover.

In contrast, others have worked with low spatial resolution (>250 m) imagery, which allows for frequent, often daily, imaging from space (Table 1). These high temporal resolution or "time-series" data have been used for studies of invasive plants and their habitats. The very common Normalized Difference Vegetation Index (NDVI) enhances the signal of photosynthetically active vegetation with a combination of visible and near-infrared spectral bands. Several phenological metrics such as vegetation onset time, duration of growth, end time, and rate of vegetation green-up and senescence can be derived from NDVI time-series data. Using the NOAA Advanced Very High Resolution Radiometer (AVHRR), Bradley and Mustard showed in 2005 that the highly invasive grass *Bromus tectorum* could be readily mapped in the Great Basin of the United States, based on its unique phenological response to precipitation relative to neighborhood

TABLE 1
Sensor Specifications and Applications as Applied to Biological Invasion of Terrestrial Ecosystems

Sensor Specifications				Species Traits and Remote Sensing Strategies	Limitations
Spatial (m)	Temporal (days)	Spectral (bands)	Examples		
Moderate Spatial/Spectral					
10–100	15+	<20	ASTER, SPOT, Landsat TM/TEM + satellite sensors	Large stands; differentiate phenology of coexisting plants; select images acquired in the right season	Coarse spatial and spectral resolution incompatible with invasive species detection among native species
High Spatial					
<10	<10	~5	Aerial photographs, Quickbird and IKONOS satellite sensors	Unique spatial patterns; flowing phenology	Inflexibility of data collection; pixel spacing still not fine enough to observe plants at the species level; cannot detect plants lacking distinct flowering pattern; impractical for large-area monitoring
High Temporal					
≥250	2	<40	AVHRR, MODIS, VEGETATION satellite sensors	Unique phenological characteristics; combination of models and time-series vegetation indices derived from the images	Insufficient spectral bands to extract non-native species from large pixels; difficult to conduct field validation due to the large pixel size; non-native species signals overwritten by climatic variations such as precipitation
Hyperspectral					
Varied (0.5–30)	Variable	>75	AVIRIS airborne and Hyperion satellite sensors	Unique spectral signatures; spectral mixture analysis; biochemical analysis	Limited geographical coverage; spectral similarity of species
Active Remote Sensing					
0.5–100	Variable	1 (3-D imaging)	LiDAR, RADAR aircraft and satellite sensors	Monitoring structural changes of stand during invasion from repeat coverage	No spectral information so species are hard to detect without other information
Image Fusion					
~0.5	Variable	50+ (3-D view)	Carnegie Airborne Observatory	Unique spectral signatures of species; detection of non-native species of different height levels (overstory and understory) at the very fine spatial scale	High performance computing required; limited spatial coverage; pushing the limits of spatial, spectral, and dimensional resolutions

vegetation. In 2006, Morisette and colleagues combined NDVI and land cover from the NASA Moderate Resolution Imaging Spectroradiometer (MODIS) sensor with thousands of field points to predict the habitat suitability for nonnative *Tamarix* spp. in the mainland United States. The analysis highlighted the Southwest as the most suitable region for these Asian trees and shrubs.

Imaging spectrometers (also called hyperspectral imagers) collect continuous data across a wide region of the visible (400–700 nm), near-infrared (700–1,200 nm), and shortwave infrared (1,200–2,500 nm) spectra, often using hundreds of spectral bands (≤10 nm resolution) (Table 1). Some of the most commonly used sensors for scientific and management purposes include the

NASA Airborne Visible/Infrared Imaging Spectrometer (AVIRIS) and the Compact Airborne Spectrographic Imager (CASI). There has been one civilian research satellite, Earth Observing-1 (EO-1), which has provided hyperspectral imagery from space. Although the history of hyperspectral remote sensing is relatively short (less than 25 years) compared to other types of remote sensing (e.g., aerial photographs, which have been around for more than 100 years; satellite images, around for about 50 years), hyperspectral images are currently the most heavily used source of remotely sensed data for invasive species research.

The main advantages of hyperspectral data are that detailed spectral profiles can be developed for native and nonnative plants, and that specific spectral regions can be analyzed that are most sensitive to the abundance of the species of interest. In 2002 Williams and Hunt used hyperspectral data from the AVIRIS sensor to map a European herbaceous perennial plant *Euphorbia esula* in the Great Plains of the United States. Similarly, Underwood and colleagues used airborne hyperspectral imagery to map invasive plants in California in 2003.

Uses of hyperspectral data are not limited to studies of the presence or abundance of alien species, but can be further extended to study the ecosystem effects of invasion. In 2005 Asner and Vitousek used AVIRIS with canopy models to investigate how the invasive nitrogen-fixing tree *Morella (Myrica) faya* and the understory herb *Hedychium gardnerianum* affect the canopy chemistry of Hawaiian montane forests. They found that *M. faya* significantly increases canopy nitrogen concentrations and water content in invaded sites. In contrast, *H. gardnerianum* substantially reduces nitrogen concentrations in the overstory tree canopies (presumably through competition for nitrogen in soils). In 2008 Asner and colleagues further studied the uniqueness of hyperspectral signatures of 43 Hawaiian native and alien trees (Fig. 1). They found that invasive trees were spectrally unique from native trees, resulting from differences in leaf pigments, nutrients, and structural and biophysical properties of invasive and native species.

Hyperspectral images from the spaceborne EO-1 Hyperion sensor are less commonly used for alien species mapping, probably due to the coarse spatial resolution (30 m) and lower instrument performance compared to airborne imaging spectrometers such as AVIRIS. Nonetheless, Hyperion provides an opportunity for systematic data collection over a remote region on a stable platform (EO-1 satellite) and at much lower cost than many airborne mapping efforts. The data have been utilized to

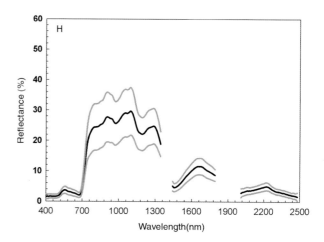

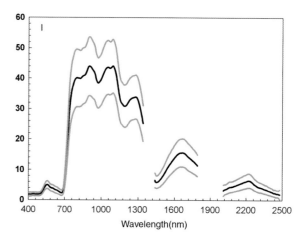

FIGURE 1 Hyperspectral reflectance signatures grouped as Hawaiian native (H) and invasive (I) tree species are distinct from one another, making it possible to map them using current airborne imaging spectrometer technology. The black and gray lines indicate mean and range of spectral reflectance in each group, respectively. The unique spectral properties of each group arise from the unique biochemical and structural properties of the species, especially those which tend to be successful invaders of Hawaiian ecosystems. (Derived from Asner et al., 2008.)

map coastal alien plants with demonstrably satisfactory accuracies (overall accuracy ≥81%).

Light detection and ranging (LiDAR) is another optical remote sensing technology that measures the distance between the sensor and a target surface, obtained by determining the elapsed time between the emission of a short duration laser pulse and the arrival of the reflection of that pulse at the LiDAR receiver. It has been widely used to estimate the three-dimensional structure of individual plants and vegetation. LiDAR has not yet proven to be an efficient tool for detecting alien plants; however, recent efforts have integrated hyperspectral and LiDAR instrument systems to improve the detection of invasive species based on their chemical

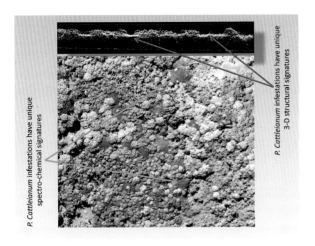

FIGURE 2 New airborne three-dimensional (3-D) remote sensing of invasive species and their canopy structural properties. This image was collected by the Carnegie Airborne Observatory (CAO) over a lowland moist tropical forest region on Hawaii Island. Spectroscopic analysis in the lower panel indicates canopy species composition of the forest (shown as different colors), while LiDAR analysis in the upper panel provides quantitative measurements of the 3-D structure of the canopies. Infestations of the highly invasive tree *Psidium cattleianum* is highlighted. (From G. Asner, unpublished.)

and structural properties. One of the only systems in the world with this capability is the Carnegie Airborne Observatory (CAO; http://cao.ciw.edu). In 2008, using a "laser-guided spectroscopy" approach, Asner and colleagues mapped individual crowns of five highly invasive tree species throughout Hawaii Island (Fig. 2). This study showed that different invasive tree species alter the fundamental structure of the forests they invade, but in different ways, either from the ground up or from the top down.

CURRENT LIMITATIONS

The information provided by moderate spatial and spectral resolution satellite images (e.g., Landsat Thematic Mapper) is usually insufficient to delineate the distribution of alien plants or their functional traits in ecosystems. Aerial photographs and some high spatial resolution satellite imagery techniques (e.g., IKONOS, GeoEye-1) have been used to assess the extent of some invasive species with distinct phenological characteristics, but not in a systematic way that might allow large-scale application. Sensors with large area coverage such as MODIS have large pixels (low spatial resolution), and thus the images collected by these sensors capture not only the species of interest but other components as well, such as untargeted plants, surface soils, and senescent vegetation; this limits the

utility of images for alien plant monitoring especially in early stages of invasion.

Data patterns in hyperspectral images can be complicated and hard to unravel. In many cases, data reduction techniques are required for generalization, and they usually involve sophisticated algorithms that "mine" for invasive species signals in the high-dimensional spectral space. An imaging spectrometer is also usually flown onboard aircraft, so the swath width of a hyperspectral image is usually narrow, which limits its ability for large-scale invasive plant mapping. Also, scheduling of aircraft and related regulations for different countries could make data acquisition inflexible.

SEE ALSO THE FOLLOWING ARTICLES

Disturbance / Early Detection and Rapid Response / Fire Regimes / Habitat Compatibility / Hawaiian Islands: Invasions / Landscape Patterns of Plant Invasions / Range Modeling

FURTHER READING

Asner, G. P., R. F. Hughes, P. M. Vitousek, D. E. Knapp, T. Kennedy-Bowdoin, J. Boardman, R. E. Martin, M. Eastwood, and R. O. Green. 2008. Invasive plants transform the three-dimensional structure of rain forests. *Proceedings of the National Academy of Sciences, USA* 105: 4519–4523.

Asner, G. P., M. O. Jones, R. E. Martin, D. E. Knapp, and R. F. Hughes. 2008. Remote sensing of native and invasive species in Hawaiian forests. *Remote Sensing of Environment* 112: 1912–1926.

Asner, G. P., and P. M. Vitousek. 2005. Remote analysis of biological invasion and biogeochemical change. *Proceedings of the National Academy of Sciences, USA* 102: 4383–4386.

Bradley, B. A., and J. F. Mustard. 2005. Identifying land cover variability distinct from land cover change: Cheatgrass in the Great Basin. *Remote Sensing of Environment* 94: 204–213.

Cuneo, P., C. R. Jacobson, and M. R. Leishman. 2009. Landscape-scale detection and mapping of invasive African Olive (*Olea europaea* L. ssp. *cuspidata* Wall ex G. Don Ciferri) in SW Sydney, Australia using satellite remote sensing. *Applied Vegetation Science* 12: 145–154.

Everitt, J. H., G. L. Anderson, D. E. Escobar, M. R. Davis, N. R. Spencer, and R. J. Andrascik. 1995. Use of remote sensing for detecting and mapping leafy spurge (*Euphorbia esula*). *Weed Technology* 9: 599–609.

Fuller, D. O. 2005. Remote detection of invasive Melaleuca trees (*Melaleuca quinquenervia*) in South Florida with multispectral IKONOS imagery. *International Journal of Remote Sensing* 26: 1057–1063.

Huang, C., and G. P. Asner. 2009. Applications of remote sensing to alien invasive plant studies. *Sensors* 9: 4869–4889.

Morisette, J. T., C. S. Jarnevich, A. Ullah, W. Cai, J. Pedelty, J. E. Gentle, T. J. Stohlgren, and J. L. Schnase. 2006. A tamarisk habitat suitability map for the continental United States. *Frontiers in Ecology and the Environment* 4: 11–17.

Müllerová, J., P. Pyšek, V. Jarošík, and J. Pergl. 2005. Aerial photographs as a tool for assessing the regional dynamics of the invasive plant species *Heracleum mantegazzianum*. *Journal of Applied Ecology* 42: 1042–1053.

Pengra, B. W., C. A. Johnston, and T. R. Loveland. 2007. Mapping an invasive plant, *Phragmites australis*, in coastal wetlands using the EO-1 Hyperion hyperspectral sensor. *Remote Sensing of Environment* 108: 74–81.

Robinson, T. P., R. D. van Klinken, and G. Metternicht. 2008. Spatial and temporal rates and patterns of mesquite (*Prosopis* species) invasion in Western Australia. *Journal of Arid Environments* 72: 175–188.

Underwood, E., S. Ustin, and D. DiPietro. 2003. Mapping nonnative plants using hyperspectral imagery. *Remote Sensing of Environment* 86: 150–161.

Williams, A. P., and E. R. Hunt Jr. 2002. Estimation of leafy spurge cover from hyperspectral imagery using mixture tuned matched filtering. *Remote Sensing of Environment* 82: 446–456.

REPRODUCTIVE SYSTEMS, PLANT

SPENCER C. H. BARRETT

University of Toronto, Ontario, Canada

The reproductive systems of plants concern the diverse strategies by which populations reproduce, including the balance between sexual and asexual reproduction, and the structures and processes governing the union of gametes and the transmission of genes between generations. The reproductive system is a particularly important life history trait for invasive species, because the quantity and genetic quality of propagules produced by reproduction are important for successful colonization and spread in novel environments. Flowering plants (angiosperms) display a wide range of reproductive systems, including a diversity of pollination and mating strategies, as well as striking variation in the relative importance of sexual versus asexual reproduction. Reproductive systems differ in their ecological and biogeographical consequences, particularly when colonizing individuals occur at low density without mates or pollinators following dispersal. Under these demographic circumstances, uniparental reproductive systems are often favored over biparental reproduction, especially for species with short life histories. Reproductive systems also determine opportunities for adaptive evolution, because they influence key population-genetic parameters such as genetic recombination, effective population size, gene flow, and the partitioning of genetic diversity within and among populations.

REPRODUCTIVE DIVERSITY IN PLANTS

Three distinctive features of plant biology complicate reproduction. First, being immobile, plants require vectors to transfer pollen (pollination) between individuals to facilitate mating. This has promoted the evolution of diverse floral adaptations associated with the agents of pollen dispersal (animals, wind, water) and has given rise to the field of pollination biology. Second,

most plants exhibit hermaphroditic sex expression; thus many are capable of self-fertilization (selfing) as well as cross-fertilization (outcrossing). For sexual species, selfing requires that plants are self-compatible, which means that they have the ability to set seed following self-pollination. In contrast, outcrossing is most commonly enforced by self-incompatibility, a genetically determined physiological mechanism that prevents self-fertilization. Lastly, due to the modular structure of plants and to the fact that they often exhibit a high degree of phenotypic plasticity, male and female gametes can be combined in diverse ways at the flower, inflorescence, and plant levels. This variation in sex allocation results in diverse gender strategies despite the basic hermaphroditic condition of most plants. These distinctive features of plant reproduction result in a high level of sexual promiscuity within plant populations, and individuals often mate with many sexual partners (multiple paternity), including themselves. Recent research in reproductive biology is concerned with using genetic markers to determine who mates with whom, and, through ecological experiments, with evaluating the fitness consequences of different levels of outcrossing and selfing in populations.

In addition to reproducing sexually, many plants also reproduce asexually. In such cases, the parents produce genetically identical copies of themselves through either vegetative reproduction or apomixis (agamospermy). Vegetative reproduction, also termed clonal growth or clonal propagation, is very common among perennial plants, and the morphological structures involved include axillary shoots, stolons, runners, bulbs, corms, and tubers. A functionally important distinction between the different forms of clonal growth is whether clonal offspring become physiologically autonomous and capable of dispersal away from the parental plant. Clonal propagation often dominates at geographical range limits in species with broad distributions and is also a characteristic feature of aquatic plants, for whom floating clonal propagules are an important means of dispersal. Rates of vegetative reproduction vary widely among species, depending on local environmental conditions, but there are very few examples, if any, in which clonal propagation is the exclusive means of reproduction throughout a species' geographical range.

Apomixis, the production of clonal seed in the absence of fertilization, is less common than vegetative reproduction. Nevertheless, it has been reported from at least 300 species from over 40 families and is especially common in the Asteraceae, Poaceae, and Rosaceae, all families in which invasive species are well represented. Apomixis can be either obligate or facultative and can be hard to detect, especially when pollination is required for seed maturation and endosperm formation. Apomixis is commonly associated with polyploidy and can explain the ability of otherwise sterile polyploids, such as triploids, to produce abundant seed. Although sharing some features in common with selfing (e.g., single plants can establish populations), apomixis has entirely different genetic consequences and is commonly associated with hybridity and heterozygosity, rather than inbreeding and homozygosity, which are typical in species with high rates of selfing.

Despite the complexity of reproductive mechanisms in angiosperms, a fundamental dichotomy is whether new offspring arise from uniparental reproduction, in which only a single parent is involved, or, alternatively, whether progeny arise from biparental reproduction. Selfing and all forms of asexual reproduction involve uniparental reproduction, while sexual outbreeders engage in biparental reproduction. This distinction is of particular importance when considering the establishment phase of the invasion process, where mates may be limiting. Two axes of continuous variation are also of particular significance when considering the reproduction of invasive populations. First, for species with dual reproductive modes involving both sexual and asexual reproduction, it is important to establish what proportion of plants in a population has originated by either mechanism, as this can influence the demographic characteristics of populations, including dispersal and rates of spread. Similarly, for sexually produced offspring, determining the fraction of seeds that have arisen in each generation from cross-fertilization versus self-fertilization is relevant for understanding the maintenance of genetic diversity and the fitness of offspring. The use of polymorphic allozyme and microsatellite markers allows for the measurement of these variables, providing a more quantitative assessment of plant reproductive systems.

UNIPARENTAL REPRODUCTION AND INVASIVENESS

Are particular reproductive traits a prerequisite for invasion success, and is it possible to identify a syndrome of life history characters that typically occur in successful invaders? These questions have been of longstanding concern to biologists and managers who are interested in predicting potential invasiveness. They have generated considerable debate and are still not fully resolved. On one hand, some argue that the attributes of successful invaders and the diverse environments that they occupy are simply too heterogeneous to draw any general

conclusions. According to this perspective, ecological context and opportunity (e.g., commercial introductions) are the primary determinants of whether a particular species becomes invasive. Certainly the wide range of life histories represented among invasive species, ranging from annuals to woody plants, would seem to limit attempts at broad generalization across all angiosperms. Nevertheless, others have sought to identify key adaptive traits that are consistently associated with invasiveness in targeted surveys of selected groups of species. Some generalizations have been possible by restricting the scope to specific taxonomic groups or lineages, geographical areas, and life forms. Traits associated with the reproductive system are often found to be associated with invasion success, with the capacity for uniparental reproduction most commonly identified. For example, about half of the species judged to be among the world's worst weeds reproduce primarily by selfing, and most of the remainder reproduce primarily by asexual means. Why should this be so?

The Role of Selfing in Plant Invasions

The benefits of selfing in promoting successful colonization were first emphasized by Herbert G. Baker (University of California, Berkeley) in a classic paper in 1955 (*Evolution* 9: 347–348). He proposed that self-compatible plants, particularly those that were capable of autonomous self-pollination (without the aid of pollinators), were more likely to be favored during colonization because of their ability to establish colonies from a single individual following long-distance dispersal. This idea, now referred to as Baker's Law, was initially formulated to explain the high incidence of self-compatible rather than self-incompatible plants on oceanic islands. However, it has been extended to include any demographic situation involving low density where the reproduction of colonizing individuals is limited by an absence of pollinators or mates. This could include conditions such as those that occur at the leading edge of an invasion front, or in a metapopulation in which recurrent colonization–extinction cycles result in episodes of low density. Under these circumstances, the Allee effect may operate, and individuals with the facility for autonomous self-pollination should therefore be favored. Indeed, uniparental reproduction, by providing reproductive assurance, may allow small populations to grow faster than if they were obligately outcrossing, and such an effect could, in principle, reduce the duration of post-introduction bottlenecks. These arguments lead to the prediction that among successful invaders there should be a high incidence of species capable of uniparental reproduction because low-density conditions commonly prevail, especially during the establishment phase of invasions.

Comparative evidence to support a general association between invasiveness and uniparental reproduction is mixed. Surveys of introduced weed species of arable crops commonly reveal a high incidence of self-compatibility, and the capacity for clonal propagation is well represented among "environmental weeds" of nonagricultural habitats. A recent North American survey demonstrated that invasive species capable of autonomous self-pollination had larger geographical ranges than those that require pollen vectors for seed production. Nevertheless, other analyses have shown a reduced incidence of self-compatibility among exotic North American weeds, a result inconsistent with Baker's Law. In the United Kingdom, there is no evidence that native and invasive species differ in the incidence of self-compatibility, but this may simply reflect the fact that postglacial colonization initially favored self-compatibility in the native flora. Part of the problem in these types of analyses is determining whether the trait of interest is overrepresented among invasives as theory predicts; for this, a control group is required for comparison. Surveys sometimes involve samples of native versus introduced species of diverse affinities. However, a more powerful approach is to compare paired samples of related (e.g., congeneric) introduced species in which one is invasive and the other is not. Here shared ancestry and equivalent opportunity are to some extent controlled. This approach recently identified the facility for autonomous selfing as an important correlate of invasiveness in paired samples of South African members of the family Iridaceae that have been exported for use in horticulture to various regions of the world.

In general, reproductive systems in angiosperms are associated with life history, especially longevity. For example, annual and short-lived species tend to have higher selfing rates than long-lived species, which are frequently self-incompatible or dioecious (with separate sexes) and are therefore predominantly outcrossing. Therefore, in contrast to regions such as California and the Mediterranean parts of Europe where selfing annuals make up a significant proportion of the native and introduced flora, this is not the case in the Cape Province and New Zealand, because most invaders are perennial, and many are woody. However, it is worth noting that there is some evidence that among woody invaders in one region of South Africa (KwaZulu-Natal), the capacity for uniparental reproduction through autonomous

self-pollination occurs very commonly, a pattern consistent with Baker's Law.

Clonal Propagation and Genetic Uniformity

Taxonomic groups that contain diverse reproductive systems are of particular interest for investigating the role of uniparental reproduction and invasiveness. *Eichhornia*, a small genus of eight species in the aquatic family Pontederiaceae, is instructive in this regard. It contains both selfing and outcrossing species as well as species exhibiting diverse strategies for clonal propagation. However, only one species (water hyacinth, *E. crassipes*) has become a serious worldwide invader; it is generally considered the world's worst aquatic weed (Fig. 1). Interestingly, the remaining species in the genus all have geographically widespread distributions in their native ranges and colonize a range of wetland habitats, including lakes, rivers, pools, drainage ditches, and rice fields. However, none

FIGURE 1 Floating mats of the invasive aquatic weed water hyacinth (*Eichhornia crassipes*). (A) Sexual population in the native range in Argentina, illustrating mass flowering. (B) Highly clonal population impeding boats in Vietnam. (Photographs courtesy of the author.)

have shown the dramatic spread exhibited by *E. crassipes* during the past century, despite the fact that several have been introduced to various regions as pond ornamentals. Water hyacinth's aggressive behavior is principally the result of its prolific powers of asexual reproduction through clonal growth, combined with its free-floating habit. Under favorable conditions, it has been estimated that ten plants of *E. crassipes* can produce a 0.4 ha mat of floating plants in eight months. These mats choke reservoirs, rivers, and drainage canals in many tropical and subtropical regions. Because daughter rosettes produced through clonal growth are easily detached from floating mats, they provide an important mechanism of long-distance dispersal in water currents.

Water hyacinth illustrates several important general principles found in invasive species with both sexual and asexual reproductive systems. The species is tristylous, and in its native range (lowland South America, primarily Brazil and Argentina), populations contain the three floral morphs (referred to as the long-, mid- and short-styled forms) that characterize this particular outcrossing system. Native populations are cross-pollinated by specialized long-tongued bees, and sexual reproduction commonly occurs when water level fluctuations allow seeds to germinate and establish on riverbanks. However, in most introduced populations, clonal growth dominates, and sexual reproduction is uncommon.

The rarity of sexual reproduction in the introduced range of *E. crassipes* is a result of three factors. First, strong founder events caused by human introductions and subsequent clonal dispersal have strongly influenced the distribution of floral morphs, interfering with the functioning of tristyly. Second, the specialized pollinators that normally service populations in the native range are absent, and cross-pollination between morphs is limited. Finally, many of the habitats occupied by *E. crassipes* in the introduced range do not exhibit the periodic water level fluctuations that are a feature of habitats in its native range. As a result, any seeds that are produced in introduced populations generally fail to establish. Most populations in the introduced range of water hyacinth are therefore composed of a single or small number of clones that multiply by vegetative reproduction.

Other examples of invasive species in which sexual reproduction is severely limited or absent altogether in the introduced range include Canadian pondweed (*Elodea canadensis*, Hydrocharitaceae) and Japanese knotweed (*Fallopia japonica*, Polygonaceae). In both species the restrictions on sexual reproduction are associated with founder events, which, as with water hyacinth, have

disrupted the functioning of the species' sexual systems. In both species, there are limited opportunities for outcrossing because of the absence of conspecific pollen parents. Canadian pondweed is dioecious (having separate female and male plants) in its native North American range, but in the United Kingdom, where it has spread through canals and river systems, only females are currently reported. Similarly, Japanese knotweed is gynodioecious (having female and hermaphrodite plants), but invasive populations in the United Kingdom, where it is considered one of the most pernicious weeds, are composed exclusively of female plants. Genetic analyses of populations of Japanese knotweed have identified a single clone in the United Kingdom and other parts of Europe, as might be expected if all reproduction is asexual.

A more complex situation is evident in the case of the European flowering rush (*Butomus umbellatus*, Butomaceae), a serious invader of wetlands along the southern Canadian and northern U.S. border. In Europe both diploid and triploid populations occur. Diploid populations reproduce by seed and asexually through pea-sized vegetative bulbils and rhizome fragmentation, while triploid populations are sterile, propagating exclusively through clonal propagation. Although both ploidy levels have been introduced to North America, the diploid cytotype is far more abundant with a much broader geographical range. Unexpectedly, despite the dominance of the diploid cytotype in North America, reproduction in invasive populations is largely, if not exclusively, asexual. This is reflected in very low levels of genetic diversity throughout North America compared to Europe, where diversity is much higher. Although abundant viable seed is produced in diploid populations, virtually all propagation and dispersal in North America is through the production of large quantities of bulbils. In contrast, North American triploid populations do not produce bulbils and can only propagate clonally via rhizome fragmentation. This is a much slower process, limiting rates of clonal propagation and the spread of the triploid cytotype. The difference between the clonal strategies of the diploid and triploid cytotypes of *B. umbellata* in North America appears to be the result of founder events associated with human introduction.

These findings of extensive clonal propagation in the introduced ranges of invasive plant species have important implications. First, the absence or low levels of sex and recombination means that populations are often genetically depauperate, and therefore that opportunities for the evolution of local adaptation to prevailing environmental conditions are restricted. Second, the genetic uniformity of some invasive clonal species indicates that genetic variation is clearly not a prerequisite for successful spread. Well-developed phenotypic plasticity can allow clonal genotypes of some species to cope with considerable environmental heterogeneity. Finally, the genetic uniformity of clonal populations can be successfully exploited in biological control programs aimed at eradicating invasions. Clonal populations are less likely to evolve resistance to introduced control agents (e.g., insects, pathogens) because of the low genetic diversity their populations contain in contrast with invaders that are sexual outbreeders.

OUTBREEDERS AS SUCCESSFUL INVADERS

Although there are obvious benefits to uniparental introduction during the establishment phase of an invasion, this has not prevented many obligate outbreeders from becoming successful invaders, especially among perennial species. Most perennial outbreeding invaders also reproduce vegetatively, and the longevity of clones reduces the risk of reproductive failure if mates arrive later in the invasion process. Obligate outcrossing enforced by self-incompatibility obviously represents a severe liability for an annual invader, because there are only opportunities for mating during one flowering season. However, even self-incompatible annuals can be found among successful invaders, raising an interesting question: what other traits do they possess that might ameliorate the effects of low density and maintain fertility during the early phases of population establishment?

Overcoming Constraints Imposed by Biparental Reproduction

Common ragweed (*Ambrosia artemisiifolia*, Asteraceae) is a wind-pollinated annual native to North America that has become invasive in several regions including Europe, Asia, and Australia and is a major pollen allergen (Fig. 2). Ragweed has been frequently reported in the literature as self-compatible and was assumed to be largely selfing, presumably because most successful annual invaders are indeed selfers. However, recent investigations of this species have demonstrated that this is not the case. Ragweed plants are strongly self-incompatible, and measurements of mating patterns using genetic markers indicate high levels of outcrossing. Several features of the reproductive biology of ragweed populations limit opportunities for Allee effects to operate. These include an enormous production of windborne pollen that reduces the likelihood of pollen limitation, prolific seed production, and the maintenance of persistent seed banks. Metapopulation models confirm that high fecundity and seed banks can offset the costs associated with low density in self-

FIGURE 2 *Ambrosia artemisiifolia* (common ragweed), a loose wind-pollinated, self-incompatible annual. This photo depicts a flowering shoot composed mostly of male flowers dispersing pollen into the wind. (Photograph courtesy of R. W. Ness.)

incompatible colonizing species. Ragweed's invasive powers are not an exception in the large family Asteraceae; other self-incompatible annual plants with prolific colonizing abilities include *Centaurea solstitialis*, *Crepis sancta*, and *Senecio squalidus*.

Not all invasive species are equally subject to Allee effects, for the influence of low density on mating and fertility depends, in part, on the dispersal biology of species, and particularly on the extent to which diaspores (dispersal units including fruits and seeds) are adapted for long-distance dispersal. *Turnera subulata* (Turneraceae) is a perennial self-incompatible species that is native to Brazil but common in disturbed habitats throughout the neotropics (Fig. 3). It has also been introduced to various

FIGURE 3 The showy flowers of *Turnera subulata*, a perennial, distylous, insect-pollinated tropical weed of disturbed roadsides with ant-dispersed seeds. The long-styled morph is on the left, and the short-styled morph is on the right. (Photograph courtesy of the author.)

FIGURE 4 Inflorescences of the perennial, tristylous, insect-pollinated, wetland invader purple loosestrife (*Lythrum salicaria*). The species is outbreeding, and populations maintain considerable genetic variation for a range of traits, including flower color, as illustrated in the photograph. (Photograph courtesy of the author.)

parts of the Old World (e.g., India, Indonesia, Malaysia) as a showy ornamental but has escaped to become a common roadside weed. Although *T. subulata* is distylous (with two cross-compatible floral morphs), this sexual system has not handicapped the species' ability to spread widely in its adventive range. Seeds of *T. subulata* possess a lipid-rich body and are dispersed over relatively short distances by ants. As a result, plant colonies tend to move gradually rather than through long-distance jumps, and local dispersal ensures that both morphs are dispersed together, facilitating outcrossing.

Outcrossing and the Evolution of Local Adaptation

One of the most successful self-incompatible invasives is purple loosestrife (*Lythrum salicaria*, Lythraceae), which was introduced approximately 200 years ago to the eastern seaboard of the United States and has since spread into Canada, becoming a noxious wetland invader (Fig. 4). There is evidence that purple loosestrife was introduced to North America both deliberately (as a popular garden ornamental, for medicinal use, and for honey production) and also accidentally. Purple loosestrife flowers prolifically and has showy floral displays that attract a wide range of bees and butterflies in both its native and introduced ranges. Reliable pollination and a predominantly outcrossed mating system provide considerable opportunities for genetic admixture in the alien range, and there is good evidence that populations contain high levels of genetic diversity. Recent studies of the spread of purple loosestrife indicate that this diversity has played a significant role

in enabling populations to adapt to latitudinal variation in climate. This has occurred through the adjustment of important life history traits (such as flowering time) to the shorter growing seasons that characterize populations at northern range limits. Large-scale range expansion during biological invasion requires that invading species adapt to geographical variation in climatic and local ecological conditions. Invaders capable of outcrossing have the capacity to harness the diversity arising from multiple introductions, and introduced populations may in some cases be more genetically diverse than those occurring in the native range.

EVOLUTION OF REPRODUCTIVE SYSTEMS DURING INVASION

Evolutionary changes to life history traits during invasion may be most significant when they involve aspects of the reproductive system itself. This is because changes to the reproductive strategies of populations feed back on demographic and genetic parameters, which in turn affect the course of evolution. The frequent occurrence in plants of inter- and intraspecific variation in reproductive traits indicates that many traits influencing reproductive mode are evolutionarily labile and hence could potentially respond to stochastic and selective forces during invasion. As of yet, there is only limited evidence for evolutionary changes to reproductive systems associated with the invasion process, but those that have been documented are instructive.

The annual *Eichhornia paniculata* is largely outcrossing in its native northeastern Brazil, where it occupies seasonal pools and is generally nonweedy. In contrast, populations infesting rice fields in Cuba and Jamaica are predominantly selfing. Genetic and ecological studies of this species demonstrate that the evolutionary transition from outcrossing to selfing has favored establishment in the Caribbean and Central America following long-distance dispersal. Similarly, native populations of *Rubus alceifolius* (Rosaceae) in Southeast Asia are largely sexual, while colonization of Madagascar and La Réunion is associated with a shift to uniparental reproduction through apomixis. Asexuality in these Indian island populations has arisen through hybridization with a native species (*R. roridus* in Madagascar), a pattern commonly found in apomictic groups. Reproductive transitions most commonly involve a shift to uniparental reproduction, but this is not always the case. In New Zealand, a common pasture weed in the south island is *Hieracium pilosella* (Asteraceae), a complex of apomictic races of varying cytotype native to Europe. Genetic markers indicate that in at least three locations, reversions from apomixis to obligate sexuality have occurred, enabling biparental reproduction because sexual individuals are self-incompatible. These examples indicate that it is unwise to treat the reproductive system of a plant species as a fixed property based on its behavior in the native range.

SEE ALSO THE FOLLOWING ARTICLES

Apomixis / Evolution of Invasive Populations / Invasiveness / Life History Strategies / Pollination / Vegetative Propagation

FURTHER READING

Barrett, S.C.H. 1989. Waterweed invasions. *Scientific American* 260: 90–97.

Barrett, S.C.H., R.I. Colautti, and C.G. Eckert. 2008. Reproductive systems and evolution during biological invasion. *Molecular Ecology* 17: 373–383.

Barrett, S.C.H., and L.D. Harder. 1996. Ecology and evolution of plant mating. *Trends in Ecology and Evolution* 11: 73–79.

Brown, A.H.D., and J.J. Burdon. 1987. Mating systems and colonizing success in plants (115–131). In A.J. Gray, M.J. Crawley, and P.J. Edwards, eds. *Colonization, Succession and Stability.* Oxford: Blackwell Scientific Publications.

Friedman, J., and S.C.H. Barrett. 2008. High outcrossing in the annual colonizing species *Ambrosia artemisiifolia* (Asteraceae). *Annals of Botany* 101: 1303–1309.

Montague, J.L., S.C.H. Barrett, and C.G. Eckert. 2008. Reestablishment of clinal variation in flowering time among introduced populations of purple loosestrife (*Lythrum salicaria*, Lythraceae). *Journal of Evolutionary Biology* 21: 234–245.

Pannel, J.R., and S.C.H. Barrett. 1998. Baker's Law revisited: Reproductive assurance in a metapopulation. *Evolution* 52: 657–668.

Richards, A.J. 1997. *Plant Breeding Systems.* London: Chapman and Hall.

Van Kleunen, M., J.C. Manning, V. Pasqualetto, and S.D. Johnson. 2008. Phylogenetically independent associations between autonomous self-fertilization and plant invasiveness. *American Naturalist* 171: 195–201.

REPTILES AND AMPHIBIANS

FRED KRAUS

Bishop Museum, Honolulu, Hawaii

There are more than 15,000 species of reptiles and amphibians in the world. Of these, at least 678 species are documented to have been introduced outside their native ranges by humans, and at least 322 of these have become established on foreign shores, resulting in more than 1,060 successfully established occurrences of alien reptiles and amphibians in the world. Most species have only a few records of introduction, but several have been introduced numerous times, and the pace of introduction has been increasing exponentially for the past century and a half, with a doubling time of a little over 27 years. Frogs, lizards, and turtles have predominated in volume of introductions, but significant numbers of snakes, salamanders, and crocodilians have also been involved. Despite the expanding magnitude of this phenomenon, study of alien reptiles and amphibians lags behind that for other taxa—such as plants, insects, or mammals—that create more obvious problems for humans. But sufficient information is now available such that we can begin to limn some of the dynamics and consequences of human introduction of these groups.

PATHWAYS OF INTRODUCTION

Herpetofaunal introductions involve 11 pathways, and six of these account for the large majority: unintentional introduction via cargo shipments or nursery trade (or their shipping vessels) and intentional introduction via the pet trade, biocontrol use, food use, or for unspecified but deliberate purposes most often associated with aesthetic enjoyment. Five additional pathways—via aquaculture, bait use, religious release, scientific research, and the zoo trade—comprise the remainder of herpetofaunal introductions, but they form a small proportion of them. The greatest number of introductions has occurred via the pet-trade and cargo pathways, and the sum of introductions for intentional purposes far outweighs that due to accidental importation via trade. But pathway importance varies taxonomically, temporally, and geographically, and knowledge of these variations is important for stemming future introductions.

Some pathways have been taxonomically limited. For example, the large majority of biocontrol introductions have involved frogs. Similarly, the food pathway has been limited to frogs, lizards, and turtles, and the nursery trade pathway has involved only frogs, lizards, and snakes. Other pathways, however, such as the pet trade and deliberate releases for personal aesthetic enjoyment, have involved all major groups of reptiles and amphibians. Frogs and lizards have been introduced via all six major pathways, snake introductions have involved all except the food pathway, and the other taxa have been moved for a more limited variety of reasons. Crocodilians have been introduced only via the pet trade and deliberate release; salamanders have been introduced primarily for these same reasons but have also had some introductions occur via cargo shipments; turtles have been introduced via the pet-trade, deliberate release, biocontrol, and food pathways.

Pathway importance has varied through time. Until the middle of the twentieth century, intentional introductions for unspecified purposes were the largest source of herpetofaunal introductions. Since the 1960s, the largest source has been the pet trade. In the past two decades, accidental introductions via cargo imports have become the second-largest source of introductions. Pathways can also decline dramatically in importance. For example, the biocontrol pathway was very important in the 1930s and 1940s but has stagnated for the past several decades.

These patterns are not uniform around the globe, but vary geographically. Most introductions have occurred to North America and Europe, although all parts of the subpolar globe are affected. Most introductions to these continents—and to Asia, South America, and the Atlantic Islands—have resulted from the pet trade. In contrast, introductions to most other regions have involved a more even distribution of pathways, although introductions to Australia have been predominantly caused by internal movement of cargo. Source regions of alien herpetofauna are also not evenly distributed: by far the largest share of alien herpetofauna has come from North America, with less than half as many introductions involving species native to Europe and Asia, and the remaining regions of the globe have contributed a relative smattering of species.

ESTABLISHMENT SUCCESS

Some families and genera are more liable to establish successfully than others and hence pose a greater risk of invasiveness. Species with a history of prior establishment are more likely to establish in a new location than are those lacking such a history, and establishment success increases with increasing number of release events. It is likely that establishment success improves as larger numbers of individuals are released in a single event, but that supposition has yet to be tested.

A variety of ecological factors is expected to correlate with establishment success of alien species. Similarity of climate between native and introduced ranges of alien herpetofauna is an important predictor of establishment success; however, the importance of other ecological factors has not yet been assessed. It is known that native range size, which might be taken as a rough measure of a species' ecological breadth, does not explain establishment success for reptiles and amphibians, so more sensitive and direct ecological variables must be investigated.

Establishment success varies by introduction pathway, with nursery-trade, biocontrol, and food-use introductions most often leading to successfully established populations. The first is successful because transportation conditions are very favorable for hitchhiking herpetofauna, with animals being moved in large blocks of equable habitat. Success of biocontrol and food introductions is because establishment is the explicit goal of such activities, and considerable effort and large numbers of animals are typically involved in such actions.

Lastly, small islands (those <6000 km^2) have proven more easily colonized by alien herpetofauna than have large islands (those >8000 km^2) or continents. This could reflect the increased contribution to small islands of introductions from pathways especially liable to successful establishment (e.g., nursery trade, biocontrol, food use), or it could be that small islands are especially prone to successful colonization because of increased availability of empty niche space.

IMPACTS

Documented impacts from alien reptiles and amphibians are diverse, even though only about two dozen of the more than 1,000 established occurrences of alien herpetofauna have been investigated. Impacts may be grouped for convenience into the categories of ecological, evolutionary, and social. The first and second groups affect native species in the recipient regions harboring alien herpetofauna, and the last affects humans directly.

Alien herpetofauna exert ecological impacts largely by disrupting native food webs. This happens because many reptiles and amphibians occur at high densities and can form high standing biomass. Thus, they can have a large effect on native ecosystems or communities by preying upon native species, poisoning native predators, or competing with native animals for food, shelter, or some other critical resource. Examples of these effects include the almost total removal of Guam's native forest-bird fauna by introduction of the brown treesnake (*Boiga irregularis*, Fig. 1); poisoning of many of Australia's predatory mammals, lizards, snakes, and crocodiles by introduction of the cane toad (*Rhinella marina* [*Bufo marinus*], Fig. 2); and restriction of the rare endemic *Nactus* geckos of Mauritius and surrounding islets to relictual patches of habitat by competitive exclusion from refugia by the introduced gecko *Hemidactylus frenatus* (Fig. 3). However, ecological effects need not be so direct. Removal of fruit-eating birds from Guam is expected to lead to loss of pollination and seed-dispersal services that these birds formerly provided to native forest trees, and extirpation of insectivorous birds seems to have led to an increase in spider numbers. Poisoning of large predatory lizards in parts of Australia has increased reproductive success of native turtles

FIGURE 1 Brown treesnake (*Boiga irregularis*) from Fergusson Island, Papua New Guinea. (Photograph courtesy of the author.)

FIGURE 2 Cane toad (*Rhinella marina* [*Bufo marinus*]) from Oahu, Hawaii. (Photograph courtesy of David Preston.)

that formerly lost most of their nests to lizard predation. Smaller species of introduced herpetofauna may serve as an inexhaustible food source to keep unnaturally high the populations of other, predatory, aliens. In this manner, brown treesnake numbers on Guam remain artificially inflated by vast numbers of a small alien skink, *Carlia ailanpalai* (Fig. 4), which provides an abundant source of food. Lastly, native species and communities can be altered by novel parasites or pathogens arriving with alien herpetofauna. As one important example, extinction of amphibian populations around the globe by the disease fungus *Batrachochytrium dendrobatidis* seems to be due in part to widespread introduction of amphibian hosts, such as North American bullfrogs (*Lithobates catesbeianus*, Fig. 5), which are resistant to the fungus but introduce it into naive amphibian populations.

Introduction of cane toads to Australia has led to longer-term evolutionary changes in the morphology, behavior, and physiology of some native snake predators. For example, the red-bellied black snake (*Pseudechis porphyriacus*) has evolved greater resistance to toad toxins, greater aversion to eating toads, relatively smaller gape size, and relatively larger body size in areas where it has suffered long exposure to toads. These changes have all been such as to decrease the threat the toads pose to the snakes. Evolved behavioral and morphological changes to better avoid predation by alien snakes have been documented in native Mallorcan frogs. Such evolutionary changes are expected to result commonly from introductions but have been little studied in alien herpetofauna. A far more common evolutionary effect is genetic pollution of native species by hybridization with closely related alien herpetofauna. This is the most widely documented

impact from herpetofaunal introductions, and in some cases, native wildlife are threatened with extinction or local extirpation. As one example, many populations of the endangered California tiger salamander (*Ambystoma californiense*) are threatened with genetic swamping by alien congeners introduced as fishing bait, with the alien genes predominating in areas already suffering habitat alteration.

Finally, some herpetofaunal introductions have directly affected humans themselves, most often by threatening human health or impacting economic activities. Bites from the mildly venomous brown treesnake have led to hundreds of emergency-room admissions on Guam, and the toll on Okinawa from introductions of two species of viper is expected to lead to much larger costs as they expand their ranges in coming decades. Public-health effects from frogs and lizards have also been proposed, but none of these has been compellingly demonstrated.

Guam has suffered significant economic damage from brown treesnakes because the snakes have induced

FIGURE 3 Common house gecko (*Hemidactylus frenatus*) from Molokai, Hawaii. (Photograph courtesy of the author.)

thousands of power outages by draping themselves over power lines, and they have dramatically impacted the local poultry industry. Alien coqui frogs (*Eleutherodactylus coqui*) have depressed property values of affected residential properties in Hawaii and have led to quarantine treatment for exported nursery plants. And cane toads have imposed costs on the honeybee industry in Australia by preying on bees. Compared to many other alien pests, though, economic costs from alien herpetofauna have generally been rather minor.

MANAGEMENT

Management of alien herpetofauna is in its infancy. These species are generally ignored in quarantine programs to prevent the ingress of alien pests. One exception is that, on Guam, quarantine is directed to outbound cargo and vehicles so as to keep brown treesnakes from colonizing other shores. Because so many herpetofaunal introductions derive from the pet trade and other deliberate importations, fauna proposed for importation could potentially be assessed for invasion risk and barred from import if necessary, but reliable methods for predicting such risk have yet to be developed for reptiles and amphibians. Development and application of reliable predictive models could greatly reduce the number of future herpetofaunal invasions.

Eradication of newly established populations has rarely been successful for alien amphibians and has never been seriously attempted for an alien reptile. The few documented successes have been against frog populations whose breeding requirements concentrated them in narrowly circumscribed aquatic habitats where they could be effectively targeted for control. Most often, attempts at "eradication" have been poorly executed, often because of insufficient planning and funding, initial misperception of the size of the targeted population, or lack of involvement by trained management staff.

Control of widespread alien herpetofauna has rarely been attempted and could not yet be considered successful in any instance. The primary effort in this vein is the program reducing numbers of brown treesnakes near the ports on Guam so as to decrease their probability of successful emigration to other jurisdictions. This has been measurably successful but does nothing to limit snake damages on Guam.

In general, control operations against alien herpetofauna have suffered from a number of constraints, making success difficult to achieve. These animals are often cryptic, have high reproductive rates, and can achieve high population densities. This means that, for many species, populations are easily overlooked initially but can quickly build to numbers impossible to control. Working against this need for rapid response are human behavioral constraints, including widespread disbelief that alien reptiles and amphibians constitute an actionable problem, possible support for the introduction by some segments of society, frequent opposition to the killing of vertebrates by other segments of society, and lack of effective control methods. Operation of one or more of these last four constraints often means that reaction to a new alien reptile or amphibian population is delayed beyond the short time during which control might have been successful.

Management of alien reptiles and amphibians is capable of improvement, but this will require removal of the social constraints identified above. Managers and other government officials need to recognize that alien herpetofauna may merit management response and that the desires of limited segments of society should not be

FIGURE 4 Admiralty Islands skink (*Carlia ailanpalai*) from Guam. (Photograph courtesy of Gordon H. Rodda.)

FIGURE 5 North American bullfrog (*Lithobates catesbeianus*) from Maui, Hawaii. (Photograph courtesy of the author.)

allowed to trump the common good. And more effective control methods must be developed for a variety of the most invasive taxa.

SEE ALSO THE FOLLOWING ARTICLES

Biological Control, of Animals / Brown Treesnake / Burmese Python and Other Giant Constrictors / Pet Trade / Predators

FURTHER READING

Bomford, M., F. Kraus, S. Barry, and E. Lawrence. 2009. Determinants of establishment success in introduced reptiles and amphibians: A role for climate matching. *Biological Invasions* 11: 713–724.

Fritts, T. H., and G. H. Rodda. 1998. The role of introduced species in the degradation of island ecosystems: A case history of Guam. *Annual Review of Ecology and Systematics* 29: 113–140.

Kraus, F. 2009. *Alien Reptiles and Amphibians: A Scientific Compendium and Analysis.* Dordrecht, The Netherlands: Springer Science and Business Media B.V.

Lever, C. 2001. *The Cane Toad: The History and Ecology of a Successful Colonist.* Otley, UK: Westbury Publishing.

Lever, C. 2003. *Naturalized Reptiles and Amphibians of the World.* New York: Oxford University Press.

Rodda, G. H., Y. Sawai, D. Chiszar, and H. Tanaka. 1999. *Problem Snake Management: The Habu and Brown Treesnake.* Ithaca, NY: Comstock Publishing.

Rodda, G. H., and C. L. Tyrrell. 2008. Introduced species that invade and species that thrive in town: Are these two groups cut from the same cloth? (327–341). In J. C. Mitchell, R. E. Jung-Brown, and B. Bartholomew, eds. *Urban Herpetology.* Herpetological Conservation, Vol. 3. Salt Lake City, UT: Society for the Study of Amphibians and Reptiles.

RESTORATION

TRUMAN P. YOUNG AND KURT J. VAUGHN

University of California, Davis

Ecological restoration is defined by the Society for Ecological Restoration as "the process of assisting the recovery of an ecosystem that has been degraded, damaged, or destroyed." Although this can mean the attempt to return the site to a "natural" or even wilderness condition, it also encompasses many kinds of ecosystem repair. Examples include revegetating mine tailings, returning forest to abandoned farmland, improving habitat for wildlife, and reestablishing historic fire or flood regimes. This work may require the elimination of degrading forces, preparation of sites, planting of appropriate species, and maintenance and monitoring of sites. At each of these stages, invasive species may impede successful restoration.

THE FACES OF RESTORATION

Ecological restoration can take many forms and appear under different names, depending on the objectives.

REVEGETATION The reestablishment of vegetation on sites from which it has been lost, often with the primary goal of erosion control. This vegetation may or may not be specifically designed to replicate the indigenous vegetation of the site.

HABITAT ENHANCEMENT Habitat "improvement," usually directed toward some desirable or threatened taxon (e.g., waterfowl). This may explicitly not include the return to predisturbance vegetation.

RECLAMATION Originally used in the opposite sense of "reclaiming" land from nature for human use, now increasingly used to designate the more general idea of reclaiming land from a less desirable or degraded state to a more desirable state, which may include the reestablishment of natural vegetation.

REHABILITATION The improvement of degraded land, which may include major restoration activities, but which may also include less intensive management techniques that favor shifts in certain aspects of the plant community, often with an extractive use in mind, such as ranching.

MITIGATION The attempt to create a community type in a site where it may or may not have occurred before, done to balance the loss of a similar community or population elsewhere (such as from development).

REMEDIATION Similar to mitigation, but with fewer legalistic overtones of one-for-one replacement. This can include the use of biodiversity, such as vegetation buffers, to increase ecosystem function or environmental quality (as in bioremediation).

GOALS

The list above comprises some of the diversity in the nature and goals of ecological restoration. Still, the most basic and general goal of restoration is the reestablishment of the predisturbed ecosystem. However, this objective may raise multiple questions when attempting to define measurable goals for restoration activities. For example, what is the proper reference state for restoration in systems with a long history of human management (e.g., meadows maintained by Native American fire practices in the Yosemite Valley, or by grazing practices in the Alps)?

More specific restoration goals may include objectives that are biological (e.g., reestablishment of native species or control of exotic species), hydrological (e.g., erosion control or reestablishment of inundation regime), geochemical (e.g., soil condition amelioration), social (e.g., reestablishment of recreation areas), or economic (e.g., reestablishment of ecosystem services). At some sites, the control of invasive species is an ultimate goal of restoration activities, but more often it is an important prerequisite for other, broader goals.

Biological goals at multiple scales can serve as quantifiable measures of restoration success. At the species and community level, such goals may include species richness (the number of species), species composition (relative abundance), guild structure (i.e., perennial grasses, nitrogen-fixers, annual forbs, woody shrubs), eradication or control of invasive species, and the restoration of threatened species and wildlife habitat.

Ecosystem-level functions may also be measurable goals of restoration. These include primary productivity, nutrient cycling and retention, carbon sequestration, hydrological function, and trophic integrity (mutualists, herbivores, carnivores). Ecosystem functions that more directly serve human needs are referred to as ecosystem services, and these include erosion control, runoff quality and quantity, pollination services, nutrient balance (soil fertility), extractive use (grazing forage, fisheries, timber), and aesthetic values (public use, tourism).

STAGES OF RESTORATION

Restoration activities comprise multiple stages (Table 1). The first stage in a restoration project is site assessment and project planning. This activity often includes the determination of restoration goals, which can include a combination of general and specific measurable objectives. Reference sites are often identified in this initial stage to represent quantifiable targets for measuring restoration success. Lists of appropriate, usually native species are generated, either for planting or as indicators of restoration success.

Restoration is unlikely to be successful unless the forces that caused the degradation of the system are identified and remediated. In some cases, the removal of degradative forces is sufficient to allow for the natural regeneration of the ecosystem. For example, the recovery of the eastern deciduous forest of the United States has proceeded well without intervention following the widespread abandonment of agriculture in the region in the nineteenth century. Other possible degradative forces include acidification, nutrient enrichment, soil compaction, mine tailings, topsoil removal, soil

TABLE 1
The Restoration Process

Stage 1: Site Assessment and Project Planning

Initial site evaluation/data collection
Selection of reference site/criteria
Determination of goals
Identification of degradative forces
Development of a restoration plan

Stage 2: Amelioration of Degradative Forces/Legacies

Remove disturbance (or reestablish disturbance regime)
Improve soil conditions
Restore hydrological function
Control invasive species

Stage 3: Active Planting

Site preparation
Seed collection/increase
Planting
Horticultural amendments (irrigation, mulch, shelters, fertilizer)

Stage 4: Maintenance/Monitoring

Maintain amendments as needed
Data collection/analysis
Adaptive management

NOTE: A simplified outline of the major stages and activities of a restoration project.

salinization, overgrazing, timber harvesting, and river damming and levees. The degradative force may also be the disruption of historic disturbance regimes, such as fire or flood regimes, and the reestablishment of these regimes on a landscape scale is often a broad goal of restoration activities.

If the degradative forces leave behind a strong or long-lasting legacy, it may be necessary to prepare the site before natural regeneration can proceed. This can mean adding topsoil, ameliorating soil conditions (including impoverishing artificially enriched soils and waters), reconstituting hydrologic function, and providing weed control. Herbicide application, hand weeding, disking, tilling, fire, and grazing are all potential weed control techniques.

If, after this site preparation, natural regeneration is not possible, or if it occurs too slowly to meet project goals, then active planting of desirable species can be carried out. This may be preceded by seed collections from reference communities and by a process of seed increase, if natural seed sources are insufficient. Often seed or seedlings are commercially produced, ideally of local genotypes.

Plants may be seeded directly or planted as seedlings. Horticultural amendments such as irrigation, tree shelters, mulching, fertilizers, and mycorrhizal inocula may also be part of this planting effort.

Ideally, the project is then monitored to assess restoration success and to determine the need for appropriate adaptive management activities.

CONCEPTUAL ECOLOGICAL UNDERPINNINGS

Restoration ecology is the science that supplies research answers to questions concerning how best to achieve restoration goals and explores the ecological concepts underlying restoration. Restoration ecology draws from a wide range of environmental sciences, including agronomy, horticulture, hydrology, biogeochemistry, and ecology (as well as the social sciences; see below). Within ecology, important aspects of restoration include genetics; ecophysiology; and population, community, ecosystem, and landscape ecology.

Disturbance describes a change in environmental conditions that interferes with ecosystem structure or function. While disturbance is a natural process in all ecosystems, and the reestablishment of natural disturbance regimes is often a restoration goal in and of itself, severe or chronic anthropogenic disturbances may alter ecosystems beyond their capacity to naturally recover.

The concept of *succession* can be simplified as the tendency for disturbed communities to recover to their predisturbed state. There are systems where natural regeneration proceeds without intervention as long as propagule pressure from outside the disturbed area is sufficient, and the soils are relatively intact. Restoration in these cases may be either completely passive or may simply consist of jumpstarting or accelerating successional processes. *Assembly* theory suggests that multiple steady states may exist for a given site, determined in part by the order of arrival of its colonists. This theory suggests that the outcome of a restoration project may depend strongly on planting decisions, and that a site may not inevitably converge to a previous or desired state. *State-transition* models take the idea of assembly one step further and seek to outline not only the different possible states of a community but also the drivers determining transitions among these states, and they specifically include the activities of restoration. It is not uncommon for a highly invaded community structure to be one of these states.

RESTORATION IN A SOCIOECONOMIC CONTEXT

Restoration is a human enterprise, and it takes place in a complex social context. The social complexity of a particular restoration project is partially a function of the spatial extent, intensity of intervention, number of landowners and managers potentially directly or indirectly affected, number

FIGURE 1 Tamarisk (*Tamarix parviflora*) has severely invaded many riparian areas in the arid and semi-arid western United States. (A) Tamarisk (the reddish shrub in the foreground) in a restoration site in Utah. (B) Mechanical removal. (C) Vegetation immediately post-removal, with remnant native cottonwoods and space for recruitment of willows and more cottonwoods. (Photographs courtesy of the Bureau of Land Management, U.S. Department of the Interior.)

of funding sources, presence of threatened species, and character of local, state, and federal laws and policies.

Restoration activities range from small, locally conceived and locally implemented "postage stamp" projects to regional-scale projects involving the coordination of hundreds of partners and multiple government agencies and nonprofits, and costing hundreds of millions of dollars. (In the United States, such regional-scale projects include the Everglades, the California Bay delta, Lake Tahoe, and the Mississippi delta.)

Community involvement and volunteerism are hallmarks of ecological restoration. These efforts often include opportunities for assessing and informing public perceptions of natives, exotics, landscape, and "wilderness." Community participation in restoration activities can offer occasion for hands-on environmental education, as well as instill a sense of local land stewardship.

A substantial proportion of restoration projects are the result of legally mandated mitigation for development that has as its (largely unsubstantiated) underlying assumption that losses of species or habitats at one site can be recouped through restoration of another site.

INVASIVE SPECIES AND RESTORATION

In many sites, invasive species are one of the major sources of ecosystem degradation, and their control is often one of the primary goals of (and challenges to) restoration efforts. Invasive species can sometimes completely preempt successional regeneration, leading to the formation of a highly invaded stable community state. Invasive plant (and animal) species can be a degradative force in their own right, whose initial control is a prerequisite for restoration, and whose long-term control often requires the restoration of a native plant community that is resistant to further invasion (Fig. 1). In highly invaded sites like island ecosystems or the western grasslands of the United Sates, it has been said that there can be no successful restoration without effective weed control, and there can be no effective weed control without successful restoration. Conversely, in some ecosystems, nonnative species can be used to assist in restoration.

SEE ALSO THE FOLLOWING ARTICLES

Endangered and Threatened Species / Fire Regimes / Grasses and Forbs / Herbicides / Hydrology / Land Use / Mechanical Control / Succession

FURTHER READING

Baker, J. P. 2005. Vegetation conservation, management and restoration (309–331). In E. van der Maarel, ed. *Vegetation Ecology*. Malden, MA: Blackwell.

Ewel, J. J., and F. E. Putz. 2004. A place for alien species in ecosystem restoration. *Frontiers in Ecology and the Environment* 2: 354–360.

Falk, D. A., M. A. Palmer, and J. B. Zedler. 2006. *Foundations of Restoration Ecology*. Washington, DC: Island Press.

Hobbs, R. J. 2007. Setting effective and realistic restoration goals: Key directions for research. *Restoration Ecology* 15: 354–357.

Society for Ecological Restoration International. www.ser.org/

Van Andel, J., and J. Aronson. 2006. *Restoration Ecology*. Oxford: Blackwell.

Young, T. P., D. A. Petersen, and J. J. Clary. 2005. The ecology of restoration: Historical links, emerging issues, and unexplored realms. *Ecology Letters* 8: 662–673.

RINDERPEST

ANDY DOBSON
Princeton University, New Jersey

RICARDO M. HOLDO AND ROBERT D. HOLT
University of Florida, Gainesville

Rinderpest is a virus in the Morbillivirus family that causes disease in cattle; it created the largest pandemic ever recorded when it was introduced into East Africa in the early 1890s. Its invasion and spread caused the deaths of around 50 to 90 percent of cattle and wild artiodactyl species (wildebeest, buffalo, giraffe) in sub-Saharan Africa. The loss of hosts for tsetse flies created a human epidemic of sleeping sickness throughout sub-Saharan Africa. Rinderpest is closely related to human measles virus and canine distemper virus; control was only achieved once methods used to develop a measles vaccine were applied to rinderpest in the early 1960s.

INTRODUCTION HISTORY IN AFRICA

Africa has been afflicted by many disasters in the relatively short period for which we have historical records: droughts, dictators, deforestation, swarms of locusts, and of course disease. Although over 20 million Africans are currently infected with the HIV virus that causes AIDS, and several thousand children die each day from malaria, arguably the worst disaster ever to have hit the African continent was the great rinderpest pandemic that started in 1889. In the ten years of its initial spread from the coast of Ethiopia in the Horn of the Africa until its arrival in Cape Town, it is estimated to have killed 80 to 90 percent of the cattle population and similar proportions of many artiodactyl species (wildebeest, giraffe, and particularly buffalo). These losses essentially devastated the protein supply for most of the human population of sub-Saharan Africa while simultaneously triggering an epidemic of sleeping sickness when starved tsetse flies switched to human hosts in the absence of their natural hosts.

Rinderpest provides detailed and important insights into the impacts that pathogens can have when they invade host populations in locations that have no prior experience of them. While the population dynamics of rinderpest are, at first sight, deceptively simple, there is ultimately a significant level of subtle complexity that sharply echoes that seen for measles. For instance, it was quickly realized that early measles models that divided the host population into susceptible, infected, and recovered classes required more details about how transmission was driven, for example by mixing rates among age classes and by spatial heterogeneity across the landscape in the host population. These pathogens thus provide an important set of insights into studies of invasive species and conservation of endangered species as pathogen dynamics are inherently those of other species distributed as metapopulations, because each host is, in effect, a habitat patch that can be colonized (causing infection) or can experience local extinction (disease clearance or host death). Ultimately, the invasion success of any pathogen and the efficacy of its control require on an understanding of how it manages to persist in the patches of habitat that are its hosts.

THE MORBILLIVIRUS FAMILY

Rinderpest (RPV) is likely the basal species in a Morbillivirus family. It is a single-stranded RNA paramyxovirus and is the most likely ancestral species for canine distemper in dogs (CDV), peste des petites ruminants in goats and gazelles (PPRV), and measles in humans (MV). The trifurcation in the evolutionary split between RPV, CDV, and MV is too poorly resolved to accurately determine the order with which these pathogens spread into their major classes of hosts. It seems likely that either rinderpest's association with cattle predates the domestication of aurochs or that CDV's association with dogs predates dog domestication. Whichever is the case, the domestication of cattle and dogs allowed the ancestral morbillivirus to jump between domestic species and humans and gave rise to three viruses that have had major impacts on the welfare and abundance of humans, canids, cattle, and goats. There is no other closely related group of pathogens with a comparative impact on humans and their domestic livestock. The symptoms of infection are initially subtle and are often missed by farmers and herders, thus allowing the pathogen to be transmitted before intervention can occur. After several days, infected animals stop feeding, become dehydrated, and have difficulty breathing. Soon thereafter, open sores appear around the mouth and nasal passages. These manifestations are rapidly followed by diarrhea, further dehydration, and eventual death. Mortality rates within herds range from 30 to 90 percent,

causing the pathogen to have devastating effects whenever outbreaks occur.

DYNAMICS OF RINDERPEST

The population dynamics of rinderpest are inherently similar to those of measles, arguably the pathogen for which we have developed the deepest quantitative understanding, so it is worth briefly considering the population dynamics of measles for insights it provides into the dynamics of rinderpest. In the text that follows, one could readily replace the word "measles" with "rinderpest," as broadly similar things will happen at both the individual and population levels when rinderpest infects cattle and measles infects humans. Indeed, were measles or rinderpest to resurge and create an epidemic, it would be perfectly possible to use the human measles vaccine to protect susceptible cattle, and vice versa! There is almost perfect cross-immunity between the vaccines, as the pathogens are essentially indistinguishable to the immune system.

Early work on the dynamics of measles identified a key epidemiological principle: measles was able to persist only in a population of greater than half a million people. This pattern was consistent in relatively isolated human populations living on oceanic islands, as well as those living in cities that were connected to other cities by emerging rail and road networks. In these larger cities, the birth rate was high enough to sustain an input of susceptible hosts into the population, thereby maintaining a constant chain of infection for the currently infected individuals in the population. Constant chains of infection are required by all morbilliviruses if they are to persist in the host population, and this will occur only if the population is larger than a "critical community size." As individuals who have recovered from measles are immunologically resistant for the rest of their lives, new susceptible hosts can enter the population only by birth or by immigration. Newborn individuals are said to have maternal immunity if their mothers had previously been infected and were placed in the resistant class of hosts that maintain antibodies protecting them from further infection. However, maternal immunity is lost after a period of around six months, so all infants will from then on become susceptible to infection. When these individuals contact infected hosts, they are likely to become infected themselves and enter a period a period of infectiousness, during which they either die from the infection or, more usually and if well nourished, recover and enter the immunologically resistant class.

The dynamics of measles, rinderpest, and all the other morbilliviruses can thus be well described by dividing the host population into four classes of hosts: susceptible, infected, resistant, and with transient maternal immunity; we designate their abundances as S, I, R, and M, respectively. This leads to a set of four coupled differential equations that describe the dynamics of a morbillivirus infection in a well-mixed host population.

$$dM/dt = bR - (\delta + d)M$$
$$dS/dt = bS + \delta M - dS - \beta SI$$
$$dI/dt = \beta SI - (\alpha + d + \sigma)I$$
$$dR/dt = \sigma I - dR$$

Here, $S + I + R = N$, the total size of the adult population, and maternal immunity is lost at a rate δ (whence newborn hosts are, on average, protected for a period of $1/\delta$ years). The population gives birth at a rate b and dies at a rate d, and it is assumed that infected individuals are too ill to give birth. Transmission occurs at a rate β between infectious and susceptible hosts, and individuals are infectious for a period of $1/\sigma$ years. To regulate the host, we could assume direct density dependence in, say, births (details are not explicitly shown in the above equations). The properties of this model have been exhaustively explored in the mathematical epidemiology literature; these properties provide a number of important insights into the dynamics of many viral pathogens. As mentioned above, a central property of pathogens whose dynamics can be described by this model structure is that the host population size needs to be large enough to allow the pathogen to persist. A slightly more subtle consequence is that the average age of infection will be coupled to the size of the population. Thus, in large and aggregated host populations, transmission rates will be high, and average age of infection will be skewed toward the younger individuals who have recently lost their maternal immunity. In contrast, in host populations that are below the critical community size, the average age of infection will be higher and will depend more on contact rates between the local small population and the external populations that are themselves large enough to maintain a constant chain of infection.

The structure of the equations also reveals a second key feature of most morbillivirus infections: the relative proportions of the host population in the different classes of infection will be roughly proportional to the time each individual host spends in each class of infection: $1/\delta$, $1/(\alpha+\sigma)$, and $1/d$. As hosts are susceptible for only a couple of years in endemic areas, and infective for, at best, a couple of weeks and then immune for life, the largest proportion of the host population will be in the recovered class. There will be significantly fewer susceptible hosts, and infected hosts will be comparatively rare, often constituting less than 0.01 percent of the population. This creates an important paradox that has confounded ecologists' perspective on the importance of pathogens until recently; although infectious individuals are comparatively rare, the presence of a pathogen can significantly reduce the abundance of a host population. However, once the pathogen has established, it holds the population at a significantly lower abundance (around the local critical community size), and although the chain of transmission is maintained in the reduced host population, infected individuals will be comparatively rare. We tend only to see large numbers of infected individuals when the pathogen first invades a host population and all hosts are susceptible. Rinderpest provides one of the best examples of this; when first introduced into sub-Saharan Africa, it produced a dramatic pandemic, as all host populations were composed only of susceptible individuals, so everyone became infected at almost the same time, causing widespread disease and mortality. It is likely that some of the plagues of Egypt recorded in the Old Testament were of similar form when measles first crossed over into human populations from cattle; similar devastating epidemics were recorded when Central American Indians were introduced to measles and smallpox following the Columbian conquest.

The equations described above can be rearranged to provide an expression for the basic reproductive number of the pathogen, commonly termed R_0. This can be defined formally as the number of secondary cases produced by the first infected individual introduced into the population. A moment's thought suggests we are highly unlikely ever to witness, let alone sample and quantify, this event, so other methods are used to measure R_0 formally. Nonetheless, deriving and inspecting an algebraic expression for R_0 provides important insight into the demographic components of the host–pathogen interaction that determine the magnitude of a disease outbreak. It has also been argued that deriving expressions for R_0 for other types of invasive species might provide insights into how best to control them. The expressions for R_0 for rinderpest, measles, or distemper are essentially identical:

$$R_0 = \frac{\beta N}{(\alpha + d + \sigma)}$$

Inspection of this expression provides a key biological insight; R_0 is the rate at which an infected host contacts and successfully transmits to susceptible hosts, times the duration of time over which it is infectious. This is arguably analogous to one measure of the fitness of a free-living organism: the number of viable offspring it produces during its reproductive lifespan (often called "lifetime reproductive success"). Conditions that create large values of R_0 will produce dramatic, large epidemics that quickly exhaust the supply of susceptible individuals and then die out. In subtle contrast, conditions that produce R_0 values that are slightly larger than unity will produce small epidemics that may persist for quite some time. One can use R_0 as a measure of pathogen fitness. However, there are tradeoffs for a virulent pathogen between optimizing its epidemic potential (a traditional perspective on "fitness," i.e., growth rate) and optimizing persistence; the latter may favor strains that are less competitive in the short run, particularly in spatially structured host and pathogen populations, permitting local extinctions and recolonizations. These processes may even interact; when a pathogen is repeatedly spilling over from a reservoir host, the initial outbreaks may be dramatic and then die out. Subsequent epidemics are less dramatic, as R_0 is reduced, owing to the presence of immunologically resistant hosts who recovered from earlier outbreaks. Persistence will now increase, and the pathogen may be able to establish a longer-term chain of transmission in the host population.

Ultimately, the principal goal of vaccination is to increase the proportion of resistant hosts in the population to levels at which the chain of transmission is broken and the pathogen dies out. Estimating the magnitude of R_0 provides an important quantitative guideline for the fraction of hosts that need to be vaccinated, quarantined, or culled in order for chains of transmission to break and cause the pathogen to die out in the host population. Vaccination and quarantine reduce the quantity β, whereas culling boosts d. The level of vaccination needs to reduce the current reproductive number for the pathogen below unity for control to be complete (i.e., the infection tends toward extirpation); these conditions are usually met when the proportion vaccinated, p_v, meets the following inequality:

$$p_v > 1 - \frac{1}{R_0}$$

Pathogens with high estimated values of R_0 require much higher levels of vaccination than do those with lower R_0 values. Two important caveats apply here: (1) this initial calculation assumes that the vaccine is administered once and lasts for the rest of the host's lifespan; this seems to be the case for measles, rinderpest, and canine distemper (although two doses of vaccine are increasingly used

to boost individual levels of protection). It is not often the case for other pathogens. These either mutate rapidly and create new variants for which new vaccines must be developed (e.g., influenza), or the vaccine produces only short-term immunity, in which case hosts must be repeatedly vaccinated if eradication is the goal, and (2) while individual shots of vaccine protect the host to whom they have been administered, increasing the numbers of immunologically resistant hosts in the population creates a secondary important effect termed "herd immunity," which entails a reduced risk of infection for hosts who have been neither vaccinated nor naturally infected. In the case of measles, cultural beliefs and misgivings about possible vaccine side-effects act to reduce levels of vaccination coverage to below levels where eradication can occur. Similar problems beset rinderpest vaccination, as herders are reluctant to have cattle vaccinated at times of stress (e.g., during droughts), or when emergency vaccination schemes in response to outbreaks lead to rumors about whether the vaccine or the pathogen led to cattle deaths.

A key potential difference between rinderpest and measles is that rinderpest may be present as a pathogen in a community of different host species; this introduces a number of significant complications in the population dynamics of outbreaks. These can be reduced to a manageable scale by recognizing that rates of between-species transmission are likely to be considerably lower than rates of within-species transmission. Furthermore, rates of between-species transmission are likely to be significantly determined by levels of spatial niche overlap between potential host species. Thus, species with very different feeding niches are unlikely to transmit aerosol pathogens between each other. Similarly, if species use similar niches sequentially, as occurs in some "grazing successions," then rates of between-species transmission will be asymmetrical and will be more likely to occur from the earlier species in the succession to the successive ones, rather than vice versa. More subtly, it is likely that some group of host species will act as a reservoir for the pathogen, while others will act as spillover hosts. A key point here is that spillover hosts may play a key role in indicating the presence of background transmission in the reservoir hosts that would go undetected owing to less pronounced pathology and etiology in the reservoir hosts. This occurs with Nipah and Hendra virus, emerging pathogens that are fairly closely related to rinderpest and other morbilliviruses. Their reservoir hosts are pteropid fruit bats, which show essentially no symptoms of infection with the virus. However, when it is transmitted from bats to horses or pigs, significant pathology and mortality are very apparent.

Rinderpest tends to show similar pathologies in most of its host species, although different strains of the virus may show significant differences in pathology that range from mild to severe. Similarly, with morbilliviruses, older individuals may exhibit much higher levels of pathology and mortality than do younger ones, who are characteristically infected in a population large enough to support the continuous presence of the pathogen. The classic studies on measles in the Faeroe Islands are instructive here; measles had been absent from this isolated group of islands for over 50 years, and when it was introduced by a passing fisherman, it created an epidemic that infected most individuals under the age of fifty; those older than this still had immunity from exposure during the previous outbreak. The mortality rates of those under fifty increased exponentially, from relatively low mortality when under the age of five to very severe mortality in those in those in their thirties and forties. The dramatic *grand mal* seizures and mass mortalities of adult humans exposed to measles observed in the Faeroes, Hawaii, and ancient Athens echo those seen when rinderpest devastated the cattle, wildebeest, and buffalo populations of sub-Saharan Africa in the 1890s. However, once the pathogen became endemic, younger animals that had survived infection created a significant level of herd immunity that slowed the rate of spread of the pathogen and lowered its average prevalence. New cases were largely restricted to young animals born in the previous 12 months; this feature led to rinderpest being known as "yearling disease" to African veterinarians and farmers. This is comparable to human populations with endemic measles, which mainly infects susceptible youngsters.

CONTROL OF RINDERPEST

Once a vaccine had been developed for measles in the 1950s, it was relatively straightforward to develop one for rinderpest, and subsequently for canine distemper. The principal motivation for developing the vaccine was to protect the cattle herds of British East Africa from recurrent outbreaks of rinderpest, which conventional wisdom dictated were the result of frequent spillovers from wildlife reservoirs in wildebeest and other wild game species. The British government even debated the seriousness of the problem, worried in part about the national cattle herd's ability to keep beef on the British Sunday dinner menu and in part about an increasing reliance on Argentinian beef, which might be cut off in a political pique if relations with Argentina were to become sufficiently strained

over ownership of the Malvinas/Falkland Islands. Indeed, it might be argued that the rise of the Argentinian beef-based economy in the early twentieth century was significantly aided by the presence of rinderpest throughout the rest of the world. The absence of the pathogen and the advent of fast-traveling steamships allowed Argentina to supply much of Europe with beef at a time when rinderpest hampered production in most other cattle-producing countries.

Once the rinderpest vaccine had been developed, it was quickly deployed to protect cattle in East Africa and other parts of the world (Fig. 1). The results were quite startling; certainly the vaccine provided excellent protection and allowed vibrant and viable cattle herds to reestablish in areas of sub-Saharan Africa from which they had been largely excluded for over 60 years. Simultaneously, the disease began to disappear from wildlife—although not a single wild animal was vaccinated. It is hard to think of a better experimental way to identify the true reservoir of a pathogen!

The effects of rinderpest removal on wild host species such as wildebeest and buffalo were spectacular. The Serengeti National Park had been established in 1959 to protect the herd of around a quarter of a million wildebeest; the eradication of "yearling disease" allowed the population to increase by a factor of around six to over 1.5 million animals. Buffalo were felt to be rare or absent from the region because of their popularity as game for "sportsmen" and poachers. Rinderpest control, however, allowed buffalo numbers to increase massively, suggesting that disease, not overhunting, may have been the culprit in keeping buffalo scarce. In like manner, buffalo had been rare in Ngorongoro Crater; by the early 1970s, there

FIGURE 1 Widespread vaccination of cattle has eradicated rinderpest in the wild. In this picture Maasai cattle are vaccinated against rinderpest during an annual control campaign in the Ngorongoro highlands of Tanzania in 1991. (Photograph courtesy of Andy Dobson.)

were over 5,000 in Ngorongoro Crater and over 20,000 by the late 1990s.

IMPACT IN THE SERENGETI

The full impact of rinderpest at the ecosystem level has only recently been fully appreciated and quantified. This can be done largely as a consequence of long-term studies in and around Serengeti National Park. Removal of rinderpest by vaccination of cattle can be seen as a form of experiment that has been replicated in other areas of East Africa where roughly similar phenomena are recorded. The differences between Serengeti and other areas most likely reflect differences in the ability to prevent poaching of wild game as their numbers recover and agricultural expansion into areas in and around the national parks. The Mara region of Kenya that forms the northern extension of the Serengeti provides the ultimate "control." The migratory wildebeest use this area as grazing habitat in the summer when the southern part of the ecosystem (in Tanzania) has dried from lack of rain. While both the resident and migratory populations of wildebeest in the Mara initially increased following the eradication of rinderpest, the resident populations have now declined by as much as 95 percent following agricultural expansion around the park and the removal of river water for agriculture. Ultimately, the mantra for all who work in Serengeti is "We need to prevent the same things from happening in Tanzania that have happened around the Kenyan Mara region!"

Removal of rinderpest from the Serengeti created cascading impacts that affected the abundance and diversity of nearly all other species in the ecosystem (Fig. 2). It is again hard to think of another invasive alien for which this has been so extensively documented. This is testimony not only to the importance of long-term ecosystem studies, but also, more ominously, to the power of viral pathogens to produce dramatic impacts on their host's abundance, and thus to significantly alter the ecosystem processes that they are involved in. The removal of rinderpest provided a blunt illustration that the large-bodied herding herbivores of the Serengeti were not limited in abundance by their predators. The six- to eightfold increase in wildebeest and the huge increase in buffalo following rinderpest release led to an increase in both lions and hyenas, the principal predators of these species. This strongly suggests that, in the absence of disease, food availability is more important than predation for regulating herbivore populations in East African savannas, at least among free-living herding species. It is noteworthy that not all grazers increased in abundance. Indeed, some

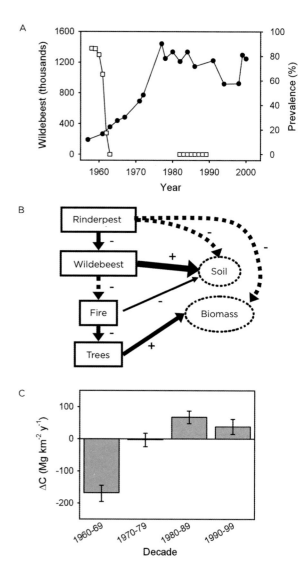

FIGURE 2 (A) Serengeti wildebeest population (•) reported for the period 1958–2000, and rinderpest seroprevalence (□) reported for the periods 1958–1963 and 1982–1989. (B) Pathways linking rinderpest with ecosystem C pools (in soils and biomass) as a result of a trophic cascade (solid and dashed lines signify direct and indirect effects, respectively). (C) Simulated annualized decadal net changes in ecosystem C balance (mean and 95% credible intervals) as a result of rinderpest eradication in the 1960s. (From Holdo et al. 2009, *PLoS Biology*.)

of the smaller grazers, particularly Thompson's gazelles, declined by around 50 percent, most likely because of increased hyena predation. There thus may be an emergent indirect interaction between rinderpest and gazelles: relaxation of rinderpest boosted wildebeest and buffalo, boosting hyena numbers, which then inflicted heavier predation upon the smaller-bodied gazelles. Curiously, the huge herds of plains zebra that comigrate with the wildebeest and exhibit considerable dietary overlap have remained at approximately the same level of abundance over at least the last 50 years. The zebra are completely resistant to rinderpest and associate with the wildebeest, as lions and hyenas significantly prefer wildebeest to zebra as prey. Increasing wildebeest numbers permitted by the decline in rinderpest may have buffered any impact of increased predator numbers on zebra mortality, and this effect might even outweigh competition for forage with the burgeoning wildebeest population.

The changes in herbivore abundance have precipitated a trophic cascade that has significantly altered the fire regime of the Serengeti. In the early 1960s, prior to rinderpest vaccination, around 80 to 90 percent of the Serengeti burned each year, following lightning strikes during the dry season in areas with large amounts of unconsumed grass. As wildebeest numbers increased in the 1970s and 1980s, fire frequency declined to less than 25 percent, and acacia-dominated woodland and bush began to colonize the central and northern regions of the park. This recovery was also partly aided by low numbers of elephants following the ivory-poaching epidemics of the 1970s. Long-term studies of tree recruitment suggest that the Serengeti ecosystem has changed from a source of atmospheric carbon when rinderpest was present to a major carbon sink, now that rinderpest has been removed. Carbon accumulated in the trees is more than matched by carbon build-up in the soil of both the wooded and grassland areas of the park; in fact, as much as 80 percent of the carbon is stored in the soils, and carbon continues to accumulate there at a significant annual rate. It is suspected that similar processes apply in other East African national parks. This ecosystem effect should permit them to be promoted as areas where airlines and businesses could offset their carbon budgets by investing in schemes that control cattle and wildlife diseases, minimize poaching, and provide health and education facilities for people living around the parks.

A SUCCESSFUL EXTINCTION?

There are increasing reports that rinderpest has been eradicated in the wild, which would place it alongside smallpox as one of only two pathogens to be eradicated in the wild; in both cases, this was achieved by a sequential combination of mass, and then targeted, vaccinations. Celebrations are still muted, as it was previously thought that rinderpest was eradicated following the JP15 (Joint Project) African project initiated in 1962. This project focused on vaccinating all cattle for each of three years and then only calves in subsequent years. In Tanzania and Kenya, its success gave rise to the eruption of wildebeest and buffalo described above. The project seemed a

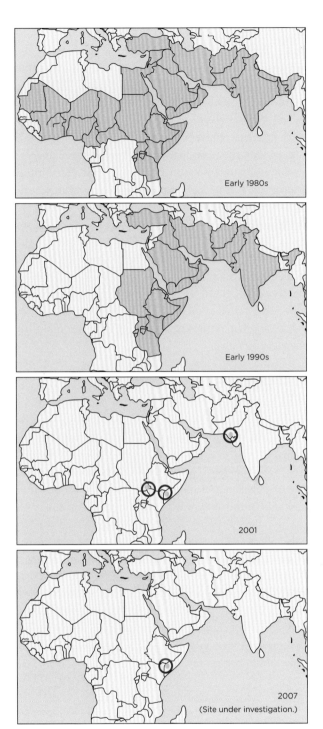

FIGURE 3 Four maps illustrating the declining geographical distribution of rinderpest in the face of mass cattle vaccination. (From Normile, 2008, *Science*.)

great success, and by the mid-1970s, rinderpest seemed to have disappeared from each of the 22 countries involved. Unfortunately, initial signs of success led to apathy and a reduction in vaccination coverage levels, which allowed the pathogen to resurge from troubled areas along the

Mali–Mauritania border in the west and from southern Sudan in the east. Losses from this new pandemic in the early 1980s were thought to have matched those in the original pandemic, with an estimated 100 million cattle deaths and an economic cost to Nigeria alone of over $2 billion. Similar problems prevailed in areas from Turkey to Bangladesh, where persistence in metapopulations of host patches allowed rinderpest to resurge. A second eradication campaign was launched in 1987 (the Pan-African Rinderpest Campaign); it was then fairly swiftly realized that attempts to eradicate rinderpest in Africa were being undermined by a constant trickle of infected cattle from the Indian subcontinent. This realization led the Food and Agriculture Organization to patch together the Global Rinderpest Eradication Campaign in 1993, which set the goal of global rinderpest eradication by 2004, followed by a period of surveillance to ensure that the pathogen was extinct. At present, it seems to have accomplished these goals (Fig. 3).

RINDERPEST AS AN INVASIVE SPECIES

Rinderpest has run the full gamut of possibilities as an invasive species: it has caused massive changes in ecosystems on several continents, has had major effects on human health, and has caused major economic disruption, yet ultimately it provides one of the very few examples of global control and eradication. Its impact on cattle and wild artiodactyls was largely driven by its ability to use domestic livestock as a reservoir and then repeatedly spill over into wildlife populations that were insufficiently large to maintain continuous chains of infection. This case study of rinderpest as an invasive species on one hand illustrates the importance of community interactions in determining the spread and control of invasions and, on the other, exemplifies the potential amplification of impacts of invasive species due to interspecific interactions and ecosystem processes.

SEE ALSO THE FOLLOWING ARTICLES

Epidemiology and Dispersal / Eradication / Flaviviruses / Influenza / Pathogens, Animal / Pathogens, Human

FURTHER READING

Anderson, R. M., and R. M. May. 1982. Directly transmitted infectious diseases: Control by vaccination. *Science* 215: 1053–1060.

Bolker, B., and B. Grenfell. 1995. Space, persistence and dynamics of measles epidemics. *Philosophical Transactions of the Royal Society of London Series B–Biological Sciences* 348: 309–320.

Ferrari, M. J., R. F. Grais, N. Bharti, A. J. K. Conlan, O. N. Bjornstad, L. J. Wolfson, P. J. Guerin, A. Djibo, and B. T. Grenfell. 2008. The dynamics of measles in sub-Saharan Africa. *Nature* 451: 679–684.

Fine, P. E. M. 1993. Herd immunity: History, theory, practice. *Epidemiologic Reviews* 15: 265–302.

Holdo, R. M., A. R. E. Sinclair, A. P. Dobson, K. L. Metzger, B. Bolker, and R. D. Holt. 2009. A disease-mediated trophic cascade in the Serengeti and its implications for ecosystem C. *PLoS Biology* 7.

Mariner, J. C., J. McDermott, J. A. P. Heesterbeek, A. Catley, and P. Roeder. 2005. A model of lineage-1 and lineage-2 rinderpest virus transmission in pastoral areas of East Africa. *Preventive Veterinary Medicine* 69: 245–263.

Normile, D. 2008. Rinderpest: Driven to extinction. *Science* 319: 1606–1609.

Plowright, W. 1982 The effects of rinderpest and rinderpest control on wildlife in Africa. *Symposia of the Zoological Society of London* 50: 1–28.

Runyoro, V. A., H. Hofer, E. B. Chausi, and P. D. Moehlman. 1995. Long-term trends in the herbivore populations of the Ngorongoro Crater, Tanzania (146–168). In A. R. E. Sinclair and P. Arcese, eds. *Serengeti II: Dynamics, Management, and Conservation of an Ecosystem*. Chicago: University of Chicago Press.

Sinclair, A. R. E. 1979. The eruption of the ruminants (82–103). In A. R. E. Sinclair and P. Arcese, eds. *Serengeti II: Dynamics, Management, and Conservation of an Ecosystem*. Chicago: University of Chicago Press.

RISK ASSESSMENT AND PRIORITIZATION

W. M. LONSDALE

CSIRO Entomology, Canberra, Australian Capital Territory, Australia

Risk assessment is a set of analytical techniques for estimating the frequency of undesired outcomes and their consequences. As applied to invasions, risk assessment typically concerns analyzing the likelihood that a particular species, if allowed into a region, will harm the environment, agriculture, or human health. Prioritization for managing invasions generally concerns the risk-based ranking of different invasive species already within a region. Important recent developments in invasion risk assessment have explored the risks posed by particular pathways (e.g., ballast water, the pet trade, shipping containers) for entry of species. The field is in its early stages, with one strand of science focusing on creating increasingly elaborate models for assessing invasion risk, and another taking a more skeptical approach, arguing that attempts to predict invasion risk are unlikely to succeed.

RISK ASSESSMENT CONTEXT

Three drivers have stimulated demand for invasion risk assessment and prioritization. The first is the trend toward risk-weighted resource allocation in government, a trend that has increasingly permeated natural resource policymaking. The second has been the policy need to minimize intersectoral conflict, especially between the economy and the environment, which has necessitated the creation of tools to assess the environmental risks of intentional species introductions such as pets, garden ornamentals, and new crops or pastures. Lastly, as global trade increases, the unintended movement of species via commodities is intensifying, and it is a requirement under the World Trade Organization that trade restrictions invoked for biosecurity purposes must be science-based.

In essence, risk assessment for an introduced species consists of establishing the likelihood that it will enter, establish, and spread in a new region, and predicting the consequences (for the environment, economy, or human health) should it do so. In some cases, these two considerations are treated separately; in others they are conflated.

A useful breakdown of the basic elements of risk assessment is the National Academy of Science's (1983) formulation of risk assessment. This was developed for chemical pollutants affecting human health and consists of a four-step procedure:

1. Hazard identification: What type of harm can the substance cause?

2. Exposure assessment: How long will a target population be exposed to how much of the substance?

3. Dose-response assessment: How does the target population respond to this exposure?

4. Risk characterization: Information from the steps above is combined to estimate the likelihood and magnitude of harm.

For assessing risks to the environment, more complex approaches are needed. The USEPA has developed such a framework. It is conceptually similar to that of the NAS, consisting of a problem formulation phase, an analysis phase (divided into characterization of exposure and characterization of ecological effects), and a risk characterization phase. It differs, however, in that

- The NAS model was developed for a single species—humans—but ecological risk assessment must address risks to a wide variety of species.
- Ecological endpoints are diverse, and there is no single set of ecological values, analogous to human fatality or injury, that can serve as the endpoint for an ecological risk assessment.
- The NAS model considers risks to the individual, while ecological effects may occur at the population, community, or ecosystem levels.

- Biological stressors can reproduce and multiply, and potentially evolve, in stark contrast to chemical stressors.
- Human health risk assessments commonly assume that there are no external feedback loops from the damaged to undamaged portion of the target population. Ecological risk assessment, however, must consider such interactions, especially where species are important determinants of community structure and function.

Clearly, the EPA formulation represents a more powerful tool than the NAS framework for considering environmental risks such as introduced species. However, much of the risk assessment carried out on invasions falls short even of the NAS model. For many intentional introductions, risk approaches generally consist of hazard identification only—step 1 in the NAS formulation above—asking essentially, "Will this species become a pest?" The information and resources required to conduct a risk assessment according to the NAS scheme above are generally not available. Furthermore, high-impact invasions are rare and are thus difficult to predict.

APPLICATIONS OF RISK ASSESSMENT TO INVASIONS

Overview

Invasion risk assessments may be broken down into the following categories:

- traditional approaches such as target lists of potential harmful species, often associated with a particular industry sector (e.g., pests and diseases of the wheat industry)
- species-focused approaches, typically for intentional introductions
- pathway-focused approaches, exploring routes whereby an array of harmful species, known or unknown, may enter

Risk assessments for invasions tend to be desk studies rather than experimental studies, except in the case of biocontrol agents, where specificity to the target pest must be tested. This is not only because of the resources that would be required, but also because potential interactions between factors, especially biotic ones, are difficult to simulate experimentally.

Target Lists

This is the most venerable approach to risk assessment for invasive species, typically consisting of the expert compilation of lists of pest species that threaten a particular agricultural sector, based on knowledge of the global panoply of pests affecting the crop or livestock species. It can be problematic in that it does not assess the risk of species previously unknown as pests, it can be hard to assess the relative risks of different species in the encyclopedic lists so produced, and it does not generally take account of the likelihood of entry. Nevertheless, if combined with a pathway analysis and modern approaches to comparing species assemblages such as self-organizing maps, this approach could be a very powerful tool for prioritizing invasion risk, at least from known invaders.

Species-Based Risk Assessments for Intentional Introductions

There has been a great expansion in the availability of risk assessment tools for species introductions over the past decade, especially for introduced plants. A well-known example is the Australian Weed Risk Assessment (WRA) System, adopted by the Australian Government in 1997 and tested in various other regions around the world, including Japan, Florida, the Czech Republic, and the Pacific Islands. The system is spreadsheet based, asking a series of questions about the attributes of the species, which contribute to a final score that indicates whether the species risks becoming a weed. The system has been found to reject an average of 90 percent of known invasive species and to accept an average of 70 percent of known noninvasive species (but see Evaluating Risk Assessment Systems). In Australia, the tool has been applied especially to the consideration of pasture and ornamental plant species, but there are moves to apply it to biofuel crop risk assessment. It is, strictly speaking, a hazard identification tool rather than a risk assessment tool, as it does not attempt to predict the geographic extent or intensity of the impact.

For vertebrates there exist tools for assessing the risks of importation of mammals, reptiles and amphibians, and freshwater fish. For invertebrates and pathogens, the main intentional introductions are as agents for biocontrol, and here the risk assessment focus is on experimental tests of agent specificity to the target pest.

Pathway-based Risk Assessments for Unintentional Introductions

Where the array of possible introduced species is large, policy focus tends to turn to assessing and managing the risk of different pathways of entry. Trade is an important vector for species movement internationally—for example, there is a close correlation between international trade volume and the establishment rates of introduced insects, plant

pathogens, and molluscs in the United States over the past century. A key focus for the science of pathway-based risk assessment has been ballast water. Ballast water is collected from the sea by ships to provide stability after they have off-loaded their cargo and then is transported across the oceans and rereleased in the process of loading with new cargo. It is estimated that 10,000 species (including fish, molluscs, plankton, and algae) are mobile in ballast water at any one time. Obviously, this constitutes a major potential vector for species translocation, and there has been a substantial international effort not just to assess the risk but also to manage it actively, culminating in the *International Convention for the Control and Management of Ships' Ballast Water and Sediments*, adopted under the International Maritime Organization in 2004. Other pathways for unintentional introductions include containers (especially for timber and stored product pests), agricultural pests, and diseases in food and cut-flower movements. While the volume of movement is likely to be far greater by sea, the rapidity of movement by air means that the latter pathway may be important in delivering living organisms to a new destination.

Intentional pathways of invasion, however, such as the pet and aquarium trades and agricultural crop introductions, are probably much more important than unintentional pathways. This is because human intervention is likely to give a significant helping hand to intentionally introduced species in the vulnerable early stages of their establishment in the new range.

COMPONENTS OF INVASION RISK

Overview

Much invasion risk assessment focuses on the attributes of the introduced species itself, but the success or failure of a species is the outcome of much more than this. It depends on the interaction between its properties, the properties of the invaded environment, and those of chance events, particularly in the early stages of establishment when the species' numbers are likely to be low.

Species Attributes

Plant scientists, in particular, have focused heavily on correlative studies to deduce predictors of species invasion potential. Various studies have analyzed suites of introduced plant species in particular regions, seeking biological generalizations about invasion success. However, these generalizations tend to apply only to the group of species from which they derive. For example, different studies have variously found that large-seeded and small-seeded species are more likely to establish, or that there is no effect of seed size at all. Other studies have found no capacity to explain invasion success using biological attributes. Some studies are also confounded by methodological flaws such as using individual species as independent data points (i.e., not controlling for taxonomic relationships between species) or ignoring the question of failure rates (how many species with the supposed invasive trait have actually vanished without trace?). Furthermore, definitions of "invasive" may vary substantially between studies (e.g., "invasive" might, for example, mean "established" in one study, but "having a negative impact" in another), thus limiting their comparability.

This is problematic, because such studies would help to directly inform weed risk assessments if they could elicit broad and robust generalizations. As it is, the key species attribute we can usually invoke safely, for plants and animals, is that of pre-adaptation to the new climate. This is a necessary, but not sufficient, condition for invasion success, however.

Environmental Factors

Climate is just one aspect of the invaded environment that must match the species' preferences. Other aspects include the biotic environment, such as the presence of facilitating organisms (e.g., pollinators), or the absence of predators and pathogens, and the occurrence of the appropriate disturbance regime (e.g., fire, flooding cycles, grazing, etc).

Collectively, the susceptibility of the environment to invasion is known as its invasibility. Whether invasibility varies spatially, with ecosystem, or over time is not known. Recent studies have explored the concept of invasional meltdown, in which interspecific facilitation by preceding invaders leads to an accelerating increase in the number of introduced species and their impact. Similarly, climate change may disrupt ecosystem structure and function and so increase invasibility. While it is not yet clear how widespread these phenomena will be, the idea that environments may differ in their susceptibility to different invaders, and over time, is plausible and emphasizes that invasion risk assessments need to take account of the invaded environment as well as of species' attributes.

One approach to risk assessment that integrates species attributes with aspects of the environment is correlation of a species' abundance in one region with its abundance in a climatically similar region elsewhere. It is common to find that abundance in one region predicts abundance in the other for a majority of test species.

Effects of Chance

Founder populations are extremely vulnerable to the effects of chance. Human intervention to protect such populations from harm (predation, disease, competition) or to increase propagule pressure is likely to have a profound impact on survival probability, and thus on the potential to establish and invade. The importance of human intervention in increasing invasion success has been demonstrated for cases as varied as birds in New Zealand and ornamental plants in Florida. Experimental work with forest understory plants has shown that propagule pressure overrides environmental factors such as disturbance regime and ecosystem diversity in affecting invasion success. One little-understood phenomenon that is hard to attribute specifically to chance, environment, or species make-up is the occurrence in many species of lag phases of decades or even centuries after introduction, before they become invasive, so that we are never sure if a diagnosed noninvader will actually remain so in perpetuity.

EVALUATING RISK ASSESSMENT SYSTEMS

Successful invasions are the outcome of the interaction between species attributes, the invaded environment, and chance. Small wonder that many ecologists doubt that it will ever be possible to predict the outcome of an introduction reliably. Yet many risk assessment systems do claim impressive levels of accuracy. One reason is that many appraisals of risk assessment systems confuse explanation with prediction: describing what has happened is not the same as prediction, yet much risk assessment modeling involves analysis of one set of past invasions to describe another set of past invasions, in order to develop rules for prediction. Another reason is that there may be methodological errors in the evaluation of risk assessment screening systems, errors well known to medical epidemiologists assessing the validity of mass health screening systems. It is rare to see truly rigorous methods such as those used in epidemiology applied to the evaluation of invasion risk assessment systems. In particular, a critique such as the following would be usefully applied. In testing the validity of a screening system,

1. The initial sample is often not appropriately taken. At the moment, a typical method for testing pest risk assessment systems is to carry out the evaluation of the system using different subsamples of the same species population, which is often composed of unnatural proportions of extreme pests and nonpests. Even if found to be accurate, it is unsafe to generalize results from such an evaluation to the wider population. The results are likely to be valid only for the original sample population.

2. The tests are run on species that are already widely known to be pests (contravening the requirement for blind evaluation). This becomes particularly

critical where the screening procedure involves value judgments (e.g., weed or not?).

3. Species may be dropped from the evaluation because they do not fit the test or because insufficient information is available to complete the test (the problem of "dropouts," in medical terms).

4. Characters are typically measured or estimated on introduced, not native, populations of the organism. Invasive species often are not as vigorous in the native range, where they are subject to depredations of natural enemies, as in the introduced range, where these enemies are often missing. In operation, a pest risk assessment system would generally be required to evaluate "pestiness" based on phenotype in the native, not the introduced, range. Questions about the vigor of an invader would often have a very different answer if asked prior to introduction.

5. Because the proportion of organisms entering a region that will actually become pests is generally low, even in a situation where the accuracy is rather high, there will be a high rate of false positives, and the predictive value will be low (Fig. 1)—the base-rate effect. This phenomenon alone could account for much of the discrepancy between skeptical ecologists, on one hand, and claimants of accurate screening systems, on the other: all are simultaneously correct.

It is very possible that measured accuracy levels would not be as high after rigorous evaluation using the same standards as those applied to medical screening tests.

APPLYING INVASION RISK ASSESSMENT IN A POLICY CONTEXT

The application of risk assessment tools to invasions is potentially problematic for policymakers. Given the role of trade in global species movement, the World Trade Organization (WTO) and the various agreements that it administers are critically important in considering how to manage invasives. The WTO gives significant latitude for countries to take strong biosecurity measures to prevent introductions, and many countries have such measures in place. Nevertheless, the emphasis on a mechanistic risk assessment approach under the WTO is hard to reconcile with the uncertain nature of invasions. The sheer numbers of species involved, the emergence of hitherto unknown harmful species, and the inherently low predictability of the phenomenon have led a number of authors to call for a more adaptive, cooperative,

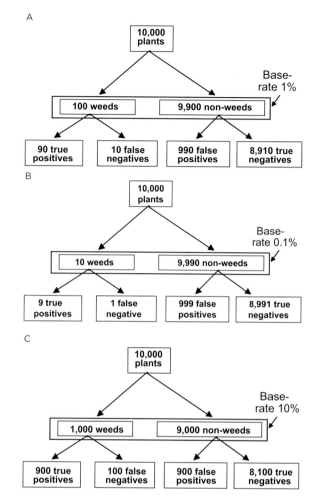

FIGURE 1 Counterintuitive effects of the base rate or prevalence of pestiness in a population of species. (A) Imagine a suite of 10,000 plant species introduced to a region, of which 100 species of unknown identity will become weeds, giving a base rate for weediness of 1 percent (i.e., 100/10,000). Imagine applying a weed risk assessment system to the 10,000 species in order to seek out the weeds. Assume this system can identify 90 percent of true weeds and 90 percent of non-weeds. The proportion of rejected species that would be weeds is about 8 percent; in other words, we would be wrong about 92 percent of the rejected species (called "false positives"). (B) As for A, but with a weed base rate of 0.1%; here, the proportion of rejected species that would be weeds is only 9/(9 + 999) = 0.009 percent. (C) As for A, but with a weed base rate much higher, at 10 percent; here, the weedy proportion of rejected species is also much higher, at 900/(900 + 900) = 50 percent.

precautionary approach to regulating the global movement of species.

Essentially, a precautionary approach, in circumstances where the base rate is low, might involve policy acceptance that, to protect the environment from invasive species, there will be a high rate of false positives (Fig. 1). Weighing up the risks and benefits of an action in this way is the domain of decision theory. For example, the decision on whether to heed a recommendation to exclude a new crop plant depends on the following factors:

- the damage that would be caused if a useful plant were excluded
- the damage that would be caused if a weed were allowed in
- the background probability that a plant would become a weed (the base rate)
- the accuracy of the system that predicts whether the plant will be a weed

The decision-theory approach to building invasion risk assessments into policy decisions holds promise but will demand more knowledge than we currently have on invasion epidemiology, particularly information on the benefits of useful species (not normally considered or questioned in invasion risk assessment) and information on how the base rate varies with type and location of introduction.

SEE ALSO THE FOLLOWING ARTICLES

Agreements, International / Invasional Meltdown / Invasibility, of Communities and Ecosystems / Invasion Economics / Lag Times / Propagule Pressure

REFERENCES

Davies, J. C. 1996. *Comparing Environmental Risks.* Washington, DC: Resources for the Future.
Lipton, J., H. Galbraith, J. Burger, and D. Wartenberg. 1996. A paradigm for ecological risk assessment. *Environmental Management* 17: 1–5.
Lonsdale, W. M., and C. S. Smith. 2001. Evaluating pest screening systems: Insights from epidemiology and ecology (52–60). In R. H. Groves, F. D. Panetta, and J. G. Virtue, eds. *Weed Risk Assessment.* Melbourne: CSIRO Publishing.
Pheloung, P. C., P. A. Williams, and S. R. Halloy. 1999. A weed risk assessment model for use as a biosecurity tool evaluating plant introductions. *Journal of Environmental Management* 57: 239–251.
Simberloff, D. 2005. The politics of assessing risk for biological invasions: The USA as a case study. *Trends in Ecology and Evolution* 20: 216–222.
Smith, C. S., W. M. Lonsdale, and J. Fortune. 1999. When to ignore advice: Invasion predictions and decision theory. *Biological Invasions* 1: 89–96.
Williamson, M. 1999. Invasions. *Ecography* 22: 5–12.

RIVERS

EMILI GARCÍA-BERTHOU

University of Girona, Spain

PETER B. MOYLE

University of California, Davis

Rivers are centers of aquatic biodiversity but have been enormously affected by invasive species. Some of the most widely introduced species in rivers are fishes such as salmonids in headwater streams or carps in large rivers, but many invertebrates (e.g., zebra mussels, various crayfish, and amphipods) and plants (e.g., water hyacinths, Eurasian milfoils) have also become widespread. The documented ecological impacts of invasive species in rivers are enormous and occur from the individual to the ecosystem level. Although environmental degradation facilitates invasion, even protected ecosystems such as headwater streams are not free from invasions (e.g., salmonids). The "island nature" of river basins for many freshwater organisms underlines the need for prevention and early eradication of new introductions. Preserving or restoring the natural flow regime usually favors native species and is essential to reduce the abundance and impacts of invasive species.

RIVERS AS ECOSYSTEMS

Rivers and lakes occupy only about 0.8 percent of the Earth's surface and hold only 0.3 percent of its freshwater (most of it is in glaciers, snow, or underground), which in turn is only 2.5 percent of the total water availability (the rest is seawater). The importance of rivers is, however, huge in terms of biodiversity and human services. Freshwater ecosystems have about 2.4 percent of the world's animal and plant species and have almost as many described fish species as the sea (about 12,000 versus 16,000), even though oceans occupy 70 percent of the Earth's surface. Many freshwater organisms, mostly those that cannot tolerate seawater or those without dispersing stages such as many insects with aquatic larvae and flying adults, cannot move naturally among river basins. This makes watersheds strong natural barriers and explains the high frequency of endemic species in river basins.

Rivers provide major services to human society, including potable water, irrigation for agriculture, fisheries, transport, and hydroelectric energy. For this reason, freshwater ecosystems are among the most altered and threatened in the world and have suffered many extinctions and many species introductions. Most large river basins are heavily regulated by either dams or water abstraction, have been polluted by fertilizers and chemical wastes, and have been interconnected to facilitate transport or water transfer. Most of these human activities facilitate invasions of freshwater organisms, which are introduced either intentionally or as byproducts of other human activities (e.g., shipping). As a result, riverine ecosystems are increasingly being invaded by alien species to the point where alien flora and fauna can dominate these systems. Alien trout (Salmonidae) and invertebrates, such as New Zealand mud snails (*Potamopyrgus antipodarum*), may be very abundant in headwaters. In large rivers, abundant alien species worldwide include common and Asian carps (Cyprinidae), Asian clams (*Corbicula* spp.; Fig. 1), various crayfish species (Astacidae, Cambaridae), and dozens of aquatic and riparian plants.

INVASIVE SPECIES IN RIVERS

Fishes

Among the best-documented invasive taxa in rivers are fishes, which are usually introduced intentionally. Most are large species that are highly desirable to humans and are capable of dispersing over long distances quickly. As Europeans invaded many regions of the world, they also spread their favorite species. This explains why the common carp *Cyprinus carpio* is the second most introduced fish species throughout the planet (after *Oreochromis mossambicus*) and why the European brown trout *Salmo trutta* has been introduced to all the continents (except Antarctica, but including sub-Antarctic islands). The ecological impact of common carp ranges from well-demonstrated macrophyte declines and water quality impairments to likely effects on some native fish and waterfowl. The brown trout has also been shown to have effects from the individual to the ecosystem level and displaces native species such as other salmonids in North America, galaxiids in Australasia, and amphibians in many regions. Increasingly, fishes that have been widely introduced into rivers are Asian carps, such as the grass carp (*Ctenopharyngodon idella*), silver carp (*Hypophthalmichthys molitrix*), and bighead carp (*Hypophthalmichthys nobilis*), because of their desirability for food and weed control. Their impacts are not well understood, although they can become extremely abundant in rivers such as those of the Mississippi basin.

Other Vertebrates

Other well-known invasive vertebrate species of rivers and riparian zones include amphibians such as the American bullfrog (*Lithobates catesbeianus*); turtles,

FIGURE 1 The Asian bivalve *Corbicula fluminea* in a drawn irrigation ditch. (Photograph courtesy of Josep M. Dacosta.)

FIGURE 2 The red swamp crayfish *Procambarus clarkii*. (Photograph courtesy of Dani Boix.)

especially the red-eared slider (*Trachemys scripta elegans*); mammals such as the American mink (*Neovison vison*), muskrat (*Ondatra zibethicus*), and nutria or coypu (*Myocastor coypus*); and waterfowl such as Canada geese (*Branta canadensis*) and mute swans (*Cygnus olor*).

Invertebrates

The best-known invertebrate river invaders are molluscs and crustaceans. Among the molluscs with strongest effects are the zebra mussel (*Dreissena polymorpha*) and the Asian clam (*Corbicula fluminea*), both of which threaten native molluscs worldwide, largely through direct displacement. They have the potential to alter, at least locally, riverine food webs through their actions as filter feeders. Among crustaceans, crayfish (e.g., *Procambarus clarkii*; Fig. 2) are among the best documented as invaders, but other species ranging from amphipods to crabs have had impacts. One such problem species is the Chinese mitten crab (*Eriocheir sinensis*), which lives as an adult in estuaries but migrates upstream into rivers in huge numbers to breed, clogging water intake structures and causing other damage.

Parasites

Parasites are little appreciated as invaders in rivers, yet their effects can be dramatic. The introduction of crayfish plague (*Aphanomyces astaci*) into Europe from North America, by North American crayfish, dramatically reduced Europe's native crayfish populations. Likewise, the introduction of whirling disease by brown trout from Europe has reduced native trout populations in some North American rivers, and the introduction of tapeworms from Asian grass carp has caused problems with endangered riverine cyprinids in the western United States.

FIGURE 3 The water fern *Azolla filiculoides*, such as these in the Ter River in Spain, is native to parts of North America but invasive in many other regions. Like other floating plants, it has ecosystem effects (through alteration of the light regime and biogeochemical cycles) and is more typical of wetlands and slow-flowing river stretches. (Photograph courtesy of Josep M. Dacosta.)

Plants

Aquatic and riparian plants can cause major changes to river ecosystems by reducing flow through stream channels, shading large areas, and dominating riparian ecosystems. They often create habitats favored by other nonnative species, especially snails and fishes, thus reducing local species diversity. Among the best-known global invasive plant species are water hyacinth *Eichhornia crassipes*, Eurasian water-milfoil *Myriophyllum spicatum*, hydrilla *Hydrilla verticillata*, the ferns *Salvinia molesta* and *Azolla filiculoides* (Fig. 3), and the giant reed *Arundo donax*. Ecosystem changes caused by alien plants include reduction of stream flows by salt cedars *Tamarix* spp. in North America (Fig. 4), nitrogen soil enrichment by black locust *Robinia pseudoacacia*, increased erosion rates by *Impatiens glandulifera*, and alteration of the light regime and biogeochemical cycles by floating plants such as water hyacinth or *Salvinia molesta*. These floating plants can also impede navigation in rivers.

INVASIBILITY OF RIVERS

The linear nature, large size, and diverse habitats of rivers make them extremely vulnerable to introductions of alien species and to their spread. The lower reaches often end in estuaries or large cities, where shipping and large

FIGURE 4 Tamarisk in bloom along a California stream. The stream has been denuded of riparian vegetation by the grazing of livestock, favoring the inedible tamarisk bushes. (Photograph courtesy of Peter Moyle.)

numbers of people increase the frequency of introductions. The rivers then become convenient corridors for dispersal. Species introduced into headwater systems likewise can spread downstream rapidly.

The lower reaches of rivers and streams in developed countries of the northern hemisphere harbor more invasive species than do headwater streams or tropical rivers. This probably reflects more frequent introductions, more intense sampling, and more environmental degradation of the former rather than a higher intrinsic invasibility. Evidence suggests that most river systems can be invaded. For example, in South America, tropical rivers with diverse fish faunas have been invaded by tilapia species and African catfish (*Clarias gariepinus*). Near-pristine headwater streams throughout the world have likewise been successfully invaded by salmonid fishes. In fact, there is growing evidence to suggest that, in temperate areas, some of the most invaded rivers are those with rich native faunas.

Invasion success is also increased by habitat degradation. For example, the Colorado River in North America is one of the most altered rivers in the world, with its waters being diverted, stored in reservoirs, or highly polluted. As a result, most of the native aquatic species are gone, and the river is dominated by alien fishes, invertebrates, and plants, including dense growths of salt cedars in the riparian zone. Many of this river's freshwater invaders, such as common carp, western mosquitofish (*Gambusia affinis*), tilapia (*Oreochromis* spp.), zebra mussels, bullfrogs, and water hyacinth, are also very resistant to pollution. Not surprisingly, species such as these increasingly have global distributions. As rivers worldwide are degraded, the shift toward alien species becomes increasingly common, resulting in the homogenization of riverine faunas worldwide, especially in temperate regions.

Many freshwater invaders are often first introduced into reservoirs behind dams on rivers; they typically are species that thrive in lakes or in backwater habitats of rivers. Thus, widespread aliens such as the common carp, mosquitofishes (*G. affinis* and *G. holbrooki*), largemouth bass (*Micropterus salmoides*), tilapia, or water hyacinth can become well established in reservoirs, which then serve as a source for continual downstream colonization, even if the species are not particularly well adapted to riverine habitats. Rivers with highly regulated flows, however, can favor these introduced species, which can then displace native species. The invasive success of introduced fish such as the rainbow trout (*Oncorhynchus mykiss*) has been shown to depend on the matching of the hydrological regime with critical reproductive events such as fry emergence.

MANAGEMENT OF INVASIVE SPECIES IN RIVERS

The management of invasive species in rivers is similar to that in other systems, with prevention of new introductions and spread of harmful species having the highest priority. Specific management measures emerge from the properties of freshwater ecosystems and their invaders. Avoiding or reversing environmental degradation is essential because many invaders thrive in altered habitats. Preserving or restoring the natural flow regime will often benefit native species adapted to the regime, especially for reproduction, and will discourage alien species, although it will rarely eliminate them entirely. In a regulated river, for example, prescribed floods can be useful to reduce the populations of introduced species such as largemouth bass, mosquitofish, and many invasive plants, especially if used in conjunction with more direct control measures.

In some instances, dams may be useful barriers to prevent upstream spread of invasive species, although more typically they are the source of invasions downstream. For some dams, screens have been developed on outlets to prevent alien fish from escaping downstream. Eradication of unwanted fishes (usually common carp) and plants using toxicants and other means has been tried in a number of rivers, but it is rarely successful because some individuals usually manage to escape in the complex habitats present in rivers.

SEE ALSO THE FOLLOWING ARTICLES

Carp, Common / Crayfish / Fishes / Freshwater Plants and Seaweeds / Hydrology / Invasibility, of Communities and Ecosystems / Mammals, Aquatic / Water Hyacinth / Zebra Mussel

FURTHER READING

Allan, J. D. 1995. *Stream Ecology: Structure and Function of Running Waters.* Dordrecht: Springer.

Claudi, R., and J. H. Leach, eds. 1999. *Nonindigenous Freshwater Organisms: Vectors, Biology, and Impacts.* Boca Raton, FL: Lewis Publishers.

Elton, C. S. 1958. *The Ecology of Invasions by Animals and Plants.* London: Methuen (reprinted 2000, Chicago: University of Chicago Press).

Gherardi, F., ed. 2007. *Biological Invaders in Inland Waters: Profiles, Distribution, and Threats.* Dordrecht: Springer.

Lockwood, J. L., M. F. Hoopes, and M. P. Marchetti. 2007. *Invasion Ecology.* Malden, MA: Blackwell.

Moyle, P. B., and T. Light. 1996. Biological invasions of fresh water: Empirical rules and assembly theory. *Biological Conservation* 78: 149–161.

Tickner, D. P., P. G. Angold, A. M. Gurnell, and J. O. Mountford. 2001. Riparian plant invasions: Hydrogeomorphological control and ecological impacts. *Progress in Physical Geography* 25: 22–52.

RODENTS (OTHER)

DANIEL SIMBERLOFF

University of Tennessee, Knoxville

Although rats have received the most attention, other rodent species have significantly affected native ecosystems, communities, and species. By far the greatest impacts have been on islands. Those rodents that change entire ecosystems, generally by affecting the physical structure of the habitat (ecosystem engineers), have caused the most far-reaching changes. These include three aquatic species widely introduced as fur-bearers: nutria (*Myocaster coypus*), North American beavers (*Castor canadensis*), and muskrats (*Ondatra zibethicus*). Many other rodents have been introduced and have persisted; some have spread. Often, damage to human enterprises (e.g., agriculture) is described, but reports of ecological damage are not as common. The house mouse (*Mus musculus*) and the North American gray squirrel (*Sciurus carolinensis*) are the most important ecologically.

AQUATIC FUR BEARERS

North American Beavers

The North American beaver (Fig. 1), by virtue of its tree-cutting and damming activities, is the paradigmatic ecosystem engineer, transforming forests into wetlands, changing stream flows, and modifying nutrient dynamics and decomposition. It has been successfully introduced to Tierra del Fuego, Kodiak Island, the Queen Charlotte Islands, Anticosti Island, Finland, and Russia.

FIGURE 1 North American beaver (*Castor canadensis*). (Photograph courtesy of Guillermo Martínez Pastur, CADIC-CONICET.)

In Tierra del Fuego, approximately 25 pairs were introduced in 1946 to the Argentinean side of Isla Grande to establish a fur industry. They now number over 100,000 and occupy not only Isla Grande but several Chilean islands as well; they have recently spread to the Chilean mainland. Their main impact is conversion of substantial areas of *Nothofagus* forest to grassy meadows that persist long after beavers abandon the sites. These sites are much more heavily invaded than are adjacent forests by introduced plant species, perhaps because of higher nitrogen levels in sediments. The feasibility of eradicating the beaver from southern South America is currently under study. In the Queen Charlotte Islands, beavers interact with introduced deer to prevent regeneration of native crabapple and willow species. As in South America, some beaver dams greatly affect stream flow in the Queen Charlottes. Impacts of the beaver have not been studied on Kodiak Island or Anticosti Island.

In Europe, increasing and spreading populations of North American beaver are, along with habitat destruction, a major impediment to a broad-scale effort to reintroduce the European beaver (*C. fiber*), which had been hunted almost to extinction. The North American beaver engages in more dam-building activity, matures earlier, and has larger litters than the European beaver and tends to replace it where they cooccur. Efforts to reintroduce the European beaver to areas from which it had been extirpated have generated some concern about its impacts on forests and hydrology, which resemble those of the North American beaver. Although the impacts of the latter would likely be greater, the growing acceptance of the reintroduction of the European beaver has lessened concern about ecological impacts of its introduced congener.

Muskrats

Muskrats (Fig. 2), native to North America, can greatly affect habitat by virtue of eating aquatic plants and by burrowing into shorelines in habitats in which they cannot construct their "houses" of mud and plant material. They have been successfully introduced to central and northern Europe, Russia, parts of China and Mongolia, Honshu Island (Japan), Tierra del Fuego, the Queen Charlotte Islands, Vancouver Island, and Anticosti Island. The population initiated by introduction of five individuals near Prague in 1905 spread throughout Europe, facilitated by subsequent deliberate releases and escapes from fur farms. They were eradicated from Great Britain after 12 years. They are well established in northern Asia and, in Japan, in the vicinity of Tokyo; although they are not as numerous as beavers in Tierra del Fuego, they have spread from Isla Grande to several other islands and into southern mainland Argentina and Chile. In the Queen Charlotte Islands, the muskrat population exploded in the 1930s and mysteriously crashed in 1944–1945 (although they still persist).

In general, introduced muskrats have not been shown to cause as much ecological damage as beavers, which occupy similar aquatic habitats, although in Russia they have locally displaced the native desman (*Desmana moschata*). None of their activities are as consequential as beaver dam-building. In the Queen Charlotte Islands and Tierra del Fuego, they are not as numerous as beavers. In Europe, their burrowing (which undermines banks) and feeding on plants are sometimes problematic. In South America, muskrats and beavers are now both invading Patagonia, where another large aquatic rodent—the nutria, which is invasive elsewhere—is native. So far, there is no research on their interaction.

FIGURE 2 Muskrat (*Ondatra zibethicus*). (Photograph courtesy of R. Town, U.S. Fish and Wildlife Service.)

Nutria

Nutria (Fig. 3), native to southern and central South America, are often seen as an ecological scourge because they damage wetlands, sometimes turning them into open water. They have been introduced successfully (usually as escapees or releases from fur farms) to many parts of Europe, Asia, and North America, as well as to Japan and East Africa. In many areas, they have achieved staggering numbers, although typically cold winters have caused abrupt temporary population declines. Their burrowing activities inflict the most ecological damage, and their impact on aquatic vegetation is exacerbated by the fact that they prefer lower, more succulent parts of plants and also dig out and eat roots. Nutria are also often substantial crop pests.

In North America, nutria are established in many Gulf Coast and eastern states, as well as in Oregon and perhaps other western states, and on Vancouver Island. In addition to escaping or being released from fur farms, nutria have spread in some areas by being sold as biological control agents for aquatic weeds (although their feeding is too generalized for them to serve well in this capacity). Ecological impact is greatest along the Gulf Coast and in the vicinity of the Chesapeake Bay. In Louisiana and Maryland, thousands of hectares of wetlands have been lost to the sea. In Great Britain, crop loss, and especially damage to wetlands, led to a successful eradication campaign.

FIGURE 3 Nutria (*Myocaster coypus*) killed by sharpshooters over two days in late 2003 as part of a population control program undertaken by the U.S. Geological Survey at the Rockefeller Wildlife Management Area, Louisiana. (Photograph courtesy of Sergio Merino, U.S. Geological Survey.)

MICE

The house mouse (*Mus musculus*) originated in Eurasia but has been introduced globally, including to many subantarctic islands (e.g., Marion Island, South Georgia, Macquarie). This species is legendary because of the staggering damage it causes to stored food; it also transmits human disease and is a household nuisance. Its ecological impact has received much less attention, possibly because the house mouse largely cooccurs with rats, which suppress the populations of the former by preying on them. However, on two islands lacking rats—Gough and Marion—intensive research suggests that their ecosystem impact equals that of rats. On Gough they prey on seabirds and almost certainly greatly affect several invertebrate populations, while on Marion the sheer biomass of invertebrates they consume affects nutrient cycles and has led to important declines in populations of several seabirds. On both of these islands, the feasibility of eradicating house mice is being studied. On several small islands from which rats have been eradicated, mouse populations have exploded.

A number of other mice have been introduced to various islands but are not known to damage ecosystems or harm native populations. The bank vole (*Clethrionomys glareolus*), native to continental Europe and Great Britain, was introduced to Ireland and is widespread in the south of that island. The Eurasian and North American redbacked mouse (*C. rutilus*) was introduced to the Commander Islands in the Russian Far East and is widespread at least on Bering Island. The common vole (*Microtus arvalis*) was probably introduced to the Orkney Islands by Neolithic settlers around 3500 BC. The Cairo spiny mouse (*Acomys cahirinus*) is a successful introduction to Cyprus and Crete.

SQUIRRELS

The North American gray squirrel (*Sciurus carolinensis*; Fig. 4) is a problematic introduction on several islands. It is now causing great concern because a project to eradicate a localized population in Italy was stymied by objections from animal rights advocates, and the species is spreading and can reach continental Europe. Introduced to Great Britain in the late nineteenth century, it is a chief cause of the rapid decline of the native European red squirrel (*S. vulgaris*) through competition for food. The plight of the red squirrel is locally exacerbated because it is adapted to conifer forests, and in some areas of Europe, these are being replaced by deciduous forests or agriculture. Gray squirrels also carry and transmit parapox virus to red squirrels, which inevitably kills the latter. Gray squirrels

FIGURE 4 North American gray squirrel (*Sciurus carolinensis*). (Photograph courtesy of Jonathan Slaght.)

in Great Britain also strip bark from trees, which can kill trees outright and subject them to lethal infection. Gray squirrels were introduced in 1913 to Ireland and present the same threat there to the red squirrel (probably an ancient introduction to Ireland, where it went extinct by the early eighteenth century until it was reintroduced in the mid-nineteenth century) as in Great Britain. The gray squirrel escaped from a game farm on Vancouver Island in 1966, and the population spread to cover the entire southern part of the island, where they are displacing native North American red squirrels (*Tamiasciurus hudsonicus*) and threatening the woodlands of garry oak (*Quercus garryana*) by stripping bark and notching acorns. The gray squirrel has also been successfully introduced to several cities on the West Coast of the United States and to the Cape Peninsula and Western Cape of South Africa. In South Africa, it has aided invasion by the European stone pine (*Pinus pinea*) by dispersing its seeds. Established gray squirrel populations in Australia died out for unknown reasons.

The fox squirrel (*S. niger*) of eastern North America was introduced further north in New York State and also in California, Oregon, Washington, and Idaho. Although they are primarily of concern in the west as a pest of agriculture and because they gnaw telephone cables, they are also implicated in the decline of the native western gray squirrel (*S. griseus*) and are also believed to compete with native Douglas squirrels (*Tamiasciurus douglasii*) and Abert's squirrels (*S. aberti*). The European red squirrel (*S. vulgaris*) has been successfully introduced to the southern Russian Federation and adjacent republics where it is not native.

Introduced populations of the North American red squirrel (*T. hudsonicus*) have established on Newfoundland, the Queen Charlotte Islands, Kodiak Island, Afognak Island, Sidney Island, and several islands in the Alexander Archipelago. In the Queen Charlottes, they prey on songbird nests and are in turn prey for martens (*Martes americana*), whose populations have thus increased. There is no other substantial research on the ecological impact of this squirrel in its introduced range.

The Barbary ground squirrel (*Atlantoxerus getulus*) was introduced around 1966 to Fuerteventura in the Canary Islands; it has spread widely and become especially numerous in mountain habitats. It heavily browses some native plants of special concern and now constitutes the bulk of the diet of an endemic buzzard (*Buteo buteo insularum*) on the island. The Central American red-bellied squirrel (*S. aureogaster*) was introduced to Elliott Key, Florida, in 1938 and thrives there, possibly threatening native plant and animal species; an eradication campaign is under way. Pallas's squirrel (*Callosciurus erythraeus*) from Asia has been introduced to Argentina, France, and Japan, where it is common. In Argentina and France, it is recorded as damaging trees as well as eating native bird eggs, but the population consequences of these activities are unknown. In all three locales, the spread of Pallas's squirrel has been associated with anthropogenic environments. Finlayson's squirrel (*C. finlaysoni*), also Asian, flourishes in an urban Italian park and strips bark from native trees. The Indian palm squirrel (*Funambulus pennanti*) was introduced to Sydney and Perth. It died out in Sydney but persisted in the vicinity of the Perth Zoological Gardens for about 75 years before spreading to suburban areas, where it is apparently restricted to areas having many exotic trees. Siberian chipmunks (*Tamias sibiricus*) have been successfully introduced to France, Germany, Austria, the Netherlands, Switzerland, and Japan. No reports of ecological impacts have been published. Finally, the arctic souslik (*Spermophilus parryii*), a widespread arctic species, was successfully introduced to several islands in the Aleutians; its impact there is unknown.

OTHER RODENTS

The great bandicoot (*Bandicota indica*) from southern Asia was introduced to Java. The Eurasian edible dormouse (*Glis glis*) was introduced to Great Britain in 1902, and the species has been slowly spreading ever since. They are a pest of orchards and also damage homes. The North African crested porcupine (*Hystrix cristata*) was introduced to southern Europe, possibly by the Romans, and persists in Italy, Sicily, and the Balkans. Ord's kangaroo rat (*Dipodomys ordii*), from the western United States, was introduced to Ohio and persists on the shores of Lake Erie. South American guinea pigs (*Cavia porcellus*) are widely kept as pets and laboratory animals; they have escaped or been released often and in many areas. However, the only introduction known to have persisted and grown is on Santa Cruz Island, in the Galápagos. No damage has been reported. The Brazilian agouti (*Dasyprocta leporina*) in the Virgin Islands and on many of the Lesser Antilles is believed to be a pre-Columbian introduction, while the Central American red agouti (*D. punctata*) was successfully introduced to the Cayman Islands in the late nineteenth century. Neither agouti is reported to cause ecological damage. The Central and South American paca (*Cuniculus paca*) is introduced and established on Cuba; its impact there is unreported.

SEE ALSO THE FOLLOWING ARTICLES

Eradication / Islands / Mammals, Aquatic / Rats / Transformers / Wetlands

FURTHER READING

Anderson, C. B., C. R. Griffith, A. D. Rosemond, R. Rozzi, and O. Dollenz. 2006. The effects of invasive North American beavers on riparian plant communities in Cape Horn, Chile. Do exotic beavers engineer differently in sub-antarctic ecosystems? *Biological Conservation* 128: 467–474.

Lever, C. 1985. *Naturalized Mammals of the World.* London: Longman.

Long, J. L. 2003. *Introduced Mammals of the World: Their History, Distribution and Influence.* Wallingford, UK: CABI Publishing.

Lowery, G. H. 1974. *The Mammals of Louisiana and Its Adjacent Waters.* Baton Rouge: Louisiana State University Press.

Palmer, G. H., J. L. Koprowski, and T. Pernas. 2007. Tree squirrels as invasive species: Conservation and management implications (273–282). In G. W. Witmer, W. C. Pitt, and K. A. Fagerstone, eds. *Managing Vertebrate Invasive Species: Proceedings of an International Symposium.* Fort Collins, CO: USDA/APHIS/WS National Wildlife Research Center.

Simberloff, D. 2009. Rats are not the only introduced rodents producing ecosystem impacts on islands. *Biological Invasions* 11: 1735–1742.

S

SCOPE PROJECT

DANIEL SIMBERLOFF

University of Tennessee, Knoxville

Although the beginning of modern invasion biology is often attributed to Charles Elton's publication in 1958 of *The Ecology of Invasions by Animals and Plants*, his monograph, though often prescient, did not directly trigger a burst of research and publicity on invasions. Rather, the main impetus was a project on the ecology of biological invasions by the Scientific Committee on Problems of the Environment (SCOPE, an arm of the International Council of Scientific Unions), initiated in mid-1982.

CONCEPTION AND INITIATION OF THE PROJECT

Many ecologists attended the Third International Conference on Mediterranean-Type Ecosystems in Stellenbosch, South Africa, in 1980. Invasions were not a major focus of ecology then, and among ecologists the prevailing view of introduced species was that problematic invasions were a feature only of certain disturbed habitats; this was the main message of Elton's monograph. During a field trip at the conference, participants were surprised to see introduced pines advancing over a hill into undisturbed fynbos vegetation. Subsequently at the Stellenbosch conference, Harold Mooney, a Stanford University ecologist, and Fred Kruger, a South African forestry researcher, discussed topics that might fruitfully be addressed through a SCOPE project using the comparative approach adopted by the conferences on Mediterranean-type ecosystems. Having just observed

the invaded fynbos, and being familiar with other invasions into Mediterranean-type ecosystems, Mooney and Kruger determined to propose to SCOPE a project on plant invasions into the five Mediterranean-type systems. An 11-member international working group to develop the project—to determine its proposed scope and goals—met in May 1982 and produced a draft proposal for presentation to the general assembly of SCOPE, recommending that Mooney chair the project's scientific advisory committee.

At the general assembly meeting in Ottawa in June 1982, Ralph Slatyer, an Australian ecologist and president of SCOPE, suggested a global project, in accord with the SCOPE mandate to advance knowledge on environmental problems of global significance, and one that would treat all habitats, not just Mediterranean-type vegetation. Thus the project was termed the Scope Programme on the Ecology of Biological Invasions, Mooney was named chair, and the main questions specified by SCOPE were the following:

1. What factors determine whether a species will become an invader?

2. What site properties determine whether an ecological system will be prone to, or resistant to, invasion?

3. How should management systems be developed to best advantage, in light of the knowledge gained from attempting to answer questions 1 and 2?

SCOPE PRODUCTS

The SCOPE project struck a responsive chord. Under its aegis, national programs started in Great Britain, Australia, the United States, the Netherlands, and South Africa, and there were regional syntheses for Europe and

617

the Mediterranean basin as well as for the tropics. An international conference in Hawai'i in 1986 synthesized and extended the national efforts; the proceedings of this conference, published in 1989, became a highly influential book in invasion biology. Although the international synthesis conference was the formal culmination of the project, many publications associated with the project continued to appear, at least through 1991. Several other projects were either begun under the aegis of the SCOPE program (e.g., on mathematical modeling of invasions [1985] and invasions into nature reserves [1985]), inspired by the SCOPE program (e.g., on biogeography of invasions into Mediterranean ecosystems [1983]), or associated themselves with particular parts of the SCOPE program (e.g., on the relationship of succession to colonization [1987]). The national reports and global synthesis alone included 145 editors and authors, while associated efforts added many more.

INFLUENCE OF THE SCOPE PROJECT

The SCOPE project and SCOPE-inspired publications had an enormous impact. Invasion biology papers were published at a very low level through the 1980s, but by the mid-1990s, they were increasing at an exponential rate (Fig. 1). This initial surge largely reflects the influence of publications of the SCOPE project, mostly from the late 1980s. The two most prominent, one on invasions of North America and Hawai'i and the other the global synthesis volume, were cited with dramatically increasing frequency in the mid-1990s (Fig. 2) at exactly the time

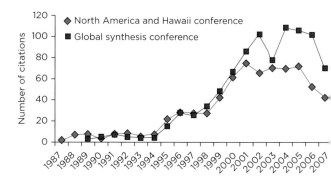

FIGURE 2 Citations of two leading SCOPE project products, publications based on the North America and Hawaii conference (Mooney and Drake, 1986, blue diamonds) and the global synthesis conference (Drake et al., 1989, red squares).

of the general surge in invasion ecology publications. In addition, many participants in SCOPE workshops and conferences did not publish papers in the associated SCOPE products but continued research in invasion biology. Also important is the fact that although various parts of the SCOPE project engaged most of the people doing prominent research on invasions in the 1980s (not a large group), they also enlisted a number of leading scientists in related disciplines (particularly ecology and mathematical modeling) who had not previously studied invasions but who subsequently conducted invasion biological research.

The SCOPE project triggered a surge in invasion research in way a way that Elton's monograph did not. Many SCOPE participants had read Elton, often years earlier. In the SCOPE national volumes and global synthesis, 33 of 100 chapters cite his monograph as a general reference for the importance of invasions as an environmental problem. But Elton's monograph did not set off an explosion of research, suggesting that the time was ripe for such a response in the 1980s as it had not been in 1958. Suggestive evidence for this fact is the appearance of an Australian book edited by Roger Kitching in 1986—*The Ecology of Exotic Animals and Plants: Some Australian Case Histories*—simultaneously with that of the Australian contribution to SCOPE (Groves and Burdon, 1986): *Ecology of Biological Invasions*. The two books deal with very similar topics, yet Kitching (personal communication) was unaware of the SCOPE project when he initiated his book in 1984, and there is only one author in common among the 20 chapters in the SCOPE volume and the 16 chapters in the other. These two independent yet strikingly similar collections arose for the same reason: by the 1980s, the burgeoning and increasingly evident number of invasions causing environmental or economic problems

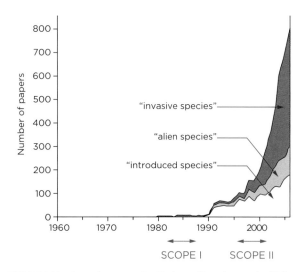

FIGURE 1 Numbers of papers using the terms "invasive species," "alien species," or "introduced species" from 1960 to 2006. (Modified from E. García-Berthou, *Journal of Fish Biology* 71(Suppl. D): 33–55. Courtesy of John Wiley and Sons Ltd.)

in many habitats—not only disturbed ones—finally led to the recognition that invasions were an integral and global problem, not just a collection of intriguing and sometimes troubling special cases. This was the raison d'être of the SCOPE program, and it is unsurprising that other scientists would have recognized the same phenomenon at roughly the same time.

A second reason the SCOPE program was timely is that the 1980s were a time of intense interest in various academic theories (especially in population and community biology) and their potential relevance to ecological problems; among these were the dynamic equilibrium theory of island biogeography, theories of limiting similarity, and theoretical explorations of the relationship between stability of biotic communities and their diversity or complexity. Data on introduced species seemed appropriate to test such ideas, and several contributions to the SCOPE volumes used them to just that end. Quantitative treatment and modeling were prominent features of many parts of the SCOPE program and have remained a key component of modern invasion biology. Elton's monograph, by contrast, was largely a collection of particular cases and interesting anecdotes.

The SCOPE project, then, was the proximate impetus behind the explosive growth of modern invasion biology beginning in the 1980s. It did not address all aspects of the current invasion biology agenda. For instance, evolutionary questions had a low profile in the SCOPE project (and virtually no profile in Elton's monograph), and evolution of both invaders and members of invaded communities became an important part of the field by the late 1990s. However, the three SCOPE questions—what makes some species invasive, what makes some systems invasible, and how information on these matters can inform management—remain major components of invasion biology.

FURTHER READING

Davis, M.A. 2006. Invasion biology 1958–2005: The pursuit of science and conservation (35–64). In M.W. Cadotte, S.M. McMahon, and T. Fukami, eds. *Conceptual Ecology and Invasion Biology*. Dordrecht, The Netherlands: Springer.

di Castri, F., A.J. Hansen, and M. Debussche, eds. 1990. *Biological Invasions in Europe and the Mediterranean Basin*. Dordrecht, The Netherlands: Kluwer.

Drake, J.A., H.A. Mooney, F. diCastri, R.H. Groves, F.J. Kruger, M. Rejmánek, and M. Williamson. 1989. Preface (xxiii–xxiv). In J.A. Drake, H.A. Mooney, F. di Castri, R.H. Groves, F.J. Kruger, M. Rejmánek, and M. Williamson, eds. *Biological Invasions: A Global Perspective*. Chichester: Wiley.

Gray, A.J., M.J. Crawley, and P.J. Edwards, eds. 1987. *Colonization, Succession and Stability*. Oxford: Blackwell Scientific Publications.

Groves, R.H., and J.J. Burdon, eds. 1986. *Ecology of Biological Invasions*. Cambridge: Cambridge University Press.

Kitching, R.L., ed. 1986. *The Ecology of Exotic Animals and Plants: Some Australian Case Histories*. Brisbane: Wiley.

Kornberg, H., and M.H. Williamson, eds. 1986. Quantitative aspects of the ecology of biological invasions. *Philosophical Transactions of the Royal Society of London, Series B* 314.

Macdonald, I.A.W., F.J. Kruger, and A.A. Ferrar, eds. 1986. *The Ecology and Management of Biological Invasions in Southern Africa*. Cape Town: Oxford University Press.

Mooney, H.A., and J.A. Drake, eds. 1986. *Ecology of Biological Invasions of North America and Hawaii*. Ecological Studies 58. New York: Springer.

Usher, M.B., F.J. Kruger, I.A.W. Macdonald, L.L. Loope, and R.E. Brockie. 1988. The ecology of biological invasions into nature reserves: An introduction. *Biological Conservation* 44: 1–8.

SEA LAMPREY

PETER W. SORENSEN

University of Minnesota, St. Paul

ROGER A. BERGSTEDT

U.S. Geological Survey, Millersburg, Michigan

The sea lamprey (*Petromyzon marinus*) is an ancient and parasitic fish that invaded the Laurentian Great Lakes from its native habitat in the coastal regions of the North Atlantic. Its complex anadromous life cycle includes both a filter-feeding larval phase, during which it lives burrowed in freshwater stream sediments, and a parasitic phase, during which it lives in saltwater, where it feeds voraciously on the blood of other fish before returning to freshwater streams to reproduce. The sea lamprey invaded the Laurentian Great Lakes in the early twentieth century, adopted a freshwater lifestyle, and within a few decades had triggered a near total collapse of this ecosystem's fisheries. Since the 1950s, it has been the focus of one of the world's only successful invasive fish control program, resulting in a substantial, albeit partial, recovery of Great Lakes fisheries. This control program relies mainly on taxon-specific toxicants and barriers but is developing new approaches, including the release of sterile males and pheromone-mediated trapping. Ironically, the sea lamprey is now threatened across much of its original range, where it is considered a delicacy.

THE SPECIES, ITS BIOLOGY AND LIFE HISTORY

The sea lamprey, *Petromyzon marinus*, is an ancient, jawless cartilaginous fish that has changed little over the past 400 million years (Fig. 1). It is native to coastal regions of both the European and North American sides of the Atlantic Ocean and is closely related to three dozen other species of northern hemisphere lampreys, none of which are invasive. The sea lamprey has a complex anadromous life history that includes a filter-feeding larval stage, during which it usually lives burrowed in sediments of freshwater streams; a juvenile parasitic stage, during which it usually lives in saltwater, where it feeds on the blood of host fishes; and an adult stage,

FIGURE 1 Drawing of a sexually mature adult sea lamprey. Image originally prepared by Ellen Edmonson and Hugh Chrisp as part of the 1927–1940 New York Biological Survey. (Permission for use granted by the New York State Department of Environmental Conservation [NYSDEC].)

when it enters freshwater streams to spawn and die (Fig. 2). In the past century, the sea lamprey has invaded a number of North American lakes including the Laurentian Great Lakes (see section below), where it appears to continue to pursue the same life history strategies with just a few minor exceptions. The sea lamprey is by far the best understood of the lampreys, especially in the Great Lakes, where it is invasive and its presence has promoted intensive study.

The life history of the sea lamprey is complex and is key to both its success and control. Adult sea lampreys migrate into freshwater streams from their oceanic or lacustrine feeding habitat in May and June each year. Mature males, which have a distinctive rope-like dorsal ridge, use their oral discs like a suction cup to move stones into horseshoe-shaped nests (*Petromyzon* means "stone sucker" in Latin). Sea lampreys are fecund: Atlantic Ocean females carry between 150,000 and 300,000 eggs, while Great Lakes fish have 20,000 to 110,000 eggs. Ripe males attract females to their nests by releasing a steroidal sex pheromone. After depositing fertilized eggs into nests, males and females die. Eggs hatch within about a week into tiny, blind, filter-feeding larvae, known as ammocoetes. Larvae burrow in stream sediments and filter-feed on biofilm sloughed from the substrate. The ecological significance of this

activity is considerable, as some Great Lakes streams have hundreds of thousands of larval lampreys of multiple species. Once larval sea lampreys reach a length of about 120 mm (in the Great Lakes, this takes between 3 and 20 years), they undergo a radical metamorphosis and develop eyes, a greatly enlarged olfactory system, and an oral disc that they later use to attach to host fish (Fig. 3). Metamorphosed sea lampreys leave streams between late fall and spring to enter the Atlantic Ocean or freshwater lakes.

Juvenile sea lampreys are voracious predators and use their well-developed oral discs with proteinacious teeth to attach to host fishes and bore a hole so they can then suck blood and body fluids. Estimates of the biomass and number of fish killed per sea lamprey vary widely but range up to 20 kg per lamprey. Parasite growth rates are greatest in late autumn, and recoveries of dead fish in the Great Lakes suggest that this is when most fish are killed. When lake trout are abundant, they bear a disproportionate number of sea lamprey attacks, but all large fish are vulnerable to attack, and apparent host preferences may simply be due to parasite and host depth and temperature distributions. When feeding, sea lampreys are typically carried far from their natal streams by their hosts and become widely dispersed.

The parasitic stage of sea lampreys is thought to last two seasons in the Atlantic Ocean but is known (through tagging) to last only one in the Great Lakes. Adult lampreys reach an average length of 50 cm in the

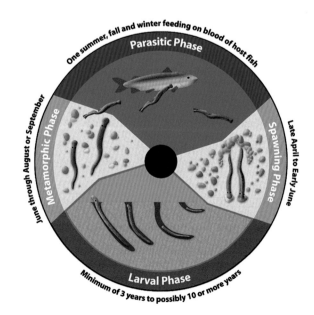

FIGURE 2 Life history of the sea lamprey. (Courtesy of the Great Lakes Fishery Commission.)

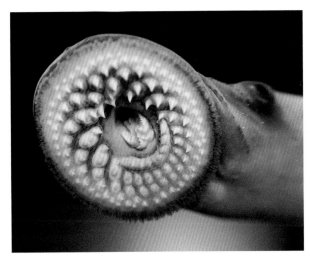

FIGURE 3 Oral disc of a parasitic-phase sea lamprey. The animal uses this sucker-like structure to attach to its prey and then cuts a hole into the prey using the sharp cusps on its "tongue." (Photograph courtesy of Ted Lawrence, Great Lakes Fishery Commission.)

Great Lakes by late winter, when they start to mature and search for spawning streams that they enter in early spring. Field studies have demonstrated that sea lampreys do not return to natal streams, but instead choose (locate) streams that contain large numbers of larvae. This feat appears attributable to a larval pheromone that is composed of unique sulfated sterols. It is the only migratory pheromone identified and synthesized in a fish and is active at concentrations below 10^{-13} Molar (a gram in 80 billion liters). Reliance on a pheromone makes evolutionary sense for a species that may be transported far from its natal stream by its host and for which the larvae odor can serve as a universal indicator of spawning and nursery habitat.

The sea lamprey has a remarkable physiology. It is an extremely robust animal and undergoes two radical metamorphoses (first to a parasitic and then to an adult phase) and has a sophisticated kidney that functions in either full strength sea water or the low-ion waters of the Great Lakes. While the sea lamprey's eyesight is relatively poor and its taste system primitive, it has one of the most developed and specialized olfactory systems of the vertebrates (its nose is about a third the size of its brain)—a trait that allows it to use pheromones to great advantage in its expansive environment. Like most ancient fishes, it also is sensitive to electric fields and possesses some unusual enzyme systems, which make it susceptible to certain toxicants (see below). It is a surprisingly persistent swimmer, has an indeterminate growth rate, and is fecund—all traits that give it enormous flexibility and contribute to its invasiveness.

INVASIVE PATHWAYS AND DAMAGE

The sea lamprey has been a valued fish since the Middle Ages in Europe, where it served as food for both commoners and royalty and seems to have caused no problems (although King Henry I of England died of eating a "surfeit of lamprey pie"). Apparently, the sea lamprey was not eaten by indigenous peoples in North America, although its close relative, the Pacific lamprey (*Entosphenus tridentate*), was (and still is). By all indications, sea lampreys played a balanced role in coastal food webs prior to settlement. This situation changed quickly at the end of the nineteenth century in North America, with the construction of shipping canals that allowed ships to enter the Great Lakes and other inland lake systems. With the exception of Lake Ontario, these lakes had been isolated since their formation at the end of the last glaciations by Niagara Falls and had evolved simple but productive ecosystems that also came to support huge inland fisheries. Within a few decades of the construction and enlargement of these canals (especially the Welland Canal that bypasses Niagara Falls), the sea lamprey was observed in Lake Erie (Fig. 4). Sea lampreys were subsequently observed in rapid succession in lakes Huron, Michigan, and Superior during the course of the 1930s, at which time self-sustaining populations developed and flourished. The sea lamprey also became prevalent (and problematic) in the Finger Lakes of New York and in Lake Champlain. Although the details of how and when it spread and whether it may have originally been native in some of these systems are unclear, the dramatic increase in its numbers is not.

At the same time that sea lamprey numbers increased in the Great Lakes, fish populations collapsed dramatically. In particular, the lake trout (*Salvelinus namaycush*),

FIGURE 4 Dates of the first documented observation of sea lampreys in each of the Great Lakes. (Courtesy of the Great Lakes Fishery Commission.)

a keystone predator in the Great Lakes, was the first to suffer severe losses, and by the late 1940s, fisheries for them had collapsed with devastating economic and ecological consequences. Although some argue that initial decreases may have been promoted by overfishing and pollution, the strikingly inverse relationship between lamprey and lake trout abundance suggests an important link. Lake trout became extinct in all of the Great Lakes except Superior by the end of the 1950s. This change caused dramatic, cascading changes in the fish communities and lake ecology. With the reduction in predators came explosions in the populations of invasive "prey" species such as smelt and alewives, followed by massive annual die-offs of many during the 1950s and 1960s (Fig. 5). These events caused tremendous, but never estimated, losses in property and recreational value. Extinctions of deepwater sculpins in Lake Ontario and several deepwater coregonine fishes across the Great Lakes also coincided with extended periods of high abundance of prey fishes. After initial success controlling the sea lamprey, government agencies attempted to stabilize these systems by stocking large numbers of native lake trout along with nonnative species of trout and Pacific salmon. Those additions (which continue at a reduced level) have resulted in viable recreational fisheries and stabilized fish communities. The Great Lakes ecosystem continues to change and remains in a state of flux (see below), as other invaders continue to enter this system and agencies continue their efforts to control

them. Similar versions of this scenario play out in the Finger Lakes and Lake Champlain.

Ironically, at the same time that the number of sea lampreys exploded in the Great Lakes, the abundance of this fish plunged across most of the Atlantic coast, especially in Europe. Most believe that dams, declining water quality, and overfishing (the sea lampreys are prized as food by the Portuguese and Finns) have been and continue to be responsible. Several reasons seem to explain why sea lamprey should be invasive in the large inland lakes of North America. Here, key waterways were opened up to shipping and not blocked by dams. Perhaps most importantly, their simple ecosystems seem both to lack predators for sea lampreys and to contain large spawning and nursery habitats previously occupied by native lampreys. Sea lampreys could quickly locate that habitat by following pheromone trails left by once abundant (and beneficial) native lampreys that also appear to excrete the same migratory pheromone. Lastly, larger fishes native to the Great Lakes had evolved no defense mechanisms for predators.

SEA LAMPREY CONTROL

Control of the sea lamprey in the Great Lakes is one of only a few success stories for an invasive fish. Efforts started in the late 1940s, when the State of Michigan began studying the biology of the sea lamprey to identify key aspects of its unusual life cycle that could be exploited. In 1950 the U.S. Fish and Wildlife Service assumed these efforts. In 1955, after a continuing deterioration in lake fisheries, the binational Great Lakes Fishery Commission (GLFC) was formed under a convention between the United States and Canada. One of the GLFC's primary duties was (and is) the control or eradication of sea lampreys, and it currently funds and oversees management of sea lamprey populations across the Great Lakes. Control is delivered through U.S. and Canadian government agents.

Control of the sea lamprey has focused on breaking its life cycle at vulnerable points: adult migration and larval survival. Early control efforts used screen weirs, often in conjunction with electrical fields, to block access to the spawning areas and trap and remove adult spawners. However, this proved only partially effective, and a screening program to find a selective toxicant was initiated in the early 1950s. It evaluated over 6,000 chemicals before identifying 3-trifluoromethyl-4-nitrophenol (TFM), which interferes with oxidative phosphorylation. Selectivity is attributable to the lamprey's lack of an enzyme to break down TFM. Starting in Lake Superior in 1958, TFM has been carefully applied as a "lampricide" to infested streams. A second,

FIGURE 5 Photograph of a mass die-off of alewives in the Great Lakes, which can be attributed to predator-prey imbalances associated with the invasion of the sea lamprey. (Photograph courtesy of the Great Lakes Fishery Commission.)

nonselective toxin, 2′, 5-dichloro-4′-nitrosalicylanilide (niclosamide) is also occasionally used in conjunction with TFM in high-alkalinity waters to enhance potency and selectivity. Only streams heavily infested with sea lamprey larvae are currently treated with lampricides, and then only when the larvae are large enough to metamorphose. These streams are identified by yearly surveys. Fortunately, of over 5,000 tributaries to the Great Lakes, only 474 have historically produced sea lampreys, and only 284 regularly require treatments. Notably, extensive toxicity studies permit the concentration of TFM to be precisely gauged relative to flow, temperature, pH, and water chemistry. Treatments rarely kill significant numbers of nontarget organisms, with the exception of native lampreys and some insects. Although niclosamide and TFM degrade rapidly with sunlight and wave action, do not bioaccumulate, and have no known effects on higher vertebrates, their use can promote controversy. Lampricide treatments have proven highly successful, leading to a decrease in sea lamprey numbers to about 10 percent of their former numbers. Reduced lamprey numbers have allowed native and stocked fish to survive and the lake trout populations to rebound. Recently, the restoration of lake trout in Lake Superior was declared a success, and most stocking stopped. A substantial, although much changed, fishery once again exists in the Great Lakes, as well as in Lake Champlain and the Finger Lakes.

New control strategies are being developed to enhance an integrated control program that relies less on expensive toxicants. Ineffective screen weirs have been replaced with low-head barriers that block sea lampreys but allow jumping fish to pass. Adjustable-height barriers and electrical barriers that use safe levels of pulsed DC current are also in use. A new technique is the release of male sea lampreys sterilized by injection with a biodegradable mutagenic chemical (bisazir) administered in a self-contained facility that eliminates environmental release. Bisazir makes sperm nonviable without affecting behavior. Sterile males are released into the last large untreated population in the Great Lakes, the St. Marys River, where it is hoped they can nullify the success of other males. Some success is indicated. Lastly, sex and migratory pheromones of the sea lamprey have now been identified and synthesized. The migratory pheromone is being explored as a potent and safe attractant to bring migratory sea lampreys into spawning systems while the sex pheromone is being examined for its ability to trap females on spawning grounds. Most recently, the sea lamprey genome has been mapped, and research is also under way to understand better which genes regulate key processes in the development and sexual maturation of sea lampreys. Many interesting challenges await those seeking to understand and manage this enigmatic, extremely damaging invasive animal.

SEE ALSO THE FOLLOWING ARTICLES

Biological Control, of Animals / Canals / Fishes / Great Lakes: Invasions / Lakes / Life History Strategies / Parasites, of Animals / Pesticides (Fish and Mollusc)

FURTHER READING

Applegate, V. C. 1950. Natural history of the sea lamprey (*Petromyzon marinus*) in Michigan. *U.S. Fish and Wildlife Service Special Scientific Report* 55.

Great Lakes Fishery Commission. 2001. Strategic vision of the Great Lakes Fishery Commission for the first decade of the new millennium. www.glfc.org/pubs/SpecialPubs/StrategicVision2001.pdf

Hardisty, M. W. 2006. *Lampreys: Life without Jaws*. Ceredigion, UK: Forrest Text.

Hardisty, M. W., and I. C. Potter. 1971. *The Biology of Lampreys*. 5 volumes. London: Academic Press.

Jones, M. L., C. H. Olver, and J. W. Peck, eds. 2003. Sea Lamprey International Symposium (SLIS II). *Journal of Great Lakes Research* 23(suppl 1).

Smith, B. R., ed. 1980. Proceedings of the Sea Lamprey International Symposium *Canadian Journal of Fisheries and Aquatic Sciences* 37: 11.

Sorensen, P. W., and T. E. Hoye. 2007. A critical review of the discovery and application of a migratory pheromone in an invasive fish, the sea lamprey, *Petromyzon marinus* L. *Journal of Fish Biology* 71(suppl D): 100–114.

SEAS AND OCEANS

EDWIN GROSHOLZ

University of California, Davis

The patterns of invasion in the ocean realm are obvious in that they clearly reflect the human footprint of global commerce and overexploitation of natural resources. However, there are significant exceptions to the expected patterns, with some regions seemingly much less invaded than others despite being subject to generally similar frequencies of ship traffic, urbanization of the coastal region, and other characteristics that are broadly associated with increased levels of invasion. Also, certain taxa, despite significant commercial interest and opportunities for intentional introductions, unlike similar taxa in terrestrial habitats, have not been the focus of introduction.

TAXONOMIC LIMITS OF INVASIONS

The extent of invasions in marine seas and oceans involves a broad array of taxa, including fishes, invertebrates, and algae. Many entries in this volume address the diversity of the many groups that are introduced species. However, introductions are glaringly absent among groups of higher vertebrates. Marine vertebrates including mammals (pinnipeds, otters, sirenians, cetaceans), reptiles (crocodiles, iguanas, turtles, snakes), and birds (shorebirds, waterbirds, seabirds) are not found on any list of introduced species. Of course, these species are unlikely to be introduced by the introduction pathways responsible for the movement of fishes, invertebrates, and algae. But despite the fact that many species of marine birds and mammals have been exploited historically, and there have been efforts to culture some of these taxa (e.g., turtle farms), these have not resulted in established species introductions.

The possibility of introducing nonnative marine vertebrates (nonfish) may be changing in the future because of shifts in conservation strategies for threatened and endangered species. One issue arising in endangered species conservation is the concept of "managed relocation," in which species are introduced to new regions where they did not exist previously. Efforts to reestablish threatened or endangered species have historically involved transplanting individuals to sites within the historical range. But with plans for managed relocation being developed for ranges of species, the possibility of an "introduced" population of marine vertebrate may become a reality in the future.

ABUNDANCES IN DIFFERENT REGIONS AND HABITATS

As with any group, the degree to which invasive species have been described is a function of the taxonomic effort in addition the magnitude and source of invasive propagules, characteristics of the recipient community, phylogenetic similarity between native species and potential invaders, and so forth. Natural dispersal of species between large regions or between ocean basins does occur naturally, but at a very low frequency, often associated with large storms or other rare events. However, on top of this very low natural rate of introduction is the vastly accelerated rate of introduction by human activities, which has resulted in very different levels of invasion among the different oceanic regions and habitats of the Earth.

Hawaii and the Central Pacific

The list of introduced species in this broad reach of the central Pacific is not well known for the most part. However, the nonnative species of Hawaii are particularly well described and represent an important window into the patterns of invasion in this region. The extensive records and accurate descriptions of nonnative species demonstrate that this is one of the most invaded regions of the

world, with more than 340 species including 287 invertebrates, 24 algae, 20 fish, and 12 flowering plants.

Northeastern Pacific

The northeastern Pacific coast is among the best-studied areas in the world in terms of introduced species. This area has been the focus of extensive ship traffic from around the Pacific and the recipient of massive shipments of oysters from the eastern United States and Asia, which have contributed significantly to the high number of invasive species in this region. These currently number at least 296 species.

Northwestern Atlantic and Gulf of Mexico

The northwestern Atlantic coast of North America is well studied. Many of the introduced species are described from urbanized bays and estuaries that have been the recipient of many centuries of ship traffic and the intentional introductions of many aquaculture and fisheries species. The number of nonnative species in this region of the Atlantic is in excess of 172 species, with at least 40 introduced species in the Gulf of Mexico alone.

Northeast Atlantic Including Baltic and North Seas

Another well-studied region is the northeastern Atlantic region of Europe, where studies of introduced species have a long history. Well-defined lists of introduced species are available individually for several marine regions including the Atlantic, Baltic, and North Sea regions. A total of at least 350 species are described from this broadly defined region, ranging from the frigid Barents Sea to the warm-temperate northeast Atlantic.

Mediterranean Sea

The Mediterranean Sea has one of the most fully described introduced biotas, with quite complete descriptions of most nonnative taxa. There are at least 325 species of nonnative aquatic and marine species. The opening of the Suez Canal, which allowed access by large numbers of species from the Red Sea and the greater northwestern Indian Ocean region, contributed greatly to the large numbers of introduced species in this region.

Australia, New Zealand, and the Southern Oceans

The introduced species in Australia and New Zealand are very well cataloged and are known from some of the most comprehensive surveys of introduced marine species anywhere on the planet. This region is one of the most highly invaded regions, although to an extent, this status reflects the extensive documentation of invasions. The number of introduced species in Australia is now determined to be at least 473 (including cryptogenic ones), while New Zealand has at least 167 introduced species (including cryptogenic ones).

Black, Aral, and Caspian Seas

This area has been the recipient as well as the source of many species introductions. At least 70 introduced species are found in this region, which includes the Black and Azov seas and the Caspian Sea. Among the most notorious invasions has been that of the comb jelly (ctenophore) *Mnemiopsis leidyi*, which contributed to crashes in the populations of anchovies in the Black and Azov seas and to crashes of a small commercial fish known as kilka in the Caspian Sea.

Sea of Japan and Northwestern Pacific

The northwestern Pacific region, including the Sea of Japan, the East China Sea, and surrounding regions, contains remarkably few introduced species. In the Sea of Japan, only 17 introduced species are currently described. At least one recent survey in the Yellow Sea region near Dalian, China, failed to yield any new introduced species. Despite significant ship traffic and substantial urban development of the coastal zone in this region, the number of introduced species is surprisingly small at present.

Polar Regions

The extreme physical conditions and the very limited opportunity for introductions both argue for a low likelihood of the successful establishment of nonnative species in both the Arctic and Antarctic regions. Among the few recorded examples of nonnative species in Antarctic waters are the Atlantic spider crab *Hyas araneus* and the galatheid crab *Munidopsis albatrossae*, both taken from Antarctic trawls. However, because there has been only one collection for each of these species, is it uncertain whether these species are truly established in Antarctic waters. Depending on how the Arctic Ocean region is defined, as many as 20 introduced species are found there.

Pelagic Open Ocean

Of course, the most extensive habitats in the world's oceans are the pelagic blue-water environments of the open ocean. There are comparatively few documented introductions in the pelagic realm of the large ocean basins. Among the most important invasions in these pelagic habitats are those of introduced gelatinous zooplankton. These have been reported from a number of

FIGURE 1 Medusae of invasive jellies *Maeotias marginata* and *Moerisia* spp. in San Francisco Bay in central California. (Photograph courtesy of Alpa Wintzer.)

regions from the Gulf of Mexico to the Hawaiian Islands. A number of more restricted seas and inland oceans have also been subject to "jelly" invasions, including the eastern Mediterranean, the Black Sea region, and the Sea of Japan.

Among the most dramatic introductions in pelagic habitats have been those of introduced cnidarians. These include the dramatic blooms in the Gulf of Mexico of the Australian spotted jellyfish *Phyllorhiza punctata*, which was introduced from the western Pacific. These blooms have substantially impacted shrimp and crab fisheries in this region and have clogging shrimp nets, resulting in millions of dollars in fishery losses. Also, the recent invasion of San Francisco Bay by several hydrozoan species, including *Maeotias marginata*, *Blackfordia virginica*, and *Moerisia* spp. (Fig. 1), is now being considered as a factor contributing to the decline of native species in this heavily invaded system.

Benthic Offshore Habitats

With increasing use of remotely operated vehicles and other devices with the ability to image benthic habitats in increasingly deeper areas, introduced species are becoming identified in increasingly deeper waters. However, few species have yet to be described from offshore areas of the continental shelf. An example of a recent introduction that has become widely established and surprisingly dominant is the invasive tunicate *Didemnum* spp., which has become established across large areas (>100 mi^2) of the commercially important fishing grounds known as George's Bank, off the southern New England coast of the United States, at depths of up to 81 m.

Deep Oceans

It goes without saying that because we know comparatively little about the deep oceans it would be difficult to verify that a certain species is nonnative. Therefore, it should not be surprising that no introduced species have been described from deep ocean (abyssal) regions. Even from the moderate depths of the continental shelf regions, very few introduced species are described, and even fewer are described from slope depths below 150 m. Two exceptions to this pattern are the well-documented introductions of the western Atlantic snow crab *Chionoecetes opilio* and the North Pacific red king crab *Paralithodes camtschatica*. Both of these species were intentionally introduced into the Russian Barents Sea and have since moved southward into Norwegian waters. Snow crabs have been found in trawls taken between depths of 180 and 350 m, and king crabs have been taken from waters as deep as 400 m.

Coral Reefs

Despite the common view that disturbed habitats are more likely to become invaded, relatively pristine regions have been rapidly invaded by introduced species. There are extensive records of species being introduced throughout coral reefs of the Pacific for commercial gain. These examples include giant clams (*Tridacna*) that were raised in hatcheries in places like Hawaii and Palau and introduced to other island regions including Fiji and American Samoa, where they did not exist previously. The same is true for topshell snails (*Trochus*), whose shells were used commercially to make pearl buttons. These snails were native to the Indian Ocean and southwestern Pacific and were introduced to islands in many other areas of the Pacific, where they became successfully established.

Kelp Forests and Sea Grasses

Among the most notorious marine plant introductions is that of the tropical alga *Caulerpa taxifolia* into offshore seagrass and benthic habitats of the Mediterranean and the California coast (Fig. 2). This normally tropical alga was bred for generations in aquaria in Germany in temperate conditions, which produced a fast-growing, chemically protected alga that could tolerate temperate winter conditions. This Mediterranean strain of *C. taxifolia* escaped from the Monaco aquarium and spread along the northern Mediterranean coast, producing one of the most damaging algal introductions to date. This strain of *Caulerpa* also became established in two sites in southern California, where it has since been eradicated. The recent introduction of *Caulerpa racemosa* var. *cylindracea*, native

FIGURE 2 The green alga *Caulerpa taxifolia* (Med. strain) invading the French coast of the Mediterranean Sea. (Photograph courtesy of Alexandre Meinesz.)

to Australia, has now resulted in a spreading scourge that has affected a broader area with greater impact than the better known *C. taxifolia*.

Introductions have also become commonplace in kelp forest habitats. Among the most extensive and well documented has been that of the European bryozoan *Membranipora membranacea*, introduced into the northwestern Atlantic. Here, it grows densely on a variety of brown algae, including kelps (*Laminaria*), to the point where the kelp blades become brittle and undergo early senescence and break-off. The resulting decline in *Laminaria* can facilitate the rapid invasion of the green alga *Codium fragile* ssp. *tomentosoides*, which has rapidly come to dominate shallow, rocky subtidal habitats in this region.

Salt Marshes and Mangroves

Salt marshes and estuarine environments are among the most invaded systems in the world. This is thought to be the result of several factors including significant levels of human disturbance, high levels of propagule pressure from ship traffic, past aquaculture activities, and so forth. Among the most damaging introductions of salt marshes worldwide are those by several species of cordgrasses in the genus *Spartina*. Nonnative cordgrasses have successfully established in North and South America, Europe, China, Australia, and New Zealand.

Mangroves are also intertidal habitats that have significant numbers of introduced species. However, the plant itself has also been the focus of intentional introductions. For example, at least six species of mangroves were introduced to Hawaii in the early part of the twentieth century. Of these, only the red mangrove (*Rhizophora mangle*) has spread successfully, resulting in significant impacts on native species including habitat loss for endangered waterbirds.

Bays and Estuaries

The most invaded aquatic habitats in the world are bays and estuaries. This is likely the result of extensive ship traffic, introductions of aquaculture species, coastal urbanization and associated habitat disturbance, and overexploitation of native species. Many of the same species can be found in multiple harbors on several different continents, resulting in an overall decline in beta diversity (among-site diversity), even as the list of species at any one site may be increasing.

PATHWAYS OF INTRODUCTIONS

Introduced species have become introduced to new regions in many ways. In many or most cases, the initial introduction is due to human activities ranging from ship traffic involving the transfer of ballast water to hull fouling between ports, which releases propagules of potential new invaders. Also, the movement of live species as fisheries products, ornamental aquarium objects, bait for fishing, and so forth provides many avenues for new introductions. Although we have begun to quantify the importance of ship-based pathways, we know little about the relative importance of these other pathways for facilitating new introductions.

CONCLUSIONS

The most heavily invaded areas are the nearshore coastal areas most influenced by ship traffic, historical aquaculture activities, urbanization, overexploitation, and other factors that have resulted in increased opportunities for nonnative species to become established. Deep ocean areas, and to a lesser extent polar and pelagic open ocean habitats, are much less invaded. The taxonomic identity of most invasive species is invertebrate, fish, or algae, with virtually no examples of introduced marine mammals, birds, or reptiles.

SEE ALSO THE FOLLOWING ARTICLES

Algae / Ballast / Crabs / Invertebrates, Marine / Mediterranean Sea: Invasions / Ponto-Caspian: Invasions / Wetlands

FURTHER READING

Allen, J. A. 1998. Mangroves as alien species: The case of Hawaii. *Global Ecology and Biogeography Letters* 7: 61–71.
CIESM. 2002. *Atlas of Exotic Species in the Mediterranean.* Monaco: CIESM Publishers. www.ciesm.org/online/atlas/intro.htm
Introduced Marine Species of Hawaii Guidebook. www2.bishopmuseum.org/HBS/invertguide/index.htm
Levin, P. S., J. A. Coyer, R. Petrik, and T. P. Good. 2002. Community-wide effects of nonindigenous species on temperate rocky reefs. *Ecology* 83: 3182–3193.
Meinesz, A. 1999. *Killer Algae.* Chicago: University of Chicago Press.
NAS Nonindigenous Aquatic Species Database. United States Geological Survey. http://nas.er.usgs.gov/
NEMESIS National Exotic Marine and Estuarine Species Information System. http://invasions.si.edu/nemesis/index.html
NOBANIS North European and Baltic Network on Invasive Alien Species. www.nobanis.org/
NIMPIS National Introduced Marine Pest Information System (Australia). www.marine.csiro.au/crimp/nimpis/
Zvyagintsev, A. Y. 2006. Introduction of species into the northwestern Sea of Japan and the problem of marine fouling. *Russian Journal of Marine Biology* 29: S10–S21.

SEAWEEDS

SEE FRESHWATER PLANTS AND SEAWEEDS

SEED ECOLOGY

MICHELLE R. LEISHMAN AND CARLA J. HARRIS

Macquarie University, Sydney, New South Wales, Australia

The size and number of seeds produced by plants is a critical factor in their ability to colonize ground and establish seedlings. There is a fundamental tradeoff within the reproductive strategy of a plant between producing more small versus fewer large seeds for a given investment in reproduction. There is always selection pressure to produce more seeds, because more seeds translate into more potential offspring. However, larger seeds have a greater chance of establishing, particularly in environments that are hazardous to seedlings, such as those in deep shade, with low soil moisture, or located beneath litter. The success of invasive plants compared to co-occurring native plants may be due to their greater reproductive output, through more and/or larger seeds. Seeds are also critical in determining the dispersal method of plants, through both space and time. Adaptations for long-distance seed dispersal and dormancy mechanisms allowing the spread of risk across multiple germination times are both potentially important contributors to invasive success in plants.

THE ROLE OF SEED PRODUCTION IN INVASION SUCCESS

The number of seeds a plant produces is an important element of plant fitness, potentially conferring benefits to invasive species if large numbers of seeds are produced. Propagule pressure—the number of propagules (seeds or other plant components capable of giving rise to new individuals) reaching a new location—is regarded as a key factor in the success of exotic plant species establishing and subsequently invading new areas. Therefore, the number of seeds an introduced plant produces may be an important factor in its invasion success, with a greater number of seeds resulting in a greater number of colonization opportunities, and hence a greater chance of successful establishment and spread into new areas.

There are several mechanisms by which plants introduced into novel environments may have greater seed production than co-occurring native species, resulting in successful invasion. First, the introduced plant may be "pre-adapted" to be a successful invader by already having a large amount of its resources allocated to reproductive output. Second, in a novel environment, it may be freed of its natural enemies, enabling a reduction in the resources allocated to defense and an increase in the resources allocated to reproductive output. Third, freedom from enemies may result in increased growth, and hence in larger plant size. As the reproductive output of a plant is often positively correlated with its size, this may result in greater seed production per individual plant.

There are surprisingly few data comparing seed production of invasive and native or noninvasive plants, or comparing seed production of invasive plants in their original and novel ranges. Anecdotal evidence suggests that successful invaders frequently have copious seed production (Fig. 1). Within-genera studies have found greater seed production in more invasive species within some genera but not all. One large cross-species study has shown that invasive species do have greater seed production than native species, producing nearly seven times more seeds per individual per year for a given seed mass (Fig. 2). The study showed that invasive plants produced more seeds on average than did natives, as well as a greater total mass of seeds, and this was not associated with species phylogeny. However, the study was not able to determine which of the mechanisms described above resulted in increased seed production of invasive compared to native plant species, suggesting a likely future area of research.

FIGURE 1 Copious seed production in the introduced invader *Ligustrum lucidum*. (Photograph courtesy of John Martyn.)

Other components of seed production, apart from number of seeds produced, may also be important contributors to invasive success. For example, studies comparing invasive with noninvasive species within the genus *Pinus* have shown that short juvenile periods and short intervals between seed crops are also characteristic of invasive species. Thus, successful invaders are likely to be those species that produce seed early, often, and in large quantities.

A number of biotic and abiotic variables also need to be considered when measuring seed production, as they can influence a plant's reproductive output. Seed

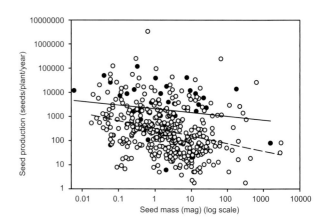

FIGURE 2 Mean seed production and mean seed mass of native (open circles) and invasive (closed circles) species compiled from the global literature and existing datasets. (From Mason et al., 2008. *Global Ecology and Biogeography* 17: 633–640.)

production may be limited by factors including resource availability (e.g., light, soil nutrients), rainfall, and seed and fruit predation. Potential seed numbers are also influenced by pollination and fertilization success as well as by the retention and development of fruits to maturity.

RELATIONSHIPS BETWEEN SEED SIZE AND INVASION SUCCESS

Seeds consist of an embryo plus endosperm (the seed reserve) plus a protective seed coat. Many seeds also have dispersal structures attached such as wings or hairs for wind dispersal, hooks or spines for adhesion dispersal, or arils and flesh for vertebrate dispersal (Fig. 3). The seed plus its dispersal appendages is called the diaspore. In this section, we consider seed size as the mass of the embryo, endosperm, and seed coat.

The size of a seed represents the amount of maternal investment in an individual offspring. Seed size is therefore intimately linked to many aspects of the ecology and life history of a species, including seedling establishment, seedling growth rate, seed number, and dispersal. There are a number of possibilities regarding how the direction of selection on seed size might act for exotic species introduced into a novel environment. Smaller seed size (and thus larger numbers of seeds) might be advantageous due to increased dispersal capacity (smaller seeds can be carried further by wind) and increased chances of colonization (more individual seeds equals more individual colonization opportunities). However, larger seeds have a greater chance of successful establishment, which may be particularly advantageous in sites with low light availability, high rates of herbivory, or high rates of competition. Thus, it is likely that the relationship between seed size and invasion success is context-dependent and that consistent patterns in seed size between native and invasive species, or between invasive species in their original and novel ranges, may not be clear.

Within the genus *Pinus*, seed mass has been identified as a useful predictor of invasiveness, with small seed mass being significantly related to invasion success. In contrast, within-genera (Asteraceae and Poaceae) studies of co-occurring native and invasive species have found that invasive species have larger average seed size. A large cross-species study has shown that small seeds are associated with invasion success among the nonnative flora of eastern Australia, but no evidence was found for a difference in seed size between native and invasive species from around the world, or between the original and novel ranges of invasive species.

The conflicting evidence exists also for studies at the local scale. However, in studies where seed size and invasion success are considered in relation to particular disturbance types, clearer patterns emerge. In sites that are physically

FIGURE 3 Seed dispersal modes of a variety of invasive species. (A) Vertebrate-dispersed fruits of African olive *Olea europea* subsp. *cuspidata*. (Photograph courtesy of Peter Cuneo.) (B) Wind-dispersed seed of *Cirsium vulgare*. (Image courtesy of Marcel Rejmánek.) (C) Spiny adhesion-dispersed seeds of *Bidens pilosa*. (Photograph courtesy of Botanic Gardens Trust, Sydney.)

disturbed, there is a clear association between invasive success and smaller seeds. Importantly, the association between small seeds and invasion success within the genus *Pinus* was specifically in relation to their invasion in physically disturbed landscapes. In contrast, in sites with high availability of soil moisture and nutrients resulting in dense vegetation, invasion is associated with larger vertebrate-dispersed seeds. In these sites, invaders may be advantaged both by long-distance dispersal capacity and by the ability to establish in a shaded, competitive environment.

In summary, although small seed size has been found to be associated with invasiveness in a range of studies, the pattern is by no means universal. Rather, the context of the invaded site is important in constraining the advantage of smaller seeds (increased long-distance dispersal, more colonization opportunities) in the invasion process.

THE ROLE OF SEED DISPERSAL IN INVASION

The ability to disperse away from the area of introduction into novel areas is critical for a plant to become a successful invader. Plants can disperse by means of vegetative propagation or dispersal of their seeds. In this section, seed dispersal is considered.

Plants employ a variety of mechanisms to disperse their seeds across the landscape (Fig. 3). This may be by means of morphological adaptations to the seed and includes wings, hairs, and pappuses for wind dispersal; hooks, barbs, or spines for adhesion dispersal; an oil-rich appendage termed an elaiosome for dispersal by ants; an aril or flesh for vertebrate dispersal; flotation devices to promote dispersal by water; and various mechanisms that eject seeds for ballistic dispersal. Up to half of the plant species within a flora have no morphological adaptations for dispersal. In these cases, smaller seeds have a greater chance of being dispersed long distances by wind. Different dispersal modes are able to achieve different dispersal distances. Dispersal distances of tens of meters to kilometers can be achieved by vertebrate, wind, and water dispersal, while ballistic and ant dispersal rarely achieve distances greater than a few meters.

Many studies have shown that successful invaders are more likely to have long-distance dispersal mechanisms compared to noninvasive or native plants. Seed dispersal by vertebrates is particularly common in invasive plant species (e.g., members from the blackberry family [genus *Rubus*], a globally invasive taxa). Vertebrate-dispersed exotic plants are often found invading along road edges (where birds have perched on fence posts, depositing seeds below) or clumped under perch trees. Rapid rates of spread can be achieved by vertebrate-dispersed species: new populations become established beyond the invading front via long-distance dispersal, and the population then spreads slowly out from these foci, eventually joining up to create a large invaded area. Dispersal by vertebrates is also particularly effective for invasion into vegetation types with closed canopies and into undisturbed habitats. Risk assessment for alien plant introductions usually includes an assessment of vertebrate dispersal as it is recognized as a significant factor in successful invasion.

Many invasive species also have adaptations for dispersal by wind. This dispersal mechanism is more effective for plants invading open habitats or physically disturbed sites. Traits of successful invaders of these habitats include small seed size and seeds adapted for wind dispersal, both of which can result in long dispersal distances. In riparian environments, successful invaders often have seeds that can float and hence be dispersed long distances by water. For example, *Mimosa pigra* (catclaw) invades floodplain ecosystems in Australia, where its buoyant seed pods are spread over long distances by floodwaters. In production landscapes, many invasive species achieve long dispersal distances by adhering to livestock. Finally, exotic plants can be transported large distances by human activities. In particular, small seeds may be transported as contaminants in soil, in mud adhering to machinery or vehicle tires, or in stock feed.

In contrast, few invasive plants have seeds adapted for dispersal by ants or are large and lacking morphological adaptations for dispersal. For example, many of the Australian *Acacia* species invasive in South Africa have arils for dispersal by birds rather than elaiosomes for dispersal by ants. An exception to this generalization is *Ricinus communis* (castor oil plant) which is a fast-growing tree native to Africa and Eurasia and is now found throughout tropical regions and the Pacific. Its seeds are covered by an elaiosome; however, it is also dispersed via a ballistic mechanism and through human activities.

In summary, the ability to disperse seeds long distances is crucial for introduced plants to successfully and rapidly invade new environments. Plants achieve long-distance seed dispersal either by being small or by having adaptations for dispersal, with dispersal by birds, wind, water, and human activities being most strongly correlated with invasion success.

THE ROLE OF SEED BANKS IN INVASION

Most plant species, particularly in areas outside the tropics, have seed dormancy mechanisms that enable the development of a soil- or canopy-stored seed bank. Seed banks are particularly advantageous when conditions for establishment are variable and unpredictable. Seed banks enable plants to spread the risk of germination and establishment over time, which is analogous to the effect of having more numerous, more widely dispersed seeds, which enables plants to spread risk across space. In general, seed banks may assist in stabilizing population dynamics and in facilitating population recovery after disturbances.

When a species is introduced to a new environment, it faces many challenges in establishing self-sustaining populations that are able to spread and invade across the landscape. Seed dormancy is a trait that confers an advantage for newly invading species by reducing the risk of population loss due to unfavorable conditions or disturbances. Many successful invaders have long-lived seeds, and soil seed banks are frequently dominated by exotic species. Long-lived seeds stored in the soil are also readily transported by animal and human activity, via soil adhering to feet, vehicle tires, and machinery. Similarly, in fire-prone regions, many successful invaders have moderate to high levels of serotiny, where seed is stored within the plant canopy in woody fruits that provide protection from fire. Serotinous species are able to maximize recruitment from seed in the relatively favorable post-fire conditions and to maximize seed dispersal, as there are few obstructions to seed movement. Good examples of successful invaders that have a serotinous strategy are *Pinus* and *Hakea* species that have invaded fire-prone fynbos in South Africa.

Knowledge of the soil seed bank of invasive species is critical for successful management of their abundance and spread. Removal of an invasive plant species at a site may require follow-up control for many years when the species has a long-lived soil seed bank. Alternatively, for invasive species with short-lived soil seed banks, control measures can be targeted at removing adults and seedlings over a relatively short time period. For invasive species that have long-lived seeds that are cued to germinate after a particular disturbance such as fire, control measures can be targeted to short periods after the disturbance.

SEE ALSO THE FOLLOWING ARTICLES

Dispersal Ability, Plant / Invasiveness / Reproductive Systems, Plant / Propagule Pressure

FURTHER READING

Buckley, Y. M., P. Downey, S. V. Fowler, R. Hill, J. Memmot, H. Norambeuna, M. Pitcairn, R. Shaw, A. W. Sheppard, C. Winks, R. Wittenberg, and M. Rees. 2003. Are invasives bigger? A global study of seed size variation in two exotic shrubs. *Ecology* 84: 1434–1440.

Daws, M. L., J. Hall, S. Flynn, and H. W. Pritchard. 2007. Do invasive species have bigger seeds? Evidence from intra- and inter-specific competition. *South African Journal of Botany* 73: 138–143.

Fenner, M., and K. Thompson. 2005. *The Ecology of Seeds*. Cambridge: Cambridge University Press.

Mason, R. A., J. Cooke, A. T. Moles, and M. R. Leishman. 2008. Reproductive output of native and exotic plants: A global comparison. *Global Ecology and Biogeography* 17: 633–640.

Myers, J. H., and D. R. Bazely. 2003. *Ecology and Control of Invasive Plants*. Cambridge: Cambridge University Press.

Rejmánek, M., D. M. Richardson, S. I. Higgins, M. J. Pitcairn, and E. Grotkopp. 2005. Ecology of invasive plants: State of the art (104–161). In H. A. Mooney, R. N. Mack, J. A. McNeely, L. E. Neville, P. J. Schei, and J. K. Waage, eds. *Invasive Alien Species: A New Synthesis*. Washington, DC: Island Press.

Richardson, D. M., and M. Rejmánek. 2004. Conifers as invasive aliens: A global survey and predictive framework. *Diversity and Distributions* 10: 321–331.

SHRUBS

SEE TREES AND SHRUBS

SLUGS

SEE SNAILS AND SLUGS

SMALL INDIAN MONGOOSE

WARREN HAYS

Hawaii Pacific University, Honolulu

The small Indian mongoose, *Herpestes auropunctatus*, was intentionally introduced to at least 45 islands (including 8 in the Pacific) and one continental mainland between 1872 and 1979, in the hope of controlling rat and snake populations. This small carnivore is now found on the mainland or islands of Asia, Africa, Europe, North America, South America, and Oceania. In many areas of introduction, it has severely reduced or extirpated native birds, small mammals, reptiles, and amphibians. The small Indian mongoose is now firmly established in dense populations on many islands around the world, creating a major barrier to preservation or reestablishment of dozens of rare or endangered species.

NATURAL HISTORY

The small Indian mongoose (Fig. 1) is a small mammalian carnivore with an adult body mass typically between 300 and 900 g. It is an entirely diurnal species that dens in shallow burrows at night and spends the day largely concealed under the cover of vegetation. Its natural range consists of South and Southeast Asia from Iraq to Myanmar, with

FIGURE 1 The small Indian mongoose, *Herpestes auropunctatus*. (Photograph courtesy of jackjeffreyphoto.com.)

the exception of southern India. The northern boundary of its range is set by the temperature isocline at which mean annual low temperature falls below 10°C. It is also found on the islands of Java and Macau, where it was presumably established by human agency in ancient times.

Populations commonly extend from sea level to 2200 m in elevation, both in the natural range and in areas of introduction. The small Indian mongoose is generally more common in dry environments, but in some areas of introduction, such as the Hawaiian Islands, even very wet forest may be densely populated with mongooses. Individuals have been reported in Maui's Kipahulu valley, which has an annual rainfall of 750 cm per year.

Small Indian mongooses are omnivorous, but primarily insectivorous. Nevertheless, they are extremely effective opportunistic predators upon small mammals, ground-nesting birds, snakes, lizards, frogs, toads, and crabs, and a dense mongoose population can place extreme stress on populations of these prey species. Although mongoose population densities have never been established with

certainty either in their natural range or in areas of introduction, the home ranges of individuals are known to be extraordinarily small in many areas of introduction, suggesting dense populations. Radio-tracking studies have shown home ranges of between 1.4 and 3.6 ha for small Indian mongooses in both Hawaiian and Caribbean studies. These home range areas are roughly 5 percent the expected size for a mammalian carnivore with this species' body mass.

Pregnant females are found between February and August in Hawaii and the Virgin Islands, while in Fiji, in the southern hemisphere, they are found between August and February. Females sometimes produce two litters per year, carrying one to five embryos per pregnancy. The young stay with their mothers until they are fully grown, at four to six months of age. Females born early in the season may produce their first litter during their first year, before their natal breeding season ends. Some degree of complex sociality exists in the species during the breeding season, but this social system has not yet been fully elucidated. For example, adult males have been found to den in groups during the breeding season.

HISTORY OF INTRODUCTIONS

The small Indian mongoose has been intentionally introduced into wild habitats more widely than any other mammal (Fig. 2). A sugar planter introduced this species to Jamaica from eastern India in 1872, hoping to control rat infestations in sugar plantations. From there, the species was eventually introduced to 29 islands throughout the Caribbean, to reduce populations of rats and of the fer-de-lance pit viper (*Bothrops atrox*). It was also successfully introduced to the South American mainland in 1900, but the population there appears to be still confined to the narrow developed coastal strip of Guyana, Suriname, and French Guiana.

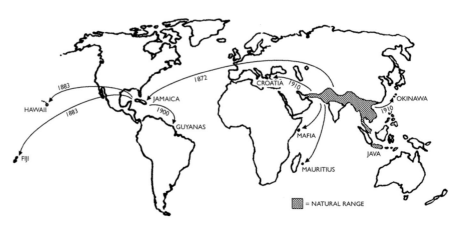

FIGURE 2 Native range and routes of introduction of the small Indian mongoose.

In the 1880s, sugar planters in Hawaii and Fiji imported the small Indian mongoose from Jamaica. It is now well established on all the largest Hawaiian and Fijian islands, with the exception of Kauai. It has also been introduced to the island of Ambon in the Indonesian archipelago, and Mauritius, Mafia, and Grande Comore in the Indian Ocean. It was introduced to Okinawa to control the habu pit viper (*Trimeresurus flavoviridis*), and later also to nearby Amami Island. It was introduced to the island of Mljet off the coast of Croatia in the Adriatic Sea to control the horned viper (*Vipera ammodytes*) in 1910, and has more recently been introduced to four other Dalmatian islands and the European mainland. The mainland population is now in Croatia, Bosnia and Herzegovina, and Montenegro as far south as the Albanian border, and it appears to still be spreading. These populations are also of special note because this is by far the coldest environment inhabited by the species anywhere, with occasional frosts in winter.

IMPACT ON OTHER SPECIES

There is a common story in Hawaii that small Indian mongooses failed to control Norway rats (themselves an unwanted introduced species), because the mongoose is diurnal and rats are primarily nocturnal. But most published accounts dispute this story, asserting that the small Indian mongoose served as an excellent cane-field ratter, though it was eventually made obsolete by the development of improved techniques of rat poisoning.

The effects of mongooses on native fauna are less contentious—particularly the impact of the species on ground-nesting birds. In Fiji, the barred-wing rail (*Nesoclopeus poecilopterus*) was common in 1875; in 1883, the mongoose was introduced, and within a few years, the bird was extinct. Similarly, the Jamaica petrel (*Pterodroma caribbaea*) was abundant in the late eighteenth century; the mongoose was introduced to Jamaica in 1872, and this bird has not been seen since 1893.

In addition to these outright extinctions, a number of island populations of ground-nesting birds were extirpated when mongooses were introduced. Four once-common bird species have been extirpated from all Fijian islands with mongooses, but persist on mongoose-free islands: the banded rail (*Rallus philippensis*), sooty rail (*Porzana tabuensis*), white-browed rail (*Poliolimnas cinereus*), and purple swamphen (*Porphyrio porphyrio*). The uniform crake (*Amaurolimnas concolor*) was extirpated from Jamaica sometime after 1881, and mongooses were probably a major factor. In the Hawaiian Islands, the mongoose was probably a major cause of extirpation or reduction of the endangered Newell's shearwater (*Puffinus newelli*)

on Oahu, Molokai, Maui, and Hawaii. The mongoose is currently considered one of the most serious threats to the Okinawan rail (*Gallirallus okinawae*), an endangered flightless bird discovered on Okinawa in 1978.

In Puerto Rico, mongooses have contributed to the population decline of five ground-nesting birds: two species of quail dove (*Geotrygon* spp.), the short-eared owl (*Asio flammeus*), the West Indian nighthawk (*Chordeiles gundlachii*), and the Puerto Rican nightjar (*Caprimulgis noctitherus*). Mongooses are suspected of preying on the nests of at least seven federally listed endangered birds in Hawaii, in addition to the Newell's shearwater: the nene or Hawaiian goose (*Branta* [*Nesochen*] *sandvicensis*), the alala or Hawaiian crow (*Corvus hawaiiensis*), the Hawaiian duck (*Anas wyvilliana*), the Hawaiian coot (*Fulica americana alai*), the Hawaiian stilt (*Himantopus mexicanus knudseni*), the Hawaiian gallinule (*Gallinula chloropus sandvicensis*), and the Hawaiian dark-rumped petrel (*Petrodroma phaeopygia sandwichensis*). Mongooses are considered a major impediment to the ongoing efforts to protect and reestablish these species.

Mongoose introductions have also had a major impact on reptile and amphibian populations on Caribbean and Pacific islands. In particular, the small Indian mongoose probably caused the extinction of the Hispaniola racer (*Alsophis melanichnus*), a colubrid snake that was endemic to Hispaniola. It has also extirpated the Antiguan racer (*Alsophis antiguae*) from Antigua, the Martinique clelia snake (*Liophis cursor*) from Martinique, and the Santa Lucia groundsnake (*L. ornatus*) from St. Lucia. Each of these species persists on mongoose-free satellite islands. Mongooses have also reduced Jamaica's black racer (*Alsophis ater*) to IUCN critically endangered status.

Census data show that lizards may be nearly 100 times more abundant on Pacific islands without mongooses than on islands that have them. Mongooses were probably largely responsible for the extirpation of the two largest skinks, *Emoia nigra* and *E. trossula*, from the two largest Fijian islands, Viti Levu and Vanua Levu. These lizard species vanished from these islands within a few years of the mongoose's arrival, and they are still found on mongoose-free Fijian islands. The mongoose was also largely responsible for the extirpation of the Saint Croix ground lizard (*Ameiva polops*) on St. Croix.

Small Indian mongooses prey upon the eggs and young of at least four endangered species of sea turtle: the hawksbill sea turtle (*Eretmochelys imbricata*), leatherback turtle (*Dermochelys coriacea*), green sea turtle (*Chelonia mydas*), and loggerhead sea turtle (*Caretta caretta*). The edible frog (*Leptodactylus pentadactylus*) has been extirpated from the three Caribbean islands of its original range to which mongooses have been introduced but persists on two mongoose-free islands.

Mongooses may also have caused, or contributed to, the extinction of the Haitian island shrew (*Nesophontes hypomicrus*), the small Haitian island shrew, (*N. zamicrus*), the large Haitian island shrew (*N. paramicrus*), and the Hispaniolan spiny rat (*Brotomys voratus*). All four of these small mammals went extinct within a few decades of the mongoose introduction to Hispaniola. Mongooses have probably been the primary cause of the reduction of the Cuban solenodon, *Solenodon cubanus*, a small mammalian insectivore, to endangered status.

OUTLOOK

In virtually all of its introduced range, the small Indian mongoose is a major impediment to the reestablishment of many extirpated or reduced vertebrate populations on dozens of islands. However, this is a role that the mongoose shares in most places with other introduced predators such as feral cats and various rat species, as well as other forms of habitat degradation. On most of the islands where mongooses are found today, managing the mongoose population is just one part of the difficult task of habitat restoration.

Almost all efforts at mongoose population management in areas of introduction have focused on the use of traps to remove individuals from ecologically sensitive areas, particularly the nesting sites of rare, ground-nesting birds. But trapping programs are of limited use, because removing the mongooses from a section of habitat creates a "population vacuum"—a region of low competition that lures more mongooses into the area.

The small Indian mongoose is strongly entrenched in most of its areas of introduction. It appears to have high population densities on many large islands, it is often found in habitat that would have been unsuitable for it within its natural range, and in some places (e.g., the Hawaiian Islands) it has no natural predators and few communicable diseases and parasites. Only the most drastic measures, such as aerial dispersion of bait pellets containing diphacinone or other toxins, would have any chance of eliminating such populations from even small islands—and the ecological costs of such efforts might outweigh their benefits in many cases. Managing the existing populations, although difficult, is likely to remain the most viable option for the foreseeable future.

SEE ALSO THE FOLLOWING ARTICLES

Biological Control, of Animals / Birds / Carnivores / Rats / Reptiles and Amphibians

FURTHER READING

Barun A., I. Budinski, and D. Simberloff. 2008. A ticking time-bomb? The small Indian mongoose in Europe. *Aliens* 26: 14–16.

Corbet, G. B., and J. E. Hill. 1992. *The Mammals of the Indomalayan Region*. Oxford: Oxford University Press.

Hays, W. S. T., and S. Conant. 2007. Impact of the small Indian mongoose (*Herpestes javanicus*) (Carnivora: Herpestidae) on native vertebrate populations in areas of introduction. *Pacific Science* 61: 3–16.

Henderson, R. W. 1992. Consequences of predator introductions and habitat destruction on amphibians and reptiles in the Post-Columbus West Indies. *Caribbean Journal of Science* 28: 1–10.

Hinton, H. E., and A. M. S. Dunn. 1967. *Mongooses: Their Natural History and Behavior*. London: Oliver and Boyd, Ltd.

Nellis, D. W., and C. O. R. Everard. 1983. The biology of the mongoose in the Caribbean. *Studies of the Fauna of Curaçao and the Caribbean Islands* 195: 1–162.

SNAILS AND SLUGS

ROBERT H. COWIE

University of Hawaii, Honolulu

Snails and slugs are gastropod molluscs (phylum Mollusca, class Gastropoda). They occur in marine, brackish, freshwater, and terrestrial habitats. Many species have been introduced both deliberately and inadvertently throughout the world. Some of them have become invasive and have caused serious environmental, agricultural, human and animal health, and commercial problems. The main problems are posed by terrestrial and freshwater species, although in notable instances, brackish-water or marine species have been introduced and have become invasive.

PATHWAYS OF INTRODUCTION

With some notable exceptions, most alien snails and slugs are transported inadvertently. Quarantine agencies around the world routinely intercept numerous species.

Deliberate Introductions

The following are the major pathways of deliberate introduction of snails. Slugs are not known to have been deliberately introduced.

FOOD (INCLUDING AQUACULTURE) Various species of European land snails (mostly Helicidae) have been deliberately introduced around the world as "escargot" (e.g., *Cornu aspersum* [formerly *Helix aspersa*; Fig. 1] to New Caledonia and Brazil, *Otala lactea* to Bermuda), becoming serious pests in many areas (e.g., California). They are

FIGURE 1 *Cornu aspersum*, the European brown garden snail, has been widely introduced globally as a species of "escargot," known in France as the "petit gris." It has probably also been introduced inadvertently. This snail has become an important pest in both agricultural and garden situations, for instance in Californian citrus groves. (Photograph courtesy of K.A. Hayes.)

often transported in personal baggage, or even sewn into clothing, to be released into backyards or vacant lots for harvest once the populations have become large enough, or to be used for large-scale captive commercial enterprises, from which they readily escape.

The South American freshwater apple snail *Pomacea canaliculata* (Ampullariidae; Fig. 2) was taken to Southeast Asia in about 1980, with the intention of developing aquaculture programs to supplement local food resources and to develop a gourmet export industry. It was also taken to Hawaii (probably from Southeast Asia) for the same reasons. The viviparids *Cipangopaludina chinensis* and *C.*

japonica, also freshwater snails, were probably also introduced to the United States from Asia as food resources; *C. chinensis* has been in Hawaii since at least 1900.

The giant African snail, *Lissachatina fulica* (formerly *Achatina fulica*; Achatinidae), has been introduced to many primarily tropical areas as a food resource, not only for humans but also for domestic ducks, as well as for other reasons (see below). Other achatinids such as *Archachatina marginata* may also have been introduced for this purpose.

AQUARIUM INDUSTRY Many species of freshwater snails are moved around the world to be sold in pet stores for use in domestic aquaria. They then either escape or are released. These include various Ampullariidae (apple snails; mostly *Pomacea* spp. and *Marisa cornuarietis*), Viviparidae (e.g., *Cipangopaludina chinensis*), Planorbidae (ramshorn snails, e.g., *Helisoma* spp., *Planorbis* spp.), Physidae, and Lymnaeidae (e.g., *Radix auricularia*). These and other species may also be transported inadvertently in association with domestic aquarium plants and fish (see below).

MEDICINAL PURPOSES Some snail species have been used for a variety of medical purposes, mostly for extracting medicinally useful compounds, although they have not been introduced widely for such purposes. For example, *Lissachatina fulica* was introduced to Hawaii for unspecified medicinal purposes, as well as for other reasons, and was much earlier introduced to La Réunion to make snail soup as a remedy for a chest infection.

BIOLOGICAL CONTROL The best publicized example is the introduction of predatory snails, notably *Euglandina rosea* (Spiraxidae) and *Gonaxis* spp. (Streptaxidae), but

FIGURE 2 A species of apple snail, *Pomacea canaliculata*, crawling under water (A), and its eggs laid on emergent vegetation (B), in a taro field in Hawaii, where it has become a major crop pest. Freshwater apple snails, notably *Pomacea canaliculata* and *P. insularum*, have also been introduced to many parts of southern and eastern Asia, where they have become major pests of rice and other aquatic crops, having been introduced deliberately for food. Originally from South America, both species, and a number of others, have also been introduced to the United States, where they threaten natural as well as agricultural ecosystems. (Photographs courtesy of K.A. Hayes.)

FIGURE 3 (A) The giant African snail, *Lissachatina fulica*, which has been introduced widely throughout the tropics and subtropics for food and other purposes. (Photograph courtesy of the author.) (B) The carnivorous snail *Euglandina rosea* (native to the southeastern United States), which has been introduced as a predator in attempts to control the giant African snail. This strategy has been unsuccessful, and even worse, *E. rosea* has caused declines and extinctions of native snail species, notably on Pacific islands. (Photograph courtesy of K. A. Hayes.)

also a large number of other species, in ill-conceived attempts to control the giant African snail (*Lissachatina fulica*) in many islands of the Pacific and Indian oceans (Fig. 3). *Euglandina rosea* has also been used against other snail pest species (e.g., the helicid *Otala lactea* in Bermuda).

The southern European facultatively predatory snail *Rumina decollata* (Subulinidae) is widely used in attempts to control other invasive snail species in the United States. Initially introduced to California to control the European *Cornu aspersum*, it has been widely introduced to other areas both deliberately as a biocontrol agent and inadvertently by the horticultural trade (see below).

A number of freshwater species (e.g., *Marisa cornuarietis*) have been introduced, especially in the Caribbean region, to eradicate or control invasive aquatic weeds such as water lettuce (*Pistia stratiotes*). Various species of

Ampullariidae and Thiaridae have also been introduced to control snail vectors of human schistosomes by either predation or competition.

PETS A major infestation of *Lissachatina fulica* in Florida in the 1960s (subsequently eradicated) resulted from the deliberate importation of snails from Hawaii by a child. Interception of *L. fulica* in personal luggage of tourists returning from Hawaii to the U.S. mainland is routine, and deliberate smuggling of *L. fulica* by the U.S. pet trade has occurred. *Helix pomatia*, intercepted coming into California, was being imported for snail races! Probably there are other instances involving these and other species.

AESTHETICS Snails have been deliberately imported and released for aesthetic reasons. One of the original introductions of *L. fulica* into Hawaii was for such ornamental purposes. The European land snail *Otala lactea* was introduced to the United States on at least one occasion to remind immigrants of their homeland, as well as for food (see above). Shell collectors and dealers may have introduced aesthetically attractive species (e.g., the European helicid *Cepaea nemoralis* to the United States). Other large or brightly colored species, especially Caribbean species of *Liguus* and *Orthalicus*, have been released for similar reasons (e.g., species from Cuba and the Bahamas into Florida) and perhaps continue to be imported.

BIOLOGICAL RESEARCH Snails imported for research purposes may have escaped or even been released into the wild. For instance, in the early twentieth century, various species of *Cerion* from the Bahamas, Puerto Rico, and Cuba were transplanted to the Florida Keys and to Hawaii to investigate the relative importance of genetic and environmental control of shell morphology; one of them became established in the Keys.

Inadvertent Introductions

Many species of snails and slugs are transported in or on a huge range of products, vehicles, containers, and so forth. It is likely that some travel as eggs or small juveniles, which are much less readily detected than adults. Reproductive ecology may facilitate colonization. Although pulmonate land (Stylommatophora) and freshwater (Basommatophora) snails and slugs are hermaphrodites, many species must nonetheless cross-fertilize in order to reproduce. However, some species can self-fertilize, such that introduction of a single individual could lead to successful invasion. European slugs introduced to North America are disproportionately represented by such selfing species. Some cross-fertilizing species are also facultative selfers in the absence of mates. Also, sperm storage, which is well known in pulmonates, would permit single individuals of cross-fertilizing species to invade successfully. Furthermore, although most "prosobranchs" (Prosobranchia is not now considered a valid phylogenetic grouping) have separate sexes and must cross-fertilize, some species (e.g., the thiarid *Melanoides tuberculata*) are predominantly parthenogenetic, which would permit colonization by a single individual. Although single individuals could lead to invasions, there may be long-term genetic consequences of reduced genetic variation in these introduced populations that influence invasion success and dynamics.

The major pathways of inadvertent introductions for snails and slugs are listed below. Other miscellaneous pathways, not discussed in detail, include association with various wood products (e.g., pallets, crating), flowerpots and other earthenware, quarry products (e.g., ornamental rocks), machinery and heavy equipment, and possibly hiking and camping equipment.

AGRICULTURAL PRODUCTS (EXCLUDING HORTICULTURE) Agriculture (products for consumption as food) is here distinguished from horticulture (predominately decorative or propagative products). Many species of snails and slugs have been found associated with shipments of agricultural produce in many countries. In the United States, for instance, numerous species are intercepted on a wide variety of fruits and vegetables, with no obvious patterns of association, except notably several species of slugs (family Arionidae) with European mushrooms.

HORTICULTURAL PRODUCTS Cut flowers, live plants, seeds, turf, leaves used for mulch, and similar products provide a ready pathway for introduction, and many species of snails and slugs are frequently intercepted in shipments of these products. The scale of the international horticultural trade is huge, and this appears to be the most important pathway for the introduction of terrestrial snails and slugs globally. Many nurseries are infested with invasive snails and slugs, especially small species (Fig. 4), which has led to their global characterization as "greenhouse aliens." Notable among these are the European slug *Lehmannia valentiana* (Limacidae) and the small North American zonitid land snails *Zonitoides arboreus* and *Hawaiia minuscula*, the latter of which was first described from invasive Hawaiian specimens without realizing that it is not native to Hawaii. Often, new records of alien species in an area are associated with nurseries, garden

FIGURE 4 *Succinea tenella*, a small snail frequently transported accidentally with horticultural products. The horticultural trade is one of the major vectors of alien snails and slugs globally. (Photograph courtesy of K. A. Hayes.)

stores, botanical gardens, or recent landscaping activities, including *Parmarion martensi* (Ariophantidae) and *Polygyra cereolus* (Polygyridae) in Hawaii and *Bulimulus guadalupensis* (Bulimulidae) in Florida.

Many of the species associated with this pathway may have been imported to a nursery from elsewhere and then exported to one or more additional regions. For example, the euconulid *Liardetia doliolum* (alien in Hawaii) has been exported to the U.S. mainland from a nursery in Hawaii, and the neotropical *Guppya gundlachi* is regularly intercepted in horticultural shipments to the United States from Thailand. Leaves used as mulch have been suggested as the pathway by which litter-dwelling snails could be introduced, and hay used as mulch facilitated the spread of *Theba pisana* (Helicidae) in California.

COMMERCIAL AND DOMESTIC SHIPMENTS Numerous commodities, usually shipped in containers, may have snails and slugs associated with them. They may be inadvertently transported from the source with the product itself, or they may attach themselves to the product, its packaging, or the shipping container en route. Particularly important may be the time the shipment or container spends on a dockside or cargo holding area. Especially notable in the United States is shipment of domestic tiles from the Mediterranean region that are frequently infested with European snails. Snails and slugs can be transported with domestic shipments of household goods just as they can with commercial shipments.

MILITARY SHIPMENTS Snails and slugs can also be transported with military supplies and other goods associated with military campaigns, as well as during routine peacetime transportation of military equipment and supplies. For example, there were 231 snail interceptions on military equipment returning from Europe to the United States in the immediate aftermath of the 1999 U.S. involvement in the Balkans.

VEHICLES, RAIL, AND AIRPLANES Snails may attach themselves to vehicles (e.g., parked cars) and then be transported to wherever the vehicle is driven. Many snails are intercepted on private cars and commercial trucks crossing the U.S.–Mexican border. Similar transport by rail is likely. Airplanes may be less important because of the extreme cold at high altitude and the low likelihood of survival on landing in the inhospitable environment of large, open runways. Nevertheless, prior to reaching the extreme cold of high altitude, attached snails may be blown off as a result of the speed of the airplane, resulting in expansion of their range by relatively short-distance dispersal.

SOIL Snails and slugs, and especially their eggs, are readily transported in soil. This could result from (1) soil that is deliberately or accidentally transported with agricultural or horticultural products, (2) soil that is to be used for landscaping or top-dressing, or (3) soil that becomes transplanted by vehicles, shoes, and so forth. The United States prohibits import of any shipment with soil for these reasons.

AQUARIUM INDUSTRY Small freshwater snails (e.g., species of Physidae, Thiaridae, Planorbidae, and Lymnaeidae) are easily transported inadvertently attached to aquarium plants. Also, snails may be associated with aquaculture of aquarium and other alien fish and have been transported with them (e.g., *Biomphalaria straminea* from Hong Kong to Australia).

AQUACULTURE Aquatic snails can be introduced accidentally (or even intentionally) along with a different species (not necessarily snails) specifically introduced for aquaculture. This may have been the mechanism of introduction of the freshwater hydrobiid snail *Potamopyrgus antipodarum* to the Snake River from elsewhere within the United States. Also, aquaculture of aquarium fish may be important in this respect, as mentioned above.

SHIPS AND BOATS Although ballast water is more often implicated in the introduction of marine species, if a ship takes on ballast in a freshwater harbor and then discharges it at the end of its voyage in another freshwater harbor, it is possible that freshwater snails (or their larvae) could be introduced (e.g., *Potamopyrgus antipodarum, Bithynia tentaculata, Radix auricularia* introduced to North America). Dry ballast is no longer used and no longer constitutes a pathway for introduction of terrestrial species. Hull-fouling may also constitute a mechanism of transport, possibly if snails or their eggs (e.g., Physidae, Lymnaeidae) are attached to hulls or to the associated algae of small boats or barges, and then moved from one body of freshwater to another, as is the case for zebra mussels, although no instances involving snails appear to have been reported in the literature.

CANALS AND OTHER MODIFIED WATERWAYS Canals joining two or more formerly unconnected bodies of water form "biotic corridors" permitting the formerly distinct faunas to mix. For instance, two pleurocerid snails (*Elimia livescens, Pleurocera acuta*) and a valvatid (*Valvata piscinalis*) have been introduced to the Hudson River basin in the northeastern United States via this route. Canals can also facilitate the spread of alien species from a focus of introduction (e.g., *Potamopyrgus antipodarum* in Britain). Dams and other modifications of existing rivers may alter the aquatic habitat in ways that facilitate the spread of alien snails. The Suez Canal has permitted the invasion of the Mediterranean by many species of Indian Ocean marine molluscs.

ROADS By acting as corridors for the introduction of alien plants, roads may create habitat more suitable for invasion by alien snails, thereby facilitating the spread of a species from a focus of introduction. However, in no instance has this been demonstrated to have taken place. Roadside ditches may act similarly as corridors along which freshwater snails can invade.

IMPACTS

The impacts of invasive snails and slugs can be categorized as ecological (including impacts on native biodiversity), agricultural and horticultural, health-related (affecting humans and livestock), and commercial. These categories are not mutually exclusive and may overlap in some cases, and some invasive snails and slugs may have multiple impacts.

Environmental Impacts

PREDATION The best-known story of introduced nonmarine snails is that of *Lissachatina fulica* and *Euglandina rosea*. In 1936, the giant African snail, *L. fulica*, was introduced to Hawaii from Japan. By the 1950s, it had become so abundant that efforts were initiated to try to control it by introducing predatory snails as biological control agents. In short, 15 predatory snail species were introduced into the wild in Hawaii; a small number of them became established, the most notable of which is *Euglandina rosea* from Florida. There is no good scientific evidence that *E. rosea* (or any of the other predatory snails) has controlled populations of *L. fulica* in Hawaii, but *E. rosea* has been heavily implicated in the decline and possible extinction of Hawaiian tree snails (Achatinellinae) and possibly other endemic Hawaiian snails. *Euglandina rosea* (and some of the other predatory species) has been taken to other Pacific islands, notably the Society Islands of French Polynesia (Tahiti, Moorea, etc.), where it has caused similar declines or extinctions, especially of the partulid tree snails, but probably also of other endemic species.

Another example of the use of nonindigenous predatory snails in attempts to control other invasive snail species is the widespread use of the southern European *Rumina decollata*. This facultative predator was introduced initially to California and subsequently to other areas to "control" the European *Cornu aspersum* (formerly *H. aspersa*), although evidence that populations of *C. asperjum* were reduced as a direct result of predation by *R. decollata* is scanty. Its impact on native species is not known.

Various species of Ampullariidae and Thiaridae have also been introduced to control snail vectors of human schistosomes by either predation or competition. Impacts on native vegetation and associated fauna are generally not discussed or are simply ignored.

Such biological control agents have frequently been introduced simply because they were reported to feed on the target, without demonstrating (1) that they are specific to the target and will not attack native species, and (2) that they can indeed reduce populations of the target. Most official biocontrol programs now make some attempt to demonstrate specificity to the target, but the much more complex problem of demonstrating potential reduction of pest populations is rarely addressed. Furthermore, putative biocontrol agents are frequently introduced unofficially with no testing at all.

COMPETITION Introduced *Pomacea canaliculata* and *Pomacea insularum* (Ampullariidae) have been implicated in the decline of native freshwater snail species in Southeast Asia and Florida. Other ampullariids have caused the decline of native species in the Caribbean islands. In North America, the introduced New Zealand mud snail, *Potamopyrgus antipodarum*, is outcompeting native stream snails. Among land snails, there are no clear cases in which competition between introduced and native species has been demonstrated, although it is possible that the decline of certain native litter-dwelling snail species in the Pacific islands (e.g., *Pleuropoma fulgora* in the Samoan Islands) has been caused by competition between these species and introduced subulinids, notably *Allopeas gracile*, *Subulina octona*, and *Paropeas achatinaceum*.

HABITAT MODIFICATION In Florida, the introduced ampullariid *Marisa cornuarietis* dramatically reduces native freshwater plants, thereby destroying habitat for the native animals associated with them. *Marisa cornuarietis* and *Pomacea glauca*, introduced to Caribbean islands to eradicate or control invasive aquatic weeds such as water lettuce (*Pistia stratiotes*), may also have impacted native vegetation. Australia is especially concerned should these apple snails be introduced to natural wetland areas such as Kakadu.

In Hawaii, a recent study showed that invasive slugs (*Deroceras laeve*, *Limacus flavus*, *Limax maximus*, *Meghimatium striatum*) are preventing rainforest regeneration by destroying newly germinated seedlings of native forest plants. The widely invasive giant African snail (*Lissachatina fulica*) and veronicellid slugs (notably *Veronicella cubensis*) may be causing similar problems in Hawaii and elsewhere, although this has not been formally documented.

AGRICULTURAL AND HORTICULTURAL IMPACTS Many species of snails and slugs have become agricultural and horticultural pests, as well as being moved around the globe in association with agricultural and horticultural products. Notable agricultural pests in North America are the slugs *Deroceras reticulatum* and various species of *Arion*, which are particularly serious pests of maize, soybean, lucerne, and other forage legumes. These slugs, as well as species of *Milax*, are also the most serious agricultural pests in Europe and New Zealand, where they attack maize and other cereals, oilseed rape, sunflower, potatoes, and other vegetable and ornamental crops. Helicid snails (*Cornu aspersum*, *Theba pisana*) are serious pests of grapevines in California, South Africa, and Australia, although *T. pisana* has been almost completely controlled in California; *Bradybaena similaris* is a grapevine pest in Taiwan. In Australia, *T. pisana* and a number of other species are extremely serious pests of cereals, as they aestivate on the stems and ears, contaminating the crop and clogging machinery during harvest (Fig. 5).

In tropical regions, *Lissachatina fulica* and other achatinids have been considered serious pests, although their

FIGURE 5 The European snail *Cernuella virgata* on stubble remaining after cereal harvest (A) and congregated on a fence post (B) in South Australia, where this species and other invasive European species, notably *Theba pisana* and *Cochlicella* spp., are serious pests, contaminating the harvested grain and clogging harvesting machinery. *Theba pisana,* in particular, has also been considered a pest elsewhere, especially in California and South Africa. (Photographs courtesy of the South Australian Research and Development Institute.)

agricultural impacts may have been exaggerated. Veronicellid slugs are also serious pests in the tropics. The North American land snail *Zonitoides arboreus* is a pest in orchid nurseries in Hawaii.

The freshwater "apple snails" *Pomacea canaliculata* and *P. insularum*, introduced to Southeast Asia, escaped, or were released, and are now serious agricultural pests, primarily of rice; they have also spread to natural ecosystems (see above). *Pomacea canaliculata* and other ampullariids were also brought to Hawaii as potential human food resources. They were released into taro fields (similar to rice paddies) and are now the most serious pests of taro. Apple snails have also become widespread in parts of the continental United States, where they threaten rice crops; Australia is especially concerned should they be introduced.

For comprehensive information on snails and slugs as crop pests, see the accounts listed under Further Reading.

HUMAN AND LIVESTOCK HEALTH AND WELL-BEING Many snails and slugs act as vectors of human and animal parasites. Most notable among these parasites are the rat lungworm, *Angiostrongylus cantonensis*, which causes eosinophilic meningoencephalitis in humans, and the cattle liver fluke *Fasciola hepatica*. Veronicellid slugs (Fig. 6), *Lissachatina fulica* (Fig. 3), the semi-slug *Parmarion martensi*, and apple snails (*Pomacea* spp.; Fig. 2) have been especially implicated as vectors of *A. cantonensis*. Freshwater lymnaeids are particularly notable as vectors of cattle liver flukes. Many of these major vectors are invasive alien species.

Some species have simply become a general nuisance. *Lissachatina fulica* became so abundant in the 1950s in Hawaii that it was almost impossible to avoid them when one walked on sidewalks. They would crawl up the sides of houses in great numbers, and there were even stories of cars skidding on the massed snails.

COMMERCE Probably the most important commercial impact of introduced snails and slugs, apart from their

FIGURE 6 Three color forms of the Caribbean slug *Veronicella cubensis*. This and other veronicellid slugs, including *Sarasinula plebeia*, *Laevicaulis alte*, and others, are becoming increasingly widespread in humid tropical and subtropical regions, where they may become serious agricultural, garden, and environmental pests, as well as being vectors of human parasitic diseases such as angiostrongyliasis. (Photograph courtesy of the author.)

direct impacts on commercial agricultural and horticultural production and the consequent economic costs of implementing control measures, is that they may often be contaminants of shipments of these products. This can lead to rejection of a shipment exported to another jurisdiction. Australia has suffered major economic losses from rejection of bulk shipments of contaminated grain to other countries. The first knowledge that *Liardetia doliolum* was present in Hawaii was its interception by quarantine authorities in California (and shortly thereafter in Arizona) on shipments of horticultural products exported from Hawaii.

WHAT CAUSES A SNAIL OR SLUG TO BE INVASIVE?

Criteria thought to promote invasiveness in snails and slugs include both biological attributes of the species and attributes related to their interaction with people, as for most invasive organisms. Evaluating these attributes may permit identification of those species most likely to become invasive if introduced.

Biological Attributes

RANGE If a species has a wide natural climatic range, it could impact a larger area when introduced.

PHYLOGENETIC RELATIONSHIPS If a species is closely related to known pests, then the likelihood of its becoming a pest is greater.

ADULT SIZE Larger snail species are favored for deliberate introductions, but for inadvertent introductions, smaller species have a greater chance of evading quarantine.

JUVENILE/EGG SIZE Production of smaller, and therefore more readily dispersed, offspring could lead to a species' more rapid and wider dispersal.

REPRODUCTIVE POTENTIAL If a species produces large numbers of young in a short period of time, then the chances of it being more invasive may be greater.

SEMELPARITY OR ITEROPARITY Semelparous species put all their reproductive effort into a single reproductive event, thereby shortening their life cycle. Semelparity is probably correlated with high reproductive potential, so semelparous species may be more invasive than iteroparous species, which spread their reproductive effort over more than one breeding season.

BREEDING SYSTEM Self-fertilizing or parthenogenetic species, rather than outcrossing species, may be better invaders.

Human Interaction Attributes

PROPAGULE PRESSURE Higher propagule pressure, often inferred from frequent interception, may increase the likelihood of establishment.

INTRODUCTION HISTORY A species introduced widely elsewhere may have a greater likelihood of becoming invasive in the region in question.

KNOWN AS A MAJOR PEST ELSEWHERE If a species is a major pest elsewhere, there is a greater likelihood that it will cause serious problems in the region in question.

ERADICATION, CONTROL, AND PREVENTION

Eradication

Eradication of newly detected infestations of snails and slugs may sometimes be possible, but eradication of well-established species, especially small ones that may go unnoticed until they are widespread, is generally difficult. *Lissachatina fulica* was eradicated from Florida in the 1960s after a lengthy and expensive effort. In California, *Theba pisana*, a citrus pest, was eradicated in the late 1940s after a long campaign that involved extensive use of flame throwers. It has since reappeared. In the Pacific, new infestations of *L. fulica* have been eradicated in Tuvalu and possibly Kosrae, and eradication of a new infestation of the freshwater ampullariid snail *Pila conica* (and perhaps of a species of *Pomacea*) was accomplished in Palau. Efforts have been made to eradicate and then exclude well-established alien predatory snails (and rats) from small (approximately 500 m^2) native snail reserves in Hawaii and French Polynesia, although with limited success largely because of the difficulty of maintaining the exclosures.

Control

Control or management of alien snails, as opposed to complete eradication, is possible in some circumstances. Molluscicides are used over large areas of agricultural crops in both Europe and North America to control slugs, although cultural methods, such as increased tilling and removal of crop residues, are also practiced. In Asia, invasive apple snails (*Pomacea* spp.) are controlled to varying degrees by using pesticides, various cultural methods, and laborious hand collection of snails and destruction of their egg clutches. In Japan, winter tilling of rice paddies reduces populations of *Pomacea canaliculata*.

Major efforts have been made to control *Lissachatina fulica* in Pacific islands and elsewhere, especially focusing on biological control with predatory snails. But these efforts have been disastrous, as noted above. In Hawaii a bait has been developed against the predatory snail *Euglandina rosea*, but it has been used only experimentally because of permitting issues.

On a small scale, baiting, most commonly with metaldehyde formulated in pellets, is moderately successful. Placing barriers of copper strips helps to prevent access to small areas. Cultural practices such as raising plant pots off the ground reduce resting places for slugs. Good nursery and garden hygiene in general seems to reduce snail and slug numbers.

Prevention of Further Introduction and Spread

Preventing entry of a harmful species is always preferable to attempting to eradicate or control it after it has been introduced. For snails and slugs, prevention is especially important because, once introduced, populations may be noticed only when they are too numerous and widespread to be eradicated. Prevention fundamentally requires aggressive quarantine regulations and actions at three basic stages, as follows.

PREENTRY REGULATIONS AND SCREENING Screening protocols, such as those developed for plants by a number of countries, would allow authorities to evaluate the likely invasiveness of an alien species of snail or slug and determine whether it should be placed on a list of prohibited species (black list) or species permitted entry (white list).

PORT-OF-ENTRY QUARANTINE This can be at the level of a country, a subdivision of a country (state, county), an island, or even small areas such as a reserve. It requires far more personnel, who must be adequately trained, than are currently employed even in wealthy countries.

POST-ENTRY RAPID RESPONSE The need to quickly eradicate new populations that result from alien species slipping through the first two stages of the quarantine net has been discussed above.

EDUCATION AND NEEDS

Finally, and although it is becoming something of a refrain, educating and influencing not only the general public but also politicians, managers, and business people about the impacts, realized and potential, of alien species is crucial. Snails and slugs are rarely mentioned as problematic invasive species in public arenas. While it is important that snail and slug specialists communicate with their

colleagues by publishing in their peer-reviewed journals, it is in the end more important to air concerns more widely. With the exception of some well-known examples, disproportionately little effort is put into prevention of snail and slug invasions compared to the huge sums put into insect, weed, and to some extent plant disease pest management. Yet apple snails (*Pomacea* spp.) are now the number one rice pest in Asia; *Theba pisana* and a number of hygromiids are major cereal pests in Australia; invasive European slugs are major crop pests in North America; *Lissachatina fulica* is often considered a major pest, disease vector, and general nuisance throughout much of the tropics; and *Euglandina rosea* has caused the extinction of endemic species on Pacific islands. Heightened awareness of the potential problems that could be caused by invasive snails and slugs should lead to greater efforts to prevent them from being introduced and, should they be introduced, greater awareness of the need to act quickly in eradicating or controlling them before they become invasive. A key need also, however, is better knowledge of the basic biology of many of these potentially invasive species, and especially increased understanding of their potential environmental, as opposed to agricultural, impacts.

SEE ALSO THE FOLLOWING ARTICLES

Agriculture / Biological Control, of Animals / Black, White, and Gray Lists / Disease Vectors, Human / Eradication / Hawaiian Islands: Invasions / Horticulture / Pesticides (Fish and Mollusc)

FURTHER READING

Barker, G. M. 1999. *Naturalised Terrestrial Stylommatophora (Mollusca: Gastropoda)*. Lincoln, Canterbury, New Zealand: Manaaki Whenua Press.

Barker, G. M., ed. 2001. *The Biology of Terrestrial Molluscs*. Wallingford, UK: CABI Publishing.

Barker, G. M., ed. 2002. *Molluscs as Crop Pests*. Wallingford, UK: CABI Publishing.

Cowie, R. H. 2001. Can snails ever be effective and safe biocontrol agents? *International Journal of Pest Management* 47: 23–40.

Cowie, R. H., ed. 2004. Non-marine alien molluscs: The future is a foreign ecosystem. *American Malacological Bulletin* 20: 87–159.

Cowie, R. H., R. T. Dillon Jr., D. G. Robinson, and J. W. Smith. 2009. Alien non-marine snails and slugs of priority quarantine importance in the United States: A preliminary risk assessment. *American Malacological Bulletin* 27: 113–132.

Cowie, R. H., K. A. Hayes, C. T. Tran, and W. M. Meyer III. 2008. The horticultural industry as a vector of alien snails and slugs: Widespread invasions in Hawaii. *International Journal of Pest Management* 54: 267–276.

Cowie, R. H., and D. G. Robinson. 2003. Pathways of introduction of nonindigenous land and freshwater snails and slugs (93–122). In G. Ruiz and J. T. Carlton, eds. *Invasive Species: Vectors and Management Strategies*. Washington, DC: Island Press.

Joshi, R. C., and L. C. Sebastian. 2006. *Global Advances in Ecology and Management of Golden Apple Snails*. Muñoz, Nueva Ecija, Philippines: Philippine Rice Research Institute.

Robinson, D. G. 1999. Alien invasions: The effects of the global economy on non-marine gastropod introductions into the United States. *Malacologia* 41: 413–438.

SOUTH AFRICA: INVASIONS

DAVID M. RICHARDSON
Stellenbosch University, Matieland, South Africa

JOHN R. WILSON
South African National Biodiversity Institute, Matieland

OLAF L. F. WEYL
South African Institute for Aquatic Biodiversity, Grahamstown

CHARLES L. GRIFFITHS
University of Cape Town, South Africa

Because of the variety of biomes present in South Africa, a substantial proportion of the planet's biota is preadapted to conditions that exist currently in the country. South Africa faces other challenges unique to the region; for example, the invasion of trees into the largely treeless fynbos biome. However, the major challenge for South Africa has been to manage biological invasions in the context of a country with scarce water resources and a pressing need for social and economic development. In the Working for Water Programme, South Africa has come up with a workable solution to integrate the management of biological invasions (plants in particular) with social development.

THE REGION: GEOGRAPHY, HISTORY, BIODIVERSITY, AND THE SOCIOECONOMIC MILIEU

South Africa covers 1.22 million km^2 at the southern tip of Africa. It has a coastline of 2,798 km, with the Indian Ocean to the east, the Atlantic to the west, and the meeting of the two oceans off the south coast. On land, South Africa is bordered to the north by Namibia, Botswana, and Zimbabwe, and to the northeast by Mozambique and Swaziland, with landlocked Lesotho being surrounded by South Africa. The Republic of South Africa (RSA) also governs the small sub-Antarctic archipelago, comprising Marion Island (290 km²) and Prince Edward Island (45 km²).

Mainland South Africa is predominantly arid, with more than half the region receiving less than 500 mm of rainfall per year. However, there is a substantial east–west rainfall gradient and strong variation in rainfall seasonality from winter rains in the southwest, through all year rains, to summer rains in the east. This rainfall variation, along with a complex topography and soils, gives rise to interesting biogeographical patterns.

Ecologically, South Africa has been called "a world in one country," with several major biome types (desert, succulent karoo, Nama karoo, fynbos, Albany thicket, grassland, and savanna) concentrated in a relatively small area (Fig. 1). As the main terrestrial biomes all have climatic and environmental analogs covering large areas of the world, a large proportion of the planet's biota is preadapted to conditions that exist currently in South Africa. The variety of biomes and their isolation from other similar areas may explain why RSA is listed by the UNEP World Conservation Monitoring Centre as being among 18 megadiverse countries. RSA is also home to three of Conservation International's biodiversity hotspots: the Cape Floristic Region, Maputaland-Pondoland-Albany, and the succulent karoo (www.biodiversityhotspots.org). The country has a long and strong tradition of biodiversity conservation with a well-managed network of protected areas.

The history of South Africa's human habitation is arguably as varied as its biodiversity. Some of the earliest paleontological records of *Homo sapiens* come from South Africa, and modern humans have inhabited the region for more than 100,000 years. Since then, there has been a succession of human migration from further north in Africa. While trade with Arab states may also have been a significant vector for the introduction of new organisms to the region, by far the most profound influences shaping the composition and arrangement of the country's nonnative biota followed the foundation of Cape Town in 1652 by the Dutch East India Company. Initially, Cape Town was a way station for trading ships, but during the 1800s, the Boers (original Dutch, Flemish, German, and Huguenot French settlers) and the British 1820 Settlers moved north and east within South Africa. The resulting colonial battles for control of the land were exacerbated by the discovery of diamonds, and later gold (e.g., the Anglo–Boer Wars). In 1948, institutionalized segregation known as apartheid was established, and the majority of South Africans, i.e. those classified as "non-Europeans," were officially disenfranchised. Racial segregation continued until the first democratic elections in 1994. Since then, all major political bodies have ascribed to the goal of ensuring equitable and sustainable development for

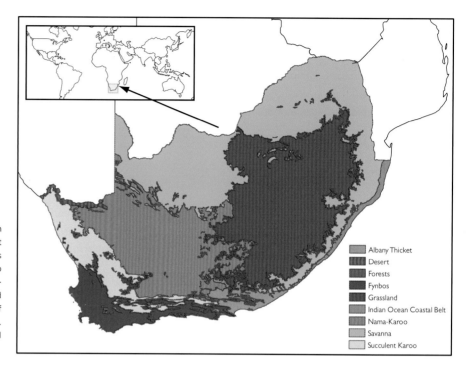

FIGURE 1 South Africa has a rich diversity of vegetation types but a similar variety of opportunities for biological invasions. The map shows the biomes present in continental South Africa (per Mucina and Rutherford, 2006, The vegetation of South Africa, Lesotho and Swaziland. *Strelitzia* 19. South African National Biodiversity Institute, Pretoria).

Legend:
- Albany Thicket
- Desert
- Forests
- Fynbos
- Grassland
- Indian Ocean Coastal Belt
- Nama-Karoo
- Savanna
- Succulent Karoo

everyone living in South Africa, but RSA remains one of the most inequitable countries in the world.

The socioeconomic challenges imposed by the demands for poverty alleviation and the provision of housing for a rapidly growing and urbanizing population have created substantial challenges for the conservation of natural capital. South Africa's biodiversity is valuable in terms of subsistence use, ranching, tourism, and sales of raw materials. Roughly 80 percent of South Africa still has semi-natural vegetation, which supports, to varying degrees, subsistence and commercial ranching. Only a tiny part of the country supports native forest, and plantation forestry using nonnative trees is an important land use in regions that previously supported grassland and shrubland vegetation. The low rainfall and dearth of perennial rivers limits primary production to 1–6 tons ha^{-1} year^{-1}. Increasing extraction from perennial rivers causes some to cease flowing for extended periods. Invasive nonnative species, especially plants, are compromising options for the conservation of natural capital and the sustained delivery of ecosystem services in many parts of the country. Interventions to provide sustained water supply to support development can create major dispersal pathways for biota, in particular interbasin water transfer schemes. Increasing international trade is also forging new (and strengthening existing) pathways for the accidental and intentional introduction of nonnative biota. Global issues similarly influence patterns in the use of and demand for a wide

range of nonnative species. For example, RSA's strategy on biofuels, if fully implemented, would result in widespread plantings of nonnative plants, including some known invasive species.

CURRENT STATUS OF NONNATIVE SPECIES

Plants

South Africa currently has about 750 nonnative tree species and around 8,000 nonnative shrubby and herbaceous species. However, only about 10 percent (roughly 850) of nonnative plant species are naturalized—permanently established without direct human support. As of December 2009, additional new legislation governing invasive alien organisms was in process, but the Conservation of Agricultural Resources Act (first passed 1983, and amended 2001) lists 199 taxa, of which 44 are legally declared or proposed for declaration as noxious weeds (i.e., their removal is required by law), and 31 have been declared or proposed for declaration as invaders (i.e., their spread must be controlled). Australian species feature especially prominently in the lists of invasive plants. Many trees and other woody plants are highly invasive in South Africa, a feature that sets the country apart from most other regions of the world, where herbaceous species generally dominate invasive floras. Certainly among Mediterranean-type ecosystems (MTEs), South Africa's fynbos biome differs markedly from the other four regions with MTEs in the dominance of trees in its invasive flora. Widespread invasions of nonnative trees from the genera *Acacia, Hakea,*

FIGURE 2 Invasive nonnative trees are a prominent feature of many South African landscapes, especially in the fynbos biome. (A) shows *Hakea sericea* (front) and *Pinus pinaster* invading fynbos near Villiersdorp in the Western Cape. (B) shows galls formed by the introduced insect *Trichilogaster acacialongifoliae* on *Acacia longifolia*. Most rivers across the country are fringed with dense stands of invasive trees, for example *Acacia mearnsii* (C). Biological control is a crucial part of the integrated management strategy for nonnative trees. (Photograph courtesy of D.M. Richardson.)

and *Pinus* in the fynbos biome are extraordinary; the rapid proliferation and spread of these invaders in fynbos are probably the best-known examples of invasive plants in South Africa (Fig. 2). *Prosopis* species have invaded large areas in the arid center of the country. Nonnative grasses are widespread: 15 percent of grass genera and 12 percent of grass species in southern Africa are naturalized nonnatives. A recent exercise defined 117 "major invaders" (invasive nonnative species that are well established and that already have a substantial impact on natural and semi-natural ecosystems) and 84 "emerging invaders" (taxa with less influence at present, but with attributes and potentially suitable habitat that could result in increased range and consequences in the next few decades). Most invasive plant species are currently confined to 10 percent or less of the region but could potentially invade up to 40 percent. Depending on the species, between 2 and 79 percent (in the case of *Arundo donax*) of quarter-degree grid cells in the region are climatically suitable for species to invade, and some areas are suitable for up to 45 members of the current set of plant invaders. Species richness of invasive plants is highest along the more mesic coastal regions and, as in other parts of the world, is positively correlated with native plant species richness. Range size for major invaders is explained by environmental variables, the traits of the species, the amount of time since the initial introduction,

and the extent and spatial distribution of potential suitable habitat. The only national-scale mapping exercise undertaken to date (in the mid-1990s) estimated that 10 million ha had been invaded (by 180 woody nonnative taxa).

Freshwater ecosystems have been invaded by a range of nonnative aquatic plant species. The five most widespread and influential species are red water fern (*Azolla filiculoides*), water hyacinth (*Eichhornia crassipes*), water lettuce (*Pistia stratiodes*), parrot's feather (*Myriophyllum aquaticum*), and salvinia (*Salvinia molesta*).

Mammals

At least 50 nonnative mammal species are known to have been introduced into natural ecosystems in South Africa, and many other species (including primates, wolves, marsupials, and bears) are known to have been introduced but held in captivity. Ungulates have been especially popular introductions for hunting, and South Africa has the second highest number of introduced ungulates globally, after the United States. Very few can be considered invasive or even established. Mammal species that are definitely invasive include the gray squirrel (*Sciurus carolinensis*; only in suburban and other human-modified areas), the house mouse (*Mus musculus*; commensal with humans), the Norway rat (*Rattus norvegicus*), the black rat (*Rattus rattus*), the fallow deer (*Dama dama*), and the red lechwe (*Kobus leche*).

Molecular studies of rats in South Africa recently revealed that *Rattus tanezumi*, until now unknown in the region, is widespread and sympatric with *R. rattus*. Two prominent invasive species that have been eradicated are cats (*Felis catus*; introduced to Marion Island and eradicated by 1991) and Himalayan tahrs (*Hemitragus jemlahicus*; introduced to Table Mountain around 1935 and eradicated by 2008). There are two known herds of feral horses (*Equus caballus*) and several populations of feral pigs (*Sus scrofa*).

Of more concern than mammals nonnative to the country are native South African mammals that have been moved by humans outside their natural range. Many species have self-sustaining, extralimital populations. Species with the smallest initial ("natural") ranges have shown the greatest increases in range extent. Extralimitals that have been the most invasive and problematic in their novel ranges are the warthog (*Phacochoerus africanus*), the impala (*Aepyceros melampus melampus*), and the nyala (*Tragelaphus angasii*). Categorization of such species as "nonnative" in parts of the country is controversial and is the subject of heated debate between conservation authorities and others, especially game ranchers.

Birds

Forty-eight bird species in South Africa owe their presence in the country to introduction by humans. Only seven species have viable feral populations, and the populations of a further 13 species are geographically restricted or decreasing. The seven species that have viable populations are all commensal with humans, and four of these (the rock pigeon *Columbia livia*, the common myna *Acridotheres tristis*, the common starling *Sturnus vulgaris*, and the house sparrow *Passer domesticus*) are widespread. All except the house crow *Corvus splendens*, which was "self-introduced," were deliberate introductions.

Reptiles and Amphibians

There are 275 species of nonnative reptiles from 30 families that have been imported to South Africa (based on records from importation and captivity permits, pet stores, zoos, and other collections). There are no invasive species in terrestrial ecosystems, although at least two species (the tropical house gecko *Hemidactylus mabouia* and the flowerpot snake *Ramphotyphlops braminus*) have established feral populations. The red-eared slider (*Trachemys scripta*), a freshwater turtle, has established several feral populations in freshwater ecosystems, although the long-term persistence of these populations has not been verified. Although the importation of amphibians is illegal in most provinces, at least 14 nonnative taxa (including five *Dendrobates* species—poison dart frogs

native to South America) are known to be in the country. Other than two species that are native to South Africa, but which are invasive outside their natural ranges (the painted reed frog *Hyperolius marmoratus* and the guttural toad *Amietophrynus gutturalis*), there are no known invasive species of amphibians in South Africa.

Terrestrial Invertebrates

Details of invasive nonnative invertebrates in South Africa are sketchy, and research has been largely restricted to agricultural pests. Of the top 40 invertebrate crop pests in the country, 42 percent are nonnative. These include the red scale (*Aonidiella aurantii*), the Mediterranean fruit fly (*Ceratitis capitata*), and the codling moth (*Cydia pomonella*). Several other species are known to be present because of their actual or potentially adverse effects. The European wasp *Vespula germanica* is still confined to a small area in the extreme southwest of the country. The Argentine ant (*Linepithema humile*) has invaded large parts of the fynbos biome, although only in human-modified areas. Its range also extends into the interior (Free State and Gauteng provinces), and it has also been recorded from localities on the periphery of the Northern Cape, but it is apparently absent from the central region of that province. Other prominent insect invaders include the wood wasp (*Sirex noctilio*), first recorded in South Africa in 1994 and now rapidly spreading northward along the eastern seaboard, and the ectoparasitic mite *Varroa destructor*, which crossed from its natural host, the eastern honeybee (*Apis cerana*), to the western honeybee, *Apis mellifera*, which is native in South Africa. In 1997 the varroa mite was found only in the Western Cape, but it has spread rapidly throughout South Africa as a result of migratory beekeeping activities and is now present in commercial honeybee colonies countrywide. A number of European species of springtails have also been introduced to South Africa, but the taxonomic status of many of them is still unclear. One undoubtedly invasive, nonnative springtail identified to date is *Hypogastrura manubrialis*, which has been observed in huge numbers in farmlands of the Western Cape. There are 34 species of nonnative terrestrial molluscs that have been recorded in South Africa; 28 are naturalized and 13 are invasive.

Freshwater Animals

All major southern African river systems are inhabited by nonnative animal species, including introduced nonnative taxa and translocated species native to the country. At least 58 fish species are known to be either nonnative in South Africa or native to part of the region but introduced by humans to other parts of the country. At least ten species of freshwater gastropods have been introduced into South Africa, mostly through the aquarium trade. Two of these, *Lymnaea columella* (Lymnaeidae) and *Physa acuta* (Physidae), have been invasive in river systems across the country for many years, probably since the 1940s or 1950s, and two recent arrivals, *Tarebia granifera* (Thiaridae) and *Aplexa marmorata* (Physidae), are spreading. The recent increase in aquaculture has resulted in the introduction, and subsequent escape, of four nonnative species of freshwater crayfish (*Cherax tenuimanus, C. destructor, C. quadricarinatus*, and *Procambarus clarkii*) in South Africa.

Marine Organisms

The first paper specifically listing introduced marine species in South African waters appeared in 1992, listing just 15 species. This number has grown rapidly to over 120 in 2009, and even this is certainly a considerable underestimate because many species were introduced centuries ago, making their origins difficult or impossible to determine. Others occur in areas that have been inadequately surveyed, and yet others belong to groups that will require considerable resources to identify. The most prominent marine invader is probably *Mytilus galloprovincialis*, which was likely introduced in the late 1970s, but which was only identified as nonnative in 1984, by which time it had spread along the entire west coast north of Cape Point.

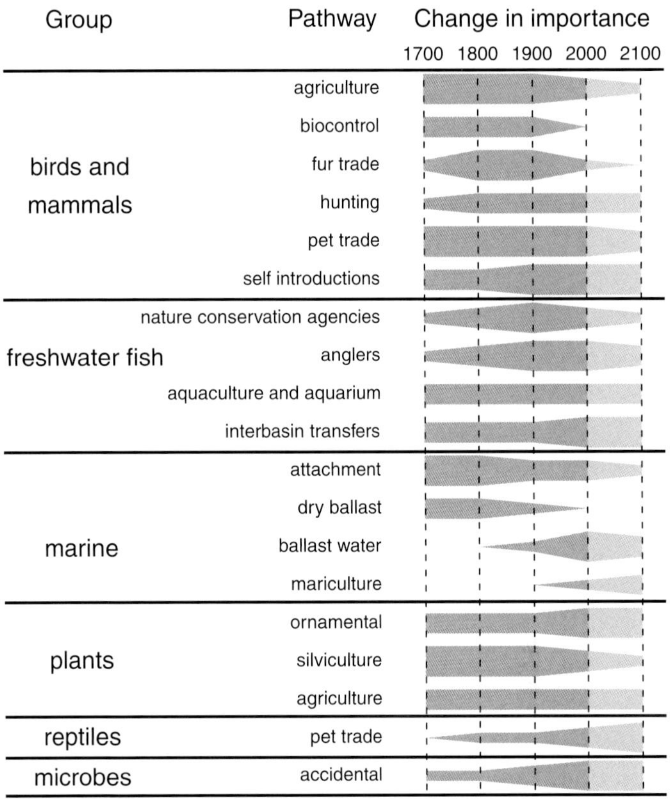

PATHWAYS OF INVASION

The numbers of invasive species in different taxonomic groups and their current extent of spread have been influenced by numerous factors, including biological traits of the species, features of the environment, residence time, propagule pressure, and interactions among these factors. Fundamentally important in all cases has been the combination of socioeconomic factors that have defined the pathways for their introduction to, and dissemination within, the country. The dimensions of these pathways have changed markedly over time, and will continue to change into the future (Fig. 3). Some examples illustrate the complexity of the phenomenon.

For plants, many of the most widespread invaders were introduced and disseminated for forestry and horticulture. These pathways were strongly shaped by changing sociopolitical forces. For example, between 1652 and the present, there have been radical changes in policies relating to the use of nonnative trees in forestry. These changes mean there were reasonably discrete phases determining how, where, and when nonnative tree species were introduced

FIGURE 3 The importance of different pathways for introducing species to South Africa has changed and keeps changing. A rough estimate of the relative importance of the pathways is indicated by the thickness of bars. Post-2000 patterns (pale shading) are speculative. (Modified from Wilson et al., 2009, *Trends in Ecology and Evolution* 24: 136–144.)

and planted, and similar discrete phases in how the problems caused were perceived, and consequently the policy and management responses. The reasons for introducing nonnative birds and mammals have also changed through time. In particular, introductions for agriculture, the fur trade, and biological control have declined markedly, while introductions for the pet trade remain important, and self-introductions have increased. However, the legal and illegal trade in nonnative amphibians and reptiles has, as in other parts of the world, escalated dramatically in recent years. For reptiles, the larger, colorful, patterned species and those that are easy to breed and handle are the most popular in the trade.

There have been clear changes in the relative importance of known pathways for the introduction of nonnative freshwater fishes into open waters. Whereas introductions by nature conservation agencies to stock water bodies have decreased, many other pathways (e.g., introductions for angling and for the aquarium trade) have remained fairly constant, while introductions via engineered interbasin transfers has increased substantially.

TABLE 1

Major Invasive Fish Species in South Africa

Family Species	Native Range	First Record	Introduction Pathway	LI	BR	OR	OL	BE	IN	GO	CR	GA	SU	GF	KE	MZ	MV	UM	TU	MF	Nonnative Occurrence (% of systems)
Centrarchidae																					
Lepomis macrochirus	NA	1938	FO	a	a	a	a	a	a	-	a	a	-	a	a	a	a	a	a	a	88
Micropterus dolomieu	NA	1937	AN	a	a	a	a	a	a	a	a	a	-	a	a	a	a	a	a	-	71
Micropterus punctulatus	NA	1939	AN	a	-	a	a	-	a	a	a	a	-	-	a	a	a	-	a	-	76
Micropterus salmoides	NA	1928	AN	a	a	a	a	a	a	a	a	a	a	a	a	a	a	a	a	a	100
Cichlidae																					
Oreochromis mossambicus	AF	Various	AN, AQ, FO, OR	n	n	a	a	a	n	a	n	a	a	n	n	n	n	n	n	n	100
Oreochromis niloticus	AF	1959	CB, AN, AQ	a	-	-	-	-	-	-	a	-	-	-	-	-	-	-	-	-	12
Tilapia rendalli	AF	VAR	CO	n	-	-	-	-	n	-	a	-	-	-	-	-	-	-	n	-	6
Tilapia sparrmanii	AF	VAR	FF	n	a	n	a	a	n	a	a	a	-	a	a	n	n	n	n	n	59
Clariidae																					
Clarias gariepinus	AF	1975	IBT, AN, AQ	n	a	n	a	a	n	a	a	a	a	a	a	-	n	n	n	n	90
Locariidae																					
Pterygoplichthys disjunctivus	SA	2006	OR	-	-	-	-	-	-	-	a	-	-	-	-	-	-	-	-	-	6
Cyprinidae																					
Carassius auratus	AS	1726	OR	-	-	-	-	-	-	-	a	a	a	a	-	a	a	a	-	-	24
Ctenopharyngodon idella	AS	1967	AQ, CO	-	-	-	-	-	-	-	a	-	-	a	-	a	a	a	-	-	12
Cyprinus carpio	AS	1859	OR	a	a	a	a	a	-	a	a	a	a	a	a	a	a	-	a	a	82
Hypophthalmichthys molitrix	AS	1975	AQ	a	-	-	-	-	-	-	-	-	a	a	-	-	-	-	-	-	6
Labeo capensis	AF	1975	IBT, AN	-	-	n	-	-	-	-	-	-	a	a	-	-	-	-	a	-	24
Labeobarbus aeneus	AF	VAR	IBT, AN	a	-	n	-	-	-	a	-	-	a	a	a	-	-	-	a	-	47
Tinca tinca	EU	1896	Unknown	-	a	-	-	-	-	-	-	-	-	-	-	-	-	-	-	-	6
Poeciliidae																					
Gambusia affinis	NA	1936	CON	a	a	a	a	a	a	a	a	a	a	a	-	a	-	-	-	-	47
Poecilia reticulata	SA	1912	ORN	-	-	-	-	-	-	-	a	-	-	-	-	a	a	-	a	-	18
Xiphophorus helleri	CA	1974	ORN	-	-	-	-	-	a	-	a	-	-	-	-	a	a	-	-	-	18
Salmonidae																					
Oncorhynchus mykiss	NA	1897	AN, AQ	a	a	a	a	a	a	a	a	a	-	a	a	a	a	a	a	a	88
Salmo trutta	EU	1892	AN, AQ	a	a	a	a	a	a	a	a	a	-	-	a	a	a	-	a	a	76

NOTE: Major invasive fish species in South African freshwater ecosystems, their native range, first record in South Africa, introduction pathway, and distribution by major river system. The number of river systems invaded by each species expressed as a percentage of the 17 assessed river systems is also provided. a = alien (nonnative), n = native, - = absent. Native range: CA = Central America, NA = North America, AF = Africa, SA = South America, EU = Europe, AS = Asia. Introduction pathways: AN = stocked for angling, OR = ornamental ponds and aquarium trade, AQ = aquaculture, CB = cross border, CO = bio-control. River system: LI = Limpopo, BR = Bree, OR = Orange, OL = Olifants/Doring, BE = Berg, IN = Incomati, GO = Gouritz, CR = coastal rivers, GA = Gamtoos, SU = Sundays, GF = Great Fish, KE = Great Kei, MZ = Mzimvubu, MV = Mvoti, UM = Umzimkulu, TU = Tugela, MF = Mfolozi.

The complexity of these changes is summarized in Table 1. For marine organisms, the most profound changes have involved the decline in the importance of dry ballast and the marked increase in the magnitude and volume of ballast water discharge.

KEY IMPACTS OF INVASIVE NONNATIVE SPECIES IN SOUTH AFRICA

Impacts on Water Resources

Invasive nonnative trees and shrubs have had a major impact on surface water resources in South Africa. Not all invasive plants use more water than the natural vegetation that they replace, but dense stands of invasive trees definitely use more water than dominant native plants in grasslands or shrublands. Most evidence for this is from catchment experiments in high-rainfall areas where shrublands or grasslands were afforested with pines or eucalypts. Invasive woody plants are thought to be using 3.3 billion m^3 $year^{-1}$ more water than what is used by native vegetation (about 7% of the runoff of the country).

Impacts on Grazing Resources

Mesquite (*Prosopis* spp.), jointed cactus (*Opuntia aurantiaca*), and nasella tussock grass (*Stipa trichotoma*) are the most important invasive species that currently affect grazing most directly, but others such as triffid weed (*Chromolaena odorata*), silverleaf nightshade (*Solanum elaeagnifolium*), and pompom weed (*Campuloclinium macrocephalum*) are rapidly becoming more important.

Impacts on Biodiversity in Terrestrial Ecosystems

The impacts of invasive species on biodiversity in South Africa are still poorly understood. Most South African research on nonnative-plant impacts has focused on small spatial scales (plots or communities), and much of this work has been in the fynbos biome. Results to date show that dense stands of nonnative trees and shrubs in fynbos can rapidly reduce abundance and diversity of native plants at the scale of small plots. Such invasions also dramatically increase biomass and change litterfall dynamics and nutrient cycling. Australian acacias have had particularly marked effects on local-scale plant diversity patterns. Tree and shrub invasions in fynbos change many aspects of faunal communities, for example by altering the abundance and composition of native ant communities, with implications for seed dispersal functions of native plants. Changed feeding behaviors of native generalist birds that disperse seeds, with likely detrimental effects on native species, have

also been documented. Invasions by *Chromolaena odorata* reduce vegetation heterogeneity in grasslands, savannas, and forests. Such changes have been reported to cause a wide range of effects on native diversity, ranging from altered native spider abundance and species diversity to changes in reproduction biology of native Nile crocodiles (*Crocodylus niloticus*), including altered sex ratios of hatchlings due to reduced temperatures due to shading and nest abandonment caused by the fibrous root mats of *C. odorata*.

Impacts on Fire Regimes and Erosion

Invasion of grasslands and shrublands by tall trees and shrubs increases the amount of available fuel by up to ten times. This leads to dramatically increased fire intensity and physical damage to the soil, sometimes leading to increased erosion after fire.

Impacts on Freshwater Ecosystems

Invasive nonnative fish are having a dramatic impact on the indigenous fish fauna. Of the 30 IUCN red-listed fish species in South Africa, 16 are threatened primarily by invasive nonnative species (almost exclusively predatory fish). A more recent threat is hybridization, which, at the very least, results in wasted reproductive effort (if hybrids are sterile) and, at worst, results in complete introgression and species loss. The translocation of the Orange River mudfish (*Labeo capensis*) from its native west-flowing Orange River to the east-flowing Sundays River through interbasin water transfers has resulted in its hybridization with the previously isolated, genetically unique *L. umbratus* population. The cichlid *Oreochromis mossambicus* is threatened in its native South African range by hybridization with the alien invasive *O. niloticus* and is now IUCN red-listed as near threatened. Nonnative fish species are now often the dominant fishes in many river systems. Besides their impact on native fishes, the nonnative species have also affected other components such as aquatic macro-invertebrates and can alter habitat structure, which in turn affects ecosystem functioning.

Impacts of Nonnative Mammals

Many native South African mammals cause impacts in areas where they are extralimital. Reported impacts include hybridization with related but previously isolated species (e.g., *Damaliscus dorcas dorcas* hybridizing with *D. d. phillipsi*); a range of herbivore-driven impacts, such as giraffes (*Giraffa camelopardalis*) driving a change in tree composition; and complex mediator-facilitated competitive effects, such as detrimental effects on *D. d. phillipsi* due to a parasite (*Bronchonema magna*) vectored by the extralimital springbok (*Antidorcas marsupialis*). Mammals introduced from outside Africa have also had noticeable impacts. Himalayan tahrs caused erosion and affected the structure and composition of fynbos vegetation on Table Mountain, while on the subantarctic Marion Island, between 1949 and 1991, introduced cats had a major impact on native birds, notably burrowing petrels and common diving petrels (which are thought to have been exterminated by cats by 1965). Both the tahrs on Table Mountain and the cats on Marion Island have been eradicated (see below).

RESPONSES

South Africa has a long history of successful management responses to biological invasions. However, the management of invasions took a giant leap forward with the launching of the Working for Water Programme (WfW; www.dwaf.gov.za/wfw/) in 1995. WfW provided the foundation for initiating and, more importantly, sustaining control programs at local, regional, and national scales. Results of scientific research on the ecology and impacts of the principal invasive plant species provided the initial rationale for WfW. This national-scale, government-sponsored initiative aims to protect water supplies and restore productive and conservation land, primarily through labor-intensive invasive-plant clearing operations. It thus fulfils environmental, economic, and social goals. The program has grown rapidly: at the end of 2008, it comprised 303 projects located in all nine provinces, employed more than 25,000 people, and had initially cleared over 1.8 million ha of invaded land and followed-up 4.1 million ha. The program has been lauded internationally for its vision in simultaneously addressing environmental and socioeconomic issues. As its name implies, the specific aim of WfW was initially to alleviate the well-documented impacts of invasive species on water resources through control measures. As such, it represents a model case of the leverage of conservation action based on a scientific evaluation of the value of ecosystem services and the threats to these from invasive species. The focus of the program has been expanded to deal with all invasive plant species, not only those with clear impacts on water resources. The program has resuscitated the national initiative to apply biological control as a key part of integrated control. WfW is also spearheading innovative initiatives to integrate a "payments for ecosystem services" mechanism to strengthen the sustainability of management operations, ensuring that programs simultaneously address poverty alleviation, ecosystem service delivery, and ecological objectives. Another facet of WfW involves facilitating enabling partnerships with key sectors. For example, the Nurseries Partnership Programme is a joint venture between WfW, the South African Nursery Association, and the National Department of Agriculture to address the problem of invasive nonnative plants within the nursery industry and among the gardening public. Similar engagement with the plantation forestry industry has greatly improved awareness of problems of invasive forestry trees and has resulted in cofunding of significant clearing operations.

As already mentioned, there have been notable successes in the control of invasive nonnative mammals. Cats were eradicated on Marion Island through an intensive campaign of biological control using viruses, trapping, and hunting that started in 1977 and resulted in the extermination of all cats by 1991. Himalayan tahrs were eradicated from Table Mountain by 2008.

South Africa also has very good legislation to support the management of biological invasions. The main legal instruments that have direct bearing on invasive species and their management are the Conservation of Agricultural Resources Act (CARA; Act 43 of 1983); the Agricultural Pests Act (36 of 1983); and the National Environmental Management: Biodiversity Act (NEMBA, Act 10 of 2004). Among many far-reaching facets, the draft regulation pertaining to invasive species under NEMBA lists 109 fishes as prohibited and categorizes nonnative fishes into those requiring compulsory control, those to be controlled as part of a nonnative invasive species program, and those to be regulated by area or activity. Other legal instruments influence, or potentially influence, the management of nonnative species indirectly, through the potential effects of invasive species on water security (National Water Act, Mountain Catchment Areas Act) or economic security (National Forests Act). Much work remains to ensure that provisions made in existing legislation are incorporated into effective and sustainable programs for managing the full suite of nonnative species in South Africa to alleviate the major threats that they pose to the country's natural capital.

An obstacle to further progress and implementation of plans is the shortage of human resources needed to drive such programs. A milestone toward solving the capacity problem for managing biological invasions was the establishment of the Centre for Invasion Biology (CIB; www.sun.ac.za/cib) in 2004. It is one of six "Centres of Excellence" created by South Africa's Department of Science and Technology in all fields of science. Creation of the CIB points to the recognition of biological invasions as a major challenge to the South African environment.

South Africa has made substantial contributions to the international scientific knowledge base on biological invasions in the past. The formation of the CIB provides a platform and critical mass for further research on key issues relating to the management of invasive species. In its first four years, the CIB produced 246 primary research publications in peer-reviewed journals, supported over 100 graduate students and post-doctoral associates, and provided high-level input to policy formulation.

SEE ALSO THE FOLLOWING ARTICLES

Fishes / Forestry and Agroforestry / Grasses and Forbs / Hydrology / Trees and Shrubs / Weeds

FURTHER READING

Dean, W. R. J. 2000. Alien birds in southern Africa: What factors determine success? *South African Journal of Science* 96: 9–14.

De Wit, M., D. Crookes, and B. W. van Wilgen. 2001. Conflicts of interest in environmental management: Estimating the costs and benefits of a tree invasion. *Biological Invasions* 3: 167–178.

Henderson, L. 2001. *Alien Weeds and Invasive Plants: A Complete Guide to Declared Weeds and Invaders in South Africa*. Pretoria: Agricultural Research Council.

Le Maitre, D. C., D. M. Richardson, and R. A. Chapman. 2004. Alien plant invasions in South Africa: Driving forces and the human dimension. *South African Journal of Science* 100: 103–112.

Macdonald, I. A. W., F. J. Kruger, and A. A. Ferrar. 1986. *The Ecology and Management of Biological Invasions in Southern Africa*. Cape Town: Oxford University Press.

Milton, S. J. 2004. Grasses as invasive alien plants in South Africa. *South African Journal of Science* 100: 69–75.

Nel, J. L., D. M. Richardson, M. Rouget, T. Mgidi, N. Mdzeke, D. C. Le Maitre, B. W. van Wilgen, L. Schonegevel, L. Henderson, and S. Neser. 2004. A proposed classification of invasive alien plant species in South Africa: Towards prioritising species and areas for management action. *South African Journal of Science* 100: 53–64.

Olckers, T., and M. P. Hill. 1999. *Biological Control of Weeds in South Africa (1990–1998)*. African Entomology Memoir No. 1: 1–182.

Richardson, D. M., W. J. Bond, W. R. J. Dean, S. I. Higgins, G. F. Midgley, S. J. Milton, L. Powrie, M. C. Rutherford, M. J. Samways, and R. E. Schulze. 2000. Invasive alien organisms and global change: A South African perspective (303–349). In H. A. Mooney and R. J. Hobbs, eds. *Invasive Species in a Changing World*. Washington, DC: Island Press.

Richardson, D. M., J. A. Cambray, R. A. Chapman, et al. 2003. Vectors and pathways of biological invasions in South Africa: Past, present and future. In G. M. Ruiz and J. T. Carlton, eds. *Invasive Species: Vectors and Management Strategies*. Washington, DC: Island Press.

Richardson, D. M., I. A. W. Macdonald, J. H. Hoffmann, and L. Henderson. 1997. Alien plant invasions (535–570). In R. M. Cowling, D. M. Richardson, and S. M. Pierce, eds. *The Vegetation of Southern Africa*. Cambridge: Cambridge University Press.

Richardson, D. M., and B. W. van Wilgen. 2004. Invasive alien plants in South Africa: How well do we understand the ecological impacts? *South African Journal of Science* 100: 45–52.

Spear, D., and S. L. Chown. 2009. Non-indigenous ungulates as a threat to biodiversity. *Journal of Zoology* 279: 1–17.

van Wilgen, B. W., W. J. Bond, and D. M. Richardson. 1992. Ecosystem management (345–371). In R. M. Cowling, ed. *The Ecology of Fynbos: Nutrients, Fire and Diversity*. Cape Town: Oxford University Press.

van Wilgen, B. W., R. M. Cowling, and C. J. Burgers. 1996. Valuation of ecosystem services: A case study from South African fynbos ecosystems. *BioScience* 46: 184–189.

Versfeld, D. B., D. C. Le Maitre, and R. A. Chapman. 1998. *Alien Invading Plants and Water Resources in South Africa: A Preliminary Assessment*. Report TT99/98. Pretoria: Water Research Commission.

SPREAD, MODELS OF

SEE EPIDEMIOLOGY AND DISPERSAL

SUCCESSION

SCOTT J. MEINERS

Eastern Illinois University, Charleston

STEWARD T. A. PICKETT

Cary Institute of Ecosystem Studies, Millbrook, New York

Succession in a strict sense refers to the recovery and revegetation of an area following a disturbance such as the cessation of agriculture, the retreat of a glacier, or an intense forest fire. Succession is a special case of vegetation dynamics, although many early ecologists referred to all vegetation change as succession. Succession includes a series of compositional and structural changes, often in a directional manner. The common occurrence of natural disturbances coupled with the extent of human activity on the planet makes succession one of the most ubiquitous ecological processes. Because invasion is a crucial feature of succession, understanding the nature and controls of community dynamics is important for the science and management of invasive species.

CONTEXT OF SUCCESSION

The study of succession began as ecologists first struggled with the idea that plant communities were not static over time but were dynamic over both long and short time scales. The long history of successional studies has generated both progress and controversy, even in the way that succession is defined. Early work described successional dynamics as both directional and with a clear endpoint. Neither of these ideas is currently accepted without caveats. In many, if not most, cases succession will generate a directional change in communities over time. For example, succession in many mesic temperate environments will generate the recovery of deciduous forests, with successional transitions

FIGURE 1 A successional sequence from the Piedmont region of New Jersey. (A) Newly abandoned agriculture land. (B) Herbaceous stage, five years after abandonment. (Photographs courtesy of Steward Pickett.) (C) Increasing dominance of trees 40 years after abandonment. (Photograph courtesy of Scott Meiners.) (D) Mature forest. (Photograph courtesy of Helen Buell.) Data from this system are used in all of the examples illustrated.

from herbaceous to shrubby to forested communities (Fig. 1). However, the ecological literature is filled with examples where successional dynamics stall at some intermediate stage or reverse direction following some change in the environment. For example, the introduction of drought and fire to mesic succession may generate savanna or grassland communities instead of a closed-canopy deciduous forest. Some systems exhibit essentially cyclical dynamics, where each successional stage is continually replaced by another, with no stage able to regenerate itself. Cyclical successions typically occur as a function of the internal patch dynamics of a community rather than in response to a disturbance. The coarse-scale community would be an aggregate of patches in all successional phases.

Opinions on the existence of a successional endpoint, or climax community, have also changed over time. While succession generally does result in a community which has the ability to regenerate itself, these communities are by no means stable. All communities are dynamic; they just exhibit turnover and compositional changes at different time scales. If conditions change, then the inherent dynamics of the community may also change and generate a new community structure.

SUCCESSIONAL DYNAMICS

The structure of communities tends to change dramatically during succession. Often, the disturbance that initiated succession produces openings within the community that are available for colonization by plants. The amount of bare substrate may range from complete, as in succession following volcanic eruptions, to very little, as in succession following hay production or forest clearing. Over successional time, this colonizable space decreases. There is also an increase in the total biomass of the community, as spaces fill in and larger plants replace smaller, disturbance-adapted species. In successions that regenerate forests, the increase in biomass will largely be due to the accumulation of woody material. Within successions to perennial grasslands or other herbaceous communities, much of this biomass may be accumulated below ground in roots and storage structures. Associated with the increase in biomass is the ability of the community to accumulate and

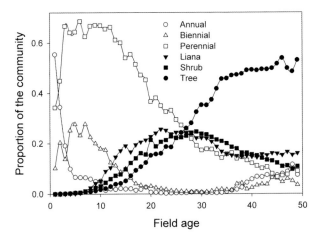

FIGURE 2 Successional transitions of life forms over time during succession from row crop agriculture to deciduous forest.

retain nutrients. While early successional communities often contain relatively low amounts of nutrients in the biomass and have little ability to retain these nutrients within the system, late successional communities tend to contain larger pools of nutrients in the biomass and have a greater capacity for retention.

The diversity of the plant community, and often the associated invertebrate, vertebrate, and microbial communities, changes during succession. In general, early successional communities are relatively low in diversity, being dominated by only those species that are able to either survive the initiating disturbance or quickly colonize following it. As succession proceeds, the community accumulates more species, many of which are not dependent on disturbance for their regeneration. This accumulation of species increases the diversity of mid-successional communities. However, as late successional species accumulate in the community, opportunities for the regeneration of early successional species decrease, eventually leading to their loss from the community. The greatest diversity in succession tends to be in mid-successional communities that contain a mixture of early and late species.

The simplest and clearest compositional transitions that occur during succession are changes in the life form of the dominant plant species (Fig. 2). As succession proceeds, the community shifts in dominance from short-lived herbaceous species to long-lived herbaceous species. In areas where the climate supports the growth of forests, these herbaceous species will be replaced by shrubs and lianas, and ultimately by trees. This does not mean that late successional communities lack other life forms. A late successional forest would also be expected to contain shade-tolerant herbaceous and woody species

in the understory, although canopy trees would clearly be the dominant life form. Succession on bare mineral substrates, which is limited by the availability of nutrients, may begin with extended periods that are dominated by cyanobacteria and mosses, which will precede dominance by herbaceous species.

Concurrent with the changes in life form would be other changes in the characteristics of the community (Fig. 3). In successions that produce forests, the community will become taller over time as short-statured herbaceous plants are overtopped and replaced by shrubs and then by canopy trees. The community will also shift over time from small-seeded species toward larger-seeded species that are more able to regenerate under the dense plant canopy. Related to this change in seed

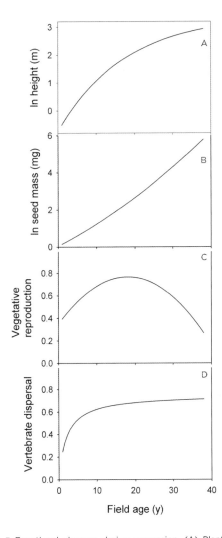

FIGURE 3 Functional changes during succession. (A) Plant height. (B) Seed mass. (C) Vegetative reproduction. (D) Vertebrate dispersal. Plant height and seed mass data are abundance-weighted means for the entire community, while vegetative reproduction and dispersal are the proportion of the community with each trait.

mass is an increased dependence on vertebrates, particularly mammals and birds, for seed dispersal. Smaller-seeded species tend to rely on abiotic mechanisms such as wind for seed dispersal. Vegetative reproduction often peaks in mid-successional communities and is strongly linked with the ability of species to take over and exploit patches of resources. Competitive ability and presumably vegetative reproduction are important in both succession and species invasion.

CAUSES OF SUCCESSION

Successional dynamics are the result of variation in three general classes of successional drivers—site availability, species availability, and species performance (Fig. 4). Much of the theoretical and experimental work on the mechanisms that generate succession focuses either on a specific mechanism within each of these general causes or on the importance of these general causes relative to each other. As specific mechanisms and their relative importance are likely to vary dramatically among locations, we will focus primarily on the general causes.

The conditions following disturbance are critical in determining the vegetation that will develop through succession. Disturbances of varying intensity or type will generate different successional trajectories. Even within the same region, disturbance from logging will generate a very different successional plant community than would be found on land retired from row crop agriculture or from grazing. Similarly, the spatial extent of the disturbance will influence the rate of succession; large disturbed areas will regenerate much more slowly and differently than smaller areas of disturbance. Of paramount importance is the condition of the soil following the disturbance. Agricultural practices often alter the fertility of soils, which may reduce the speed of recovery when fertility is low or may generate novel plant communities when fertilizer residues generate high fertility in normally poor soils. One of the major dichotomies in succession, that of primary and secondary succession, is largely based on the quality of soil available for regeneration. Primary succession typically occurs quite slowly as community dynamics are limited by fertility while the mineral soils accumulate

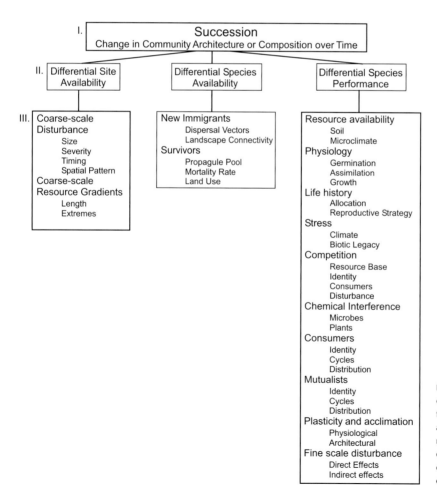

FIGURE 4 Hierarchy of successional drivers ranging from the broad classes of differential site availability, species availability, and species performance to the detailed mechanisms within each class. Successional dynamics within a site can result from one or many of these specific mechanisms within each broad class.

organic matter during succession. All of the site and disturbance variables function within the constraints of local climatic conditions. Succession in relatively moist areas tends to generate forests relatively quickly, while succession in dry habitats slowly produces more open communities composed of shrubs and grasses.

Differential availability of species to a disturbed area can have large effects on the successional processes that occur. Plant species vary dramatically in their dispersal ability and in their ability to survive a given disturbance. Many species may be present immediately following a disturbance, persisting as vegetative fragments or as seeds dormant in the soil. These species often dominate the earliest successional communities. Other species must disperse into the disturbed area. Which species arrive, and their relative abundances, will be determined by the spatial location of reproductive individuals in the surrounding landscape, the efficiency and behavior of dispersal vectors, and the landscape connectivity of the disturbed habitat to source populations. As the seed production of some plant species is temporally sporadic, the composition of successional communities may vary stochastically based on which species reproduced when the site became available. The abundance of an invasive species in the surrounding landscape may determine the initial importance of the species in the successional community, but once the species becomes reproductive, it may spread throughout the community.

Finally, differential performance generates much of the ecological sorting of species seen during succession. Species performance is determined by a variety of characteristics of a species. Factors such as life form and longevity determine the physical constraints on a species and how long it may potentially occupy a particular spot within the community. Other factors include physiological characteristics such as growth rate, shade tolerance, response to herbivory, and competitive ability, to list a few. These characteristics will together determine the outcome of individual plant–plant interactions at local scales. At broader spatial scales, compositional changes within the community will be the sum of all of the individual interactions. Of course, all interactions occur within the constraints of the original site conditions and may vary dramatically from place to place. For example, annuals are typically rapidly displaced by longer-lived species. However, when soil fertility is maintained at artificially high levels with fertilizers, annuals may persist and even dominate the community for years.

While the individual successional drivers are discussed separately above, a plant species must integrate across all of these factors to be successful in a plant community. As

no species can be perfectly suited to all conditions and interactions, tradeoffs must exist that allow species to succeed within particular types of environments experienced during their evolutionary history. These tradeoffs generate the ecological strategies employed by the species, which in turn determine the ability of species to capitalize on a specific successional phase. From a plant strategy perspective, early successional communities will be dominated by short-lived species, which are able to either survive the disturbance or quickly disperse into the site. These species maximize their growth rates but are largely dependent on disturbances to maintain their populations. Early colonizers are replaced by mid-successional species, which employ a competitive strategy of rapidly expanding and capturing resources. Competitive species tend to disperse relatively well but are initially slower to grow and reproduce. Many mid-successional species expand vegetatively via stolons or rhizomes and generate large patches where they dominate local resources. Finally, late successional species expand most slowly of all but can regenerate under earlier successional species or replace them as they die. These species are the longest lived, but they often disperse poorly owing to their large seed mass. Of course, due to spatial heterogeneity within the habitat and stochastic effects on populations, natural communities contain species with strategies adapted for all positions along the successional gradient, not just the phases described here for convenience.

PLANT INVASIONS IN SUCCESSION

Early successional communities worldwide are typically heavily invaded by a diversity of nonnative species. The dominance of nonnative species, particularly in secondary succession, is probably driven by a combination of two factors. First, the disturbances that initiate succession also tend to generate conditions favorable for nonnative species. As disturbance is clearly linked with invasion, it is not surprising that disturbance-adapted nonnative species are also important components of successional systems. Second, the most abundant types of successional habitats are the result of agricultural activities. These activities, whether focused on grazing, row crops, or forestry practices, all support associated suites of weedy species, many of which are nonnative. Agricultural activities increase both the abundance of these species within the landscape and their contribution to the successional flora.

While early successional habitats are often heavily invaded, the relative abundance of nonnative species typically decreases during succession (Fig. 5). While early

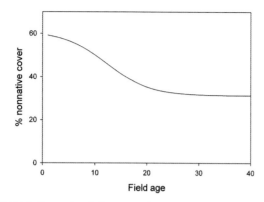

FIGURE 5 Change in relative cover of nonnative species during succession. While the relative abundance decreases sharply early in succession, nonnative species persist in later successional communities.

successional environments are dominated by disturbance-adapted species associated with agricultural practices, these species tend not to regenerate as the influences of agriculture decrease after abandonment. However, the successional replacement of nonnative species is expected only when the pool of nonnative species is dominated by early successional species. In contrast to agriculture, horticultural introductions may spread species from a variety of successional stages throughout the landscape. In communities where sufficient late successional species have also been introduced, the dominance of nonnative species may persist through succession.

Besides being important components of successional systems, invasive nonnative species may have direct impacts on successional processes. As many invaders are aggressive colonizers of disturbed areas, they may alter the direction of successional trajectories or reduce the rate of succession. This is particularly true if the invasive species is capable of regenerating in late successional communities or represents a life form new to the community. For example, woody invasive species in coastal areas may generate a persistent forested community in an area previously dominated by graminoids. Species-rich shrubby communities in South Africa (fynbos) are often replaced by near monocultures of nonnative acacias or pines. Similarly, the introduction of shade-tolerant shrubs into disturbed forests may generate much denser shrub layers that inhibit tree regeneration and the herbaceous layer. Most invasive species that are abundant in early successional communities will probably not become problematic, as they will be replaced by later successional species. However, any species that inhibits a successional transition may delay or even stall succession. In mesic environments that allow succession to forest, tree establishment is a key successional transition. If competition

with invasive species reduces or prevents the growth of trees, then a new, potentially persistent community may develop.

As successional areas often border areas targeted for conservation, they may also represent important propagule sources of invasive species. Invasive species may become abundant in the disturbed habitat and then disperse to the edges of the conservation area or colonize gaps within the community. The presence of successional communities may effectively increase the propagule pressure on the conservation area and increase local rates of invasion.

MANAGEMENT OF SUCCESSION

Much of ecological restoration involves the manipulation of ecological succession to achieve management goals. Typically, management involves increasing the rate of succession to achieve late successional communities more quickly, but this is not always the case. It may be necessary to retard succession in an earlier phase if an herbaceous species or its pollinator, or a grassland bird species, is the management target. Similarly, herbaceous communities in many climates are replaced by woody species during normal successional processes. If a meadow or other herbaceous community is the desired target, then either fire or mowing must be employed to inhibit the establishment of woody species. When management goals require the establishment of late successional communities or species, the successional drivers suggest several potential interventions. As late successional species are typically poor dispersers, seeds or seedlings may be installed to alleviate dispersal limitation. A complementary approach is to assess the factors limiting the growth of the species of interest. Competition and herbivory are two common processes that limit tree regeneration. Mechanical or chemical removal of competing vegetation and protection from herbivores may increase the growth rate of trees and hasten the transition from herbaceous to forested community.

Management of invasive species comes into play in two main ways. If the invasive species within a region are primarily early successional species, then succession, and management practices that increase its rate, will be sufficient to reduce the abundance of the invasive species. In contrast, if an invasive species inhibits successional transitions, then management intervention may be necessary to allow succession to proceed. For invasions involving species capable of regenerating within late successional communities, management of successional processes will have little influence. In these cases, management of the invasion must focus directly on the removal of the species itself. However, following removal of the invader, it may be necessary to further restore the community to install species that may prevent recolonization of the site. For example, in forests with invasive shrub understories, invasive species should be replaced with suitable native shrubs that may compete with the invader. These management goals and activities are examples of the importance of understanding the processes of succession and the successional conditions that can affect invasive species.

SEE ALSO THE FOLLOWING ARTICLES

Agriculture / Disturbance / Fire Regimes / Forestry and Agroforestry / Horticulture / Restoration

FURTHER READING

Bazzaz, F. A. 1996. *Plants in Changing Environments*. Cambridge: Cambridge University Press.

Cramer, V. A., and R. J. Hobbs, eds. 2007. *Old Fields: Dynamics and Restoration of Abandoned Farmland*. Washington, DC: Island Press.

Grime, J. P. 2001. *Plant Strategies, Vegetation Processes, and Ecosystem Properties*. Chichester: John Wiley and Sons.

Luken, J. O. 1990. *Directing Ecological Succession*. London: Chapman and Hall.

Meiners, S. J., T. A. Rye, and J. R. Klass. 2008. On a level field: The utility of studying native and non-native species in successional system. *Applied Vegetation Science* 12: 45–53.

Pickett, S. T. A., and M. L. Cadenasso. 2005. Vegetation dynamics (172–198). In E. van der Maarel, ed. *Vegetation Ecology*. Malden, MA: Blackwell.

Walker, L. R., and R. del Moral. 2003. *Primary Succession and Ecosystem Rehabilitation*. Cambridge: Cambridge University Press.

Walker, L. R., J. Walker, and R. J. Hobbs. 2007. *Linking Restoration and Ecological Succession*. New York: Springer.

T

TAXONOMIC PATTERNS

JULIE L. LOCKWOOD AND THOMAS VIRZI

Rutgers University, New Brunswick, New Jersey

Evolution is a conservative process, such that closely related species tend to share sets of life history traits, and this pattern is often reflected in their taxonomic affiliation. For example, species in the same taxonomic family tend to have similar reproductive strategies, life spans, and habitat requirements. If any of these traits is important in determining invasiveness, then it follows that taxonomically closely related species are more likely to become invasive given the opportunity. Many plant and animal higher taxa are more common in lists of invasive species than their overall numbers would warrant. What is more controversial is how these taxonomic patterns are produced. We suggest that two broad mechanisms produce taxonomic patterns among lists of invaders. The first is taxonomically nonrandom patterns in what species are transported out of their native range as exotics. The second is taxonomically nonrandom patterns in which species, of those transported and released, establish and go on to cause ecological or economic impacts.

TRANSPORT PATTERNS

The first step on the invasion pathway is for individuals of a species to be picked up within their native ranges and transported far away via human actions. It is the suite of species represented in this group that are then released into a nonnative environment and thus have the potential to establish a self-sustaining exotic population. The range of human actions that serve to transport species is incredibly diverse, while at the same time, each of these activities is very taxonomically selective in the species that it moves. A convenient way to make sense of this variation, including understanding which higher taxa are more often transported, is to impose a simple division between human actions that intentionally transport exotic species and those that transport them accidentally.

Intentional Processes

Intentional transport of exotic species includes primarily activities related to the pet and horticultural trades. For example, for most countries, the list of invasive plants is dominated by species that were intentionally transported there by humans for cultivation. Often, these plants go on to establish self-sustaining populations outside of cultivated fields, thereby becoming invasive. Other taxa that people have favored for planting include those used as forage, fuel, lumber, and medicine. Because of their fast growth rates and tendency to have straight trunks, which makes them good lumber trees, pines (genus *Pinus*) are quite commonly established as exotics worldwide. Similarly, many species of grass (Poaceae) and amaranths (Amaranthaceae) are widespread as nonnatives, in part because of their usefulness to humans as forage and food crops. Many more plants have been transported and released as exotics because of their perceived aesthetic value. Europeans who colonized the "neo-Europes" of North America, Australia, and New Zealand tended to favor the importation and planting of species that they were familiar with from their home countries, and for this reason there are many species of exotic plants in these countries that are in common plant families of Europe. Gardeners, however, have a taste for the unfamiliar, and several plant families (such as lilies, Lilaceae) are found worldwide as exotics because of their attractive flowers.

Some portion of these garden plants will "jump the fence" and become established as exotic species.

The pet trade is equally particular about what species it transports and releases as exotics, and there are apparently very specific traits that people find appealing in their pets. It is the "exotic" pets that constitute the majority of the species that end up as invasive species. For birds, the prime example of a higher taxon that is much more commonly traded as a pet than should be expected based on its overall richness is the parrot family (Psittacidae). Millions of individual parrots are captured from their native ranges (usually as young nestlings) and placed for sale in the international market, presumably because of their bright plumage and ability to mimic human voices. Some of these individuals will escape from their cages, or their owners will grow tired of their loud calls and messy habits and will purposefully release them. No matter the means of escape, large urban areas like Los Angeles, London, and Miami now have healthy and diverse populations of exotic parrots due to the intentional movement of a range of parrot species for sale (Fig. 1). Just like the gardener, the

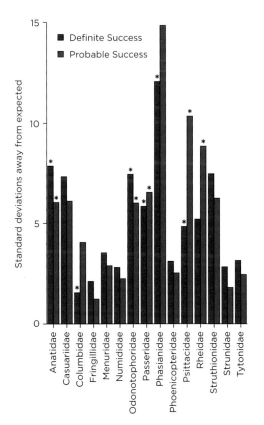

FIGURE 1 The number of standard deviations the observed value (number of successfully established exotic birds) lies away from the number expected by chance within several bird families. The asterisk indicates a difference between the observed and the expected that is statistically significant. (Reprinted from Lockwood, 1999.)

avid aquarist strives to have the beautiful and unfamiliar fish in his or her tank. For this reason, many fish families (such as cichlids, Cichlidae) have relatively many species established as exotics worldwide because they are commonly traded in the pet fish market and then dumped from fish tanks when their owners (or shopkeepers) can no longer care for them.

Unintentional Processes

Because intentional introductions reflect human desires, taxonomic patterns for these species reflect what humans find pleasing. It is not as clear how taxonomic patterns can arise when the species are being unintentionally transported (i.e., when they are hitchhikers). Good examples of hitchhiking species are those transported as seeds within the pots or soil stock used in the nursery plant trade. These species tend to have very small seeds, quick generation times, and adaptations that allow them to grow in urban habitats. To the extent that these traits are characteristic of some plant families (e.g., knotweeds, Polygonaceae), we tend to see many more hitchhiking species from these groups. In the animal world, species that burrow, hide, or nest in cavities are very likely to hitchhike on ships and airplanes, or within the cargo carried by these transportation modes. Once the ship or plane has docked, these hitchhikers decamp, and many establish as exotics (e.g., rodents, Rodentia).

Often, unintentionally introduced animals and plants are quite small, and for this reason many thousands of individuals can be moved in one ship or airplane. For example, many thousands of individuals of marine plants and animals are accidentally transported via ballast water on ships. Marine fish and mussel taxa that have pelagic juvenile stages are particularly likely to be transported in this way. Insects are also often transported accidentally as hitchhikers, often along with the plants they feed on or within soil used as ship ballast. Good examples of insect families that hold relatively many exotic species established in the United States are mole crickets (Gryllotalpidae), likely transported unintentionally in soil ballast, and thrips (Thripidae), likely transported with the economically important plants they feed on.

PATTERNS IN ESTABLISHMENT AND IMPACT

Once species are transported and established within a nonnative environment, many will expand their numbers and geographic ranges, potentially causing economic or ecological harm. There are two ways to look at how taxonomic identity can influence the probability of establishment or impact. Both of these ways were aptly described by Charles Darwin in *On the Origin of Species.*

> It might have been expected that the plants which have succeeded in becoming naturalized [exotic] in any land would generally have been closely allied to the indigenes [native species]; for these are commonly looked at as specially created and adapted for their own country. It might, also, perhaps have been expected that naturalized plants would have belonged to a few groups more specially adapted to certain stations in their own homes.

Darwin is proposing two sides of a taxonomic argument as to how the degree of relatedness between the native and exotic species affects the exotic's ability to establish and cause impact. On one hand, exotics in the same taxonomic group as the natives should share many of the traits that are clearly beneficial to living in the environment of the natives: a phenomenon often called pre-adaptation. On the other hand, exotic species that have traits very different from the natives (in Darwin's quote, have a "different station") may have an advantage because they will not be competing with the natives for scarce resources. This latter supposition is often called Darwin's naturalization hypothesis. It is based on the idea that exotics are successful, and sometimes harmful, because they are taxonomically (and by proxy, ecologically) distinct from the natives. Both sides of this argument have been tested in the modern ecological literature, and both find some level of support. Indeed, some authors argue that the two hypotheses are not mutually exclusive, and that the same data can generate support for either hypothesis depending on the spatial scale of analysis and the geographic location under consideration.

Pre-Adaptation

If pre-adaptation holds, then we should see that exotics that have a close relative in the native system will be more likely to succeed than exotics that are taxonomically distinct from the natives. We see just this pattern among the exotic plants of New Zealand. Exotic plant species were far more likely to succeed at establishing self-sustaining populations in New Zealand if they came from a genus that had one or more native New Zealander relatives within it. Similarly, we see that some plant families common to both Australia and southern Florida hold relatively many invasive species in each region (e.g., sedges, Cyperaceae), and at least some of this crossover is due to the species' sharing the ability to exist in the climates each location affords. There is similar evidence that birds tend to succeed in establishing in climates that are very

similar to the climate of their native range in Australia and New Zealand. It is not clear, however, how well this tendency is captured by the species' taxonomic affiliation. Conversely, the invasive plants of Mediterranean islands tend not to show any taxonomic pattern, thus suggesting that whether these invaders are, or are not, closely related to the natives does not matter.

Distinctiveness

Under Darwin's naturalization hypothesis, exotic species are less likely to establish self-sustaining populations in places with congeneric species due to competition or because they may be more likely to be attacked by native predators or pathogens. In plants, there is some empirical support for this hypothesis, such as the successful establishment of highly invasive grasses in California that are significantly less related to native grasses than are exotic but noninvasive grasses. Plant families that contain many species that are aquatic or semi-aquatic, that are nitrogen-fixers, that are climbers, and that reproduce clonally are likely to become invaders. Each of these traits may confer a competitive advantage over native species because the natives do not have similar traits or because their abilities in this regard are sufficiently different from that of the exotics as to present little resistance to the exotic's dominance. Plant species that commonly invade disturbed habitats (e.g., urban areas) are more likely to flower at an early age and to have high seed production, rapid vegetative growth, and wide environmental tolerance. In many instances, the species native to these disturbed areas do not also have these traits, thus leaving an open door for the exotics to establish and flourish.

Only a few studies directly test Darwin's naturalization hypothesis among animals. There is no evidence for taxonomic distinctiveness influencing success among exotic freshwater fishes; however, the data also did not support a role for pre-adaptation. More broadly, and somewhat contradictorily, there is good evidence that high-impact aquatic invaders come from genera that are not already present within the native system. Thus, fish, algae, invertebrates, and vascular plants that are distinct from the native species in such places as the Laurentian Great Lakes, the Hudson River, and the New Zealand coast are more likely to cause declines in one or more native species.

SEE ALSO THE FOLLOWING ARTICLES

Darwin, Charles / Evolution of Invasive Populations / Geographic Origins and Introduction Dynamics / Horticulture / Pet Trade

FURTHER READING

Daehler, C. C. 1998. The taxonomic distribution of invasive angiosperm plants: Ecological insights and comparison to agricultural weeds. *Biological Conservation* 84: 167–180.

Duncan, R. P., and P. A. Williams 2002. Darwin's naturalization hypothesis challenged. *Nature* 417: 608–609.

Lockwood, J. L. 1999. Using taxonomy to predict success among introduced avifauna: Relative importance of transport and establishment. *Conservation Biology* 13: 560–567.

Mack, R. N., and W. M. Lonsdale 2001. Humans as global plant dispersers: Getting more than we bargained for. *BioScience* 51: 95–102.

Proches, S., J. R. U. Wilson, D. M. Richardson, and M. Rejmánek. 2008. Searching for phylogenetic pattern in biological invasions. *Global Ecology and Biogeography* 17: 5–10.

Ricciardi, A., and S. K. Atkinson 2004. Distinctiveness magnifies the impact of biological invaders in aquatic ecosystems. *Ecology Letters* 7: 781–784.

Ricciardi, A., and M. Mottiar 2006. Does Darwin's naturalization hypothesis explain fish invasions? *Biological Invasions* 8: 1403–1407.

Smith, K. F., M. Behrens, L. M. Schloegel, N. Marano, S. Burgiel, and P. Daszak. 2009. Reducing the risks of the wildlife trade. *Science* 324: 594–595.

THREATENED SPECIES

SEE ENDANGERED AND THREATENED SPECIES

TOLERANCE LIMITS, ANIMAL

ALBERTO JIMÉNEZ-VALVERDE
Universidad de Málaga, Spain

JORGE M. LOBO
National Museum of Natural Sciences, Madrid, Spain

Ecophysiological constraints prevent species from occupying the entirety of abiotic gradients present in nature and restrict them to just a portion lying between certain environmental bounds (i.e., their tolerance limits) beyond which they cannot survive. The fact that every species shows specific environmental adaptations that allow it to grow and reproduce is the basis of the niche concept. These environmental restrictions are the first factor that demarcates the geographic regions that a species can inhabit. Therefore, knowledge of tolerance limits is crucially important in order to assess the risk that a species will become invasive in a new site. Are environmental tolerance limits fixed? What role is played by these environmental restrictions in delimiting geographic range limits? Are invasive species characterized by broader tolerances than native ones? To what extent is their survival and spread limited by tolerances? These and related questions are essential to understanding the potential invasiveness of species as well as their abundance and distribution.

PHYSIOLOGICAL LIMITS

Temperature is probably the most important environmental factor influencing the performance of species, especially in ectotherms, the majority of species on Earth. Thermal performance curves are typically asymmetrically bell-shaped, with a progressive increase in fitness from the lowest limit of temperature tolerance to an optimum, after which a sharp decline occurs (Fig. 1A). Denaturation of proteins and degradation of enzymatic processes are believed to be among the main determinants of maximum

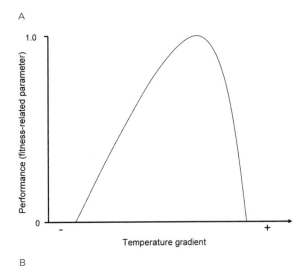

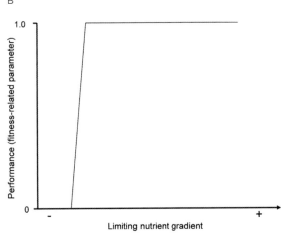

FIGURE 1 (A) Theoretical thermal performance curve in which the performance of the organism increases slowly with temperature and eventually reaches an optimum from which the performance sharply decreases. (B) Theoretical performance curve of an organism in relation to limiting nutrient availability. Below a certain threshold, the organism has a severe deficit of that nutrient and is unable to survive.

lethal limits. The need for oxygen when an increase in the metabolic rate occurs may also be an important additional determinant of these maximum lethal limits. Insufficient aerobic capacity of mitochondria also plays a significant role in setting low thermal tolerance limits, although the freezing of body fluids is probably the main determinant. Physiological adaptations to low temperatures have been extensively studied, and we know that several strategies in response to freezing have evolved in animals. Some amphibians, reptiles, and invertebrates are freeze-tolerant (i.e., they can survive a certain amount of extracellular ice formation in their tissues), while freeze-avoiding species synthesize antifreezing molecules that decrease the temperature at which the formation of ice starts.

Besides temperature, many other factors such as humidity, salinity, and limiting nutrient availability influence the fitness of species, with each factor producing a different shape of performance curve (Fig. 1B). Laboratory experiments provide the underlying explanation for physiologically imposed tolerance limits and the mechanisms behind their respective curve shapes. However, laboratory experiments are usually performed over short periods, and their results are often difficult to extrapolate to real-world situations. Moreover, the mismatch between laboratory theory and field observations—species living above or below reported tolerance limits—highlights the importance of taking into account interactions between different factors and demonstrates the complexity of the phenomenon.

PHYSIOLOGY AND GEOGRAPHIC DISTRIBUTION

Physiological tolerances clearly influence species distributions, and the match between the geographic ranges of species and climate has long been recognized by naturalists and biogeographers. Also, observed range shifts in response to past and present climate changes indicate the undeniable influence of climate on species distributions. However, as explained above, critical lethal limits observed in the laboratory are not likely to limit species ranges. We now know that a slight departure from optimal conditions can cause important adverse effects on the populations of species even before lethal limits are reached. For example, many species die if exposed repeatedly to moderately low temperatures. A decrease in the activity rate can lead to fasting periods that eventually may kill the organism by the indirect effect of temperature on starvation. Similarly, a rise in temperature increases the amount of energy required to maintain a given metabolic rate, compromising growth and reproduction. Low temperatures may also

affect dispersal rates, while high temperatures can increase the activity of predators, competitors, and parasites, jeopardizing the long-term persistence of populations.

The interaction between different climatic factors also has a decisive influence on organisms' survival. For example, the response to winter temperatures depends on the temperature of the former season. The capacity to survive low temperatures is also highly dependent on the rate of thermal change—rapid changes offer fewer opportunities for acclimatization—or on the length of the cold period. Snow may enhance the opportunities to survive cold extremes by creating an insulating surface, and the presence of water in the environment may promote tissue-freezing by inoculation. However, behavior is probably the most important and flexible factor that modifies tolerance limits. For instance, selection of overwintering places with higher temperatures than the surrounding area and aggregation of individuals are strategies to avoid exposure to harmful cold conditions. Researchers must also consider that tolerance limits and functions describing the relationship between performance and environmental factors can change between the different life stages of the same species, and so the distribution will then be determined by the most constrained stage.

PHENOTYPIC PLASTICITY AND EVOLUTION

Physiological limitations are not fixed but rather vary among individuals and populations of the same species, giving phenotypic plasticity an important role in the processes of adaptation to new environments. The capacity to change the phenotype may differ between life stages of the same species. Similarly, not all traits are equally plastic. For example, upper thermal tolerances are usually quite constant; on the contrary, lower thermal limits are quite flexible, and individuals can more readily adapt to decreases in temperature. Interestingly, it has been shown that even specialist species that are at present restricted to very specific climatic conditions retain the capacity to respond to environmental changes. Changes due to phenotypic plasticity can take place very quickly; a few days of acclimation in insects may be enough to alter their thermal limits. These phenotypic changes can promote genetic adaptation, which, together with the effect of either genetic drift owing to bottlenecks or genetic diversity caused by multiple source populations, makes rapid physiological evolution an important process in colonization events. In fact, the available evidence is sometimes contrary to what would be expected, and there is no general rule that invasion success is the direct consequence of having higher environmental tolerances or wider phenotypic plasticity.

The capacity to adapt to new environments allows both generalist and specialist species to increase their geographic ranges. Invasiveness should thus be viewed as a species-specific characteristic affected by many different (unknown) factors that interact in complex ways.

PREDICTING THE GEOGRAPHIC DISTRIBUTION OF INVADERS

Because the control of populations of already-established invasive species can be extremely costly and difficult, prevention is the most important strategy. Risk maps showing invasiveness potential are thus a valuable tool for environmental managers. As we have seen, the complexity of the responses of species to the environment, interactions among factors, and the ability to adapt make the prediction of the potential areas of establishment challenging. Prediction of areas susceptible to invasion is usually done using distribution data of the focal species, several bioclimatic variables, and correlative techniques that establish a relationship between both kinds of data: the so-called species distribution models (or ecological niche models). However, these techniques have been criticized because they do not rely on any mechanistic basis for the response of the species to the environment. Thus, for exercises such as predicting the potential distribution of a species in a new territory, models that incorporate physiological and biological knowledge of the species are recommended, and ideally, a combination of both approaches (mechanistic and correlative—which may not be mutually exclusive but complementary) is preferred when possible. Unfortunately, physiological information is lacking for most species, so correlative models will continue to be a necessary tool. Correlative models must be parameterized, and this step should avoid, as much as possible, blind procedures that rely on automatic variable-selection techniques that ignore the basic knowledge about the response of organisms to environmental gradients. Similarly, because the distribution data do not reflect the fundamental niche of a species but only a part, and because interactions between environmental factors are not stationary but vary geographically, simple techniques that account for extreme and rare data should be preferred over complex overfitting methods.

SEE ALSO THE FOLLOWING ARTICLES

Climate Change / Evolution of Invasive Populations / Genotypes, Invasive / Range Modeling / Tolerance Limits, Plant

FURTHER READING

Angilletta, M. J. 2009. *Thermal Adaptation: A Theoretical and Empirical Synthesis*. Oxford: Oxford University Press.

Bale, J. S. 2002. Insects and low temperatures: From molecular biology to distributions and abundance. *Philosophical Transactions of the Royal Society B* 357: 849–862.

Chown, S. L., and K. J. Gaston. 2008. Macrophysiology for a changing world. *Proceedings of the Royal Society B* 275: 1469–1478.

Cox, G. W. 2004. *Alien Species and Evolution*. Washington, DC: Island Press.

Ghalambor, C. K., J. K. McKay, S. P. Carroll, and D. N. Reznick. 2007. Adaptive versus non-adaptive phenotypic plasticity and the potential for contemporary adaptation in new environments. *Functional Ecology* 21: 394–407.

Jeffree, E. P., and C. E. Jeffree. 1994. Temperature and the biogeographical distributions of species. *Functional Ecology* 8: 640–650.

Kearney, M., and W. Porter. 2009. Mechanistic niche modelling: Combining physiological and spatial data to predict species' ranges. *Ecology Letters* 12: 334–350.

McNab, B. K. 2002. *The Physiological Ecology of Vertebrates: A View from Energetics*. New York: Cornell University Press.

Peck, L. S., M. S. Clark, S. A. Morley, A. Massey, and H. Rossetti. 2009. Animal temperature limits and ecological relevance: Effects of size, activity and rates of change. *Functional Ecology* 23: 248–256.

Terblanche, J. S., C. J. Klok, E. S. Krafsur, and S. L. Chown. 2006. Phenotypic plasticity and geographic variation in thermal tolerance and water loss of the tsetse *Glossina pallidipes* (Diptera: Glossinidae): Implications for distribution modelling. *American Journal of Tropical Medicine and Hygiene* 74: 786–794.

TOLERANCE LIMITS, PLANT

JAMES H. RICHARDS AND BENJAMIN R. JANES

University of California, Davis

Plants, including invasive species, vary immensely in tolerance of the extremes of abiotic factors such as temperature, moisture, light, nutrient availability, and toxic substances. Such tolerance sets the ultimate limits of species' distributions, and the reaction of plants to the complex of abiotic factors present in any habitat strongly affects productivity and abundance. Different combinations of physiological mechanisms, developmental patterns, and morphology at a variety of scales allow plants to exploit habitats that differ in the combination of stresses present. The ability of plants to adjust to variable conditions during growth (i.e., to acclimate) is critical for tolerance of abiotic extremes and for maximizing growth and reproduction under different growing conditions. Usually growth and reproduction are very limited when abiotic conditions are near tolerance limits, yet some examples of exceptional tolerance can explain invasive success.

PLANT DISTRIBUTION AND ABUNDANCE

Tolerance of plants to key abiotic factors such as freezing and drought interacts with climatic extremes to determine the distribution of plants globally and among habitats on regional or local landscape scales. These interactions also affect the potential distribution of invasive species. Some examples of these interactions include how freezing tolerance may determine species' northern latitudinal limits (Fig. 1A), how chilling sensitivity of species with tropical evolutionary background limits their presence in temperate climates, and how drought sensitivity of seedlings can be related to tree distribution. Rare climatic events such as severe frosts or extended droughts can occur frequently enough to limit distribution of long-lived species.

When abiotic environmental variables are less extreme, productivity and abundance of plants are determined by the shapes of plant response curves to temperature, moisture, light, nutrient availability, and toxic substances, as well as by interactions with biotic factors. In nearly every environment, plants that avoid the extremes of abiotic factors coexist with plants that tolerate the extremes. These two strategies of tolerance and avoidance result from combinations of traits in plants that affect phenology, morphology, physiology, and metabolism.

For plants with either avoidance or tolerance traits, the approximate mapping of growth and reproductive output to physiological responses means that when abiotic conditions are not optimal, growth and potential population growth rates are also limited (Fig. 1B). Extensive information from agronomic, horticultural, and forest species shows that there is a rapid decline in seed, fruit, or biomass yield as nutrient availability declines below a broad zone of sufficiency (Fig. 1C). Similarly, soil water availability below threshold levels or salinity above threshold levels can cause steep declines in growth, yield, or survival. When one considers environmental characteristics of a habitat and the ability of a species to become invasive, thresholds for the beginning of decline in plant performance are likely more important determinants of successful invasion than are absolute tolerance limits. Even if plants have great tolerance, both individual growth and population growth rates are too low for invasion success when plants are growing near their physiological limits. Interspecific competition and other biotic interactions can have large impacts on species distribution, and thus physiological tolerance limits would normally be outside the realized niche of plants. Nevertheless, there are some examples where greater physiological tolerance allows invasive species to be successful in habitats where natives are not.

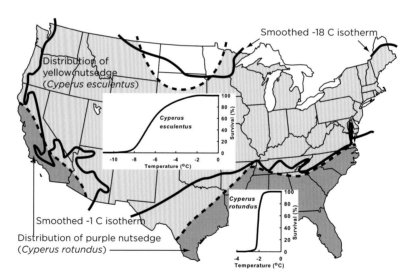

FIGURE 1 (A) Distribution of purple nutsedge (*Cyperus rotundus*) is limited to areas where average minimum air temperatures of the coldest month (January) are not very extreme (i.e., warmer than approximately –1 °C). This correlates with the limited freezing tolerance of tubers of this species determined in a laboratory trial (lower inset). By contrast, tubers of yellow nutsedge (*Cyperus esculentus*) tolerate much lower temperatures (upper inset) than purple nutsedge, and this greater tolerance of freezing contributes to the former's much wider distribution. Differences between laboratory and field temperature limits may be due to insulation of tubers by soil, litter, or snow, to genetic differences among populations, or to different acclimation periods and conditions. Based on the laboratory results and the –1 °C isotherm location (1931–1960), the predicted distribution of purple nutsedge extends north of the distribution shown for 1973. Indeed, current data from the USDA Plants database (http://plants.usda.gov/java/profile?symbol=CYRO) show the occurrence of this species in Oregon and also that it is present in many locations somewhat north of the 1973 distribution shown here (e.g., into Oklahoma, Missouri, Kentucky, Maryland, Delaware, New Jersey, Pennsylvania, and New York). Whether this northern spread is due to climatic warming, to genetic changes, or simply to spread into local habitats where soil temperatures remain above the freezing tolerance limit of purple nutsedge tubers is not clear. (Map and insets derived from Stoller, 1973. Effect of minimum soil temperature on differential distribution of *Cyperus rotundus* and *C. esculentus* in the United States. *Weed Research* 13: 209–217.)

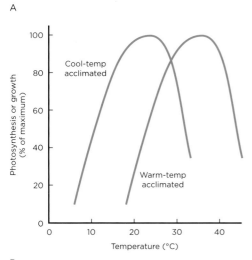

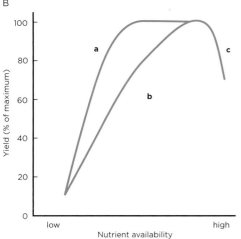

FIGURE 2 (A) Typical temperature response curves of photosynthesis or of growth for a species that acclimates to cool or to warm temperatures. (B) Typical threshold responses of two different species (a, b) to nutrient availability. At moderate nutrient availability, species b is much less productive than a, although their limits of tolerance and reduced productivity at high, toxic nutrient levels (c) are the same in this example.

SCALES OF STRESS TOLERANCE OR AVOIDANCE

Characterizing plants as tolerators or avoiders is based on mechanistic understanding of responses to individual factors. Understanding mechanisms responsible for tolerance or avoidance requires appreciation of the scale at which each mechanism operates. Tolerance at one scale (e.g., the whole organism) may result from avoidance mechanisms at another scale (e.g., cellular or tissue level). For example, halophytic plants, including some invasive species that persist in saline environments (e.g., *Halogeton glomeratus*, *Tamarix* spp.), may have high leaf salt concentrations, yet this "tolerance" requires that toxic sodium ions be compartmentalized in cell vacuoles with

concomitant accumulation of balancing osmotic concentrations of compatible solutes in the cytoplasm. Tolerance at the whole plant or even whole leaf level is achieved by efficient avoidance of high salt concentrations in the cytoplasm. In higher plants, leaf cell metabolism is not tolerant of salt, but halophytes are very efficient at compartmentalizing the toxic ions to avoid damage to essential enzymatic and other cellular-level processes. Contrast this with the stress of leaf tissue dehydration where water stress tolerators must have cytoplasmic tolerance, not avoidance, of low water content and activity.

Many annual invasive species in temperate climates are drought avoiders, with phenological patterns that allow them to grow during periods of high moisture availability and physiological responses that prevent significant dehydration of tissues (e.g., *Avena fatua*, *Brassica* spp., *Erodium cicutarium*, etc.). Corresponding to these drought avoidance characteristics, the plants often exhibit tolerance of cool or even freezing temperatures that are typical of periods with high moisture availability. Tolerance and avoidance of multiple abiotic factors are not necessarily correlated.

MECHANISMS OF TOLERANCE

Tolerance limits at the high and low ends of abiotic environmental gradients are usually determined by very different mechanisms. Plant tolerance of too much water (flooding) depends on physiological acclimation that balances limited energy production capacity with simultaneously reduced demand while minimizing cytoplasmic acidification and on morphological characteristics that increase oxygen supply to roots and root tips. In contrast to these mechanisms, tolerance of too little water (drought) is characterized by a complex of mechanisms that link plant water uptake, transport, loss, and carbon assimilation. Drought tolerators maintain some stomatal opening and continue photosynthetic activity even as leaves dehydrate to some extent. Maintaining photosynthetic activity provides substrate for continued root growth and synthesis of osmotically active compounds like sugars and other compatible solutes. Increasing the concentration of osmotically active compounds in the leaves reduces leaf osmotic potential, maintaining a gradient for continued water uptake from drying soil. Continued water loss associated with stomatal opening under water-limited conditions leads to greater tension on water in xylem and may lead to xylem cavitation and loss of the ability to conduct water to sites of evaporation in leaves. Stomatal control of water loss is tuned such that xylem tension is constrained below the threshold of complete cavitation. The safety margins between stomatal operational points and thresholds for cavitation are typically greater for more tolerant plants. Under severe or extended droughts, less tolerant plants with smaller safety margins may die because xylem cavitates, but the most tolerant plants appear to be well enough protected that this does not often happen. As drought conditions continue, tolerant plants that avoid cavitation can die because of carbon starvation caused by continued stomatal limitation of photosynthetic carbon assimilation. The most dehydration-tolerant plants must have metabolic characteristics such as accumulation of compatible solutes and protection of macromolecules and membranes from dehydration damage that allow continued functioning despite tissue dehydration, xylem resistance to cavitation, and leaves that can avoid heat and high light damage when stomata are closed. Although there may be a few points of overlap, the complex of mechanisms that characterizes drought tolerators is very different from the mechanisms found in flooding-tolerant plants.

Interactions among drought, heat, and light stress are critical to understanding actual limits in natural environments. The threshold for heat damage to leaves is generally higher for plants in low light or for downward-facing leaf surfaces than for plants in high light. Damage to the light reactions of photosynthesis, specifically photosystem II (PS II), and not protein denaturation, is the major negative outcome of heat or excess light. Plants have exquisite mechanisms to avoid excess light absorption by leaves or chloroplasts as well as enzymatic and non-enzymatic mechanisms to channel excess energy away from PS II and prevent generation or to regulate the amount of damaging reactive oxygen species (ROS).

Tissue type and development strongly affect tolerance limits. Growing or immature tissues are seldom able to tolerate freezing or dehydration to the same extent as mature or dormant tissues. Moreover, differences between leaves, reproductive tissues, stems, and roots may be pronounced, with reproductive tissues and roots often being more sensitive to frost or dehydration, for example, than leaf or stem tissues. Because the warmest and coldest temperatures occur at the soil surface, seedlings are usually exposed to more extreme temperatures than larger plants. Freezing tolerance in seedlings is essential for fall and spring germinating annuals in temperate and mountain climates, while heat tolerance limitations may be important in locations such as deserts, wind-protected microsites in high mountains, and open areas where high midday insolation leads to high surface temperatures. These microclimate–stress tolerance interactions can prevent successful colonization of locations that are otherwise suitable for mature plants.

Plant physiognomy also especially affects freezing tolerance. In temperate or colder climates, low-growing plants with buds near the ground that are often covered with an insulating snow layer in winter usually are less freeze-tolerant than are trees in the same community. Changes in snow cover, spatially and temporally, affect the distribution of such species.

ACCLIMATION TO ABIOTIC STRESS

A species' evolutionary history affects not only its absolute stress tolerance limits but also its acclimation capacity. Plants with tropical origins are commonly chill-sensitive (i.e., unable to tolerate temperatures less than about 10–15 °C). Even with slow exposure to cool conditions, they cannot acclimate to these temperatures very well. By contrast, plants from areas with larger variation in temperatures during the growing season usually have the capacity to acclimate photosynthesis and growth so that temperature optima for these processes typically shift by 10–15 °C and the temperature at which freezing damage occurs may shift over 30 °C from winter to summer. Ecotypes of a species with different evolutionary histories may also differ greatly in ability to acclimate, as shown by classic comparisons of Arctic and alpine populations' different abilities to acclimate photosynthesis and respiration to variable growth temperatures. Tolerance limits shift in parallel with acclimation of optimal temperatures. Similar patterns exist for tolerance of and acclimation capacity to other abiotic factors such as moisture availability, freezing, salinity, and light.

Prevention of cell lysis during freeze–thaw cycles results from induction of desaturase enzymes and changes in plasma membrane composition during acclimation. Increased proportions of unsaturated fatty acids in plasma membrane phospholipids allow cells to shrink with freeze-induced dehydration and then to reexpand upon thawing by incorporating plasma membrane material still attached to the inner side of the membrane, thus preventing expansion-induced lysis. Such expansion is not possible for unacclimated cells, and they are lysed by swelling that exceeds their capacity to expand cell membrane area. Cellular dehydration accompanies freezing, and full acclimation to minimum temperature tolerance requires the induction of numerous changes that include compatible solute accumulation and synthesis of many proteins that protect macromolecules, prevent membrane–membrane fusion, and regulate rates of extracellular ice formation. Acclimation of plants to freezing temperatures requires exposure to cool temperature and short days. Full acclimation may change the minimum temperature tolerated by leaves and shoots by as much as 30 °C or more for temperate plants native to areas with very cold winter temperatures. Species from more constant climates may be unable to acclimate significantly.

Tolerance of even colder temperatures, especially below levels where water may supercool and remain liquid (down to the homogeneous nucleation point of approximately −40 °C) requires anatomical or morphological features that allow space for intercellular or extra-organ ice formation and avoidance of freeze-induced xylem cavitation. Such cavitation is greater in large-volume xylem vessels because the air bubbles that form during freezing are larger and more susceptible to expansion upon thawing than are the bubbles in smaller conduits. This mechanism of cavitation is very different from air seeding through pit membrane pores, which causes cavitation under drought-induced xylem tensions.

Because different physiological mechanisms determine tolerance at upper and lower limits, it is seldom possible for a single species to be most tolerant at both ends of the environmental spectrum. An excellent example of this is for acclimation to cold versus warm temperatures. Biochemical membrane characteristics required for maximum cold and freeze tolerance, including a high degree of fatty acid desaturation, are nearly polar opposites of key characteristics needed for tolerance of high temperatures (i.e., high levels of saturated lipids in thylakoid membranes).

TOLERANCE LIMITS AND INVASIVE SUCCESS

Small differences in limits of tolerance may be important in the success of invasive species in some cases. Greater drought and salinity tolerance of *Tamarix* spp. compared to natives contributes to its dominance in many riparian areas of semi-arid western North America. The similarly extensive invasion of *Bromus tectorum* in western North America can be partially attributed to its lower minimum critical value for root growth (about 3 °C) than that of native grass seedlings (about 8 °C). This root cold tolerance difference, while not the only factor that results in the invasive success of *B. tectorum*, contributes to its competitive success in at least two ways. First, greater root growth rate in cold soils allows *B. tectorum* to position root tips deeper in the soil more quickly after fall germination so that the growing roots inhabit soil that is slightly warmer than surface soils, where competing grass seedlings roots are located. Second, by late winter or early spring, *B. tectorum* roots will have penetrated to greater depths, allowing them to preemptively exploit nitrogen and water resources in soil horizons with few growing roots of competing native grass seedlings. *Taeniatherum caput-medusae* is another invasive species that appears to have similar root growth cold tolerance.

Understanding not only the limits of tolerance but the shape of plant response curves to environmental factors will be essential to predict invasion success. Moreover, not only aboveground but also belowground organs must be considered. Understanding relationships of evolutionary history to potential mechanisms and limits of acclimation to abiotic factors will be critical for determining levels of preadaptation to new environment–species combinations.

SEE ALSO THE FOLLOWING ARTICLES

Cheatgrass / Climate Change / Evolution of Invasive Populations / Invasibility, of Communities and Ecosystems / Invasiveness / Tolerance Limits, Animal

FURTHER READING

Alpert, P., E. Bone, and C. Holzapfel. 2000. Invasiveness, invasibility and the role of environmental stress in the spread of non-native plants. *Perspectives in Plant Ecology, Evolution and Systematics* 3: 52–66.

Bray, E. A., J. Bailey-Serres, and E. Weretilnyk. 2000. Responses to abiotic stresses (1175–1177). In B. B. Buchanan, W. Gruissem, and R. L. Jones, eds. *Biochemistry and Molecular Biology of Plants*. Rockville, MD: American Society of Plant Physiology.

Chaves, M. M., J. P. Maroco, and J. S. Pereira. 2003. Understanding plant responses to drought: From genes to the whole plant. *Functional Plant Biology* 30: 239–264.

Crawford, R. M. M. 2008. *Plants at the Margin: Ecological Limits and Climate Change*. Cambridge: Cambridge University Press.

Drew, M. C. 1997. Oxygen deficiency and root metabolism: Injury and acclimation under hypoxia and anoxia. *Annual Reviews Plant Physiology and Plant Molecular Biology* 48: 223–250.

Lambers, H., F. S. Chapin III, and T. L. Pons. 2008. *Plant Physiological Ecology*, 2nd ed. Berlin: Springer.

McDowell, N., W. T. Pockman, C. D. Allen, D. D. Breshears, N. Cobb, T. Kolb, J. Plaut, J. Sperry, A. West, D. G. Williams, and E. A. Yepez. 2008. Mechanisms of plant survival and mortality during drought: Why do some plants survive while others succumb to drought? *New Phytologist* 178: 719–739.

Nilsen, E. T., and D. M. Orcutt. 1996. *The Physiology of Plants under Stress: Abiotic Factors*. New York: Wiley.

Schulze, E-D., E. Beck, and K. Müller-Hohenstein. 2002. *Plant Ecology*. Berlin: Springer.

Woodward, F. I. 1987. *Climate and Plant Distribution*. Cambridge: Cambridge University Press.

TRANSFORMERS

JOAN G. EHRENFELD

Rutgers University, New Brunswick, New Jersey

An invasive species is termed a "transformer" when it causes a substantial change to the structure or function of the ecosystem of which it has become a part, relative to the structure and function of the uninvaded ecosystem, and when these changes have taken place over a large area. Transformers may change the physical or chemical characteristics of the environment, they may change the composition of and connections within the food web of the system, they may change ecosystem process rates, or they may create some combination of these three major types of change. There are often important economic and practical ramifications of ecosystem change. A well-known example is the invasion of several species of trees into the scrub shrub ecosystems of South Africa. The widespread invasion resulted in a large loss of water, a scarce resource, to evaporation, necessitating an expensive program to remove the trees; the invasion also resulted in increased fire frequencies, in the outcompeting of grasses and damaging of important grazing lands, in reduced bird diversity, and in changes to stream biotas through the reduction of stream flow. Ecosystem transformation can establish an alternate stable state, in which the structure of the transformed ecosystem itself creates conditions that are self-reinforcing, making it difficult to restore the system to its preinvasion state. While there are no generally accepted quantitative criteria for determining when a change in structure or function is "substantial," the concept implies that more than one, and usually several, of the components of an ecosystem, such as species abundances, nutrient flux rates or pool sizes, food web connections, or physical structure, are significantly altered by the presence of the transforming species.

MECHANISMS DRIVING ECOSYSTEM TRANSFORMATION

Individual species transform whole ecosystems through the introduction of some physical, physiological, or behavioral trait that is unlike those of any native species in the given ecosystem. The trait or traits that have been associated with ecosystem transformation vary greatly; the key feature is that they represent a unique and novel feature in the invaded ecosystem. In some cases of plant invasions, ecosystems may be transformed merely by the size or growth form of the novel species. Animals can similarly transform ecosystems through the introduction of a behavior novel to the invaded system; the most infamous example is the introduction of the brown tree snake (*Boiga irregularis*) into Guam and other South Pacific islands, where its ability to climb trees resulted in extensive predation of species that had never experienced predators within the tree canopy before. Invasive species can also transform ecosystems by significantly reducing the abundance of native species that had been the dominant

organisms in terms of either biomass or numbers. Finally, species transform ecosystems when their traits lead to changes in the physical distribution of resources and structures, the frequency of disturbances, or the chemical composition of soil or water.

The concept of a species being a transformer is related to several other ecological concepts. Keystone species are those that have a large impact on community structure and function that is disproportionate to their own density or biomass. The concept of an ecosystem engineer refers to species that modify the physical environment; while it is recognized that virtually all organisms affect their environment in some way, the concept specifically targets species that cause large changes to some component of the physical environment in ways that affect many other species. Another key concept is that of a foundation species, defined as a species that contributes most of the density or biomass to an ecosystem, and thereby has a controlling effect on many other aspects of structure and function. Invasive species can act as transformers either by acting as keystone, engineering, or foundation species, or by affecting the abundance of native keystone, engineering, or foundation species. In all cases, the most important feature is that the species alters many aspects of the physical environment or affects the abundance, distribution, and interactions of many co-occurring species (usually it does both).

The following are examples of the various pathways and mechanisms by which invasive species can transform ecosystems.

Transformations Mediated by Alterations to the Physical Environment

Among the most notable environmental transformations wrought by invasive species are those in which a species acts to trap moving sand or sediments, thereby altering the geomorphology of the land surface. Well-documented examples include *Ammophila arenaria*, *Spartina alterniflora*, and *Spartina anglica*. These coastal grasses have morphological and growth form characteristics, notably dense swards of stiff stems, rapid vegetative spread through runners, and the ability to emerge from under deposited sediments that are very effective in trapping sediments. The result is the construction of tall continuous dunes (*Ammophila*) in environments that had supported only small, discontinuous dunes and the creation of densely vegetated marsh surfaces (*Spartina* spp.) in estuaries that had previously had expansive open mud flats (Fig. 1). The conversion of mud flats to vegetated marshes transforms the composition of the

FIGURE 1 Extensive invasion of *Spartina alterniflora* in Yacheng, Zhejiang Province, China. The estuarine area is completely filled with vegetated mud flats. (Photograph courtesy of the author.)

estuarine bird community, as mudflat-feeding species lose habitat and marsh-feeding species gain habitat. The spread of *S. alterniflora* also results in altered hydrologic patterns, altered water chemistry, soil pools and fluxes of carbon and nitrogen, and the shifting of food webs from webs based on consumption of surface algae to webs based on buried plant detritus, with a concomitant change in the species composition of the benthic invertebrate community. Other well-documented physically mediated transformations include the action of nonnative earthworms in incorporating surface organic matter into the mineral soil in temperate and boreal forests, which causes changes to soil nutrient cycles, plant community structure, the abundance and food web relationships of litter-dwelling or litter-feeding organisms, and the ability of some invasive annual grasses to promote very frequent fire regimes, with a similar range of physical, chemical, and biotic consequences. A final physical mechanism by which species transform ecosystems is simply through a change in spatial distribution of biomass, as a result of invasive plant growth form. The canopy height and density, the density of understory strata, and the morphology of plant structures that enable animals to utilize them in novel ways are all pathways of transformation, largely because many animals depend more heavily on the three-dimensional structure of the habitat than on species identity for breeding and refuge sites. For example, the presence of *Centaurea maculosa* in the grasslands of western North America allows some species of spider to become up to 74 times as abundant as they previously had been, because the stiff plant stems of the invasive species not only support but also enhance the size of their webs in comparison with webs on native plants; as a result, invertebrate predation increases 89-fold.

Transformations Mediated by Changes to Biogeochemical Cycling Processes

Invasive species can alter biogeochemical cycling processes through numerous mechanisms. Some invasive species take up from the soil and translocate to their leaves large amounts of salts, which, when deposited on the soil surface, can increase soil salinity enough to prevent other species from growing; the change in species and habitat structure causes cascading effects on other organisms. Nitrogen-fixing plants similarly transform communities by increasing the supply of a limiting resource, thereby stimulating changes in plant community composition and associated animals. A third biogeochemical mechanism acts through the introduction of plant tissues that decompose faster or more slowly than those of the uninvaded community because of differences in leaf tissue chemistry (concentrations of nitrogen, lignin, phenolics, alkaloids, and other secondary compounds). Rapidly decomposing litter in forests results in a loss of the leaf litter layer, an important habitat for many animal species and a source of food for ground-foraging birds. Invasive species in riparian areas can also affect stream invertebrate communities when litter falling into the stream provides poor substrate for the key invertebrate shredders and decomposers of plant materials. Invasives also transform ecosystems through their effects on the hydrologic cycle. Species with much higher rates of water withdrawal and transpiration (for example, *Eucalyptus* spp., *Tamarix* spp., *Acacia* spp.) alter water availability for other plants and alter stream flows, affecting aquatic organisms as well as terrestrial species.

TRANSFORMATIONS MEDIATED BY ANIMAL BEHAVIOR

Just as plant traits can transform ecosystems through differences in physical structure, tissue chemistry, or biogeochemical behavior, animal behavioral traits can transform ecosystems through cascading effects on food webs. Herbivory can reduce or remove foundation species, thus altering the environment for all other components of the ecosystem. Prominent examples of such a cascade of change are the decline of hemlocks (*Tsuga canadensis*) from herbivory by the hemlock woolly adelgid (*Adelges tsugae*), which alters plant, animal, soil, and stream communities, and the effects of the golden apple snail (*Pomacea canaliculata*) on aquatic ecosystems in tropical wetlands. In the latter case, herbivory results in a loss of aquatic macrophytes and an increase in algal biomass, turbidity, and phosphorus concentrations. Patterns of predation can also cause ecosystem transformations in cases in which the invasive predator alters the abundance of a keystone species, usually through the introduction of a novel predation pathway. On islands near New Zealand, rat predation reduces the abundance of burrow-nesting sea birds; the reduction in nutrient import from oceanic food webs via the birds results in large shifts in the aboveground vegetation of the islands. On Christmas Island, in the Indian Ocean, the introduction of crazy ants (*Anoplolepis gracilipes*) resulted in high mortality of red land crabs; without the crabs, litter was not incorporated into the soil to decompose, but rather accumulated on the forest floor. Furthermore, seedling densities increased, and both native and introduced scale insects increased, leading to fungal pathogens infecting the canopy trees. Behaviors that result in novel patterns of intraguild competition can also transform ecosystems; for example, introduced rainbow trout (*Oncorhynchus mykiss*) in Japan are better competitors for surface invertebrate prey than are the native fish, which shift their feeding to benthic algal grazers, resulting in an increase in algal biomass. As a result, fewer insects that form an important part of the adjacent forest food web emerge from the stream, with a resulting decrease in spider density

CONCLUSION

Ecosystem transformation by invasive species occurs through numerous pathways; the common feature is the introduction of a structure, process, or behavior that is novel to the invaded system and that has the potential to affect multiple connected ecosystem components. Transformation of the abiotic as well as the biotic environment may create alternate stable states, making reversal and restoration difficult at best. Thus, when an invader either functions as an engineer or a keystone species, or strongly alters the abundance of such species or foundation species, there is a high potential for transformations to occur, with profound implications for both ecological health and human use of the invaded system.

SEE ALSO THE FOLLOWING ARTICLES

Brown Treesnake / Earthworms / Eutrophication, Aquatic / Fire Regimes / Hemlock Woolly Adelgid / Herbivory / Invasional Meltdown / Nitrogen Enrichment

FURTHER READING

Ellison, A.M., M.S. Bank, B.D. Clinton, E.A. Colburn, K. Elliott, C.R. Ford, D.R. Foster, B.D. Kloeppel, J.D. Knoepp, G.M. Lovett, J. Mohan, D.A. Orwig, N.L. Rodenhouse, W.V. Sobczak, K.A. Stinson, J.K. Stone, C.M. Swan, J. Thompson, B. Von Holle, and J.R. Webster. 2005. Loss of foundation species: Consequences for the structure and dynamics of forested ecosystems. *Frontiers in Ecology and the Environment* 3: 479–486.

Grosholz, E.D., L.A. Levin, A.C. Tyler, and C. Neira. 2009. Changes in community structure and ecosystem function following *Spartina alternifolia* invasion of Pacific estuaries (23–40). In B.R. Sillian, E.D. Grosholz, and M.D. Bertness, eds. *Human Impacts on Salt Marshes: A Global Perspective.* Berkeley: University of California Press.
Kenis, M., M.-A. Auger-Rozenberg, A. Roques, L. Timms, C. Péré, M.J.W. Cock, J. Settele, S. Augustin, and C. Lopez-Vaamonde. 2009. Ecological effects of invasive alien insects. *Biological Invasions* 11: 21–45.
Liao, C.Z., R.H. Peng, Y.Q. Luo, X.H. Zhou, X.W. Wu, C.M. Fang, J.K. Chen, and B. Li. 2008. Altered ecosystem carbon and nitrogen cycles by plant invasion: A meta-analysis. *New Phytologist* 177: 706–714.
Lovett, G.M., C.D. Canham, M.A. Arthur, K.C. Weathers, and R.D. Fizhugh. 2006. Forest ecosystem responses to exotic pests and pathogens in eastern North America. *BioScience* 56: 395–407.
Richardson, D.M., P. Pyšek, M. Rejmánek, M.T. Barbour, F.D. Panetta, and C.J. West. 2000. Naturalization and invasion of alien plants: Concepts and definitions. *Diversity and Distributions* 6: 93–107.
Strayer, D.L. 2009. Twenty years of zebra mussels: Lessons from the mollusk that made headlines. *Frontiers in Ecology and the Environment* 7: 135–141.
Strayer, D.L., V.T. Eviner, J.M. Jeschke, and M.L. Pace. 2006. Understanding the long-term effects of species invasions. *Trends in Ecology and Evolution* 21: 645–651.

TREES AND SHRUBS

DAVID M. RICHARDSON

Stellenbosch University, Matieland, South Africa

Until fairly recently, tall woody plants were not widely recognized as major invasive species. In the past few decades, however, humans have moved thousands of tree and shrub species out of their natural ranges for many purposes, and hundreds of species have become naturalized or invasive. In many parts of the world, trees and shrubs now feature prominently on lists of invasive alien plants, and in some areas, nonnative species in these life forms are now among the most conspicuous and damaging of invasive species. Invasive trees invade many habitats, have a wide range of impacts, and create many types of conflicts of interest. Issues related to invasiveness are now influencing how people perceive, use, and manage alien trees and shrubs.

DEFINING TREES AND SHRUBS

What constitutes a tree? Here, perennial woody plants that have many secondary branches supported clear of the ground on a single main stem or trunk with clear apical dominance are considered trees. Setting a minimum height specification at maturity for separating trees from shrubs is difficult, but here we consider species that meet the aforementioned criteria and that regularly attain a height of at least 3 m to qualify as trees. Woody plants that do not meet these criteria by having multiple stems or small size are shrubs. Species typically classified as shrubby herbs or forbs (e.g., *Catharanthus roseus*, *Crotalaria* spp., *Hypericum* species, many *Solanum* species such as *S. sisymbriifolium* and *S. viarum*) are excluded. Many species of climbers and lianas are sometimes called shrubs. For example, the genus *Rubus* has many species that are called shrubs, but some are really lianas; the same applies to *Colubrina asiatica* and *Hiptage benghalensis*. In some genera, some species could be called shrubs, while others could not; for example, in *Opuntia*, *O. ficus-indica* could qualify as a shrub whereas many taxa could not. The approach taken here was to exclude groups in which most taxa are not woody. The treatment thus deals with *woody* plants.

WHICH SPECIES TO INCLUDE?

This article aims to provide a global snapshot of the current situation regarding nonnative woody plants as invaders. There are many sources of information on invasive trees and shrubs, including books and monographs, peer-reviewed papers, sundry reports and articles in the gray literature, and countless contributions on the Internet. There are also several databases of invasive woody plants, notably the Global Invasive Species Database (www.issg.org/database/welcome/), the Provisional List of Invasive Woody Plants in Tropical and Sub-Tropical Regions created and maintained by Pierre Bingelli (www .bangor.ac.uk/~afs101/iwpt/web1-99.pdf), the Invasive and Introduced Tree Species Database of the Food and Agriculture Organization of the United Nations (www .fao.org/forestry/24107/en/), and the Hawaiian Ecosystems at Risk project (HEAR; www.hear.org/). These databases use different criteria for categorizing alien species, greatly complicating the task of collating a single list on which to base a global review. Consequently, a new list was compiled, drawing on all the above sources and many others. The list compiled for this article includes only trees and shrubs that are clearly invasive (sensu Richardson et al., 2000). It includes taxa that feature on regional lists as "major invaders" and "transformers" (e.g., in South Africa), "EPPC category 1" species (in the United States), "weeds of national significance" (in Australia), "widespread" invaders (e.g., for New Zealand), and lists of "worst weeds" from various sources (always checked). The list includes only the most invasive species, not those just naturalized or established only in highly disturbed areas

such as roadsides or in heavily human-modified landscapes. It is not complete, but it is deemed suitable for the purpose of this global review.

A GLOBAL LIST OF INVASIVE ALIEN TREES AND SHRUBS: EMERGENT PROPERTIES

The list of invasive tree and shrub taxa assembled for this article comprises 350 taxa (250 trees, 100 shrubs). The list is available as supplementary material at http://academic .sun.ac.za/cib/supplementary.asp. The distribution of taxa in classes, orders, and families is summarized in Table 1. Among the features of the list are the large number of taxa in the class Pinopsida, order Pinales (4 families, 12 genera, 38 species), and in the Magnoliopsida, large numbers of taxa in the orders Fabales (1 family, 39 genera, 87 species), Rosales (6 families, 23 genera, 39 species), Myrtales (6 families, 18 genera, 29 species), Malpighiales (6 families, 12 genera, 25 species), Sapindales (5 families, 15 genera, 23 species), and Lamiales (6 families, 17 genera, 22 species). These seven orders make up 75 percent of the list (Table 1). Two families and two genera stand out as particularly important: the Fabaceae, and in particular the genus *Acacia* (22 species; most widespread is *A. mearnsii*, which is invasive in at least eight regions), and Pinaceae, particularly the genus *Pinus* (22 species; most widespread are *P. pinaster*, *P. radiata* and *P. elliottii*, all of which are invasive in five or more regions). Other genera with five or more species on the list are *Senna* (Fabaceae; 13), *Salix* (Salicaceae; 10), *Ligustrum* (Oleaceae; 7), *Eucalyptus* (Myrtaceae; 5, but see below), *Rubus* (Rosaceae; 5), *Populus* (Salicaceae; 5), and *Rosa* (Rosaceae; 5).

The list represents a tiny proportion of the global woody plant flora, comprising probably around 60,000 species of "trees" and somewhere around the same number of "shrubs." The list is hugely biased in favor of temperate species with obvious usefulness to humans. Tropical species are markedly underrepresented.

Families with many trees and shrubs that have not (yet) contributed many invasive species including the following: Anacardiaceae (850 species, including about 200 *Rhus* spp.), Betulaceae (140 species, including 35 species in *Alnus* and 35 in *Betula*), Dipteracarpaceae (535 species), Ericaeae (3,850 species, including about 1,000 *Rhododendron* [650 in China] and 860 *Erica* species), Fagaceae (970 species, including 34 in *Nothofagus* and 530 in *Quercus*), Lauraceae (2,550 species), Magnoliaceae (221 species), Moraceae (1,150 species; 850 Ficus spp.), Proteaceae (1,755 species, including 77 *Banksia*, 149 *Hakea*, and 103 *Protea* species), Sapindaceae (1,450 species, including 114 *Acer* species), and Ulmaceae (50 species, including 25–30

TABLE 1
Abridged List of Invasive Alien Trees and Shrubs

Coniferae

Order Pinales: 4 families, 12 genera, 38 species
Pinaceae (2 *Abies*, 2 *Larix*, 2 *Picea*, 22 *Pinus*, *Pseudotsuga*, *Tsuga*); Araucariaceae (*Araucaria*); Podocarpaceae (*Afrocarpus*); Cupressaceae (*Cryptomeria*, 3 *Juniperus*, *Tetraclinis*, *Thuja*)

Magnoliidae

Order Laurales: 1 family, 2 genera, 3 species
Lauraceae (2 *Cinnamomum*, *Litsea*)

Order Magnoliales: 2 families, 2 genera, 2 species
Annonaceae (*Annona*); Magnoliaceae (*Magnolia*)

Order Piperales: 1 family, 1 genus, 1 species
Piperaceae (*Piper*)

Commelinidae

Order Arecales: 1 family, 12 genera, 15 species
Arecaceae (*Arenga*, 2 *Archontophoenix*, *Areca*, *Cocos*, *Elaeis*, *Heterospathe*, *Livistona*, *Nypa*, 2 *Phoenix*, *Ptychosperma*, *Roystonea*, 2 *Washingtonia*)

Eudicotyledoneae

Order Proteales: 1 family, 2 genera, 6 species
Proteaceae (2 *Grevillea*, 4 *Hakea*)

Order Ranunculales: 1 family, 2 genera, 3 species
Berberidaceae (*Mahonia*, 2 *Berberis*)

Pentapetalae (Core Eudicots)

Order Caryophyllales: 4 families, 5 genera, 7 species
Cactaceae (*Pereskia*); Phytolaccaceae (*Phytolacca*); Polygonaceae (*Fallopia*, *Triplaris*); Tamaricaceae (3 *Tamarix*)

Rosidae

Order Myrtales: 6 families, 18 genera, 29 species
Combretaceae (*Guiera*); Lythraceae (*Punica*); Melastomataceae (*Clidemia*, *Melastoma*; *Miconia*; *Ossaea*; *Tibouchina*); Myrtaceae (5 *Eucalyptus*, 2 *Eugenia*, *Kunzea*, 2 *Leptospermum*, *Melaleuca*, *Metrosideros*, 3 *Psidium*, 2 *Rhodomyrtus*, 3 *Syzygium*); Onagraceae (*Fuchsia*); Vochysiaceae (*Vochysia*)

Fabidae (Eurosids I)

Order Celastrales: 1 family, 2 genera, 2 species
Celastraceae (*Celastrus*, *Euonymus*)

Order Fabales: 1 family, 38 genera, 86 species
Fabaceae (*Abrus*, 22 *Acacia* [*sensu lato*], *Adenanthera*, 3 *Albizia*, *Amorpha*, 2 *Bauhinia*, *Caesalpinia*, *Calicotome*, *Calliandra*, *Caragana*, *Cassia*, 3 *Cytisus*, *Dalbergia*, *Delonix*, *Dichrostachys*, *Falcataria*, 2 *Genista*, *Gleditsia*, *Gliricidia*, *Leucaena*, *Lupinus*, *Medicago*, 4 *Mimosa*, *Myroxylon*, 2 *Paraserianthes*, *Parkinsonia*, *Pithecellobium*, *Polygola*, 4 *Prosopis*, *Psoralea*, *Robinia*, 13 *Senna*, 3 *Sesbania*, *Spartium*, *Tamarindus*, *Teline*, *Tipuana*, *Ulex*)

Order Fagales: 4 families, 6 genera, 9 species
Betulaceae (*Alnus*, *Betula*); Casuarinaceae (3 *Casuarina*); Fagaceae (*Castanea*, 2 *Quercus*); Myricaceae (*Morella*)

Order Malpighiales: 6 families, 12 genera, 25 species
Chrysobalanaceae (*Chrysobalanus*); Clusiaceae (*Calophyllum*, *Clusia*); Euphorbiaceae (*Homolanthus*, *Macaranga*, *Ricinus*, *Triadica*); Phyllanthaceae (*Bischofia*); Rhizophoraceae (*Rhizophora*); Salicaceae (*Flacourtia*, 5 *Populus*, 10 *Salix*)

Order Rosales: 6 families, 23 genera, 39 species
Cannabaceae (*Trema*); Elaeagnaceae (3 *Elaegnus*); Moraceae (*Artocarpus*, *Broussonetia*, *Castilla*, *Ficus*, *Morus*); Rhamnaceae (*Colubrina*, *Frangula*, *Hovenia*, *Maesopis*, 3 *Rhamnus*, 4 *Rhus*, *Zizyphus*); Rosaceae (*Aronia*, *Cotoneaster*, *Crataegus*, *Eriobotrya*, 2 *Prunus*, 2 *Pyracantha*, *Pyrus*, 5 *Rosa*, 5 *Rubus*); Ulmaceae (*Celtis*); Urticaceae (2 *Cecropia*)

Malvidae (Eurosids II)

Order Malvales: 2 families, 4 genera, 4 species
Malvaceae (*Brachychiton*, *Lavatera*, *Malvastrum*); Muntingiaceae (*Muntingia*)

Order Sapindales: 5 families, 15 genera, 23 species
Anacardiaceae (*Mangifera*, *Persea*, 2 *Rhus*, 2 *Schinus*); Meliaceae (*Azadirachta*, 2 *Cedrela*, *Melia*, *Swietenia*, *Toona*); Rutaceae (3 *Citrus*, *Triphasia*, *Zanthoxylum*); Sapindaceae (4 *Acer*, *Cupaniopsis*); Simaroubaceae (*Ailanthus*)

Asteridae

Order Ericales: 3 families, 5 genera, 6 species
Ericaceae (*Calluna*, *Erica*, *Rhododendron*); Myrsinaceae (2 *Ardisia*); Sapotaceae (*Chysophyllum*)

Lamiidae (Euasterids I)

Order Gentianales: 2 families, 6 genera, 6 species
Apocynaceae (*Nerium*, *Thevetia*, *Calotropis*, *Cryptostegia*); Rubiaceae (*Cinchona*, *Coprosma*)

Order Lamiales: 6 families, 17 genera, 25 species
Bignoniaceae (*Cordia*, *Jacaranda*, *Spathodea*, *Tecoma*); Lamiaceae (*Clerodendron*, *Plectranthus*, *Vitex*); Oleaceae (3 *Fraxinus*, *Jasminum*, 7 *Ligustrum*, *Olea*); Paulowniaceae (*Paulownia*); Scrophulariaceae (*Buddleja*, *Myoporum*); Verbenaceae (*Duranta*, *Lantana*, *Citharexylum*)

Order Solanales: 2 families, 4 genera, 6 species
Convolvulaceae (*Ipomoea*); Solanaceae (2 *Cestrum*, *Nicotiana*, *Solanum*)

Campanulidae (Euasterids II)

Order Apiales: 2 families, 2 genera, 2 species
Araliaceae (*Schefflera*); Pittosporaceae (*Pittosporum*)

Order Aquifoliales: 1 family, 1 genus, 1 species
Aquifoliaceae (*Ilex*)

Order Asterales: 1 family, 5 genera, 6 species
Asteraceae (*Baccharis*, *Cassinia*, *Chromolaena*, *Chrysanthemoides*, 2 *Pluchea*)

Order Dipsacales: 1 family, 1 genus, 1 species
Caprifoliaceae (*Lonicera*)

Ulmus species). Why are these taxa underrepresented in the global list of major woody invaders? This question cannot be answered definitively yet, because species from some of these families have not been sufficiently widely transported and disseminated around the world for a long enough period to give them a chance to invade. Very few woody plant groups have been surveyed in enough detail to assess levels of invasiveness in relation to the degree of transport and dissemination outside their natural ranges. This complicates the quest to comment meaningfully on levels of invasiveness in different groups—the natural experiment of translocations of trees and shrubs across the planet has been uneven and is ongoing. An exception is the class Pinopsida, for which enough evidence is available to allow for reasonably robust conclusions to be drawn on the determinants of invasiveness, taking into account life history traits, propagule pressure, and facets of invasibility. For this group, a syndrome of life history traits (small seed mass [<50 mg], a short juvenile period [<10 years], and short intervals between large seed crops) separates the most invasive species from others with less potential to invade.

The list reveals a marked overrepresentation of species used for ornamentation (224 species) and for "forestry" (70 species). Other prominent reasons for introduction and dissemination are agroforestry (35 taxa) and food (31 taxa). It is worth noting that many trees and shrubs on the list were introduced and propagated for purposes for which they proved to be poorly suited (e.g., *Bischofia javanica* was used for ornamentation in Florida in the early 1900s but was soon discovered to be unsuitable because it grew too big, had large surface roots, had foliage that was prone to diseases, and regenerated freely).

Numbers of invasive alien trees and shrubs vary considerably between regions of the world, although it is difficult to determine whether the pattern is real or is simply due to different levels of reporting in different regions. Nonetheless, southern Africa seems to top the list, with at least 126 species of invasive alien trees and shrubs, followed by the Pacific Islands and North America (both with 83 species), Australia (with 72 species), South America (with 56), and Europe (with 52). Again, the pattern reflects predominantly the magnitude of introductions and plantings, rather than a clear propensity for regions to accommodate invasive nonnative woody plants.

FIGURE 1 Examples of invasive trees and shrubs. (A) *Prunus serotina* (Rosaceae; black cherry), native to Central and North America, was introduced to Europe as an ornamental tree in the seventeenth century and is now a widespread invasive species in many parts of Western Europe. It is seen here invading a plantation of *Pinus sylvestris* in northern France. (Photograph courtesy of Guillaume Decocq.) (B) Common broom (*Cytisus scoparius*; Fabaceae) is native to western and central Europe, and is invasive in many parts of the world. It is seen here invading on New Zealand's South Island. (Photograph courtesy of Yvonne Buckley.) (C) *Euonymus alata* (Celastraceae; burning bush) invading a sugar maple forest in the mid-Hudson region of New York State. The species is native to central and northern China, Japan, and Korea and was widely planted as an ornamental. (Photograph courtesy of Patrick Martin.)

FIGURE 2 More invasive trees and shrubs. (A) Many species in the genera *Cotoneaster*, *Crataegus*, *Pyracantha*, *Rosa*, and *Rubus* (family Rosaceae) have small, brightly colored, fleshy fruits that are readily dispersed by many animal species. The picture shows *Pyracantha angustifolia* invading secondary shrubland in the Córdoba mountains of central Argentina. (Photograph courtesy of Paula Tecco.) (B) *Fallopia japonica* (Polygonaceae; Japanese knotweed), a native of eastern Asia, is invasive in many parts of Europe (seen here in southern Poland) and North America. (Photograph courtesy of the author.) (C) A stand of *Rhododendron ponticum* (Ericaceae; common rhododendron) in the Galtee Mountains, Ireland. The species is native to the area south of the Black Sea and was first introduced into the British Isles for cultivation in the mid-1700s; it is now a major invader. (Photograph courtesy of Alexandra Erfmeier.)

INVASION PROCESSES

The taxa represented on the list possess a good cross-section of life history strategies evident in tree and shrub life forms worldwide (Figs. 1–3). Consequently, invasions have been reported from all major biomes and most conceivable habitat types. Invaded vegetation types span semi-arid shrublands to tropical forests. All species on the list were intentionally introduced and planted or disseminated to some extent in new habitats; considerable propagule pressure or the creation of multiple foci have aided naturalization and invasion in many, probably most, cases. Marked lag phases between introduction and naturalization or invasion have been noted for several species. Such delays are informative regarding the barriers that mediate invasion success for introduced trees and shrubs. Introduced pines in the southern hemisphere are a classic example; the absence of appropriate mycorrhizal fungal symbionts initially precluded establishment and spread. Only once inoculated soil was introduced from the native range of pines was establishment, and later widespread naturalization and invasion, possible. Other mutualisms have mediated the performance of introduced trees and shrubs in many cases. As is the case for invasive plants in general, those invasive woody plants that require animal-mediated pollination or propagule dispersal have usually managed to infiltrate prevailing networks in the invaded regions.

Many alien tree and shrub species have invaded very quickly. Among the fastest linear, local-scale dispersal rates reported for invasive plants have been those for trees and shrubs, including *Mimosa pigra* (87.3 m year^{-1}); *Pinus radiata* (31.0 m year^{-1}); *Prunus serotina* (14.6 m year^{-1}); *Frangula alnus* (6.7 m year^{-1}); and *Rhododendron ponticum* (5.0 m year^{-1}). Such dispersal rates have been achieved through a wide range of dispersal syndromes, involving numerous types and permutations of disturbance, and into diverse habitats. For example, many introduced trees and shrubs have invaded intact vegetation in many parts of the world. Perhaps the most dramatic and well studied has been the large-scale invasion of natural fynbos vegetation in South Africa by a suite of alien trees and shrubs, especially taxa in the genera *Acacia*, *Hakea*, and *Pinus*. These taxa possess life history traits that enable them to establish in the nutrient-poor fynbos soils, to deal with the prevailing fire regime either by resprouting or regenerating from seed after fire and also by achieving reproductive maturity earlier than most native taxa, and to disperse rapidly. Together with these attributes, their capacity to produce massive numbers of seeds in the absence of key

FIGURE 3 More invasive trees and shrubs. (A) *Rubus niveous* (Rosaceae; mora, mysore, or hill raspberry), a thorny, perennial shrub of Asiatic origin, is one of the most serious plant invaders in the Galapagos Islands. It arrived only in the early 1980s and has spread rapidly; it is now present on three islands and is distributed over at least 10,000 ha. (Photograph courtesy of the author.) (B) *Hakea sericea* (Proteaceae; silky hakea), an Australian shrub, invading fynbos vegetation in the Western Cape, South Africa. (Photograph courtesy of the author.) (C) *Acer negundo* (Aceraceae; box elder) invading natural vegetation near Bordeaux, France. (Photograph courtesy of Annabel Porté.)

natural enemies affords them great resilience to fluctuating inter-fire intervals. Since their introduction and, in most cases, substantial human-aided dissemination, which predominantly occurred more than 150 years ago, these species have invaded entire landscapes, forming either monospecific or mixed-species stands that pose a major threat to biodiversity and ecosystem services in this biodiversity hotspot.

IMPACTS

Invasive alien trees and shrubs have affected the ecosystems they have invaded in many ways. Many trees act as ecosystem engineers, and invasive trees that reach very high densities and substantially increase biomass or change the type and arrangement of aboveground material have the potential to change many facets of ecosystem functioning. In some cases, the impacts of tree and shrub invasions are obvious and dramatic. Some invasions have radically transformed entire ecosystems. For example, invasive *Acacia*, *Hakea*, and *Pinus* species rapidly transform species-rich fynbos shrublands in South Africa into species-poor, alien-dominated forests or woodlands. Invasion of *Melaleuca quinquenervia* in Florida's Everglades has changed large areas of open grassy marshes to closed-canopy swamp forests. The net effect of invasive trees and shrubs is determined by the product of the per capita effect, the abundance they achieve (reflected by numbers of stems per area, the total biomass added, or total leaf area per ground area), and their geographical range. The taxa listed in Table 1 span the range from causing localized impacts to massive ecosystem-level transformations. In their impacts, invaders may also be categorized as "discrete-trait invaders" (DTIs) that add one or more new functions an ecosystem or "continuous-trait invaders" (CTIs), which differ only quantitatively from native species in providing one or more functions. Some of the most prominent characteristics of invasive alien trees and shrubs are summarized below.

Excessive Users of Resources

Many invasive trees invade riparian ecosystems, where they achieve dominance and huge abundance and thus consume more water than would the native species that normally frequent these ecosystems. This impact is due primarily to increased biomass, and therefore increased water use (a characteristic of CTIs). Prominent examples are *Tamarix* spp. in southwestern North America and *Acacia* species (e.g., *A. longifolia* and *A. mearnsii*) in South Africa.

Donors of Limiting Resources

Many invasive alien species of woody legumes impact invaded ecosystems primarily via their addition of nitrogen (the classic example of DTIs). Well-studied examples are *Morella (Myrica) faya*, which doubles canopy nitrogen as it replaces native forest species in Hawaii, and Australian *Acacia* species in South Africa.

Fire Promoters and Suppressors

The best-studied example of an invasive tree that brings fire to a previously fire-free system is that of *Melaleuca quinquenervia*, which invades wetland habitats in Florida, where a massive increase in flammable material leads to very intense fires. Examples of tree and shrub invasions suppressing fire frequency are *Mimosa pigra* in northern Australia and *Schinus terebinthifolius* and *Triadica sebifera* in North America; in all cases, tree invasions have resulted in reduced horizontal continuity of fuel, which has reduced fire frequency and intensity. Tree and shrub invasions in South African fynbos may promote or suppress fire frequency in fynbos vegetation, depending on the density of invasive stands.

Sand Stabilizers

Acacia cyclops was widely planted along coastal dunes in western South Africa to stabilize sand movement. Planted and self-sown stands of this species perform this function very well; in some areas, dune stabilization has resulted in massive beach erosion.

Colonizers of Intertidal Mudflats and Sediment Stabilizers

Invasive mangrove species, notably *Rhizophora mangle*, have had a significant impact in Hawaii, altering coastline hydrodynamics and nearshore sedimentation.

Litter Accumulators

Pinus strobus invades native *P. sylvestris* forests in the northern Czech Republic, where its major impact is attributable to increased litter production and consequent effects on soil acidity.

Transformers

Many invasive alien trees and shrubs qualify as "transformers" (sensu Richardson et al., 2000). Examples of shrubs include *Calluna vulgaris* (in New Zealand), *Chrysanthemoides monilifera* (in Australia), *Clidemia hirta* (on many Pacific islands), *Cytisus scoparius* (syn. *Sarothamnus scoparius*) (in many parts of the world), and *Hakea sericea* (in Portugal and South Africa). Prominent examples of transformers among invasive alien trees are *Cinchona pubescens* (in the

Galapagos Islands), *Ligustrum robustum* var. *walkeri* (on La Réunion Island), *Melaleuca quinquenervia* (in Florida), *Miconia calvesens* (in Tahiti), *Mimosa pigra* (in northern Australia and Zambia), *Morella (Myrica) faya* (in Hawaii), *Pinus pinaster* (in South Africa), and *Triadica sebifera* (syn. *Sapium sebiferum*) (in North America).

Besides the effects mentioned above that are attributable to effects on physical resources, either due to large size and biomass or to impacts on resource availability, many tree and shrub invasions affect resident biota in rather more subtle ways, for example by altering the habitat for other organisms. For example, American robins nesting in two invasive shrubs in Illinois, *Lonicera maackii* and *Rhamnus cathartica*, experience higher predation than robins in nests built in comparable native shrubs. In Hawaii, the spread of introduced mangroves has led to habitat loss for wetland birds, including the endemic Hawaiian stilt (*Himantopus mexicanus knudseni*), the Hawaiian coot (*Fulica americana alai*), and the Hawaiian duck (*Anas wyvilliana*). The new mangrove habitats also provide refugia for shorebird predators, which include invasive rats (*Rattus* spp.) and mongooses (*Herpestes auropunctatus*), as well as alien marine species such as the mangrove crab (*Scylla serrata*). Several species have a major impact by creating impenetrable thorny thickets that limit the passage of animals (e.g., *Caesalpinia decapetala*, *Lantana camara*, several *Rosa* and *Rubus* species). In Australia *Annona glabra* grows in estuaries and chokes mangrove swamps, where its seedlings carpet the banks and prevent other species from germinating or thriving. An interesting impact is attributed to the invasive shrub *Chromolaena odorata* in eastern South Africa; greater shading created by this species leads to changes in sex ratios of crocodiles.

Besides the many and diverse ecological effects discussed above, tree and shrub invasions have many complex, and mostly negative, effects on human livelihoods. These have been clearly documented in South Africa (especially for Australian acacias and *Chromolaena odorata*) and in Papua New Guinea (due to invasion of *Piper aduncum*). *Prosopis* invasions in sub-Saharan Africa have led to considerable rangeland degradation, causing many problems for human societies, especially those relying on subsistence agriculture. Tree and shrub invasions have huge financial costs in many regions. For example, the annual budget of the Working for Water Programme in South Africa (for clearing invasive alien plants, mainly trees and shrubs, throughout the country) was $65 million for 2008. At least $3 million is spent annually to manage *Melaleuca quinquenervia* in Florida.

MANAGEMENT ISSUES

Management efforts are under way in many parts of the world to reduce problems associated with invasive alien trees and shrubs. These range from ad hoc local-scale efforts to control invasions and mitigate their effects to national-scale, systematic strategies that integrate all potential options for reducing current problems and reducing the risk of future problems.

In devising sustainable strategies for managing problems arising from invasions of introduced trees and shrubs, managers and planners must confront several complex challenges. As discussed above, most widespread tree and shrub invaders were intentionally introduced to the regions where they now cause problems, and most are still useful in parts of the regions where they occur. Conflicts of interest abound. Especially for forestry and agroforestry, replacing invasive alien species with native or less invasive nonnative alternatives has limited potential. For commercial forestry, eucalypts and pines will remain the foundation of exotic forestry enterprises, and options must be sought to reduce invasiveness and to mitigate negative impacts of the key taxa. There is more scope for finding acceptable alternatives for invasive nonnative ornamental species, but the nursery trade has substantial financial investments in many countries. The demand for popular ornamentals also has strong cultural ties, and it is thus difficult to change quickly. There are also other challenges for managing invasiveness in ornamental plants. In many taxa, different cultivars and other subspecific entities show different levels of invasiveness (e.g., the cases of *Lantana* spp. and *Pyrus calleryana*). This complicates the implementation of clear policies. In forestry, invasiveness may change substantially in hybrids and transgenics, with scope for both enhanced and reduced invasiveness. Biotechnology has the potential to reduce invasiveness of useful trees by producing sterile trees. Although technologically feasible, important barriers exist. For example, the Forestry Stewardship Council prohibits the use of transgenic species. Interventions must consider that invasive alien trees and shrubs (other than the economically important taxa discussed above) may serve useful purposes in some situations. For instance, many alien trees and shrubs have strong value as nurse plants for the restoration of degraded natural forests. Increasing land degradation in many parts of the world will increase the need for stabilization and rehabilitation efforts, including the controlled use of nonnative species, even those with known or predicted invasive potential. "Weediness" is often welcomed in such cases, and this is difficult to reconcile with biodiversity conservation concerns. Management strategies for invasive trees and shrubs must accommodate such issues.

Another factor that must be taken into account is the rapidly changing global market for products from trees and shrubs, including new uses. For example, many alien trees and shrubs are being proposed for wide-scale planting for the production of biofuels, among them known invasive taxa such as *Azadirachta indica*, *Calotropis procera*, *Olea europaea*, *Populus* spp., *Ricinus communis*, *Salix* spp., *Triadica sebifera*, and *Zizyphus mauritiana*, as well as many other species that are very likely to be invasive. Altered planting configurations, including massive increases in propagule pressure and the number of planting sites and thus foci for launching invasions, to accommodate biofuel production will radically change conditions. In many cases, this will favor invasions, leading to more invasions of more species over a greater area. Consideration must be given to potential invasions when deciding on strategies for biofuels and other emerging markets for wood-based products in different parts of the world.

SEE ALSO THE FOLLOWING ARTICLES

Eucalypts / Forestry and Agroforestry / Horticulture / *Lantana camara* / Melastomes / Transformers

FURTHER READING

Brooks, M. L., C. M. D'Antonio, D. M. Richardson, J. B. Grace, J. E. Keeley, J. M. Di Tomaso, R. J. Hobbs, M. Pellant, and D. Pyke. 2004. Effects of invasive alien plants on fire regimes. *BioScience* 54: 677–688.

Ewel, J. J., et al. 1999. Deliberate introductions of species: Research needs. *BioScience* 49: 619–630.

Kowarik, I. 1995. Time lags in biological invasions with regard to success and failure of alien species (15–38). In P. Pyšek, K. Prach, M. Rejmánek, and P. M. Wade, eds. *Plant Invasions: General Aspects and Special Problems*. Amsterdam: SPB Academic Publishing.

Pyšek, P., and P. E. Hulme. 2005. Spatio-temporal dynamics of plant invasions: Linking pattern to process. *Ecoscience* 12: 302–315.

Pyšek, P., M. Kivánek, and V. Jarošík. 2009. Planting intensity, residence time, and species traits determine invasion success of alien woody species. *Ecology* 90: 2734–2744.

Rejmánek, M., and D. M. Richardson. 1996. What attributes make some plant species more invasive? *Ecology* 77: 1655–1661.

Richardson, D. M., N. Allsopp, C. M. D'Antonio, S. J. Milton, and M. Rejmánek. 2000. Plant invasions: The role of mutualisms. *Biological Reviews* 75: 65–93.

Richardson, D. M., I. A. W. Macdonald, P. M. Holmes, and R. M. Cowling. 1992. Plant and animal invasions (271–308). In R. M. Cowling, ed. *The Ecology of Fynbos: Nutrients, Fire and Diversity*. Cape Town: Oxford University Press.

Richardson, D. M., P. Pyšek, M. Rejmánek, M. G. Barbour, D. F. Panetta, and C. J. West. 2000. Naturalization and invasion of alien plants: Concepts and definitions. *Diversity and Distributions* 6: 93–107.

Richardson, D. M., P. A. Williams, and R. J. Hobbs. 1994. Pine invasions in the southern hemisphere: Determinants of spread and invadability. *Journal of Biogeography* 21: 511–527.

UNGULATES

SEE GAME ANIMALS; GRAZERS

V

VEGETATIVE PROPAGATION

JITKA KLIMEŠOVÁ

Czech Academy of Sciences, Třeboň

Vegetative propagation (= clonal growth) is a form of plant reproduction in which vegetative growth produces new, physically independent individuals. Vegetatively derived offspring are genetically identical and usually remain interconnected for a longer or shorter period. Vegetative offspring are usually less numerous than generative offspring (seeds), but because of support from the parent plant, they are much more likely to establish. The distance to which vegetative offspring spread depends on how they spread. When offspring spread by the lateral growth of the plant, they do not usually disperse more than 25 cm per year. On the other hand, aquatic plants spread by water currents or plants dispersed by humans may disperse enormous distances. Clonal growth may be triggered or enhanced by plant injury caused by natural means or by humans.

MORPHOLOGY AND TAXONOMY

Ability to grow clonally is not evenly distributed among plant taxa and growth forms. Monocotyledonous plants are more often clonal than dicotyledonous plants because they lack secondary thickening, have short-lived tissues, and readily undergo adventitious rooting. The majority of clonal plants are perennial herbs; among trees, clonal growth is far less common. Morphological diversity of clonal growth is great. A necessary attribute of clonality, the existence of many connections between belowground and aboveground parts of the plant, may be achieved by virtue of stems producing adventitious roots, by virtue of roots producing adventitious shoots, or by both means. Stem-derived clonality is generally more common and prevails among herbs and, particularly, among monocots (Fig. 1). Root-derived clonality is less common and is prevalent among dicot trees. Root-derived clonality by means of adventitious sprouting from roots is often induced (or at least enhanced) by injury to the plant.

GEOGRAPHY

Clonal growth is more important at higher latitudes than at lower ones, owing to the increasing importance of the herbaceous habit in seasonal climates. Another factor affecting clonal growth on a large geographical scale is

FIGURE 1 *Cynodon dactylon* (Bermudagrass) on ash fields under Paricutin Volcano, Mexico. Due to vigorous vegetative propagation by stolons, Bermudagrass is an excellent colonizer. This is an example of stem-derived clonality. In spite of its common name, this species is native in Africa and has become one of the most difficult weeds in all tropical, all subtropical, and many temperate countries. (Photograph courtesy of M. Rejmánek.)

recurrent disturbances like fire and herbivory, leading to sheltering of bud-bearing organs from which new shoots are produced to compensate for those that were damaged. Also, regularly anthropogenically disturbed ecosystems such as arable fields or hay meadows host many plants with remarkable abilities to grow clonally or to regenerate from a bud bank after injury.

THE ROLE OF VEGETATIVE PROPAGATION IN PLANT INVASION

Owing to large size and sensitivity to aridity, the role of vegetative propagules in dispersal of plants to new areas is restricted to cases in which successful introduction of an alien plant is enabled by either deliberate or unintentional dispersal by humans. On the other hand, establishment of an introduced plant is enhanced by clonality, as there is no risk of failed generative reproduction caused by lack of stimuli for flowering or lack of sexual partners or pollen vectors in the invaded area. Clonal plants, although they are less common among invasive plants, are more common in undisturbed vegetation than in anthropogenic habitats, which are invaded predominantly by nonclonal plants. From the geographical pattern described above, it follows that invaders in temperate zones are more often clonal plants than are invaders in tropical regions. Nevertheless, clonal plants are among the most problematic invaders in the tropics (e.g., *Cynodon dactylon*, *Cyperus* spp., *Opuntia* spp., *Pennisetum purpureum*, *Salvinia molesta*).

Water and human activities play an important role in spreading clonal plants. Among freely floating clonal aquatic plants are such detrimental invaders as *Elodea canadensis* in Europe and *Eichhornia crassipes* in Africa. Among ornamentals, perennial weeds and plants used for rehabilitation of eroded areas include invaders that spread through transport of vegetative diaspores by humans. For example, *Ammophila arenaria* is widespread in America, where it has been used to stabilize coastal dunes, while *Oxalis pes-caprae* (Fig. 2), a deliberately introduced ornamental species, is invading arable land and pastures in the Mediterranean, and *Fallopia* species, planted as ornamentals, are spreading especially along rivers in Europe.

RISKS

Clonal growth allows for the spread of plants with very poor or no generative reproduction in their invaded ranges.

Clonal growth organs provide a bud bank for vegetative regeneration after a winter or summer resting period,

FIGURE 2 (A) Invasion by the pseudoannual geophyte *Oxalis pescaprae* in Monterey County, California. (B) Each plant can produce 25 viable bulbs per year, all of which have the potential to germinate the following season. (Photographs courtesy of J. M. DiTomaso.)

and also where there is severe disturbance. This ability of clonal plants to spread hinders eradication efforts, and this is especially true for plants capable of adventitious sprouting from roots, even though they show no or limited clonal growth without injury.

SEE ALSO THE FOLLOWING ARTICLES

Apomixis / Dispersal Ability, Plants / Disturbance / Freshwater Plants and Seaweeds / Life History Strategies / Water Hyacinth / Weeds / Wetlands

REFERENCES

Klimešová, J., and L. Klimeš. 2007. Bud banks and their role in vegetative regeneration: A literature review and proposal for simple classification and assessment. *Perspectives in Plant Ecology, Evolution and Systematics* 8: 115–129.

Pyšek, P. 1997. Clonality and plant invasions: Can a trait make a difference? (405–427). In H. de Kroon and J. van Groenendael, eds. *The Ecology and Evolution of Clonal Plants*. Leiden, The Netherlands: Backhuys Publishers.

van Groenendael, J. M., L. Klimeš, J. Klimešová, and R. J. J. Hendriks. 1996. Comparative ecology of clonal plants. *Philosophical Transactions of the Royal Society of London Series B* 351: 1331–1339.

VINES AND LIANAS

CARLA J. HARRIS AND RACHAEL GALLAGHER

Macquarie University, Sydney, New South Wales, Australia

Invasive vines and lianas are one of the most destructive plant functional groups that successfully invade new areas, with climbing plants producing many unique and deleterious effects in the ecosystems they invade. Invasive vines and lianas are particularly problematic, as their climbing habit results in the smothering of vegetation that they use as support (Fig. 1).

CLASSIFICATION OF CLIMBING PLANTS

Climbing plants are a group of plants possessing long and flexible stems that are not strong enough to carry their foliage to a height where there is adequate light. Consequently, climbing plants such as vines and lianas use surrounding objects including trees and shrubs to obtain sufficient height within the ecosystems in which they live.

Vines climb in a variety of ways, with a number of different subtypes that can be grouped according to the different mechanisms with which they climb. Stem twiners, which form the largest group of vines, can be both woody and nonwoody, and use various mechanisms, including modified stems, branches, or petioles, to twine around the stems and branches of host plants. Tendril climbers (e.g., *Passiflora* species) use specialized structures to wrap around and hook onto the stems of host plants. Root climbers (e.g., *Hedera helix*) use modified roots to adhere to the branches of trees and shrubs. Scramblers (e.g., *Rubus* species) are the least specialized, with their weak, flexible stems sprawling through the foliage of surrounding vegetation.

FIGURE 1 *Ipomoea purpurea* smothering host vegetation in New South Wales, Australia. (Photograph courtesy of R. Gallagher.)

Lianas are woody vines that may begin life as self-supporting terrestrial seedlings and go on to rely on external structural support as they grow. Liana climbing habits range from draping over hosts or structures to using branched tendrils or adhesive adventitious roots to secure themselves on surrounding vegetation.

The climbing habit has arisen multiple times in the evolutionary history of plants and is considered a key strategic innovation. Following this evolutionary trajectory, exotic vines and climbers come from a wide array of lineages, with almost half (48%) of invasive vines coming from outside the six most abundant families (Fig. 2). There are also seedless vascular plants such as the climbing ferns *Lygodium microphyllum* and *L. japonicum* that have become invasive when introduced to novel regions. These ferns climb using creeping rhizomes that attach to host plants and can spread rapidly through windborne spores.

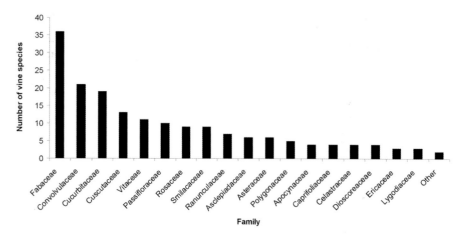

FIGURE 2 Global taxonomic distribution of invasive vine and liana species. (Compiled from information obtained at www.invasive.org.)

ENVIRONMENTAL PREFERENCES OF EXOTIC VINES

Exotic vines, like their native counterparts, are generally early successional species that exploit high-light environments such as canopy gaps and forest edges. For this reason, they are often described as "shade-intolerant" or "light-demanding" species. Their abundance reaches a peak on or near the edges of vegetation boundaries, yet the most problematic species have the ability to penetrate deep into established vegetation, often through the colonization of shade-tolerant seedlings that lie in wait for a canopy disturbance. For instance, *Celastrus orbiculatus* (Oriental bittersweet) can successfully establish seedlings under very low-light environments, but plants grow rapidly when exposed to high light conditions.

Exotic vine infestations tend to be localized in habitats characterized by high water or light availability. This tendency to colonize canopy gaps and edges makes them particularly problematic invaders of closed forests such as rainforest and riparian zones. Consequently, the abundance of exotic climbers in riparian areas may also be a result of a lack of water limitation in these environments. Woody vines, such as *Mafadyena unguis-cati* (cat's claw creeper), are characterized by large xylem vessel elements that facilitate the movement of water to the leaf surface for physiological processes such as photosynthesis. Water often needs to be transported over long distances in vines that loop through the canopy, and the maintenance of the transpiration stream within the xylem is critical for avoiding potentially irreversible cavitation. The role of high soil water content in maintaining water relations may contribute to the abundance of vines in environments where water is not limiting, as exemplified by the case of the exotic twiner *Mikania micrantha* (mile-a-minute).

PROBLEMS CAUSED BY INVASIVE VINES AND LIANAS

Climbing plants are often described as structural parasites. Their capacity to use surrounding vegetation for support underpins their ability to impact upon plant communities in very specific ways. Introduced and invasive vines may adversely affect their novel environments in a variety of ways that may be direct or indirect, such as through competing for light, nutrients, and water or by directly by causing mechanical damage (Fig. 3). An example of an indirect mechanism is a significant reduction in the leaf area and fecundity of trees that occurs when introduced vines become abundant and trees have failed to deter them from invading the canopy. The climbing mechanism of

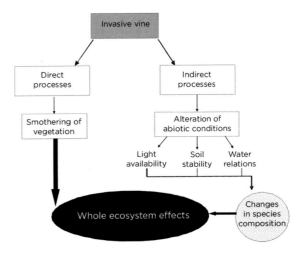

FIGURE 3 Conceptual diagram of direct and indirect processes by which invasive vines and lianas affect their invaded ecosystems.

a vine or liana may also become an important element in determining the nature of the ecosystem impact. For example, twiners may interfere with downward translocation of nutrients in the phloem by constricting the bole of the host.

The ecological effects of invasive vines may manifest across a range of spatial scales, and the nature of the damage may also vary across different spatial resolutions. At a global scale, reported increases in tropical forest turnover rates, possibly linked to increases in atmospheric carbon dioxide concentrations, have been paralleled by increases in native liana density, a change that may be exacerbated by the increasing vine and liana abundance arising from anthropogenic activities. At a smaller scale, invasive vines have the potential to suppress tree growth, encourage tree mortality, and decrease the abundance of native animal populations by providing a suboptimal habitat within natural communities.

While the abundance of native vines reaches an equilibrium state with competing vegetation, exotic vines are often observed to arrest or delay the process of succession. For example, the formation of nonnative "vine blankets" in the vegetation of Florida's Everglades region in the years following Hurricane Andrew severely diminished native plant recruitment. Similarly, species richness declined with increasing cover of the exotic twiner *Lonicera japonica* through the reduced local colonization of native species over time. This ability for exotic vines to transform the composition of native vegetation results from the pronounced shift in the abiotic conditions available for both recruitment and growth associated with dense vine infestation. This shift is often catalyzed by a disturbance event, such as the formation of a canopy gap, which increases the availability of light and the colonization opportunities for foreign propagules.

Once established in the canopy, vines may monopolize the available light resources by forming a dense layer of leaves atop surrounding host vegetation. This dense mantle of leaves reduces the amount of photosynthetically active radiation reaching both host vegetation and vegetation on the forest floor. The weight of dense infestations of exotic vines on host trees may lead to the breakage of limbs and, in extreme cases, to an increased incidence of tree mortality. The destruction of mature trees not only promotes the formation of new canopy gaps, but may also reduce rates of seed production of secondary successional and climax community species. This cycle of canopy gap creation, vine infestation, and tree breakage has the potential to significantly alter vegetation composition in invaded sites through the transformation of abiotic conditions and competitive interactions.

ATTRIBUTES RELATING TO EXOTIC VINE AND LIANA INVASION SUCCESS

One of the most important steps for the successful management of invasive plants, including vines, is the identification of factors that enable an introduced species to become invasive. Not all introduced plants become invasive, and thus it is desirable to determine the traits, or suite of traits, common to invasive plants that distinguish them from noninvasive exotic species. The identification of characteristics of highly invasive vine and liana species provides a greater understanding of factors contributing to their invasion success and contributes to their effective management. Broadly, such attributes can be divided into two categories, abiotic and biotic.

Abiotic Attributes

Introduction history attributes are variables that relate to the introduction of a species into a new area, such as when it was introduced (residence time), why it was introduced (reason for introduction—ornamental, agricultural, or accidental or unknown), and to which area it is native (origin).

In an Australian study, factors relating to the introduction of exotic vines and lianas into Australia showed that introduction for ornamental reasons was the most common explanation for importation for the 179 exotic species of climbing plants that have established persistent populations in native vegetation across Australia. This finding was in agreement with a range of other studies on invasive plants. Additionally, there were significant differences in numbers of exotic vines arriving in Australia from different continents, with a higher number of exotic vines coming from South America than from any other continent. Notably, it was found that the longer a species had been present in Australia, the more abundant it was, with a positive relationship existing between minimum residence time and exotic vine abundance.

While clear patterns exist between the occurrence of exotic vines and their introduction history attributes, neither the continent of origin nor the reason for introduction appears to contribute to their distribution or success. Hence, apart from minimum residence time, this Australian example indicates that using introduction history attributes when forming predictive frameworks to target potentially invasive climbing species does not appear to offer useful predictive information for vines and lianas. Attributes such as biological properties and life history traits of a species may offer more effective predictive tools.

Biotic Traits

An understanding of the life history traits of exotic vines and their relationships to invasion success is integral to their management. The way in which vines reproduce and grow may determine the invasion success of a species and may also inform choices about the most effective method of control.

REPRODUCTIVE TRAITS Asexual reproduction is a common adaptation in exotic vines. Many species are capable of clonal reproduction via rooting stem fragments or tuberous storage organs formed either underground or aerially. Underground tubers are a particular feature of the genus *Asparagus* and are commonly associated with both reinfestation of cleared sites as well as spread to novel areas. Some exotic vines, such as species in the genus *Dioscorea*, produce stem-borne tubers that, once mature, fall to the ground to form new plants, thus increasing the density of an infestation.

A lack of seed production does not appear to be a barrier to invasion. For example, some exotic species in the genus *Ipomoea* as well as *Anredera cordifolia* (balloon vine) produce no viable seed in their exotic Australian range but are among the most widespread and destructive exotic vines in that country.

Exotic vines also spread through the production of seeds. There is substantial variation in the seed and dispersal traits of exotic vines that reflects their phylogenetic diversity. For example, species from the family Bignoniaceae, such as *Macfadyena* and *Campsis*, generally possess small seeds with adaptations for wind dispersal, while leguminous species such as kudzu (*Pueraria montana* var. *lobata*) have larger seeds with no dispersal adaptations. A number of problematic vines are dispersed across the landscape by birds and bats, which facilitates the colonization of new sites. For example, members of the Solanaceae and Asparagaceae families possess brightly colored, fleshy fruits that are attractive to vertebrate dispersers.

A large knowledge gap still remains surrounding the reproductive ecology of invasive vines; however, some studies have been conducted for individual vine species enabling some general conclusions to be drawn regarding the role that reproductive traits play in their invasion success. For example, *Cryptostegia grandiflora* (rubber vine), one of Australia's worst invasive species, has a number of dispersal agents including wind, water, birds, and mammals. Over 95 percent of its seeds are viable, with each seed pod being capable of producing over 800 seeds. This high reproductive output means that one hectare of rubber vine can produce millions of seeds every year. It may be that this extraordinarily high level of fecundity, combined with their many dispersal mechanisms, may enhance the invasion potential of introduced vines, although there are currently no comparative studies to highlight such relationships.

LEAF-LEVEL TRAITS Once exotic vines have established, their prolific growth rate facilitates their dominance over competing native vegetation. For example, *Pueraria montana* var. *lobata* (kudzu), one of the most invasive exotic vines, has been reported as spreading up to 50,000 ha per year in the United States alone. Kudzu now covers more than 3 million ha in the United States. Specific leaf area (SLA) approximates the relative growth rate of a species and is often used as a means of evaluating the ecological strategies employed by a species. Exotic and native vines exhibit a high SLA when compared to surrounding host vegetation. SLA exhibits a positive scaling relationship with other important characteristics such as leaf nitrogen content on a mass basis (Nmass) and photosynthetic gas exchange (Amass). A high SLA indicates the species or functional group possesses an ecological strategy of producing thin, short-lived leaves that are relatively cheap to construct. Many exotic species exhibit this strategy, and it has been associated with the ability to rapidly dominate a community. Exotic vines are no exception, with some of the most problematic species such as *Celastrus orbiculatus* (Oriental bittersweet) and *Lonicera japonica* (Japanese honeysuckle) possessing a very high SLA (>250 cm^2 g^{-1}) or extended periods of photosynthetic activity.

GLOBAL CHANGE AND EXOTIC VINES

Habitat fragmentation is the leading cause of biodiversity loss across the globe and has significant implications for exotic vine management. Given the affinity of exotic vines for disturbance, particularly in edge and gap environments, it is not surprising that habitat fragmentation can facilitate vine invasion. The fragmentation may be directly attributable to human activities, such as forest clearing, or indirectly associated with human actions, such as anthropogenically induced climatic changes that increase the frequency or severity of disturbance events such as fires and cyclones.

Atmospheric carbon dioxide concentration (CO_2) has been steadily increasing since the 1850s and now is around 35 percent higher than in preindustrial times. Exotic vines, like most plant groups, exotic or native, respond positively to experimental increases in elevated CO_2. Although this is not a surprising result, it is important to note that exotic vines have emerged as a particularly responsive functional group in this experimental work. For example, in a comparative study between native and exotic *Lonicera* species, the invasive species *L. japonica* showed a total biomass increase of 135 percent at 675 μL/L CO_2, compared to a 40 percent increase in the native *L. sempervirens* at similar CO_2 concentrations. *L. japonica* also exhibited similar increases in annual net primary production when grown in a natural competitive environment in a free-air carbon enrichment experiment.

The production of phytotoxins has also been shown to be altered in problematic exotic vine species grown under elevated CO_2 levels. The production of a more potent form of urushiol, the dermatological irritant that causes a reaction in around 80 percent of all those who make contact with *Toxicodendron radicans* (poison ivy), was associated with a doubling of ambient CO_2 in a natural community setting. These findings have potentially important implications for human health as carbon dioxide concentrations continue to rise.

Native climbing plant distribution is heavily skewed toward the tropics. Vines have undergone large evolutionary radiations in tropical regions, and the most problematic exotic vines generally arise from tropical environments. For example, South America contributes the majority of exotic vine species to Australia (see above) and is the continent with the highest concentration of tropical forests. Climate warming projections indicate the onset of warmer conditions in habitats in more poleward areas in the future. The effect that this warming may have on exotic vine distributions is largely unknown, but it has been hypothesized that species that emanate from warmer climes will be favored under future climates.

CONCLUSION

Vines are a unique and complex group, whose growth form presents a novel set of ecosystem impacts and management challenges to new environments when they become invasive. While invasive vines share a range of characteristics in common with other functional groups, such as high SLA and asexual reproduction, they also impose specific impacts on surrounding vegetation, such as their ability to smother hosts. There is still a great deal of research required in order to obtain a deeper understanding of the characteristics of successfully invasive vines. Essential areas for study include determining the suite of life history traits that relate to the invasion success of invasive vines in a comparative framework, investigating the population genetics of invasive vines, investigating mutualistic associations with dispersers and soil symbionts, and determining the effect of changing climate regimes and CO_2 levels on vines and their hosts. It is only then that we can begin to comprehend their responses to a changing climate and be able to manage them successfully in our current one.

SEE ALSO THE FOLLOWING ARTICLES

Australia: Invasions / Climate Change / Kudzu / Life History Strategies / Seed Ecology / Succession / Transformers

FURTHER READING

Baars, R., and D. Kelly. 1996. Survival and growth responses of native and introduced vines in New Zealand to light availability. *New Zealand Journal of Botany* 34: 389–400.

Gianoli, E. 2004. Evolution of a climbing habit promotes diversification in flowering plants. *Proceedings of the Royal Society of London B* 271: 2011–2015.

Harris, C. J., B. R. Murray, G. C. Hose, and M. A. Hamilton. 2007. Introduction history and invasion success in exotic vines introduced to Australia. *Diversity and Distributions* 13: 467–475.

Horvitz, C. C., and A. Koop. 2001. Removal of nonnative vines and post-hurricane recruitment in tropical hardwood forests of Florida. *Biotropica* 33: 268–281.

Horvitz, C. C., J. B. Pascarella, S. McMann, A. Freedman, and R. H. Hofstetter. 1998. Functional roles of invasive non-indigenous plants in hurricane-affected subtropical hardwood forests. *Ecological Applications* 8: 947–974.

Putz, F. E., and H. A. Mooney, eds. 1991. *The Biology of Vines*. Cambridge: Cambridge University Press.

Wu, Y., K. Rurchey, N. Wang, and J. Godin. 2006. The spatial pattern and dispersion of *Lygodium microphyllum* in the Everglades wetland ecosystem. *Biological Invasions* 8: 1483–1493.

Yurkonis, K. A., and S. J. Meiners. 2004. Invasion impacts local species turnover in a successional system. *Ecology Letters* 7: 764–769.

W

WASPS

JACQUELINE BEGGS

University of Auckland, New Zealand

Several species of social wasps have become invasive in many parts of the world. *Vespula* wasps can reach extremely high densities in their introduced range. They have restructured ecological communities; disrupted some industries such as forestry, grape growing, and tourism; and created a human health hazard from the risk of stings. Other social wasps are also pests, such as two species of paper wasp (*Polistes*). There is currently no technique for delivering widespread, effective control of social wasps, although experimental toxic baiting of *Vespula* has been used successfully in several countries. Trapping and direct poisoning of nests provide limited control.

DISTRIBUTION AND ABUNDANCE

At least three *Vespula* and two *Polistes* species are considered invasive (Table 1). However, these two genera belong to a family (Vespidae) containing thousands of species, so not all members of this family are necessarily potential invaders. *Vespula germanica* is now particularly widespread, but there is potential for the other invasive wasps to increase their worldwide ranges. Most *Vespula* species are difficult to distinguish, making it more difficult to detect and prevent new invasions, particularly in regions with high wasp densities. There is still some taxonomic uncertainty for some species; it is likely that the U.S. species *V. vulgaris* is not the same as the U.K. or New Zealand *V. vulgaris* (on the basis of molecular work), which could lead to confusion for biosecurity managers.

New Zealand and Hawaii are unusual in that they have no native social bees or wasps, which may partly explain why *Vespula* and *Polistes* have invaded these areas so successfully. New Zealand now has four species of social wasp (two *Vespula*, two *Polistes*), and Hawaii has seven (one *Vespula*, six *Polistes*). Although many species overlap in range, in New Zealand honeydew beech (*Nothofagus*) forests, *V. vulgaris* displaces *V. germanica*, probably because they are more efficient than *V. germanica* at collecting honeydew (a sugary exudate produced by scale insects). In other habitats, both species coexist.

The density of *Vespula* wasps varies considerably from year to year and site to site as a result of strong density

TABLE 1

Distribution of Invasive Social Wasps

Species	Origin	Introduced Range
Vespula germanica	Palearctic	New Zealand, Australia, Ascension Island, South Africa, the United States, Canada, Chile, Argentina
Vespula vulgaris	Holarctic	New Zealand, Australia, Argentina
Vespula pensylvanica	Western half of temperate North America	Hawaiian Islands
Polistes dominulus	Palearctic	Canada, United States, Chile, Argentina, Australia
Polistes chinensis antennalis	Eastern Asia	New Zealand, Norfolk Island

FIGURE 1 Wasp warning sign in honeydew beech forest, New Zealand. Extensive road works through some of the highest-density *Vespula* habitat in the world posed a serious risk from angry wasps whose colonies were disturbed. (Photograph courtesy of the author.)

dependence (probably through queen competition in spring) or as a negative effect of wet, cold spring weather. In New Zealand honeydew beech forest, the mean density of *V. vulgaris* measured at six sites over 13 years at the peak of the wasp season (early autumn) was 12 nests per ha; this equated to about 10,000 workers per ha. However, on occasion, they can exceed 30 nests per ha, and such high densities can pose a serious threat to humans and other biota (Fig. 1). This high abundance is associated with copious quantities of honeydew excreted by endemic scale insects (Fig. 2). There are similar associations between other invasive social insects and honeydew, such as for yellow crazy ants (*Anoplolepis gracilipes*) on Christmas Island. In contrast, *Vespula* density in habitats without honeydew is much lower: for example, around 0.4 nests per ha in the United Kingdom, and between 0.06 and 0.33 nests per ha in New Zealand scrubland and pasture (without honeydew). In Patagonia, the abundance of introduced *Vespula germanica* was estimated to be three to eight times lower than in New Zealand honeydew beech forest.

Polistes can also reach relatively high densities in their introduced range. In New Zealand, introduced *P. chinensis antennalis* is particularly abundant in warm, lowland areas of open habitat, such as shrublands, swamps, and salt meadows, rather than in forest. Densities of more than 200 nests per ha (6,300 wasps per ha) have been recorded in northern shrubland. In some areas of the United States, *P. dominulus* also is relatively abundant. These introduced populations are relatively genetically diverse, suggesting that there have been multiple introductions of this species.

BIOLOGY

The invasion ability of social wasps is facilitated by their reproductive and dispersal strategies. They form colonies of related individuals with a division of labor between separate castes that enables them to cooperate to collect food, raise offspring, defend against predators and competitors, and cope with abiotic stress. Furthermore, insects can store sperm for long periods, so a female who has mated can disperse and establish a new sexually reproducing population without having to be accompanied by a male. The species of *Vespula* and *Polistes* that have successfully invaded new areas all mate at the end of their annual cycle. The fertilized queen then finds a safe location to enter diapause—sometimes hidden away in human goods that are then transported to a new location.

Vespula wasps typically form annual monogynous colonies that die at the end of each wasp season (late autumn or early winter). However, in warmer margins of a species' range, and in its introduced range, perennial polygynous colonies can be formed. In some areas, about 10 percent of colonies are perennial, but due to their large size (Fig. 3), they can make a major contribution to the total wasp population. Perennial colonies have been estimated to contain up to 500,000 cells per nest, while an annual colony contains around 10,000 cells per nest.

IMPACTS

Social wasps potentially cause a range of impacts in natural ecosystems, from predation of invertebrates to alteration of nutrient cycling or competition with native species

FIGURE 2 *Vespula vulgaris* feeding on a droplet of honeydew. Copious quantities of honeydew produced by endemic scale insects (*Ultracoelostoma* spp.) in New Zealand beech forest are associated with extremely high densities of *Vespula* wasps. (Photograph courtesy of Bob Brown, University of Auckland.)

FIGURE 3 Perennial *Vespula germanica* nest in beech forest, New Zealand. Such nests are polygynous and can become extremely large. (Photograph courtesy of Landcare Research.)

for resources. However, the impact of wasps is variable, dependent on the species involved, their abundance, and the habitat invaded.

In New Zealand honeydew beech forests where wasps are extremely abundant, the wasps have altered the structure, connectedness, and interaction strengths of the food web at several trophic levels. First, they consume large quantities of honeydew (more than 90% of the standing crop for about four months of the year), thus competing with a range of native species that also feed on honeydew (e.g., kaka, a forest parrot, and tui and bellbirds [honeyeaters]). As generalist predators (Fig. 4), *Vespula* also have a direct impact on many native invertebrates and can kill nestling birds. Calculations show that *Vespula* foraging rates are so high on vulnerable species of spiders and caterpillars that they have virtually no chance of surviving the wasp season. This has led to a restructuring of the invertebrate community. These impacts are likely to have flow-on effects, such as reducing the flow of honeydew carbon to microorganisms in the soil and hence altering nutrient cycling. Similarly, reducing the abundance of some invertebrate taxa may have consequences for nutrient cycling and pollination.

There is also concern about the ecological impact of *V. pensylvanica* in upper elevations of Hawaii, particularly as predators of native insects. In Australia, there is dietary

overlap between native *P. humilis* and introduced *V. germanica*, and so competition with native species for food resources may be a problem in addition to direct predation. In contrast, research on *V. germanica* in Patagonia did not detect any difference in the abundance, species richness, or composition of the invertebrate community (305 species) between control areas and areas where the abundance of introduced wasps was experimentally reduced. However, a likely explanation for this apparent contradiction in impacts is the relatively low abundance of wasps in Patagonia compared to New Zealand honeydew beech forest.

Polistes also prey on invertebrates, particularly caterpillars (Lepidoptera). It has been estimated that *P. chinensis antennalis* consume almost 1 kg per ha per season of invertebrate biomass in northern New Zealand shrubland, where they reach high densities. Similarly, *P. dominulus* may detrimentally affect invertebrates in its introduced range. It is also possible that *Polistes* will disrupt pollination by competing with other nectar feeders at flowers.

In some areas of North America, *P. dominulus* displaces native *Polistes* species as a result of indirect competition. For example, the proportion of *P. fuscatus* nests has substantially declined in the northeastern United States, probably because of the superior productivity and survivorship of the introduced species. However, most studies have been done in human-altered habitat where colonies are more easily studied, so it is not clear that the displacement will occur in other habitats. *Polistes dominulus* may also compete for nest sites with cavity-nesting birds such as the eastern bluebird. In contrast, *Polistes* may be

FIGURE 4 *Vespula* wasps are generalist feeders; they collect protein from a range of dead and live sources. This worker is scavenging prosciutto to take back and share with her colony, but such trophallaxis makes *Vespula* vulnerable to toxic baiting. (Photograph courtesy of the author.)

viewed as a beneficial organism in some modified habitats because it can act as a biological control agent for caterpillar pests. However, it remains to be seen whether these benefits outweigh the disadvantages.

CONTROL

There are three main strategies for controlling social wasps: direct poisoning or trapping, toxic baiting, and biological control. Wasps are susceptible to a range of contact toxins, so locating a nest and sprinkling an insecticide dust (e.g., pirimiphos-methyl or pyrethrum) at the entrance of an underground nest is one option. The smaller and generally less aggressive *Polistes* colonies can be carefully clipped into a bag and frozen or treated with a contact insecticide. However, it can be difficult to locate nests, or the nests may be too widespread to make this feasible. Trapping workers away from the nest has also been tried, but it is difficult to achieve widespread control with this technique unless early-season queens are caught. An experiment that removed 75 percent of *Polistes* adults from nests to simulate trapping did not reduce the growth rate of nests compared to that in intact colonies. Nevertheless, a range of lure and kill traps are commercially available, and in some cases, that will at least reduce the abundance of workers in a specific location.

Another option is to exploit the trophallaxis behavior of social wasps and lure the workers to bait that is mixed with a toxin they cannot detect. Worker wasps collect the bait and return to their nest to share the food, inadvertently passing toxin around the colony and often killing the queen. This technique has been used against *Vespula* very successfully in several countries, achieving a greater than 90 percent kill of colonies over large areas (approximately 1,000 ha in New Zealand). These control operations use a protein-based bait to avoid attracting honeybees to the bait because the toxins used are not wasp-specific, and honeybee repellants also deter wasps. Toxic baiting is difficult for *Polistes* because they do not scavenge from dead animals, so they will not collect protein baits such as fish or chicken. A carbohydrate bait that does not attract honeybees would be useful in many situations. Experiments have shown that isobutanol, a volatile component of fermenting molasses, mixed with acetic acid is attractive to workers in at least four wasp genera (*Vespula, Dolichovespula, Vespa, Polistes*), but several species in these taxa are not attracted, including *Vespula vulgaris*.

A range of toxins (e.g., fipronil, microencapsulated diazinon, hydramethylnon, sulfluramid, and sodium monofluoroacetate [1080]) have been used experimentally for toxic baiting of *Vespula*, although to date a commercial

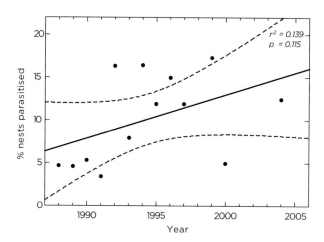

FIGURE 5 The biological control agent *Sphecophaga vesparum* was introduced to New Zealand to control *Vespula* wasps, but relatively few nests are parasitized. There is no evidence that *S. vesparum* has significantly reduced *Vespula* populations. (Graph courtesy of Elsevier, 2008, *Biological Control* 44[3].)

product is not available. Even a low concentration of fipronil (0.1% w/w) mixed with protein bait such as canned sardine cat food, minced chicken, or minced beef can provide effective control up to 400 m from a bait station. However, because queen wasps can disperse 30 km in a year, toxic baiting usually provides control for only one season, and wasps will reinvade the following year.

The third strategy, biological control, has been tried in New Zealand by introducing tiny specialist parasitoid wasps (*Sphecophaga vesparum*) (Fig. 5). The adult winged female parasitoid enters a *Vespula* wasp nest, lays eggs on late fifth-instar larvae or on pupae, and each developing parasitoid larva kills its host. The high fecundity, short generation time, and host specificity of the parasitoid meant that it had the potential to be an effective and safe control agent. The parasitoid established at least three populations in New Zealand. However, at the only site that was systematically monitored, the proportion of nests attacked remained relatively low (between 3% and 17%), the number of parasitoids produced exponentially declined to less than 20 per nest, and there was no net effect on *Vespula* at the population level. After almost 20 years since establishment, there was no evidence that *S. vesparum* was a successful biological control agent.

A range of other organisms are associated with social wasp colonies, including more than 80 microbes associated with *Vespula, Vespa,* and *Dolichovespula*. However, few if any of these are known to be both host-specific and capable of effectively controlling populations, two key requirements for classical biological control. Several fungi, bacteria, and nematodes have been confirmed as

FIGURE 6 Paper wasps (*Polistes dominulus*) infested with the endo-parasite *Xenos vesparum* (Strepsiptera, Stylopidae) desert their colonies to form extranidal aggregations. Arrows indicate the parasites. (Photograph courtesy of the author.)

pathogenic to wasps and could potentially be developed as biopesticides. Biological control of *Polistes* may be feasible with specialized parasites, such as *Xenos vesparum* (Strepsiptera, Stylopidae) that attack *Polistes dominulus* (Fig. 6).

SEE ALSO THE FOLLOWING ARTICLES

Ants / Bees / Biological Control, of Ánimals / Competition, Animal / Dispersal Ability, Animal / New Zealand: Invasions / Predators

FURTHER READING

Beggs, J. 2001. The ecological consequences of social wasps (*Vespula* spp.) invading an ecosystem that has an abundant carbohydrate resource. *Biological Conservation* 99: 17–28.

Beggs, J. R., J. S. Rees, R. J. Toft, T. E. Dennis, and N. D. Barlow. 2008. Evaluating the impact of a biological control parasitoid on invasive *Vespula* wasps in a natural forest ecosystem. *Biological Control* 44: 399–407.

Beggs, J. R., and D. A. Wardle. 2006. Keystone Species: Competition for honeydew among exotic and indigenous species (281–294). In R. B. Allen and W. G. Lee, eds. *Biological Invasions in New Zealand.* Berlin: Springer-Verlag.

Goodisman, M. A. D., R. W. Matthews, J. P. Spradbery, M. E. Carew, and R. H. Crozier. 2001. Reproduction and recruitment in perennial colonies of the introduced wasp *Vespula germanica. Journal of Heredity* 92: 346–349.

Kasper, M. L., A. F. Reeson, S. J. B. Cooper, K. D. Perry, and A. D. Austin. 2004. Assessment of prey overlap between a native (*Polistes humilis*) and an introduced (*Vespula germanica*) social wasp using morphology and phylogenetic analyses of 16S rDNA. *Molecular Ecology* 13: 2037–2048.

Liebert, A. E., G. J. Gamboa, N. E. Stamp, T. R. Curtis, K. M. Monnet, S. Turillazzi, and P. T. Starks. 2006. Genetics, behavior and ecology of a paper wasp invasion: *Polistes dominulus* in North America. *Annales Zoologici Fennici* 43: 595–624.

Sackmann, P., and J. C. Corley. 2007. Control of *Vespula germanica* (Hym. Vespidae) populations using toxic baits: Bait attractiveness and pesticide efficacy. *Journal of Applied Entomology* 131: 630–636.

Toft, R. J., and R. J. Harris. 2004. Can trapping control Asian paper wasp (*Polistes chinensis antennalis*) populations? *New Zealand Journal of Ecology* 28: 279–282.

WATER HYACINTH

MARTIN HILL AND JULIE COETZEE
Rhodes University, Grahamstown, South Africa

MIC JULIEN
Commonwealth Scientific and Industrial Research Organisation (CSIRO), Indooroopillay, Queensland, Australia

TED CENTER
U.S. Department of Agriculture, Ft. Lauderdale, Florida

Water hyacinth, *Eichhornia crassipes* Martius (Solms-Laubach) (Pontederiaceae), is a perennial free-floating aquatic plant that originated in the Amazon basin of South America. This plant has been spread around the world by humans since the late 1800s as an ornamental pond plant. It has established in most tropical and subtropical countries as well as in many warm-temperate regions between 40° N and 40° S. In the absence of natural enemies, water hyacinth forms large mats on still and slow-moving water bodies, where it severely degrades aquatic ecosystems and limits their utilization.

TAXONOMY

Eichhornia crassipes is in the family Pontederiaceae, which has been included in the order Commelinales. There are eight genera within the family and eight species in the genus *Eichhornia*, all but one of which are South American in origin. The exception is *E. natans* (P. Breauv.), which is native to tropical Africa.

DESCRIPTION AND BIOLOGY

Water hyacinth is a perennial plant, generally free floating in open water but often rooted in moist mud along the banks of rivers, lakes, and impoundments. Each plant consists of a rosette of six to ten leaves attached to a rhizome with a well-developed fibrous root system (Fig. 1). The growth form of this plant is variable. When growing at the edge of an infestation or otherwise not in dense mats, they are short (<30 cm tall) with inflated bulbous petioles, while in dense mats, the plants characteristically have elongated petioles up to 1.5 m in length. Circular leaf blades are supported by the petioles and, on large plants, these may be saucer sized.

Reproduction is both sexual and asexual. The plant produces inflorescences with up to 23 flowers. After pollination, the flower spike bends to position the seed capsule below the surface of the water, where the seeds are released.

FIGURE 1 Morphology of water hyacinth. (Photograph courtesy of M. P. Hill.)

Each seed capsule contains up to 50 seeds, which sink and remain viable in the sediment for 15 to 20 years. Seed germination occurs in moist sediments or in warm shallow water. However, the main mode of reproduction is asexual through the production of daughter plants or ramets, which are produced on stolons from the mother plant. *Eichhornia crassipes* populations increase rapidly, doubling every 11–18 days under suitable uncrowded conditions and in warm, tropical climates and eutrophic waters.

DISTRIBUTION

The center of origin of *Eichhornia crassipes* is the Amazon basin in Brazil and Peru. It has been spread to most of South America and the Caribbean islands and was first recorded in the United States in 1884 in New Orleans. By the end of the nineteenth century, the plant was recorded in Egypt, India, Australia, and Java. The main mode of spread of water hyacinth throughout the world has been through anthropogenic means, via the horticultural and aquarium trades, due to the appeal of its attractive smooth, green foliage and beautiful purple flowers (Fig. 2). It continues to be spread in this fashion.

UTILIZATION

There has been considerable research into the utilization of water hyacinth. Uses include in biogas production, animal fodder, fertilizer, mulch, the manufacture of paper and furniture, and water quality management. The main factor arguing against successful utilization is that water

hyacinth has a 95 percent water content, which makes most utilization projects commercially inviable on anything other than a cottage-industry level. Furthermore, no utilization program has been shown to control water hyacinth to acceptable levels, and creating a reliance on the weed as a resource could lead to its increased spread and to possible conflicts of interest between groups utilizing the plant and those wanting it controlled.

IMPACT

Extensive mats of water hyacinth have a negative impact on aquatic ecosystems and cause problems for all aspects of water resource utilization. Socioeconomic impacts include reduction in the quality and quantity of drinking water caused by bad odors, color, taste, and turbidity; increased incidence of waterborne, water-based, and water-related disease (e.g., malaria, encephalitis, and filariasis); increase in siltation and sedimentation of rivers, lakes, and impoundments; reduction of useful water surface area for fishing, recreation, and water transport (Fig. 3); clogging of irrigation canals and pumps; drowning of livestock; interruption of hydroelectric power generation; and enhanced flood damage to road and rail bridges and impoundment walls.

Dense mats of water hyacinth reduce light penetration into the water column, which negatively impacts submerged vegetation, reducing oxygen production in aquatic communities. The lack of phytoplankton production alters the composition of invertebrate communities,

FIGURE 2 Water hyacinth flower. (Photograph courtesy of M. P. Hill.)

FIGURE 3 A 7-km plug of water hyacinth on the Vaal River, South Africa. (Photograph courtesy of J. A. Coetzee.)

ultimately affecting fish and other higher-order vertebrate populations. Water hyacinth also outcompetes indigenous aquatic macrophytes.

CONTROL

Traditionally the control of water hyacinth has fallen into one of three broad categories: physical control (manual and mechanical removal), herbicidal control, and biological control. More recently, the emphasis has moved to an integration of these three methods.

Physical Control

Manual removal through hand-pulling or through the use of other handheld tools such as pitchforks is employed in a number of developing countries such as Guyana, South Africa, and China. This method is very labor-intensive, is effective only for small infestations, and essentially is used as an employment-creation exercise. Mechanical control using custom-designed machinery has also been implemented, for example, on Lake Victoria, with limited success. However, the amount of biomass that has to be removed and the growth rate of water hyacinth render this method ineffective, except for very small infestations or to keep specific areas such as the waters around hydro intakes free of the weed. Furthermore, the remoteness of many infestations makes mechanical control impractical. Booms and cables have also been used to prevent water hyacinth from entering water abstraction pumps and hydropower coolant intakes (e.g., the Kafue Gorge Dam, Zambia). Cables have also been used to concentrate weed infestations behind them, making physical removal and herbicide applications more efficient.

Herbicide Application

Herbicide control has been successfully used against infestations of water hyacinth since the early 1900s. Although this control method is relatively expensive, it has the advantage of being quick and temporarily effective. *Eichhornia crassipes* is susceptible to herbicides such as 2,4-dichlorophenoxyacetic acid (2,4-D), diquat, paraquat, and glyphosate, which have resulted in successful control in small (<1 ha), single-purpose water systems such as irrigation canals and impoundments. Herbicide control provides only short-term control and requires regular follow-up applications. It is often the lack of commitment to follow-up that has led to the failure of chemical control programs. In developing countries of the world, many water hyacinth–infested sites are used for drinking water, for washing, and for fishing, and so the use of chemical sprays contaminates these sites and can threaten human health. The herbicide control of water hyacinth is often not appropriate in developing countries as it is expensive and requires highly skilled personnel, and as herbicides are often perceived as poisons. Herbicide application is not permitted near or on waterways in some countries, thus reducing the management options for water hyacinth.

Biological Control

Mechanical and herbicidal controls are viewed as short-term or immediate control options; biological control is perceived as the long-term or sustainable control option for this weed. The first agent for water hyacinth was the weevil, *Neochetina eichhorniae* Warner (Coleoptera: Curculionidae), which was released in Florida in 1972. Since then, biological control agents for water hyacinth have been introduced to over 30 countries around the world. The most widely used agents are the weevils, *N. eichhorniae* and *N. bruchi* Hustache, and the moth, *Niphograpta albiguttalis* (Warren) (Lepidoptera: Pyralidae). Other agents, released as classical biological control agents, are the moth, *Xubida infusella* (Walker) (Lepidoptera: Pyralidae), the water hyacinth bug, *Eccritotarsus catarinensis* (Carvalho) (Hemiptera: Miridae), the galumnid mite, *Orthogalumna terebrantis* Wallwork, and the fungal pathogen, *Cercospora piaropi* Tharp. Most recently, the leafhopper, *Megamelus scutellaris* Berg (Hemiptera: Delphacidae), has been released in the United States. Several other insects are currently being screened for release on water hyacinth in regions of the world, where the other agents have not achieved the acceptable level of control or where the time taken to achieve control (1.5 to 3 years in the tropics) is perceived to be too long.

Biological control has met with varied success across the world. It has been highly successful in tropical areas such as Southeast Asia; Papua New Guinea; and western, central, and southern Africa. However, biological control has been less successful in some areas and this has been ascribed to variable climatic conditions (biological control takes longer in temperate climates [in excess of three years]), eutrophication of aquatic ecosystems (the plants grow so quickly in polluted systems that they are not easily suppressed by the introduced agents), interference from other control options such as herbicide application and mechanical control, and stochastic events such as floods and droughts, all of which remove the plants along with the agents. The plants then reinvade from seed and flourish in the absence of the biological control agents.

CONCLUSION

Water hyacinth remains the world's most invasive and damaging aquatic plant despite the fact that there are a number of effective ways to control it. Even though good progress has been made in controlling water hyacinth around the world, it still poses a threat to aquatic ecosystems and to human activities. The long-term control of this plant will require an integrated management approach utilizing all appropriate control methods, but with special emphasis on the need to reduce the inflow of nitrate and phosphate pollutants into aquatic environments, as water hyacinth infestations are largely symptoms of eutrophication.

SEE ALSO THE FOLLOWING ARTICLES

Eutrophication, Aquatic / Freshwater Plants and Seaweeds / Herbicides / Lakes / Reproductive Systems, Plant / Vegetative Propagation

FURTHER READING

Barrett, S. C. H. 1988. Evolution of breeding systems in *Eichhornia*: A review. *Annals of the Missouri Botanical Garden* 75: 741–760.

Center, T. D., M. P. Hill, H. A. Cordo, and M. H. Julien. 2002. Water hyacinth (41–64). In R. van Driesche, B. Blossey, M. Hoddle, S. Lyon, and R. Reardon, eds. *Biological Control of Invasive Plants in the Eastern United States*. West Virginia: Forest Health and Technology Enterprises Team.

Cilliers, C. J., M. P. Hill, J. A. Ogwang, and O. Ajuonu. 2003. Aquatic weeds in Africa and their control (161–178). In P. Neuenschwander, C. Borgemeister, and J. Langewald, eds. *Biological Control in IPM Systems in Africa*. Wallingford, UK: CAB International.

Coetzee, J. A., M. P. Hill, M. H. Julien, T. D. Center, and H. A. Cordo. 2009. *Eichhornia crassipes* (183–210). In R. Muniappan, G. V. P. Reddy, A. Raman, and V. P. Gandhi, eds. *Weed Biological Control with Arthropods in the Tropics*. Cambridge: Cambridge University Press.

Gopal, B. 1987. *Water Hyacinth*. Amsterdam: Elsevier.

Hill, M. P. 2003. The impact and control of alien aquatic vegetation in South African aquatic ecosystems. *African Journal of Aquatic Science* 28: 19–24.

Julien, M. H. 2008. Plant biology and other issues that relate to the management of water hyacinth: A global perspective with focus on Europe. *Bulletin OEPP/EPPO Bulletin* 38: 477–486.

WEEDS

JODIE S. HOLT

University of California, Riverside

A universally accepted definition of a weed is a plant growing where it is not wanted. In addition to this anthropocentric definition, which is based on human needs and perceptions, biological definitions have been suggested that attempt to distinguish weeds from other plants and explain their worldwide success. These definitions intersect and overlap to some extent with definitions of invasive plant species, which also vary. Because of the role of human perception in defining unwanted plants, there will likely never be a universally accepted set of biological or ecological criteria that define weedy taxa. Nevertheless, weeds, as a subset of pests, are widely recognized around the world for their close associations with human activities and their undesirable impacts on agricultural, natural, and even urban landscapes.

DEFINITIONS AND DESCRIPTIONS OF WEEDS

Weeds are universally recognized as plants growing out of place (Table 1). While this definition implies undesirability, a weed may not be undesirable to all people at all times and places. The concept of a weed has been recognized since the beginnings of agriculture: a plant that interferes with human activities. Just as crops were domesticated from wild plants, some weeds are thought to have evolved from wild plants during the process of crop domestication, while others are believed to have escaped from domestication. Invasive plants, in contrast, are defined in biological terms as plants introduced to a particular region that establish, naturalize, and spread to new habitats away from the point of introduction. When invasive plants are recognized as having negative impacts on desirable species or ecosystems, they are also considered weeds. However, many weeds are native or are fairly restricted to human-disturbed habitats or agricultural ecosystems, so not all weeds are invasive.

Most weeds possess biological characteristics that enable them to establish, grow, and flourish in habitats where they were not planted. A list of the "ideal characteristics

TABLE 1
Definitions of Weeds

Definition	Source
A plant whose virtues have not yet been discovered	R. W. Emerson 1878
Persistent plants that are difficult to control	A. Gray 1879
A plant that grows without being deliberately sown or cultivated	W. E. Brenchley 1920
A plant that grows plentifully and crowds out more valuable plants	W. E. Brenchley 1920
Unwanted, undesirable plants that should be destroyed	L. H. Bailey and E. Z. Bailey 1941
A plant that grows spontaneously in habitats highly disturbed by humans	R. M. Harper 1944
A plant growing out of place or in an undesirable location	Weed Science Society of America 1956 (www.wssa.net/index.htm)

NOTE: A chronological list of definitions of weeds that have appeared throughout the published literature for over a century, all of which show human value or judgment. Sources are shown in the table. Adapted from Radosevich, S. R., J. S. Holt, and C. M. Ghersa. 2007. *Ecology of Weeds and Invasive Plants: Relationship to Agriculture and Natural Resource Management*, 3rd ed. New York: John Wiley and Sons, Inc.

of weeds" was proposed by a botanist, Herbert Baker, in 1974 (Table 2). Baker suggested that most weeds would possess a combination of several characteristics on the list, while plants with many of these characteristics would be highly successful weeds.

A more comprehensive definition of weeds that combines anthropocentric and biological viewpoints

TABLE 2
"Ideal" Characteristics of Weeds

Germination requirements fulfilled in many environments

Discontinuous germination (internally controlled) and great longevity of seed

Rapid growth through vegetative phase to flowering

Continuous seed production for as long as growing conditions permit

Self-compatibility but not complete autogamy or apomixy

Cross-pollination, when it occurs, by unspecialized visitors or wind

Very high seed output in favorable environmental circumstances

Production of some seed in a wide range of environmental conditions; tolerance and plasticity

Adaptations for short-distance dispersal and long-distance dispersal

If perennial, vigorous vegetative reproduction or regeneration from fragments

If perennial, brittleness, so as not to be drawn from the ground easily

Ability to compete interspecifically by special means (rosettes, choking growth, allelochemicals)

NOTE: From Baker, H. G. 1974. The evolution of weeds. *Annual Review of Ecology and Systematics* 5: 1–24. Reprinted, with permission, from the *Annual Review of Ecology and Systematics*, Volume 5 © 1974 by Annual Reviews www.annualreviews.org.

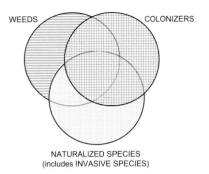

FIGURE 1 The relationship of weeds, colonizers, and naturalized species, which includes invasive species. In this view, the three categories of plants are not identical, although large areas of overlap occur. (Adapted from Rejmánek, M. 2000. Invasive plants: Approaches and predictions. *Austral Ecology* 25: 497–506.)

is shown in Figure 1 Here, weeds are defined as undesirable plants, while invasive plants are defined biogeographically as a subset of naturalized plants that establish in new habitats and spread to areas where they are not native. These categories overlap with each other and with colonizers, defined ecologically as early successional species. Thus, while weediness could be a transient condition (e.g., a volunteer corn plant in a soybean field), certain plants possess suites of traits that result in their widespread recognition as weeds nearly everywhere that they occur.

CLASSIFICATION OF WEEDS

In order to manage or control weeds, it is necessary to know their identity and understand something about their characteristics and impacts, because most methods of weed control target certain groups of weeds and not others. Weeds are classified in many ways for various purposes: according to taxonomy, life history, habitat, undesirability, and ecological behavior.

Taxonomic Classification

Nearly any plant could be a weed to someone at some time; however, most weeds are flowering plants in the phylum Anthophyta. About 250 plants are major agricultural weeds around the world, although many others are considered local weeds with more restricted distributions. Certain plant families contain the majority of agricultural weeds. The grass family (Poaceae) and sunflower family (Asteraceae) are particularly noteworthy for their large numbers of weedy members. Poaceae also contains the most important crops worldwide, including corn and many grain crops, which reinforces the view that weeds evolved from a close association with human agricultural activities.

Classification by Life History

Weeds are often classified by the type of life cycle they possess. Many weeds are annual plants that complete their life cycle in one year or less, while others are perennials that live for two or more years. Some biennial weeds occur, as well, in which vegetative growth occurs in the first year, and flowering, seed production, and senescence occur in the second year. Regardless of the length of the life cycle, weeds can persist year to year as seeds in the soil seedbank. Herbaceous perennial weeds also have continuous vegetative growth from meristems, often belowground, over a few to many years, which makes them particularly difficult to control. There are also many woody weeds (trees, shrubs, and vines) that interfere with forests, rangeland, and other managed ecosystems.

Classification by Habitat

Weeds occur in terrestrial habitats including agricultural fields, rangelands, forests, urban landscapes, and wildlands. Some weeds are restricted to aquatic habitats, including ponds, lakes, streams, and rivers. Some terrestrial weeds grow predominantly along waterways, although they are not classified as aquatic. Methods of weed control are typically specific to a particular habitat and are usually not directly transferable to dissimilar habitats. For example, some herbicides used in terrestrial locations are not registered for use near water, while herbicides used in or near waterways would not be suitable for terrestrial habitats. Likewise, physical methods of weed control are usually specific to the type of habitat where the weeds occur.

Classification by Undesirability

Collectively, weeds have undesirable effects on many aspects of human life and are often classified according to these effects. In the United States and many other countries, weeds may be legally classified at the national, state, and county level as noxious and may be placed on lists whereby their import and transport can be prohibited and regulated. The U.S. Federal Noxious Weed Act, enacted in 1974, provides crop inspection at ports of entry for seeds of listed noxious weeds in order to prevent their introduction into the country. As of 2006, 96 species were listed across a range of taxa, including an aquatic alga caulerpa (*Caulerpa taxifolia*), weedy crop relatives (red rice, *Oryza* spp.), and 25 species of the woody plant mesquite (*Prosopis* spp.). Some weeds are poisonous and pose threats to the health of humans and livestock. While these may not necessarily be regulated legally, they are often listed with public health officials who would have

FIGURE 2 Poison oak (*Toxicodendron diversilobum*) growing in a forest floor in northern California. (Photograph courtesy of the author.)

to deal with their impacts. Poison ivy (*Toxicodendron radicans*) and poison oak (*Toxicodendron diversilobum*; Fig. 2) are two examples of plants native to the United States that cause contact dermatitis in many people.

Ecological and Evolutionary Classification

As noted above, in an ecological context, weeds are considered to be colonizers of disturbed land that readily establish and persist in association with crop cultivation in agroecosystems and other human-caused land disturbances. Weeds have also been classified in an evolutionary context in terms of their adaptations for carbon allocation and response to environmental stress and disturbance. By this scheme, most weeds are considered to fall into one of two categories. Most herbaceous plants are competitive ruderals, with rapid early growth rates and early onset of competition, which is favored by fertile soils and regular disturbance such as is found in agroecosystems. Others, including trees and shrubs, are stress-tolerant competitors that occur primarily in sites characterized by low disturbance.

WEED IMPACTS

In agroecosystems, weeds can reduce crop yields and quality, interfere with harvesting operations, reduce utility and value of land for raising livestock or other productive uses, and pose threats to livestock through toxicity, indigestibility, and low nutritional content, as well as by having spines, thorns, or other detrimental physical features. Weeds also interfere with use of land and waterways for recreation, lower property values, and pose public health hazards through toxicity and allergic reactions. In wildlands, weeds compete with and sometimes displace desirable native species, alter fire regimes and hydrology,

FIGURE 3 Giant reed (*Arundo donax*) in the Santa Ana riverbed in Riverside, California. In the 1990s, a huge mat of giant reed washed downstream and damaged the Van Buren Bridge; it cost the city $5 million to replace the damaged bridge with the one in the photo. (Photograph courtesy of M. A. Rauterkus.)

and alter biodiversity and ecosystem services. Worldwide, weeds cause more crop yield loss than any other type of pest and cost billions of dollars annually in loss of productivity and cost of removal. Because of their significant impacts on human activities, including on agriculture, natural resources, public health, and property, weeds are often subject to legal regulations, as noted above.

Some weeds have spread so widely that they are known around the world. In agroecosystems, purple nutsedge (*Cyperus rotundus*) and bermudagrass (*Cynodon dactylon*) are two of the world's worst weeds. In wildlands, giant reed (*Arundo donax*; Fig. 3) has invaded many riparian habitats worldwide, while leafy spurge (*Euphorbia esula*) is an aggressive invader in rangelands throughout the world except Australia. Other weeds such as alang alang or cogon grass (*Imperata cylindrica*) are widespread in both agroecosystems and wildlands around the world. Most weeds, however, are important primarily at the regional or local level. For example, cheatgrass (*Bromus tectorum*) is a rangeland weed in the western United States that has displaced nearly all the native plants in this ecosystem, while kudzu (*Pueraria lobata*) is an invasive vine that continues to spread in the southeastern United States and has earned the name "the vine that ate the South."

WEED MANAGEMENT

Weed management encompasses all the activities pertaining to the prevention, eradication, and control of unwanted plants (Table 3). Management typically also includes growing or fostering desired vegetation, such as crop plants. The most successful management programs

TABLE 3
Weed Management in Agroecosystems

Weed Management Method	Tool or Technique
Prevention	Sanitation, use of certified weed-free crop seed, use of weed-free animal feed, use of clean nursery stock, adherence to weed laws and quarantines
Eradication	Total weed removal by any means
Control	Reducing weed presence to an economically acceptable level
Physical weed control	Hand-pulling, hand-hoeing, controlled burning, flaming, tillage/cultivation, mowing and shredding, chaining, dredging, flooding, mulching, soil solarizing
Cultural weed control	Planting competitive crops, managing crop planting times and densities, rotating crops, planting cover crops or living mulches, planting smother crops
Biological weed control	Using living organisms to reduce weed density or presence
Chemical weed control	Using phytotoxic chemicals to suppress or kill weeds
Integrated weed management	Using a combination of methods and tools to foster desired vegetation and to control weeds

NOTE: A list of commonly used methods of weed management in agroecosystems and the tools and techniques commonly used to accomplish them.

are based on understanding weeds and their characteristics, including identity, taxonomy, life history, habitat, and relationships with other organisms. The approaches, methods, and tools of weed management in agroecosystems are well established and have been developed over many decades. Many if not most of the approaches are also applicable to invasive plants in wildlands, although the specific tools are not necessarily appropriate or transferable to nonagricultural habitats, particularly when sensitive native species are present. Nevertheless, it is important that managers of land impacted by invasive plant species be familiar with the concepts used in weed management in agroecosystems in order to maximize efficiency and avoid spending time reinventing techniques that are already well established in other ecosystems.

Prevention, Eradication, and Control

Prevention is acting to keep weeds from establishing in areas where they do not already occur. Thus, preventive measures attempt to stop the introduction, establishment, and spread of new weeds in an area. One of the most important methods of prevention is sanitation so that weed seeds and propagules are not inadvertently moved from place to place on equipment, animals, vehicles, or even shoes. The use of certified weed-free crop seed and animal feed and clean nursery stock is also a type of prevention. In some cases, noxious or other serious weeds can be restricted through weed laws (as noted above) or quarantines.

Eradication is the total removal of a weed from a specified area so that no parts remain, including propagules. Although eradication is often a goal of weed removal, it is seldom achieved without great expense. For this reason, it is rarely attempted except in very high value crops or vulnerable landscapes. The best chance of eradicating a weed is when the infestation is new and small and before it has expanded in size or spread beyond the point of introduction. For example, kochia (*Kochia scoparia*) was introduced intentionally into Australia in the 1990s but soon began to spread widely. A national eradication program was developed quickly, and this invader was eradicated from Australia by 2000. Witchweed (*Striga* spp.) is a noxious weed in many countries and is the focus in the eastern United States of an active quarantine and eradication program, which has been successful in local areas at great expense.

Weed control in agroecosystems is defined as the reduction of weed presence to an economically acceptable level. The level of control desired depends on the situation; in most agricultural fields, it is not economically feasible to attempt 100 percent control (eradication) because the value of the crop would not exceed the cost of weed removal. Thus, the effectiveness of weed control will depend on the weeds present and the ease of controlling them, the methods and tools available for control, and the value of the crop or land infested with the weed. The four methods of weed control are physical, cultural, biological, and chemical.

Physical Methods of Weed Control

Physical weed control is a method that removes weeds by physical means, either by using hand labor or by using various types of mechanical equipment designed for that purpose. Through physical means, weeds can be uprooted and pulled from the ground, shoots can be cut off from their root systems, or whole plants can be buried, smothered, or burned. Among these methods, hand-pulling and hoeing using hand tools are the oldest yet still most common methods of weed control around the world. Removal by hand is especially effective against weeds that cannot resprout from vegetative propagules that might be left in the ground. This method is equally effective on weeds in agroecosystems and on invasive weeds in wildlands.

Fire is used as another physical method of weed control. Agricultural fields are sometimes burned to clear off vegetation prior to planting, and noncrop areas such as roadsides and ditches are sometimes burned to remove weeds. Prescribed burns are used in rangelands, shrublands, and forests in some areas to reduce fuel loads and decrease the potential for catastrophic wildfire. In some agricultural situations, directed flame is used to heat and desiccate weeds in a crop field without damaging the crop.

Tillage and cultivation are physical methods of controlling weeds using mechanical equipment powered by animals or engines. Typically an agricultural field will be tilled before crop planting to loosen the soil, prepare the seedbed, and remove weeds. After crop planting, tillage is also practiced as a method of weed control. Many different types of equipment are available to perform tillage, depending on the situation. Tillage kills weeds by breaking or pulling them loose from the soil, exposing roots to the air where they desiccate, or covering and smothering them with soil. If weeds are able to resprout from vegetative propagules, then repeated tillage is used to kill resprouts and deplete the soil of propagules. Repeated tillage combined with irrigation is also used to bring weed seeds to the surface, promote germination, and then uproot and kill weed seedlings.

Another physical method of weed control is mowing or shredding unwanted plants. This method is used to reduce the biomass of vegetation without killing and removing it in cases where ground cover is desired and bare soil would be detrimental. In areas where weedy shrubs are present, large chains are dragged between two tractors to uproot large woody weeds. A similar approach is used to dredge waterways for removal of aquatic weeds. These methods are very destructive, so they cannot be used in habitats where sensitive native species or wildlife are present.

Several physical methods of weed control entail smothering weeds. In some specific situations, flooding for several weeks can be used to control weeds by depriving their roots of oxygen. A more common approach to smother weeds is by covering them with opaque mulches to block light, which prevents photosynthesis and starves the covered plants. Many kinds of mulches are used in agricultural and landscape settings, including straw, hay, grass clippings, composted manure, rice hulls, and even paper and plastic. Effective mulches are those that exclude light but permit water and air to pass through. Soil solarization is similar to mulching, but it uses clear plastic. In this case, light penetration through the plastic is desirable and results in the build-up and trapping of heat under the plastic, which kills weed seedlings as well as many microorganisms in the soil. This method is being tested for use in restoration sites where weed removal is necessary before native species can be planted.

Cultural Methods of Weed Control

Cultural weed control involves practices that promote a healthy, competitive crop and discourage weeds. Cultural methods of controlling weeds include planting competitive crop varieties, manipulating crop planting times and densities, rotating different crops to avoid a build-up of particular weeds, planting cover crops or living mulches along with the primary crop to minimize bare soil and resource availability for weeds, and planting smother crops (which grow fast and outcompete weeds) between or alternated with the primary crop. The many methods of weed prevention are also considered cultural methods of weed control because they reduce the presence of weeds in an area and stop their movement into other areas.

Biological Weed Control

Biological weed control uses living organisms to reduce the density or incidence of a weed population. Ideally, the biological control agent is a natural enemy of the weed that uses it as a food source or host. This method is not suitable for annual cropping systems because of the time required to establish populations of the biological control agent. Furthermore, biological control agents rarely kill weeds or eliminate them completely, but instead reduce their growth, competitiveness, or reproduction. A successful biological control program is one in which populations of the weed and control agent reach equilibrium such that the weed is no longer a significant problem. One advantage of this method is its relatively low cost once an agent has been identified. Good control agents are also selective and host-specific and rarely damage other species besides the targeted weed.

A well-known biological control success story was the introduction of the cactus moth (*Cactoblastis cactorum*) into Australia in 1925 for control of prickly pear cactus (*Opuntia stricta*), which by that time had infested over 60 million acres of rangeland. Today, the cactus occurs there only at low levels, which are maintained by cactus moth grazing. Biological control was also highly successful on St. Johns wort (*Hypericum perforatum*) in western U.S. rangelands in the 1940s by the introduction of the St. Johns wort beetle (*Chrysolina quadrigemina*), which reduced a widespread infestation to about 1 percent of its earlier extent. The grass carp (*Ctenopharyngodon idella*) is an herbivorous fish that is highly effective for biological control of submersed aquatic weeds. However, this species will sometimes consume native plant species and remove valuable habitat for wildlife, so it has been banned in some states.

Typically, biological control programs begin with exploration to locate a suitable natural enemy for use as a control agent for a specific weed. In some countries such as the United States, the processes of importation, testing, and release of a biological control agent are carefully regulated by federal agencies to ensure that the agent is host-specific, is able to survive in the new habitat, and poses no threat to crops or native species. Besides this classical method of biological control, in which natural enemies are imported and released, augmentation is sometimes practiced. In this method, biological control agents are mass-reared and released repeatedly for control of a particular weed.

Grazing is another method of biological control in which mammals, birds, insects, and even fish are used to eat weeds. For example, large animals such as sheep and goats are used for weed control in perennial crops or on rangeland, while some fish have been employed to eat aquatic weeds, as described above. Goats are commonly used for controlling weedy shrubs, spiny weeds, and allergenic plants such as poison ivy or oak in wildlands. In agroecosystems, geese are used for biological control of grasses and grass-like plants in orchards, vineyards, and many perennial and annual crop fields. Another type of biological control is the use of plant pathogens as bioherbicides or fungi as mycoherbicides. As their name suggests, these organisms can be cultured, packaged, and then applied to weeds using herbicide application tools and techniques. Finally, some plants can be employed as biological control agents when they produce allelopathic chemicals that are released into the soil. Some smother crops and living mulch crops are chosen because they are allelopathic to other plants. Wheat and rye, for example, can suppress germination of some weed seedlings when they are grown in a field and then allowed to decompose on the land before another crop is grown. Allelopathic chemicals occur in soil through decomposition of plant components or exudation or volatilization from plant parts. Plant-to-plant biological control remains poorly understood, and for this reason is not yet well developed as a weed control method.

Chemical Weed Control

Chemical weed control involves the use of phytotoxic chemicals to suppress or kill weeds. Inorganic chemicals including salts, acids, and oils have been used for over a century to kill vegetation in agricultural fields, rangelands, forests, and noncrop areas. Herbicides are organic, synthetic chemicals used for the same purpose. The many benefits of herbicide use include control of weeds that are difficult to kill by any other means, control of weeds in inaccessible areas, reduced need for tillage, greater flexibility in cropping and management systems, and reduction in human labor required for weeding. However, there are potential problems caused by herbicides, including injury to nontarget plants, animals, or microorganisms; residues in the environment; evolution of herbicide resistance; and impacts on human health and safety. For these and other reasons, the registration and use of herbicides is tightly regulated by government agencies in some countries.

More herbicides are used in the United States than any other type of pesticide, including insecticides, fungicides, and rodenticides. Major advances in crop production have occurred as a result of the use of herbicides, and they are increasingly being used in wildland areas to control invasive weeds. Invasions of some wildland weeds are so extensive that chemical control is an essential part of a comprehensive management plan. For example, melaleuca (*Melaleuca quinquenervia*), a desirable native tree in Australia, has invaded vast expanses of the Florida Everglades of the United States, and chemicals are a necessary tool for its control. However, over the past few decades, there has been increased awareness of unintended effects of herbicides on the environment, including nontarget impacts, and an increased emphasis on reducing herbicide use in favor of more integrated and ecological approaches.

Integrated pest management employs a combination of methods to suppress pests. These methods include prevention, monitoring, and a combination of control techniques. As environmental concerns have escalated, many researchers have scaled up from studying multiple methods for weed control to examining entire cropping systems to understand the ecological basis for crop productivity and weed impacts. As weeds increasingly move into nonagricultural wildland ecosystems, it has become even more important that weed control be viewed in an ecological context.

SEE ALSO THE FOLLOWING ARTICLES

Agriculture / Allelopathy / Biological Control, of Plants / Cheatgrass / Eradication / Herbicides / Kudzu / Mechanical Control / "Native Invaders" / Seed Ecology

FURTHER READING

Anderson, W. P. 2007. *Weed Science: Principles and Applications*, 3rd ed. Long Grove, IL: Waveland Press, Inc.

Baker, H. G. 1974. The evolution of weeds. *Annual Review of Ecology and Systematics* 5: 1–24.

De Wet, J. M. J., and J. R. Harlan. 1975. Weeds and domesticates: Evolution in the man-made habitat. *Economic Botany* 29: 99–107.

Grime, J. P. 1979. *Plant Strategies and Vegetation Processes*. New York: John Wiley and Sons, Inc.

Holm, L. G., J. Doll, E. Holm, J. Pancho, and J. Herberger. 1997. *World Weeds: Natural Histories and Distribution*. New York: John Wiley and Sons, Inc.

Radosevich, S. R., J. S. Holt, and C. M. Ghersa. 2007. *Ecology of Weeds and Invasive Plants: Relationship to Agriculture and Natural Resource Management*, 3rd ed. New York: John Wiley and Sons, Inc.

Randall, J. M. 1997. Defining weeds of natural areas (18–25). In J. O. Luken and J. W. Thieret, eds. *Assessment and Management of Plant Invasions*. New York: Springer-Verlag.

Rejmánek, M. 2000. Invasive plants: Approaches and predictions. *Austral Ecology* 25: 497–506.

Ross, M. A., and C. A. Lembi. 2008. *Applied Weed Science: Including the Ecology and Management of Invasive Plants*, 3rd ed. Upper Saddle River, NJ: Prentice Hall.

Zimdahl, R. L. 2004. *Weed-Crop Competition: A Review*, 2nd ed. Hoboken, NJ: Wiley-Blackwell.

WEST NILE VIRUS

SEE FLAVIVIRUSES

WETLANDS

JOY B. ZEDLER

University of Wisconsin, Madison

Wetlands are shallow-water ecosystems that depend on continuous or recurring inundation. While the maximum depth of "shallow water" is debated, the 6 m limit developed by the 1971 Ramsar Convention on Wetlands is appropriate here, because it includes estuaries and ports, where many invaders arrive or depart. The full spectrum of wetlands spans reefs and aquatic beds at the deepest extreme and seasonal wetlands at the driest extreme. Attention to wetland invasions is important because wetlands perform critical ecosystem services to a greater extent than predicted by the small portion of the Earth they occupy. Shallow waters cover less than 6 percent of global area, and wetlands cover less than 9 percent of the Earth's land area, yet they provide a much greater proportion of global, annually renewable ecosystem services. Valued functions include high primary productivity, support of high animal diversity, ability to improve water quality, storage of carbon, and hydrologic functions. Effects of invasive species on these services are largely unquantified, but many of the impacts are negative.

INVASIBILITY OF WETLANDS

Wetlands are especially vulnerable to biological invasions, first because many highly invasive species are wetland plants and animals, and second because many wetland invaders rapidly dominate their adopted community. Plants often displace native species by forming monotypic vegetation, and several invasive animals dramatically impact large areas through habitat disruption or consumption of native species.

Numerous Invaders

Of the world's 100 most invasive alien species (see Appendix), 19 are shallow-water animals (eight fish, three molluscs, two crustaceans, two insects—both mosquitoes, two amphibians, one flatworm, and one sea star), and 13 are plants (one fungus, five woody plants, two grasses, two herbaceous plants, one aquatic plant, and two algae). Of the world's 33 most invasive plants, eight are wetland species. In both lists, the proportion of wetland species is large, given that wetlands cover a small minority of global area.

Wetland invaders are often dispersed via flowing water. Seeds of 61 percent of 441 wetland plant species listed by Beth Middleton are water-dispersed. Seeds can remain dormant in natural waters, on fur and feathers, in the guts of granivores, in the ballast water of ships, on the mud of tires and machines, and in the crevices of shoes. Dormancy facilitates their transport to new areas with shallow water and wet soil.

Estuaries are easily invaded by aquatic animals, largely due to shipping and discharge of ballast water. California's San Francisco Bay, for example, has more than 210 invasive species from multiple origins. Freshwater systems are also easily invaded. The majority of plant and animal introductions (two-thirds of those in the tropics and one-half of those in temperate regions) have led to invasions, according to the Millennium Ecosystem Assessment. Notorious animal invaders include the cane toad (*Rhinella marina [Bufo marinus]*, native to the Americas and invasive in Australia); the bullfrog (*Rana catesbeiana*, native to eastern North America and now living worldwide), carp (four genera from Asia now widespread in the United States), nutria (*Myocastor coypus*, from South America, now invasive in wetlands along the Gulf of Mexico), and the zebra mussel.

A few wetland species are relatively rare in their native regions, yet are invasive elsewhere. For example, spike rush (*Juncus acutus*) is a rare brackish-marsh species in its native southern California coastal marshes but is highly invasive in wet pastures of Western Australia.

The Propensity of Wetland Invasive Plants to Form Monotypes

Wetlands tend to allow plant invaders to form monotypes that displace native vegetation and allow very few cooccurring plants to persist. Examples are giant reed (*Phragmites australis*), especially in the northeastern United States; invasive cattails (*Typha* spp.), reed canary grass (*Phalaris arundinacea*), and purple loosestrife (*Lythrum salicaria*) in temperate North America; paperbark (*Melaleuca quinquenervia*) and Brazilian pepper (*Schinus terebinthefolius*) in Florida; and catclaw mimosa (*Mimosa pigra*) and aligatorweed (*Alternanthera philoxeroides*) in Asia and Australia.

Clonal grasses are especially effective at forming monotypes. In mesocosms, reed canary grass rapidly invaded wet prairies that were given high nutrient loading, prolonged flooding (15 cm deep), and an influx of topsoil. Strong synergisms between flooding and nutrients were responsible for the most rapid conversions of diverse wet prairie to monotypes of reed canary grass. The simulated conditions mimic those of landscape sinks, where water and nutrients accumulate. As monotypes develop, they displace native species. In Wisconsin, sedge meadow plots had 16 species per square meter where uninvaded and six species where reed canary grass had 80 percent canopy cover. Similarly, in marshes, two-thirds of the native species were absent where invasive cattails had 80 percent canopy cover.

A Conceptual Model of Wetland Invasibility

The formation of wetlands in landscape sinks explains their vulnerability to invasion (Fig. 1). Wetlands form

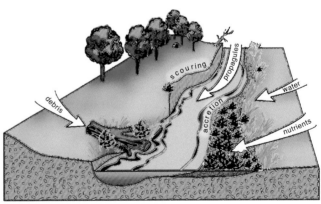

Landscape sink:
- Propagules arrive and accumulate
- Floods and debris create canopy gaps
- Ample moisture and nutrients

→ • Opportunists establish and capitalize on the opportunity to form monotypes

FIGURE 1 Most wetlands are "landscape sinks" that receive water, nutrients, and propagules from upstream and upslope. These features and the associated disturbances that create canopy gaps provide the opportunity for establishment. The opportunists that establish are often introduced alien species.

where water, sediments, nutrients, and organic matter accumulate, and these same processes facilitate invasion. Plant propagules float into wetlands and germinate or sprout where moisture is abundant, sediments are exposed by disturbance, and nutrients are abundant. After establishment, the same conditions facilitate rapid vegetative spread of clonal species.

Disturbances and altered hydrological conditions typically increase invasibility. Flowing water scours banks and deposits sediment, providing bare substrate for plants to establish. Any debris that accumulates can smother existing vegetation and create canopy openings. Human structures also play a role. Behind dams and levees, water levels are unnaturally stabilized, soils develop anoxia, and soil chemistry changes in ways that can lead to internal eutrophication (release of phosphorus under waterlogged conditions). In Wisconsin, hybrid cattails (*Typha x glauca*) spread an average of 4 m per year behind a dam, but native cattails (*T. latifolia*) remained stable where hydroperiods were naturally fluctuating.

INVASIVENESS OF WETLAND PLANTS

Invasive Traits

While no single attribute explains invasiveness, water dispersal (hydrochory) and vegetative reproduction confer definite advantages. Seeds, rhizome fragments, stem fragments with nodes, and turions (Fig. 2) can float along waterways. Where viable plant fragments land on wet soils, the accumulated debris acts as a mulch, facilitating sprouting and establishment. Runners and rhizomes that remain attached to a parent are particularly effective at colonization, because the parent can subsidize moisture and nutrients until leaves of the new ramet reach sunlight.

Once a wetland plant has invaded, it will likely grow tall rapidly through efficient growth. Stems and leaves of many species have large volumes of air tissue (aerenchyma; Fig. 3). Dense rhizomes and roots compete strongly for nutrients. And some species extend the growing season beyond that of native species (e.g., *Rhamnus cathartica, R. frangula*). Reed canary grass remains alive through October, and its green canopies are visible on satellite imagery, allowing it to be mapped as dominant in 200,000 ha of Wisconsin wetlands.

Allelopathic chemicals secreted by an invader can effectively kill native competitors. The thorough study of allelopathy in *Phragmites australis* by Rudrappa and colleagues showed that its roots exude gallic acid and cause acute rhizotoxicity to other species, such as the native cordgrass *Spartina alterniflora*, while not harming

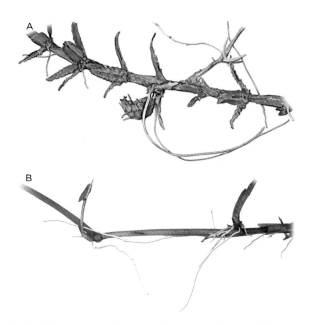

FIGURE 2 Means of vegetative reproduction. (A) Turion of *Hydrilla verticillata*. (Photograph courtesy of W. Michael Dennis at http://tenn.bio .utk.edu/vascular/photoD/Monocots/Hydrocharitaceae/hy_vert1.jpg.) (B) Rooting stem nodes of *Phalaris arundinacea*. (Photograph courtesy of the author.)

itself. The persistence of gallic acid in soil suggests that an invader can leave a legacy even after it is eradicated, thereby hindering restoration efforts.

Genetic Concerns

The evolutionary consequences of introducing invasive species are becoming increasingly understood for wetland plants. For example, reed canary grass from various European regions is evolving rapidly in eastern North America. Genetic shuffling has produced novel genotypes with phenotypic plasticity, rapid selection, and increased vegetative colonization. Similarly, invasive *Typha* in Laurentian Great Lakes wetlands was once thought to have two parents (the native *T. latifolia* and the alien *T. angustifolia*) and a sterile F_1 generation (*T. x glauca*). It is now clear that the invader represents a hybrid swarm that includes F_2 generations resulting from the occasional fertile seed, as well as many backcrosses.

Hybrid vigor is best known in cordgrass (genus *Spartina*). *Spartina alterniflora* is native to eastern North American coasts; when brought to the United Kingdom, it hybridized with the native *S. maritima* and formed a highly aggressive, invasive hybrid, *S. townsendii*. After chromosome doubling, the newly created *S. anglica* continued to expand and dominate British salt marshes. Introductions to Asia, Australia, and New Zealand led to invasiveness

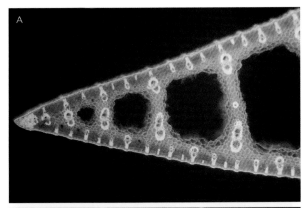

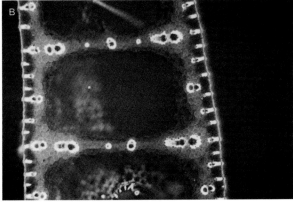

FIGURE 3 *Typha glauca* leaf edge and center cross-sections showing aerenchyma (large air spaces between "I-beams"). (Photographs courtesy of Hilary McManus, University of Connecticut.)

there as well. Deliberate introductions of *S. alterniflora* to the U.S. Pacific Northwest negatively affected oyster harvesting and shorebird watching. Introductions to San Francisco Bay were more problematic, because *S. alterniflora* hybridized with the native *S. foliosa*.

IMPACTS

The introduction of invasive species is second only to habitat loss in causing the extinction of freshwater species and changing wetland food webs. Invasive species have altered every component and function of wetland ecosystems.

Impacts on Ecosystem Structure and Function

Above ground, woody invaders often overtop low-growing herbaceous wetlands. In the Florida Everglades, for example, the "river of grass" is displaced by the Brazilian pepper tree and the Australian paperbark. When trees replace nonwoody plants, they can deplete soil water. The paperbark covers over 160,000 ha of Florida wetlands and transpires so much additional water that it makes the vegetation more fire-prone than native sawgrass (*Cladium jamaicense*).

Below ground, nitrogen-fixing invaders alter soil nutrient status. The tropical American shrub, catclaw mimosa (*Mimosa pigra*), is among the world's 100 most invasive plants. As a nitrogen fixer, it alters nutrient cycling and enriches soil of some 80,000 ha of invaded tropical wetlands in Australia. Once enriched, eutrophic soils are even better able to support invasive plants. Hydrological functioning of arid region riparian zones is altered by deep-rooting invaders (*Tamarix* spp.), which can lower water tables and increase surface soil salinity. Where watersheds are heavily populated, entire streams are dewatered.

In shallow tropical waters, floating invasive plants usurp light and cause anoxia in deeper water, ultimately suffocating invertebrates and fish. Hypoxic sediments can also release contaminants and phosphorus to the water column. Examples of mat-forming invasives are water lettuce (*Pistia*), of uncertain origin, and water fern (*Salvinia*) and water hyacinth (*Eichhornia*), both from South America. Humans are inconvenienced and livelihoods threatened when floating aquatic plants clog waterways and hinder boat traffic. Sometimes the impacts are life threatening; for example, floating mats of red water fern (*Azolla filiculoides*) provide habitat for snails that carry bilharzia, which infects millions of people in Africa.

Food Webs

Both plant and animal invaders alter wetland food webs: invasive plants can change the quality of food available, and invasive omnivores can directly modify both plant and animal populations. In 1973, the North American red swamp crayfish (*Procambarus clarkii*) was deliberately introduced to Spain, where there were no native crayfish but where suitable wetland habitat was widespread. Thirty years and many translocations later, it occurred throughout the Mediterranean region and Europe. Attributes that facilitate its invasiveness include omnivory, short time to reproduction, high reproductive output, and mobile larvae. Once abundant, it altered and simplified historical food webs (Fig. 4). Through grazing and shredding of macrophytes, it removed plants and created turbidity, converting marshes to open water and shifting the food base toward phytoplankton, filamentous algae, and detritus (which crayfish also consumed). Juvenile crayfish fed on invertebrates (arthropods, gastropods) and outcompeted native carnivores. Adults fed on fish and other abundant animals. Additional impacts resulted from disease organisms carried by the crayfish.

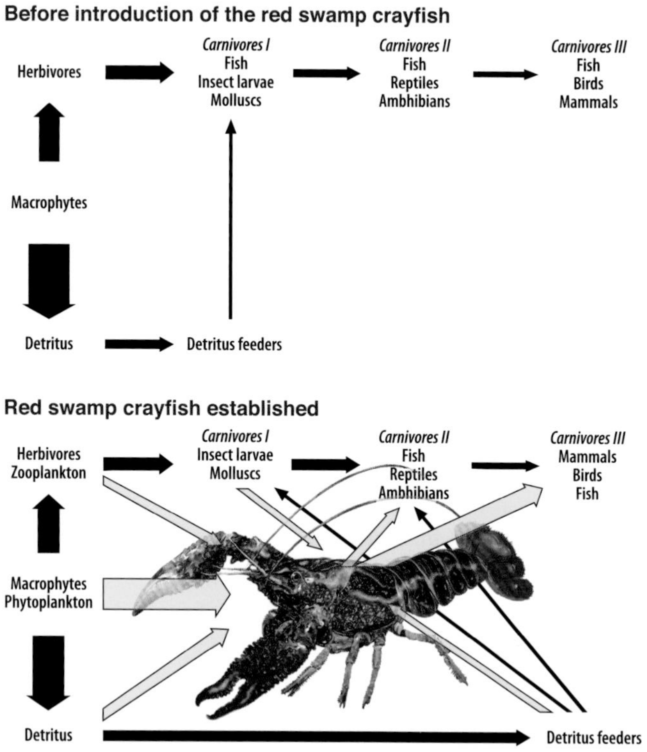

Before introduction of the red swamp crayfish

Red swamp crayfish established

FIGURE 4 The wetland food web had a five-step food chain before the introduction of the invasive red swamp crayfish, *Procambarus clarkii*, and a two-to-three-step food chain after (based on information in Geiger et al., 2005).

Invasional Meltdowns

In many cases, the effects of invaders are direct and independent of other aliens. Invasive *Typha* was the most common dominant in 74 wetlands along the shores of the five Laurentian Great Lakes, with no indication of facilitation by other species. In Australia, thickets of the invasive catclaw mimosa (*Mimosa pigra*) had direct negative effects on the abundance of birds and lizards within three to five years. In other cases, however, the effect on one species has depended on that of another (leading to a meltdown).

In Bodega Bay, the introduced amethyst gem clam (*Gemma gemma*, native to the east coast of the United States) did not displace the native clam species (*Nutricola* spp.) until the European green crab (*Carcinus maenas*) was introduced and began to prey preferentially on native clams. With fewer native clams, the introduced clam expanded. Early evidence of invasional meltdowns among vertebrates came from created ponds in Oregon, where permanent ponds are not a natural type of wetland. The alien bullfrog (*Rana catesbeiana*) is attracted to ponds, where it can consume native frogs, salamanders, turtles,

and even ducklings. Native dragonfly nymphs, however, feed on bullfrog tadpoles. But a nonnative sunfish can reduce dragonfly density and increase bullfrog tadpole survival. Thus, the second invader indirectly benefits the first by reducing the effect of a native predator.

CONCERNS ABOUT THE FUTURE

Human Impacts

As the human population grows in numbers, in area of land occupied, and in impacts on the land, wetlands will experience greater invasibility (see the first section of this article). More runoff and more sewage will enrich downstream water bodies, even following treatment. Eutrophic water favors the growth of invasive species and allows native vegetation to be outcompeted. At the same time as the potential for invasion and establishment are enhanced, increased globalization will foster introductions of more alien species.

Climate Change

A model of the potential future distribution invasive species in the western United States predicts that tamarisk (salt cedar, *Tamarix* spp.) will expand considerably, because most of the riparian lands of the western United States would be climatically suitable for it by 2100, while only 2 percent of the currently invaded land would become unsuitable. Projections for other wetland and riparian invaders are needed using models that include more estimates of future environments. Anticipated are greater extremes of precipitation, increased storminess, and land use intensification.

Novel Ecosystems

Given that human- and climate-change-induced impacts to the land will eventually be felt in low-lying areas, the world's wetlands will increasingly become novel ecosystems. At a minimum, the soils will be more enriched with nutrients, sediments, and contaminants, and the waters will have more altered chemistry. There will also be more roads, increased habitat fragmentation, and altered fire regimes. Native biota will form new assemblages, and new alien species will invade.

Novel environments with novel species will require novel management approaches. Jack Ewel argues that noninvasive alien species can play important roles in novel ecosystems—for example, as nurse crops to reduce colonization by invasives, trees that can attract birds to disperse native plant seeds, grasses that can fuel control burns, nitrogen fixers that will enrich barren sites, remediators of toxic soil, and substitutes for services previously performed by extinct species. Land stewards might need to use alien species to build resilience into restored and constructed wetlands.

RESTORATION OF INVADED WETLANDS

Controlling Invaders

Herbicides, mechanical removal, fire, and biological control agents are widely used to manage wetland invasions, but none is problem-free. A constraint in wetlands is that herbicides or their surfactants can be toxic to valued aquatic invertebrates or fish or birds. Mechanical approaches (pulling, mowing, plowing) are difficult in wet soils, which tend to mire both people and machines. In temperate climates, machines are less damaging when used in winter, over frozen soil. Prescribed burning is effective for managing some woody species, especially nonresprouters, but controlling fires in wetlands can be hazardous to both people and fire-control equipment. Biocontrol measures are allowed for purple loosestrife in the United States, but not for invasive strains of native species (e.g., common reed and reed canary grass) due to the risk of affecting native strains.

In Michoacán, Mexico, *Typha domingensis* is harvested for fodder and for basket-making and other crafts. Four harvests per year are possible during the long growing season. In field experiments, frequent shoot removal effectively depleted starch reserves in rhizomes and reduced *Typha* growth, to the benefit of native plant diversity. Harvesting of invasive wetland plants for biofuels is often suggested as a control measure in the United States, but mechanical harvesting is not compatible with wet soil.

For species that have become dominant by escaping native predators, it is possible to gain control by introducing predators (after extensive research to rule out side effects). Release of multiple predators to control purple loosestrife (*Lythrum salicaria*) substantially diminished invasive populations in the midwesern United States.

Restoring Native Biota

Wetland restoration is challenging because efforts tend to disturb the site and create conditions that are suitable for invasion. Allowing species to colonize on their own (self-restoration) can also allow invasion. The largest self-restoration of wetlands under way is the rewetting of the Central Marshes in Iraq, along the Iran border. Deliberate drainage of over 20,000 km² in the 1990s caused salts and contaminants to concentrate in the surface soil. Local residents began rewetting the marshes in 2003 by breaching levees, but without planting. Two salt-tolerant native species quickly reinvaded (common reed and *Typha domingensis*), but other colonizers were new to the region

(e.g., *Hydrilla verticillata*). Novel conditions (e.g., saline soils) tend to attract invasive alien species.

Efforts to replace the invasive species typically aim to restore predisturbance vegetation, when such targets are known. Often, native species of plants are introduced, rather than relying on self-restoration. But even where aggressive, competitive native species are planted, reinvasion remains a threat. Historical levels of diversity are rarely achieved. A broader goal (e.g., keeping all the native species somewhere in the region) is more achievable.

Restoring Functions

The dollar value of functions provided by restored wetlands exceeds that of all other ecosystems by a factor of ten, according to Dodds and colleagues. Disturbance regulation (storm protection, flood control, and drought recovery) and nutrient cycling (storage, cycling, and processing of nutrients; nitrogen fixation; and the nitrogen and phosphorus cycles) are at the top of the list of services provided by wetlands. It is unclear how invasive plants affect the magnitude of these functions. While invasive species tend to be more productive than natives, and potentially more effective at erosion control and more resilient to disturbance, the deliberate use of invasive species to restore such ecosystem services is incompatible with the goal of restoring the biodiversity-support function.

Adaptive Restoration

Efforts to reverse even small-scale invasions can require large, long-term investments in resources. Even where invasive monotypes have been eradicated and the native species restored, the methods might not be effective in another place or time, because both the environmental conditions and the invaders will likely differ. Adaptive restoration can restore native vegetation while learning how best to do so. Restoration of wetlands dominated by invasive plants becomes adaptive when planners begin by listing the unknowns that need to be known to control invaders and restore native vegetation. The next step is to prioritize the questions and develop experimental approaches to answer each question. Next, one designs an overall project that can accommodate a series of modules, so that early experimental results can inform later restoration and experimentation. The first module is often small and focused, perhaps comparing methods that might control the invader. The next module involves a larger field experiment using the control mechanism selected from the first module combined with new tests of which native species can tolerate current site conditions. A third module could then test alternatives for native vegetation reestablishment, assemblage comparison, planting methods, plant spacing, or soil amendments. Throughout the process, an adaptive restoration team would use new information to decide what to do next, modifying restoration in response to the outcomes of previous modules.

Because wetlands produce ecosystem services of extraordinary value, and because they are highly vulnerable to invasions of both plants and animals, it is justifiable that wetland enhancement and restoration be given high priority. Restrictions on discharges to wetlands (e.g., under the U.S. Clean Water Act) and requirements for buffers around wetlands provide additional protection. Given that wetland environments are continually changing and the native biota are challenged by physical and chemical variations as well as invasive species, wetland restoration sites offer major opportunities for research to control invasive species and sustain biodiversity and ecosystem functions.

SEE ALSO THE FOLLOWING ARTICLES

Competition, Plant / Dispersal Ability, Plant / Genotypes, Invasive / Hydrology / Invasional Meltdown / Land Use / Vegetative Propagation / Water Hyacinth / Zebra Mussel

FURTHER READING

Braithwaite, R. W., W. M. Lonsdale, and J. A. Estbergs. 1989. Alien vegetation and native biota in tropical Australia: The spread and impact of "*Mimosa pigra.*" *Biological Conservation* 48: 189–210.

Dodds, W. K., K. G. Wilson, R. L. Rehmeier, G. L. Knight, S. Wiggam, J. A. Falke, H. J. Dalgleish, and K. N. Bertrand. 2008. Comparing ecosystem goods and services provided by restored and native lands. *BioScience* 58: 837–845.

Geiger, W., P. Alcorlo, A. Baltanás, and C. Montes. 2005. Impacts of an introduced crustacean on the trophic webs of Mediterranean wetlands. *Biological Invasions* 7: 49–73.

Lavergne, S., and J. Molofsky. 2007. Increased genetic variation and evolutionary potential drive the success of an invasive grass. *Proceedings of the National Academy of Sciences, USA* 104: 3883–3888.

Millennium Ecosystem Assessment. 2005. *Ecosystems and Human Well-Being: Wetlands and Water Synthesis.* Washington, DC: World Resources Institute.

Tickner, D. P., P. G. Angold, A. M. Gurnell, and J. O. Mountford. 2001. Riparian plant invasions: Hydrogeomorphological control and ecological impacts. *Progress in Physical Geography* 25: 22–52.

van der Velde, G., S. Rajogopal, M. Kuyper-Kollenaar, A. bij de Vaate, D. W. Thieltges, and H. J. MacIsaac. 2006. Biological invasions: Concepts to understand and predict a global threat (61–90). In R. Bobbink, B. Beltman, J. T. A. Verhoeven, and D. F. Whigham, eds. *Wetlands: Functioning, Biodiversity Conservation, and Restoration.* Ecological Studies, Vol. 191. Berlin: Springer-Verlag.

Zedler, J. B., and S. Kercher. 2004. Causes and consequences of invasive plants in wetlands: Opportunities, opportunists, and outcomes. *Critical Reviews in Plant Sciences* 23: 431–452.

Zedler, J. B., and S. Kercher. 2005. Wetland resources: Status, ecosystem services, degradation, and restorability. *Annual Review of Environment and Resources* 30: 39–74.

WHITE LISTS

SEE BLACK, WHITE, AND GRAY LISTS

XENOPHOBIA

PETER COATES
University of Bristol, United Kingdom

Discussion of invasive introduced species is not confined to biologists and natural resource managers. Nor does the debate take place in a social and cultural vacuum. The subject attracts widespread attention because it is often treated as part of a larger debate, scholarly and public, about the impact of immigration. Those who regard scientific and popular concern over invasive introductions as exaggerated have a particular tendency to draw connections between attitudes to human and nonhuman immigrants. A typical strategy is to level the charge of xenophobia against those advocating action to control "problem" species. Xenophobia, a virulent fear and pathological dislike of foreigners and things foreign or alien, can take the form of extreme acts of violence such as pogroms, genocide, and "ethnic cleansing." Xenophobia may also be expressed in more trivial ways. Anti-German sentiment in the United States during World War I caused sauerkraut to be renamed "liberty cabbage," and resentment at French noncooperation during the invasion of Iraq in 2003 was vented in the reinvention of French fries as "freedom fries."

NATIVISM: A RELATED CONCEPT

A related concept, often used interchangeably with xenophobia, is nativism: an acute, defensive-aggressive manifestation of nationalism stemming from perception of certain immigrant or national groups as threats to the existing ethnoracial composition of a nation-state and its socioeconomic, political, and cultural condition. One early outbreak of nativism in the United States was an Anglo-American Protestant backlash against Irish Catholics in northeastern U.S. cities in the 1840s. However, the term is more usually associated with late nineteenth and early twentieth century opposition of "old stock" Americans to the "new" immigration of Catholics and Jews from southern and eastern Europe, as well as, on the west coast, efforts to exclude newcomers from China and Japan. Xenophobic and nativist sentiments and activities pertaining to immigration have not been confined to the United States. Britain's Aliens Act of 1905 sought to exclude and expel unwanted elements among its own "new" immigrants, while the "White Australia" immigration policy, formalized in 1901, barred Asians on racial grounds.

HUMAN AND OTHER IMMIGRANTS

The story of human immigration is just one aspect of the process whereby North America and other so-called neo-Europes (South Africa, southern South America, Australia, and New Zealand) were populated by nonnatives. Europeans were accompanied and often supported, directly and indirectly, by a "biotic portmanteau" (Alfred Crosby's phrase, in his seminal study of species transfer, *Ecological Imperialism: The Biological Expansion of Europe, 900–1900* [1986]) of flora and fauna, ranging from livestock, crops, and game animals to insect pests, weeds, and plant pathogens. In fact, the two eras in U.S. history that experienced the greatest volume of floral and faunal introductions since the initial Columbian Exchange—the 1850s to the 1920s and the period since the 1960s—coincide with the peak periods of human immigration. Given the parallel and complementary nature of these migrations, and the vocabulary of colonization, transplantation, invasion, and naturalization that those who talk about immigration share with commentators on biological introductions, it is not surprising that perception and discussion of human arrivals both "desirable" and "undesirable" and of their nonhuman counterparts frequently overlap. The long-established common names of many invasive species highlight this fusion between human and nonhuman realms and underscore their foreign character: the Argentine ant, the English sparrow, German ivy, the Hessian fly, Japanese knotweed, Russian thistle.

Our tendency to express ourselves through metaphors reflects and intensifies this ingrained habit (Fig. 1). A metaphor is a figure of speech comparing two ostensibly unrelated things; it is a linguistic device whose force relies on the established connotations of the comparative entity (the "English" sparrow is a dirty foreigner, to adopt the parlance of the bird's enemies in the late nineteenth century American "Sparrow War"). The simile, another comparative technique, preserves the nominal distinction between the two subjects (the "English" sparrow is *like* a dirty foreigner), but the

effect, in practice, is the same. Despite its pervasiveness in debates about invasive introductions, the precise role of analogizing through metaphor and simile is less clear.

CRITIQUES OF INVASIVE INTRODUCED SPECIES AND XENOPHOBIA: HISTORICAL CONNECTIONS

The function of metaphor and the process of conflation are most transparent in responses to invasive introductions in the United States between the 1850s and the 1920s. During this period of heightened anxiety over mass immigration from nontraditional sources, attitudes to nonhuman introductions were shaped by cultural impulses as well as ecological understandings. The standard critical ploy humanized a perceived natural threat through analogy with the supposedly low-living, high-breeding behavior of the unwelcome and injurious human stranger who menaced the "native"-born American (Fig. 2). One native bird conservationist reviled the "English" sparrow (*Passer domesticus*, introduced in 1852) as the "feathered embodiment of those instincts and passions which belong to the lowest class of foreign immigrants" (1892), and a Chicago journalist characterized the Japanese beetle (*Popillia japonica*), a notorious 1920s plant pest, as "the real yellow peril."

Nonetheless, this practice of ethnic stigmatization served to intensify rather than create hostility. The English sparrow (and its "Cockney cousin," the starling, *Sturnus vulgaris*, introduced in 1890) may have been unduly demonized; there were additional reasons for native bird decline. But this strategy of vilification through negative humanization tied up with notions of national identity was supplementary and was never adopted in the absence of palpable material foundations of objection, not even in xenophobia's heyday. An historian of biology (Philip Pauly) has emphasized the close proximity in time between legislation to curb invasive introductions, such as the Lacey Act (1900) and various state and federal plant quarantine laws, and the imposition of the first numerical quotas on selected groups of immigrants in 1921 and 1924. He points out, for example, that the Chinese Exclusion Act of 1882 followed hot on the heels of California's plant quarantine law of 1881. This is circumstantial evidence. Measures to protect U.S. agriculture and native birds would have been enacted even if there had not been a single Italian, Polish, or Jewish immigrant in the United States. Likewise, the imposition of immigrant quotas did not require or rely on the presence of English sparrows, gypsy moths, chestnut blight, or citrus canker.

NOT WANTED

Zebra Mussel Outlaws

Threats to the West ~ Why Be Concerned?

Zebra mussels cause devastating impacts on municipal water systems, recreation and fisheries. Currently, they are widespread in Eastern USA and as far west as Oklahoma. We don't want these outlaws in California where they would rapidly reproduce and cause millions of dollars in damage to our water resources and recreation. We need your help to stop these mussels from entering our lakes, rivers and streams.

HOW COULD THESE OUTLAWS 'RIDE' HERE?

Attach to boat hulls and motors.

On infested recreational boats and commercial boat haulers from infested waters like the Mississippi River and Great Lakes.

HOW CAN WE ARREST THE SPREAD?

Learn how to identify zebra mussels (see sidebar).
Remove all aquatic plants and animals from boat, motor, trailer, and equipment.
Drain water from livewells, bilge, and motor.
Dispose of unwanted live minnows and worms in the trash.
Rinse boat and equipment with high pressure or hot water, especially if moored for more than a day, OR
Dry everything for at least 5 days.
Never launch watercraft with a suspected infestation.
Report sightings on watercraft or in a lake or river – note location, place mussel in a sealed container with rubbing (isopropyl) alcohol, and call the Zebra Mussel Watch Hotline, 1-888-840-8917.

Found only in freshwater. Small barnacle-like clams with dark and light colored stripes.

Cost millions of dollars each year to control in power plants and water delivery systems.

VOLUNTEER FOR A POSSE

Early detection is key to preventing and mitigating impacts of zebra mussels. If you would like to help as a volunteer monitor to protect your lake or river, please contact:

Zebra Mussel Watch Program
1 (888) 840-8917 (toll free)
mussel@water.ca.gov

Cover crayfish and clams, and outcompete native species for food and habitat.

 CALIFORNIA BAY-DELTA AUTHORITY

FIGURE 1 To raise public awareness of the threat this species presents, California's Department of Water Resources (2005) humanizes the zebra mussel by deploying the familiar mythology of the western frontier.

FIGURE 2 In a drawing accompanying one of his best-known animal stories, about the adventures in Manhattan's urban wilderness of two English sparrows, Randy and Biddy, Ernest Thompson Seton evokes popular hostility toward this successful immigrant species. (From *Lives of the Hunted*, New York: Charles Scribner's, 1901, p. 125.)

Pauly also claims, with reference to incidents such as the rejection of the original gift of flowering cherry trees from Japan to Washington, D.C., in 1910, that the personal racial bias of decision makers such as the federal government entomologist Charles Marlatt clouded their scientific judgment. Yet when the Japanese authorities closely followed his advice for rigorous selection, inspection, and fumigation to exclude root gall worm, fungus mycelium, and a host of other harmful pests, the second shipment of trees was readily admitted.

CRITIQUES OF INVASIVE INTRODUCED SPECIES AND XENOPHOBIA: RECENT AND CONTEMPORARY CONNECTIONS

Since the early 1990s, a revival of nativism (although barely comparable in scale to the previous wave) has accompanied an upsurge in immigration to North America and western Europe, producing measures such as California's Proposition 187 (1994) and growing support for anti-immigrant political parties in France, Germany, and Italy. The past two decades have also been marked by growing concern for the future of native plants. An assortment of commentators have responded to these two phenomena by assailing critics of invasive introductions with assorted charges of botanical xenophobia, biological nativism, plant racism, and plant Nazism. They also question whether a rigorous scientific case against invasive introductions can be made and explain support for natives as a matter of ideological preference. These assailants include

Philip Pauly, bestselling writer Michael Pollan, philosopher Mark Sagoff, sociologists of science Jonah Peretti and Banu Subramaniam, and German landscape historians Gert Gröning and Joachim Wolschke-Bulmahn.

Grafted onto Nazi ideology was a sinister plant sociology, stemming from landscape architects Willy Lange and Alwin Seifert, that aimed to purify German nature by uprooting non-Nordic plants. Today, along the banks of British rivers, volunteer working parties of native plant enthusiasts hack away at the riotous growth of Himalayan balsam (*Impatiens glandulifera*, a horticultural import from the 1830s), while their American counterparts rip out Scotch broom (*Cytisus scoparius*), another escaped ornamental, from California grasslands. Tangible evidence that native flora advocates in Britain, the United States, or Germany are the Nazis' ideological heirs, wittingly or unwittingly, is lacking, however. "Weed whacking" is not a natural extension of immigrant bashing (or a surrogate form). Headline-grabbing accusations that fall short of invoking fascism are equally strained and unsubstantiated. Immigration restriction organizations and far right groups have no views or policies on invasive introductions. And native plant societies have no immigration policy. The frequently articulated argument that concern about invasive species is underpinned by a belief that all introduced species are innately undesirable holds no water either. The implications that nationalism and nativism are synonymous and that eco-nativism provides a medium for expressing discrimination in officially postracist and multicultural societies withers under scrutiny, too.

Self-styled "multihorticulturalists" like Pollan overemphasize rhetorical parallels; there is no unsavory hidden agenda. And some feel that references to hordes of illegal aliens flooding in and menacing innocent and well-behaved native species that occupy their rightful place have more or less been reduced to banal figures of speech and toothless journalistic cliché. Others claim that the human–nonhuman comparison is a vital tool for raising public awareness of a pressing conservation problem. Metaphors and analogies are useful for transferring meaning from the specialist to the lay sphere. Yet they inevitably simplify and exude ambiguity, and they are often saturated with connotations from other contexts in which their use is more established, associations that the popular media magnifies and dramatizes. Provocative book titles like *Alien Invasion* and vernacular terms such as "killer bee" (Africanized honeybee, *Apis mellifera scutellata*) and "killer algae" (*Caulerpa taxifolia*) certainly inflame current debates over the impact of invasive introductions.

With crossover language so rife, allegations of biological xenophobia persist. Some commentators further contend that, because unreflective use of metaphor has such a strong ability to embed patterns of thinking, the deeply resonant historical analogy with human immigration has left an ingrained "cognitive residue" that perpetuates (however unintentionally) a climate of xenophobia and nativism. These observers also worry that casual, ill-considered use of strongly flavored language—which includes military terminology—by scientists and natural resource managers who ought to know better, as well as by glib journalists, impedes dispassionate discussion and the public's ability to grasp the ecological, economic, and public health implications of invasive introductions, even eroding public respect for invasion biology. What invasion biologists and conservation managers need, they assert, is a nontendentious terminology that does not compromise the perfectly legitimate case for acting against invasive introductions—or even an alternative set of metaphors that acknowledges how humans have largely created the conditions for invasive successes and how much invasive species reflect our own priorities.

In defending themselves against the continuing charge of xenophobia and backward-looking parochialism that clings to an outmoded vision of what (and who) belongs, those who seek to mitigate the effects of invasive species have adopted a fresh emphasis. They present their cause as part of a socially and culturally progressive and ethically sound desire to preserve regional distinctiveness and embattled minorities in a world of rampant globalization in which we face the threat of suffocating sameness. In Britain, to combat the gray squirrel (*Sciurus carolinensis*, a nineteenth century American import) in favor of its numerically much smaller red native counterpart (*Sciurus vulgaris*) involves the same commitment to cultural survival, community identity, and tasty diversity that fuels the championing of local cheeses and apples against the tasteless universalism of international agribusiness's products. Welcoming the tea shop's brave reappearance in London amid the sprawling American empire of a coffee shop chain, and casting around for an evocative analogy, a journalist described Starbucks as "the Yankee grey squirrel of the high street, which has reduced our native tea and coffee shops to a few redoubts in the North" (George Trefgarne, "Is It Time for Tea?," *The Spectator*, 17 January 2007).

Britain's ongoing "squirrel war" is a salutary reminder that not all invasive species are "old world" in origin and that the problem of invasive species is not confined to "new worlds" that Europeans and their descendants colonized. A number of the most vigorously contested invasive introductions in the United Kingdom are deliberate transplantations from North America: the gray squirrel (1870s), muskrat (1920s), mink (1940s), ruddy duck (1940s), signal crayfish (1970s), and American bullfrog (1990s). The list of *Top Ten Most Wanted Foreign Species* ("that have overstayed their environmental visa") compiled by the U.K. government's Environment Agency (August 2006) features the American signal crayfish in second place, and the American mink ranks at number three. Unable to resist the famous British (male) characterization of American troops stationed in Britain during and after World War II as "overpaid, overfed, oversexed, and over here," the agency warned that, from the standpoint of the native white-clawed crayfish and the water vole, the signal crayfish and mink were "oversized, oversexed, and over here." No offense to Britain's closest ally was intended, and none was taken.

SEE ALSO THE FOLLOWING ARTICLES

Birds / Laws, Federal and State / Rodents / Zebra Mussel

FURTHER READING

Coates, P. 2006. *American Perceptions of Immigrant and Invasive Species: Strangers on the Land*. Berkeley: University of California Press.
Gröning, G., and J. Wolschke-Bulmahn. 2003. The native plant enthusiasm: Ecological panacea or xenophobia? *Landscape Research* 28: 75–88.
Hettinger, N. 2001. Exotic species, naturalisation, and biological nativism. *Environmental Values* 10: 193–224.
O'Brien, W. 2006. Exotic invasions, nativism, and ecological restoration: On the persistence of a contentious debate. *Ethics, Place and Environment* 9: 63–77.
Pauly, P. J. 1996. The beauty and menace of the Japanese cherry trees. *Isis* 87: 51–73.
Peretti, J. H. 1998. Nativism and nature: Rethinking biological invasion. *Environmental Values* 7: 183–192.
Pollan, M. 1994. Against nativism. *New York Times Magazine* (15 May): 52–55.
Simberloff, D. 2003. Confronting introduced species: A form of xenophobia? *Biological Invasions* 5: 179–192.
Simberloff, D. 2006. Invasional meltdown 6 years later: Important phenomenon, unfortunate metaphor, or both? *Ecology Letters* 9: 912–919.
Subramaniam, B. 2001. The aliens have landed! Reflections on the rhetoric of biological invasions. *Meridians: Feminism, Race, Transnationalism* 2(1): 26–40.

YELLOW FEVER

SEE FLAVIVIRUSES

Z

ZEBRA MUSSEL

LADD ERIK JOHNSON

Université Laval, Québec, Canada

The zebra mussel, *Dreissena polymorpha*, is a small (1–3 cm shell length), freshwater, filter-feeding bivalve (Fig. 1) whose overall appearance belies its iconic status among aquatic invasive species, especially in North America. Its appearance there, along with a congener, the quagga mussel (*D. rostriformis bugensis*; see below), in the late 1980s was followed by such dramatic ecological and economic impacts that an act of the U.S. Congress was passed to study and control "aquatic nonindigenous nuisance species" and prevent their establishment and spread. Although this legislation and the dramatic infusion of funding that accompanied it were directed at all aquatic invaders, the reality was an impressive undertaking of basic and applied research on the zebra mussel itself, as evidenced by over a decade of international conferences dedicated wholly or partly to the subject (now a more general conference known as ICAIS, the International Conference on Aquatic Invasive Species). It is arguably the most well-studied invasive introduced species in the world, and certainly in aquatic ecosystems.

HISTORICAL OVERVIEW

All this attention on the North American invasion has neglected the long invasive history of this species in Europe, including many countries of the former Soviet Union, over the past 200 years. Native to the Ponto-Caspian region of Eurasia, the zebra mussel began its long

march west in the early 1800s due largely to the extensive construction of canals that linked Europe's large rivers and lakes and promoted commerce. By the 1820s, populations were already established in England, and the invasion continued throughout the continent with more recent invasions of Ireland (1994) and Spain (2001), further demonstrating that this is an ongoing process. Although it is a long-established and well-studied feature of the European aquatic fauna, its reputation as an invader and nuisance species was, however, possibly underappreciated, although in light of the general rule that "the best predictor of invasiveness is a history of invasiveness elsewhere," predictions of the eventual appearance of the zebra mussel in North America had periodically been made over the twentieth century before its arrival. These came true in the late 1980s when a population was discovered in Lake St. Clair, likely the result of the release of larval stages

FIGURE 1 The zebra mussel showing its characteristic striped shell and gregarious nature. (Photograph by Scott Camazine, courtesy of New York Sea Grant.)

in the ballast of international ships or possibly of adults taken up by anchors.

A WINNING FORMULA

The success of this invader can be attributed principally to two characteristics it shares with its distant marine relatives: the ability to attach to solid surfaces with fine hair-like filaments (known as byssal threads) and a complex life cycle that includes a microscopic larval stage. The first characteristic gives zebra mussels the ability to form multilayered beds on almost any surface, and thus to create large populations, which can then affect both directly and indirectly a wide range of species and ecological processes (see below). This "biofouling" nature, ubiquitous in marine systems, is almost nonexistent in freshwater systems, being limited mostly to microscopic organisms such as diatoms. The ability to attach to objects, including other zebra mussels, allows it to create a matrix of living and dead material up to several centimeters thick—an extraordinary example of ecosystem engineering. Moreover, zebra mussels can exploit shell debris, aquatic plants, and even litter (Fig. 2) as footholds from which they can spread out over soft-sediment habitats. While the physical matrix thus created can serve as a habitat for other creatures, it can also cause ecological and economic problems (see below). The other "marine" characteristic is the planktotrophic life cycle, which includes a free-living larval stage that spends weeks in the water column feeding on microscopic algae and detritus. This strategy of producing self-sufficient offspring not only increases the number of gametes produced per adult (>10^6/female/year in some cases) but

FIGURE 2 Zebra mussels take the "Pepsi Challenge." Biofouling of litter found in a midwestern lake in the United States. (Photograph courtesy of C. Kraft.)

also permits widespread dispersal within a water body. While high fecundity itself does not assure rapid population growth and range expansion (mortality rates for planktotrophic larvae are exceedingly high), it creates this potential if environmental conditions are right. This combination of high dispersal and high reproductive rates thus provides the potential for impressive invasions: for example, in just three years after its initial detection in North America, the zebra mussel had spread west to Lake Superior and east to the Hudson River, although there may have been more than one initial introduction involved.

ZEBRAS VS. QUAGGAS

While the zebra mussel has grabbed public attention in North America, a parallel invasion by a congener, the quagga mussel, has been simultaneously occurring there. Despite arriving at around the same time and having the same general characteristics described above, subtle differences in its morphology and ecology have allowed it to exploit certain habitats better, especially those colder and calmer habitats of deeper waters; indeed, it is now outcompeting the zebra mussel in many places, especially in soft-sediment habitats.

DISPERSAL AND RANGE EXPANSION

Given the planktonic nature of the larval stage, the rapid colonization of a given water body is not surprising, as the larvae can be transported great distances during their time in the water column; over the long term, however, the maintenance of populations in rivers or streams is problematic, as the larvae are unable to move upstream by their own means. Therefore, unless there is a headwater lake that can "feed" downstream populations with propagules, such populations generally cannot be established or maintained. This problem has been largely circumvented by humankind's predilection for damming rivers, thereby creating reservoirs or impoundments that can serve the same purpose. For this reason, large rivers in both Europe and North America are able to support substantial populations of zebra mussels. In addition to natural dissemination, range expansion within a body of water can be accelerated by human activities, especially the movement of ships, boats, and barges. Indeed, the fouling nature of the zebra mussel permits it to "hitchhike" on any object that might be transported by currents (e.g., wood, aquatic plants) or human beings. Such human-mediated dispersal is certainly responsible for its rapid range expansion in the contiguous waterways of Europe and North America

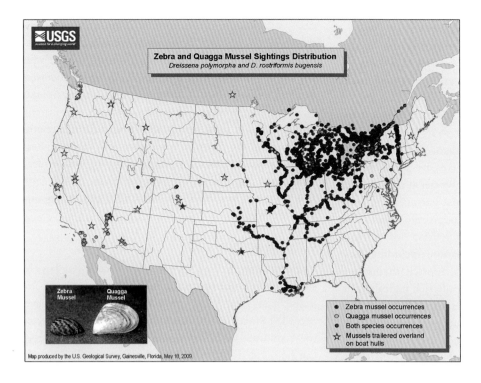

FIGURE 3 Distribution of sightings of zebra and quagga mussels across North America through 2009. (Map courtesy of the U.S. Geological Service.)

(e.g., the Laurentian Great Lakes; Fig. 3), which has been furthered by the construction of canals that connected adjacent watersheds (e.g., the Erie Canal in New York and the Dnieper–Bug Canal in Belarus).

This propensity for hitchhiking is also the key to the overland dispersal of zebra mussels. Unlike some other freshwater organisms, zebra mussels have no specific adaptations for overland dispersal (e.g., the ephippia stage of cladocerans) and thus must rely on the chance transport of individuals or submerged fouled objects. While natural mechanisms do exist (e.g., waterfowl flying between lakes), these vectors are dwarfed by the traffic of boats and the movement of other material between lakes and rivers resulting from human activities. Of the many possibilities, the transfer of objects or structures that have been left in water for long periods of time (e.g., docks, barges) is the most likely way in which a large founding population can be established; however, even things that are only in the water for a short time can pick up mussels hitchhiking on other objects—for example, growing on aquatic macrophytes that easily become entangled in boat trailers (Fig. 4) and may even be nonindigenous species themselves! Certainly, the scale and magnitude of recreational boating and associated fishing activity makes it the leading contender for the principal vector of secondary spread of this invader. The extensive databases on boat use and the discrete nature of

aquatic habitats have permitted scientists to develop and apply gravity dispersal models to predict the spread of the zebra mussel across regional "landscapes," especially when combined with appropriate habitat suitability models. Rare events and unlikely vectors must, however, still be taken into consideration and may be responsible for some of the long-distance jumps that have greatly expanded the extent of invasions, for example the recent establishment of populations of quagga and zebra mussels in California—thousands of kilometers from the nearest population.

FIGURE 4 The entanglement of aquatic macrophytes on boat trailers—one means of overland "hitchhiking" for zebra mussels and other aquatic organisms. (Photograph courtesy of the author.)

ECOLOGICAL IMPACTS

As with many invasive introduced species, the impact of the zebra mussel on natural communities and ecosystems depends largely on its abundance, which can reach more than 20,000 m^{-2}, or even more than 100,000 m^{-2} if recent recruits are included. This is particularly true for local impacts of the zebra mussel, as its high reproductive potential and gregarious settlement permit it, like its marine cousins, to establish extensive beds, thereby forming a biological and physical barrier at the interface between the water column and the bottom. The mussels' mere physical presence can negatively affect other benthic organisms through competition for primary space or by increasing sedimentation, including feces and pseudofeces produced by the mussels themselves. The accumulation of this organically rich material can, however, benefit certain organisms (e.g., detritivores), as can the matrix of shells, which can provide microhabitats for mobile invertebrates such as snails and amphipods and secondary space for other sessile organisms (e.g., sponges, benthic diatoms). Unfortunately, one of the clearest and most important impacts of zebra mussels has been achieved through their use of indigenous unionids, a family of freshwater mussels, as secondary space for settlement. Unlike dreissenid mussels, unionids usually inhabit soft-sediment environments and typically have just a small portion of their shells exposed to the water; this foothold is, however, sufficient to support a biomass of zebra mussels nearly as great as the unionid itself. Either indirectly by competition for food or directly by smothering, the consequences are usually fatal for the native species, leading to a bottom littered with empty shells that then ironically provide additional substrata for zebra mussel colonization.

Beyond these immediate, more local impacts, zebra mussels can have system-wide effects on the entire body of water through their ceaseless filtration, as they extract most particles greater than 1–2 μm from the surrounding water. Adult mussels can filter from 1–2 L of water per day, a rate that leads to estimated times for filtering the entire body of water ranging from days (Hudson River) to weeks (Great Lakes). Although these rates do not take into account the fact that most estimates of densities include very small juvenile stages and that most water in or adjacent to mussel beds is simply being refiltered, the overall impact is clear: water transparency dramatically increases in areas invaded by the mussels. While initially perceived by the public as a positive outcome of the invasion, clearer does not mean cleaner, as pollution remains within the ecosystem, and in some cases, exposure to or transfer of particular pollutants may even be increased by the presence of zebra mussels. Certainly, the overall impact on the aquatic food web has been a shift from planktonic to benthic processes, as food availability for zooplankton decreases, to the benefit of the benthos. Moreover, increases in nutrient availability and light penetration have led to dramatic increases in the biomass and extent of macrophytes. Subtler changes include shifts in composition of phytoplankton assemblages toward species that are less consumable, such as colonial cyanobacteria, and changes in nutrient dynamics.

ECONOMIC IMPACTS

The attention focused on the zebra mussel has largely stemmed from the perceived impacts it has had on economic activities, which have been estimated to have exceeded $100 million. Parallel to its ecological impacts, the economic problems associated with zebra mussels can be divided into two general categories. The first is associated with the mechanical or physical effects of biofouling—in other words, the rapid accumulation of material capable of interfering with machinery or controlled water flow. This latter aspect is particularly critical for municipal drinking water and industrial cooling systems, which depend heavily on the large freshwater sources that are the natural habitat of the zebra mussel. Indeed, the provision of abundant solid surfaces and a continuous supply of water often make such facilities an ideal habitat for zebra mussels, thereby exacerbating the problem. The second area, more difficult to assess against the rhythm of natural variation, involves changes in the environment caused by the invasion and the associated impacts on the recreational or commercial exploitation of resources (e.g., fisheries). Certainly, declines or shifts in fish assemblages or the accumulation of shell debris (or worse, macrophytes) on beaches has the potential to discourage the recreational use of lakes and rivers, but humans have proven remarkably adaptable, or perhaps simply tolerant, of environmental degradation caused by their own activities, including, but not limited to, the introduction of invasive species.

CONTROL

As with most invasive species, prevention is the key to minimizing the impacts of zebra mussels, and considerable outreach efforts have been made by numerous governmental and nongovernmental organizations to this effect. Most efforts have been directed at changing the behavior of boaters and fishers who might inadvertently transfer mussels between bodies of water, and a plethora of brochures and signage has been produced and distributed to this effect. The efficacy of these efforts remains largely unassessed, but there is no doubt that between these efforts and the general publicity of the invasion, the public is much more aware of the problem of aquatic invasive species in general.

Eradication of existing populations is impossible in most cases, but owing to the isolation of many aquatic systems, early detection and aggressive rapid responses offer some hope for areas where invasion has not yet occurred. Where the mussel is already well established, control of populations in natural settings is largely inconceivable because of the high connectivity between distant populations provided by the larval stage. Predators and parasites do exist (some nonindigenous themselves), but zebra mussel populations appear principally controlled by ecological processes that determine the availability of food resources.

More success has been achieved in controlling populations affecting the artificial habitats created by municipal and industrial facilities. Whereas a creative panoply of ideas (e.g., ultrasound, cayenne pepper) has been assessed, it appears that chlorine, the old water treatment standby, is still the most effective and manageable solution; substantial progress has been made in targeting the more vulnerable life stages with appropriate chemical levels that minimize collateral environmental problems (e.g., dechlorinization).

SUMMARY

While the zebra mussel has, over the past two centuries, fundamentally altered ecosystems on two continents, its arrival and rapid spread across the North American landscape came at a time when the importance of invasive introduced species was just beginning to gain substantial attention from both the scientific community and the general public. Its timing could not have been better, and it has arguably become the "poster child" of aquatic invasions. Although the invasion continues, it must remembered that the zebra mussel is but one in a long line of aquatic invaders (the Asian clam, *Corbicula fluminea*, and the sea lamprey, *Petromyzon marinus*, being two more historic cases in North American freshwaters) that will continue to arrive if the lessons learned from this invasion are not well applied.

SEE ALSO THE FOLLOWING ARTICLES

Ballast / Canals / Great Lakes: Invasions / Invasion Economics / Invertebrates, Marine / Lakes / Ponto-Caspian: Invasions / Rivers

FURTHER READING

D'Itri, F. M., ed. 1997. *Zebra Mussels and Aquatic Nuisance Species*. Baton Rouge, LA: CRC Press.

Johnson, L. E., and J. T. Carlton. 1996. Post-establishment spread in large-scale invasions: Dispersal mechanisms of the zebra mussel *Dreissena polymorpha*. *Ecology* 77: 1686–1690.

Karatayev, A. Y., L. E. Burlakova, and D. K. Padilla. 2002. Impacts of zebra mussels on aquatic communities and their role as ecosystem engineers (433–446). In E. Leppäkoski, S. Gollasch, and S. Olenin, eds. *Invasive Aquatic Species of Europe: Distribution, Impacts and Management*. Dordrecht: Kluwer Academic Publishers.

Leung, B., J. M. Bossenbroek, and D. M. Lodge. 2006. Boats, pathways, and aquatic biological invasions: Estimating dispersal potential with gravity models. *Biological Invasions* 8: 241–254.

Nalepa, T. F., and D. W. Schlosser, eds. 1993. *Zebra Mussels: Biology, Impacts and Control*. Boca Raton, FL: Lewis Publishers.

Roberts, L. 1990. Zebra mussel invasion threatens United States waters. *Science* 249: 1370–1372.

Strayer, D. L. 2009. Twenty years of zebra mussels: Lessons from the mollusk that made headlines. *Frontiers in Ecology and the Environment* 7: 135–141.

Ward, J. M., and A. Ricciardi. 2007. Impacts of *Dreissena* invasions on benthic macroinvertebrate communities: A meta-analysis. *Diversity and Distributions* 13: 155–165.

100 OF THE WORLD'S WORST INVASIVE ALIEN SPECIES
A Selection From The Global Invasive Species Database

Note: Species were selected for the list using two criteria: their serious impact on biological diversity or human activities, and their illustration of important issues of biological invasion. To ensure a wide variety of examples, only one species from each genus was selected. Absence from the list does not imply that a species poses a lesser threat. Development of the *100 of the World's Worst Invasive Alien Species* list has been made possible by the support of the Fondation d'Entreprise TOTAL. For further information on these and other invasive alien species, consult the *Global Invasive Species Database* (1998–2000) at www.issg.org/database.

MICROORGANISMS

Banana bunchy top virus (BBTV)
Plasmodium relictum (avian malaria)
Rinderpest virus (cattle plague)

FUNGI AND OOMYCETES

Aphanomyces astaci (crayfish plague)
Batrachochytrium dendrobatidis (frog chytrid fungus)
Cryphonectria parasitica (chestnut blight)
Ophiostoma ulmi (Dutch elm disease)
Phytophthora cinnamomi (phytophthora root rot)

AQUATIC PLANTS

Caulerpa taxifolia (caulerpa seaweed)
Eichhornia crassipes (water hyacinth)
Spartina anglica (common cordgrass)
Undaria pinnatifida (wakame seaweed)

LAND PLANTS

Acacia mearnsii (black wattle)
Ardisia elliptica (shoebutton ardisia)
Arundo donax (giant reed)
Cecropia peltata (pumpwood)
Chromolaena odorata (Siam weed)
Cinchona pubescens (quinine tree)
Clidemia hirta (Koster's curse)
Euphorbia esula (leafy spurge)
Fallopia japonica (Japanese knotweed)
Hedychium gardnerianum (Kahili ginger)
Hiptage benghalensis (hiptage)
Imperata cylindrica (cogon grass)
Lantana camara (lantana)
Leucaena leucocephala (leucaena)
Ligustrum robustum (tree privet)
Lythrum salicaria (purple loosestrife)
Melaleuca quinquenervia (melaleuca)
Miconia calvescens (miconia)
Mikania micrantha (mile-a-minute weed)
Mimosa pigra (mimosa)
Morella (Myrica) faya (fire tree)
Opuntia stricta (erect prickly pear)
Pinus pinaster (cluster pine)
Prosopis glandulosa (mesquite)
Psidium cattleianum (strawberry guava)
Pueraria montana var. *lobata* (kudzu)
Rubus ellipticus (yellow Himalayan raspberry)
Schinus terebinthifolius (Brazilian pepper tree)
Spathodea campanulata (African tulip tree)
Sphagneticola trilobata (wedelia)
Tamarix ramosissima (salt cedar)
Ulex europaeus (gorse)

(Continued on next page)

(Continued from previous page)

AQUATIC INVERTEBRATES

Asterias amurensis (northern Pacific seastar)
Carcinus maenas (green crab)
Cercopagis pengoi (fish hook flea)
Dreissena polymorpha (zebra mussel)
Eriocheir sinensis (Chinese mitten crab)
Mnemiopsis leidyi (comb jelly)
Mytilus galloprovincialis (Mediterranean mussel)
Pomacea canaliculata (golden apple snail)
Potamocorbula amurensis (marine clam)

LAND INVERTEBRATES

Aedes albopictus (Asian tiger mosquito)
Anopheles quadrimaculatus (common malaria mosquito)
Anoplolepis gracilipes (crazy ant)
Anoplophora glabripennis (Asian longhorned beetle)
Bemisia tabaci (sweet potato whitefly)
Cinara cupressi (cypress aphid)
Coptotermes formosanus shiraki (Formosan subterranean termite)
Euglandina rosea (rosy wolf snail)
Linepithema humile (Argentine ant)
Lissachatina fulica (formerly *Achatina fulica*) (giant African snail)
Lymantria dispar (gypsy moth)
Pheidole megacephala (big-headed ant)
Platydemus manokwari (flatworm)
Solenopsis invicta (red imported fire ant)
Trogoderma granarium (khapra beetle)
Vespula vulgaris (common wasp)
Wasmannia auropunctata (little fire ant)

AMPHIBIANS

Eleutherodactylus coqui (Caribbean tree frog)
Rana catesbeiana (bullfrog)
Rhinella marina (*Bufo marinus*) (cane toad)

FISHES

Clarias batrachus (walking catfish)
Cyprinus carpio (carp)
Gambusia affinis (Western mosquito fish)
Lates niloticus (Nile perch)
Micropterus salmoides (large-mouth bass)
Oncorhynchus mykiss (rainbow trout)
Oreochromis mossambicus (Mozambique tilapia)
Salmo trutta (brown trout)

BIRDS

Acridotheres tristis (Indian myna)
Pycnonotus cafer (red-vented bulbul)
Sturnus vulgaris (starling)

REPTILES

Boiga irregularis (brown treesnake)
Trachemys scripta (red-eared slider)

MAMMALS

Capra hircus (goat)
Cervus elaphus (red deer)
Felis catus (domestic cat)
Herpestes auropunctatus (small Indian mongoose)
Macaca fascicularis (macaque monkey)
Mus musculus (mouse)
Mustela erminea (short-tailed weasel)
Myocastor coypus (nutria)
Oryctolagus cuniculus (rabbit)
Rattus rattus (ship rat)
Sciurus carolinensis (gray squirrel)
Sus scrofa (pig)
Trichosurus vulpecula (brushtail possum)
Vulpes vulpes (red fox)

The glossary that follows defines more than 600 specialized terms that appear in the text of this encyclopedia. Included are a number of terms that may be familiar to the lay reader in their common sense but that have a distinctive meaning within these fields of study. Definitions have been provided by the encyclopedia authors so that these terms can be understood in the context of the articles in which they appear.

abiotic resistance The lack of environmental compatibility between an ecosystem and introduced organisms, irrespective of resident biota identity and abundance.

acclimation The process of adjusting physiological or morphological characteristics in response to an environmental change.

acclimatization The establishment of a species in the wild outside its natural range with the support of humans.

adaptation 1. Genetic change in a population resulting from natural selection, whereby the average state of a character becomes improved with respect to a specific function or a population becomes better suited to its environment. **2.** A property or trait that is the result of adaptation.

adaptive radiation The more or less simultaneous divergence of several new species from a common ancestor when that ancestor enters a novel environment.

adaptive restoration Strategic, phased experimentation developed so as to learn how best to restore an ecosystem.

additive genetic variance Genetic variance attributed to the effects of substituting one allele for another at a given locus, or at the multiple loci governing a polygenic trait.

admixture The mixing of previously isolated populations within a species.

advected energy Energy transferred by horizontal movement of an air mass; e.g., energy supplied by the movement of hot, dry air into the moist environment from the adjacent dryland areas. A classic example is energy transfer in an oasis.

adventitious roots Roots that grow from stems, usually from those creeping on the soil (called *stolons*) or in the soil (called *rhizomes*).

adventitious shoots Shoots produced from roots or hypocotyl; i.e., from organs that usually bear no buds. This ability is restricted to 10% of herbs and 30% of trees in temperate regions.

aggregate species A named species that represents a group of closely related, usually morphologically very similar, species.

agrobioterrorism Bioterrorism targeting agriculture.

agroecosystem A system composed of a farm or other unit of agricultural activity and its interactions with its biotic and physical environment.

algae (*singular,* **alga**) A large, diverse group of uni- or multi-cellular autotrophic eukaryotic organisms that lack true tissues and organs and that are thus not considered true plants.

algaecides Chemicals that are marketed and approved for controlling algae.

Allee effect Decreased reproduction or survival of individuals at low population densities, caused by a variety of factors, such as difficulty finding mates or pollen limitation. (Identified by Warder C. Allee, American zoologist and ecologist.)

allele In diploid individuals, including humans, an alternative form of a gene.

allelochemical A secondary metabolite that hinders the growth of other species.

allelopathy The process whereby chemicals produced by one species kill or inhibit the growth of other species.

allopatric Occurring in separate, disjunct geographic areas.

allopolyploid A hybrid between different species in which two or more complete sets of chromosomes derived from both parental species are retained; common in plants but rare in animals.

allotetraploid See AMPHIDIPLOID.

allozyme markers Simply inherited allelic variants of genes controlling enzymes and proteins that can be used to analyze mating patterns and determine genetic relationships among species.

alternate host An organism other than a primary crop host that harbors a pathogen or pest until it can attack the crop.

alternate stable state A combination of environmental and biotic conditions having self-reinforcing, positive feedback mechanisms that resist transition to a different combination of environmental and biotic conditions. Ecological systems may shift from one stable state to another when perturbed.

altricial Born featherless or otherwise naked, and needing care by parents.

ammocoete A larval lamprey.

ammonification Conversion of organically bound nitrogen (N) to ammonia (NH_3) during the decomposition of organic matter; the most common form of nitrogen mineralization.

amphidiploid An individual having a diploid set of chromosomes derived from each parent. Also, ALLO-TETRAPLOID.

amphidromous Having a life cycle with adults living in fresh water and larvae in marine waters.

anadromous Having a life history in which an organism migrates into fresh water to reproduce but spends most of its growth phase at sea.

anekeitaxonomy Classification of plants and animals according to where they are thought to "belong" as a function of their exposure to human agency (e.g., native and introduced species).

anemochory Seed dispersal by wind.

annual plant A plant that completes its life cycle from seed to seed within a single year.

anoxia Lack of oxygen.

anthropogenic Caused or influenced by humans.

anthropophilic Literally, *man-loving;* of blood-sucking insects, preferring human hosts.

anticoagulant A substance that prevents coagulation of blood (e.g., a compound that inhibits the synthesis of vitamin K–dependent clotting factors in the liver).

apomict A plant that reproduces asexually by seeds, without meiotic reduction and fertilization. In *autonomous apomicts* pollination is not required for endosperm formation, while in *pseudogamous apomicts* it is a requirement.

apoplasm The nonliving components of a plant, including the cell walls, interstitial space, and xylem.

apparent competition Indirect suppression (by a host or prey population) of another host or prey population by subsidizing a shared natural enemy (pathogen or predator) population.

aquaculture The farming of aquatic organisms in inland and coastal areas, involving intervention in the rearing process to enhance production, often entailing individual or corporate ownership of the stock being cultivated.

aquaria (*singular,* **aquarium**) Containers of various sizes and shapes in which aquatic plants, algae, fish, and other organisms can be grown and maintained.

aquatic Living in marine or freshwater environments.

aquatic herbicides Chemicals that are marketed and approved for controlling aquatic weeds.

aquifer Weathered or fractured rock and sediment deposits whose structure or texture holds water or permits appreciable water movement.

arbovirus Acronym for arthropod-borne virus.

arbuscular mycorrhizal fungi Mycorrhizal fungi in the order Glomeromycota whose diagnostic structures are formed inside cortex cells of a plant root.

ardeids Birds in the family Ardeidae, including herons, egrets, and bitterns.

arthropod-borne viruses Viruses transmitted by ticks, mosquitoes, midges, and sandflies that replicate in the arthropod prior to transmission.

asexual reproduction See VEGETATIVE PROPAGATION.

assortative mating Preferential mating among individuals with similar (or dissimilar) phenotypes.

autochory Self-dispersal of seeds by such means as ballistic dispersal from the parent, or seeds' creeping across the surface while hygroscopic awns contract and extend with changes in humidity.

autogenous Literally, *self-generating;* of blood-sucking insects, capable of producing eggs without a blood meal.

ballast water Water carried by ships in special tanks to increase stability when not carrying a load of cargo.

barrier zone A geographic area designed to slow or prevent the spatial spread of a nonnative or pest species into or beyond the zone.

basal meristem A tissue consisting of small, unspecialized, quickly dividing cells located at the base of the organ it produces.

base-rate The rate at which a given phenomenon occurs. Successful invasions generally have a low base-rate, in that few species arriving in a new range will successfully establish, naturalize, and become harmful.

benthic Describing or referring to the sea bed, river bed, or lake floor.

benthophagous Feeding on benthic organisms.

benthos The organisms associated with the bottom of a body of water, including both sedimentary and rocky substrates.

bilateral agreements Agreements between two governments.

bilharziosis See SCHISTOSOMIASIS.

biocontrol agent A natural enemy deliberately used by humans to control the population of a pest species.

biofouling The accumulation of living organisms and associated debris on submerged objects.

biofuels Liquid fuels derived from plant materials.

biogeochemistry The chemical transformations of nutrients and other materials that are carried out by organisms, and the fluxes of such substances among different components of the ecosystem.

biogeography The study of the distribution patterns of living organisms and of the processes that contribute to those patterns.

biological control/biocontrol The deliberate introduction of a natural predator, parasite, or pathogen to reduce the abundance of a pest.

biosecurity Preventive measures taken to exclude, eradicate, and control unwanted introduced organisms in order to protect the economy, the environment, and the health of humans, other animals, and plants.

biota All living organisms in a given area.

bioterrorism Terrorism that uses biological weapons.

biotic resistance The ability of resident biological communities to deter potential invasions of introduced organisms. Also, ECOLOGICAL RESISTANCE.

bivalve Any member of a class (Mollusca; Bivalvia) of invertebrate organisms having two shells joined by a hinge. Examples include oysters, mussels, scallops, and clams.

black list 1. A policy response to harmful nonindigenous species that prohibits the introduction of only those species that have been determined to be harmful. **2.** A list of such species determined to be harmful.

bottleneck See GENETIC BOTTLENECK.

bridge vector A vector that transmits a pathogen from its enzootic cycle to a host outside that cycle.

broad-spectrum insecticide A chemical substance used to control insect pests that can affect many different species, including nontarget species. Often acts by affecting the nervous system; its effects depend on the dose and duration of exposure.

bryophytes Nonvascular embryophytes with dominance of the haploid gametophyte; they do not form a monophyletic group but are split into liverworts (Marchantiophyta), mosses (Bryophyta), and hornworts (Anthocerotophyta).

bud bank A pool of reserve meristems on a plant that are usually activated only when aboveground shoots are lost; i.e., after winter or summer inactivity or after injury.

carbohydrate Any of a group of organic compounds that includes sugars, starches, celluloses, and gums and that serve as a major energy source in animals or plants.

carnivore 1. A member of the order Carnivora, including dogs, bears, raccoons, weasels, mongooses, hyenas, and cats. **2.** A colloquial term for any primarily animal-eating organism, especially a vertebrate.

carp, Asian A collective term for several large Asian fishes in the family Cyprinidae, especially grass carp (*Ctenopharyngodon idella*), silver carp (*Hypophthalmichthys molitrix*), bighead carp (*Hypophthalmichthys nobilis*), and black carp (*Mylopharyngodon piceus*).

carp, common A large species of minnow, *Cyprinus carpio*.

catechin A flavonoid exuded through the roots of *Centaurea maculosa*, an aggressive exotic invasive in North America.

cation exchange capacity A measure of the capacity of soil to hold cations.

Ciconiiformes An avian order of marsh birds including herons, egrets, bitterns, storks, ibises, and spoonbills.

classical biological control The intentional introduction of natural enemies from the native range of a pest species to the invasive range of the pest species in order to control the pest.

climate envelope A specified set of climatic conditions, typically describing the climatic conditions required by a species for survival and reproduction.

climax community The local community in which dynamics no longer result from the disturbance that initiated succession; the successional endpoint.

clonal growth/clonality Growth through vegetative spread and production of genetically uniform individuals; in plants, usually via rhizomes or stolons (rooted stems) that can each grow independently of the original parent plant, though typically they are attached to it.

co-adapted gene complex A set of genes in an individual that have evolved together and that, as a group, confer a fitness advantage on the individual.

coevolution Mutual evolutionary influences (either negative or positive) that exist between species that exert selection pressure on one another.

coexistence The ability of species to cohabit in an area through time.

collinearity High correlation among explanatory variables.

colonizer An early successional organism that readily establishes and spreads in open, disturbed areas.

columbids Doves and dovelike birds such as pigeons in the family Columbidae.

commensal An organism that benefits from another without having any impact upon it.

commensal rodents Rats and mice that are generally found living in close association with humans.

common garden experiment An experiment in which populations from one or more locations are reared in a common environment to distinguish between heritable (genetically based) traits and those that are environmentally induced (plastic). If traits that differ between populations or species in the wild still show significant difference when reared under common-garden conditions, those trait differences are interpreted as being genetically based rather than induced by the environment.

common law Principles of law derived through case-by-case adjudication in the courts. Common law is the foundation for many basic legal principles in property law, contract law, tort (injury) law, and so forth, but has largely been supplanted by statutory law in many jurisdictions.

community The suite of populations of different species coexisting in a given area.

compatible solutes Small molecules that are compatible with enzymatic and other metabolic activity even when present in high concentrations in the cytoplasm. They include compounds such as proline, glycine betaine, and mannitol, whose accumulation can be induced by exposure to osmotic stress as a result of drought, salinity, or freezing.

competition The simultaneous reduction in fitness (survivorship, growth, and reproduction) of two or more organisms owing to shared use of resources in limited supply.

conditions Intrinsic abiotic environmental factors of a location that can affect suitability for survival, growth, and reproduction, such as soil texture, temperature, pH, and relative humidity. These factors usually fluctuate temporally and spatially.

congeneric Belonging to the same genus; in invasion studies, generally refers to two species originally from two different parts of the world, currently co-occurring.

connectance The fraction of all possible links that are realized in networks, such as food webs.

consumer An ecological term for all nonautotrophic organisms that must consume food.

control Activities related to suppression of an invasive species within a defined geographic area.

cordillera A large range of high mountains that spans the entire length of a continent.

coregonine Any of a group of fishes native to the Great Lakes from the genus *Coregonus* (e.g., the ciscoes).

cosmopolitan Having a worldwide distribution, at least in suitable habitats.

cotyledon A part of a plant embryo within the seed. Upon germination, the cotyledons usually become the first leaves of a seedling.

cover crop A crop grown in solid stands along with the primary crop in order to protect and improve soil between crop rows or between periods of regular crop production and to suppress weeds. Also, GREEN MANURE, LIVING MULCH.

crayfish Freshwater crustaceans related to lobsters; phylum Arthropoda, subphylum Crustacea, order Decapoda, infraorder Astacidea.

crayfish plague A disease that is lethal to Australian and European crayfish species; caused by the oömycete *Aphanomyces astaci*, a funguslike organism endemic to North America.

crevicolous Describing or referring to species living preferentially in crevices, holes, and burrows.

critical community size The size of host population required to maintain a constant flow of new susceptible hosts so that a pathogen population can persist; a concept that originated in human epidemiology literature, in which *community* means population.

cross-fertilization See OUTBREEDING.

cross-validation A method of assessing the validity of a statistical model. The available data set is divided into two parts. Modeling of the data uses one part only. The model derived from this part is then used to predict the values in the other part of the data. A valid model should show high predictive accuracy.

cryptic Describing or referring to an animal or plant that is very difficult to see because it blends into the background (camouflage) or because it uses inaccessible or thickly vegetated habitats.

cryptic diversity A property of genetically distinctive evolutionary lineages or species that are morphologically indistinguishable, and thus hidden from human view.

cryptogamic crust The thin layer of mosses, lichens, liverworts, free-living algae, and cyanobacteria that comprise the nonvascular plant matrix on the soil surface in steppe habitats.

cryptogenic Of uncertain geographic origin; describing or referring to species not yet resolved as introduced or native.

cultch Shell pieces on which young oysters (spat) settle.

cultivar A plant variety that has been produced in cultivation by selective breeding (short for "cultivated variety").

cuticle In plants, the protective waxy covering on the epidermal cells; most prominent on leaves.

cyanobacteria Photosynthetic bacteria not considered to be true algae (formerly known as *blue-green algae*).

cytoplasm/cytosol The substance found inside the cell membrane; the material that fills the cell.

damage costs The costs incurred by people in the host country as a result of an invasion. They include both direct costs (the cost of illness, lost output, and the like) and indirect costs (the loss of ecosystem services owing to, e.g., the extinction or extirpation of native species).

damping off Different pathogen-caused ailments that can kill seeds or seedlings.

Darwin's naturalization hypothesis The hypothesis that new regions proportionally gain more new genera than species, meaning that species more phylogenetically distinct from native species are more likely to be successful invaders.

daughter plant A plant that results from vegetative growth from another plant.

decision theory A mathematical field concerned with identifying the values and uncertainties relevant to optimizing decisions. Also, GAME THEORY.

decision tree A decision support tool that uses a treelike graph or model of decisions and their possible consequences, including chance event outcomes, resource costs, and measures of satisfaction.

decomposer organisms Organisms that break down dead or decaying organisms and their parts. The main examples are fungi, bacteria, and soil fauna (worms, springtails, and the like).

decomposition The breakdown of organic matter into simpler compounds, which results in the release of nutrients for plant growth.

demersal Living in or near the deepest part of a water body. *Demersal plankton* are invertebrates that live near the bottom but that migrate to upper layers at night.

demographic stochasticity Variation in population growth rates arising from random differences among individuals in survival and reproduction.

detection The use of population surveys, such as the deployment of pheromone-baited traps, to identify the arrival of nonnative species.

detection probability The chance of successfully finding an individual of a particular taxon.

detritivore An organism that derives its nutrition by consuming detritus.

detritus Nonliving particulate organic matter derived from plant and animal remains.

diapause A state of dormancy in which an organism minimizes its activity and physiological functions in order to persist through unfavorable environmental conditions.

diatom A specific group of single-celled algae that possess a silica-based cell wall. Diatoms can grow as single cells or in colonies; both freshwater and marine species exist.

diffuse competition The combined effect, on one species, of competition from two or more other species.

diploid 1. Describing or referring to the life cycle stage in which there are two complete sets of chromosomes in the cell nucleus, one from each parent. **2.** A species in which each individual has two complete sets of chromosomes in the cell nuclei.

disease agent An organism that causes or contributes to the development of a disease. Also, ETIOLOGICAL AGENT.

dispersal The spread of mature individuals, their offspring, or propagules away from the place of their birth to another region.

dispersal kernel 1. Description of the movement of individuals in discrete time as the distribution of final locations relative to initial location. **2.** The frequency distribution of seedfall as a function of distance from the source.

dispersal mode The mechanism of dispersal of propagules (seeds, tubers, viable fragments, and the like).

disturbance Any relatively discrete event in time that disrupts ecosystem, community, or population structure and changes resources, substrate availability, or the physical environment. Examples of disturbances include fires, floods, storms, pathogen outbreaks, and human activities (construction, deforestation, agricultural cultivation, and the like).

disturbance regime The different types and characteristics of disturbance experienced in an area, including their intensity, frequency, extent, duration, and seasonality.

diversity-invasibility hypothesis The theory that communities of greater species diversity are less invasible than species-poor communities.

domestication The taming process in which a population of animals, through a process of selective breeding and conditioning, becomes accustomed to human provision and control.

donor region A geographical area that is the source of a nonnative species.

dunnage Material used to pack, stabilize, or protect cargo transported on ships or airplanes.

early successional communities Plant communities that occur following large-scale disturbances on open ground.

ecological filter Dispersal, environmental, or biotic factors that limit which species are present in a community.

ecological niche 1. The position occupied by an organism (taxon), as defined by the ranges of biotic and abiotic factors within which it is able to survive and reproduce. The *fundamental ecological niche* is the largest possible ecological niche in the absence of competitors, predators, and pathogens. The *realized ecological niche* is the ecological niche as realized in nature; i.e., in the presence of other organisms. Typically, the realized ecological niche is smaller than the fundamental ecological niche. 2. The precisely formulated description of the natural history of an organism (taxon); its functional role in the environment or biotic community. Also, NICHE.

ecological resistance See BIOTIC RESISTANCE.

ecological plasticity The ability of an organism to change its ecology in response to changes in the environment.

economic threshold (ET) The pest population density at which a control tactic should be initiated in order to stop the pest density from reaching the economic injury level.

ecophysiological traits Adaptations of an organism's physiology that allow it to respond to environmental conditions, such as the production of hollow vessels in plant stems (aerenchyma) that allow diffusion of oxygen into roots when the soil is waterlogged.

ecosystem A system composed of all the organisms (animal, plant, microbe) and the abiotic environment, and all interactions among and between these components.

ecosystem engineer An organism that creates or modifies its habitat (especially the physical structure of its habitat), thereby affecting other organisms and the environment.

ecosystem functioning Key ecosystem processes that maintain the functioning of a system, such as productivity, decomposition, and nutrient cycling.

ecosystem services The products provided by functioning ecosystems that benefit humans as well as other organisms.

ecoterrorism Terrorism targeting ecosystems and ecosystem services.

ectomycorrhizae Fungi living as symbionts with roots of certain plant species with which they form sheaths around the root tip and nets of hyphae surrounding the plant cells.

ectotherm A species in which thermoregulation depends on the gain of heat from the environment.

efficacy The ability of an insecticide or other action to control the target pest.

elaiosomes Fleshy lipid- and protein-rich structures attached to seeds of some plant species. Depending on their size, such seeds are dispersed by either ants or birds.

emergent properties Complex processes and patterns that emerge at a higher level of organization (e.g., the community level) derived from a number of processes and patterns at a lower level of organization (e.g., the population level).

empty niche An environmental condition or mode of living unexploited by resident species that may be exploited by an invader, sometimes with little or no impact on native populations.

encephalitis Inflammation of the brain.

endangered species A species that has a high risk of extinction in the wild in the near future; a category in the IUCN system and under the U.S. Endangered Species Act.

Endangered Species Act (ESA) An important U.S. law passed to protect endangered species and the ecosystems in which they live.

endemic Found in and restricted to a particular geographical area. *Endemic species* are found in one place and nowhere else; for example, lemur species are found only on the island of Madagascar.

endemic disease A disease that does not need external inputs in order to persist in a population.

endoparasite A parasite that lives in the internal organs or tissues of its host.

endozoochory Dispersal of seeds by their consumption, often but not always in a fruit, and regurgitation or defecation.

enemy release Decreased enemy attack on a host or prey species resulting from the introduction of the host or prey to a new geographic range lacking the host's or prey's natural enemies.

enemy release hypothesis The hypothesis that the release from enemies (e.g., pathogens, parasites, predators) constitutes the fitness advantage that allows a nonindigenous species to thrive.

entomopathogenic fungus A fungus that acts as a parasite of insects, killing or seriously disabling them.

environmental stochasticity Variation in population growth rates caused by environmental fluctuations that affect survival and reproduction of all individuals in a population.

enzootic A disease occurring in a particular animal species within a limited geographical area.

epibenthic Living on or near the surface of the submerged substrate.

epidemic In humans, an outbreak of an infection that substantially exceeds the expected level based on recent experience.

epifauna Animals that live attached to firm surfaces, usually in fresh water or in the sea.

epigenetics The study of heritable changes in phenotype without changes in the nucleotide sequence of the genetic code (DNA or RNA). Most epigenetic changes occur only within the lifetime of an individual organism (with heritability only among cell lineages), but some epigenetic changes are inherited from one generation to the next. *Epigenetic mechanisms* include DNA methylation, histone modification, chromatin remodeling, and the action of microRNAs.

epizoochory Dispersal of seeds that are attached externally (e.g., to fur) on animals or that are carried in the mouth without being swallowed.

epizootic The occurrence of disease cases significantly in excess of the normal or endemic level in a nonhuman animal population. The analog in humans is termed an *epidemic.*

eradication Total elimination of an invasive species from a defined geographic area.

Erythrean introduced species A species that has entered the Mediterranean from the Red Sea via the Suez Canal and that has established a population, at least temporarily, in the Mediterranean Sea.

ethylene A phytohormone that significantly affects the growth, development, and stress responses of plants.

etiological agent See DISEASE AGENT.

exotic species See INTRODUCED SPECIES.

euryhaline Tolerating a wide range of salinities.

eurythermal Tolerating a wide range of temperatures.

eutrophic Describing or referring to the aquatic environment with highest natural or anthropogenic primary productivity of the ecosystem (>10 mg/m^3 chlorophyll a).

eutrophication A process, or several processes, whereby nutrients such as nitrogen and phosphorus accumulate in aquatic ecosystems, usually from external sources.

evaporation The loss of moisture from any surface in the form of water vapor.

evolutionarily conserved traits Genetically preserved ancestral similarity of species traits owing to retention of a high proportion of common ancestral alleles.

evolutionary tradeoff A phenomenon in which an increase in fitness caused by a change in one trait is opposed by a decrease in fitness caused by a concomitant change in a second trait, limiting the first trait's response to selection. Tradeoffs limit the number of traits that can be maximized simultaneously.

evolution of increased competitive ability (EICA) The hypothesis that introduced plants can redirect resources away from defense and toward increased competitive ability.

evolvability 1. The ability of genomes within a population to produce adaptive variants, such that the population can respond to selection. **2.** The ability of a population to respond to selection owing to its standing genetic variance, as quantified by the genetic coefficient of variation ($CV_A = 100\sqrt{V_A} / \bar{X}$, where V_A is the additive genetic variation and $\bar{X}$ is the trait mean).

explanatory (or predictor) variable A variable used to explain or predict changes in the values of another (dependent) variable.

exploitative competition Competition arising among organisms when they use and deplete the same pool of resources (e.g., food).

exponential growth A pattern of population increase in which growth rates are proportional to population size, leading to initially slow but ever-increasing rates.

externalities Those effects of market transactions that are not taken into account by the parties engaged in those transactions.

extinction Global elimination of all individuals of a species.

extinction by hybridization The extinction of at least one species as a genetically distinct entity by repeated hybridization events.

extirpation Local extinction of a species that continues to persist elsewhere.

extralimital Describing or referring to organisms or populations found outside a defined area (such as dolphins outside a marine reserve). In the invasive species literature, the term typically refers to organisms found outside of their native geographical ranges.

extra-nidal Of insects, located or occurring in a place away from the colony.

facilitation Positive interaction between species, leading to an increase in the fitness or population size of one or both species.

facultative mutualism A mutualism in which a species can persist without its partner, though its performance is greater in the presence of the partner.

febrile Relating to or affected by fever.

fecundity The number of gametes or propagules (e.g., seeds or eggs) produced; the potential reproductive capacity of an organism or population.

federal appropriations legislation Laws that provide annual funding for authorized federal efforts, originating in the U.S. House of Representatives and then sent to the Senate, with detailed consideration within their respective Appropriations Committees.

federal authorizing legislation Laws passed by the U.S. Congress, originating in either the Senate's or the House of Representatives' committees with jurisdiction over a given topic (e.g., the House Committee on Natural Resources).

female reproductive success The fraction of female eggs that are fertilized.

feral Describing or referring to formerly domestic animals now existing in a wild state (e.g., pigs, goats, and sheep).

fertility The actual reproductive performance of an organism or population, measured as the actual number of viable offspring produced per unit of time.

fertility control Temporary or permanent contraception.

fertilization Fusion of male and female gametes. In angiosperms, an egg cell fused with a sperm cell gives rise to an embryo; a central cell fused with a sperm cell results in formation of endosperm.

filariasis An animal or human infection caused by parasitic nematodes transmitted by vectors.

finite rate of population increase (λ) The number of times a population multiplies itself each generation. $\lambda > 1$ means that population is growing, and $\lambda < 1$ means that population is declining.

fire behavior Characteristics of a single fire event as it responds to weather, fuels, and topography.

fire frequency The number of fires per unit of time.

fire intensity The rate of heat release at the flaming front of a fire.

fire regime The characteristics of fires that are typical for a region. Important features include timing, frequency, intensity, heterogeneity, and size.

fire return interval A measure of fire frequency defined as the average number of years before fire reburns a given area.

fitness The extent to which an individual or population contributes genes to future generations; usually quantified as finite rate of population increase (λ).

fitness homeostasis The ability to maintain consistent fitness across a range of conditions. *Individual fitness homeostasis* depends on adaptive phenotypic plasticity. *Population fitness homeostasis* results from both individual fitness homeostasis and genetic polymorphism.

fixed heterosis The maintenance of hybrid vigor in a fixed manner in a population, common in allopolyploid

hybrids. The individuals reproduce asexually and favorable genotypes do not recombine.

fluvial Belonging to or found in flowing water, usually rivers.

food web The network of feeding relationships among the organisms of a community, including the consumption of both live organisms and dead (detrital) materials.

foundation species A species responsible for ecosystem structure because of its abundance and influence.

founder effect or **event** A special case of genetic drift in which loss of genetic variation occurs when a new population is established by a small number of individuals from a larger population.

founder (or founding) population size The number of individuals introduced or first arriving into a new territory that subsequently establish a population.

frugivorous Describing or referring to animals that consume fleshy fruit and frequently disperse intact seeds.

fumigant Any volatile, toxic material used to kill animals, plants, or other organisms.

functional types Organisms that have the same general morphology and that function in a similar manner ecologically, such as trees, or epiphytes, or vines.

fur farming The practice of breeding or raising certain types of animals for their fur.

galliforms Chickens and chickenlike birds such as quails, francolins, turkeys, and grouse, in the order Galliformes.

game theory See DECISION THEORY.

gastropod Literally, *stomach-foot;* a member of the phylum Mollusca, class Gastropoda (snails and slugs) that does not have legs but crawls on the extended body ("foot").

gene flow· The transfer of genes from one population to another.

generalist A species that can survive and reproduce in a wide range of environmental conditions using a wide variety of resources to do so.

generalist pathogen A pathogen that infects more than one host.

general purpose genotype A genotype responsible for high individual fitness homeostasis.

genet A clonal colony or group of individuals that are genetically identical and that have arisen by vegetative reproduction from the same ancestral individual.

genetic algorithm An algorithm that finds solutions to optimization and search problems using techniques inspired by evolutionary biology, such as inheritance, mutation, recombination, and selection.

genetic architecture A term used in the field of quantitative genetics to refer to the number of loci and the number and frequency of alleles at those loci that affect a particular trait, their allelic and genic interactions, and their relationship to other traits (e.g., genetic correlations, pleiotropy).

genetic assimilation 1. A process by which a phenotype that is expressed under stressful conditions is then favored by natural selection such that the organism becomes "canalized" for that particular phenotype. The extreme trait is then no longer inducible but becomes fixed in the population. 2. Disappearance (or assimilation) of a species as its genes are diluted through hybridization with a closely related species.

genetic bottleneck The reduction in gene frequencies in a population owing to a rapid reduction in population size for several generations. May lead to lower genetic diversity. Also, BOTTLENECK, POPULATION BOTTLENECK.

genetic correlation Association of different traits caused by the correlation between the two loci controlling the two traits; often used in breeding programs as a measure of the strength of the impact of selection on one character on the change in another unselected character.

genetic diversity The total amount of genetic variation in the genetic makeup of a species.

genetic drift Random changes in genetic diversity of a population over time; typically more pronounced in smaller populations.

genetic load The decrease in the average fitness of individuals in a population owing to the presence of deleterious alleles.

genetic marker A polymorphic neutral gene (e.g., microsatellite or allozyme) that can be used to analyze various population genetic processes such as mating and gene flow.

genetic polymorphism The persistent occurrence in a population of two or more alleles of the same gene, all in frequencies too great to be maintained by recurrent mutation.

genetic structure The degree to which the genetic diversity of a species is partitioned within and among populations.

genetic swamping Repeated hybridization events primarily in the direction of one species into other that result in the eventual dilution and loss of genetic material of the recipient species.

genetic variance The degree of phenotypic variance within populations that is due to genetic differences among individuals (as opposed to environmental or chance factors), including dominance, additive genetic, and epistatic genetic variance.

genomic architecture The structure, content, and organization of a genome, including the location and order of genes and of noncoding regions of the genome.

genospecies A species defined by its unique genetic makeup, rather than by morphology.

genotype The combination of alleles carried by an individual, usually with reference to a certain gene region or character under consideration; an inherited character that, in combination with environmental conditions, contributes to the phenotype, or appearance, of an individual.

geographic origins Regions (or ancestral populations) in the native range from which individuals are sampled and introduced into a new territory, or range. Also, SOURCE POPULATIONS.

geographic range Landscape- or larger-scale locations where a species is found in the wild.

gonotrophic cycle In blood-sucking insects, the process of egg production through oviposition, initiated by blood consumption.

granivore A seed-eater.

greenhouse gases Gases in the atmosphere that absorb and emit infrared radiation and that can trap heat in the Earth's atmosphere. The principal greenhouse gases in the Earth's atmosphere are carbon dioxide, methane, and nitrous oxide.

green manure See COVER CROP.

gross domestic product (GDP) The standard measure of economic performance: the market value of all final goods and services made within a country in a year.

groundwater Water occurring below the land surface in materials (e.g., soil, fractured rock) that is in a saturated state and able to move to discharge areas; this term generally refers to water contained in aquifers.

habitat The type of environment that a species inhabits or can inhabit.

halophytes Plants that persist in saline environments.

haploid Describing or referring to the number of non-homologous chromosomes in a single set (i.e., in a gamete).

haplotype A set of closely linked genetic markers (or a DNA sequence) that is inherited as a unit and that does not recombine; a form of genotype.

hatchery A place where people foster the reproduction of an aquatic species to overcome growth and survival constraints in the wild and provide a consistent, reliable flow of young organisms to be used in aquaculture or for sport fishing.

helminth A worm that is parasitic on the intestines of vertebrates, especially roundworms, tapeworms, and flukes.

hemagglutinin A rod-shaped glycoprotein radiating from the virus surface; the cell-binding protein initiating cell infection. There are 15 (H_1–H_{15}) distinct forms of this glycoprotein in nature.

hemorrhagic Caused by, or resulting in, loss of blood or bleeding.

hepatic See LIVERWORT.

herbivore An animal that consumes mostly plants.

herd immunity Reduction in the numbers of susceptible hosts that become infected, caused by the vaccination of other hosts, breaking the chains of transmission that lead to new infections.

herpetofaunal Pertaining to reptiles and amphibians.

heterosis A phenomenon resulting from hybridization, in which hybrid offspring display greater fitness, size, or growth rate than the parental lines. *Fixed heterosis* refers to cases in which heterosis does not become disrupted by recombination; it occurs often in allopolyploid hybrids that reproduce asexually. Also, HYBRID VIGOR.

heteroxenous parasites Parasites having life cycles involving multiple host species (intermediate and final hosts).

heterozygosity The state of having two different alleles at the same locus (gene) in an individual. *Average heterozygosity* (*H*) is calculated as the sum of the proportions of heterozygotes at all loci ÷ total number of loci sampled and has a value ranging from 0 to 1.0; it is used as a measure of genetic diversity.

higher taxon A group that includes animal or plant species that have a natural relationship, such as a phylum, order, family, or genus.

high spatial resolution remote sensing Acquisition of surface physical properties at a small (e.g., < 10m) ground sampling distance.

high temporal resolution remote sensing Frequent data collection (e.g., every 1–2 days for the same location) of surface physical properties.

Holocene The geologic period since the end of the last ice age (approximately 10,000 BP).

holoplankton Organisms that spend all of their life cycle in the water column.

home range The region of habitat used by an individual animal.

honeydew A sugary waste product excreted by scale insects.

host 1. The organism from which a parasite obtains its nutrition or shelter and within which it may replicate. **2.** Usually, the larger or sessile organism participating in a mutualism.

host shift Change by an herbivore or pathogen of the host species on which it can develop or reproduce.

hotspot 1. In geology, a relatively fixed location on the Earth's surface, not near a plate boundary, where active volcanism has occurred for a long period of time. **2.** In ecology, an area with exceptional endemism of plants or animals as well as significant levels of habitat loss.

hot spring A spring that is thermally influenced by, or that is composed of, geothermally heated groundwater.

HPAIV The highly pathogenic avian influenza virus strain of influenza A(H_5); lethal for many bird species and in some 50% of human infections.

hybrid An offspring that results from the mating of two closely related species.

hybridization The interbreeding of genetically differentiated groups. *Intraspecific hybridization* is interbreeding between members of the same species, whereas *interspecific hybridization* is interbreeding between different species.

hybrid vigor See HETEROSIS.

hydrochory Propagule dispersal by water.

hydroelectric Generating electricity by use of water energy.

hydrological cycle The continuous circulation of water in vapor, liquid, and solid form between the atmosphere, the ocean, and the land, including surface and groundwater, driven by energy from the sun.

hydrology The field of science that studies movement, distribution, and quality of water on and in the Earth.

hydroperiod Timing of wet conditions; a graph of water level over time.

hyperspectral imaging Measurement of solar reflected light in hundreds of contiguous spectral channels.

hypoxia Lack of oxygen, typically owing to either ice cover or high temperature.

ichthyoplankton Eggs and larvae of pelagic and benthic fish in the water column.

import The bringing of a species into a country from another country.

incipient population A population that has not yet become firmly established but that has the required ingredients (i.e., males and females) for becoming established.

incrimination Definitive identification of a vector species as transmitter of a pathogen.

indigenous species A species occurring naturally in an area, whose presence does not result from human activity. An indigenous species need not be endemic, as it may be found in a broader area than the one under consideration. Also, NATIVE SPECIES.

indirect effects Effects of one species on another via its actions on a third species.

infection The presence of a disease agent within a host; infected hosts are not necessarily diseased.

inoculation 1. The placement of organisms that will grow or reproduce into the body of a human or animal. **2.** Penetration of external ice into the body of an organism, causing tissue to freeze.

integral projection model A discrete time model for dynamics of population size and structure wherein birth, growth, and death rates depend upon a continuous-state variable such as size.

integrated pest management (IPM) An approach to pest management that takes advantage of all appropriate options, including, but not limited to, the judicious use of pesticides, and that relies on current, comprehensive information on the life cycles of pests and their interaction with the environment.

internal nutrient loading The amount of nitrogen and phosphorus that is present and recycled within a body of water.

interspecific pollen transfer The transport of pollen from anthers of one species to the stigma of other species.

intraguild predation Predation among potential competitors.

intrinsic rate of population increase (r) The potential rate of growth of a population in an infinite environment, calculated as $r = b - d$, where b is the instantaneous birth rate and d is the instantaneous death rate. The dimension is *individuals/individual/time unit*.

introduced species Individuals of a species moved from their native range to a new location outside of their native range. Colloquially, EXOTIC SPECIES, NONNATIVE SPECIES.

introduction Release of a species outside of its native range, either accidentally or deliberately.

introduction dynamics The specific details associated with an alien species' introduction into a new territory or range.

introduction history Attributes such as reason for origin and minimum residence time that relate to the introduction of a species into an area.

introgression The repeated transfer of genetic material from one species into another via hybridization.

invasibility The susceptibility of a community or ecosystem to the establishment and spread of one or more introduced species.

invasional meltdown The process wherein two or more introduced species increase each other's likelihood of establishment or impact on native species.

invasion hub A source (e.g., lake or garden) that serves as the predominant source of nonindigenous species spread to other locations.

invasion pathway A route, such as a commercial shipping route, through which a species is moved from an area in which it is established to one in which it is not.

invasion potential The characteristics of a species that render it capable of successfully invading.

invasion stages Variously defined sequential states during the process of invasion, from initial introduction to full invasion.

invasion window An opportunity for invasion created by a change in ecosystem properties, such as the creation of a canopy gap by disturbance.

invasiveness The ability of a species to reproduce, spread away from places where it is introduced, and establish in new locations.

invasive population A population that has become invasive. Some species contain both invasive and noninvasive populations.

invasive species 1. A naturalized species that produces reproductive offspring, often in very large numbers, and that spreads over large areas. This definition is usually used by ecologists. **2.** A nonindigenous species that spreads rapidly, causing environmental or economic damage. This definition (equivalent to "nonnative pest species") is often used by managers, particularly in the United States.

IPM Integrated pest management.

irruption A drastic and rapid increase in the density of a species population.

isocline 1. A line on a map that connects points on the Earth's surface that all share the same property (e.g., the same annualized temperature conditions). **2.** In a plot of resource uptake versus resource supply, the line representing the equality *resource uptake = resource supply.*

isomers Chemical compounds that have the same atomic composition but different structures.

IUCN The World Conservation Union, a major international conservation organization; previously known as the International Union for the Conservation of Nature.

Judas animal An animal that is marked with a tracking device and used to locate other animals of the same species.

jump dispersal A pattern of species spread characterized by movement of individuals across relatively long distances, rather than the progressive expansion of a contiguous population.

keystone species A species that has a large effect on community structure and function (high overall importance value) that is highly disproportionate to its own abundance or biomass, and whose actions are not carried out by any other organism in the ecosystem.

Lacey Act A U.S. federal statute that regulates imports and interstate commerce for some types of animals, in particular species obtained, transported, or sold in violation of a state, national, or foreign law. The "injurious species" section of the Lacey Act created a listing process by the Secretary of the Interior under which about 25 taxa of nonnative animals are prohibited.

lag The delayed onset or relatively slow rate of an event or process.

lag period/time/phase A time of relative inactivity preceding rapid population growth and range expansion of an invasive population. The term is typically used when this period is relatively long, sometimes lasting decades or more.

lampricide A toxin used to kill larval lamprey.

land use The use to which any particular area or ecosystem is put by humans. This may range from no use or use for conservation, through use of particular ecosystem components, to removal or complete transformation to a new system.

larva (*plural*, **larvae**) An early phase of the life cycle of some animals, different from adults in morphology, diet, and, often, habitat. For example, oyster larvae spend at least some of their time swimming and feeding in the water, whereas adults are entirely immobile on the bottom.

lateral roots More or less horizontal, generally shallow roots.

LC50 The concentration of material in medium to which test organisms are exposed that is estimated to be lethal to 50% of the organisms.

legally binding agreements Agreements that must be observed and met in good faith; generally referred to as *treaties* or *conventions*.

lentic Describing or referring to bodies of water that lack a strong current, such as lakes, ponds, and swamps.

liana A woody vine or climbing plant.

lichens Composite organisms arising from a strong mutualistic association between a saprophytic fungus and a photosynthetic alga or bacterium.

LiDAR Light Detection and Ranging, an optical remote sensing technology that measures the distance between a sensor and a target surface.

life form A combination of the longevity and architecture of a plant species; e.g., annuals, shrubs, lianas.

life history traits Characteristics of a species that relate to important life history stages such as growth and reproduction.

limiting similarity The hypothesis that species with similar traits (e.g., niches or morphology) for resource capture will compete strongly for resources, likely leading to dominance of one species or to an invader's being limited or excluded by other "similar" species.

lineage A group of individuals related by descent from a common ancestor.

linear model Any parametric model with one response variable, including generalized linear models with other than normal distribution (e.g., logistic regression).

lipophilic herbicides Herbicide compounds that are soluble in oils but not in water; generally formulated as emulsifiable concentrates.

litter quality The quality of litter for decomposer organisms. High-quality litter has high concentrations of nutrients and low levels of defensive and complex structural compounds, and decomposes more quickly, while the reverse is true for low-quality litter.

liverwort A division of bryophytes with an estimated total of 6,000 to 8,000 species globally. Also, HEPATIC.

living mulch See COVER CROP.

locus The position that a gene occupies on a chromosome or within a segment of genomic DNA.

logistic population growth model A model of population growth based on intrinsic growth rate, r, and limited by carrying capacity, K.

log phase A period in the numerical growth of a population in which growth is rapid and accelerating; in a plot of growth over time, the slope of the plot is strongly positive.

long-distance dispersal A dispersal event at the extreme far end of the distribution of recorded distances from the source population.

LTRE Life Table Response Experiment, a type of analysis for a structured population in which the experimental unit is the whole population and the entire set of vital rates constitutes the dependent variable.

macroalgae Plantlike, multicellular algae that are identifiable without a microscope.

macrophytes Freshwater algae and vascular plants that can be easily seen by the unaided eye.

male reproductive success 1. The production of fertile offspring by a male relative to that of other males in the population. **2.** The success of a plant in siring seeds by its own pollen.

management The wide range of activities undertaken to exclude, detect, contain, eradicate, or otherwise control invasive species.

marginal costs and benefits The costs and benefits of a small change in the provision of some service. A basic condition for the efficient use of resources is that marginal costs and benefits are equal.

marine macroalgae Marine algae that are typically large and easily seen with the unaided eye; may be known as *kelp* or *seaweed*.

meiofauna Animals that live between grains of sand and other finer sediments.

meiosis A process of reductional cell division in which the number of chromosomes per cell is cut in half.

meromictic Describing or referring to a body of water having no vertical mixing between upper and deeper layers, so that oxygen does not penetrate to the deeper layers.

meroplankton Organisms that spend part of their life cycle in the water column; the pelagic larvae of benthic animals.

mesotrophic Describing or referring to an aquatic environment with low primary productivity (ca. 1 mg/m^3 chlorophyll *a*).

meta-analysis A statistical analysis that combines results from different studies to obtain a quantitative estimate of the overall effect of a particular intervention or variable on a defined outcome. This combination may produce a statistically more powerful conclusion than that provided by any individual study.

metabolism The chemical reactions that occur in an organism to maintain life.

metapopulation A population of populations more or less connected by gene flow and characterized by colonization–extinction dynamics.

microflora/microbes Bacteria, small algae, and small fungi.

microsatellites A class of highly polymorphic genetic markers present in nuclear and organellar DNA that consist of repeating units of 1–6 base pairs in length and that can be used to investigate paternity and gene flow. Also, SIMPLE SEQUENCE REPEATS (SSRS).

minimum generation time Duration of the juvenile period, from birth or germination to first reproduction.

minimum residence time An estimate of the minimum length of time (in years) that an exotic species has been present in its new area.

mitigation An attempt, often mandated by law, to offset the loss of value (e.g., death of individuals of a protected species).

mixed effect and nested statistical design A model combining fixed and random effects in which the random effects can be hierarchically nested; sometimes used on a taxonomic hierarchy treating genera, families, orders, and other taxonomic levels as nested random effects.

mode of action The sequence of events of an herbicide or other pesticide from the time it contacts a plant to the time the plant dies, including absorption, translocation, metabolism, and mechanism of activity at the target site.

MODIS Moderate Resolution Imaging Spectroradiometer, a remote sensing technology with large-area coverage.

molecular markers Physical sites in the genome of a species at which the nucleic acid sequence varies among individuals; often used in various molecular techniques to reconstruct the demographic history of populations and to determine the influence of these events on the genetics of populations.

monoculture The exclusive cultivation of a single crop over wide areas; a stand of a single species.

monogynous Of insects, having only one queen per colony.

monotype 1. A taxonomic group that is virtually a single species. **2.** A monodominant stand of vegetation (any associates contribute very little cover).

monoxenous parasites Parasites having life cycles with a single host species.

Morbillivirus A family of single-stranded RNA viruses that cause acute illness in their hosts.

mosses The most species-rich division of bryophytes, with approximately 12,000 species globally, characterized by their multicellular rhizoids.

motile Capable of self-propelled movement.

multilateral agreements Agreements made by three or more governments.

multiple introductions An introduction of a species in a new territory characterized by the release of individuals from multiple source populations in the native range.

multivariate statistics A type of statistical analysis wherein two or more variables are treated simultaneously; especially, statistical analysis in which there is more than one response variable.

multivoltine Having multiple generations per year.

mutation Evolution through alterations in the genetic code; e.g., nucleotide substitutions (point mutations), insertions, deletions, gene duplications, translocations, chromosomal inversions, and chromosomal duplications. Mutations can occur through DNA damage, errors during DNA replication or meiosis, or the action of transposable elements.

mutualism A positive interaction between two species, in which each species aids the other.

mutualist An organism that enters a mutually positive relationship with another organism; examples include associations of plant roots with mycorrhizal fungi and with nitrogen-fixing bacteria.

mycorrhizae/mycorrhizas Symbioses between mycorrhizal fungi and plant roots. From the Latin *myco* (fungi) and Greek *rhiza* (root).

mycorrhizal fungi Fungi that form mutualistic associations with plant roots. The plant provides the fungus with carbon, and the fungus takes up soil nutrients that are in turn provided to the plant.

myrmechory Seed dispersal by ants.

native species See INDIGENOUS SPECIES.

natural enemy An organism (predator, parasite, parasitoid, pathogen, or competitor) that diminishes populations of a species in nature, even if only (as with parasites) by weakening individuals rather than killing them. Some natural enemies may be used by humans as biocontrol agents.

natural flow regime The natural pattern of temporal variation in river flow features, including frequency, magnitude, timing, and duration of floods and droughts.

natural history A study of living organisms relying more on observational than experimental methods; especially applied to observation of whole organisms in nature.

naturalization The establishment of a species in the wild outside its natural range without the support of humans.

naturalized Describing or referring to an introduced organism that reproduces without human help and that may or may not spread further.

natural selection The process by which individuals having particular heritable traits are more likely to survive and reproduce, so that the traits become more common in a population over successive generations.

NDVI Normalized difference vegetation index.

nekton Actively swimming animals.

neritic Occurring in shallow coastal waters over the continental shelf.

net reproduction rate (R_0) The average number of female offspring that would be born to each member of a birth cohort of females during her lifetime if the population experienced a fixed pattern of age-specific birth and death rates.

neurologic Pertaining to the nervous system, including the brain, spinal cord, and peripheral nerves.

neutral loci Genetic loci (specific locations of a gene or DNA sequence on a chromosome) that are evolving not in response to natural selection but rather through neutral processes of random genetic drift, mutation, and migration.

niche See ECOLOGICAL NICHE.

NIP Nonindigenous parasite.

NIS Nonindigenous species.

nitrification Biological oxidation of ammonia (NH_3) first to nitrite (NO_2^-) and then to nitrate (NO_3^-) in soils, sediments, or water columns.

nitrogen fixation A process in which atmospheric nitrogen is assimilated by certain bacteria into compounds available for use by fungi and plants, frequently in symbiotic associations.

nitrogen-fixer A plant species that interacts with bacteria to incorporate atmospheric nitrogen (N_2) into compounds such as ammonia and nitrate that are useful for various chemical processes, including physiological processes of plants.

nitrophilic species A species that responds to nitrogen enrichment through increased growth or survival.

non-binding agreements Agreements that provide guidance but that are not enforceable; generally referred to as "soft law."

nonindigenous Occurring in an area beyond the native range.

nonindigenous parasite (NIP) A successfully introduced and established parasite.

nonindigenous species (NIS) A successfully introduced and established species.

nonnative species See INTRODUCED SPECIES.

nonparametric model A model independent of any particular statistical distribution defined by parameters.

nontarget species An organism that is not the intended object of a management tactic.

normalized difference vegetation index (NDVI) A spectral metric to enhance the signal of photosynthetically active vegetation with a combination of visible and near-infrared spectral bands.

noxious weed A weed regulated under federal, state, or local laws or regulations.

nutrient limitation Scarcity of essential elements needed for plant growth.

obligate *By necessity;* as, for example, a type of symbiosis in which an organism cannot persist without its partner species. (Antonym of *facultative*.)

occurrence The presence of a particular species in a particular geographically defined entity (e.g., *Chamaeleo jacksonii* in Hawaii and *C. jacksonii* in California comprise two occurrences, although each may consist of many biological populations); used as a measure of the extent of alien introductions globally.

ornithophilic Of blood-sucking insects, preferring birds as hosts.

outbreaks Explosive increases in the abundance of a particular species.

outbreeding Mating between different genotypes within a population. Also, CROSS-FERTILIZATION.

oviposition Deliberate placement of eggs, usually on a firm surface, as by monarch butterflies.

pandemic Worldwide outbreak arising from a single area and caused by a pathogen (usually, a virus) unrelated to previous infecting pathogens.

parasite An organism that lives in or on, and takes nourishment from, another organism and that has a negative effect on its host.

parasitoid An organism that develops in or on a single host individual, eventually killing it; refers to some invertebrate animals (e.g., some groups of flies and wasps) whose larvae show this trait.

parthenogenesis A form of asexual reproduction in which females produce eggs that develop without fertilization.

pathogen Any disease-causing agent; most commonly refers to infectious bacteria, viruses, and other microorganisms.

pathogenic Causing disease, or capable of doing so.

pathway The means by which a species is introduced.

pelagic Referring to or occurring in open waters of lakes and seas.

peregrine Having established, though not necessarily invasive, population(s) outside the natural range of the species.

perennial Persisting for two or more years.

performance curve A curve illustrating the effect of a variable on a species' fitness-related trait.

persistence The ability of a species to avoid extirpation.

pest A species possessing characteristics that are considered hazardous or unwanted by humans.

pesticide Any chemical used to kill pests. Often specific terms are used for chemicals targeting plant pests (herbicides), insects (insecticides), fungi (fungicides), and the like.

pesticide resistance Adaptation of a pest population resulting in decreased susceptibility to some pesticides.

pest management strategy or **tactic** An overall plan for alleviating a pest problem.

petiole The slender stalk that attaches a leaf to a stem.

phenology The timing of a species' life history events in response to seasons or climatic events.

phenotype External manifestation of the genotype.

phenotypic plasticity The capacity of a genotype to produce different phenotypes in different environments.

pheromone A chemical emitted by one organism to communicate with another organism of the same species.

phloem A thin layer of tissue between the outer bark and the sapwood on trees; transports nutrients within trees.

phosphites or **phosphonates** Systemic fungicides that have a high level of environmental safety and very low nontarget toxicity.

photosystem II (PSII) A protein complex in the chloroplast thylakoid membranes that is responsible for initiating photosynthetic electron transport and splitting water molecules, releasing protons and oxygen.

phylogenetic approach An analysis using a hypothesis describing evolutionary relationships among the species involved.

phylogeography The study of processes that control the geographic distributions of lineages by constructing the genealogies of populations and genes.

phytotelmata Small bodies of water impounded by terrestrial plants that harbor aquatic metazoan communities.

piscivorous Fish-eating.

planktivore A species that feeds primarily on zooplankton.

plankton Organisms living in the water column that drift with currents.

planktotrophic Feeding in the water column.

plant propagule Any viable structure with the capacity to give rise to a new plant (e.g., a seed, spore, or vegetative part capable of independent growth if detached from the parent plant).

plant traits Attributes of plants that differ among species: for example, leaf size, nutrient concentration, growth form, leaf toughness. Key traits are often important in determining how plant species influence ecosystem functioning.

plastic Able to express different phenotypes in different environmental conditions.

ploidy A multiple of the basic number of sets of chromosomes in a cell.

pollen limitation A reduction in potential seed set caused when some ovules remain unfertilized owing to insufficient pollination or to the receipt of poor quality pollen.

pollination The process by which pollen is moved from the male part of a flower (anther) to the female part (stigma). If the pollen is chemically suitable for the stigma, a pollen tube will grow and fertilize an ovule, producing a seed.

pollinator An animal vector that carries pollen from anthers to stigmas, usually of different flowers.

pollinator limitation Limitation of seed production by insufficient pollinator visitation.

polygynous Of insects, having more than one queen per colony.

polylectic Having a broad diet.

polyphagous Of insects, feeding on many species and multiple genera of host plants.

polyploid Having more than two complete sets of chromosomes in the cell nucleus. This condition can arise by doubling within one species or by hybridization between two species.

polyvectic Capable of being introduced by two or more vectors.

Ponto-Caspian species Species originating either from the lower reaches of rivers along the margins of the Black, Caspian, and Azov seas, or from these seas.

population A collection of individuals belonging to the same species in one location.

population bottleneck See GENETIC BOTTLENECK.

population growth rate The rate of change in the number of individuals in a population over time.

population projection matrix model A discrete time model for dynamics of population size and structure in which birth, growth, and death rates are structured by categorical states such as age or stage.

population rate of spread The change in the spatial extent of a population over time.

positive feedback A phenomenon or process resulting from species interactions that builds on itself in an accelerating fashion, eventually becoming unstoppable.

post-immigration evolution Phenotypic or life-history trait evolution that takes place following the introduction of a species into a new territory and that can occur through either stochastic processes or natural selection.

pre-adaptation A trait inherited from an ancestor and used by an animal or plant for a purpose unrelated to that for which the trait was used by the ancestor.

precipitation The deposition of moisture in liquid form on the surface of the Earth, including through rainfall, snow, and dew, and as intercepted fog and mist.

prevalence In epidemiology, the percentage of infected individuals in a host population.

prey The organism that serves as food for a predator and that is killed by the act of predation.

primary or **secondary space** The place occupied by sessile organisms directly on an abiotic surface (primary) or on other organisms (secondary).

principal coordinate axes Mutually noncorrelated axes derived by principal coordinates analysis, a method for building sequence ordinations using dissimilarity matrices derived from measurements on objects of interest.

producers In ecology, autotrophic green plants that form the basis of the food chain.

propagule A dispersal unit for reproduction of plants or other organisms by sexual or asexual means, including seeds, fruits, bulbs, and other structures.

propagule pressure The extent to which an environment is exposed to the introduction of new organisms; a combination of the number of viable propagules (e.g., seeds, eggs) in individual dispersal episodes and the number of dispersal episodes.

prosobranch A term applied to groups of snails possessing an operculum (a trap door–like structure that closes the shell aperture when the snail withdraws); includes most marine and some land and freshwater snails.

protocols Supplementary, often more specific, guidance within the context of legally binding agreements.

provisional measure A temporary regulatory action taken to limit introduction of a particular species when there is insufficient scientific evidence available to evaluate its potential risk.

pseudofeces Mucus-bound packets of detrital material rejected as being unsuitable food.

pseudoindigenous Describing or referring to an introduced species incorrectly thought to be native.

public goods Goods that are nonrival (consumption by one person does not preclude consumption by others) and non-exclusive (no one can be prevented from accessing the good).

pulmonate Describing or referring to snails and slugs that lack an operculum and possess a lung with which to breathe air. Most land and freshwater snails are pulmonates.

pycnohalocline Density and salinity gradient in the vertical stratification of the sea.

pyrene The seed and the innermost layer of the wall of a fruit.

ramet An individual member of a clone; e.g., the mother plant and each daughter plant separately.

rangeland A "natural" habitat whose primary use is for livestock grazing. Rangelands include mainly deserts, montane meadows, grasslands, and savannas.

rapid responders Personnel who can be mobilized to respond, usually as part of a team, to reports of invasive organisms.

Raunkiaer's scheme A life-form classification based on the location of the plant's buds during unfavorable growing conditions. (Proposed by Danish botanist Christen Raunkiaer.)

reactive oxygen species (ROS) Oxygen-containing free radicals (singlet oxygen, superoxide, peroxide, and hydroxyl radical) that rapidly react with nucleic acids, proteins, and lipids, causing undesirable oxidative damage.

reassortment The mixing of the genetic material of a species into new combinations in different individuals; e.g., a process by which new influenza viruses can be generated through combinations of genes in one individual that had previously existed only in separate individuals.

recruitment 1. The addition of new individuals to a population through reproduction or immigration. **2.** Survival of young fish or eggs to adulthood.

recurved Curved backward.

reduction-oxidation (or redox) potential Affinity of chemical species for electrons, and thus the tendency of the species to be reduced.

reference sites or **communities** Less disturbed (even "original") plant communities that can serve as targets for restoration efforts, being located nearby or at least having similar environmental traits (e.g., soils, elevation, rainfall).

reflex bleeding An insect's production of droplets of hemolymph or noxious chemicals at leg joints.

regional agreements Agreements made among neighboring countries; may include the distant protectorates of those neighboring countries.

relative growth rate (RGR) A measure of biomass productivity, defined as the increase in mass per unit of organism mass over a time unit (for plants, usually *g/g/day*).

removed In epidemiology, describing or referring to an individual that is either dead or recovered and thus no longer able to infect others.

reproductive isolation The inability of a species to breed successfully with a related species, usually as a result of geographic, behavioral, morphological, or genetic barriers.

reproductive number In epidemiology, the mean number of new infections produced by a single individual while it is infective.

reproductive output The production of offspring by an individual.

resource availability The availability of resources, such as water, nutrients, and space. High resource availability in an environment usually means that it is more invasible.

resources Abiotic environmental factors necessary for survival, growth, and reproduction, such as water, light, mineral nutrients, and various forms of fixed carbon (e.g., plant material or meat).

response variable Dependent (*y*), or outcome, variable.

restoration The repairing of damage caused to an ecosystem by human use or degradation.

rhizomatous species Plant species able to reproduce by underground stems known as *rhizomes*.

rhizome A characteristically horizontal plant stem that is usually found underground, often sending out roots and shoots from its nodes. Rhizomes can store carbohydrates, allowing for continued resprouting from a "bud bank" after removal of aboveground portions of the plant.

rhizosphere The soil zone in which plant roots are embedded that receives root exudates.

risk A state of uncertainty in which some possible outcomes have an undesired effect or entail a significant loss.

risk assessment The evaluation of the likelihood of entry, establishment, or spread of a pest or disease within the territory of an importing country, and the associated potential biological, health, and economic consequences of such a spread.

root exudate Any of a suite of different chemicals or plant products released from roots into the environment.

ruderal Describing or referring to a plant species that colonizes frequently disturbed and nutrient-rich areas.

saturation A condition of an area in which resource limitations prevent the establishment of additional species.

scale insect A member of the Coccoidea (insect order Hemiptera): small, sessile, plant-feeding insects having sucking mouthparts. Many of these species are worldwide pests.

scarification The breaking, scratching, or softening of a seed coat, thus allowing water to penetrate.

schistosomiasis The second most socioeconomically devastating parasitic disease after malaria; caused by trematodes of the genus *Schistosoma*, intermediate hosts of which are several species of freshwater snails. Also, BILHARZIOSIS, SNAIL FEVER.

screening A systematic examination or assessment, performed especially to detect an unwanted substance or attribute. In the context of biological invasion and disease risks, the term refers to *risk assessment*.

sea chest A space inside the hull of a vessel into which water is drawn to be used for ballast and other purposes.

seaweed Large, usually multicellular marine alga.

secondary metabolites Organic compounds not directly involved in primary metabolic processes such as photosynthesis, cell respiration, cell division, or cell growth.

secondary poisoning The poisoning of an animal that consumes the tissues or organs of another poisoned animal.

secondary spread Spread of a nonindigenous species to adjacent areas following initial invasion.

seed 1. The ripened ovule of a plant consisting of the embryo plus a protective seed coat; often endosperm (a nutritive tissue) is present. **2.** Young postlarval bivalves that can be planted to grow into a cultured crop.

seed bank The population of living seeds that occurs in the soil. Also, SOIL SEED BANK.

seed dispersal Dispersal of a seed away from the vicinity of the mother plant.

seed dormancy A state in which viable seeds are prevented from germination even though environmental conditions are favorable for germination.

seed production The number of seeds produced by a plant; may be expressed on a per individual or per unit biomass or canopy area or canopy volume basis.

selection Any process, natural or artificial, by which certain individuals or characters are favored or perpetuated in preference to others.

selective insecticide A substance that is used to control a pest insect but that is typically less toxic to some groups of insects than a broad-spectrum insecticide is.

sensitivity The degree to which a summary variable (e.g., population growth rate) changes when there is a small change of a particular parameter: e.g., vital rate (technically a partial derivative).

serotiny The storage of seeds by plants in aerial structures (e.g., cones) rather than in the soil seed bank, with seed release triggered in response to an environmental factor (e.g., fire).

shallow lakes Lakes that are shallow enough that they typically mix well.

simple sequence repeats See MICROSATELLITES.

siphonous Describing or referring to organization of algae that usually consists of hollow bags or filaments (siphons) in which cells are lacking. There are no cross walls separating nuclei, and the whole thallus consists of a continuous mass of cytoplasm contained only by the outer walls of the filament or sac.

slug A gastropod mollusc lacking, or almost lacking, a shell. *Semislugs* possess a rudimentary shell into which they cannot fully retract the body. Slugs have evolved from snails many times independently.

small world In epidemiology, a network characterized by connectedness of all individuals through a small number of other individuals.

smother crop A fast-growing crop grown at high densities or in closely spaced rows in order to suppress weed growth.

snail A gastropod mollusc with a shell, permanently attached to the shell by a single muscle, into which it can generally fully retract its body.

snail fever See SCHISTOSOMIASIS.

soil moisture or **soil water** Water held in the soil; includes both liquid and vapor phases.

soil seed bank See SEED BANK.

source populations See GEOGRAPHIC ORIGINS.

spatfall Natural recruitment, appearance of juvenile oysters (spat) on hard surfaces.

specialist A species that can survive and reproduce only within a narrow range of environmental conditions using specific resources (e.g., a specific diet).

specialist pathogen A pathogen that infects a single host species.

species complex A group of species that satisfy the biological definition of species (i.e., that are reproductively isolated from each other) but that are very similar (if not identical) morphologically.

specific leaf area (SLA) The ratio of one-sided fresh leaf area and dry leaf mass; generally measured in $cm^2\ g^{-1}$. Many invasive plant species have high SLA (> 150 $cm^2\ g^{-1}$).

spillover Increased pathogen transmission to a host population resulting from the presence of another host population that serves as a reservoir for the pathogen population.

spore A reproductive structure, usually haploid, that is produced by meiosis and that is adapted for dispersal and survival for extended periods of time in unfavorable conditions. Spores form part of the life cycles of several taxonomic groups (e.g., bryophytes, ferns and fern allies, fungi).

SPS Agreement The Sanitary and Phytosanitary Agreement, an international treaty under the World Trade Organization that sets out the rights and obligations of member states for establishing national regulations that govern animal and plant health as well as food safety.

standing biomass The total mass (measured as volume or weight) of a species occupying a given area.

standing genetic variation Allelic variation that is currently present and segregating within a population.

statutory laws Positive and authoritative legal statements of responsibilities, duties, obligations, and liability by a legislature.

steppe Temperate grassland dominated by perennial grasses and devoid of trees on zonal soils.

stepping stone The sequence of spread of nonindigenous species in a series of unconnected areas.

sterile male release technique (SMRT) A method used to control nuisance populations, mostly of insect species, whereby sterile males are released to the wild; females that mate with a sterile male thus have no offspring, reducing the population size of the next generations.

stigma The part of a female portion of a flower that receives pollen.

stocking The release of hatchery-produced organisms in order to create added value in a fishery. Methods include *restocking* (release of cultured juveniles to restore spawning biomass), *stock enhancement* (release of cultured juveniles to overcome recruitment limitations), and *ranching* (release of cultured juveniles or adults to be taken in a fishery).

stolon A modified stem that grows horizontally and produces new individuals (ramets) capable of independent existence.

stratified dispersal A common mode of dispersal in invading species in which local growth and dispersal are coupled with long-range movement.

substrate Any firm surface, usually terrestrial soil or mud, sand, or gravel at the bottom of rivers, lakes, and the sea.

succession The change in species composition of ecological communities over time, as when an abandoned field becomes a forest. *Primary succession* occurs on sites having mineral substrate only, such as that following the retreat of glaciers or production of lava flows. In *secondary succession,* the soil and some survivors, seeds, or clonal fragments remain after removal of all or part of the previous plant community.

surface water Water in liquid form that emerges onto or flows over the land surface, including into rivers, wetlands, marshes, lakes, and manmade dams. Flows over land and into rivers are also called *runoff.*

surfactant Generally, a wetting agent that reduces the surface tension of water on a leaf surface; added to, or formulated with, an herbicide to cause spreading of the spray solution on the surface. (Derived from "surface-acting agent.")

surveillance Monitoring to determine pest presence, density, dispersal, and dynamics.

susceptible Describing or referring to an individual that could potentially be infected.

sylvatic Affecting only wild animals.

symbiont An organism that benefits from an interaction with another organism; usually, the smaller, or mobile, organism participating in a mutualism.

sympatric speciation The process whereby populations of originally interbreeding individuals cease to interbreed and become distinct without undergoing geographical separation.

symplasm The living components of plants; includes all cells containing a nucleus.

synergism Interaction among agents (e.g., chemical compounds or biological species) such that the total effect is greater than the sum of the individual effects.

systemic herbicides Herbicides that translocate in the living tissues of a plant, particularly the phloem, and accumulate in the growing points or the storage organs.

taxa (*singular,*** taxon)** Taxanomic gropus of any rang (e.g., family, genus, or species), including all the subordinate groups (e.g., genera within families).

taxonomy The classification of animals and plants in an ordered system that indicates natural relationships.

temporal and **spatial autocorrelation** Temporal or spatial dependence among samples. Such dependence usually violates statistical assumption of random independent sampling.

teratogenesis The formation of congenital defects during embryonic development.

terminal velocity The speed of a falling object when constant owing to the restraining force of the still air through which it is moving.

thicket A thick or dense growth of shrubs.

threatened species Species that fall into the *endangered* or *vulnerable to extinction* categories in the IUCN system; refers, under the U.S. Endangered Species Act, to species at risk of extinction, though at lower risk than *endangered species.*

throughfall Precipitation that passes through a tree canopy.

toxicant An agent or material capable of producing an adverse response in a biological system.

toxicology The study of toxic effects of drugs and other harmful substances.

transformation The modification of an existing ecosystem through altered land use that results in new ecosystem configurations or replacement with crops or other intensively managed vegetation.

transformer species An introduced species that has established self-replacing populations and that has traits that make it able to change the structure and functioning of at least some of the ecosystems it invades.

transgressive segregation The expression of phenotypes observed in segregating hybrid populations that are extreme compared to phenotypes in either of the parental lines.

translocation The movement (by humans) of a species from a country where it occurs to another part of the same country.

transpiration Evaporation of water from plants; occurs primarily in the leaves through the stomata. The rate of water loss is regulated by the plant and occurs primarily during daylight hours when the stomata are open.

trematodes An invertebrate class that includes the liver fluke.

tristyly A genetic polymorphism in which plant populations are composed of three floral morphs that differ in the length and position of their female (style) and male (anther) sexual organs. The floral polymorphism promotes animal-mediated cross-pollination among the floral morphs.

trophic cascade A chain reaction within food webs that results when altered population densities at higher trophic levels cause shifts in the dominance and impact of consumers at lower levels.

trophallaxis The sharing of food among social insects.

tuber A swollen stem that serves as a vegetative propagule. It is produced belowground in some terrestrial and aquatic plants and can remain viable for up to 12 years.

turion A vegetative propagule (dormant bud) that is produced by some aquatic plants either aboveground or belowground on shoots and that can serve as a dispersal propagule.

tychoplankton Benthic organisms that are swept up into the water column.

type locality The location where a new species is first described (although not necessarily where it is native).

uncertainty In risk assessment, a state of having limited knowledge such that it is impossible to describe exactly the existing state or the future outcome.

ungulate A large herbivorous land mammal that is either odd- (e.g., horse) or even- (e.g., deer) toed.

unionids Freshwater mussels (family Unionidae) that normally inhabit soft sediment habitats.

univoltine Having one generation per year.

variety A taxonomic category below that of species and subspecies that differentiates variable populations.

vector 1. A mechanism or pathway (e.g., transoceanic shipping or canal) that is responsible for the introduction of nonindigenous species. **2.** An organism that does not cause or experience disease itself but that instead serves as a means of transmission between host individuals.

vector-borne disease An infectious disease caused by a pathogen transmitted from one host to another by another organism, most often a blood-sucking arthropod vector (insect or tick).

vegetative propagation Generally, reproduction by nonsexual means; the reproduction and dispersal of a plant or alga without exchange of genetic material. Also, ASEXUAL REPRODUCTION.

veliger The larval stage in bivalves, characterized by two tiny hinged shells holding ciliated lobes of tissue that enable the larva to swim (slowly) and to capture food particles.

velvet Soft tissue that covers antlers during growth.

venison Meat from deer or elk.

vernal pools Small, isolated wetlands that retain water on a seasonal basis but that dry completely at other times of the year.

viable population A population that can replace itself through birth and the survival of young.

water quality Overall suitability of water for human and animal use as reflected by its clarity, purity, nutrient load, and possible presence of algae.

water resources The field of science and management that assesses the quantity and quality of water available for various purposes, mainly with a focus on human use.

watershed The area of land that contributes water (typically) to a stream or river tributary; called a *catchment* or *river basin* when the area covers an entire river system.

water-use The total use of water by a plant, including transpiration and interception losses.

weed An unwanted plant, usually perceived as detrimental to human interests.

weed control Reduction of weed presence to an economically acceptable level.

weed prevention Actions taken to keep weeds from becoming established where they do not already occur.

white list 1. A policy response to harmful nonindigenous species that prohibits the introduction of all or many species and that allows the introduction of only those species that have been approved, either because they have been found to present little or no threat of harm or because the benefits of allowing further introduction are believed to outweigh the costs. **2.** A list of species that are approved for introduction.

wool gardens Stands of plants growing from seeds washed out of imported fleeces.

xylem cavitation Loss of continuity of the water in the water-conducting tissue of plants as a result of formation of bubbles that block conducting elements. When extreme, the loss of conductivity usually cannot be reversed, and distal tissues dehydrate and die.

xylem The woody tissue that supports and transports water in trees.

zoobenthic Describing or referring to animals (zoobenthos) living on or near the bottom of a body of water.

zoonosis Disease transmitted from animals to humans.

Below is a short list of the most important book/volume references on biological invasions. This is not a complete bibliography of publications on invasions. Items listed under both General References and Regional Volumes are intentionally arranged chronologically, with the aim of illustrating historical development of invasion biology as a science. Volumes on more specialized topics are arranged alphabetically by author or editor name(s). Authors and titles of textbooks are printed in bold.

General References (Listed Chronologically)

Elton, C. S. 1958. *The Ecology of Invasions by Animals and Plants.* London: Methuen and Co. LTD. [Reprinted edition 2000. Chicago: University of Chicago Press.]

Baker, H. G., and G. L. Stebbins, eds. 1965. *The Genetics of Colonizing Species.* New York: Academic Press.

Crosby, A. 1986. *Ecological Imperialism: The Biological Expansion of Europe, 900–1900.* Cambridge, UK: Cambridge University Press. [Revised edition 2004.]

Kronberg, H., and M. H. Williamson. 1987. *Quantitative Aspects of the Ecology of Biological Invasions.* London: The Royal Society.

Drake, J. A., H. A. Mooney, F. di Castri, R. H. Groves, F. J. Kruger, M. Rejmánek, and M. Williamson, eds. 1989. *Biological Invasions: A Global Perspective.* Chichester, UK: John Wiley and Sons.

Hengeveld, R. 1989. *Dynamics of Biological Invasions.* London: Chapman and Hall.

McKnight, B. N., ed. 1993. *Biological Pollution: The Control and Impact of Invasive Exotic Species.* Indianapolis: Indiana Academy of Science.

Carey, J. R., P. Moyle, M. Rejmánek, and G. Vermeij, eds. 1996. *Invasion Biology.* Special issue of *Biological Conservation* 78: 1–214.

Williamson, M. 1996. *Biological Invasions.* London: Chapman and Hall.

Shigesada, N., and K. Kawasaki 1997. *Biological Invasions: Theory and Practice.* Oxford: Oxford University Press.

Sandlund, O. T., P. J. Schei, and A. Viken, eds. 1999. *Invasive Species and Biodiversity Management.* Dordrecht, The Netherlands: Kluwer Academic Publishers.

Yano, E., K. Matsuno, M. Shiyomi, and D. A. Andow, eds. 1999. *Biological Invasions of Ecosystems by Pests and Beneficial Organisms.* Tukuba, Japan: National Institute of Agro-Environmental Sciences.

Mooney, H. A., and R. J. Hobbs, eds. 2000. *Invasive Species in a Changing World.* Washington, DC: Island Press.

Perrings, C., M. Williamson, and S. Dalmazzone, eds. 2000. *The Economics of Biological Invasions.* Cheltenham, UK: Edward Elgar.

National Research Council. 2001. *Predicting Invasions of Nonindigenous Plants and Plant Pests.* Washington, DC: National Academy Press.

Baskin, Y. 2002. *A Plague of Rats and Rubbervines.* Washington, DC: Island Press/Shearwater Books.

Pimentel, D., ed. 2002. *Biological Invasions: Economic and Environmental Costs of Alien Plant, Animal, and Microbe Species.* Boca Raton, FL: CRC Press.

Veitch, C. R., and M. N. Clout, eds. 2002. *Turning the Tide: The Eradication of Invasive Species.* Gland, Switzerland/Cambridge, UK: IUCN, The World Conservation Union.

Kowarik, I. 2003. *Biologische Invasionen: Neophyten und Neozoen in Mitteleuropa.* Stuttgart, Germany: Verlag Eugen Ulmer.

Ruiz, G. M., and J. T. Carlton, eds. 2003. *Invasive Species: Vectors and Management Strategies.* Washington, DC: Island Press.

Summer, D. A. 2003. *Exotic Pests and Diseases: Biology and Economics for Biosecurity.* Ames: Iowa State Press.

Alimov, A. F., and N. G. Bogutskaya, eds. 2004. *Biological Invasions in Aquatic and Terrestrial Ecosystems.* Moscow: KMK Scientific Press Ltd. [In Russian]

Cox, G. W. 2004. *Alien Species and Evolution.* Washington, DC: Island Press.

Mooney, H. A., R. N. Mack, J. A. McNeely, L. E. Neville, P. J. Schei, and J. K. Waage, eds. 2005. *Invasive Alien Species: A New Synthesis.* Washington, DC: Island Press.

Sax, D. F., J. J. Stachowitz, and S. D. Gaines, eds. 2005. *Species Invasions: Insights into Ecology, Evolution, and Biogeography.* Sunderland, MA: Sinauer.

Cadotte, M. W., S. M. McMahon, and T. Fukami, eds. 2006. *Conceptual Ecology and Invasion Biology: Reciprocal Approaches to Nature.* Dordrecht, The Netherlands: Springer.

Lockwood, J. L., M. F. Hoopes, and M. P. Marchetti. 2007. *Invasion Biology.* Malden, MA: Blackwell Publishing.

Nentwig, W., ed. 2007. *Biological Invasions.* Dordrecht, The Netherlands: Springer.

van der Weijden, W., R. Leewis, and P. Bol. 2007. *Biological Globalisation.* Utrecht, The Netherlands: KNNV Publishing.

Clout, M. N., and P. A. Williams. 2009. *Invasive Species Management: A Handbook of Principles and Techniques.* Oxford: Oxford University Press.

Davis, M. A. 2009. *Invasion Biology.* Oxford: Oxford University Press.

Keller, R. P., D. M. Lodge, M. A. Lewis, and J. F. Shogren, eds. 2009. *Bioeconomics of Invasive Species.* Oxford: Oxford University Press.

Barbault, R., and M. Atramentowicz, eds. 2010. *Les Invasions Biologiques, une Question de Natures et de Sociétés.* Antony Cedex, France: Quae.

Franklin, J. 2010. *Mapping Species Distributions: Spatial Inference and Prediction.* Cambridge, UK: Cambridge University Press.

Perrings, C., H. Mooney, and M. Williamson, eds. 2010. *Bioinvasions and Globalization.* Oxford: Oxford University Press.

Richardson, D. M., ed. 2010. *Fifty Years of Invasion Ecology.* Oxford: Wiley-Blackwell.

Regional Volumes (Listed Chronologically)

Thomson, G. M. 1922. *The Naturalisation of Animals and Plants in New Zealand.* Cambridge, UK: Cambridge University Press.

Groves, R. H., and J. J. Burdon, eds. 1986. *Ecology of Biological Invasions: An Australian Perspective.* Canberra: Australian Academy of Science.

Macdonald, I. A. W., F. J. Kruger, and A. A. Ferrar, eds. 1986. *The Ecology and Management of Biological Invasions in Southern Africa.* Cape Town: Oxford University Press.

Mooney, H. A., and J. A. Drake, eds. 1986. *Ecology of Biological Invasions of North America and Hawaii.* New York: Springer-Verlag.

di Castri, F., A. J. Hansen, and M. Debussche, eds. 1990. *Biological Invasions in Europe and the Mediterranean Basin.* Dordrecht, The Netherlands: Kluwer Academic Publishers.

Groves, R. H., and F. di Castri, eds. 1991. *Biogeography of Mediterranean Invasions.* Cambridge, UK: Cambridge University Press.

Ramakrishnan, P. S., ed. 1991. *Ecology of Biological Invasions in the Tropics.* New Delhi: International Scientific Publications.

Office of Technology Assessment. 1993. *Harmful Non-Indigenous Species in the United States.* Washington, DC: Office of Technology Assessment, United States Congress.

Simberloff, D., D. C. Schmitz, and T. C. Brown, eds. 1997. *Strangers in Paradise: Impact and Management of Non-Indigenous Species in Florida.* Washington, DC: Island Press.

Smith, I. M., D. G. McNamara, P. R. Scott, and M. Holderness, eds. 1997. *Quarantine Pests for Europe.* Wallingford, UK: CAB International.

Cox, G. W. 1999. *Alien Species in North America and Hawaii: Impacts on Natural Ecosystems.* Washington, DC: Island Press.

Low, T. 1999. *Feral Future.* Victoria, Australia: Viking, Ringwood. [Also 2002. Chicago: University of Chicago Press.]

Claudi, R., P. Nantel, and E. Muckle-Jeffs, eds. 2002. *Alien Invaders in Canada's Waters, Wetlands, and Forests.* Ottawa: Natural Resources Canada, Canadian Forest Service.

Essl, F., and W. Rabitsch, eds. 2002. *Neobiota in Österreich.* Vienna: Umweltbundesamt GmbH.

Tellman, B., ed. 2002. *Invasive Exotic Species in the Sonoran Region.* Tucson: University of Arizona Press.

Weidema, I. R., ed. 2002. *Introduced Species in the Nordic Countries.* Copenhagen: Nordic Council of Ministers.

Capdevilla-Argüelles, L., B. Zilletti, and N. Pérez, eds. 2003. *Contribuciones al Conocimiento de las Especies Exóticas Invasoras en España.* León, Spain: Grupo Especies Invasoras.

Mihaly, B., and Z. Botta-Dukát, eds. 2004. *Biológiai Inváziók Magyarországon Özönnövények.* Budapest: TermészetBÚVAR Alapítvány Kiadó.

de la Vega, S. G. 2005. *Invasión en Patagonia.* Buenos Aires: Contacto Silvestre.

Allen, R. B., and W. G. Lee, eds. 2006. *Biological Invasions in New Zealand.* Berlin: Springer-Verlag.

Beauvais, M.-L., A. Coléno, and H. Jourdan, eds. 2006. *Les Espèces Envahissantes dans l'Archipel Néo-Calédonien.* Paris: IRD.

Boersma, P. D., Reichard, S. R., and A. N. Van Buren. 2006. *Invasive Species in the Pacific Northwest.* Seattle: University of Washington Press.

Coates, P. 2006. *American Perceptions of Immigrant and Invasive Species.* Berkeley: University of California Press.

Mlíkovsky, J., and P. Styblo. 2006. *Nepuvodní Druhy Fauny a Flóry Ceské Republiky.* Prague: Cesky Svaz Ochrancu Pririrody.

Pascal, M., O. Lorvelec, and J.-D. Vigne. 2006. *Invasions Biologiques et Extinctions.* Paris: Belin.

DAISIE. 2009. *Handbook of Alien Species in Europe.* Dordrecht, The Netherlands: Springer.

Wan, F.-H., J.-Y. Guo, and F. Zhang. 2009. *Research on Biological Invasions in China.* Beijing: Science Press.

Plants

AKEPIC—Alaska Exotic Plant Information Clearinghouse. 2005. *Invasive Plants of Alaska.* Anchorage: Alaska Association of Conservation Districts Publication.

Booth, B. D., S. D. Murphy, and C. J. Swanton. 2003. *Weed Ecology in Natural and Agricultural Systems.* Wallingford, UK: CABI Publishing.

Bossard, C. C., J. M. Randall, and M. C. Hoshovsky, eds. 2000. *Invasive Plants of California's Wildlands.* Berkeley: University of California Press.

Botta-Dukát, Z., and L. Balogh. 2008. *The Most Important Invasive Plants in Hungary.* Vácrátót: Institute of Ecology and Botany, Hungarian Academy of Sciences.

Brock, J. H., M. Wade, P. Pyšek, and D. Green, eds. 1997. *Plant Invasions: Studies from America and Europe.* Leiden: Backhuys Publishers.

Brundu, G., J. Brock, I. Camarda, L. Child, and M. Wade., eds. 2001. *Plant Invasions: Species Ecology and Ecosystem Management.* Leiden: Backhuys Publishers.

Celesti-Grapow, L., F. Pretto, G. Brundu, E. Carli, and C. Blasi, eds. 2009. *Plant Invasions in Italy: An Overview.* Rome: Palombi and Partner S.r.l.

Child, L., J. H. Brock, G. Brundu, K. Prach, P. Pyšek, M. Wade, and M. Williamson, eds. 2003. *Plant Invasions: Ecological Threats and Management Solutions.* Leiden: Backhuys Publishers.

Clement, E. J., and M. C. Foster 1994. *Alien Plants of the British Isles.* London: Botanical Society of the British Isles.

Coombs, E. M., et al., eds. 2004. *Biological Control of Invasive Plants in the United States.* Corvalis: Oregon State University Press.

Cronk, Q. C. B., and J. L. Fuller. 1995. *Plant Invaders.* London: Chapman and Hall.

De Candolle, A. P. 1855. *Géographie Botanique Raisonné,* Vol. 2. Paris: V. Masson.

Groves, R. H., F. D. Panetta, and J. G. Virtue, eds. 2001. *Weed Risk Assessment.* Collingwood, Australia: CSIRO.

Groves, R. H., R. C. H. Shepherd, and R. G. Richardson. 1995. *The Biology of Australian Weeds,* Vol. 1. Melbourne: R. G. and F. J. Richardson.

Henderson, L. 2001. *Alien Weeds and Invasive Plants: A Complete Guide to Declared Weeds and Invaders in South Africa.* Pretoria: Agricultural Research Council.

Holm, L., J. Doll, E. Holm, J. Pancho, and J. Herberger. 1997. *World Weeds: Natural Histories and Distribution.* New York: John Wiley and Sons.

Holm, L. G., D. Plucknett, J. V. Pancho, and J. P. Herberger. 1977. *The World's Worst Weeds: Distribution and Biology.* Honolulu: The University of Hawaii Press.

Inderjit, ed. 2005. *Invasive Plants: Ecological and Agricultural Aspects.* Basel: Birkhäuser.

Inderjit, ed. 2009. *Management of Invasive Weeds.* Dordrecht, The Netherlands: Springer.

Institute of Pacific Islands Forestry. 2010. *Pacific Island Ecosystems at Risk (PIER): Plant Threats to Pacific Ecosystems.* Honolulu: USDA Forest Service. http://www.hear.org/pier

Kaufman, S. R., and W. Kaufman 2007. *Invasive Plants. Guide to Identification and the Impacts and Control of Common North American Species.* Mechanicsburg, PA: Stackpole Books.

Kohli, R. K., S. Jose, H. P. Singh, and D. R. Batish, eds. 2009. *Invasive Plants and Forest Ecosystems.* Boca Raton, FL: CRC Press.

Lazarides, M., K. Cowley, and P. Hohmen. 1997. *CSIRO Handbook of Australian Weeds.* Collingwood, Australia: CSIRO Publishing.

Luken, J. O., and J. W. Thieret, eds. 1997. *Assessment and Management of Plant Invasions.* New York: Springer.

Matthei, O. 1995. *Manual de las Malezas que Crecen en Chile.* Santiago, Chile: Alfabeta Impresores.

Muller, S., ed. 2004. *Plantes Invasives en France.* Paris: Publications Scientifiques du Muséum National d'Histoire Naturelle.

Myers, J. H., and D. R. Bazely 2003. *Ecology and Control of Introduced Plants.* Cambridge, UK: Cambridge University Press.

Panetta, F. D., R. H. Groves, and R. C. H. Shepherd. 1998. *The Biology of Australian Weeds*, Vol. 2. Melbourne: R. G. and F. J. Richardson.

Parsons, W. T., and E. G. Cuthbertson. 2001. *Noxious Weeds of Australia*, 2nd ed. Collingwood, Australia: CSIRO Publishing.

Pyšek, P., K. Prach, M. Rejmánek, and M. Wade, eds. 1995. *Plant Invasions: General Aspects and Special Problems*. Amsterdam: SPB Academic Publishing.

Radosevich, S. R., J. S. Holt, and C. M. Ghersa. 2007. *Ecology of Weeds and Invasive Plants*. Hoboken, NJ: Wiley-Interscience.

Randall, R. P. 2002. *A Global Compendium of Weeds*. Meredith, Victoria, Australia: R.G. and F.J. Richardson. http://www.hear.org/gcw/

Reed, C. C. 1977. *Economically Important Foreign Weeds*. Agriculture Handbook No. 498. Washington, DC: USDA, U.S. Government Printing Office.

Reynolds, S. C. P. 2002. *A Catalogue of Alien Plants in Ireland*. Glasnevin, Ireland: National Botanic Gardens.

Rice, P. M. 2010. *INVADERS Database System*. Missoula: Division of Biological Sciences, University of Montana. http://invader.dbs.umt.edu

Ridley, H. N. 1930. *The Dispersal of Plants throughout the World*. Ashford, UK: Reeve and Co.

Ryves, T. B., E. J. Clement, and M. C. Foster. 1996. *Alien Grasses of the British Isles*. London: Botanical Society of the British Isles.

Salisbury, E. 1961. *Weeds and Aliens*. London: Collins.

Sauer, J. D. 1988. *Plant Migration: The Dynamics of Geographic Patterning in Seed Plant Species*. Berkeley: University of California Press.

Sheley, R. L., and J. K. Petroff, eds. 1999. *Biology and Management of Noxious Rangeland Weeds*. Corvallis: Oregon State University Press.

Starfinger, U., K. Edwards, I. Kowarik, and M. Williamson, eds. 1998. *Plant Invasions: Ecological Mechanisms and Human Responses*. Leiden: Backhuys Publishers.

Stewart, C. N. 2009. *Weedy and Invasive Plant Genomics*. Ames, Iowa: Wiley-Blackwell.

Stone, C. P., C. W. Smith, and J. T. Tunison, eds. 1992. *Alien Plant Invasions in Native Ecosystems of Hawaii*. Honolulu: University of Hawaii Press.

Tokarska-Guzik, B., J. H. Brock, G. Brundu, L. Child, C. C. Daehler, and P. Pyšek, eds. 2008. *Plant Invasions: Human Perception, Ecological Impacts and Management*. Leiden: Backhuys Publishers.

Villasenor Rios, J. L., and F. J. Espinosa Garcia. 1998. *Catálogo de Malezas de México*. Mexico City: Universidad Nacional Autonoma de México.

Weber, E. 2003. *Invasive Plant Species of the World*. Wallingford, UK: CABI Publ.

White, D. J., E. Haber, and C. Keddy. 1993. *Invasive Plants of Natural Habitats in Canada*. Ottawa: Canadian Wildlife Service.

Animals

Blackburn, T. M., J. L. Lockwood, and P. Cassey. 2009. *Avian Invasions: The Ecology and Evolution of Exotic Birds*. Oxford: Oxford University Press.

Dahlsten, D. L., and R. Garcia, eds. 1989. *Eradication of Exotic Pests*. New Haven, CT/London: Yale University Press.

Fuller, P. L., L. G. Nico, and J. D. Williams. 1999. *Nonindigenous Fishes Introduced into Inland Waters of the United States*. Bethesda, MD: American Fisheries Society.

Hendrix, P. F., ed. 2006. *Biological Invasions Belowground: Earthworms as Invasive Species*. Dordrecht, The Netherlands: Springer.

Kraus, F. 2009. *Alien Reptiles and Amphibians*. Dordrecht, The Netherlands: Springer.

Kapuscinski, A. R., ed. 2007. *Environmental Risk Assessment of Genetically Modified Organisms: Methodologies for Transgenic Fish*. Cambridge, MA: CABI Publ.

Langor, D. W., and J. Sweeney, eds. 2009. *Ecological Impacts of Non-Native Invertebrates and Fungi on Terrestrial Ecosystems*. Dordrecht, The Netherlands: Springer.

Laycock, G. 1966. *The Alien Animals*. Garden City, NY: Natural History Press.

Lever, C. 1985. *Naturalized Mammals of the World*. London: Longman.

Lever, C. 1987. *Naturalized Birds of the World*. Burnt Mill, UK: Longman Scientific and Technical. [2nd ed. 2005. London: T. and A. D. Poyser.]

Lever, C. 1994. *Naturalized Animals*. London: T. and A. D. Poyser.

Lever, C. 1996. *Naturalized Fishes of the World*. San Diego: Academic Press.

Lever, C. 2003. *Naturalized Reptiles and Amphibians of the World*. Oxford: Oxford University Press.

Long, J. L. 1981. *Introduced Birds of the World*. New York: Universe Books.

Long, J. L. 2003. *Introduced Mammals of the World*. Wallingford, UK: CABI Publishing.

Mattson, W. J., P. Niemela, I. Millers, and Y. Inguanzo. 1994. *Immigrant Phytophagous Insects on Woody Plants in the United States and Canada: An Annotated List*. USDA Forest Service, General Technical Report NC-169.

Roots, C. 1976. *Animal Invaders*. New York: Universe Books.

Williams, D. F., ed. 1994. *Exotic Ants*. Boulder: Westview.

Microorganisms and Pathogens

Agrios, G. N. 2005. *Plant Pathology*, 5th ed. Burlington, MA: Elsevier Academic Press.

Brachman, P. S., and E. Abrutyn, eds. 2009. *Bacterial Infections of Humans: Epidemiology and Control*, 4th ed. New York: Springer.

Cooke, B. M., D. G. Jones, and B. Kaye, eds. 2006. *The Epidemiology of Plant Diseases*, 2nd ed. Dordrecht, The Netherlands: Springer.

Callahan, J. R. 2010. *The Emerging Biological Threats: A Reference Guide*. Santa Barbara, CA: Greenwood Press.

Fong, I. W., and K. Alibek, eds. 2005. *Bioterrorism and Infectious Agents*. New York: Springer.

Fong, I. W., and K. Alibek, eds. 2007. *New and Evolving Infections of the 21st Century*. New York: Springer.

Frantzen, J. 2007. *Epidemiology and Plant Ecology*. Singapore: World Scientific Publishing.

Hudson, P. J., A. Rizzoli, B. T. Grenfell, H. Heesterbeek, and A. P. Dobson, eds. 2001. *The Ecology of Wildlife Diseases*. Oxford: Oxford University Press.

Huisman, J., H. C. P. Matthijs, and P. M. Visser, eds. 2005. *Harmful Cyanobacteria*. Dordrecht, The Netherlands: Springer.

Mahy, B. W. J., and M. H. van Regenmortel, eds. 2010. *Desk Encyclopedia of Animal and Bacterial Virology*. Amsterdam: Elsevier Academic Press.

Mahy, B. W. J., and M. H. van Regenmortel, eds. 2010. *Desk Encyclopedia of Human and Medical Virology*. Amsterdam: Elsevier Academic Press.

Mahy, B. W. J., and M. H. van Regenmortel, eds. 2010. *Desk Encyclopedia of Plant and Fungal Virology*. Amsterdam: Elsevier Academic Press.

Maier, R. M., I. L. Pepper, and C. P. Gerba. 2009. *Environmental Microbiology*. Amsterdam: Elsevier Academic Press.

Thomas, F., J.-F. Guégan, and F. Renaud. 2007. *Ecology and Evolution of Parasitism*. Oxford: Oxford University Press.

Vidhyasekaran, P. 2004. *Concise Encyclopedia of Plant Pathology*. New York: Food Products Press.

Marine and Freshwater Invasions

Claudi, R., and J. H. Leach, eds. 2000. *Nonindigenous Freshwater Organisms: Vectors, Biology and Impacts*. Boca Raton, FL: Lewis Publishers.

DeVoe, M. R., ed. 1992. *Introductions and Transfers of Marine Species*. Charleston: South Carolina Sea Grant Consortium.

Dumont, H. J., T. A. Shiganova, and U. Nierman, eds. 2004. *Aquatic Invasions in the Black, Caspian, and Mediterranean Seas*. Dordrecht, The Netherlands: Kluwer Academic.

Gherardi, F., ed. 2007. *Biological Invasions in Inland Waters: Profiles, Distribution and Threats*. Dordrecht, The Netherlands: Springer.

Gollasch, S., B. S. Galil, and A. N. Cohen, eds. 2006. *Bridging Divides: Maritime Canals as Invasion Corridors*. Dordrecht: The Netherlands: Springer.

Leppäkoski, E., S. Gollasch, S. Olenin, eds. 2002. *Invasive Aquatic Species of Europe: Distribution, Impacts, and Management.* Dordrecht, The Netherlands: Kluwer Academic Publishers.

National Research Council (U.S.). 2008. *Great Lakes Shipping, Trade, and Aquatic Invasive Species.* Washington, DC: Transportation Research Board.

Pederson, J., ed. 2000. *Marine Bioinvasions.* Cambridge, MA: MIT Sea Grant College Program.

Rilov, G., and J. Crooks, eds. 2009. *Biological Invasions in Marine Ecosystems.* Berlin: Springer-Verlag.

Rosenfield, A., and R. Mann, eds. 1992. *Dispersal of Living Organisms into Aquatic Ecosystems.* College Park, MD: Maryland Sea Grant College.

Welcomme, R. L. 1988. *International Introductions of Inland Aquatic Species.* FAO Fisheries Technical Paper 294.

Noteworthy Invasions

Buhs, J. B. 2004. *The Fire Ant Wars.* Chicago: The University of Chicago Press.

Ejeta, G., and J. Gressel, eds. 2007. *Integrating New Technologies for Striga Control: Towards Ending the Witch-hunt.* Singapore: World Scientific Publishing Co.

Freinkel, S. 2007. *American Chestnut: The Life, Death, and Rebirth of a Perfect Tree.* Berkeley: University of California Press.

Goldschmidt, T. 1996. *Darwin's Dreampond.* Cambridge, MA: MIT Press.

Jaffe, M. 1994. *And No Birds Sing.* New York: Simon and Schuster.

Leader-Williams, N. 1988. *Reindeer on South Georgia.* Cambridge: Cambridge University Press.

Meinesz, A. 1999. *Killer Algae.* Chicago: The University of Chicago Press.

Montague, T. L., ed. 2000. *The Brushtail Possum.* Lincoln, New Zealand: Manaaki Whenua Press.

Pyšek, P., M. J. W. Cock, W. Nentwig, and H. P. Ravn, eds. 2007. *Ecology and Management of Giant Hogweed (Heracleum mantegazzianum).* Wallingford, UK: CAB International.

Spear, R. J. 2005. *The Great Gypsy Moth War.* Amherst: University of Massachusetts Press.

Thompson, H. V., and C. M. King, eds. 1994. *The European Rabbit: The History and Biology of a Successful Colonizer.* Oxford: Oxford University Press.

Winston, M. L. 1992. *Killer Bees: The Africanized Honey Bee in the Americas.* Cambridge, MA: Harvard University Press.

Young, J. A., and C. D. Clements. 2009. *Cheatgrass: Fire and Forage on the Range.* Reno: University of Nevada Press.

Periodicals

Aquatic Invasions [electronic resource]. (2006–). Helsinki, Finland: Regional Euro-Asian Biological Invasions Centre (REABIC).

Biological Control. (1991–). San Diego, CA: Elsevier.

Biological Invasions. (1999–). Dordrecht, The Netherlands: Springer.

Diversity and Distributions. (1998–). Oxford, UK: Wiley-Blackwell.

Emerging Infectious Diseases. (1995–). Atlanta, GA: Centers for Disease Control and Prevention.

Invasive Plant Science and Management. (2008–). Champaign, IL: Weed Science Society of America.

Journal of Aquatic Plant Management. (1976–). Vicksburg, MS: Aquatic Plant Management Society, Inc.

Neobiota. (2002–). Berlin: European Group on Biological Invasions, Institute of Ecology of the Technische Universität Berlin.

Boldface indicates main articles. *Italics* indicate illustrations, text boxes, and tables.

carp, common (Cyprinidae), 27, 38, **100–104**, 229, 232, *411*, 414, 419, 610
    Asian, *27*, 452, 578
    biological control and management, 230, 256–257, 415, 419, 432, 578, 697
    diseases, 421
    distribution, worldwide, 27, 611
    grass carp, 28
    habitat alteration and, 31
    impacts, worldwide, 102, 103, 296, 414, 415, 610, 611, 612, 699
Carpobrotus spp., *174*, 361–362, 450, 521
CART (classification and regression trees), **104–108**
Caspian Sea. *See* Ponto-Caspian region
Castor canadensis. See beavers
catclaw mimosa (*Mimosa pigra*), 171, 630, 675, 676, 699, 701, 702
cats (*Felis* spp.), 95, 96, 98, 483, 558
    biological control and, 62, 96
    dispersers, as, 362
    eradication, 99, 100, 198, 199, 202, 294, 395, 481, 646, 650
    hybridization, 345
    impact on
        Australia, 36, 41, 42, 332, 565
        Hawaii, 312
        islands, 41, 96, 99, 294, 393, 395, 481, 558
        New Zealand, 192, 481, 559
        South Africa, 646, 650
    invasional meltdown and, 362
    synergies, 99, 294, 332, 395, 481, 559
cattle, 37, 342, 476, 479, 484, 512, 597, 598, 603–604. *See also* grazers/grazing
caulerpa seaweed (*Caulerpa taxifolia*), 12, *13*, 14, 15, 35, 199, 251, 254, 256, 455, 490, 578, 626, 694, 708
Cecropia peltata (pumpwood), 41, 244
Celastrus orbiculatus (Oriental bittersweet), *124*, 321, 681, 683
Centaurea spp. (knapweeds), 165, *285*, 489, 492, 669
Central America, 420–421, 473–474. *See also* New World; *specific organisms*
Cercopagis pengoi (fish hook flea), 136, 217, 412
Cervus elaphus (red deer), 37, 266, 290, 331, 345, 361
chance, effects of, 607
cheatgrass (*Bromus tectorum*), **108–113**, 275, 282, 424, 695
    costs, 289
    dispersal, 160, 162
    fire and, 109, 112, 224, *225*, 227, 286–287, 306, 491, 565
    habitat compatibility and, 307
    impacts, worldwide, 108–112, 276, *276*, 565, 667, 695
    invasiveness, 110–111, 384
    protected areas and, 565
    remote sensing and, 580–581
    tolerance limits, 667
    transport and establishment, 282–283
chemical control, 509, 622–623, 697–698, 713–714. *See also* herbicides; insecticides; pesticides

cherry, black (*Prunus serotina*), 55, *160*, 164, 521, *673*, 674
chestnut blight (*Cryphonectria parasitica*), 55, 194, 259, 523
Chile, 223, 226. *See also specific organisms*
China, *32*, 60, 92, 494, 625, 691. *See also specific organisms*
Chinese mitten crab (*Eriocheir sinensis*), 46, 127, 128, 135, 552, 610
Chinese tallow (*Triadica sebifera*), 164, 226, *383*, 451, 676, 677
Christmas Island, 22, 24. *See also specific organisms*
Chromolaena odorata (Siam weed), 67, 244, 649, 676
Chrysanthemoides monilifera (bitou bush), 41, 69, 382, 676
cichlids. *See* tilapias and other cichlids (Cichlidae)
Cinchona pubescens (quinine tree), 383, 676
clams
    crabs and, 127
    dispersal, 31, 377
    extinction, 297
    green crabs and, 127
    impacts, worldwide, 297, 387, 456, 610, 626, 702, 714
    invasional meltdown and, 361
    marine invertebrates and, 386
    mutualism, 467
Clarias batrachus (walking catfish), 29, 231
classification and regression trees (CART), **104–108**
classifications, 249–251, 364, *366*, 372
Clathrus archeri (octopus stinkhorn), 261, *263*
Clidemia hirta (Koster's curse), 315, 380, 458, 460, 461, 676
climate and weather, 356–357
    biological control and, 692
    diseases and, 513
    disturbance and, 166
    evolution and, 220
    evolutionary responses and, 214
    genetics and, 367
    habitat compatibility and, 307–308
    Hawaii and, 310, 314
    hemlock woolly adelgid and, 321, 322–323
    herbicides and, 330
    invasibility and, 607, 660
    management and, 703
    mosquitoes and, 464
    nitrogen and, 488
    plant dispersal and, 162
    protected areas and, 564
    range models and, 568, 569
    reptiles and amphibians and, 591
    weevils and, 335
    See also tolerance limits, animal; tolerance limits, plant
climate change, **113–117**
    agreements, international, and, 6
    apomixis and, 26
    birds and, 70
    bryophytes/lichens and, 85
    butterflies and, 156
    crayfish and, 129

diseases and, 510, 513, 518
dispersal and, 411
genetics and, 271
habitat compatibility and, 305–306
herbivory and, 334
impacts, worldwide, 298, 452, 457–458, 702–703
invasibility and, 607
lag times and, 409
land use and, 426
management and, 116
marine invertebrates and, 390
oysters and, 498
plants and, 217, 559, 683–684
species shift and, 369
See also acid rain
climate matching, 72, 90. *See also* habitat compatibility; habitat matching
climbing ferns (*Lygodium* spp.), 226, 680
clonal growth and clonality, 284, 584, 586–587, **678–679**, 682, 699. *See also* apomixis
Coccinella septempunctata (seven-spot ladybug), *401*, 402, 403
Codium fragile (codium), 13, 14, 40, 141, 250, *386*, 626
coexistence models, 357–358
cogon grass (*Imperata cylindrica*), 124, 281, *282*, 283, 290, 695
comb jellyfish (*Mnemiopsis leidyi*), 119, 417, 551, 552, 554, 555, 556, 625
common reed (*Phragmites australis*), 210, 271–272, 275, 412, 474, *596*, 699, 700, 703
competition
    Darwin on, 144, 372
    habitat compatibility and, 308
    invasibility and, 358, 359, 383–384, 440
    lag times and, 409
    range models and, 569
    taxonomic patterns and, 660
    See also limiting similarity hypothesis
competition, animal, **117–122**
    aquatic organisms, 28–29, 30, 31, 35, 132–133, 232
    birds and, 72, 313
    carnivores and, 98, 100
    climate and, 115
    diseases and, 31
    earthworms, 181–182
    insects, 19–20, 22, 39, 50–51, 121, 403, 686
    snails, land, and, 639
    See also competition
competition, plant, **122–125**
    biological control and, 63, 64
    bryophytes and lichens, 84
    cheatgrass and, 109, 111
    climate control and, 115
    eutrophication and, 210
    evolution and, 221
    fire and, 227
    grazers and, 291
    Hawaii, 315, 316
    islands and, 393
    nitrogen and, *489*

diatoms, 13, 251
  effects of, 14, 553, 556
  habitat, 388, 550, 712
  mutualism, 362
  North Sea, 46
  vectors, 31, 46, 413, 552, 555, 712
  See also Didymosphenia geminata (didymo;
    rock snot)
Didemnum spp., 625
Didymosphenia geminata (didymo; rock snot),
  250, 483
dieback "fungus" (*Phytophthora cinnamomi*),
  41, 42, 57, 259, 543
dingo, 36, 37, 42, 98, 99, 558
diseases, **150–154**, 519
  aquaculture and, 30–31
  climate change and, 115
  costs, 375
  crayfish and, 701
  dispersal and vectors, 11, **150–154**, 365, 512,
    614
  ecoterrorism and, 183, 184, 185
  impact on
    Africa, 45, 151, 153, 512, 513, 516, 517, 518,
      519, 526, 597, 601–602
    Asia, 151, 153, 516, 517, 526
    Australasia, 512, 545
    Australia, 513
    Central America, 154, 512, 599
    Europe, 151–152, 197, 421, 445, 512, 513,
      515–516, 517, 519, 544–555
    Hawaii, 510
    India, 516, 517, 518
    islands, 151, 153, 391, 443, 444, 465
    lakes, 421
    Mediterranean region, 444, 513
    New Zealand, 479, 512
    North America, 151–152, 153, 236, 237,
      259, 300, 421, 512, 513, 516, 518–519, 545
    South Africa, 259
    South America, 153, 236, 444, 512, 518
    United Kingdom, 197, 512, 517
  management, 152, 153, 432, 434, 513, 546
  meltdown and, 394
  pet trade and, 539
  plant resistance, 194
  rats and, 571
  risk assessments and, 606
  threats from, 192
  trade and, 376–377
  vectors, human, **150–154**
  water hyacinth and, 690
  weeds and, 694
  See also epidemiology; pathogens, animal;
    pathogens, human; *specific diseases;*
    specific locales; specific organisms;
    specific vectors
dispersal ability, animal, **154–159**
dispersal ability, plant, **159–165**, 584, 588
dispersal and vectors, 373, 584, 606
  agriculture and, 8
  climate change and, 114
  DAISIE data, 141
  demography and, 149
  epidemiology and, 196–198

eradication and, 200
invasibility and, 359
invasion economics and, 376–378
lag times and, 407–408
lakes and, 410–411, 412–413, *414*
pollen and, 584
range model assumption, 570
risk and, 15
risk assessments and, 606
rivers and, 611
seas and oceans and, 624
See also disease vectors, human; *main*
  articles for specific organisms; spread
  (dispersal) (stage of invasion);
  transport (arrival) (stage of invasion);
  specific organisms; specific activities
distribution/impacts, world wide, 270, 466,
  664. *See also specific organisms*
disturbance, **165–168**, 305, 306–307, 596, 617
  agriculture and, 8
  apomicts and, 25
  aquatic plants and, 252–253
  biotic resistance and, 188
  clonal growth and, 676
  competition and, 123
  control of, 703
  earthworms and, 179, 181
  eutrophication and, 210
  forestry and, 248
  grasses and forbs and, 281, 285
  herbivory and, 333–335
  hybridization and, 344, 345
  impacts, worldwide, 476–477, 554–555, 556,
    700
  invasibility and, 20, 147, 358–359, 409,
    440–441
  land use and, 425
  life histories and, 438
  plants and, 422–423
  reservoirs and, 418
  restoration and, 703
  seeds and, 629–630
  soil, 56
  succession and, 654–655, 656
  taxonomic patterns and, 660
  vines and lianas and, 681–682, 683
  weeds and, 694
  See also fragmentation; habitat alteration;
    restoration; *specific disturbances*
diversity. *See* biodiversity (species richness)
dogs (*Canis* spp.), 95, 98, 192, 312, 342, 475–
  476, 537, 558, 598. *See also* dingo
domestication, 30, 598. *See also specific*
  domesticated organisms
dominance, ecological, 19, 20
Douglas fir (*Pseudotsuga menziesii*), 55, 111, 361,
  383, 473
Dreissena spp. (zebra and quagga mussels),
  710–714
drivers of invasion, 5, 490. *See also specific drivers*
drought, 665–666, 692
dunes and beaches, 40, 44, 306, 359–360, *388*,
  449, 676, 679, 713. *See also* intertidal
  habitats and species
Dutch elm disease. See *Ophiostoma ulmi*

early detection and rapid response (EDRR),
  169–177, 186, 202, 240, *247*, 567,
  593, 642
earthworms, 57, **177–183**, *178*, 312, 363, 380,
  491, 669
  Pontoscolex corethrurus, 38, *178*, 179, 180
East Africa, 377, 444
East China Sea, 625
Echinochloa crus-galli (barnyard grass), 217, 219
economic factors, 367–368
  agreements, international, and, 5, 6
  algae and, 14–15
  animals and, 79, 80, 98, 264, 292, 640–641
  apomixis and, 26
  aquatic organisms and, 133, 447, 487, 494,
    622, 690, 713
  assessments, 141
  biological control and, 67–68, 69, 195
  databases and, 146
  diseases and, 515, 545, 601
  early detection and rapid response and, 169
  eradication and, 201–202
  forestry and, 246
  fungi and, 258
  grasses and forbs and, 289
  grazers and, 293
  gypsy moth and, 300
  impacts, worldwide, 456–457, 479, 484, 644
  insects and, 23, 50, 53
  management and, 354
  plants and, 112, 170
  restoration and, 595, 596–597
  risk assessment and, 605
  See also costs of invasive species;
    ecoterrorism; invasion economics;
    trade, international; *specific industries*
ecosystem engineering, 233, 292, 498, 612, 668,
  675, 711
ecosystem services, 67, 68, 73, 129, 195, 212,
  595, 695, 698, 703
ecoterrorism, **183–186**, 514. *See also* biosecurity
Egeria densa (Brazilian waterweed), 35, 249,
  253, 416, 480
egrets, 149, 512
Egypt, 442, 454, 456, 599
EICA (evolution of increased competitive
  ability), 221, 278, 409, 440
Eichhornia crassipes (water hyacinth). *See* water
  hyacinth (*Eichhornia crassipes*)
Elaeagnus angustifolia (Russian olive), 160, *161*,
  432, 449
Eleutherodactylus coqui (Caribbean tree frog),
  313, 593
Elodea canadensis (Canadian waterweed), 249,
  372, *411*, 416, 480, 587, 679
Elton, Charles S., **187–189**, 369, 373, 617, 619
  Bates, Marston and, 374
  biotic resistance, on, 360
  citations of, *365*
  invasibility, on, 356, 358
  "native invaders," on, 472
  oysters, on, 494
  plant competition, on, 122
  terminology, 373
  textbooks, 374–375

emerald ash borer (*Agrilus planipennis*), 194, 238, *239*, 240, 241, 268, 533, 567, 575, 579
endangered/extinct/protected/threatened species, **189–193**
  amphibians, 261
  animals and, 87, 95, 96, 98, 99–100, 264, 266, 292, 293, 315, 331
  aquatic organisms, 129, 132, 133, 232, 233, 297
  bees, 52
  birds and, 72, 73–74, 96, 310, 549, 633
  competition and, 120–121, 125
  hybridization and, 29, 98, 422
  impacts, worldwide, 36, 37, 38, 41, 42, 72, 297, 309–310, 565
  plants, 549
  See also extinction
endemic arboviral encephalitides, 152
endozoochory, 413
enemy release
  animal parasites and, 500, 502–503
  biological control of animals and, 148–149
  eutrophication and, 210–211
  habitat compatibility and, 308
  invasibility and, 440
  "native invaders" and, 472
  origin of invasives and, 276, 279
  plant pathogens and, 520–523
  See also evolution of increased competitive ability (EICA)
enemy release hypothesis (ERH), 63, 72, 148, 149, **193–196**, 221
England. *See* United Kingdom
environmental niche models (ENMs), 18, 305, 306
environmental shifts, 220, 221
epidemiology, **196–198**, 350, 518, 598, 599, 607, 609
eradication, **198–203**, 272, 354, 378, 528. *See also main articles for specific organisms; management methods*
Erharta calycina (perennial veldt grass), *167*
Eriocheir sinensis (Chinese mitten crab), 46, 127, 128, 135, 552, 610
erosion, 37, 460, 461, 611, 649, 703
establishment (naturalization) (stage of invasion), *245*, 365, 379
  animals and, 231, 498, 500–501, 541, 591
  chance and, 607
  Galápagos Islands and, 392
  historical precedents, 364
  humans and, 606
  invasional meltdown and, 361
  invasiveness and, 382
  lag times and, 404, 408–409
  life histories and, 438
  melastomes and, 461
  mutualism and, 466
  plants and, 206, 207, 283–284, 548, 586, 627, 629
  propagule pressure and, 561–563
  taxonomic patterns and, 659–660
eucalypts (*Angophora, Corymbia,* and *Eucalyptus* spp.), **203–209**, 671
  fire and, 203, 208
  forest insects and, 240

forestry and, 242, 243–244, 677
fungi and, 469, *470*, 471
impacts, 307, 308
impacts, worldwide, 39, *40*, 41, 203, 205, 206, *207*, 208
invasiveness, 206, 383
pathogens and, 318
Euglandina rosea (rosy wolf snail), 62, 315, 635, 636, 638–639, 642
Euphorbia esula (leafy spurge), 67, 282, *283*, 288, 289, 580, 582, 695
Eurasian watermilfoil (*Myriophyllum spicatum*), 249, 256, *260*, 412, 415–416, 417, 420, 451, 609, 611
Europe
  agreements, international, and, 6, 373
  canals, 92, 417
  competition and, 118
  dunes, 359–360
  eradication, 198
  forestry, 242
  horticulture and, 336
  hybridization, 345
  invasibility of forests, 359
  lakes, 412
  "native invaders," 473, 474
  natural enemies of plants, 521
  nitrogen and, 491
  novel weapons and, 493
  scientific projects, 617
  xenophobia and, 705
  See also European Union; *specific countries; specific organisms*
European rabbit (*Oryctolagus cuniculus*), 290
  biological control and, 62, 96, 197, 290, 293, 558
  competition and, 118–119, 121–122
  costs, 291
  dispersers, as, 361
  Elton, Charles S. on, 187
  endangerment by, 191
  evolution and, 220
  herbivory and, 332, 335
  impact on
    Australia, 10, 36–37, 41, 62, 118–119, 185, 291
    islands, 42, 395
    subantarctic, *42*
  management, 294, 537
  plants and, 332
  See also rabbits
European Union (EU), 134, 185, 339. *See also* DAISIE Project
eutrophication, aquatic, **209–213**
  aquatic organisms and, 419, 486, 488
  biological control and, 692
  herbivory and, 332–333
  human activities and, 305
  impact on Ponto-Caspian region, 550, 556
  kudzu and, 398
  wetlands and, 701, 702
evolution, **213–215, 215–222**
  competition and, 120–121
  diseases and, 545
  fishes, 488

habitat compatibility and, 308
herbicides and, 9
hybridization and, 342
invasibility and, 522
invasives, of, **215–222**
plant reproductive systems and, 589–590
predators and, 560
rapid, 213
response of natives, as, **213–215**
snakes and, 592
See also adaptation; genetics
evolution of increased competitive ability (EICA), 221, 278, 409, 440
expansion, population, 220–221
extinction, 367–368
  diversity and, 368
  freshwater species, 701
  hybridization and, 343–344, 422
  islands and, 392, 393, 395
  kudzu and, 398
  New Zealand, 475, 476, 481
  nitrogen and, 489
  predators and, 557
  propagule pressure and, 561
  viruses, 517
  See also endangered/extinct/protected/ threatened species; *specific organisms*

Fallopia spp. (knotweeds), 281, *282*, 288, 451, 587, *673*, 679
Fasciola spp., 526
  cattle liver fluke, 640
  and fascioliasis, 526
  gigantica, 527
Felis spp. *See* cats (*Felis* spp.)
fellow travelers (hitchhikers), 5, 9, 13, 14, 15, 31, 35, 416, 659
ferns, climbing (*Lygodium* spp.), 226, 680
Fiji, 151, 460
fir, Douglas (*Pseudotsuga menziesii*), 55, 111, 361, *383*, 473
fire, **223–228**, 306
  clonal growth and, 679
  costs, 289
  dams and, 307
  diseases and, 546
  dispersal and, 155
  disturbance and, 166–167
  ecoterrorism and, 184
  impacts, worldwide, 225, 227, 286, 289 (*see also specific geographical locations*)
  land use and, 427
  management tool, as, 223, 228, 430, 491, 695, 703
  "native invaders" and, 473
  niches and, 149
  restoration and, 594
  seed dormancy and, 631
  See also specific organisms
fire ants. *See* little fire ant (*Wasmannia auropunctata*); red imported fire ant (*Solenopsis invicta*); tropical fire ant (*Solenopsis geminata*)
fire tree (*Morella* [*Myrica*] *faya*), 54, 182, 226, 565, 582, 675–676

fisheries. *See* fishing and fisheries

fishes, **229–234, 525–528**
  algae and, 455
  biodiversity and, 229, 231, 232
  biological control and, 257, 415, 418–419,
    610, 697
  climate change and, 114, 116
  competition and, 118, 121
  crayfish and, 133
  crustaceans and, 136
  diseases and, 232–233, 575
  dispersal and vectors, 33, 34, 146, 155, 158,
    411, 606
  dispersers, as, 412, 413
  extinction, 343, 420–421, 479, 484, 486, 622
  habitat alteration and, 31
  hybridization and, 343, 345
  impact on
    Africa, 103, 158, 231, 232, 420, 484–488
    Asia, 100, 103, 155, *229*, 670
    Australasia, 70, 610
    Australia, 2, 38, 42, 100, 101, 102, 158,
      231, 422
    Caribbean, 542
    Europe, 1, 100, 101, 103, 229, 231, 232,
      233, 415
    Great Lakes (U.S.), 232, 296–297, 415,
      416
    Hawaii, 231, 232, 310, 319
    islands, 392
    lakes, 410, *411*, 412
    Mediterranean Sea, 453, 455
    New Zealand, 35, 100, 101, 102, 158, 230,
      231, 232
    North America, 35, 100, 101, 158, 230,
      231, 233, 415, 420, 421
    Ponto-Caspian region, 100, 101, 119, 552,
      553, 554, 555, 556, 625
    rivers, 609, 610
    Russia and former USSR, 231, 415
    South Africa, 232, 646, 647–649
    South America, 158, 230, 231, 232
    United Kingdom, 103, 415
    United States (continental), 4, 34, 192,
      229, 230, 231, 232, 233, 541, 542
    wetlands, 699
  impact on humans, 35, 487, 488 (*see also*
    main article on fishes)
  invasiveness, 34, 101, 231–232, 438, 440
  light and, 701
  management, 100, 452, 619
  mussels and, 420
  pesticides, **525–528**
  plants and, 420, 611
  predators, 621
  taxonomic patterns and, 659, 660
  threats to, 191
  water clarity and, 418–419
  See also aquaculture; aquaria; fishing and
    fisheries; *specific habitats*; *specific*
    organisms; *specific locations*
fish hook flea (*Cercopagis pengoi*), 136, 217, 412
fishing and fisheries, 191, 230, 539
  agreements, international, and, 6
  algae and, 12, 13

bait, 231
brown trout and, 158
carp and, 103, 610
competition and, 119
crabs and, 127
crustaceans and, 135
dispersal and, 413–414
earthworms and, 180, 182
ecosystem improvement and, 230
fishes and, 232
herbivory and, 333
hybridization and, 192, 214–215
impact on
  Africa, 484, 485, 487–488
  Mediterranean Sea, 453, 456
  New Zealand, 479
  Ponto-Caspian region, 550, 551
  United States, 296, 297, 593, 621, 622
insects and, 418
jellyfish and, 625
marine invertebrates and, 386
pathogens and, 513
stocking and, 28, 29, 30, *158*
water hyacinth and, 690
zebra mussel and, 713
  See also aquaculture; *specific introduced*
    organisms
fitness, population, 380–381, 600
flatworms, 57, 182, 388, 412, 417, 558, 699
flaviviruses, **234–238**
flexibility, 19
flies, 45, 144, 153, 199, 201, 202, 220, 256, 646.
  See also specific flies
flooding, 165, 252, 347, 348
  biological control and, 692
  dispersal and, 162
  habitat compatibility and, 305, 306, 307
  herbivory and, 332, 333
  impacts, worldwide, 37, 41, 553
  invasibility and, 205, 217, 357, 607, 630, 699
  management and, 167, 207, 355, 452, 594,
    595, 612, 696, 703
  phosphorus and, 211
  plants and, 112, 393, 422, 423, 690, 692
  tolerance of, 665–666
  wetlands and, 699
flora. *See* plants
Florida
  animals, 85–91, *87, 89, 90*, 219, 231, 266
  fire and, 286
  insects, 409, 462
  invasibility and, 380
  laws, 431, 436
  "native invaders," 474
  nitrogen and, 490
food webs
  aquatic, 251
  biological control and, 68, 69
  carnivores and, 99
  cordgrass (*Spartina* spp.) and, 669
  crayfish and, 129, 133
  crustaceans and, 135, 136
  herbivory and, 332
  insecticides and, 532
  island, 395

lakes and, 421
management and, 481
molluscs and, 610
parasitoids and, 315
predators and, *557–558, 559–560*
rabbits and, 332
wetland, 701–702
zebra mussel and, 713
  See also nitrogen; nutrients and nutrient
    cycling
forbs. *See* grasses and forbs
forest insects, **238–241**, 243, 315, 322
forestry and agroforestry, 203, **241–248**, 425, 426
  eucalypts, 205
  forest insects and, 240
  fungi and, *259, 261*
  gypsy moth and, 300–301
  impact on South Africa, 647
  invasibility and, 607
  woody plants and, 677
  worst plants and, 673
  See also logging
forests, 356
  beavers and, 612, 613
  biocontrol of plants, 461
  competition and, 123
  grazers and, 292, 331
  gypsy moth and, 300
  impact on
    Americas, 56, 322, 545–546
    Australia, 545
    Guam, 591
    Hawaii, 315, 316, 317
    New Zealand, 240, *244*, 245, 476, 477,
      483, 559
    South Africa, 649
  invasibility, 359
  "native invaders," 473
  Nile perch and, 559
  novel weapons and, 493
  predators and, 559
  remote sensing and, 582
  slugs and, 639
  succession and, 657
  transformers and, 669
  weeds, 460, 482
  See also forestry and agroforestry;
    succession; *specific forest organisms*
foundation species, 668
founder events, 7–8, 25, 587. *See also* propagule
    pressure
foxes
  biological control by, 97, 294
  diseases and, 513
  disturbance and, 167
  grazers and, 292
  impact on
    Arctic, 56, 97, 98, 99, 100, 559–560, 565
    Australia, 36, 42, 565
    islands, 393, 394
  parasites and, 503
  pesticides and, 537
  predators, 468
  rabbits and, 332, 335
  See also red fox (*Vulpes vulpes*)

fragmentation, 225, 426, 683, 703

France, 1, 344, 361, 445, 454, 474, 707. *See also specific organisms*

French Guiana, 51, 632

freshwater habitats and organisms, **248–258**, 298, 305, 318–319, 553, 555, 609, 646–647, 649. *See also* lakes; rivers; wetlands; *specific organisms*

freshwater plants and seaweeds, **248–258**

freshwater snail (*Melanoides tuberculata*), 277, 278, 637

frog chytrid fungus (*Batrachochytrium dendrobatidis*), 41, 192, 261, 262, 592

frogs and toads

    biological control and, 61–62, 590

    cane toad (*Rhinella marina* [*Bufo marinus*]), 38, 42, 591, *592*, 699

    competition and, 119–120

    diseases and, 192, 541

    distribution, worldwide, 611

    endangered/extinct, 38, 41, 42

    extinction, 480, 633–634

    impact on

        Americas, 192, 611

        Australia, 38, 41, 42, 192

        Europe, 119–120

        Hawaii, 313

        New Zealand, 476

        rivers, 610

        South Africa, 646

        United Kingdom, 709

        wetlands, 699

    invasional meltdown and, 361, 702

    See also amphibians; bullfrog

fungi, 54, 56–57, 140, 146, **258–263**

    biodiversity and, 261, 262–263, 469

    biological control by, 259, 321, 461, 524, 688–689, 691

    diseases, 513

    dispersal, 52, 411, 415

    earthworms and, 181

    ecoterrorism and, 184–185

    eucalypts and, 206

    forest insects and, 238

    frogs and, 192, 480

    genetic variation of, 220

    grasses and forbs and, 288

    gypsy moth and, 300

    impact on

        Americas, *259*, *260*, 261, 263, 469, 470, 471, 520, 523, 592

        Australia, 57, 206, 261, 469–470

        Europe, 259, *260*, 261, 263, *470*

        Hawaii, 461, *470*

        India, 261

        New Zealand, 57, *259*, *260*, 470, 480

        South Africa, 259, *259*, *260*

        United Kingdom, 471, 523

        wetlands, 699

        worldwide, *470*

    invasiveness, 262, 263

    mutualism, 57, 466

    novel weapons and, 492, 493

    parasitic plants and, 504

    pathogens, 10–11

plants and, 55, 521, 523

predators of, 400

See also mycorrhizal fungi

fur bearers, 95, 290

    biological control by, 268–269

    diseases and, 515

    farm escapes, 447

    impacts, worldwide, 3, 290, 647

    parasites, 501

    plant dispersal and, 160–161

    See also game animals; *specific fur bearers*

Galápagos Islands, 198, 267, 344, 383, 392, 475–476. *See also specific organisms*

Gambusia affinis and *G. holbrooki* (mosquitofish), 38, 229, *230*, 415, 552, 611

game animals, 3, **264–269**, 373

gardens. *See* horticulture

garlic mustard (*Allaria petiolata*), 288

gastropods, 30, 133, 379, 416, 526, 552, 646–647. *See also specific gastropods*

generalist species, 61, 69, 101, 217–218, 300, 308–309, 402, 558–559. *See also main articles for specific organisms; specific organisms*

genetic drift, 219–220

genetic engineering, 62, 248, 354

genetic/molecular information

    aquatic mammals, 448

    diseases and, 444

    earthworms, 179

    fungi, 263

    genotypes, 270–272

    invasives and, 216

    ladybugs and, 403

    lakes and, 412

    marine invertebrates and, 386

    mosquitoes and, 465

    origins and, 273–274

    oysters and, 494, 497

    plant reproduction and, 584, 589–590

    rats and, 646

    sea lamprey and, 623

    viruses, 235

    wasps and, 685

    See also evolution

genetics

    aquaculture and, 29

    horticulture and, 340

    invasibility and, 219

    lag times and, 408

    mutation, 219, 474

    pathogens and, 510

    range models and, 569–570

    rats and, 571

    wetlands and, 701

    See also evolution; hybridization

genetic variation, 219–221

genotypes, 29–30, **270–272**, 380, 398, 412, 440, 474

geographic origins. *See* origins, geographic

geography, plant, 366, 367

Germany, 4, 404, 408, 707–708

giant African snail (*Lissachatina fulica*; *Achatina fulica*), 38, 558, 635, 636, 638, 639, 640, 641, 642

giant hogweed (*Heracleum mantegazzianum*), 288, 580

giant reed (*Arundo donax*), *8*, *9*, 64, 114, 249, 282, *286*, 287, 289, 307, 383, 423, 424, 611, 645, *695*

    biological control, 64

    costs, 289

    disturbance and, 281

    habitat compatibility and, 307

    impacts, *286*, 289

    impacts, worldwide, 645, 695, 699

    invasiveness, 380

    rivers and, 611

    transport and establishment, 282–283

giant salvinia (*Salvinia molesta*), 40, 67, 172, 252, *253*, 256, *257*, 335, 382, 415, 611, 645, 679

glassy-winged sharpshooter (*Homalodisca vitripennis*), 59, 524

Global Invasive Species Programme (GISP), 365

globalization, 319, 337, 377–378, 407, 443, 486, 512. *See also* Internet

global warming. *See* climate change

goat (*Capra hircus*), 290

    biological control by, 697

    eradication, 312, 481

    game animals, 267

    impacts, worldwide, 37, 42, 476, 480

    invasiveness, 357

    management, 198, 201, 294

    mutualism, 56

    vines and, 294

gobies (Gobiidae)

    bacteria and, 297

    diseases and, 421

    dispersal and vectors, 416

    fellow travelers, 297, 420, 421

    impacts, worldwide, 229, 553

    lakes and, 412

    mussels and, 420

    mutualism, 298

    round (*Neogobius melanostomus*), 231, 297, *411*, 412

    vectors, 34, 156, 231, 416, 421

golden apple snail (*Pomacea canaliculata*), 332, 635, 639, 641, 669

goldfish (*Carassius auratus*), 32, 38, 101, 419

gorse (*Ulex europaeus*), 41, 53, *160*, 161, 176, 393, *477*

    benefits of, 480

    invasiveness, 715

grass carp (white amur; *Ctenopharyngodon idella*), 28, 31, 172, 230, 256, 415, 452, 610, 697

government, 65, 76. *See also* laws, non-U.S.; laws, U.S.; regulation

grasses and forbs, **280–290**, 373

    biodiversity and, 287, 288

    biological control of, 697

    climate change and, 217

    competition and, 124, 332, 335

    diversity and, 424

    evolution/genetics and, 219, 270, 271

    fire and, 39, 54–55, 167, 224–226, 227, *286*, 316, 491

    hydrology and, 347–348

impact on
    Africa, 108
    Aleutian Islands, 559–560
    Asia, 109
    Australia, 39–40, 42, 54–55, 108, *167*, 281,
        284, 289
    California, *167*, 285, 286, 287, 289, 491
    China, 282
    Europe, 491
    Hawaii, 282, 287
    Indonesia, 286, 289
    New Zealand, 288, 476–477
    North America, 108–109, 219, 281, 288, *678*
    South Africa, 645, 649, 668
    South America, 108
    United Kingdom, 450
    United States (continental), *167*, 286,
        287, 288, 289, 451, 491
    wetlands, 699, 703
    worldwide, 356, 695
invasiveness, 384
litter and, 287
management and, 289, 449, 450, 491
"native invaders," 473
nitrogen and, 488–489, 490, 491
protected areas and, 566
range models, 569
taxonomic patterns and, 660
transformers, as, 669
transport (stage of invasion) and, 658
weeds and, 693
See also specific grasses and forbs
gray squirrel (*Sciurus carolinensis*), 290, 292,
    612, *613*, 614, 645, 708
grazers/grazing, **290–294**, 331
    benefits of, 335
    biological control and, 64, 293, 697
    birds and, 70
    diseases and, 290, 291
    disturbance and, 166, 167
    earthworms and, 181
    eradication, 294
    fire and, 225–226, 227
    grasses and forbs and, 110, 111, 289
    habitat compatibility and, 307
    impacts, ecological, 293
    impact on
        Americas, 282, 331
        Australia, 291, 293, *293*
        Europe, 110, 111, 289, 331, 332
        Hawaii, 292
        islands, 79, 292
        New Zealand, 290, 291, 292, 293
        South Africa, 649
        United Kingdom, 292
    land use and, 427
    nitrogen and, 491
    over-, 69
    plants and, 191
    rinderpest and, 602–603
    See also specific grazers
Great Britain, 187, 188, 336, 339, 474, 617. *See
    also* United Kingdom; *main articles
    for specific organisms; countries; locales;
    organisms*

Great Lakes (U.S.) (Laurentian), **294–298**
    dispersal, 415, 416
    evolution and, 219, *219*
    hybridization and, 422
    invasibility and, 381
    invasional meltdown, 363
    laws and, 432, 433
    management, 298
    mutualism, 361
    "native invaders," 474
    naturalization hypothesis and, 660
    nitrogen and, 491
    regulations and, 579
green crab (*Carcinus maenas*), 39, 61, 127, 128,
    135, 270, 361, 452, 503, 570
Guam, 151, 265. *See also* brown treesnake
    (*Boiga irregularis*)
Gulf of Mexico, 624, 625, 699
gypsy moth (*Lymantria dispar*), 10, **298–304**,
    307, 528–530, 707

habitat alteration (modification)
    aquaculture and, 31
    Australian, 36
    crustaceans and, 135
    deer and, 473
    diseases and, 192, 510, 513
    evolution and, 217–218
    fishes and, 233
    islands and, 395
    ladybugs and, 403
    lag times and, 409
    mutualism and, 467
    reptiles and amphibians and, 593
    rivers and, 612
    seed dispersal and, 165
    snails and, 639
    woody plants and, 676
    See also disturbance
habitat compatibility, **305–309**
habitat matching, 128. *See also* climate matching
habitats, 119–120, 136, 140, *190*. *See also* habitat
    alteration (modification); *specific
    habitats*
Haiti, 151, 152, 393, 519, 634
Harmonia axyridis (multicolored Asian
    ladybug), *8, 10, 401*
Hawaii, **309–319**, 624
    acclimatization societies, 4
    aquaria and, 416
    biological control, 65–66, 68, 194
    climate change and, 115
    competition and, 123
    diseases, 151, 444, 601
    Egler, Frank and, 374
    evolutionary responses, 215
    fire, 223, 225, *225*, 226, 286, 316
    forests, 639
    herbivory and, 334
    hybridization, 344
    invasibility and, 357, 381
    laws, 430, 433, 436
    nitrogen and, 490–491
    See also main articles for specific organisms;
        specific organisms

hawkweed (*Hieracium* spp.), 25, 380
health impacts, human, 367
    ants, of, 23–24
    biological control and, 484
    crayfish, 134
    diseases and, 510, 527
    fungi and, 258, *470*
    globalization and, 378
    herbicides and, 698
    hosts, 526
    hydrology and, 348–349
    parasites and, 503
    policies, global, 514
    risk assessment and, 605
    weeds and, 694
    See also diseases; pathogens
Hedychium gardnerianum (kahili ginger), 282,
    282, 289, 315, *316*, 582
hemlock (*Tsuga* spp.), 15, 56, 61, 115, 334, 669
hemlock woolly adelgid (*Adelges tsugae*)
    (HWA), 61, 115, **320–323**, 334, 669
Heracleum mantegazzianum (giant hogweed),
    288, 580
herbicides, 64, **323–331**, 430, 698
    adaptation to, 492
    aquatic plants and, 256, 691
    biological control and, 692
    ecoterrorism and, 184
    invasibility and, 440
    parasitic plants and, 506
    protected areas and, 566
    weeds and, 698
    wetlands and, 703
herbivores/herbivory, 54, **331–336**
    aquaculture and, 28
    belowground phenomena and, 55
    clonal growth and, 679
    demography and, 148
    disturbance and, 167
    enemy release and, 193–196, 211
    eradication and, 201, 335
    evolutionary responses, 213
    grasses and forbs and, 285
    habitat compatibility and, 307
    impacts, worldwide, 36, 56, 310, 334, 649
    invasional meltdown and, 361, 362
    mutualism, 56
    pollination and, 549
    predation and, 307, 602–603
    transformers and, 669
    See also grazers/grazing; *specific herbivores*
Herpestes auropunctatus. See small Indian
    mongoose
Hieracium spp. (hawkweed), 25, 380
hitchhikers (fellow travelers), 5, 9, 13, 14, 15, 31,
    35, 416, 659
HIV and AIDS, 187, 375, 376, 487, 518–519,
    597
hogweed, giant (*Heracleum mantegazzianum*),
    288, 580
Homalodisca vitripennis (glassy-winged
    sharpshooter), 59, 524
homogenization. *See* biodiversity
honeybee (*Apis mellifera*), 39, 50, 119, 646
honeycreepers, 74, 115, 310, *311*, 317, 394

introductions. *See specific organisms*

introgression, 52, **342–346**

invasibility, **356–360**, 365–366, 606–607
   pollination and, 547–548
   propagule pressure and, 563, 607
   reservoirs and, 418
   resource availability and, 307
   rivers, of, 611–612
   wetlands, of, 699–700
   See also invasiveness (success of invasives);
      SCOPE (Scientific Committee on
      Problems of the Environment); *main
      articles for locations*

invasional meltdown, 307, **360–364**, 394, 552,
      607, 702

invasion biology, 19, 147, 155, **364–369**,
      369–375, 466
   biogeography and, 44
   conservation biology and, 319
   historical precedents, **369–375**
   invasibility and, 356, 357
   lag times and, 404–405
   reproduction and, 196
   See also Elton, Charles S.; *main articles for
      specific organisms*

invasion economics, **375–378**

invasiveness (success of invasives), 215, **379–
      385**, 483, 710
   climate change and, 114–115
   clonal growth and, 679
   competition and, 122–124
   Darwin on, 143–144
   defined, 607
   dispersal and, 156
   disturbance, 409
   diversity and, 271
   Elton, Charles on, 188
   establishment (stage of invasion) and, 591
   eutrophication and, 210, 212
   evolution and, 216–217, 217–218, 221
   fire and, 223–224, 226–227
   forestry and, 242, 243, *245*, 246–247
   genetic, 270
   habitat compatibility and, 306
   horticulture and, 339
   impact versus, 144, 146, 283
   islands and, 185
   landscapes and, 423
   land use and, 426
   life histories and, 437–441
   mutualism and, 466
   natural enemies and, 193–194
   nitrogen and, 489–490
   novel weapons and, 493
   origin of invasives and, 276–279
   pathogens, of, 513
   plant pathogens and, 521–522, 524
   propagule pressure and, 561–563
   reproductive systems and, 220, 381, 438,
      584–588, 589
   rivers and, 612
   seeds and, 627–631
   succession and, 655–656
   trees and shrubs, of, 673–674
   wetlands and, 700

See also demography; invasibility;
      main articles for specific organisms;
      range modeling; taxonomic
      patterns; tolerance limits, animal;
      tolerance limits, plant; *specific
      animals*; *specific factors*; *specific
      organisms*

invasive species, 8, 44, 147, 438. *See also*
      invasiveness; most invasive species;
      specific organisms

invertebrates, 140, 141, **385–390**
   aquatic habitat alteration and, 31
   crayfish and, 133
   dispersal and vectors, 33, 35, 412, 413, 417
   eradication, 198–199, 560
   impact on
      Australia, 38
      Hawaii, 313–314, 318, 319
      islands, 41, 79, 392
      lakes, 410, *411*, 420
      New Zealand, 479, 483
      North America, 467, 669
      rivers, 610
      South Africa, 646
   invasibility and, 382
   lakes and, 420
   light and, 701
   management, 449
   marine, 39, **385–390**
   mice and, 614
   mutualism, 56
   plants and, 288, 669
   predators, 558–559, 701
   risk assessment and, 606
   soil fertility and, *57*
   tolerance limits, 662
   wasps and, 686, 687
   zebra mussel and, 712
   *See also specific invertebrates; specific marine
      invertebrates*

Iraq, 157, 703

Ireland, 83, 129, 133, 415, 417, 450

islands, 291, 292, 306, **391–395**
   beavers and, 612–613
   biological control and, 558
   biotic resistance and, 188
   climate and weather and, 564
   Darwin on, 143, 157
   ecological impacts, 394
   eradication and, 100, 198
   invasibility, 357
   management and, 409
   nitrogen and, 490–491
   pesticides and, 535
   predator control, 560
   quarantine and, 314–315
   rats and, 571, 573
   reptiles and amphibians and, 591
   rodents and, 612
   selfing and, 585
   small Indian mongoose and, 631
   threats to, 191
   See also specific islands; specific organisms

Israel, 120, 130, 455, 509, 539

Italy, 4, 151, 201, 455, 456, 501–502

Japan
   acclimatization societies, 4
   algae and, 12
   carnivores, 100
   crayfish, 129, 130, 134
   ecoterrorism and, 185
   fishes, 101
   insects, 377
   marine invertebrates, 387
   snails and, 35
   weeds, 275

Japanese beetle (*Popillia japonica*), 533, *534*, 706

Johnsongrass (*Sorghum halepense*), 8, 9

kahili ginger (*Hedychium gardnerianum*), 282,
      282, 289, 315, *316*, 582

Kalm, Peter, 371

Kenya, 72, 133, 348

keystone species, 395, 498, 668, 670

knapweeds (*Centaurea* spp.), 165, *285*, 489,
      492, 669

knotweeds (*Fallopia* spp.), 282, *282*, 288, 451,
      587, *673*, 679

Koster's curse (*Clidemia hirta*), 315, 380, 458,
      460, 461, 676

kudzu (*Pueraria montana*), 124, 164, 226,
      396–399, 683

Lacey Act, 76, 431, 542, 543, 577, 706

ladybugs, *8*, *10*, **400–404**

lag times, **404–410**
   brown treesnake, 78
   horticulture and, 340
   import bans and, 542
   invasibility and, 607
   Nile perch, 486
   oysters, 498
   plant pathogens and, 522
   seed banks and, 630–631
   trees and shrubs and, 675

lakes, 211, 212, 251, *254*, 258, **410–422**, 452,
      467, 484. *See also* Great Lakes (U.S.)
      (Laurentian)

lamprey, sea (*Petromyzon marinus*), 232, 296–
      297, *297*, 298, 432, **619–623**, 620, 714

landscape patterns of plant invasions, **422–425**

land use, *7*, 357, **425–427**, 433, 473–474, 513,
      546, 703

lantana (*Lantana* spp.), 65–66, 194, 344,
      428–430, 677

La Réunion Island, 443, 461, 635

largemouth bass (*Micropterus salmoides*), 31,
      229, 414, 612

Lates niloticus (Nile perch), 28, 191, 232, 233,
      413, 420, **484–488**, 559

laws, non-U.S., 476, 481, 482–483, 575–576,
      577, 578, 597, 644, 646, 650. *See also*
      agreements, international; import
      restrictions; policy; regulation

laws, U.S., 170, 171, 295–296, 312, 313,
      396, **430–437**, 704, 707. *See also*
      agreements, international; import
      restrictions; policy; regulation

leafy spurge (*Euphorbia esula*), 67, 282, *283*,
      288, 289, 580, 582, 695

nematodes (*continued*)
  impacts, worldwide, 233, 483, 501–502, 553
  mysids and, 136
  plant pathogens and, 520, 521–522
  plants and, 54, 55
  rats and, 394
Neogobius melanostomus (round goby), 231, 297, *411*, 412
Neovison vison (mink). See mink (*Neovison vison*)
Netherlands, 185, 211, 405, 417, 445, 489, 617
Nevada, 265, 287, 417, 418, 534
New Guinea, 266
New Jersey, 357, *467*
New Mexico, 264, 436
New World, 463, 465, 474
New Zealand, 372, 373, 374, **475–484**
  acclimatization societies, 3
  biological control, 60, 62, 67, *68*, 558, 688
  bioterrorism, 514
  ecoterrorism and, 183, 185
  eradication, 198, 199, 200, 201, 481–482
  extinction, 343
  forests, 686, *687*
  horticulture and, 341
  import bans, 539
  invasibility and, 379
  islands, 394
  lakes, 417, 420
  marine invaders, 624–626
  naturalization hypothesis and, 660
  pesticides, 535, 537
  water clarity, 419
  See also specific organisms
New Zealand flatworm (*Arthurdendyus triangulatus*), 182
New Zealand mud snail (*Potamopyrgus antipodarum*), *48*, 156–157, *157*, 380, 382, 609, 638, 639
niches, 217, 305, 661
  ants and, 18
  Darwin on, 144
  diversity and, 308
  invasibility and, 359, 380, 384
  invasion resistance and, 285
  Mediterranean Sea, 455–456
  models, 663
  overlap, 51, 182, 384, 446
  plant, 149
  range models and, 569, 570
Nile perch (*Lates niloticus*), 28, 191, 232, 233, 413, 420, **484–488**, 559
Nile tilapia (*Oreochromis niloticus niloticus*), 28, 29, *32*, 414–415, 486–487
nitrogen, 210, **488–492**
  bryophytes/lichens and, 84, 85
  carbon and, 491
  crayfish and, 419
  earthworms and, 182
  forestry and, 244, 246
  grasses and forbs and, 287, 289
  habitat compatibility and, 307
  herbivory and, 334
  impact on Hawaii, 582
  invasibility and, 440

invasional meltdown and, 363
island invasions and, 393
kudzu and, 398
mutualism and, 467
"native invaders" and, 473
pathogens and, 521
plants and, 307, 316, 675–676
transformers and, 669
wetlands and, 703
See also nutrients and nutrient cycling
North Africa, 138, 377, *470*
North America
  biological control, 60, 64
  climate change and, 114, 115
  dispersal to, 46, 92, 155, 411
  18th-20th centuries, 364
  fire regimes, 227
  forests, 240, 322
  hybridization, 343, 344, 345
  invasibility, 357, 359
  invasional meltdown, 363
  lakes, 412 (*see also* Great Lakes (U.S.) (Laurentian))
  "native invaders," 474
  natural enemies of plants, 521
  nitrogen and, 491
  scientific projects, 618
  soil, 55
  See also main articles for specific organisms; New World; *specific countries*; *specific locales*; *specific organisms*
North American islands, 97, 98
northern Pacific seastar (*Asterias amurensis*), 39
Norway, 127–128, 626
Norway maple (*Acer platanoides*), 124, 147, 383, 424
novel weapons hypothesis (NWH), 17, 277–278, **492–493**
nursery trade. *See* horticulture (gardens, nursery trade)
nutria (coypu; *Myocastor coypus*), 187, 290, 333–334, 432, 446–448, 503, 610, 612, 614, 699
nutrients and nutrient cycling
  agriculture and, 8, 486
  algae and, 319
  animals and, 56, 99, 181–182, 332, 612, 713
  disturbance and, 167
  extinction and, 558
  fire and, 227
  fungi and, 55, 258
  impacts, worldwide, 394, 649
  insects and, 302, 322, 686, *687*
  invasional meltdown and, 363
  plants and, 56, 284, 289, 297, 316, 440, 630
  transformers and, 669
  wetlands and, 322, 699, 701, 703
  See also eutrophication, aquatic; soils; *specific nutrients*

oaks, 300, 355, 473, 523, 545
Oceania, 108, 458
oceans, 113, 609, **623–627**. *See also* pelagic habitats and organisms; seas; *specific oceans*

octopus stinkhorn (*Clathrus archeri*), 261, *263*
omnivores/omnivory, 557
  ants, 19, 531, 558
  aquatic organisms, 27, 28, 29, 101, 388
  carnivores, 95, 312, 632
  invasiveness and, 380
  land crabs, 362
  rodents, 446
  wetlands and, 701
Oncorhynchus mykiss (rainbow trout), 29, 192, 229–230, *229*, 233, 343, 414, 421, 612, 670
  Australia, 38
oomycetes. See *Phytophthora*
Ophiostoma ulmi (Dutch elm disease), 197, 261, 344, 523, 524
Opuntia spp., 65, 66, 68, 69, 194, 195, 649, 671, 679. *See also* prickly pear (*Opuntia stricta*)
Oregon
  acclimatization societies, 4
  animals, *46*, 265, 269, 361, 614, 615
  insects, 534
  invasional meltdown and, 361
  invertebrates, marine, 390
  plants, 110, 171, 282, 449, 523, 524, 545
  wetlands, 702
Oreochromis mossambicus (Mozambique tilapia), 27, 29, 38, 229, *648*, 649
Oreochromis niloticus niloticus (Nile tilapia), 28, 29, 32, 414–415, 486–487
Orgyia thyellina (white-spotted tussock moth), 530–531
Oriental bittersweet (*Celastrus orbiculatus*), *124*, 321, 681, 683
origins, geographic, **273–279**, 281, 282
ornamental animals, 101, 103, 134, 230, 414, 416, 466, 636
ornamental plants, 149, 282, 426
  biological control and, 63, 65, 69, 415
  climate change and, 114, 116
  diseases and, 544
  dispersal, 9, 149, 164, 240, 282, 339, 344, 627, 679
  dispersal by, 81, 83, 85, 240, 544
  distributions, 589
  hybridization and, 341, 344
  impact on
    Australia, 39, 40, 205, 682
    Europe, 141, *673*, 679
    Mediterranean region, 679
    New Zealand, 480, *673*
    United States, 607, 673, *673*
  invasiveness, 339, 341, 589
  management, 175, 255, 605, 606, 677
  nutrients and, 255
  reproduction and, 589
  threats to, 377, 524, 533, 639
  See also specific ornamental plants
Orobanchaceae (parasitic plants), 508–509
Oryctolagus cuniculus (European rabbit). *See* European rabbit (*Oryctolagus cuniculus*)
ostriculture, **494–499**
outcrossing, 179, 273, 584
  eucalypts and, 203
  invasibility and, 381, 586, 587, 588, 589, 641

soil and, 56
See also waterbirds; waterfowl
sea lamprey (*Petromyzon marinus*), 232, 296–297, *297*, 298, 432, **619–623**, *620*, 714
Sea of Japan, 625
seas, **623–627**. *See also* oceans; pelagic habitats and organisms; *specific seas*
sea turtles, 23, 633
seaweeds, 12, 13–15, **248–258**, 455, 467, 490
seeds, **627–631**
    aquatic plants, 252–253
    dispersal, 159–164, 207, 366, 413, 477, 591
    dispersal and vectors, 23, 362
    eucalypts, 206, 207, 208–209
    fire and, 224, 227
    forestry and, 242
    fungi and, *260*
    grasses and forbs and, 288
    herbivory and, 332, 334
    horticulture and, 337, 341
    human activities and, 374
    invasional meltdown and, 361–362
    invasiveness, 163, 381, 382, 383, 440, 588, 606–607, 630, 682–683
    life history strategies, 629
    plant reproductive systems and, 589
    rats and, 573
    succession and, 653–654
    taxonomic patterns and, 659
    water levels and, 587
    weeds and, 460–461, 694
    wetlands and, 699, 700, 703
selfing, 220, 275, 381, 397, 438, 584–588, 637
seven-spot ladybug (*Coccinella septempunctata*), *401*, 402, 403
sexual reproduction, *26*, 271, 381, 412. *See also* reproductive systems, plant
Seychelles Archipelago, 443–444, 458, 572
shade. *See* light
sheep, 267, 476, 527
shellfish, 113, 388. *See also* crustaceans; molluscs; *specific shellfish*
shipping. *See* transport (air, land, sea)
short-tailed weasel (stoat; *Mustela erminea*), 95, 96, 478, 483, 558
shrimps, 135
    diseases and, 30–31
    dispersal, 27–28, 416
    effects of alien species on, 319
    grazing and, 335
    impacts, worldwide, 136–137, 415, 486, 555
    lakes and, 411, 412
    tilapia and, 29
shrubs, **671–677**
    biological control by, 697
    competition and, 123
    earthworms and, 182
    fire and, 565, 675, 701
    fire regime and, 225, 227
    herbicides and, 326
    herbivory and, 331–332
    impact on
        Africa, *674*
        Australia, 40–41, 53, 676
        Hawaii, 54

islands, 559–560, 676
    New Zealand, 480, *673*, 676
    South Africa, 649, 675, 676
    United States, 182, *673*
    worldwide, 356
    invasional meltdown and, 361
    invasiveness, 359
    management, 449
    "native invaders," 475
    succession and, 657
    See also woody plants; *specific shrubs*
Siam weed (*Chromolaena odorata*), 67, 244, 649, 676
signal crayfish (*Pacifastacus leniusculus*), 31, 129, *130*, 132
skinks, 78, *78*, 96, 393, 481, 483, 572, 592, 633
skunks, 292, 393, 394, 468
slugs, 61, 315, **634–643**. *See also* snails
small Indian mongoose (*Herpestes auropunctatus*), 62, **631–634**
    biological control by, 393, 558
    control and, 62, 80, 95, 99
    endangerment/extinction by, 98
    eradication of, 100
    impacts, worldwide, 312, 313, 393, 558, 633, 676
smallpox, 184, 197, 509, 515, 516–517, 539, 599
snails, **634–643**
    aquatic plants and, 611
    Australia, 35, 44, 62, 157, 388, 640
    biological control and, 558, 635–636, 639, 642
    biological control by/of, 61, 62, 129, 134, 210, 315, 415, 558
    corals, 626
    crabs and, 127
    dispersal, 35, 44, 156–157, 387
    diversity and, 380
    eradication, 636, 641
    extinction, 133, 309, 639
    freshwater (*Melanoides tuberculata*), 277, *278*, 637
    Hawaii, 315, 558
    impact on
        Africa, 701
        Asia, 28, 35, 332, *333*
        Australia, 35, 44, 62, 157, 388, 640
        corals, 626
        Europe, 157
        Great Britain, 157
        Hawaii, 315, 558
        Mediterranean Sea region, 456
        New Zealand, 35, 156–157, 573
        North America, 35, 157, 387, 388
        Pacific region, 626
        Tahiti, 558
    invasiveness, 157, 637
    mutualism, 467, 468
    New Zealand mud snail (*Potamopyrgus antipodarum*), *48*, 156–157, *157*, 380, 382, 609, 638, 639
    periwinkle (*Littorina littorea*), 45, 387, 388, *389*
    pesticides, 525–530
    rivers and, 609
    transformers, as, 669

zebra mussel and, 712
    See also specific snails
snakes
    biological control of, 631, 632
    evolutionary responses of, 214
    impact on
        Australia, 592
        Caribbean, 86
        islands, 87, 88–89 (*see also* brown treesnake (*Boiga irregularis*))
        South Africa, 646
        United States, 89, 540, 541
        world, 88
    impacts on humans, 78–79, 87, 89, 559
    management, 61, 90–91, 312
    pathways, 590
    predators, 633
    See also specific snakes
social factors
    birds and, 542
    eradication and, 201–202
    grazers, 291
    insects and, 401, 529–530
    management and, 293
    New Zealand, 483–484
    Nile perch, 487
    ornamental plants and, 677
    reptiles and amphibians and, 593–594
    restoration and, 595, 596–597
    snails and, 642
    South Africa and, 643, 650
    squirrels and, 614
    weeds and, 692
    See also xenophobia
social species, 71, 686
soils, 54, 308
    benefits of invasions on, 368
    chemicals in, 17
    diseases and, 544
    disturbance, 306
    earthworms and, 57, 178–182, *178*
    fox impacts on, 560
    fungi and, 469
    grazers and, 292
    hemlock woolly adelgid and, 322
    herbivory and, 333–334
    horticulture and, 337
    insecticides and, 321
    invasibility and, 490
    invasional meltdown and, 363
    kudzu and, 396
    management tool, as, 696
    pathogens and, 514–515
    plants and, 112, 287, 288, 307, 611, 654–655
    protected areas, 565
    snails and, 638
    wetlands and, 703
    See also belowground phenomena; litter (soil cover); novel weapons hypothesis; nutrients and nutrient cycling
Solenopsis geminata (tropical fire ant), 19, 21, 22, 39
Solenopsis invicta. See red imported fire ant
Sorghum halepense (Johnsongrass), 8, 9

South Africa and, 644
See also canals; globalization; Internet;
    transport (air, land, sea); *specific
    industries*
transformer species, 282, **667–670**, 675, 676, 681
transport (air, land, sea), 141, 145–146, 365
    algae and, 12, 13, 15
    animal pathogens and, 512–513
    animals and, 95, 96, 149, 178, 179, 180, 573,
        590, 591, 637–638
    aquatic organisms and, 32–33, 34, 129, 133,
        135, 155, 496, 622, 711, 712
        marine organisms and, 385–386, 624, 627
    brown treesnake and, 80
    climate change and, 114
    diseases and, 150, 152, 153, 367–368, 443,
        444–445, 510, 517–518, 519
    fungi and, 259
    genetic information and, 270–271
    habitat alteration and, 31
    impact on
        lakes, 295, 417
        Mediterranean Sea, 453, 454
        oceans, 623
        Ponto-Caspian region, 550–552, 555, 556
        rivers, 609
    insects and, 240, 314, 402, 403, 462, 463
    international agreements, 6
    lag times and, 404, 407
    laws, 436
    plant dispersal and, 165
    predators and, 557
    protected areas and, 564
    taxonomic patterns and, 659
    viruses and, 234, 236
    See also ballast; boats, pleasure; canals;
        trade, international
transport (arrival) (stage of invasion), 407–
    408, 438, 498, 500–501, 658–659, 664.
    *See also main articles for specific species;
    specific species (dispersal)*
trapping (management technique), 634, 688
tree-of-heaven (*Ailanthus altissima*), 147, 160,
    321
trees, **671–677**
    aquatic plants and, 257
    beavers and, 56, 333
    beetles and, 532–533
    benefits of invasions, 368
    biocontrol and, 61
    climate change and, 115
    competition and, 124
    demography and, 147
    diseases, 214, *259*
    fire and, 226, 565, 675, 701
    Florida, 565
    fungi and, 259, 469, 470, 471
    gypsy moth and, 298, 299, 300
    hybridization, 345
    hydrology and, 347
    impact on
        Africa, *673*
        Australia, 41
        Germany, 404, 408
        Hawaii, 490–491

islands, 393, 394, 676
    New Zealand, 480
    North America, 55–56, 548, 565, 611, 676
    Pacific region, 630
    South Africa, 357, 644, 647, 649, 668, 675
insects and, 528–529
kudzu and, 397
ladybugs and, 401, 402
lag times, 404
laws, 432
life histories and, 424
management, 449, 451
mutualism, 466
nitrogen and, 490–491
pathogens, 317–318, 521
protected areas and, 565
remote sensing and, 580, 581, 582
squirrels and, 615
succession and, 656
wetlands and, 701, 703
See also forest insects; forestry and
    agroforestry; forests; woody plants;
    specific tree species
Triadica sebifera (Chinese tallow), 164, 226,
    383, 451, 676, 677
Trichosurus vulpecula (brush tail possum), 62, 119,
    290, *477*, 535, 549
tropical fire ant (*Solenopsis geminata*), 19, 21,
    22, 39
trout, 115, 296, 298, 414, 609, 670
    brook (brook char; *Salvelinus fontinalis*),
        115, *121*, 141, 421
    See also salmon and trout (Salmonidae);
        specific trout species
Tsuga spp. (hemlock), 15, 56, 61, 115, 334, 669
Turkey, 133, 179
turtles, 37, *38*, 540, 590, 591–592, 610, 624,
    646. *See also* sea turtles

Ulex europaeus (gorse), 41, 53, *160, 161*, 176,
    393, *477*
    benefits of, 480
    invasiveness, 715
Undaria pinnatifida (wakame seaweed), 12, *13*,
    40, 141, 250
ungulates, 37, 42, 311–312, 513
unicoloniality, 20
unintentional introductions, 7. *See also specific
    organisms*
United Kingdom, 185, 201. *See also main
    articles for specific organisms; specific
    countries; specific organisms*
United Nations, 6, 49, 248, 671
United States
    acclimatization societies, 3–4
    biological control of plants, 67–69, 194, 703
    climate change and, 115, 116
    codes of conduct, 339
    competition and, 123, 124
    databases, 146
    dispersal, 49
    ecoterrorism and, 185
    18th–19th centuries, 365–366
    eradication, 173, 177, 198, 199
    eutrophication, 211, 212

extinction by hybridization, 343
fire regime, 224, 226, 227, 286
hemlock woolly adelgid and, 321–322
horticulture and, 336
hybridization, 345
hydrology and, 348
invasibility, 357
invasional meltdown and, 361
invasion economics, 376, 378
lakes and, 417
management, 169, 355
"native invaders," 473, 475
nitrogen enrichment, 489, 490
novel weapons and, 493
pesticides and, 527, 698
plant diversity, 423
protected areas, 564
risk assessment and, 606
rodents, 333–334
xenophobia and, 705, 707
See also laws, U.S.; North America; *main
    articles for specific organisms; specific
    locales; specific organisms; specific states*

vectors. *See* dispersal and vectors
vegetation. *See* plants
vegetative propagation, *16*, 24–27, 271, 438,
    678–679, 700
vertebrates, 139, 140, 141, 372
    agriculture and, 10
    ants and, 21, 22–23
    aquatic, *27*
    diseases, 442
    dispersers, as, 683
    eradication, 199
    extinction, 557, 558
    fire regime and, 224
    grasses and forbs and, 288
    hemlock woolly adelgid and, 322
    impacts, worldwide, 38, 310–311, 476, 541, *541*
    invasiveness, 383, 438
    management, 354, 449, 452, 482
    marine, 624
    parasites, 134
    pesticides, 535, 537
    risk assessment and, 606
    vectors, as, 630
    viruses and, 234, 236
    See also specific vertebrates
Vespula spp. *See* wasps
Vespula vulgaris (common wasp), 558, 688. *See
    also* wasps
vines and lianas, 40, 294, 393, 653, 671, **680–
    684**. *See also* kudzu
viruses, 146
    aphids and, 314
    aquaculture and, 31
    ballast and, 46
    bees and, 52–53
    biocontrol and, 62, 332, 395, 559
    bird invasions and, 73
    dispersal, 411, 524
    eradication and, 200
    genetics and, 510
    grazers and, 292

viruses (*continued*)
    impacts, worldwide, 236, 237, 297, 510, 516, 520, 539
    invasiveness, 234, 237–238
    lag times, 408
    multiple invasions, 214
    See also pathogens, animal; *specific diseases*; *specific viruses*
Viscaceae (parasitic plants), 506–507
Vitex rotundifolia (beach vitex), *175*
    on North Carolina coast, 175
    symposium, 175
    Task Force, *174*, *175*, 177

wakame seaweed (*Undaria pinnatifida*), 12, *13*, 40, 141, 250
walking catfish (*Clarias batrachus*), 29, 231
wallabies, 478, 549
Wallace, Alfred Russell, 372
warfare, biological, **183–187**
Washington state, 362, 462, 512, 524
Wasmannia auropunctata (little fire ant), 18, 19, 20, 21, 23, 24, 38
wasps, 588, **685–689**
    biocontrol and, *60*, 314
    competition and, 119
    Compsilura concinnata, 191, 301
    forests and, 238
    hosts, 52
    impacts, worldwide, 119, 315, 409, 480, 559, 646
    ladybugs and, 402
    lag times and, 409
    mutualism, 466
    Vespula vulgaris (common wasp), 558, 688
water
    abiotic resistance and, 308
    clarity, 418–419
    climate change and, 116
    clonal growth and, 679
    competition and, 124
    diseases and, 518
    eucalypts and, 208
    forestry and, 246
    grasses and forbs and, 289
    grazing and, 331
    impact on
        Hawaii, 582
        South Africa, 124, 348, 378, 649, 650, 676
    invasibility and, 675
    invasion economics and, 378
    kudzu and, 397
    mosquitoes and, 151
    pathogens and, 521
    plant dispersal and, 164
    rivers and, 609
    tolerance limits and, 665
    vines and lianas and, 681, 683
    wetlands and, 698, 699, 700, 701, 703
    See also aquatic habitats; aquatic organisms; ballast; *bodies of water*; eutrophication, aquatic; hydrology; marine habitats; precipitation; water levels; water quality

waterbirds, 412–413, 420, 624, 627, 633, 676. *See also* seabirds; waterfowl
waterfleas, 136
waterfowl, 103, 344, 476, 480, 610, 708–709. *See also* seabirds; waterbirds; *specific waterfowl*
water hyacinth (*Eichhornia crassipes*), 586, **689–692**
    biological control of, 415
    impacts, worldwide, 176, 177, 611, 645
    invasiveness, 382
    management, 451
    rivers and, 611, 612
    wetlands and, 701
water levels, 587
water molds (*Phytophthora*), **543–546**
water quality
    agriculture and, 486
    aquatic organisms and, 102, 103, 133, 610, 622, 690, 691, 713
    biomanipulation of, 418–419
    eutrophication and, 212
    Great Lakes and, 298
    plants and, 242, 346–347, 348–349, 690, 691
Watson, Hewett Cottrell, 371
weasels, 478. *See also Mustela erminea* (short-tailed weasel)
weather. *See* climate and weather
wedelia (*Sphagneticola trilobata*), 281, *282*
weeds, 284, 437, **692–698**
    apomicts, 25
    aquatic, 40, 230
    assessments, 76
    biological control and, 63–70, *65*, *66*, 199, 230, 335, 677, 697
    climate change and, 114
    demography and, 148
    distribution, 280
    eradication, 173, 177, 199, 201, 481, 482, 695
    evolution and, 217
    fire and, 694
    forestry and, 246
    grazing and, 331
    horticulture and, 337
    impact on
        Africa, 280, 281, 429, 458
        Asia, 281, 429, 458, 547, 588
        Australia, 37, 39, 53, 172–173, 176, 246, 281, 282, 284, 372–373, 429, 458, 588
        Central America, 460
        Chile, *505*, 547
        Europe, 280, 281, 588
        Hawaii, 65–66, 315, 458, 460
        Indian Ocean region, 429
        islands, 41, *460*
        Mediterranean region, 429
        New Zealand, 25, 173, 429, 480–481, 548, 589–590
        North America, 279, 282, 284, 586
        South America, 281, 460, *505*, 547
        United Kingdom, 286
        United States, 148, 169, 170–172, 284, 429
        United States (continental), 65, 66, 171, 199, 282, 284, 547, 564, 597

impacts on humans, 429
invasiveness, 379, 437–441, *440*
Leopold on, 373
management, 169–177, *171*, 353, 354, 449–450, 482 (*see also* herbicides)
pathogens and, 524
pollination and, 53
range models, 569
regulations and, 577, 579
risk assessments and, 606, 607
selfing and, 586
succession and, 655
See also parasitic plants; seaweeds; *specific weeds*
weevils
    biological control and, 61, 256, 335, 415, 416, 691
    boll weevil (*Anthonomus grandis grandis*), 396, 532, 534
    evolutionary responses to, 214
    extinction, 481
    impacts, worldwide, 61, 351, 531
    lag times and, 409
West Nile fever, 151–152, 236–238, 314, 367, 376, 465, 510, 512–513
wetlands, 306, **698–704**
    aquatic mammals and, 446–447
    beavers and, 612
    biological control and, 703
    clonality and, 586–587
    dispersal, 413
    fire and, 703
    grasses and forbs and, 286–287, 288–289
    herbivory and, 332, 335
    impacts, worldwide, 37, 481, *596*
    ladybugs and, 402
    management, 452
    "native invaders," 473
    nitrogen and, 491
    nutria and, 614
    plant reproduction and, 587
    protected areas, 564
    restoration, *596*
    transformers and, 669
    woody plants and, 676
whirling disease (*Myxobolus cerebralis*), 232–233, 421
white-eye, Japanese (*Zosterops japonicus*), 4, 72, *313*, 460
whiteflies, 61, 524, *525*
white-spotted tussock moth (*Orgyia thyellina*), 530–531
Willdenow, Karl, 371
wind, 164, 411, 412, 442, 444, 477, 480, 630, 683
Wisconsin, 10, 133, 135, 211, 299, *302*, 473, 523, 699, 700
witchweed (*Striga* spp.), 170, 171, 173, 504, 508, 696
wolves (*Canis* spp.), 98, 268, 342, 343
woody plants, 671
    biological control and, 677
    fire and, 676
    grazers and, 331
    herbicides and, 326

horticulture and, 336
hydrology and, 347–348
impacts, worldwide, 586, 644, 671, 674, 676
import restrictions, 694
invasiveness, 382, *383*
lag times and, *408*
succession and, 656
wetlands and, 699, 701
See also shrubs; trees; *specific woody plants*
worms, 39, 44, *46*, 363, 386, 465, 639
marine, 552, 553, 554
See also earthworms
worst invasive species. *See* most invasive species

xenophobia, **705–709**

yellow crazy ant (*Anoplolepis gracilipes*), 18, 19,
20, 21, 22, 24, 558
impact on
Australia, 39

Christmas Island, 41, 56, 394, 670,
686
invasional meltdown and, 362, 364
yellow fever, 150–151, 236, 462–463, 519, 531

zebra mussel (*Dreissena polymorpha*) and quagga
mussel (*D. bugensis*), **710–714**
beneficial effects, 212
climate change and, 116
competition and, 118
crustaceans and, 137
dispersal and vectors, 31, 155, 156, 377, 411,
416, 417, 567
distribution, worldwide, 611
impacts, 118, 191, 233, 296, 297, 420, 611
invasiveness, 711
lag times, 404
lakes and, 412
management, 418, 432, 452
mutualism, 298, 467, 468

quagga mussel versus zebra mussel, 711
range models and, 568
rivers and, 610
water quality and, 419
zooplankton
ballast and, *45*
biomanipulated, 419
Black Sea and, 119, 551, 552, 553, 554, 555,
556
crab, 127, 128
crustaceans and, 136
fishes and, 230, 232, 233, 419
lakes and, 410
oceans and, 625
predators, *413*, 415, 420
zebra mussel and, 711, 713
zoos. *See* pets and zoos
Zosterops japonicus (Japanese white-eye), 4, 72,
313, 460